CONCEPTS OF GENETICS

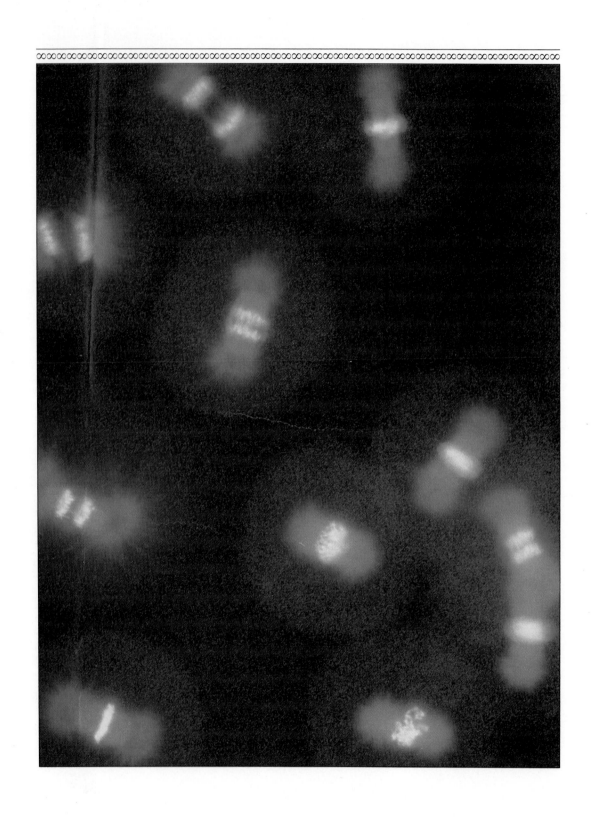

THIRD EDITION

CONCEPTS OF GENETICS

William S. Klug
Trenton State College at Hillwood Lakes

Michael R. Cummings
University of Illinois, Chicago

Macmillan Publishing Company
New York

Collier Macmillan Canada, Inc.
Toronto

Maxwell Macmillan International Publishing Group
New York   Oxford   Singapore   Sydney

Senior Editor: Robert L. Rogers
Production Editor: JoEllen Gohr
Art Coordinator: Mark D. Garrett
Photo Editor: Chris Migdol
Photo Researcher: Yvonne Gerin
Text Designer: Cynthia Brunk
Cover Designer: Russ Maselli
Production Buyer: Pamela D. Bennett

This book was set in Palatino.

Macmillan Publishing Company
866 Third Avenue, New York, New York 10022

Collier Macmillan Canada, Inc.

Library of Congress Cataloging-in-Publication Data
Klug, William S.
    Concepts of genetics/William S. Klug, Michael R.
  Cummings.—3rd ed.
        p.   cm.
    Includes bibliographical references and index.
    ISBN 0-02-364793-0 (hard)
    1. Genetics. I. Cummings, Michael R.   II.   Title.
QH430.K574   1991
575.1—dc20   90-13407
                                              CIP

Printing: 3   4   5   6   7   8   9   Year: 1   2   3   4

Cover Illustration: Supplied by Lionel I. Rebhun at the
University of Virginia, this micrograph was derived from
fertilized eggs of the sea urchin *Strongylocentrotus
purpuratus*. The eggs were processed after spindle
formation had occurred with a buffer that preserves the
cytoskeleton; fixed; and mounted on coverslips. The
preparation was then exposed to an antibody which
recognizes beta-tubulin and, thus, reveals microtubules.
This antibody was visualized with a rhodamine-labelled
secondary antibody. The chromosomes were stained with
a DNA intercalating dye (Hoechst 33342) and the whole
preparation photographed with a fluorescence
microscope.

Text Illustrations: Hans & Cassady, Inc.

To those who mean the very most,
Cindy, Brad, and Dori
Lee Ann, Brendan, and Kerry

Captains and kings may rule the world,
but it is only the presence and memory of those we love
and with whom we share our intellectual and genetic heritage
that bring beauty to living and justify their good deeds

And to the memory of Bill Gray,
who understood this more so than most of us.
Rest easy, good friend.

WILLIAM S. KLUG is currently Professor and Chairman of the Department of Biology at Trenton State College at Hillwood Lakes, New Jersey. He was previously a member of the faculty of Wabash College.

MICHAEL R. CUMMINGS, currently Associate Professor of Biological Sciences and Research Associate Professor at the Institute for the Study of Developmental Disabilities, University of Illinois, Chicago, has also served on the faculties of Northwestern University and Florida State University.

Preface

While planning the third edition, it became clear that our goals for the third edition have not changed from the previous editions, namely to:

- Emphasize concepts rather than excess detail
- Write clearly and directly, to students rather than to researchers
- Provide a clear organizational framework within and between chapters
- Design meaningful illustrations
- Create extensive sets of challenging problems
- Propagate the rich history of genetics that so beautifully illustrates how a discipline develops and how science works

Feedback from a variety of sources has confirmed that we have been reasonably successful in achieving these goals, and that they should once again serve as the text's cornerstone.

The field of genetics has moved much too rapidly to allow a revision with only minor polishing and updating. The third edition reflects substantial reorganization; a considerable amount of new information; and many new features to enhance the pedagogical value of the book. As was true for the first two editions, this text will accommodate many different organizational approaches and lecture formats.

Chapters are written so that they are as independent of others as possible, allowing instructors to use them in various sequences. Within each chapter, central concepts are emphasized that will support a variety of lecture formats. We believe that this focus has contributed most to the success of *Concepts of Genetics*.

FEATURES

The major features that distinguish this edition from the second edition include

- A revised chapter organization
- A new, full-color art program
- Several new sections in each chapter which further emphasize problem solving as well as conceptual and analytical thinking
- Updating to include the most recent topics in the field of genetics, as well as expansion of several existing sections
- A new supplement package, which includes a *Student's Handbook: A Guide to Concepts and Problem Solving*, an *Instructor's Sourcebook*, and a package of full-color overhead transparencies

Organization

In response to comments from users and reviewers of the text over the past five years, we have rearranged the sequence of chapters.

Part One, which emphasizes transmission genetics and phenotypic variation, now includes the chapter entitled "Variations in Chromosome Number and Arrangement" (Chapter 7). In addition, while the topic of polygenic inheritance has been retained in Chapter 4 as it relates to "Modification of Mendelian Ratios," the remaining quantitative genetics topics have been moved and expanded upon in the new Chapter 23.

Part Two again begins with three chapters on DNA structure, analysis, and replication, but these are now followed immediately by the three chapters introducing the genetic code and its transcription and translation into proteins. This part concludes with the chapter discussing the process of mutation.

Part Three has been reorganzied around the central theme of the organization and regulation of genetic information. Following the introduction of bacterial and viral genetic systems and the technology of recombinant DNA, the ensuing chapters pursue the topics of the structure and organization of the genome and the genes within it, as well as their regulation. The latter topic considers bacterial-viral modes separately from those of eukaryotes.

Part Four emphasizes the genetic basis of the development of organisms, as well as their behavior and organization into populations. The text concludes with the considerations of evolutionary genetics.

We have retained an appendix on experimental methodology, a glossary, and answers to selected problems.

Art Program

Redoing all artwork in a full-color format and adding a large number of color photographs and micrographs have undoubtedly enhanced the pedagogical value of the illustrations. The student will find a consistent assignment of color to each of the essential genetic molecules and components throughout the text (e.g., DNA, replicated DNA, the various RNAs, proteins, ribosomes, plasmids, etc.). Whenever possible, we have continued to use the "flow-chart diagrams" that were praised in the first two editions. We also have included a color photograph of nearly every plant that is discussed in the text.

Problem Solving and Conceptual and Analytic Thinking

Several new features added to each chapter capitalize on the way in which genetics lends itself to learning how to solve problems and to think conceptually and analytically. Initiating each chapter is a short section called "Chapter Concepts." In a few sentences, this section summarizes the most important ideas presented. At the conclusion of each chapter, a "Chapter Summary" enumerates the major points that have been discussed. Following the summary is a section entitled "Insights and Solutions," which presents students with solved problems and responses to analytical inquiries. This section emphasizes problem solving in those chapters lending themselves to quantitative analysis, and stresses analytical thinking and experimental rationale in other chapters. "Insights and Solutions" will teach students how to approach and solve problems, will stimulate analytical thinking, and will facilitate their success in the ensuing "Problems and Discussion Questions" section, which has been expanded with more challenging entries.

Updating

Every year new techniques are developed and new findings elevate certain topics to greater prominence. Some topics and techniques that have been introduced for the first time, or have been given greater attention, in this edition include

- DNA supercoiling
- gene conversion
- oncogenes
- transposable elements
- directed mutation
- retroviruses
- site-directed mutagenesis
- DNA fingerprinting
- polymerase chain reaction (PCR)
- restriction fragment length polymorphisms
- eukaryotic transcription factors
- centromere and telomere sequences
- human mitochondrial disorders

In addition, many other areas have been expanded, reflecting our growing knowledge of the genetics of eukaryotes. There are also several new sections—one in Chapter 1 that details the early history of genetics. A new section on probability has been added in Chapter 3.

Supplement Package

Student's Handbook: A Guide to Concepts and Problem Solving by Harry Nickla of Creighton University reviews vocabulary, concepts, and problem solving chapter by chapter. In addition, it provides a very detailed solution or lengthy discussion for *every* problem and question in the text, and supplies additional problems to be used for practice by students.

The *Instructor's Sourcebook*, also by Harry Nickla, contains questions and problems an instructor can use to prepare exams.

Overhead transparencies of approximately 100 illustrations from the text will be available to adopters of the text.

ACKNOWLEDGMENTS

No text of this breadth and depth could be the sole work of its authors. The field is just too extensive to be mastered by only two of us. While we take complete responsibility for any errors herein, we gratefully acknowledge the advice and contributions made by the reviewers of all three editions, particularly those who were involved in this edition:

Sidney L. Beck	De Paul University
Elliot S. Goldstein	Arizona State University
George Haughn	University of Saskatchewan
Dennis Hynes	California Polytechnic University, SLO
Keith K. Klein	Mankato State University
Harry Nickla	Creighton University
W. Stuart Riggsby	University of Tennessee
Robert M. Zarcaro	Providence College

One reviewer, Elliot Goldstein, has advised us during the preparation of all three editions. Another, Harry Nickla, has written the *Student's Handbook* and the *Instructor's Sourcebook*. We thank them both for efforts far beyond reasonable expectations. Gratitude is also extended to Allan Gotthelf and James G. Lennox for reviewing and editing the new historical section in Chapter 1.

We also thank our colleagues and our secretarial staff, who together have bolstered our efforts with encouragement, specific discussions, and endless technical support. In particular, Dr. Jim Bricker, Ms. Bette Baier, and Mrs. Monica Zrada have provided invaluable assistance at Trenton State College. And to Linda Burroughs and Lisa Andrasz, who helped structure the Chapter Summaries.

At Macmillan Publishing, we express our appreciation to Cindy Brunk, whose design efforts have greatly enhanced the book; to JoEllen Gohr, who so skillfully guided the text through production; to Dick Morel and to the entire staff at Hans & Cassady, whose creative efforts are reflected in the many figures; to Mark Garrett, who coordinated the art program; to Yvonne Gerin and Chris Migdol, who researched the color photographs; and to Vickie Brewster, whose experience in genetics has enhanced the index, which she produced. Particular thanks are due to Bob Rogers, who has become a good friend as he coordinated all of these efforts in his role as Senior Editor. More than any other factor, it has been his belief in and support of this project that has brought the third edition to fruition. It has been a real pleasure working with all of the above individuals, who deserve to share in any success that this text enjoys. Our gratitude to them goes well beyond these written words.

William S. Klug
Michael R. Cummings

Brief Contents

PART FOUR
GENETICS OF ORGANISMS
AND POPULATIONS

Contents

PART TWO
MOLECULAR BASIS OF HEREDITY

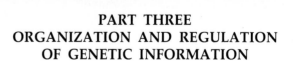

**PART THREE
ORGANIZATION AND REGULATION
OF GENETIC INFORMATION**

PART FOUR
GENETICS OF ORGANISMS
AND POPULATIONS

CONCEPTS OF GENETICS

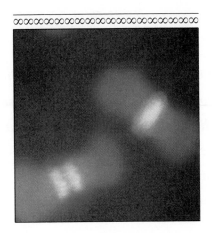

1

An Introduction to Genetics

THE HISTORICAL CONTEXT OF GENETICS
Prehistoric Times: Domesticated Animals and Cultivated Plants | The Greek Influence:
The Origin of Pangenesis and Inheritance of Acquired Characteristics | 1600–1850: The
Dawn of Modern Biology | Charles Darwin and Evolution

BASIC CONCEPTS OF GENETICS

INVESTIGATIVE APPROACHES IN GENETICS

THE SIGNIFICANCE OF GENETICS IN SOCIETY
Soviet Genetics: Science and Politics | Genetic Advances in Agriculture and Medicine

CHAPTER CONCEPTS

Genetics is the science of heredity. The discipline has a rich history and involves investigations of molecules, cells, organisms, and populations, utilizing many different experimental approaches. Because of the significance of genetic information and its expression at all levels of the function of individuals and during evolution, genetics has unified the study of biology and had a profound impact on human affairs.

Welcome to the study of the discipline of genetics. You are about to embark on the exploration of a subject that many students before you have found to be the most interesting and fascinating in the field of biology. This is not surprising because an understanding of genetic processes is fundamental to the comprehension of life itself. Genetic information directs cellular function, determines an organism's external appearance, and serves as the link between generations in every species. Knowing how these processes occur is important to understanding the living world. Knowledge of genetic concepts also helps us to understand the other disciplines of biology. The topics studied in genetics overlap directly with molecular biology, cell biology, physiology, evolution, ecology, systematics, and behavior. Usually, the study of each of these disciplines is incomplete without the knowledge of the genetic component underlying each of them. Genetics is therefore said to unify biology and serve as its "core."

Interest in and fascination with this discipline further stem from the fact that, in genetics, many initially vague and abstract concepts have been investigated in a logical fashion until they have become clearly and definitively understood. As a result, genetics has a rich history that exemplifies the nature of scientific discovery and the analytical approach used to acquire information. Scientific analysis, moving from the unknown to the known, is one of the major forces that attract students to biology.

But there is still another reason why the study of genetics is so appealing. Since its establishment, this field of study has expanded continuously. Every year large numbers of new findings are made. While it has been said that scientific knowledge doubles every ten years, one estimate holds that the doubling time in genetics is less than five years. Certainly, over the

past four decades, no five-year period has passed without some of the newly acquired information causing significant excitement for geneticists and for biologists in general. And each advance becomes part of an ever-expanding cornerstone upon which further progress is based. It is particularly stimulating to be in the midst of these developments, whether you are studying or teaching genetics.

THE HISTORICAL CONTEXT OF GENETICS

In the chapters that follow, we will discuss the behavior of chromosomes, the way in which genes present on them are transmitted from one generation to the next, and the way in which genetic information is stored, altered, expressed, and regulated. The most significant scientific findings that serve as the foundation for this information were obtained in the nineteenth century. As the twentieth century dawned, a period of integration of this knowledge occurred, clarifying the physical basis of living organisms and their relationship to one another. For example, the ideas that matter is composed of atoms, that cells are the fundamental units of living organisms, that cells contain nuclei that house threadlike structures called chromosomes, and that chromosomes are constant in number in a species and thus might be important in heredity together provided the basis for an important synthesis of ideas. When these ideas were combined with the newly rediscovered findings of Gregor Mendel and integrated with Darwin's theory of evolution and the origin of species, a more complete picture of life at the level of the individual and of the population emerged. On this foundation, the era of modern-day biology was initiated. And, beyond this introductory chapter, it is largely at this point in history that we will begin our coverage of genetics.

But what of the many important ideas and hypotheses which preceded this period in history and which, valid or not, served as forerunners of nineteenth-century thought? In the following short section, we shall consider some of these important ideas, two of which persisted incorrectly for over 2000 years! Nevertheless, they served to propel forward the quest to better understand the basis for the existence of all living things.

Prehistoric Times: Domesticated Animals and Cultivated Plants

It may never be known when people first recognized the existence of heredity. However, a variety of archeological evidence, such as primitive art, preserved bones and skulls, and dried seeds, have provided many insights. Such evidence documents human efforts that led to the successful domestication of animals and cultivation of plants thousands of years ago. Such efforts represent artificial selection of genetic variants within populations which then bred true.

Between 8000 and 1000 B.C., horses, camels, oxen, and even various breeds of dogs (derived from the wolf family) were domesticated to serve various roles. In Egypt, herds of gazelles were maintained for food, to be replaced eventually by sheep and then later by goats, which also provided milk. In contrast to the ancestral forms present in nature, pigeons were bred and maintained in captivity. Even insects were selected for useful purposes, including bees in the Nile Valley and silk moths in China.

FIGURE 1.1
Relief carving depicting artificial pollination of date palms during the reign of Assyrian King Assurnasirpal II (883–859 B.C.).

AN INTRODUCTION TO GENETICS   3

Cultivation of plants paralleled these developments. Several primitive cereals, including maize, wheat, and rice, are thought to have been developed around 5000 B.C. Large volumes of maize remains dating to this period have been recovered in caves in the Tehucan Valley of Mexico. Assyrian art depicts artificial pollination of the date palm, thought to originate in Babylonia during this same time. Such cultivation very likely influenced the types of modern-day palms found in the region. Today, there are over 400 kinds of date palm in just four oases in the Sahara, differing from one another in various traits such as fruit taste.

The presence of cultivated plants and domestic animals during prehistoric times documents people's successful attempts, perhaps unconscious at first, to manipulate the genetic composition of organisms. There seems little doubt that it was soon learned that desirable and undesirable traits were passed to successive generations. This recognition of diversity of forms on which selection could be performed to produce new breeds and varieties of animals and plants documents human awareness of heredity during prehistoric times.

The Greek Influence: The Origin of Pangenesis and Inheritance of Acquired Characteristics

While few, if any, ideas attempted to seriously explain heredity during prehistoric times, much more is known about such attempts during the Golden Age of Greek Culture. Considerable attention was devoted to the subjects of reproduction and heredity, particularly as they related to the origin of humans.

This is most evident in the writings of the Hippocratic school of medicine (5th–4th century B.C.) and subsequently of the philosopher and naturalist Aristotle (384–322 B.C.), as well as in the writings of their many followers. Even before this time, pre-Socratic philosophers (600–450 B.C.) were shifting their interest from religious mythology to philosophical and proto-scientific inquiries regarding the basis of human life. Indeed, their thinking influenced every area of Greek life, including poetry, music, art, and politics.

Central to an explanation of the hereditary basis of procreation of animals is: (1) the source of the physical substance of the offspring, and (2) the generative force behind the direction taken by that substance as it materializes into a living organism. Ancient accounts suggested that the male semen was somehow involved, an idea much refined by Hippocrates and his followers. As such, they are credited

with the earliest thinking leading to the **theory of pangenesis,** later proposed more formally by Charles Darwin.

The Hippocratic treatise *On the Seed* argues that male semen is formed in numerous parts of the body and is transported through blood vessels to the testicles. Active "humors," as the bearer of hereditary traits, are drawn from various parts of the body. These humors could be healthy or diseased, the latter condition accounting for the appearance of newborns exhibiting congenital disorders or deformities. A second idea, which like pangenesis was to survive for over 2000 years, was central to this hypothesis: **the inheritance of acquired characteristics**. This idea, today mostly associated with Jean Baptiste Lamarck, suggested that the performance of various physical and intellectual activities by parents could predispose their offspring with similar abilities or characteristics. Humors could thus be altered in parents, and offspring could inherit traits that their parents had acquired in their "environment."

Aristotle, who had studied under Plato for some twenty years, was more critical and more expansive than Hippocrates in his analysis of the origin of human individuals and their heritage. Better known as a philosopher, Aristotle was also a keen naturalist, having identified and classified hundreds of organisms and having dissected and studied many of these. His observations and reflections led him to be critical of pangenesis, which he argued did not explain the appearance of features inherited from generations preceding the parents of an individual, such as grandparents.

Aristotle proposed instead that male semen was formed from blood rather than each organ and that its generative power resided in a vital heat it contained. This vital heat had the capacity to produce an offspring of the same "form" (i.e., basic structure and capacities) as the parent. It generated the offspring by cooking and shaping the menstrual blood produced by the female, which was the "matter" for the offspring. The properties of the offspring were determined primarily by the heat supplied by the male but partly by the material supplied by the female. The embryo would develop from the initial "setting" of the menstrual blood by the semen into a mature offspring, not because it already contained the parts in miniature (as some Hippocratics had thought), but because of the shaping power of that vital heat. These ideas constituted but one part of Aristotelian philosophy of order in the living world and in the universe in general.

While the ideas of Hippocrates and Aristotle may sound primitive and naive today, we should recall that neither sperm nor eggs had yet to be observed in any mammal, let alone humans. In fact, eggs were not discovered until 1827! Thus, in their own right, these explanations were worthy ones in their time and, in fact, for centuries to come. Often in the history of science, incorrect or inadequate theories have served as important stepping stones for acquiring more accurate knowledge. For embodied in them is much insightful information which, even if subsequently refuted, influences future thought.

In the case of the ideas of Hippocrates and Aristotle, we can see this very clearly. They suggested that inheritance was biparental and controlled by genetic information present in discrete units. Further, Aristotle's "setting" is the same event that, even today, has evaded a complete scientific explanation: **procreation**. What precisely occurs following the fusion of two gametes that endow the zygote with the capability to develop into an adult, while neither gamete alone seems to possess such qualities? We still cannot accurately describe that creative force, which at one moment before an egg and a sperm fuse, is absent, and then after conception leads to the development of an individual. Thus, while the theories of pangenesis and of inheritance of acquired characters survived well into the nineteenth century, only to be shown to be incorrect, we should not lose sight of the importance of the influence of Greek thought throughout the history of science and particularly of genetics.

1600–1850: The Dawn of Modern Biology

During the ensuing 1900 years (300 B.C. to A.D. 1600), the theoretical understanding of genetics was not extended by significant new ideas, but interest in applied genetics remained strong. As early as Roman times, plant grafting and animal breeding were greatly emphasized. By the Middle Ages, naturalists, well aware of the impact of heredity on organisms they studied, were faced with reconciling their findings with religious beliefs. The naturalist Albert Magnus (1193–1280), his student Saint Thomas Aquinas (1225–1274), and Roger Bacon (1214–1294) attempted to meld scientific findings and philosophies with the current doctrines of the Christian Church. The theories of Hippocrates and Aristotle still prevailed at that time and in the case of humans, they no doubt conflicted with the religious doctrine of **special creation**.

However, beginning in the seventeenth century and continuing through the nineteenth century, major strides were made in experimental biology, which provided much greater insights into the basis of life. The many findings set the scene for the revolutionary work and principles put forward by

Charles Darwin and by Gregor Mendel. A brief review of some of the most significant observations and interpretations will illustrate this point.

It was in the 1600s when the English anatomist William Harvey (1578–1657) studied early development. Better known for his experiments demonstrating that the blood was pumped by the heart through a circulatory system made up of the arteries and veins, he also wrote a treatise on reproduction and development, patterned after Aristotle's work on the same topic. In it he is credited with the earliest statement of the **theory of epigenesis**—that an organism is derived from substances present in the egg which must first be assembled and which then differentiate during embryonic development. Modifying the ideas of pangenesis somewhat, epigenesis holds that new structures such as the many organs are not present initially, but instead arise during development—the so-called *de novo* appearance of new body components. This theory is consistent with Aristotle's ideas describing procreation. Indeed, Harvey had studied Aristotle and was well aware of his ideas.

The theory of epigenesis conflicted directly with the **theory of preformationism,** first put forward in

FIGURE 1.2
Gregor Johann Mendel, who in 1865 put forward the major postulates of transmission genetics as a result of experiments with the garden pea.

the seventeenth century. Stating that sex cells contain a complete miniature adult called the **homunculus,** perfect in every form, preformationism was popular well into the eighteenth century. Some, called ovists, believed it was the ovum, while others (spermists) believed it was the sperm that housed the homunculus. However, work by the embryologist Casper Wolff (1733–1794) and others clearly disproved this theory and strongly favored epigenesis. For example, Wolff was quite convinced that several structures, such as the alimentary canal, were not present in the earliest embryos he studied.

During this same period, there were other significant chemical and biological findings and proposals. In 1808, John Dalton expounded his atomic theory, which stated that all matter is composed of small invisible units called atoms. Improved microscopes were available, and around 1830, Matthias Schleiden and Theodor Schwann proposed that all organisms are composed of basic visible units called cells. By this time, the **idea of spontaneous generation,** the creation of living organisms from nonliving components, had clearly been disproved by the experiments of Francesco Redi (1621–1697), Lazzaro Spallanzani (1729–1799), and Louis Pasteur (1822–1895), among others. Thus, all living organisms were considered to be derived from preexisting organisms and to consist of cells made up of atoms.

Another prevailing notion had a major influence on nineteenth-century thinking: **fixity of species.** Embraced particularly by those also adhering to a belief in special creation, this doctrine was attributed to the Swedish physician and plant taxonomist, Carolus Linnaeus (1707–1778), among others. Linnaeus is best known today for devising the binomial system of classification. According to the doctrine of fixity of species, animal and plant groups remained unchanged from that moment of their appearance on earth. Had such a view been fervently adhered to by plant hybridizers such as Mendel, the principles that govern inheritance may never have been established. Linnaeus did become convinced late in his life that species were not fixed.

The influence of this doctrine is illustrated by considering the work of the German plant breeder Joseph Gottlieb Kolreuter (1733–1806), who was considered to be the first to successfully hybridize a variety of plants. Several of his findings were potentially quite far-reaching. In work with tobacco, he cross-bred two groups and derived a new hybrid form, which he then converted back to one of the parental types by repeated backcrosses. In other breeding experiments, using carnations, he clearly observed segregation of traits, which was to become one of Mendel's principles of genetics. Kolreuter's belief in both special creation and the fixity of species caused him to be ambivalent about the importance of his own work, and he failed to recognize its real significance.

Neither was the prevailing scientific climate kind to plant hybridizers and undoubtedly worked against the acceptance of Kolreuter's findings. This influence was the result of the German school of thought known as *Naturphilosophie.* The beliefs that nature was unity and that only the study of the whole organism was significant framed this doctrine. As proponents of this approach, several plant hybridizers chose not to emphasize the study of individual traits, but instead focused holistically on organisms. Like Kolreuter, Karl Friedrich Gartner (1772–1850) performed experiments with peas that led to results similar to Mendel's monohybrid cross, upon which the principles of dominance/recessiveness and segregation are based. However, Gartner did not concentrate on the analysis of individual traits and failed to grasp the significance of his own work.

Unfortunately, proponents of the *Naturphilosophie* actively discouraged this type of research, creating an atmosphere that made it difficult for the work of Kolreuter, Gartner, and others to find acceptance. Undoubtedly, this influence also adversely affected the acceptance of Mendel's work half a century later.

Charles Darwin and Evolution

With the above information as background, we conclude our coverage of the historical context of genetics with a brief discussion of the work of Charles Darwin, who in 1859 published the book-length statement of his evolutionary theory, *The Origin of Species.* His many geological, geographical, and biological observations convinced Darwin that existing species arose by descent from other ancestral species. He embarked on two years of intensive inquiry, which culminated in his formulation of the **theory of natural selection,** a theory of the causes of evolutionary change. This theory, formulated and proposed independently by Alfred Russell Wallace, hypothesized that populations tend to leave more offspring than the environment can support, leading to a struggle for existence among them. In such a struggle, those organisms with heritable traits that better adapt them to their environment are better able to survive and reproduce than those with less adaptive traits. Over a long period of time, numerous slight, advantageous variations will accumulate. Once reproductive isolation results, a new species is formed and change in form through evolution continues.

The primary gap in the theory was a lack of understanding of the genetic basis of variation and inheritance, a gap that left Darwin's theory open to reasonable criticism well into the twentieth century. Aware of this weakness in his theory of evolution, in 1868 Darwin published a second book, *Variations in Animals and Plants under Domestication,* in which he attempted to provide a more definitive explanation. He seized on two ideas that we know were not novel ones: pangenesis and the inheritance of acquired characteristics. Together, these ideas bolstered his thought that heritable variation arises gradually over time. In his **provisional hypothesis of pangenesis,** Darwin coined the term *gemmules* to describe the units representing each body part that were gathered by the blood into the semen. We need not dwell on this further since we have described pangenesis before. However, Darwin felt that it was the gemmules that respond to the environment in an adaptive way, which allows for the inheritance of acquired characteristics. This theory, central to Greek thought, was formalized by Lamarck in his 1809 treatise, *Philosophie Zoologique.* Lamarck's theory, which became known as the **doctrine of use and disuse,** proposes that organisms acquire or lose characteristics which gradually become heritable.

On this backdrop, the experiments of Gregor Johann Mendel were performed between 1856 and 1863, forming the basis for his 1865 paper. In it, Mendel demonstrated a number of statistical patterns underlying hybrid inheritance and developed a theory involving hereditary factors in the germ cells that explained these patterns. His research was virtually ignored until it was partially duplicated, and then cited, by Carl Correns, Hugo de Vries, and Eric Von Tschermak in 1900, and championed by William Bateson. During this interval, chromosomes were discovered and support for the epigenetic interpretation of development grew considerably. Thus, it gradually became clear that heredity and development were dependent on "information" contained in these bodies that were contributed by gametes to each individual.

As we have seen, a rich history of scientific endeavor and thinking preceded and surrounded Mendel's work. In Chapter 3, we will return to a thorough analysis of his findings, which have served to this day as the foundation of genetics. And his work was but one important part of the body of knowledge that would initiate the era of modern biological thought of the twentieth century.

BASIC CONCEPTS OF GENETICS

In this introductory chapter we would like to review some of the simple but basic concepts in genetics which you have undoubtedly already studied. By reviewing them at the outset, we can establish an initial vocabulary and proceed through the text with a common foundation. We shall approach these basic concepts by asking and answering a series of questions. You may wish to write or think through an answer before reading the explanation of each question. Throughout the text, the answers to these questions will be expanded as more detailed information is presented.

What does genetics mean?
Genetics is the branch of biology concerned with heredity and variation. This discipline involves the study of cells, individuals, their offspring, and the populations within which organisms live. Geneticists investigate all forms of inherited variation and the nature of the underlying genetic basis of such characteristics.

What is the center of heredity in a cell?
In eukaryotic organisms, the **nucleus** contains the genetic material. In prokaryotes such as bacteria, the genetic material exists in an unenclosed area of the cell called the **nucleoid region**. In viruses, which are not true cells, the genetic material is ensheathed in the protein coat referred to as the **viral head**.

What is the genetic material?
In eukaryotes and prokaryotes, **DNA** serves as the molecule storing genetic information. In viruses, either DNA or **RNA** serves this function.

What do DNA and RNA stand for?
DNA and RNA are abbreviations for **deoxyribonucleic acid** and **ribonucleic acid,** respectively. These are the two types of nucleic acids found in organisms. Nucleic acids, along with carbohydrates, lipids, and proteins, compose the four major classes of organic biomolecules found in living things.

How is DNA organized to serve as the genetic material?
DNA, while single-stranded in a few viruses, is usually a double-stranded molecule organized as a double helix. Contained within each DNA molecule are hereditary units called **genes,** which are part of a larger element, the **chromosome**.

What is a gene?
In simplest terms, the gene is the functional unit of heredity. In chemical terms, it is a linear array of nucleotides—the chemical building blocks of DNA and RNA. A more sophisticated approach is to consider it as an informational storage unit capable of undergoing replication, mutation, and expression.

As investigations have progressed, the gene has been found to be a very complex genetic element.

What is a chromosome?

In viruses and bacteria, the chromosome is most simply visualized as a long, usually circular DNA molecule organized into genes. In eukaryotes, the chromosome is more complex. It is composed of a linear DNA molecule complexed with proteins. In addition to genes, the chromosome contains many regions that do not store genetic information. It is not yet clear what role, if any, is played by some of these regions of chromosomes. Our knowledge of the chromosome, like the gene, is continually expanding.

When and how can chromosomes be visualized?

If the chromosomes are released from the viral head or the bacterial cell, they can be visualized under the electron microscope. In eukaryotes, chromosomes are most easily visualized under the light microscope when they are undergoing **mitosis** or **meiosis**. In these division processes, the chromosomes are tightly coiled and condensed. Following division, they uncoil and exist as **chromatin fibers** during interphase, when they can be studied under the electron microscope.

How many chromosomes does an organism have?

While there are many exceptions, members of most species have a specific number of chromosomes present in each somatic cell called the **diploid number (2n)**. Upon close analysis, these chromosomes are found to occur in pairs, each member of which shares a nearly identical appearance, when visible during cell division. Called **homologous chromosomes,** the members of each pair are identical in their length and in the location of the centromere, the point of spindle fiber attachment during division. They also contain the same sequence of gene sites, or loci, and pair with one another during meiosis. Thus, the number of different *types* of chromosomes in any diploid species is equal to half the diploid number and called the **haploid (n)** number. Some organisms, such as yeast, are haploid and contain only one "set" of chromosomes. Other organisms, notably plants, are sometimes characterized by more than two sets of chromosomes and are said to be **polyploid**.

What is accomplished during the processes of mitosis and meiosis?

Mitosis is the process by which the genetic material of eukaryotic cells is duplicated and distributed during cell division. **Meiosis** is the process whereby cell division produces gametes in animals and spores in most plants. While mitosis occurs in somatic tissue and yields two progeny cells with an amount of genetic material identical to the progenitor cell, meiosis creates cells with precisely one-half of the genetic material. Each gamete receives one member of each homologous pair of chromosomes and is haploid. This accomplishment is essential if offspring arising from two gametes are to maintain a constant number of chromosomes characteristic of their parents and other members of the species.

FIGURE 1.3
The DNA of the chromosome of a bacterial virus viewed under the electron microscope.

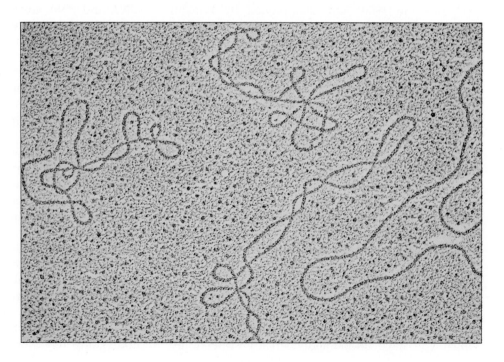

What are the sources of genetic variation?

Classically, there are two sources of genetic variation: **chromosomal** and **gene mutations**. The former, also called **chromosomal aberrations,** includes duplication, deletion, or rearrangement of chromosome segments. Gene mutations result from a change in the stored chemical information in DNA. Such a change may include substitution, duplication, or deletion of nucleotides, which compose this chemical information. Alternate forms of the gene which result from mutation are called **alleles**. Genetic variation frequently, but not always, results in a change in some characteristic of an organism.

How does DNA store genetic information, and how is it altered to produce variation?

The sequence of nucleotides in a segment of DNA constituting a gene is present in the form of a **genetic code**. This code specifies the chemical nature of proteins, which are the end product of genetic expression. Mutations are produced when the genetic code is altered.

How is the genetic code organized?

There are four different nucleotides in DNA, each varying in one of its components, the **nitrogenous base**. The genetic code is a **triplet;** therefore, each combination of three nucleotides constitutes a code word. Almost all possible codes specify one of twenty **amino acids,** the chemical building blocks of proteins.

How is the genetic code expressed?

The coded information in DNA is first transferred during a process called **transcription** into a **messenger RNA (mRNA)** molecule. The mRNA subsequently associates with the cellular organelle, the **ribosome,** where it is **translated** into a protein molecule.

Are there exceptions where proteins are not the end product of a gene?

Yes. For example, genes coding for **ribosomal RNA (rRNA),** which is part of the ribosome, and for **transfer RNA (tRNA),** which is involved in the translation process, are transcribed but not translated. Therefore, RNA is sometimes the end product of stored genetic information.

Why are proteins so important to living organisms that they serve as the end product of the vast majority of genes?

Many proteins serve as highly specific biological catalysts, or **enzymes**. In this role, these proteins control cellular metabolism, determining which carbohydrates, lipids, nucleic acids, and other proteins are present in the cell. Many other proteins perform nonenzymatic roles. For example, hemoglobin, collagen, immunoglobulins, and some hormones are proteins.

Why are enzymes necessary in living organisms?

As biological catalysts, enzymes lower the activation energy required for most biochemical reactions and speed the attainment of equilibrium. Otherwise, these reactions would proceed so slowly as to be ineffectual in organisms living under the conditions on earth. Thus, some genes control the variety of enzymes present in any cell type, which in turn dictates its overall biochemical composition.

INVESTIGATIVE APPROACHES IN GENETICS

The scope of topics encompassed in the field of genetics is enormous. Studies have involved viruses, bacteria, and a wide variety of plants and animals and have spanned all levels of biological organization, from molecules to populations. It is helpful, before we embark on a detailed study of genetics, to know the types of investigations that have been used most often in this field. With some overlap, most have used one of four basic approaches.

The most classical investigative approach is the study of **transmission genetics,** in which the patterns of inheritance of traits are examined. Experiments are designed so that the transmission of traits from parents to offspring can be analyzed through several generations, and patterns of inheritance are sought that will provide insights into more general genetic principles. The first significant experimentation of this kind to have a major impact on understanding heredity was performed by Gregor Mendel in the middle of the nineteenth century. Information derived from his work serves today as the foundation of transmission genetics. In human studies, where designed matings are neither possible nor desirable, **pedigree analysis** is used. In pedigree analysis, patterns of inheritance are traced through as many generations as possible, leading to predictions of the mode of inheritance of the trait under investigation.

The second approach involves **cytological investigations** of the genetic material. The earliest studies used the light microscope. The initial discovery in the early twentieth century of chromosome behavior during mitosis and meiosis was a critical event in the history of genetics. In addition to playing an important role in the rediscovery and acceptance of Mendelian principles, these observations served as the basis of the **chromosomal theory of inheritance**. This

FIGURE 1.4
Visualization of DNA under ultraviolet light. The bands were produced using recombinant DNA technology.

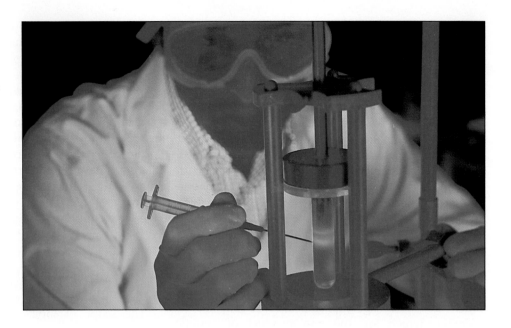

theory, which viewed the chromosome as the carrier of genes and the functional unit of transmission of genetic information, was the cornerstone for further studies in genetics throughout the first half of this century.

The light microscope continues to be an important research tool. It is useful in the investigation of chromosome structure and abnormalities and is instrumental in preparing **karyotypes,** which illustrate the chromosomes characteristic of any species arranged in a standard sequence.

With the advent of electron microscopy, the repertoire of investigative approaches in genetics has grown. In high-resolution microscopy genetic molecules and their behavior during gene expression can be directly visualized.

The third general approach, **molecular and biochemical analysis,** has had the greatest impact on the growth of genetic information. Molecular studies beginning in the early 1940s have consistently expanded our knowledge of the role of genetics in life processes. Although experimental sources were initially bacteria and viruses, extensive information is now available concerning the nature, expression, replication, and regulation of the genetic information in eukaryotes as well. The precise nucleotide sequence of many genes has been determined and cloned in the laboratory. **Recombinant DNA studies,** where genes from any organism are literally spliced into bacterial or viral DNA and cloned *en masse,* is the most significant and far-reaching research technology used in molecular genetic investigations. As a result, it is now possible to probe gene structure and function with a resolution heretofore impossible. This

technology also has profound implications in medicine and agriculture.

The final approach involves the study of the **genetic structure of populations**. In these investigations scientists attempt to define how and why certain genetic variation is preserved in populations while other variation diminishes or is lost with time. Such information is critical to the understanding of the evolutionary process. Population genetics also allows us to predict gene frequencies in future generations.

Together these approaches have transformed a subject poorly understood in 1900 into one of the most advanced disciplines in biology today. As such, the impact of genetics on society has been immense. We shall discuss many examples of the applications of genetics in the following section and throughout the text.

THE SIGNIFICANCE OF GENETICS IN SOCIETY

In addition to acquiring information for the sake of extending knowledge in any discipline of science—an experimental approach called **basic research**—scientists conduct investigations to solve problems facing society or simply to improve the well-being of members of that society—an approach called **applied research**. The history of applied research shows that it is usually possible only as an extension of prior basic research. In order for the two types of research to be efficiently executed and ultimately to benefit society, they must be allowed to proceed in a reasonably uninhibited and rational manner. Scientists must be allowed to examine existing knowledge, determine the direction of their efforts, and freely exchange their ideas and findings with other scientists, as has been the case in most countries with respect to the field of genetics.

However, in one era in one country—the USSR under the government of Joseph Stalin—this was not the case. In order to illustrate the consequences of the loss of free scientific inquiry and the imposition of controlled research, we will recount this era. Then we will turn to a brief survey of some of the benefits society has enjoyed as the result of unrestrained research in genetics.

Soviet Genetics: Science and Politics

One of the most famous examples of an interaction between science and politics occurred in the Soviet Union. This instance happened to involve genetics. In particular, it had to do with the genetic and evolutionary theory of the inheritance of acquired characteristics. As we know from our earlier discussion, the theory originated with the ancient Greeks, was formalized by Lamarck in 1809, and considered by Darwin. However, experimental evidence early in the twentieth century argued against it. Nevertheless, in the 1930s, Trofim Denisovich Lysenko, a young plant breeder from the Ukraine, espoused the idea that plant development and agricultural productivity could be improved by manipulating environmental conditions. Lysenko further stated that such improvements would be incorporated into the genetic material and thus passed on to future generations of plants.

The political climate in the Soviet Union at that time provided a hospitable environment for the revival of Lamarck's ideas. First, the theory that

FIGURE 1.5
A bumblebee pollinates an aster flower, achieving cross-fertilization and enhancing genetic variability within the species.

environmental change could produce permanent genetic change was compatible with the Marxist political thesis that the proper social conditions would induce permanent changes in human behavior. Second, Ivan Pavlov, the Soviet scientist known for his work on the learned or conditioned reflex, emphasized the importance of environmental stimuli. Pavlov claimed, although he later retreated from this statement, to have found an example of a conditioned reflex that was inherited in mice.

Perhaps most significant in the acceptance of Lysenko's ideas was the poor condition of Soviet agriculture during this period. The Soviet government was desperate to improve agricultural production, since the traditional methods of selective breeding were not perceived as producing adequate harvests.

Facing a difficult situation, Stalin was impressed with Lysenko's ideas. Lysenko proposed that germination of the winter wheat crop could be speeded up by **vernalization,** the practice of subjecting plants to an artificial cold period to shorten the dormant period of the seeds, which are usually shed in autumn. The early shedding of the seeds permitted the planting of an additional crop that could be harvested before autumn. Lysenko claimed that because the changes induced by vernalization were permanent ones, this technique need be applied only once.

In fact, this practice did not lead to a permanent increase in agricultural output, but by the time this became apparent, Lysenko had become director of the Institute of Genetics of the USSR Academy of Science. Appointed to this position in 1940, he managed to suppress throughout the Soviet Union all work in genetics that ran contrary to his own views. He was responsible for destroying research records, laboratory supplies, and experimental material of those who opposed him, and in some cases, he had his opponents arrested and sentenced to prison.

One of Russia's leading geneticists, Nikolai Ivanovich Vavilov was a favorite target of Lysenko. Director of the Lenin Academy of Agriculture, Vavilov was one of the world's experts on wheat. He was sentenced to prison for agricultural espionage in 1941. He died from malnutrition in 1943.

During this time great advances were made in genetics, particularly in the United States. Scientifically based selective breeding programs had developed new varieties of hybrid grains whose yields were much greater than those of the old varieties. Yet none of this knowledge was made available to Soviet geneticists until 1964, when Lysenko finally fell from power and lost his stranglehold on Soviet genetics. One American botanist calculated that USSR corn production would have been increased by six million

FIGURE 1.6
A spider protects her egg case, ensuring genetic continuity between generations.

tons between 1947 and 1957 if only half the country's acreage had been planted with the new hybrid strains.

Unfortunately, Soviet genetics was held back for nearly a generation at a time when numerous significant advances were being made elsewhere in the world. Adherence to theories that were politically rather than scientifically acceptable was indeed costly to Soviet society.

Genetic Advances in Agriculture and Medicine

The major benefit to society as a result of genetic study has been in the areas of agriculture and medicine. Even though cultivation of plants and domestication of animals had begun long before, the rediscovery of Mendel's work in the early twentieth century spurred scientists to apply genetic principles to these processes. In both cases, the use of selective breeding and hybridization techniques has had the most significant impact.

In plants, four major categories of improvements have been possible: (1) more efficient energy utilization during photosynthesis, resulting in more vigorous growth and increased yields; (2) increased resistance to natural predators and pests, including insects and disease-causing microorganisms; (3) pro-

FIGURE 1.7
A coral reef environment demonstrates extensive genetic variation among organisms.

duction of hybrids exhibiting a combination of superior traits derived from two different strains or even two different species; and (4) selection of genetic variants with increased protein value or an increased content of limiting amino acids that are essential in the human diet.

These improvements have resulted in a tremendous increase in yield and nutrient value in such crops as barley, beans, corn, oats, rice, rye, and wheat, among others. It is estimated that in the United States the use of improved genetic strains has led to a threefold increase in crop yield per acre. In Mexico, where corn is the staple crop, a significant increase in protein content and yield has occurred. A substantial effort has also been made to improve the growth of Mexican wheat. Led by Norman Borlaug, a team of researchers was able to develop a strain of wheat that incorporated favorable genes from other strains found in various parts of the world, creating a superior variety that is now grown in many underdeveloped countries besides Mexico. Because of this effort, which led to the well-publicized "Green Revolution," Borlaug received the Nobel Peace Prize in 1970. There is little question that this application of genetics, which is now prevalent worldwide, has contributed to the well-being of our own species and to the maintenance of world peace.

Similarly, applied research in genetics has developed superior breeds of animals. Enormous increases in usable meat supplies produced per unit of food intake have occurred. For example, selective breeding has produced chickens that grow faster, produce more high-quality meat per chicken, and lay greater numbers of larger eggs. In larger animals, including pigs and cows, the use of artificial insemination has been particularly important. Sperm samples derived from a single male with superior genetic traits may now be used to fertilize thousands of females located in all parts of the world.

Equivalent strides have been made in medicine as a result of advances in genetics, particularly since 1950. Numerous disorders in humans have been discovered to result from either a single mutation or a specific chromosomal abnormality. For example, the genetic basis of sickle-cell anemia, erythroblastosis fetalis, cystic fibrosis, hemophilia, muscular dystrophy, Tay-Sachs disease, Down syndrome, and countless metabolic disorders is now well documented. It is estimated that over ten million children or adults in the United States suffer from some form of genetic affliction.

Recognition of the genetic basis of these disorders has provided direction for the development of detection and treatment. **Genetic counseling** provides parents with objective information upon which they can base rational decisions about parenting. It is estimated that every childbearing couple stands an approximately three percent risk of having a child with some form of genetic anomaly. There are currently thousands of documented genetic disorders.

Applied research in genetics has provided other medical benefits. Increased knowledge in **immunogenetics** has made possible compatible blood transfusions as well as organ transplants. The discovery of tissuebound antigens has led to the important concepts of **histocompatibility** and tissue typing. In

FIGURE 1.8
Triticale, a hybrid grain derived from wheat and rye, produced as a result of genetic research.

conjunction with immunosuppressive drugs, transplant operations involving human organs, including the heart, liver, and kidney, are increasing annually.

Recombinant DNA technology is also an important part of applied genetics. Cloned human genes that code for such medically important molecules as insulin and interferon serve as the source of mass production of many essential molecules. Recombinant DNA techniques will also play an increasing and essential role in **human genetic engineering,** which involves the direct manipulation of the genetic material. In the not-so-distant future, human genetic engineering will undoubtedly be used to alter the genetic constitution of individuals harboring genetic defects. While such a process presents ethical questions, it may correct serious genetic errors in members of our species.

In later chapters, these and other examples in agriculture and medicine will be discussed in greater detail. While other scientific disciplines are also expanding in knowledge, none have paralleled the growth of information that is occurring annually in genetics. By the end of this course, we are confident you will agree that the present truly represents the "Age of Genetics."

CHAPTER SUMMARY

1 Genetics, which emerged as a fundamental discipline of biology early in the twentieth century, has a rich history dating back to prehistoric times.
2 Numerous concepts and a basic vocabulary essential to the study of genetics have been presented.
3 Four investigative approaches are most often used in the study of genetics, including transmission genetic studies, cytogenetic analyses, molecular-biochemical experimentation, and inquiries into the genetic structure of populations.

4 Genetic research can be either basic or applied. Basic genetic research extends our knowledge of the discipline; the objective of applied genetics research is to solve specific problems affecting members of our species.

5 Genetic research has been critical to several agricultural successes that have increased world food production. In Russia during the 1940s, however, Lysenko's agricultural policies coupled with political factors greatly diminished crop production.

6 Recombinant DNA technology has greatly enhanced our research capability. It also has had a profound impact in elucidating the genetic basis of inherited diseases as well as in the mass production of medically important gene products.

PROBLEMS AND DISCUSSION QUESTIONS

1 Describe and contrast the ideas of Hippocrates and Aristotle related to the genetic basis of life.

2 Define pangenesis, epigenesis, and preformationism.

3 Which ideas and doctrines that preceded Darwin were central to his thinking?

4 Describe Darwin's and Wallace's theory of natural selection. What information was lacking from it, i.e., what gap remained in it?

5 Describe how DNA is organized into genes and how it encodes information leading to the synthesis of proteins.

6 Describe the four major investigative approaches used in studying genetics.

7 Contrast basic vs. applied research.

8 How did Lysenko's genetic views differ from those held by plant breeders in the United States in 1940?

9 Norman Borlaug received the Nobel Peace Prize for his work in genetics. Why do you think he was awarded this prize?

10 How has genetic research been applied to the field of medicine?

SELECTED READINGS

ANDERSON, W. F., and DIRCUMAKOS, E. G. 1981. Genetic engineering in mammalian cells. *Scient. Amer.* (July) 245: 106–21.

BORLAUG, N. E. 1983. Contributions of conventional plant breeding to food production. *Science* 219: 689–93.

BOWLER, P. J. 1989. *The Mendelian revolution: The emergence of hereditarian concepts in modern science and society.* London: Athlone.

COCKING, E. C., DAVEY, M. R., PENTAL, D., and POWER, J. B. 1981. Aspects of plant genetic manipulation. *Nature* 293: 265–70.

DAY, P. R. 1977. Plant genetics: Increasing crop yield. *Science* 197: 1334–39.

DUNN, L. C. 1965. *A short history of genetics.* New York: McGraw-Hill.

GARDNER, E. J. 1972. *History of biology.* 3rd ed. New York: Macmillan.

GARFIELD, E. 1981. Medical genetics: The new preventive medicine. *Current Contents—Life Sciences*, vol. 24, no. 36, pp. 5–20.

HOPWOOD, D. A. 1981. The genetic programming of industrial microorganisms. *Scient. Amer.* (Sept.) 245: 91–102.

HUTTON, R. 1978. *Bio-revolution: DNA and the ethics of man-made life.* New York: New American Library.

JORAUSKY, D. 1970. *The Lysenko affair.* Cambridge, Mass.: Harvard University Press.

KING, R. C., and STANSFIELD, W. D. 1990. *A dictionary of genetics.* 4th ed. New York: Oxford University Press.

LERNER, I. M., and LIBBY, W. J. 1976. *Heredity, evolution, and society.* 2nd ed. New York: W. H. Freeman.

MEDVEDEV, Z. A. 1969. *The rise and fall of T. D. Lysenko.* New York: Columbia University Press.

MOORE, J. A. 1985. Science as a way of knowing—Genetics. *Amer. Zool.* 25: 1–165.

OLBY, R. C. 1966. *Origins of Mendelism.* London: Constable.

POPOVSKY, M. 1984. *The Vavilov affair.* Hamden, Conn.: Archon.

SILBERNER, J. 1982. Superchicken. *Science Digest* 90: 30–32.

SOYFER, V. N. 1989. New light on the Lysenko era. *Nature* 339: 415–20.

STUBBE, H. 1972. *History of genetics: From prehistoric times to the rediscovery of Mendel.* (Translated by T. R. W. Waters.) Cambridge, MA: MIT Press.

TORREY, J. G. 1985. The development of plant biotechnology. *Amer. Scient.* 73: 354–63.

WEINBERG, R. A. 1985. The molecules of life. *Scient. Amer.* (Oct.) 253: 48–57.

PART ONE

Heredity and Phenotype

2

Mitosis and Meiosis

CHAPTER CONCEPTS

Genetic continuity between cells and organisms of any sexually reproducing species is maintained by the processes of mitosis and meiosis. The processes are orderly and efficient, serving to produce diploid somatic cells and haploid gametes, respectively.

In every living thing there exists a substance referred to as the **genetic material**. Except in certain viruses, this material is composed of the nucleic acid, DNA. A molecule of DNA is organized into units called **genes** that direct all metabolic activities of cells. In this chapter, we consider the topic of genetic continuity between cells and organisms. The manner in which the genetic material is transmitted from one generation of cells to the next, and from organisms to their descendants, is exceedingly precise.

Two major processes are involved in the transmission of the genetic material in eukaryotes: **mitosis** and **meiosis**. While the mechanisms of the two processes are similar in many ways, the outcomes are quite different. Mitosis leads to the production of two cells with an identical amount and type of genetic information. Meiosis, on the other hand, reduces the genetic content by precisely half. This reduction is essential if sexual reproduction is to occur without doubling the amount of genetic material at each generation. Strictly speaking, mitosis is that portion of cell division in which the hereditary components are precisely and equally divided into daughter cells. Meiosis is part of a special type of cell division leading to the production of sex cells: gametes and spores. This process is an essential step leading to the transmission of genetic information from an organism to its offspring.

CELL STRUCTURE

Before describing mitosis and meiosis, we will briefly review the structure of cells. As we shall see, many components, such as the nucleolus, ribosome, and centriole, are involved directly or indirectly with the genetic material as it functions. Other components, including the mitochondria and chloroplasts, contain their own unique genetic information. It is also useful for us to compare the structural differences of the prokaryotic bacterial cell with the animal and plant eukaryotic cell. Variation in cell structure is dependent on the specific genetic expression by each cell type.

Before 1940, knowledge of cell structure was based on information obtained with the **light microscope**. Around 1940 the **transmission electron microscope** was still in its early stages of development, and by 1950 many details of cell ultrastructure had been unveiled. Under the electron microscope cells were seen as highly organized, precise structures. A new world of whorling membranes, miniature organelles, microtubules, granules, and filaments was revealed. These discoveries revolutionized thinking in the entire field of biology. We will be concerned with those aspects of cell structure relating to genetic study. As the parts of the cell are described, refer to Figure 2.1.

The entire cell is surrounded by a **plasma membrane,** an outer covering that defines the cell boundary and protects the cell from its immediate environment. This membrane is not lifeless and passive, but rather it actively controls the movement of materials such as gases, nutrients, and waste products into and out of the cell. In addition to this membrane, plant cells have an outer covering called the **cell wall**. This rigid structure is primarily composed of a polysaccharide called cellulose. Bacterial cells also have a cell wall, but its chemical composition is quite different from that of the plant cell wall. The major component is a complex macromolecule called a **peptidoglycan**. As its name suggests, the molecule consists of peptide and sugar units. Long polysaccharide chains are cross-linked with short peptides, which impart great strength and rigidity to the bacterial cell. Some

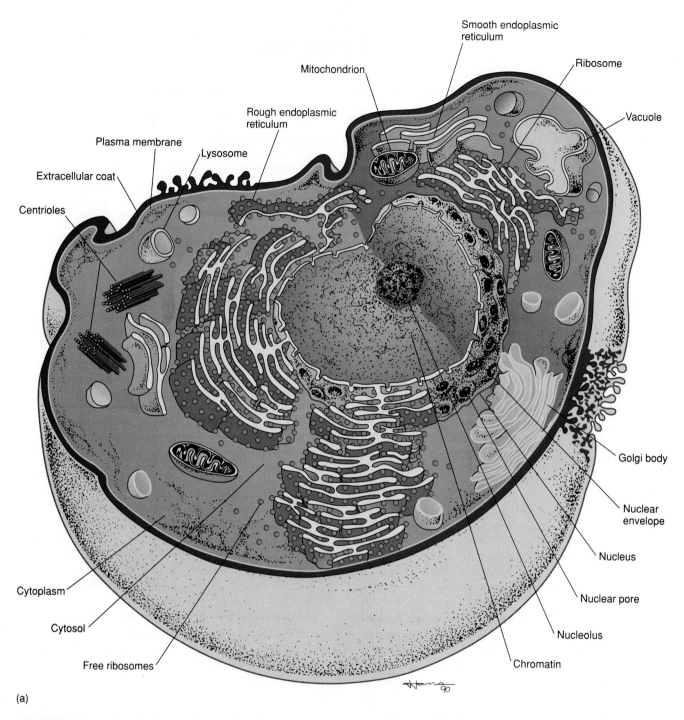

Smooth endoplasmic
reticulum

Ribosome

Mitochondrion

Vacuole

Rough endoplasmic
reticulum

Plasma membrane

Lysosome

Extracellular coat

Centrioles

Golgi body

Nuclear
envelope

Nucleus

Nuclear pore

Cytoplasm

Cytosol

Nucleolus

Free ribosomes

Chromatin

(a)

FIGURE 2.1
(a) Drawing of a generalized animal cell as seen under the electron microscope.
Emphasis has been placed on the cellular components discussed in the text. (b)
Drawing of a generalized plant cell. Note the presence of the cell wall, large vacuoles,
and chloroplasts, and the absence of centrioles and lysosomes in the plant cell
compared with the animal cell.

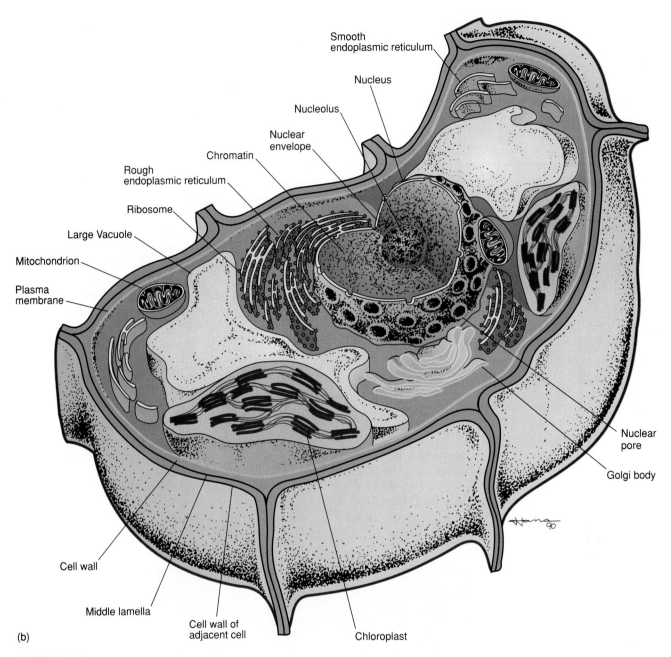

Smooth
endoplasmic reticulum

Nucleus

Nucleolus

Nuclear
envelope

Chromatin

Rough
endoplasmic reticulum

Ribosome

Large Vacuole

Mitochondrion

Plasma
membrane

Nuclear
pore

Golgi body

Cell wall

Middle lamella

Cell wall of
adjacent cell

Chloroplast

(b)

FIGURE 2.1
(continued)

bacterial cells have still another covering, a **capsule**. It is a mucus-like polysaccharide that protects certain bacteria from phagocytic activity by the host during their pathogenic invasion of eukaryotic organisms.

Many, if not most, animal cells have a covering over the plasma membrane called a **cell coat**. Consisting of glycoproteins and polysaccharides, the chemical composition of the cell coat differs from comparable structures in either plants or bacteria. In addition to protecting the cell membrane, the cell coat provides biochemical identity at the surface of cells. Among other forms of molecular recognition, various

antigenic determinants are part of the cell coat. For example, the **ABO and MN antigens** are found on the surface of red blood cells. In other cells, the **histocompatibility antigens,** which elicit an immune response during tissue and organ transplants, are part of the cell coat. All forms of biochemical identity at the cell surface are under genetic control, and many have been thoroughly investigated (see Chapter 4).

The presence of the **nucleus** and other membranous organelles characterizes eukaryotic cells and organisms. The nucleus houses the genetic material, DNA, which is found in association with large numbers of acidic and basic proteins. During nondivisional phases of the cell cycle this DNA/protein complex exists in an uncoiled, dispersed state called **chromatin**. As we will soon discuss, during mitosis and meiosis this material coils up and condenses into structures called **chromosomes**.

The lack of a nuclear envelope and membraneous organelles is characteristic of the kingdom Monera (eubacteria and cyanobacteria). In bacteria such as *E. coli* the genetic material is present as a long, circular DNA molecule that is compacted into a region referred to as the **nucleoid**. Part of the DNA may be attached to the cell membrane, but in general, the nucleoid constitutes a large area throughout the cell. While the DNA is compacted, it does not undergo the extensive coiling characteristic of the stages of mitosis where, in eukaryotes, chromosomes are visible. Nor is the DNA in these organisms associated as extensively with proteins as is eukaryotic DNA. Figure 2.2 shows two bacteria undergoing cell division and illustrates the bacterial chromosomes compacted into nucleoids.

An amorphous structure called the **nucleolus,** which is composed of RNA and protein, is also contained within the eukaryotic nucleus. This organelle is a processing center for the production of ribosomes. The nucleolus is generally associated with a specific chromosomal region called the **nucleolar organizer region (NOR)**. This region contains DNA that codes for ribosomal RNA (rRNA). Following its synthesis, rRNA is processed and combined with many specific proteins to form the mature ribosome.

Prokaryotic cells do not have a distinct nucleolus, but do contain genes that specify rRNA molecules.

The remainder of the cell enclosed by the plasma membrane, and excluding the nucleus, is composed of **cytoplasm**. Cytoplasm consists of a nonparticulate, colloidal material referred to as the **cytosol** that surrounds and encompasses the numerous types of cellular organelles. Beyond these components, an extensive system of tubules and filaments comprising the cytoskeleton provides a lattice of support structures within the cytoplasm. Consisting primarily of tubulin-derived microtubules and actin-derived microfilaments, this structural framework maintains cell shape, facilitates cell mobility, and anchors the various organelles. Tubulin and actin are both proteins found abundantly in eukaryotic cells.

One such organelle, the membranous **endoplasmic reticulum (ER),** compartmentalizes the cytoplasm, greatly increasing the surface area available for biochemical synthesis. ER may be smooth, in which case it serves as the site for synthesis of fatty acids and phospholipids. Or, the ER may be rough because it is studded with ribosomes. The amount of endoplasmic reticulum in any cell is correlated with the degree of synthetic activity exhibited by the cell. Ribosomes, which will be discussed in detail in Part 4, serve as nonspecific workbenches for the translation of genetic information contained in messenger RNA (mRNA) into proteins.

Three other cytoplasmic structures are very important in the eukaryotic cell's activities: **mitochondria, chloroplasts,** and **centrioles.** Mitochondria are found

FIGURE 2.2
Color-enhanced electron micrograph of *E. coli* undergoing cell division. Particularly prominent are the two chromosomal areas that have been partitioned into the daughter cells.

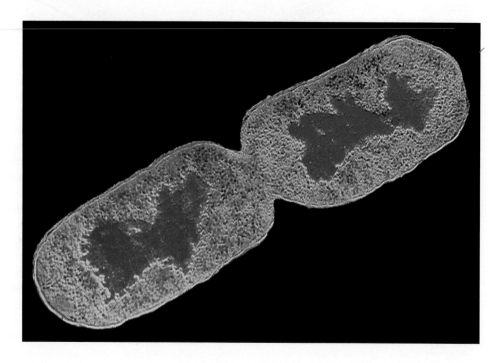

in both animal and plant cells and are the sites of the **oxidative phases of cell respiration**. These chemical reactions generate large amounts of ATP, an energy-rich molecule. Chloroplasts are found in plants and some protozoans. This organelle is associated with **photosynthesis,** the major energy-trapping process on earth. Both mitochondria and chloroplasts contain a type of DNA distinct from that found in the nucleus. Furthermore, these organelles can duplicate themselves and transcribe and translate their genetic information. It is interesting to note that the genetic machinery of mitochondria and chloroplasts closely resembles that of prokaryotic cells. This observation has led to speculation concerning the evolutionary origin of these organelles.

Animal and lower plant cells also contain a pair of complex structures called the centrioles. These cytoplasmic bodies, contained within a specialized region called the **centrosome,** are associated with the organization of those spindle fibers that function in mitosis and meiosis. In some organisms, the centriole is derived from another structure, the **basal body,** which is associated with the formation of cilia and flagella. Over the years, there have been many reports which have suggested that centrioles and basal bodies contain DNA that is involved in the replication of these structures. Recently, very convincing evidence has been presented that this is most certainly the case in the single-celled green alga, *Chlamydomonas*. Several genes are contained within a circular DNA structure associated with the basal body and centriole.

The organization of **spindle fibers** by the centrioles occurs during the early phases of mitosis and meiosis. Composed of arrays of microtubules, these fibers play an important role in the movement of chromosomes as they separate during cell division. The microtubules consist of polymers of alpha and beta subunits of the protein tubulin. The interaction of the chromosomes and spindle fibers will be considered later in this chapter.

HOMOLOGOUS CHROMOSOMES, HAPLOIDY, AND DIPLOIDY

To describe mitosis and meiosis, we must employ the concept of **homologous chromosomes**. An understanding of this concept will also be of critical importance in the following chapters on transmission genetics.

Chromosomes are most easily visualized during mitosis; when they are examined carefully, they are seen to take on distinctive lengths and shapes. Each contains a condensed or constricted region called the **centromere,** which establishes the general appearance of each chromosome. Figure 2.3 illustrates chromosomes with centromere placements at different points along their lengths. Extending from either side of the centromere are the arms of the chromosome. Depending on the position of the centromere, different arm ratios are produced. As Figure 2.3 illustrates, chromosomes are classified as **metacentric, submetacentric, acrocentric,** or **telocentric** on the basis of the centromere location.

When studying mitosis, we may make several other important observations. First, the number of chromosomes is species-specific. Generally, each somatic cell within members of the same species contains an identical number of chromosomes. This is called the **diploid number (2n)**. When the lengths and centromere placements of all such chromosomes are examined, a second general feature is apparent. Nearly all of the chromosomes exist in pairs with regard to these two criteria. The members of each pair are called **homologous chromosomes**. For each chromosome exhibiting a specific length and centromere placement, another exists with identical features.

Figure 2.4 illustrates the identical physical appearance of members of homologous chromosome pairs. There, the mitotic chromosomes of a human female have been photographed, cut out of the print, and matched up, creating a **karyotype**. As you can see, human chromosomes have a $2n$ value of 46 and exhibit a diversity of sizes and centromere placements.

Prior to being photographed, the chromosome preparation was treated with the enzyme trypsin and then stained. This has caused the banded appearance, the significance of which will be discussed later in the text. Note also that each of the 46 chromosomes is actually a double structure. Had these chromosomes been allowed to continue dividing, the two parts of each chromosome—called *chromatids*—would have separated into the two new cells as division continued.

Collectively, the genes contained on one member of each homologous pair of chromosomes constitute the **haploid genome** of the species. The **haploid number (n)** of chromosomes is one half of the diploid number. Table 2.1 demonstrates the wide range of n values found in a variety of plants and animals.

Homologous pairs of chromosomes have important genetic similarity. They contain identical gene sites, or **loci,** along their lengths. Thus, they have identical genetic potential. In sexually reproducing organisms, one member of each pair is derived from the maternal parent (through the ovum) and one from the paternal parent (through the sperm). Thus, each diploid organism contains two copies of each

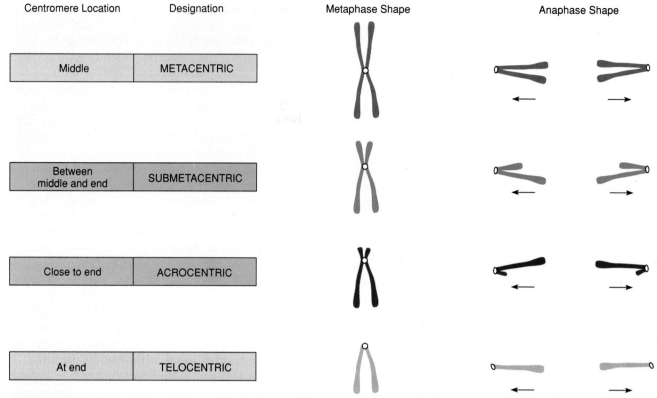

Centromere Location	Designation	Metaphase Shape	Anaphase Shape
Middle	METACENTRIC		
Between middle and end	SUBMETACENTRIC		
Close to end	ACROCENTRIC		
At end	TELOCENTRIC		

FIGURE 2.3
Centromere locations and descriptions of chromosomes based on that location. Note that the shape of the chromosome during anaphase is determined by the position of the centromere.

gene as a consequence of biparental inheritance. As will be seen in the following chapters on transmission genetics, the members of each pair of genes, while influencing the same characteristic or trait, need not be identical. Alternate forms of the same gene are called **alleles**. It is an important point to remember that in a population of members of the same species, many alleles of the same gene may exist.

The concepts of haploid number, diploid number, and homologous chromosomes may be related to the process of meiosis. During the formation of gametes or spores, meiosis converts the diploid number of chromosomes to the haploid number. As a result, haploid gametes or spores contain precisely one member of each homologous pair of chromosomes. Following fusion of two gametes in fertilization, the diploid number is reestablished, thus maintaining the constancy of genetic material from generation to generation.

There is one important exception to the concept of homologous pairs of chromosomes. In many species, one pair, the **sex-determining chromosomes,** may not be homologous in size, centromere placement, arm ratio, or genetic potential. For example, in humans, males contain the Y chromosome in addi-

tion to one X chromosome, while females carry two homologous X chromosomes. The X and Y chromosomes are not strictly homologous. The Y is considerably smaller and lacks most of the gene sites contained on the X. Nevertheless, in meiosis they behave as homologues so that gametes produced by males receive either the X or Y chromosome.

MITOSIS AND CELL DIVISION

The process of mitosis is critical to all eukaryotic organisms. In single-celled organisms such as protozoans, fungi, and algae, mitosis is a part of cell division, which provides the basis of asexual reproduction. Multicellular diploid organisms begin life as single-celled fertilized eggs or **zygotes.** The mitotic activity of the zygote and the subsequent daughter cells is the foundation for development and growth of the organism. In adult organisms, mitotic activity is prominent in wound healing and other forms of cell replacement in certain tissues. For example, the epidermal skin cells of humans are continuously being sloughed off and replaced. Cell division also results in a continuous production of reticulocytes,

FIGURE 2.4

Mitotic chromosomes constituting the standard human female karyotype. Chromosomes are present in homologous pairs. Each member of a pair can be seen to be a double structure.

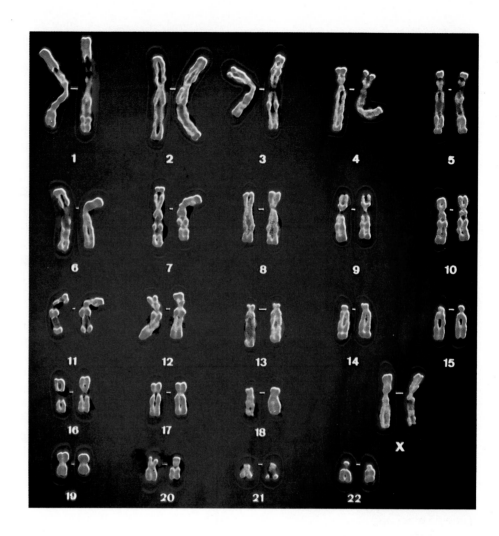

TABLE 2.1

The haploid number of chromosomes for representative organisms.

Common Name	Genus-Species	Haploid No.	Common Name	Genus-Species	Haploid No.
Black bread mold	*Aspergillus nidulans*	8	House fly	*Musca domestica*	6
Broad bean	*Vicia faba*	6	House mouse	*Mus musculus*	20
Cat	*Felis domesticus*	19	Human	*Homo sapiens*	23
Cattle	*Bos taurus*	30	Jimson weed	*Datura stramonium*	12
Chicken	*Gallus domesticus*	39	Mosquito	*Culex pipiens*	3
Chimpanzee	*Pan troglodytes*	24	Pink bread mold	*Neurospora crassa*	7
Corn	*Zea mays*	10	Potato	*Solanum tuberosum*	24
Cotton	*Gossypium hirsutum*	26	Rhesus monkey	*Macaca mulatta*	21
Dog	*Canis familiaris*	39	Roundworm	*Caenorhabditis elegans*	6
Evening primrose	*Oenothera biennis*	7	Silkworm	*Bombyx mori*	28
Frog	*Rana pipiens*	13	Slime mold	*Dictyostelium discoidium*	7
Fruit fly	*Drosophila melanogaster*	4	Snapdragon	*Antirrhinum majus*	8
Garden onion	*Allium cepa*	8	Tobacco	*Nicotiana tabacum*	24
Garden pea	*Pisum sativum*	7	Tomato	*Lycopersicon esculentum*	12
Grasshopper	*Melanoplus differentialis*	12	Water fly	*Nymphaea alba*	80
Green alga	*Chlamydomonas reinhardi*	18	Wheat	*Triticum aestivum*	21
Horse	*Equus caballus*	32	Yeast	*Saccharomyces cerevisiae*	17

which eventually shed their nuclei and replenish the supply of red blood cells in vertebrates. In abnormal situations, adult somatic cells may exhibit uncontrolled cell divisions, resulting in cancer.

It is generally observed that following cell division, the initial size of each new daughter cell is approximately one-half the size of its parent. However, the nucleus of each new cell is not appreciably smaller than the nucleus of the original cell. Quantitative measurements of DNA confirm that there are equivalent amounts of genetic material in the daughter nuclei.

The process of cytoplasmic division is called **cytokinesis**. The division of cytoplasm seems to require only a mechanism that results in a partitioning of the volume into two parts, followed by the enclosure of both new cells within a distinct plasma membrane. Cytoplasmic organelles either replicate themselves, arise from existing membrane structures, or are synthesized *de novo* (anew) in each cell. The subsequent proliferation of these structures is a reasonable and adequate mechanism for reconstituting the cytoplasm in daughter cells.

The division of the genetic material into daughter cells is more complex than cytokinesis and requires more precision. The chromosomes must first be exactly replicated and then accurately distributed into daughter cells. The end result is the production of two daughter cells, each with a chromosome composition identical to the parent cell.

Interphase and the Cell Cycle

Many cells undergo a continuous alternation between division and nondivision. The interval between each mitotic division is called **interphase** (Figure 2.5). It was once thought that the biochemical activity during interphase was devoted solely to the cell's growth and primary function, whatever it might be. However, we now know that another biochemical step critical to the next mitosis occurs during interphase: **the replication of the DNA of each chromosome**. Occurring midway through interphase, this period during which DNA is synthesized is called the **S phase**. The initiation and completion of synthesis can be detected using radioactive DNA precursors such as ^{3}H-thymidine. Their incorporation into DNA can be monitored using the technique of **autoradiography** (see Appendix A).

Investigations of this nature have demonstrated two periods during interphase, before and after S, when no DNA synthesis occurs. These are designated **G_1** (gap I) and **G_2** (gap II), respectively. During both of these periods, as well as during S, intensive metabolic activity, cell growth, and cell differentiation occur. By the end of G_2, the volume of the cell has roughly doubled, DNA has been replicated, and mitosis (M) is initiated. Continuously dividing cells then repeat this cycle (G_1, S, G_2, M) over and over. This concept of such a **cell cycle** is illustrated in Figure 2.5.

FIGURE 2.5
Diagrammatic representation of the stages that compose an arbitrary cell cycle. Following mitosis (M), cells initiate a new cycle (G_1). Cells may become nondividing (G_0) or continue through the restriction point (R) where they become committed to begin DNA synthesis (S) and complete the cycle (G_2 and M). Following mitosis, two daughter cells are produced.

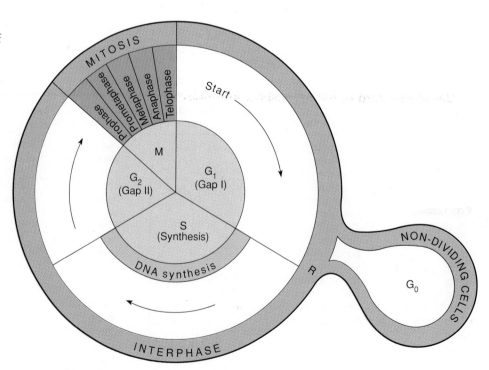

Much is known about the cell cycle based on *in vitro* (test tube) studies. When grown in culture, most cell types traverse the complete cycle in about 20 hours. The actual process of mitosis occupies only a small part of the cycle. The length of the S and G_2 stages of interphase are fairly consistent among different cell types. Most variation is seen in the length of time spent in the G_1 stage. Figure 2.6 illustrates the relative length of these periods in a typical cell.

G_1 is of great interest in the study of cell proliferation and its control. At a point late in G_1, all cells follow one of two paths. They either withdraw from the cycle and become quiescent in the so-called **G_0 stage,** or they become committed to initiate DNA synthesis and complete the cycle. The time when this decision is made has been referred to as the restriction or **R point**. Cells which enter G_0 remain viable and metabolically active but are nonproliferative. Cancer cells seem to lose the ability to achieve this status. Some cells enter G_0 and never reenter the cell cycle. Others can remain quiescent in G_0, but they may be stimulated to reenter the cycle.

Prophase

Once G_1, S, and G_2 are completed, mitosis is initiated. Mitosis is a dynamic period of vigorous and continual activity. For discussion purposes, the entire process is subdivided into discrete phases, and specific events are assigned to each stage. These stages, in order of occurrence, are **prophase, prometaphase, metaphase, anaphase,** and **telophase.** Each of these stages is depicted in a drawing in Figure 2.7A. Many are shown as they actually occur in Figure 2.7B.

Almost one-third of mitosis is spent in **prophase**; significant activities occur during this stage. One of the early events in animal cell and lower plant prophase involves the migration of two pairs of centrioles to opposite ends of the cell. These structures are found just outside the nuclear envelope in an area of differentiated cytoplasm called the **centrosome**. It is thought that each pair of centrioles consists of one mature unit and a smaller, newly formed centriole.

The direction of migration of the centrioles is such that two poles are established at opposite ends of the cell. This creates an axis along which chromosomal separation occurs. Following their migration, the centrioles are responsible for the organization of cytoplasmic microtubules into a series of **spindle fibers**. Even though the cells of plants, fungi, and certain algae seem to lack centrioles, spindle fibers

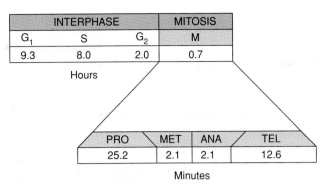

INTERPHASE			MITOSIS
G_1	S	G_2	M
9.3	8.0	2.0	0.7

Hours

PRO	MET	ANA	TEL
25.2	2.1	2.1	12.6

Minutes

FIGURE 2.6
The time spent in each phase of one complete cell cycle of a human cell in culture. Times vary according to cell types and conditions.

are nevertheless apparent during mitosis. If some other center that organizes microtubules into spindle fibers exists in these cells, it has yet to be discovered.

As the centrioles migrate, the nuclear envelope begins to break down and gradually disappears. In a similar fashion, the nucleolus disintegrates within the nucleus. While these events are taking place, the diffuse chromatin—the characteristic uncoiled form of the genetic material during interphase—condenses until distinct threadlike structures, or chromosomes, are visible. As condensation continues it becomes apparent near the end of prophase that each chromosome is a double structure split longitudinally except at a single point of constriction, the centromere.* The two parts of each chromosome are called **chromatids**. The DNA contained in each pair of chromatids constituting a chromosome was derived from a replication event during the S phase of the previous interphase. Thus both members of each chromosome are genetically identical and called **sister chromatids**. In humans, with a diploid number of 46, a cytological preparation of late prophase will reveal 46 such chromosomal structures (Figure 2.4). In the cell, these are found randomly distributed in the area formerly occupied by the nucleus.

By the completion of prophase, spindle fibers have been laid down between the centrioles, which are now at opposite ends of the cell. The nucleolus and nuclear envelope are no longer visible, and sister

---

*You may sometimes see the term *kinetochore* used synonymously with *centromere*. The kinetochore is actually a granule within the centromere that attaches to spindle fibers during mitosis. Two kinetochores form on opposite faces of the centromere, each attaching one or the other member of a pair of sister chromatids to the spindle fibers.

MITOSIS

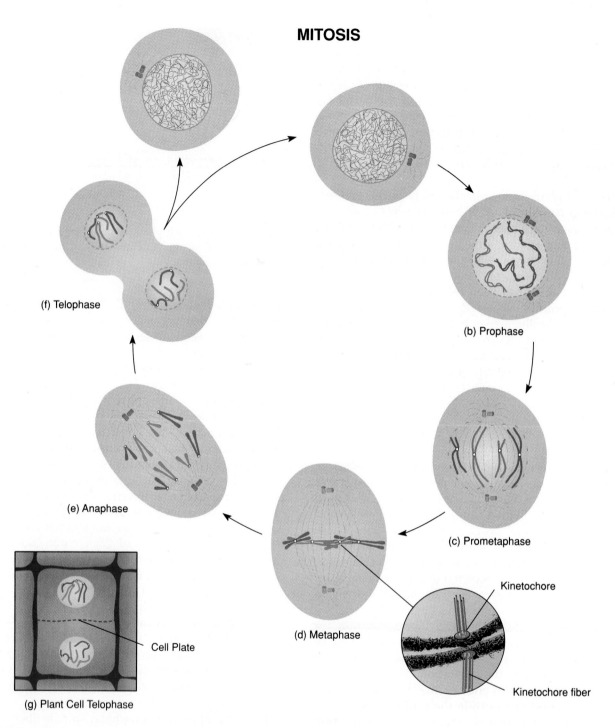

(f) Telophase

(b) Prophase

(e) Anaphase

(c) Prometaphase

(d) Metaphase

Kinetochore

Kinetochore fiber

Cell Plate

(g) Plant Cell Telophase

FIGURE 2.7A
Mitosis in an animal cell with a diploid number of 4. The events occurring in each
stage are described in the text. Of the two homologous pairs of chromosomes, one
contains longer, metacentric members and the other shorter, submetacentric members.
The maternal chromosomes and the paternal chromosomes are shown in different
colors. The insert (g), showing the telophase stage in a plant cell, illustrates the
formation of the cell plate and lack of centrioles.

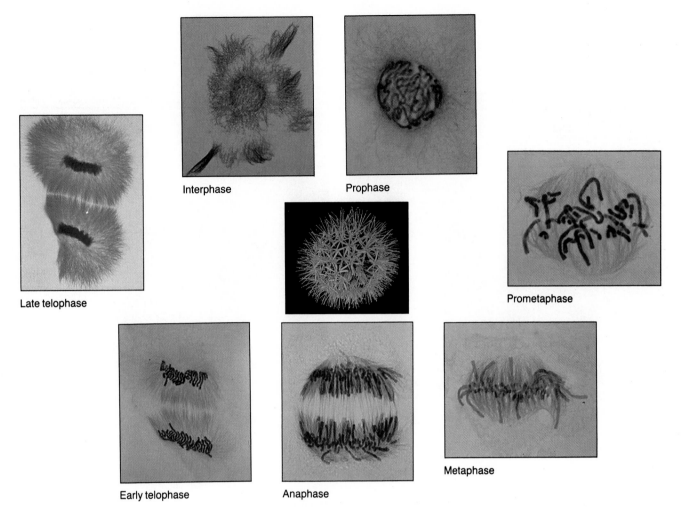

Interphase

Prophase

Prometaphase

Late telophase

Metaphase

Early telophase

Anaphase

FIGURE 2.7B
Light micrographs illustrating the stages of mitosis depicted in Figure 2.7A. These
stages were derived from the flower of *Haemanthus,* shown in the center of the figure.

chromatids are apparent. Parts (a) and (b) of Figure
2.7A illustrate interphase and prophase.

Prometaphase and Metaphase

The distinguishing event of the ensuing stages is the
migration of the centromeric region of each chromosome to the equatorial plane. In some descriptions
the term **metaphase** is applied strictly to the chromosome configuration following this movement. In such
descriptions, **prometaphase** refers to the period of
chromosome movement, as depicted in part (c) of
Figure 2.7A. The equatorial plane, also referred to as
the **metaphase plate,** is the midline region of the cell,
a plane that lies perpendicular to the axis established
by the spindle fibers.

Migration is made possible by the binding of one
or more spindle fibers to the kinetochore contained
within the centromere of each chromosome. There

are two major classes of spindle microtubules. Those
called kinetochore microtubules have one end near
the centrosome region and the other anchored in the
kinetochore. The nonkinetochore microtubules grow
from the centrioles of the centrosome, but have their
other ends free. These sometimes interdigitate with
one another, providing a framework to the spindle
and maintaining the separation of the two poles
during chromosome separation. At the completion of
metaphase, each centromere is now aligned at the
plate with the chromosome arms extending outward
in a random array. This configuration is shown in
part (d) of Figure 2.7A.

Anaphase

Events critical to chromosome distribution during
mitosis occur during its shortest stage, **anaphase**. It is
during this phase that sister chromatids of each

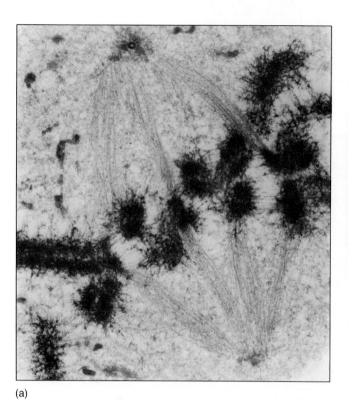

(a)

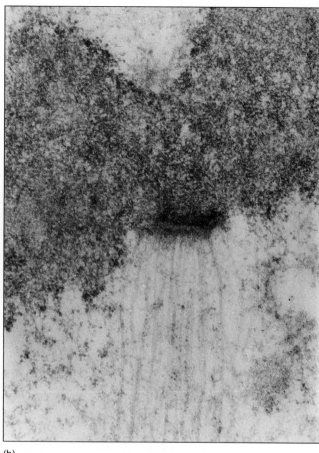

(b)

FIGURE 2.8
Electron micrographs revealing the attachment of spindle fibers to chromosomes during mitosis. In (a), spindle fibers emanating from both poles are apparent. In (b), spindle attachment to the kinetochore of the centromere can be seen. Photo (a), magnification = 12,000×. Photo (b), magnification = 55,000×.

double chromosomal structure separate from each other and migrate to opposite ends of the cell. In order for complete separation to occur, each centromeric region must be divided into two. Once this has occurred, each chromatid is now referred to as a daughter chromosome. Movement of the chromosomes to the opposite poles of the cell is dependent upon the centromere-spindle fiber attachment (Figure 2.8). While chromosome movement is dependent on a still undefined mechanism involving the spindle fibers, the centromeres appear to lead the way during migration, with the chromosome arms trailing behind. Depending on the location of the centromere along the chromosome, different shapes are assumed during separation (review Figure 2.3).

As the chromosomes begin their migration, the spindle elongates, extending the distance between the two centrosome regions. At the completion of anaphase, the chromosomes have migrated to the opposite poles of the cell. The steps occurring during anaphase are critical in providing each subsequent daughter cell with an identical set of chromosomes. In human cells there would now be 46 chromosomes at each end, one from each original sister pair. Part (e) of Figure 2.7A illustrates anaphase prior to its completion.

Telophase

Telophase is the final stage of mitosis. The most significant event is cytokinesis, the division or partitioning of the cytoplasm. Cytokinesis is essential if two new cells are to be produced from one. The mechanism differs greatly in plant and animal cells.

In plant cells, a **cell plate** is synthesized and laid down across the region of the metaphase plate. Animal cells, however, undergo a constriction of the cytoplasm, in much the same way a loop might be tightened around the middle of a balloon. The end result is the same: Two distinct cells are formed.

It is not surprising that the process of cytokinesis varies among cells of different organisms. Plant cells, which are more regularly shaped and structurally rigid, require a mechanism for the deposition of new cell wall material around the plasma membrane. The cell plate, laid down during telophase, becomes the **middle lamella**. Subsequently, the primary and secondary layers of the cell wall are deposited between the cell membrane and middle lamella on both sides of the boundary between the two daughter cells. In animals, complete constriction of the cell membrane produces the **cell furrow** characteristic of newly divided cells.

Other events necessary for the transition from mitosis to interphase are also initiated during late telophase. They represent a general reversal of those that occurred during prophase. In each new cell, the chromosomes begin to uncoil and become diffuse chromatin once again, while the nuclear envelope reforms around them. The nucleolus gradually reforms and is completely visible in the nucleus during early interphase. The spindle fibers also disappear. Telophase in animal and plant cells is illustrated in part (f) and part (g), respectively, of Figure 2.7A.

MEIOSIS AND SEXUAL REPRODUCTION

The process of meiosis, unlike mitosis, reduces the amount of genetic information. While mitosis produces daughter cells with a full diploid complement, meiosis produces gametes or spores with exactly half the number of chromosomes. During sexual reproduction, gametes then combine in fertilization to reconstitute the diploid complement found in parental cells. The process itself must be very specific, since it is insufficient to produce gametes with a random array of one-half of the total number of chromosomes. Instead, each gamete must receive one member of each homologous pair of chromosomes, ensuring genetic continuity from generation to generation.

The process of sexual reproduction also ensures genetic variety among members of a species. Each offspring receives one copy of every gene on every chromosome from one parent as well as a second complete set of copies from the other parent. For any given gene site on a chromosome, called a **locus,** alternate forms of that gene may exist. These alternate forms are called **alleles** and have arisen from genetic changes or mutations. Sexual reproduction, therefore, reshuffles the combinations of alleles, producing an offspring that is never identical to either parent. This process constitutes one form of genetic recombination within species. A second type of genetic recombination—**crossing over,** or genetic exchange between homologous chromosomes during meiosis—may also occur. Crossing over produces an even greater potential for genetic variability among individuals.

An Overview of Meiosis

We have already established what must be accomplished during meiosis. Before systematically considering the stages of this process, we will briefly describe how diploid cells are converted to haploid gametes or spores. Unlike mitosis, in which each paternally and maternally derived member of any given homologous pair of chromosomes behaves autonomously during division, in meiosis homologous chromosomes pair together, or **synapse**. Each synapsed structure, called a **bivalent,** gives rise to a unit, the **tetrad,** consisting of four chromatids; this demonstrates that both chromosomes have duplicated. In order to achieve haploidy, two divisions are necessary. In the first division, described as **reductional,** each tetrad separates and yields two **dyads,** each of which contains two sister chromatids joined at a common centromere. During the second division, described as **equational,** each dyad splits into two **monads** of one chromosome each. Thus, the two divisions may potentially produce four haploid cells.

The First Meiotic Division

During the first stage of meiosis, three significant events occur:

1 Homologous chromosomes pair up, or synapse, and take on the tetrad configuration, demonstrating that each chromosome has replicated.
2 The chromatid arms within a tetrad may overlap at one or more places, forming a single **chiasma** (or several **chiasmata**). Each of these crosslike configurations is thought to result from genetic exchange or crossing over between homologues.
3 Following crossing over, each tetrad divides, yielding two dyads.

We will now examine the actual stages leading to these events. As you read the following descriptions, you should identify each stage in Figure 2.9.

MEIOSIS I

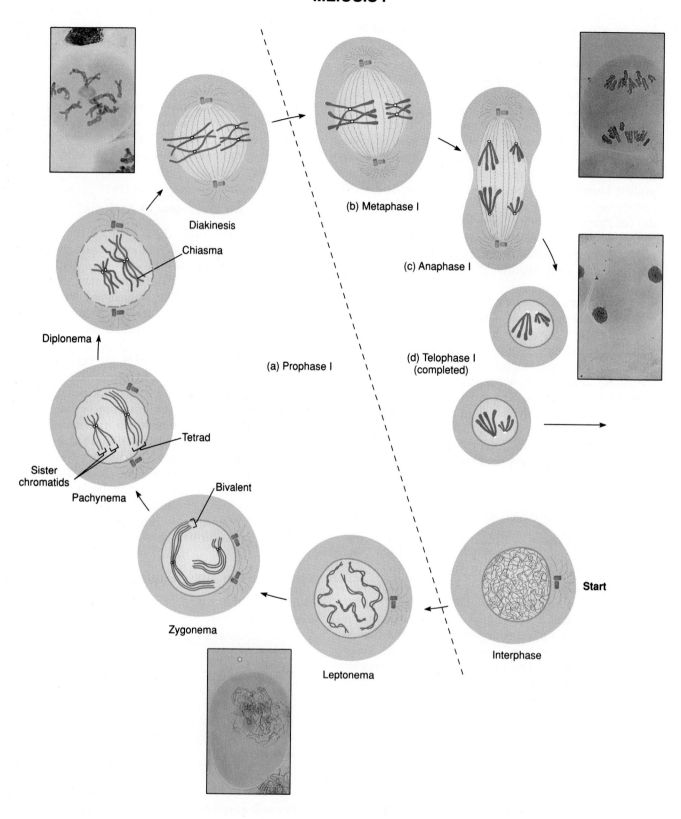

FIGURE 2.9

A diagrammatic representation of the major events occurring during meiosis in a male animal with a diploid number of 4. The same chromosomes described in Figure 2.7 are followed, as described in the legend there. Note that the combination of chromosomes contained in the cells produced following telophase II is dependent on the random

MEIOSIS II

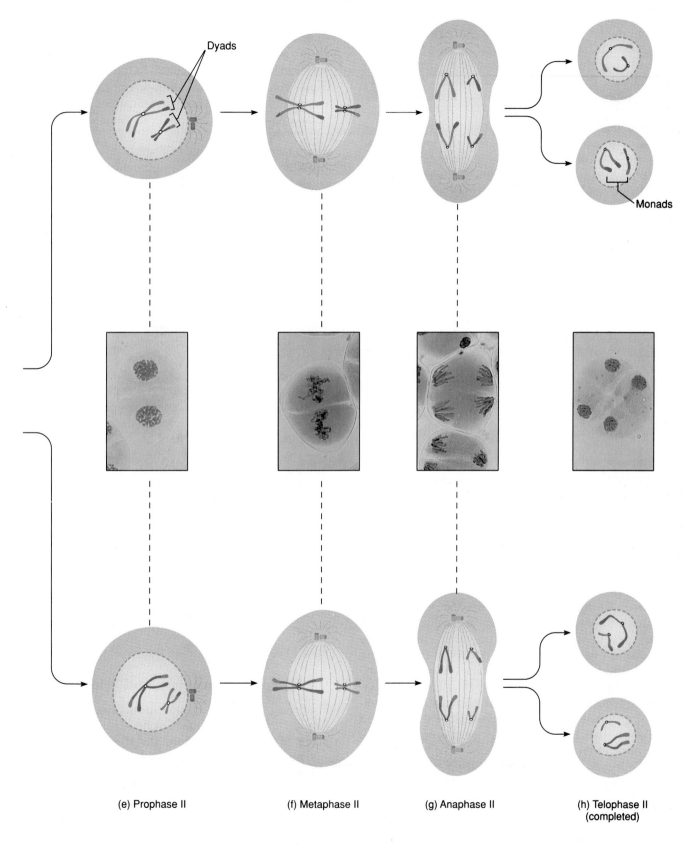

(e) Prophase II   (f) Metaphase II   (g) Anaphase II   (h) Telophase II (completed)

alignment of each tetrad and dyad on the equatorial plate during metaphase I and metaphase II. Several other combinations, which are not shown, can be formed. The events depicted in this figure should be correlated with the description in the text. Light micrographs illustrate many of the stages of meiosis depicted in the diagrams.

Prophase I

As in mitosis, DNA synthesis precedes meiosis, although the products of replication are not visible until the first prophase. The first meiotic prophase is a fairly lengthy one, and it has been further subdivided into five substages: leptonema,* zygonema,* pachynema,* diplonema,* and diakinesis.

Leptonema During the **leptotene stage,** the interphase chromatin material begins to condense, and the chromosomes, although still extended, become visible. Along each chromosome are **chromomeres,** localized condensations that resemble beads on a string.

Zygonema The chromosomes continue to shorten and thicken in the **zygotene stage.** Homologous chromosomes are attracted to each other and begin to undergo zipperlike pairing or synapsis. Pairing is accompanied by the appearance of a unique ultrastructural cell component, the **synaptonemal complex** (Figure 2.10 on page 35). This structure looks like a tripartite ribbon, and as prophase continues, it is found in association with the synapsed homologues. Thus, it is speculated that the function of the synaptonemal complex is associated with chromosome pairing. At the completion of zygonema, the number of threadlike chromosomes is reduced by one-half because of synapsis. The paired structures are sometimes referred to as **bivalents.** While each member of any pair has already replicated, this is not visually apparent until the following pachytene stage.

Pachynema In the **pachytene stage,** coiling and shortening of the synapsed chromosomes continue. While such configurations are not emphasized in Figure 2.9, cytological preparations reveal the chromosomes to overlap and be twisted around one another. During this stage, it first becomes evident that each chromosome is really a double structure, thus providing visual evidence that chromosome replication has already occurred. As in mitosis, the two members of each chromosome are connected by a common centromere and are called **sister chromatids.** Thus, each bivalent now contains four members, (two pairs of sister chromatids) and is referred to as a **tetrad.** The number of tetrads is equivalent to the haploid number of the species.

While not visually evident during pachynema, a major meiotic event occurs that is of great significance in genetics: **crossing over.** Crossing over is the reciprocal exchange of parts of chromosome arms between nonsister chromatids of the tetrad. Within each tetrad, one or more such events occur. The points of exchange become apparent as crosslike structures called **chiasmata** in the next stage, diplonema. Biochemical studies confirm that a small amount of DNA synthesis accompanies the chromosome exchanges during the pachytene stage.

Diplonema During visualization of the ensuing **diplotene stage** it is readily apparent that each tetrad consists of two pairs of sister chromatids. Within each tetrad, each pair of sister chromatids begins to separate. However, one or more areas remain in contact where chromatids are intertwined. Each such area, called a **chiasma,** is thought to represent a point where nonsister chromatids have undergone crossing over. Although the physical exchange between chromosome areas occurs during the previous pachytene stage, the evidence of crossing over is visible only when the duplicated chromosomes begin to separate.

Crossing over is an important source of genetic variability gained during gamete formation. New combinations of genetic material are formed during this process, mixing alleles from both members of a homologous chromosome pair.

Diakinesis Further shortening and condensation of the chromatids take place during the final stage of the first meiotic prophase, **diakinesis.** The chromosomes pull farther apart, but sister chromatids on either side of the chiasmata remain loosely associated. As this separation proceeds, the chiasmata move laterally toward the ends of the tetrad. This process, called **terminalization,** begins in late diplonema, and is completed during diakinesis. During this final period of prophase I, the nucleolus and nuclear envelope break down, and the two centromeres of each tetrad become attached to the recently formed spindle fibers.

Metaphase, Anaphase, and Telophase I

Following the first meiotic prophase stage, steps similar to those of mitosis occur. In the **metaphase stage of the first division,** the chromosomes have maximally shortened and thickened. The terminal chiasmata of each tetrad are visible and appear to be the only factor holding the nonsister chromosomes together. Each tetrad interacts with spindle fibers, facilitating movement to the metaphase plate.

---

*These are the noun forms of these stages. The adjective forms (leptotene, zygotene, pachytene, and diplotene) are also used in the text of this chapter.

During the first division, a single centromere holds each pair of sister chromatids together. It does *not* divide. At the **first anaphase,** therefore, one-half of each tetrad (one pair of sister chromatids) is pulled toward each pole of the dividing cell. The products of the separation, or **disjunction,** of these homologous chromosomes are two **dyads.** At the completion of anaphase I, there is a series of dyads equal to the haploid number present at each pole.

If no crossing over had occurred in the first meiotic prophase, each dyad at each pole would consist of either paternal or maternal chromatids. However, the exchanges produced by crossing over create mosaic chromatids of paternal and maternal origin. The alignment of each tetrad prior to this first anaphase stage is random. One-half of each tetrad will be pulled to one or the other pole at random, and the other half will move to the opposite pole. This random **segregation** of dyads is the basis for the Mendelian principle of **independent assortment,** which we will discuss in Chapter 3. You may wish to return to this discussion when you study this principle.

In many organisms, **telophase of the first meiotic division** reveals a nuclear membrane forming around the dyads. Then, the nucleus enters into a short interphase period. In other cases, the cells go directly from the first anaphase into the second meiotic division. In any event, telophase is much shorter than the corresponding stage in mitosis.

The Second Meiotic Division

A second division of the sister chromatids is essential to achieve haploidy in the meiotic products. During **prophase II,** each dyad is composed of one pair of sister chromatids attached by a common centromere. During **metaphase II,** the centromeres are directed to the equatorial plate. Then, the centromere divides, and during **anaphase II** the sister chromatids of each dyad are pulled to opposite poles. Since the number of dyads is equal to the haploid number, **telophase II** reveals one member of each homologous chromosome pair present at each pole. Each chromosome is referred to as a **monad.** Not only has the haploid state been achieved, but, if crossing over has occurred, each monad is a combination of maternal and paternal genetic information. As a result, the offspring produced by any gamete will receive from it a mixture of genetic information originally present in his or her grandparents. Following cytokinesis in telophase II, potentially four haploid gametes may result from a single meiotic event.

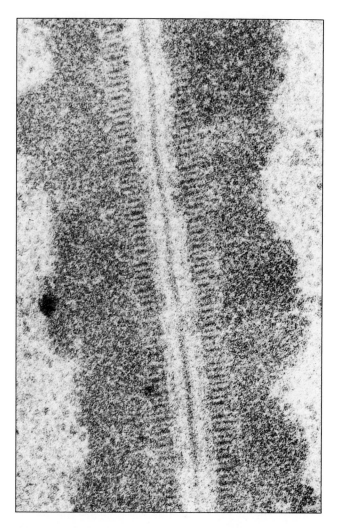

FIGURE 2.10
Electron micrograph of a portion of a synaptonemal complex derived from a pachytene bivalent of *Neotiella rutilans.*

SPERMATOGENESIS AND OOGENESIS

Although events that occur during the meiotic divisions are similar in all cells that participate in gametogenesis, there are certain differences between the production of a male gamete (spermatogenesis) and a female gamete (oogenesis) in most animal species.

Spermatogenesis takes place in the testes, the male reproductive organs. The process begins with the expanded growth of an undifferentiated diploid germ cell called a **spermatogonium.** This cell enlarges to become a **primary spermatocyte,** which undergoes the first meiotic division. The products of this division are called **secondary spermatocytes.** Each secondary spermatocyte contains a haploid number of dyads. The secondary spermatocytes then undergo the second meiotic division, and each of these cells produces two haploid **spermatids.** Spermatids go through a series of developmental changes, **spermiogenesis,** and become highly specialized, motile **spermatozoa** or **sperm.** All sperm cells produced during spermatogenesis receive equal amounts of genetic material and cytoplasm. Figure 2.11 summarizes these steps.

Spermatogenesis may be continuous or occur periodically in mature male animals, with its onset determined by the nature of the species' reproductive cycle. Animals that reproduce year-round produce sperm continuously, while those whose breeding period is confined to a particular season produce sperm only during that time.

In animal **oogenesis,** the formation of **ova** (singular: **ovum**), or eggs, occurs in the ovaries, the female reproductive organs. The daughter cells resulting from the two meiotic divisions receive equal amounts of genetic material, but they do *not* receive equal amounts of cytoplasm. Instead, during each division, almost all the cytoplasm of the **primary oocyte,** itself derived from the **oogonium,** is concentrated in one of the two daughter cells. The concentration of cytoplasm is necessary because the function of the mature ovum is to nourish the developing embryo following fertilization.

During the first meiotic anaphase in oogenesis, the tetrads of the primary oocyte separate, and the dyads move toward opposite poles. During the first telophase, the dyads present at one pole are pinched off with very little surrounding cytoplasm to form the **first polar body**. The other daughter cell produced by this first meiotic division contains most of the cytoplasm and is called the **secondary oocyte.** The first polar body may or may not divide again to produce two small haploid cells. The mature ovum will be produced from the secondary oocyte during the second meiotic division. During this division, the cytoplasm of the secondary oocyte again divides unequally, producing an **ootid** and a **second polar body.** The ootid then differentiates into the mature ovum. Figure 2.11 illustrates the steps leading to formation of the mature ovum and polar bodies.

Unlike the divisions of spermatogenesis, the two meiotic divisions of oogenesis may not be continuous. In some animal species the two divisions may directly follow each other. In others, including the human species, the first division of all oocytes begins in the embryonic ovary, but arrests in prophase I. Many years later, the first division is reinitiated in each oocyte upon its ovulation. The second division is completed only after fertilization, if it occurs.

THE SIGNIFICANCE OF MEIOSIS

The process of meiosis is critical to the successful sexual reproduction of all diploid organisms. It is the mechanism by which the diploid amount of genetic information is reduced to the haploid amount. In animals, meiosis leads to the formation of gametes, while in plants haploid spores are produced, which in turn lead to the formation of haploid gametes.

Each diploid organism contains its genetic information in the form of homologous pairs of chromosomes. Each pair consists of one member derived from the maternal parent and one from the paternal parent. Following meiosis, haploid cells contain a mixture of either the paternal or maternal representative of each homologous pair of chromosomes. Thus, meiosis results in a vast amount of genetic variation. Additionally, crossing over, which occurs in the first meiotic prophase stage, further reshuffles the genetic information. Crossing over occurs between the maternal and paternal members of each homologous pair, resulting in the production of even greater amounts of genetic variation in gametes.

Thus, the two most significant points about meiosis are that the process leads to:

1 the maintenance of a constant amount of genetic information between generations.
2 extensive genetic variation following the partitioning of the genetic material.

It is important that we also briefly elaborate on the important role that meiosis plays in the life cycle of fungi and plants. In many single-celled organisms in these groups, the predominant vegetative cells are haploid. They arise through meiosis and proliferate by cell division, including mitosis. In multicellular plants, the life cycle alternates between the diploid **sporophyte** stage and the haploid **gametophyte** stage. While one or the other predominates in different species, the processes of meiosis and fertilization bridge the sporophyte and gametophyte generations.

Finally, it is important to know what happens when meiosis fails to achieve the expected outcome. Rarely, at either the first or second division, separation or disjunction of the chromatids of a tetrad or dyad *does not* occur. Instead, both members move to the same pole during anaphase. Such an event is called **nondisjunction,** because the two members fail to disjoin. If nondisjunction occurs at the first division stage, it is said to be a primary event; at the second division stage, it is called a secondary event.

The results of primary and secondary nondisjunction for just one chromosome of a diploid genome are

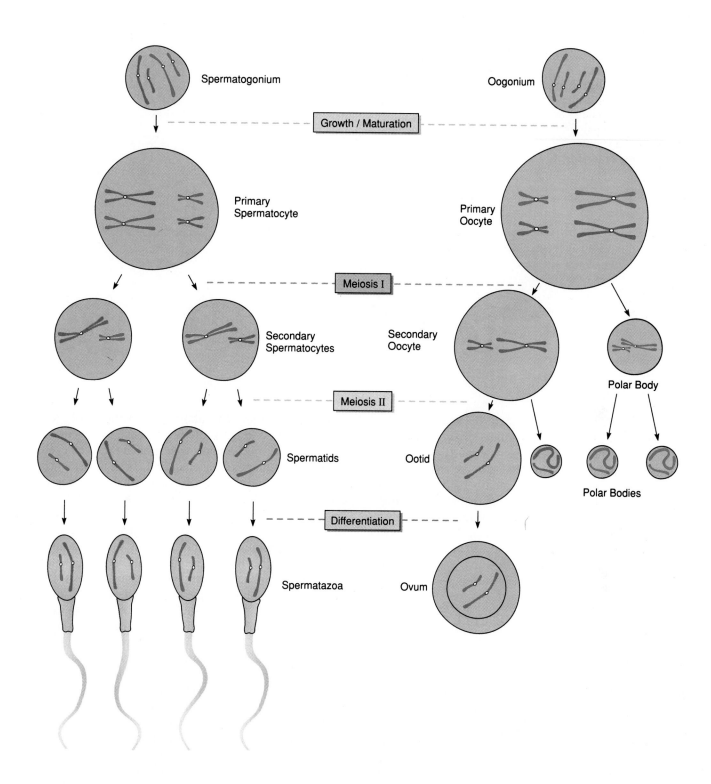

FIGURE 2.11
Oogenesis and spermatogenesis in animal cells.

FIGURE 2.12
Diagram illustrating
nondisjunction during the first
and second meiotic divisions.
In both cases, gametes are
formed either containing two
members of a specific
chromosome or lacking it
altogether.

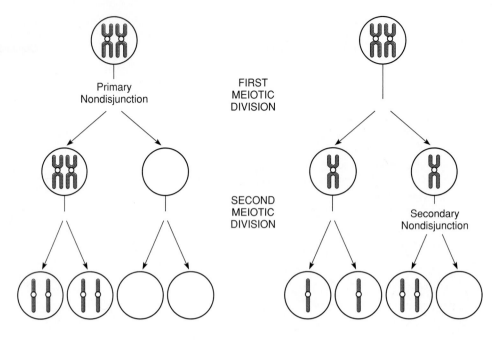

shown in Figure 2.12. As can be seen, for the chromosome in question, gametes may be formed containing either two members or none at all. Following fertilization with a normal gamete, a zygote is produced with either three members (**trisomy**) or only one member (**monosomy**) of this chromosome. While these conditions are more frequently tolerated in plants, they usually have severe or lethal effects in animals. Trisomy and monosomy will be described in greater detail in Chapter 7.

REGULATION OF THE CELL CYCLE

We conclude this chapter by returning to a discussion of the cell cycle in order to consider a most important question: "How is the cell cycle regulated?" Many cells alternate between mitosis and interphase, passing continuously from one cell cycle to another. For example, meristematic cells in plants and certain epithelial cells in animals exhibit this behavior. Other cells withdraw from the cell cycle, entering the G_0 stage. Some of these cells never divide again, while others can be induced to reenter the cycle. Different cell types, even in the same organism, vary widely in the lengths of their cell cycles. For example, animal cells exhibit *in vivo* cycles ranging from a few hours to several months. When examined carefully, it is apparent that most of this variation can be traced to the time spent in G_1. The total time necessary to complete S and G_2 is relatively constant in different cells.

Taken together, the above observations suggest that the cell cycle is tightly regulated. The discovery of the control mechanisms that regulate the cell cycle is significant for two reasons. First, such information extends our knowledge of the normal behavior of cells. Second, this information may be critical to our understanding of cancer, a disease process in which the regulation of the cell cycle is disrupted and cells divide continuously and more rapidly than is normal. We will briefly discuss several observations which suggest that the control mechanisms are at the molecular level and are genetically based.

One line of supporting evidence comes from studies of fusion of cells which are in different stages of the cell cycle. If a cell in any part of interphase (G_1, S, or G_2) is fused with one undergoing mitosis, the chromatin of the interphase nucleus begins to condense into chromosomes prematurely. If cells in G_1 and S are fused, the G_1 nucleus is prompted to initiate DNA synthesis prematurely.

These and other observations strongly suggest that some factors present in the cytoplasm influence the sequence of events in the cell cycle. One factor appears to promote uncoiled chromatin to condense into chromosomes. It is also probable that the cytoplasm contains molecular information that triggers DNA synthesis, thus initiating the S phase of the cycle. Once a cell enters the S phase and DNA replication is initiated, the normal sequence of events leading to mitosis will occur in the time sequence characteristic of that cell. Inhibitors of DNA synthesis will completely block the occurrence of these preliminary events and mitosis. These observations suggest that the most critical signal in the cell cycle is the one responsible for the initiation of DNA synthesis. Once synthesis is initiated, the cycle is self-perpetuating, leading to cell division.

A second type of investigation has sought to identify genes that affect the reproduction of cells and influence the cell cycle. One such approach is illustrated by the efforts of Lee Hartwell and others, leading to the isolation of mutations in yeast that affect the cell division cycle (*cdc* **mutations**). These investigations have involved the study of over 50 mutant genes that interfere with the ability to synthesize DNA, to complete nuclear division, or to undergo cytokinesis, among other defects. It is assumed that if a mutant gene causes the cell to arrest at different points in the yeast cell cycle, then the nonmutant or normal gene plays an important role in the cell cycle.

Another interesting group of genes that affect cell reproduction are the **oncogenes**. Originally discovered in RNA-containing viruses (retroviruses), oncogenes have the ability to transform normal eukaryotic cells into malignant cells. Surprisingly, the nearly identical counterparts of these genes have been found in the DNA of normal cells. It is believed that these cellular genes, sometimes called **proto-oncogenes,** play a variety of specific roles in normal cell proliferation and thus in the cell cycle. The viral versions have apparently been derived from the host cell and undergone various types of mutation, imparting them with transforming ability. We will have much more to say about oncogenes in Chapters 7 and 20.

The observations of *cdc* genes and proto-oncogenes suggest that the cell cycle is under strict genetic regulation. It appears that many genes thus far analyzed play a supporting role in the normal cell cycle. What causes such a cell to behave normally or to lose the ability to control proliferation remains an active area of research investigation.

CHAPTER SUMMARY

1 Mitosis and meiosis allow for the precise transmission of genetic information from cells or organisms to their descendants.

2 Mitosis is part of cell division, which is the basis of growth and replacement of nongametic tissues. In mitosis, each new daughter cell has the total hereditary complement replicated.

3 Meiosis results in the formation of haploid gametes or spores in the reproductive tissues of animals and plants. In diploid organisms, nuclei are produced that contain a haploid complement consisting of one member of each homologous pair of chromosomes.

4 Centrioles organize microtubules into spindle fibers, which together serve to separate chromosome pairs during mitosis and meiosis and facilitate their movement to the poles.

5 During cell division, cytoplasmic elements such as the plasma membrane, nucleolus, endoplasmic reticulum, ribosomes, mitochondria, and chloroplasts are equally proportioned into new daughter cells. It is essential that both the genetic components of the nucleus and the cytoplasmic elements are proportioned equally into the new daughter cells.

6 Knowledge of the nature of homologous chromosomes is essential in understanding inheritance. Each pair of homologous chromosomes shares the same size, centromere placement, and arm ratio. Each chromosome pair shares corresponding genetic information in the form of alleles at each gene site, or locus—one maternally derived, one paternally derived.

7 The cell cycle consists of the time needed for each cell to replicate its DNA, differentiate, grow, and divide. Interphase is the time during the cell cycle when DNA is replicated and cell products are made. Mitosis is the period of nuclear division during which chromosomes are partitioned into daughter cells.

8 Mitosis is subdivided into discrete phases: prophase, prometaphase, metaphase, anaphase, and telophase. Condensation of chromatin into chromosome structures occurs during prophase, and the nuclear membrane begins to disappear. During prometaphase, chromosomes take on the appearance of double structures, each represented by a pair of sister chromatids. In metaphase, chromosomes line up on the equatorial plane of the cell, and spindle fibers attach to their kinetochores. Spindle fibers function during anaphase, pulling the chromosome pairs apart and toward the poles. Telophase completes daughter cell formation, and is characterized by the division of the cytoplasm.

9 During cytokinesis in animal cells, the cytoplasm is pinched into two cells. In plants, a cell plate is synthesized and laid down, dividing the original cell into two daughter cells.

10 Meiosis provides the basis for sexual reproduction. Additionally, the process results in extensive genetic variation by virtue of the exchange between homologous chromosomes during crossing over.

11 A major difference exists between meiosis in males and females. Spermatogenesis partitions cytoplasmic volume equally and produces four haploid sperm cells. Oogenesis, on the other hand, accumulates the cytoplasm around one egg cell and reduces the other haploid sets of genetic material to polar bodies. The extra cytoplasm contributes to zygote development following fertilization.

12 Evidence is accumulating that the cell cycle is tightly regulated and under genetic control. Disruption of this control by mutations may cause the cycle to arrest or to proceed uncontrollably. Because of the significance of malignancy, this area is under intense investigation.

Insights and Solutions

With this initial appearance of "Insights and Solutions," it is appropriate to provide a brief description of its value to you as a student. This section will precede the "Problems and Discussion Questions" in each chapter. One or more examples will be provided, illustrating approaches that are useful in solving genetics problems. Our initial emphasis will be on insights that will help you arrive at correct solutions to ensuing problems.

1 In an organism with a haploid number of 3, how many individual chromosomal structures will align on the metaphase plate during (a) mitosis; (b) meiosis I; and (c) meiosis II? Describe each configuration.

ANSWER: (a) In mitosis, where homologous chromosomes do not synapse, there will be 6 double structures, each consisting of a pair of sister chromatids. The number of structures is equivalent to the diploid number.

(b) In meiosis I, the homologues have synapsed, reducing the number of structures to 3. Each is called a tetrad and consists of two pairs of sister chromatids.

(c) In meiosis II, the same number of structures exist (3), but in this case, they are called dyads. Each consists of a pair of sister chromatids. When crossing over has occurred, each chromatid may contain part of one of its nonsister chromatids obtained during exchange in prophase I.

2 For the chromosomes illustrated in Figure 2.9, draw all possible alignment configurations that may occur during metaphase of meiosis I. How many different configurations can occur with three pairs of chromosomes ($n = 3$)?

ANSWER: If $n = 3$, then 8 different configurations would be possible. (See illustration on page 41.) The formula 2^n, where n equals the haploid number, will allow you to calculate the number of potential alignment patterns. As we will see in the next chapter, these patterns serve as the physical basis of the Mendelian postulate of independent assortment.

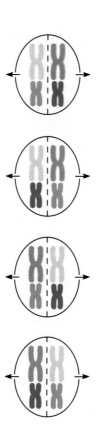

1   What role do the following cellular components play in the storage, expression, or transmission of genetic information: (a) chromatin, (b) nucleolus, (c) ribosome, (d) mitochondrion, (e) centriole, (f) centromere?

2   Discuss the concepts of homologous chromosomes, diploidy, and haploidy. What characteristics are shared between two chromosomes considered to be homologous?

3   If two chromosomes of a species are the same length and have similar centromere placements yet are *not* homologous, what *is* different about them?

4   Describe the events that characterize each stage of mitosis.

5   If an organism has a diploid number of 16, how many chromatids are visible at the end of mitotic prophase? How many chromosomes are moving to each pole during anaphase of mitosis?

6   How are chromosomes generally named on the basis of centromere placement?

7   Contrast telophase in plant and animal mitosis.

8   Outline and discuss the events of the cell cycle. What experimental technique was used to demonstrate the existence of the S phase?

9   Contrast the end results of meiosis with those of mitosis.

10 Define and discuss the following terms: (a) synapsis, (b) bivalents, (c) chiasmata, (d) crossing over, (e) chromomeres, (f) sister chromatids, (g) tetrads, (h) dyads, (i) monads, and (j) synaptonemal complex.

11 If an organism has a diploid number of 16 in an oocyte,
    (a)   How many tetrads are present in the first meiotic prophase?
    (b)   How many dyads are present in the second meiotic prophase?
    (c)   How many monads migrate to each pole during the second meiotic anaphase?

12 Contrast spermatogenesis and oogenesis. What is the significance of the formation of polar bodies?

13 Explain why meiosis leads to significant genetic variation while mitosis does not.

14 In Figure 2.9 the four spermatids formed as a result of meiosis each contain a unique combination of paternally and maternally derived chromosome segments (as represented by the different colors). These four different combinations of genetic information resulted from the random alignment of both sets of chromosomes on the equatorial plate during metaphase I and metaphase II. Had they become aligned differently, many other chromosome combinations might have been produced in the spermatids. With reference to Figure 2.9, how many other combinations are possible? Draw them.

15 During oogenesis in an animal species with a haploid number of 6, one dyad undergoes secondary nondisjunction. Following the second meiotic division, the involved dyad ends up intact in the ovum. How many chromosomes are present in: (a) the mature ovum, and (b) the second polar body? (c) Following fertilization by a normal sperm, what chromosome condition is created?

16 Assume that a diploid cell contains three pairs of homologous chromosomes designated C^1C^2, M^1M^2, and S^1S^2 and that no crossing over occurs. What possible configurations of chromosomes will be present in: (a) two daughter cells following mitosis?; (b) the first meiotic metaphase?; and (c) haploid cells following meiosis?

17 During the first meiotic prophase, (a) when does crossing over occur?; (b) when does synapsis occur?; (c) during which stage are the chromosomes least condensed?; and (d) when are chiasmata first visible?

18 What is the role of meiosis in the life cycle of a higher plant such as an angiosperm?

19 What is known about the regulation of the cell cycle?

20 Once DNA synthesis is initiated in the S phase of the cell cycle, the events leading to and including cell division continue automatically. What is the significance of this observation to investigations of cancer?

SELECTED READINGS

ALBERTS, B., et al. 1989. *Molecular biology of the cell.* 2nd ed. New York: Garland.

BAKER, B. A., et al. 1976. The genetic control of meiosis. *Ann. Rev. Genet.* 10: 53–134.

BASERGA, R., and KISIELESKI, W. 1963. Autobiographies of cells. *Scient. Amer.* (Aug.) 209: 103–110.

BRACHET, J., and MIRSKY, A. E. 1961. *The cell: Meiosis and mitosis.* Vol. 3. Orlando: Academic Press.

DUPRAW, E. 1968. *Cell and molecular biology.* Orlando: Academic Press.

FERRARI, S. and BASERGA, R. 1987. Oncogenes and cell cycle genes. *BioEssays* 7: 9–13.

HALL, J. L., RAMANIS, Z., and LUCK, D. J. 1989. Basal body/centriolar DNA: Molecular genetic studies in *Chlamydomonas. Cell* 59: 121–32.

HARTWELL, L. H. 1978. Cell division from a genetic perspective. *J. Cell Biol.* 77: 627–37.

HARTWELL, L. H., et al. 1974. Genetic control of the cell division cycle in yeast. *Science* 183: 46–51.

MAZIA, D. 1961. How cells divide. *Scient. Amer.* (Jan.) 205: 101–120.

———. 1974. The cell cycle. *Scient. Amer.* (Jan.) 235: 54–64.

McINTOSH, J. R., and McDONALD, K. L. 1989 The mitotic spindle. *Scient. Amer.* (Oct.) 261: 48–56.

MOENS, P. B. 1973. Mechanisms of chromosome synapsis at meiotic prophase. *Int. Rev. Cytol.* 35: 117–134.

PARDEE, A. B., et al. 1978. Animal cell cycle. *Ann. Rev. Biochem.* 47: 715–50.

PRESCOTT, D. M. and FLEXER, A. S. 1986. *Cancer, the misguided cell.* 2nd ed. Sunderland, MA: Sinauer.

———. 1976. *Reproduction of eukaryotic cells.* Orlando: Academic Press.

PRESCOTT, D. M., and GOLDSTEIN, L. 1967. Nuclear-cytoplasmic interaction in DNA synthesis. *Science* 155: 469–70.

RAO, P. N., and JOHNSON, R. T. 1970. Mammalian cell fusion: Studies on the regulation of DNA synthesis and mitochondria. *Nature* 225: 159–64.

————. 1974. Induction of chromosome condensation in interphase cells. *Adv. Cell Mol. Biol.* 32: 327–41.

SIMCHEN, G. 1978. Cell cycle mutants. *Ann. Rev. Genet.* 12: 161–191.

SWANSON, C. P.; MERZ, T.; and YOUNG, W. J. 1981. *Cytogenetics, the chromosome in division, inheritance, and evolution.* 2nd ed. Englewood Cliffs, N. J.: Prentice-Hall.

WANG, E. 1985. A 57000-mol-wt protein uniquely present in nonproliferating and senescent human fibroblasts. *J. Cell Biol.* 100: 545–51.

WESTERGAARD, M., and vonWETTSTEIN, D. 1972. The synaptinemal complex. *Ann. Rev. Genet.* 6: 71–110.

WHEATLEY, D. N. 1982. *The centriole: A central enigma of cell biology.* New York: Elsevier/North-Holland Biomedical.

WISSINGER, W. L., and WANG, R. J. 1983. Cell cycle mutants. *Int. Rev. Cytol.* 15 (supplement): 91–113.

YANISHEVSKY, R. M., and STEIN, G. M. 1981. Regulation of the cell cycle in eukaryotic cells. *Int. Rev. Cytol.* 69: 223–59.

YUNIS, J. J., and CHANDLER, M. E. 1979. Cytogenetics. In *Clinical diagnosis and management by laboratory methods*, ed. J. B. Henry, vol. 1. Philadelphia: W. B. Saunders.

ZIMMERMAN, A. M., and FORER, A., eds. 1981. *Mitosis/Cytokinesis.* New York: Academic Press.

3

Mendelian Genetics

∞∞∞∞∞∞∞∞∞∞∞∞∞∞∞∞∞∞∞∞∞∞∞∞∞∞∞∞∞∞∞∞∞∞

CHAPTER CONCEPTS

Inherited characteristics are the result of particulate factors
called genes that are transmitted from generation to generation
on vehicles called chromosomes, according to rules
first described by Gregor Mendel.

Even though inheritance has been recognized for thousands of years, the first significant insights into the mechanisms involved occurred only a little over a century ago. In 1866, Gregor Mendel published the results of a series of experiments that would lay the foundation for the formal discipline of genetics. In the ensuing years the concept of the gene as a distinct hereditary unit was established, and the ways in which genes are transmitted to offspring and control traits were clarified. Research in these areas was accelerated in the first half of the twentieth century, generating the interest so important to the acquisition of the knowledge derived in this field since about 1950. It is safe to say that studies in genetics, particularly at the molecular level, have remained continually at the forefront of biological research since that time.

In this chapter we will focus on the development of the principles established by Mendel, now referred to as **Mendelian genetics**. These principles describe the transmission of genes from parents to offspring and were derived directly from his experimentation.

When Mendel began his studies of inheritance using *Pisum sativum*, the garden pea, there was little or no knowledge of chromosomes and the role and mechanism of meiosis. Nevertheless, he was able to determine that distinct **units of inheritance** exist. He also predicted their behavior during the formation of gametes. Subsequent investigators, with access to cytological data, were able to relate their observations of chromosome behavior during meiosis to Mendel's principles of inheritance. Once this correlation had been made, Mendel's postulates were accepted as the basis for the study of transmission genetics. Even today, they serve as the cornerstone of the study of inheritance.

GREGOR MENDEL

In 1822, Gregor Johann Mendel was born of a peasant family in the village of Heinzendorf, now part of Czechoslovakia. An excellent student in high school, Mendel studied philosophy for several years afterward and was admitted to the Augustinian Monastery in Brno (previously called Brünn) in 1843, where he received support for his studies and research throughout the rest of his life. In 1849, he was relieved of pastoral duties and received a teaching appointment that lasted several years. From 1851 to 1853 he attended the University of Vienna, where he studied physics and botany. In 1854 he returned to Brno, where for the next sixteen years he taught physics and natural science.

In 1856 Mendel performed the first set of hybridization experiments with the garden pea. The research phase of his career lasted until 1868, when he was elected abbot of the monastery. While his interest in genetics remained, his new responsibilities demanded most of his time. In 1884 Mendel died of a kidney disorder. The local newspaper paid him the following tribute: "His death deprives the poor of a benefactor, and mankind at large of a man of the noblest character, one who was a warm friend, a promoter of the natural sciences, and an exemplary priest. . . ."

Mendel's Experimental Approach

In 1865, Mendel first reported the results of some simple genetic crosses between certain strains of the garden pea. Although, as we saw in Chapter 1, his was not the first attempt to provide experimental evidence pertaining to inheritance, Mendel's work is an elegant model of practical experimental design, insightful analysis, and data interpretation.

Mendel showed remarkable insight into the methodology necessary for good experimental biology. He chose an organism that was easy to grow and interbreed. The pea plant is self-fertilizing in nature, but is easily crossbred in designed experiments. It reproduces well and grows to maturity in a single season. Mendel worked with seven unit characters, visible features that were each represented by two contrasting forms or traits. For the character stem height, for example, he experimented with the traits *tall* and *dwarf*. He selected six other contrasting pairs of traits involving seed shape and color, pod shape and color, and pod and flower arrangement. True-breeding strains were available from local seed merchants. Each trait appeared generation after generation in self-fertilizing plants; that is, the strains exhibiting them "bred true."

Mendel's success in an area where others had failed may be attributed to several factors in addition to the choice of a suitable organism. He restricted his examination to one or very few pairs of contrasting traits in each experiment. He also kept accurate quantitative records, a necessity in genetic experiments. From the analysis of his data, Mendel derived certain postulates that have become the principles of transmission genetics.

The significance of Mendel's experiments was not realized until the early twentieth century, well after his death. Once Mendel's publications were rediscovered by geneticists investigating the function and behavior of chromosomes, the implications of his postulates were immediately apparent. He had discovered the basis for the transmission of hereditary traits!

THE MONOHYBRID CROSS

The simplest crosses performed by Mendel involved only one pair of contrasting traits. Each such breeding experiment is called a **monohybrid cross**. A monohybrid cross is made by mating individuals from two parent strains, each of which exhibits one of the two contrasting forms of the character under study. Initially we will examine the first generation of offspring of such a cross, and then we will consider the offspring of **selfing** or **self-fertilizing** individuals from this first generation. The original parents are called the P_1 or **parental generation,** their offspring are the F_1 or **first filial generation,** and the individuals resulting from the selfing of the F_1 generation are

called the F_2 or **second filial generation**. We can, of course, continue to follow the F_3, F_4, F_5, and subsequent generations, if desirable.

The cross between peas with tall stems and dwarf stems is representative of Mendel's monohybrid crosses. *Tall* and *dwarf* represent contrasting forms or traits of the character of stem height. Unless tall or dwarf plants are crossed together or with another strain, they will undergo self-fertilization and breed true, producing their respective trait generation after generation. However, when Mendel crossed tall plants with dwarf plants, the resulting F_1 generation consisted only of tall plants. When members of the F_1 generation were selfed, Mendel observed that 787 of 1064 F_2 plants were tall, while 277 of 1064 were dwarf. Note that in this cross (Figure 3.1) the dwarf trait disappears in the F_1, only to reappear in the F_2 generation.

Genetic data are usually expressed and analyzed as ratios. In this particular example, many identical P_1 crosses were made and many F_1 plants—all tall—were produced. When these F_1 offspring were self-fertilized, 787 F_2 progeny were tall and 277 were dwarf—a ratio of approximately 2.8:1.0, or about 3:1.

Mendel made similar crosses between pea plants exhibiting each of the other pairs of contrasting traits. The results of these crosses are also shown in Figure 3.1. In every case, the outcome was similar to the tall/dwarf cross just described. All F_1 offspring were identical to one of the parents. In the F_2, an approximate ratio of 3:1 was obtained. Three-fourths appeared like the F_1 plants, while one-fourth exhibited the contrasting trait, which had disappeared in the F_1 generation.

It is appropriate to point out one further aspect of the monohybrid crosses. In each, the F_1 and F_2 patterns of inheritance were similar regardless of which P_1 plant served as the source of pollen, or sperm, and which served as the source of the ovum, or egg. The crosses could be made either way—that is, pollen from the tall plant pollinating dwarf plants, or vice versa. These are called **reciprocal crosses**. Therefore, the results of Mendel's monohybrid crosses were not sex-dependent.

To explain these results, Mendel proposed the existence of particulate **unit factors** for each trait. He suggested that these factors serve as the basic units of heredity and are passed unchanged from generation to generation, determining various traits expressed by each individual plant. Using these general ideas, Mendel proceeded to hypothesize precisely how such factors could account for the results of the monohybrid crosses.

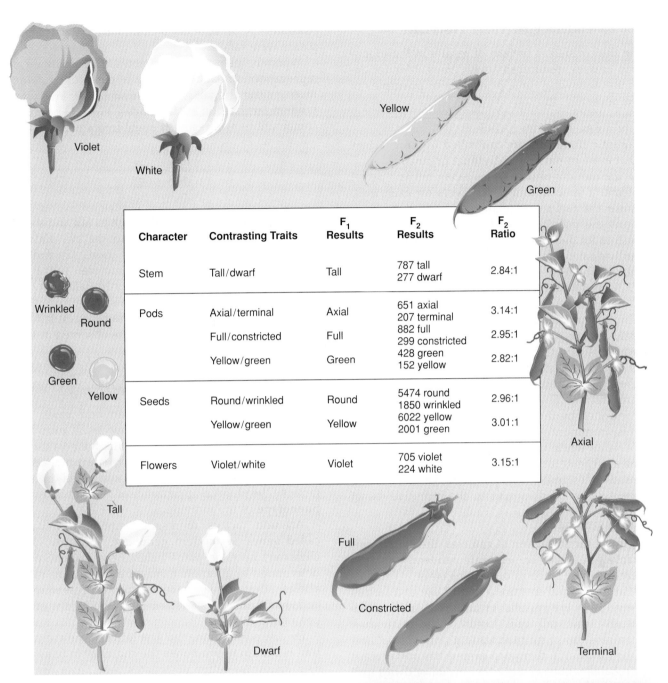

Character	Contrasting Traits	F_1 Results	F_2 Results	F_2 Ratio
Stem	Tall/dwarf	Tall	787 tall 277 dwarf	2.84:1
Pods	Axial/terminal	Axial	651 axial 207 terminal	3.14:1
	Full/constricted	Full	882 full 299 constricted	2.95:1
	Yellow/green	Green	428 green 152 yellow	2.82:1
Seeds	Round/wrinkled	Round	5474 round 1850 wrinkled	2.96:1
	Yellow/green	Yellow	6022 yellow 2001 green	3.01:1
Flowers	Violet/white	Violet	705 violet 224 white	3.15:1

FIGURE 3.1

A summary of the seven pairs of contrasting traits and the results of Mendel's seven monohybrid crosses. In each case, pollen derived from plants exhibiting one contrasting trait was used to fertilize the ova of plants exhibiting the other contrasting trait. In the F_1 generation one of the two traits, referred to as dominant, was exhibited by all plants. The contrasting trait, referred to as recessive, then reappeared in approximately one-fourth of the F_2 plants. The garden pea (*Pisum sativum*) is also shown.

Mendel's First Three Postulates

Using the consistent pattern of results in the monohybrid crosses, reinforced seven times, Mendel derived the following three postulates or principles of inheritance.

1 UNIT FACTORS IN PAIRS
   Genetic characters are controlled by unit factors that exist in pairs in individual organisms.
   In the monohybrid cross involving tall and dwarf stems, a specific unit factor exists for each trait. Since the factors occur in pairs, three combinations are possible: two factors for tallness, two factors for dwarfness, or one of each factor. Every individual contains one of these three combinations, which determines stem height.
2 DOMINANCE/RECESSIVENESS
   When two unlike unit factors responsible for a single character are present in a single individual, one unit factor is dominant to the other, which is said to be recessive.
   In each monohybrid cross, the trait expressed in the F_1 generation is controlled by the dominant unit factor. The trait that is not expressed is controlled by the recessive unit factor. Note that this dominance/recessiveness relationship only pertains when unlike unit factors are present in pairs. The terms *dominant* and *recessive* are also used to designate the traits. In this case, tall stems are said to be dominant to the recessive dwarf stems.
3 SEGREGATION
   During the formation of gametes, the paired unit factors separate or segregate randomly so that each gamete receives one or the other.
   If an individual contains a pair of like unit factors (e.g., both are specific for tall), then all gametes receive one tall unit factor. If an individual contains unlike unit factors (e.g., one for tall and one for dwarf), then each gamete has a 50 percent probability of receiving either the tall *or* dwarf unit factor.

These postulates provide a suitable explanation for the results of the monohybrid crosses. The tall/dwarf cross will be used to illustrate this explanation. Mendel reasoned that P_1 tall plants contained identical paired unit factors, as did the P_1 dwarf plants. The gametes of tall plants all received one tall unit factor as a result of segregation. Likewise, the gametes of dwarf plants all received one dwarf unit factor. Following fertilization, all F_1 plants received one unit factor from each parent, a tall factor from one and a dwarf factor from the other, reestablishing the paired relationship. Because tall is dominant to dwarf, all F_1 plants were tall.

When F_1 plants form gametes, the postulate of segregation demands that each gamete randomly receive *either* the tall *or* dwarf unit factor. Following random fertilization events during F_1 selfing, four F_2 combinations will result in equal frequency:

> (1) tall/tall
> (2) tall/dwarf
> (3) dwarf/tall
> (4) dwarf/dwarf

Combinations (1) and (4) will result in tall and dwarf plants, respectively. According to the postulate of dominance/recessiveness, combinations (2) and (3) will both yield tall plants. Therefore, the F_2 is predicted to consist of three-fourths tall and one-fourth dwarf, or a ratio of 3:1. This is approximately what Mendel observed in the cross between tall and dwarf plants. A similar pattern was observed in each of the other monohybrid crosses.

Modern Genetic Terminology

In order to illustrate the monohybrid cross and Mendel's first three postulates, we must introduce several new terms as well as a set of symbols for the unit factors. Traits such as tall or dwarf are visible expressions of the information contained in unit factors. The physical appearance of a trait is called the **phenotype** of the individual.

All unit factors represent units of inheritance called **genes** by modern geneticists. For any given character, such as plant height, the phenotype is determined by alternate forms of a single gene called **alleles**. For example, the unit factors representing tall and dwarf are alleles determining the height of the pea plant.

By one convention, the first letter of the recessive trait is chosen to symbolize the character in question. The lowercase letter designates that allele for the recessive trait, and the uppercase letter designates the allele for the dominant trait. Therefore, we let *d* stand for the dwarf allele and *D* represent the tall allele. When alleles are written in pairs to represent the two unit factors present in any individual (*DD*, *Dd*, or *dd*), these symbols are referred to as the **genotype**. This term reflects the genetic makeup of an individual whether it is haploid or diploid. By reading the genotype, it is possible to know the phenotype of the individual: *DD* and *Dd* are tall, while *dd* is dwarf. When both alleles are the same (*DD* or *dd*), the individual is said to be **homozygous** or a **homozygote;** when alleles are different (*Dd*), we use the term **heterozygous** or **heterozygote**. These symbols and terms are used in Figure 3.2 to illustrate the complete monohybrid cross. Given this information, we may explore the possible rationale used by Mendel to arrive at his postulates, keeping in mind that hindsight makes this task much easier.

What led Mendel to deduce unit factors in pairs? Since there were two contrasting traits for each character, it seemed logical that two distinct factors must exist. However, why does one of the two traits or phenotypes disappear in the F_1 generation? Observation of the F_2 generation helps to answer this question. The recessive trait and its unit factor do not actually disappear in the F_1; they are merely hidden or masked, only to reappear in one-fourth of the F_2 offspring. Therefore, Mendel concluded that one unit factor for tall and one for dwarf were transmitted to each F_1 individual; but because the tall factor or allele is dominant to the dwarf factor or allele, all F_1 plants are tall. Finally, how is the $3:1$ F_2 ratio explained? As shown in Figure 3.2, if the tall and dwarf alleles of the F_1 heterozygote segregate randomly into gametes, and if fertilization is random, this ratio is the natural outcome of the cross.

Because he operated without the hindsight that modern geneticists enjoy, Mendel's analytical reasoning must be considered a truly outstanding scientific achievement. On the basis of rather simple, but precisely executed breeding experiments, he proposed that discrete particulate units of heredity exist, and he explained how they are transmitted from one generation to the next!

While today we know these to be genes, of particular significance in Mendel's work was the evidence favoring these **particulate units of inheritance**. This concept was unique in 1866. The most

FIGURE 3.2
An explanation of the monohybrid cross between tall and dwarf pea plants. The symbols *D* and *d* are used to designate the tall and dwarf unit factors, respectively, in the genotypes of mature plants and gametes. All individuals are shown in rectangles. All gametes are shown in circles.

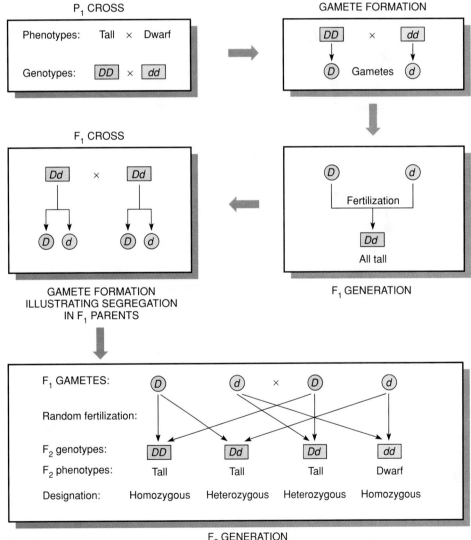

prevalent theory at that time suggested that offspring were a blend or mixture of their parents' traits. If this were true, then all of Mendel's F_1 plants would have exhibited an intermediate expression. In the tall/dwarf cross, the F_1 generation would all have been taller than the P_1 dwarf plants, but not so tall as the P_1 tall plants. The fact that Mendel's proposals for inheritance ran counter to the most widely accepted genetic theories has been cited as one reason why his work went "undiscovered" for over three decades.

Punnett Squares

The genotypes and phenotypes resulting from the recombination of gametes during fertilization can be easily visualized by constructing a **Punnett square,** so named after the person who first devised this approach, Reginald C. Punnett. Figure 3.3 illustrates

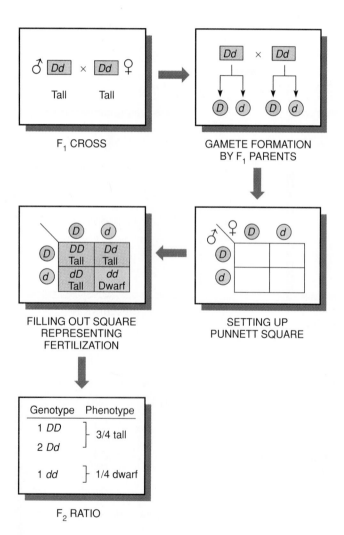

FILLING OUT SQUARE
REPRESENTING
FERTILIZATION

F$_2$ RATIO

FIGURE 3.3
The use of a Punnett square in generating the F_2 ratio of the $F_1 \times F_1$ cross shown in Figure 3.2.

this method of analysis for the $F_1 \times F_1$ monohybrid cross. All possible gametes are assigned to a column or a row, with the vertical column representing those of the male parent and the horizontal row those of the female parent. After entering the gametes in rows and columns, the new generation is predicted by combining the male and female gametic information for each combination and entering the resulting genotypes in the boxes. This process represents all possible random fertilization events. The genotypes and phenotypes of all potential offspring are ascertained by reading the entries in the boxes. The Punnett square method is particularly useful when one is first learning about genetics and how to solve problems.

The Test Cross: One Character

Tall plants produced in the F_2 generation are predicted to be of either the *DD* or *Dd* genotypes. We might ask whether there is a way to distinguish the genotype. Mendel devised a rather simple method that is still used today in breeding procedures of plants and animals: the **test cross**. The organism of the dominant phenotype but unknown genotype is crossed to a homozygous recessive individual. For example, if a tall plant of genotype *DD* is test-crossed to a dwarf plant, which must have the *dd* genotype, all offspring will be tall phenotypically and *Dd* genotypically. However, if a tall plant is *Dd* and is crossed to a dwarf plant, one-half of the offspring will be tall (*Dd*) and the other half will be dwarf (*dd*). Therefore, a 1:1 tall/dwarf ratio demonstrates the heterozygous nature of the tall plant of unknown genotype. The basis for these conclusions is illustrated in Figure 3.4. The test cross reinforced Mendel's conclusion that separate unit factors control the tall and dwarf traits.

THE DIHYBRID CROSS

A natural extension of performing monohybrid crosses was for Mendel to design experiments where two characters were examined simultaneously. We will refer to such a cross, involving two pairs of contrasting traits, as a **dihybrid cross**. It is also called a two-factor cross. For example, if pea plants having yellow seeds that are also round were bred with those having green seeds that are also wrinkled, the results shown in Figure 3.5 will occur. The F_1 offspring will be all yellow and round. It is therefore apparent that yellow is dominant to green, and that round is dominant to wrinkled. When the F_1 individuals are selfed, approximately 9/16 of the F_2 plants express

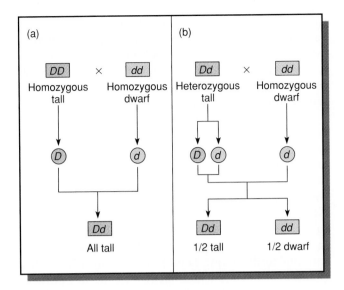

FIGURE 3.4
The test cross illustrated with a single character. In (a), the tall parent is homozygous. In (b), the tall parent is heterozygous. The genotypes of each tall parent are determined by examining the offspring when each is crossed to the homozygous recessive dwarf plant.

yellow and round, 3/16 express yellow and wrinkled, 3/16 express green and round, and 1/16 express green and wrinkled.

A variation of this cross is also shown in Figure 3.5. If, instead of having one P_1 parent with both dominant traits (yellow, round) and one with both recessive traits (green, wrinkled), plants with yellow wrinkled seeds can be bred with those having green, round seeds in a P_1 cross. In spite of the change in parental phenotypes, both the F_1 and F_2 results

remain unchanged. It will become clear in the next section why this is so.

Mendel's Fourth Postulate: Independent Assortment

We can most easily understand the results of a dihybrid cross if we consider it theoretically as consisting of two monohybrid crosses conducted separately. Think of the two sets of traits as being inherited independently of each other; that is, the chance of any plant becoming tall or dwarf is not at all influenced by the chance that this plant will have round or wrinkled seeds. Thus, because yellow is dominant to green, all F_1 plants in the first theoretical cross would have yellow seeds. In the second theoretical cross, all F_1 plants would have round seeds because round is dominant to wrinkled. When Mendel examined the F_1 plants of the dihybrid cross, all were yellow and round, as predicted.

The predicted F_2 results of the first cross are 3/4 yellow and 1/4 green. Similarly, the second cross would yield 3/4 round and 1/4 wrinkled. Figure 3.5 shows that in the dihybrid cross, 12/16 F_2 plants are yellow while 4/16 are green, exhibiting the 3:1 ratio. Similarly, 12/16 F_2 plants have round seeds while 4/16 have wrinkled seeds, again revealing the 3:1 ratio.

Since it is evident that the two pairs of contrasting traits are inherited independently, we can predict the frequencies of all possible F_2 phenotypes by applying the "product law" of probabilities: **When two independent events occur simultaneously, their combined probability is equal to the product of their individual probabilities of occurrence.** For example, the probability of an F_2 plant having yellow *and* round

FIGURE 3.5
The F_1 and F_2 results of Mendel's dihybrid crosses between yellow, round and green, wrinkled pea plants, and between yellow, wrinkled and green, round pea plants.

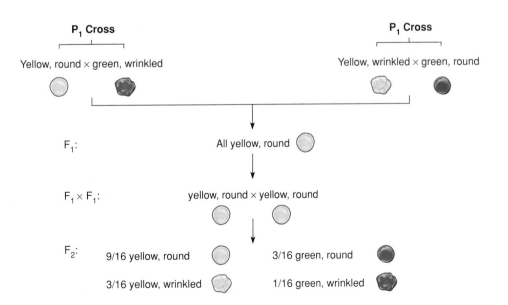

seeds is (3/4)(3/4), or 9/16, since 3/4 of all F_2 plants should be yellow and 3/4 of all F_2 plants should be round.

In a like way, the probabilities of the other three F_2 phenotypes may be calculated: yellow (3/4) *and* wrinkled (1/4) are predicted to be present together 3/16 of the time; green (1/4) *and* round (3/4) are predicted 3/16 of the time, and green (1/4) *and* wrinkled (1/4) are predicted 1/16 of the time. These calculations are illustrated in Figure 3.6. It is now apparent why the F_1 and F_2 results are identical whether the initial cross is yellow, round bred with green, wrinkled or if yellow, wrinkled are bred with green, round. In both crosses, the F_1 genotype of all plants is identical. Each plant is heterozygous for both gene pairs. As a result, the F_2 generation is also identical in both crosses.

On the basis of similar results in numerous dihybrid crosses, Mendel proposed a fourth postulate:

4 INDEPENDENT ASSORTMENT
During gamete formation, segregating pairs of unit factors assort independently of each other.

This postulate stipulates that segregation of any pair of unit factors occurs independently of all others. As a result of segregation, each gamete receives one member of every pair of unit factors. For one pair, whichever unit factor is received does not influence the outcome of segregation of any other pair. Thus, according to the postulate of independent assortment, all possible combinations of gametes will be formed in equal frequency.

Independent assortment is illustrated in the formation of the F_2 generation, shown in the Punnett square in Figure 3.7. Examine the formation of gametes by the F_1 plant. Segregation prescribes that every gamete receives either a G or g allele *and* a W or w allele. Independent assortment stipulates that all combinations (GW, Gw, gW, and gw) will be formed with equal probabilities.

In every $F_1 \times F_1$ fertilization event, each zygote has an equal probability of receiving one of the four combinations from each parent. If a large number of offspring is produced, 9/16 are yellow and round, 3/16 are yellow and wrinkled, 3/16 are green and round, and 1/16 are green and wrinkled, yielding what is designated as **Mendel's 9:3:3:1 dihybrid ratio**. This ratio is based on probability events involving segregation, independent assortment, and random fertilization. Therefore, it is an ideal ratio. Because of deviation due strictly to chance, particularly if small numbers of offspring are produced, the perfect ratio will seldom be approached.

The Test Cross: Two Characters

The test cross may also be applied to individuals that express two dominant traits, but whose genotypes are unknown. For example, the expression of the yellow, round phenotype in the F_2 generation just described may result from the $GGWW$, $GGWw$, $GgWW$, and $GgWw$ genotypes. If an F_2 yellow, round plant is crossed with the homozygous recessive green, wrinkled plant ($ggww$), analysis of the offspring will indicate the correct genotype of that yellow, round plant. Figure 3.8 illustrates the use of the test cross with offspring resulting from a dihybrid cross.

FIGURE 3.6
The determination of the combined probabilities of each F_2 phenotype for two independently inherited characters. The probability of each plant being yellow or green is independent of the probability of it being round or wrinkled.

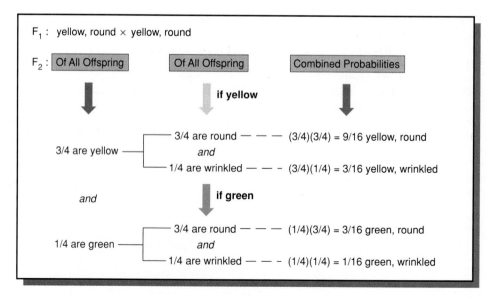

F_1: yellow, round × yellow, round

F_2: [Of All Offspring]   [Of All Offspring]   [Combined Probabilities]

if yellow

3/4 are yellow ———— 3/4 are round — — — (3/4)(3/4) = 9/16 yellow, round
and
1/4 are wrinkled — — — (3/4)(1/4) = 3/16 yellow, wrinkled

and

if green

1/4 are green ———— 3/4 are round — — — (1/4)(3/4) = 3/16 green, round
and
1/4 are wrinkled — — — (1/4)(1/4) = 1/16 green, wrinkled

FIGURE 3.7
Diagram of the dihybrid crosses shown in Figure 3.5. The F_1 heterozygous plants are self-fertilized to produce an F_2 generation, which is computed using a Punnett square. Both the phenotypic and genotypic F_2 ratios are shown.

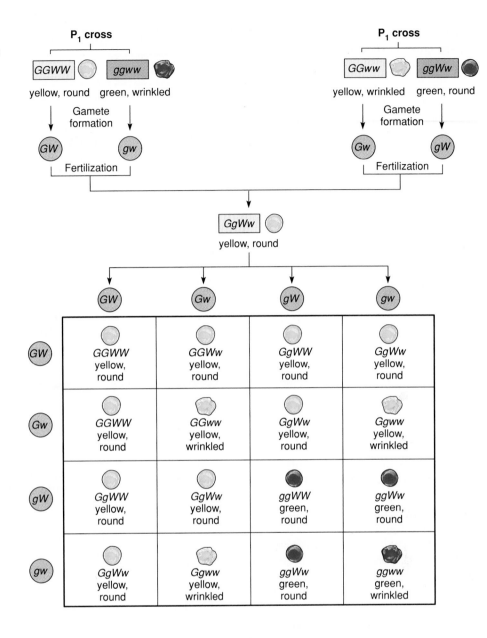

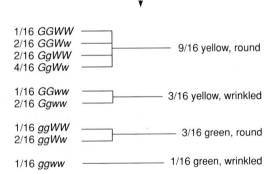

FIGURE 3.8
The test cross illustrated with two independent characters.

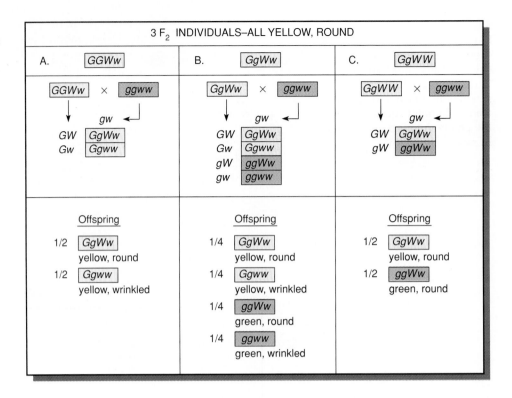

THE TRIHYBRID CROSS

We have thus far considered inheritance by individuals of up to two pairs of contrasting traits. Mendel demonstrated that the identical processes of segregation and independent assortment apply to three pairs of contrasting traits in what is called a **trihybrid cross,** also referred to as a three-factor cross.

While a trihybrid cross is somewhat more complex than a dihybrid cross, its results are easily calculated if the principles of segregation and independent assortment are followed. For example, consider the cross shown in Figure 3.9, where the gene pairs representing theoretical contrasting traits are symbolized A/a, B/b, and C/c. In the cross between $AABBCC$ and $aabbcc$ individuals, all F_1 individuals are heterozygous for all three gene pairs. Their genotype, $AaBbCc$, results in the phenotypic expression of the dominant A, B, and C traits. When F_1 individuals are parents, they all produce 8 different gametes in equal frequencies. At this point, we could construct a Punnett square with 64 separate boxes and read out the phenotypes. Because such a method is cumbersome in a cross involving so many factors, another method has been devised to calculate the predicted ratio.

The Forked-Line Method, or Branch Diagram

It is much less exacting to consider each contrasting pair of traits separately and then to combine these results using the **forked-line method,** which was first illustrated in Figure 3.6. This method, also called a **branch diagram,** relies on the simple application of the laws of probability established for the dihybrid cross. Each gene pair is assumed to behave independently during gamete formation.

When the monohybrid cross AA × aa is made, we know that:

1. All F_1 individuals have the genotype Aa and demonstrate the phenotype represented by the A allele, which is called the A phenotype in the following discussion.
2. The F_2 generation consists of individuals with either the A phenotype or the a phenotype in the ratio of 3:1, respectively.

The same generalizations may be made for the BB × bb and CC × cc crosses. Thus, in the F_2 generation, 3/4 of all organisms will have phenotype A, 3/4 will have B, and 3/4 will have C. Similarly, 1/4 of all organisms will have phenotype a, 1/4 will have b, and 1/4 will have c. The proportions of organisms that express each phenotypic combination may be predicted by assuming that fertilization, following the independent assortment of these three gene pairs during gamete formation, is a random process. We must simply apply once again the product law of probabilities.

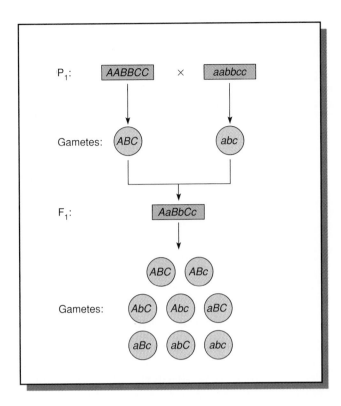

FIGURE 3.9
The formation of P_1 and F_1 gametes in a trihybrid cross.

The phenotypes of the F_2 generation calculated in this way using the forked-line method are illustrated in Figure 3.10. They fall into the **trihybrid ratio of 27:9:9:9:3:3:3:1**. The same method may be applied when solving crosses involving any number of gene pairs, *provided* that all gene pairs assort independently from each other. We will see later that this is not always the case. However, it appeared to be true for all of Mendel's characters.

Note in Figure 3.10 that only phenotypic ratios of the F_2 generation have been derived. It is possible to generate genotypic ratios as well. To do so, we again consider the A/a, B/b, and C/c gene pairs separately. For example, for the A/a pair the F_1 cross is $Aa \times Aa$. Phenotypically, an F_2 ratio of 3/4 A:1/4 a is produced. Genotypically, however, the F_2 ratio is different; 1/4 AA:1/2 Aa:1/4 aa will result. Using Figure 3.10 as a model, we would enter these genotypic frequencies on the left side of the calculation. Each would be connected by three bars to 1/4 BB, 1/2 Bb, and 1/4 bb, respectively. From each of these nine designations, three more bars would extend to the 1/4 CC, 1/2 Cc, and 1/4 cc genotypes. Thus, on the right side of the completed diagram, 27 genotypes and their frequencies of occurrence would appear. One of the problems at the end of this chapter asks you to use the forked-line or branch diagram method to determine the genotypic ratios generated in a trihybrid cross (see Problem 16).

In crosses involving two or more gene pairs, the calculation of gametes and genotypic and phenotypic results is quite complex. There are several simple mathematical rules that will enable you to check the accuracy of various steps required in working genetics problems. First, you must determine the number of heterozygous gene pairs (n) involved in the cross. For example, where $AaBb \times AaBb$ represents the cross, $n = 2$; for $AaBbCc \times AaBbCc$, $n = 3$; for $AaBBCcDd \times AaBBCcDd$, $n = 3$ (because the B genes are not heterozygous). Once n is determined, 2^n is the number of different gametes that can be formed by each parent; 3^n is the number of different genotypes that result following fertilization; and 2^n is the number of different phenotypes that are produced from these genotypes. Table 3.1 summarizes these rules, which may be applied to crosses involving any number of gene pairs, provided that the gene pairs assort independently from one another.

FIGURE 3.10
The generation of the F_2 trihybrid phenotypic ratio using the forked-line, or branch diagram, method.

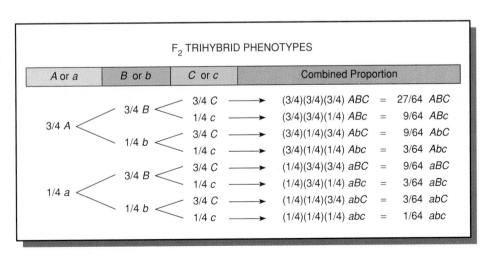

A or a	B or b	C or c	Combined Proportion		
		3/4 C	(3/4)(3/4)(3/4) ABC	=	27/64 ABC
	3/4 B	1/4 c	(3/4)(3/4)(1/4) ABc	=	9/64 ABc
3/4 A		3/4 C	(3/4)(1/4)(3/4) AbC	=	9/64 AbC
	1/4 b	1/4 c	(3/4)(1/4)(1/4) Abc	=	3/64 Abc
		3/4 C	(1/4)(3/4)(3/4) aBC	=	9/64 aBC
	3/4 B	1/4 c	(1/4)(3/4)(1/4) aBc	=	3/64 aBc
1/4 a		3/4 C	(1/4)(1/4)(3/4) abC	=	3/64 abC
	1/4 b	1/4 c	(1/4)(1/4)(1/4) abc	=	1/64 abc

TABLE 3.1
Simple mathematical rules useful in working genetics problems. The fourth column assumes dominance and recessiveness is operational for all gene pairs.

CROSSES BETWEEN ORGANISMS HETEROZYGOUS FOR GENES EXHIBITING INDEPENDENT ASSORTMENT			
Number of Heterozygous Gene Pairs	Number of Different Types of Gametes Formed	Number of Different Genotypes Produced	Number of Different Phenotypes Produced
n	2^n	3^n	2^n
1	2	3	2
2	4	9	4
3	8	27	8
4	16	81	16

THE REDISCOVERY OF MENDEL'S WORK

Mendel's work, initiated in 1856, was presented to the Brünn Society of Natural Science in 1865 and published one year later. However, his findings went largely unnoticed for about thirty-five years! Many reasons have been suggested to explain why the significance of his research was not immediately recognized.

First of all, Mendel's adherence to mathematical analysis of probability events was quite an unusual approach in biological studies. Perhaps his approach seemed foreign to his contemporaries. More importantly, his conclusions drawn from such analyses did not fit well with the existing theories involving the cause of variation between organisms. The source of natural variation intrigued students of evolutionary theory. These individuals, stimulated by the proposal developed by Charles Darwin and Alfred Russell Wallace, believed that variation was of a **continuous nature** and that offspring were a blend of their parents' phenotypes. As we have mentioned earlier, Mendel theorized that variation was due to discrete or particulate units and was therefore of a **discontinuous nature**. For example, Mendel proposed that the F_2 offspring of a dihybrid cross were merely expressing traits produced by new combinations of previously existing unit factors. As a result, Mendel's theories did not fit well with the evolutionists' preconceptions about causes of variation.

Beyond this interpretation is still a further speculation as to why Mendel's contemporaries failed to grasp the significance of his findings. Perhaps they did not realize that Mendel's postulates explained *how* variation was transmitted to offspring. Instead, they may have attempted to interpret his work in a way that addressed the issue of *why* certain phenotypes survive preferentially. It was this latter question that had been addressed in the **theory of natural selection**. Thus, it may well be that the collective vision of Mendel's scientific colleagues was obscured by the impact of this extraordinary theory of organic evolution.

The Rebirth of Mendel's Work

However, near the end of the nineteenth century, a remarkable discovery set the scene for the rebirth of Mendel's work—**The discovery of chromosomes**, which has influenced our understanding of living organisms to this very day. In 1879, Walther Flemming discovered threadlike structures that he called "**chromatin**" in the nuclei of salamanders. He was further able to describe the behavior of these chromosomes during cell division. As a result of the findings of Flemming and many other cytologists, the presence of a nuclear component soon became an integral part of ideas surrounding inheritance. In 1885, August Weismann, a proponent of the doctrine of epigenesis, pronounced that "the nuclear substance alone can be the bearer of hereditary tendencies." Weismann also stated that "a determinant is nothing more than a living element of the germ substance, and its presence in the germ conditions the appearance of and specific development of a particular part of the body." It was in this setting that scientists were able to reexamine Mendel's findings.

In the very early twentieth century, two types of research led to the rebirth of Mendel's work. Hybridization experiments similar to Mendel's were independently performed by three botanists, Hugo DeVries, Karl Correns, and Erich Tschermak. DeVries' work, for example, had focused on unit characters, and he demonstrated the principle of segregation in his experiments with several plant species. Apparently, he had searched the existing literature and found that Mendel's work had anticipated his own conclusions! Correns and Tschermak also drew conclusions similar to those of Mendel.

In 1902 two cytologists, Walter Sutton and Theodor Boveri, independently published papers, linking their discoveries of the behavior of chromosomes during meiosis to the Mendelian principles of segregation and independent assortment. They pointed out that the separation of chromosomes during meiosis could serve as the cytological basis of these two postulates. While they thought that Mendel's unit factors were probably chromosomes rather than genes on chromosomes, their findings made Mendel's work the foundation of ensuing genetic investigations.

Unit Factors, Genes, and Homologous Chromosomes

Because the correlation between Sutton's and Boveri's observations and Mendelian principles is the foundation for the modern interpretation of transmission genetics, we will examine this correlation before moving to the more complex topics of the next several chapters.

As pointed out in Chapter 2, each species possesses a specific number of chromosomes in each somatic (body) cell nucleus. For diploid organisms, this number is called the **diploid number** ($2n$) and is characteristic of that species. During the formation of gametes, this number is precisely halved (n), and when two gametes combine during fertilization, the diploid number is reestablished. The chromosome number is not reduced in a random manner, however. It was apparent to early cytologists that the diploid number of chromosomes is composed of homologous pairs identifiable by their morphological appearance. The gametes contain one member of each pair. The chromosome complement of a gamete is thus quite specific, and the number of chromosomes in each gamete is equal to the haploid number.

With this basic information, we can see the correlation between the behavior of unit factors and chromosomes and genes. Figure 3.11 shows three of Mendel's postulates and the accepted explanation of each. Unit factors are really genes located on homologous pairs of chromosomes [Figure 3.11(a)]. Members of each pair of homologues separate, or segregate, during gamete formation [Figure 3.11(b)].

To illustrate the principle of independent assortment, it is important to distinguish between members of any given homologous pair of chromosomes. One member of each pair is derived from the **maternal parent,** while the other comes from the **paternal parent**. We represent their different origins by different colors. In Figure 3.11(c) two pairs of segregating homologues are considered. They behave independently during gamete formation, with each gamete always receiving one of each pair. All possible combinations are shown. If we add the symbols used

in Mendel's dihybrid cross (D, d and W, w) to the diagram, we see why equal numbers of the four types of gametes are formed. The independent behavior of Mendel's pairs of unit factors was due to the fact that they were on separate pairs of homologous chromosomes.

From observations of the phenotypic diversity of living organisms, we see that it is logical to assume that there are many more genes than chromosomes. Therefore, each homologue must carry genetic information for more than one trait. The currently accepted concept is that a chromosome is composed of a large number of linearly ordered, information-containing units called **genes**. Thus, Mendel's unit factors (which determine tall or dwarf stems, for example) actually constitute a pair of genes located on one pair of homologous chromosomes. The location on a given chromosome where any particular gene occurs is called its **locus**. The different forms taken by a given gene, called **alleles** (D or d), contain slightly different genetic information that determines the same character (stem length). Alleles are alternate forms of the same gene. Although we have only discussed genes with two alternative alleles, most genes have *more* than two allelic forms. We will discuss the concept of **multiple alleles** in Chapter 4.

We conclude this section by reviewing the criteria necessary to classify two chromosomes as a homologous pair:

1 During mitosis and meiosis, when chromosomes are visible as distinct figures, both members of a homologous pair are the same size and exhibit identical centromere locations.
2 During early stages of meiosis, homologous chromosomes pair together, or synapse.
3 Although not generally microscopically visible, homologues contain identical, linearly ordered, gene loci.

INDEPENDENT ASSORTMENT AND GENETIC VARIATION

One of the major consequences of independent assortment is the production by individuals of genetically dissimilar gametes. Genetic variation results because the two members of any homologous pair of chromosomes are rarely, if ever, genetically identical. For them to be identical, homozygosity at every locus along both homologous chromosomes would be required. Various methods used to estimate the degree of heterozygosity show that on the average, 10 to 40 percent of the loci are occupied by different alleles. However, some geneticists feel that organ-

FIGURE 3.11
The correlation between the
Mendelian postulates of (a) unit
factors in pairs, (b) segregation,
and (c) independent
assortment, and the presence
of genes located on
homologous chromosomes and
their behavior during meiosis.

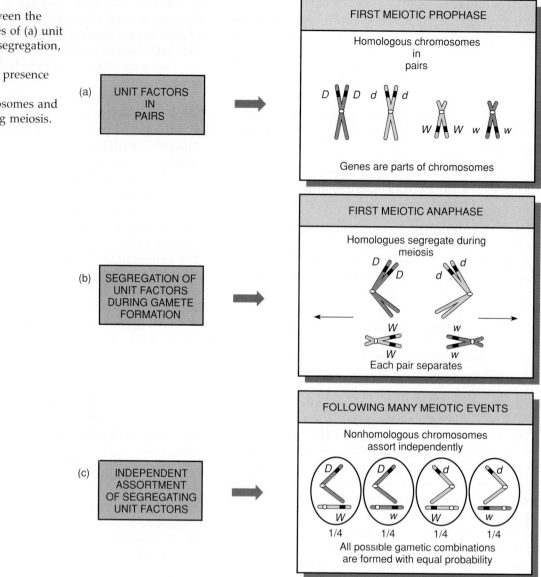

isms that are not inbred display heterozygosity at most, if not all, of their loci. Therefore, since independent assortment leads to the production of all possible chromosome combinations, extensive genetic diversity results.

The number of possible gametes, each with different chromosome compositions, is 2^n, where n equals the haploid number. Thus, if a species has a haploid number of 4, then 2^4 or 16 different gamete combinations can be formed as a result of independent assortment. While this number is not great, consider the human species, where $n = 23$. If 2^{23} is calculated, we find that in excess of 8×10^6, or over 8 million,

different types of gametes are represented. Since fertilization represents an event involving only one of approximately 8×10^6 possible gametes from each of two parents, each offspring represents only one of $(8 \times 10^6)^2$, or 64×10^{12}, potential genetic combinations! It is no wonder that, except for identical twins, each member of the human species demonstrates a distinctive appearance and individuality. This number of combinations is far greater than the number of humans who have ever lived on earth! Genetic variation resulting from independent assortment has been extremely important to the process of organic evolution in all organisms.

PROBABILITY AND GENETIC EVENTS

Genetic ratios are most properly expressed as probabilities, e.g., 3/4 tall:1/4 dwarf. These values predict the outcome of each fertilization event, such that the probability of each zygote having the genetic potential for becoming tall is 3/4, while the potential for becoming dwarf is 1/4. Since probabilities can always be expressed as fractions, the values range from 0, where an event is certain *not to* occur, to 1.0, where an event is certain *to* occur.

When two or more events occur independently of one another, but at the same time, we can calculate the probability that they will indeed occur together. This is accomplished by applying the **product law**. It states that the probability of two or more events occurring simultaneously is equal to the *product* of their individual probabilities. Two or more events are independent of one another if the outcome of each one does not affect the outcome of any of the others under consideration.

To illustrate the use of the product law, consider the outcomes if you were to toss a penny (P) and a nickel (N) at the same time and examined all combinations of heads (H) and tails (T) that can occur. There are four possible outcomes:

$$(P_H : N_H) = (1/2)(1/2) = 1/4$$
$$(P_T : N_H) = (1/2)(1/2) = 1/4$$
$$(P_H : N_T) = (1/2)(1/2) = 1/4$$
$$(P_T : N_T) = (1/2)(1/2) = 1/4$$

The probability of obtaining a head or a tail in the toss of either coin is 1/2 and unrelated to the outcome of the toss of the other coin. All four events are predicted to occur with equal probability.

If we were interested in calculating the probability of an event where the possible outcomes are independent of one another, but can be accomplished in more than one way, we would apply the **sum law**. For example, we can ask, what is the probability of tossing our penny and nickel and obtaining one head and one tail? In such a case, we do not care whether it is the penny or nickel that comes up heads, provided that the other coin has the alternate outcome. As we can see above, there are two ways in which the desired outcome can be accomplished, each with a probability of 1/2. Thus, according to the sum law, the overall probability is equal to:

$$(1/4) + (1/4) = 1/2$$

One-half of all such tosses are predicted to yield the desired outcome.

These simple probability laws will be useful throughout our discussions of transmission genetics and as you solve genetics problems. In fact, we have already applied the product law earlier when the forked-line method was used to calculate the phenotypic outcome of Mendel's trihybrid cross. When we wish to know the overall outcome of a cross, we need only calculate the probability of each possible outcome. The results of this calculation then allow us to predict the proportion of offspring that will parallel each possible outcome.

There is a very important point to remember when dealing with probability. Predictions of possible outcomes are based on large sample sizes. If we predict that 9/16 of the offspring of a dihybrid cross will express both dominant traits, it is very unlikely that 9 out of 16 offspring will express this phenotype. Instead, our prediction is that, of a large number of offspring, approximately 9/16 of them will do so. The deviation from the predicted ratio in smaller sample sizes is attributed to deviation due to chance, a subject we shall deal with in our discussion of statistics in the next section. As we will see, the impact of deviation due strictly to chance is diminished as the sample size increases.

EVALUATING GENETIC DATA: CHI-SQUARE ANALYSIS

Mendel's 3:1 monohybrid and 9:3:3:1 dihybrid ratios are hypothetical predictions based on the following assumptions: (1) dominance/recessiveness; (2) segregation; (3) independent assortment; and (4) random fertilization. The last three processes are influenced by chance events and therefore are subject to normal deviation. This concept, **chance deviation,** is most easily illustrated by tossing a coin numerous times and recording the number of heads and tails observed. In each toss, there is a probability of 1/2 that a head will occur and a probability of 1/2 that a tail will occur. Therefore, the expected ratio of many tosses is 1:1. If a coin were tossed 1000 times, usually *about* 500 heads and 500 tails would be observed. Any reasonable fluctuation from this hypothetical ratio (e.g., 486 heads and 514 tails) would be attributed to chance deviation.

As the total number of tosses is reduced, the impact of chance deviation increases. For example, if a coin were tossed only 4 times, the hypothetical ratio would predict 2 heads and 2 tails. However, you wouldn't be too surprised if all 4 tosses resulted in only heads or only tails. But, for 1000 tosses, 1000 heads or 1000 tails would be most unexpected. In fact, you might believe that such a result would be impossible. Actually, all heads or all tails in 1000 tosses would be predicted to occur with a probability of only $(1/2)^{1000}$. Since $(1/2)^{20}$ is equivalent to less than 1 in 1 million times, an event occurring with a probability of only $(1/2)^{1000}$ would be virtually impossible to achieve.

Two major points are significant here:

1 The outcomes of segregation, independent assortment, and fertilization, like coin tossing, are subject to random fluctuations from their predicted occurrences as a result of chance deviation.
2 As the sample size increases, the average deviation from the expected decreases. Therefore, a larger sample size diminishes the impact of chance deviation on the final outcome.

It is important in genetics to be able to evaluate observed deviation. When we assume that data will fit a given ratio such as 1:1, 3:1, or 9:3:3:1, we establish what is called the **null hypothesis**. It is so named because the hypothesis assumes that there is no real difference between the **measured values** (or ratio) and the **predicted values** (or ratio). Evaluation of the null hypothesis is accomplished by statistical analysis. On this basis, the null hypothesis may either: (1) be rejected, or (2) fail to be rejected. If it is rejected, any observed deviation from the expected cannot be attributed to chance alone. The null hypothesis and the underlying assumptions leading to it must be reexamined. If the null hypothesis fails

to be rejected, any observed deviations are attributed to chance.

Thus, statistical analysis provides a mathematical basis for examining how well observed data fit or differ from predicted or expected occurrences, testing what is called the **goodness of fit**. Assuming that the data do not "fit" exactly, just how much deviation can be allowed before the null hypothesis is rejected? One of the simplest statistical tests devised to answer this question is **chi-square analysis (χ^2)**. This test takes into account the observed deviation in each component of an expected ratio as well as the sample size and reduces them to a single numerical value. This value (χ^2) is then used to estimate how frequently the observed deviation can be expected to occur strictly as a result of chance. The formula used in chi-square analysis is

$$\chi^2 = \Sigma \frac{(o - e)^2}{e}$$

where o is the observed value for a given category and e is the expected value for that category. Σ (sigma) represents the sum of the calculated values for each category of the ratio. Because $(o - e)$ is the deviation (d) in each case, the equation can be reduced to

$$\chi^2 = \Sigma \frac{d^2}{e}$$

TABLE 3.2
Chi-square analysis.

(a) Hypothetical monohybrid cross.

Expected Ratio	Observed (o)	Expected (e)	Deviation ($o - e$)	Deviation^2 (d)^2	Deviation^2/Expected (d^2/e)
3/4	740	3/4 (1000) = 750	740 − 750 = −10	$(-10)^2$ = 100	100/750 = 0.13
1/4	260	1/4 (1000) = 250	260 − 250 = +10	$(+10)^2$ = 100	100/250 = 0.40
TOTAL = 1000					$\Sigma\chi^2$ = 0.53
					p = 0.48

(b) Hypothetical dihybrid cross.

Expected Ratio	o	e	$o - e$	d^2	d^2/e
9/16	587	567	+20	400	0.71
3/16	197	189	+8	64	0.34
3/16	168	189	−21	441	2.33
1/16	56	63	−7	49	0.78
TOTAL = 1008				χ^2 = 4.16	
				p = 0.26	

Table 3.2(a) illustrates the step-by-step procedure necessary to make the χ^2 calculation for the F_2 results of a hypothetical monohybrid cross. If you were analyzing these data, you would work from left to right, calculating and entering the appropriate numbers in each column. Regardless of whether the calculated deviation $(o - e)$ is initially positive or negative, it becomes positive after the number is squared. Table 3.2(b) illustrates the analysis of the F_2 results of a hypothetical dihybrid cross. Based on your study of the calculations involved in the monohybrid cross, check to make certain that you understand how each number was calculated in the dihybrid example.

The final step in the chi-square analysis is to interpret the χ^2 value. To do so, you must initially determine the value of the **degrees of freedom (df)**, which is equal to $n - 1$ where n is the number of different categories into which each datum point may fall. For the $3:1$ ratio, $n = 2$, so $df = 2 - 1 = 1$. For the $9:3:3:1$ ratio, $df = 3$. Degrees of freedom must be taken into account because the greater the number of categories, the more deviation is expected as a result of chance.

With this accomplished, the χ^2 value must now be interpreted in terms of a corresponding **probability value (p)**. Because this calculation is complex, the p value is usually located on a table or graph. Figure 3.12 shows the wide range of χ^2 and p values for numerous degrees of freedom in both forms. We will use the graph to determine the p value. The caption for Figure 3.12(b) explains how to use the table.

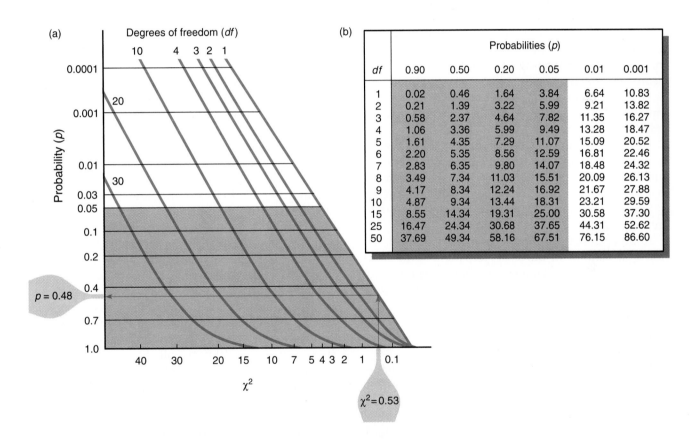

FIGURE 3.12
(a) A graph used to convert χ^2 values to p values. The conversion of a χ^2 value of 0.53 with 1 degree of freedom to an estimated probability value of 0.48 is illustrated. (b) A table showing χ^2 values for a variety of combinations of df and p. Any χ^2 value greater than that shown at the $p = 0.05$ level for a particular df serves as the basis to reject the null hypothesis in question. In our example, a χ^2 value of 0.53 for a df of 1 is converted to a probability value between 0.20 and 0.50. From our graph in (a), the more precise value ($p = 0.48$) was estimated by interpolation. All values that serve to fail to reject the null hypothesis are shaded in both the graph and the chart.

These simple steps must be followed to determine p:

1 Locate the χ^2 value on the abscissa.
2 Draw a vertical line from this point to the line representing the appropriate *df*.
3 Extend a horizontal line from this point to the left until it intersects the ordinate.
4 Estimate, by interpolation, the corresponding p value.

For our first example (the monohybrid cross) in Table 3.2, the p value of 0.48 may be estimated in this way and is illustrated in Figure 3.12(a). For the dihybrid cross, use this method to see if you can determine the p value. χ^2 is 4.16 and *df* equal 3. A p value of 0.26 is the approximate value. Use of the table rather than the graph confirms that both p values are between 0.20 and 0.50. Examine the table to confirm this.

So far, we have been concerned only with the determination of p. The most important aspect of χ^2 analysis is understanding what the p value actually means. We will use the example of the dihybrid cross ($p = 0.26$) to illustrate. In these discussions, it is simplest to think of the p value as a percentage (e.g., $0.26 = 26$ percent).

The p value indicates the probability of obtaining as great or even greater deviation by chance alone. In our dihybrid example, the probability is 26 percent that the observed deviation or more will occur owing to chance. The larger the p value, the closer the data are to the predicted or ideal ratio.

Another description provides more specific statistical information. In our example, the p value indicates that, were the same experiment repeated many times, 26 percent of the trials would be expected to exhibit chance deviation as great or greater than that seen in the initial trial. Conversely, 74 percent of the repeats would show less deviation as a result of chance than initially observed.

These interpretations of the p value reveal that a hypothesis (a 9:3:3:1 ratio in this case) is never proved or disproved absolutely. Instead, a relative standard must be set to serve as the basis for either supporting or rejecting the hypothesis. This standard is often a probability value of 0.05. In chi-square analysis, a value less than 0.05 makes it unlikely that the observed results could be obtained by chance alone. Instead, such a p value indicates substantial deviation from the predicted results and thus serves as the basis for rejecting the null hypothesis.

On the other hand, p values of 0.05 or greater ($1.0 - 0.05$) fail to reject the null hypothesis. In our example where $p = 0.26$, the hypothesis of independent assortment is supported by the experimental data. That is, the data do not provide any reason to reject the hypothesis.

HUMAN PEDIGREES

In all crosses discussed so far, one of the two traits for each character has been dominant to the other. Based on this observation, two significant questions may be asked:

1 Does the expression of all genes occur in this fashion?
2 Is it possible to ascertain the mode of inheritance of genes in organisms where designed crosses and the production of large numbers of offspring are impossible?

The answer to the first question is no. As we will see in Chapters 4 and 5, many modes of inheritance exist that modify the monohybrid and dihybrid ratios observed by Mendel.

The answer to the second question is yes. Even in humans the pattern of inheritance of a specific phenotype can be studied.

The simplest way to study this pattern is to construct a family tree indicating the phenotype of the trait in question for each member. Such a family tree is called a **pedigree**. By analyzing the pedigree, we may be able to determine how the gene controlling the trait is inherited.

Figure 3.13 shows the conventions used in constructing a pedigree. Circles represent females, and squares designate males. If the sex is unknown, a diamond is used. If a pedigree traces only a single trait, as Figure 3.13 does, the circles, squares, and diamonds are shaded if the phenotype being considered is expressed. If two traits are considered, one way to follow them is to divide the square or circle into an upper and lower half. For example, a circle might be unshaded, completely shaded, or have either half shaded, depending on the phenotype.

The parents are connected by a horizontal line, and a vertical line leads to their offspring. All such offspring are called **sibs** and are connected by a horizontal **sibship line**. Sibs are placed from left to right according to birth order and are labeled with Arabic numerals. Each generation is indicated by a Roman numeral.

Twins are indicated by connected diagonal lines. **Monozygotic** or **identical twins** stem from a single line itself connected to the sibship line (see III-5,6 in Figure 3.13). **Dizygotic** or **fraternal twins** are connected directly to the sibship line (see III-8,9). A number within one of the symbols (II-10–13) repre-

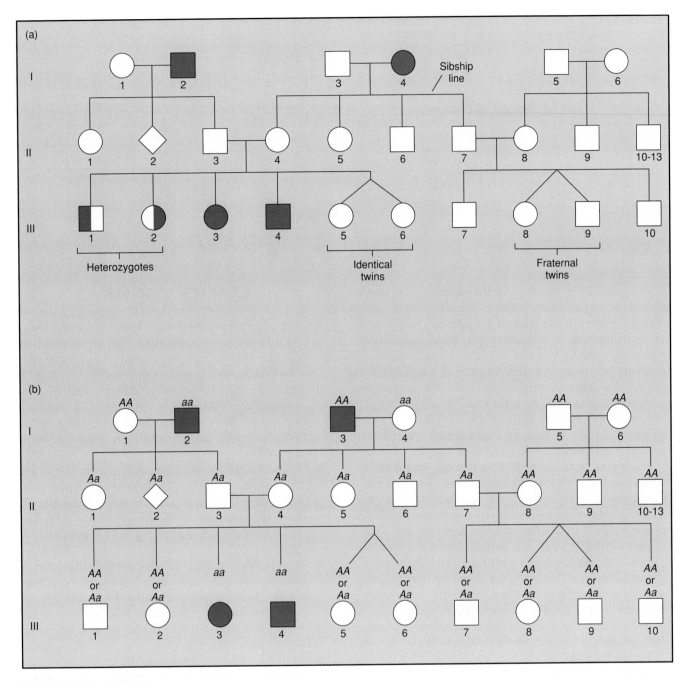

FIGURE 3.13
(a) A representative pedigree for a single character through three generations. (b) The most probable genotypes of each individual in the pedigree.

sents numerous sibs of the same or unknown phenotypes. A male whose phenotype drew the attention of a physician or geneticist is called the **propositus** (a female is a **proposita**) or sometimes the **proband** and is indicated by an arrow (see III-4).

The pedigree shown in Figure 3.13 traces the theoretical pattern of inheritance of the human trait **albinism**. By analyzing the pedigree, we will see that albinism is inherited as a recessive trait.

Two of the parents of the first generation, I-2 and I-4, are affected. Because none of their offspring show the disorder, it is reasonable to conclude that the unaffected parents (I-1 and I-3) were homozygous normal individuals. Had they been heterozygous, one half of their offspring would be expected to exhibit albinism.

An unaffected second generation is characteristic of a rare recessive trait. If albinism were inherited as

TABLE 3.3
Representative recessive and dominant human traits.

Recessive Traits	Dominant Traits
Albinism	Achondroplasia
Alkaptonuria	Brachydactyly
Ataxia telangiectasia	Congenital stationary night blindness
Color blindness	Ehler-Danlos syndrome
Cystic fibrosis	Fascio-scapulo-humeral muscular dystrophy
Duchenne's muscular dystrophy	Huntington's disease
Galactosemia	Hypercholesterolemia
Hemophilia	Marfan syndrome
Lesch-Nyhan syndrome	Neurofibromatosis
Phenylketonuria	Phenylthiocarbamide tasting (PTC)
Sickle-cell anemia	Widow's peak
Tay-Sachs disease	

a dominant trait, one-half of the second generation would be expected to exhibit the disorder in the crosses involving the I-2 and I-4 parents. Inspection of the offspring constituting the third generation provides further support for the hypothesis that albinism is a recessive trait. Parents II-3 and II-4 are apparently both heterozygous, and approximately one-fourth of their offspring should be affected. Two of the six offspring do show albinism. This deviation from the expected ratio is characteristic of crosses with few offspring.

Individual II-7 is undoubtedly heterozygous, while II-8 is most likely homozygous normal, given the low frequency of the *a* allele in the population. If so, we would predict that none of their offspring (III-7,8,9, and 10) would be albino, and this is borne out.

Based on this pedigree analysis, and the conclusion that albinism is a recessive trait, the genotypes of all individuals can be predicted. For both the first and second generations, this can be done with some certainty. In the third generation, for normal individuals, we can only guess whether they are homozygous or heterozygous. These predictions are shown in Figure 3.13(b).

Pedigree analysis of many traits has been an extremely valuable research technique in human genetic studies. However, this approach does not usually provide the certainty in drawing conclusions that is afforded by designed crosses yielding large numbers of offspring. Nevertheless, when many independent pedigrees of the same trait or disorder are analyzed, consistent conclusions can often be drawn. Table 3.3 lists numerous human traits and classifies them according to their recessive or dominant expression. As we will see in Chapter 5, the genes controlling some of these traits are located on the sex-determining chromosomes.

CHAPTER SUMMARY

1 The cornerstone for the discipline of genetics was established over a century ago as a result of Gregor Mendel's experimentation with the garden pea.

2 Mendel studied inheritance patterns exhibited in crosses involving one or more pairs of contrasting traits. From the analyses of these results, the principles of transmission genetics were derived.

3 Mendel's postulates help describe the basis for the genetic control of phenotypic expression. He showed that unit factors, later called alleles, exist in pairs and exhibit a dominant/recessive relationship in determining the expression of traits.

4 Succeeding geneticists discovered that alleles occupy identical loci on homologous chromosomes. Mendel postulated that these alleles must segregate during meiosis so that individual gametes receive only one of the two factors with equal probability.

5 Mendel's final postulate of independent assortment states that pairs of segregating unit factors do so independently of other such pairs. As a result, all possible combinations of gametes will be formed with equal probability.

6 The discovery of chromosomes in the late 1800s and subsequent studies of their behavior during meiosis led to the rebirth of Mendel's work.

7 The Punnett Square and the forked-line methods are used to predict the probabilities of genotypes and phenotypes from crosses of one or more gene pairs.

8 Genetic ratios are expressed as probabilities. Thus, understanding and applying the laws of

probability are essential to understanding genetics and working genetics problems.

9 Statistical analysis is used to test the validity of experimental outcomes. In genetics, variations from the expected ratios are anticipated owing to chance deviation. A chi-square analysis tests the probability of these variations being generated from chance alone. It provides the basis for assessing the null hypothesis.

10 Pedigree analysis provides a method for studying the inheritance pattern of human traits over several generations. This often provides the basis for determining the mode of inheritance of human characteristics and disorders.

Insights and Solutions

Students demonstrate their knowledge of transmission genetics by solving genetics problems. Success at this task represents not only comprehension of theory but its application to more practical genetic situations. Most students find problem solving in genetics to be challenging but rewarding. This section is designed to provide basic insights into the reasoning essential to this process.

Genetics problems are in many ways similar to algebraic word problems. The approach taken should be identical: (1) analyze the problem carefully; (2) translate words into symbols, defining each one first; and (3) solve the problem. The first two steps are most critical. The third step is largely mechanical.

The simplest problems are those that state all necessary information about the P_1 generation and ask you to find the expected ratios of the F_1 and F_2 genotypes and/or phenotypes. The following steps should always be followed when you encounter this type of problem:

1 Determine the genotypes of the P_1 generation.

2 Determine what gametes may be formed by the P_1 generation.

3 Recombine gametes either by the Punnett square method, the forked-line method, or, if the situation is very simple, by inspection. Read the F_1 phenotypes directly.

4 Repeat the process to obtain information about the F_2 generation.

Determining the genotypes from the given information requires an understanding of the basic theory of transmission genetics. For example, consider the following problem: A recessive mutant allele, *black*, causes a very dark body in *Drosophila* when homozygous. The wild-type color is described as gray. What F_1 phenotypic ratio is predicted when a black female is crossed to a gray male whose father was black?

To work this problem, you must understand dominance and recessiveness as well as the principle of segregation. Further, you must use the information about the male parent's father. You can work out the problem as follows:

1 Since the female parent is black, she must be homozygous for the mutant allele (*bb*).

2 The male parent is gray and thus he must have at least one dominant allele (*B*). Since his father was black (*bb*), and since he received one of the chromosomes bearing these alleles, the male parent must be heterozygous (*Bb*).

From here, the problem is simple:

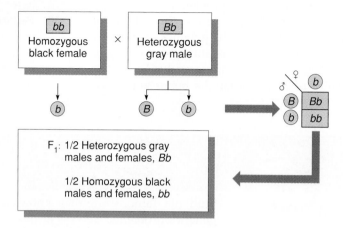

Apply the above approach to the following problems:

1 In Mendel's work, he found that full pods are dominant to constricted pods while round seeds are dominant to wrinkled seeds. One of his crosses was between full, round plants and constricted, wrinkled plants. From this cross, he obtained an F_1 that was all full and round. In the F_2, Mendel obtained his classic 9:3:3:1 ratio. Using the above information, determine the expected F_1 and F_2 results of a cross between constricted, round and full, wrinkled plants.

ANSWER: First of all, define gene symbols for each pair of contrasting traits. Select the lowercase forms of the first letter of the recessive traits to designate those phenotypes, and use the uppercase forms to designate the dominant traits. Thus, use C and c to indicate full and constricted, and use W and w to indicate the round and wrinkled phenotypes, respectively.

Now, determine the genotypes of the P_1 generation, form gametes, reconstitute the F_1 generation, and read off the phenotype(s):

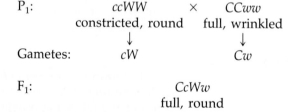

You can see immediately that the F_1 generation expresses both dominant phenotypes and is heterozygous for both gene pairs. Thus, we can expect that the F_2 generation will yield the classic Mendelian ratio of 9:3:3:1. Let's work it out anyway, just to confirm this, using the forked-line method. Since both gene pairs are heterozygous and can be expected to assort independently, we can predict the F_2 outcomes from each gene pair separately and then proceed with the forked-line method.

Every F_2 offspring is subject to the following probabilities:

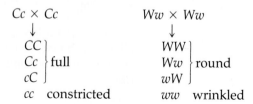

The forked-line method then allows us to confirm the 9:3:3:1 phenotypic ratio. Remember that this represents proportions of 9/16:3/16:3/16:1/16. Note that we are applying the product law as we compute the final probabilities:

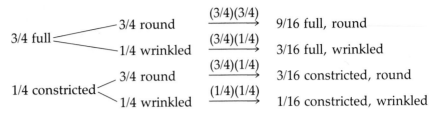

2 Determine the probability that a plant of genotype *CcWw* will be produced from parental plants of the genotypes *CcWw* and *Ccww*.

ANSWER: Since the two gene pairs demonstrate straightforward dominance and recessiveness and they independently assort during gamete formation, we need only calculate the individual probabilities of the two separate events (*Cc* and *Ww*) and apply the product law to calculate the final probability:

$$Cc \times Cc \rightarrow 1/4\ CC:1/2\ Cc:1/4\ cc$$

$$Ww \times ww \rightarrow 1/2\ Ww:1/2\ ww$$

$$p = (1/2\ Cc)(1/2\ Ww) = 1/4\ CcWw$$

3 In another cross, involving parent plants of unknown genotype and phenotype, the offspring shown below were obtained. Determine the genotypes and phenotypes of the parents.

Offspring: 3/8 full, round
3/8 full, wrinkled
1/8 constricted, round
1/8 constricted, wrinkled

ANSWER: This problem is more difficult and requires keener insights since you must work backwards. The best approach is to consider the outcomes of pod shape separately from those of seed texture.

Of all plants, 6/8 (3/4) are full and 2/8 (1/4) are constricted. Of the various genotypic combinations that can serve as parents, which will give rise to a ratio of 3/4:1/4? Since this ratio is identical to Mendel's monohybrid F_2 results, we can propose that both unknown parents share the same genetic characteristic as the monohybrid F_1 parents: they must both be heterozygous for the genes controlling pod color and thus are

Cc

Before accepting this hypothesis, let's consider the possible genotypic combinations that control seed texture. If we consider this characteristic alone, we can see that the traits are expressed in a ratio of 4/8 (1/2) round:4/8 (1/2) wrinkled. In order to generate such a ratio, the parents *cannot* both be heterozygous or their offspring would yield a 3/4:1/4 phenotypic ratio. They *cannot* both be homozygous or all offspring would express a single phenotype. Thus, we are left with testing the hypothesis that one parent is homozygous and one is heterozygous for the alleles controlling texture. The potential case of *WW* × *Ww* will not work since it also would yield only a single phenotype. This leaves us with the potential case of *Ww* ×

ww. Offspring in such a mating will yield 1/2 *Ww* (round) : 1/2 *ww* (wrinkled), exactly the outcome we are seeking.

Now, let's combine our hypotheses and predict the outcome of crossing:

$$CcWw \times Ccww$$

3/4 C–	1/2 *Ww*	$\xrightarrow{(3/4)(1/2)}$	3/8 C–*Ww*	full, round
	1/2 *ww*	$\xrightarrow{(3/4)(1/2)}$	3/8 C–*ww*	full, wrinkled
1/4 *cc*	1/2 *Ww*	$\xrightarrow{(1/4)(1/2)}$	1/8 *ccWw*	constricted, round
	1/2 *ww*	$\xrightarrow{(1/4)(1/2)}$	1/8 *ccww*	constricted, wrinkled

As we can see, this cross produces offspring according to our initial information. Thus, we have solved the problem. Note that in this solution, we have used *genotypes* in the forked-line method, in contrast to the use of *phenotypes* in the earlier solution. The use of the dash (as in C–*Ww*, for example) indicates that either allele can be at this position.

PROBLEMS AND DISCUSSION QUESTIONS

When working genetics problems in this and succeeding chapters, always assume that members of the P_1 generation are homozygous, unless the information given indicates or requires otherwise.

1 In a cross between a black and a white guinea pig, all members of the F_1 generation are black. The F_2 generation is made up of approximately 3/4 black and 1/4 white guinea pigs.
(a) Diagram this cross, showing the genotypes and phenotypes.
(b) What will the offspring be like if two F_2 white guinea pigs are mated?
(c) Two different matings were made between black members of the F_2 generation with the results shown below. Diagram each of the crosses.

Cross	Offspring
Cross 1	All black
Cross 2	3/4 black, 1/4 white

2 Albinism in humans is inherited as a simple recessive trait. For the following families, determine the genotypes of the parents and offspring. When two alternative genotypes are possible, list both.
(a) Two nonalbino (normal) parents have five children, four normal and one albino.
(b) A normal male and an albino female have six children, all normal.
(c) A normal male and an albino female have six children, three normal and three albino.
(d) Construct a pedigree of the families in (b) and (c). Assume that one of the normal children in (b) marries one of the albino children in (c) and that they have eight children.

3 Which of Mendel's postulates are illustrated by the pedigree in Problem 2? List and define these postulates.

4 Discuss the rationale used by Mendel in relating his monohybrid results to his postulates.

5 What advantages were provided by Mendel's choice of the garden pea in his experiments?

6 Pigeons may exhibit a checkered or plain pattern. In a series of controlled matings, the following data were obtained:

P$_1$ Cross	F$_1$ Progeny	
	Checkered	Plain
(a) checkered × checkered	36	0
(b) checkered × plain	38	0
(c) plain × plain	0	35

Then, F$_1$ offspring were selectively mated with the following results. The P$_1$ cross giving rise to each F$_1$ pigeon is indicated in parentheses.

F$_1$ × F$_1$ Crosses	F$_1$ Progeny	
	Checkered	Plain
(d) checkered (a) × plain (c)	34	0
(e) checkered (b) × plain (c)	17	14
(f) checkered (b) × checkered (b)	28	9
(g) checkered (a) × checkered (b)	39	0

How are the checkered and plain patterns inherited? Select and define symbols for the genes involved and determine the genotypes of the parents and offspring in each cross.

7 Mendel crossed peas having round seeds and yellow cotyledons with peas having wrinkled seeds and green cotyledons. All the F$_1$ plants had round seeds with yellow cotyledons. Diagram this cross through the F$_2$ generation using both the Punnett square and forked-line, or branch diagram, methods.

8 Determine the genotypes of the F$_2$ plants given here by analyzing the phenotypes of the offspring of these crosses.

F$_2$ Plants	Offspring
(a) round, yellow × round, yellow	3/4 round, yellow 1/4 wrinkled, yellow
(b) wrinkled, yellow × round, yellow	6/16 wrinkled, yellow 2/16 wrinkled, green 6/16 round, yellow 2/16 round, green
(c) round, yellow × round, yellow	9/16 round, yellow 3/16 round, green 3/16 wrinkled, yellow 1/16 wrinkled, green
(d) round, yellow × wrinkled, green	1/4 round, yellow 1/4 round, green 1/4 wrinkled, yellow 1/4 wrinkled, green

9 Which of the crosses in Problem 8 is a test cross?
10 Which of Mendel's postulates can only be demonstrated in crosses involving at least two pairs of traits? Define it.
11 Correlate Mendel's four postulates with what is now known about homologous chromosomes, genes, alleles, and the process of meiosis.
12 What is the basis for homology among chromosomes?
13 Distinguish between homozygosity and heterozygosity.
14 In *Drosophila*, *gray* body color is dominant to *ebony* body color, while *long* wings are dominant to *vestigial* wings. Work the following crosses through the F$_2$ generation and determine the genotypic and phenotypic ratios for each generation. Assume the P$_1$ individuals are homozygous.
(a) gray, long × ebony, vestigial
(b) gray, vestigial × ebony, long
(c) gray, long × gray, vestigial

15 How many different types of gametes can be formed by individuals of the following genotypes: (a) *AaBb*, (b) *AaBB*, (c) *AaBbCc*, (d) *AaBBcc*, (e) *AaBbcc*, and (f) *AaBbCcDdEe*? What are they in each case?

16 Using the forked-line, or branch diagram, method, determine the genotypic and phenotypic ratios of the trihybrid crosses (a) *AaBbCc × AaBBCC*, (b) *AaBBCc × aaBBCc*, and (c) *AaBbCc × AaBbCc*.

17 Mendel crossed peas with green seeds with those of yellow seeds. The F_1 generation produced only yellow seeds. In the F_2, the progeny consisted of 6022 plants with yellow seeds and 2001 plants with green seeds. Of the F_2 yellow-seeded plants, 519 were self-fertilized with the following results: 166 bred true for yellow and 353 produced a 3:1 ratio of yellow:green. Explain these results by diagraming the crosses.

18 In a study of black and white guinea pigs, 100 black animals were crossed individually to white animals and each cross was carried to an F_2 generation. In 94 of the cases, the F_1 individuals were all black, and an F_2 ratio of 3 black:1 white was obtained. In the other 6 cases, half of the F_1 animals were black and the other half were white. Why? Predict the results of crossing the black and white F_2 guinea pigs from the 6 exceptional cases.

19 Mendel crossed peas with round, green seeds to ones with wrinkled, yellow seeds. All F_1 plants had seeds that were round and yellow. Predict the results of test-crossing these F_1 plants.

20 Thalassemia is an inherited anemic disorder in humans. Individuals can be completely normal, they can exhibit a "minor" anemia, or they can exhibit a "major" anemia. Assuming that only a single gene pair and two alleles are involved in the inheritance of these conditions, which phenotype is recessive?

21 Below are shown F_2 results of two of Mendel's monohybrid crosses. Calculate the χ^2 value and determine the p value for both. Which of the two shows a greater amount of deviation?

(a) Full pods	882
Constricted pods	299
(b) Violet flowers	705
White flowers	224

22 In one of Mendel's dihybrid crosses, he observed 315 smooth, yellow, 108 smooth, green, 101 wrinkled, yellow, and 32 wrinkled, green F_2 plants. Analyze these data using the chi-square test to see if
(a) It fits a 9:3:3:1 ratio
(b) The smooth:wrinkled traits fit a 3:1 ratio
(c) The yellow:green traits fit a 3:1 ratio

23 A geneticist, in assessing data that fell into two phenotypic classes, observed values of 250:150. She decided to perform chi-square analysis using two different null hypotheses: (a) the data fit a 3:1 ratio; and (b) the data fit a 1:1 ratio. Calculate the χ^2 values for each hypothesis. What can be concluded about each hypothesis?

24 The basis for rejection of any null hypothesis is arbitrary. The researcher can set more or less stringent standards by deciding to raise or lower the p value used to reject or fail to reject the hypothesis. In the case of chi-square analysis of genetic crosses, would the use of a standard of $p = 0.10$ be more or less stringent in failing to reject the null hypothesis? Explain.

25 For the following pedigree, predict the mode of inheritance and the resulting genotypes of each individual. Assume that the alleles A and *a* control the expression of the trait.

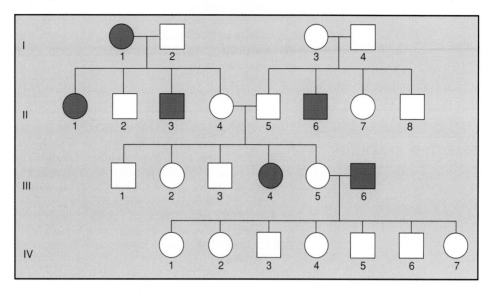

26 The following pedigree is for myopia in humans. Predict if the disorder is inherited as the result of a dominant or recessive allele. Determine the most probable genotype for each individual based on your prediction.

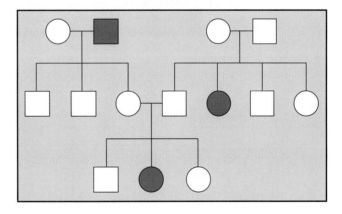

27 Draw all possible conclusions concerning the mode of inheritance of the trait denoted in the following limited pedigree:

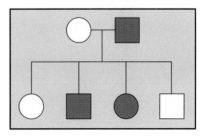

28 Consider three independently assorting gene pairs, *A/a*, *B/b* and *C/c* where each demonstrates typical dominance (*A–*, *B–*, *C–*) and recessiveness (*aa, bb, cc*). What is the *probability* of obtaining an offspring that is *AABbCc* from parents that are *AaBbCC* and *AABbCc*?

29 What is the probability of obtaining a triply recessive individual from the parents shown in Problem 28?

30 Of all offspring of the parents in Problem 28, what proportion will exhibit all three dominant traits?

SELECTED READINGS

CUMMINGS, M. R. 1991. *Human heredity: principles and issues.* 2nd ed. St. Paul: West.

DUNN, L. C. 1965. *A short history of genetics.* New York: McGraw-Hill.

OLBY, R. C. 1966. *Origins of Mendelism.* London: Constable.

PETERS, J., ed. 1959. *Classic papers in genetics.* Englewood Cliffs, N.J.: Prentice-Hall.

SNEDECOR, G. W., and COCHRAN, W. G. 1980. *Statistical methods.* 7th ed. Ames, Iowa: Iowa State University Press.

SOKAL, R. R., and ROHLF, F. J. 1981. *Biometry—The principles and practice of statistics in biological research.* 2nd ed. New York: W. H. Freeman.

SOUDEK, D. 1984. Gregor Mendel and the people around him. *Am. J. Hum. Genet.* 36: 495–98.

STERN, C., 1950. *The birth of genetics.* (Supplement to *Genetics* 35.)

STERN, C., and SHERWOOD, E. 1966. *The origins of genetics: A Mendel source book.* San Francisco: W.H. Freeman.

STUBBE, H. 1972. *History of genetics: From prehistoric times to the rediscovery of Mendel's laws.* Cambridge: MIT Press.

STURTEVANT, A. H. 1965. *A history of genetics.* New York: Harper & Row.

VOELLER, B. R., ed. 1968. *The chromosome theory of inheritance: Classical papers in development and heredity.* New York: Appleton-Century-Crofts.

4

Modification of Mendelian Ratios

POTENTIAL FUNCTION OF AN ALLELE

SYMBOLS FOR ALLELES

INCOMPLETE, OR PARTIAL, DOMINANCE

CODOMINANCE

MULTIPLE ALLELES
The ABO Antigens | The Bombay Phenotype | The Secretor Locus | The
Rh Antigens | Multiple Alleles in *Drosophila:* The *white* Locus

LETHAL ALLELES

COMBINATIONS OF TWO GENE PAIRS

GENE INTERACTION: DISCONTINUOUS VARIATION
Epistasis | Novel Phenotypes | Other Modified Dihybrid Ratios

GENE INTERACTION: CONTINUOUS VARIATION
Quantitative Inheritance: Polygenes

CHAPTER CONCEPTS

Specific phenotypes are often controlled by one or more gene pairs whose alleles exhibit modes of expression other than dominance and recessiveness. In all such cases, however, the Mendelian principles of segregation and independent assortment are operative during the distribution of the alleles into gametes.

In Chapter 3, we discussed the simplest principles of transmission genetics. We saw that genes are present on homologous chromosomes and that these chromosomes **segregate** from each other and **independently assort** with other segregating chromosomes during gamete formation. These two postulates are the fundamental principles of gene transmission from parent to offspring. However, once an offspring has received the total set of genes, it is the expression of genes that determines the organism's phenotype. If gene expression does not adhere to a simple dominant/recessive mode, or if more than one pair of genes influences the expression of a single trait, the classic 3:1 and 9:3:3:1 F_2 ratios may be modified. Although in this and the next several chapters we consider more complex modes of inheritance, the fundamental principles set down by Mendel hold true in these situations as well.

In this chapter, our discussion will initially be restricted to the inheritance of traits that are under the control of only one gene. In diploid organisms, where homologous pairs of chromosomes exist, two copies of each gene influence such traits. The copies need not be identical since alternate forms of genes, or **alleles,** occur within populations. How alleles act to influence a given phenotype will be our major consideration. Then we will proceed to consider how a single phenotype may be controlled by more than one gene. This general phenomenon is referred to as **gene interaction,** indicating that phenotypes are frequently under the influence of more than one gene product. Numerous examples will be presented to illustrate a variety of heritable patterns observed in such situations.

POTENTIAL FUNCTION OF AN ALLELE

Following the rediscovery of Mendel's work in the early 1900s, research focused on the many ways in which genes can influence an individual's phenotype. This course of investigation, stemming from Mendel's findings, is called **neo-Mendelian genetics** (*neo* from the Greek word meaning since or new).

Each type of inheritance described in this chapter was investigated when observations of genetic data did not precisely conform to the expected Mendelian ratios. Hypotheses that modified and extended the Mendelian principles were proposed and tested with specifically designed crosses. The explanations for these observations were in accordance with the principle that a phenotype is under the control of one or more genes located at specific loci on one or more pairs of homologous chromosomes. If we adhere to the principles of segregation and independent assortment, we can predict accurately the transmission of any number of allele pairs.

To understand the various modes of inheritance, we must first examine the potential function of an allele. Alleles are alternate forms of the same gene; therefore, they contain modified genetic information and specify an alteration in the original gene product. For example, there are well over 100 alleles of the genes that specify the protein portion of human hemoglobin. Even though each allele specifies a modification of one of the components of hemoglobin, they all store information necessary for this molecule's chemical synthesis. Once manufactured, however, the product of an allele may or may not be functional. The allele that occurs most frequently in a population or the one that is arbitrarily designated as normal is called **wild type**. This common allele is usually dominant, such as the allele for tall plants in the garden pea, and its product is functional in the cell. Wild-type alleles thus, serve as standards for comparison against all mutations occurring at a particular locus.

The process of **mutation** is the source of new alleles. Each allele may be recognized by a change in the phenotype. A new phenotype results from a change in functional activity of the cellular product controlled by that gene. Usually, the alteration or mutation is expressed as a loss of the specific wild-type function. For example, if a gene is responsible for the synthesis of a specific enzyme, a mutation in the gene may change the conformation of this enzyme, thus eliminating its affinity for the substrate. This mutation results in a total loss of function. On the other hand, another organism may have a different mutation in this gene, which results in an enzyme with a reduced or increased affinity for binding the substrate. This mutation may reduce or enhance rather than eliminate the functional capacity of the gene product. In either case, the phenotype may or may not be altered in a discernible way.

Although phenotypic traits may be affected by a single mutation, traits are often the result of many gene products. In the case of enzymatic reactions, most are part of complex metabolic pathways. Therefore, phenotypic traits may be under the control of more than one gene and the allelic forms of each gene involved. In the initial part of this chapter, examples will be restricted to only one gene pair and the alleles associated with it. Thus, in these cases we see a modification of the 3:1 monohybrid ratio. Then we will consider traits controlled by two genes and the accompanying modification of the 9:3:3:1 dihybrid ratio.

SYMBOLS FOR ALLELES

In Chapter 3, we learned to symbolize alleles for very simple Mendelian traits. We used the lowercase form of the initial letter of the name of a recessive trait to denote the recessive allele and the same letter in uppercase form to refer to the dominant allele. Thus, for tall and dwarf, where dwarf is recessive, D and d represent the alleles responsible for these respective traits.

As more complex inheritance patterns were investigated, another useful system was developed to discriminate between wild-type and mutant traits. In this system, the initial letter of the name of the mutant trait is selected. If the trait is recessive, the lowercase form is used; if it is dominant, the uppercase form is used. The contrasting wild-type trait is denoted by the same letter, but with a + as a superscript.

For example, *ebony* is a recessive body color mutation in the fruit fly, *Drosophila melanogaster*. The normal wild type body color is gray. Using the above system, *ebony* is denoted by the symbol e while gray is denoted by e^+. If we focus on the *ebony* mutation, the responsible locus may be occupied by either the wild type allele (e^+) or the mutant allele (e). A diploid fly may thus exhibit three possible genotypes:

e^+/e^+: gray homozygote (wild type)
e^+/e: gray heterozygote (wild type)
e/e: ebony homozygote

The slash is used to indicate that the two allele designations represent the same locus on two homologous chromosomes. If we were instead considering a dominant mutation such as *Wrinkled (Wr)*, the three possible designations would be Wr^+/Wr^+, Wr^+/Wr, and Wr/Wr. The latter two genotypes express the wrinkled-wing phenotype.

One advantage of this system is that further abbreviation may be used when convenient: the wild type allele may simply be denoted by the + symbol. Using *ebony* as an example under consideration in a cross, the designations of the three possible genotypes become:

$+/+$: gray homozygote (wild type)
$+/e$: gray heterozygote (wild type)
e/e: ebony homozygote (mutant)

As we will see in Chapter 6, this abbreviation is particularly useful when two or three genes linked together on the same chromosome are considered simultaneously.

Still other allele designations are sometimes useful. The system just described works well with alleles which are either dominant or recessive to one another. However, if no dominance exists, we may simply use uppercase letters and superscripts to denote alleles, e.g., R^1 and R^2, L^M and L^N, I^A and I^B. Their use will become apparent in the ensuing three sections.

Finally, note that in each of the many crosses discussed in the next few chapters, only one or a few gene pairs are involved. It may be useful for you to remember that in each cross, all genes not under consideration are assumed to be normal, or wild type.

INCOMPLETE, OR PARTIAL, DOMINANCE

Incomplete, or **partial, dominance** in the offspring is based on the observation of intermediate phenotypes generated by a cross between parents with contrasting traits. For example, if plants such as four-o'clocks or snapdragons with red flowers are crossed with plants with white flowers, offspring may have pink flowers. It appears that neither red nor white flower color is dominant. Since some red pigment is produced in the F_1 intermediate-colored pink flowers, dominance appears to be incomplete or partial.

If this phenotype is under the control of a single pair of alleles where neither is dominant, the results of the F_1 (pink) $\times$ F_1 (pink) cross can be predicted. The resulting F_2 generation is shown in Figure 4.1, confirming the hypothesis that only one pair of alleles determines these phenotypes. The genotypic ratio (1:2:1) of the F_2 generation is identical to that of Mendel's monohybrid cross. Because there is no dominance, however, the phenotypic ratio is identical to the genotypic ratio. Note here that since neither of the alleles is recessive, we have chosen not to use upper and lowercase letters. Instead, we have chosen R^1 and R^2 to denote the red and white alleles.

Incomplete dominance, which results in an intermediate expression of the overt phenotype, is relatively rare. However, even when complete dominance is evident, careful examination of the level of the gene product, rather than the phenotype, shows some intermediate gene expression. For example, in human biochemical disorders such as **Tay-Sachs disease,** homozygous recessive individuals are severely affected while heterozygotes are phenotypically normal. In affected individuals there is almost no activity of the enzyme **hexosaminidase**. Heterozygotes, on the other hand, express only about 50 percent of the enzyme activity found in homozygous normal individuals. Fortunately, this level of enzyme activity is sufficient to allow normalcy. This situation is not uncommon in enzyme disorders. It illustrates the somewhat arbitrary nature of the terms *dominance* and *recessiveness.*

CODOMINANCE

If one pair of alleles is responsible for the production of two distinct and detectable gene products, a situation unlike that of incomplete dominance or dominance/recessiveness arises. The distinct genetic expression of both alleles is called **codominance**. For example, the **MN blood groups** in humans are characterized by certain molecules called glycoproteins found on the surface of red blood cells. Discovered by Karl Landsteiner and Philip Levine, these molecules are **native antigens** that provide immunological identity to individuals. Each native antigen can elicit an antibody response if it is present on tissue transfused or transplanted to another individual lacking the antigen.

The MN system, under the control of a locus on chromosome 4, results in persons having blood group M, MN, or N. The MN phenotypes of offspring of parents with the various blood group combinations are as follows:

Parental Phenotypes	Offspring Phenotypes
M $\times$ M	All M
N $\times$ N	All N
M $\times$ N	All MN
M $\times$ MN	1/2 M : 1/2 MN
N $\times$ MN	1/2 N : 1/2 MN
MN $\times$ MN	1/4 M : 1/2 MN : 1/4 N

Examination of these results confirms that only a single pair of alleles is involved in MN inheritance. The results are consistent with the hypothesis that two alleles (L^M and L^N) determine the three phenotypes: $L^M L^M$ results in blood group M; $L^M L^N$ results in group MN; and $L^N L^N$ results in group N. The offspring of two heterozygous MN parents appear in a 1:2:1 ratio, the same as that found for the incomplete dominance mode of inheritance. However, while incomplete dominance produces an intermediate, blending effect in heterozygous individuals, codominant inheritance results in distinct evidence of the gene products of both alleles. Once again, the mode of inheritance can only be discerned by analyzing the specific gene products.

MULTIPLE ALLELES

Because the information stored in any gene is extensive, mutations may modify this information in many ways. Each change has the potential for producing a different allele. Therefore, at any given locus on the chromosome, the number of alleles within a population of individuals need not be restricted to only two. When three or more alleles are found for any particular gene, the mode of inheritance is called **multiple allelism.**

The concept of multiple alleles can only be studied in populations. Any individual diploid organism has, at most, two homologous gene loci which may be occupied by different alleles. However, among members of a species, many alternative forms of the same gene may exist. The following examples

FIGURE 4.1
Incomplete dominance illustrated by flower color. The photograph illustrates not only red, white, and pink snapdragons, but other colors produced as a result of the effect of other genes.

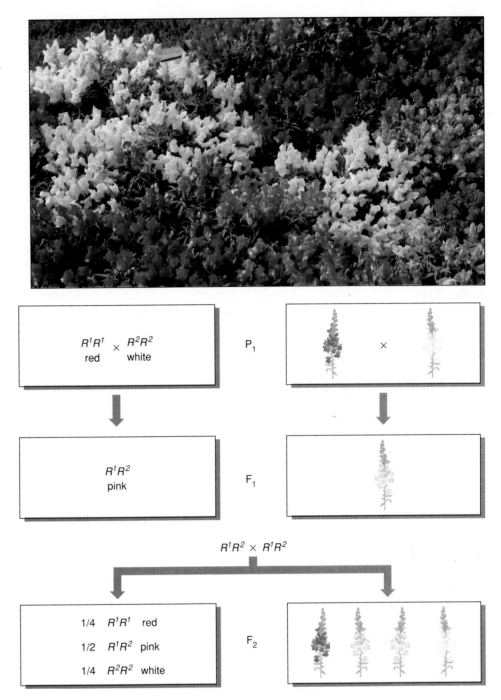

R^1R^1 × R^2R^2
red   white

P₁

R^1R^2
pink

F₁

R^1R^2 × R^1R^2

1/4 R^1R^1 red

1/2 R^1R^2 pink

1/4 R^2R^2 white

F₂

illustrate the concept of multiple alleles. In several of the examples the relationship between genetics, immunology, and medicine is shown.

The ABO Antigens

The simplest possible case of multiple alleles is that in which there are three alleles of one gene. This situation exists in the inheritance of the **ABO blood types** in humans, discovered by Landsteiner in the early 1900s. The ABO system, like the MN blood types, is characterized by the presence of native antigens on the surface of red blood cells. The ABO antigens are distinct from the MN antigens and are under the control of a different gene, located at a distinct locus on chromosome 9. As in the MN system, one combination of alleles in the ABO system exhibits a codominant mode of inheritance.

The ABO phenotype of any individual is ascertained by mixing a blood sample with antiserum containing type A or type B antibodies. If the antigen is present in the blood, it will react with an antibody that was formed against it and cause clumping or agglutination of the red blood cells. When individuals are tested in this way, four phenotypes are revealed. Each individual has either the A antigen (A phenotype), the B antigen (B phenotype), the A and B antigens (AB phenotype), or neither antigen (O phenotype). In 1924, it was hypothesized that these phenotypes were inherited as the result of three alleles of a single gene. This hypothesis was based on studies of the blood types of many different families.

Although different designations may be used, we will use the symbols I^A, I^B, and I^O for the three alleles. The I designation stands for **isoagglutinogen,** another term for antigen. If we assume that the I^A and I^B alleles are responsible for the production of their respective A and B antigens and that I^O is an allele that does not produce any detectable A or B antigens, the various genotypic possibilities can be listed and the appropriate phenotype assigned to each:

Genotype	Antigen	Phenotype
$I^A I^A$	A	
$I^A I^O$	A	A
$I^B I^B$	B	
$I^B I^O$	B	B
$I^A I^B$	A, B	AB
$I^O I^O$	Neither	O

Note that in these assignments the I^A and I^B alleles behave dominantly to the I^O allele, but codominantly to each other. We can test the hypothesis that three alleles control ABO blood types by examining potential offspring from all possible matings, as shown in Table 4.1. If we assume heterozygosity wherever possible, we can predict which phenotypes can occur. These theoretical predictions have been upheld in numerous studies examining the blood types of children of parents with all possible phenotypic combinations. The hypothesis that three alleles control ABO blood types in the human population is now universally accepted.

Our knowledge of human blood types has several practical applications. Compatible blood transfusions can be achieved and decisions about disputed parentage more accurately made. The latter cases can occur when newborns are inadvertently mixed up in hospitals, or when an unmarried male is suspected of, but denies, fathering a child out of wedlock. In both cases, an examination of the ABO phenotypes as well as other inherited antigens of the possible parents and the child may help to resolve the situation. Table 4.1 demonstrates numerous cases where it is impossible for a parent of a particular ABO phenotype to produce a child of a certain phenotype. In fact, the only mating that can result in offspring of all four phenotypes is between two heterozygous individuals, one showing the A phenotype and the other showing the B phenotype. On genetic grounds alone, a male or female may be unequivocally ruled out as the parent of a certain child. On the other hand, it should be obvious that this type of genetic evidence *never proves* parenthood.

The Bombay Phenotype

The biochemical basis of the ABO blood type system has now been carefully worked out. The A and B

TABLE 4.1
Potential phenotypes in the offspring of parents with all possible ABO blood type combinations, assuming heterozygosity whenever possible.

P₁ Generation		F₁ Phenotype			
Phenotypes	Genotypes	A	B	AB	O
A × A	$I^A I^O \times I^A I^O$	3/4	—	—	1/4
B × B	$I^B I^O \times I^B I^O$	—	3/4	—	1/4
O × O	$I^O I^O \times I^O I^O$	—	—	—	all
A × B	$I^A I^O \times I^B I^O$	1/4	1/4	1/4	1/4
A × AB	$I^A I^O \times I^A I^B$	1/2	1/4	1/4	—
A × O	$I^A I^O \times I^O I^O$	1/2	—	—	1/2
B × AB	$I^B I^O \times I^A I^B$	1/4	1/2	1/4	—
B × O	$I^B I^O \times I^O I^O$	—	1/2	—	1/2
AB × O	$I^A I^B \times I^O I^O$	1/2	1/2	—	—
AB × AB	$I^A I^B \times I^A I^B$	1/4	1/4	1/2	—

antigens are actually carbohydrate groups (sugars) that are bound to lipid molecules (fatty acids) protruding from the membrane of the red blood cell. The specificity of the A and B antigens is based on the terminal sugar of the carbohydrate group.

Almost all individuals possess what is called the **H substance,** to which a terminal sugar is added. As shown in Figure 4.2, the H substance itself consists of three sugar molecules, N-acetylglucosamine, galac-

tose, and fucose, linked together. The I^A allele is responsible for an enzyme that can add the terminal sugar N-acetylgalactosamine to the H substance. The I^B allele is responsible for a modified enzyme that cannot add N-acetylgalactosamine, but instead can add the terminal sugar galactose. Heterozygotes ($I^A I^B$) add either one or the other entity at the many sites available. This latter phenomenon illustrates the biochemical basis of codominance in individuals of

FIGURE 4.2
The biochemical basis of the ABO blood groups. The *H* allele, present in almost all humans, directs the conversion of a precursor molecule to the H substance by adding a molecule of fucose. Failure to do so results in the Bombay phenotype. The I^A and I^B alleles are then able to direct the addition of terminal sugar residues to the H substance. The I^O allele is unable to direct either of these terminal additions. Gal: galactose; AcGLuNH: N-acetyl-D-glucosamine; AcGa1NH: N-acetylgalactosamine.

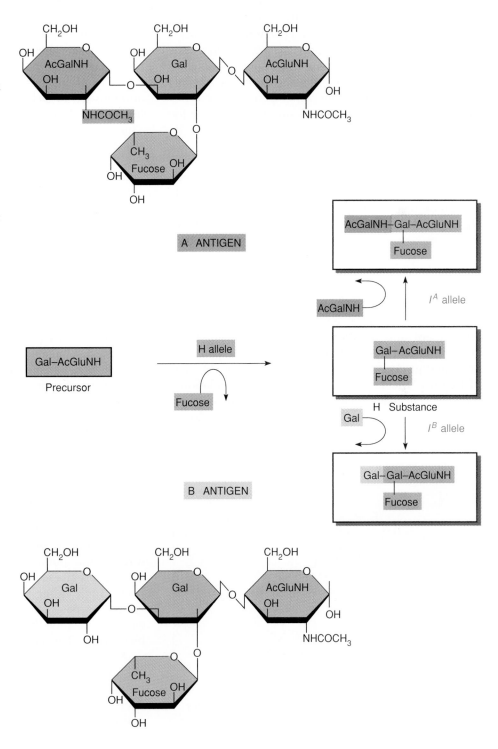

the AB blood type. Persons of type O (I^OI^O) cannot add either terminal sugar, possessing only the H substance protruding from the surface of their red blood cells.

The molecular basis of the mutations leading to the I^A, I^B, and I^O alleles has recently been elucidated. We shall return to this topic in Chapter 14 when we discuss mutation and mutagenesis.

In 1952, a most interesting situation was observed in a woman in Bombay which provided information concerning the genetic basis of the H substance. She displayed a unique genetic history that was inconsistent with her blood type. In need of a transfusion, she was found to lack both the A and B antigens and was thus typed as O. However, as shown in the partial pedigree in Figure 4.3, one of her parents was type AB and she was the obvious donor of an I^B allele to two of her offspring. Thus she was genetically type B but functionally type O!

It was subsequently shown that this woman was homozygous for a rare recessive mutation, h, which prevented her from synthesizing the complete H substance. The terminal portion of the carbohydrate chain protruding from the red cell membrane was shown to lack fucose. In the absence of fucose, the enzymes specified by the I^A and I^B alleles are apparently unable to recognize the incomplete H substance as a proper substrate. Thus, neither the terminal galactose or N-acetylgalactosamine can be added, even though the enzymes capable of doing so are present and functional. As a result, the ABO system genotype cannot be expressed in individuals of genotype hh, and they appear as type O. To distinguish them from the rest of the population, they are said to demonstrate the **Bombay phenotype**. The frequency of the h allele is exceedingly low.

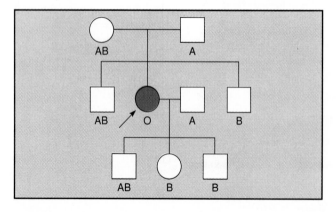

FIGURE 4.3
A partial pedigree of a woman displaying the Bombay phenotype. Functionally, her ABO blood group behaves as type O. Genetically, she is type B.

Thus, the vast majority of the human population is of the HH genotype and can synthesize the H substance.

The Secretor Locus

Still a third gene is known to affect the expression of the ABO blood type system: the **secretor locus**. In most people, the A and B antigens are present in various body secretions as well as on the membrane of red blood cells. The ability to secrete these antigens in body fluids such as saliva is under the influence of the dominant allele, *Se*, carried by more than three-fourths of the human population (*Se/Se* or *Se/se*). The minority of the population who do not secrete the antigens (*se/se*) lack an enzyme that normally modifies the H substance, rendering it water soluble. Thus, nonsecretors make the antigens as specified by the ABO loci but do not secrete them.

The Rh Antigens

Another set of antigens thought by some geneticists to illustrate multiple allelism includes those designated Rh. Discovered by Landsteiner, Levine, and others around 1940, the Rh antigens have received a great deal of attention because of their direct involvement in the disorder **erythroblastosis fetalis**. The initial investigations led to the belief that in the human population only two alleles controlled the presence or absence of the antigen. It was thought that the Rh^+ allele determined the presence of the antigen and behaved as a dominant gene. The Rh^- allele seemed to result in the absence of the antigen.

Erythroblastosis fetalis, also referred to as **hemolytic disease of the newborn (HDN)**, is a form of anemia. It occurs in an Rh-positive fetus whose mother is Rh-negative and whose father is Rh-positive, contributing that allele to the fetus. Such a genetic combination results in a potential immunological incompatibility between the mother and fetus. If fetal blood passes through the ruptured placenta at birth and enters the maternal circulation, the mother's immune system recognizes the Rh antigen as foreign and builds antibodies against it. During a second pregnancy, the antibody concentration becomes high enough that when maternal antibodies, which can pass across the placenta, enter the fetus's circulation, they begin to destroy the fetus's red blood cells. This causes the hemolytic anemia.

About 10 percent of all human pregnancies demonstrate Rh incompatibility. However, for numerous reasons, less than 0.5 percent actually result in anemia. Currently, incompatible mothers are given anti-Rh antisera immediately after giving birth to an

Rh-positive baby. This destroys any Rh-positive cells that have entered the mother's circulation so that she does not produce her own anti-Rh antibodies. Before this treatment was developed, many fetuses failed to survive to term. For those that did, complete blood transfusions were often necessary. Even then, many newborns failed to survive.

With the development of more refined antisera to test for the presence of the antigen, it became apparent that the genetic control of Rh antigens was much more complex than originally thought. For example, some presumed Rh-negative blood was found to contain the antigen, but in a different chemical form. Alexander Wiener has proposed the existence of at least eight alleles at a single Rh locus. Other workers, including Ronald A. Fisher, Robert R. Race, and Ruth Sanger, have proposed an alternative type of inheritance. These workers feel that there are three closely linked genes, each with two alleles involved in the inheritance of Rh factors. The term **linkage** is used to describe genes located on the same chromosome. We will discuss this concept in detail in Chapter 6.

The two systems of nomenclature designating the alleles are contrasted in Table 4.2. In the Wiener system, the presence of at least one of the four dominant alleles is sufficient to yield the Rh-positive blood type. Examination of the Fisher-Race system shows that the genetic locus bearing the set of *D* and *d* alleles is most critical. The presence of at least one dominant *D* allele results in the Rh-positive phenotype, while the *dd* genotype ensures the Rh-negative blood type. While the *C, c, E,* and *e* alleles specify distinguishable antigens, they are not immunologically significant. Because of the complexity of the antigenic patterns, it is difficult to favor one system over the other.

TABLE 4.2
A comparison of the alleles involved in the Wiener system with those of the Fisher-Race system to explain the genetic basis of the Rh blood groups.

Wiener Nomenclature	Fisher-Race Nomenclature	Phenotype
R^1	CDE	
R^2	CDe	
R^0	cDE	RH$^+$
R^z	cDe	
r	CdE	
r'	Cde	
r''	cdE	Rh$^-$
r^y	cde	

Multiple Alleles in *Drosophila:* The *white* Locus

Many other phenotypes in plants and animals are known to be controlled by multiple allelic inheritance. In *Drosophila*, for example, where the induction of mutations has been used extensively as a method of investigation, many alleles are known at practically every locus. The *white* eye mutation, discovered by Thomas H. Morgan and Calvin Bridges in 1912, is only one of over 100 alleles that may occupy this locus. In this allelic series, eye colors range from complete absence of pigment in the *white* allele, to deep ruby in the *white-satsuma* allele, to orange in the *white-apricot* allele, to a buff color in the *white-buff* allele. These alleles are designated w, w^{sat}, w^a and w^{bf}, respectively (Table 4.3). In each of these cases, the total amount of pigment in these mutant eyes is reduced to less than 20 percent of that found in the brick red wild-type eye.

LETHAL ALLELES

Many gene products are essential to an organism's survival. Mutations resulting in the synthesis of a gene product that is nonfunctional can sometimes be tolerated in the heterozygous state; that is, one wild-type allele may be sufficient to produce enough of the essential product to allow survival. However, such a mutation behaves as a **recessive lethal allele,** and homozygous recessive individuals will not survive. The time of death will depend upon when the product is needed during development. In instances where even one copy of the wild gene is not sufficient for normal development, the heterozygote will not survive. In this case, the mutation is behaving as a **dominant lethal allele** because its presence somehow overrides the expression of the wild-type product, or the amount of wild type product is simply insufficient to support its essential function.

In some cases, the allele responsible for a lethal effect when homozygous may result in a distinctive mutant phenotype when present heterozygously. Such an allele is behaving as a recessive lethal but is dominant with respect to the phenotype. For example, a mutation causing a yellow coat in mice was discovered in the early part of this century. The yellow coat varied from the normal agouti-coat phenotype, as illustrated in Figure 4.4. Crosses between the various combinations of the two strains yielded unusual results:

Cross A: agouti × agouti → all agouti
Cross B: yellow × yellow → 2/3 yellow: 1/3 agouti
Cross C: agouti × yellow → 1/2 yellow: 1/2 agouti

TABLE 4.3
Some of the alleles present at the white locus of *Drosophila melanogaster* and their eye color phenotype.

Allele	Name	Eye Color
w	white	pure white
w^a	white-apricot	yellowish
w^{bf}	white-buff	light buff
w^{bl}	white-blood	yellowish ruby
w^{cf}	white-coffee	deep ruby
w^e	white-eosin	yellowish pink
w^{mo}	white-mottled orange	light mottled orange
w^{sat}	white-satsuma	deep ruby
w^{sp}	white-spotted	fine grain, yellow mottling
w^t	white-tinged	light pink

These results are explained on the basis of a single pair of alleles, one of which behaves as an autosomal recessive lethal. The mutant *yellow* allele (A^Y) is dominant to the wild-type *agouti* allele (A), so heterozygous mice will have yellow coats. However, the yellow allele also behaves as a recessive lethal; thus no homozygous yellow mice are ever recovered. The genetic basis for these three crosses is provided in Figure 4.4.

Many genes are known to behave similarly in other organisms. In *Drosophila, Curly* wing (*Cy*), *Plum* eye (*Pm*), *Dichaete* wing (*D*), *Stubble* bristle (*Sb*), and *Lyra* wing (*Ly*) behave as recessive lethals but are dominant with respect to the expression of the mutant phenotype when heterozygous.

Other alleles are known to behave as dominant lethals. In humans, a disorder called **Huntington disease** (also referred to as Huntington chorea) is due to a dominant allele that behaves quite differently from the alleles just described. While individuals homozygous for the lethal gene apparently never survive through fetal development, heterozygotes develop normally well into adulthood. Affected individuals then undergo gradual nervous and motor degeneration until they die. This lethal disorder is particularly tragic because it has such a late onset, typically at about age 40. By that time, the affected individual may have produced a family. Each child has a 50% probability of also developing the disorder and passing the lethal gene to his or her offspring. The American folk singer and composer Woody Guthrie died from this disease.

Dominant lethal alleles are rarely observed. In order for them to exist in a population, the affected individual must reproduce before the allele's lethality is expressed. If all affected individuals die before reaching the reproductive age, the mutant gene will

FIGURE 4.4
Inheritance patterns in three crosses involving the mutant *yellow* allele (A^Y) in the mouse. Note that the mutant allele behaves dominantly to the wild-type agouti (A) allele in controlling coat color, but it also behaves as a recessive lethal allele. The genotype $A^Y A^Y$ does not survive.

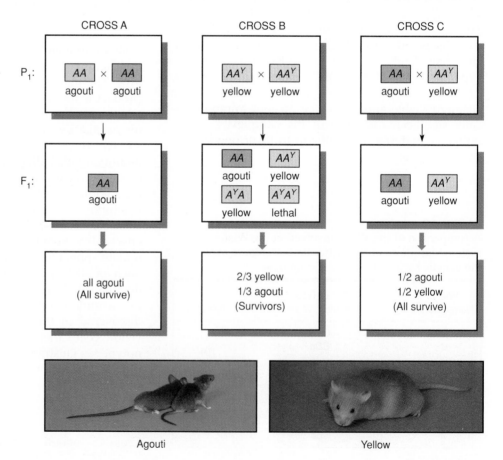

not be passed to future generations. The mutation will thus disappear from the population unless it recurs as a result of mutation.

COMBINATIONS OF TWO GENE PAIRS

Each example discussed so far modifies Mendel's $3:1$ F_2 monohybrid ratio. Therefore, combining any two of these modes of inheritance in a dihybrid cross will likewise modify the classical $9:3:3:1$ ratio. Having established the foundation for the modes of inheritance of incomplete dominance, codominance, multi-

ple alleles, and lethal genes, we can now deal with the situation of two modes of inheritance occurring simultaneously. Mendel's principle of independent assortment applies to these situations, provided that the genes controlling each character are not located on the same chromosome.

Suppose, for example, that a mating occurs between two humans who are both heterozygous for the autosomal recessive gene that causes albinism and who are both of blood type AB. What is the probability of any particular phenotypic combination occurring in each of their children? Albinism is

FIGURE 4.5
A theoretical modified dihybrid cross involving the ABO blood type and albinism in humans. The probabilities of each phenotype in the offspring of such a mating are calculated using a Punnett square.

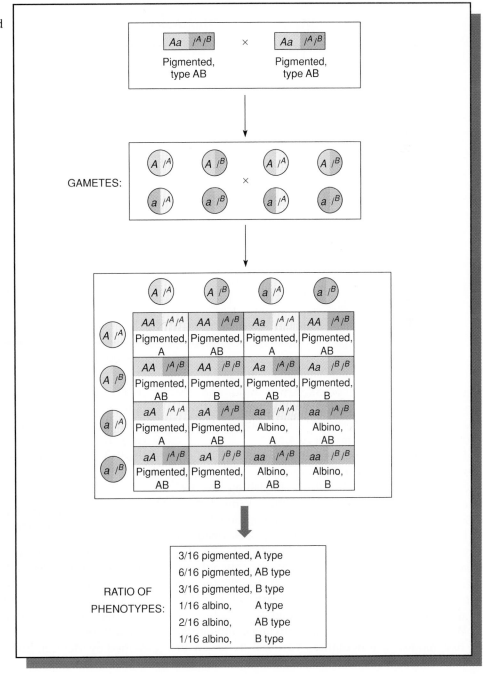

FIGURE 4.6
The calculation of the probabilities in the mating illustrated in Figure 4.5 using the forked-line method.

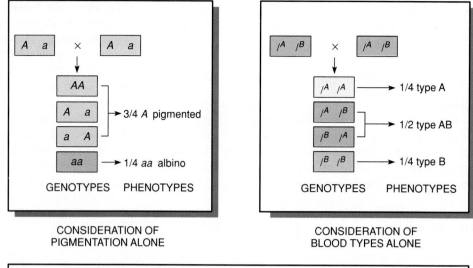

CONSIDERATION OF
PIGMENTATION ALONE

CONSIDERATION OF
BLOOD TYPES ALONE

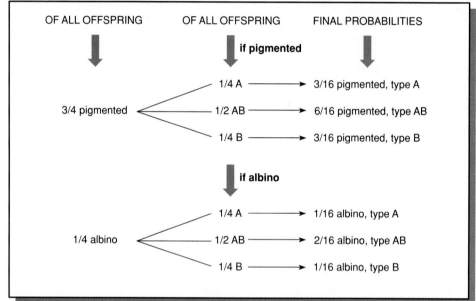

CONSIDERATION OF BOTH TRAITS TOGETHER

inherited in the simple Mendelian fashion, and the blood types are determined by the series of three multiple alleles, I^A, I^B, and I^O. The solution to this problem is diagramed in Figure 4.5.

Instead of this dihybrid cross yielding the classical four phenotypes in a 9:3:3:1 ratio, six phenotypes occur in a 3:6:3:1:2:1 ratio, establishing the expected probability for each phenotype. While Figure 4.5 solves the problem using the conventional Punnett square, the forked-line method described in Chapter 3 simplifies this task (Figure 4.6). Recall that this method simply requires that the phenotypic ratios for each trait be computed individually (which can usually be done by inspection); all possible combinations can then be calculated.

We can deal in a similar way with any combination of two modes of inheritance. You will be asked to

determine the phenotypes and their expected probabilities for many of these combinations when you solve the problems at the end of the chapter. In each case, the final phenotypic ratio is a modification of the 9:3:3:1 dihybrid ratio.

GENE INTERACTION: DISCONTINUOUS VARIATION

Soon after the rediscovery of Mendel's work, experimentation revealed that phenotypic characters were often under the control of more than one gene pair. This was a significant discovery because it revealed for the first time that genetic influence on the phenotype is much more sophisticated than envisioned by Mendel. Instead of single genes controlling the development of individual parts of the plant and animal body, it soon became clear that each phenotypic character is influenced by many gene products.

The concept of **gene interaction** does not mean that two or more genes, or their products, necessarily interact directly to influence a particular phenotype. Instead, this concept implies that the cellular function of numerous gene products is related to the development of a common phenotype. To clarify this point, we will present several examples that directly illustrate gene interaction at the biochemical level.

In the following discussion, we will make several assumptions and adopt certain conventions:

1 In each case, distinct phenotypic classes are produced, each clearly discernible from all others. Such traits illustrate **discontinuous variation,** where discrete phenotypic categories exist.
2 The genes considered in each cross are unlinked and therefore assort independently of one another during gamete formation.
3 If complete dominance exists between the alleles of any gene pair, such that AA and Aa or BB and Bb are equivalent in their genetic effects, the designations $A-$ or $B-$ will be used for both combinations. Therefore, the dash $(-)$ indicates that either allele may be present, without consequence to the phenotype.
4 All P_1 crosses will involve homozygous individuals (e.g., $AABB \times aabb$, $AAbb \times aaBB$, or $aaBB \times AAbb$). Therefore, each F_1 generation will consist of only heterozygotes of genotype $AaBb$.
5 In each example, the F_2 generation produced from these heterozygous parents will be the main focus of analysis. When two genes are involved (Figure 4.7, p. 86), the F_2 genotypes fall into four categories: $9/16$ $A-B-$, $3/16$ $A-bb$, $3/16$ $aaB-$, and $1/16$ $aabb$. Because of dominance, all genotypes in each category are equivalent in their effect on the phenotype.

The study of gene interaction has revealed inheritance patterns in which these four categories are grouped together in various ways, such that each grouping yields a different phenotype. Thus, the $9:3:3:1$ ratio, characteristic of Mendel's dihybrid cross, is modified in several ways (Figure 4.8, p. 87). In the next several sections, we shall proceed to discuss a number of examples from Figure 4.8.

Epistasis

Perhaps the best examples of gene interaction are those illustrating the phenomenon of **epistasis**. De-

rived from the Greek word meaning "stoppage," epistasis can in genetics be equated with the word *masking*. The phenomenon occurs when the expression of one gene pair masks or modifies the expression of another gene pair. This masking may occur under different conditions. For example, the presence of two recessive alleles at one locus may prevent or override the expression of alleles at a second locus (or several other loci). Or, a single dominant allele at the first locus may influence the expression of the alleles at a second gene locus. In a third case, two gene pairs may complement one another such that at least one dominant allele at each locus is required to express a particular phenotype. Each of these three forms of epistasis will be examined in more detail.

An example of the homozygous recessive condition at one locus masking the expression of a second locus has been examined earlier in this chapter when we discussed the Bombay phenotype. There, the homozygous condition (hh) masked the expression of the I^A and I^B alleles. Another example is seen in the inheritance of coat color in mice (Case 1 of Figure 4.8). Wild-type coat color is agouti, a grayish pattern formed by alternating bands of pigment on each hair. Agouti is dominant to nonagouti (black) hair caused by a recessive mutation, b. Thus, $B-$ results in agouti, while bb yields nonagouti coat color. When homozygous, a recessive mutation, a, at a separate locus, eliminates pigmentation altogether, yielding albino mice.

In a cross between agouti ($AABB$) and albino ($aabb$), members of the F_1 are all $AaBb$ and have agouti coat color. In the F_2 progeny of a cross between two F_1 double heterozygotes, the following genotypes and phenotypes are observed:

F_1: $AaBb \times AaBb$

F_2 Ratio	Genotype	Phenotype	Final Phenotypic Ratio
9/16	$A-B-$	agouti	9/16 agouti
3/16	$A-bb$	black	3/16 black
3/16	aa $B-$	albino	4/16 albino
1/16	aa bb	albino	

Gene interaction yielding the observed $9:3:4$ F_2 ratio can be envisioned as a two-step process:

	Gene A		Gene B	
Precursor	↓	Black	↓	Agouti
Molecule	⟶	Pigment	⟶	Pattern
(colorless)	$A-$		$B-$	

FIGURE 4.7
Generation of the four
genotypic groupings from the
nine genotypes produced in a
cross between double
heterozygotes.

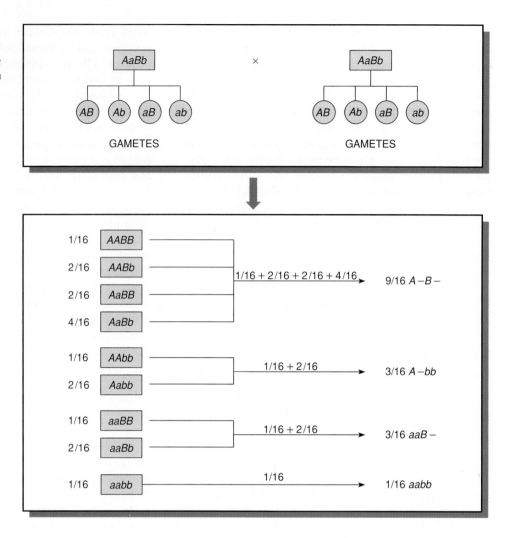

In the presence of an *A* allele, black pigment can be
made from a colorless substance. In the presence of a
B allele, the black pigment is deposited during the
development of hair in a pattern producing the agouti
phenotype. If the *bb* genotype occurs, all of the hair
remains black. If the *aa* genotype occurs, no black
pigment is produced, regardless of the presence of
the *B* or *b* alleles, and the mouse is albino. Therefore,
the *aa* genotype masks or suppresses the expression
of the *B* gene.

A second type of epistasis occurs when a dominant allele at one genetic locus masks the expression of the alleles of a second locus. For instance,
Case 2 of Figure 4.8 deals with the inheritance of
fruit color in summer squash. Here, the dominant
allele *A* results in white fruit color regardless of the
genotype at a second locus, *B*. In the absence of the
dominant *A* allele (the *aa* genotype), *BB* or *Bb*
results in yellow color while *bb* results in green
color. Therefore, if two white-colored double het-

erozygotes (*AaBb*) are crossed together, an interesting genetic ratio occurs because of this type of
epistasis.

$$F_1: AaBb \times AaBb$$

F_2 Ratio	Genotype	Phenotype	Final Phenotypic Ratio
9/16	*A– B–*	white ⎫	
3/16	*A– bb*	white ⎭	12/16 white
3/16	*aa B–*	yellow	3/16 yellow
1/16	*aa bb*	green	1/16 green

Of the offspring, 9/16 are *A–B–* and are thus white.
The 3/16 bearing the genotypes *A–bb* are also white.
Of the remaining squash, 3/16 are yellow (*aaB–*),
while 1/16 are green (*aabb*). Thus, the modified ratio
of 12:3:1 occurs.

Our third example (Case 3 of Figure 4.8) is
demonstrated in a cross between two strains of

Case	Organism	Character	AABB 1/16	AABb 2/16	AaBB 2/16	AaBb 4/16	AAbb 1/16	Aabb 2/16	aaBB 1/16	aaBb 2/16	aabb 1/16	Final Phenotypic Ratio
			F₁: AaBb × AaBb → F₂ Genotypes									
	Pea	Mendel's dihybrid	9/16				3/16		3/16		1/16	9:3:3:1
1	Mouse	Coat color	Agouti				Black		Albino			9:3:4
2	Squash	Color	White						Yellow		Green	12:3:1
3	Pea	Flower color	Purple				White					9:7
4	Squash	Fruit shape	Disc				Sphere				Long	9:6:1
5	Chicken	Color	White						Colored		White	13:3
6	Mouse	Color	White-spotted				White		Colored		White-spotted	10:3:3
7	Shepherd's purse	Seed capsule	Triangular								Ovoid	15:1
8	Flour beetle	Color	Red	Sooty	Red	Sooty	Black		Jet		Black	6:3:3:4

FIGURE 4.8

The basis of modified dihybrid F₂ phenotypic ratios resulting from crosses between doubly heterozygous F₁ individuals. The four groupings shown in Figure 4.7 are combined in various ways to produce these ratios.

white-flowered sweet peas. Unexpectedly, the F₁ plants were all purple, and the F₂ occurred in a ratio of 9/16 purple to 7/16 white. The proposed explanation for these results suggests that the presence of at least one dominant allele of each of two gene pairs is essential in order for flowers to be purple. All other genotype combinations yield white flowers because the homozygous condition of either recessive allele masks the expression of the dominant allele at the other locus.

The cross is shown as follows:

$$P_1: \quad AAbb \times aaBB$$
$$\text{white} \quad \text{white}$$
$$F_1: \quad \text{All } AaBb \text{ (purple)}$$

F₂ Ratio	Genotype	Phenotype	Final Phenotypic Ratio
9/16	A– B–	purple	9/16 purple
3/16	A– bb	white	
3/16	aa B–	white	7/16 white
1/16	aa bb	white	

We can envision the way in which two gene pairs might yield such results:

	Gene A			Gene B		
Precursor Substance (colorless)	↓ A–	→	Intermediate Product (colorless)	↓ B–	→	Final Product (purple)

At least one dominant allele from each pair of genes is necessary to ensure both biochemical conversions to the final product, yielding purple flowers. In the cross above, this will occur in 9/16 of the F₂ offspring. All other plants have flowers that remain white.

These three examples illustrate in a simple way how two gene products interact to influence a common phenotype. In most, if not all, instances, many more than two genes interact to produce each phenotypic character. Provided that each such gene is homozygous for the wild-type alleles, these additional genes would not influence the outcome of the above crosses.

Novel Phenotypes

Other cases of gene interaction yield novel, or new, phenotypes in the F₂ generation, in addition to producing modified dihybrid ratios. Case 4 in Figure 4.8 depicts the inheritance of fruit shape in the summer squash *Cucurbita pepo*. When plants with *disc*-shaped fruit (*AABB*) are crossed to plants with *long* fruit (*aabb*), the F_1 generation all have *disc* fruit. However, in the F_2 progeny, fruit with a novel shape—*sphere*—appear as well as fruit exhibiting the parental phenotypes. The F_2 generation, with a modified 9:6:1 ratio, is as follows:

$$F_1: \quad AaBb \times AaBb$$
$$\text{disc} \qquad \text{disc}$$

F_2 Ratio	Genotype	Phenotype	Final Phenotypic Ratio
9/16	$A- B-$	disc	9/16 disc
3/16	$A- bb$	sphere	6/16 sphere
3/16	$aa\ B-$	sphere	
1/16	$aa\ bb$	long	1/16 long

In this example of gene interaction, both gene pairs influence fruit shape equivalently. A dominant allele at either locus ensures a sphere-shaped fruit. In the absence of dominant alleles, the fruit is long. However, if both dominant alleles (*A* and *B*) are present, the fruit is flattened into a disc shape.

Another interesting example of an unexpected phenotype arising in the F_2 generation is the inheritance of eye color in *Drosophila melanogaster*. The wild-type eye color is brick red. When two autosomal recessive mutants, *brown* and *scarlet*, are crossed, the F_1 generation consists of flies with wild-type eye color. In the F_2 generation wild, scarlet, brown, and white-eyed flies are found in a 9:3:3:1 ratio. While this ratio is numerically the same as Mendel's dihybrid ratio, the *Drosophila* cross involves only one character, eye color. The diagram of this cross uses the gene symbols *A* to *B* to maintain consistency in this section. (The actual symbols for the two genes are *bw* and *st*.)

$$P_1: \quad AAbb \times aaBB$$
$$\text{brown} \qquad \text{scarlet}$$
$$F_1: \quad AaBb \text{ (wild type)}$$

F_2 Ratio	Genotype	Phenotype	Final Phenotypic Ratio
9/16	$A- B-$	wild type	9/16 wild
3/16	$A- bb$	brown	3/16 brown
3/16	$aa\ B-$	scarlet	3/16 scarlet
1/16	$aa\ bb$	white	1/16 white

This cross is an excellent example of gene interaction because the biochemical basis of eye color in this organism has been determined. *Drosophila*, as a typical arthropod, has compound eyes made up of individual visual units called **ommatidia**.

The wild-type eye color is due to the deposition and mixing of two separate pigments in each ommatidium. These include the bright red pigment **drosopterin** and the brown pigment **xanthommatin**. Each pigment is produced by separate biosynthetic pathways. Each step of each pathway is catalyzed by a separate enzyme and is thus under the control of a separate gene. As shown in Figure 4.9, the *brown* mutation, when homozygous, interrupts the synthesis of the bright red pigment. Thus, the eye contains only xanthommatin pigments and is brown. The recessive mutation *scarlet*, located on a separate autosome, interrupts the synthesis of the brown xanthommatins and renders the eye color bright red in homozygous mutant flies. Each mutation apparently causes the production of a nonfunctional enzyme. Flies that are double mutants and thus homozygous for both *brown* and *scarlet* lack both functional enzymes and can make neither of the pigments; they represent the novel white-eyed flies appearing in 1/16 of the F_2 generation.

Other Modified Dihybrid Ratios

The remaining cases (5–8) in Figure 4.8 illustrate additional modifications of the dihybrid ratio and provide still other examples of gene interactions. All cases (1–8) have two things in common. First, in arriving at a suitable explanation of the inheritance pattern of each one, we have not violated the principles of segregation and independent assortment. Therefore, the added complexity of inheritance in these examples does not detract from the validity of Mendel's conclusions. Second, the F_2 phenotypic ratio in each example has been expressed in sixteenths. When a similar observation is made in crosses where the inheritance pattern is unknown, it suggests to geneticists that two gene pairs are controlling the observed phenotypes. You should make the same inference in the analysis of genetics problems. Other insights into solving genetics problems are provided in the Insights and Solutions section at the conclusion of this chapter.

GENE INTERACTION: CONTINUOUS VARIATION

In the preceding section, examples illustrated gene interaction leading to phenotypic variation that is easily classified into distinct traits. Pea plants may be tall or dwarf; squash shape may be spherical, disc-

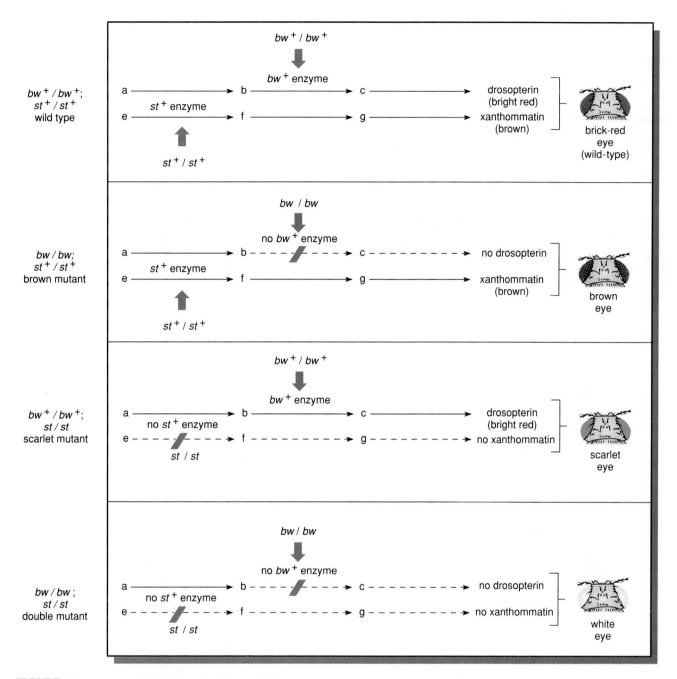

FIGURE 4.9

A theoretical explanation of the biochemical basis of the four eye-color phenotypes produced in a cross between flies with *brown* and *scarlet* eyes in *Drosophila melanogaster*. In the presence of at least one wild-type *bw*⁺ allele, an enzyme is produced that converts substance b to c, and drosopterins are synthesized. In the presence of at least one wild-type *st*⁺ allele, substance e is converted to f, and xanthommatins are synthesized. The homozygous presence of the recessive *bw* and *st* mutant alleles blocks the synthesis of these respective pigment molecules. Either none, one, or both of these pathways can be blocked, depending on the genotype.

shaped, or elongated; and fruit-fly eye color red, brown, scarlet, or white. These phenotypes are examples of discontinuous variation where discrete phenotypic categories exist. There are many other traits in a population that demonstrate considerably more variation and are not easily categorized into distinct classes. Such phenotypes represent **continuous variation**. For example, in addition to his work with sweet peas, Mendel experimented with beans. In a cross between a purple- and a white-flowered strain, the F_1 was purple. However, the F_2 contained not only purple- and white-flowered bean plants, but ones with numerous intermediate shades. He was unable to explain these results satisfactorily, but recognized that they were inconsistent with most of his data derived from sweet peas. It was not until more than 50 years later that the inheritance of characters exhibiting continuous variation was explained.

It is now known that traits exhibiting continuous variation are often controlled by two or more genes and are termed **polygenic**. In cases where the products of several genes interact to make additive contributions to the phenotype, the trait is said to exhibit **quantitative or continuous variation**. We will examine such patterns of inheritance in which a phenotypic trait is controlled by genes at two or more loci. Later in the text (Chapter 23), we will return to this general topic and outline the statistical tools used by geneticists to study traits that exhibit continuous variation. In addition, we will consider the effects of nongenetic factors on gene expression. These include various types of environmental effects and may lead to an estimate of **heritability,** a concept used by geneticists to calculate the degree of genetic influence on the expression of traits controlled by genes at many loci.

Quantitative Inheritance: Polygenes

In the late eighteenth century, Josef Gottlieb Kolreuter showed that when tall and dwarf tobacco plants were crossed, the F_1 generation was not all tall, as Mendel found with garden peas. Instead, the individual plants were all intermediate in height. When the F_2 generation was examined, individuals showed

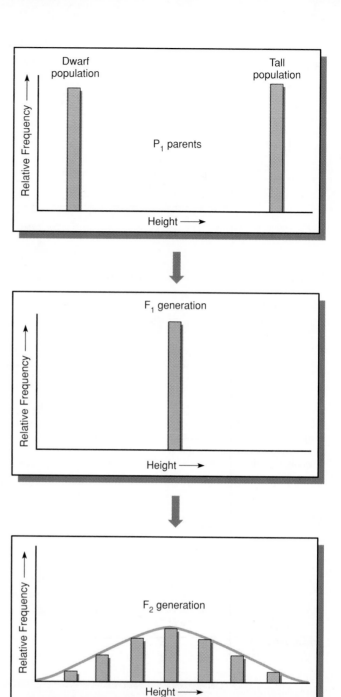

FIGURE 4.10
Histograms showing the relative frequency of individuals expressing various height phenotypes derived from Kolreuter's cross between dwarf and tall tobacco plants carried to an F_2 generation. The photograph shows a field of tobacco plants.

continuous variation in height, ranging from tall to dwarf. The majority of the F_2 plants were intermediate like the F_1, but a few were as tall or dwarf as the P_1 parents. These distributions are depicted in histograms in Figure 4.10. Note that the F_2 data have assumed a normal distribution, as evidenced by the bell-shaped curve drawn over the top of the bars in the histogram.

At the beginning of the twentieth century, geneticists noted that many characters in different species had similar patterns of inheritance, such as height and stature in humans, seed size in the broad bean, grain color in wheat, and kernel number and ear length in corn. In each case, offspring in the succeeding generation seemed to be a blend of their parents' characteristics.

The issue of whether continuous variation could be accounted for in Mendelian terms caused considerable controversy in the early 1900s. Those such as William Bateson and Gudny Yule, who adhered to the Mendelian explanation of inheritance, suggested that a large number of factors or genes were responsible for the observed patterns. This proposal, called the **multiple-factor** or **multiple-gene hypothesis,** implied that many factors or genes contributed to the phenotype in a cumulative or quantitative way. However, some geneticists argued that Mendel's unit factors could not account for the blending of parental phenotypes characteristic of these patterns of inheritance. As a result, geneticists became skeptical of Mendel's ideas and did not accept them for some time.

By 1920, the conclusions of several critical sets of experiments had been disclosed. These findings largely resolved the controversy and demonstrated that Mendelian factors could account for continuous variation. In one experiment, Edward M. East performed crosses between two strains of the tobacco plant *Nicotiana longiflora*. The fused inner petals of the flower, or corollas, of strain A were decidedly shorter than the corollas of strain B. With only minor variation, each strain was true breeding. Thus, the differences between them were clearly under genetic control.

When plants from the two strains were crossed, the F_1, F_2, and selected F_3 data (Figure 4.11) demonstrated a very distinct pattern. The F_1 generation contained corollas that were intermediate in length, compared with the P_1 varieties, and showed only minor variability among individuals. When the cross was carried to the F_2 generation, however, the range of corolla lengths increased considerably. While corolla lengths of the P_1 plants were about 40 mm and 94 mm, the F_1 generation contained plants with corollas that were all about 64 mm. In the F_2 generation, lengths ranged from 52 mm to 82 mm. Most individuals were similar to their F_1 parents, and as the deviation from this average increased, fewer and fewer plants were observed. When the data are plotted graphically (number vs. length), a bell-shaped curve results.

East further experimented with this population by selecting F_2 plants of various corolla lengths and allowing them to produce separate F_3 generations. Several of these F_3 generations are illustrated in Figure 4.11. In each case, a bell-shaped distribution was observed in the F_3, with most individuals similar in height to the F_2 parents, but with considerable variation around this value.

East's experiments thus demonstrated that although the variation in corolla length seemed continuous, experimental crosses resulted in the segregation of distinct phenotypic classes as observed in the three independent F_3 categories. This finding strongly suggested that the multiple-factor hypothesis could account for traits that deviate considerably in their expression.

The multiple-factor hypothesis, suggested by the observations of East and others, embodies the following major points:

1 Characters under such control can usually be quantified by measuring, weighing, counting, etc.
2 Two or more pairs of genes, located throughout the genome, account for the hereditary influence on the phenotype in an **additive way**. Because many genes may be involved, inheritance of this type is often called **polygenic**.
3 Each gene locus may be occupied by either an additive allele, which contributes a set amount to the phenotype, or by a nonadditive allele, which does not contribute quantitatively to the phenotype.
4 The total effect of each additive allele at each gene, while small, is approximately equivalent to all other additive alleles at other gene sites.
5 Together, the genes controlling a single character produce substantial phenotypic variation.
6 This genetic variation is enhanced by environmental factors. For example, despite their genetic basis, height in humans is influenced by individual diets, and plant characters are influenced by rainfall and soil nutrients.
7 Analysis of polygenic traits requires the study of large numbers of progeny from a population of organisms.

These points, as well as an explanation of the multiple-factor hypothesis, can be illustrated by ex-

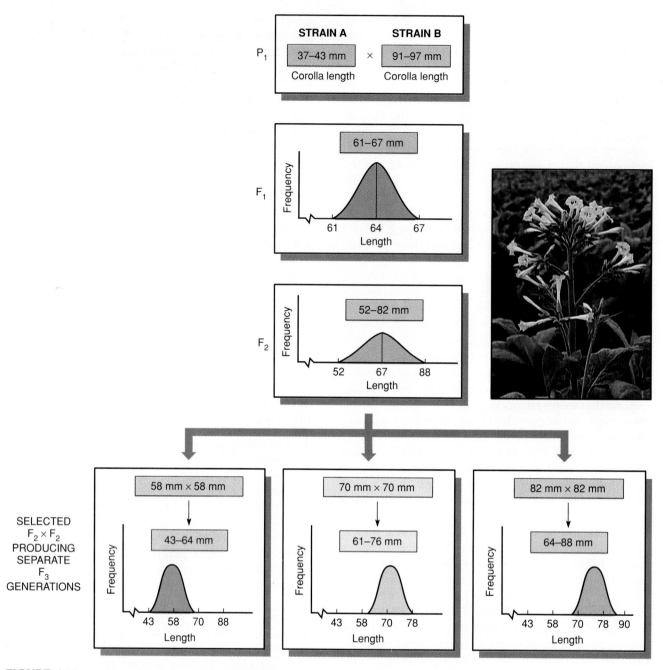

FIGURE 4.11
The F_1, F_2, and selected F_3 results of East's cross between two strains of *Nicotiana* with different corolla lengths. Plants of strain A vary from 37 to 43 mm, while plants of strain B vary from 91 to 97 mm. The photograph illustrates the flower and corolla of a tobacco plant.

amining Herman Nilsson-Ehle's experiments involving grain color in wheat performed in the early twentieth century. In one set of experiments, wheat with red grain was crossed to wheat with white grain. The F_1 generation demonstrated an intermediate color. In the F_2, approximately 15/16 of the plants showed some degree of red grain while 1/16 of the

plants showed white grain. Since the ratio occurred in sixteenths, we can hypothesize that two gene pairs control the phenotype and, if so, they segregate independently from one another in a Mendelian fashion.

Upon careful examination of the F_2, grain with color could be classified into four different shades of

FIGURE 4.12
An illustration of how the multiple factor hypothesis can account for the 1:4:6:1 phenotype ratio of grain color when all alleles designated by an uppercase letter are additive and contribute an equal amount of pigment to the phenotype.

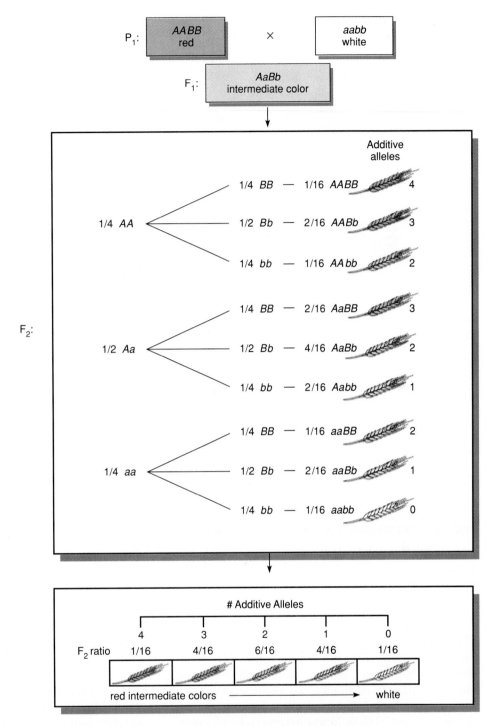

red. If two gene pairs were operating, each with one potential additive allele and one potential nonadditive allele, we can envision in Figure 4.12 how the multiple factor hypothesis could account for this variation. In the P_1, both parents were homozygous; the red parent contains only additive alleles (uppercase) while the white parent contains only nonadditive alleles (lowercase). The F_1, being heterozygous, contains only two additive alleles and expresses an intermediate phenotype. In the F_2, each offspring has either 4, 3, 2, 1, or 0 additive alleles. Wheat with no additive alleles (1/16) is white like the P_1 parent, while wheat with 4 additive alleles is red like the other P_1 parent. Plants with 3, 2, or 1 additive alleles constitute the other three categories of red color observed in the F_2, with most (6/16) having 2 additive alleles like the F_1 parent.

Thus, it appeared that if examined carefully, even apparent continuous variation can be explained in a Mendelian fashion. This was confirmed by examining

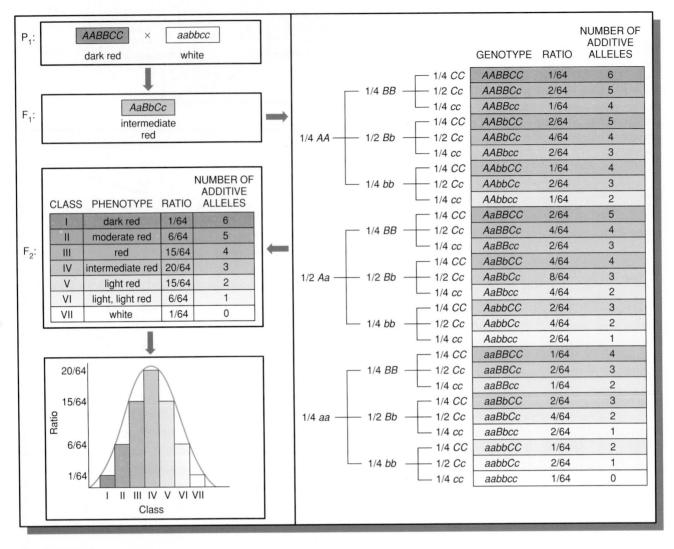

FIGURE 4.13
A detailed explanation of the F_2 results of Nilsson-Ehle's cross involving polygenic inheritance, where three gene pairs are contributing additively to grain color in wheat.

other wheat crosses made by Nilsson-Ehle where three gene pairs affected grain color. He again crossed red-grained wheat with white-grained wheat. In this case, the P_1 red plant expressed a particularly deep (or dark) red color. From this cross, he obtained an F_1 generation intermediate in color. Members of the F_1 generation were then crossed, and as shown in Figure 4.13, the F_2 phenotypes were classified into seven shades: dark red, moderate red, red, intermediate red, light red, light-light red, and white in a ratio of approximately $1:6:15:20:15:6:1$, respectively. When the components of this ratio are totaled, the sum of 64 is obtained. Because this is the same as that found in the F_2 of trihybrid crosses, it was proposed that three gene pairs controlled the observed pattern of grain-color inheritance.

Nilsson-Ehle assumed that in the F_1 generation each of the three gene pairs was heterozygous, being the product of two homozygous P_1 plants. He proposed that each heterozygous gene pair contained one additive and one nonadditive allele. Following independent assortment and fertilization, an array of F_2 genotypes are produced (Figure 4.13). If each additive allele, designated by an uppercase letter (A, B, and C) contributes equally to grain color, seven different gradations are generated. Only one of 64 genotypes contains either all six additive alleles or all six nonadditive alleles; these F_2 genotypes are identical to those of the P_1 dark red and white parents, respectively. Six of the 64 genotypes contain either five additive alleles or one additive allele, accounting for the moderate red and light-light red phenotypes,

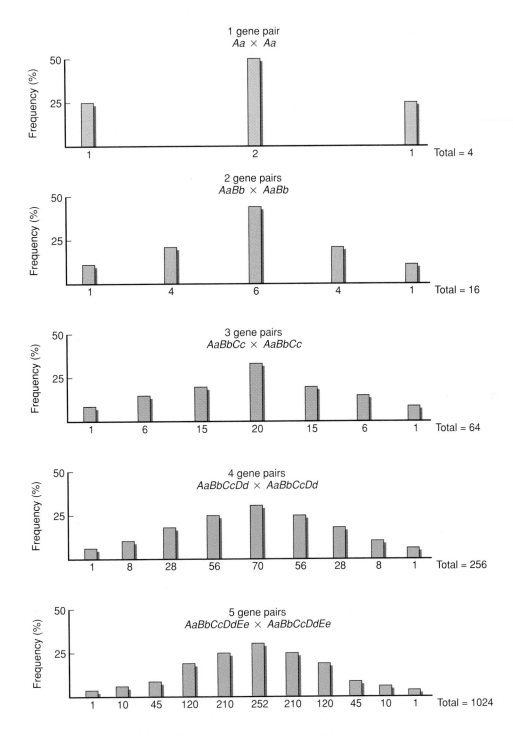

FIGURE 4.14
The results of crossing two heterozygotes where polygenic inheritance is in operation with one to five gene pairs. Each histogram bar indicates a distinct phenotypic class from one extreme (left end) to the other extreme (right end). Each phenotype results from a different number of additive alleles.

respectively. Fifteen of the 64 genotypes contain either four or two additive alleles, combinations producing the red and light red phenotypes, respectively. Finally, 20 of the 64 combinations contain some combination of three additive alleles and are responsible for the most prevalent phenotype, intermediate red.

This explanation fits nicely with the data. Multiple-factor inheritance as just described is now an acceptable mechanism to account for phenotypes displaying continuous variation. This type of inheritance, in which alleles contribute additively to a phenotype, is different from any other mode of inheritance discussed thus far. Although the Nilsson-Ehle experiments involved two or three gene pairs, there is no reason why three, four, or more gene pairs cannot function in controlling various phenotypes. As we saw in Nilsson-Ehle's initial cross, if two gene pairs were involved, only five F_2 phenotypic categories, in a 1:4:6:4:1 ratio, would be expected. On the other hand, as three, four, five, or more gene pairs become involved, greater and greater numbers of classes would be expected to appear in more complex ratios. The number of phenotypes and the expected F_2 ratios of crosses involving up to five gene pairs are illustrated in Figure 4.14.

If the ratio of F_2 individuals resembling *either* of the two most extreme phenotypes can be determined, then the number of gene pairs involved (n) may be calculated using the following simple formula:

$$\frac{1}{4^n} = \text{ratio of } F_2 \text{ individuals expressing either extreme phenotype}$$

In our past example, the P_1 phenotypes represent these two extremes. In Figure 4.13, 1/64 of the F_2 are either dark red *or* white like the P_1 classes; this ratio can be substituted on the right side of the equation prior to solving for n:

$$\frac{1}{4^n} = \frac{1}{64}$$
$$\frac{1}{4^3} = \frac{1}{64}$$
$$n = 3$$

Table 4.4 lists the ratio and the number of F_2 phenotypic classes produced in crosses involving up to five gene pairs.

For low numbers of gene pairs, it is sometimes easier to use the $(2n + 1)$ rule. If n equals the number of gene pairs, $2n + 1$ will determine the total number of categories of possible phenotypes. Where $n = 3$, $2n + 1 = 7$, since there may be 6, 5, 4, 3, 2, 1, or 0 additive alleles. For low numbers of gene pairs, it is fairly simple to work backwards from the F_2 data.

Polygenic control is a significant concept because it is believed to be the mode of inheritance for a vast number of traits. For example, height, weight, and stature in all plants and animals, grain yield in crops, beef and milk production in cattle, egg production in chickens, as well as skin color in humans are thought

TABLE 4.4
Determination of number of gene pairs from polygenic inheritance patterns.

n	Ratio from Cross between Heterozygous Parents	Number of Distinct F_2 Phenotypic Classes
1	1/4	3
2	1/16	5
3	1/64	7
4	1/256	9
5	1/1024	11

to be under polygenic control. Thus, knowledge of this mode of inheritance is of prime importance in animal breeding, agriculture, and sociological studies of humans. In each case of polygenic inheritance, the genotype establishes the potential range in which a particular phenotype may fall, while environmental factors determine exactly how much of the potential will be realized. In the crosses described in this section we have assumed an optimal environment, which minimizes variation resulting from that source.

CHAPTER SUMMARY

1 Since Mendel's work was rediscovered, the study of transmission genetics has been expanded to include modes of inheritance controlling phenotypes that are often influenced by two or more genes in a variety of ways.

2 Incomplete or partial dominance prevails over the more classical dominant/recessive relationship when an intermediate phenotypic expression of a trait occurs in an organism heterozygous for two alleles.

3 Codominance reflects the situation in which joint expression of two alleles occurs in a heterozygous organism.

4 The concept of multiple allelism applies to populations, since a diploid organism may host only two alleles at any given locus. Within a population, however, many alleles of the same gene often occur.

5 Lethal mutations usually result in the inactivation or the lack of synthesis of gene products essential during development. Such mutations may be recessive or dominant. Some lethal traits, such as Huntington disease, are expressed during adulthood.

6 Mendel's classic F_2 ratio is often modified in instances where gene interaction results in either continuous or discontinuous variation.

7 Epistasis involves discontinuous variation, where two or more genes influence a single characteristic. Usually, the expression of one of the genes masks the expression of the other gene or genes.

8 Polygenic inheritance results in continuous variation and is illustrated when products of several genes make additive contributions to such characteristics as height or pigment deposition.

Insights and Solutions

Genetic problems take on an added complexity if they involve two independent characters and multiple alleles, incomplete dominance, or epistasis. The most difficult type of problems are those that were faced by pioneering geneticists in the laboratory. In these problems they had to determine the mode of inheritance by working backwards from the observations of offspring to parents of unknown genotype. For example, consider the problem of comb shape inheritance in chickens, where walnut, rose, pea, and single are the observed distinct phenotypes. *How is comb shape inherited and what are the genotypes of the P_1 generation of each cross? Use the following data to answer these questions.*

Cross 1: single × single   ⟶ all single
Cross 2: walnut × walnut ⟶ all walnut
Cross 3: rose × pea   ⟶ all walnut
Cross 4: F_1 × F_1 of Cross 3
          walnut × walnut ⟶ 93 walnut
                                28 rose
                                32 pea
                                10 single

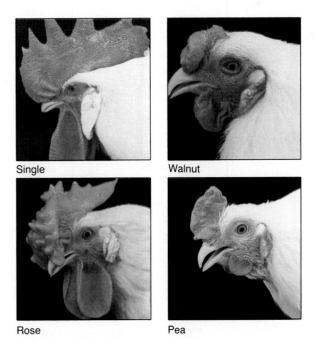

Single   Walnut

Rose   Pea

At first glance, this problem may appear quite difficult. The approach used in solving it must involve two steps. First, carefully analyze the data for any useful information. Once you determine something concrete, follow an empirical approach; that is, formulate a hypothesis and, in a sense, test it against the given data. Look for a pattern of inheritance that is consistent with all cases.

(a) In this problem there are two immediately useful facts. First, P_1 singles breed true. Second, while P_1 walnut breeds true (Cross 2), a walnut phenotype is also produced in crosses between rose and pea (Cross 3). When these F_1 walnuts are crossed together (Cross 4), all four comb shapes are produced in a ratio that approximates a $9:3:3:1$. This observation should immediately suggest a cross involving two gene pairs because the resulting data closely resemble the ratio of Mendel's dihybrid crosses. Since only one trait is involved (comb shape), epistasis must be occurring. This may serve as a working hypothesis, and we must now propose how the two gene pairs interact to produce each phenotype.

(b) If we call the allele pairs A, a and B, b, we might predict that since walnut represents 9/16 in Cross 4, $A-B-$ will produce walnut. We might also hypothesize that in the case of Cross 2, the genotypes were $AABB \times AABB$ where walnut was seen to breed true. (Recall that $A-$ and $B-$ mean AA or Aa and BB or Bb, respectively.)

(c) Since single is the phenotype representing 1/16 of the offspring of Cross 4, we could predict that this phenotype is the result of the $aabb$ genotype. This is consistent with Cross 1.

(d) Now we have only to determine the genotypes for rose and pea. The most logical prediction would be that at least one dominant A or B allele combined with the double recessive condition of the other allele pair can account for these phenotypes. For example,

$$A-bb \rightarrow \text{rose}$$
$$aa\ B- \rightarrow \text{pea}$$

If, in Cross 3, $AAbb$ (rose) were crossed with $aaBB$ (pea), all offspring would be $AaBb$ (walnut). This is consistent with the data, and we must now only look at Cross 4. We predict these walnut genotypes to be $AaBb$ (as above), and from the cross

$$\frac{AaBb}{\text{(walnut)}} \times \frac{AaBb}{\text{(walnut)}}$$

we expect

9/16 $A-\ B-$ (walnut)
3/16 $A-\ bb$ (rose)
3/16 $aa\ B-$ (pea)
1/16 $aa\ bb$ (single)

Our prediction is consistent with the information we were given. The initial hypothesis of the epistatic interaction of two gene pairs proves consistent throughout, and the problem has been solved.

This example illustrates the need to have a basic theoretical knowledge of transmission genetics. Then, you must search for the appropriate clues so that you can proceed in a stepwise fashion toward a solution. Mastering problem solving requires practice, but provides a great deal of satisfaction.

Apply this general approach to the following problems:

1 In radishes, flower color may be red, purple, or white. The edible portion of the radish may be long or oval. When only flower color is studied, red × white yields all purple. If these F_1 purples are interbred, no dominance is evident and the F_2 generation consists of 1/4 red:1/2 purple:1/4 white. Regarding radish shape, long is dominant to oval in a normal Mendelian fashion.

(a) Determine the F_1 and F_2 phenotypes from a cross between a true-breeding red long radish and one that is white oval. Be sure to define all gene symbols initially.

ANSWER: This is a modified dihybrid cross where the gene pair controlling color exhibits incomplete dominance. Shape is controlled conventionally. First, establish gene symbols:

$$RR = \text{red} \qquad Rr = \text{purple} \qquad rr = \text{white}$$
$$O- = \text{long} \qquad oo = \text{oval}$$

P_1: $RROO$ × $rroo$
     (red long)   (white oval)

F_1:   all $RrOo$ (purple long)

$F_1 \times F_1$: $RrOo \times RrOo$

F_2:

1/4 RR	3/4 $O-$	3/16 RR $O-$	red long
	1/4 oo	1/16 RR oo	red oval
2/4 Rr	3/4 $O-$	6/16 Rr $O-$	purple long
	1/4 oo	2/16 Rr oo	purple oval
1/4 rr	3/4 $O-$	3/16 rr $O-$	white long
	1/4 oo	1/16 rr oo	white oval

Note above that to generate the F_2 results, we have used the forked-line method. First, the outcome of crossing F_1 parents for the color genes is considered ($Rr \times Rr$). Then the outcome of shape is considered ($Oo \times Oo$).

(b) A red oval plant was crossed with a plant of unknown genotype and phenotype, yielding the data shown below. Determine the genotype and phenotype of the unknown plant.

Offspring: 103 red long:101 red oval:98 purple long:100 purple oval

ANSWER: Since the two characters are inherited independently, consider them separately. The data indicate a 1/4:1/4:1/4:1/4 proportion. First, consider color:

P_1:   red × ?
F_1:   204 red (1/2)
        198 purple (1/2)

Since the red parent *must* be RR, the unknown must have a genotype of Rr to produce these results. It is thus purple. Now, consider shape:

P_1:   oval × ?
F_1:   201 long (1/2)
        201 oval (1/2)

Since the oval plant *must* be *oo*, the unknown plant must have a genotype of *Oo* to produce these results. It is thus long. The unknown plant is thus:

$$RrOo \quad \text{purple long}$$

2 Plants may be 10, 20, 30, 40, 50, 60, or 70 cm high where plant height is under polygenic control. A true-breeding plant that is 10 cm is crossed to another true-breeding plant that is 50 cm high. How many gene pairs are involved? What F_1 and F_2 results can be predicted?

ANSWER: Since there are seven groups of phenotypes, using our 2n + 1 rule, we can propose that there are three pairs of genes that control height additively. Each additive allele adds 10 cm to a base of 10 cm. The 10 cm plant contains no additive alleles. The 50 cm plant, since it is true-breeding, must be homozygous for additive alleles at two of the three gene sites:

$$P_1: \quad aabbcc \times AABBcc$$
$$F_1: \quad \text{all } AaBbcc \text{ (30 cm)}$$

It is simplest to determine the F_2 ratio by using the forked-line method. Since each allele influences plant height additively, we must compute the genotypic ratio. All F_2 plants will be *cc*, so we can add the *c* alleles to the *B* group:

			# Additive Alleles	Height
F_2:		1/4 BB cc — 1/16 AA BB cc	4	50
	1/4 AA	2/4 Bb cc — 2/16 AA Bb cc	3	40
		1/4 bb cc — 1/16 AA bb cc	2	30
		1/4 BB cc — 2/16 Aa BB cc	3	40
	2/4 Aa	2/4 Bb cc — 4/16 Aa Bb cc	2	30
		1/4 bb cc — 2/16 Aa bb cc	1	20
		1/4 BB cc — 1/16 aa BB cc	2	30
	1/4 aa	2/4 Bb cc — 2/16 aa Bb cc	1	20
		1/4 bb cc — 1/16 aa bb cc	0	10

Final Ratio: 1/16 10 cm
4/16 20 cm
6/16 30 cm
4/16 40 cm
1/16 50 cm

PROBLEMS AND DISCUSSION QUESTIONS

1 In shorthorn cattle, coat color may be red, white, or roan. Roan is an intermediate phenotype expressed as a mixture of red and white hairs. The following data were obtained from various crosses:

red × red $\longrightarrow$ all red
white × white $\longrightarrow$ all white
red × white $\longrightarrow$ all roan
roan × roan $\longrightarrow$ 1/4 red: 1/2 roan: 1/4 white

How is coat color inherited? What are the genotypes of parents and offspring for each cross?

2 Contrast incomplete dominance and codominance.

3 In foxes, two alleles of a single gene, *P* and *p*, may result in lethality (*PP*), platinum coat (*Pp*), or silver coat (*pp*). What ratio is obtained when platinum foxes are interbred? Is the *P* allele behaving dominantly or recessively in causing lethality? in causing platinum coat color?

4 In mice, a short-tailed mutant was discovered. When it was crossed to a normal long-tailed mouse, 4 offspring were short-tailed and 3 were long-tailed. Two short-tailed mice from the F_1 generation were selected and crossed. They produced 6 short-tailed and 3 long-tailed mice. These genetic experiments were repeated three times with approximately the same results. What genetic ratios are illustrated? Hypothesize the mode of inheritance and diagram the crosses.

5 List all possible genotypes for the A, B, AB, and O phenotypes. Is the mode of inheritance of the ABO blood types representative of dominance? of recessiveness? of codominance?

6 With regard to the ABO blood types in humans, determine the genotypes of the male parent and female parent below:

Male Parent: Blood type B; mother type O
Female Parent: Blood type A; father type B

Predict the blood types of the offspring that this couple may have and the expected proportion of each.

7 In a disputed parentage case, the child is blood type O while the mother is blood type A. What blood type would exclude a male from being the father? Would the other blood types prove that a particular male was the father?

8 The A and B antigens in humans may be found in water-soluble form in secretions, including saliva, of some individuals but not in others. The population thus contains "secretors" and "nonsecretors". The following inheritance patterns have established that this trait is inherited:

secretor × secretor	$\longrightarrow$	all secretors
nonsecretor × nonsecretor	$\longrightarrow$	all nonsecretors
secretor × nonsecretor	$\longrightarrow$	all secretors
F_1 secretors	$\longrightarrow$	F_2 = 3/4 secretors: 1/4 nonsecretors

How is the trait inherited? Determine the genotypes of all parents and offspring.

9 Distinguish between epistasis and polygenic inheritance, and between discontinuous and continuous variation.

10 In rabbits, a series of multiple alleles controls coat color in the following way: *C* is dominant to all other alleles and causes full color. The chinchilla phenotype is due to the c^{ch} allele, which is dominant to all alleles other than *C*. The c^h allele, dominant only to c^a (albino), results in the Himalayan coat color. Thus, the order of dominance is $C > c^{ch} > c^h > c^a$. For each of the three cases below, the phenotypes of the P_1 generations of two crosses are shown, as well as the phenotype of one member of the F_1 generation. For each case, determine the genotypes of the P_1 generation and the F_1 offspring and predict the results of making each cross between F_1 individuals as shown.

	P_1 Phenotypes		F_1 Phenotypes
(a)	Himalayan × Himalayan	$\longrightarrow$	albino
			× $\longrightarrow$??
	full color × albino	$\longrightarrow$	chinchilla
(b)	albino × chinchilla	$\longrightarrow$	albino
			× $\longrightarrow$??
	full color × albino	$\longrightarrow$	full color
(c)	chinchilla × albino	$\longrightarrow$	Himalayan
			× $\longrightarrow$??
	full color × albino	$\longrightarrow$	Himalayan

11 In the guinea pig, one locus involved in the control of coat color may be occupied by any of four alleles: C (black), c^k (sepia), c^d (cream), or c^a (albino). Like coat color in rabbits (Problem 10), an order of dominance exists: $C > c^k > c^d > c^a$. In the following crosses, write the parental genotypes and predict the phenotypic ratios that would result.
    (a) sepia × cream, where both guinea pigs had an albino parent
    (b) sepia × cream, where the sepia guinea pig had an albino parent and the cream guinea pig had two sepia parents
    (c) sepia × cream, where the sepia guinea pig had two full color parents and the cream guinea pig had two sepia parents
    (d) sepia × cream, where the sepia guinea pig had a full color parent and an albino parent and the cream guinea pig had two full color parents.

12 Three gene pairs located on separate autosomes determine flower color and shape as well as plant height. The first pair exhibits incomplete dominance where color can be red, pink (the heterozygote) or white. The second pair leads to personate (dominant) or peloric (recessive) flower shape, while the third gene pair produces either the dominant tall trait or the recessive dwarf trait. Homozygous plants that are red, personate, and tall are crossed to those that are white, peloric, and dwarf. Determine the F_1 genotype(s) and phenotype(s). If the F_1 plants are interbred, what proportion of the offspring will exhibit the same phenotype as the F_1 plants?

13 As in Problem 12, color may be *red, white,* or *pink,* and flower shape may be *personate* or *peloric.* For the following crosses, determine the P_1 and F_1 genotypes. What phenotypic ratios would result from crossing the F_1 of (a) to the F_1 of (b)?

(a) red peloric × white personate ⟶ F_1 = all pink personate
(b) red personate × white peloric ⟶ F_1 = all pink personate
(c) pink personate × red peloric   ⟶ F_1 = 1/4 red personate
                                          1/4 red peloric
                                          1/4 pink personate
                                          1/4 pink peloric
(d) pink personate × white peloric ⟶ F_1 = 1/4 white personate
                                          1/4 white peloric
                                          1/4 pink personate
                                          1/4 pink peloric

14 With reference to the eye color phenotypes produced by the recessive, autosomal, unlinked *brown* and *scarlet* loci in *Drosophila* (see Figure 4.9), predict the F_1 and F_2 results of the following P_1 crosses. Recall that when both the *brown* and *scarlet* alleles are homozygous, no pigment is produced, and the eyes are white.
    (a) wild type × white
    (b) wild type × scarlet
    (c) brown × white

15 Pigment in the mouse is only produced when the C allele is present. Individuals of the cc genotype have no color. If color is present, it may be determined by the A, a alleles. AA or Aa results in gray (agouti) color, while aa results in black coats.
    (a) What F_1 and F_2 genotypic and phenotypic ratios are obtained from a cross between $AACC$ and $aacc$ mice?
    (b) In three crosses between gray females whose genotypes were unknown and males of the $aacc$ genotype, the following phenotypic ratios were obtained:

    (1) 8 gray   (2) 9 gray   (3) 4 gray
        8 colorless   10 black   5 black
                                            10 colorless

What are the genotypes of these female parents?

16 In some plants a red pigment, cyanidin, is synthesized from a colorless precursor. The addition of a hydroxyl group (OH^-) to the cyanidin molecule causes it to become purple. In a cross between two randomly selected purple plants, the following results were obtained:

> 94 purple
> 31 red
> 43 colorless

How many genes are involved in the determination of these flower colors? Which genotypic combinations produce which phenotypes? Diagram the purple × purple cross.

17 In rats, the following genotypes of two independently assorting autosomal genes determine coat color:

> $A- B-$ (gray)
> $A- bb$ (yellow)
> aa $B-$ (black)
> aa bb (cream)

A third gene pair on a separate autosome determines whether or not any color will be produced. The CC and Cc genotypes allow color according to the expression of the A and B alleles. However, the cc genotype results in albino rats regardless of the A and B alleles present. Determine the F_1 phenotypic ratio of the following crosses:
(a) $AA\ bb\ CC \times aa\ BB\ cc$
(b) $Aa\ BB\ CC \times AA\ Bb\ cc$
(c) $Aa\ Bb\ Cc \times Aa\ Bb\ cc$
(d) $Aa\ BB\ Cc \times Aa\ BB\ Cc$
(e) $AA\ Bb\ Cc \times AA\ Bb\ cc$

18 Given the inheritance pattern of coat color in rats, as described in Problem 17, predict the genotype and phenotype of the parents who produced the following F_1 offspring:
(a) 9/16 gray:3/16 yellow:3/16 black:1/16 cream
(b) 9/16 gray:3/16 yellow:4/16 albino
(c) 27/64 gray:16/64 albino:9/64 yellow:9/64 black:3/64 cream
(d) 3/8 black:3/8 cream:2/8 albino
(e) 3/8 black:4/8 albino:1/8 cream

19 Consider the genes and alleles involved in the Bombay phenotype and the ABO blood groups. How would you describe the genetic interaction between these alleles?

20 In a species of the cat family, eye color can be gray, blue, green, or brown, and each trait is true-breeding. In separate crosses involving homozygous parents, the following data were obtained:

Cross	P_1	F_1	F_2
A	green × gray	all green	3/4 green: 1/4 gray
B	green × brown	all green	3/4 green: 1/4 brown
C	gray × brown	all green	9/16 green: 3/16 brown: 3/16 gray: 1/16 blue

(a) Analyze the data: How many genes are involved? Define gene symbols and indicate which genotypes yield each phenotype.
(b) In a cross between a gray-eyed cat and one of unknown genotype and phenotype, the F_1 generation was not observed. However, the F_2 resulted in the same F_2 ratio as in cross C. Determine the genotypes and phenotypes of the unknown P_1 and F_1 cats.

21 In a plant, a tall variety was crossed with a dwarf variety. All F_1 plants were tall. When $F_1 \times F_1$ plants were interbred, 9/16 of the F_2 were tall and 7/16 were dwarf.

(a) Explain the inheritance of height by indicating the number of gene pairs involved and by designating which genotypes yield tall and which yield dwarf (use dashes where appropriate).

(b) Of the F_2 dwarf plants, *what proportion* of them will be true-breeding if self-fertilized? *List* these genotypes.

22 In a unique species of plants, flowers may be yellow, blue, red, or mauve. All colors may be true-breeding. If plants with blue flowers are crossed to red-flowered plants, all F_1 plants have yellow flowers. When carried to an F_2 generation, the following ratio was observed:

9/16 yellow: 3/16 blue: 3/16 red: 1/16 mauve

In still another cross using true-breeding parents, yellow-flowered plants are crossed with mauve-flowered plants. Again, all F_1 plants had yellow flowers and the F_2 showed a 9:3:3:1 ratio, as underlined above.

(a) Describe the inheritance of flower color by defining gene symbols and designating which genotypes give rise to each of the four phenotypes.

(b) Determine the F_1 and F_2 results of a cross between true-breeding red and true-breeding mauve-flowered plants.

23 What types of offspring (and in what ratio) can be expected from two parents of blood type AB who are both heterozygous at the H locus (controlling the Bombay phenotype)?

24 Shown below are five maternal and paternal phenotypes (1–5), each designating the ABO, MN, and Rh blood group antigens. Each combination resulted in one of the five offspring shown to the right (a–e). Arrange the offspring with the correct parents such that all five cases are consistent. Is there more than one set of correct answers?

Parental Phenotypes		Offspring
(1) A, M, Rh^-	$\times$ A, N, Rh^-	(a) A, N, Rh^-
(2) B, M, Rh^-	$\times$ B, M, Rh^+	(b) O, N, Rh^+
(3) O, N, Rh^+	$\times$ B,N,Rh^+	(c) O, MN, Rh^-
(4) AB, M, Rh^-	$\times$ O, N, Rh^+	(d) B, M, Rh^+
(5) AB, MN, Rh^-	$\times$ AB, MN, Rh^-	(e) B, MN, Rh^+

25 An invading alien geneticist from a planet where genetic research was prohibited brought with him to earth two pure-breeding lines of pet frogs. One line croaked by uttering rib-it rib-it, and had purple eyes. The other line croaked more softly by muttering knee-deep knee-deep, and had green eyes. With a new-found sense of inquiry, he mated the two types of frogs. In the F_1, all frogs had blue eyes and uttered rib-it rib-it. He proceeded to make many $F_1 \times F_1$ crosses, and when he fully analyzed the F_2 data, he realized they could be reduced to the following ratio:

27/64 blue-eyed, rib-it utterer
12/64 green-eyed, rib-it utterer
9/64 blue-eyed, knee-deep mutterer
9/64 purple-eyed, rib-it utterer
4/64 green-eyed, knee-deep mutterer
3/64 purple-eyed, knee-deep mutterer

(a) How many total gene pairs are involved in the inheritance of both traits? Support your answer.

(b) Of these, how many are controlling eye color? How can you tell? How many are controlling croaking?

(c) Assign gene symbols for all phenotypes and indicate the genotypes of the P_1 and F_1 frogs.

(d) Indicate the genotypes of the six F_2 phenotypes.

(e)   After years of experiments, the geneticist isolated pure-breeding strains of all six F_2 phenotypes. Indicate the F_1 and F_2 phenotypic ratios of the following cross using these pure breeding strains:

blue-eyed, knee-deep mutterer × purple-eyed, rib-it utterer

(f)   One set of crosses with his true-breeding lines initially caused the geneticist some confusion. When he crossed true-breeding purple-eyed, knee-deep mutterers with true-breeding green-eyed, knee-deep mutterers, he often got different results. In some matings, all offspring were blue-eyed, knee-deep mutterers but in other matings, all offspring were purple-eyed, knee-deep mutterers. In still a third mating, 1/2 blue-eyed, knee-deep mutterers and 1/2 purple-eyed, knee-deep mutterers were observed. Explain why the results differed.

(g)   In another experiment, the geneticist crossed two purple-eyed, rib-it utterers together with the results shown below. What were the genotypes of the two parents?

9/16 purple-eyed, rib-it utterer
3/16 purple-eyed, knee-deep mutterer
3/16 green-eyed, rib-it utterer
1/16 green-eyed, knee-deep mutterer

26   Labrador retrievers may be black, brown, or golden in color. While each color may breed true, many different outcomes occur if many litters are examined from a variety of matings, where the parents are not necessarily true breeding. Shown below are just some of the many possibilities. Propose a mode of inheritance that is consistent with these data, and indicate the corresponding genotypes of the parents in each mating. Indicate as well the genotypes of dogs that breed true for each color.

(a) black × brown ⟶ all black
(b) black × brown ⟶ 1/2 black
                    1/2 brown
(c) black × brown ⟶ 3/4 black
                    1/4 golden
(d) black × golden ⟶ all black
(e) black × golden ⟶ 4/8 golden
                     3/8 black
                     1/8 brown
(f) black × golden ⟶ 2/4 golden
                     1/4 black
                     1/4 brown

(g) brown × brown ⟶ 3/4 brown
                    1/4 golden
(h) black × black ⟶ 9/16 black
                    4/16 golden
                    3/16 brown

27   Assume that height in a plant is controlled by two gene pairs and that each additive allele contributes 5 cm to a base height of 20 cm (i.e., *aabb* is 20 cm).
(a)   What is the height of an *AABB* plant?
(b)   Predict the phenotypic ratios of F_1 and F_2 plants in a cross between *aabb* and *AABB*.
(c)   List all genotypes that give rise to plants which are 25 and 35 cm in height.

28   In a cross where three gene pairs determine weight in squash, what proportion of individuals from the cross *AaBbCC* × *AABbcc* will contain only two additive alleles? Which genotype or genotypes fall into this category?

29 An inbred strain of plants has a mean height of 24 cm. A second strain of the same species from a different geographical region also has a mean height of 24 cm. When plants from the two strains are crossed together, the F_1 plants are the same height as the parent plants. However, the F_2 generation shows a wide range of heights; the majority are like the P_1 and F_1 plants, but approximately 4 of 1000 are only 12 cm high, and about 4 of 1000 are 36 cm high.

(a) What mode of inheritance is occurring here?
(b) How many gene pairs are involved?
(c) How much does each gene contribute to plant height?
(d) Indicate one possible set of genotypes for the original P_1 parents and the F_1 plants that could account for these results.
(e) Indicate three possible genotypes that could account for F_2 plants that are 18 cm high and F_2 plants that are 33 cm high.

30 In a series of crosses between plants of various heights to a 20-inch plant, the following results were obtained in the F_1 generations:

(a) $4'' \times 20'' \rightarrow$ All $12''$
(b) $8'' \times 20'' \rightarrow$ All $14''$
(c) $12'' \times 20'' \rightarrow$ All $16''$
(d) $16'' \times 20'' \rightarrow$ All $18''$

Propose an explanation for the inheritance of height in the above plant. Under the constraints of your explanation, predict the genotypes of plants of each height.

31 Members of a strain of inbred skunks whose stripes are uniformly 20 cm long are crossed to those of another strain whose stripes are uniformly 24 cm. The F_1 hybrids all have stripes that are 22 cm. When the F_1 skunks were crossed, the F_2 generation produced animals with stripes of 16, 18, 20, 22, 24, 26, and 28 cm. The 22 cm variety was most frequent and the 16 and 28 cm variety least frequent. Of 1000 F_2 skunks, 15 were 16 cm and 16 were 28 cm.

(a) What is the mode of inheritance illustrated above?
(b) How many gene pairs are involved?
(c) What is the contribution of each additive allele to stripe length?
(d) What are the genotypes of the P_1 parents and F_1 offspring?
(e) List four genotypes which give rise to skunks whose stripes are 20 cm long.

32 Erma and Harvey were a compatible barnyard pair, but a curious sight. Harvey's tail was only 6 cm while Erma's was 30 cm. Their F_1 piglet offspring all grew tails that were 18 cm. When inbred, an F_2 generation resulted in many piglets (Erma and Harvey's grandpigs) whose tails ranged in 4 cm intervals from 6 to 30 cm (6, 10, 14, 18, 22, 26, 30). Most had 18 cm tails, while 1/64 had 6 cm and 1/64 had 30 cm tails.

(a) Explain how tail length is inherited by describing the mode of inheritance, indicating how many gene pairs are at work, and designating the genotypes of Harvey, Erma, and their 18 cm offspring.
(b) If one of the 18 cm F_1 pigs were mated with the 6 cm F_2 pigs, what phenotypic ratio would be predicted if many offspring resulted? Diagram the cross.

∞∞∞∞∞∞∞∞∞∞∞∞∞∞∞∞∞∞∞∞∞∞∞

SELECTED READINGS

BRINK, R. A., ed. 1967. *Heritage from Mendel*. Madison: University of Wisconsin Press.
CHAPMAN, A. B. 1985. *General and quantitative genetics*. Amsterdam: Elsevier.
CLARKE, C. A. 1968. The prevention of "Rhesus" babies. *Scient. Amer.* (Nov.) 219: 46–52.
CORWIN, H. O., and JENKINS, J. B. 1976. *Conceptual foundations of genetics: Selected readings*. Boston: Houghton-Mifflin.
CROW, J. F. 1983. *Genetics notes*. 8th ed. New York: Macmillan.
DUNN, L. C. 1966. *A short history of genetics*. New York: McGraw-Hill.

EAST, E. M. 1910. A Mendelian interpretation of variation that is apparently continuous. *Amer. Naturalist* 44: 65–82.

———. 1916. Studies on size inheritance in *Nicotiana*. *Genetics* 1: 164–76.

FALCONER, D. S. 1981. *Introduction to quantitative genetics*. 2nd ed. New York: Longman.

FOSTER, H. L., et al., eds. 1981. *The mouse in biomedical research. Vol. 1: History, genetics, and wild mice*. Orlando: Academic Press.

FOSTER, M. 1965. Mammalian pigment genetics. *Adv. in Genet*. 13: 311–39.

GRANT, V. 1975. *Genetics of flowering plants*. New York: Columbia University Press.

LINDSLEY, D. C., and GRELL, E. H. 1967. *Genetic variations of* Drosophila melanogaster. Washington, D.C.: Carnegie Institute of Washington.

NOLTE, D. J. 1959. The eye-pigmentary system of *Drosophila*. *Heredity* 13: 233–41.

PAWELEK, J. M., and KÖRNER, A. M. 1982. The biosynthesis of mammalian melanin. *Amer. Scient*. 70: 136–45.

PETERS, J. A., ed. 1959. *Classic papers in genetics*. Englewood Cliffs, N.J.: Prentice-Hall.

RACE, R. R., and SANGER, R. 1975. *Blood groups in man*. 6th ed. Oxford, England: Blackwell Scientific Publishers.

RANSON, R., ed. 1982. *A handbook of* Drosophila *development*. New York: Elsevier Biomedical Press.

STERN, C. 1973. *Principles of human genetics*. 3rd ed. New York: W. H. Freeman.

TEARLE, R. G., et al. 1989. Cloning and characterization of the *scarlet* gene of *Drosophila melanogaster*. *Genetics* 122: 595–606.

VOELLER, B. R., ed. 1968. *The chromosome theory of inheritance—Classic papers in development and heredity*. New York: Appleton-Century-Crofts.

VOGEL, F., and MOTULSKY, A. G. 1986. 2nd ed. *Human genetics: Problems and approaches*. New York: Springer-Verlag.

WATKINS, M. W. 1966. Blood group substances. *Science* 152: 172–81.

WIENER, A. S., ed. 1970. *Advances in blood groupings, III*. New York: Grune and Stratton.

YOSHIDA, A. 1982. Biochemical genetics of the human blood group ABO system. *Amer. J. Human Genet*. 34: 1–14.

ZIEGLER, I. 1961. Genetic aspects of ommochrome and pterin pigments. *Adv. in Genet*. 10: 349–403.

5

Sex Determination
and Sex Linkage

CHAPTER CONCEPTS

Genetic mechanisms have evolved that result in various forms of sexual

differentiation in plants and animals. The mechanisms that determine

maleness and femaleness most often involve a unique pair of

chromosomes that differ in the two sexes. Transmission of genetic

information present on these chromosomes leads to the unique

inheritance patterns referred to as sex linkage.

Thus far in the text, we have discussed how, in many animals and plants, different alleles and different genes and their products can interact to influence phenotypic variation. The orderly transmission of these genetic units from parents to offspring relies on: (1) the production of haploid gametes through the process of meiosis; and (2) subsequent fertilization events which reestablish the diploid number and ensure genetic constancy within members of a species. Highly effective fertilization, in turn, depends on some form of sexual differentiation in organisms. Some hint of this differentiation occurs as low on the evolutionary scale as bacteria and single-celled eukaryotic algae. In higher forms of life, the differentiation of the sexes is often much more apparent as phenotypic dimorphism in the male and female members of each species. The shield and spear (♂), the ancient symbols of iron and Mars, and the mirror (♀), the symbol of copper and Venus, represent the maleness and femaleness so acquired by individuals.

In this chapter we will first review several representative modes of **sexual differentiation** by examining the life cycle of three organisms often studied in genetics: the green alga, *Chlamydomonas;* the bread mold, *Neurospora;* and the corn plant, *Zea mays.* Then, we will investigate what is known about the genetic basis for the **determination of sexual differences**. While dissimilar, or heteromorphic, chromosomes such as the X-Y pair are often involved, they are not always the direct basis of sex determination. Following this discussion we will turn to the direct influence of sexual differentiation and sex chromosomes on patterns of inheritance and genetic expression. The topics of sex linkage, as well as sex-limited and sex-influenced inheritance, will be discussed. Sex linkage, the inheritance of traits under the control of

genes located on the X chromosome, is another example of neo-Mendelian genetics. This section of the chapter thus extends the discussion of this topic initiated in Chapter 4. In the course of the chapter we will also consider numerous related topics involving humans.

REPRESENTATIVE MODES OF SEXUAL DIFFERENTIATION

In the biological world, a wide range of reproductive modes is recognized. Asexual organisms are those for which no evidence of sexual reproduction is known. Other species alternate between short periods of sexual reproduction and prolonged periods of asexual reproduction. In most diploid eukaryotes, however, sexual reproduction is the only natural mechanism resulting in new members of a species.

In multicellular organisms, it is important to distinguish between **primary sexual differentiation,** which involves only the gonads where gametes are produced, and **secondary sexual differentiation,** which involves other organs, such as mammary glands and external genitalia. In plants and animals, the terms **unisexual, dioecious,** and **gonochoric** are equivalent; they all refer to individuals containing only male or female reproductive organs. Conversely, the terms **bisexual, monoecious,** and **hermaphroditic** refer to individuals containing both male and female reproductive organs, an occurrence known in both the plant and animal kingdoms. This latter group of organisms can produce fertile gametes of both sexes. The term **intersex** is usually reserved for individuals of intermediate sexual differentiation, who are most often sterile.

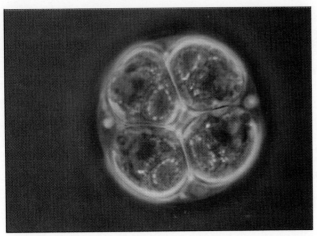

Meiotic products

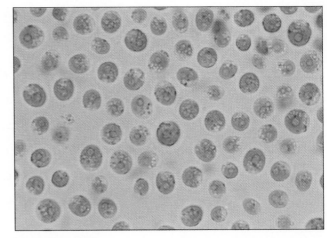

Vegetative colonies

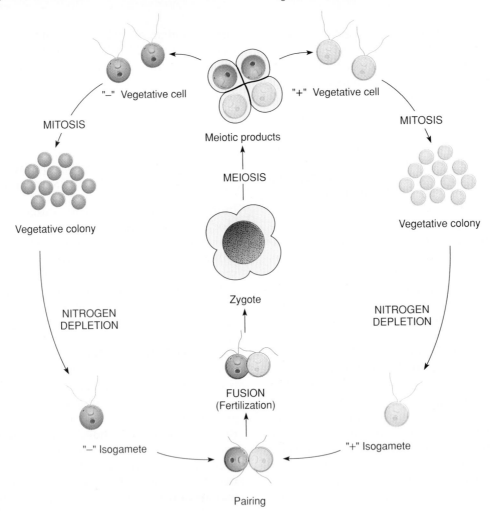

"−" Vegetative cell

MITOSIS

Meiotic products

"+" Vegetative cell

MITOSIS

Vegetative colony

MEIOSIS

Vegetative colony

NITROGEN
DEPLETION

Zygote

NITROGEN
DEPLETION

FUSION
(Fertilization)

"−" Isogamete

"+" Isogamete

Pairing

FIGURE 5.1
The life cycle of *Chlamydomonas*. The diploid zygote (in the center) undergoes meiosis, producing "+" or "−" haploid cells that undergo mitosis, yielding vegetative colonies. Unfavorable conditions stimulate them to form isogametes, which fuse in fertilization, producing a zygote, and repeating the cycle. Several of these stages are illustrated photographically.

Chlamydomonas

The life cycle of the green alga *Chlamydomonas*, shown in Figure 5.1, is representative of organisms that exhibit only infrequent periods of sexual reproduction. Such organisms spend most of their life cycle in the haploid phase. Asexually reproducing organisms produce daughter cells by mitotic divisions. However, under unfavorable nutrient conditions, such as nitrogen depletion, certain daughter cells function as gametes. Following fertilization, a diploid zygote, which can withstand the unfavorable environment, is formed. When conditions become more suitable, meiosis ensues and haploid vegetative cells are produced.

In species such as these, there appears to be little visible difference between the haploid vegetative cells that reproduce asexually and the haploid gametes that are involved in sexual reproduction. The two gametes that fuse together during mating are not usually morphologically distinguishable. Such gametes are called **isogametes,** and species producing them are said to be **isogamous.**

In 1954, Ruth Sager and Sam Granik demonstrated that gametes in *Chlamydomonas* could be subdivided into two mating types. Producing clones derived from single haploid cells, they showed that cells from a given clone would mate with cells from some other clones but not all of them. When they tested mating abilities of large numbers of clones, all could be placed into either a **plus (+)** or a **minus (−)** mating category. Plus cells would mate only with minus cells, and vice versa, as represented in Figure 5.2. Following fertilization and meiosis, the four haploid cells (**zoospores**) produced were found to consist of two plus types and two minus types. Therefore, in this alga, a primitive means of sex differentiation exists even though there is no morphological indication that such differentiation has occurred.

If chemical extracts are prepared from cloned *Chlamydomonas* cells or their flagella, and these extracts are then added to cells of the opposite mating type, clumping or agglutination occurs. This observation suggests that despite the morphological similarity between isogametes, a chemical differentiation has occurred between them, influencing sex determination.

Neurospora

A more complex form of sexual differentiation occurs in Ascomycetes. One ascomycete, the bread mold *Neurospora crassa*, has been used in numerous biochemical and genetic studies. Its life cycle is illus-

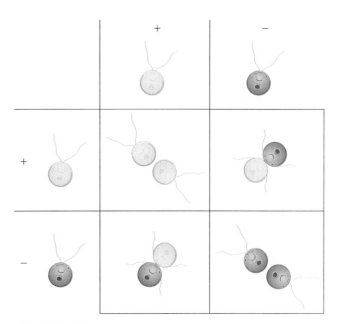

FIGURE 5.2
Mating types in *Chlamydomonas*.

trated in Figure 5.3. The predominant vegetative phase is represented by a spongy mass, the **mycelium**. The mycelium is composed of intertwined branching filaments called **hyphae**. The nuclei found within the segmented hyphae are haploid. Most reproduction is asexual and occurs in one of two ways. First, hyphae may grow by cell division and mitosis and fragment, increasing the mass of the mycelium. Second, special haploid spores called **conidia** may be formed mitotically. These spores may become airborne, and when present on a suitable medium, such as bread, may originate new mycelia through mitosis.

Two mating types, designated *A* and *a*, have been recognized in *Neurospora*. Under proper nutrient conditions, mycelial masses of either type will form fruiting bodies, or **protoperithecia,** within which the **ascogonium** develops. Extending out of the fruiting body are receptive hyphae called **trichogynes**. Mating occurs when a conidium of the opposite mating type contacts a trichogyne. When this occurs, multiple fertilization events soon follow. The conidial nucleus enters the trichogyne, dividing mitotically as it moves inward. These conidial nuclei fuse with the ascogonial cells. This female structure buds off numerous hyphae or proasci containing binucleate cells of opposite mating types. Nuclei then fuse to form a diploid zygote, and the cell containing it enlarges, forming a saclike structure called the **ascus**. Within one fruiting body, many such structures are formed.

FIGURE 5.3

Sexual reproduction during the life cycle of *Neurospora* is initiated following fusion of conidia of opposite mating types. Following fertilization, each diploid zygote becomes enclosed in an ascus where meiosis occurs, leading to four haploid cells, two of each mating type. A mitotic division then occurs and the eight haploid ascospores are released. Upon germination, the cycle may be repeated. Many asci, as shown in the photograph, may form in a single structure. We have illustrated the events occurring in only one ascus.

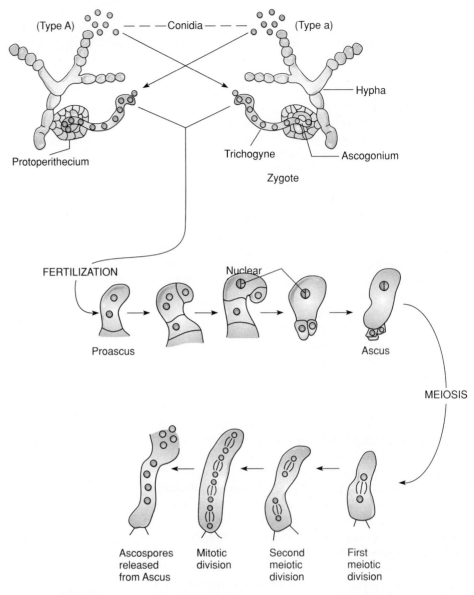

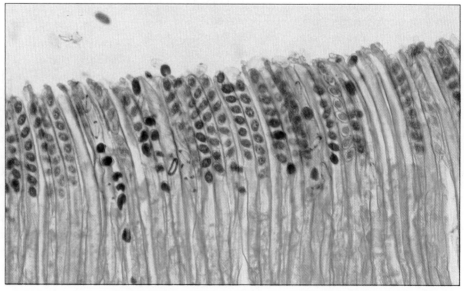

The zygote immediately undergoes meiosis, and each of the four haploid products divides mitotically, yielding eight **ascospores**. The meiotic and mitotic products are retained in the ascus in the precise linear order of their formation. Thus, each half of the ascus contains separate products resulting from the first meiotic division. This observation has been very useful in examining products of genetic crossing-over because each ascospore can be dissected from the ascus and germinated experimentally (see Chapter 6). The haploid ascospores are released from the ascus and may form new mycelia through mitosis and branching.

Like *Chlamydomonas,* the diploid phase of *Neurospora* is short-lived. The existence of distinct mating types ensures that sexual reproduction does not occur between conidia derived from the same fruiting body or mycelium. This restriction promotes greater genetic variability through sexual reproduction.

Zea mays

In many plants, life cycles alternate between the haploid **gametophyte stage** and the diploid **sporophyte stage**. The processes of meiosis and fertilization link the two phases during the life cycle. The relative amount of time spent in the two phases varies between the major plant groups. In some nonseed plants, such as mosses, the haploid gametophyte phase and the morphological structures representing this stage predominate. The reverse is true in all seed plants.

Maize (*Zea mays*) is an example of a monoecious seed plant that bears both male and female structures on a single diploid **sporophyte**. The life cycle of *Zea mays* is illustrated in Figure 5.4. **Microgametophytes,** housing the male gametes, arise by meiosis from small haploid microspores in the **stamen** or **tassels**. Each of four products of meiosis, the microspores, develops into a mature male gametophyte—the pollen grain—which contains two sperm nuclei.

Equivalent female diploid cells exist in the **pistil** of the sporophyte. Following meiosis, only one of the four haploid megaspores survives. It usually divides mitotically three times, producing a total of eight haploid nuclei enclosed in the embryo sac. Two of these nuclei unite near the center of the embryo sac to form the diploid **endosperm nucleus**. At the micropyle end of the sac where the sperm enters, three nuclei remain: the **egg nucleus** and two **synergids**. The other three **antipodal nuclei** are clustered at the opposite end of the embryo sac.

Pollination occurs when pollen grains make contact with the silks (or *stigma*) of the pistil and develop extensive pollen tubes that grow toward the embryo sac. When contact is made at the micropyle, the two sperm nuclei enter the embryo sac. **Double fertilization** occurs when one sperm nucleus unites with the haploid oocyte nucleus and the other sperm nucleus unites with two endosperm nuclei. This process results in a diploid zygote and a triploid endosperm, respectively. Each ear of corn may contain as many as 1000 such structures, each of which develops into a single kernel. Although the majority of kernels serve as nutrition for animal populations, including humans, each structure, if allowed to germinate, will give rise to a new sporophyte.

The mechanism of sex determination and differentiation in a monoecious plant like *Zea mays,* where the tissues that form both male and female gametes are of the same genetic constitution, was difficult to comprehend at first. However, the discovery of a large number of mutant genes that disrupt normal tassel and pistil formation supports the concept that gene products play an important role in sex determination and differentiation. Table 5.1 lists eleven mutations representing separate genes on a number of chromosomes of *Zea mays*. These mutant genes affect the differentiation of male or female tissue in several ways.

Those mutant genes that cause a sex reversal provide the most information. All mutations classified as *tassel seed* (*ts*) interfere significantly with tassel production when homozygous and induce the formation of female structures. Thus, it is possible for a single gene to cause a normally monoecious plant to become functionally female. On the other hand, the recessive mutations *silkless* (*sk*) and *barren stalk* (*ba*) interfere with the development of the pistil, resulting in plants with only functional male reproductive organs.

By manipulating these mutations in certain strains of corn, geneticists can produce dioecious, or single sex, strains. A plant of genotype $sk/sk; ts_2/ts_2$ has no functional pistil along the stem, but has the tassels converted to pistils. This plant is female. The plant of genotype $sk/sk; ts_2^+/ts_2$ also lacks pistils along the stem, but the tassels function normally to produce pollen. This is a male plant. If these two strains are interbred, both parental genotypes are produced, perpetuating the dioecious condition.

Data gathered from studies of these and other mutants suggest that the wild-type alleles of these genes interact in the sex determination of certain cells. They subsequently undergo sexual differentiation into either male or female structures, which in turn produce male or female gametes.

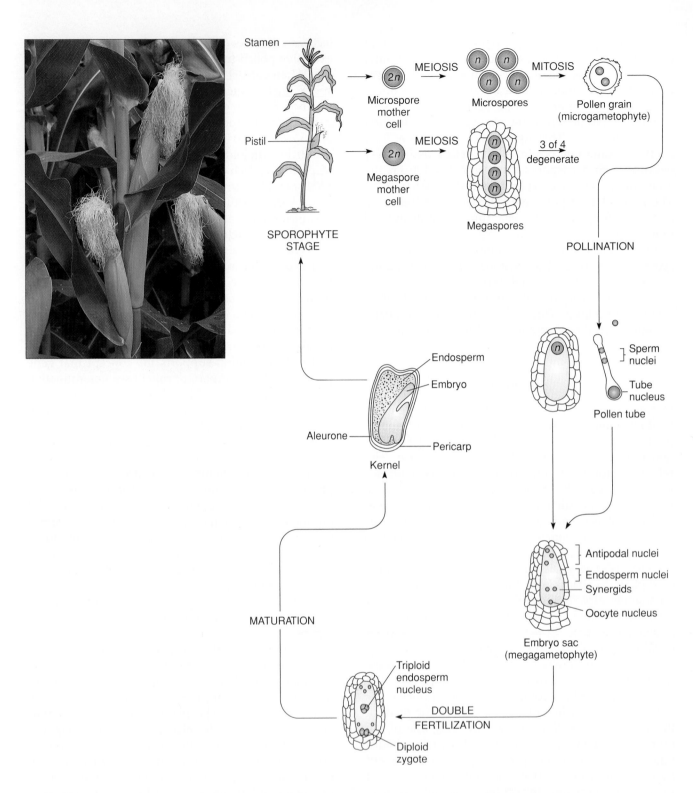

FIGURE 5.4

The life cycle of maize *(Zea mays).* The diploid sporophyte bears stamens and pistils, which give rise to haploid microspores and megaspores. These develop into the microgametophyte and the megagametophyte stages. Following pollination, two sperm nuclei fertilize the endosperm and egg nuclei housed in the embryo sac. The fertilized oocyte is the diploid zygote, which divides and develops into the embryo. The embryo develops within the kernel (or seed) and is nourished by the developing endosperm. Upon germination, a new plant, the diploid sporophyte stage, develops and the cycle may repeat itself. The photograph illustrates the mature sporophyte stage, including the ears that bear the many kernels.

TABLE 5.1
Mutants of *Zea mays* affecting the development of the tassel and pistil.

Chromosome	Locus	Mutant Symbol	Mutant Name	Mutant Phenotype
1	24	ts_2	Tassel seed	Tassel completely pistillate
1	119	ts_3	Tassel seed	Tassel with both pistillate and staminate flowers
1	158	ts_6	Tassel seed	Tassel completely pistillate
2	56	sk	Silkless	Pistils abort; no silks produced
3	55	ts_4	Silkless	Tassel with pistillate and staminate flowers
3	72	ba_1	Barren stalk	Ear shoots missing
4	56	ts_5	Tassel seed	Tassel with both silks and anthers
6	4	po_1	Polymitotic	Microspores divide rapidly without chromosome division; tassel sterile
6	17	ms_1	Male sterile	Anthers shriveled
8	14	ms_8	Male sterile	No anthers exserted; microsporophytes usually degenerate
9	67	ms_2	Male sterile	No anthers exserted

HETEROMORPHIC CHROMOSOMES AND SEX DETERMINATION

The earliest observations directly linking genetics and sex determination occurred around 1900. The research involved the cytological examination of sperm cells derived from insects. Variation in chromosome number between different sperm cells and between male and female somatic cells was observed and correlated with sex differences. From these observations geneticists postulated a chromosomal basis for sex determination and the 1:1 sex ratio observed in unisexual species. Hypotheses concerning special **sex chromosomes** and their role in sexual differentiation conformed nicely with the rediscovery early in this century of Mendelian concepts.

Modes of Heterogametic Sex Determination

In 1891, H. Henking identified a nuclear structure, which he labeled the **X-body,** in the sperm of certain insects. Several years later, Clarance McClung showed that some grasshopper sperm contain a genetic structure, which he called a **heterochromosome,** but the remainder do not. He mistakenly associated the presence of the heterochromosome with the production of male progeny. In 1906, Edmund B. Wilson clarified Henking and McClung's findings when he demonstrated that female somatic cells in the insect *Protenor* contain 14 chromosomes,

including 2 X chromosomes. During oogenesis, an even reduction occurs, producing gametes with 7 chromosomes, including one X. Male somatic cells, on the other hand, contain only 13 chromosomes, including a single X chromosome. During spermatogenesis, gametes are produced containing either 6 or 7 chromosomes, without or with an X, respectively. Fertilization by X-bearing sperm results in female offspring, and fertilization by X-deficient sperm result in male offspring (Figure 5.5).

The presence or absence of the X chromosome in male gametes provides an efficient mechanism for sex determination in this species and also produces a 1:1 sex ratio in the resulting offspring. This mechanism, now called the **XX/XO** or **Protenor mode of sex determination,** depends on the random distribution of the X chromosome into one-half of the male gametes during segregation.

Wilson also experimented with the hemipteran insect *Lygaeus turicus*, in which both sexes have 14 chromosomes. Twelve of these are autosomes. Additionally, the females have 2 X chromosomes, and the males have only a single X and a smaller heterochromosome labeled the **Y chromosome.** Females in this species produce only gametes of the (6A + X) constitution, but males produce two types of gametes in equal proportions: (6A + X) and (6A + Y). Therefore, following random fertilization, equal numbers of male and female progeny will be produced with distinct chromosome complements. This

FIGURE 5.5
The XX/XO or Protenor mode
of sex determination.

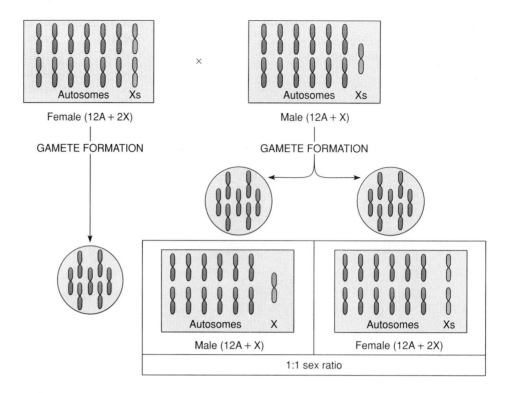

Female (12A + 2X) × Male (12A + X)

GAMETE FORMATION   GAMETE FORMATION

Male (12A + X)   Female (12A + 2X)

1:1 sex ratio

mode of sex determination is called the **Lygaeus** or **XX/XY type**.

The examples presented so far have implied that the male always has unlike sex chromosomes and produces two types of gametes. In these cases, the male is described as the **heterogametic sex;** the female, who has like sex chromosomes and produces uniform gametes with regard to chromosome numbers, is called the **homogametic sex**. However, in other organisms, the female is the heterogametic sex and exhibits either the Protenor (XX/XO) or Lygaeus (XX/XY) mode of sex determination. To immediately distinguish situations in which the female is the heterogametic sex, some workers use the notation **ZZ/ZW,** where ZW is the heterogamous female, instead of the XX/XY notation.

Studies have provided strong evidence that the female is the heterogametic sex for many organisms. For example, all moths and butterflies, most birds, some fish, reptiles, amphibians, and at least one species of plants (*Fragaria orientalis*) demonstrate such a pattern.

The situation with fowl (chickens) illustrates the difficulty in establishing which sex is heterogametic and whether the Protenor or Lygaeus mode is operable. While genetic evidence supported the hypothesis that the female is the heterogametic sex, the cytological identification of the sex chromosomes was not accomplished until 1961. Their identification was difficult to achieve because of the large number of

chromosomes (78). When the sex chromosomes were finally identified, the female was shown to contain an unlike chromosome pair, including a Y (or W) chromosome. Thus, in fowl, the female is indeed heterogametic and is characterized by the Lygaeus type of sex determination.

Sex Determination in *Melandrium*

Investigations with the dioecious angiosperm *Melandrium** have enhanced our knowledge of sex determination in heterogametic organisms. In 1923 it was discovered that the chromosome composition varied between male and female plants of these species. One set of chromosomes, designated X and Y, has been shown to determine maleness or femaleness. The male contains four pairs of autosomes plus an X and a Y chromosome. The Y chromosome in these plants is larger than the X chromosome. Female plants contain the same number of autosomes as the male but have two X chromosomes.

In 1953, Mogens Westergaard correlated abnormal chromosome compositions in plants with the sex of the individual, as shown in Table 5.2. He concluded that the Y chromosome has a strong masculinizing influence because a male plant is always produced when it is present. The X chromosome has a feminizing influence, but this influence is masked by the

*Melandrium is now subsumed under the *Silene* genus.

TABLE 5.2
A comparison of chromosome composition and sex in *Melandrium*.

Chromosome Composition		Sex
Number of Each Autosome	Sex Chromosomes	
2	XX	Normal female
2	XY	Normal male
2	XXX	Female
3	XX	Female
4	XXXX	Female
2	XYY	Male
2	XXY	Male
3	XY	Male
3	XXXY	Male
4	XXXY	Male

SOURCE: After Westergaard, 1953.

action of the Y chromosome. He based this conclusion on the observation that XXY and XXXY plants, for example, are male despite the presence of two or three X chromosomes.

Westergaard also investigated plants that contained the normal number of autosomes and one X chromosome, but only fragments of the Y chromosome. These studies allowed him to identify specific sex-determining functions for certain areas of the sex chromosome. Figure 5.6 shows these subdivisions for the sex chromosomes of a male *Melandrium*.

As shown in Figure 5.6, regions I and IV of the Y chromosome suppress female development, some-

how counteracting the influence of region V of the X chromosome, which promotes female development. If either region I or IV is missing, bisexual development occurs. Region II promotes male development, and if male tissue develops, region III is essential for male fertility. In its absence, male tissue develops, but the plant is sterile. A region of the X chromosome is also designated IV because it has been identified as the only part of the chromosome that synapses with the Y chromosome during meiosis.

In summary, genes localized on the Y chromosome initiate or trigger the development of male sex organs, and at the same time suppress the expression of the female-determining genetic information localized on the X chromosomes. However, the mechanisms involved in the interaction and expression of genetic information resulting in maleness or femaleness are unknown. It should be noted that these studies do not rule out the possible influence of autosomal genes in the process of sexual differentiation.

CHROMOSOME RATIOS IN *DROSOPHILA*: ANOTHER MODE OF SEX DETERMINATION

Studies of the fruit fly *Drosophila melanogaster* demonstrate yet another mode of sex determination. Cytological evidence reveals that the male *Drosophila* is the heterogametic (XY) sex (Figure 5.7). As in *Melandrium* we might assume that the Y chromosome determines maleness in this species were it not for the elegant work of Calvin Bridges. Bridges' work on sex determination in *Drosophila*, initiated in 1916, can be

Female   Male

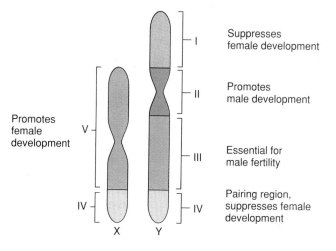

FIGURE 5.6
The subdivisions of the X and Y chromosomes of *Melandrium* and the proposed role of each chromosomal area. A female flower (left) is compared with the male flower (right) in the photographic insert.

MALE   FEMALE

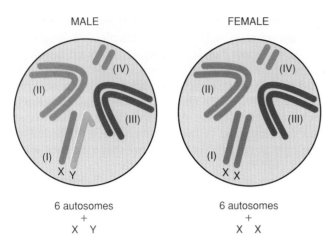

6 autosomes
+
X   Y

6 autosomes
+
X   X

FIGURE 5.7
The chromosomal composition of male and female
Drosophila melanogaster.

divided into two phases: a study of offspring resulting from nondisjunction of the X chromosomes in
females, and subsequent work with progeny of
triploid (3*n*) females.

Nondisjunction is the failure of paired chromosomes to segregate or separate during the anaphase
stage of the first or second meiotic divisions. The
result of nondisjunction is the production of two
abnormal gametes, one of which contains an extra
chromosome ($n + 1$) and the other that lacks a
chromosome ($n - 1$). Fertilization of such gametes
produces ($2n + 1$) or ($2n - 1$) **aneuploid** zygotes.
Bridges studied flies that were the product of gametes
formed by nondisjunction of the X chromosome. In
addition to the normal complement of six autosomes,
these individuals had either an XXY or an XO
chromosome composition. (The zero signifies that the
second chromosome is absent.) The XXY flies were
normal females, and the XO flies were sterile males.
The presence of the Y chromosome in the XXY flies
did not cause maleness, and its absence in the XO
flies did not produce femaleness. From these data
Bridges concluded that the Y chromosome in *Drosophila* lacks male-determining factors, but apparently
contains genetic information essential to male fertility
since the XO males were sterile.

Bridges was able to clarify the mode of sex
determination in *Drosophila* by studying the progeny
of triploid females (3*n*), which have *three* copies each
of the haploid complement of chromosomes. These
females apparently originate from rare diploid eggs
fertilized by normal haploid sperm. Triploid females
have heavy-set bodies, coarse bristles, and coarse
eyes, and may be fertile. During meiosis a wide range
of chromosome complements is distributed into gametes, and these gametes give rise to offspring with

a variety of abnormal genetic constitutions. A correlation between the sexual morphology, chromosome
composition, and Bridges' interpretation is shown in
Figure 5.8.

Bridges realized that the critical factor in determining sex is the **ratio of X chromosomes to the number
of haploid sets of autosomes present**. Normal
(2X:2A) and triploid (3X:3A) females each have a
ratio equal to 1.0, and both are fertile. As the ratio
exceeds unity (3X:2A, or 1.5, for example), what was
originally called a **superfemale** is produced. Since
this female is rather weak and infertile and has
lowered viability, this type is now more appropriately
called a **metafemale**.

Normal (XY:2A) and sterile (X:2A) males each
have a ratio of 1:2, or 0.5. When the ratio decreases to
1:3, or 0.33, as in the case of an XY:3A male, infertile
metamales result. Other flies recovered by Bridges in
these studies contained an X:A ratio intermediate
between 0.5 and 1.0. These flies were generally
larger, and they exhibited a variety of morphological
abnormalities and rudimentary bisexual gonads and
genitalia. They were invariably sterile and were
designated as **intersexes**.

These results indicate that in *Drosophila*, male-
determining factors are not localized on the sex
chromosomes, but are instead found on the autosomes. Some female-determining factors, however,
are localized on the X chromosomes. Thus, with
respect to primary sex determination, male gametes
containing one of each autosome plus a Y chromosome result in male offspring *not* because of the
presence of the Y chromosome but because of the lack
of an X chromosome—quite a different situation
from that previously described in *Melandrium*. This
mode of sex determination is explained by the **genic
balance theory**. Bridges proposed that a threshold for
maleness is reached when the X:A ratio is 1:2
(X:2A), but that the presence of an additional X
(XX:2A) alters this balance and produces female
differentiation.

Bridges' conclusion received further support from
the experiments of Jack Schultz and Theodosius
Dobzhansky. They were able to obtain males, females, and intersexes that had gained or lost vari-
ously sized fragments of the X chromosome. Some of
their results are shown in Table 5.3. The effects were
proportional to the extent of the addition or subtraction of the X and were consistent with Bridges' genic
balance interpretation of sex determination. This
finding led to the conclusion that multiple factors
critical to sex determination exist on the X chromosome.

Several mutant genes, which behave in a manner
analogous to the sex-determining genes of *Zea mays*,

FIGURE 5.8

Chromosome compositions, the ratios of X chromosomes to sets of autosomes, and the resultant sexes in *Drosophila melanogaster*.

CHROMOSOME COMPOSITION	CHROMO-SOME FORMU-LATION	RATIO OF X CHROMOSOMES TO AUTOSOME SETS	SEX
	$3X/2A$	1.5	Metafemale
	$3X/3A$	1.0	Female
	$2X/2A$	1.0	Female
	$2X/3A$	0.67	Intersex
	$3X/4A$	0.75	Intersex
	$X/2A$	0.50	Male
	$XY/2A$	0.50	Male
	$XY/3A$	0.33	Metamale

have been located on the autosomes of *Drosophila*. The recessive autosomal gene *transformer* (*tra*), discovered by Alfred H. Sturtevant, is especially interesting. Homozygous mutant females are transformed

TABLE 5.3

The effect of the addition or subtraction of fragments of the *Drosophila* X chromosome.

Chromosome Composition	Sex	Addition or Subtraction	Effect
2X/3A	Intersex	+	Toward female
2X/2A	Female	+	Toward metafemale
2X/3A	Intersex	−	Toward male

into sterile males, but normal males are unaffected when homozygous for *tra*. Experiments by Robert C. King and Dietrich Bodenstein showed that the transplantation of normal but immature female gonadal tissue into the abdomens of *tra/tra* sterile males results in normal ovarian development. Thus, it appears that the abdominal environment can support the production of eggs, but that the *transformer* gene has interfered with primary sex determination in these flies.

More recently, another gene, *Sex-lethal* (*Sxl*), has been shown to play a critical role in sex determination by responding to the products of a group of at least four separate regulatory genes. Activation of the *Sxl* gene, located on the X chromosome, is essential to subsequent female development, and relies on a ratio of X chromosomes to sets of autosomes that equals 1.0. In the absence of activation, resulting from an

X:A ratio of 0.5, male development occurs. While it is not yet clear how this ratio influences either the group of regulatory genes or the *Sxl* locus, nor how the product of this locus influences sex determination, some insights are available. The *Sxl* locus is part of a hierarchy of gene expression involved in this process, and itself exerts control over still other genes, including the *tra* gene discussed above. The wild-type allele of *tra* functions only in females and is required for their somatic sexual differentiation. We have already seen that the effect of mutations in *tra* is to cause maleness in chromosomally female flies. Mutations in *Sxl* in such female flies are usually lethal.

Based on this information, it is evident that many genes play important roles in sex determination as well as sexual differentiation in *Drosophila*. This is an area of active investigation, so we can soon expect the entire genetic system to be elucidated.

SEX DETERMINATION IN HUMANS

From the time dividing cells of humans were first observed, geneticists tried to determine accurately the chromosome number of our own species. The first significant attempt was made in 1912, when H. von Winiwarter counted 47 chromosomes in a spermatogonial metaphase. At that time, geneticists believed that the sex-determining mechanism in humans was based on the presence of an extra chromosome in females; that is, females were thought to have 48 chromosomes. In the 1920s, Theophilus Painter observed between 45 and 48 chromosomes in testicular tissue and also discovered the small Y chromosome, which is now known to occur only in males. In his original paper, Painter favored 46 as the diploid number in humans, but he later concluded that 48 was the correct number in both males and females.

For thirty years, 48 was accepted as the correct number of human chromosomes. Then, in 1956, Joe Hin Tjio and Albert Levan introduced improvements in chromosome preparation techniques. These improved techniques led to a strikingly clear demonstration of metaphase stages showing that 46 was indeed the human diploid number. Later that same year, C. E. Ford and John L. Hamerton, also working with testicular tissue, confirmed this finding.

Within the normal 23 pairs of human chromosomes, one pair was shown to vary in configuration in males and females. These two chromosomes were designated the sex chromosomes. The human female has two X chromosomes, and the human male has one X and one Y chromosome. As with *Drosophila*, where XX and XY constitutions also yield females and

males, respectively, these observations are insufficient to conclude that the Y chromosome determines maleness.

Klinefelter and Turner Syndromes

Around 1940 it was observed that two human abnormalities, the **Klinefelter** and **Turner syndromes,** * caused aberrant sexual development. The study of these genetic disorders has been instrumental in sex determination studies with humans. Individuals with Klinefelter syndrome [Figure 5.9(a)] have genitalia and internal ducts that are usually male, but their testes are underdeveloped and fail to produce sperm. Although masculine development occurs, feminine sexual development is not entirely suppressed. Slight enlargement of the breasts is common, for example. Intersexuality may lead to abnormal social development.

In Turner syndrome [Figure 5.9(b)], the affected individual has female external genitalia and internal ducts, but the ovaries are rudimentary. Other characteristic abnormalities include short stature (usually under five feet); a short, webbed neck; and a broad, shieldlike chest.

In 1959, the karyotypes of individuals with these syndromes were independently determined to be abnormal with respect to the sex chromosomes. Individuals with Klinefelter syndrome most often have an XXY complement in addition to 44 autosomes. People with this karyotype are designated **47,XXY**. Individuals with Turner syndrome have only 45 chromosomes, including just a single X chromosome, and are designated **45,X**. Both conditions result from nondisjunction of the sex chromosomes during meiosis. Note the convention used in designating chromosome compositions. The number indicates how many chromosomes are present, and the information after the comma designates the deviation from the normal diploid content.

These karyotypes and their corresponding sexual phenotypes allow us to conclude that, as in *Melandrium*, the Y chromosome determines maleness in humans. In its absence, the sex of the individual is female. The presence of the Y chromosome in the individual with Klinefelter syndrome is sufficient to determine maleness, even though its expression is not complete. Similarly, in the absence of a Y chromosome, as in the case of individuals with Turner syndrome, no masculinization occurs.

---

*Although the possessive form of the names of most syndromes (eponyms) is often used (e.g., Klinefelter's), the current preference is to use the nonpossessive form for syndromes, as recommended in McKusick, 1975.

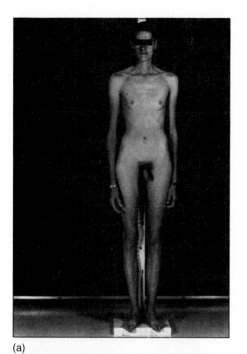

(a)

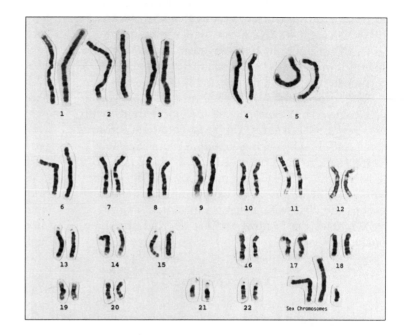

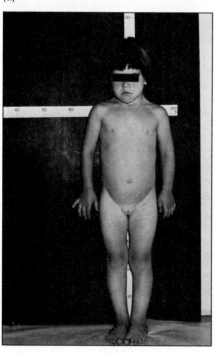

(b)

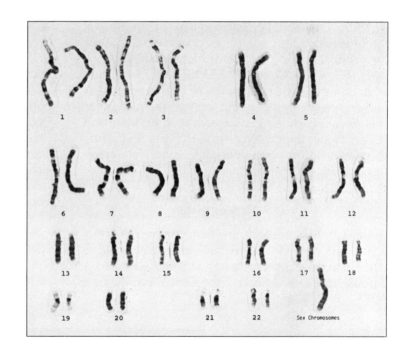

FIGURE 5.9
(a) The karyotype and a photograph of an individual with Klinefelter syndrome. In the karyotype, two X chromosomes and one Y chromosome are visible, creating the 47,XXY condition. (b) The karyotype and a photograph of an individual with Turner syndrome. In the karyotype, only one X chromosome is visible, creating the 45,X condition.

Klinefelter syndrome occurs in about 2 of every 1000 male births. The karyotypes **48,XXXY; 48,XXYY; 49,XXXXY;** and **49,XXXYY** are similar phenotypically to **47,XXY.** Generally, more severe manifestations are seen in individuals with a greater number of X chromosomes.

Karyotypes other than 45,X also lead to Turner syndrome. These include mosaic individuals with two apparent cell lines, the most common chromosome combinations being **45,X/46,XY** and **45,X/46,XX.** Turner syndrome is observed in only about 1 in 3000 female births, a frequency much lower than that for Klinefelter syndrome. One explanation for this difference is the observation that a substantial majority of **45,X** fetuses die *in utero* and are aborted spontaneously.

47,XXX Syndrome

The presence of three X chromosomes (**47,XXX**) results in female differentiation. This syndrome, which is estimated to occur in about 1 of 1200 female births, is highly variable in expression. Frequently, 47,XXX women are perfectly normal. In other cases, underdeveloped secondary sex characteristics, sterility, and mental retardation may occur. In rare instances, **48,XXXX** and **49,XXXXX** karyotypes have been reported. The syndromes associated with these karyotypes are similar to but more pronounced than the 47,XXX. Thus, the presence of additional X chromosomes appears to disrupt the delicate balance of genetic information essential to normal female development.

The XYY Condition

A third human trisomy involving the sex chromosomes has been discovered and intensively investigated. This case involves males with one X and two Y chromosomes in addition to the normal complement of 44 autosomes (**47,XYY**). Many studies have attempted to clarify the effect of this chromosome composition.

In 1965, Patricia Jacobs discovered 9 of 315 males in a Scottish maximum security prison to have the 47,XYY karyotype. These males were significantly above average in height and were involved in criminal acts of serious social consequence. Of the 9, 7 were of subnormal intelligence, and all suffered personality disorders. In several other studies, similar findings were obtained.

Because of these investigations, the phenotype and frequency of the 47,XYY condition in criminal and noncriminal populations have been examined more extensively. Above average height and subnormal intelligence have been generally substantiated, and the frequency of males displaying this karyotype is indeed higher in penal and mental institutions compared with unincarcerated males. Table 5.4 compares the frequency of XYY individuals in newborns with the frequency found in various mental-penal settings. Each value represents data compiled from many individual studies. Note that when men are selected for above average height, the frequency of XYY cases increases dramatically.

The possible correlation between this chromosome composition and antisocial and criminal behavior was of considerable interest, leading to a particularly controversial long-term study initiated in 1968. This investigation was undertaken with the hope of establishing or dispelling any possible link between the presence of an extra Y chromosome and specific behavior. Many questions were unresolved. For example, is excessive height a predisposing factor in abnormal social development of these individuals? Does their subnormal intelligence result in a higher likelihood that they will be apprehended, elevating their frequency in penal settings? And most importantly, what characteristics are displayed by XYY individuals who are not in penal settings? The only

TABLE 5.4
Frequency of XYY individuals in various settings.

Setting	Restriction	Number Studied	Number XYY	Frequency XYY
Control population	Newborns	28,366	29	0.10%
Mental-penal	No height restriction	4,239	82	1.93%
Penal	No height restriction	5,805	26	0.44%
Mental	No height restriction	2,562	8	0.31%
Mental-penal	Height restriction	1,048	48	4.61%
Penal	Height restriction	1,683	31	1.84%
Mental	Height restriction	649	9	1.38%

SOURCE: Compiled from data presented in Hook, 1973, Tables 1–8. Copyright 1973 by the American Association for the Advancement of Science.

constant association noted was that XYY males are almost always over six feet tall!

The study was initiated to identify 47,XYY individuals at birth and to follow their behavioral patterns during preadult and adult development. By 1974 the two primary investigators, Stanley Walzer and Park Gerald, had identified about 20 XYY newborns in 15,000 births at Boston Hospital for Women. However, they soon came under great pressure to abandon their research. Those opposed to the study argued that the investigation could not be justified and might cause great harm to those individuals who displayed this karyotype. They argued that (1) no association between the additional Y chromosome and abnormal behavior had been previously established in the population at large, and (2) by "labeling" these individuals, a self-fulfilling prophecy might be created. That is, as a result of participation in the study, parents, relatives, and friends might treat those identified as 47,XYY differently, leading to the type of antisocial behavior expected of them. Despite the support of a government funding agency and the faculty at Harvard Medical School, Walzer and Gerald abandoned the investigation in 1975.

Since it is now clear that many XYY males do not exhibit any form of antisocial behavior and lead normal lives, we must conclude that there is no direct correlation between the extra Y chromosome and behavior. In retrospect, the objections to Walzer and Gerald's work were justified.

Sexual Differentiation in Humans

Once researchers had established that in humans, it is the Y chromosome that houses genetic information necessary for maleness, efforts were made to pinpoint a specific gene or genes capable of determining sex. Before pursuing a discussion of this topic, it is important to distinguish between **sex determination** and **sexual differentiation** to comprehend how adult male and female humans arise. During early development, every human embryo undergoes a period when it is potentially hermaphroditic or bisexual. Gonadal primordia arise as a pair of ridges associated with each embryonic kidney. Primordial germ cells migrate to these ridges, where an outer cortex and inner medulla form. The **cortex** is capable of developing into an ovary, while the inner **medulla** may develop into a testis. In addition, two sets of undifferentiated male (Wolffian) and female (Mullerian) ducts exist in each embryo.

The genital ridge is present after five weeks of gestation, and if these cells have the XY constitution, male development of the medullary region can be detected by the seventh week. In the absence of the Y

chromosome, no male development is initiated, and the genital ridge is destined to subsequently form ovarian tissue. Parallel development of the appropriate male or female duct system then occurs, and the other duct system degenerates. There is a substantial amount of evidence that once testes differentiation is initiated, the embryonic testicular tissue secretes a hormone that supports continued sexual differentiation. For example, in rabbits, if castration occurs after testes development has begun but before the duct system has developed, embryos develop female characteristics regardless of the presence of the Y chromosome.

As the twelfth week of human female development approaches, the oogonia within the ovaries begin meiosis and primary oocytes can be detected (see Figure 2.10). By the twenty-fifth week of gestation, all oocytes become arrested in meiosis, where all will remain dormant until puberty is reached, some ten to fifteen years later. Primary spermatocytes, on the other hand, are not formed in males until puberty is reached.

We can conclude that genetic information on the sex chromosomes is responsible for the primary sex determination event. Under normal conditions, once testicular or ovarian development has begun, subsequent early differentiation occurs under the influence of the male or female sex hormones. These hormones support, if not determine, expression of other secondary sexual characteristics during development.

Sex-Determining Function of the X and Y Chromosomes

The sex chromosome abnormalities previously described demonstrate that both the X and Y chromosomes play a critical role in human sex determination. Indeed, the Y chromosome seems most decisive. When it is present, the sexual phenotype is male; and in its absence, female development occurs. The genetic function of the Y chromosome appears to be similar in all mammalian species. In the mouse, rabbit, rat, horse, cow, sheep, goat, pig, cat, and dog, all males have the XY karyotype and all females are XX.

Even though it is known that the presence of a Y chromosome leads to male development and that its absence leads to female development, just how the genetic information induces sexual differentiation is not yet clear. Some insights have been gained as a result of observations of partial loss of an X or Y chromosome in individuals with otherwise normal karyotypes. Partial deletion of one X in an apparent 46,XX female results in a variable expression of the Turner syndrome. If the short arm (called the p arm) of the X is missing (**46,XXp-**), the complete syndrome results. If the long arm (called the q arm) is missing (**46,XXq-**), incomplete ovarian development is usually the only consequence.

The significance of the Y chromosome is even more enigmatic because it is largely heterochromatic. As such, it remains condensed in interphase cells and in spermatozoa and can be identified by fluorescent staining techniques. Heterochromatic chromosomal material is thought to be genetically inert in contrast to euchromatin, which contains potentially active genes. This interpretation is consistent with the observation that the Y chromosome lacks information homologous to the X chromosome. Nevertheless, the Y chromosome must contain genetic information essential for sex determination in humans. Any such genes are not heterochromatic.

Data have been collected on individuals who have one X chromosome and only part of a Y chromosome. When these deletions involve the short arm of the Y (**46,XYp-**), a sexual phenotype similar to that of individuals with Turner syndrome is seen. Thus, genetic information located on the short arm of the Y chromosome appears to be important for male development. The short arm of the Y chromosome also contains a region that synapses with the terminal portion of the short arm of the X chromosome during meiosis. This partial pairing region is essential to male spermatogenesis during segregation of the sex chromosomes.

A Holandric Gene Product: The H-Y Antigen

For many years, geneticists have sought to document the presence of genes on the human Y chromosome. In addition to identification of those genetic factors causing male sex determination, investigators have attempted to show that other genes reside on the Y chromosome. Such genes must exhibit a **holandric** or **Y-linked** inheritance pattern, with the trait in question being passed directly from the father and expressed in all his sons. While the evidence is questionable, one such trait, **hairy pinna** (hair developing on the cartilagenous projection of the external ear) has often been cited as one holandric trait in humans.

Another gene, responsible for the production of the **H-Y antigen**, clearly fits the criteria for residence on the Y chromosome. Discovered in 1955 as a transplantation antigen in mice, it has been shown to be common to all mammalian species studied. In mice, the H-Y antigen is thought to be responsible for the females' rejection of male tissue transplants between members of highly inbred strains. In these strains, male-to-male, female-to-female, and female-to-male transplants are not rejected because of their genetic compatibility.

Recent studies, particularly by Stephen Wachtel, have suggested that the expression of this histocompatibility antigen is associated with a gene located on the Y chromosome. The strongest evidence has been provided by dose-related studies. If anti-H-Y antibodies are added to white blood cells, the H-Y antigen on the surface of the cells will bind the antibody. Human males with 47,XYY and 48,XXYY karyotypes bind about twice as much antibody as do males with only one Y chromosome (46,XY).

In the absence of contradictory evidence, some geneticists, including Wachtel and Susumo Ohno, postulated that the H-Y antigen might be the gene product responsible for male development. It was hypothesized that the embryonic gonad somehow responds to this antigen, being stimulated to undergo testicular development. In the absence of this antigen, ovarian development subsequently occurs. While various types of circumstantial evidence supported this idea, no direct evidence in its favor has been forthcoming. As we will see in the next section, more recent information has provided still another, more promising candidate.

The Primary Signal: Testis-Determining Factor (TDF)

Work completed by David Page and his colleagues in 1987 soon altered the notion that the H-Y antigen might determine maleness in humans. Using molecular genetic techniques, they were able to subdivide the Y chromosome into a number of regions that could be identified as being present or absent in the genetic make-up of individuals. They investigated two unique groups who were ''**sex-reversed**.'' First, there were normal-appearing, but sterile, males who displayed the female karyotype of 46,XX. Second, there were normal-appearing females who displayed the male karyotype of 46,XY.

Research proceeded by first asking whether or not the XX males might contain a small piece of the Y chromosome hooked to one of the X chromosomes, thereby causing maleness. And might there be a small piece of the Y chromosome missing in the XY females that prevented them from becoming males?

Attention soon focused on one particular region of the Y chromosome called **1A2**. This region was determined to encode a protein containing repeating amino acid sequences that can form domains called ''fingers.'' Together with an atom of zinc, these fingers can bind to nucleic acids. Because of these characteristics and the association with the Y chromosome, the protein, encoded by at least part of the 1A2 region, was named **ZFY**.

The question was then asked whether the ZFY gene was present or absent in the XX males and XY females. In an initial sample of 90 sex-reversed patients, a perfect correlation was found. The genetic information, normally present on the Y chromosome, was present in XX males and absent in XY females! In the most extreme cases, one XX male contained a fragment of the Y that was only 0.5 percent of the total chromosome, while one XY female had lost less than 1.0 percent of the Y chromosome.

Thus, it appeared that the ZFY protein, encoded by part of the 1A2 region of the Y chromosome, might be the **testis-determining factor** (or **TDF** for short). Since region 1A2 is located on the distal part of the short arm (the p arm) of the Y chromosome, TDF could not be the H-Y antigen because the location of the H-Y gene is known to be near the centromere on the long arm (the q arm).

Since the initial work of Page, more recent findings have caused us to reexamine whether ZFY is indeed the testis-determining factor. First, another group of XX males have been examined, some of whom *lack* the ZFY gene sequences of the 1A2 region. And, XX true hermaphrodites, who develop both testicular and ovarian tissue, have been studied. They are expected to contain the ZFY sequences, but do not. Furthermore, while ZFY appears to have been conserved on the Y chromosome for over a million years of mammalian evolution, one group fails to demonstrate this region. Marsupials, where males are clearly XY, do not have ZFY. One other investigation, involving a mutant strain of mice, has sealed the case of ZFY, so to speak. This strain forms testicular tissue, but the testes do not contain the germ cells essential to sperm formation. Such mice lack ZFY sequences. Thus, it is possible that the gene sequences of region 1A2 are essential for fertility, but not testis formation. If so, we must conclude that ZFY is not the TDF.

In spite of this setback in resolving the issue, findings such as these continue to generate a great deal of excitement in genetics. With each new finding, we continue to move closer to the ultimate identification of the TDF in humans and all mammals. Many researchers believe that TDF will prove to be a gene that behaves as a central switch, whose product activates a group of other genes involved in influencing the development of the undifferentiated gonad. Two genes in this group are very likely encoding ZFY and the H-Y antigen.

Sex Ratio in Humans

The presence of heteromorphic sex chromosomes in one sex of a species but not the other provides a potential mechanism for producing equal proportions of male and female offspring. The actual proportion of male to female offspring is called the **sex ratio**. It can be assessed in two ways. The **primary sex ratio** reflects the proportion of males and females *conceived* in a population. The **secondary sex ratio** reflects the proportion that are *born*. The secondary sex ratio is much easier to determine, but has the disadvantage of not accounting for disproportionate embryonic or fetal mortality, should it occur.

When the secondary sex ratio in the human population is determined using worldwide census data, it does not equal 1.0. For example, in the Caucasian population in the United States, the secondary ratio is 1.06, reflecting 106 males born for each 100 females. In the black population in the United States, the ratio is 1.025. In other countries the excess of male births is even higher. In Korea, the secondary sex ratio is 1.15.

To account for this discrepancy, it has been suggested that prenatal female mortality might be greater than prenatal male mortality. If so, it is possible that the primary sex ratio is 1.0 and that it is altered before birth. However, this hypothesis has been shown to be false. In fact, just the opposite occurs. In a Carnegie Institute study, reported in 1948, the sex of approximately 6000 embryos and fetuses recovered from miscarriages and abortions was determined. On the basis of the data derived from this study, the primary sex ratio was estimated to be 1.079. More recent data have estimated that this figure is even higher—between 1.20 and 1.60! Therefore, many more males than females are conceived in the human population.

It is not clear why such a radical departure from the expected 1.0 primary sex ratio occurs. A suitable explanation can be derived only from examining the assumptions upon which the theoretical ratio is based:

1 Because of segregation, males produce equal numbers of X- and Y-bearing sperm.
2 Each type of sperm has equivalent viability and motility in the female reproductive tract.
3 The egg surface is equally receptive to both types of sperm.

There is no strong experimental evidence to suggest that any of these assumptions are invalid. However, it has been speculated that since the human Y chromosome is smaller than the X chromosome,

Y-bearing sperm are of less mass and therefore more motile. If this is true, then the probability of a fertilization event leading to a male zygote is increased. Other explanations must also be considered.

GENES ON THE X CHROMOSOME: SEX LINKAGE

By 1920, researchers had gained clear insights into the genetic basis of sex determination in several organisms. The discovery of X and Y chromosomes is a particularly significant chapter in the story of neo-Mendelian genetics. With the role of these so-called sex chromosomes in sex determination established, research ensued involving the inheritance of traits controlled by genes located on the X and Y. Many questions were open to investigation. Two pertinent ones involved the genetic constitution of these chromosomes:

1 Are the X and Y chromosomes in a species homologous with regard to genetic loci?
2 Are there genes other than those involved in sex determination residing on the X and Y chromosomes?

The answer to the first question is a resounding *no*. The Y chromosome bears very few gene sites. Overwhelmingly, the answer to the second question is yes. Most genes on the X are unrelated to sex determination. As we will see in the remainder of this chapter, the answers to many other related questions were also soon forthcoming.

Sex-Linked Genes in *Drosophila*

The first clue that genes reside on the X chromosome became evident during Thomas H. Morgan's genetic analysis of the *white* mutation in *Drosophila*. Recall from Chapter 4 that the *white* locus may be occupied by any number of a large series of multiple alleles. Morgan's work established that the inheritance pattern of the white-eye trait was clearly related to the sex of the parent carrying the mutant allele. Unlike the outcome of the typical monohybrid cross, Morgan's reciprocal crosses between white- and red-eyed flies did not yield identical results. In contrast, in all of Mendel's monohybrid crosses, F_1 and F_2 data were very similar in reciprocal crosses regardless of which P_1 parent exhibited the recessive mutant trait. Morgan's analysis led to the conclusion that the *white* locus is present on the X chromosome rather than one

of the autosomes. As such, both the gene and the trait are said to be **sex linked**.*

Results of reciprocal crosses between white-eyed and red-eyed flies are shown in Figure 5.10. The obvious differences in phenotypic ratios in both the F_1 and F_2 generations are dependent on whether the P_1 white-eyed parent was male or female.

Morgan was able to correlate these observations with the difference found in chromosome composition between male and female *Drosophila*. As we have seen, one of the four pairs of chromosomes varies between sexes. This chromosome pair is involved in the sex determination mechanism and constitutes the **sex chromosomes**. All other chromosomes are called **autosomes**. Females possess two rod-shaped homologues called the X chromosomes and designated XX. Males possess a single X chromosome and a single Y chromosome, together designated XY. The complex chromosome composition of both sexes of *Drosophila* is shown in Figure 5.7.

On the basis of this correlation, the postulate was put forth that the allele for white eye is found on the X chromosome but its locus is absent from the Y chromosome. Females thus have two available gene sites, one on each X chromosome, while males have only one available gene site on their single X chromosome. This explanation supposes that the Y chromosome lacks loci homologous to those on the X chromosome. However, the X and Y chromosomes still behave as homologues in that they do partially synapse with each other and segregate into gametes during meiosis.

Morgan's interpretation of sex-linked inheritance, shown in Figure 5.11, provides a suitable theoretical explanation for his results. Since the Y chromosome lacks homology with most genes on the X chromosome, whatever alleles are present on the X chromosome of males will be expressed in the phenotype. Since males cannot be either homozygous or heterozygous for sex-linked genes, this condition is called **hemizygous**. In such cases, no alternative alleles are present, and the concept of dominance and recessiveness is irrelevant.

One result of sex linkage is the so-called **crisscross pattern** of inheritance. Phenotypic traits controlled by recessive sex-linked genes are passed from the homogametic sex to the heterogametic sex in each generation. In *Drosophila* and other species where the

---

*Traditionally, the term *sex linked* has been used to describe genes located on the X chromosome. Thus, we will use this term. However, **X-linkage** is a more accurate description of this phenomenon.

FIGURE 5.10
The F₁ and F₂ results of T. H.
Morgan's reciprocal crosses
involving the sex-linked white-
eye mutation in *Drosophila
melanogaster*. The photograph
contrasts white eyes with the
brick red wild-type eye color.

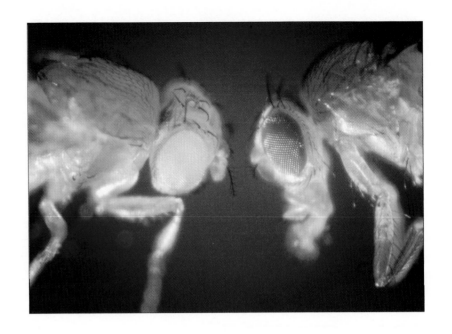

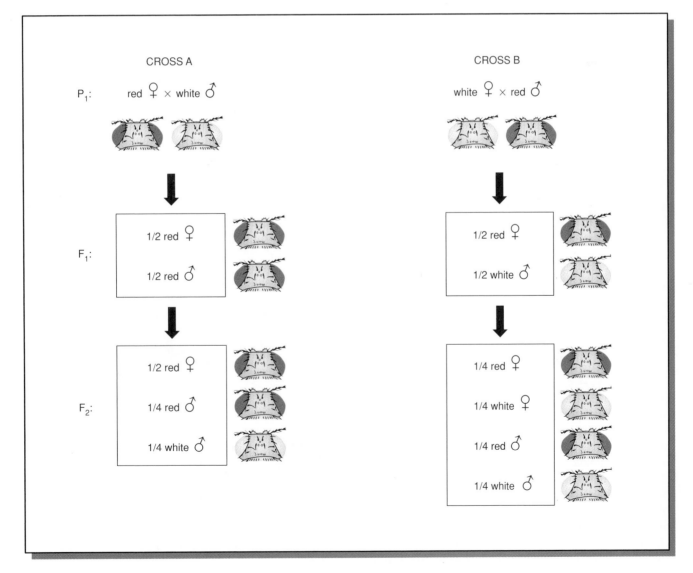

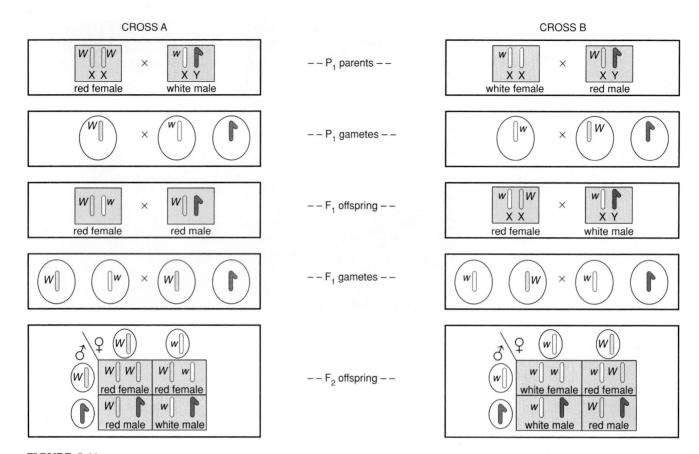

FIGURE 5.11
The chromosomal explanation of the results of the sex-linked crosses shown in Figure 5.10.

male is the heterogametic sex, such traits are passed from mother to son. In order for a female to exhibit a recessive trait, both of her X chromosomes must contain the mutant allele. Since all male offspring receive one of their mother's two X chromosomes and are hemizygous for all alleles present on that X, they will express all the recessive sex-linked traits she expresses.

Drosophila Mosaics

The mode of sex determination and our knowledge of sex linkage in *Drosophila* help to explain the appearance of the extremely unusual fruit fly shown in Figure 5.12. This fly was recovered from a stock where all other females were heterozygous for the sex-linked genes *white* eye (*w*) and *miniature* wing (*m*). It is a **bilateral gynandromorph,** which means that one-half of its body has developed as a male and the other as a female. If a zygote heterozygous for *white* eye and *miniature* wing were to lose one of the X chromosomes during the first mitotic division, the two cells would be of the XX and XO constitution,

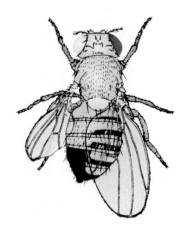

FIGURE 5.12
A bilateral gynandromorph of *Drosophila melanogaster* formed following the loss of one X chromosome in one of the two cells during the first mitotic division. The left side of the fly, composed of male cells containing a single X, expresses the mutant white-eye miniature-wing alleles. The right side is composed of female cells containing two X chromosomes heterozygous for the two recessive alleles.

respectively. Thus, one cell would be female and the other would be male.

In the case of the bilateral gynandromorph, each of these cells is responsible for producing all progeny cells that make up either the right half *or* the left half of the body during embryogenesis. The cell of XO constitution apparently produced only identical progeny cells and gave rise to the left half of the fly, which was male. Since the male half demonstrated the *white, miniature* phenotype, the X chromosome bearing the w^+, m^+ alleles was lost, while the w, m-bearing homologue was retained. All female cells of the right side of the body remained heterozygous for both mutant genes and therefore contained normal eye-wing phenotypes. Depending on the orientation of the spindle during the first mitotic division,

gynandromorphs can be produced where the "line" demarcating male versus female development occurs at almost any place along or across the fly's body.

Sex-Linked Inheritance in Humans

In humans, many genes and the respective traits controlled by them are recognized as being linked to the X chromosome. These sex-linked traits may be easily identified in pedigrees because of the crisscross pattern of inheritance. A pedigree for one form of human color blindness is shown in Figure 5.13. The mother in generation I passes the trait to all her sons but to none of her daughters. If the offspring in generation II marry normal individuals, the color-blind sons will produce all normal male and female

FIGURE 5.13
(a) A human pedigree of the sex-linked color blindness trait.
(b) The most probable genotypes of each individual in the pedigree.

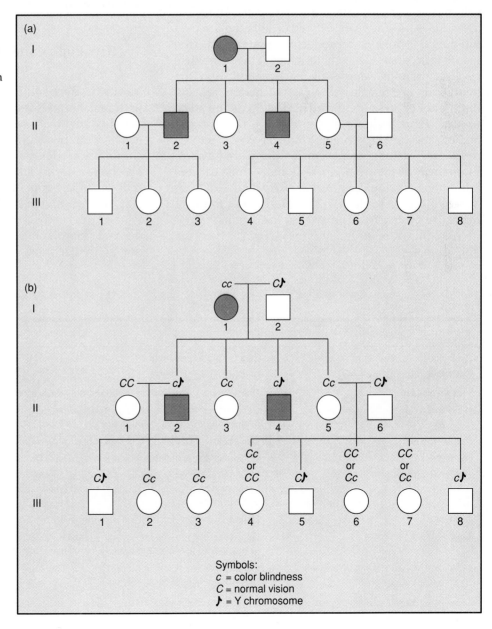

Symbols:
c = color blindness
C = normal vision
♪ = Y chromosome

offspring (III-1, 2, and 3); the normal-visioned daughters will produce normal-visioned female offspring (III-4, 6, and 7), as well as color-blind (III-8) and normal-visioned (III-5) male offspring.

Many sex-linked human genes have now been identified, as shown in Table 5.5. For example, the genes controlling two forms of hemophilia and one form of muscular dystrophy are located on the X chromosome. Additionally, numerous genes whose expression yields enzymes are sex-linked. Glucose-6-phosphate dehydrogenase and hypoxanthine-guanine-phosphoribosyl transferase are two examples. In the latter case, the **Lesch-Nyhan syndrome** results from the mutant form of the X-linked gene product.

Because of the way in which sex-linked genes are transmitted, unusual circumstances may be associated with recessive sex-linked disorders, in comparison to recessive autosomal disorders. For example, if a sex-linked disorder debilitates or kills the affected individual prior to reproductive maturation, the disorder occurs exclusively in males. This is the case because the only sources of the lethal allele in the population are females who carry it heterozygously but do not express the disorder. They pass the allele to one-half of their sons, who develop the disorder because they are hemizygous, but who rarely, if ever, reproduce. Heterozygous females also pass the allele to one-half of their daughters, who become carriers but do not develop the disorder. An example of such a sex-linked disorder is the Duchenne form of muscular dystrophy. The disease has an onset prior to age 6 and is often lethal prior to age 20. It occurs only in males.

THE X CHROMOSOME AND DOSAGE COMPENSATION

The presence of two X chromosomes in normal human females and only one X in normal human males is unique compared with the equal numbers of autosomes present in the cells of both sexes. On theoretical grounds alone, it is possible to speculate that this situation should create a genetic dosage problem between males and females for all X-linked genes. Females have two copies and males only one. The additional X chromosomes in both males and females exhibiting the various syndromes discussed earlier in this chapter should compound this dosage problem. In this section, we will describe certain research findings regarding X-linked gene expression which demonstrate that a **genetic dosage compensation mechanism** does indeed exist.

Barr Bodies and Dosage Compensation in Mammals

Murray L. Barr and Ewart G. Bertram's experiments with female cats, and Keith Moore and Barr's subsequent study with humans, demonstrate a genetic mechanism in mammals that compensates for X chromosome dosage disparities. Barr and Bertram observed a darkly staining body in interphase nerve cells of female cats. They found that this structure was absent in similar cells of males. In human females, this body can be easily demonstrated in cells derived from the buccal mucosa or in fibroblasts but not in similar male cells (Figure 5.14). This highly

TABLE 5.5
Human sex-linked traits.

Condition	Characteristics
Color blindness, deutan type	Insensitivity to green light
Color blindness, protan type	Insensitivity to red light
Fabry's disease	Deficiency of galactosidase A; heart and kidney defects, early death
G-6-PD deficiency	Deficiency of glucose-6-phosphate dehydrogenase; severe anemic reaction following intake of primaquines in drugs and certain foods including fava beans
Hemophilia A	Classical form of clotting deficiency; lack of clotting factor VIII
Hemophilia B	Christmas disease; deficiency of clotting factor IX.
Hunter syndrome	Mucopolysaccharide storage disease resulting from iduronate sulfatase enzyme deficiency; short stature, clawlike fingers, coarse facial features, slow mental deterioration, and deafness
Ichthyosis	Deficiency of steroid sulfatase enzyme; scaly dry skin, particularly on extremities
Lesch-Nyhan syndrome	Deficiency of hypoxanthine-guanine phosphoribosyl transferase enzyme (HGPRT), leading to motor and mental retardation, self-mutilation, and early death
Muscular dystrophy (Duchenne type)	Progressive, life-shortening disorder characterized by muscle degeneration and weakness; sometimes associated with mental retardation. Deficiency of the protein dystrophin

FIGURE 5.14
Photomicrographs of numerous human fibroblast nuclei. Cases of one, two, and three Barr bodies are illustrated.

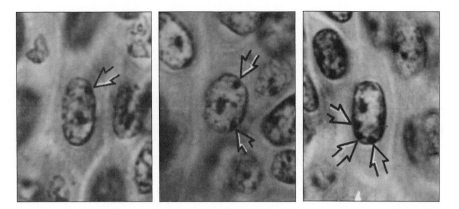

condensed structure, about 1 μm in diameter, lies against the nuclear envelope of interphase cells. It stains positively in the Feulgen reaction for DNA.

Current experimental evidence strongly suggests that this body, called a **sex chromatin body** or simply a **Barr body,** is an inactivated X chromosome. Ohno was the first to suggest that the Barr body arises from one of the two X chromosomes. This hypothesis is attractive because it provides a mechanism for dosage compensation. If one of the two X chromosomes is inactive in the cells of females, the dosage of genetic information that may be expressed in males and females is equivalent. Convincing but indirect evidence for this hypothesis comes from the study of the sex chromosome syndromes described earlier in this chapter. Regardless of how many X chromosomes exist, all but one of them appear to be inactivated and can be seen as Barr bodies. For example, none is seen in Turner 45,X females; one is seen in Klinefelter 47,XXY males; two in 47,XXX females; three in 48,XXXX females; and so on (Figure 5.15). Therefore, the number of Barr bodies follows an $N - 1$ rule, where N is the total number of X chromosomes present.

While this mechanism of inactivation of all but one X chromosome increases our understanding of dosage compensation, it further complicates our perception of other matters. Since one of the two X chromosomes is inactivated in normal human females, why then is the Turner 45,X individual not entirely normal? Why aren't females with the triplo-X and tetra-X karyotypes (47,XXX and 48,XXXX) normal? Further, in Klinefelter syndrome (47,XXY), X chromosome inactivation effectively renders such persons 46,XY. Why aren't they unaffected by the additional X chromosome in their nuclei?

One possible explanation is that chromosome inactivation does not normally occur in the very early stages of development of those cells destined to form gonadal tissues. Another possible explanation is that not all of each X chromosome forming a Barr body is

inactivated. As a result, excessive expression of certain X-linked genes might still occur despite apparent inactivation of additional X chromosomes. In spite of these possible explanations, the above observations remain enigmatic.

The Lyon Hypothesis

In mammalian females one X chromosome is of maternal origin, and the other is of paternal origin. Which one is inactivated? Is the inactivation random? Is the same chromosome inactive in all somatic cells? In 1961, Mary Lyon and Liane Russell independently proposed a hypothesis that answers these questions.

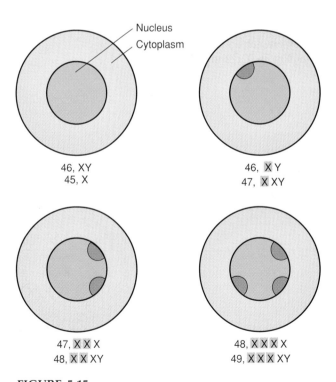

FIGURE 5.15
Diagrammatic representation of Barr body occurrence in various human karyotypes.

They postulated that the inactivation of X chromosomes occurs randomly in somatic cells at a point early in embryonic development and that once inactivation has occurred, all progeny cells have the same X chromosome inactivated.

This explanation, which has come to be called the **Lyon hypothesis,** was initially based on observations of female mice heterozygous for sex-linked coat color genes (Figure 5.16). The pigmentation of these heterozygous females was mottled, with large patches of skin expressing the color allele on one X and other patches expressing the allele on the other X. Indeed, if one or the other of the two X chromosomes was inactive in adjacent patches of cells, such a phenotypic pattern would result. Similar mottling occurs in the black and orange patches of female tortoise-shell (or calico) cats. Such sex-linked coat color patterns do not occur in male cats since all cells are hemizygous

for only one sex-linked coat color allele. Two such cats are compared in Figure 5.17.

The most direct evidence in support of the Lyon hypothesis comes from studies of gene expression in clones of human fibroblast cells. Individual cells may be isolated following biopsy and cultured *in vitro*. If each culture is derived from a single cell, it is referred to as a **clone.** The synthesis of the enzyme **glucose-6-phosphate dehydrogenase (G-6-PD)** is controlled by a sex-linked gene. Numerous mutant alleles of this gene have been detected, and their gene products can be differentiated from the wild-type enzyme by their migration pattern in an electrophoretic field. Fibroblasts have been taken from females heterozygous for different allelic forms of G-6-PD and studied.

The Lyon hypothesis predicts the results of the examination of numerous clones derived from a female heterozygous for two G-6-PD alleles. If inac-

FIGURE 5.16
Diagrammatic representation of the Lyon hypothesis in the mouse. Following random inactivation of one of the two X chromosomes in female nuclei, all progeny cells inactivate the same chromosome. The adult is a mosaic, with some groups of cells expressing the alleles on one X chromosome and the other groups expressing the alleles present on the homologous X chromosome.

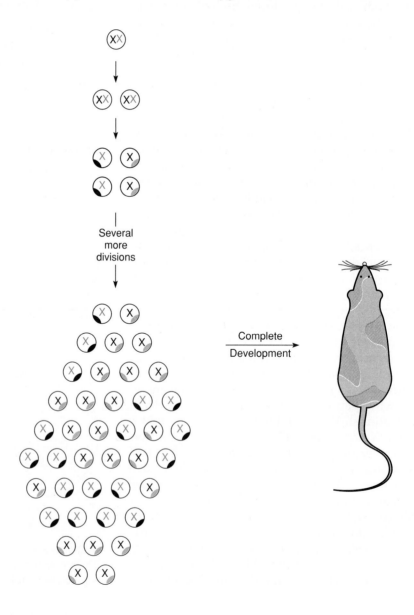

FIGURE 5.17
A male and female cat from the same litter, illustrating the Lyon hypothesis. The female (on the left) is a calico with orange and black patches resulting from the inactivation of one or the other X chromosome early in development. The male (on the right) is hemizygous for the orange allele. The white patches are due to still another gene.

tivation of an X chromosome occurs randomly early in development and is permanent in all progeny cells, such a female should show two types of clones in approximately equal proportion. Each clone shows one electrophoretic variant of G-6-PD or the other, but never both. In 1963, Ronald Davidson and colleagues performed an experiment involving 14 clones from a single heterozygous female. Seven showed only one form of the enzyme, and seven showed only the other form. What was most important was that none of the 14 showed both forms of the enzyme. Studies of G-6-PD mutants thus provide strong support for the random permanent inactivation of either the maternal or paternal X chromosome.

Although not all of the existing evidence is as clear-cut as the preceding experiment, the Lyon hypothesis is generally accepted as valid. One extension of the hypothesis is that mammalian females are mosaics for all heterozygous X-linked alleles. Some areas of the body express only the maternally derived alleles, and others express only the paternally derived alleles. Two especially interesting examples involve **red-green color blindness** and **anhidrotic ectodermal dysplasia,** both X-linked recessive disorders. In the former case, hemizygous males are fully color-blind in all retinal cells. However, heterozygous females display mosaic retinas with patches of defective color perception and surrounding areas with

normal color perception. In the latter disorder, hemizygous males show absence of teeth, sparse hair growth, and lack of sweat glands. The skin of heterozygous females reveals patterns of tissue with and without sweat glands (Figure 5.18). In both examples, random inactivation of one or the other X chromosome early in the development of heterozygous females has led to these occurrences.

The least understood aspect of the Lyon hypothesis concerns how inactivation occurs. While studies on the mechanisms that inactivate an X chromosome are just beginning, evidence suggests that DNA in inactivated chromosome regions has been chemically modified. One line of work has shown an association between the addition of methyl groups to cytosine residues and the condensation of chromosomes. Perhaps this process of methylation is part of the mechanism responsible for X chromosome inactivation.

SEX-LIMITED AND SEX-INFLUENCED INHERITANCE

There are numerous examples in different organisms where the sex of the individual plays a determining role in the expression of certain phenotypes. In some cases, the expression of a specific phenotype is absolutely limited to one sex; in others, the sex of an individual influences the expression of a phenotype

	Phenotype	
Genotype	♀	♂
HH	Hen-feathered	Hen-feathered
Hh	Hen-feathered	Hen-feathered
hh	Hen-feathered	Cock-feathered

In the development of certain breeds of fowl, one allele or the other has become fixed in the population. In the Leghorn breed, all individuals are of the *hh* genotype, and thus all males and females show distinctive plumage. Sebright bantams are all *HH*, showing no sexual distinction in feathering.

Cases of sex-influenced inheritance include pattern baldness in humans, horn formation in sheep, and certain coat patterns in cattle. In such cases, autosomal genes are responsible for the contrasting phenotypes displayed by both males and females. However, the heterozygous genotype may exhibit one phenotype in one sex and the contrasting one in the other. For example, **pattern baldness** in humans, where the hair is very thin on the top of the head (Figure 5.20), is inherited in the following way:

	Phenotype	
Genotype	♀	♂
BB	Bald	Bald
Bb	Not bald	Bald
bb	Not bald	Not bald

Even though females may display pattern baldness, this phenotype is much more prevalent in males. When females do inherit the *BB* genotype, the phenotype is much less pronounced than in males and is expressed later in life.

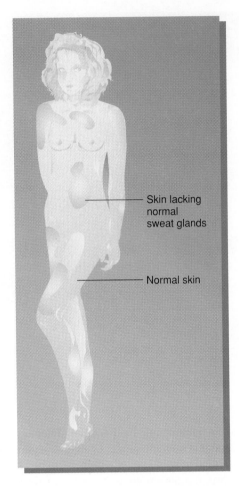

FIGURE 5.18
The hypothetical pattern of gene expression in a woman heterozygous for the sex-linked disorder anhidrotic ectodermal dysplasia. The patches of skin lacking sweat glands consist of cells where the X chromosome carrying the normal allele has been inactivated. Patches of normal skin consist of cells where the X chromosome carrying the mutant allele has been inactivated.

Skin lacking normal sweat glands

Normal skin

that is not limited to one sex or the other. This distinction differentiates **sex-limited inheritance** from **sex-influenced inheritance**.

In domestic fowl, tail and neck plumage is often distinctly different in males and females (Figure 5.19), demonstrating sex-limited inheritance. Cock-feathering is longer, more curved, and pointed, while hen-feathering is shorter and more rounded. The inheritance of feather type is due to a single pair of autosomal alleles whose expression is modified by the individual's sex hormones.

As shown in the following chart, hen-feathering is due to a dominant allele, *H*; but regardless of the homozygous presence of the recessive *h* allele, all females remain hen-feathered. Only in males does the *hh* genotype result in cock-feathering.

CHAPTER SUMMARY

1   The genetic variability of a species resulting from sexual reproduction is an integral part of the evolutionary process. Sexual differentiation between members of a species is important for efficient sexual reproduction. This is evident morphologically in corn, *Drosophila*, and humans, and physiologically in *Neurospora* and *Chlamydomonas*.

2   The genetic basis of sexual differentiation is often related to the presence of different chromosome compositions. The heterogametic sex either lacks one chromosome or contains a unique chromosome, usually called the Y chromosome. Genes on autosomes also play an important role in

FIGURE 5.19
Hen-feathering (left) versus cock-feathering (right) in domestic fowl. The feathers in the hen are shorter and less curved.

determining sex. In *Drosophila*, a balance between genetic expression of the X chromosome and the autosomes is critical to sex determination.

3 In humans, studies of individuals showing abnormal karyotypes provide evidence for the

FIGURE 5.20
Pattern baldness, a sex-influenced autosomal trait in humans.

strong influence of the Y chromosome on male development. In the absence of the Y, female development is initiated. Once gonad development has been initiated, secondary sexual differentiation occurs under the influence of hormones.

4 Testis-determining factor (TDF) is a holandric, Y-linked gene product involved in the initial event leading to maleness. In humans this gene is located on the short arm of the Y chromosome, but has yet to be identified.

5 Sex (or X) linkage leads to a unique mode of inheritance, since members of the heterogametic sex lack a chromosome strictly homologous to the X chromosome. Such individuals express alleles on the X chromosome directly, even if they are recessive. Such expression is the basis of hemizygosity.

6 When males are the heterogametic sex, females homozygous for a recessive allele pass the trait to all of their sons. Daughters become heterozygous carriers and pass the trait to half of their sons.

7 The primary sex ratio in humans substantially favors males at conception. During embryonic and fetal development, male mortality is higher than that of females, but the secondary sex ratio at birth still favors males by a small margin.

8 Dosage compensation mechanisms limit the expression of sex-linked genes in females, who

SEX DETERMINATION AND SEX LINKAGE   135

have two X chromosomes, as compared to males, who have only one X. In mammals, compensation is achieved as a result of the inactivation of either the maternal or paternal X early in development. This process results in the formation of Barr bodies in female somatic cells.

9 The Lyon hypothesis states that, early in development, inactivation is random between the maternal and paternal X. All subsequent prog-

eny cells inactivate the same X as their progenitor cell. Mammalian females thus develop as genetic mosaics with respect to their expression of heterozygous sex-linked alleles.

10 Hormones can substantially modify gene expression, leading to sex-influenced and sex-limited inheritance, including cases of pattern baldness in humans and feather patterns in fowl.

Insights and Solutions

1 In humans, colorblindness is inherited as a sex-linked recessive trait. A woman with normal vision, but whose father is colorblind, marries a male who has normal vision. Predict the color vision of their male and female offspring.

   ANSWER: The female is heterozygous since she inherited an X chromosome with the mutant allele from her father. Her husband is normal. Therefore, the parental genotypes are

$$Cc \times C\uparrow$$

   All female offspring are normal (CC or Cc). One half of the male children will be colorblind ($c\uparrow$), and the other half will have normal vision ($C\uparrow$).

2 For the pedigree shown below, which depicts the inheritance pattern of a recessive sex-linked trait in humans (*a*), determine all genotypes that can be predicted with absolute assurance.

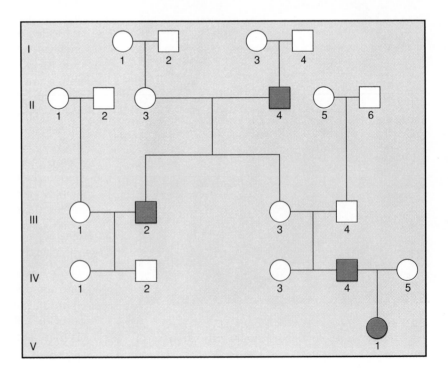

ANSWER: First, the genotypes of *all* males can be predicted. Individuals II-4, III-2, and IV-4 express the trait and must be hemizygous ($a \uparrow$). All other males must carry a normal allele and are therefore $A \uparrow$.

The only female to express the trait (V-1) must be homozygous (*aa*), carrying the mutant allele on both X chromosomes. Many of the remaining females can be determined absolutely to be carriers (*Aa*) since they are the only possible source of the mutant allele that is eventually passed to either a son or grandson: I-1, I-3, II-3, III-3. Two other females are known to be heterozygous by virtue of receiving the allele from their father, who was affected: III-3 (we already knew about her) and IV-1. Finally, female IV-5 is heterozygous because she passed the allele to an affected daughter, who also had to receive an allele from her affected father. In all other females, the genotype cannot be determined with certainty, although we do know that they are not homozygous recessive (*aa*).

3 In *Drosophila*, all members of a strain of flies express a jagged wing margin due to a recessive mutation. How can you determine if the mutation is sex-linked or autosomal?

ANSWER: Cross a mutant female to a normal male. If the mutant allele is autosomal, all male and female offspring will have normal wings. If the allele is sex-linked, all male offspring will express the trait and all females will be normal.

∞∞∞∞∞∞∞∞∞∞∞∞∞∞∞∞∞∞∞∞

**PROBLEMS AND
DISCUSSION QUESTIONS**

1 Define the following terms: (a) heterogamy, (b) heteromorphic chromosome, (c) nondisjunction, (d) Barr body, and (e) Lyon hypothesis.

2 An insect species is discovered in which the heterogametic sex is unknown. A sex-linked recessive mutation for *reduced wing* size (*rw*) is found in this species. Contrast the F_1 and F_2 generations from a cross between a female with reduced wings and a normal male when:
   (a) the female is the heterogametic sex.
   (b) the male is the heterogametic sex.
   (c) Is it possible to distinguish between the Protenor and Lygaeus type of sex determination based on these crosses?

3 Contrast the Protenor and Lygaeus modes of sex determination.

4 Contrast the evidence leading to the explanation of the different modes of sex determination in *Drosophila* and humans. Does *Melandrium* show a mode of sex determination that parallels either of these species?

5 In *Drosophila*, females homozygous for the recessive sex-linked gene causing *singed* bristles (*sn*) are crossed to wild-type males. Both the males and females are heterozygous for the gene *transformer* (*tra*), which is located on chromosome 3. What phenotypic ratios in the male and female offspring will result from this cross? The *tra* mutation is discussed on page 119.

6 Devise a method of nondisjunction in human female gametes that would give rise to Klinefelter and Turner syndrome offspring following fertilization by a normal male gamete.

7 A sex-linked dominant mutation in the mouse, *Testicular feminization* (*Tfm*), eliminates the normal response to the testicular hormone testosterone during sexual differentiation. An XY animal bearing the *Tfm* allele on the X chromosome develops testes, but no further male differentiation occurs. The external genitalia of such an animal are female. From this information, what might you conclude about the role of the X and Y chromosomes in sex determination and differentiation in mammals?

8 Discuss the possible reasons why the primary sex ratio in humans is as high as 1.40 to 1.60.

9   It has been suggested that any male-determining genes contained on the Y chromosome in humans should not be located in the limited region that synapses with the X chromosome during meiosis. What might be the outcome if such genes were located in this region?

10 What does the apparent need for dosage compensation mechanisms suggest about the expression of genetic information in normal diploid individuals?

11 Indicate the expected number of Barr bodies in interphase cells of the following individuals: Klinefelter syndrome; Turner syndrome; and karyotypes 47,XYY, 47,XXX, and 48,XXXX.

12 Cat breeders are aware that kittens with the calico coat pattern are almost invariably females. Why?

13 It is not uncommon to encounter self-sterile plant species. For example, in *Nicotiana*, pollen that falls on the stigma of the same plant will not develop in some cases, although the same pollen will fertilize another plant. Speculate on the evolutionary advantages of self-sterility.

14 When cows have twin calves of unlike sex (fraternal twins), the female twin is usually sterile and has masculinized reproductive organs. This calf is referred to as a **free martin**. In cows, twins may share a common placenta and fetal circulation. Based on the information in this chapter concerning sexual differentiation, predict why a free martin develops.

15 In certain *Drosophila* strains, the X chromosomes are attached to each other. Some flies contain attached-X chromosomes and a separate Y chromosome. What sex would this fly be? If a similar situation existed in the human species, what sex would the individual be? What syndrome would the individual exhibit?

16 An attached-X plus Y fly, as described in Problem 15, with *white* eyes is crossed to a fly of the opposite sex which has *miniature* wings (both are sex-linked, recessive genes). What proportion of phenotypes in male and female offspring is expected? Assume that flies with three or no X chromosomes do not develop to adulthood.

17 *Parthenogenesis* is the process in which unfertilized eggs initiate development and give rise to offspring. In birds such as the turkey, females are the heterogametic sex and usually of the Protenor type (X0). How could a parthenogenic male offspring arise?

18 In the wasp *Bracon hebetor*, a form of parthenogenesis resulting in haploid organisms is common. All haploids are males. When fertilization occurs, diploid individuals are almost always female. P. W. Whiting has shown that an X-linked gene with nine multiple alleles (*Xa, Xb, Xc*, etc.) controls sex determination. Any homozygous or hemizygous condition results in males, and any heterozygous condition results in females. If an *Xa/Xb* female mates with an *Xa* male and lays 50 percent fertilized and 50 percent unfertilized eggs, what proportion of male and female offspring will result?

19 The marine echiurid worm *Bonellia viridis* is an extreme example of the environment's influence on sex determination. Undifferentiated larvae either remain free-swimming and differentiate into females or they settle on the proboscis of an adult female and become males. If larvae that have been on a female proboscis for a short period are removed and placed in sea water, they develop as intersexes. If larvae are forced to develop in an aquarium where pieces of proboscises have been placed, they develop into males. Contrast this mode of sexual differentiation with that of mammals. Suggest further experimentation to elucidate the mechanism of sex determination in *Bonellia*.

20 In cats, yellow coat color is determined by the *b* allele and black coat color by the *B* allele. The heterozygous condition results in a color known as tortoise shell. These genes are sex-linked. What kinds of offspring would be expected from a cross of a black male and a tortoise-shell female? What are the chances of getting a tortoise-shell male?

21 A husband and wife have normal vision, although both of their fathers are red-green color-blind, which is inherited as a sex-linked recessive condition. What is the probability that their first child will be
(a) a normal son?
(b) a normal daughter?

(c) a color-blind son?

(d) a color-blind daughter?

22 In humans, the ABO blood type is under the control of autosomal multiple alleles. Color blindness is a recessive sex-linked trait. If two parents who are both type A and have normal vision produce a son who is color blind and is type O, what is the probability that their next child will be a female who is normal visioned and type O?

23 In *Drosophila*, a sex-linked recessive mutation, *scalloped* (*sd*), causes irregular wing margins. Diagram the F_1 and F_2 results if:

(a) A *scalloped* female is crossed with a normal male.

(b) A *scalloped* male is crossed with a normal female.

Compare these results to those that would be obtained if *scalloped* were not sex-linked.

24 Another recessive mutation in *Drosophila*, *ebony* (*e*), is on an autosome (chromosome 3) and causes darkening of the body compared with wild-type flies. What phenotypic F_1 and F_2 male and female ratios will result if a *scalloped*-winged female with normal body color is crossed with a normal-winged *ebony* male? Work this problem by both the Punnett square method and the forked-line method.

25 In *Drosophila* the recessive gene for *white* eye (*w*) is sex linked, while the recessive gene for *dumpy* wing (*dp*) is located on chromosome 2. A female who was both red-eyed (homozygous) and dumpy-winged was crossed to a male whose mother was white-eyed and whose mother and father both had homozygously normal wings. Show the F_1 and F_2 phenotypic ratios of the sex, eye color, and wing shape.

26 In *Drosophila*, the sex-linked recessive mutation *vermilion* (*v*) causes bright red eyes, which is in contrast to brick red eyes of wild type. A separate autosomal recessive mutation, *suppressor of vermilion* (*su-v*), causes flies homozygous or hemizygous for *v* to have wild-type eyes. In the absence of vermilion alleles, *su-v* has no effect on eye color. Determine the F_1 and F_2 phenotypic ratios from a cross between a female with wild-type alleles at the *vermilion* locus, but who is homozygous for *su-v*, with a *vermilion* male who has wild-type alleles at the *su-v* locus.

27 While *vermilion* is sex linked and brightens the eye color, *brown* is an autosomal recessive mutation that darkens the eye. Flies carrying both mutations lose all pigmentation and are white eyed. Predict the F_1 and F_2 results of the following crosses:

(a) vermilion females × brown males

(b) brown females × vermilion males

(c) white females × wild males.

28 In spotted cattle, the colored regions may be mahogany or red. If a red female and a mahogany male, both derived from separate true-breeding lines, are mated and the cross carried to an F_2 generation, the following results are obtained:

F_1: 1/2 mahogany males

1/2 red females

F_2: 3/8 mahogany males

1/8 red males

1/8 mahogany females

3/8 red females

When the reciprocal of the initial cross is performed (mahogany female and red male), identical results are obtained. Explain these results by postulating how the color is genetically determined. Diagram the crosses.

29 Predict the F_1 and F_2 results of crossing a male fowl that is cock-feathered with a true-breeding hen-feathered female fowl. Recall that these traits are sex-limited, as discussed in the text.

30 Shown below are four graphs that plot the percentage of males that occur in various reptile groups vs. the temperature fertilized eggs encounter during early development.

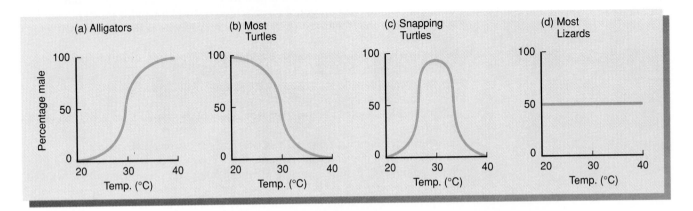

Interpret these data as they relate to sex determination in reptiles.

31 Analyze the following pedigree. What is the most likely mode of inheritance? Defend your answer by stating what information *supports* your answer and what information *rules out* other potential modes. You may rule out spontaneous mutation.

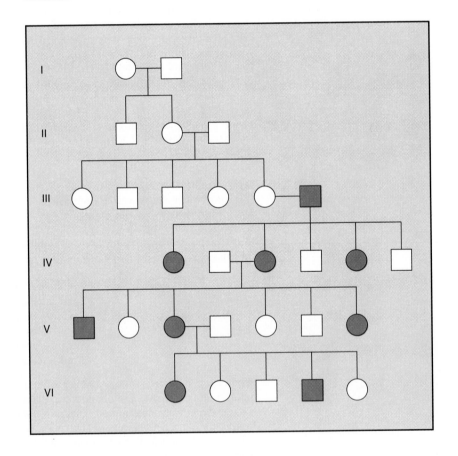

SELECTED READINGS

BACCI, G. 1965. *Sex determination*. Orlando: Academic Press.

BAKER, B. S. 1989. Sex in flies: The splice of life. *Nature* 340: 521–24.

BARR, M. L. 1966. The significance of sex chromatin. *Int. Rev. Cytol.* 19: 35–39.

BULL, J. T. 1983. *The evolution of sex-determining mechanisms*. Menlo Park, Calif.: Benjamin Cummings.

CARR, D. H. 1971. Chromosomes and abortion. *Adv. Hum. Genet.* 2: 201–57.

COURT-BROWN, W. M. 1968. Males with an XYY sex chromosome complement. *J. Med. Genet.* 5: 341–59.

DAVIDSON, R., NITOWSKI, H., and CHILDS, B. 1963. Demonstration of two populations of cells in human females heterozygous for glucose-6-phosphate dehydrogenase variants. *Proc. Natl. Acad. Sci.* 50: 481–85.

DRAYNA, D., and WHITE, R. 1985. The genetic linkage map of the human X chromosome. *Science* 230: 753–58.

ERICKSON, J. D. 1976. The secondary sex ratio of the United States 1969–71: Association with race, parental ages, birth order, paternal education and legitimacy. *Ann. Hum. Genet. Lond.* 40: 205–12.

ERICKSON, R. P., and VERGA, V. 1989. Is zinc-finger Y the sex determining gene? *Am. J. Hum. Genet.* 45: 671–74.

GORDON, J. W., and RUDDLE, F. H. 1981. Mammalian gonadal determination and gametogenesis. *Science* 211: 1265–78.

HASELTINE, F. P., and OHNO, S. 1981. Mechanisms of gonadal differentiation. *Science* 211: 1272–78.

HEAD, G., MAY, R., and PENDLETON, L. 1987. Environmental determination of sex in the reptiles. *Nature* 329: 198–99.

HODGKIN, J. 1990. Sex determination compared in *Drosophila* and *Caenorhabditis*. *Nature* 344: 721–28.

HOOK, E. B. 1973. Behavioral implications of the human XYY genotype. *Science* 179: 139–50.

JACOBS, P. A., et al. 1974. A cytogenetic survey of 11,680 newborn infants. *Ann. Hum. Genet.* 37: 359–76.

LUCCHESI, J. 1983. The relationship between gene dosage, gene expression, and sex in *Drosophila*. *Dev. Genet.* 3: 275–82.

LYON, M. F. 1961. Gene action in the X-chromosome of the mouse (*Mus musculus* L.). *Nature* 190: 372–73.

———. 1962. Sex chromatin and gene action in the mammalian X chromosome. *Am. J. Hum. Genet.* 14: 135–48.

———. 1972. X-chromosome inactivation and developmental patterns in mammals. *Biol. Rev.* 47: 1–35.

———. 1988. X-chromosome inactivation and the location and expression of X-linked genes. *Am. J. Hum. Genet.* 42: 8–16.

McKUSICK, V. A. 1962. On the X chromosome of man. *Quart. Rev. Biol.* 37: 69–175.

———. 1990. *Mendelian inheritance in man*. 9th ed. Baltimore: Johns Hopkins University Press.

McLAREN, A. 1988. Sex determination in mammals. *TIG* 4: 153–57.

McMILLEN, M. M. 1979. Differential mortality by sex in fetal and neonatal deaths. *Science* 204: 89–91.

MITTWOCH, U. 1967. *Sex chromosomes*. Orlando: Academic Press.

———. 1973. *Genetics of sex differentiation*. Orlando: Academic Press.

MORGAN, T. H. 1910. Sex limited inheritance in *Drosophila*. *Science* 32: 120–22.

OHNO, S. 1967. *Sex chromosomes and sex-linked genes*. New York: Springer-Verlag.

———. 1978. *Major sex determining genes*. New York: Springer-Verlag.

PAGE, D. C., et al. 1987. The sex-determining region of the human Y chromosome encodes a finger protein. *Cell* 51: 1091–1104.

POLANI, P. E. 1982. Pairing of X and Y chromosomes, non-inactivation of X-linked genes, and the maleness factor. *Hum. Genet.* 60: 207–11.

RUSSELL, L. B. 1961. Genetics of mammalian sex chromosomes. *Science* 133: 1795–1803.

SILVERS, W. K., and WACHTEL, S. S. 1977. H-Y antigen: Behavior and function. *Science* 195: 956–60.

SIMPSON, J. L. 1982. Abnormal sexual differentiation in humans. *Ann. Rev. Genet.* 16: 193–224.

WACHTEL, S. S. 1977. H-Y antigen and the genetics of sex determination. *Science* 198: 797–99.

———. 1983. *H-Y antigen and the biology of sex determination*. New York: Grune and Stratton.

WESTERGAARD, M. 1958. The mechanism of sex determination in dioecious flowering plants. *Adv. in Genet.* 9: 217–81.

WHITING, P. W. 1939. Multiple alleles in sex determination of *Habrobracon*, *J. Morph.* 66: 323–55.

WIBERG, U. H., 1987. Facts and considerations about sex-specific antigens. *Hum. Genet.* 76: 207–19.

WITKIN, H. A., et al., 1976. Criminality in XYY and XXY men. *Science* 193: 547–55.

6

Linkage, Crossing Over, and Chromosome Mapping

CHAPTER CONCEPTS

Many genes reside on a single chromosome. Unless separated by crossing over, alleles present at the many loci on each homologue segregate as a unit during gamete formation. Recombinant gametes resulting from crossing over enhance genetic variability within a species and serve as the basis for constructing chromosomal maps.

As early as 1903, Walter Sutton, who along with Theodor Boveri united the fields of cytology and genetics, pointed out the likelihood that in organisms there are many more "unit factors" than chromosomes. Soon thereafter, genetic investigations with several organisms revealed that certain genes were not transmitted according to the law of independent assortment. When studied together in matings, these genes seemed to segregate as if they were somehow joined or linked together. Further investigations showed that such genes were part of the same chromosome, and they were indeed transmitted as a single unit.

We now know that most chromosomes consist of very large numbers of genes and, in fact, contain sufficient DNA to encode thousands of these units. Genes that are part of the same chromosome are said to be **linked** and to demonstrate **linkage** in genetic crosses.

Since the chromosome, not the gene, is the unit of transmission during meiosis, linked genes are not free to undergo independent assortment. Instead, the alleles at all loci of one chromosome should, in theory, be transmitted as a unit during gamete formation. However, in many instances this does not occur. During the first meiotic prophase, when homologues are paired or synapsed, a reciprocal exchange of chromosome segments may take place. This event, called **crossing over,** results in the reshuffling or **recombination** of the alleles between homologues.

The degree of crossing over between any two loci on a single chromosome is proportionate to the distance between them. Thus, the percentage of recombinant gametes varies, depending on which loci are being considered. This correlation serves as the basis for the construction of **chromosome maps**.

Crossing over is currently viewed as an actual physical breaking and rejoining process that occurs during meiosis. This exchange of chromosome segments provides for an enormous potential variation in the gametes formed by any individual. This type of variation, in combination with that resulting from independent assortment, ensures that all offspring will contain a diverse mixture of both maternal and paternal alleles. In addition to producing individual diversity, genetic variability is of paramount importance to the process of organic evolution.

In this chapter, we will discuss linkage, crossing over, and chromosome mapping. We will also consider a variety of topics involving the exchange of genetic information. We will conclude the chapter by entertaining the rather intriguing question of why Mendel, who studied seven genes, did not encounter linkage. Or did he?

LINKAGE VS. INDEPENDENT ASSORTMENT

In order to provide a simplified overview of the major theme of this chapter, Figure 6.1 illustrates and contrasts the meiotic consequences of independent assortment, linkage without crossing over, and linkage with crossing over. Two homologous pairs of chromosomes are considered in the case of independent assortment and one homologous pair in the two cases of linkage.

Figure 6.1(a) illustrates the results of independent assortment of two pairs of nonhomologous chromosomes, each containing one heterozygous gene pair. When a large number of meiotic events are observed, four genetically different gametes are formed in equal proportions.

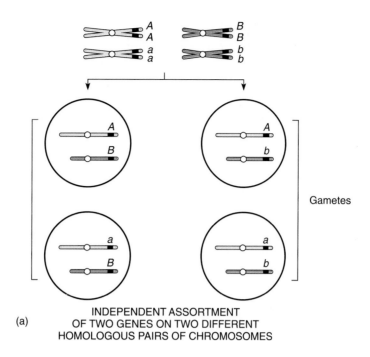

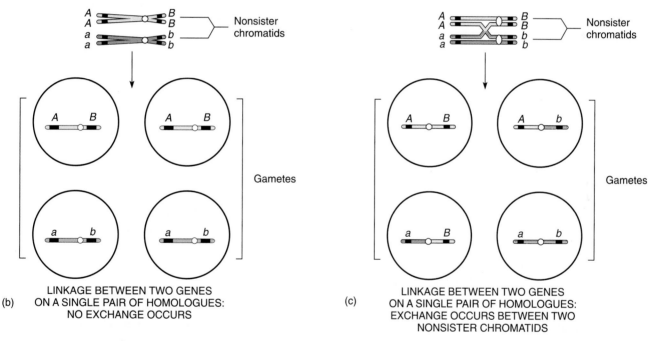

FIGURE 6.1

A comparison of the results of gamete formation where two heterozygous genes are (a) on two different pairs of chromosomes; (b) on the same pair of homologues, but with no exchange occurring between them; and (c) on the same pair of homologues, with an exchange between two nonsister chromatids.

We can compare these results with those that occur if the same genes are instead linked on the same chromosome. If no crossing over occurs between the two genes [Figure 6.1(b)], only two genet-ically different gametes are formed. Each gamete receives the alleles present on one homologue or the other, which has been transmitted intact as the result of segregation. This case illustrates **complete linkage,** which is said to result in the production of only **parental** or **noncrossover gametes**. The two parental gametes are formed in equal proportions.

Figure 6.1(c) illustrates the results when crossing over occurs between two linked genes. As you will note, the crossover involves only two nonsister chromatids of the four chromatids present in the tetrad. This exchange generates two new allele combinations, called **recombinant** or **crossover gametes**. The two chromatids not involved in the exchange result in noncrossover gametes [as those in part (b) of this figure].

The frequency with which crossing over occurs between any two linked genes is proportional to the distance separating the respective loci along the chromosome. While it is rare, it is possible for two randomly selected genes to be so close to each other that an undetectable number of crossover events occurs between them. This circumstance, complete linkage, results in the production of only parental gametes. If a distinct but small distance separates two genes, few recombinant and many parental gametes will be formed. As the distance between two genes increases, the proportion of recombinant gametes increases and that of the parental gametes decreases.

As will be explored later in this chapter, when two linked genes whose loci are far apart are considered, the number of recombinant gametes approaches, but never exceeds, 50 percent. If the maximal number were produced, a 1:1:1:1 ratio of the four types (two parental and two recombinant gametes) would result. In such a case, transmission of two linked genes would be indistinguishable from that of two unlinked, independently assorting genes.

The Linkage Ratio

If complete linkage exists between two genes because of their proximity, and organisms with mutant alleles representing these genes are mated, a unique F_2 phenotypic ratio results that is characteristic of linkage. To illustrate this ratio, we will consider a cross involving the closely linked recessive mutant genes *brown* (*bw*) eye and *heavy* (*hv*) wing vein in *Drosophila melanogaster* (Figure 6.2). The normal, wild-type alleles bw^+ and hv^+ are both dominant and result in red eyes and thin wing veins, respectively.

In this cross, flies with mutant brown eyes and normal thin veins are mated to flies with normal red eyes and mutant heavy veins. In more concise terms, brown-eyed flies are crossed with heavy-veined flies. If we extend the system of using genetic symbols established in Chapter 4, linked genes may be represented by placing their allele designations above and below a single or double horizontal line. Those

placed above the line are located at loci on one homologue and those placed below at the homologous loci on the other homologue. Thus, we may represent the P_1 generation as follows:

$$P_1: \quad \frac{bw \;\; hv^+}{bw \;\; hv^+} \quad \times \quad \frac{bw^+ \; hv}{bw^+ \; hv}$$

brown, thin   red, heavy

Since the genes are located on an autosome, no distinction for male and female is necessary.

In the F_1 generation each fly receives one chromosome of each pair from each parent; all flies are heterozygous for both gene pairs and exhibit the dominant traits of red eyes and thin veins:

$$F_1: \quad \frac{bw \;\; hv^+}{bw^+ \; hv}$$

red, thin

As shown in Figure 6.2, when the F_1 generation is interbred, because of complete linkage, each F_1 individual forms only parental gametes. Following fertilization, the F_2 generation will be produced in a 1:2:1 phenotypic and genotypic ratio. One-fourth of this generation will show brown eyes and thin veins; one-half will show both wild-type traits, red eyes and thin veins; and one-fourth will show red eyes and thick veins. In more concise terms, the ratio is 1 brown:2 wild:1 thick. Such a ratio is characteristic of complete linkage.

Figure 6.2 also demonstrates the results of a test cross with the F_1 flies. Such a cross produces a 1:1 ratio of brown, thin and red, thick flies. Had the genes controlling these traits been incompletely linked or located on separate autosomes, four phenotypes rather than two would have been produced.

When large numbers of mutant genes present in any given species are investigated in crosses similar to those just described, the genes located on the same chromosome will show evidence of linkage to one another. As a result, **linkage groups** can be established, one for each chromosome. Hence, the number of linkage groups should correspond to the haploid number of chromosomes. In organisms where large numbers of mutant genes are available for genetic study, this correlation has been upheld.

INCOMPLETE LINKAGE, CROSSING OVER, AND CHROMOSOME MAPPING

If two genes linked on the same chromosome are selected randomly it is unlikely that their respective loci will be contiguous to each other. Therefore, because crossing over will occur between them,

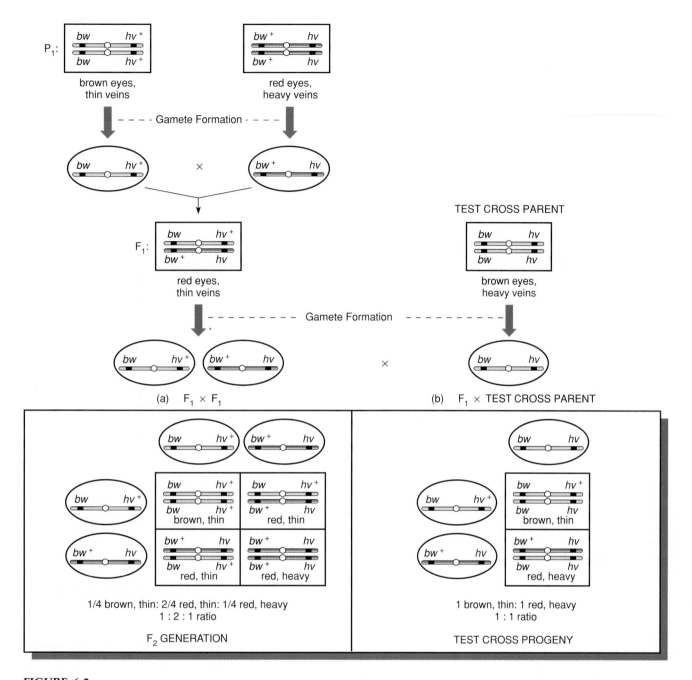

FIGURE 6.2
(a) The results of a cross between two linked genes demonstrating complete linkage. The cross is carried to the F₂ generation. (b) The results of a test cross with an F₁ individual are also illustrated.

complete linkage is rarely achieved. Instead, crosses involving two linked genes usually produce a percentage of offspring resulting from recombinant gametes. This percentage is variable, depending upon which two genes are involved in the cross. This phenomenon was first studied and explained by two *Drosophila* geneticists, Thomas H. Morgan and his undergraduate student, Alfred H. Sturtevant.

Morgan and Crossing Over

As you may recall from our earlier discussion of sex-linked genes in Chapter 5, Morgan first discovered the phenomenon of sex linkage. In his studies, he investigated numerous *Drosophila* mutations located on the X chromosome. When he analyzed crosses involving only one trait, he was able to deduce the mode of sex-linked inheritance. However, when he made crosses involving two sex-linked genes, his results were at first puzzling. For example, as shown in Cross A of Figure 6.3, he crossed mutant *yellow* body (*y*) and *white* eyes (*w*) females with wild-type males (gray bodies and red eyes). The F₁ females were wild-type, while the F₁ males expressed both mutant traits. In the F₂, 98.7 percent of the offspring showed the parental phenotypes—yellow-bodied, white-eyed flies and wild type flies. The remaining 1.3 percent of the flies were either yellow-bodied with red eyes or gray-

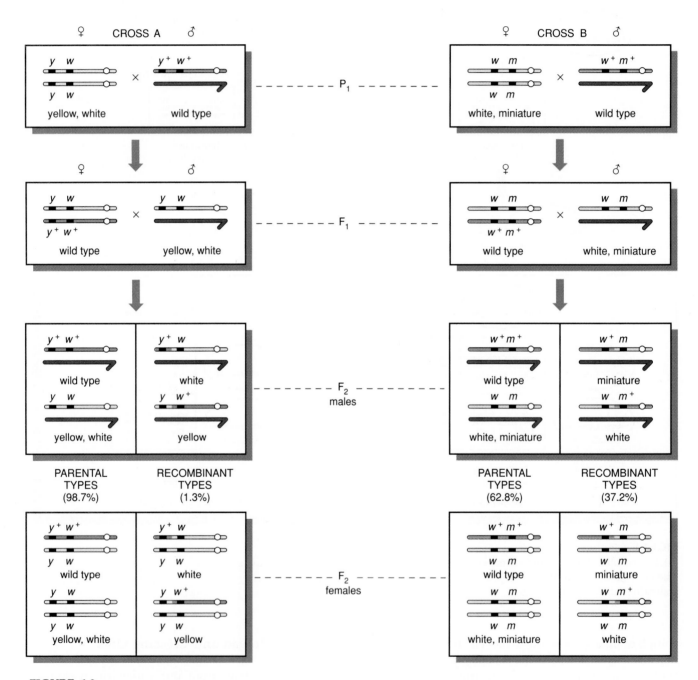

FIGURE 6.3
The F₁ and F₂ results of crosses involving the yellow-body, white-eye mutations and the white-eye, miniature-wing mutations. In the F₂ generation of Cross A, 1.3 percent of the flies demonstrate recombinant phenotypes, which are either *white* or *yellow*. In the F₂ generation of Cross B, 37.2 percent of the flies demonstrate recombinant phenotypes, which are either *miniature* or *white*.

bodied with white eyes. It was as if the genes had somehow separated from each other during gamete formation in the F_1 flies.

When he made crosses involving other sex-linked genes, the results were even more puzzling (Cross B of Figure 6.3). The same basic pattern was observed, but the proportion of F_2 phenotypes differed; for example, in a cross involving white-eye, miniature-wing mutants, only 62.8 percent of the F_2 showed the parental phenotypes, while 37.2 percent of the offspring appeared as if the mutant genes had been separated during gamete formation.

In 1911, Morgan was faced with two questions: (1) What was the source of gene separation? and (2) Why did the frequency of the apparent separation vary depending on the genes being studied? The proposed answer to the first question was based on his knowledge of earlier cytological observations made by F. Janssens and others. Janssens had observed that synapsed homologous chromosomes in meiosis wrapped around each other, forming points of **chiasmata** [sing.: **chiasma**]. Morgan proposed that these chiasmata could represent points of genetic exchange, the crux of the so-called **chiasmatype theory**.

In the crosses shown in Figure 6.3, Morgan postulated that if an exchange occurred between the mutant genes on the two X chromosomes of the F_1 females, it would lead to the observed results. He suggested that such exchanges led to 1.3 percent recombinant gametes in the *yellow-white* cross and 37.2 percent in the *white-miniature* cross. On the basis of this and other experimentation, Morgan concluded that linked genes exist in a linear order along the chromosome, and that a variable amount of exchange occurs between any two genes.

As an answer to the second question, Morgan proposed that two genes located relatively close to each other along a chromosome are less likely to have a chiasma form between them than if the two genes are farther apart on the chromosome. Thus, the closer two genes are, the less likely it is that a genetic exchange will occur between them. Morgan proposed the term **crossing over** to describe the physical exchange leading to recombination. Although there still is some doubt concerning the validity of the chiasmatype theory, it did provide an adequate basis for Morgan's theory of genetic recombination and the subsequent work on crossing over and mapping.

Sturtevant and Mapping

Morgan's student, Alfred H. Sturtevant, was the first to realize that his mentor's proposal could be used to map the sequence of and distance between linked genes. Sturtevant argued that if the frequency of

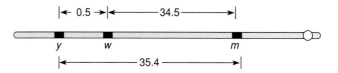

FIGURE 6.4
A simple map of the *yellow* (y), *white* (w), and *miniature* (m) genes on the X chromosome of *Drosophila melanogaster*. Each number represents the percentage of recombinant offspring produced in the three crosses, each involving two different genes.

crossing over between genes is related to the distance between them, the recombination frequencies between a series of linked genes must be additive.

For example, Sturtevant compiled data on recombination between the genes represented by the *yellow*, *white*, and *miniature* mutants initially studied by Morgan. Frequencies of crossing over between each pair of these three genes were observed to be

  (1) *yellow, white*   0.5%
  (2) *white, miniature* 34.5%
  (3) *yellow, miniature* 35.4%

Since the sum of (1) and (2) is approximately equal to (3), Sturtevant suggested that the order of the genes on the X chromosome was *yellow-white-miniature*. The *yellow* and *white* genes are apparently close to each other because the recombination frequency is low. However, both of these genes are quite far apart from *miniature* because the *white, miniature* and *yellow, miniature* combinations show large recombination frequencies. Since *miniature* shows more recombination with *yellow* than with *white* (35.4 vs. 34.5), it follows that *white* is between the other two genes.

Sturtevant suggested that the frequency of exchange could be taken as an estimate of the distance between two genes or loci along the chromosome. He constructed a map of the three genes on the X chromosome, with one map unit being equated with one percent recombination between two genes.* In the preceding example, the distance between *yellow* and *white* would be 0.5 map unit, and between *yellow* and *miniature*, 35.4 map units. It follows that the distance between *white* and *miniature* should be (35.4 − 0.5) or 34.9. This estimate is close to the actual frequency of recombination between *white* and *miniature* (34.5). The simple map for these three genes is shown in Figure 6.4.

In addition to these three genes, Sturtevant considered three other genes on the X chromosome and produced a more extensive map including all six

_____

*In honor of Morgan's work, one map unit is often referred to as a **centimorgan (cM)**.

genes. He and a colleague, Calvin Bridges, soon began a search for autosomal linkage in *Drosophila*. By 1923, they had clearly shown that linkage and crossing over were not restricted to sex-linked genes. Rather, they had discovered linked genes, on autosomes between which crossing over occurred.

During this work, they made another interesting observation. In *Drosophila*, crossing over was shown to occur only in females. The fact that no crossing over occurs in *Drosophila* males made the genetic analysis of autosomal inheritance and mapping much simpler to perform. However, crossing over does occur in both sexes in most other organisms.

Although many refinements in chromosome mapping have developed since Sturtevant's initial work, his basic principles are accepted as correct. They have been used to produce detailed chromosome maps of organisms for which large numbers of linked mutant genes are known. In addition to providing the basis for chromosome mapping, Sturtevant's findings were historically significant to the field of genetics. In 1910, the **chromosomal theory of inheritance** was still being widely disputed. Even Morgan was skeptical of the theory prior to the time he conducted the bulk of his experimentation. Research has now firmly established that chromosomes contain genes in a linear order and that these genes are the equivalent of Mendel's unit factors.

Single Crossovers

Why should the relative distance between two loci influence the amount of recombination and crossing

over observed between them? The basis for this variation is explained in the following analysis.

During meiosis, a limited number of crossover events occurs in each tetrad. These recombinant events occur randomly along the length of the tetrad. Therefore, the closer two loci reside along the axis of the chromosome, the less likely it is that a crossover event will occur between them. The same reasoning suggests that the farther apart two linked loci are, the more likely it is that a random crossover event will occur between them.

In Figure 6.5(a), a single crossover occurs between two nonsister chromatids, but not between the two loci; therefore, the crossover goes undetected because no recombinant gametes are produced. In (b), where two loci are quite far apart, the crossover occurs between them, yielding recombinant gametes.

The evidence supporting the concept that crossing over occurs in the four-strand tetrad stage will be discussed later in this chapter. However, we must point out certain consequences of this fact. If only one single crossover occurs between two nonsister chromatids, the other two strands of the tetrad will not be involved in this exchange and may enter a gamete unchanged. Thus, even if a single crossover occurs 100 percent of the time between two linked genes, this recombinant event will subsequently be observed in only 50 percent of the potential gametes formed. This concept is diagramed in Figure 6.6. For example, when only singles exchanges are considered, and when 20 percent recombinant gametes are observed, crossing over actually occurred between these two loci in 40 percent of the tetrads. Therefore, the upper

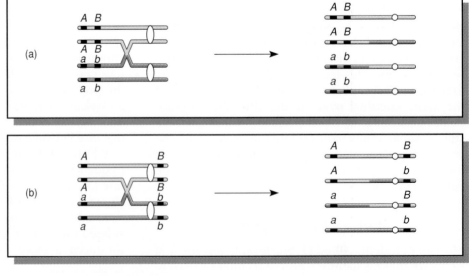

FIGURE 6.5
Two cases of exchange between nonsister chromatids and the gametes subsequently produced. In (a) the exchange does not separate the alleles of the two genes, only parental gametes are formed, and the exchange goes undetected. In (b) the exchange separates the alleles, resulting in recombinant gametes.

Exchange between
nonsister chromatids

Gametes

FIGURE 6.6
The consequences of a single exchange between two nonsister chromatids occurring in the tetrad stage. Two noncrossover (parental) and two crossover (recombinant) gametes are produced.

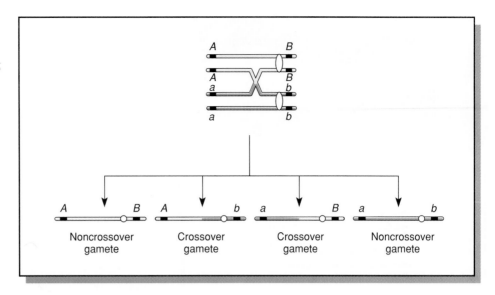

Noncrossover gamete · Crossover gamete · Crossover gamete · Noncrossover gamete

limit of observed crossing over is 50 percent. The general rule is that under these conditions, the percentage of tetrads involved in an exchange between two genes is twice as great as the percentage of recombinant gametes produced.

When two linked genes are greater than 50 map units apart, a crossover can theoretically be expected to occur between them in 100 percent of the tetrads. If this theoretical prediction were to be achieved, each tetrad would yield equal proportions of the four gametes shown in Figure 6.6, just as if the genes were on different chromosomes and assorting independently. For a variety of reasons, this theoretical limit is never achieved.

Multiple Crossovers

It is possible that in a single tetrad, two, three, or more exchanges will occur between nonsister chromatids as a result of several crossing over events. Double exchanges of genetic material result from **double crossovers,** as shown in Figure 6.7. For a double exchange to be studied, three genes must be investigated, each heterozygous for two alleles. Before we can determine the frequency of recombination between all three loci, we must review some simple probability calculations.

The probabilities of a single exchange occurring between the A and B or the B and C genes are directly related to the physical distance between each locus. The closer A is to B and B is to C, the less likely it is that a single exchange will occur between either of the two sets of loci. In the case of a double crossover, two separate and independent events or exchanges must occur simultaneously. The mathematical probability of two independent events occurring simulta-

neously is equal to the product of the individual probabilities. This is the product law previously introduced in Chapter 3 (p. 59).

Suppose that crossover gametes resulting from single exchanges between A and B are recovered 20 percent of the time ($p = 0.20$) and between B and C 30 percent of the time ($p = 0.30$). The probability of recovering a double-crossover gamete arising from two exchanges, between A and B and B and C, is predicted to be $(0.20) \cdot (0.30) = 0.06$, or 6 percent. It is apparent from this calculation that the frequency of double-crossover gametes is always expected to be much lower than that of either single-crossover class of gametes.

If three genes are relatively close together along one chromosome, the expected frequency of double-crossover gametes is extremely low. For example, consider the $A–B$ distance in Figure 6.7 to be 3 map units and the $B–C$ distance in that figure to be 2 map units. The expected double-crossover frequency would be $(0.03) \cdot (0.02) = 0.0006$, or 0.06 percent. This translates to only 6 events in 10,000. Thus, in a mapping experiment involving closely linked genes, very large numbers of offspring are required in order to detect double-crossover events. In this example, it would be unlikely that a double crossover would be observed even if 1000 offspring were examined. If these probability considerations are extended, it is evident that if four or five genes were being mapped, even fewer triple and quadruple crossovers can be expected to occur.

Three-Point Mapping in *Drosophila*

The information presented in the previous section serves as the basis for simultaneously mapping three

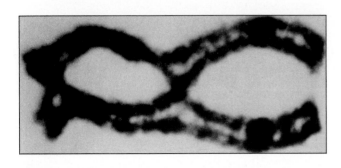

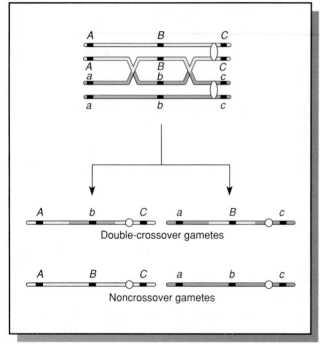

FIGURE 6.7
The results of a double exchange occurring between nonsister chromatids. Because the exchanges involve only two strands, two noncrossover gametes and two double crossover gametes are produced. The accompanying photograph was derived from a salamander spermatocyte.

or more linked genes. To illustrate this, we will examine a situation involving three linked genes.

Three criteria must be met for a successful mapping cross:

1 The genotype of the organism producing the crossover gametes must be heterozygous at all loci under consideration.
2 The cross must be constructed so that the genotypes of all gametes can be accurately determined by observing the phenotypes of the resulting offspring. This is necessary because the gametes and their genotypes can never be observed directly. Thus, each phenotypic class must reflect the genotype of the gametes of the parents producing it.
3 A sufficient number of offspring must be produced in the mapping experiment to recover a representative sample of all crossover classes.

These criteria are met in the three-point mapping cross from *Drosophila melanogaster* shown in Figure 6.8. In this cross, three sex-linked recessive mutant genes—*yellow* body color, *white* eye color, and *echinus* eye shape—are considered.

In the P_1 generation, males hemizygous for all three wild-type alleles are crossed to females that are homozygous for all three recessive mutant alleles. Therefore, the P_1 males are wild type with respect to body color, eye color, and eye shape. They are said to have a **wild-type** phenotype. The females, on the other hand, exhibit the three mutant phenotypes—*yellow* body color, *white* eyes, and *echinus* eye shape.

This cross produces an F_1 generation consisting of females heterozygous at all three loci and males that, because of the Y chromosome, are hemizygous for the three mutant alleles. Phenotypically, all F_1 females are wild type, while all F_1 males are *yellow*, *white*, and *echinus*. The genotype of the F_1 females fulfills the first criterion for constructing a map of the three linked genes; that is, it is heterozygous at the three loci and may serve as the source of recombinant gametes generated by crossing over. Note that because of the genotypes of the P_1 parents, all three mutant alleles are on one homologue and all three wild-type alleles are on the other homologue. *Other arrangements are possible.* For example, the heterozygous F_1 female might have the *y* and *ec* mutant alleles on one homologue and the *w* allele on the other. This would occur if, in the P_1 cross, one parent was *yellow* and *echinus* and the other parent was *white*.

Now, returning to our original cross, the second criterion is met by virtue of the gametes formed by the F_1 males. Every gamete will contain either an X chromosome bearing the three mutant alleles or a Y chromosome, which is genetically inert for the three loci being considered. Whichever type participates in fertilization, the genotype of the F_1 female gamete will be expressed phenotypically in the F_2 male and female offspring. As a result, all F_1 noncrossover and crossover gametes can be detected by observing the F_2 phenotypes.

With these two criteria met, we can construct a chromosome map from the crosses illustrated in Figure 6.8. First, we must determine which F_2 phenotypes correspond to the various noncrossovers and crossover categories. Two of these can be determined immediately.

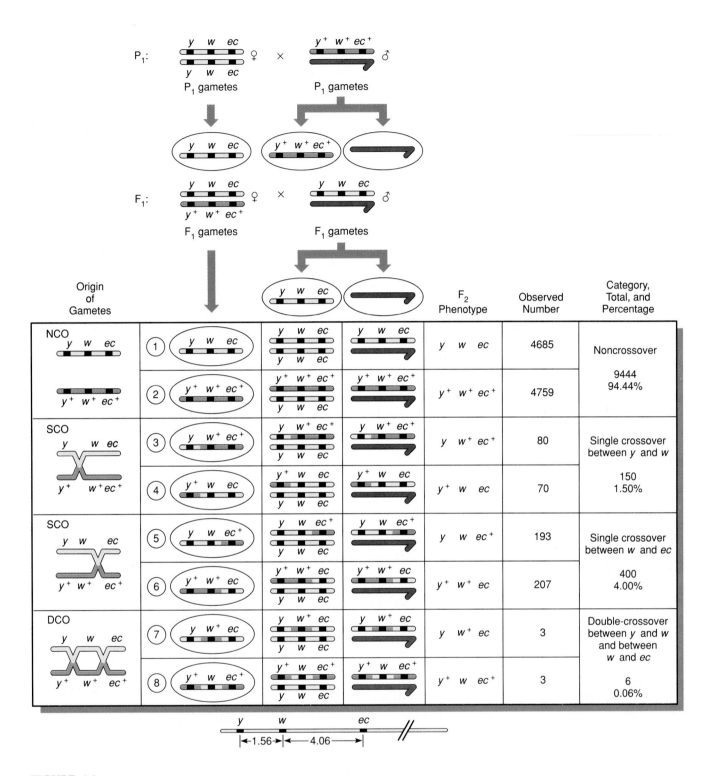

FIGURE 6.8
A three-point mapping cross involving the *yellow* (*y* or *y⁺*), *white* (*w* or *w⁺*), and *echinus* (*ec* or *ec⁺*) genes in *Drosophila melanogaster*. NCO, SCO, and DCO refer to noncrossover, single crossover, and double crossover groups, respectively. Note that because of the complexity of this and several of the ensuing figures, centromeres have not been included on the chromosomes.

The first is the **noncrossover category** of F_2 phenotypes. These phenotypes are determined by the combination of alleles present in the parental gametes formed by the F_1 female. Each such gamete contains one or the other of the X chromosomes unaffected by crossing over. As a result of segregation, equal proportions of the two types of gametes, and subsequently the F_2 phenotypes, are produced. Because they are derived from a heterozygote, the genotypes of the two parental gametes and the phenotypes of the two F_2 phenotypes complement one another. For example, if one is wild type, the other is completely mutant. This is the case in the cross under consideration. In other situations, if one chromosome shows one mutant allele or trait, the other shows the other two mutant traits, and so on. They are therefore called **reciprocal classes** of gametes and phenotypes.

The two noncrossover phenotypes are most easily recognized because *they exist in the greatest proportion.* Figure 6.8 shows that classes (1) and (2) are present in the greatest numbers. Therefore, flies that are *yellow, white,* and *echinus* and those that are normal or wild type for all three characters constitute the noncrossover category and represent 94.44 percent of the F_2 offspring.

The second category that can be easily detected is represented by the double-crossover phenotypes. Because of their probability of occurrence, *they must be present in the least numbers.* Remember that this group represents two independent but simultaneous single-crossover events. Two reciprocal phenotypes can be identified: class (7), which shows the mutant traits *yellow* and *echinus* but normal eye color; and class (8), which shows the mutant trait *white* but normal body color and eye shape. Together these double-crossover phenotypes constitute only 0.06 percent of the F_2 offspring.

The remaining four phenotypic classes represent two categories resulting from single crossovers. Classes (3) and (4), reciprocal phenotypes produced by single-crossover events occurring between the *yellow* and *white* loci, are equal to 1.50 percent of the F_2 offspring. Classes (5) and (6), constituting 4.00 percent of the F_2 offspring, represent the reciprocal phenotypes resulting from single-crossover events occurring between the *white* and *echinus* loci.

The map distances between the three loci can now be calculated. The distance between y and w, or between w and ec, is equal to the percentage of all detectable exchanges occurring between them. For any two genes under consideration, this will include all appropriate single crossovers as well as all double crossovers. The latter are included because they represent two simultaneous single crossovers. For the y and w genes this includes classes (3), (4), (7),

and (8), totaling 1.50% + 0.06%, or 1.56 map units. Similarly, the distance between w and ec is equal to the percentage of offspring resulting from an exchange between these two loci: classes (5), (6), (7), and (8), totaling 4.00% + 0.06%, or 4.06 map units. The map of these three loci on the X chromosome, based on these data, is shown at the bottom of Figure 6.8.

Determining the Gene Sequence

In the preceding example, the order or sequence of the three genes along the chromosome was assumed to be $y–w–ec$. In most mapping experiments, the gene sequence is not known, and this constitutes another variable in the analysis. Had the gene order been unknown in this example, it could have been determined in a straightforward way. There are two methods by which to accomplish this. You should select one or the other method and adhere to its use in your own analysis.

Method I This method is based on the fact that there are only three possible orders, each containing one of the three genes in between the other two. Note that the right-to-left direction along the chromosome is meaningless. While we assumed $y–w–ec$, $ec–w–y$ is fully equivalent because the data provide no basis for knowing which is correct. Therefore, only three possibilities exist, and one of them must be correct:

> **(I)** w-y-ec (y is in the middle)
> **(II)** y-ec-w (ec is in the middle)
> **(III)** y-w-ec (w is in the middle)

If you use the following steps during your analysis, you will be able to determine gene order:

1 Assuming any of the three orders, first determine the **arrangement** of alleles along each homologue of the heterozygous parent giving rise to noncrossover and crossover gametes (the F_1 female in our example).
2 Determine whether a double-crossover event occurring within that arrangement will produce the **observed** double-crossover phenotypes. Remember that these phenotypes occur least frequently and can be easily identified.
3 If this order does not produce the correct phenotypes, try each of the other two orders. One must work!

These steps will be illustrated for the cross shown in Figure 6.8. The three possible orders are labeled **(I)**, **(II)**, and **(III)**, as just shown. Either y, ec, or w must be in the middle.

1 Assuming y is between w and ec, the arrangement of alleles along the homologues of the F_1 heterozygote is

$$\textbf{(I)} \quad \frac{w \quad y \quad ec}{w^+ \quad y^+ \quad ec^+}$$

We know this because of the way in which the P_1 generation was crossed. The P_1 female contributed an X chromosome bearing the w, y, and ec alleles, while the P_1 male contributed an X chromosome bearing the w^+, y^+, and ec^+ alleles.

2 A double crossover within the above arrangement would yield the following gametes:

$$\underline{w \quad y^+ \quad ec} \quad \text{and} \quad \underline{w^+ \quad y \quad ec^+}$$

Following fertilization, the F_2 double-crossover phenotypes would correspond to these gametic genotypes, yielding offspring that are *white, echinus* and offspring that are *yellow*. Determination of the actual double crossovers reveals them to be *yellow, echinus* flies and *white* flies. Therefore, our assumed order is incorrect.

3 If we try the other two orders, one with the ec/ec^+ alleles in the middle (II) and one with the w/w^+ alleles in the middle (III):

$$\textbf{(II)} \quad \frac{y \quad ec \quad w}{y^+ \quad ec^+ \quad w^+} \quad \text{and} \quad \textbf{(III)} \quad \frac{y \quad w \quad ec}{y^+ \quad w^+ \quad ec^+}$$

we see that arrangement II again provides **predicted** double-crossover phenotypes that do not correspond to the **actual** double-crossover phenotypes. The predicted phenotypes are *yellow, white* flies and *echinus* flies in the F_2 generation. Therefore, this order is also incorrect. However, arrangement III **will** produce the **observed** phenotypes, *yellow, echinus* flies and *white* flies. Therefore, this order, where the w gene is in the middle, is correct.

To summarize, utilizing this method is rather straightforward. Determine the arrangement of alleles on the homologues of the heterozygote yielding the crossover gametes. Then, test each of three possible orders to determine which one yields the observed double-crossover phenotypes. Whichever of the three does so represents the correct order. Testing the three possibilities in our example is summarized in Figure 6.9.

Method II This method again requires that we determine the arrangement of alleles along each homologue of the heterozygous parent. It also as-

sumes that one of the three genes must be in the middle. It requires one further assumption:

Following a double-crossover event, the allele representing the middle gene will find itself present with the outside or flanking alleles present on the opposite parental homologue.

To illustrate, assume order (I), $w–y–ec$, in the arrangement

$$\textbf{(I)} \quad \frac{w \quad y \quad ec}{w^+ \quad y^+ \quad ec^+}$$

Following a double-crossover event, the y and y^+ alleles would find themselves switched to the arrangement

$$\frac{w \quad y^+ \quad ec}{w^+ \quad y \quad ec^+}$$

Following segregation, two gametes would be formed:

$$\underline{w \quad y^+ \quad ec} \quad \text{and} \quad \underline{w^+ \quad y \quad ec^+}$$

Since the genotype of the gamete will be expressed directly in the phenotype following fertilization, the double-crossover phenotypes will be

white, echinus flies   and   *yellow* flies

Note that the *yellow* allele, assumed to be in the middle, is now associated with the two outside markers of the other homologue, w^+ and ec^+. However, these predicted phenotypes do not coincide with the observed double-crossover phenotypes. Therefore, the *yellow* gene is not in the middle.

This same reasoning can be applied to the assumption that the *echinus* gene or the *white* gene is in the middle. In the former case, a negative conclusion will be reached. If we assume that the *white* gene is in the middle, the predicted and actual double crossovers coincide. Therefore, we conclude that the *white* gene is located between the *yellow* and *echinus* genes.

To summarize, this method is also straightforward. Determine the arrangement of alleles on the homologues of the heterozygote yielding crossover gametes. Then determine the actual double-crossover phenotypes. Simply select the single allele that has been switched so that it is now associated with two other alleles that have not been separated by crossing over.

In our example above, y, ec, and w are together in the F_1 heterozygote, as are y^+, ec^+, and w^+. In the F_2 double-crossover classes, it is w and w^+ that have been switched. The w allele is now associated with

Three Theoretical Sequences	Double-crossover Gametes	Phenotypes

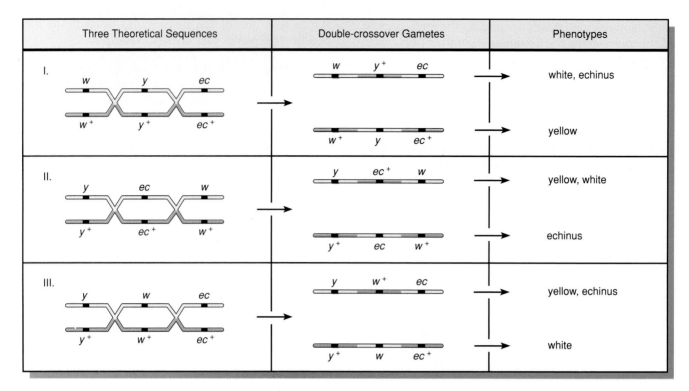

I.
| | w y^+ ec | white, echinus |
| | w^+ y ec^+ | yellow |

II.
| | y ec^+ w | yellow, white |
| | y^+ ec w^+ | echinus |

III.
| | y w^+ ec | yellow, echinus |
| | y^+ w ec^+ | white |

FIGURE 6.9

A summary of the three possible orders of the *white, yellow,* and *echinus* genes, the results of a double crossover in each case, and the resulting phenotypes produced. Note that this and subsequent figures use the conventions of w^+ or +, y^+ or +, etc., to denote the dominant allele, in contrast to the use of *W* or *Y*.

y^+ and ec^+, while the w^+ allele is now associated with the y and ec alleles. Therefore, the *white* gene is in the middle, and the *yellow* and *echinus* genes are the flanking markers.

A Mapping Problem in Maize

Having established the basic principles of chromosome mapping, we will now consider a related problem in maize (corn) where the gene sequence and interlocus distances are unknown.

This analysis differs from the preceding discussion in several ways and therefore will expand your knowledge of mapping procedures:

1 In the previous discussion we initially knew the gene sequence, and we used this information to explain how to determine an unknown sequence. In this analysis of maize, the sequence is *initially* unknown.
2 The previous mapping cross involved sex-linked genes. Here, autosomal genes are considered.
3 The previous cross was discussed in bits and pieces as each principle was established. Here, the

analysis proceeds uninterrupted from beginning to end.
4 In the discussion of this cross we will make a transition in the use of symbols, as first suggested in Chapter 4. Instead of using the symbols bm^+, v^+, and pr^+, we will simply use + to denote each wild-type allele. This symbol is less complex to manipulate, but requires a better understanding of mapping procedures.

When we consider three autosomally linked genes in maize, the experimental cross must still meet the same three criteria established for the X-linked genes in *Drosophila*: (1) one parent must be heterozygous for all traits under consideration; (2) the gametic genotypes produced by the heterozygote must be apparent from observing the phenotypes of the offspring; and (3) a sufficient sample size must be available for complete analysis.

In maize, the recessive mutant genes *bm* (brown midrib), *v* (virescent seedling), and *pr* (purple aleurone) are linked on chromosome 5. Assume that a female plant is known to be heterozygous for all three traits. Nothing is known about (1) the arrangement of the mutant alleles on the maternal and paternal homologues; (2) the sequence of genes; or (3) the map distances between the genes. What genotype must the male plant have to allow successful mapping? In order to meet the second criterion, the male must be homozygous for all three recessive mutant alleles. Otherwise, offspring of this cross showing a given phenotype might represent more than one genotype, making accurate mapping impossible.

Figure 6.10 diagrams this cross. As shown, we do not know either the arrangement of alleles or the sequence of loci in the heterozygous female. Various possibilities are shown, each followed by a question mark. In the test cross male parent, *we do not know the sequence,* but must write it down initially as a random selection.

The offspring have been arranged in groups of two for each pair of reciprocal phenotypic classes. The two members of each reciprocal class are derived from either no crossing over (**NCO**), one single-crossover event (**SCO**), or a double crossover (**DCO**).

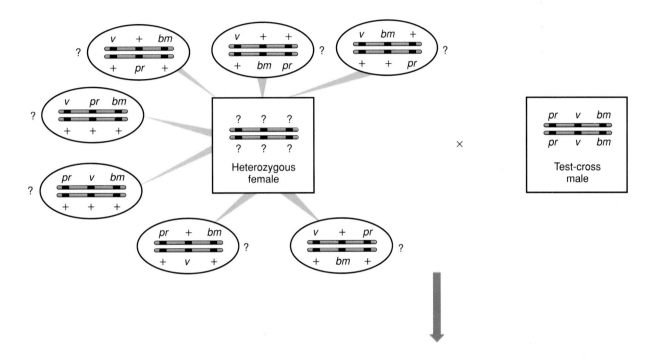

Offspring Phenotype			Number	Total and Percentage	Exchange Classification
+	v	bm	230	467	Noncrossover
pr	+	+	237	42.1%	(NCO)
pr	v	+	82	161	Single-crossover
+	+	bm	79	14.5%	(SCO)
+	v	+	200	395	Single-crossover
pr	+	bm	195	35.6%	(SCO)
pr	v	bm	44	86	Double-crossover
+	+	+	42	7.8%	(DCO)

FIGURE 6.10
The results of a three-point mapping cross in maize where the arrangement of alleles, the order of the genes, and the distance between them are all unknown.

To solve this problem, answer the following questions. It will be helpful to refer to Figures 6.10 and 6.11 while you consider them.

1 *What is the correct heterozygous arrangement of alleles in the female parent?*

Determine the two noncrossover classes, those that occur in the highest frequency. In this case, they are $+ \ v \ bm$ and $pr \ + \ +$. Therefore, the arrangement of alleles on the homologues of the female parent must be as shown in Figure 6.11(a). These homologues segregate into gametes, unaffected by any recombination event. Any other arrangement of alleles could not yield the observed noncrossover classes. (Remember that $+ \ v \ bm$ is equivalent to $pr^+ \ v \ bm$, and that $pr \ + \ +$ is equivalent to $pr \ v^+ \ bm^+$.)

2 *What is the correct sequence of genes?*

The reasoning established in Method I will first be used to answer this question. We know that the arrangement of alleles is

$$\overline{\overline{+ \ v \ bm}} \\ \overline{pr \ + \ +}$$

But, is the assumed sequence correct? That is, will a double-crossover event yield the observed double-crossover phenotypes following fertilization? Simple observation shows that it will not [Figure 6.11(b)]. Try the other two orders [Figure 6.11(c) and (d)], keeping the same arrangement:

$$\overline{\overline{+ \ bm \ v}} \over {pr \ + \ +}} \quad \text{and} \quad {\overline{\overline{v \ + \ bm}} \over {+ \ pr \ +}}$$

Only the latter case will yield the observed double-crossover classes [Figure 6.11(d)]. Therefore, the *pr* gene is in the middle.

The same conclusion is reached if the problem is analyzed using Method II. In this case, no assumption of gene sequence is necessary. The arrangement of alleles in the heterozygous parent is

$$\overline{\overline{+ \ v \ bm}} \\ \overline{pr \ + \ +}$$

The double-crossover gametes are also known:

$$\overline{pr \ v \ bm} \quad \text{and} \quad \overline{+ \ + \ +}$$

It is obvious that the *pr* allele has shifted so as to be associated with *v* and *bm* following a double crossover. The latter two alleles were present together on one homologue, and they stayed together. Therefore, *pr* is the odd gene, so to speak, and it is in the middle.

3 *What is the distance between each pair of genes?*

Having established the sequence of loci as $v-pr-bm$, we can now determine the distance between *v* and *pr* and between *pr* and *bm*. Remember that the map distance between two genes is calculated on the basis of all detectable recombinational events occurring between them. This includes both the single- and double-crossover events involving the two genes being considered.

Figure 6.11(e) shows that the phenotypes $v \ pr \ +$ and $+ \ + \ bm$ result from single crossovers between *v* and *pr*, accounting for 14.5 percent of the offspring. By adding the percentage of double crossovers (7.8%) to the number obtained for single crossovers, the total distance between *v* and *pr* is calculated to be 22.3 map units.

Figure 6.11(f) shows that the phenotypes $v \ + \ +$ and $+ \ pr \ bm$ result from single crossovers between *pr* and *bm*, totaling 35.6 percent. With the addition of the double-crossover classes (7.8%) the distance between *pr* and *bm* is calculated to be 43.4 map units.

The final map for all three genes in this example is shown in Figure 6.11(g).

The Accuracy of Mapping Experiments

Until now, we have considered crossover frequencies to be directly proportional to the distance between any two loci along the chromosome. However, it is not always possible to detect all crossover events. A case in point is a double exchange that occurs between the two loci in question. As shown in Figure 6.12(a), if a double exchange occurs, the original arrangement of alleles on each nonsister homologue is recovered. Therefore, even though crossing over occurs, it is impossible to detect in these cases. This factor holds for all even-numbered exchanges between two loci.

As a result of this and other types of multiple exchanges, mapping determinations will usually slightly underestimate the actual distance between two genes. The farther apart two genes are, the greater the probability that undetected crossovers will occur. Therefore, the discrepancy is minimal for two genes that are relatively close together, but increases as the distance becomes greater. This fact is reflected in Figure 6.12(b), where the relationship between recombination frequency and map distance is graphed.

Accurate linkage distances between widely separated genes must therefore be determined by many different experiments. These experiments should utilize numerous genes in between the two in

FIGURE 6.11
The steps used in producing a map of the three genes involved in the cross shown in Figure 6.10.

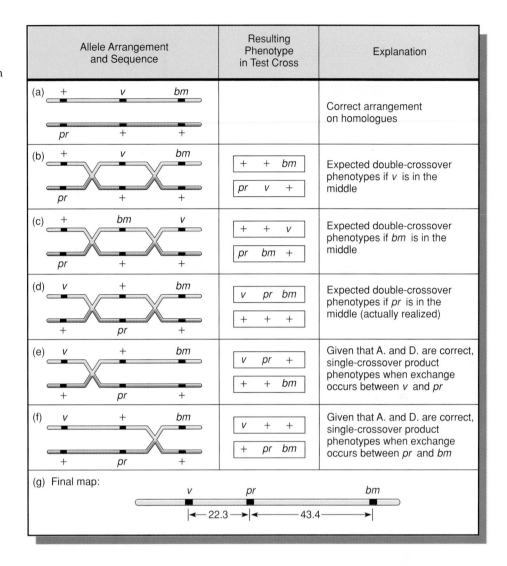

Allele Arrangement and Sequence	Resulting Phenotype in Test Cross	Explanation
(a) + v bm / pr + +		Correct arrangement on homologues
(b) + v bm / pr + +	+ + bm / pr v +	Expected double-crossover phenotypes if v is in the middle
(c) + bm v / pr + +	+ + v / pr bm +	Expected double-crossover phenotypes if bm is in the middle
(d) v + bm / + pr +	v pr bm / + + +	Expected double-crossover phenotypes if pr is in the middle (actually realized)
(e) v + bm / + pr +	v pr + / + + bm	Given that A. and D. are correct, single-crossover product phenotypes when exchange occurs between v and pr
(f) v + bm / + pr +	v + + / + pr bm	Given that A. and D. are correct, single-crossover product phenotypes when exchange occurs between pr and bm
(g) Final map: v —— 22.3 —— pr —— 43.4 —— bm		

question. For example, if genes *A* and *K* are widely separated, mapping experiments using genes *B, C, D, E, F, G, H, I,* and *J* may be necessary, assuming these to be between *A* and *K*. If each distance is determined (e.g., *A–B, B–C, C–D. . .J–K*), then the distance between *A* and *K* is additive. That is, by adding the *A–B, B–C, C–D. . .J–K* distances together, we obtain a more accurate measurement of the distance between *A* and *K*.

Interference and the Coefficient of Coincidence

Still another factor tends to limit the accuracy of mapping data. This factor involves the actual reduction of the number of expected double crossovers when genes are reasonably close to one another along the chromosome. This reduction, called **interference,** will be illustrated for three-point mapping.

We have already considered the probability relationships between single- and double-crossover events. In theory, the percentage of **expected double crossovers** is predicted by multiplying the percentage of the total crossovers between each pair of genes. Remember that a double crossover (DCO) represents two single-crossover events. For example, the expected double-crossover frequency of the cross illustrated in Figures 6.10 and 6.11 may be calculated in the following manner:

$$DCO_{exp} = (0.223) \times (0.434) = 0.097 = 9.7\%$$

Frequently this predicted figure does not correspond precisely with the observed DCO frequency. Generally, there are fewer DCOs observed than predicted. In the maize cross, only 7.8 percent DCOs were observed. In some cases there are more DCOs than expected. These disparities are explained by the

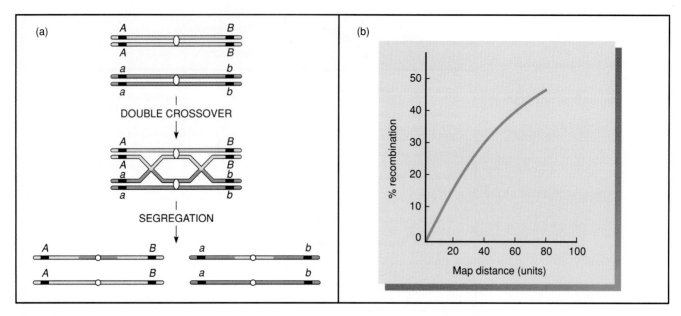

FIGURE 6.12
(a) An illustration of a double crossover that goes undetected because no rearrangement of alleles occurs. (b) A comparison of the theoretical and actual relationships between recombination and map distance, as studied in *Drosophila, Neurospora,* and maize.

concept of interference, which is quantified by calculating the **coefficient of coincidence (C):**

$$C = \frac{\text{Observed DCO}}{\text{Expected DCO}}$$

In the maize cross, we have

$$C = \frac{0.078}{0.097} = 0.804$$

Once C is calculated, interference (I) may be quantified using the simple equation

$$I = 1 - C$$

In the maize cross, we have

$$I = 1.000 - 0.804 = 0.196$$

If interference is complete and no double crossovers occur, $I = 1.0$. If fewer DCOs than expected occur, I is a positive number (as above) and **positive interference** has occurred. If more DCOs than expected occur, I is a negative number and **negative interference** has occurred.

In eukaryotic systems, positive interference is most often observed. It appears that a crossover event in one region of a chromosome inhibits a second crossover in neighboring regions of the chromosome. In general, the closer genes are to one another along the chromosome, the more positive interference is observed and the lower the C value. In *Drosophila,* when three genes are clustered within 10 map units, interference is often complete and no double-crossover classes are recovered. This observation suggests that interference may be explained by physical constraints that prohibit the formation of closely aligned chiasmata. Perhaps a mechanical stress is imposed on chromatids during crossing over such that one chiasma inhibits the formation of a second chiasma in the neighboring region. This interpretation is consistent with the finding that the impact of interference decreases as the genes in question are located farther apart. In the maize cross illustrated in Figure 6.11, the three genes are relatively far apart, and 80 percent of the expected double crossovers are observed.

The Genetic Map of *Drosophila*

In organisms such as *Drosophila,* maize, and the mouse, where large numbers of mutants have been discovered and experimental crosses are easy to perform, extensive maps of each chromosome have been made. Illustrated in Figure 6.13 are partial maps of the four chromosomes of *Drosophila.* Virtually

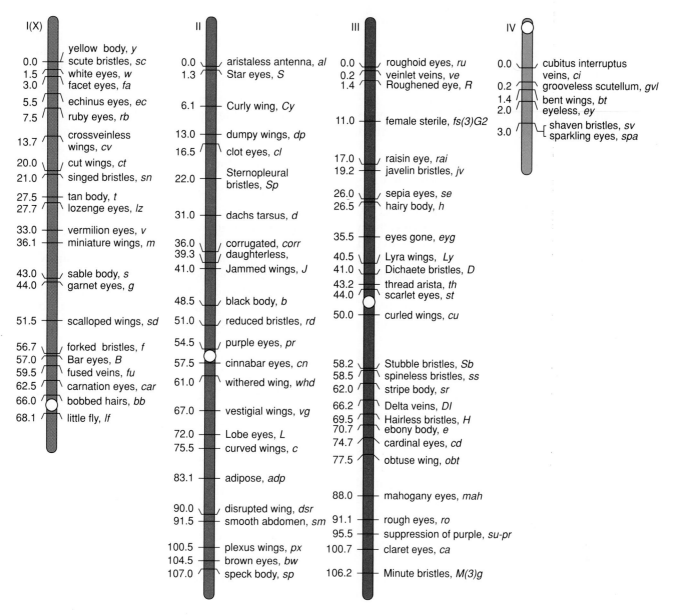

FIGURE 6.13
A partial genetic map of the four chromosomes of *Drosophila melanogaster*. The circle on each chromosome represents the position of the centromere.

every morphological feature of the fruit fly has been observed to be subject to mutation. The gene locus involved in determining an altered phenotype is first localized to one of the four chromosomes, or linkage groups, and then mapped in relation to other linked genes of that group. As can be seen, the genetic map of the X chromosome is somewhat less extensive than that of autosome 2 or 3. In comparison to these three, autosome 4 is minuscule. Based on cytological evidence, the relative lengths of the genetic maps have been found to correlate with the relative physical lengths of these chromosomes.

Somatic Cell Hybridization and the Human Gene Map

In humans, where neither designed matings nor large numbers of offspring are available, the earliest linkage studies were based on pedigree analysis. Attempts were made to establish whether a trait was sex-linked or autosomal. All of the traits shown to be sex-linked were obviously shown to be linked to the X chromosome. For autosomal traits, geneticists tried to distinguish clearly whether pairs of traits demonstrated linkage or independent assortment. In this way, it was hoped that a human gene map could be created.

The results, even as recently as the 1960s, were discouraging because of the limitations of this approach and because of the relatively high haploid number of human chromosomes (23). Relatively few cases of either X-linkage or autosomal linkage were established and almost no mapping information became available.

However, in the 1960s, a new technique, somatic cell hybridization, was developed which aided immensely in assigning human genes to their respective chromosomes. This technique, first discovered by Georges Barsky, relies on the fact that two cells in culture can be induced to fuse into a single hybrid cell. While Barsky used two mouse cell lines, it soon became evident that cells from different organisms will also fuse together. When this event occurs, an initial cell type called a **heterokaryon** is produced. The hybrid cell contains two nuclei in a common cytoplasm. Using the proper techniques, it is possible to fuse human and mouse cells, for example, and isolate the hybrids from the parental cells.

As the heterokaryons are cultured *in vitro*, two interesting changes occur. Eventually, the nuclei fuse together, creating what is termed a **synkaryon**. Then, as culturing is continued for many generations, chromosomes from one of the two parental species are gradually lost. In the case of the human-mouse hybrid, human chromosomes are lost randomly until eventually, the synkaryon has a full complement of mouse chromosomes and only a few human chromosomes. It is the result of this event that allows the assignment of human genes to the chromosomes upon which they reside.

The experimental rationale is straightforward. If a specific human gene product is synthesized in a synkaryon containing one to three human chromosomes, then the gene responsible for that product must reside on one of the one to three human chromosomes remaining in the hybrid cell. Or, if the human gene product is absent, the responsible gene cannot be present on any of the remaining human chromosomes. Ideally, a panel of 23 hybrid cell lines, each with but one unique human chromosome, would allow the immediate assignment of any human gene for which the product could be characterized.

In practice, a panel of cell lines, each with several remaining human chromosomes is most often utilized. The correlation of the presence or absence of each chromosome with the presence or absence of each gene product is called **synteny testing**. Consider, for example, the hypothetical data provided in Figure 6.14, where four gene products (A, B, C, and D) are tested in relationship to eight human chromosomes. Let us carefully analyze the gene which produces product A:

1 Product A is not produced by cell line 23, but chromosomes 1, 2, 3, and 4 are present in cell line 23. Therefore, we can rule out the presence of gene A on those four chromosomes and conclude that it must be on chromosome 5, 6, 7, or 8.
2 Product A is produced by cell line 34, which contains chromosomes 5 and 6, but not 7 and 8. Therefore, gene A is on chromosome 5 or 6.
3 Product A is also produced by cell line 41 which contains chromosome 5 but not chromosome 6. Therefore, gene A is on chromosome 5, according to this analysis.

Using a similar approach, gene B can be assigned to chromosome 3. You should perform this analysis to demonstrate for yourself that this is correct. Gene

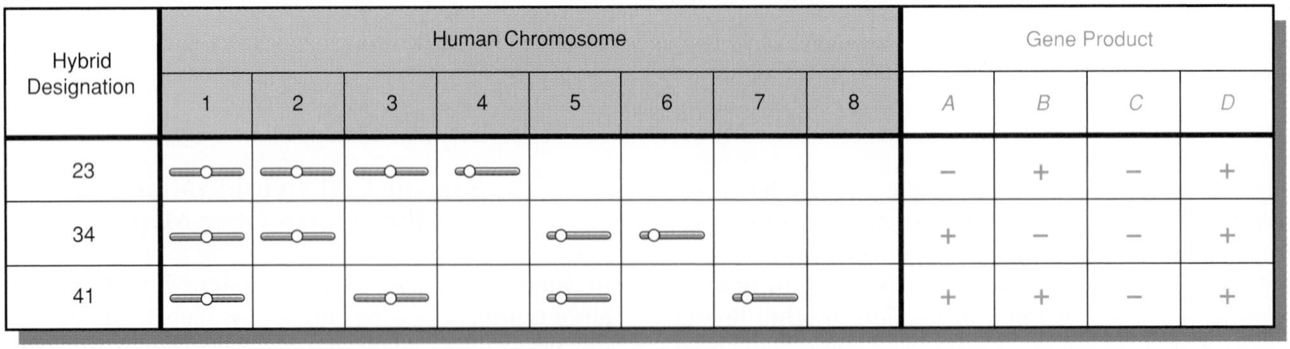

Hybrid Designation	Human Chromosome								Gene Product			
	1	2	3	4	5	6	7	8	A	B	C	D
23	⊟	⊟	⊟	⊟					−	+	−	+
34	⊟	⊟			⊟	⊟			+	−	−	+
41	⊟		⊟		⊟		⊟		+	+	−	+

FIGURE 6.14
A hypothetical grid of data used in synteny testing to assign genes to their appropriate human chromosomes. Three somatic cell hybrid lines, designated 23, 34, and 41, have each been scored for the presence or absence of human chromosomes 1 through 8, as well as for their ability to produce the hypothetical human gene products A through D.

C presents a unique situation. The data indicate that it is not present on any of the first seven chromosomes (1–7). While it might be on chromosome 8, no direct evidence supports this conclusion. Other panels are needed. We shall leave gene *D* for you to analyze. Upon what chromosome does it reside?

Using the approach described above, literally hundreds of human genes have been assigned to one chromosome or another. Many of the assignments shown in Table 6.1 were derived in this way. For mapping still other genes, where the products have yet to be discovered, researchers have had to rely on other linkage techniques. For example, by combining a rather sophisticated approach using recombinant DNA technology with pedigree analysis, it has been possible to assign the genes responsible for **Huntington disease**, **cystic fibrosis**, and **neurofibromatosis** to their respective chromosomes, 4, 7, and 17. This technology will be discussed in Chapter 16.

Finally, we might ask how human genes can be assigned to different regions of a given chromosome. Sometimes in hybrid cell lines, fragments of a particular chromosome become transferred to another chromosome, resulting in a **translocation**. It is possible using chromosome banding techniques to identify the exact origin of the translocation and correlate its

TABLE 6.1
Some human genes and their chromosome assignments.

Symbol	Gene Name	Chromosome
ABO	ABO blood group	9
ACP1	Acid phosphatase-1	2
ADA	Adenosine deaminase	20
AK-2	Adenylate kinase 2	1
CBD	Color blindness (deutan)	X
CBP	Color blindness (protan)	X
CC	Congenital cataracts	7
CF	Cystic fibrosis	7
COLIA1	Collagen I alpha-1 chain	17
DMD	Duchenne muscular dystrophy	X
F7	Blood clotting factor VII	8
Fy	Duffy blood group	1
GALK	Galactokinase	17
GLB1	B-galactosidase-1	3
GPD	Glucose phosphate dehydrogenase	X
H1	Histone 1	7
HBA	Hemoglobin, alpha chains	16
HBB	Hemoglobin, beta chain	11
HD	Huntington disease	4
HEMA/(FBC)	Classic hemophilia	X
HEXA	Hexosaminidase A	15
HEXB	Hexosaminidase B	5
HLA, A, B, C, D	Human leucocyte antigens	6
HPA	Hpa restriction endonuclease polymorphism	16
HPRT	Hypoxanthine phosphoribosyl transferase	X
IDH-1	Isocitrate dehydrogenase	2
IF-1	Interferon Type 1	2
IGH	Immunoglobulin, heavy chain	14
INS	Insulin	11
JK	Kidd blood group	2
LDHA	Lactate dehydrogenose A	11
LDHB	Lactate dehydrogenose B	12
MDH	Malate dehydrogenase	2
MN	MN Blood Group	4
NF	Neurofibromatosis	17
PAH	Phenylalanine hydroxylase (PKU)	12
PEP-B	Peptidase B	12
PEP-C	Peptidase C	1
PGD	Phosphoglycerate dehydrogenase	1
PGK	Phosphoglycerate kinase	X
PGM-1	Phosphoglucomutase-1	1
PI	α-1-Antitrypsin	14
RH	Rhesus blood group	1
TK1	Thymidine kinase	17
XG	Xg blood group	X

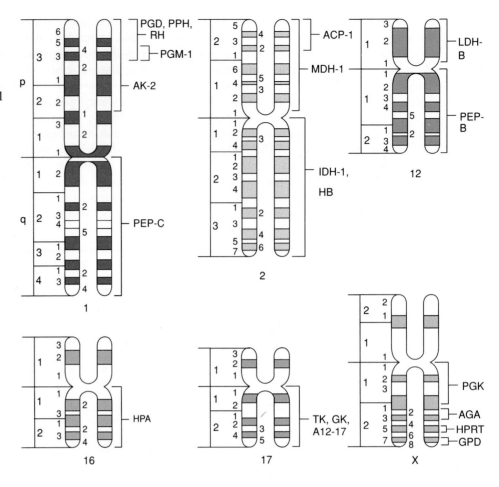

FIGURE 6.15
Representative regional gene assignments for human chromosomes 1, 2, 12, 16, 17, and the X. These assignments were derived using somatic cell hybridization techniques.

presence in hybrid cells with specific gene expression. In this way, gene maps of human chromosomes may be compiled. Partial maps of chromosomes 1, 2, 12, 16, 17 and the X are shown in Figure 6.15 to illustrate this point. While these maps are not as specific as the genetic map of *Drosophila*, we are beginning to learn a great deal about the chromosome locations of a multitude of human genes.

THE USE OF HAPLOID ORGANISMS IN RECOMBINATIONAL STUDIES

Many of the single-celled eukaryotes are haploid during the vegetative stages of their life cycle. The alga *Chlamydomonas* and the mold *Neurospora* illustrate this genetic condition. These organisms do form reproductive cells that fuse during fertilization, producing a diploid zygote. However, this structure soon undergoes meiosis, resulting in haploid vegetative cells that are then propagated by mitotic divisions. The life cycles of *Chlamydomonas* and *Neurospora* were presented in Figures 5.1 and 5.3, respectively.

Haploid organisms have several important advantages in genetic studies compared with more ad-

vanced diploid eukaryotes. They can be cultured and manipulated in genetic crosses much more easily. In addition, a haploid organism contains only a single allele of each gene, which is expressed directly in the phenotype. This fact greatly simplifies genetic analysis. As a result, organisms such as *Chlamydomonas* and *Neurospora* have served as subjects of research investigations in many areas of genetics, including linkage and mapping studies.

In order to perform genetic experiments with such organisms, crosses are made, and following fertilization, the meiotic structures are isolated. Because in each of these structures all four meiotic products give rise to spores, each structure is called a **tetrad**. *This term has a different meaning here than when it was used earlier to describe a precise chromatid configuration in meiosis.* In any case, individual tetrads are isolated, and the resultant cells are grown and analyzed separately from those of other tetrads. In the results to be described, the data will reflect the proportion of tetrads that showed one combination of genotypes, the proportion that showed another combination, and so on. Thus, such experimentation is called **tetrad analysis**.

In the following sections, we will examine the use of such data to: (1) demonstrate that crossing over occurs in the four-strand stage; (2) determine the map distance between a gene and the centromere; (3) distinguish between independent assortment and linkage of two genes; and (4) map two linked genes.

Crossing Over in the Four-Strand Stage

The question of when crossing over occurs during meiosis is critical to understanding the process and consequence of genetic exchange in eukaryotes. There are two alternative times at which crossing over might occur. First, exchange could take place before the chromosomes have duplicated in the **two-strand stage**. Alternatively, exchange could occur at the **four-strand stage** after the chromosomes have duplicated.

If crossing over occurs at the two-strand stage, all four products of a single meiotic event will be recombinant gametes because each pair of sister chromatids in the tetrad is derived from one of the members of the two-strand stage. If, on the other hand, crossing over occurs between two nonsister chromatids in the four-strand stage, two parental

(noncrossover) chromatids and two recombinant chromatids will be formed. These alternative possibilities are compared in Figure 6.16.

To decide between these two alternatives, we need to examine the results of an experiment with an organism from which all four products of single meiotic events may be recovered and observed. The organism used in this experiment is the ascomycete *Neurospora*, a haploid mold. Fertilization occurs, and meiosis results in the formation of a four-celled tetrad. Then, in *Neurospora*, a mitotic division of each product occurs, producing eight haploid cells called **ascospores**. Most important, the entire process takes place in a thin ascus sac that retains the haploid ascospore products in the order in which they are formed. These events are depicted in Figure 5.3.

Carl L. Lindegren's observations of meiotic segregation strongly support the theory that crossing over occurs in the four-strand stage. He examined the various possibilities of ascospore formation resulting from a cross between an *albino* mutant strain (*a*) and one with normal pigmentation (+). His work focused on the results of a crossover that occurred in the region between the mutant *albino* locus and the

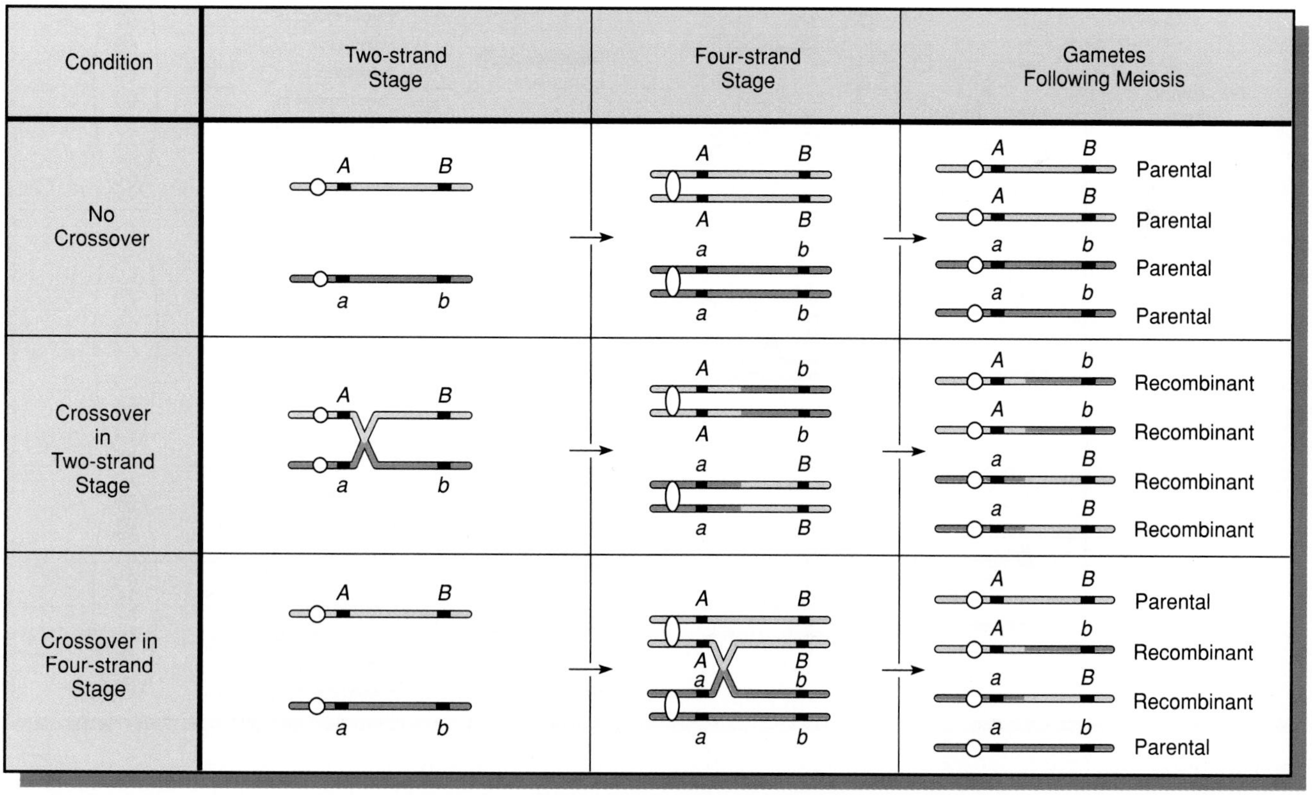

FIGURE 6.16
Comparison of the genotypes of gametes formed as a result of crossing over in the two-strand and four-strand stages with those formed when no crossing over occurs.

centromere. Figure 6.17 illustrates the theoretical results for various alternatives of an exchange in this region for both the two- and four-strand stages. In case (a) no exchange occurs. In case (b) the results of crossing over in the two-strand stage are predicted. With or without a crossover event, the resulting ascus will always contain four pigmented ascospores and four unpigmented ascospores, in that sequence. The asci produced in cases (a) and (b) cannot be distinguished from each other.

If, on the other hand, a crossover occurs in this region during the four-strand stage, an alternate arrangement of ascospores in the ascus is predicted,

as seen in case (c). Case (d) shows still another arrangement that occurs as a result of a slightly different exchange during the four-strand stage.

Lindegren observed asci with arrangements found in cases (a), (b), (c), and (d). His findings are consistent with the conclusion that crossing over indeed occurs in the four-strand stage and exclude crossing over in the two-strand stage, which cannot generate arrangements (c) and (d).

Similar findings have been drawn from studies of other organisms, including *Drosophila*. To date, no experiments have been reported that seriously dispute crossing over in the four-strand stage in eukary-

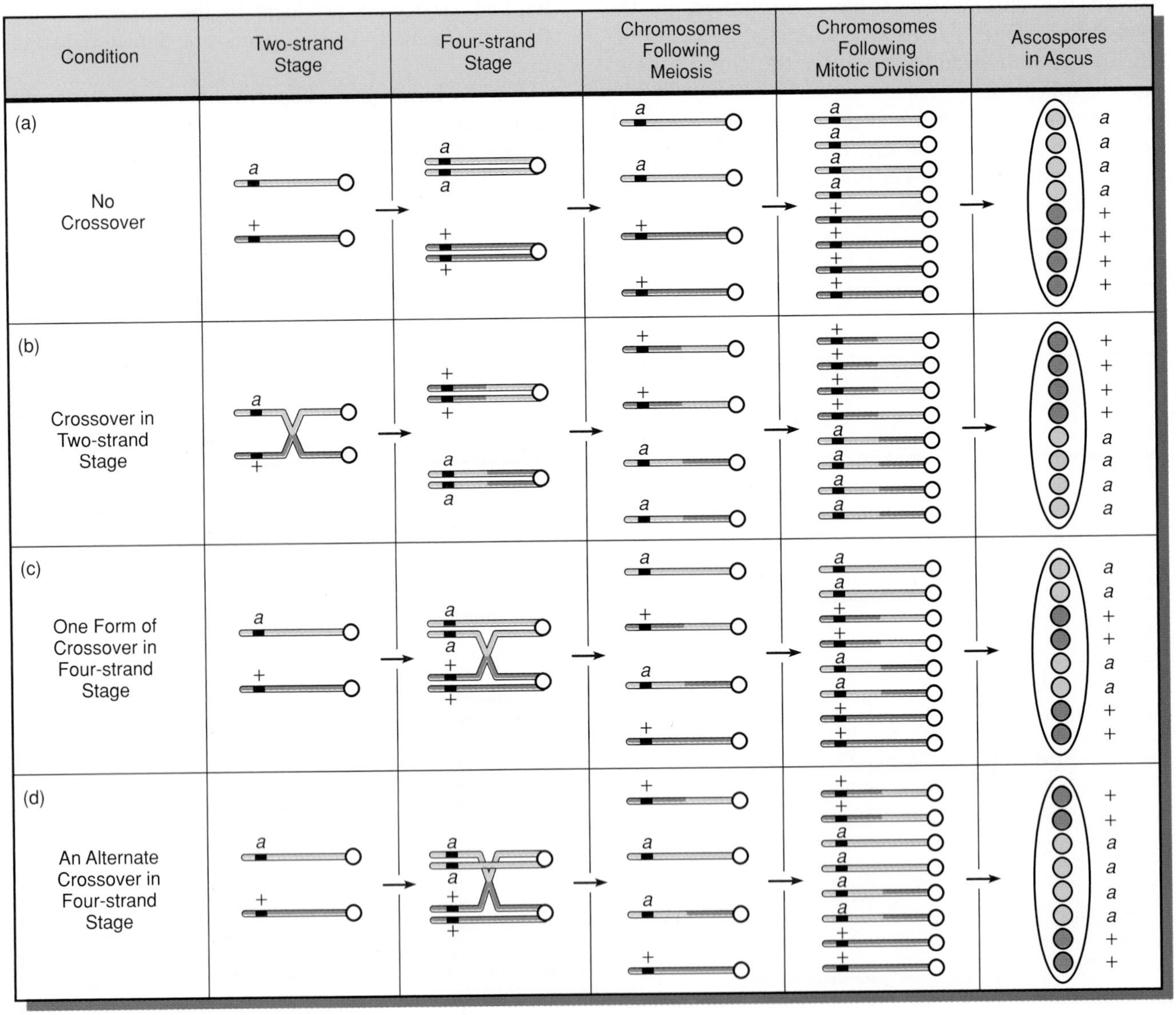

FIGURE 6.17
Four ways in which ascospore patterns can be generated in the asci of *Neurospora* as a result of genetic events. While the patterns produced in (a) and (b) cannot be distinguished from one another, these and the patterns produced in (c) and (d) were observed, leading to the conclusion that crossing over occurs in the four-strand stage.

otic organisms. The interpretation of these results is that crossing over occurs between any two nonsister chromatids of the four-strand stage. When such an event occurs, two of the four strands will be uninvolved in the exchange. Therefore, the theoretical upper limit of recombination between any two linked loci, however distantly located, is 50 percent. Crossing over also occurs between sister chromatids, but since each contains identical alleles, no new genotypic combinations are produced. Therefore, such crossing over is undetectable during genetic analysis.

Gene to Centromere Mapping

When a single gene is analyzed in *Neurospora*, as diagramed in Figure 6.17, the data can be used to calculate the map distance between the gene and the centromere. This process is sometimes referred to as **mapping the centromere**. It may be accomplished by experimentally determining the frequency of recombination using tetrad data.

If no crossover event between the gene under study and the centromere occurs, the pattern of ascospores in the ascus appears as shown in Figure 6.17(a) (*aaaa+++*). This pattern represents **first division segregation** because the two alleles are separated during the first meiotic division. However, a crossover event will alter this pattern, as shown in Figure 6.17(c) (*aa++aa++*) and 6.17(d) (*++aaaa++*). Actually, two recombinant patterns may occur, depending on the chromatid orientation during the second meiotic division: *++aa++aa* and *aa+++aa*. All four of the latter patterns reflect **second division segregation** because the two alleles are not separated until the second meiotic division. Usually, the ordered tetrad data are condensed to reflect the genotypes of the identical ascospore pairs. Thus, five combinations are possible:

First Division Segregation

aa + +

Second Division Segregation

a + a +
+ a + a
+ a a +
a + + a

In order to calculate the distance between the gene and the centromere, a large number of asci resulting from a controlled cross must be scored. Using these data, the distance is calculated:

1/2(second division segregant asci)
———————————————
total asci scored

The recombination percentage is only one-half the number of second division segregants because crossing over in each of them has occurred in only two of the four strands during meiosis.

To illustrate, assume that *a* represents albino and + represents wild type in *Neurospora*. In crosses between the two genetic types, suppose the following data were observed:

65 first-division segregants
70 second-division segregants

The distance between *a* and the centromere is thus:

$$\frac{(1/2)\ (70)}{135} = 0.259$$

or about 26 map units.

As the distance increases up to 50 units, in theory, all asci should reflect second division segregation. However, numerous factors prevent this. As in diploid organisms, accuracy is greatest when the gene and centromere are relatively close together.

Ordered vs. Unordered Tetrad Analysis

In the previous two sections we have generally assumed that the genotype of each ascospore and its position in the tetrad can be determined. To perform such an **ordered tetrad analysis,** individual asci must be dissected and each ascospore must be tracked as it germinates. This is a tedious process, but it is essential for two types of analysis:

1 To distinguish between first division segregation and second division segregation of alleles in meiosis.
2 To determine whether recombinational events are reciprocal or not.

In the first case, such information is essential to "map the centromere" as we have just discussed. Thus, ordered tetrad analysis must be performed in order to map the distance between a gene and the centromere.

In the second case, ordered tetrad analysis has revealed that recombinational events are not always reciprocal, particularly when closely linked genes are studied in *Ascomycetes.* This observation has led to the investigation of the phenomenon called **gene conversion**. Since its discussion requires a background in DNA structure and analysis, we will return to this topic in Chapter 10.

It is much less tedious to isolate individual asci, allow them to mature, and then determine the genotypes of each ascospore, but not in any particular order. This approach is referred to as **unordered** tetrad analysis. As we will see in the next section, this type of analysis can be used to determine whether two genes are linked on the same chromosome or not, and if so, to determine the map distance between them.

Linkage and Gene Mapping in Haploid Organisms

Analysis of haploid organisms can also be performed in order to distinguish between linkage and independent assortment of two genes; it further allows mapping distances to be calculated between gene loci once linkage is established. In the following discussion, we will consider tetrad analysis in the alga, *Chlamydomonas*. With the exception that the four meiotic products are not ordered and *do not* undergo a mitotic division following the completion of meiosis, the general principles discussed for *Neurospora* also apply to *Chlamydomonas*.

To compare independent assortment and linkage, we will consider two theoretical mutant alleles, *a* and *b*, representing two distinct loci in *Chlamydomonas*. Suppose that 100 tetrads derived from the cross *ab* × ++ yield the data shown in Table 6.2. As you can see, all tetrads produce one of three patterns. For example, all tetrads in category I produce two ++ cells and two *ab* cells and are designated as **parental ditypes (P)**. Category II tetrads produce two *a*+ cells and two +*b* cells and are called **nonparental ditypes (NP)**. Category III tetrads produce one cell each of the four possible genotypes and are thus termed **tetratypes (T)**.

These data support the hypothesis that the genes represented by the *a* and *b* alleles are located on separate chromosomes. In order to understand why, you must refer to Figure 6.18. In parts a (category I)

and b (category II) of this figure, the origin of parental and nonparental ditypes is demonstrated for two unlinked genes. According to the Mendelian principle of independent assortment of unlinked genes, approximately equal proportions of these tetrad types are predicted. Thus, when the parental ditypes are equal to the nonparental ditypes, then the two genes are not linked. The data in Table 6.2 confirm this prediction. Since independent assortment has occurred, it can be concluded that the two genes are located on separate chromosomes.

The origin of category III, the **tetratypes**, is diagramed in Figure 6.18(c) and (d). The genotypes of tetrads in this category can be generated in two possible ways. Both involve a crossover event between one of the genes and the centromere. In Figure 6.16(c) the exchange involves one of the two chromosomes and occurs between gene *a* and the centromere; in Figure 6.18(d), the other chromosome is involved and the exchange occurs between gene *b* and the centromere.

The production of tetratype tetrads does not alter the final ratio of the four genotypes present in all meiotic products. If the genotypes from 100 tetrads (which yield 400 cells) are computed, 100 of each genotype are found. This 1:1:1:1 ratio is predicted according to the expectation of independent assortment.

Now consider the case where the genes *a* and *b* are linked. The same categories of tetrads will be produced. However, parental and nonparental ditypes will not necessarily occur in equal proportions; nor will the four genotypic combinations be found in equal numbers if the genotypes of all meiotic products are computed. For example, the following data might be encountered:

Category I (Parental Ditype)	Category II (Nonparental Ditype)	Category III (Tetratype)
64	6	30

Since the parental and nonparental categories are not produced in equal proportion, we can predict that independent assortment is not in operation. If not, we can conclude that the two genes are linked and proceed to determine the map distance between the two genes.

In the analysis of these data, we are concerned with the determination of which tetrad types represent genetic exchanges between the two genes. The **parental ditype tetrads** arise only when no crossing over occurs between the two genes. The **nonparental ditype tetrads** arise only when a double exchange

TABLE 6.2
Tetrad analysis in *Chlamydomonas*.

Category	I	II	III
Tetrad Type	Parental (P)	Nonparental (NP)	Tetratypes (T)
Genotypes Present	+ +	a +	+ +
	+ +	a +	a +
	a b	+ b	+ b
	a b	+ b	a b
Number of Tetrads	43	43	14

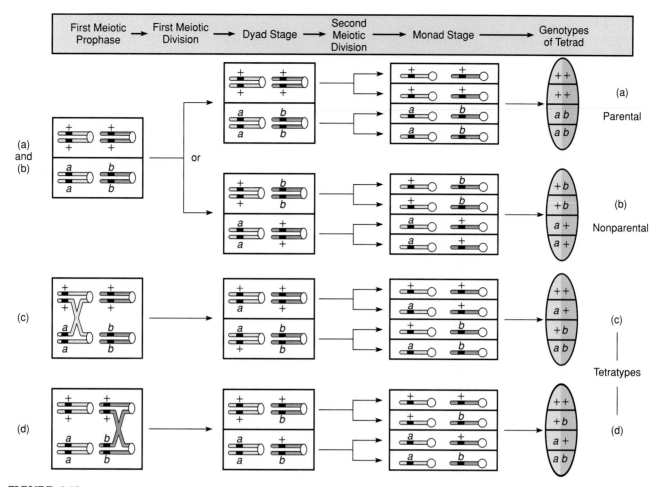

FIGURE 6.18
The origin of various genotypes found in tetrads in *Chlamydomonas* when two genes located on separate chromosomes are considered.

involving all four chromatids occurs between two genes. The **tetratype tetrads** arise when either a single crossover occurs or when an alternate type of double exchange occurs between the two genes. The various types of exchanges described here are diagramed in Figure 6.19.

When the proportion of the three tetrad types has been determined, it is possible to calculate the map distance between the two linked genes. The following formula computes the exchange frequency, which is proportional to the map distance between the two genes:

$$\text{Exchange Frequency (\%)} = \frac{\text{NP} + 1/2(\text{T})}{\begin{array}{c}\text{Total Number}\\\text{of Tetrads}\end{array}} \times 100$$

In this formula NP represents the nonparental tetrads; all meiotic products represent an exchange. The

tetratype tetrads are represented by T; one-half of the meiotic products represents exchanges. The sum of the scored tetrads that fall into these categories is then divided by the total number of tetrads examined. If this calculated number is multiplied by 100, it is converted to a percentage, which is directly equivalent to the map distance between the genes.

In our example, the calculation reveals that genes *a* and *b* are separated by 21 map units:

$$\frac{6 + 1/2(30)}{100} = \frac{6 + 15}{100} = \frac{21}{100} = 0.21 \times 100 = 21\%$$

This discussion introduces linkage analysis and chromosome mapping of two genes in haploid organisms. We can extend such investigations to the consideration of three or more genes. In these cases, the gene sequence as well as map distances can be determined.

FIGURE 6.19

The origin of various genotypes found in tetrads in *Chlamydomonas* when two genes located on the same chromosome are considered.

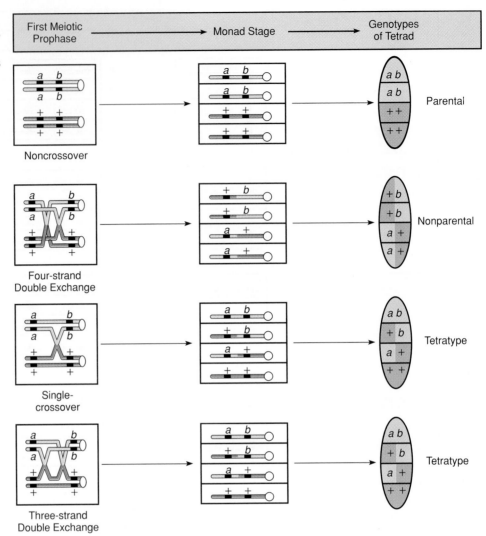

OTHER ASPECTS OF GENETIC EXCHANGE

We have thus far established that careful analysis of crossing over during gamete formation can serve as the basis for the construction of chromosome maps in both diploid and haploid organisms. We should not, however, lose sight of the real biological significance of the process, which is to generate genetic variation in gametes, and subsequently, in the offspring derived from the resultant eggs and sperm. Because of the critical role of crossing over in generating variation, the study of genetic exchange has remained an important topic for study in genetics. Many questions have been addressed. For example, does crossing over involve an actual exchange of chromosome arms? What is the mechanism by which exchange occurs? And, does exchange occur between paired sister chromatids during mitosis? We will conclude this chapter by considering observations that attempt to answer these questions.

Cytological Evidence for Crossing Over

Visual proof that genetic crossing over in higher organisms is accompanied by an actual physical exchange between homologous chromosomes has been demonstrated independently by Curt Stern in *Drosophila* and by Harriet Creighton and Barbara McClintock in *Zea mays* (corn). Since the experiments are similar, we will consider only the work with corn. In Creighton and McClintock's work, two linked genes on chromosome 9 were studied. At one locus, the alleles *colorless* (c) and *colored* (C) control endosperm coloration. At the other locus, the alleles *starchy* (Wx) and *waxy* (wx) control the carbohydrate characteristics of the endosperm.

A corn plant was obtained that was heterozygous at both loci and contained two unique cytological markers on one of the homologues. The markers consisted of a densely stained knob at one end of the chromosome and a translocated piece of another chromosome (8) at the other end. The arrangement of alleles and cytological markers in this plant is shown in Figure 6.20. Creighton and McClintock crossed this plant to one homozygous for the color alleles and heterozygous for the endosperm alleles. They obtained several different phenotypes in the offspring, one of which could arise only as a result of crossing over. The chromosomes of this plant, with the colorless waxy phenotype (Case I in Figure 6.20) were examined for the presence of the cytological markers. As expected if genetic crossing over is accompanied by a physical exchange between homologues, the translocated chromosome was still present, but the knob was not. In a second plant (Case II), the phenotype colored starchy was potentially the result of crossing over. If so, chromosomes from it should contain the dense knob but not the translocated chromosome. This was the case, and again the findings supported the conclusion that a physical exchange took place. Similar findings by Stern using *Drosophila* leave no doubt about this conclusion involving the cytological basis of crossing over.

The Mechanism of Crossing Over

It has long been of interest to determine how and when during meiosis crossing over occurs. If we accept the evidence just discussed, then crossing over must be considered to be the result of an actual physical exchange between DNA molecules of two homologous chromosomes. Of particular interest is the relationship between the chiasmata observed cytologically during prophase I of meiosis and the breakage and reunions presumed to occur during genetic crossing over. Are chiasmata the cytological manifestations of crossover events?

Two main theories have been proposed, both of which involve chiasmata, but in quite different ways. The **classical theory** holds that crossing over events are the result of physical strains imposed by chiasmata, the presence and location of which occur randomly. While there is not necessarily a one-to-one relationship—a chiasma may or may not induce the breakage and rejoining that leads to crossing over—this theory holds that chiasmata are responsible for crossover events and clearly precede them. Since we know that chiasmata are first seen rather late in meiotic prophase I, during the diplotene stage, the classical theory predicts that crossing over occurs sometime after diplonema but before the chromosomes separate at meiotic anaphase I. The order of events presumed to occur and the resultant pairing relationships created according to the classical theory are illustrated in Figure 6.21(a).

The other proposal, called the **chiasmatype theory,** was first set forth by F. A. Janssens in the early twentieth century and later modified by John Belling and C. D. Darlington around 1930. It predicts that crossing over precedes chiasma formation and occurs early in meiotic prophase I, presumably in pachytene. Chiasmata are subsequently formed at

FIGURE 6.20
The phenotypes and chromosome composition observed in Creighton and McClintock's demonstration in maize that crossing over involves a breakage and rejoining process.

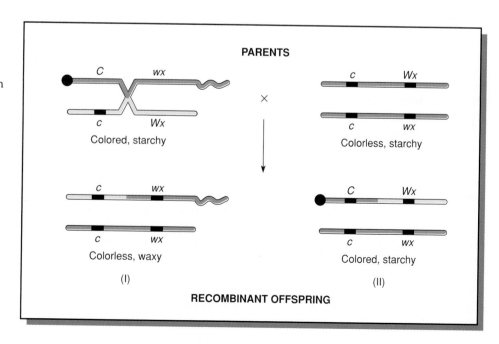

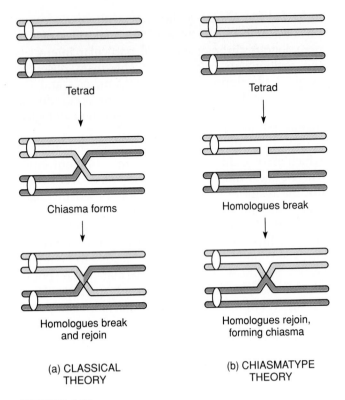

Tetrad

Chiasma forms

Homologues break
and rejoin

(a) CLASSICAL
THEORY

Tetrad

Homologues break

Homologues rejoin,
forming chiasma

(b) CHIASMATYPE
THEORY

FIGURE 6.21
Two main theories proposed to account for the
mechanism of crossing over.

points of genetic exchange. Thus, a chiasma present in diplotene is a cytological manifestation of one crossing over event. Chromosomes become wrapped around one another as a result of the previous exchange between homologues. These events are illustrated in Figure 6.21(b).

The chiasmatype theory has gained much support over the years, even though it does not account for all related observations. Several findings derived from the use of modern research techniques are pertinent to this subject. The first involves the discovery of DNA synthesis during meiosis. Yasuo Hotta and Herbert Stern have performed careful analysis using *Lilium* (lily) anthers, where the stages of meiotic prophase of microsporophytes can be easily identified and studied.

They have shown that a small but measurable amount of DNA synthesis occurs during the zygotene stage of meiotic prophase I. Recall from our earlier discussion of mitosis and meiosis in Chapter 2 that DNA replication occurs during the S phase of the cell cycle, well before mitosis and meiosis begin. Thus, this finding has generated much interest. What is the role of this late-replicating DNA? It amounts to 0.3 percent of the total nuclear DNA; it is distributed generally among all chromosomes; and it is not

strongly bonded covalently to the DNA already present. Importantly, if this DNA synthesis is inhibited, then chromosomal synapsis is also inhibited and meiosis arrests. These latter findings suggest that the newly synthesized DNA plays an important role in chromosome alignment during meiosis.

Coincidental with this DNA synthesis is the formation of the **synaptonemal complex (SC)** during the zygotene stage. The SC is found between synapsed homologues (see Figure 2.10) and is essential to chromosome alignment. Perhaps the DNA synthesis occurring during zygonema is directly related to SC formation between homologues. Interestingly, there is a possibility that the SC is also involved in crossing over. Several observations are particularly significant here. In male *Drosophila*, no crossing over occurs, and no SCs are found when spermatocytes are viewed under the electron microscope. In the silk moth, *Bombyx*, no crossing over occurs in females, and no synaptonemal complexes are observed in the oocytes. Because of the constant association between this structure and crossing over, the participation of the SC in this event must be considered seriously.

Many geneticists feel that these events, characteristic of the zygotene stage, are essential to genetic crossing over which occurs in the subsequent pachytene stage. Just how crossing over occurs, however, is still a matter of debate and speculation. Some interesting biochemical information derived from Stern's studies of the lily bears on this issue.

It is generally held that the physical exchange occurring during crossing over involves breakage and rejoining of the DNA strands between homologues. Stern and coworkers have demonstrated the presence of an enzyme system during zygonema and pachynema capable of breaking and resealing DNA strands. For example, certain **endonucleases** are capable of "nicking" one of the two strands of the DNA double helix. Other enzymes called **ligases** are capable of rejoining broken ends of DNA strands. Additionally, a small but discernible amount of DNA synthesis occurs during the pachytene stage in *Lilium* and other organisms, such as wheat. This synthesis, unlike that occurring during zygonema, is of the repair type. It does not result in a net increase in DNA and thus does not represent replication of DNA. Inhibitors of premeiotic S-phase and zygonema DNA synthesis are not effective in preventing the DNA synthesis that occurs during pachynema.

With good reason, these discoveries have generated much enthusiasm and the hope that we will soon come to understand the precise molecular mechanisms underlying crossing over. Several popular models have been put forth. Because we have yet to discuss the specific chemistry of DNA structure, replication, and synthesis, we will postpone their presentation. Those already versed in DNA chemistry may wish to jump ahead to Figure 10.16 and the accompanying discussion. These models provide a suitable explanation for how crossing over might occur.

Mitotic Recombination

In 1936, Curt Stern demonstrated that exchanges similar to crossing over occur during mitosis in *Drosophila*. This finding was considered unusual because homologues do not normally pair up during mitosis in most organisms. However, such synapsis is apparently the rule in *Drosophila*. Since Stern's discovery, genetic exchange during mitosis has been shown to be a general event in certain fungi as well.

Stern observed small patches of mutant tissue in females heterozygous for the sex-linked recessive mutations *yellow* body and *singed* bristles. Under normal circumstances, a heterozygous female is completely wild type (gray-bodied with straight, long bristles). He explained these observations by postulating that, during mitosis in certain cells during development, homologous exchanges could occur between the loci for *yellow* and *singed* or between *singed* and the *centromere,* or that a double exchange could occur. These three possibilities are diagrammed in Figure 6.22. After these three types of exchanges occur, tissues derived from the progeny cells are produced with *yellow* patches, adjacent *yellow* and *singed* patches **(twin spots),** and *singed* patches, respectively. The last type of tissue, which represents the double exchange, was found in the lowest frequency, similar to the less frequent double-crossover frequency seen in meiosis. The frequency of twin spots argues strongly against the appearance of *yellow* or *singed* tissue being due to two spontaneous but independent mutational events. If the twin spots arose in such a way, their frequency would occur in only one in many million flies.

Table 6.3 on page 175 compares the relative frequency of occurrence for each of the three spotting types. These frequencies are correlated with the known genetic distances between the *yellow* and *singed* loci and the centromere, which effectively serves as an additional genetic marker. The data parallel the predicted frequencies if the exchanges occur according to the rules we established earlier for meiotic crossing over. This is additional evidence that the occurrence of mutant tissue spots is due to mitotic exchange in somatic cells.

In 1958 George Pontecorvo and others described a similar phenomenon in the fungus *Aspergillus*. Al-though the vegetative stage is normally haploid, some cells and their nuclei fuse, producing diploid cells that divide mitotically. Occasionally, crossing over occurs between linked genes during mitosis in this diploid stage so that resulting cells are recombinant. Pontecorvo referred to these events that produce genetic variability as the **parasexual cycle**. On the basis of such exchanges, genes can be mapped by estimating the frequency of recombinant classes.

As a rule, if mitotic recombination occurs at all in an organism, it does so at a much lower frequency than meiotic crossing over. While it is assumed that there is always at least one exchange per meiotic tetrad, mitotic exchange occurs in 1 percent or less of mitotic divisions in organisms that demonstrate it. Some researchers feel that the low frequency of exchange may be explained by a pairing of only some portions of homologous chromosomes.

Sister Chromatid Exchanges

While homologous chromosomes do not usually pair up or synapse in somatic cells, each individual chromosome in prophase and metaphase of mitosis consists of two identical sister chromatids, joined at a common centromere. Surprisingly, several experimental approaches have demonstrated that reciprocal exchanges similar to crossing over occur between sister chromatids. While these **sister chromatid exchanges** (SCE) do not produce new allelic combinations, evidence is accumulating which attaches significance to these events.

Identification and study of sister chromatid exchanges (SCEs) are facilitated by several modern staining techniques. In one approach, if cells are allowed to replicate for several generations in the presence of the thymidine analogue bromodeoxyuridine (BUdR), and the cells are then shifted to medium lacking the analogue, chromatids with or without BUdR are then distinguishable. Chromatids with both strands containing BUdR stain less brightly than chromatids with only one strand of the double helix containing the analogue. In Figure 6.23 numerous instances of SCE events may be detected.

While the significance of sister chromatid exchanges is still uncertain, several observations have led to great interest in this phenomenon. It is known, for example, that agents that induce chromosome damage (such as viruses, X-rays, ultraviolet light, and certain chemical mutagens) also increase the frequency of sister chromatid exchanges. This frequency of SCEs is also elevated in **Bloom syndrome,** an autosomal recessive disorder in humans. This rare genetic disease is characterized by retardation of growth, a great sensitivity of the facial skin to the

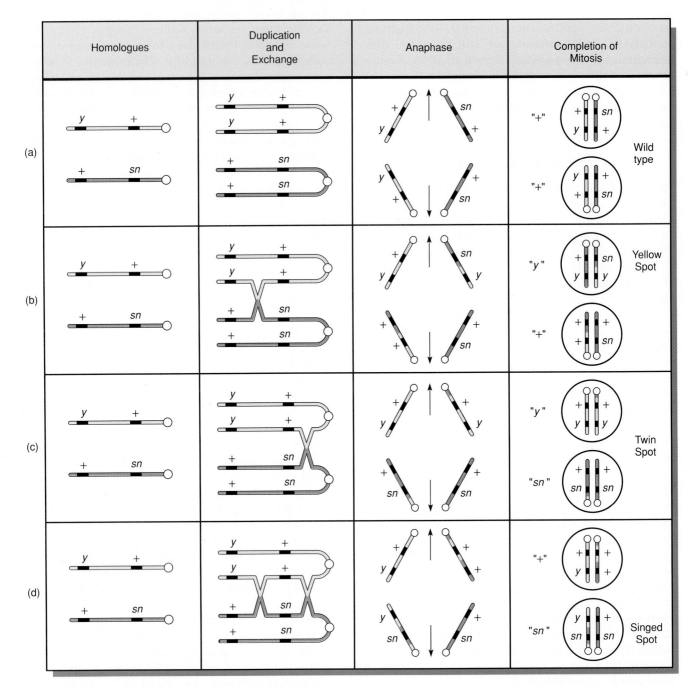

FIGURE 6.22
The production of mutant tissue in a female heterozygous for the recessive *yellow* and *singed* alleles as a result of mitotic recombination in *Drosophila*.

sun, abnormal immunological function, and a predisposition to cancer. The chromosomes from cultured leucocytes, bone marrow cells, and fibroblasts derived from homozygotes (*bl/bl*) are very fragile and unstable when compared to those derived from unaffected (+/+) and heterozygous (*bl/*+) individuals. Increased breaks and rearrangements between nonhomologous chromosomes are observed in addition to excessive amounts of sister chromatid exchanges.

The mechanisms of exchange between nonhomologous chromosomes and between sister chromatids may prove to be similar because the frequency of both events increases substantially in individuals with genetic disorders. These findings suggest that further study of sister chromatid exchange may contribute to

TABLE 6.3
Comparison of the various parameters related to mitotic recombination studies in *Drosophila.*

Mutant Tissue Type	Region of Exchange	Relative Frequency of Exchange	Type of Exchange	Relative Distance
Yellow	*y-sn*	High	Single	21 map units
Yellow-singed (twin spot)	*sn*-centromere	Highest	Single	49 map units
Singed	*y-sn* and *sn*-centromere	Lowest	Double	——

an increased understanding of recombination mechanisms and the relative stability of normal and genetically abnormal chromosomes. We shall encounter still another demonstration of SCEs in Chapter 10 when we consider replication of DNA (see Figure 10.5).

DID MENDEL ENCOUNTER LINKAGE?

We conclude this chapter by examining a modern-day interpretation of the experiments that serve as the cornerstone of transmission genetics—the crosses with garden peas performed by Mendel.

It has often been said that Mendel had extremely good fortune in his classical experiments with the garden pea. In none of his crosses did he encounter apparent linkage relationships between any of the seven mutant characters. Had Mendel obtained highly variable data characteristic of linkage and crossing over, these unorthodox ratios might have hindered his successful analysis and interpretation.

The simplest explanation for the absence of linkage is that each of the seven genes was located on a different linkage group or chromosome. Since *Pisum sativum* has a haploid number of 7, this speculation has been widely publicized. An article by Stig Blixt, reprinted on page 176 in its entirety from *Nature,*

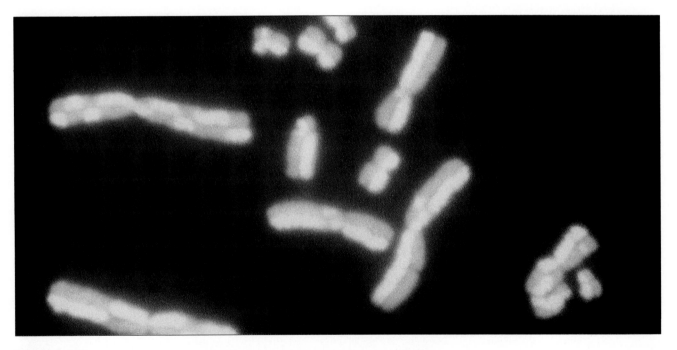

FIGURE 6.23
Demonstration of sister chromatid exchanges (SCEs) in mitotic chromosomes. Sometimes called harlequin chromosomes because of their patchlike appearance, chromatids containing the thymidine analogue BUdR in both DNA strands fluoresce *less* brightly than those with the analogue in only one strand. These chromosomes were stained with 33258-Hoechst reagent and viewed under fluorescence microscopy.

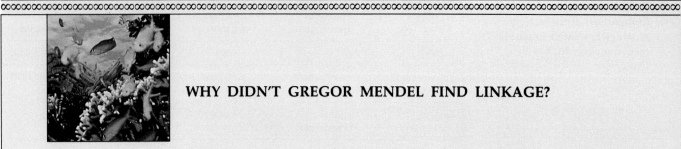

WHY DIDN'T GREGOR MENDEL FIND LINKAGE?

It is quite often said that Mendel was very fortunate not to run into the complication of linkage during his experiments. He used seven genes and the pea has only seven chromosomes. Some have said that had he taken just one more, he would have had problems. This, however, is a gross oversimplification. The actual situation, most probably, is shown in Table 1. This shows that Mendel worked with three genes in chromosome 4, two genes in chromosome 1, and one gene in each of chromosome[s] 5 and 7. It seems at first glance that, out of the 21 dihybrid combinations Mendel theoretically could have studied, no less than four (that is, *a-i*, *v-fa*, *v-le*, *fa-le*) ought to have resulted in linkages. As found, however, in hundreds of crosses and shown by the genetic map of the pea[1], *a* and *i* in chromosome 1 are so distantly located on the chromosome that no linkage is normally detected. The same is true for *v* and *le* on the one hand, and *fa* on the other, in chromosome 4. This leaves *v-le*, which ought to have shown linkage.

TABLE 1
Relationship between modern genetic terminology and character pairs used by Mendel.

Character Pair Used by Mendel	Alleles in Modern Terminology	Located in Chromosome
Seed colour, yellow-green	*I-i*	1
Seed coat and flowers, coloured-white	*A-a*	1
Mature pods, smooth expanded-wrinkled indented	*V-v*	4
Inflorescences, from leaf axils—umbellate in top of plant	*Fa-fa*	4
Plant height, 1 m-around 0.5 m	*Le-le*	4
Unripe pods, green-yellow	*Gp-gp*	5
Mature seeds, smooth-wrinkled	*R-r*	7

Mendel, however, seems not to have published this particular combination and thus, presumably, never made the appropriate cross to obtain both genes segregating simultaneously. It is therefore not so astonishing that Mendel did not run into the complication of linkage, although he did not avoid it by choosing one gene from each chromosome.

Weibullsholm Plant Breeding Institute, Landskrona, Sweden and Centro Energia Nuclear na Agricultura, Piracicaba, SP, Brazil

STIG BLIXT

Received March 5; accepted June 4, 1975.
[1]Blixt, S. 1974. In *Handbook of genetics*, ed. R. C. King. New York: Plenum Press.
SOURCE: Reprinted by permission from *Nature*, Vol. 256, p. 206. Copyright © 1975 Macmillan Journals Limited.

demonstrates the inadequacy of this hypothesis and explains why Mendel did not encounter ratios characteristic of linkage and crossing over. At this point in your studies of genetics, you have acquired a sophistication quite sufficient to both comprehend and appreciate this information.

CHAPTER SUMMARY

1 Genes located on the same chromosome are said to be linked. Alleles located on the same homologue, therefore, can be transmitted together during gamete formation. However, the mechanism of crossing over between homologues during meiosis results in the reshuffling of alleles, thereby contributing to genetic variability within gametes.

2 Early in this century, geneticists realized that crossing over could provide an experimental basis for mapping the location of linked genes relative to one another. As a result, chromosomal maps have been produced for organisms such as corn, the mouse, and *Drosophila*. Somatic cell hybridiza-

tion techniques have made possible mapping of human chromosomes.

3 Mapping studies are also possible in haploid organisms such as *Chlamydomonas* and *Neurospora*. Studies using *Neurospora* indicate that crossing over occurs during the four-strand stage of meiosis.

4 Cytological investigations of both corn and *Drosophila* have revealed that crossing over requires actual breaking and rejoining of nonsister chromatids. The exact relationships between chiasmata and crossing over, and between crossing over and the synaptonemal complex, are not yet resolved.

5 An exchange of genetic material between sister chromatids may occur during mitosis as well. These events are referred to as sister chromatid exchanges (SCEs). An elevated frequency of such events is seen in the human disorder Bloom syndrome.

6 Evidence now suggests that several of Mendel's seven genetic characters were, in fact, linked. However, those that were linked were sufficiently far apart to prevent them from being distinguished from unlinked genes that undergo independent assortment.

Insights and Solutions

1 In rabbits, *black* (B) is dominant to *brown* (b), while *full color* (C) is dominant to *chinchilla* (C^ch). The genes controlling these traits are linked. Rabbits that are heterozygous for both traits and express *black, full color*, were crossed to rabbits that are *brown, chinchilla*, with the following results:

> 31 *brown, chinchilla*
> 35 *black, full*
> 16 *brown, full*
> 19 *black, chinchilla*

Determine the arrangement of alleles in the heterozygous parents and the map distance between the two genes.

ANSWER: This is a two-point map problem, where the two reciprocal phenotypes most prevalent are the noncrossovers. The less frequent reciprocal phenotypes arise from a single crossover. The arrangement of alleles is derived from the noncrossover phenotypes because they enter gametes intact. The cross is shown on the following page.

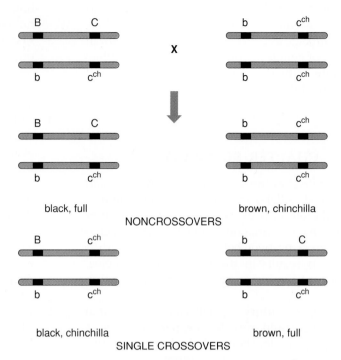

B C   b c^ch

X

b c^ch   b c^ch

B C   b c^ch

b c^ch   b c^ch

black, full   brown, chinchilla

NONCROSSOVERS

B c^ch   b C

b c^ch   b c^ch

black, chinchilla   brown, full

SINGLE CROSSOVERS

The single crossovers give rise to 35/100 offspring (35%). Therefore, the distance between the two genes is 35 map units (mu).

2 In *Drosophila, Lyra* (*Ly*) and *Stubble* (*Sb*) are dominant mutations located at locus 40 and 58, respectively, on chromosome 3. A recessive mutation with bright red eyes was discovered and shown also to be on chromosome 3. A map was obtained by crossing a female who was heterozygous for all three mutations to a male homozygous for the bright red mutation (which we will temporarily call *br*). The following data were obtained:

	Phenotype	Number
(1)	*Ly Sb br*	404
(2)	+ + +	422
(3)	*Ly* + +	18
(4)	+ *Sb br*	16
(5)	*Ly* + *br*	75
(6)	+ *Sb* +	59
(7)	*Ly Sb* +	4
(8)	+ + *br*	2
		1000

Determine the location of the bright red mutation on chromosome 3. By referring to Figure 6.13, predict what mutation has been discovered. How could you be sure?

ANSWER: First, determine the **arrangement** of the alleles on the homologues of the heterozygous crossover parent (the female in this case). This is done by locating the most frequent reciprocal phenotypes, which arise from the noncrossover gametes. These are phenotypes (1) and (2). Each one represents the arrangement of alleles on one of the homologues. Therefore, the arrangement is:

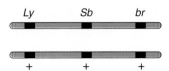

Second, determine the correct **sequence** of the three loci along the chromosome. This is done by determining which sequence will yield the observed double-crossover phenotypes, which are the least frequent reciprocal phenotypes (7 and 8).

If the sequence is correct as written, then a double crossover

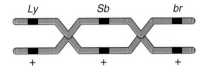

will yield $Ly + br$ and $+ Sb +$ as phenotypes. Inspection shows that these categories (5 and 6) are actually single crossovers, not double crossovers. Therefore, the sequence, as written, is incorrect. There are only two other possible sequences. The br gene is either to the left of Ly, or it is between Ly and Sb:

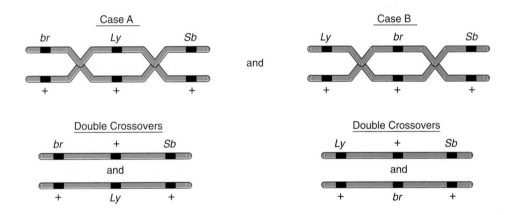

Comparison with the actual data shows that Case B is correct. The double-crossover gametes yield flies that express Ly and Sb, but not br, or express br, but not Ly and Sb. Therefore, the correct arrangement *and* sequence is:

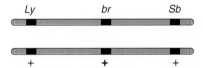

Once this has been determined, it is possible to determine the location of br relative to Ly and Sb. A single crossover between Ly and br

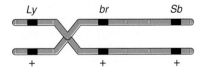

yields flies that are *Ly* + + and + *br Sb* (categories 3 and 4). Therefore, the distance between the *Ly* and *br* loci is equal to:

$$\frac{18 + 16 + 4 + 2}{1000} = \frac{40}{1000} = 0.04 = 4 \text{ map units}$$

Remember that we must add in the double crossovers, since they represent two single crossovers occurring simultaneously. Because we need to know the frequency of all crossovers between *Ly* and *br*, they must be included.

Similarly, the distance between the *br* and *Sb* loci is derived mainly from single crossovers between them:

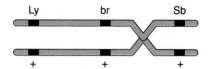

This event yields *Ly br* + and + + *Sb* phenotypes (categories 5 and 6). Therefore, the distance equals:

$$\frac{75 + 59 + 4 + 2}{1000} = \frac{140}{1000} = 0.14 = 14 \text{ map units}$$

The final map shows that *br* is located at locus 44, since *Lyra* and *Stubble* are known:

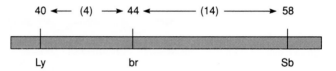

Inspection of Figure 6.13 reveals that the mutation *scarlet*, which has bright red eyes, is known to exist at locus 44, so it is reasonable to hypothesize that the bright red eye mutation is an allele of *scarlet*. To test this hypothesis, we could cross females of our bright red mutant with known *scarlet* males. If they are alleles, all progeny will reveal a bright mutant eye. If not, all progeny will show normal brick red wild type eyes, since our mutant and scarlet are actually at slightly different loci (very near 44.0). In such a case, all progeny will be heterozygous and will not show a bright red eye. This cross represents an **allelism test**.

∞∞∞∞∞∞∞∞∞∞∞∞∞∞∞∞∞∞∞∞∞∞

PROBLEMS AND DISCUSSION QUESTIONS

1 What is the significance of genetic exchange and recombination to the process of evolution?
2 Describe the cytological events that suggest that crossing over occurs during the first meiotic prophase.
3 Why does more crossing over occur between two distantly linked genes than between two genes that are very close together on the same chromosome?
4 Why is a 50 percent recovery of single-crossover products the upper limit, even when crossing over *always* occurs between two linked genes?
5 Why are double-crossover events expected in lower frequency than single-crossover events?
6 What is the proposed basis for positive interference?
7 What essential criteria must be met in order to execute a successful mapping cross?

8 The genes *dumpy wing* (*dp*), *clot eye* (*cl*), and *apterous wing* (*ap*) are linked on chromosome 2 of *Drosophila*. In a series of two-point mapping crosses, the following genetic distances were determined:

$$
\begin{aligned}
dp\text{-}ap &\quad 42 \\
dp\text{-}cl &\quad 3 \\
ap\text{-}cl &\quad 39
\end{aligned}
$$

What is the sequence of the three genes?

9 Consider two hypothetical recessive autosomal genes *a* and *b*. Where a heterozygote is test-crossed to a double homozygous mutant, predict the phenotypic ratios under the following conditions:
   (a) *a* and *b* are located on separate autosomes.
   (b) *a* and *b* are linked on the same autosome but are so far apart that a crossover always occurs between them.
   (c) *a* and *b* are linked on the same autosome but are so close together that a crossover almost never occurs.
   (d) *a* and *b* are linked on the same autosome about 10 map units apart.

10 In corn, colored aleurone (in the kernels) is due to the dominant allele *R*. The recessive allele *r*, when homozygous, produces colorless aleurone. The plant color (not the kernel color) is controlled by the gene pair *Y* and *y*. The dominant *Y* gene results in green color, while the homozygous presence of the recessive *y* gene causes the plant to appear yellow. In a test cross between a plant of unknown genotype and phenotype and a plant that is homozygous recessive for both traits, the following progeny were obtained:

Colored green	88
Colored yellow	12
Colorless green	8
Colorless yellow	92

Explain how these results were obtained by determining the exact genotype and phenotype of the unknown plant, including the precise association of the two genes on the homologues.

11 In the cross shown below involving two linked genes, *ebony* (*e*) and *claret* (*ca*), in *Drosophila*, where crossing over does not occur in males,

$$
\frac{e \quad ca^{+}}{e^{+} \quad ca} \overset{\female}{} \times \frac{e \quad ca^{+}}{e^{+} \quad ca} \overset{\male}{}
$$

offspring were produced in a 2 +: 1 *ca*: 1 *e* phenotypic ratio. These genes are 30 units apart on chromosome 3. What contribution did crossing over in the female make to these phenotypes?

12 With two pairs of genes involved (*P/p* and *Z/z*), a test cross (*ppzz*) with an organism of unknown genotype indicated that the gametes produced were in the following proportions:

PZ, 42.4%; *Pz*, 6.9%; *pZ*, 7.1%; and *pz*, 43.6%

Draw all possible conclusions from these data.

13 Two different female *Drosophila* were isolated, each heterozygous for the autosomally linked genes *b* (*black body*), *d* (*dachs tarsus*), and *c* (*curved wings*). These genes are in the order *d—b—c*, with *b* being closer to *d* than to *c*. Shown below is the genotypic arrangement for each female along with the various gametes formed by both. Identify which categories are noncrossovers (NCO), single crossovers (SCO), and double crossovers (DCO) in each case. Then, indicate the relative frequency in which each will be produced.

	Female A						Female B		

$$\frac{d \quad b \quad +}{+ \quad + \quad c}$$ $$\frac{d \quad + \quad +}{+ \quad b \quad c}$$

⇩ ------Gametes------ ⇩

(1) *d* *b* *c*	(5) *d* + +	(1) *d* *b* +	(5) *d* *b* *c*
(2) + + +	(6) + *b* *c*	(2) + + *c*	(6) + + +
(3) + + *c*	(7) *d* + *c*	(3) *d* + *c*	(7) *d* + +
(4) *d* *b* +	(8) + *b* +	(4) + *b* +	(8) + *b* *c*

14 In *Drosophila*, a cross was made between females expressing the three sex-linked recessive traits, *scute* (*sc*) *bristles*, *sable body* (*s*), and *vermilion eyes* (*v*), and wild-type males. In the F_1, all females were wild type, while all males expressed all three mutant traits. The cross was carried to the F_2 generation and 1000 offspring were counted with the results shown below. No determination of sex was made in the F_2 data.

Phenotype			Offspring
sc	*s*	*v*	314
+	+	+	280
+	*s*	*v*	150
sc	+	+	156
sc	+	*v*	46
+	*s*	+	30
sc	*s*	+	10
+	+	*v*	14

(a) Determine the genotypes of the P_1 and F_1 parents, using proper nomenclature.
(b) Determine the sequence of the three genes and the map distance between them.
(c) Are there more or fewer double crossovers than expected? Calculate the coefficient of coincidence. Does this represent positive or negative interference?

15 Another cross in *Drosophila* involved the recessive, sex-linked genes *yellow* (*y*), *white* (*w*), and *cut* (*ct*). A female that was yellow-bodied and white-eyed with normal wings was crossed to a male whose eyes and body were normal but whose wings were cut. The F_1 females were wild type for all three traits, while the F_1 males expressed the yellow-body, white-eye traits. The cross was carried to an F_2, and only male offspring were tallied. On the basis of the data shown below, a genetic map was constructed.

Phenotype			Male Offspring
y	+	*ct*	9
+	*w*	+	6
y	*w*	*ct*	90
+	+	+	95
+	+	*ct*	424
y	*w*	+	376
y	+	+	0
+	*w*	*ct*	0

(a) Diagram the genotypes of the F_1 parents.
(b) Construct a map, assuming that *white* is at locus 1.5 on the X chromosome.
(c) Were any double-crossover offspring expected?
(d) Could the F_2 female offspring be used to construct the map? Why or why not?

16 In *Drosophila*, *Dichaete* (*D*) is a chromosome 3 mutation with a dominant effect on wing shape. It is lethal when homozygous. The genes *ebony* (*e*) and *pink* (*p*) are chromosome 3 recessive mutations affecting the body and eye color, respectively.

Flies from a *Dichaete* stock were crossed to homozygous *ebony, pink* flies, and the F_1 progeny, with a *Dichaete* phenotype, were back-crossed to the *ebony, pink* homozygotes. The results of this back cross were

Phenotype	Number
Dichaete	401
ebony, pink	389
Dichaete, ebony	84
pink	96
Dichaete, pink	2
ebony	3
Dichaete, ebony, pink	12
wild type	13

(a) Diagram this cross, showing the genotypes of the parents and offspring of both crosses.

(b) What is the sequence and interlocus distance between these three genes?

17 *Drosophila* females homozygous for the third chromosomal genes *pink* and *ebony* (the same genes from the previous problem) were crossed with males homozygous for the second chromosomal gene *dumpy*. Because these genes are recessive, all offspring were wild type (normal). F_1 females were back-crossed to triply recessive males. If we assume that the two linked genes, *pink* and *ebony*, are 20 map units apart, predict the results of this cross. If the reciprocal cross were made (F_1 males—where no crossing over occurs—with triply recessive females), how would the results vary, if at all?

18 In *Drosophila*, two mutations, *Stubble* (*Sb*) and *curled* (*cu*), are linked on chromosome 3. *Stubble* is a dominant gene that is lethal in a homozygous state, and *curled* is a recessive gene. If a female of the genotype

$$\frac{Sb\ cu}{+\ +}$$

is to be mated to detect recombinants among her offspring, what male genotype would you choose as a mate?

19 In *Drosophila*, a heterozygous female for the sex-linked recessive traits *a*, *b*, and *c* was crossed to a male that was phenotypically *a b c*. The offspring occurred in the following phenotypic ratios:

+	b	c	460
a	+	+	450
a	b	c	32
+	+	+	38
a	+	c	11
+	b	+	9

No other phenotypes were observed.

(a) What is the genotypic arrangement of the alleles of these genes on the X chromosomes of the female?

(b) Determine the correct sequence and construct a map of these genes on the X chromosome.

(c) What progeny phenotypes are missing? Why?

20 *Drosophila melanogaster* has one pair of sex chromosomes (XX or XY) and three autosomes, referred to as chromosomes 2, 3, and 4. A male fly with very short legs was discovered by a genetics student. Using this male, the student was able to establish a pure breeding stock of this mutant and found that it was recessive. This mutant was then incorporated into a stock containing the recessive gene *black* (body color located on chromosome 2) and the recessive gene *pink* (eye color located on chromosome 3). A female from the homozygous *black, pink, short* stock was then mated to a wild-type male. The F_1 males of this cross were all wild type and were then back-crossed to the homozygous *b p sh* females. The F_2 results appeared as shown in the following table. No other phenotypes were observed.

	Wild	Pink*	Black, short	Black, pink, short
Females	63	58	55	69
Males	59	65	51	60

*Pink indicates that the other two traits are wild type, and so on.

(a) Based on these results, the student was able to assign *short* to a linkage group (a chromosome). Which one was it? Include a step-by-step reasoning.

(b) The experiment was subsequently repeated making the reciprocal cross, F_1 females back-crossed to homozygous $b\ p\ sh$ males. It was observed that 85 percent of the offspring fell into the above classes, but that 15 percent of the offspring were equally divided among $b + p$, $b + +$, $+ sh\ p$, and $+ sh +$ phenotypic males and females. How can these results be explained and what information can be derived from the data?

21 In a cross in *Neurospora* involving two alleles B and b, the following tetrad patterns were observed. Calculate the distance between the locus and the centromere.

Tetrad Pattern	Number
BBbb	36
bbBB	44
BbBb	4
bBbB	6
BbbB	3
bBBb	7

22 In *Neurospora*, the cross $a+ \times +b$ yielded only two types of ordered tetrads in approximately equal numbers:

	Spore Pair			
	1-2	3-4	5-6	7-8
Tetrad Type 1	a+	a+	+b	+b
Tetrad Type 2	+ +	+ +	ab	ab

What can be concluded?

23 Below are two sets of data derived from crosses in *Chlamydomonas*, involving three genes represented by the mutant alleles *a*, *b*, and *c*. Determine as much as you can concerning genetic arrangement of these three genes relative to one another. Describe the expected results of Cross 3.

	Genes	Tetrads		
Cross	Involved	P	NP	T
1	a & b	36	36	28
2	b & c	79	3	18
3	a & c			

24 Describe the rationale of the experiment which demonstrates that crossing over occurs in the four-strand stage in *Neurospora*.

25 In *Chlamydomonas*, a cross $ab \times ++$ yielded the following unordered tetrad data where *a* and *b* are linked.

(1)	+ + + + a b a b	38	(4)	a b a + + b + +	17
(2)	+ + a b + + a b	5	(5)	a b + + + b a +	2
(3)	a + a + + b + b	6	(6)	a b + b a + + +	3

(a) Identify the categories representing parental ditypes (P), nonparental ditypes (NP), and tetratypes (T).

(b) Explain the origin of category (2).

(c) Determine the map distance between a and b.

26 Why were there more "twin spots" observed by Stern than *singed* spots in his study of somatic crossing over? If he had been studying *tan* body color (locus 27.5) and *forked* bristles (locus 56.7) on the X chromosome of heterozygous females, what relative frequencies of tan spots, forked spots, and "twin spots" would you predict might occur?

27 Are mitotic recombinations and sister chromatid exchanges effective in producing genetic variability in an individual? In the offspring of individuals?

28 What possible conclusions can be drawn from the observations that no synaptonemal complexes are observed in male *Drosophila* and female *Bombyx* and that no crossing over occurs in these respective organisms?

29 A female of genotype

$$\frac{a \quad b \quad c}{+ \quad + \quad +}$$

produces 100 meiotic tetrads. Of these, 68 show no crossover events. Of the remaining 32, 20 show a crossover between a and b, 10 show a crossover between b and c, and 2 show a double-crossover between a and b and between b and c. Of the 400 gametes produced, how many of each of the 8 different genotypes will be produced? Assuming the order a–b–c and the allele arrangement shown above, what is the map distance between these loci?

30 The following results are ordered tetrad pairs from a cross between strain cd and strain c^+d^+. They are summarized by tetrad classes.

(1) c +	(2) c +	(3) c d	(4) + d	(5) c +	(6) c d	(7) c +
c +	c d	c d	c +	+ +	+ +	+ d
+ d	+ +	+ +	c +	c d	c d	c d
+ d	+ d	+ +	+ d	+ d	+ +	+ +
1	17	41	1	5	3	2

(a) Name the ascus type of each class from 1 to 7 (P, NP or T).

(b) The above data support the conclusion that the c and d loci are linked. State the evidence in support of this conclusion.

(c) Calculate the gene-centromere distance for each locus.

(d) Calculate the distance between the two linked loci.

(e) Draw a linkage map including the centromere and explain the discrepancy between the distances determined by the two different methods in parts c and d.

(f) Describe the arrangement of crossovers needed to produce the ascus class 6 above.

SELECTED READINGS

ALLEN, G. E. 1978. *Thomas Hunt Morgan: The man and his science.* Princeton, NJ: Princeton University Press.

BAKER, W. K. 1965. *Genetic analysis.* Boston: Houghton Mifflin.

BARRATT, R. W., et al. 1954. Map construction in *Neurospora crassa. Adv. in Genet.* 6: 1–93.

BLIXT, S. 1975. Why didn't Gregor Mendel find linkage? *Nature* 256: 206.

CATCHESIDE, D. G. 1977. *The genetics of recombination.* Baltimore: University Park Press.

CHAGANTI, R., SCHONBERG, S., and GERMAN, J. 1974. A manyfold increase in sister chromatid exchange in Bloom's syndrome lymphocytes. *Proc. Natl. Acad. Sci.* 71: 4508–12.

CREIGHTON, H. S., and McCLINTOCK, B. 1931. A correlation of cytological and genetical crossing over in *Zea mays. Proc. Natl. Acad. Sci.* 17: 492–97.

DOUGLAS, L., and NOVITSKI, E. 1977. What chance did Mendel's experiments give him of noticing linkage? *Heredity* 38: 253–57.

DRESSLER, D., and POTTER, H. 1982. Molecular mechanisms in genetic recombination. *Ann. Rev. Biochem.* 51: 727–61.

EPHRUSSI, B., and WEISS, M. C. 1969. Hybrid somatic cells. *Scient. Amer.* (April) 220: 26–35.

GARCIA-BELLIDO, A. 1972. Some parameters of mitotic recombination in *Drosophila melanogaster. Molec. Genet.* 115: 54–72.

GRELL, R. F., ed. 1974. *Mechanisms in recombination.* New York: Plenum Press.

HOTTA, Y., TABATA, S., and STERN, H. 1984. Replication and nicking of zygotene DNA sequences: Control by a meiosis-specific protein. *Chromosoma* 90: 243–53.

KING, R. C. 1970. The meiotic behavior of the *Drosophila* oocyte. *Int. Rev. Cytol.* 28: 125–68.

LATT, S. A. 1981. Sister chromatid exchange formation. *Ann. Rev. Genet.* 15: 11–56.

LINDSLEY, D. L., and GRELL, E. H. 1972. *Genetic variations of Drosophila melanogaster.* Washington, DC: Carnegie Institute of Washington.

MOENS, P. B. 1977. The onset of meiosis. In *Cell biology—A comprehensive treatise,* vol. 1, *Genetic mechanisms of cells,* ed. L. Goldstein and D. M. Prescott, pp. 93–109. Orlando: Academic Press.

MORGAN, T. H. 1911. An attempt to analyze the constitution of the chromosomes on the basis of sex-linked inheritance in *Drosophila. J. Exp. Zool.* 11: 365–414.

NEUFFER, M. G., JONES, L., and ZOBER, M. 1968. *The mutants of maize.* Madison, Wis.: Crop Science Society of America.

PERKINS, D. 1962. Crossing-over and interference in a multiply marked chromosome arm of *Neurospora. Genetics* 47: 1253–74.

PONTECORVO, G. 1958. *Trends in genetic analysis.* New York: Columbia University Press.

RUDDLE, F. H., and KUCHERLAPATI, R. S. 1974. Hybrid cells and human genes. *Scient. Amer.* (July) 231: 36–49.

STAHL, F. W. 1979. *Genetic recombination.* New York: W. H. Freeman.

STERN, C. 1936. Somatic crossing over and segregation in *Drosophila melanogaster. Genetics* 21: 625–31.

STERN, H., and HOTTA, Y. 1973. Biochemical controls in meiosis. *Ann. Rev. Genet.* 7: 37–66.

———. 1974. DNA metabolism during pachytene in relation to crossing over. *Genetics* 78: 227–35.

STURTEVANT, A. H. 1913. The linear arrangement of six sex-linked factors in *Drosophila,* as shown by their mode of association. *J. Exp. Zool.* 14: 43–59.

———. 1965. *A history of genetics.* New York: Harper & Row.

TAYLOR, J. H., ed. 1965. *Selected papers on molecular genetics.* Orlando: Academic Press.

VOELLER, B. R., ed. 1968. *The chromosome theory of inheritance: Classical papers in development and heredity.* New York: Appleton-Century-Crofts.

von WETTSTEIN, D., RASMUSSEN, S. W., and HOLM, P. B. 1984. The synaptonemal complex in genetic segregation. *Ann. Rev. Genet.* 18: 331–414.

WOLFF, S., ed. 1982. *Sister chromatid exchange.* New York: Wiley-Interscience.

7

Variations in Chromosome Number and Arrangement

CHAPTER CONCEPTS

Genetic information contained within all chromosomes of a diploid organism is delicately balanced in both content and location within the genome. A change in chromosome number or in the arrangement of a chromosome region often results in phenotypic variation or disruption of development of an organism. Because the chromosome is the unit of transmission in meiosis, such variations are passed to offspring in a predictable manner, resulting in many interesting genetic situations.

In this, the concluding chapter in Part 1, we shall examine quite a different way in which phenotypic variation may result. Up to this point in the text, we have emphasized how mutations and the resulting alleles affect an organism's phenotype, and how alleles are transmitted through gametes to offspring according to Mendelian principles. We shall look now at phenotypic variation that occurs as a result of changes in the genetic material which are more substantial than alterations of individual genes. These involve modifications at the level of the chromosome.

While members of diploid species normally contain precisely two haploid chromosome sets, many cases are known in which some deviation from this standard occurs. Modifications include variations in the number of individual chromosomes as well as rearrangements of the genetic material either within or among chromosomes. Taken together, such changes are called **chromosome mutations** or **chromosome aberrations,** to distinguish such genetic alterations from gene mutations. Since it is the chromosome, and not the gene, that is transmitted according to Mendelian laws, chromosome aberrations are transmitted to offspring in a predictable manner, resulting in many interesting examples of "heredity and phenotype," the title we have chosen for Part 1.

In this chapter, we shall consider the many types of aberrations, their consequences, and their role in the evolutionary process. We will see that the genetic component of a diploid organism is delicately balanced in content and location within the genome. Even minor alterations of content or location may result in some form of phenotypic variation, while more substantial changes may cause lethality, particularly in animal species.

VARIATION IN CHROMOSOME NUMBER

Variation in chromosome number ranges from the addition or loss of one or more chromosomes to the addition of one or more haploid sets of chromosomes. When an organism gains or loses one or more chromosomes, but not a complete set, the condition of **aneuploidy** is created. This is contrasted with the condition of **euploidy,** where three or more complete haploid sets of chromosomes are found. Those with three sets are **triploid;** and when more than two sets exist, an organism is said to be **polyploid**. We shall discuss each of these categories and subsets within them in turn. Table 7.1 provides an organizational framework for the categories and subsets as well as a brief indication of what each term means.

ANEUPLOIDY

The most common examples of **aneuploidy,** where an organism has a chromosome number other than an exact multiple of the haploid set, are cases in which a single chromosome is either added to or lost from a normal diploid set. Such circumstances can arise as a result of primary or secondary nondisjunction (see Figure 2.12). The loss of one chromosome produces a $2n - 1$ complement and is called **monosomy;** the gain of one chromosome produces a $2n + 1$ complement and is described as **trisomy**. The $2n + 2$ and $2n + 3$ conditions are called **tetrasomy** and

TABLE 7.1
Terminology for variation in
chromosome numbers.

Term	Explanation
Aneuploidy	2*n* plus or minus chromosomes
Monosomy	2*n* − 1
Trisomy	2*n* + 1
Tetrasomy, pentasomy, etc.	2*n* + 2, 2*n* + 3, etc.
Euploidy	Multiples of *n*
Diploid	2*n*
Polyploidy	3*n*, 4*n*, 5*n*, ...
Triploidy	3*n*
Tetraploidy, pentaploidy, etc.	4*n*, 5*n*, etc.
Autopolyploidy	Multiples of the same genome
Allopolyploidy	Multiples of different genomes

pentasomy, indicating that an individual has four or five copies of a chromosome in an otherwise diploid genome.

Monosomy

Monosomy for one of the sex chromosomes is fairly common. In fact, it characterizes the **Protenor mode** of sex determination **(XX/XO)**. When sex is determined in the **Lygaeus (XX/XY) mode,** XO individuals may also occur, provided that such individuals have a normal complement of autosomes. In *Drosophila*, the XO chromosome complement results in a normal appearing, but sterile male. In humans, this monosomy results in females with Turner syndrome (see Chapter 5).

Monosomy for one of the autosomes is not so easily tolerated, particularly in animals. In *Drosophila*, flies monosomic for the very small chromosome 4—a condition referred to as **Haplo-IV**—develop more slowly, exhibit a reduced body size, and have impaired viability. Monosomy for the larger chromosomes 2 and 3 is apparently lethal because such flies have never been recovered.

The failure of monosomic individuals to survive in many animal species is at first quite puzzling, since at least a single copy of every gene is present in the remaining homologue. However, if just one of those genes is represented by a lethal allele, the unpaired chromosome condition leads to the death of the organism. This occurs because monosomy unmasks recessive lethals that are tolerated in heterozygotes carrying the corresponding wild-type alleles.

Another explanation is that the expression of genetic information during early development is very delicately regulated so that a sensitive equilibrium of gene products is required to ensure normal development. This requirement does not appear to be so stringent in the plant kingdom. Monosomy for auto-

somal chromosomes has been observed in maize, tobacco, the evening primrose *Oenothera*, and the Jimson weed *Datura*, among other plants. Nevertheless, such monosomic plants are usually less viable than their diploid derivatives. Haploid pollen grains, which undergo extensive development before participating in fertilization, are particularly sensitive to the lack of one chromosome and are often sterile.

Partial Monosomy: Cri-du-Chat Syndrome

In humans, autosomal monosomy has not been reported beyond birth. Individuals with such chromosome complements are undoubtedly conceived, but none apparently survive embryonic and fetal development. There are, however, examples of survivors with **partial monosomy,** where only part of one chromosome is lost. These cases are also referred to as **segmental deletions**. One such case was first reported by Jérôme LeJeune in 1963 when he described the clinical symptoms of the **cri-du-chat syndrome**. This syndrome is associated with the loss of about one-half the short arm of chromosome 5 (Figure 7.1). Thus, the genetic constitution may be designated as **46,5p−,** meaning that such an individual has all 46 chromosomes but that some of the p arm (the short arm) is missing. Infants with this syndrome exhibit multiple anatomic malformations, including gastrointestinal and cardiac complications, and are often severely mentally retarded. Abnormal development of the glottis and larynx is characteristic of individuals with this syndrome. An infant afflicted with this syndrome has a cry similar to that of the meowing of a cat, thus giving the syndrome its name.

Since 1963, over 300 cases of cri-du-chat syndrome have been reported worldwide. An incidence of 1 in 50,000 live births has been estimated. The size of the deletion appears to influence the physical, psycho-

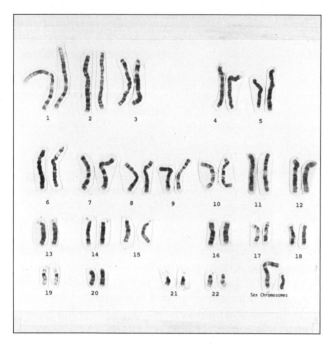

(a)

(b)

FIGURE 7.1

(a) A representative karyotype and (b) photographs of two children exhibiting cri-du-chat syndrome (46,5p−). In the karyotype, the arrow identifies the nearly complete absence of the short arm of one member of the chromosome 5 homologues.

motor, and mental skill levels of those children who survive. While the effects of the syndrome are severe, many individuals achieve a level of social development in the trainable range. Those who receive home care and early special schooling may be ambulatory, develop self-care skills, and learn to communicate verbally.

Trisomy

In general, the effects of trisomy parallel those of monosomy. However, the addition of an extra chromosome produces somewhat more viable individuals

in both animal and plant species than does the loss of a chromosome.

As in monosomy, the sex chromosome variation of the trisomic type has a less dramatic effect on the phenotype than autosomal variation. *Drosophila* females with three X chromosomes and a normal complement of two sets of autosomes (3X:2A) are much less viable than normal 2X:2A females. In humans, the addition of an extra X or Y chromosome to an otherwise normal male or female chromosome constitution (47,XXY, 47,XYY, and 47,XXX) leads to viable individuals. However, the addition of a large autosome to the diploid complement in both *Drosophila* and humans has severe effects and is usually lethal during development.

In plants, trisomic individuals are viable, but their phenotype may be altered. A classical example involves the Jimson weed *Datura*, whose diploid number is 24. Twelve different primary trisomic conditions are possible, and examples of each one have been recovered. Each trisomy alters the phenotype of the capsule sufficiently (Figure 7.2) to produce a unique phenotype. These capsule phenotypes were first thought to be caused by point mutations.

In plants as well as animals, trisomy may be detected during cytological observations of meiotic divisions. Since three copies of one of the chromosomes are present, pairing configurations are irregular. At any particular region along the chromosome length, only two of the three homologues may synapse. At various regions, however, different members of the trio may be paired. One such example, called a **trivalent**, is diagrammed in Figure 7.3. The trivalent is usually arranged on the spindle so that during anaphase one member moves to one pole, and two go to the opposite pole. In some cases, one bivalent and one univalent (an unpaired chromosome) may be present instead of a trivalent during the first metaphase stage. Meiosis thus produces gametes that can perpetuate the trisomic condition.

Down Syndrome

The only human autosomal trisomy in which a significant number of individuals survive longer than a year past birth was discovered in 1866 by Langdon Down. The condition is now known to result from trisomy of chromosome 21, one of the G group* (Figure 7.4), and is called **Down syndrome** or simply **trisomy 21** (and is designated **47,21+**). This trisomy is

---

*On the basis of size and centromere placement, human chromosomes are divided into seven groups: A (1–3); B (4–5); C (6–12); D (13–15); E (16–18); F (19–20); and G (21–22).

FIGURE 7.2
Drawings of capsule phenotypes of the fruits of *Datura stramonium*. In comparison with wild type, each phenotype is the result of trisomy of one of the twelve chromosomes characteristic of the haploid genome. The photograph illustrates the entire plant.

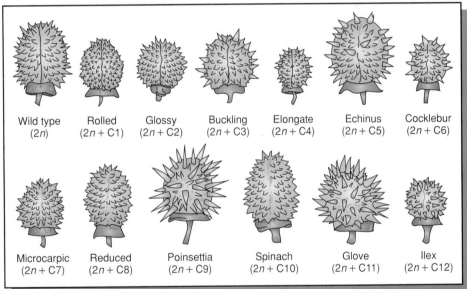

| Wild type (2n) | Rolled (2n + C1) | Glossy (2n + C2) | Buckling (2n + C3) | Elongate (2n + C4) | Echinus (2n + C5) | Cocklebur (2n + C6) |

| Microcarpic (2n + C7) | Reduced (2n + C8) | Poinsettia (2n + C9) | Spinach (2n + C10) | Glove (2n + C11) | Ilex (2n + C12) |

found in approximately 3 infants in every 2000 live births.

The external phenotype of these individuals is so similar that many of them might appear to be siblings at first glance. They display a prominent epicanthic fold in the corner of the eye and are characteristically short. They may have small, round heads; protruding, furrowed tongues, which cause the mouth to remain partially open; and short, broad hands with fingers with characteristic dermatoglyphic palm and fingerprint patterns. Physical, psychomotor, and mental development is retarded and the IQ is seldom above 70. Their life expectancy is shortened, and few individuals survive to be 50. A significant number of

Down infants do not survive the first year after birth. They are prone to respiratory disease and heart malformations and show an incidence of leukemia approximately 15 times higher than that of the normal population. However, close medical scrutiny throughout the lives of Down individuals has extended their survival significantly.

One way in which this trisomic condition may originate is through **nondisjunction** of chromosome 21 during meiosis. Failure of paired homologues to disjoin during anaphase I or II can result in male or female gametes with the $n + 1$ chromosome composition. Following fertilization with a normal gamete, the trisomic condition is created. Chromosome anal-

FIGURE 7.3
One possible pairing arrangement and segregation pattern of members of a trivalent chromosome during the first meiotic division. Two chiasmata are visible in the diplotene stage.

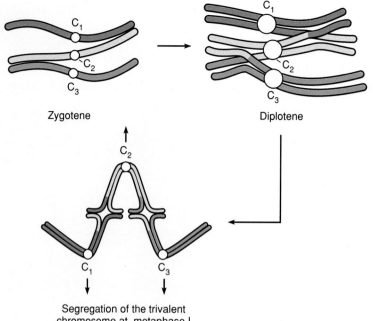

Zygotene

Diplotene

Segregation of the trivalent chromosome at metaphase I

ysis has shown that while the additional chromosome may be derived from either the mother or father, the ovum is most often the source.

Before the development of banding techniques that distinguish paternal from maternal homologues (Chapter 17), this conclusion was supported by other indirect evidence derived from studies of the age of mothers giving birth to Down infants. Figure 7.5 shows the analysis of the distribution of maternal age and the incidence of Down syndrome newborns. The frequency of Down births dramatically increases as the age of the mother increases. While the frequency

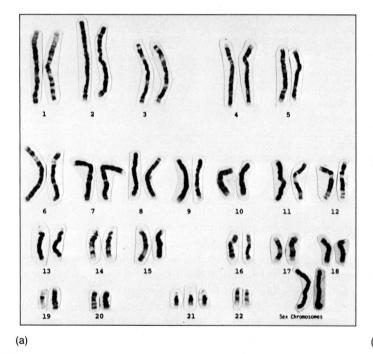

(a)

(b)

FIGURE 7.4
(a) The karyotype and (b) a photograph of a child with Down syndrome. In the karyotype, three members of the G-group chromosome 21 are present, creating the 47,21+ condition.

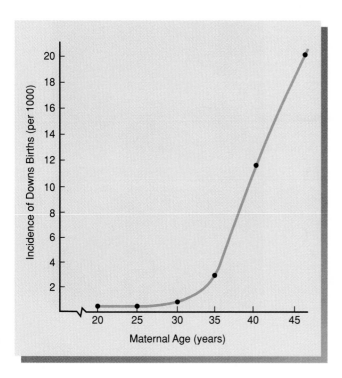

FIGURE 7.5
Incidence of Down syndrome births contrasted with maternal age.

is about 1 in 1000 at maternal age 30, a tenfold increase to a frequency of 1 in 100 is noted at age 40. The frequency increases still further to about 1 in 50 at age 45.

While the nondisjunctional event that produces Down syndrome seems more likely to occur during oogenesis in women between the ages of 35 and 45, we do not know with certainty why this is so. In human females, all primary oocytes have been formed by birth. Therefore, once ovulation begins, each succeeding ovum has been arrested in meiosis for about a month longer than the one preceding it. Thus, women 30 or 40 years old produce ova that are significantly older and arrested longer than those they ovulated 10 or 20 years previously. However, it is not yet known whether ovum age is the cause of the increased incidence of nondisjunction leading to Down syndrome.

These statistics pose a serious problem for the woman who becomes pregnant late in her reproductive years. Genetic counseling early in the pregnancy serves two purposes. First, it informs the parents about the probability that their child will be affected and educates them about Down syndrome. Although some individuals with Down syndrome must be institutionalized, others are trainable and may be cared for at home. Further, they are noted as being affectionate, loving children. Second, a genetic coun-

selor may recommend a prenatal diagnostic technique such as **amniocentesis** or **chorionic villus sampling (CVS)**. These techniques require the removal and culture of fetal cells. The karyotype of the fetus may then be determined by cytogenetic analysis. If the fetus is diagnosed as having Down syndrome, a therapeutic abortion is one option the parents may consider.

Since Down syndrome appears to be caused by a random error—nondisjunction of chromosome 21 during maternal or paternal meiosis—the disorder is not expected to be inherited. Nevertheless, a similar condition called **familial Down syndrome** does run in families and has been reported many times. Familial Down syndrome involves a **translocation** of chromosome 21, another type of chromosomal aberration which we will discuss later in this chapter.

Patau Syndrome

In 1960, Klaus Patau and his associates observed an infant with gross developmental malformations with a karyotype of 47 chromosomes (Figure 7.6). The additional chromosome was medium-sized, one of the acrocentric D group. It is now designated as chromosome 13. The trisomy 13 condition has since been described in many newborns and is called **Patau syndrome (47,13+)**. Affected infants are not mentally alert, are thought to be deaf, and characteristically have a harelip, cleft palate, and polydactyly. Autopsy has revealed congenital malformation of most organ systems, a condition indicative of abnormal developmental events occurring as early as five to six weeks of gestation. The mean survival of these infants is less than six months.

The average maternal and paternal ages of parents of affected infants are higher than the ages of parents of normal children, but are not as high as the average maternal age in cases of Down syndrome. Both male and female parents average about 32 years of age when the affected child is born. Because the condition is so rare, occurring as infrequently as 1 in 20,000 live births, it is not known whether the origin of the extra chromosome is more often maternal or paternal, or whether it arises equally from either parent.

Edwards Syndrome

In 1960, John H. Edwards and his colleagues reported on an infant trisomic for a chromosome in the E group, now known to be chromosome 18 (Figure 7.7). This aberration has been named **Edwards syndrome (47,18+)**. These infants are smaller than the average newborn. Their skulls are elongated in an anterior-

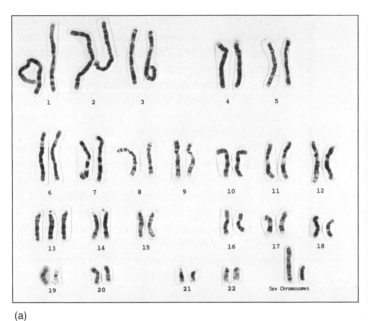
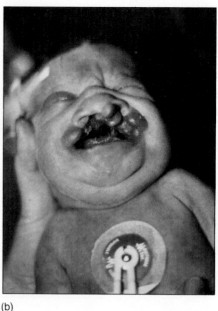

(a)   (b)

FIGURE 7.6
(a) The karyotype and (b) a photograph of an infant with Patau syndrome. In the karyotype, three members of the D-group chromosome 13 are present, creating the 47,13+ condition.

posterior direction and their ears set low and malformed. A webbed neck, congenital dislocation of the hips, and a receding chin are often characteristic of such individuals. Although the frequency of trisomy 18 is somewhat greater than that of trisomy 13, the survival time is about the same, less than four months on the average. Death is usually caused by pneumonia or heart failure. The phenotype of this child, like that of individuals with Down and Patau syndromes, illustrates that the presence of an extra autosome produces congenital malformations and reduced life expectancy.

Again, the average maternal age is high—34.7 years by one calculation. One unusual aspect of the syndrome is the preponderance of afflicted female infants. In one set of observations based on 143 cases, 80 percent were female. Overall, about 1 in 8,000 live births exhibits this malady.

Viability in Human Aneuploid Conditions

The reduced viability of individuals with recognized monosomic and trisomic conditions leads us to believe that many other aneuploid conditions may arise but that the affected fetuses do not survive to term. This observation has been confirmed by karyotypic analysis of spontaneously aborted fetuses. These studies have revealed some striking statistics. At least 15 to 20 percent of all conceptions are

terminated in spontaneous abortion (some estimates are considerably higher). About 30 percent of all spontaneous abortuses demonstrate some form of chromosomal anomaly, and approximately 90 percent of all chromosomal anomalies are terminated prior to birth as a result of spontaneous abortion.

The majority of spontaneous abortuses demonstrating chromosomal abnormalities are aneuploids. The aneuploid with highest incidence among abortuses is the 45,X condition, which produces an infant with Turner syndrome if the fetus survives to term.

An extensive review of this subject by David H. Carr has also revealed that a significant percentage of abortuses are trisomic for one of the chromosome groups. Trisomies for every human chromosome have been recovered. Monosomies were almost never found, however, even though nondisjunction should produce $n - 1$ gametes with a frequency equal to $n + 1$ gametes. This finding suggests that gametes lacking a single chromosome are functionally impaired to a serious degree or that the embryo dies so early in its development that recovery does not occur. Various forms of polyploidy and other miscellaneous chromosomal anomalies were also found in Carr's study.

These observations support the hypothesis that normal embryonic development requires a precise diploid complement of chromosomes to maintain a delicate equilibrium in the expression of genetic

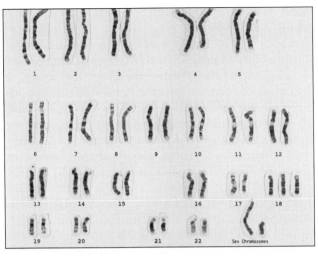

(a)

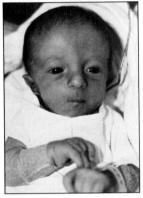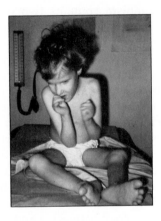

(b)

FIGURE 7.7
(a) The karyotype and (b) photographs of children with Edwards syndrome. In the karyotype, three members of the E-group chromosome 18 are present, creating the 47,18+ condition.

information. The prenatal mortality of most aneuploids does provide a barrier against the introduction of a general form of genetic anomaly into the human population.

POLYPLOIDY AND ITS ORIGINS

The term **polyploidy** describes instances where more than two multiples of the haploid chromosome set are found. The naming of polyploids is based on the number of sets of chromosomes found: a **triploid** has $3n$ chromosomes; a **tetraploid** has $4n$; a **pentaploid**, $5n$; and so forth. Several general statements may be made about polyploidy. This condition is relatively infrequent in most animal species, but is well known in lizards, amphibians, and fish. It is much more common in plant species. Odd numbers of chromosome sets are not usually maintained reliably from generation to generation because a polyploid organism with an uneven number of homologues usually does not produce genetically balanced gametes. For this reason, triploids, pentaploids, etc., are not usually found in species that depend solely upon sexual reproduction for propagation.

Polyploidy can originate in two ways: (1) the addition of one or more extra sets of chromosomes, identical to the normal haploid complement of the same species, results in **autopolyploidy,** and (2) the combination of chromosome sets from different species may occur as a consequence of hybridization and results in **allopolyploidy**. Thus, the distinction between auto- and allopolyploidy is based on the genetic origin of the extra chromosome sets.

In our discussion of polyploidy, we will use the following symbols to clarify the origin of additional chromosome sets. For example, if A represents the haploid set of chromosomes of any organism, then

$$A = a_1 + a_2 + a_3 + a_4 \ldots a_n$$

where a_1, a_2, and so on represent individual chromosomes, and where n is the haploid number. Thus, a normal diploid organism might be represented simply as AA.

Autopolyploidy

In **autopolyploidy,** each additional set of chromosomes is identical to the parent species. Therefore, **triploids** are represented as AAA, **tetraploids** are $AAAA$, and so forth.

Autotriploids may arise in several ways. A failure of all chromosomes to segregate during meiotic divisions may produce a diploid gamete. If such a gamete is fertilized by a haploid gamete, a zygote with three sets of chromosomes is produced. Or, occasionally two sperm may fertilize an ovum, resulting in a triploid zygote. Triploids may also be produced under experimental conditions by crossing diploids with tetraploids. Diploid organisms produce gametes with n chromosomes while tetraploids produce $2n$ gametes. Upon fertilization, the desired triploid is produced.

Because they have an even number of chromosomes, **autotetraploids** ($4n$) are theoretically more likely to be found in nature than autotriploids. Unlike triploids, which often produce genetically unbalanced gametes, tetraploids are more likely to produce balanced gametes, when involved in sexual reproduction. Tetraploid cells may be produced experimentally from diploid cells by applying cold or heat shock during meiosis or by applying colchicine to somatic cells undergoing mitosis. **Colchicine,** an alkyloid derived from the autumn crocus, interferes with spindle formation, and thus replicated chromosomes cannot enter anaphase and migrate to the poles. When colchicine is removed, the cell may reenter interphase. When the paired sister chromatids separate and uncoil, the nucleus will contain twice the diploid number of chromosomes and is effectively $4n$. This process is illustrated in Figure 7.8.

In general, autopolyploids are larger than their diploid relatives. Often, the flower and fruit of plants are increased in size. This increase seems to be due to larger cell size rather than greater cell number. Although autopolyploids do not contain new or unique information compared with the diploid relative, such varieties may be of greater commercial value. Economically important triploid plants include several potato species of the genus *Solanum,* Winesap apples, commercial bananas, seedless watermelons, and the cultivated tiger lily *Lilium tigrinum.* These plants are propagated asexually. Diploid bananas contain hard seeds, but the commercial, triploid, "seedless" variety has edible seeds. Tetraploid alfalfa, coffee, peanuts, and McIntosh apples are also of economic value because they are either larger or grow more vigorously than their diploid or triploid counterparts. The commercial strawberry is an octoploid. These observations attest to the importance of autopolyploidy in domesticated plants.

Allopolyploidy

Polyploidy may also result from hybridization of two closely related species. Thus, if a haploid ovum from a species with chromosome sets AA is fertilized by sperm from a species with sets BB, the resulting hybrid is AB, where $A = a_1, a_2, a_3, \ldots, a_n$ and $B = b_1, b_2, b_3, \ldots, b_n$. The hybrid may be sterile because of its inability to produce viable gametes. This occurs because, most often, some or all of the a and b chromosomes cannot synapse in meiosis and unbalanced genetic conditions result. If, however, the new AB genetic combination undergoes a natural or induced chromosomal doubling, a fertile $AABB$ tetraploid is produced. These events are illustrated in Figure 7.9. Since this polyploid contains the equivalent of four haploid genomes and since the hybrid contains unique genetic information compared with either parent, such an organism is called an **allotetraploid** (from the Greek word *allo,* meaning other or different). An equivalent term, **amphidiploid,** is also used to describe this situation in which a hybrid organism contains two complete diploid genomes. This term is preferred when the original species are known.

Allopolyploidy is the most common natural form of polyploidy in plants because the chance of forming balanced gametes is much greater than in other types of polyploidy. Since two homologues of each specific chromosome are present, meiosis may occur normally (Figure 7.9), and fertilization may successfully propagate the tetraploid sexually. This discussion assumes the simplest situation, where none of the chromosomes in set A are homologous to those in set B. In hybrids of very closely related species, some homology between a and b chromosomes is likely. In this case, meiotic pairing is more complex. Multivalents, which are more complex than the trivalents shown in Figure 7.3, may be formed, resulting in the production of unbalanced gametes. In such cases, aneuploid varieties of allotetraploids may arise. Allopolyploids are rare in most animals because mating

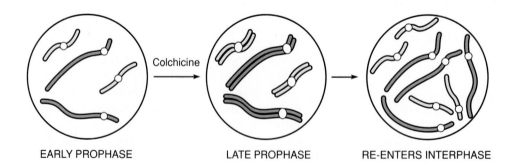

EARLY PROPHASE   LATE PROPHASE   RE-ENTERS INTERPHASE

FIGURE 7.8
Illustration of the potential involvement of colchicine in the production of an autotetraploid. The figure follows two pairs of homologous chromosomes. Colchicine interferes with the formation of spindle fibers and thus inhibits the movement of chromosomes to the poles during anaphase. Paired sister chromatids may then separate, effectively doubling the number of chromosomes.

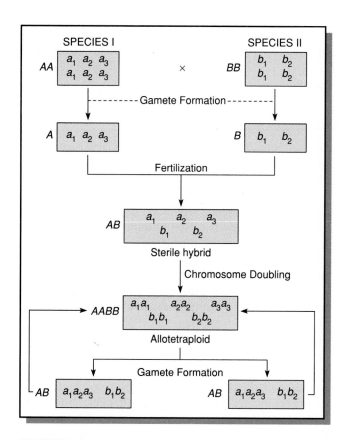

FIGURE 7.9

The origin and propagation of an allotetraploid. Species I contains genome (A) consisting of three distinct chromosomes, a_1, a_2, and a_3. Species II contains genome (B) consisting of three distinct chromosomes, b_1, b_2, and b_3. Following fertilization between members of the two species and chromosome doubling, a fertile allotetraploid containing four complete haploid genomes ($AABB$) is formed. If the original species are known, this can appropriately be called an amphidiploid as well. The autotetraploid form of *Gossypium*, which is believed to have arisen in this way, is illustrated.

behavior is most often species-specific, and thus the initial step in hybridization is unlikely to occur.

A classical example of allotetraploidy in plants is the cultivated species of American cotton, *Gossypium* (Figure 7.9). This species has 26 pairs of chromosomes: 13 are large and 13 are much smaller. When it was discovered that Old World cotton had only 13 pairs of large chromosomes, allopolyploidy was suspected. After an examination of wild American cotton revealed 13 pairs of small chromosomes, this speculation was strengthened. J. O. Beasley was able to reconstruct the origin of cultivated cotton experimentally. He crossed the Old World strain with the wild American strain and then treated the hybrid with colchicine to double the chromosome number. The result of these treatments was a fertile allotetraploid variety of cotton. It contained 26 pairs of chromosomes and characteristics similar to the cultivated variety.

Often allotetraploids exhibit characteristics of both parental species. An interesting example, but one with no practical economic importance, is that of the hybrid formed between the radish *Raphanus sativus* and the cabbage *Brassica oleracea*. Both species have a haploid number of 9. The hybrid consists of 9 *Raphanus* and 9 *Brassica* chromosomes (9R + 9B). While hybrids are almost always sterile, some fertile allotetraploids (18R + 18B) have been produced. Unfortunately, the root of this allotetraploid is more like the cabbage and its shoot resembles the radish. Had the converse occurred, the hybrid might have been of economic importance.

A much more successful commercial hybridization has been performed using the grasses wheat and rye. Wheat (genus *Triticum*) has a basic genome of 7 chromosomes. The diploid species have 14 chromosomes. Cultivated autopolyploids exist, including tetraploid ($4n = 28$) and hexaploid ($6n = 42$) varieties.

Rye (genus *Secale*) also has a genome consisting of 7 chromosomes. The only cultivated species is diploid ($2n = 14$).

Using the technique outlined in Figure 7.9, including colchicine treatment of the F_1 hybrid seedlings, geneticists have produced various allopolyploids. When tetraploid wheat is crossed with diploid rye, and the F_1 treated with colchicine, a hexaploid hybrid ($6n = 42$) is derived. The hybrid, designated *Triticale* (see Figure 1.8), represents a new genus. It contains 14 pairs of wheat chromosomes and 7 pairs of rye chromosomes. Octoploids may also be produced from crosses between hexaploid wheat and diploid rye. The hexaploid hybrid, initially described, is more stable. Fertile hybrid varieties derived from various wheat and rye species may be crossed together or back-crossed to parental strains. These crosses have created a wide variety of members of the genus *Triticale*.

The hybrid plants demonstrate characteristics of both wheat and rye. For example, certain hybrids combine the high protein content of wheat with the high content of the amino acid lysine of rye. The lysine content is low in wheat and thus is a limiting nutritional factor. Wheat is considered a high-yielding grain, while rye is noted for its versatility of growth in unfavorable environments. *Triticale* species combining both traits have the potential of significantly increasing grain production. Programs designed to improve crops through hybridization have long been underway in several underdeveloped countries of the world, as discussed in Chapter 1.

Recall that in the previous chapter, we discussed the use of somatic cell hybrids to map human genes. This technique has also been applied to the production of allopolyploid plants. Cells from the developing leaves of plants can be treated to remove their cell wall, resulting in **protoplasts**. These altered cells can be maintained in culture and stimulated to fuse with other protoplasts, producing somatic cell hybrids.

If cells from different plant species are fused in this way, hybrid cells can be produced with unique chromosome combinations. Since protoplasts can be induced to divide and differentiate into stems which develop leaves, the potential for producing allopolyploids is available in the research laboratory. In some cases, entire plants may be derived from cultured protoplasts. If only stems and leaves are produced, these may be grafted onto the stem of another plant. If flowers are formed, fertilization may yield mature seeds, which upon germination yield an allopolyploid plant.

The general process is illustrated in Figure 7.10. There are now many examples of allopolyploids which have been created commercially using the above approach. While most of them are not true amphidiploids since one or more chromosomes are missing, precise allotetraploids are sometimes produced.

Endopolyploidy

Certain cells in an otherwise diploid organism have been observed to be polyploid. The term **endopolyploidy** describes this general phenomenon. In such cases, replication and separation of chromosomes occur without nuclear division. The process leading to endopolyploidy is called **endomitosis**.

Vertebrate liver cell nuclei, including human ones, often contain $4n$, $8n$, or $16n$ chromosome sets. The stem and parenchyma tissue of apical regions of flowering plants are also often endopolyploid. Cells lining the gut of mosquito larvae attain a $16n$ ploidy, but during the pupal stages such cells undergo very quick reduction divisions, giving rise to smaller diploid cells. In the water strider *Gerris*, wide variations in chromosome numbers are found in different tissues, with as many as 1024 to 2048 copies of each chromosome in the salivary gland cells. Since the diploid number in this organism is 22, the nuclei of these cells may contain over 40,000 chromosomes!

Although the role of endopolyploidy is not clear, the proliferation of chromosome copies often occurs in cells where multiples of certain gene products are required. In fact, it is well established that certain genes, whose product is in high demand in *every* cell, exist naturally in multiple copies in the genome. Ribosomal and transfer RNA genes are examples of multiple-copy genes. In certain cells of organisms, it may be necessary to replicate the entire genome, allowing an even greater rate of expression of various genes.

VARIATION IN CHROMOSOME STRUCTURE AND ARRANGEMENT

The second general class of chromosome aberrations includes structural changes that delete, add, or rearrange substantial portions of one or more chromosomes. Included in this category are **deletions** and **duplications** of genes or part of a chromosome and **rearrangements** of genetic material in which a chromosome segment is either inverted, exchanged with a segment of a nonhomologous chromosome, or merely transferred to one. Before discussing these aberrations, we will present several general statements pertaining to them.

In most instances, these changes are due to one or more breaks along the axis of a chromosome, fol-

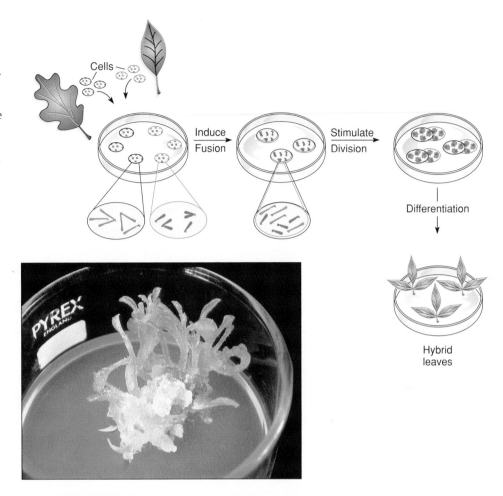

FIGURE 7.10
Application of somatic cell hybridization technique in the production of an amphidiploid. Cells from the leaves of two species of plants are removed and cultured. The cell walls are digested away and the resultant protoplasts are induced to undergo cell fusion. The hybrid cell is selected and stimulated to divide and differentiate, as illustrated in the photograph. An amphidiploid has a complete set of chromosomes from each parental cell type and displays phenotypic characteristics of each. Two pairs of chromosomes from each species are depicted.

lowed by either the loss or rearrangement of some genetic material. Chromosomes can break spontaneously, but the rate of breakage can be increased by the use of chemicals or radiation. Although the ends of chromosomes, the **telomeres,** do not readily fuse with ends of "broken" chromosomes or with other telomeres, the ends produced at points of breakage are "sticky" and can rejoin other such ends. If breakage-rejoining does not merely reestablish the original relationship, and if the alteration occurs in germ plasm, the gametes will contain the structural rearrangement, which will be heritable.

If the aberration is found in one homologue but not the other, unusual but characteristic pairing configurations are formed during meiotic synapsis. These patterns are useful in identifying the type of change that has occurred. If no loss or gain of genetic material occurs, individuals bearing the aberration "heterozygously," so to speak, will likely be unaffected phenotypically. However, the unusual pairing arrangements often lead to gametes which are duplicated or deficient for chromosomal regions. Thus, the offspring of "carriers" of certain aberrations often

have an increased probability of demonstrating phenotypic manifestations of the aberration.

DELETIONS

When a portion of a chromosome is lost, the missing piece is referred to as a **deletion** (or a **deficiency**). The deletion may occur either at one end or from the interior of the chromosome. These are called **terminal** or **intercalary deletions,** respectively. Both result from one or more breaks in the chromosome. The portion retaining the centromere region will usually be maintained and that segment without the centromere will eventually be lost in progeny cells following mitosis or meiosis. During synapsis between a chromosome with a large intercalary deficiency and a normal complete homologue, the unpaired region of the normal homologue must "buckle out." Such a configuration is called a **deficiency loop** or **compensation loop**. Such a loop can be visualized if the large polytene chromosomes derived from insect salivary gland cells are examined. Polytene

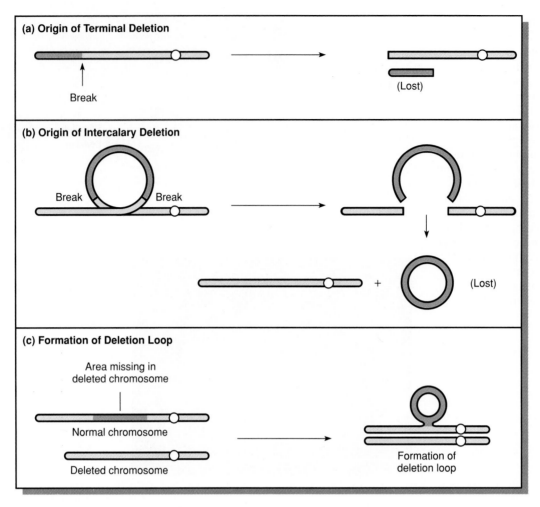

FIGURE 7.11
Origins of a terminal (a) and intercalary (b) deletion. In part (c), pairing occurs between a normal chromosome and one with an intercalary deletion by looping out the undeleted portion.

chromosomes, to be discussed in detail in Chapter 17, display a banding pattern characteristic of each chromosome, making them useful in cytological studies. The origins of both types of deletion and the formation of such a loop in *Drosophila* polytene chromosomes are diagrammed in Figure 7.11.

If too much genetic information is lost as a result of a deletion, the aberration may be lethal and never becomes available for study. As seen in the cri-du-chat syndrome, where only part of the short arm of chromosome 5 is lost, a deficiency need not be very great before the effects become severe.

A final consequence of deletions may be noted in organisms heterozygous for a deficiency. Consider the mutant *Notch* phenotype in *Drosophila*. In these flies, the wings are notched out on the posterior and lateral margins. Data from breeding studies seem to indicate that the phenotype is controlled by a sex-linked dominant mutation since heterozygous females have notched wings and transmit this allele to one-half of their male progeny. Since the mutation also appears to behave as a recessive lethal, homozygous females and hemizygous males are never recovered. It has also been noted that if notched-winged females are also heterozygous for the closely linked recessive *white*-eye, *facet*-eye, or *split*-bristle mutations, they express these mutant phenotypes as well as *Notch*. Because these mutations are recessive, heterozygotes should express the normal, wild-type phenotypes. These genotypes and phenotypes are summarized in Table 7.2.

These observations, which identified *Notch* as a deletion, were explained through a cytological examination of the polytene X chromosome of heterozygous *Notch* females. A deficiency loop was found along the X chromosome from band 3C2 through

TABLE 7.2
Notch genotypes and phenotypes.

Genotype	Phenotype
$\dfrac{N^+}{N}$	*Notch* female
$\dfrac{N}{N}$ $\dfrac{N}{\Rightarrow}$	Lethal
$\dfrac{N^+w}{N\ w^+}$	*Notch, white* female
$\dfrac{N^+fa\ spl}{N\ fa^+spl^+}$	*Notch, facet, split* female

band 3C11, as shown in Figure 7.12. These bands had previously been shown to include the *white, facet,* and *split* loci, among others. On the genetic map, this region corresponds to loci 1.5 to 3.0. Thus, this region's deficiency in one of the two homologous X chromosomes has two distinct effects. First, it results in the *Notch* phenotype. Second, by deleting the loci for genes whose mutant alleles are present on the other X chromosome, the deficiency creates a hemizygous condition so that the recessive *white, facet,* or *split* alleles are expressed. This type of phenotypic expression of recessive genes in association with a deletion is called **pseudodominance**.

Many independently arising *Notch* phenotypes have been investigated. The common deficient band for all *Notch* phenotypes has now been designated as 3C7. In every case that *white* was also expressed pseudodominantly, the common deficient band was 3C2. The bands that cytologically distinguish the

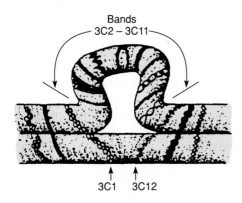

FIGURE 7.12
Deletion loop formed in salivary chromosomes heterozygous for the *Notch* deletion on the X chromosome of *Drosophila melanogaster*. The deficiency encompasses bands 3C2 through 3C11.

white locus from the *Notch* locus have been confirmed in this manner.

DUPLICATIONS

When any part of the genetic material—a single locus or a large piece of a chromosome—is present more than once in the genome, it is called a **duplication**. As in deletions, pairing in heterozygotes may produce a compensation loop. Duplications may arise as the result of unequal crossing over between synapsed chromosomes during meiosis (Figure 7.13) or through a replication error prior to meiosis. In the former case, both a duplication and a deficiency are produced.

Three interesting aspects of duplications will be considered. First, they may result in gene redundancy. Second, as with deletions, duplications may produce phenotypic variation. Third, according to one convincing theory, duplications have also been an important source of genetic variability during evolution.

Gene Redundancy and Amplification—Ribosomal RNA Genes

Although many gene products are not needed in every cell of an organism, other gene products are known to be essential components of all cells. For example, ribosomal RNA must be present in abundance in order for protein synthesis to occur. The more metabolically active a cell is, the higher is the demand for this molecule. In theory, a single copy of the gene encoding rRNA would be inadequate in most cells. Studies using the technique of molecular hybridization, which allows the determination of the percentage of the genome coding for specific RNA sequences, have shown that there are multiple copies of the genes coding for rRNA. Such DNA is called **rDNA**, and the general phenomenon is called **gene redundancy**. For example, in the common intestinal bacterium, *Escherichia coli (E. coli)*, about 0.4 percent of the haploid genome consists of rDNA. This is equivalent to 5 to 10 copies of the gene. In *Drosophila melanogaster*, 0.3 percent of the haploid genome, equivalent to 130 copies, consists of rDNA. While the presence of multiple copies of the same gene is not restricted to those coding for rRNA, we will focus on them in this section.

Studies of *Drosophila* have documented the need for the extensive amounts of rRNA and ribosomes made possible by multiple copies of these genes. In this organism, the X-linked mutation *bobbed* has, in fact, been shown to be due to a deletion of a variable

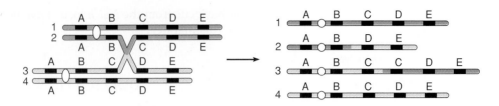

FIGURE 7.13

The origin of duplicated and deficient chromosomes as a result of unequal crossing over. The tetrad at the left is mispaired during synapsis. A single crossover between strands 2 and 3 results in a deficient and a duplicated chromosome. The two chromatids uninvolved in the crossover event remain normal in their gene sequence.

number of genes coding for rRNA. Mutant flies have low viability, are underdeveloped, and have bristles reduced in size. General development as well as bristle formation, which occurs very rapidly during normal pupal development, apparently depend on a great number of ribosomes to support translation. Many *bobbed* alleles have been studied, and each has been shown to involve a deletion, often of a unique size. The extent to which both viability and bristle length are decreased correlates well with the relative number of RNA genes deleted. Thus, we may conclude that the normal rDNA redundancy observed in wild-type *Drosophila* is the minimum required for adequate ribosome production during normal development.

In some cells, particularly oocytes, even the normal redundancy of rDNA may be insufficient to provide sufficient amounts of rRNA and ribosomes. Oocytes store abundant nutrients in the ooplasm for use by the embryo during early development. In fact, more ribosomes are included in the oocytes than in any other cell type. By considering how the amphibian *Xenopus laevis* acquires this abundance of ribosomes, we will see a second way in which the amount of rRNA is increased. This phenomenon is referred to as **gene amplification**.

The genes that code for rRNA are located in an area of the chromosome known as the **nucleolar organizer region (NOR)**. The NOR is intimately associated with the nucleolus, which is a processing center for ribosome production. Molecular hybridization analysis has shown that each NOR in *Xenopus* contains the equivalent of 400 redundant gene copies coding for the rRNA precursor molecule. Even this number of genes is apparently inadequate to synthesize the vast amount of ribosomes which must accumulate in the amphibian oocyte to support development following fertilization. To further amplify the number of rRNA genes, the rDNA is selectively replicated, and each new set of genes is released from its template. Since each new copy is equivalent to the NOR, multiple small nucleoli are formed around each NOR in the oocyte. As many as

1500 of these "micronucleoli" have been observed in a single oocyte. If we multiply the number of micronucleoli (1500) by the number of gene copies in each NOR (400), we see that amplification in *Xenopus* oocytes can result in over half a million gene copies! If each copy is transcribed only 20 times during the maturation of the oocyte, in theory, sufficient copies of rRNA may be produced to result in well over one million ribosomes.

The *Bar* Eye Mutation in *Drosophila*

Duplications may cause phenotypic variation that at first might appear to be caused by a simple gene mutation. The *Bar* eye phenotype in *Drosophila* is a classical example. Instead of the normal oval eye shape, *Bar*-eyed flies have narrow, slitlike eyes. This phenotype appears to be inherited as a dominant sex-linked mutation. However, since both heterozygous females and hemizygous males exhibit the trait, but homozygous females show a more pronounced phenotype than either of these two cases, the inheritance is more accurately described as **semidominant**.

In the early 1920s Alfred H. Sturtevant and Thomas H. Morgan discovered and investigated this "mutation." As illustrated in Figure 7.14, normal wild-type females (B^+/B^+) have about 800 facets in each eye. Heterozygous females (B/B^+) have about 350 facets, while homozygous females (B/B) average only about 50 facets. Females are occasionally recovered with even fewer facets and are designated as *double Bar* (B^+/B^D).

About ten years later, Calvin Bridges and Herman J. Muller compared the polytene X chromosome banding pattern of the *Bar* fly with that of the wild-type fly. Such chromosomes contain specific banding patterns that have been well categorized into regions. Their studies revealed that one copy of region 16A of the X chromosome was present in wild-type flies and that this region was duplicated in *Bar* flies and triplicated in *double Bar* flies. These observations provided evidence that the *Bar* phenotype is not the result of a simple chemical change in

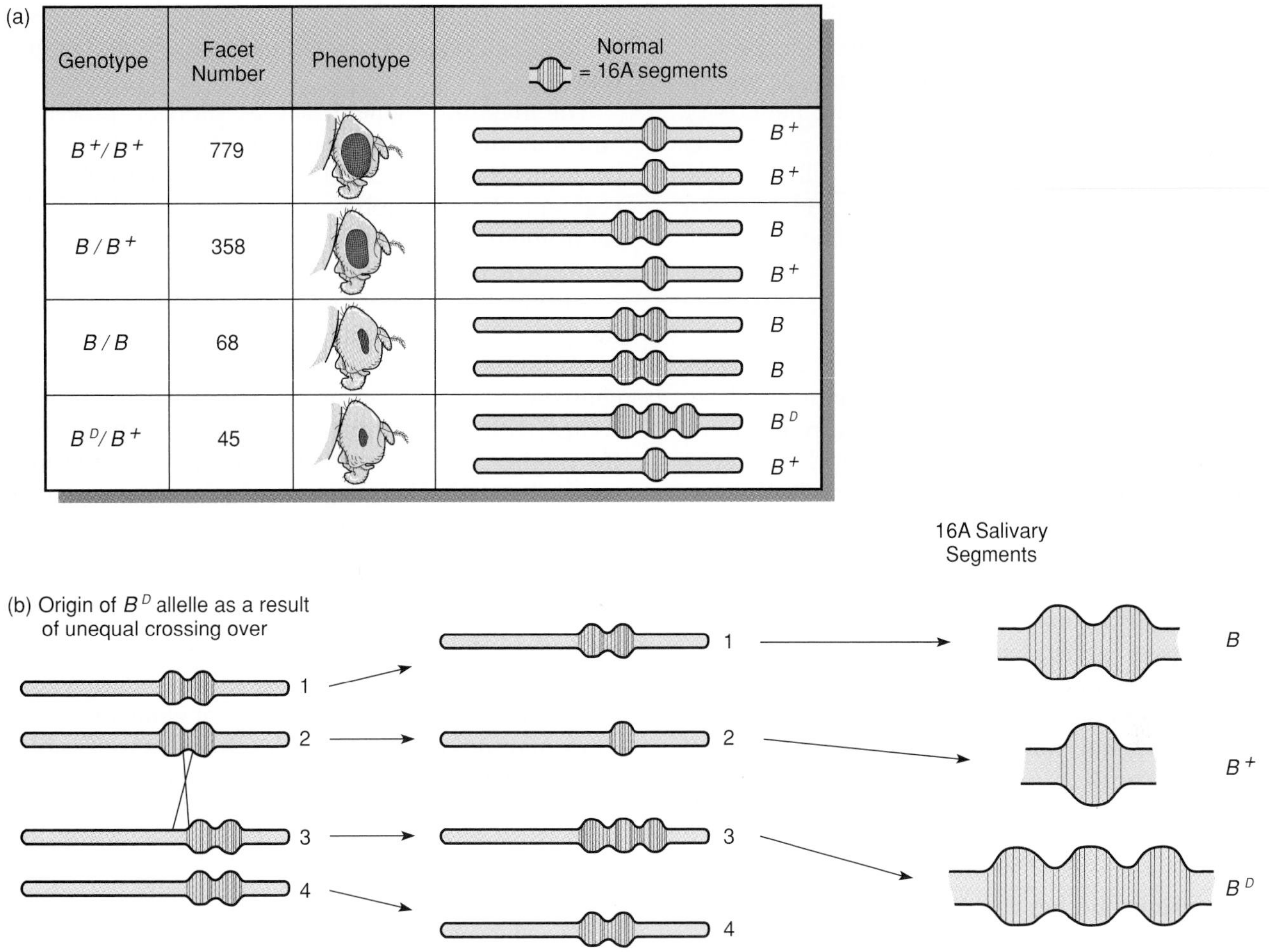

(a)

Genotype	Facet Number	Phenotype	Normal ⟨⟩ = 16A segments
B^+/B^+	779		B^+ / B^+
B/B^+	358		B / B^+
B/B	68		B / B
B^D/B^+	45		B^D / B^+

16A Salivary Segments

(b) Origin of B^D allele as a result of unequal crossing over

FIGURE 7.14
Summary of the *Bar* duplication in *Drosophila melanogaster*.

the gene, but is instead a duplication. The *double Bar* condition originates as a result of unequal crossing over, which produces the triplicated 16A region [Figure 7.14(b)].

This figure also illustrates what is referred to as a **position effect**. When the eye facet phenotypes of B/B and B^+/B^D flies are compared, an average of 68 and 45 facets are found, respectively. In both cases, there are two extra 16A regions. When the repeated segments are distributed on the same homologue instead of on two homologues, the phenotype is more pronounced. Thus, the same amount of genetic information produces a slightly different phenotype depending on the position of the genes.

The Role of Gene Duplication in Evolution

One of the most intriguing aspects of the study of evolution is the consideration of the mechanisms for genetic variation. The origin of unique gene products present in phylogenetically advanced organisms but absent in less advanced, ancestral forms is a topic of particular interest. In other words, how do "new" genes arise?

In 1970, Susumo Ohno published a provocative monograph entitled *Evolution by Gene Duplication* in which he suggests that gene duplication has been essential to the origin of new genes during evolution. Ohno's thesis is based on the supposition that the gene products of unique genes, present as only a single copy in the genome, are indispensable to the survival of members of any species during evolution. Therefore, unique genes are not free to accumulate mutations sufficient to alter their primary function and give rise to new genes.

However, if an essential gene were to become duplicated, the new copy would be inherited in all future generations. Since it is an extra copy, so to speak, major mutational changes in it are tolerated because the original gene still provides the genetic information for its essential function. The duplicated copy is now free to undergo large numbers of mutational changes within the organisms of any species over long periods of time. Over short intervals, the new genetic information may be of no practical advantage. However, over long periods of time, the gene may change sufficiently so that its product assumes a divergent role in the cell. The new function may impart an adaptability and survival advantage to organisms carrying such unique genetic information. Thus, Ohno has outlined a mechanism through which sustained genetic variability may have originated.

Ohno's thesis is based on the discovery of proteins with divergent function but whose amino acid sequence is very similar. According to current knowledge of the genetic code, amino acid sequence homology suggests nucleotide sequence homology. Similarity at the gene level suggests a common origin. So far, many pairs of molecules with related but distinct functions that share amino acid homology have been discovered. These include trypsin and chymotrypsin, myoglobin and hemoglobin, hemoglobin subunits, and the light and heavy immunoglobulin chains, to name but a few. As more and more genes and protein molecules are sequenced, the list continues to grow.

INVERSIONS

The **inversion,** another class of structural variation, is a type of chromosomal aberration in which a segment of a chromosome is turned around 180° within a chromosome. An inversion does not involve a loss of genetic information but simply rearranges the linear gene sequence. An inversion requires two breaks along the length of the chromosome and subsequent reinsertion of the inverted segment. Figure 7.15 illustrates how an inversion might arise. By forming a

chromosomal loop prior to breakage, the newly created "sticky ends" are brought close together and rejoined.

The inverted segment may be short or quite long and may or may not include the centromere. If the centromere is not part of the rearranged chromosome segment, the inversion is said to be **paracentric**. If the centromere is a part of the inverted segment, the term **pericentric** describes the inversion (Figure 7.15).

Although the gene sequence has been reversed in the paracentric inversion, the ratio of arm lengths extending from the centromere is unchanged. In contrast, some pericentric inversions create chromosomes with arms of different lengths than those of the noninverted chromosome. Thus, the **arm ratio** is often changed when a pericentric inversion is produced (Figure 7.16). The change in arm lengths may sometimes be detected during the metaphase stage of mitotic or meiotic divisions.

Although inversions may seem to have a minimal impact on the genetic material, their consequences are of interest to geneticists. Organisms heterozygous for inversions may produce aberrant gametes. Inversions may also result in position effects as well as play an important role in the evolutionary process.

Consequences of Inversions During Gamete Formation

If only one member of a homologous pair of chromosomes has an inverted segment, normal linear synapsis during meiosis is not possible. Organisms with one inverted chromosome and one noninverted homologue are called **inversion heterozygotes**. As shown in Figure 7.17, pairing between two such chromosomes in meiosis may be accomplished only if they form an **inversion loop**. In other cases, no loop

FIGURE 7.15
One possible origin of a pericentric inversion.

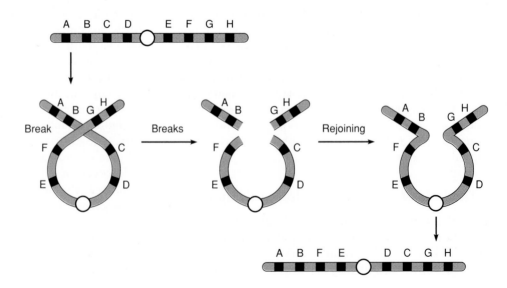

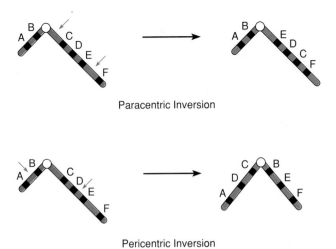

Paracentric Inversion

Pericentric Inversion

FIGURE 7.16
A comparison of the arm ratios before and after the occurrence of a paracentric and pericentric inversion.

can be formed and the homologues are seen to synapse everywhere but along the length of the inversion, where they appear separated.

If crossing over does not occur within the inverted segment of the inversion heterozygote, the homologues will segregate and result in two normal and two inverted chromatids which are distributed into gametes. However, if crossing over occurs within the inversion loop, abnormal chromatids are produced. The effects of single exchange events within paracentric and pericentric inversions are diagramed in Figure 7.17.

As in any meiotic tetrad, a single crossover between nonsister chromatids produces two parental chromatids and two recombinant chromatids. In the case of a paracentric inversion, one recombinant chromatid is **dicentric** (two centromeres) and one recombinant chromatid is **acentric** (lacking a centromere). Both contain duplications and deletions of chromosome segments as well. During anaphase, an acentric chromatid moves randomly to one pole or the other or may be lost, while a dicentric chromatid is pulled in two directions. This polarized movement produces **dicentric bridges** that are cytologically recognizable. A dicentric chromatid will usually break at some point so that part of the chromatid goes into one gamete and part into another gamete during the reduction divisions. Therefore, gametes containing either recombinant chromatid are deficient in genetic material. When such a gamete participates in fertilization, the zygote most often develops abnormally, if at all.

A similar chromosomal imbalance is produced as a result of a crossover event between a chromatid bearing a pericentric inversion and its noninverted homologue. Following meiotic divisions, each tetrad yields two parental chromatids containing the complete chromosome complement of genes. The recombinant chromatids that are directly involved in the exchange have duplications and deletions. However, no acentric or dicentric chromatids are produced. Gametes receiving these chromatids also produce inviable embryos following their participation in fertilization.

Because fertilization events involving these aberrant chromosomes do not produce viable offspring, it appears as if the inversion suppresses crossing over since crossover gametes are not recovered in the offspring. Actually, in inversion heterozygotes, the inversion has the effect of **suppressing the recovery of crossover products when chromosome exchange occurs within the inverted region**. If crossing over always occurred within a paracentric or pericentric inversion, 50 percent of the gametes would be ineffective. The viability of the resulting zygotes is therefore greatly diminished. Furthermore, up to one-half of the viable gametes have the inverted chromosome, and the inversion will be perpetuated within the species. The cycle will be repeated continuously during meiosis in future generations.

Position Effects of Inversions

Another consequence of inversions involves the new positioning of genes relative to other genes and particularly to areas of the chromosome which do not contain genes, such as the centromere. If the expression of the gene is altered as a result of its relocation, a change in phenotype may result. Such a change is an example of what is called a **position effect**, as introduced in our earlier discussion of the *Bar* duplication.

In *Drosophila* females heterozygous for the sex-linked recessive mutation *white* eye (w^+/w), the X-chromosome bearing the wild-type allele (w^+) may be inverted such that the *white* locus is moved to a point adjacent to the centromere. If the inversion is not present, a heterozygous female has wild-type red eyes, since the *white* allele is recessive. Females with the X chromosome inversion have eyes that are mottled or variegated (i.e., with red and white patches). Placement of the w^+ allele next to the centromere apparently causes the loss of complete dominance over the w allele. Other genes, also located on the X chromosome, behave in a similar manner when they are relocated. Reversion to wild-type expression has sometimes been noted. When this has occurred, cytological examination has shown that the inversion has been reversed to give the normal gene sequence.

FIGURE 7.17
The effects of a single-crossover event between nonsister chromatids at a point within a paracentric (a) and pericentric (b) inversion loop.

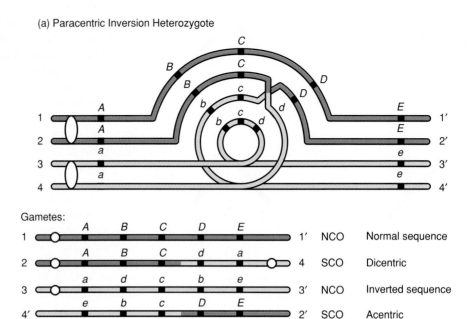

(a) Paracentric Inversion Heterozygote

Gametes:

1	A B C D E	1'	NCO	Normal sequence
2	A B C d a	4	SCO	Dicentric
3	a d c b e	3'	NCO	Inverted sequence
4'	e b c D E	2'	SCO	Acentric

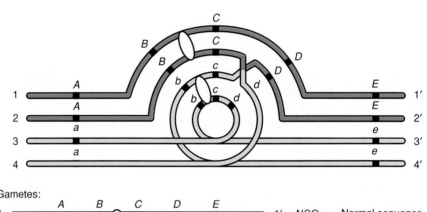

(b) Pericentric Inversion Heterozygote

Gametes:

1	A B C D E	1'	NCO	Normal sequence
2	A B C d a	4	SCO	Duplication and deletion
3	a d c b e	3'	NCO	Inverted sequence
2'	E D c b e	4'	SCO	Duplication and deletion

Evolutionary Consequences of Inversions

One major effect of an inversion is the maintenance of a set of specific alleles at a series of adjacent loci, provided they are contained within the inversion. Because the recovery of crossover products is suppressed in inversion heterozygotes, a particular gene sequence is preserved intact in the viable gametes. If this gene order provides a survival advantage to organisms maintaining it, the inversion is beneficial to the evolutionary survival of the species.

For example, if the set of alleles *ABcDef* is more adaptive than the sets *AbCdeF* or *abcdEF*, the favorable set will not be disrupted by crossing over if it is maintained within a heterozygous inversion. As we will see in Chapter 26, there are documented examples where inversions are adaptive in this way. Specifically, Theodosius Dobzhansky has shown that the maintenance of different inversions on chromosome 3 of *Drosophila pseudoobscura* through many generations has been highly adaptive to this species. Certain inversions seem to be characteristic of enhanced survival under specific environmental conditions.

TRANSLOCATIONS

Translocation, as the name implies, involves the movement of a segment of a chromosome to a new place in the genome. Translocation may occur within a single chromosome or between nonhomologous chromosomes. The exchange of segments between two nonhomologous chromosomes is a type of structural variation called a **reciprocal translocation**. The origin of a relatively simple reciprocal exchange is illustrated in Figure 7.18(a). The least complex way

for this event to occur is for two nonhomologous chromosome arms to come close to each other so that an exchange is facilitated. For this type of translocation, only two breaks are required. If the exchange includes internal chromosome segments, four breaks are required, two on each chromosome.

The genetic consequences of reciprocal translocations are, in several instances, similar to those of inversions. For example, genetic information is not lost or gained. Rather, there is only a rearrangement of genetic material. The presence of a translocation does not, therefore, directly alter viability of individuals bearing it. Like an inversion, a translocation may also produce a position effect, because it may realign certain genes in relation to other genes. This ex-

FIGURE 7.18
Origin, synapsis, and gamete formation of a simple reciprocal translocation.

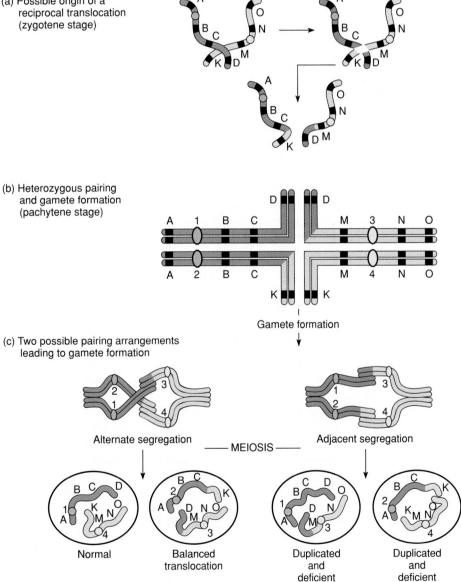

(a) Possible origin of a reciprocal translocation (zygotene stage)

(b) Heterozygous pairing and gamete formation (pachytene stage)

Gamete formation

(c) Two possible pairing arrangements leading to gamete formation

Alternate segregation —— MEIOSIS —— Adjacent segregation

Normal   Balanced translocation   Duplicated and deficient   Duplicated and deficient

change may, however, create a new genetic linkage relationship that can be detected experimentally.

In Figure 7.18(b), homologues heterozygous for a reciprocal translocation are shown to undergo unorthodox synapsis during meiosis, resulting in a cross-like configuration. As with inversions, genetically unbalanced gametes are also produced as a result of this unusual alignment during meiosis. In the case of translocations, however, aberrant gametes are not necessarily the result of crossing over. To see how unbalanced gametes are produced, focus on the homologous centromeres in Figure 7.18(b). According to the principle of independent assortment, the chromosome containing centromere 1 will migrate randomly toward one pole of the spindle during the first meiotic anaphase; it will travel with **either** the chromosome having centromere 3 **or** centromere 4. The chromosome with centromere 2 will move to the other pole along with **either** the chromosome containing centromere 3 **or** centromere 4. This results in four potential meiotic products. The 1,4 combination contains chromosomes uninvolved in the translocation. The 2,3 combination, however, contains translocated chromosomes. These contain a complete complement of genetic information and are balanced. The other two potential products, the 1,3 and 2,4 combinations, will contain chromosomes displaying duplicated and deleted segments. When incorporated into gametes, the resultant meiotic products are genetically unbalanced. If they participate in fertilization, lethality is the usual result. Therefore, only about 50 percent of the progeny of parents heterozygous for a reciprocal translocation survive. This condition is called **semisterility**.

Translocations in Humans: Familial Down Syndrome

Research performed since 1959 has revealed numerous translocations in members of the human population. One common type of translocation involves breaks at the extreme ends of the short arms of two nonhomologous acrocentric chromosomes. These small segments are lost, and the larger segments fuse at their centromeric region (Figure 7.19). This type of translocation produces a new, large submetacentric or metacentric chromosome and is called a **centric fusion translocation**. Such an occurrence in other organisms is called a **Robertsonian fusion**.

Just such a translocation accounts for cases in which Down syndrome is inherited or familial. Earlier in this chapter we pointed out that most instances of Down syndrome are due to trisomy 21. This chromosome composition results from nondisjunction during meiosis of one parent. Trisomy

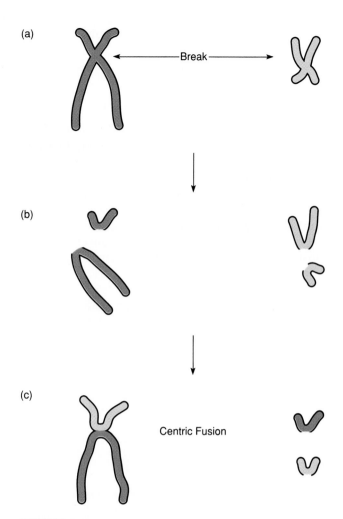

FIGURE 7.19
The possible origin of a Robertsonian translocation.
(a) Two independent breaks occur at the centromeric region on two nonhomologous chromosomes.
(b) Fragments produced. (c) Centric fusion of the long arms of the two acrocentric chromosomes.

accounts for over 95 percent of all cases of Down syndrome. In such cases, the chance of the same parents producing a second afflicted child is extremely low. However, in the remaining families with a Down child, the syndrome occurs in a much higher frequency over several generations.

Cytogenetic studies of the parents and their offspring from these unusual cases have explained the cause of **familial Down syndrome**. Analysis has revealed that one of the parents contains a **14/21 D/G translocation**. That is, one parent has the majority of the G-group chromosome 21 translocated to one end of the D-group chromosome 14. This individual has only 45 chromosomes, but apparently has a nearly complete diploid content of genetic information. During meiosis (Figure 7.20), one-fourth of the individual's gametes will have two copies of chromosome

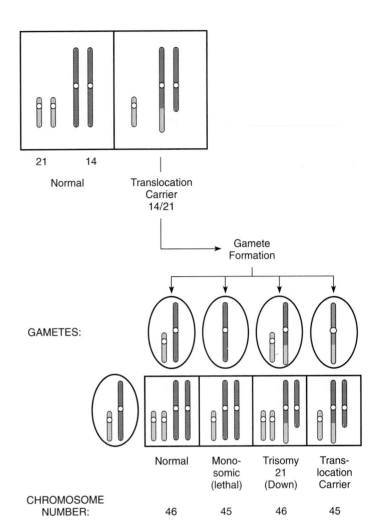

FIGURE 7.20
The origin of a D/G 14/21 translocation leading to familial Down syndrome.

21: a normal chromosome and a second copy translocated to chromosome 14. When such a gamete is fertilized by a standard haploid gamete, the resulting zygote has 46 chromosomes but three copies of chromosome 21. These individuals exhibit Down syndrome. Other potential surviving offspring contain either the standard diploid genome without a translocation or the balanced translocation like the parent. Both cases result in normal individuals. Knowledge of translocations has allowed geneticists to resolve the seeming paradox of an inherited trisomic phenotype in an individual with an apparent diploid number of chromosomes.

FRAGILE SITES IN HUMANS

We conclude this chapter by briefly discussing the results of an interesting discovery made around 1970 during observations of metaphase chromosomes pre-

pared following human cell culture. In cells derived from certain individuals, a specific area along one of the chromosomes failed to stain, giving the appearance of a gap. In different individuals whose chromosomes displayed or expressed such morphology, the gap appeared in different positions within the set of chromosomes. Such areas eventually become known as **fragile sites,** since they were considered susceptible to chromosome breakage when cultured in the absence of certain chemicals such as folic acid, which is normally present in the medium. Fragile sites were at first considered curiosities until a strong association was subsequently shown to exist between one of the sites and a form of mental retardation.

The molecular nature of fragile sites is unknown. Because they appear to represent points along the chromosome susceptible to breakage, these sites may indicate regions that are not tightly coiled or compacted. It should be noted that almost all studies of fragile sites have been carried out in mitotically dividing cells, and it is unknown whether such sites are also expressed in meiotic cells.

Most fragile sites do not appear to be associated with any clinical syndrome except for the folate-sensitive site on the X chromosome (Xq27) shown in Figure 7.21. Such a chromosome is now called the **fragile X chromosome**. This site is associated with an X-linked form of mental retardation known as **fragile X-linked mental retardation** or the **Martin-Bell syndrome (MBS)**. Males afflicted with this condition exhibit long narrow faces with protruding chins, large ears, increased testicular volume, and varying degrees of mental retardation. About 3 to 5 percent of males institutionalized for mental retardation have a fragile X chromosome. Female carriers who have one fragile X and one normal X chromosome show no clearcut physical characteristics, but have a higher incidence of mental retardation than normal individuals.

It is uncertain whether the fragile site on the X chromosome is itself the cause of the physical and mental characteristics seen in the syndrome, or whether the site serves as a marker for a closely linked gene that causes retardation. Answers to this important question must await further investigation.

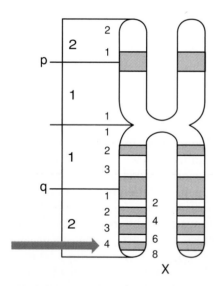

FIGURE 7.21
An illustration of the region of the human X chromosome associated with fragile X-linked mental retardation. The arrow identifies the fragile site (Xq27) of the G-banded chromosome (see Chapter 17).

CHAPTER SUMMARY

1 Investigations into the uniqueness of each organism's chromosomal constitution have further enhanced our understanding of genetic variation. Alterations of the precise diploid content of chromosomes are referred to as chromosomal aberrations or chromosomal mutations.

2 Any deviations from the expected chromosomal number, or mutations in the structure of the chromosome, are inherited in predictable Mendelian fashion; they often result in inviable organisms or substantial changes in the phenotype.

3 Aneuploidy is the gain or loss of one or more chromosomes from the diploid content, resulting in conditions of monosomy, trisomy, tetrasomy, etc. Studies of monosomy in Turner syndrome, and trisomic disorders in the Down, Patau, and Edwards syndromes have increased our understanding of the delicate genetic balance that must exist in order for normal development to occur.

4 When complete sets of chromosomes are added to the diploid genome, polyploidy is created. These sets may have identical or diverse genetic origin, creating either autopolyploidy or allopolyploidy, respectively. Polyploidy is not tolerated well in animals, but it has been especially important in the evolution and domestication of plants.

5 Large segments of the chromosome may be modified by deletions, duplications, inversions, and translocations. Deletions may produce serious conditions such as the cri-du-chat syndrome in humans, while duplications may be particularly important as a source of redundant or unique genes. Inversions and translocations, while altering the gene order, cause no loss of genetic information or deleterious effects initially. However, heterozygous combinations may cause genetically abnormal gametes following meiosis, reducing the fecundity of organisms.

6 Fragile sites in human mitotic chromosomes have sparked research interest because one such site on the X chromosome is associated with mental retardation. The significance of this and other sites is not yet clear.

Insights and Solutions

1 In a cross using maize involving three linked genes, *a*, *b*, and *c*, a heterozygote (*abc/+++*) was test-crossed (to *abc/abc*). Even though the three genes were separated by at best five map units, only two phenotypes were recovered: *abc* and *+++*. Additionally, the cross produced significantly fewer viable plants than expected. Can you propose why no other phenotypes were recovered and why the viability was reduced?

ANSWER: One of the two chromosomes contains an inversion that overlaps all three genes, effectively precluding the recovery of any "crossover" offspring. If this is a pericentric inversion and the genes are clearly separated (assuring that a significant number of crossovers will occur between them), then numerous acentric and dicentric chromosomes will be formed, resulting in the observed reduction in viability.

2 If a haplo-IV female *Drosophila* (containing only one chromosome 4, but an otherwise normal set of chromosomes) that has white eyes (a sex-linked trait) and normal bristles is crossed with a male with a diploid set of chromosomes and normal red eye color, but who is homozygous for the recessive chromosome 4 mutation, *shaven*, (*sv*), what F_1 phenotypic ratio might be expected?

ANSWER: Let's first consider only the eye color phenotypes. This is a straightforward sex-linked cross. Offspring will appear as 1/2 red females : 1/2 white males as shown below:

P_1:   *ww*   × w^+/ ⌃
         white female   red male

F_1:   1/2 ww^+ :   1/2 *w*/ ⌃
         red female   white male

The bristle phenotypes will be governed by the fact that the normal-bristled P_1 female produces gametes, one-half of which contain a chromosome 4 (sv^+) and one-half that have no chromosome 4. Following fertilization by sperm from the *shaven* male, one-half of the offspring will receive two members of chromosome 4 and be heterozygous for *sv*, expressing normal bristles. The other half will have only one copy of chromosome 4. Since its origin is from the male parent, where the chromosome bears the *sv* allele, these flies will express *shaven* since there is no wild-type allele present to mask this recessive allele.

P_1: sv^+/—   ×   *sv/sv*
normal females   shaven male

F_1:   1/2 sv^+/*sv* males and females
         1/2 *sv*/— males and females

Using the forked-line method, we can consider both eye color and bristle phenotypes together:

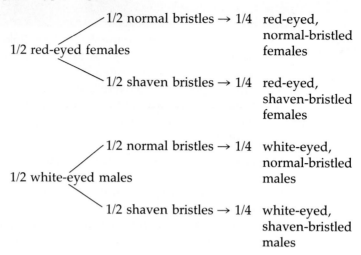

PROBLEMS AND DISCUSSION QUESTIONS

1 Define and distinguish between the following pairs of terms:

 > aneuploidy/euploidy
 > monosomy/trisomy
 > Patau syndrome/Edwards syndrome
 > autopolyploidy/allopolyploidy
 > polyteny/endopolyploidy
 > paracentric inversion/pericentric inversion

2 Contrast the relative survival times of individuals with Down syndrome, Patau syndrome, and Edwards syndrome. Why do you think such differences exist?

3 What evidence suggests that Down syndrome is more often the result of nondisjunction during oogenesis rather than during spermatogenesis?

4 Why are humans with aneuploid karyotypes usually inviable?

5 What conclusions have been drawn about human aneuploidy as a result of karyotypic analyses of abortuses?

6 Compare the fertility of allopolyploids and autopolyploids.

7 Discuss the role of polyploidy in evolution. How is its role different in animal and plant species?

8 When two plants belonging to the same genus but different species were crossed together, the F_1 hybrid was more viable and had more ornate flowers. Unfortunately, this hybrid was sterile and could only be propagated by vegetative cuttings. Explain the sterility of the hybrid. How might a horticulturist attempt to reverse its sterility?

9 What experimental techniques may be used to convert diploid plants to tetraploids?

10 In Chapter 5, we discussed Calvin Bridges' work on sex determination in *Drosophila*. Based on the information on euploidy in this chapter, describe how Bridges used triploid females to obtain the flies in Figure 5.8?

11 Discuss the origin of cultivated American cotton.

12 For a species with a diploid number of 18, indicate how many chromosomes will be present in the somatic nuclei of individuals who are haploid, triploid, tetraploid, trisomic, and monosomic.

13 Discuss the possible mechanisms involved in the production of deletions, duplications, inversions, and translocations.

14 Contrast the synaptic configurations of homologous pairs of chromosomes in which one member is normal and the other member has sustained a deletion or duplication.

15 Contrast the synaptic configurations in an inversion heterozygote and a translocation heterozygote.

16 Inversions are said to "suppress crossing over." Is this terminology technically correct?

17 Contrast the genetic composition of gametes derived from tetrads of inversion heterozygotes where crossing over occurs with a paracentric and pericentric inversion.

18 In a cross in *Drosophila*, a female heterozygous for the autosomally linked genes *a*, *b*, *c*, *d*, and *e* (*abcde/+++++*) was test-crossed to a male homozygous for all recessive alleles. Even though the distance between each of the above loci was at least 3 map units, only four phenotypes were recovered:

Phenotype	No. of Flies
+ + + + +	440
a b c d e	460
+ + + + e	48
a b c d +	52
Total	1000

Why are many expected crossover phenotypes missing? Can any of these loci be mapped from the data given here? If so, determine map distances.

19 Contrast the *Notch* locus with the *Bar* locus in *Drosophila*. What phenotypic ratios would be produced in a cross between *Notch* females and *Bar* males?

20 Under what circumstances can gene duplication be essential to an organism's survival?

21 Discuss Ohno's hypothesis on the role of gene duplication in the process of evolution.

22 What roles have inversions and translocations played in the evolutionary process?

23 A human female with Turner syndrome also expresses the sex-linked trait hemophilia, as her father did. Which parent underwent nondisjunction during meiosis, giving rise to the gamete responsible for the syndrome?

24 The primrose, *Primula kewensis*, has 36 chromosomes that are similar in appearance to the chromosomes in two related species, *Primula floribunda* ($2n = 18$) and *Primula verticillata* ($2n = 18$). How could *P. kewensis* arise from these species? How would you describe *P. kewensis* in genetic terms?

25 Varieties of chrysanthemums are known that contain 18, 36, 54, 72, and 90 chromosomes; all are multiples of a basic set of 9 chromosomes. How would you describe these varieties genetically? What feature is shared by the karyotypes of each variety? A variety with 27 chromosomes was discovered, but it was sterile. Why?

26 What is the effect of a rare double crossover within either a pericentric or paracentric inversion present heterozygously?

27 If the balanced reciprocal translocation shown below occurred in a gametic cell, would it still be possible for the translocated chromosomes to synapse with their normal homologues during meiosis? If so, draw the synaptic configuration.

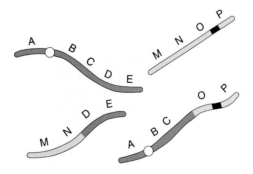

28 Many individuals who are heterozygous for a balanced reciprocal translocation are semisterile since 50 percent of their gametes contain duplications and deletions.

The other 50 percent of their gametes are either normal or balanced like the parent (see Figure 7.18). Will this be the case in the gametes formed by the individual described in the previous problem (7.27)? Defend your answer.

29 *Drosophila* may be monosomic or disomic for chromosome 4 and remain fertile. Contrast the F_1 and F_2 results of the following crosses involving the recessive chromosome 4 trait, *bent* bristles:
   (a) monosomic bent × disomic normal
   (b) monosomic normal × disomic bent

30 *Drosophila* may also be trisomic for chromosome 4 and remain fertile. Predict the F_1 and results of crossing:

trisomic bent (*b/b/b*) × normal disomic (*B/B*)

∞∞∞∞∞∞∞∞∞∞∞∞∞∞∞∞∞∞∞∞∞

SELECTED READINGS

BAKER, W. K. 1968. Position effect variegation. *Adv. in Genet.* 16: 133–69.

BEASLEY, J. O. 1942. Meiotic chromosome behavior in species, species hybrids, haploids, and induced polyploids of *Gossypium. Genetics* 27: 25–54.

BLAKESLEE, A. F. 1934. New jimson weeds from old chromosomes. *J. Hered.* 25: 80–108.

BORGAONKER, D. S. 1989. *Chromosome variation in man: A catalogue of chromosomal variants and anomalies.* 5th ed. New York: Alan R. Liss.

BOUE, A. 1985. Cytogenetics of pregnancy wastage. *Adv. Hum. Genet.* 14: 1–58.

BURGIO, G. R., et al., eds. 1981. *Trisomy 21.* New York: Springer-Verlag.

CARR, D. H. 1971. Genetic basis of abortion. *Ann. Rev. Genet.* 5: 65–80.

CUMMINGS, M. R. 1991. *Human heredity: Principles and issues.* 2nd ed. St. Paul: West.

DeARCE, M. A., and KEARNS, A. 1984. The fragile X syndrome: The patients and their chromosomes. *J. Med. Genet.* 21: 84–91.

ERICKSON, J. D. 1979. Paternal age and Down syndrome. *Amer. J. Hum. Genet.* 31: 489–97.

FELDMAN, M., and SEARS, E. R. 1981. The wild gene resources of wheat. *Scient. Amer.* (Jan.) 244: 102–12.

GUPTA, P. K., and PRIYADARSHAN, P. M. 1982. *Triticale:* Present status and future prospects. *Adv. in Genet.* 21: 256–346.

HASSOLD, T., et al. 1980. Effect of maternal age on autosomal trisomies. *Ann. Hum. Genet. Lond.* 44: 29–36.

HECHT, F. 1988. Enigmatic fragile sites on human chromosomes. *Trends in Genetics* 4: 121–22

HSU, T. C. 1979. *Human and mammalian cytogenetics: A historical perspective.* New York: Springer-Verlag.

HULSE, J. H., and SPURGEON, D. 1974. Triticale. *Scient. Amer.* (Aug.) 231: 72–81.

KAISER, P. 1984. Pericentric inversions: Problems and significance for clinical genetics. *Hum. Genet.* 68: 1–47.

KHUSH, G. S. 1973. *Cytogenetics of aneuploids.* Orlando: Academic Press.

LEWIS, E. B. 1950. The phenomenon of position effect. *Adv. in Genet.* 3: 73–115.

LEWIS, W. H., ed. 1980. *Polyploidy: Biological relevance.* New York: Plenum Press.

MANNING, C. H., and GOODMAN, H. O. 1981. Parental origin of chromosomes in Down's syndrome. *Hum. Genet.* 59: 101–3.

MANTELL, S. H., MATHEWS, J. A. and McKEE, R. A. 1985. *Principles of plant biotechnology: An introduction to genetic engineering in plants.* Oxford: Blackwell.

OBE, G., and BASLER, A. 1987. *Cytogenetics: Basic and applied aspects.* New York: Springer-Verlag.

OHNO, S. 1970. *Evolution by gene duplication.* New York: Springer-Verlag.

ROONEY, D. E., and CZEPULKOWSKI, B. H., eds. 1986. *Human cytogenetics: A practical approach.* Oxford: IRL Press.

SCHIMKE, R. T., ed. 1982. *Gene amplification.* Cold Spring Harbor, NY: Cold Spring Harbor Laboratory.

SERRA, J. A. 1968. *Modern genetics.* Orlando: Academic Press.

SHEPARD, J., et al. 1983. Genetic transfer in plants through interspecific protoplast fusion. *Science* 21: 683–88.

SIMMONDS, N. W., ed. 1976. *Evolution of crop plants.* London: Longman.

SMITH, G. F., ed. 1984. *Molecular structure of the number 21 Chromosome and Down syndrome.* New York: New York Academy of Sciences.

STEBBINS, G. L. 1966. Chromosome variation and evolution. *Science* 152: 1463–69.

SUTHERLAND, G. 1984. The fragile X chromosome. *Int. Rev. Cytol.* 81: 107–43.

———. 1985. The enigma of the fragile X chromosome. *Trends in Genetics* 1: 108–11.

SWANSON, C. P., MERZ, T., and YOUNG, W. J. 1981. *Cytogenetics: The chromosome in division, inheritance, and evolution.* 2nd ed. Englewood Cliffs, NJ: Prentice-Hall.

TAYLOR, A. I. 1968. Autosomal trisomy syndromes: A detailed study of 27 cases of Edwards syndrome and 27 cases of Patau syndrome. *J. Med. Genet.* 5: 227–52.

THERMAN, E. 1980. *Human chromosomes.* New York: Springer-Verlag.

TJIO, J. H., and LEVAN, A. 1956. The chromosome number of man. *Hereditas* 42: 1–6.

TURNER, G., and JACOBS, P. 1983. Marker(X) linked mental retardation. *Adv. Hum. Genet.* 13: 53–112.

TURPIN, R., and LeJEUNE, J. 1969. *Human afflictions and chromosomal aberrations.* Oxford, England: Pergamon Press.

WHARTON, K. A., et al. 1985. *opa:* A novel family of transcribed repeats shared by the *Notch* locus and other developmentally regulated loci in *D. melanogaster. Cell* 40: 55–62.

WILKINS, L. E., BROWN, J. A., and WOLF, B. 1980. Psychomotor development in 65 home-reared children with cri-du-chat syndrome. *J. Pediatr.* 97: 401–5.

YUNIS, J. J., ed. 1977. *New chromosomal syndromes.* Orlando: Academic Press.

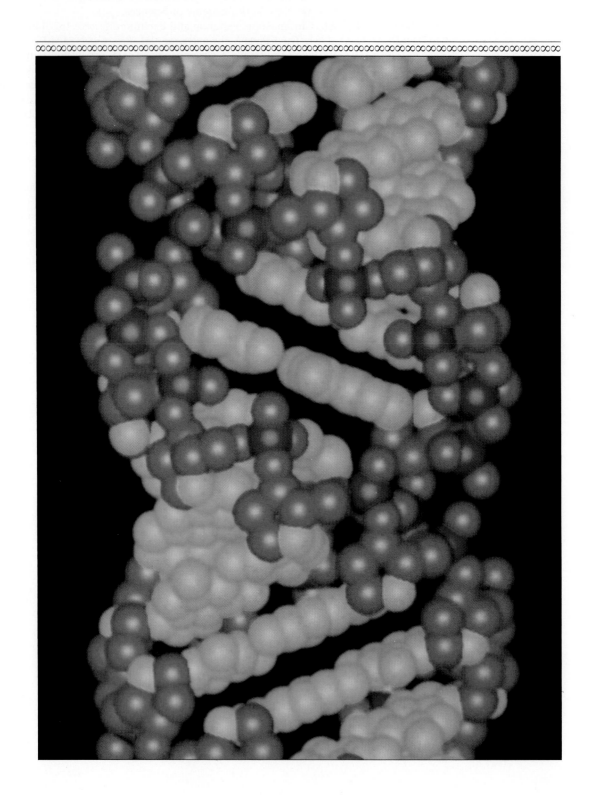

PART TWO

Molecular Basis of Heredity

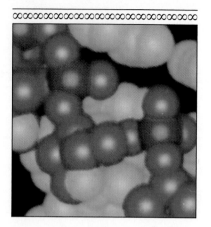

8

DNA—The Genetic Material

∞∞∞∞∞∞∞∞∞∞∞∞∞∞∞∞∞∞∞∞∞∞∞∞∞

CHAPTER CONCEPTS

With few exceptions, the nucleic acid DNA serves as the genetic material in every living thing. DNA provides the chemical basis for storing and expressing genetic information within the cell, as well as transmitting it to future generations. In some viruses, RNA plays this role.

In Part 1 of this text, we discussed the presence of genes on chromosomes that control phenotypic traits and the way in which the chromosomes are transmitted through gametes to future offspring. Logically, there must be some form of information contained in genes which, when passed to a new generation, influences the form and characteristics of the offspring; this is called the genetic information. We might also conclude that this same information in some way directs the many complex processes leading to the adult form.

Until 1944 it was not clear what chemical component of the cell constitutes the genetic material, which in turn contains the genetic information. This component was believed to be associated with chromosomes and their individual units, the genes. Chromosomes were known to have both a nucleic acid and a protein component. In 1944, however, there emerged the first direct experimental evidence that the nucleic acid, DNA, serves as the informational basis for the process of heredity. In this chapter we will discuss the era in which DNA was confirmed as the genetic material for almost every living thing.

CHARACTERISTICS OF THE GENETIC MATERIAL

The genetic material has several characteristics or functions: **replication, storage of information, expression of that information,** and **variation by mutation**. Replication of the genetic material is one facet of cell division, a fundamental property of all living organisms. Once the genetic material of somatic cells has been replicated, it may be partitioned equally into daughter cells during the process of mitosis. During the formation of gametes, the genetic material is also replicated but is partitioned so that each cell gets only

one-half of the original amount of genetic material. This process is called meiosis. Although the products of mitosis and meiosis are different, these processes are both part of the more general phenomenon of cellular reproduction.

The characteristic of storage may be interpreted as unexpressed genetic information. When is genetic information not expressed? Information contained in sperm cells is a good example. Sperm cells contain a complete haploid set of genetic information, but during their formation many genes are not expressed. For example, they do not contain numerous gene products: hemoglobin, the oxygen-carrying molecule; trypsin and chymotrypsin, the digestive enzymes; or melanin, the major pigment molecule in mammals. However, these cells do exhibit highly specialized structures and contain numerous molecules related to the fertilization process and subsequent development. These structures and molecules are dependent on the expression of numerous genes. Unexpressed versus expressed genetic information is a characteristic of all cells. The study of this aspect is an area of active research in both molecular and developmental genetics.

Expression of the information stored in the genetic material is a complex process and is the basis for the concept of **information flow** within the cell. Figure 8.1 shows a simplified illustration of this concept. The initial event is the **transcription** of genetic information stored in DNA. Transcription results in the synthesis of three types of RNA molecules: **messenger RNA (mRNA), transfer RNA (tRNA),** and **ribosomal RNA (rRNA)**. Of these, mRNAs are translated into proteins. Each type of mRNA is the product of a specific gene and leads to the synthesis of a different protein.

Translation, or protein synthesis, involves many molecular components, a supply of energy, and the

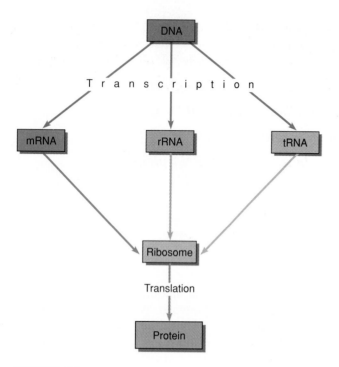

FIGURE 8.1
A simplified view of information flow involving DNA, RNA, and proteins within cells.

cellular organelle, the **ribosome**. The ribosome consists of several types of rRNA plus a variety of individual proteins. The role of tRNA is to adapt the information present in mRNA to the correct amino acids during translation. Amino acids are the building blocks of proteins. In eukaryotic cells, transcription occurs in the nucleus and translation occurs in the cytoplasm.

The genetic material is also responsible for newly arising variability among organisms through the process of mutation. If a change in the chemical composition of DNA occurs, the alteration will be reflected during transcription and translation, perhaps affecting the specified protein. If a mutation is present in gametes, it will be passed to future generations and, with time, may become distributed in the population. Genetic variation, which also includes rearrangements within and between chromosomes, provides the raw material for the process of evolution.

THE GENETIC MATERIAL: 1900–1944

The idea that genetic material is physically transmitted from parent to offspring has been accepted for as long as the concept of inheritance has existed. Beginning in the late nineteenth century, research into the structure of biomolecules progressed consid-

erably. Thus, the stage was set for the description of the genetic material in chemical terms. Although proteins and nucleic acid were both considered major candidates for the role of the genetic material, many geneticists, until the 1940s, favored proteins. Three factors contributed to this belief.

First, proteins are abundant in cells. Although the protein content may vary considerably, these molecules compose over 50 percent of the dry weight of cells. Since cells contain such a large amount and variety of proteins, it is not surprising that early geneticists believed that some of this protein could function as the genetic material.

The second factor was the accepted proposal for the chemical structure of nucleic acids during the early to mid-1900s. DNA was first studied in 1868 by Friedrich Miescher, a Swiss chemist. He was able to separate nuclei from the cytoplasm of cells and then isolate from them an acid substance that he called **nuclein**. Miescher showed that nuclein contained large amounts of phosphorus and no sulfur, characteristics that differentiate it from proteins.

As analytical techniques were improved, nucleic acids, including DNA, were shown to be composed of four similar molecules called nucleotides. Around 1910, Phoebus A. Levene proposed the **tetranucleotide hypothesis** to explain the chemical arrangement of these nucleotides in nucleic acids. He proposed a very simple four-nucleotide unit as shown in Figure 8.2. Levene based his proposal on studies of the composition of the four types of nucleotides. Although his actual data revealed proportions of the four that varied considerably, he assumed a 1:1:1:1

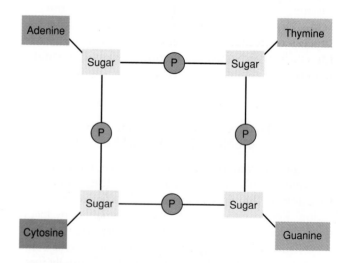

FIGURE 8.2
Levene's proposed structure of a DNA tetranucleotide containing one molecule each of the four nitrogenous bases: adenine, cytosine, guanine, and thymine.

ratio. The discrepancy was ascribed to inadequate analytical technique.

Since a single covalently bound tetranucleotide structure was relatively simple, geneticists believed nucleic acids could not provide the large amount of chemical variation expected for the genetic material. Proteins, on the other hand, are made up of 20 different amino acids, thus providing the basis for substantial variation. Thus, attention was directed away from nucleic acids as important genetic biomolecules, strengthening the speculation that proteins served as the genetic material.

The third contributing factor simply concerned the areas of most active research in genetics. Before 1940, most geneticists were engaged in the study of transmission genetics and mutation. The excitement generated in these areas undoubtedly diluted the concern for finding the precise molecule that serves as the genetic material. Thus, proteins were the most promising candidate and were accepted rather passively.

Between 1910 and 1930, other proposals for the structure of nucleic acids were advanced, but they were generally overturned in favor of the tetranucleotide hypothesis. It was not until the 1940s that the work of Erwin Chargaff led to the realization that Levene's hypothesis was incorrect. Chargaff showed that, for most organisms, the 1:1:1:1 ratio was indeed inaccurate, thus discrediting Levene's hypothesis.

EXPERIMENTAL EVIDENCE FOR DNA: EXPERIMENTS WITH PROKARYOTES AND VIRUSES

The 1944 publication by Oswald Avery, Colin Mac-Leod, and Maclyn McCarty concerning the chemical nature of a "transforming principle" in bacteria marked the initial event leading to the acceptance of DNA as the genetic material. Along with the subsequent findings of other research teams, this work constituted direct experimental proof that, in the organisms studied, DNA, and not protein, is the biomolecule responsible for heredity. This period marked the beginning of an era of discovery unprecedented in the history of biology. In the ensuing years numerous discoveries have revolutionized our understanding of the molecular basis of life on earth. Their impact on biology parallels the work that followed the publication of Darwin's theory of evolution and that following the rediscovery of Mendel's postulates of transmission genetics. Together, these constituted three great revolutions in biology.

The initial evidence implicating DNA as the genetic material was derived from studies of prokaryotic bacteria and viruses that infect them. The reasons for their use will become apparent as the experiments are studied. Primarily, bacteria and viruses are capable of rapid growth because they complete life cycles in hours. They may also be experimentally manipulated and mutations may be easily induced and selected. Thus, they are ideal for experimentation of this sort.

Transformation Studies

The research that provided the foundation for Avery, MacLeod, and McCarty's work was initiated in 1927 by Frederick Griffith, a medical officer in the British Ministry of Health. He performed experiments with several different strains of the bacterium *Diplococcus pneumoniae*. Some were **virulent** strains, which cause pneumonia in certain vertebrates (notably humans and mice), while others were **avirulent** strains, which do not cause illness.

The difference in virulence is related to the polysaccharide capsule of the bacterium. Virulent strains have this capsule, whereas avirulent strains do not. The nonencapsulated bacteria are readily engulfed and destroyed by phagocytic cells in the animal's circulatory system. Virulent bacteria, which possess the polysaccharide coat, are not easily engulfed; they multiply and cause pneumonia.

The presence or absence of the capsule is the basis for another characteristic difference between virulent and avirulent strains. Encapsulated bacteria form a **smooth**, shiny-surfaced colony (*S*) when grown on an agar culture plate; nonencapsulated strains produce **rough** colonies (*R*) (Figure 8.3). Thus, virulent and avirulent strains may be distinguished easily by standard microbiological culture techniques. Further, the smooth colonies are large compared with the colonies of the rough strains.

Each strain of *Diplococcus* may be one of dozens of different types called **serotypes**. The specificity of the serotype is due to the detailed chemical structure of the polysaccharide constituent of the thick, slimy capsule. Serotypes are identified by immunological techniques and are usually designated by Roman numerals. In the United States, types I and II are most common in causing pneumonia. Griffith used types II and III in the critical experiments that led to new concepts about the genetic material. Table 8.1 summarizes the characteristics of Griffith's two strains.

Griffith knew from the work of others that only living virulent cells would produce pneumonia in mice. If heat-killed virulent bacteria are injected into

TABLE 8.1

Strains of *Diplococcus pneumoniae* used by Frederick Griffith in his original transformation experiments.

Serotype	Colony Morphology	Capsule	Virulence
IIR	Rough	Absent	Avirulent
IIIS	Smooth	Present	Virulent

mice, no pneumonia results, just as living avirulent bacteria fail to produce the disease. Griffith's critical experiment (Figure 8.3) involved an injection into mice of living IIR (avirulent) cells combined with heat-killed IIIS (virulent) cells. Neither cell type caused death in mice when injected alone. Griffith expected that the double injection would not kill the mice, but after five days all mice receiving double injections were dead. Analysis of the blood of the dead mice revealed large numbers of living type IIIS (virulent) bacteria! As far as could be determined, these IIIS bacteria were identical to the IIIS strain from which the heat-killed cell preparation had been made. The control mice, injected only with living avirulent IIR bacteria for this set of experiments, did not develop pneumonia and remained healthy. This finding strongly suggested that the occurrence of the living IIIS bacteria in the dead mice was not caused by faulty technique or contamination, but that some interaction between the two types of injected bacteria had occurred.

Griffith suggested that the heat-killed IIIS bacteria were responsible for converting live avirulent IIR cells into virulent IIIS ones. Calling the phenomenon **transformation,** he suggested that the **transforming principle** might be some part of the polysaccharide capsule *or* some compound required for capsule synthesis, although the capsule alone did not cause pneumonia. To use Griffith's term, the transforming principle from the dead IIIS cells served as a "pabulum" for the IIR cells.

Griffith's work led other physicians and bacteriologists to research the phenomenon of transformation. By 1931, Henry Dawson, at the Rockefeller Institute, had confirmed Griffith's observations and extended his work one step further. Dawson and his coworkers showed that transformation could occur *in vitro* (in a test tube containing only bacterial cells); that is, injection into mice was not necessary for transformation to occur. By 1933, Lionel J. Alloway had refined the *in vitro* system by using crude extracts of *S* cells and living *R* cells. The soluble filtrate from the heat-killed *S* cells was as effective in inducing transformation as were the intact cells! Alloway and

others did not view transformation as a genetic event, but rather as a physiological modification of some sort. Nevertheless, the experimental evidence that a chemical substance was responsible for transformation was quite convincing.

Then, in 1944, after ten years of work, Avery, MacLeod, and McCarty published their results in what is now regarded as a classic paper in the field of molecular genetics. They reported that they had obtained the transforming principle in a highly purified state, and that beyond reasonable doubt it was DNA. The details of their work are outlined in Figure 8.4.

These researchers began their isolation procedure with large quantities (50–75 liters) of liquid cultures of type IIIS virulent cells. The cells were centrifuged, collected, and heat-killed. Following washing and several extractions with the detergent deoxycholate (DOC), they obtained a soluble filtrate which, when tested, still contained the transforming principle. Protein was removed from the active filtrate by several chloroform extractions, and polysaccharides were enzymatically digested and removed. Finally, precipitation with ethanol yielded a fibrous nucleic acid that still retained the ability to induce transformation of type IIR avirulent cells.

Further testing established beyond doubt that the transforming principle was DNA. Treatment was performed with proteolytic (protein-digesting) enzymes and an RNA-digesting enzyme, **ribonuclease**. Such treatment destroyed the activity of any remaining protein and RNA. Nevertheless, transforming activity was not diminished. Chemical testing of the final product gave strong positive reactions for DNA. The final confirmation came with experiments using crude samples of the DNA-digesting enzyme **deoxyribonuclease,** which was isolated from dog and rabbit sera. This digestion was shown to destroy transforming activity. There could be no doubt that the active transforming principle was DNA!

The great amount of work, the confirmation and reconfirmation of the conclusions drawn, and the brilliant logic involved in the research of these three scientists are truly impressive. The conclusion to the 1944 publication was, however, very simply stated: "The evidence presented supports the belief that a nucleic acid of the desoxyribose* type is the fundamental unit of the transforming principle of *Pneumococcus* type III."

Avery and his coworkers recognized the genetic and biochemical implications of their work. They

---

*Desoxyribose is now spelled *deoxyribose,* and the genus *Pneumococcus* is now referred to as *Diplococcus.*

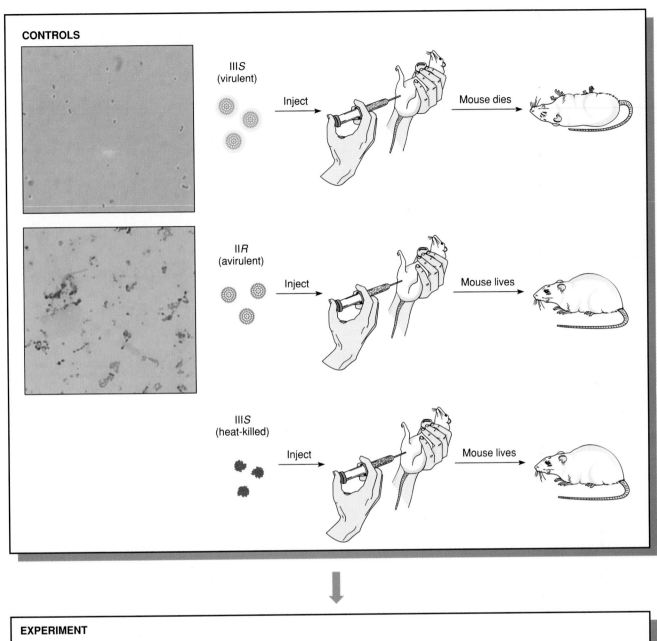

CONTROLS

IIIS
(virulent)

Inject → Mouse dies

IIR
(avirulent)

Inject → Mouse lives

IIIS
(heat-killed)

Inject → Mouse lives

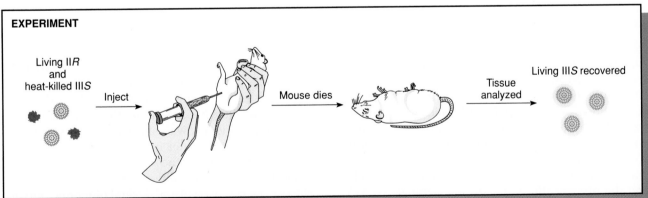

EXPERIMENT

Living IIR
and
heat-killed IIIS

Inject → Mouse dies → Tissue analyzed → Living IIIS recovered

FIGURE 8.3
Summary of Griffith's transformation experiment. The photographs show bacterial colonies containing cells with capsules (type IIIS) and without capsules (type IIR).

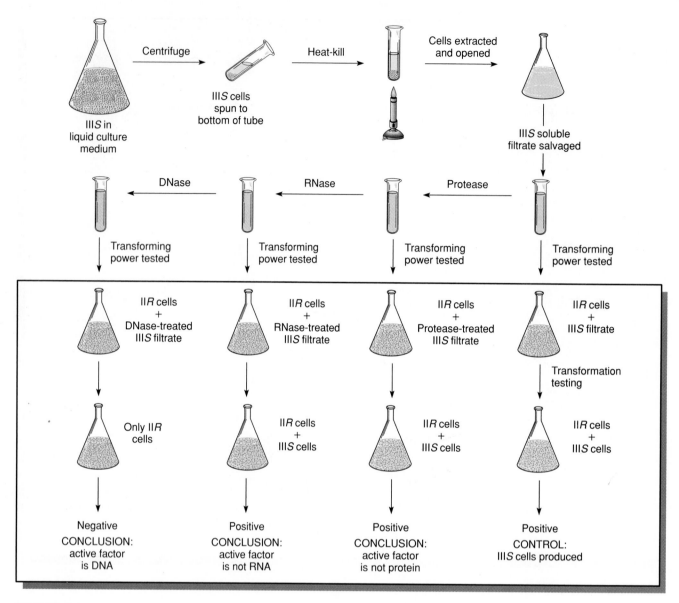

FIGURE 8.4
Summary of Avery, MacLeod, and McCarty's experiment demonstrating that DNA is
the transforming principle.

suggested that the transforming principle interacts with the *R* cell and gives rise to a coordinated series of enzymatic reactions that culminates in the synthesis of the type III capsular polysaccharide. They emphasized that, once transformation occurs, the capsular polysaccharide is produced in successive generations and that the transforming principle is replicated in daughter cells. Transformation is therefore heritable, and the process affects the genetic material.

Immediately after the publication of the report, several investigators turned to or intensified their studies of transformation in order to clarify the role of DNA in genetic mechanisms. In 1949, Harriet Taylor isolated an **extremely rough (ER)** mutant strain from

a rough (*R*) strain. This *ER* strain produced colonies that were more irregular than *R*. DNA from *R* accomplished the transformation of *ER* to *R*. Thus, the *R* strain, which served as the recipient in the Avery experiments, was shown also to be able to serve as the DNA donor in transformation.

Transformation has now been shown to occur in *Hemophilus influenzae, Bacillus subtilis, Shigella paradysenteriae, Escherichia coli,* and several other microorganisms. One line of work pursued transformation of genetic traits other than colony morphology. Many traits were found to be transformable, particularly ones involving resistance to antibiotics and the ability to metabolize various nutrients. Thus, geneticists believed that most, if not all, traits could be transformed under suitable experimental conditions.

The Hershey-Chase Experiment

The second major piece of evidence supporting DNA as the genetic material was provided by the study of the bacterial virus T2. This virus, also called a **bacteriophage** or just a **phage,** has as its host the bacterium *Escherichia coli* and consists of a protein coat surrounding a core of DNA. Electron micrographs have revealed the phage's external structure to be composed of a hexagonal head plus a tail. The life cycle of bacteriophages such as T2 is shown in Figure 8.5.

In 1952, Alfred Hershey and Martha Chase published the results of experiments designed to clarify the events leading to phage reproduction. Several of the experiments clearly established the independent functions of phage protein and nucleic acid in the reproduction process associated with the bacterial cell. Hershey and Chase knew from existing data that

1 T2 phages consist of approximately 50 percent protein and 50 percent DNA.
2 Infection is initiated by adsorption of the phage by its tail fibers to the bacterial cell.
3 The production of new viruses occurs within the bacterial cell.

It would appear that some molecular component of the phage, DNA and/or protein, enters the bacterial cell and directs viral reproduction. Which was it?

Hershey and Chase used radioisotopes to follow the molecular components of phages during infection. Both ^{32}P and ^{35}S, radioactive forms of phosphorus and sulfur, were used. Since DNA contains phosphorus but not sulfur, ^{32}P effectively labels DNA. Since proteins contain sulfur but not phosphorus, ^{35}S labels protein. **This is a key point in the experiment.** If *E. coli* cells are first grown in the presence of ^{32}P **or** ^{35}S and then infected with T2 viruses, the progeny phage will have either a labeled DNA core **or** a labeled protein coat, respectively. These radioactive phages may be isolated and used to infect unlabeled bacteria (Figure 8.6).

When labeled phage and unlabeled bacteria are mixed, an adsorption complex is formed as the phages attach their tail fibers to the bacterial wall. These complexes were isolated and subjected to a high shear force by placing them in a blender. This force strips off the attached phages, which may then be analyzed separately (Figure 8.6). By tracing the radioisotopes, Hershey and Chase were able to demonstrate that most of the ^{32}P-labeled DNA had

been transferred into the bacterial cell following adsorption; on the other hand, most of the ^{35}S-labeled protein remained outside the bacterial cell and was recovered in the phage "ghosts" (empty phage coats) after the blender treatment. Following this separation, the bacterial cells, which now contained viral DNA, were eventually lysed as new phages were produced.

Hershey and Chase interpreted these results to indicate that the protein of the phage coat remains outside the host cell and is not involved in the production of new phage. On the other hand, and most importantly, phage DNA enters the host cell and directs phage multiplication. Thus, they had demonstrated that in phage T2, DNA, not protein, is the genetic material.

This experimental work, along with that of Avery and his colleagues, provided convincing evidence to most geneticists that DNA was the molecule responsible for heredity. Since then, many significant findings have been based on this supposition. These many findings, constituting the field of molecular genetics, are discussed in detail in subsequent chapters.

Transfection Experiments

During the eight years following the publication of the Hershey-Chase experiment, additional research provided even more solid proof that DNA is the genetic material. These studies involved the same organisms used by Hershey and Chase.

In 1957, several reports demonstrated that if *E. coli* were treated with the enzyme **lysozyme,** the outer wall of the cell could be removed without destroying the bacterium. Enzymatically treated cells are naked, so to speak, and contain only the cell membrane as the outer boundary of the cell. Such structures are called **protoplasts,** or **spheroplasts**. John Spizizen and Dean Fraser independently reported that by using protoplasts, they were able to initiate phage multiplication with disrupted T2 particles. That is, provided protoplasts are used, it is not necessary for a virus to be intact in order for infection to occur.

Similar but refined experiments were reported in 1960 by George Guthrie and Robert Sinsheimer. DNA was purified from bacteriophage φX-174, a small phage that contains a single-stranded, circular DNA molecule of some 5386 nucleotides. When added to *E. coli* protoplasts, the purified DNA resulted in the production of complete φX-174 bacteriophages. This process of infection by only the viral nucleic acid, called **transfection,** proves conclusively that phage DNA alone contains all the necessary information for production of mature viruses. Thus, the evidence

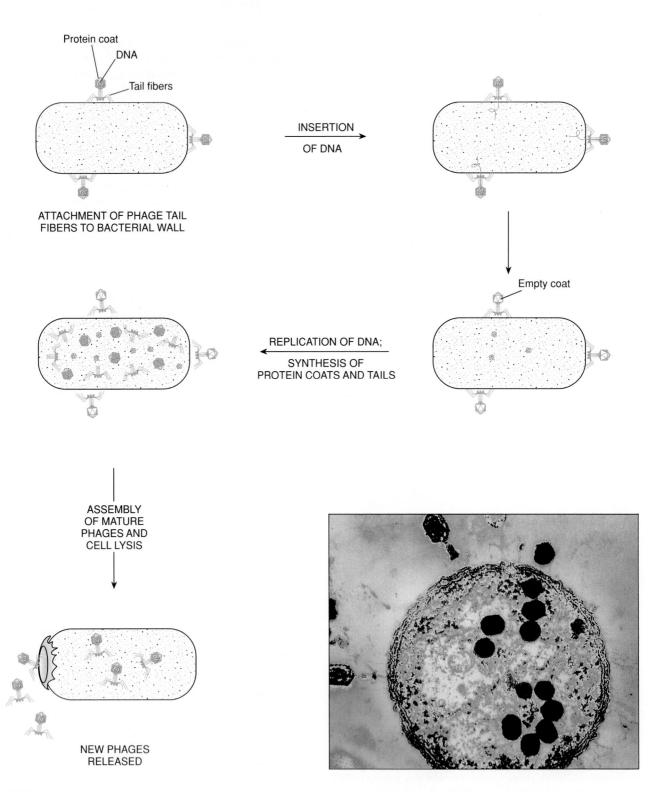

FIGURE 8.5
The life cycle of a T-even bacteriophage. The electron micrograph shows *E. coli* during infection by phage T2.

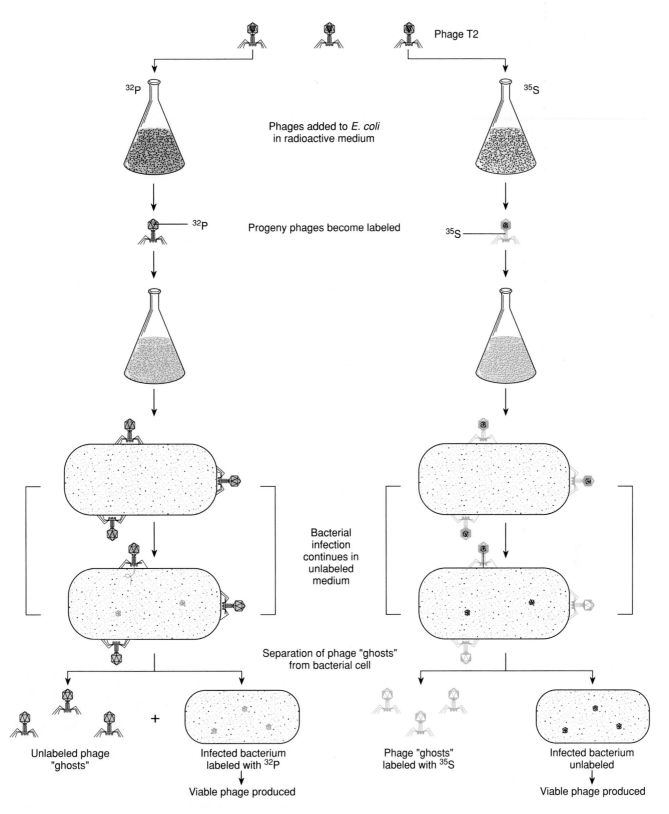

Phage T2

^{32}P

Phages added to *E. coli* in radioactive medium

^{35}S

^{32}P

Progeny phages become labeled

^{35}S

Bacterial infection continues in unlabeled medium

Separation of phage "ghosts" from bacterial cell

Unlabeled phage "ghosts"

+

Infected bacterium labeled with ^{32}P

Viable phage produced

Phage "ghosts" labeled with ^{35}S

Infected bacterium unlabeled

Viable phage produced

FIGURE 8.6
Summary of the Hershey-Chase experiment demonstrating that DNA and not protein is responsible for directing the reproduction of phage T2 during the infection of *E. coli*.

that DNA serves as the genetic material was further strengthened, even though all direct evidence had been obtained from bacterial and viral studies.

CIRCUMSTANTIAL EVIDENCE FOR DNA: EUKARYOTIC DATA

Most eukaryotic organisms are not amenable to the types of experiments performed to demonstrate that DNA is the genetic material in bacteria and viruses. Therefore, initial support for this concept in eukaryotes was based only on circumstantial evidence. Such evidence is indirect and derived from observations incidental to the hypothesis in question. While any single item of circumstantial evidence taken alone is insufficient to support the hypothesis, many different sorts of independent observations may lead to a common conclusion.

Distribution of DNA

The genetic material should be found where it functions—in the nucleus as part of chromosomes. Both DNA and protein fit this criterion. However, protein is also abundant in the cytoplasm, while DNA is not. Both mitochondria and chloroplasts are known to perform genetic functions, and DNA is also present in these organelles. Thus, DNA is only found where primary genetic function is known to occur. Protein, however, is found in all parts of the cell. These observations are consistent with an interpretation favoring DNA over protein as the genetic material.

Since it is generally accepted that chromosomes within the nucleus contain the genetic material, a correlation should exist between the ploidy of cells and the molecule that functions as the genetic material. Meaningful comparisons may be made between gametes (sperm and eggs) and somatic or body cells. The latter are recognized as being **diploid** ($2n$) and containing twice the number of chromosomes as gametes, which are **haploid** (n).

Table 8.2 compares the amount of DNA found in sperm and the nucleated precursor of erythrocytes from a variety of organisms. There is a close correlation between the amount of DNA (in picograms) and the number of sets of chromosomes.

No such correlation may be made between gametes and somatic cells for the other major classes of organic biomolecules, proteins, lipids, and carbohydrates. These data thus provide circumstantial or indirect evidence that again favors DNA over protein as the genetic material.

TABLE 8.2
DNA content of cells of various species (in picograms).

Species	Sperm (n)	Normablast ($2n$)
Human	3.25	7.30
Chicken	1.26	2.49
Trout	2.67	5.79
Carp	1.65	3.49
Shad	0.91	1.97

Mutagenesis

Ultraviolet light (UV) is one of a number of agents capable of inducing mutations in the genetic material. Bacteria and other simple organisms may be irradiated with various wavelengths of ultraviolet light, and the effectiveness of each wavelength measured by the number of mutations it induces. When the data are plotted, an **action spectrum** of ultraviolet light as a mutagenic agent is obtained. This action spectrum may then be compared with the **absorption spectrum** of any molecule suspected to be the genetic material (Figure 8.7). The molecule serving as the genetic material is expected to absorb at the wavelengths found to be mutagenic.

UV light is most mutagenic at the wavelength (λ) of 260 nanometers (nm). Both DNA and RNA absorb UV light most strongly at 260 nm. On the other hand, protein absorbs most strongly at 280 nm. Thus, this

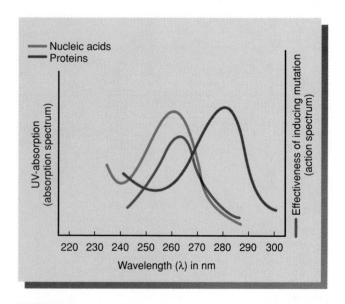

FIGURE 8.7
The absorption spectrum of nucleic acids and proteins when subjected to ultraviolet light compared to the action spectrum of ultraviolet light in inducing mutation.

indirect evidence supports both DNA and RNA as the genetic material and tends to exclude protein.

DIRECT EVIDENCE FOR DNA: EUKARYOTIC DATA

While the circumstantial evidence just described does not constitute direct proof that DNA is the genetic material in eukaryotes, these observations spurred researchers to forge ahead under this assumption. Today, there is no doubt of the validity of this conclusion; nor was there any reason to suspect otherwise.

The strongest evidence has been provided by a current experimental procedure called **recombinant DNA research**. In this procedure, segments of eukaryotic DNA corresponding to specific genes are isolated and literally spliced into bacterial DNA. Such a complex can be inserted into a bacterial cell and its genetic expression monitored. If a eukaryotic gene is selected that is absent in bacterial genetic information, the presence of the corresponding eukaryotic protein product demonstrates directly that this DNA is functional in expressing genetic information. This has been shown to be the case in numerous instances. For example, the human genes specifying the hormone insulin and the immunologically important molecule interferon are produced by bacteria following recombinant DNA procedures.

As the bacterium divides, the eukaryotic DNA is replicated along with the host DNA and is distributed to daughter cells. As divisions continue, the eukaryotic genes are also replicated, with each new bacterial cell containing an identical copy of the eukaryotic genetic information. As a result, large amounts of DNA containing specific eukaryotic genes may be isolated and studied in depth. This technique has paved the way for detailed analysis of the nucleotide sequence of specific genes, among other types of information.

The availability of vast amounts of DNA coding for specific genes has led to a second form of documentation that DNA serves as the genetic material. In the work of John Gurdon and others, billions of copies of specific genes were injected into oocytes of the frog *Xenopus laevis*. In these experiments, the egg served as a living test tube, and the transcription and translation of this genetic information were analyzed. Through such analysis, the products of these genes have been identified, confirming the informational role of DNA in genetic processes.

More recent work in the laboratory of Beatrice Mintz and others has further strengthened this evidence. This research has demonstrated that DNA encoding the human β-globin gene, when microinjected into a fertilized mouse egg, is later found in the adult mouse tissue and can be transmitted to that mouse's progeny! Such mice are referred to as **transgenic**. More recent work has introduced DNA representing the growth hormone gene from rats into fertilized mouse eggs. About one-third of the resultant animals grew to twice the size of normal mice, indicating that the foreign DNA was present *and* functional. Subsequent generations receive the gene and display the trait that it governs.

We will pursue the topic of recombinant DNA again later in the text (see Chapter 16). The point to be made here is that in eukaryotes DNA has been shown directly to meet the requirement of expression of genetic information. Later we will see how DNA is stored, replicated, expressed, and mutated.

RNA AS THE GENETIC MATERIAL

Some viruses contain an RNA core rather than one composed of DNA. In these viruses, it would thus appear that RNA must serve as the genetic material—an exception to the general rule that DNA performs this function. In 1956, it was demonstrated that when purified RNA from **tobacco mosaic virus (TMV)** was spread on tobacco leaves, the characteristic lesions caused by TMV would appear later on the leaves (Figure 8.8). It was concluded that RNA is the genetic material of this virus.

Soon afterwards, another type of experiment with TMV was reported by Heinz Fraenkel-Conrat and B. Singer, as illustrated in Figure 8.8. These scientists discovered that the RNA core and the protein coat from wild-type TMV and other viral strains could be isolated separately. In their work, RNA and coat proteins were separated and isolated from TMV and a second viral strain, **Holmes ribgrass (HR)**. Then, mixed viruses were reconstituted from the RNA of one strain and the protein of the other. When this "hybrid" virus was spread on tobacco leaves, the lesions that developed corresponded to the type of RNA in the reconstituted virus; that is, viruses with wild-type TMV RNA and HR protein coats produced TMV lesions and vice versa. Again, it was concluded that RNA serves as the genetic material in these viruses.

In 1965 and 1966, Norman R. Pace and Sol Spiegelman further demonstrated that RNA from the phage Qβ could be isolated and replicated *in vitro*. Replication was dependent on an enzyme, **Qβ RNA**

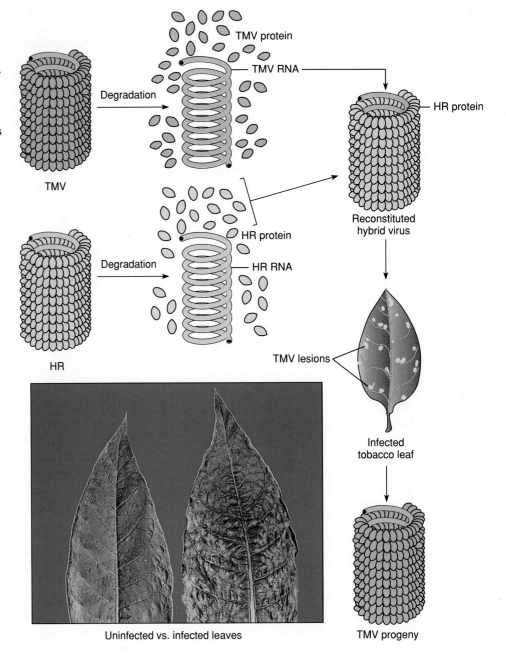

FIGURE 8.8
Reconstitution of hybrid tobacco mosaic viruses. In the hybrid, RNA is derived from the wild-type TMV virus, while the protein subunits are derived from the HR strain. Following infection, viruses are produced with protein subunits characteristic of the wild-type TMV strain and not those of the HR strain. The photograph shows TMV lesions on a tobacco leaf compared to an uninfected leaf.

TMV protein

TMV RNA

HR protein

TMV

Degradation

Reconstituted hybrid virus

HR protein

HR RNA

Degradation

HR

TMV lesions

Infected tobacco leaf

Uninfected vs. infected leaves

TMV progeny

replicase, which was isolated from host *E. coli* cells following normal infection. When the RNA replicated *in vitro* was added to *E. coli* protoplasts, infection and viral multiplication occurred. Thus, RNA synthesized in a test tube can amply serve as the genetic material in these phages.

Finally, one other group of RNA-containing viruses bears mentioning. These are the so-called **retroviruses** which replicate in an unusual way. Their RNA serves as a template for the synthesis of the complementary DNA molecule! The process, desig-

nated as reverse transcription, occurs under the direction of an RNA-dependent DNA polymerase enzyme called **reverse transcriptase**. Because the genetic material can be represented by this DNA intermediate, it may be incorporated into the genome of the host cell. Once present, if the DNA is expressed, transcription yields retroviral RNA chromosomes. Retroviruses, including the human immunodeficiency virus (HIV) that causes AIDS, will be discussed at greater length in Chapter 15.

CHAPTER SUMMARY

1   The existence of a genetic material capable of replication, storage, expression, and mutation is deducible from the observed patterns of inheritance in organisms.

2   Both proteins and nucleic acids were initially considered as the possible candidates for genetic material. Proteins are more functionally diverse than nucleic acids, a requirement for the genetic material, and were favored due to the advances being made in protein chemistry at the time. Additionally, Levene's tetranucleotide hypothesis had underestimated the magnitude of chemical diversity inherent in nucleic acid interactions.

3   By 1944, experimental evidence had invalidated Levene's hypothesis. Transformation studies, as well as experiments using bacteria infected with bacteriophages, strongly suggested that DNA was the genetic material for bacteria and most viruses.

4   Initially, only circumstantial evidence supported the concept of DNA controlling inheritance in eukaryotes. This included the distribution of DNA in the cell, quantitative analysis of DNA, and the UV-induced mutagenesis. Recent recombinant DNA techniques, as well as experiments with transgenic mice, have provided direct experimental evidence that the eukaryote genetic material is DNA.

5   Numerous viruses provide an important exception to this general rule, because many of them use RNA as their genetic material. These include bacteriophages as well as some plant and animal viruses.

6   The establishment of DNA as the genetic material paved the way for the expansion of molecular genetics research, and has served as the cornerstone for further important studies for nearly half a century.

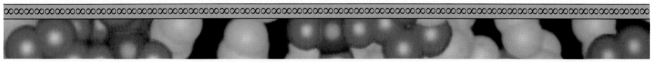

Insights and Solutions

In contrast to the preceding chapters, this one does not emphasize genetic problem solving. Instead, it recounts some of the initial experimental analysis that served as the cornerstone of modern genetics. Quite fittingly, then, our Insights and Solutions section shifts its emphasis to experimental rationale and analytical thinking, an approach that will continue through the remainder of the text whenever appropriate.

1   Based strictly on the transformation analysis of Avery, MacLeod, and McCarty, what objection might be made to the conclusion that DNA is the genetic material? What other conclusion might be considered?

    ANSWER: Based solely on their results, it may be concluded that DNA is essential for transformation. However, DNA might have been a substance that caused capsular formation by *directly* converting nonencapsulated cells to ones with a capsule. That is, DNA may simply have played a catalytic role in capsular synthesis, leading to cells displaying smooth type III colonies.

2   What observations argue against this objection?

    ANSWER: First, transformed cells pass the trait on to their progeny cells, thus supporting the conclusion that DNA is responsible for heredity, not for the direct production of polysaccharide coats. Second, subsequent transformation studies over the next five years showed that other traits, such as antibiotic resistance, could be transformed. Therefore, the transforming factor has a broad general effect, not one specific to polysaccharide synthesis. This observation is more in keeping with the conclusion that DNA is the genetic material.

3 If RNA was the universal genetic material, how would this have affected the Avery experiment and the Hershey-Chase experiment?

ANSWER: In the Avery experiment, RNase rather than DNase would have eliminated transformation. Had this occurred, Avery and his colleagues would have concluded that RNA was the transforming factor. Hershey and Chase would have received identical results, since ^{32}P would also label RNA, but not protein. Had they been using a bacteriophage with RNA as its nucleic acid, and had they known this, they would have concluded that RNA was responsible for directing the reproduction of their bacteriophage.

PROBLEMS AND DISCUSSION QUESTIONS

1 The functions ascribed to the genetic material are replication, expression, storage, and mutation. What does each of these terms mean?

2 Discuss the reasons why proteins were generally favored over DNA as the genetic material before 1940. What was the role of the tetranucleotide hypothesis in this controversy?

3 Contrast the various contributions made to an understanding of transformation by Griffith, Avery and his coworkers, and Taylor.

4 When Avery and his colleagues had obtained what was concluded to be purified DNA from the IIIS virulent cells, they treated the fraction with proteases, RNase, and DNase, followed by the assay for retention or loss of transforming ability. What were the purpose and results of these experiments? What conclusions were drawn?

5 Based on the information on transformation presented in this chapter, what other aspects of the process might you investigate to understand it more fully?

6 Why were ^{32}P and ^{35}S chosen for use in the Hershey-Chase experiment? Discuss the rationale and conclusions of this experiment.

7 Does the design of the Hershey-Chase experiment distinguish between DNA and RNA as the molecule serving as the genetic material? Why or why not?

8 Would an experiment similar to that performed by Hershey and Chase work if the basic design were applied to the phenomenon of transformation? Explain why or why not.

9 Why is the early evidence that DNA serves as the genetic material in eukaryotes called circumstantial? List and discuss these evidences.

10 What are the exceptions to the general rule that DNA is the genetic material in all organisms? What evidence supports these exceptions?

SELECTED READINGS

ALLOWAY, J. L. 1933. Further observations on the use of pneumococcus extracts in effecting transformation of type *in vitro*. *J. Exp. Med.* 57: 265–78.

AVERY, O. T., MacLEOD, C. M., and McCARTY, M. 1944. Studies on the chemical nature of the substance inducing transformation of pneumococcal types. Induction of transformation by a desoxyribonucleic acid fraction isolated from pneumococcus type III. *J. Exp. Med.* 79: 137–58. (Reprinted in Taylor, J. H. 1965. *Selected papers in molecular genetics.* Orlando: Academic Press.)

CAIRNS, J., STENT, G. S., and WATSON, J. D. 1966. *Phage and the origins of molecular biology.* Cold Spring Harbor, NY: Cold Spring Harbor Laboratory.

DAWSON, M. H. 1930. The transformation of pneumococcal types. I. The interconvertibility of type-specific S pneumococci. *J. Exp. Med.* 51: 123–47.

DeROBERTIS, E. M., and GURDON, J. B. 1979. Gene transplantation and the analysis of development. *Scient. Amer.* (Dec.) 241: 74–82.

DUBOS, R. J. 1976. *The professor, the Institute and DNA: Osward T. Avery, his life and scientific achievements.* New York: Rockefeller University Press.

FRAENKEL-CONRAT, H., and SINGER, B. 1957. Virus reconstruction. II, Combination of protein and nucleic acid from different strains. *Biochem. Biophys. Acta* 24: 530–48. (Reprinted in Taylor, J. H. 1965. *Selected papers in molecular genetics*. Orlando: Academic Press.)

GRIFFITH, F. 1928. The significance of pneumococcal types. *J. Hyg.* 27: 113–59.

GUTHRIE, G. D., and SINSHEIMER, R. L. 1960. Infection of protoplasts of *Escherichia coli* by subviral particles. *J. Mol. Biol.* 2: 297–305.

HAYES, W. 1968. *The genetics of bacteria and their viruses*. 2nd ed. New York: Wiley.

HERSHEY, A. D., and CHASE, M. 1952. Independent functions of viral protein and nucleic acid in growth of bacteriophage. *J. Gen. Phys.* 36: 39–56. (Reprinted in Taylor, J. H. 1965. *Selected papers in molecular genetics*. Orlando: Academic Press.)

HOTCHKISS, R. D. 1951. Transfer of penicillin resistance in pneumococci by the desoxyribonucleate derived from resistant cultures. *Cold Spr. Harb. Symp.* 16: 457–61. (Reprinted in Adelberg, E. A. 1960. *Papers on bacterial genetics*. Boston: Little, Brown.)

LEVENE, P. A., and SIMMS, H. S. 1926. Nucleic acid structure as determined by electrometric titration data. *J. Biol. Chem.* 70: 327–41.

McCARTY, M. 1980. Reminiscences of the early days of transformation. *Ann. Rev. Genet.* 14: 1–16.

———. 1985. *The transforming principle: Discovering that genes are made of DNA*. New York: W. W. Norton.

PALMITER, R. D., and BRINSTER, R. L. 1985. Transgenic mice. *Cell* 41: 343–45.

RAVIN, A. W. 1961. The genetics of transformation. *Adv. in Genet.* 10: 62–163.

SPIZIZEN, J. 1957. Infection of protoplasts by disrupted T2 viruses. *Proc. Natl. Acad. Sci.* 43: 694–701.

STENT, G. S., and CALENDAR, R. 1978. *Molecular genetics: An introductive narrative*. 2nd ed. New York: W. H. Freeman.

STEWART, T. A., WAGNER, E. F., and MINTZ, B. 1982. Human β-globin gene sequences injected into mouse eggs, retained in adults, and transmitted to progeny. *Science* 217: 1046–48.

VARMUS, H. 1988. Retroviruses. *Science* 240: 1427–35.

WEINBERG, R. A. 1985. The molecules of life. *Scient. Amer.* (Oct.) 253: 48–57.

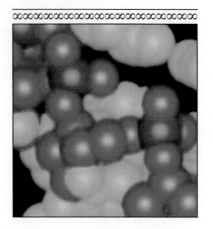

9

Nucleic Acids: Structure and Analysis

CHAPTER CONCEPTS

The structure of DNA has been determined to be a double-stranded helix held together by hydrogen bonds between complementary nitrogenous bases. One of the most exciting aspects of this structure is that it immediately suggests how DNA functions during the replication, storage, and expression of genetic information.

Chapter 8 provided evidence supporting the hypothesis that DNA is the genetic material in all organisms except certain viruses, where RNA serves this function. While this evidence was accumulating, other research was being performed to determine the chemical structure of nucleic acids. When the importance of DNA in genetic processes was realized, this work was intensified, with the hope of discerning not only the structural basis of this molecule but also the relationship of its structure to the functional characteristics of the genetic material (i.e., replication, storage, expression, and mutation).

From 1940 to 1953, many scientists were interested in DNA structure, including Erwin Chargaff, Maurice Wilkins, Rosalind Franklin, Linus Pauling, Francis Crick, and James Watson, among others. They sought information that might answer the most significant and intriguing question in the history of biology: How does DNA serve as the genetic basis for the living process? The answer was believed to depend strongly on the chemical structure of the DNA molecule, given the complex but orderly functions ascribed to it.

These efforts were rewarded in 1953 when Watson and Crick set forth their hypothesis for the double-helical nature of DNA. The assumption that the molecule's functions would be more easily clarified once its general structure was determined proved to be correct. In this chapter we are concerned with nucleic acid structure and its analysis. In subsequent chapters, we will consider in greater detail how DNA functions in the direction of life processes.

THE NUCLEOTIDE: THE BASIC UNIT

Nucleotides are the building blocks of all nucleic acid molecules. Sometimes called mononucleotides, these structural units consist of three essential components: a **nitrogenous base,** a **pentose sugar** (5 carbons), and a **phosphate group**. There are two kinds of nitrogenous bases: the nine-membered double-ringed **purines** and the six-membered single-ringed **pyrimidines**. Two types of purines and three types of pyrimidines are found commonly in nucleic acids. The two purines are **adenine** and **guanine,** abbreviated **A** and **G**. The three pyrimidines are **cytosine, thymine,** and **uracil,** abbreviated **C, T,** and **U**. The chemical structures of A, G, C, T, and U are shown in Figure 9.1. Both DNA and RNA contain A, C, and G; only DNA contains the base T, whereas only RNA contains the base U. Each nitrogen or carbon atom of the ring structures of purines and pyrimidines is designated by a number without a prime sign. Note that corresponding atoms in the two rings are numbered differently in most cases.

The pentose sugars found in nucleic acids give them their names. Ribonucleic acids (RNA) contain **ribose,** while deoxyribonucleic acids (DNA) contain **deoxyribose**. Figure 9.2 shows the straight-chain (Fisher projection formula) and ring (Haworth formula) structures for these two pentose sugars. Each carbon atom is distinguished by a number with a prime sign (e.g., C-1', C-2', etc.). As you can see, deoxyribose is missing one hydroxyl group at the C-2' position compared with ribose. This is the only difference between the two sugars.

If a molecule is composed of a purine or pyrimidine base and a ribose or deoxyribose sugar, the chemical unit is called a **nucleoside**. If a phosphate group is added to the nucleoside, the molecule is now called a **nucleotide**. Nucleosides and nucleotides are named according to the specific nitrogenous base (A, T, G, C, or U) that is part of the building block. The nomenclature and general structure of the nucleosides and nucleotides are given in Figure 9.3.

FIGURE 9.1
Chemical structures of the
pyrimidines and purines that
serve as the nitrogenous bases
in RNA and DNA.

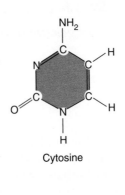

PYRIMIDINE RING

Cytosine

Uracil

Thymine

PURINE RING

Guanine

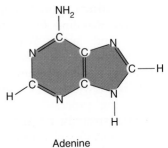

Adenine

The bonding between the three components of a nucleotide is highly specific. The C-1′ atom of the sugar is involved in the chemical linkage to the nitrogenous base. If the base is a purine, the N-9 atom is covalently bonded to the sugar. If the base is a pyrimidine, the bonding involves the N-1 atom. In a nucleotide, the phosphate group may be bonded to the C-2′, C-3′, or C-5′ atom of the sugar. The C-5′—phosphate configuration is shown in Figure 9.3 on page 238. It is by far the most prevalent one in biological systems.

NUCLEOSIDE DIPHOSPHATES AND TRIPHOSPHATES

Mononucleotides are also described by the term **nucleoside monophosphate**. The addition of one or two phosphate groups result in **nucleoside diphosphates** and **triphosphates,** as illustrated in Figure 9.4. The triphosphate form is significant because it serves as the precursor molecule during nucleic acid synthesis within the cell. Additionally, the triphosphates **adenosine triphosphate (ATP)** and **guanosine**

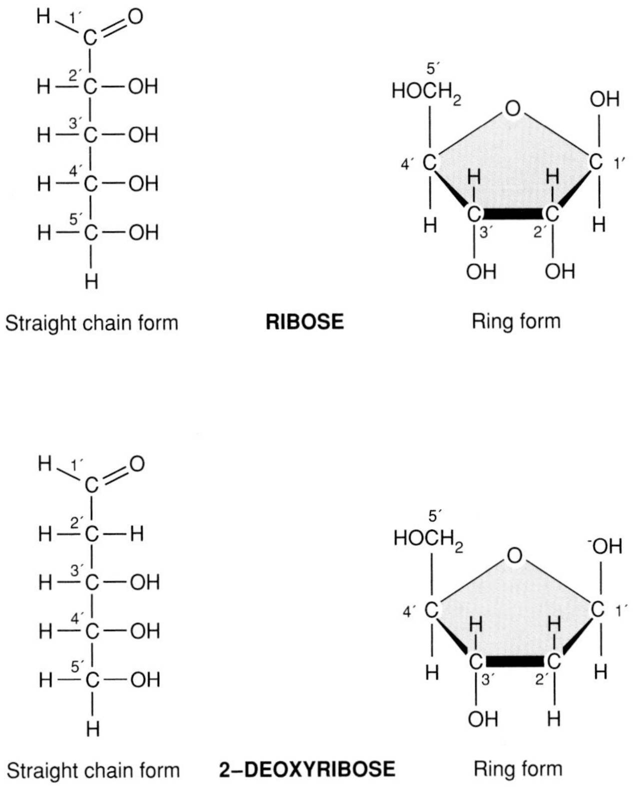

Straight chain form   **RIBOSE**   Ring form

Straight chain form   **2–DEOXYRIBOSE**   Ring form

FIGURE 9.2

Chemical structures of ribose and 2-deoxyribose, which serve as the pentose sugars in RNA and DNA, respectively. Both the straight-chain and ring forms are shown.

triphosphate **(GTP)** are important in the cell's bioenergetics because of the large amount of energy involved in the addition or removal of the terminal phosphate group. The hydrolysis of ATP or GTP to ADP or GDP and inorganic phosphate (P_i) is accompanied by the release of a large amount of energy in the cell. When these chemical conversions are coupled to energy-requiring reactions, the energy produced may be used to drive them to equilibrium. As a result, ATP and GTP are involved in many cellular activities.

POLYNUCLEOTIDES

The linkage between two mononucleotides consists of a phosphate group linked to two sugars. A **phosphodiester** bond is formed, because phosphoric acid has been joined to two alcohols (the hydroxyl groups on the two sugars) by an ester linkage on both sides. Figure 9.5(a) on page 240 shows the resultant

phosphodiester bonds in DNA and RNA. Each structure has a C-5′ end and a C-3′ end. The joining of two nucleotides forms a dinucleotide; of three nucleotides, a trinucleotide; and so forth. When long chains of nucleotides are formed, the structure is called a **polynucleotide**.

Since drawing the structures in Figure 9.5(a) is time consuming and complex, a schematic shorthand method has been devised [Figure 9.5(b)]. The nearly vertical line represents the carbons of the pentose sugar; the nitrogenous base is attached at the top, or the C-1′ position. The diagonal line, with the ⓅP in the middle of it, is attached to the C-3′ position of one sugar and the C-5′ position of the neighboring sugar; it represents the phosphodiester bond. Several modifications of this shorthand method are in use, and they can be understood in terms of these guidelines.

While Levene's tetranucleotide hypothesis (see Chapter 8) was generally accepted before 1940, research in subsequent decades revealed it to be incorrect. It was shown that DNA does not necessarily contain equimolar quantities of the four bases. Additionally, the molecular weight of DNA molecules was determined to be in the range of 10^6 to 10^9 daltons, far in excess of that of a tetranucleotide. The current view of DNA is that it consists of exceedingly long polynucleotide chains.

Long polynucleotide chains would account for the observed molecular weight and would explain the most important property of DNA—genetic variation. If each nucleotide position in this long chain may be occupied by any one of four nucleotides, extraordinary variation is possible. For example, a polynucleotide that is only 1000 nucleotides in length may be arranged 4^{1000} different ways, each one different from all other possible sequences. This potential variation in molecular structure is essential if DNA is to serve the function of storing the vast amounts of chemical information necessary to direct cellular activities.

THE STRUCTURE OF DNA

In 1953, James Watson and Francis Crick proposed that the structure of DNA is in the form of a **double helix**. Their proposal was published in a short paper in *Nature*, which is reproduced in its entirety on pages 244–46. In a sense, this publication constituted the finish line in a highly competitive scientific race to obtain what some consider to be the most significant finding in the history of biology. This "race," as recounted in Watson's book *The Double Helix*, demonstrates the human interaction,

FIGURE 9.3

The structure and names of the nucleosides and nucleotides of RNA and DNA.

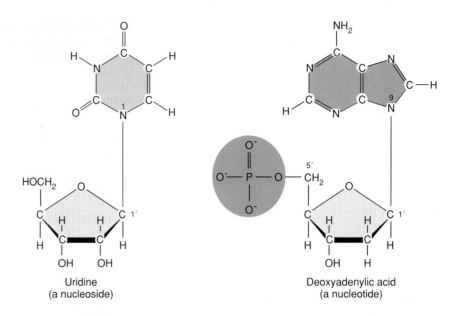

Uridine
(a nucleoside)

Deoxyadenylic acid
(a nucleotide)

Ribonucleosides	**Ribonucleotides**
Adenosine	Adenylic acid
Cytidine	Cytidylic acid
Guanosine	Guanylic acid
Uridine	Uridylic acid

Deoxyribonucleosides	**Deoxyribonucleotides**
Deoxyadenosine	Deoxyadenylic acid
Deoxycytidine	Deoxycytidylic acid
Deoxyguanosine	Deoxyguanylic acid
Deoxythymidine	Deoxythymidylic acid

genius, frailty, and intensity involved in the scientific effort that eventually led to the elucidation of DNA structure.

The available data, crucial to the development of the proposal, came primarily from two sources: base composition analysis of hydrolyzed samples of DNA, and X-ray diffraction studies of DNA. The analytical success of Watson and Crick may be attributed to model building that conformed to the above types of existing data. If the structure of DNA may be analogized by a puzzle, Watson and Crick, working in the Cavendish Laboratory in Cambridge, England, were the first to put together all of the pieces successfully.

Base Composition Studies

Between 1949 and 1953, Erwin Chargaff and his colleagues used chromatographic methods to separate the four bases in DNA samples from various organisms. Quantitative methods were then used to determine the amounts of the four bases from each source. Table 9.1(a) provides some of Chargaff's original data. Parts (b) and (c) of this table show more recently derived base composition information from various organisms which reinforce Chargaff's findings.

On the basis of these data, which you should examine, the following conclusions may be drawn:

FIGURE 9.4

The basic structures of nucleoside diphosphates and triphosphates, as illustrated by thymidine diphosphate and adenosine triphosphate.

NUCLEOSIDE DIPHOSPHATE (NDP)

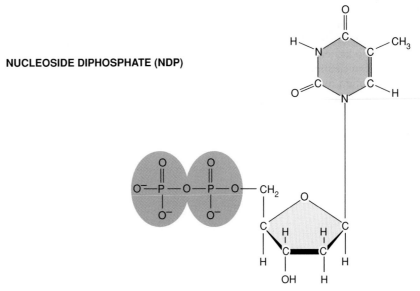

Thymidine diphosphate

NUCLEOSIDE TRIPHOSPHATE (NTP)

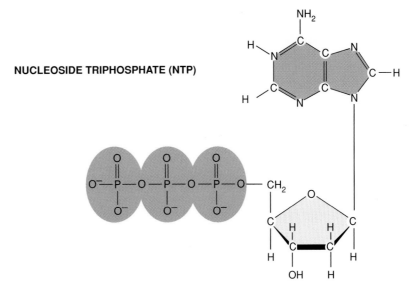

Adenosine triphosphate (ATP)

1 The number of adenine residues equals the number of thymine residues in the DNA of any species (columns 1, 2, and 5). Also, the number of guanine residues is equivalent to the number of cytosine residues (columns 3, 4, and 6).

2 The sum of the purines (A + G) equals the sum of the pyrimidines (C + T), as shown in column 7.

3 The ratio of (A + T)/(C + G) does not necessarily equal one; further, this ratio varies greatly among species, as shown in column 8, and as is apparent in Table 9.1(c).

These conclusions indicate a definite pattern of base composition of DNA molecules. These data served as the initial clue to "the puzzle." Additionally, they directly refute the tetranucleotide hypothesis, which stated that all four bases are present in equal amounts.

X-Ray Diffraction Analysis

When fibers of a DNA molecule are subjected to X-ray bombardment, these rays are scattered accord-

FIGURE 9.5
(a) The linkage of nucleotides by the formation of C-3′–C-5′ (3′–5′) phosphodiester bonds, producing a dinucleotide. (b) A shorthand notation for a polynucleotide chain.

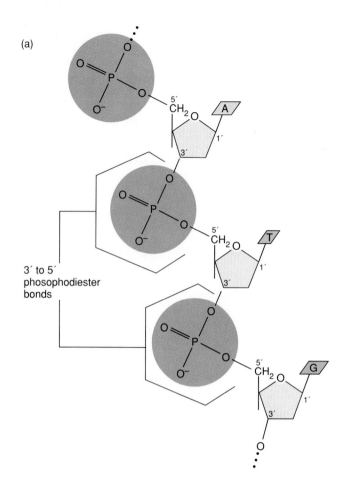

3′ to 5′ phosophodiester bonds

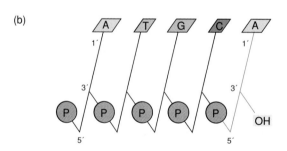

ing to the molecule's atomic structure. The pattern of scatter may be captured as spots on photographic film and analyzed, particularly for the overall shape of and regularities within the molecule. This process, X-ray diffraction analysis, was successfully applied to the study of protein structure by Linus Pauling and other chemists. The technique had been attempted on DNA as early as 1938 by William Astbury. By 1947, he had detected a periodicity of 3.4 Å that suggested

to him that the bases were stacked like pennies on top of one another.

Between 1950 and 1953, Rosalind Franklin, working in the laboratory of Maurice Wilkins, obtained improved X-ray data from more purified samples of DNA (Figure 9.6). Her work confirmed the 3.4 Å periodicity seen by Astbury and suggested that the structure of DNA was some sort of helix. However, she did not propose a definitive model. Pauling had

TABLE 9.1
DNA base composition data.
(a) Chargaff's data.

| | Approximate Percent* | | | |
| | 1 | 2 | 3 | 4 |
Source	A	T	G	C
Ox thymus	26	25	21	16
Ox spleen	25	24	20	15
Yeast	24	25	14	13
Avian tubercle bacilli	12	11	28	26
Human sperm	29	31	18	18

SOURCE: From Chargaff, 1950.
*Moles of nitrogenous constituent per mole of P (often, the recovery was less than 100 percent).

(b) Base compositions of DNAs from various sources.

| | Base Composition | | | | Base Ratio | | | Asymmetry Ratio |
| | 1 | 2 | 3 | 4 | 5 | 6 | 7 | 8 |
Source	A	T	G	C	A/T	G/C	(A + G)/(C + T)	(A + T)/(C + G)
Human	30.9	29.4	19.9	19.8	1.05	1.00	1.04	1.52
Sea urchin	32.8	32.1	17.7	17.3	1.02	1.02	1.02	1.58
E. coli	24.7	23.6	26.0	25.7	1.04	1.01	1.03	0.93
Sarcina lutea	13.4	12.4	37.1	37.1	1.08	1.00	1.04	0.35
T7 bacteriophage	26.0	26.0	24.0	24.0	1.00	1.00	1.00	1.08

(c) G + C content in several organisms.

Organism	% G + C
Phage T2	36.0
Drosophila	45.0
Maize	49.1
Euglena	53.5
Neurospora	53.7

analyzed the work of Astbury and others and incorrectly proposed that DNA was a triple helix.

The Watson-Crick Model

Watson and Crick published their analysis of DNA structure in 1953 (see pp. 244–46). By building models under the constraints of the information just discussed, they proposed the double-helical form of DNA as shown in Figure 9.7. This model has the following major features:

1   Two right-handed helical polynucleotide chains are coiled around a central axis; the coiling is **plectonic,** meaning that the two coils can only be separated by completely unwinding them.
2   The two chains are **antiparallel;** that is, their C-5'-to-C-3' orientations run in opposite directions.
3   The bases of both chains are flat structures, lying perpendicular to the axis; they are "stacked" on one another, 3.4 Å (0.34 nm) apart and are located on the inside of the structure.

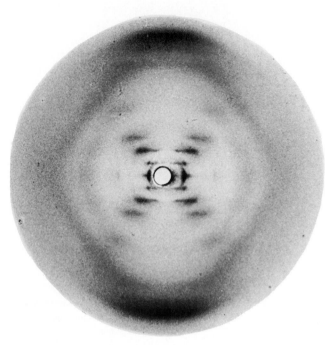

FIGURE 9.6
An X-ray diffraction photograph of the B form of crystallized DNA. The dark patterns at the top and bottom provide an estimate of the periodicity of nitrogenous bases, which are 3.4 Å apart. The central pattern is indicative of the molecule's helical structure.

4 The nitrogenous bases of opposite chains are paired to one another as the result of the formation of **hydrogen bonds** (described in the following discussion): specifically, only A-T and G-C pairs are allowed.

5 Each complete turn of the helix is 34 Å (3.4 nm) long; thus, 10 bases exist in each chain per turn.

6 In any segment of the molecule, alternating larger **major grooves** and smaller **minor grooves** are apparent along the axis.

7 The double helix measures 20 Å (2.0 nm) in diameter.

A more recent and accurate analysis of the form of DNA that served as the basis for the Watson-Crick model has revealed a minor structural difference. A precise measurement of the number of base pairs (bp) per turn has demonstrated a value of 10.4 rather than the 10.0 predicted by Watson and Crick. Where, in the classical model, each base pair is rotated around the helical axis 36°, relative to the adjacent base pair, the new finding requires a rotation of 34.6°. Thus, there are slightly more than 10 base pairs per 360° turn.

Point 2 in the model requires special emphasis. Antiparallelity means that from one end of the helix

one chain is in the 5'-to-3' orientation while the other chain is in the 3'-to-5' orientation. Given the constraints of bond angles, a double helix of the nature described by Watson and Crick could not easily be constructed if both chains were in the same orientation.

One of the most significant features of the structure proposed by Watson and Crick is the specificity of base pairing. Chargaff's data had suggested that the amounts of A equaled T and that G equaled C. Watson and Crick realized that when placed opposite each other in the model, the members of each such base pair formed hydrogen bonds, providing the chemical stability necessary to hold the two chains together. Arranged in this way, both major and minor grooves became apparent along the axis. Further, with one purine (A or G) opposite one pyrimidine (T or C) as each "rung of the spiral staircase" of the proposed helix, the Watson-Crick model conformed to the 20Å (2 nm) diameter suggested by X-ray diffraction studies.

The specific A-T and G-C base pairing is the basis for the concept of **complementarity**. This term is used to describe the chemical affinity provided by the hydrogen bonds between the bases. As we will see, this concept is very important in the processes of DNA replication and gene expression.

Two questions are particularly worthy of discussion. First, why aren't other base pairs possible? Watson and Crick discounted the A-G and C-T pairs because these represent purine-purine and pyrimidine-pyrimidine pairings, respectively. These pairings would lead to alternating diameters of more than and less than 20 nm because of the respective sizes of the purine and pyrimidine rings; additionally, the three-dimensional configurations formed by such pairings do not produce the proper alignment leading to sufficient hydrogen bond formations. It is for this latter reason that A-C and G-T pairings were also discounted, even though these pairs each consist of one purine and one pyrimidine.

The second question concerns hydrogen bonds. Just what is the nature of such a bond, and is it strong enough to stabilize the helix? A **hydrogen bond** is a very weak electrostatic attraction between a covalently bonded hydrogen atom and an atom with an unshared electron pair. The hydrogen atom assumes a partial positive charge, while the unshared electron pair—characteristic of covalently bonded oxygen and nitrogen atoms—assumes a partial negative charge. These opposite charges are responsible for the weak chemical attraction. As oriented in the double helix, adenine forms two hydrogen bonds with thymine, and guanine forms three hydrogen bonds with

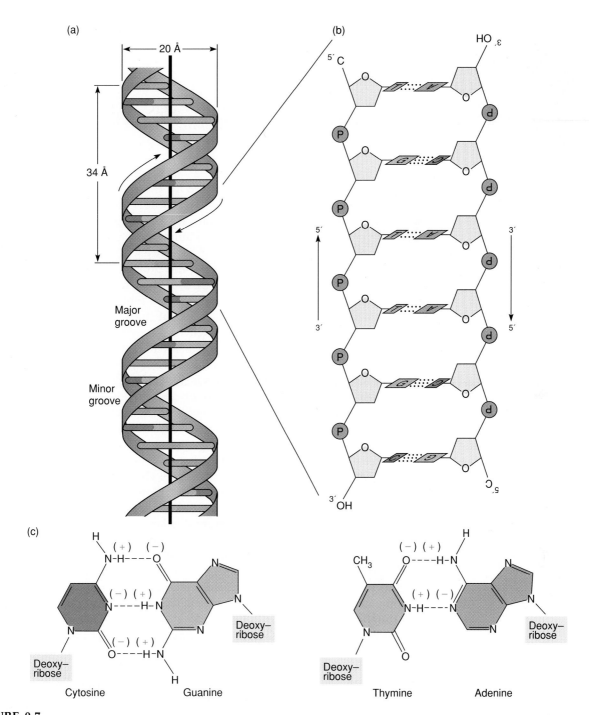

FIGURE 9.7
(a) A schematic representation of the DNA double helix as proposed by Watson and Crick. The ribbonlike strands constitute the sugar-phosphate backbones, and the horizontal rungs constitute the nitrogenous base pairs, of which there are 10 per complete turn. The major and minor grooves are apparent. The solid vertical rod has been placed through the center of the helix. (b) A representation of the antiparallel nature of the two strands of the helix. (c) The hydrogen bonds formed between cytosine and guanine and between thymine and adenine.

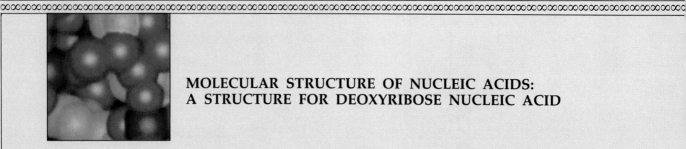

MOLECULAR STRUCTURE OF NUCLEIC ACIDS:
A STRUCTURE FOR DEOXYRIBOSE NUCLEIC ACID

We wish to suggest a structure for the salt of deoxyribose nucleic acid (D.N.A.). This structure has novel features which are of considerable biological interest. A structure for nucleic acid has already been proposed by Pauling and Corey.[1] They kindly made their manuscript available to us in advance of publication. Their model consists of three intertwined chains, with the phosphates near the fibre axis, and the bases on the outside. In our opinion, this structure is unsatisfactory for two reasons: (1) We believe that the material which gives the X-ray diagrams is the salt, not the free acid. Without the acidic hydrogen atoms it is not clear what forces would hold the structure together, especially as the negatively charged phosphates near the axis will repel each other. (2) Some of the van der Waals distances appear to be too small.

Another three-chain structure has also been suggested by Fraser (in the press). In his model the phosphates are on the outside and the bases on the inside, linked together by hydrogen bonds. This structure as described is rather ill-defined, and for this reason we shall not comment on it.

We wish to put forward a radically different structure for the salt of deoxyribose nucleic acid. This structure has two helical chains each coiled round the same axis. We have made the usual chemical assumptions, namely, that each chain consists of phosphate diester groups joining β-D-deoxyribofuranose residues with 3′,5′ linkages. The two chains (but not their bases) are related by a dyad perpendicular to the fibre axis. Both chains follow right-handed helices, but owing to the dyad the sequences of the atoms in the two chains run in opposite directions. Each chain loosely resembles Furberg's[2] model No. 1; that is, the bases are on the inside of the helix and the phosphates on the outside. The configuration of the sugar and the atoms near it is close to Furberg's "standard configuration," the sugar being roughly perpendicular to the attached base. There is a residue on each chain every 3.4 Å in the z-direction. We have assumed an angle of 36° between adjacent residues in the same chain, so that the structure repeats after 10 residues on each chain, that is, after 34 Å. The distance of a phosphorus atom from the fibre axis is 10 Å. As the phosphates are on the outside, cations have easy access to them.

cytosine. Although two or three hydrogen bonds taken alone are very weak, two or three thousand bonds in tandem (which would be found in two long polynucleotide chains) are capable of providing great stability to the helix.

Still another stabilizing factor is the arrangement of sugars and bases along the axis. In the Watson-Crick model, the hydrophobic or "water-fearing" nitrogenous bases are stacked almost horizontally on the interior of the axis, thus shielded from water. The hydrophilic sugar-phosphate backbone is on the outside of the axis, where both components may interact with water. These molecular arrangements provide significant chemical stabilization to the helix.

The Watson-Crick model had an immediate effect on the emerging discipline of molecular biology. Even in their initial 1953 article, the authors noted, "It has not escaped our notice that the specific pairing

The structure is an open one, and its water content is rather high. At lower water contents we would expect the bases to tilt so that the structure could become more compact.

The novel feature of the structure is the manner in which the two chains are held together by the purine and pyrimidine bases. The planes of the bases are perpendicular to the fibre axis. They are joined together in pairs, a single base from one chain being hydrogen-bonded to a single base from the other chain, so that the two lie side by side with identical z-coordinates. One of the pair must be a purine and the other a pyrimidine for bonding to occur. The hydrogen bonds are made as follows: purine position 1 to pyrimidine position 1; purine position 6 to pyrimidine position 6.

If it is assumed that the bases only occur in the structure in the most plausible tautomeric forms (that is, with the keto rather than the enol configurations) it is found that only specific pairs of bases can bond together. These pairs are: adenine (purine) with thymine (pyrimidine), and guanine (purine) with cytosine (pyrimidine).

In other words, if an adenine forms one member of a pair, on either chain, then on these assumptions the other member must be thymine; similarly for guanine and cytosine. The sequence of bases on a single chain does not appear to be restricted in any way. However, if only specific pairs of bases can be formed, it follows that if the sequence of bases on one chain is given, then the sequence on the other chain is automatically determined.

It has been found experimentally[3,4] that the ratio of the amounts of adenine to thymine, and the ratio of guanine to cytosine, are always very close to unity for deoxyribose nucleic acid.

It is probably impossible to build this structure with a ribose sugar in place of the deoxyribose, as the extra oxygen atom would make too close a van der Waals contact.

The previously published X-ray data[5,6] on deoxyribose nucleic acid are insufficient for a rigorous test of our structure. So far as we can tell, it is roughly compatible with the experimental data, but it must be regarded as unproved until it has been checked against more exact results. Some of these are given in the following communications. We were not aware of the details

we have postulated immediately suggests a possible copying mechanism for the genetic material." Two months later, in a second article in *Nature,* Watson and Crick pursued this idea, suggesting a specific mode of replication of DNA—the semiconservative model. The second article also alluded to two new concepts: the storage of genetic information in the sequence of the bases, and the mutation or genetic change that would result from alteration of the bases.

These ideas have received vast amounts of experimental support since 1953 and are now universally accepted. Thus, the "synthesis" of ideas by Watson and Crick was a remarkable feat and highly significant in the history of genetics and biology. For their work, they, along with Wilkins, received the Nobel Prize in Physiology and Medicine in 1962. This was to be one of many such awards bestowed for work in the field of molecular genetics.

of the results presented there when we devised our structure, which rests mainly though not entirely on published experimental data and stereochemical arguments.

It has not escaped our notice that the specific pairing we have postulated immediately suggests a possible copying mechanism for the genetic material.

Full details of the structure, including the conditions assumed in building it, together with a set of co-ordinates for the atoms, will be published elsewhere.

We are much indebted to Dr. Jerry Donohue for constant advice and criticism, especially on interatomic distances. We have also been stimulated by a knowledge of the general nature of the unpublished experimental results and ideas of Dr. M. H. F. Wilkins, Dr. R. E. Franklin and their coworkers at King's College, London. One of us (J. D. W.) has been aided by a fellowship from the National Foundation for Infantile Paralysis.

J. D. Watson
F. H. C. Crick

*Medical Research Council Unit for the Study of the
Molecular Structure of Biological Systems
Cavendish Laboratory, Cambridge.
April 2.*

[1]Pauling, L., and Corey, R. B., *Nature*, 171, 346 (1953); *Proc. U.S. Nat. Acad. Sci.*, 39, 84 (1953).

[2]Furberg, S., *Acta Chem. Scand.*, 6, 634 (1952).

[3]Chargaff, E., for references see Zamenhof, S., Brawerman, G., and Chargaff, E., *Biochim. et Biophys. Acta*, 9, 402 (1952).

[4]Wyatt, G. R., *J. Gen. Physiol*, 36, 201 (1952).

[5]Astbury, W. T., Symp. Soc. Exp. Biol. 1, Nucleic Acid, 66 (Camb. Univ. Press, 1947).

[6]Wilkins, M. H. F., and Randall, J. T., *Biochim. et Biophys. Acta*, 10, 192 (1953).

A-, B-, C-, D-, AND E-DNA

Under different conditions of isolation, purification, and crystallization, several forms of DNA have been recognized. At the time Watson and Crick performed their analysis, two forms—**A-DNA** and **B-DNA**—were known. Watson and Crick's analysis was based on X-ray studies of the B form by Franklin, which is present under a more hydrated set of conditions and is believed to be the biologically significant conformation.

While DNA studies around 1950 relied on the use of diffraction of fibers, more recent investigations have been performed using **single crystal X-ray analysis**. The earlier studies achieved limited resolution of about 5 Å, but single crystals diffract X-rays at about 1 Å, near atomic resolution. As a result, every atom is "visible" and much greater structural detail is available during analysis.

Using these modern techniques, the A form of DNA, which received minimal attention for over 25 years, has now been scrutinized. In comparison to B-DNA (Figure 9.8), A-DNA is slightly more compact, with 11 base pairs in each complete turn of the helix, which is 23 Å in diameter. While it is also a right-handed helix, the orientation of the bases is somewhat different. They are tilted and displaced laterally in relation to the axis of the helix (Figure 9.9). As a result of these differences, the appearance of the major and minor grooves is modified compared with those in B-DNA.

It is not yet clear to what extent A-DNA occurs under biological conditions. However, under certain conditions, short synthetic double-stranded fragments (e.g., CCGG, GGCCGGCC, and GGTATACC along each strand) show a preference for assuming the A configuration. A-T–rich polymers show a preference for the B form. Since A-DNA occurs under conditions of decreased hydration, it has been suggested that it may be formed physiologically as a result of interaction with hydrophobic molecules or changing cellular conditions. If so, it is possible that

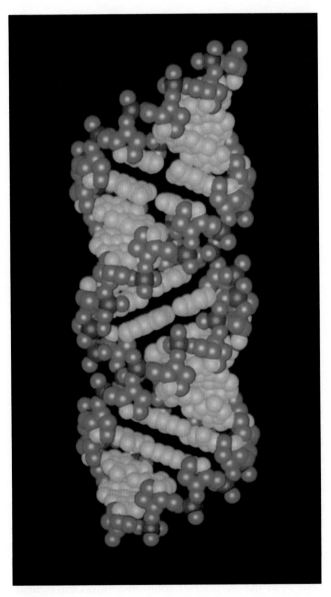 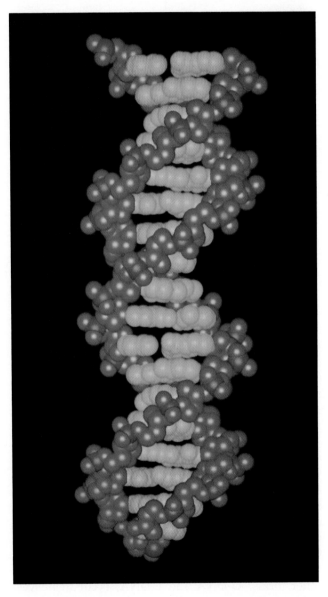

FIGURE 9.8
Space-filling models comparing the structure of the helix of A-DNA (left) and B-DNA (right). Both are right-handed helices.

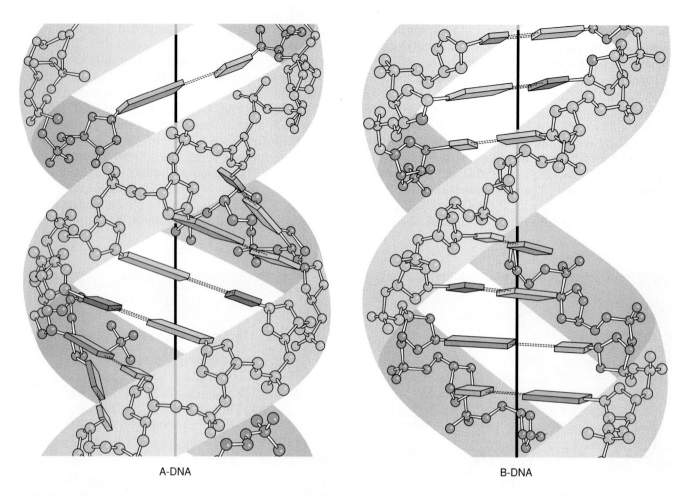

A-DNA

B-DNA

FIGURE 9.9
A comparison of the orientation of base pairs within the right-handed double helix of A-DNA (shown on the left) and B-DNA (shown on the right). In A-DNA, base pairs are tilted and pulled away from the helical axis. In B-DNA, base pairs are perpendicular to the helical axis.

DNA structure is a dynamic one, being altered under different conditions. While there is no direct evidence in support of this possibility, it is a particularly attractive idea because of the need for recognition sites along the helix during the various stages of gene activity.

Three other right-handed forms of DNA helices have been discovered. These have been designated C-, D-, and E-DNA. **C-DNA** is found to occur under even greater dehydration conditions than those observed during the isolation of A- and B-DNA. It has only 9.3 base pairs per turn and is, thus, less compact. Its helical diameter is 19 Å. Like A-DNA, C-DNA does not have its base pairs lying flat; instead they are tilted relative to the axis of the helix. Two other forms, **D-DNA** and **E-DNA,** occur in helices lacking guanine in their base composition. They have even fewer base pairs per turn: 8 and 7½, respectively.

Z-DNA

Still another form of DNA was discovered by Andrew Wang, Alexander Rich, and their colleagues in 1979 when they examined a DNA fragment of the hexanucleoside pentaphosphate d(CpGpCpGpCpG). Called **Z-DNA,** it takes on the rather remarkable configuration of a **left-handed double helix**. It had been known as early as 1972 that a dramatic shift in symmetry of the helix occurred under the conditions of high salt concentration. It has now been confirmed that this shift is the result of the conversion from a right-handed to a left-handed helix. The two configurations are illustrated and compared in Figure 9.10.

The difference between a right- and left-handed helix is in the manner of rotation around the helical axis. A right-handed helix rotates clockwise (CW) as it proceeds away from an observer looking down either axis. A left-handed helix rotates counterclock-

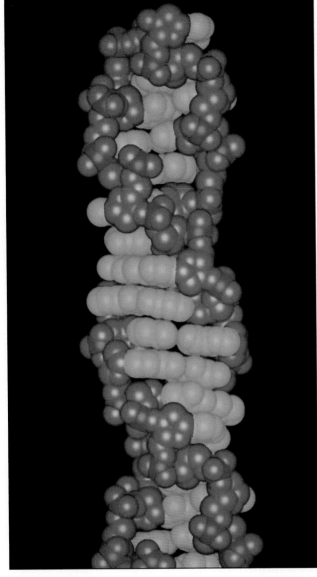

FIGURE 9.10
A space-filling model of the Z form of DNA, which is a left-handed helix. Note the near elimination of the grooves in this molecule compared to B-DNA shown in Figure 9.8.

in diameter, contains 12 base pairs per turn, and assumes a zigzag conformation (hence its name). The major groove present in B-DNA is nearly eliminated in Z-DNA.

Z-DNA has now been the subject of extensive analysis designed to demonstrate its presence in equilibrium with B-DNA under cellular conditions. Because of its more compact form, the phosphate groups on opposite strands of Z-DNA are closer together, creating a greater electrostatic repulsion than occurs in B-DNA. As a result, Z-DNA is energetically less stable and is not usually favored in an equilibrium existing between the two forms. However, DNA sequences consisting of alternating purines and pyrimidines have an affinity for forming Z-DNA. Of the various types of DNA structure that differ from B-DNA, the discovery of Z form has been the most exciting to geneticists.

THE STRUCTURE OF RNA

The second category of nucleic acids is the ribonucleic acids, or RNA. The structure of these molecules is similar to DNA, with several important exceptions. While RNA also has as its building blocks nucleotides linked into polynucleotide chains, ribose replaces deoxyribose and uracil replaces thymine. Another important difference is that most RNA is thought of as being single stranded. However, RNA molecules sometimes fold back on themselves following their synthesis; such a configuration results when regions of complementarity occur in positions that allow base pairs to form. Furthermore, some animal viruses that have RNA as their genetic material contain it in the form of double-stranded helices. Thus, there are several instances where RNA does not exist strictly as a linear, single-stranded molecule.

At least three classes of cellular RNA molecules function during the expression of genetic information: **ribosomal RNA (rRNA), messenger RNA (mRNA),** and **transfer RNA (tRNA).** These molecules all originate as complementary copies of one of the two strands of DNA segments during the process of transcription. That is, their nucleotide sequence is complementary to the deoxyribonucleotide sequence of DNA, which served as the template for their synthesis. Since uracil replaces thymine in RNA, uracil is complementary to adenine during transcription.

Each class of RNA may be characterized by its size, sedimentation behavior in a centrifugal field, and genetic function. Sedimentation behavior depends on a molecule's density, mass, and shape, and its measure is called the **Svedberg coefficient (S).** Table

wise (CCW). When viewing a picture or a model of a helix that is running vertically, one sees a coil that turns about its axis in the direction that the figures of your right or left hand would assume if they were curled around an object and held out in front of you. In DNA, the type of helix that is formed appears to be a function of the nucleotide sequence and the chemical environment in which it is studied.

Z-DNA, like A- and B-DNA, consists of two antiparallel chains held together by Watson-Crick base pairs. Beyond these characteristics, Z-DNA is quite different. The left-handed helix is 18 Å (1.8 nm)

TABLE 9.2
Sedimentation coefficients, molecular weights, and number of nucleotides for various RNAs.

RNA Type	Abbreviation	Sedimentation Coefficient	Molecular Weight	Number of Nucleotides
Ribosomal RNA	rRNA	5S	35,000	105
		18S	700,000	1740
		28S	1,800,000	4850
Transfer RNA	tRNA	4S	23,000–30,000	75–90
Messenger RNA	mRNA	6–50S	25,000–1,000,000	100–10,000

9.2 relates the S values, molecular weights, and approximate number of nucleotides of the major forms of RNA. As you can see, there is a wide variation in size of the three classes of RNA.

Ribosomal RNA is the largest of these molecules and usually constitutes about 80 percent of all RNA in the cell. The various forms of rRNA found in prokaryotes and eukaryotes differ distinctly in size. These molecules constitute an important structural component of **ribosomes,** which function in the synthesis of proteins, the process called **translation**.

Messenger RNA molecules carry genetic information from the DNA of the gene to the ribosome, where translation into protein occurs. They vary considerably in length, which is partially a reflection of the variation in the size of the gene serving as the template for transcription of mRNA species.

Transfer RNA, the smallest class of RNA molecules, carries amino acids to the ribosome during translation. Since more than one tRNA molecule interacts simultaneously with the ribosome, the molecule's smaller size facilitates these interactions.

We will discuss the functions of the three classes of RNA in much greater detail later in the text (see Chapter 12). Our purpose in this section has been to contrast the structure of DNA, which stores genetic information, with that of RNA, which functions in the expression of that information.

ANALYSIS OF NUCLEIC ACIDS

Since 1953, the role of DNA as the genetic material and the role of RNA in transcription and translation have been clarified through detailed analysis of nucleic acids. We will consider several methods of analysis of these molecules in this chapter. Some of these, as well as other research procedures, are presented in greater detail in Appendix A.

Absorption of Ultraviolet Light (UV)

Nucleic acids absorb ultraviolet light most strongly at wavelengths of 254 to 260 nm. This is the result of the interaction between UV light at these wavelengths and the ring systems of the purines and pyrimidines. Thus, any molecule containing nitrogenous bases (i.e., nucleosides, nucleotides, and polynucleotides) can be analyzed using UV light. This technique is especially important in the localization, isolation, and characterization of nucleic acids.

UV analysis is used in conjunction with many standard procedures that separate molecules. For example, compounds containing a nitrogenous base can be separated by a technique called **paper chromatography**. As illustrated in Figure 9.11, a mixture of nucleotides or short polynucleotides can be separated on the basis of the solubility of each component in various solvents. The molecules are first allowed to migrate up the paper in a specific solvent, effecting the primary separation of the mixture. Molecules of greatest solubility migrate more quickly and farther than those of lesser solubility. Then, the paper is dried, turned 90°, and placed in a second solvent, which further separates the molecules. The paper may then be exposed to UV light. The compounds containing nitrogenous bases will appear as dark spots against the fluorescing paper. Each spot may then be cut out, eluted from the paper, and investigated further.

Sedimentation Behavior

Another way that molecules of nucleic acid mixtures can be separated is by subjecting them to one of several possible **gradient centrifugation** procedures (Figure 9.12). The mixture may be loaded on top of a solution prepared so that a concentration gradient has been formed from top to bottom. Then the entire mixture is centrifuged at high speeds in an ultracentrifuge. The mixture of molecules will migrate downward, with each component moving at a different rate. Centrifugation may be stopped and the gradient eluted from the tube. Each fraction can then be measured spectrophotometrically for absorption at 260 nm. In this way, the position of a nucleic acid fraction can be located along the gradient and the fraction isolated and studied further.

FIGURE 9.11
Separation of a mixture of nucleotides and short polynucleotides using two-dimensional paper chromatography and analysis under ultraviolet light.

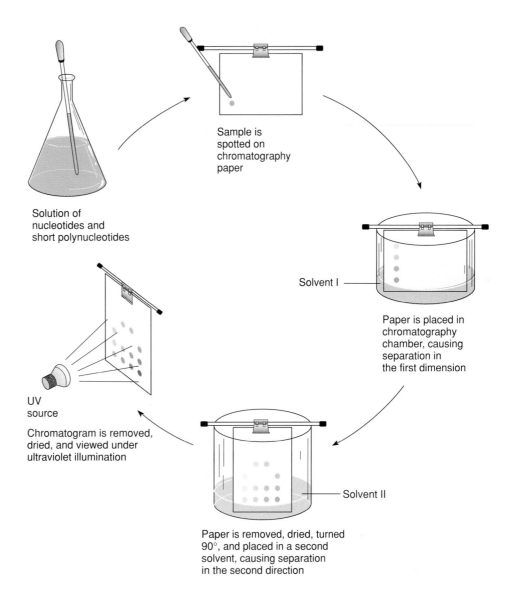

Solution of nucleotides and short polynucleotides

Sample is spotted on chromatography paper

Solvent I

Paper is placed in chromatography chamber, causing separation in the first dimension

Solvent II

Paper is removed, dried, turned 90°, and placed in a second solvent, causing separation in the second direction

UV source

Chromatogram is removed, dried, and viewed under ultraviolet illumination

The gradient centrifugations described rely on the sedimentation behavior of molecules in solution. There are two major types of sedimentation techniques employed in the analysis of nucleic acids: **sedimentation velocity** and **sedimentation equilibrium**. Both require the use of high-speed centrifugation to create large centrifugal forces upon molecules in a gradient solution.

Sedimentation velocity analysis employs an analytical centrifuge, which enables the migration of the molecules during centrifugation to be monitored with ultraviolet absorption optics. Thus, the "velocity of sedimentation" may be determined. This velocity has been standardized in units called Svedberg coefficients (S), as mentioned earlier.

In this technique, the molecules are loaded on top of the gradient, and the gravitational forces created by centrifugation drive them toward the bottom of the tube. Two forces work against this downward movement: (1) the viscosity of the solution creates a frictional resistance, and (2) part of the force of diffusion is directed upward. Under these conditions, the key variables are the mass and shape of the molecules being examined. In general, the greater the mass, the greater is the sedimentation velocity. However, the molecule's shape affects the frictional resistance. Therefore, two molecules of equal mass but different shape will sediment differently.

One use of this technique is the determination of **molecular weight (MW)**. If certain physical chemical properties of a molecule under study are also known, the MW can be calculated based on the sedimentation velocity. S values increase with molecular weight, but are not directly proportional to it.

In the second technique, sedimentation equilibrium centrifugation (sometimes called **density gradient centrifugation**), a density gradient is created that overlaps the densities of the individual components

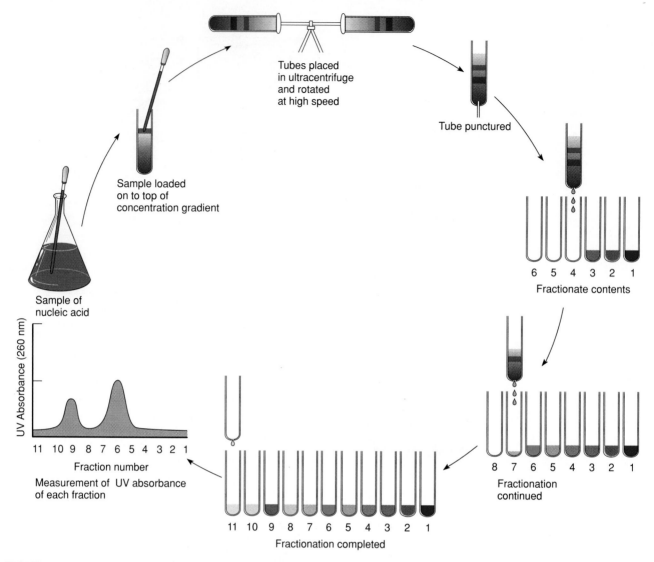

FIGURE 9.12

Separation of a mixture of two types of nucleic acid using gradient centrifugation. In order to fractionate the gradient, successive samples are eluted from the bottom of the tube. Each is measured for absorbance of ultraviolet light at 260 nm, producing a profile of the sample in graphic form.

of a mixture of molecules. Usually, the gradient is made of a heavy metal salt such as cesium chloride (CsCl). During centrifugation the molecules migrate until they reach a point of neutral buoyant density. At this point, the centrifugal force on them is equal and opposite to the upward diffusion force and no further migration occurs. If DNAs of different densities are used, they will separate as the molecules of each density reach equilibrium with the corresponding density of CsCl. The gradient may be fractionated

and the components isolated (Figure 9.12). When properly executed, this technique provides high resolution in separating mixtures of molecules varying only slightly in density.

Sedimentation equilibrium centrifugation studies may also be used to generate data on the base composition of double-stranded DNA. G-C base pairs, compared with A-T pairs, are more compact and dense. As shown in Figure 9.13, the percentage of G-C pairs in DNA is directly proportional to the

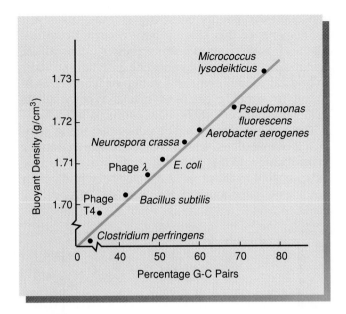

FIGURE 9.13
Percentage of guanine-cytosine (G-C) base pairs in DNA plotted against buoyant density for a variety of microorganisms.

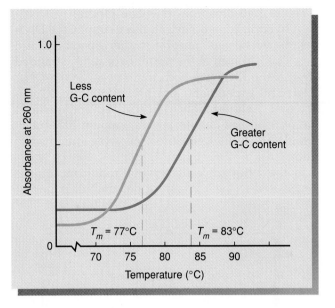

FIGURE 9.14
Comparison of the increase in UV absorbance with an increase in temperature for two DNA molecules with differing G-C contents. The molecule with a melting point (T_m) of 83°C has a greater G-C content than the molecule with a T_m of 77°C.

buoyant density of the DNA molecule. Thus, by using this technique, we can make a useful molecular characterization of DNAs from different sources.

Denaturation and Renaturation of Nucleic Acids

When **denaturation** of double-stranded DNA occurs, the hydrogen bonds of the duplex structure break, the duplex unwinds, and the strands separate. However, no covalent bonds break. During strand separation, which can be induced by heat or chemical treatment, the viscosity of DNA decreases, and both the UV absorption and the buoyant density increase. The increase in UV absorption of heated DNA in solution, called the **hyperchromic effect,** is easiest to measure. Denaturation as a result of heating is sometimes referred to as **melting.**

Since G-C base pairs have one more hydrogen bond than A-T pairs, they are more stable to heat treatment. Thus, DNA with a greater proportion of G-C pairs than A-T pairs requires more heat to denature completely. When absorption at 260 nm is monitored and plotted against temperature during heating, a **melting profile** of DNA is obtained. The midpoint of this profile, or curve, is called the **melting temperature (T_m)** and represents the point at which 50 percent of the strands are unwound or

denatured (Figure 9.14). When the curve plateaus at its maximum optical density, denaturation is complete and only single strands exist. Incubation conditions can then be changed such that these single strands can reassociate through random collisions of complementary strands. Analysis of melting profiles provides a characterization of DNA and an alternate method of estimating the base composition of DNA.

Molecular Hybridization Techniques

The property of denaturation/renaturation of nucleic acids is the basis for one of the most powerful and useful techniques in molecular genetics—**molecular hybridization.**

Denaturation occurs when double-stranded nucleic acid molecules are separated into single strands. Renaturation occurs when they return to their original duplex state. Provided that a reasonable degree of base complementarity exists, two nucleic acid strands from different sources will undergo molecular hybridization under the proper conditions. As a result, hybridization is possible between DNA strands from different species, and between DNA and RNA strands. For example, an RNA molecule will hybridize with the segment of DNA from which it was transcribed.

Figure 9.15 illustrates how the process of DNA-RNA hybridization occurs. In this example, the DNA strands are heated, causing strand separation, and then slowly cooled in the presence of single-stranded RNA. If the RNA has been transcribed on the DNA used in the experiment, and is therefore complementary to it, hybridization will occur. Several methods are available for monitoring the amount of double-stranded molecules produced following strand separation.

In the 1960s molecular hybridization techniques contributed to the increase in our understanding of transcriptional events occurring at the gene level. Refinements of this process have occurred continually and have been the forerunners of work in evolutionary homology and gene isolation studies. Hybridization can occur in solution or when DNA is bound to either a gel or a special type of filter, facilitating recovery of the newly formed hybrids.

The technique can even be performed using cytological preparations, which is called *in situ* **molecular hybridization**. In this procedure mitotic or interphase cells are fixed to slides and subjected to hybridization conditions. Radioactive single-stranded DNA or RNA

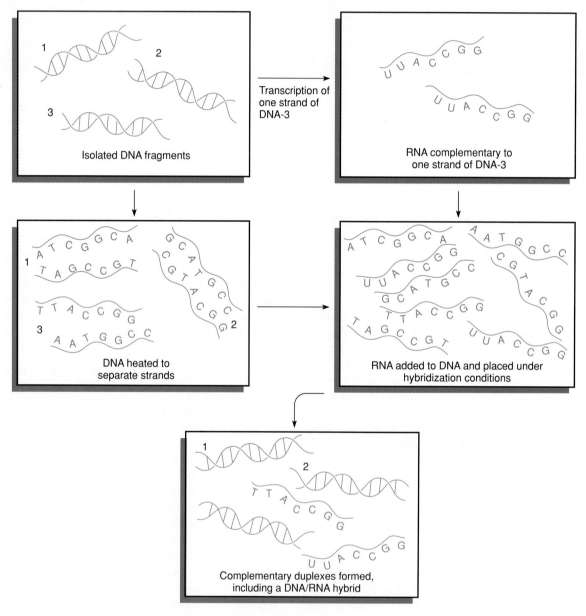

FIGURE 9.15
Diagrammatic representation of the process of molecular hybridization between DNA fragments and RNA that has been transcribed on one of the fragments.

is added, and hybridization is monitored. The nucleic acid that is added may be either radioactive or contain a fluorescent label to allow its detection. In the former case, the technique of autoradiography may be used (see Appendix A). In Figure 9.16, the use of fluorescence is illustrated. A DNA probe specific for DNA present in the centromere regions has been hybridized with human chromosomes. The use of this technique in identification of chromosomal locations housing specific genetic information has been a valuable addition to the repertoire of experimental geneticists.

Reassociation Kinetics and Repetitive DNA

One extension of molecular hybridization procedures is the technique that measures the **rate** of reassociation of complementary strands of DNA derived from a single source. This technique, called **reassociation kinetics,** was first refined and studied by Roy Britten and David Kohne.

The DNA used in such studies is first fragmented into small pieces several hundred base pairs long. Fragments are produced as a result of shearing forces introduced during isolation. After the DNA has been dissociated into single strands by heating, the temperature is lowered and reassociation is monitored. During reassociation, pieces of single-stranded DNA collide randomly. If they are complementary, a stable double strand is formed; if not, they separate and are free to encounter other DNA fragments. The process continues until all matches are made.

The results of such an experiment are presented in Figure 9.17. The percentage of reassociation of DNA fragments is plotted against a logarithmic scale of normalized time, C_0t. In this term, C_0 is equal to the initial DNA concentration in moles per liter of nucleotides, and t is equal to time in seconds. The derivation of C_0t is described in Appendix A. The shape of the curve obtained is an ideal second-order reaction.

A great deal of information can be obtained from studies comparing the reassociation of DNA of different organisms. For example, for different organisms we may compare the point in the reaction when one-half of the DNA is present as double-stranded fragments. This point is called the $C_0t_{1/2}$, or **half reaction time.** Provided that all DNA fragments contain unique nucleotide sequences and all are about the same size, $C_0t_{1/2}$ varies directly with the

FIGURE 9.16

In situ hybridization of human metaphase chromosomes using an alpha satellite DNA probe. The probe, specific to centromeric DNA, produces a yellow fluorescence signal indicating hybridization. The red fluorescence is produced by propidium iodide counterstaining of chromosomal DNA.

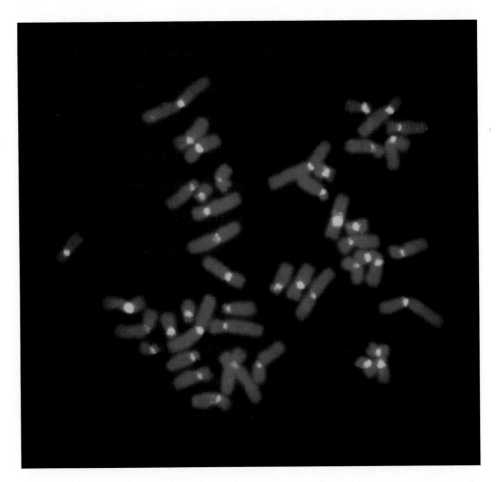

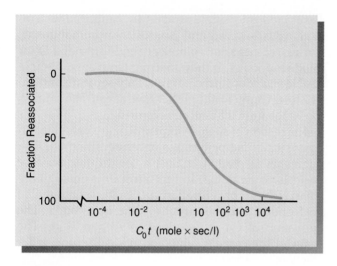

FIGURE 9.17
The time course of DNA reassociation when all fragments are unique. The curve is an ideal second-order reaction. Note that the abscissa is a logarithmic scale of C_0t.

complexity of the DNA. Designated as *X*, complexity represents the length in nucleotide pairs of all unique DNA fragments laid end to end. If the DNA used in an experiment represents the entire genome, and if all DNA sequences are different from one another, then *X* is equal to the size of the haploid genome.

Figure 9.18 illustrates what is found when DNAs from various sources are compared. As can be seen, as genome size increases, the curves obtained are shifted farther and farther to the right, indicative of an increased reassociation time. Reassociation occurs more slowly because it takes longer for matches to be made if there are initially greater numbers of unique DNA fragments. This is so because collisions are random; more sequences present will result in greater numbers of mismatches before the correct match is made.

As shown in Figure 9.19, $C_0t_{1/2}$ is directly proportional to the size of the genome. If the entire genome consists of unique DNA sequences, reassociation experiments can be used to determine the genome size of organisms. This method has been particularly useful in studying viruses and bacteria.

However, when reassociation kinetics of DNA from eukaryotic organisms were first studied, a surprising observation was made. The data showed that some of the DNA segments reassociated even more rapidly than those derived from *E. coli*! The remainder, as expected because of its greater complexity, took longer to reassociate. For example, Britten and Kohne examined DNA derived from calf thymus tissue (Figure 9.20). Based on these observations, they hypothesized that the rapidly reassociating fraction must represent repetitive sequences present many times in the calf genome. This interpretation would explain why these DNA segments reassociate so rapidly. On the other hand, they hypothesized that the remaining DNA segments consist of unique nucleotide sequences present only once in the genome; because there are more of these unique sequences, increasing the DNA complexity in calf thymus (compared with *E. coli*), their reassocia-

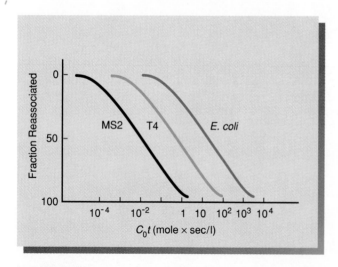

FIGURE 9.18
Comparison of the reassociation rate of DNA derived from phage MS-2, phage T4, and *E. coli*. The genome of T4 is larger than MS-2, and that of *E. coli* is larger than T4.

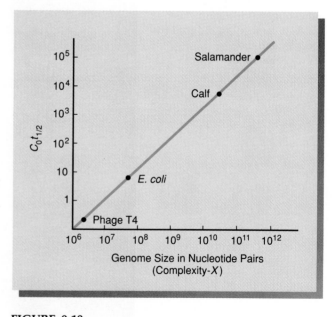

FIGURE 9.19
Comparison of $C_0t_{1/2}$ and genome size for phage T4, *E. coli*, calf, and salamander.

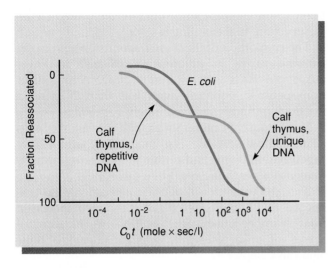

FIGURE 9.20
C_0t curve of calf thymus DNA compared with *E. coli*. The repetitive fraction of calf DNA reassociates more quickly than that of *E. coli*, while the more complex unique calf DNA takes longer to reassociate than that of *E. coli*.

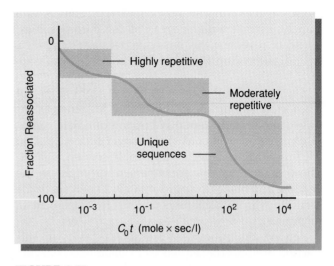

FIGURE 9.21
Reassociation analysis of eukaryotic DNA containing highly repetitive, moderately repetitive, and unique sequences. This pattern is similar to that found in mammals, including humans.

tion takes longer. The *E. coli* curve has been added to Figure 9.20 for the sake of comparison.

The various categories of DNA defined by reassociation kinetic analysis are summarized in Table 9.3. The copies present many times in the genome are collectively referred to as **repetitive DNA**. Repetitive DNA is characteristic of eukaryotic genomes and is important to our understanding of how genetic information is organized in chromosomes. Careful study has shown that, while there is a continuum of reassociation characteristics displayed by eukaryotic DNA, there are several categories of repetitive DNA, as illustrated in Figure 9.21. The first reassociates very rapidly (much more rapidly than *E. coli* DNA)

and is called **highly repetitive** or **simple sequence DNA**.

This fraction constitutes 10% to 15% of the genome of mammals, and even more in seed plants. It reassociates so rapidly because it consists of several sets of short repeating nucleotide sequences. Most of these sequences are 5 to 10 base pairs (bp) in length, and they may be present up to a million times in the genome. Most organisms have various types of simple sequence DNA which are usually similar but not identical in sequence. In humans, for example, at least ten varieties exist.

These rapidly reassociating DNA sequences are thought not to be transcribed into RNA and are most

TABLE 9.3
Types of DNA based on reassociation kinetic studies.

Descriptive Terminology		Size	Association and Frequency
Highly Repetitive or Simple Sequence DNA		Most often 5–15 bp	Tandem repeats up to 10^6/genome
Moderately Repetitive or Intermediate Repeat DNA	Short period interspersion	Most often 150–300 bp	Interspersed with single copy DNA 5×10^5/genome
	Long period interspersion	Most often 5000–10,000 bp	Interspersed with single copy DNA 10^4/genome
Unique Sequence or Single Copy DNA		Most often 1000–5000 bp	Interspersed with intermediate repeat DNA 10^4–10^5/genome

often clustered in the regions of centromeres and sometimes in telomeric areas of the chromosomes. Evidence revealing this has been derived from *in situ* hybridization studies and will be presented in Chapter 17.

The second category of repetitive DNA does not reassociate as quickly as the simple sequence DNA. Called **moderately repetitive** or **intermediate repeat DNA,** this fraction consists of longer sequences that are present fewer times than simple sequence DNA. These sequences are interspersed throughout the genome and make up approximately 20 percent of the total DNA in invertebrates and up to 40 percent in mammals.

Among the prominently studied classes in this fraction are the *Alu* sequences in mammals. Varying between 150 and 300 base pairs in length, they are characterized by a sequence that is recognized as a substrate by the highly specific restriction endonuclease enzyme called *Alu* I. In humans, *Alu* sequences are 300 base pairs long and present in the genome over half a million times. While they fail to share the exact nucleotide sequence, sufficient homology exists so that they reassociate as complementary strands. This group of sequences constitutes up to 10 percent of the human genome.

The pattern of intermediate repeat DNA within the genome takes on two forms. The 300 base pair *Alu* sequences most often alternate with unique DNA sequences of 1000 to 2000 base pairs. This pattern is called **short-period interspersion**. It is characteristic of mammals, sea urchins, and toads, among other organisms. On the other hand, **long-period interspersion** contains intermediate repeat DNA that is 5000 to 10,000 base pairs (5 to 10 kb) long, alternating with unique sequence DNA of 10 to 20 kb. Present up to 10,000 times in mammalian genomes, this pattern is also characteristic of *Drosophila* and birds.

Although we will explore the significance of repetitive DNA sequences in Chapter 17, a short discussion is appropriate here. Highly repetitive sequences are usually much too short to serve as genes. Much of moderately repetitive DNA is also nongenic and not transcribed. However, some of this fraction consists of sequences that are transcribed and constitute repeat or duplicate copies of various genes. Included in this fraction are genes coding for ribosomal RNA, ribosomal proteins, and histones (positively charged proteins bound to DNA in eukaryotes). On the other hand, most genes are present only once within the genome and are included as part of what is called **single copy** or **unique sequence DNA.** Estimates are that most eukaryotic organisms contain between 5,000 and 50,000 genes. *Drosophila* is likely to be closer to the lower estimate while humans are thought to be closer to the higher estimate.

The proportion of the three fractions present in the genome varies in different organisms. In lower eukaryotes such as yeast, almost all DNA consists of nonrepetitive sequences with less than 20 percent existing as highly or moderately repetitive. In the cells of most animals about half of the DNA is nonrepetitive and the rest moderately and highly repetitive. In plants and certain amphibians, up to 80 percent falls into the repetitive categories.

The discovery of repetitive sequences has extended our knowledge of and generated increased interest in the organization of DNA in the chromosomes of eukaryotes. We will consider this topic again in Chapter 17.

Supercoiling and Circular DNA

One major insight into the way in which DNA is organized and packaged has come from the discovery of **supercoiled DNA,** characteristic of covalently closed circular molecules and chromosomal loops. Since no discussion of the organization of DNA would be complete without a consideration of supercoiled DNA, we shall conclude this chapter by introducing this topic in some depth. The information presented here is relevant to other parts of the text as well, including those sections on the replication, expression, and regulation of genetic information and the organization of DNA into chromosomes.

Supercoiled DNA was first proposed as a result of a study of double-stranded DNA molecules derived from the **polyoma virus,** which causes tumors in mice. In 1963, it was observed that when such DNA was subjected to high-speed centrifugation, it was resolved into three distinct components, each of different density and compactness. That which was least compact, and thus least dense, demonstrated a decreased sedimentation velocity; the other two fractions each showed increasing velocity due to their greater compaction and density. All three were of identical molecular weight.

In 1965 Jerome Vinograd proposed an explanation for the preceding observation. He postulated that the two fractions of greatest sedimentation velocity both consisted of polyoma DNA molecules that were circular, while the fraction of lower sedimentation velocity contained polyoma DNA molecules that were linear rather than closed circles. It is logical that circular molecules are more compact and sediment more rapidly than linear molecules of the same length and molecular weight.

Vinograd further proposed that of the two fractions of circular molecules, the more dense consisted of covalently closed DNA helices that were slightly *underwound* in comparison to the less dense circular molecules. The energetic force stabilizing the double helix resists this underwinding, causing it to supercoil in order to retain normal base pairing. Vinograd proposed that it was the supercoiled shape that caused tighter packing and thus the increase in sedimentation velocity.

Consider a double-stranded linear molecule that exists in the normal Watson-Crick right-handed helix [Figure 9.22(a)]. As you know, this helix contains about 10 base pairs per complete turn. If the ends of the molecule are sealed, a closed circle is formed [Figure 9.22(b)] which is said to be *relaxed*. Now, if both strands of this molecule are cut, a linear molecule is once again created. Suppose that it is purposely underwound by several full turns [Figure 9.22(c)] and then resealed [Figure 9.22(d)].

Such a structure is no longer energetically "relaxed"; as a result, it exists only temporarily in this unstable physical conformation. In order to assume a more energetically favorable form, the structure must undergo an interesting change in molecular conformation. The circular molecule "coils" in the direction opposite to the helix, forming a more compact structure [Figure 9.22(e)]. In doing so, the original number of turns of the helix is reestablished, and the physical stability of the molecule is enhanced by virtue of the negative supercoils that are spontaneously formed.

Some descriptive terminology follows which will help clarify the phenomenon of supercoiling:

1   The **Linking Number (L)** is defined as the number of times the strands of a DNA double helix wind completely around one another. In a relaxed closed circle, derived from a linear helix, L equals the total number of nucleotide pairs divided by 10.4 because there are 10.4 base pairs per turn. In Figure 9.22 L is equal to 20 in part (b), but is diminished by two full turns in parts (d) and (e).

By convention, if the helix is right-handed, as it is in B-DNA, L is a positive value; it is always a positive integer. Thus, in parts (d) and (e), L has been diminished to +18 compared with the L value of +20 of the relaxed circular helix in part (b).

2   When two otherwise identical molecules demonstrate a difference in their linking number, they are said to be **topoisomers** of one another. In our example above, the conversion in Linking Number from +20 to +18 creates a topoisomer that is not relaxed and thus is unstable. To achieve an energetically stable configuration, the topoisomer undergoes supercoiling in the opposite direction (left-handed). By convention, left-handed supercoils are designated as negative. In the preceding example, two such coils will form. The number of supercoils is sometimes denoted as the **Writhing Number (W)**. In our example, W = −2.

3   It is apparent that the only way in which these two topoisomers can be interconverted is by cutting one or both of the strands and winding or unwinding the helix before resealing the ends. Biologically, this may be accomplished by one of a group of enzymes, appropriately called **topoisomerases**. First discovered by Martin Gellert and James Wang, these catalytic molecules are either Type I or II, depending on whether they cleave one or both strands in the helix, respectively. In *E. coli*, topoisomerase I serves to *relax* negative supercoils. The enzyme is a monomer that covalently bonds to the cleaved DNA strand. In the process of catalysis, the Linking Number is increased by one (+1). **Topoisomerase II,** also named **DNA gyrase,** is a tetramer that *introduces* negative supercoils into DNA. Each catalytic event requires energy and reduces L by two (−2). This enzyme is thought to bind to DNA, twist it, cleave both strands, and then pass them through the "loop" that it has created. Reestablishing the phosphodiester bonds creates the new supercoil.

The DNA in both viruses and bacteria exhibits extensive supercoiling. The virus **SV40** contains double-stranded DNA consisting of 5200 base pairs. In a relaxed, circular form, L = 5200/10.4, or +500. However, when SV40 DNA is isolated and analyzed, L is found to be only +475. As you should now be able to predict, there are 25 negative supercoils present. In *E. coli* the nucleoid region contains enormous numbers of supercoils, greatly facilitating the condensation of the chromosome and its ability to exist in the cell. Thus, there is little doubt that supercoiling solves one problem faced by viruses and bacteria—how to pack enormously long DNA molecules into relatively small spaces.

Other genetic functions, including replication and gene expression, are also thought to rely on supercoiling. And, as we shall see later in the text, supercoiled DNA and topoisomerases are also characteristic of eukaryotes, even though DNA is linear in this vast group of organisms. While chromatin and chromosomes are not circular, supercoils are made possible because areas of DNA are embedded in the lattice of proteins associated with the chromatin

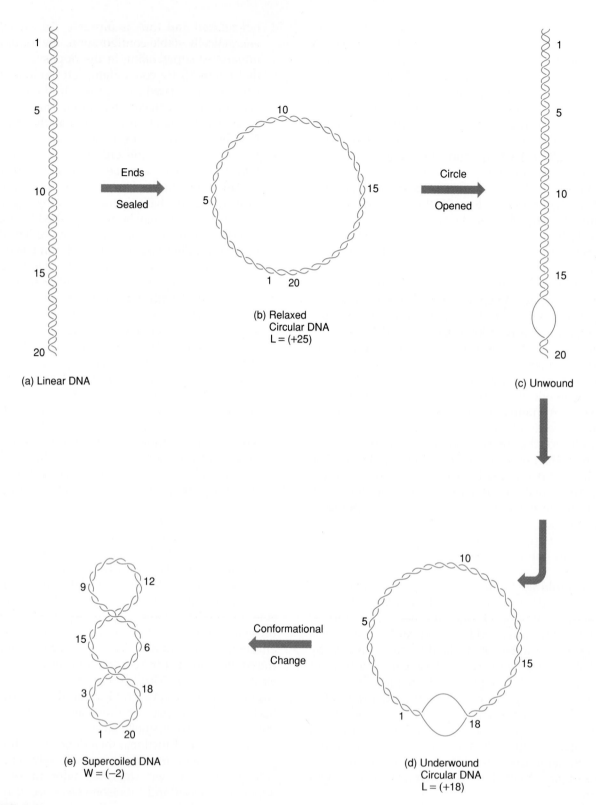

(a) Linear DNA

Ends
Sealed

(b) Relaxed
Circular DNA
L = (+25)

Circle
Opened

(c) Unwound

Conformational

Change

(e) Supercoiled DNA
W = (−2)

(d) Underwound
Circular DNA
L = (+18)

FIGURE 9.22
Transformations leading to the supercoiling of circular DNA. L designates the Linking
Number, while W designates the Writhing Number, as described in the text.

fibers. This association creates "anchored" ends, which provide the stability for the maintenance of supercoils once they are introduced by topoisomerases.

CHAPTER SUMMARY

1 During the late 1940s and early 1950s, a considerable effort was made to integrate accumulated information on the chemical structure of nucleic acids into a model of the molecular structure of DNA.

2 The X-ray crystallography data of Franklin and Wilkins confirmed the suggestion that DNA was some sort of helix. In 1953, Watson and Crick were able to assemble a model of the double helical DNA structure based on these X-ray diffraction studies as well as on Chargaff's analysis of base composition of DNA.

3 The DNA molecule exhibits antiparallel orientation and adenine-thymine and guanine-cytosine base-pairing complementarity along the polynucleotide chains. This model of DNA presents an obvious straightforward mechanism for its replication.

4 The structure assumed by the helix appears to be a function of the nucleotide sequence and its chemical environment. Several alternative forms of the DNA helical structure exist. Watson and Crick described a B configuration, one of several right-handed helices, including A, C, D, and E

forms. Wang and Rich discovered the left-handed Z-DNA currently being investigated for its physiological and genetic significance.

5 The second category of nucleic acids important in genetic function is RNA. RNA is similar to DNA with the exceptions that it is usually single-stranded, the sugar ribose replaces the deoxyribose and the pyrimidine uracil replaces thymine. Classes of RNA—ribosomal, transfer, and messenger—facilitate the flow of information from DNA to RNA to proteins, which are the end products of most genes.

6 The structure of DNA lends itself to various forms of analysis, which have in turn led to studies of the functional aspects of the genetic machinery. Absorption of UV light, sedimentation properties, and denaturation-reassociation procedures are among the important tools for the study of nucleic acids.

7 Reassociation kinetics analysis has enabled geneticists to subdivide DNA into various categories based on the frequency of existing sequences. These categories include highly repetitive and moderately repetitive sequences, in addition to the unique sequences that represent single-copy genes. Repetitive DNA is found in centromeric and telomeric regions, and is also interspersed throughout the genome.

8 Circular DNA, prominent in viruses and bacteria, is characterized by supercoiling, which alters the compaction of the molecule and achieves energetic stability.

Insights and Solutions

1 Sea urchin DNA, which is double-stranded, was shown to contain 17.5 percent of its bases in the form of cytosine (C). What percentages of the other three bases are present in this DNA?

ANSWER: The amount of C = G, so guanine is also present as 17.5 percent. The remaining bases, A and T, are present in equal amounts and together they represent the rest of the bases $(100 - 35)$. Therefore, A = T = 65/2 = 32.5 percent.

2 The quest to isolate an important disease-causing organism was successful and the molecular biologists were hard at work. The organism contained as its genetic material a remarkable nucleic acid with a base composition of A = 21 percent, C = 29 percent, G = 29 percent, U = 21 percent. When heated, it showed a major hyperchromic effect, and when kinetics were studied, the nucleic acid of this organism provided the C_0t curve shown below, in contrast to that of phage T4 and E coli. T4 contains 10^5 nucleotide pairs and exhibits a $C_0t_{1/2}$ of 0.5. The unknown organism

produced a $C_0t_{1/2}$ of 20. Analyze this information carefully and draw *all* possible conclusions about the genetic material of this organism, based strictly on the above observations. What important, straightforward information is missing and needed to confirm your hypothesis about the nature of this molecule?

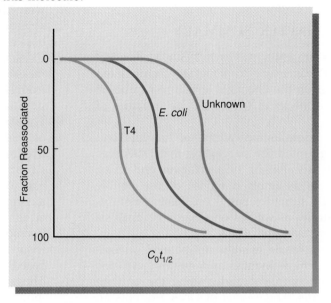

ANSWER: First of all, because of the presence of uracil (U), the molecule appears to be RNA. Since A/U = G/C = 1, the molecule may be a double helix. The hyperchromic shift and reassociation kinetics confirms this hypothesis. The kinetic study demonstrates several things. First, the shape of the C_0t curve reveals that there is no repetitive sequence DNA. Further, the complexity (X), or total length of unique sequence DNA, is greater than that of either phage T4 or *E. coli*. X can, in fact, be calculated, since a direct proportionality between $C_0t_{1/2}$ and number of base pairs exists when there are only unique sequences present:

$$\frac{0.5}{10^5} = \frac{20}{X}$$
$$0.5X = 20(10^5)$$
$$X = 40(10^5)$$
$$= 4 \times 10^6 \text{ base pairs}$$

There *may* be a greater number of genes present compared to T4 or *E. coli*, but the excessive unique sequence DNA may serve some other role, or simply play no genetic role. The missing information concerns the sugars. Our model predicts that ribose rather than d-ribose should be present. If not, the organism contains a very unusual molecule as its genetic material.

1 Draw the chemical structure of the three components of a nucleotide and then link the three together. What atoms are removed from the structures when the linkages are formed?

2 How are the carbon and nitrogen atoms of the sugars, purines, and pyrimidines numbered?

3 Adenine may also be named 6-amino purine. How would you name the other four nitrogenous bases using this alternate system? (=O is oxy, and $-CH_3$ is methyl.)

4 Draw the chemical structure of a dinucleotide composed of A and G. Opposite this structure, draw the dinucleotide TC in an antiparallel (or upside down) fashion. Form the possible hydrogen bonds.

5 Describe the various characteristics of the Watson-Crick double-helix model for DNA.

6 What evidence did Watson and Crick have at their disposal in 1953? What was their approach in arriving at the structure of DNA?

7 Had Chargaff's data from a single source indicated the following, what might Watson and Crick have concluded?

A	T	C	G
% 29	19	21	31

Why would this conclusion be contradictory to Wilkins and Franklin's data?

8 How do covalent bonds differ from hydrogen bonds? Define base complementarity.

9 List three main differences between DNA and RNA.

10 What are the three types of RNA molecules? How is each related to the concept of information flow?

11 What component of the nucleotide is responsible for the absorption of ultraviolet light? How is this technique important in the analysis of nucleic acids?

12 Distinguish between sedimentation velocity and sedimentation equilibrium centrifugation.

13 What chemical characteristics determine a molecule's Svedberg coefficient?

14 What is the basis for determining base composition using density gradient centrifugation?

15 What is the state of DNA following denaturation?

16 What is the hyperchromic effect? How is it measured? What does T_m imply?

17 Why is T_m related to base composition?

18 Which of the following terms mean approximately the same thing: denaturation, renaturation, reassociation, melting, hybridization?

19 What is the chemical basis of molecular hybridization?

20 What did the Watson-Crick model suggest about the replication of DNA?

21 Compare the following curves representing reassociation kinetics. What can be said about the DNAs represented by each set of data compared with *E. coli*?

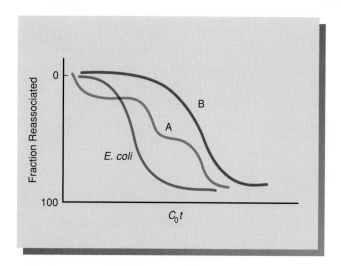

22 Draw the predicted $C_0 t$ curve if the following mixture of nucleic acid molecules were analyzed in a single experiment: 10^6 copies of a short, identical sequence; 10^3 copies of a much longer DNA sequence; and 10^2 unique sequences of DNA.

23 A genetics student was asked to draw the chemical structure of an adenine- and thymine-containing dinucleotide derived from DNA. The student made six major errors. His answer is shown below. One of them is circled, numbered ①, and explained. Find the other five. Circle them, number them ②—⑥, and briefly explain each, following the example below.

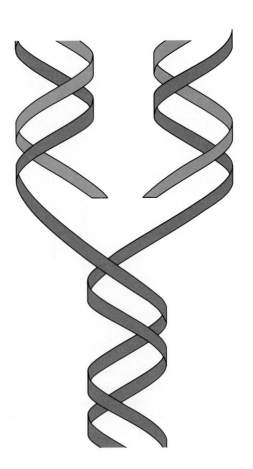

EXPLANATIONS

① Extra phosphate should not be present.

24 Considering the information in this chapter on B- and Z-DNA and right- and left-handed helices, carefully analyze the structure below and draw conclusions about its helical nature. What process is occurring?

25 The DNA of the bacterial virus T4 produces a $C_0t_{1/2}$ of about 0.5 and contains 10^5 nucleotide pairs in its genome. How many nucleotide pairs are present in the genome of the virus MS2 and the bacterium *E. coli*, whose respective DNAs produce $C_0t_{1/2}$ values of 0.001 and 10.0?

26 *Newsdate: March 1, 2005.* A unique creature has been discovered during exploration of outer space. Recently, its genetic material has been isolated and analyzed. This material is similar in some ways to DNA in its chemical makeup. It contains in abundance the 4-carbon sugar erythrose and a molar equivalent of phosphate groups. Additionally, it contains six nitrogenous bases: adenine (A), guanine (G), thymine (T), cytosine (C), hypoxanthine (H), and xanthine (X). These bases exist in the following relative proportions:

$$A = T = H \quad \text{and} \quad C = G = X$$

X-ray diffraction studies have established a regularity to the molecule and a constant diameter of about 30 Å.

Together, these data have suggested a model for the structure of this molecule.

(a) Propose a general model of this molecule. Describe it briefly.

(b) What base-pairing properties must exist for H and for X in the model?

(c) Given the constant diameter of 30 Å, do you think that H and X are *either* (1) both purines or both pyrimidines, *or* (2) one is a purine and one is a pyrimidine?

∞∞∞∞∞∞∞∞∞∞∞∞∞∞∞∞∞∞∞∞∞∞∞∞∞∞∞∞

SELECTED READINGS

AZORIN, F., and RICH, A. 1985. Isolation of Z-DNA binding proteins from SV40 minichromosomes: Evidence for binding to the viral control region. *Cell* 41: 365–74.

BAVER, W. R., CRICK, F. H. C., and WHITE, J. H. 1980. Supercoiled DNA. *Scient. Amer.* (July) 243: 118–33.

BRITTEN, R. J., and KOHNE, D. E. 1968. Repeated sequences in DNA. *Science* 161: 529–40.

———. 1970. Repeated segments of DNA. *Scient. Amer.* (April) 222: 24–31.

CANTOR, C. R. 1981. DNA choreography. *Cell* 25: 293–95.

CHARGAFF, E. 1950. Chemical specificity of nucleic acids and mechanism for their enzymatic degradation. *Experientia* 6: 201–9.

CRICK, F. H. C., WANG, J. C., and BAUER, W. R. 1979. Is DNA really a double helix? *J. Mol. Biol.* 129: 449–61.

DAVIDSON, J. N. 1976. *The biochemistry of the nucleic acids.* 8th ed. Orlando: Academic Press.

DICKERSON, R. E., et al. 1982. The anatomy of A-, B-, and Z-DNA. *Science* 216: 475–85.

DICKERSON, R. E. 1983. The DNA helix and how it is read. *Scient. Amer.* (June) 249: 94–111.

FELSENFELD, G. 1985. DNA. *Scient. Amer.* (Oct.) 253: 58–78.

FRANKLIN, R. E., and GOSLING, R. G. 1953. Molecular configuration in sodium thymonucleate. *Nature* 171: 740–41.

GEIS, I. 1983. Visualizing the anatomy of A, B and Z-DNAs. *J. Biomol. Struc. Dynam.* 1: 581–91.

JUDSON, H. 1979. *The eighth day of creation: Makers of the revolution in biology.* New York: Simon & Schuster.

LONG, E., and DAWID, I. 1980. Repeated genes in eukaryotes. *Ann. Rev. Biochem.* 49: 727–64.

OLBY, R. 1974. *The path to the double helix.* Seattle: University of Washington Press.

PAULING, L., and COREY, R. B. 1953. A proposed structure for the nucleic acids. *Proc. Natl. Acad. Sci.* 39: 84–97.

RICH, A., NORDHEIM, A., and WANG, A. H.-J. 1984. The chemistry and biology of left-handed Z-DNA. *Ann. Rev. Biochem.* 53: 791–846.

SCHILDKRAUT, C. L., MARMUR, J., and DOTY, P. 1962. Determination of the base composition of deoxyribonucleic acid from its buoyant density in CsCl. *J. Mol. Biol.* 4: 430–43.

SCHMIDTKE, J., and EPPLEN, J. T. 1980. Sequence organization of animal nuclear DNA. *Hum. Genet.* 55: 1–18.

SINGER, M. F. 1982. Sines and lines: Highly repeated short and long interspersed sequences in mammalian genomes. *Cell* 28: 433–34.

STENT, G. S., ed. 1981. *The double helix: Text, commentary, review, and original papers.* New York: W. W. Norton.

WATSON, J. D. 1968. *The double helix.* New York: Atheneum.

WATSON, J. D., and CRICK, F. C. 1953a. Molecular structure of nucleic acids. A structure for deoxyribose nucleic acids. *Nature* 171: 737–38.

———. 1953b. Genetic implications of the structure of deoxyribose nucleic acid. *Nature* 171: 964.

WILKINS, M. H. F., STOKES, A. R., and WILSON, H. R. 1953. Molecular structure of desoxypentose nucleic acids. *Nature* 171: 738–40.

ZIMMERMAN, B. 1982. The three-dimensional structure of DNA. *Ann. Rev. Biochem.* 51: 395–428.

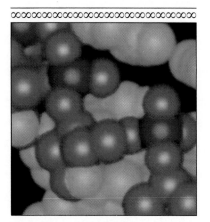

10

DNA Replication, Synthesis, and Recombination

CHAPTER CONCEPTS

Genetic continuity between parental and progeny cells is made possible by semiconservative replication of DNA, as predicted by the Watson-Crick model. Each strand of the parent helix serves as a template for the production of its complement. Synthesis of DNA is a complex but orderly process orchestrated by a myriad of enzymes and other molecules. Together, they function with great fidelity to polymerize nucleotides into polynucleotide chains. Other enzymes interact with DNA, leading to genetic recombination.

Following Watson and Crick's proposal for the structure of DNA, scientists focused their attention on how this molecule is replicated or duplicated. This process is an essential function of the genetic material and must be executed precisely if genetic continuity between cells is to be maintained following cell division. Consider for a second that in the human genome, some three billion base pairs exist. Even an error rate of 10^{-6} (one in a million) would create an excessive and unacceptable number of errors (3000) during each replication cycle. While not error-free, an extremely accurate system of DNA replication has evolved in all organisms.

As discussed in Watson and Crick's early papers in 1953, the model of the double helix gave them the initial insight into how replication could occur. This mode, called **semiconservative replication,** has since received strong experimental support from studies of viruses, prokaryotes, and eukaryotes.

Once the general mode of replication was made clear, research was intensified to determine the precise details of DNA synthesis. What has since been discovered is that numerous enzymes and other proteins are needed to copy a DNA helix. Because of the complexity of the chemical events during synthesis, this subject remains an extremely active area of research.

In this chapter, we will discuss the mode of replication as well as the chemical synthesis of DNA. Then we will discuss how DNA molecules recombine with one another. The research leading to this knowledge is still another link in our understanding of life processes at the molecular level.

THE MODE OF DNA REPLICATION

It was apparent to Watson and Crick that because of the arrangement and nature of the nitrogenous bases,

each strand of a DNA double helix could serve as a template for the synthesis of its complement. They proposed that if the helix were unwound, each nucleotide along the two parent strands would have an affinity for its complementary nucleotide. As we learned in Chapter 9, the complementarity is due to the potential hydrogen bonds that can be formed. If thymidylic acid were present, it would "attract"

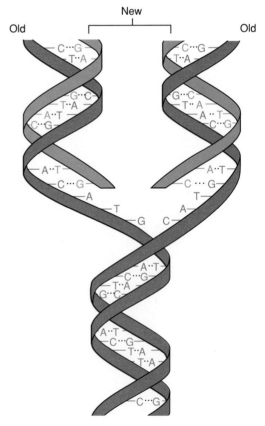

FIGURE 10.1
General model of semiconservative replication of DNA.

adenylic acid; if guanidylic acid were present, it would "attract" cytidylic acid; and so on. If these nucleotides were then covalently linked into polynucleotide chains along both templates, the result would be the production of two new but identical double strands of DNA. This concept is illustrated in Figure 10.1. Each replicated DNA molecule would consist of one "old" and one "new" strand. Therefore, this mechanism is called **semiconservative replication**.

There are two other possible modes of replication (Figure 10.2) that also rely on the parental strands as a template. In **conservative replication,** synthesis of complementary polynucleotide chains occurs as described above; following synthesis, however, the two newly created strands are brought back together, and the parental strands reassociate. The original helix is thus "conserved."

In the second alternative mode, called **dispersive replication,** the parental strands are seen to be dispersed into two new double helices following replication. This mode would involve cleavage of the parental strands during replication. Therefore, it is the most complex of the three possibilities and, as such, is least likely. It could not, however, be ruled out as an experimental model. The theoretical results of two generations of replication by the conservative and dispersive modes are compared with those of the semiconservative mode in Figure 10.2.

The Meselson-Stahl Experiment

In 1958, Matthew Meselson and Franklin Stahl published the results of an experiment providing strong evidence that semiconservative replication is the mode used by cells to produce new DNA molecules. *E. coli* cells were grown for many generations in a medium where $^{15}NH_4Cl$ (ammonium chloride) was the only nitrogen source. A "heavy" isotope of nitrogen, ^{15}N, contains one more neutron than the naturally occurring ^{14}N isotope. After many generations, all nitrogen-containing molecules, including the nitrogenous bases of DNA, contained the heavy isotope in the *E. coli* cells. DNA containing ^{15}N may be distinguished from ^{14}N-containing DNA by the use of **sedimentation equilibrium centrifugation** in a

FIGURE 10.2
The results of two rounds of replication of DNA for each of the three possible modes of replication. Each round of synthesis is shown in a different color.

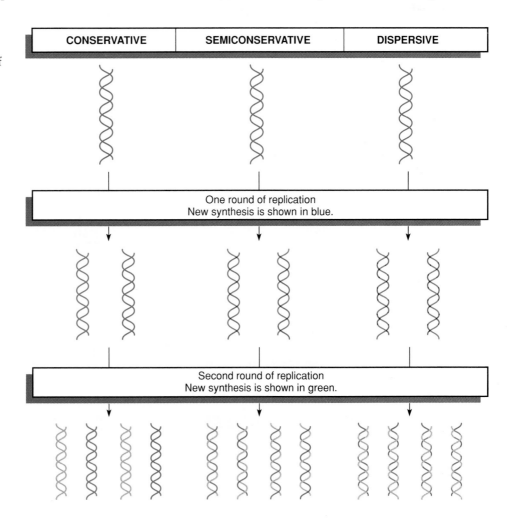

CONSERVATIVE	SEMICONSERVATIVE	DISPERSIVE

One round of replication
New synthesis is shown in blue.

Second round of replication
New synthesis is shown in green.

cesium chloride gradient (see Chapter 9). The dense ^{15}N-DNA will reach equilibrium in the gradient at a point closer to the bottom than does ^{14}N-DNA.

In this experiment, uniformly labeled ^{15}N cells were transferred to a medium containing only $^{14}NH_4Cl$. Thus, all subsequent synthesis of DNA during replication contained the "lighter" isotope of nitrogen. The time of transfer was taken as time zero ($t = 0$). The *E. coli* cells were allowed to replicate during several generations with cell samples removed at various intervals. From each sample, DNA was isolated and subjected to sedimentation equilib-

rium analysis. The actual results are depicted in Figure 10.3.

After one generation, the isolated DNA was all present in a single band of intermediate density—the expected result for semiconservative replication. Each replicated molecule would have been composed of one new ^{14}N-strand and one old ^{15}N-strand, as seen in Figure 10.4. This result was not consistent with the conservative replication mode, in which two distinct bands would have been predicted to occur.

After two cell divisions, DNA samples showed two density bands: one was intermediate and the

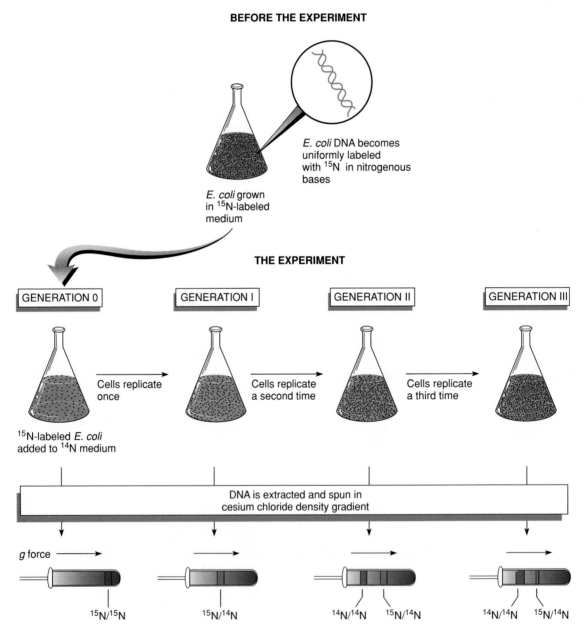

FIGURE 10.3
The Meselson-Stahl experiment.

FIGURE 10.4

The expected results of two
generations of semiconservative
replication in the Meselson-
Stahl experiment.

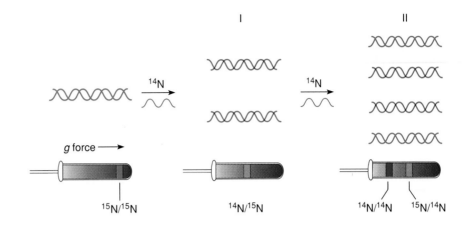

other was lighter, corresponding to the ^{14}N position in the gradient. Similar results occurred after a third generation, except that the proportion of the ^{14}N-band increased. If replication were dispersive, all subsequent generations after $t = 0$ would demonstrate DNA of an intermediate density. In each subsequent generation the ratio ^{14}N/^{15}N would increase, and the hybrid band would become lighter and lighter, eventually approaching the ^{14}N-band. This result, of course, was not observed. Thus, the results of the Meselson-Stahl experiment provided strong support for the semiconservative mode of DNA replication, as postulated by Watson and Crick.

Semiconservative Replication in Eukaryotes

In 1957, the year before the work of Meselson and his colleagues was published, evidence was also presented that supported semiconservative replication in a eukaryotic organism. J. Herbert Taylor, Philip Woods, and Walter Hughes experimented with root tips of the broad bean *Vicia faba*. Root tips are an excellent source of dividing cells. These researchers examined the chromosomes of these cells following replication of DNA. They were able to monitor the process of replication by labeling DNA with ^{3}H-thymidine, a radioactive precursor of DNA, and performing autoradiography. In this experiment, labeled thymidine is found only in association with chromosomes that contain newly synthesized DNA.

The technique of autoradiography, discussed in detail in Appendix A, is a cytological procedure that allows the isotope to be localized within the cell. In this procedure, a photographic emulsion is placed over a section of cellular material (root tips in this experiment), and the preparation is stored in the dark. The slide is then developed, much as photographic film is processed. Since the radioisotope emits energy, the emulsion turns black at the approximate point of emission following development. The end result is the presence of dark spots or grains on the surface of the section, localizing newly synthesized DNA within the cell.

Root tips were grown for approximately one generation in the presence of the radioisotope and then placed in unlabeled medium, where cell division continued. At the conclusion of each generation, cultures were arrested at metaphase by the addition of colchicine, and chromosomes were examined by autoradiography. Figure 10.5 illustrates replication of a single chromosome over two division cycles as well as the distribution of grains. The results are compatible with the semiconservative mode of replication. After the first replication cycle, radioactivity is detected over both sister chromatids. This finding is expected because each chromatid will contain one "new" radioactive DNA strand and one "old" unlabeled strand. After the second replication cycle, which also takes place in unlabeled medium, only one of the two new sister chromatids should be radioactive because half of the parent strands are unlabeled. With only the minor exception of the demonstration of **sister chromatid exchanges** (see Chapter 6), this result was observed.

Together, the Meselson-Stahl experiment and the experiment by Taylor, Woods, and Hughes soon led to the general acceptance of the semiconservative mode of replication. The same conclusion has been reached in studies with other organisms. Since this mode is suggested by the double-helix model of DNA, these experiments also strongly supported Watson and Crick's proposal for DNA structure.

FIGURE 10.5
Depiction of the experiment by Taylor, Woods, and Hughes demonstrating the semiconservative mode of replication of DNA in root tips of *Vicia faba*. The plant is shown in the photograph. In (a) an unlabeled chromatid proceeds through the cell cycle in the presence of ^{3}H-thymidine. As it enters mitosis, both sister chromatids of each chromosome are labeled, as shown by autoradiography. After a second generation of replication, this time in the absence of ^{3}H-thymidine, only one chromatid of each chromosome is expected to be surrounded by grains (b). In all cases, except where a reciprocal exchange has occurred between sister chromatids (c), the expectation was upheld.

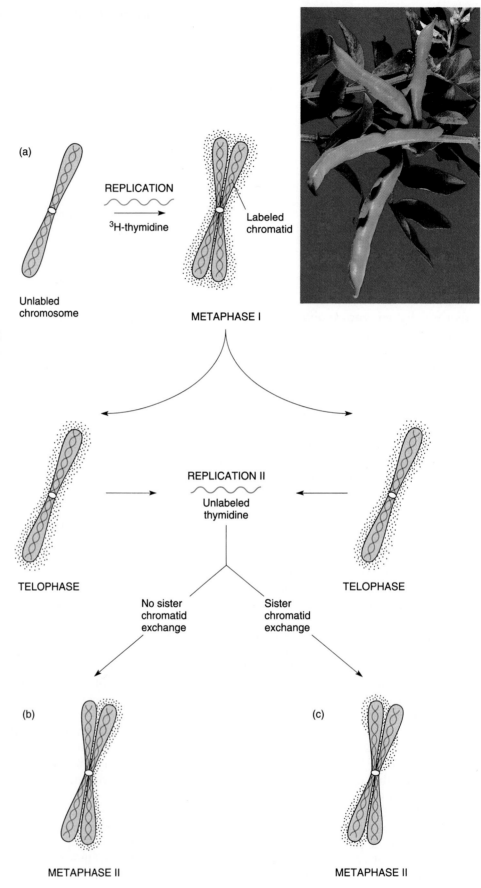

(a)

REPLICATION

^{3}H-thymidine

Labeled chromatid

Unlabled chromosome

METAPHASE I

REPLICATION II
Unlabeled thymidine

TELOPHASE

TELOPHASE

No sister chromatid exchange

Sister chromatid exchange

(b)

(c)

METAPHASE II

METAPHASE II

Replication Origins and Forks

The mode of replication just established represents the general pattern by which DNA is duplicated. Before turning to the details of the biosynthesis of DNA, we will mention several other topics relevant to the complete description of semiconservative replication. The first concerns the **origin of replication**. Along any chromosome, is there but a single origin or more than one where DNA synthesis begins? And, is an origin random or located in a specific region along the chromosome? As we address these questions, the length of DNA that is replicated following the initiation of synthesis at a single origin will be defined as a unit called the **replicon**.

A second topic involves the direction of replication. Once it begins, does it move in a single direction or in both directions away from the origin? This consideration distinguishes between **unidirectional** and **bidirectional replication**, respectively. At each point of replication, the strands of the helix must unwind, creating what is called a **replication fork**. Bidirectional replication creates two such forks, which move apart in opposite directions away from the origin.

The evidence is reasonably clear regarding these topics. In bacteria and most bacterial viruses, which have only a single circular chromosome, there is a specific region where replication is initiated. In *E. coli* this region, called *ori*, has been located. It consists of 245 base pairs, but only a small number are actually essential to the initiation of DNA synthesis. Thus, in bacteriophages and bacteria, the entire chromosome constitutes one replicon. In *E. coli*, replication is bidirectional from *ori* and proceeds until the entire chromosome is replicated, as illustrated in Figure 10.6. Both strands of the parent helix are copied at the two replication forks that are established. These forks move away from the origin in opposite directions and eventually merge as semiconservative replication of the entire chromosome is completed.

Compared to bacteria, as described above, a major difference in replication exists in eukaryotes. While replication is bidirectional, creating two replication forks during each replication cycle, there are multiple origins along each chromosome. Thus, during the S phase of interphase, there are numerous replicating events occurring along each chromosome. Eventually, the numerous replication forks merge, completing replication of the entire chromosome. The presence of multiple replicons is undoubtedly related to the much greater length of a single eukaryotic chromosome compared to one from a bacterium. As a result, replication can be completed in a reasonable period of time.

SYNTHESIS OF DNA IN MICROORGANISMS

The determination that replication is semiconservative and bidirectional indicates only the pattern of DNA duplication and the association of finished strands with one another once synthesis is completed. A much more complex issue is how the **synthesis** of long complementary polynucleotide chains occurs on a DNA template. As in most studies of molecular biology, this question was first approached by using microorganisms. Research began about the same time as the Meselson-Stahl work, and even today this topic is an active area of investigation. What is most apparent in this research is the tremendous chemical complexity of the biological synthesis of DNA.

DNA Polymerase I

Studies of the enzymology of DNA replication were first reported by Arthur Kornberg and colleagues in 1957. They isolated an enzyme from *E. coli* that was able to direct DNA synthesis in a cell-free (*in vitro*) system. The enzyme is now called **DNA polymerase I**, since it was the first of several to be isolated. Kornberg determined the following requirements for *in vitro* DNA synthesis under the direction of the enzyme:

1 All four deoxyribonucleoside triphosphates (dATP, dCTP, dGTP, dTTP = dNTP[*])
2 Mg^{++} ions
3 Template DNA

If any one of the four deoxyribonucleoside triphosphates was omitted from the reaction, no synthesis occurred. If derivatives of these precursor molecules other than the nucleoside triphosphate were used (nucleotides or nucleoside diphosphates), synthesis did not occur. If no template DNA was added, greatly reduced synthesis of DNA occurred, but only after prolonged reaction times. Thus, most synthesis directed by Kornberg's enzyme appeared to be exactly the type required for semiconservative replication. The reaction is summarized in Figure 10.7. The enzyme is known to consist of a single polypeptide containing 928 amino acids.

---

[*]dNTP designates the deoxyribose forms of the four nucleoside triphosphates: in a similar way, dNMP refers to the monophosphate forms.

FIGURE 10.6
Replication of the *E. coli*
chromosome.

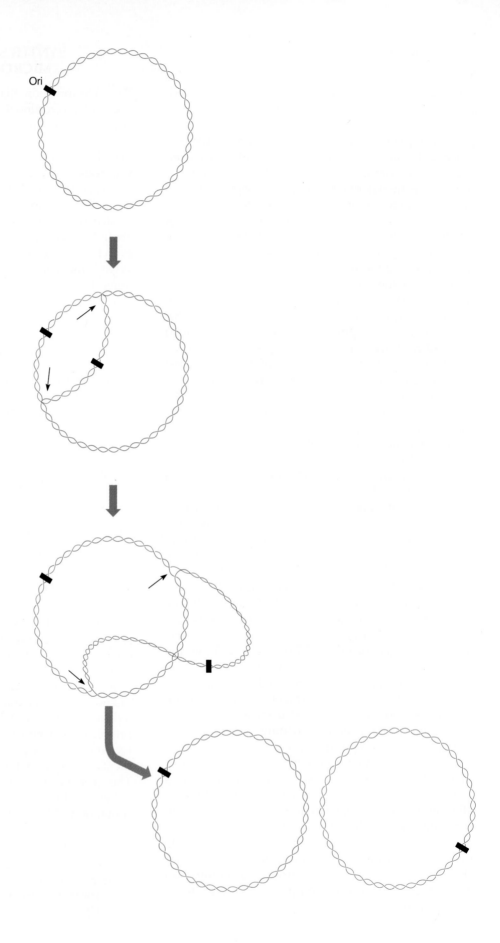

Ori

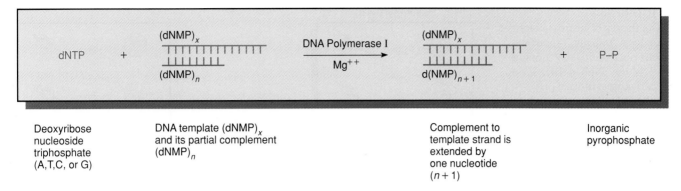

| dNTP | + | (dNMP)$_x$ —————— ‖‖‖‖‖‖‖‖‖‖ —————— (dNMP)$_n$ | $\xrightarrow[\text{Mg}^{++}]{\text{DNA Polymerase I}}$ | (dNMP)$_x$ —————— ‖‖‖‖‖‖‖‖‖‖ —————— d(NMP)$_{n+1}$ | + | P–P |

| Deoxyribose nucleoside triphosphate (A,T,C, or G) | DNA template (dNMP)$_x$ and its partial complement (dNMP)$_n$ | Complement to template strand is extended by one nucleotide ($n+1$) | Inorganic pyrophosphate |

FIGURE 10.7

Chemical reaction catalyzed by DNA polymerase I. During each step, a single nucleotide is added to the growing complement of the DNA template, using a nucleoside triphosphate as the substrate. The release and hydrolysis of inorganic pyrophosphate drives the reaction energetically.

Fidelity of Synthesis

Having shown how DNA was synthesized, Kornberg sought to demonstrate the accuracy, or fidelity, with which the enzyme had replicated the DNA template. Since the nucleotide sequences of the template and the product could not be determined in 1957, he had to rely initially on several indirect methods.

One of Kornberg's approaches was to compare the nitrogenous base compositions of the template added to the reaction mixture with those of the recovered product. Table 10.1 shows Kornberg's base composition analysis of three different DNA templates. These may be compared with the product synthesized in each case. Within experimental error, the base composition of each product agreed with the template DNAs used. These data suggested that the templates were replicated faithfully.

Kornberg also used the **nearest-neighbor frequency test** to test the fidelity of copying. With this technique, he determined the frequency with which

TABLE 10.1

Base composition of DNA template and product in Kornberg's early work.

Organism	Template/ Product	%A	%T	%G	%C
T2	Template	32.7	33.0	16.8	17.5
	Product	33.2	32.1	17.2	17.5
E. coli	Template	25.0	24.3	24.5	26.2
	Product	26.1	25.1	24.3	24.5
Calf	Template	28.9	26.7	22.8	21.6
	Product	28.7	27.7	21.8	21.8

SOURCE: From Kornberg, 1960. Copyright 1960 by the American Association for the Advancement of Science.

any two bases occur adjacent to each other along the polynucleotide chain. As illustrated in Figure 10.8, this test relies on the enzyme **spleen phosphodiesterase,** which cleaves the polynucleotide chain differently from the way in which the chain was assembled. During synthesis of DNA, 5'-nucleotides are inserted; that is, each nucleotide is added with the phosphate on the C-5' of deoxyribose. The enzyme cleaves between the phosphate and the C-5' atom, thereby producing 3'-nucleotides. If the phosphates on only one of the four nucleotides (cytidylic acid, for example) are radioactive during DNA synthesis, then after enzymatic cleavage a radioactive phosphate will be attached to the base that is the "nearest neighbor" of all cytidylic acid nucleotides. Following four separate experiments, where in each case only one of the four nucleotide types is made radioactive, the frequency of all sixteen possible nearest neighbors can be calculated. An example of such data is presented in Problem 24 at the end of this chapter.

When this technique was applied to the DNA template and the resultant product from a variety of experiments, Kornberg found general agreement between the nearest-neighbor frequencies of the two. This type of analysis is a more stringent measure of the fidelity of copying than the base composition analysis. Thus, DNA polymerase I seemed the likely candidate for the synthesis of DNA during replication within the cell, even though Kornberg's experiments were, by necessity, performed *in vitro*.

Synthesis of Biologically Active DNA

Despite Kornberg's extensive work, not all researchers were convinced that DNA polymerase I was the enzyme that replicates DNA *in vivo*. The primary

FIGURE 10.8
The theory of nearest-neighbor analysis. Initially, one of the four nucleotides (C) contains ^{32}P in the 5'-phosphate group. Following synthesis of a polynucleotide chain and enzymatic treatment with phosphodiesterase, the radioactive phosphate group is transferred to the nearest neighbors (G and T). Subsequent analysis reveals the percentage of time each of the four nucleotides (A, T, C, and G) have become radioactive.

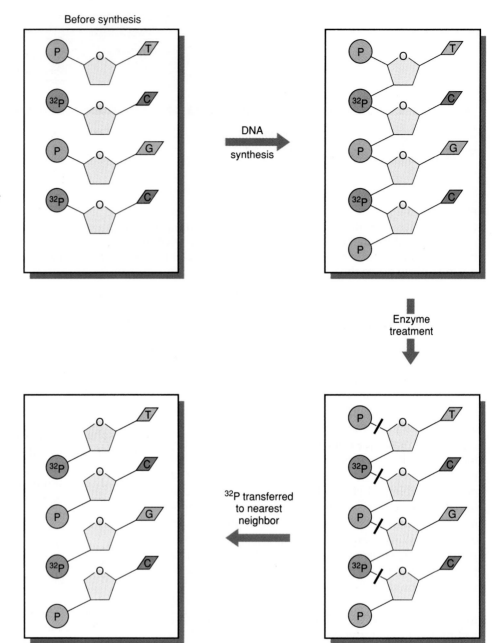

Before synthesis

DNA synthesis

Enzyme treatment

^{32}P transferred to nearest neighbor

reservations involved observations that the *in vitro* rate of synthesis was very slow, that the enzyme was much more effective replicating single-stranded DNA than double-stranded DNA, and that the enzyme appeared to be able to *degrade* DNA as well as to *synthesize* it.

Faced with the uncertainty of the true cellular function of DNA polymerase I, Kornberg pursued another approach. He reasoned that if the enzyme could be used to synthesize **biologically active DNA** *in vitro*, then DNA polymerase I must be the major catalyzing force for DNA synthesis within the cell. The term *biological activity* implies that the DNA synthesized is capable of supporting metabolic activ-

ities and directing reproduction of the organism from which it was originally derived.

In 1967, Mehran Goulian, Kornberg, and Robert Sinsheimer showed that the DNA of the small bacteriophage φX174 could be completely copied by DNA polymerase I *in vitro*, and that the new product could be isolated and used to successfully transfect *E. coli* protoplasts. This process produced mature phages from the synthetic DNA, thus demonstrating biological activity! The ingenious experimental design is outlined in Figure 10.9.

The phage φX174 provided an ideal experimental system because it contains a very small (5386 nucleotides), circular, single-stranded DNA molecule as its

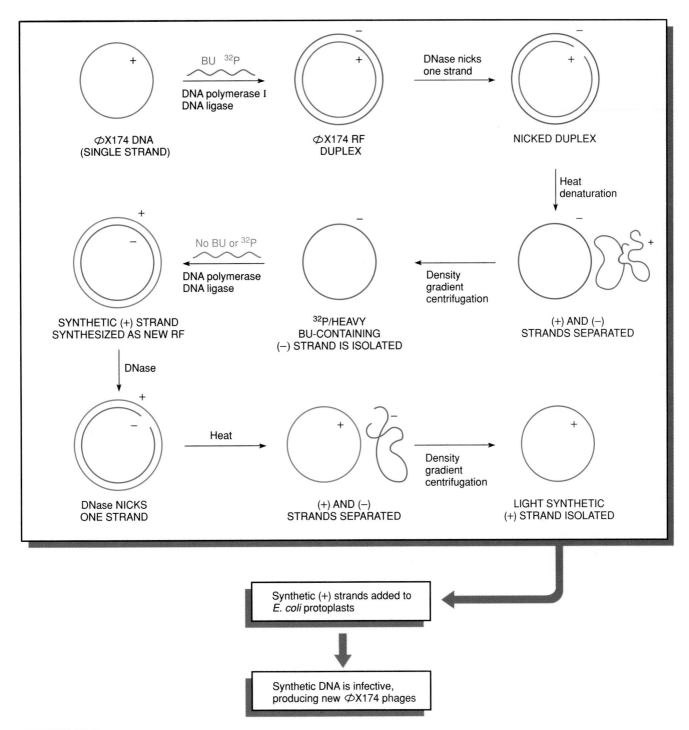

FIGURE 10.9
Schematic representation of Goulian, Kornberg, and Sinsheimer's experiment
involving *in vitro* replication and isolation of synthetic (+) DNA strands of phage
φX174. The DNA synthesized *in vitro* under the direction of DNA polymerase I
successfully transfected *E. coli* protoplasts, thus demonstrating its biological activity.

genetic material. Since the molecule is a closed circle, the experiment depended on the isolation of a second enzyme, **DNA ligase** (also called the **polynucleotide joining enzyme**), which joins the two ends of the linear molecule following replication.

In the normal course of ϕX174 infection, the circular, single-stranded DNA referred to as the **(+) strand** enters the *E. coli* cell and serves as a template for the synthesis of the complementary **(−) strand**. The two strands (+ and −) remain together in a circular double helix called the **replicative form (RF)**. The RF serves as the template for its own replication, and subsequently only (+) strands are produced. These strands are then packaged into viral coat proteins to form mature virus particles.

As shown in Figure 10.9, Goulian, Kornberg, and Sinsheimer's experiment began with ^{3}H-labeled (+) strands. Replication using DNA polymerase I and DNA ligase occurred; the strands were denatured; the (−) strand was isolated; and the process repeated with the (−) strand serving as the template. Isolation of the (−) strand depended on the use of the base analogue **5-bromodeoxyuridine,** which may be successfully incorporated in place of thymidine. In the chemical structure of this analogue, a bromine atom is substituted for a carbon atom in the methyl group at the C-5 position of the pyrimidine ring, increasing the mass and therefore the density. As a result, the (−) strands can be isolated from the (+) strands using sedimentation equilibrium centrifugation. Eventually, newly synthesized infectious (+) strands were synthesized and isolated in a similar way. The protocol of the experiment dictates that these (+) strands must have been synthesized *in vitro* under the direction of Kornberg's enzyme. When added to the bacterial protoplasts, transfection occurred, resulting in the production of mature phages. Therefore, the synthetic DNA had successfully directed reproduction!

This demonstration of biological activity was viewed as a precise assessment of faithful copying. If even a single error in one of the 5386 nucleotides had occurred, the change would most likely have constituted a lethal mutation, precluding the production of viable phages.

DNA Polymerase II and III

Although the synthesis of DNA under the direction of polymerase I did demonstrate biological activity, a more serious reservation about the enzyme's true biological role was raised in 1969. Peter DeLucia and John Cairns reported the discovery of a mutant strain of *E. coli* that was deficient in polymerase I activity. The mutation was designated *polA1*. In the absence of the functional enzyme, this mutant strain of *E. coli* still replicated its DNA and successfully reproduced! Other properties of the mutation led DeLucia and Cairns to conclude that in the absence of polymerase I, these cells were highly deficient in their ability to "repair" DNA. For example, the mutant strain was highly sensitive to ultraviolet light and radiation, both of which damage DNA and are mutagenic. Nonmutant bacteria are able to repair a great deal of UV-induced damage.

These observations led to two conclusions:

1 There must be at least one other enzyme present in *E. coli* cells that is responsible for replicating DNA *in vivo*.
2 DNA polymerase I may serve only a secondary function *in vivo*. This function is now believed by Kornberg and others to occur during the normal course of replication and to be critical to the fidelity of DNA synthesis.

To date, two unique DNA polymerases have been isolated from cells lacking polymerase I activity. These two enzymes have also been isolated from normal cells that also contain polymerase I.

The characteristics of these two enzymes, called DNA polymerase II and III, are contrasted with DNA polymerase I in Table 10.2. As is evident from that information, all three share several characteristics.

TABLE 10.2
Comparative properties of the three bacterial DNA polymerases.

Properties	I	II	III
Initiation of chain synthesis	−	−	−
5′ to 3′ elongation of an existing primer	+	+	+
3′ to 5′ exonuclease activity	+	+	+
5′ to 3′ exonuclease activity	+	−	+
Molecular weight	190,000	120,000	380,000
In vitro rate of synthesis (nucleotides/second)	10	?	500
Molecules/cell	400	75	15

While none can *initiate* DNA synthesis on a template, all can *elongate* an existing DNA strand, called a **primer**. Elongation occurs by polymerizing nucleotides in the 5′-to-3′ mode. That is, each new nucleotide is added at its 5′-phosphate end to the 3′-C of the growing polynucleotide. Each addition creates a new exposed 3′-OH group on the sugar, which can then participate in the next reaction. Synthesis occurring in this way is illustrated in Figure 10.10. As we shall see, the primer used in initiation of synthesis is a small strand of RNA.

These enzymes are all large, complex proteins exhibiting a molecular weight in excess of 100,000 daltons. All three possess 3′-to-5′ exonuclease activity. This means that they can polymerize in one direction and then reverse directions and excise nucleotides just added. As we will see, this activity provides a capacity to proofread and remove an incorrect nucleotide pair.

For two of the three (I and III), 5′-to-3′ exonuclease activity also is characteristic. This potentially allows the enzyme to excise nucleotides from the end where initial synthesis occurred and in the same direction. Thus, they have the potential ability to remove the primer, which is essential for synthesis. We will return to this topic momentarily. The final characteristics probably explain why Kornberg isolated polymerase I and not polymerase III. The former is present in much greater amounts than the latter, which is also less stable.

What then are the roles of the three polymerases *in vivo*? Polymerase III is considered to be the enzyme responsible for the polymerization essential to replication. Its 3′-to-5′ exonuclease activity also allows it to proofread and excise any incorrect base pairs created in error during polymerization. As we will soon see, gaps are a natural occurrence on one of the two strands during replication as RNA primers are removed. It is believed that polymerase I, originally studied by Kornberg, is responsible for removing the primer as well as for the synthesis that fills these gaps. Its exonuclease activity also allows proofreading to occur during this process. The role of polymerase II remains unknown.

We end this section by emphasizing the complexity of the DNA polymerase III molecule. Its active form, called a **holoenzyme,** consists of seven separate polypeptide chains. The largest, the α subunit, has a molecular weight of 140,000 daltons and is responsible for both the polymerizing and 5′-to-3′ exonuclease activity of the holoenzyme. The ε subunit possesses the 3′-to-5′ exonuclease activity. While the function of the other five subunits is not precisely clear, one binds ATP, a step essential to the initial polymerization step. It is also believed that two

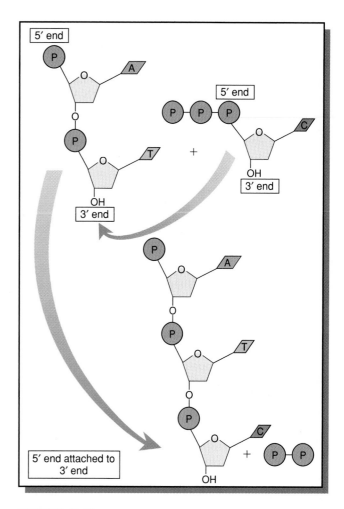

FIGURE 10.10
Demonstration of 5′-to-3′ synthesis of DNA.

holoenzymes may function together as a dimer at the replication fork, creating the possibility that a molecule of over 750,000 daltons is essential to DNA replication. Together with several other proteins at the replication fork, a complex nearly as large as a ribosome is present. It has been called a **replisome**.

DNA SYNTHESIS: A MODEL

We have thus far established that replication is semiconservative and bidirectional along a single replicon in bacteria and many viruses. And, we know that synthesis is in the 5′-to-3′ mode primarily under the direction of DNA polymerase III, creating two replication forks. These move in opposite directions away from the origin of synthesis. We are now ready to pursue several other aspects of DNA synthesis. We will combine this new information with that previously established into a coherent model. It takes into account the following additional points:

1   A mechanism must exist by which the helix is initially denatured and stabilized in this "open" configuration so that synthesis may proceed along both strands. Several types of proteins, including the **helicase** and **single-stranded DNA-binding** proteins, accomplish this.

2   As unwinding proceeds, a tension is created farther down the helix. This must be reduced and it is accomplished by the enzymatic action of the enzyme **DNA gyrase**.

3   A primer of some sort must be synthesized so that polymerization can commence under the direction of DNA polymerase III. The primer, surprisingly, is RNA!

4   Once the RNA primers have been synthesized, DNA polymerase III commences synthesis of the complement of both strands of the parent molecule. Since the strands are antiparallel, continuous synthesis in the direction that the replication fork moves is possible along only one of the two strands. On the other strand, synthesis is discontinuous in the opposite direction.

5   The RNA primers must be removed prior to completion of replication. The gaps that are temporarily created must be filled with DNA complementary to the template at each location. Both roles are believed to be accomplished by DNA polymerase I.

6   The newly synthesized DNA strand that fills each temporary gap must be ligated to the adjacent strand of DNA synthesized by DNA polymerase III. This is accomplished by the enzyme DNA ligase.

As we consider these points, two figures (Figures 10.11 and 10.12) will be used to illustrate each step. The discussion then culminates with the presentation of a complete model of DNA synthesis, illustrated in Figure 10.13.

Unwinding the DNA Helix

As discussed earlier, there is but a single origin along the circular chromosome of most bacteria and viruses. This region of the *E. coli* chromosome has been particularly well studied. Called *oriC*, the origin of replication consists of about 250 base pairs, which are characterized by the presence of repeating sequences of 9 and 13 bases (called 9mers and 13mers). One particular protein, called **dnaA** (because it is encoded by the gene called *dnaA*,) is responsible for the initial step in unwinding the helix. About 25 subunits of the dnaA protein bind to several of the 9mers. This step is essential in facilitating the subsequent binding of dnaB and dnaC helicase proteins that further open

and destabilize the helix. These interactions require the energy normally supplied by the hydrolysis of ATP.

As unwinding proceeds, another group of proteins, called single-stranded binding proteins **(SSBPs)** stabilize this conformation. As unwinding continues, supercoiling is created farther down the helix. **DNA gyrase,** a member of a larger group of enzymes called **DNA topoisomerases,** can function to reduce the tension produced. The enzyme accomplishes this by causing a break in both strands and a topological manipulation of the DNA. Then, in a more relaxed conformation, the gaps created in both strands of the DNA are resealed. These reactions also require the hydrolysis of ATP to drive them.

In combination with the polymerase complex, these proteins create an array of molecules that participate in DNA synthesis and are part of what we have previously called the replisome.

Initiation of Synthesis

Once a small portion of the helix is unwound, initiation of synthesis may occur. As previously mentioned, DNA polymerase III requires a free 3' end in order to elongate a polynucleotide chain. This prompted researchers to investigate how the first nucleotide can be added, since no free 3'-hydroxyl group is present. There is now evidence that, at least on one strand, RNA is involved in initiating DNA synthesis. It is thought that a short segment of RNA, complementary to DNA, is first synthesized on the DNA template. The RNA is made under the direction of a form of the enzyme RNA polymerase called **primase**. At least two other proteins are involved. The short segment of RNA that is synthesized serves as a **primer,** and it is to this primer that DNA polymerase III begins to add 5'-deoxyribonucleotides. The RNA polymerase does not require a free 3' end to initiate synthesis. After DNA synthesis occurs at an area adjacent to the RNA primer, the RNA segment is clipped off by an exonuclease and the gap is then filled by the action of DNA polymerase I. Recall that this enzyme also functions as an exonuclease, so it is probably also responsible for removing the segment of RNA.

RNA priming has been recognized in viruses, bacteria, and several eukaryotic organisms and is thought to be a universal phenomenon.

Continuous and Discontinuous DNA Synthesis

We must now reconsider the fact that the two strands of a double helix are antiparallel to each other. One

FIGURE 10.11
Helical unwinding of DNA during replication as accomplished by dnaA, dnaB, and dnaC proteins. Initial binding of many monomers of dnaA occurs at DNA sites containing repeating sequences of 9 nucleotides, called 9mers. Not illustrated are 13mers that are also involved.

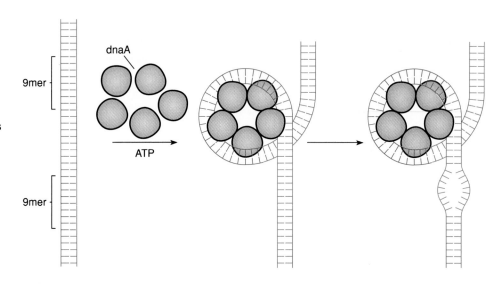

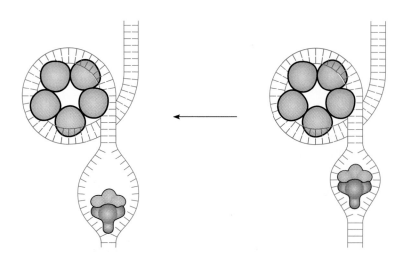

runs in the 5′-to-3′ direction, while the other has the opposite 3′-to-5′ polarity. This structural characteristic is responsible for a feature of DNA synthesis, which is illustrated in Figure 10.13. Since DNA polymerase III synthesizes DNA in only the 5′-to-3′ direction, simultaneous synthesis of antiparallel strands along an advancing replication fork must occur in one direction along one strand and in the opposite direction on the other.

Thus, as the strands unwind and the replication fork progresses down the helix, only one strand can serve as a template for **continuous synthesis**. This strand is called the **leading strand**. As the fork progresses, many points of initiation are necessary on

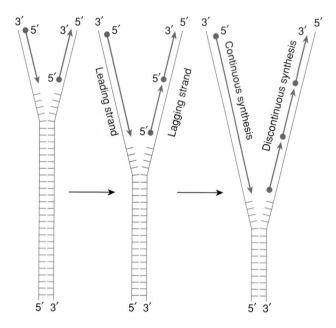

FIGURE 10.12
Illustration of the opposite polarity of synthesis along the two strands of DNA necessitated by the requirement of 5'-to-3' synthesis of DNA polymerase III. On the lagging strand, synthesis must be discontinuous, resulting in the production of Okazaki fragments. On the leading strand, synthesis is continuous.

the opposite, or **lagging, strand** resulting in **discontinuous synthesis**.

Evidence in support of discontinuous synthesis was first provided by Reiji and Tuneko Okazaki and their colleagues. They discovered that when bacteriophage DNA is replicated in *E. coli*, some of the newly formed DNA is found as small fragments containing 1000 to 2000 nucleotides. These pieces, called **Okazaki fragments,** must then be enzymatically joined to create longer DNA strands. It is quite clear that RNA primers are part of each such fragment. As synthesis proceeds, the Okazaki fragments of low molecular weight are indeed converted into longer and longer DNA strands of higher molecular weight.

Discontinuous synthesis of DNA requires the removal of the RNA primer as well as an enzyme that can unite the smaller products into the longer continuous molecule. **DNA ligase** has been shown to be capable of performing this latter function. The evidence that DNA ligase does perform this function during DNA synthesis is strengthened by the observation of a **ligase-deficient mutant** strain of *E. coli*. In

this strain, Okazaki fragments accumulate in particularly large amounts. Apparently, they are not joined adequately.

Discontinuous synthesis on the lagging strand, as described above, is characteristic of bacteria as well as eukaryotic cells.

Summary of DNA Synthesis

A summary of DNA synthesis, as depicted in the model shown in Figure 10.13, takes into account the following points:

1 Synthesis is initiated at a specific origin within each replicon. In *E. coli*, this region is called *oriC*.
2 Unwinding proteins (called helicases) denature the helix at the origin, while other proteins (SSBPs) stabilize the denatured helix.
3 Synthesis is bidirectional, creating two replication forks which move in opposite directions away from the origin.
4 Initiation of synthesis involves, at least on one strand, an RNA primer synthesized under the direction of a unique RNA polymerase called primase. The resultant RNA is complementary to its DNA template.
5 DNA polymerase III polymerizes complementary DNA strands by elongating an existing chain in the 5'-to-3' direction.
6 As the replication fork moves away from the origin, synthesis is continuous on the leading strand, but is discontinuous on the lagging strand, producing short polynucleotides called Okazaki fragments.
7 The RNA primers are removed and the resulting gaps are filled with DNA under the direction of DNA polymerase I.
8 Along the lagging strand, the Okazaki fragments are joined by DNA ligase.
9 As unwinding proceeds, the helix becomes more tightly coiled downstream from the advancing replication fork. This creates a tension which is relieved by DNA gyrase, which cuts and then reseals the strands.
10 As this process proceeds along the length of the replicon, semiconservative replication is achieved.

Since the investigation of DNA synthesis is still an extremely active area of research, this model will no doubt be extended in the future. It further provides a summary of DNA synthesis against which genetic phenomena may be interpreted.

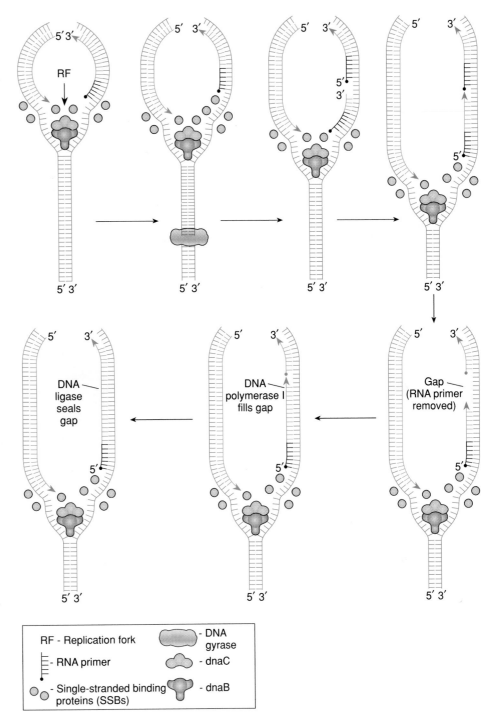

FIGURE 10.13
A model of DNA synthesis involving the many enzymes and proteins essential to the process, as described in the text and indicated in the key.

RF
RF - Replication fork
- DNA gyrase
- RNA primer
- dnaC
- Single-stranded binding proteins (SSBs)
- dnaB

DNA ligase seals gap

DNA polymerase I fills gap

Gap (RNA primer removed)

THETA (Θ) STRUCTURES AND ROLLING CIRCLES

Thus far, we have not confronted in any detail the replication of circular DNA molecules such as those found in viruses and bacteria. Bidirectional DNA synthesis at a single origin along a linear molecule creates two replication forks and results in what is called a **replication eye** or **replication bubble**. When the replicon is a circular DNA molecule, the "eye" which is formed creates a **Θ structure,** as illustrated in Figure 10.14(a).

Some bacteriophages have evolved still another mode of replication, initially involving only one of the

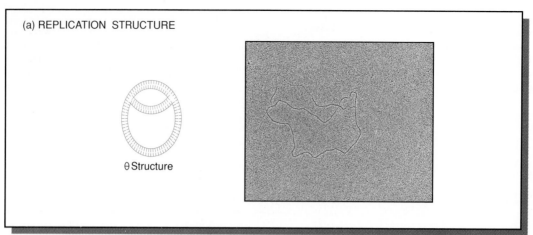

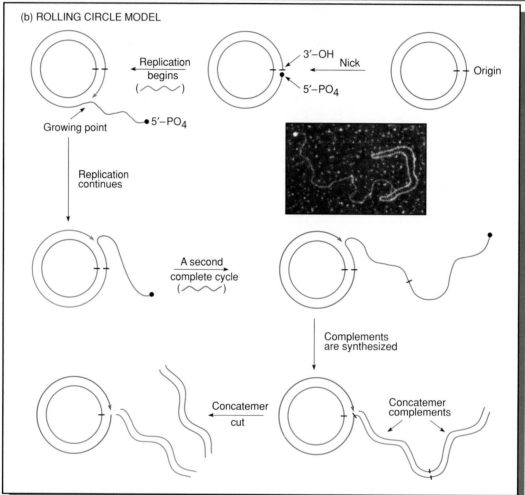

FIGURE 10.14

(a) Semiconservative replication of a circular DNA molecule, where both strands of the parent molecule are used as a template. This creates a Θ structure, illustrated in the accompanying photograph. (b) The rolling circle model, characteristic of some bacteriophages. The newly synthesized single strands serve as templates for the synthesis of their complements. Subsequent cleavage into unit chromosomes then occurs. The micrographs illustrate these structures as visualized under the electron microscope.

two strands. This mode is called the **rolling circle model** which is depicted in Figure 10.14(b).

The initial step involves enzymatic nicking of one of the two strands, creating a free 3'-OH group. DNA polymerase then extends this nicked chain by polymerizing the complement of the strand remaining intact. As replication proceeds, the original nicked strand is displaced at its 5'-PO$_4$ end. After one round of replication is completed, this DNA molecule, representing a unit genome, is totally displaced. In theory, the cycle can be repeated n number of times creating n unit genomes covalently bonded together. Each single-stranded DNA molecule can be converted to duplex form by the synthesis of the complementary strands. Such a structure is called a **concatemer** which is subsequently cleaved by an endonuclease into complete chromosomes.

This mode of replication is utilized by numerous bacterial viruses, including phage λ and φX174. In λ, the linear molecules are packaged into viral heads. Following infection of a host bacterium, the linear molecule is converted into a circular molecule to accommodate the rolling circle mode of replication.

GENETIC CONTROL OF REPLICATION

Much of what we know and have outlined in the previous section concerning the details of DNA replication in viruses and bacteria has been based on the genetic analysis of the process. For example, we have already discussed the *polA1* mutation, the study of which revealed that DNA polymerase I is not the major enzyme responsible for replication. Many other mutations have been isolated that interrupt or seriously impair some aspect of replication. These can be studied if they are **conditional mutations** that express the mutant condition at a restrictive temperature but function normally at a permissive temperature. Investigation of such temperature-sensitive mutants provides insights into the product and the associated function of the normal, nonmutant gene.

As shown in Table 10.3, the enzyme product or its general role in replication has been ascertained for a variety of genes in *E. coli*. For example, numerous mutations in genes specifying the subunits of polymerases I, II, and III have been isolated. Genes have also been identified that encode products involved in specification of the origin of synthesis, helix-unwinding and stabilization, initiation and priming, relaxation of supercoiling, repair, and ligation. The discovery of such a large group of genes attests to the complexity of the process of replication, even in the relatively simple prokaryote. This complexity is not

TABLE 10.3
A list of various *E. coli* mutant genes and their products or functions.

Mutant Gene	Enzyme or Gene Function
polA	DNA polymerase I
polB	DNA polymerase II
polC (dnaE)	DNA polymerase III α subunit
dnaX,Z	DNA polymerase III subunits
dnaG	Primase
dnaA,I,P	Initiation
dnaB,C	Helicase at *oriC*
oriC	Origin of replication
gyrA,B	Gyrase subunits
lig	Ligase
rep	Helicase
ssb	Single-stranded binding proteins
rpoB	RNA polymerase subunit

unexpected, given the enormous quantity of protein-bound DNA that must be unerringly replicated in a very brief time. As we will see, the process is even more difficult to investigate in eukaryotes.

EUKARYOTIC DNA SYNTHESIS

Most features of DNA synthesis found in microorganisms are now thought to apply also to eukaryotic systems. However, the DNA of eukaryotes is complexed with a variety of proteins, some of which maintain the molecule's structural integrity. The presence of these proteins and the general complexity of the eukaryotic cell compared with the prokaryotic cell have made analysis much more difficult.

Eukaryotic cells contain four different kinds of DNA polymerases, called α, β, γ, and δ. It appears the α form is the major enzyme involved in the replication of nuclear DNA. The β form may be involved in DNA repair, while the γ form is the only DNA polymerase found in mitochondria. Presumably, it is unique in its function within that organelle. The δ form, like α, appears to function in nuclear DNA replication. It appears that the DNA polymerases of eukaryotes have the same fundamental requirements for DNA synthesis as bacterial and viral systems: four deoxyribonucleoside triphosphates, a divalent cation (Mg^{++} or Mn^{++}), a template, and a primer.

Data and observations derived from autoradiographic and electron microscopic studies have provided many other insights. As mentioned earlier, eukaryotic DNA synthesis appears to be bidirectional, creating two replication forks from each point of origin. As one would expect because of the increased amount of DNA in eukaryotes compared to prokaryotes, many more points of origin are present. In mammals, there are about 25,000 replicons present in the genome, each consisting of an average of 100,000 to 200,000 base pairs (100–200 kb). In *Drosophila*, there are about 3500 replicons per genome of an average size of 40 kb. The many points of origin are thought not to be random along each chromosome, but instead to represent specific points of initiation of DNA synthesis. Not all origins are activated at the same time during the S phase of the cell cycle. Most DNA regions that represent genes are replicated preferentially early in the S phase, while the nongenic DNA areas are replicated later during that phase. These areas, called **euchromatin** and **heterochromatin,** respectively, will be discussed in greater detail in Chapter 17.

The *in vivo* synthesis of eukaryotic DNA is five to ten times slower than that of *E. coli*. However, most general aspects of chain elongation are thought to be similar. The actions of a variety of proteins—DNA helicase and SSBPs—are believed to modulate strand separation that precedes RNA priming by a primase enzyme. On the lagging strand, synthesis is discontinuous, resulting in Okazaki fragments that are linked together by DNA ligase. These fragments are about ten times smaller (100–150 nucleotides) than in prokaryotes. This size happens to coincide roughly with that of the basic repeating structural unit of eukaryotic chromatin, the **nucleosome,** but it is not yet clear whether there is a relationship between these two observations. Nucleosomes, consisting of about 150 base pairs of DNA and an octamer of histone proteins, will also be discussed extensively in Chapter 17. While synthesis may occur continuously on the leading strand, it is not clear whether it can proceed uninterrupted for the full length of a replicon. Thus, synthesis on the leading strand may be **semidiscontinuous.**

DNA RECOMBINATION

We conclude this chapter by returning to a topic discussed in Chapter 6—**genetic recombination**. There, it was pointed out that the process of crossing over depends on breakage and rejoining of the DNA strands between homologues. Now that we have introduced the chemistry and replication of DNA, it is appropriate to consider how recombination occurs at the molecular level. In general, the following information pertains to genetic exchange between any two homologous double-stranded DNA molecules, whether they be viral or bacterial chromosomes or eukaryotic homologues during meiosis.

While there are several models available to explain crossing over, they all share certain common features. First, all are based on the initial proposals put forth independently by Robin Holliday and Harold L. K. Whitehouse in 1964. They also depend on the complementarity between DNA strands for their precision of exchange. Finally, each model relies on a series of enzymatic processes in order to accomplish genetic recombination.

One such model is illustrated in Figure 10.15. It begins (a) with two paired DNA duplexes or homologues, each of which has a single-stranded nick introduced (b) at an identical position by an endonuclease. The ends of the strands produced by these cuts are then displaced and subsequently pair with their complements on the opposite duplex (c). A ligase then seals the loose ends (d), creating hybrid duplexes called **heteroduplex DNA molecules**. The exchange creates a cross-bridged or **Holliday structure**. The position of this cross-bridge can then move down the chromosomes by branch migration (e). This occurs as a result of a zipperlike action as hydrogen bonds are broken and then reformed between complementary bases of the displaced strands of each duplex. This migration yields an increased length of heteroduplex DNA on both homologues.

If the duplexes now separate (f), and the bottom portions rotate 180° (g), an intermediate planar structure called a **chi form** is created. If the two strands on opposite homologues previously uninvolved in the exchange are now nicked by an endonuclease (h), and ligation occurs (i), recombinant duplexes are created. Note that the arrangement of alleles has been altered as a result of the crossover that has occurred.

If the original nicks (b) do not occur at the same point on each duplex, the same end result can still be achieved. The overlapping single strands must first be clipped by an exonuclease and the gaps filled in by DNA polymerase. This tailoring of the duplexes must be accomplished before ligation (d) can occur.

Evidence supporting the above model includes the electron microscopic visualization of chi-form planar molecules from bacteria where four duplex arms are joined at a single point of exchange (Figure 10.15). Additionally, the discovery in *E. coli* of the *recA protein* has provided important evidence. This molecule promotes the exchange of reciprocal single-stranded DNA molecules as must occur in step (c) of the model. Further, *recA* enhances the hydrogen bond formation during strand displacement, thus initiating heteroduplex formation. Finally, all enzymes essential to the nicking and ligation process

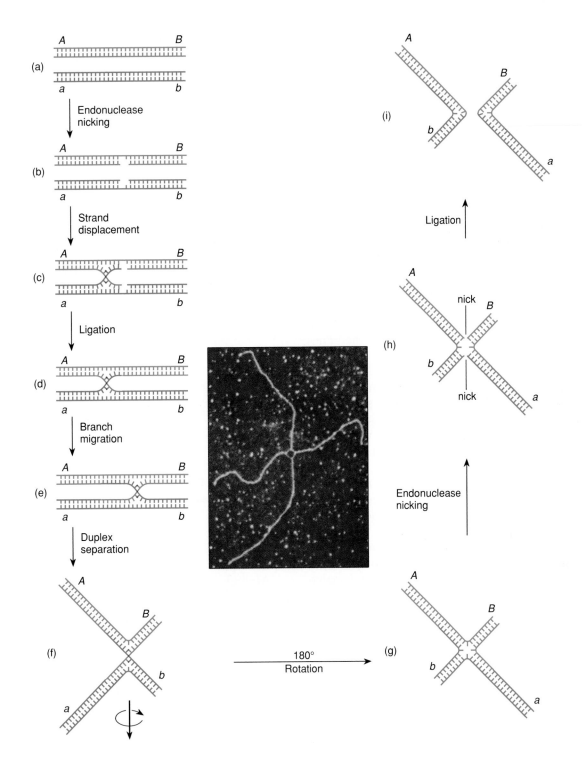

FIGURE 10.15
A possible molecular sequence depicting how genetic recombination occurs as a result of the breakage and rejoining of heterologous DNA strands. Each stage is described in the text. The electron micrograph shows DNA in a chi-form structure similar to the diagram in (g). The DNA is from the Col E1 plasmid of *E. coli.*

have been discovered and investigated. Mutations that prevent genetic recombination have been found in a number of genes in viruses and bacteria. Thus, the products of these genes play an essential role in that process. Recall from Chapter 6 that an enzyme system that can accomplish the nicking and resealing essential to crossing over has been isolated from *Lilium*.

Gene Conversion

A modification of the above model has helped to better understand a unique genetic phenomenon known as **gene conversion**. Initially found in yeast by Carl Lindegren and in *Neurospora* by Mary Mitchell, gene conversion is characterized by a genetic exchange ratio involving two closely linked genes that is nonreciprocal. If one were to cross two *Neurospora* strains each bearing a separate mutation ($a+ \times + b$), a reciprocal recombination event between the genes would yield spore pairs of the $++$ and ab genotypes. However, a nonreciprocal exchange yields one pair without the other. Working with pyridoxine mutants, Mitchell observed such a ratio. She found several asci with the $++$ genotype, but not the reciprocal product (ab). Since the frequency of these events was higher than the predicted mutation rate, and thus could not be accounted for by that phenomenon, they were called gene conversions. They were so named because it appeared that one allele had somehow been converted to another during an event where genetic exchange also occurred. Similar findings have been apparent in the study of other fungi as well.

One possible explanation of this genetic phenomenon interprets the conversion event as a mismatch of base pairs during heteroduplex formation involved in genetic recombination (Figure 10.16). Mismatched regions of hybrid strands may be created during recombination, but they can be repaired by excision of one of the strands and synthesis of the correct complement using the remaining strand as a template. Excision of either one of the strands may occur in order to accomplish the repair, yielding two

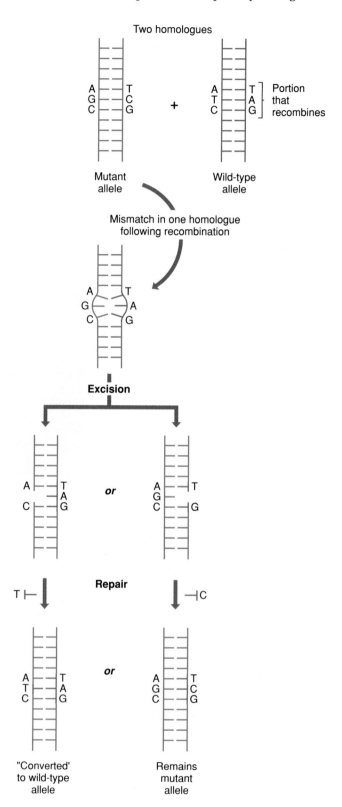

FIGURE 10.16
Illustration of a proposed mechanism that accounts for the phenomenon of gene conversion. A base pair mismatch occurs in one of the two homologues (bearing the mutant allele) during heteroduplex formation, which accompanies recombination in meiosis. During excision repair, one of the two mismatches is removed and the complement is synthesized. In one case, the mutant base pair is preserved. When it is subsequently included in a recombinant spore, the mutant genotype will be maintained. In the other case, the mutant base pair is converted to the wild-type sequence. When included in a recombinant spore, the wild-type genotype will be expressed, leading to a nonreciprocal exchange ratio.

possible "corrections" (Figure 10.16). One repairs the mismatched base pair and restores the original sequence. The other corrects the mismatch but creates a base pair substitution. This conversion may have the effect of creating identical alleles on the two homologues that were different initially.

In our example in Figure 10.16, suppose the G≡C pair on one of the two homologues was responsible for the mutant allele while the A=T pair was part of the wild type gene sequence on the other homologue. Conversion of the G≡C pair to A=T would have the effect of changing the mutant allele to wild type, just as Mitchell originally observed!

Gene conversion events have helped to explain other puzzling genetic phenomena in fungi. For example, when mutant and wild type alleles of a single gene are studied in a cross, asci should yield equal numbers of mutant and wild type spores. However, exceptional asci with ratios of 3:1 or 1:3 are sometimes observed. These ratios are more easily understood when interpreted in terms of gene conversion. The phenomenon has also been detected during mitotic events in fungi as well as during the study of unique compound chromosomes in *Drosophila*.

CHAPTER SUMMARY

1   The mechanism of replication and synthesis of DNA has constituted a fundamental issue for molecular genetics. The process resulting in the duplication of DNA is an essential function of the genetic material.

2   In theory, three modes of replication are possible: semiconservative, conservative, and dispersive. Though all three rely on base complementarity, semiconservative replication is the most straightforward and was predicted.

3   In 1958, Meselson and Stahl resolved this problem in favor of semiconservative replication by labeling DNA from *E. coli* with a heavy isotope of nitrogen. Their findings showed that newly synthesized DNA consists of one old strand and one new strand. Taylor, Woods, and Hughes used the root tips of the broad bean in autoradiographic studies to demonstrate semiconservative replication in eukaryotes.

4   During the same period, Kornberg isolated DNA polymerase I from *E. coli* and demonstrated it to be an enzyme capable of *in vitro* DNA synthesis, provided that a template and precursor nucleoside triphosphates were supplied.

5   The subsequent discovery of a mutant strain of *E. coli* capable of DNA replication, in spite of its lack of polymerase I activity, cast doubt on this enzyme's *in vivo* replicative function. DNA polymerases II and III were then isolated. While the function of polymerase II remains unresolved, polymerase III has been identified as the enzyme responsible for DNA replication *in vivo*.

6   During the process of DNA synthesis, the double helix unwinds, forming a replication fork where synthesis begins. Proteins stabilize the unwound helix and assist in relaxing the supercoils created ahead of the replication activity.

7   Synthesis is initiated at specific sites along each template strand by RNA primase, which results in a short segment of RNA that provides a suitable 3' end, upon which DNA polymerase III can begin polymerization.

8   Due to the antiparallel nature of the double helix, polymerase III synthesizes DNA continuously on the leading strand in a 5'-to-3' direction. On the opposite strand, called the lagging strand, synthesis results in short Okazaki fragments that are later joined by DNA ligase.

9   DNA polymerase I removes and replaces the RNA primer with DNA, which is joined to the adjacent polynucleotide by DNA ligase.

10   The isolation of numerous viral and bacterial mutant genes affecting many of the molecules involved in the replication of DNA attests to the complex genetic control of the entire process.

11   The entire chromosome of viruses and bacteria constitutes the unit of replication, called a replicon. In eukaryotes many replicons exist along each chromosome, and DNA synthesis proceeds five to ten times more slowly.

12   Recombination between genetic molecules relies on a series of enzymes that can cut, realign, and reseal DNA strands. The phenomenon of gene conversion may be best explained in terms of mismatch repair synthesis during these exchanges.

Insights and Solutions

1 Predict the theoretical results of conservative and dispersive models of DNA synthesis using the conditions of the Meselson-Stahl experiment. Follow the results through two generations of replication after cells have been shifted to ^{14}N-containing medium, using the following migration standards:

Density ⟶

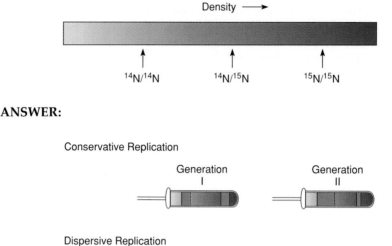

$^{14}N/^{14}N$ $^{14}N/^{15}N$ $^{15}N/^{15}N$

ANSWER:

Conservative Replication

Generation I   Generation II

Dispersive Replication

Generation I   Generation II

2 Mutations in the *dnaA* gene of *E. coli* are lethal and can only be studied following the isolation of conditional, temperature-sensitive mutations. Such mutant strains grow nicely and replicate their DNA at the permissive temperature of 18°C, but they do not grow or replicate their DNA at the restrictive temperature of 37°C. Two observations were useful in determining the function of the *dnaA* gene product. First, *in vitro* studies using DNA templates that have been nicked (opened) do not require the *dnaA* protein. Second, if intact cells are grown at 18°C and then shifted to 37°C, DNA synthesis continues at this temperature until one round of replication is completed, and DNA synthesis stops. What do these observations suggest about the role of the *dnaA* gene product?

ANSWER: These observations suggest that *in vivo* the *dnaA* protein is essential to the initiation of DNA synthesis. At 18°C (the permissive temperature) the mutation is not expressed and DNA synthesis begins. Following the shift to the restrictive temperature, DNA synthesis already initiated continues, but no new synthesis can begin. Since the *dnaA* protein is not required for synthesis of "nicked" DNA, this observation suggests that the protein functions during initiation by interacting with the intact helix and somehow facilitating the localized denaturing necessary for synthesis to proceed. In fact, both conclusions are valid.

1 Compare conservative, semiconservative, and dispersive modes of DNA replication.

2 Review the design, results, and conclusions of the Meselson-Stahl experiment.

3 In the Meselson-Stahl experiment, which of the three modes of replication could be ruled out after one round of replication? After two rounds?

4 Predict the results of the experiment by Taylor, Woods, and Hughes if replication were (a) conservative and (b) dispersive.

5 Reconsider Problem 26 in Chapter 9. In the model you proposed, could the molecule be replicated semiconservatively? Why? Would other modes of replication work?

6 What are the requirements for the *in vitro* synthesis of DNA under the direction of DNA polymerase I?

7 In Kornberg's initial experiments, he actually grew *E. coli* in Anheuser-Busch beer vats. Why do you think this was necessary?

8 How did Kornberg test the fidelity of copying DNA by polymerase I?

9 Which of Kornberg's tests is the more stringent assay? Why?

10 Which characteristics of DNA polymerase I led to doubts that its *in vivo* function is the synthesis of DNA leading to complete replication?

11 Explain the theory of nearest-neighbor frequency.

12 What is meant by "biologically active" DNA?

13 Why was the phage φX174 chosen for the experiment demonstrating biological activity?

14 Outline the experimental design of Kornberg's biological activity demonstration.

15 What was the significance of the *polA1* mutation?

16 Summarize the properties of polymerase I, II, and III.

17 Distinguish between (a) unidirectional and bidirectional synthesis and (b) continuous and discontinuous synthesis of DNA.

18 List the proteins that unwind DNA during *in vivo* DNA synthesis. How do they function?

19 Define and indicate the significance of (a) Okazaki fragments, (b) DNA ligase, and (c) primer RNA.

20 Outline the current model for DNA synthesis.

21 Why should DNA synthesis be more complex in eukaryotes than in bacteria? How is DNA synthesis similar in the two types of organisms?

22 If the analysis of DNA from two different microorganisms demonstrated very similar nearest-neighbor frequencies, is the DNA of the two organisms identical in: (a) amount, (b) base composition, and (c) nucleotide sequences?

23 Analysis of nearest-neighbor data led Josse, Kaiser, and Kornberg in 1961 (*J. Biol. Chem.* 236: 864–75) to conclude that the two strands of the double helix are in opposite polarity to one another. Demonstrate your understanding of the nearest-neighbor technique by determining the outcome of such an analysis if the strands of the following molecule are antiparallel (a) versus parallel (b):

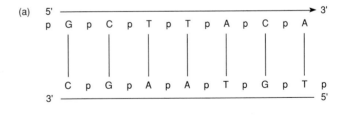

DNA REPLICATION, SYNTHESIS, AND RECOMBINATION   291

∞∞∞∞∞∞∞∞∞∞∞∞∞∞∞∞∞∞∞∞∞∞∞∞

SELECTED READINGS

BRAMHILL, D., and KORNBERG, A. 1988. A model for initiation at origins of DNA replication. *Cell* 54: 915–18.

CATCHESIDE, D. G. 1977. *The genetics of recombination.* Baltimore: University Park Press.

COZZARELLI, N. R. 1980. DNA gyrase and the supercoiling of DNA. *Science* 207: 953–60.

DARNELL, J., LODISH, H., and BALTIMORE, D. 1990. *Molecular cell biology.* 2nd ed. New York: Scient. Amer. Books.

DAVIDSON, J. N. 1976. *The biochemistry of nucleic acids.* 8th ed. Orlando: Academic Press.

DeLUCIA, P., and CAIRNS, J. 1969. Isolation of an *E. coli* strain with a mutation affecting DNA polymerase. *Nature* 224: 1164–66.

DENHARDT, D. T., and FAUST, E. A. 1985. Eukaryotic DNA replication. *BioEssays* 2: 148–53.

DRESLER, D., and POTTER, H. 1982. Molecular mechanisms in genetic recombination. *Ann. Rev. Biochem.* 51: 727–61.

GELLERT, M. 1981. DNA topoisomerases. *Ann. Rev. Biochem.* 50: 879–910.

HOLLIDAY, R. 1964. A mechanism for gene conversion in fungi. *Genet. Res.* 5: 282–304.

HUBERMAN, J. C. 1987. Eukaryotic DNA replication: A complex picture partially clarified. *Cell* 48: 7–8.

JOSSE, J., KAISER, A. D., and KORNBERG, A. 1961. Enzymatic synthesis of deoxyribonucleic acid. VIII. Frequencies of nearest neighbor base sequences in deoxyribonucleic acid. *J. Biol. Chem.* 236: 864–75.

KORNBERG, A. 1960. Biological synthesis of DNA. *Science* 131: 1503–8.

———. 1969. Active center of DNA polymerase. *Science* 163: 1410.

———. 1974. *DNA synthesis.* New York: W. H. Freeman.

———. 1979. Aspects of DNA replication. *Cold Spr. Harb. Symp.* 43: 1–10.

———. 1980. *DNA replication.* New York: W. H. Freeman.

———. 1982. *DNA replication: 1982 Supplement.* New York: W. H. Freeman.

LEHMAN, I. R. 1974. DNA ligase: Structure, mechanism, and function. *Science* 186: 790–97.

LILLEY, D. M. J. 1988. DNA opens up—supercoiling and heavy breathing. *Trends in Genetics* 4: 111–14.

LINDEGREN, C. C. 1953. Gene conversion in *Saccharomyces. J. Genet.* 51: 625–37.

MESELSON, M. S., and RADDING, C. M. 1975. A general model for genetic recombination. *Proc. Natl. Acad. Sci.* 72: 358–61.

MESELSON, M., and STAHL, F. W. 1958. The replication of DNA in *Escherichia coli. Proc. Natl. Acad. Sci.* 44: 671–82.

MITCHELL, M. B. 1955. Aberrant recombination of pyridoxine mutants of *Neurospora. Proc. Natl. Acad. Sci.* 41: 215–20.

OGAWA, T., and OKAZAKI, T. 1980. Discontinuous DNA synthesis. *Ann. Rev. Biochem.* 49: 421–57.

OKAZAKI, T., et al. 1979. Structure and metabolism of the RNA primer in the discontinuous replication of prokaryotic DNA. *Cold Spr. Harb. Symp.* 43: 203–22.

POTTER, H., and DRESSLER, D. 1979. DNA recombination: *In vivo* and *in vitro* studies. *Cold Spr. Harb. Symp.* 48: 969–85.

RADDING, C. M. 1978. Genetic recombination: Strand transfer and mismatch repair. *Ann. Rev. Biochem.* 47: 847–80.

RADMAN, M., and WAGNER, R. 1988. The high fidelity of DNA duplication. *Scient. Amer.* (August) 259: 40–46.

SCHEKMAN, R., WEINER, A., and KORNBERG, A. 1974. Multienzymes systems of DNA replication. *Science* 186: 987.

STAHL, F. W. 1979. *Genetic recombination: Thinking about it in phage and fungi.* New York: W. H. Freeman.

———. 1987. Genetic recombination. *Scient. Amer.* (Feb.) 256: 90–101.

SZOSTAK, J. W., ORR-WEAVER, T. L., and ROTHSTEIN, R. J. 1983. The double-strand-break repair model for recombination. *Cell* 33: 25–35.

TAYLOR, J. H., WOODS, P. S., and HUGHES, W. C. 1957. The organization and duplication of chromosomes revealed by autoradiographic studies using tritium-labeled thymidine. *Proc. Natl. Acad. Sci.* 48: 122–28.

TOMIZAWA, J., and SELZER, G. 1979. Initiation of DNA synthesis in *E. coli. Ann. Rev. Biochem.* 48: 999–1034.

WANG, J. C. 1982. DNA topoisomerases. *Scient. Amer.* (July) 247: 94–108.

———. 1987. Recent studies of DNA topoisomerases. *Biochimica Biophysica Acta* 909: 1–9.

WATSON, J. D., et al. 1987. *Molecular biology of the gene.* Vol. 1, *General principles.* 4th ed. Menlo Park, CA: Benjamin/Cummings.

WHITEHOUSE, H. L. K. 1982. *Genetic recombination: Understanding the mechanisms.* New York: Wiley.

WICKNER, S. H. 1978. DNA replication proteins of *E. coli. Ann. Rev. Biochem.* 47: 1163–91.

WOOD, W. B., WILSON, J. H., BENBOW, R. M., and HOOD, L. E. 1975. *Biochemistry—A problems approach.* Menlo Park, CA: Benjamin/Cummings.

ZUBAY, G. L., and MARMUR, J., eds. 1973. *Papers in biochemical genetics.* 2nd ed. New York: Holt, Rinehart and Winston.

ZYSKIND, J. W. and SMITH, D. W. 1986. The bacterial origin of replication, *oriC. Cell* 46: 489–90.

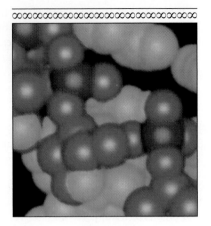

11

The Genetic Code

CHAPTER CONCEPTS

Genetic information, stored in DNA and transferred to RNA during the process of transcription, is present as three-letter codes. Using four different letters, corresponding to the four ribonucleotides in RNA, the 64 triplet codons specify the 20 different amino acids found in proteins, as well as providing signals that initiate and terminate protein synthesis. This unique language is the basis of life as we know it.

In the previous chapters in Part 2 of the text, we have established that DNA is the genetic material and elucidated the molecular structure of DNA. Together, these factors provide the physical and chemical basis for storage of genetic information. While we have yet to discuss the details of the processes, we know that the DNA of an active gene is transcribed into an RNA complement which is ultimately translated into the end product of the gene—a protein. Thus, the sequence of deoxyribonucleotides in DNA is first transferred to the complementary sequence of ribonucleotides in RNA, which somehow specifies the insertion of amino acids into protein. The end products of various genes are responsible for the normal and mutant phenotypes of organisms.

The fundamental question, then, is how an RNA molecule consisting of only four different types of nucleotides (A, U, C, and G) can specify 20 different amino acids. This question poses an intriguing theoretical problem. While the earliest proposals were imaginative, it was not until ingenious analytical research was applied that a hypothesis was shown to be correct. It was shown that information is stored in a **genetic code** which is triplet in nature. Code words, or **codons,** consisting of three ribonucleotides direct the insertion of a single amino acid into a polypeptide chain during its synthesis.

In this chapter, we will describe the way in which the code was deciphered, the coding assignments, and the general properties exhibited by the code. The work leading to these discoveries occurred most intensively in the late 1950s and early 1960s, one of the most exciting periods in the study of molecular genetics. This research revealed the intricacies of the specific chemical language that serves as the basis of all life on earth.

AN OVERVIEW OF THE GENETIC CODE

Before considering the various analytical approaches used in arriving at our current understanding of the genetic code, we shall provide a summary of the general features that characterize it:

1   The code is written in linear form using the ribonucleotide bases that compose mRNA molecules as the letters. The ribonucleotide sequence is, of course, derived from its complement in DNA.
2   Each word within the mRNA contains three letters. Thus, the code is a **triplet**. Called a **codon,** each group of *three* ribonucleotides specifies *one* amino acid.
3   The code is **unambiguous,** meaning that each triplet specifies only a single amino acid.
4   The code is **degenerate,** meaning that more than one triplet specifies a given amino acid. This is the case for 18 of the 20 amino acids.
5   The code contains "start" and "stop" punctuation signals. Certain triplets are necessary to initiate and to terminate translation.
6   No commas (or internal punctuation) are used in the code. Thus, the code is said to be **commaless**. Once translation of mRNA begins, each three ribonucleotides are read in turn, one after the other.
7   The code is **nonoverlapping**. Once translation commences, any single ribonucleotide at a specific location within the mRNA is part of only one triplet.
8   The code is almost **universal**. With only minor exceptions, a single coding dictionary is used by almost all viruses, prokaryotes, and eukaryotes.

EARLY PROPOSALS: GAMOW'S CODE

The earliest proposals for the genetic code occurred before scientists knew that mRNA is an intermediate in information transfer. As a result, these proposals assumed that DNA served directly as the template for protein synthesis. While anticipated by P. C. Caldwell and Sir Cyril N. Hinshelwood in 1950 and by A. L. Dounce in 1952, the **diamond code** proposed by George Gamow in 1954 generated the most interest. Gamow, a physicist and cosmologist, addressed directly the question of how DNA, written in four letters, could specify the 20 amino acids constituting proteins.

As illustrated in Figure 11.1, Gamow proposed that groups of nucleotides of both strands of the helix are arranged to form a series of "diamonds." These diamonds form geometric pockets, each specifying an amino acid. Notice that this model leads to an **overlapping** code, since any given nucleotide position is shared by two code words. Each diamond consists of a nucleotide on one DNA strand, the adjacent nucleotide pair, and the adjoining nucleotide on the opposite strand. Any of the four possible nucleotide pairs may occupy the middle of any diamond (A-T, T-A, C-G, and G-C). Each of these may be combined with any of the four nucleotides at either adjacent position, creating 64 possible codes. However, when geometric equivalents are taken into account, the number of codes is reduced to 20—precisely the number of amino acids that must be accounted for! Therefore, Gamow speculated that amino acids from the surrounding medium would have a chemical affinity for the pockets formed along the DNA duplex. Once sequestered in the correct sequence, they would unite to form a linear polypeptide chain.

Objections to the Diamond Code

Central to Gamow's proposal were the overlapping nature of the code and the use of a direct DNA template. Theoretical objections were quickly raised on both points. First, we will address the arguments against an overlapping code.

The assumption that the code was a triplet, with each amino acid specified by three nucleotides, was based on considerations set forth by Sidney Brenner. He argued that a triplet code represents the minimal use of four nucleotides to specify 20 amino acids. For example, four nucleotides, taken two at a time in all possible permutations, provide only 16 unique code words (4^2). While a triplet code provides 64 possibilities (4^3)—clearly more than the 20 needed—it seemed much more likely to Brenner

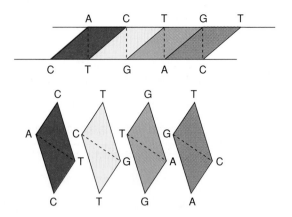

FIGURE 11.1
Gamow's diamond code that proposed that groups of four nucleotides from both strands of DNA directly encode amino acids. Note that pairs of adjoining bases are shared by adjacent codes.

than a four-letter code, where 256 code words (4^4) would exist.

Assuming a triplet code, Brenner considered restrictions that might be placed on it if it were overlapping. For example, he considered sequences within a protein consisting of three consecutive amino acids. In the nucleotide sequence GTACA, parts of the central triplet, TAC, are shared by the outer triplets, GTA and ACA. He reasoned that if this were the case, then only certain amino acids should be found adjacent to the one encoded by the central triplet, as shown below:

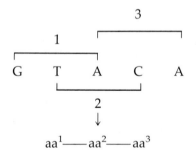

In fact, for any given central amino acid, only 16 combinations (2^4) of adjacent amino acids are theoretically possible.

Therefore, Brenner concluded that if the code were overlapping, tripeptide sequences of proteins should be somewhat limited. Looking at the available amino acid sequences of proteins that had been studied, he found no such restriction in tripeptide sequences. For any central amino acid, many more than 16 different tripeptides were found.

A second major argument against an overlapping code involved the effect of a single nucleotide change characteristic of a point mutation. With an overlapping code, two adjacent amino acids would be affected. But, mutations in the genes coding for the protein coat of tobacco mosaic virus (TMV), human hemoglobin, and the bacterial enzyme tryptophan synthetase invariably revealed only single amino acid changes.

The third argument against an overlapping code was presented by Francis Crick in 1957. He also argued against DNA serving as a direct template for the formation of proteins. Crick reasoned that any affinity between nucleotides and an amino acid would require hydrogen bonding. Chemically, however, such affinities seemed unlikely. Instead, he proposed that there must be an **adaptor molecule** that could covalently bind to the amino acid, yet be capable of hydrogen bonding to a nucleotide sequence. As we will see in the next chapter, Crick's prediction was correct; transfer RNA (tRNA) serves as the adaptor in protein synthesis. Because various adaptors would somehow have to overlap one another at nucleotide sites, Crick reasoned that such a chemical interaction would be highly unlikely. Therefore, it seemed doubtful that the code was overlapping.

Crick's and Brenner's arguments, taken together, strongly suggested that the genetic code is **nonoverlapping**. Without exception, this concept has been upheld.

THE CODE: FURTHER DEVELOPMENTS

Between 1958 and 1960, information related to the genetic code continued to accumulate. In addition to his adaptor proposal, Crick hypothesized, on the basis of genetic evidence, that the code is **comma-free;** that is, he believed no punctuation occurs along the reading frame. He also speculated that only 20 of the 64 possible triplets specify an amino acid and that the remaining 44 carry no coding assignment.

At the time, however, there was as yet no experimental evidence that the code was indeed a triplet; nor had the concept of messenger RNA, as an intermediate between DNA and protein, been established. Thus, since ribosomes had already been identified, the current thinking was that information in DNA is transferred in the nucleus to the RNA of the ribosome, which serves as the template for protein synthesis in the cytoplasm.

This concept soon became untenable as accumulating evidence demonstrated that the template intermediate was unstable. The RNA of ribosomes, on the other hand, was found to be extremely stable. As a result, in 1961 François Jacob and Jacques Monod postulated the existence of **messenger RNA (mRNA).**

The scene was thus set to demonstrate the triplet nature of the code, as contained in the intermediate mRNA, and to decipher the specific codon assignments.

Many questions beyond the triplet assignments also remained. Are there start-stop **punctuation signals**? Is the code **ambiguous,** with one triplet specifying more than one amino acid? Was Crick wrong with respect to the 44 "blank" codes? That is, is the code **degenerate,** with more than one triplet assignment for each amino acid? Is the code **universal**? As we shall see, these and other questions were answered in the next decade.

The Study of Frameshift Mutations

Before discussing the experimentation in which specific codon assignments were deciphered, we will consider the ingenious experimental work of Crick, Leslie Barnett, Brenner, and R. J. Watts-Tobin. Their work represented the first solid evidence for the triplet nature of the code.

In their work, these researchers induced insertion and deletion mutations in the B cistron of the rII locus of phage T4. The B cistron is one of two functional sites in a locus which, in mutant form, causes rapid lysis and distinctive plaques. Mutants in the rII locus will successfully infect strain B of *E. coli* but cannot reproduce on strain K12 of *E. coli*. Crick and his colleagues used the acridine dye **proflavin** to induce mutations (see Figure 14.14). This mutagenic agent intercalates within the double helix of DNA, often causing the insertion of an extra nucleotide or the deletion of one nucleotide during replication. As shown in Figure 11.2, the frame of reading will therefore shift, changing all subsequent triplets to the right of the insertion or deletion. Upon translation, the protein subsequently synthesized will be garbled, so to speak, from the point of change. These mutations are called **frameshifts**. When they are present at the rII locus, T4 will not reproduce on *E. coli* K12.

They reasoned that if phages with these induced mutations were treated again with proflavin, still other insertions or deletions would occur. This second change might result in a revertant phage, which would behave like wild type and successfully infect *E. coli* K12. For example, if the original mutant contained an insertion (+), a second event causing a deletion (−) close to the insertion would restore the original reading frame (Figure 11.2). In the same way, an event resulting in an insertion (+) might correct an original deletion (−).

In studying many mutations of this type, these researchers were able to compare various mutant combinations together on the same DNA molecule,

FIGURE 11.2
Schematic diagram of frameshift mutations caused by the addition (+) or deletion (−) of a single nucleotide in the middle of the reading frame. At the point of insertion or loss, all subsequent triplet code words are changed. A second change may reverse the frameshift and restore the correct frame of reading, as shown.

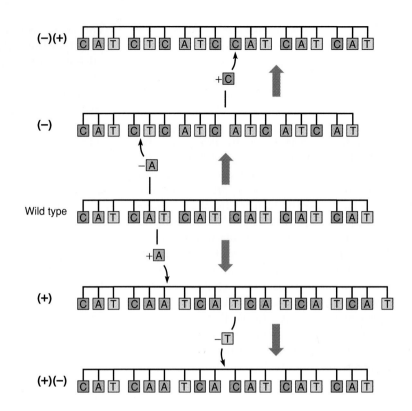

as shown in Figure 11.3. They found that various combinations of one (+) and one (−) indeed caused reversion to wild-type behavior. Still other observations shed light on the number of nucleotides constituting the genetic code. When two (+)'s were together or when two (−)'s were together, the correct reading frame was not reestablished. This argued against a doublet (two-letter) code. However, when three (+)'s or three (−)'s were present together, the original frame was reestablished. These observations strongly supported the triplet nature of the code.

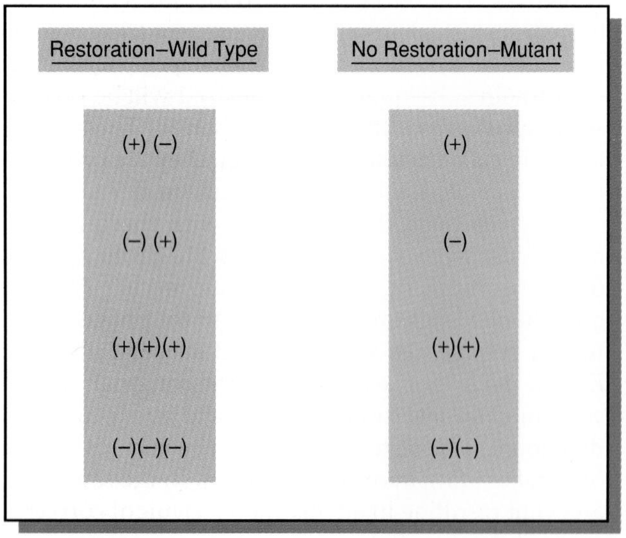

These data further suggested that the code is degenerate, which was contrary to Crick's earlier proposal. A degenerate code is one in which more than one codon specifies the same amino acid. The reasoning leading to this conclusion is as follows. In the cases where wild-type function is restored—(+) and (−), (+++), and (−−−)—the original frame of reading is also restored. However, there may be numerous triplets between the various additions and deletions that would still be out of frame. If 44 of the 64 possible triplets were blank and did not specify an amino acid, one of these so-called **nonsense triplets** would very likely occur in the length of nucleotides still out of frame. If a nonsense code were encountered during protein synthesis, it was reasoned that the process would stop or be terminated at that point. If so, the product of the rII-B locus would not be made, and restoration would not occur. Since the various mutant combinations were able to reproduce on *E. coli* K12, Crick and his colleagues concluded that, in all likelihood, most if not all of the remaining 44 codes were not blank. It follows that the genetic code is **degenerate**. As we shall see, this reasoning

Restoration–Wild Type	No Restoration–Mutant
(+) (−)	(+)
(−) (+)	(−)
(+)(+)(+)	(+)(+)
(−)(−)(−)	(−)(−)

FIGURE 11.3
Effects of multiple additions (+) and deletions (−) on the reading frame in phage T4.

proved correct and was soon established on the basis of more direct experimentation.

DECIPHERING THE CODE: INITIAL STUDIES

In 1961 Marshall Nirenberg and J. Heinrich Matthaei published results characterizing the first specific coding sequences. These results served as a cornerstone for the complete analysis of the code. Their success was dependent on the use of two experimental tools: **a cell-free protein-synthesizing system,** and the use of an enzyme, **polynucleotide phosphorylase,** which allowed the production of synthetic mRNAs. These mRNAs served as templates for polypeptide synthesis in the cell-free system.

In the cell-free (*in vitro*) system, amino acids can be incorporated into polypeptide chains. This *in vitro* mixture, as might be expected, must contain the essential factors for protein synthesis in the cell: ribosomes, tRNAs, amino acids, and other molecules essential to translation. In order to follow protein synthesis, one or more of the amino acids must be radioactive. Finally, an mRNA must be added to the system. This step is essential if a known template is to be translated.

In 1961 mRNA had yet to be isolated. However, the use of the enzyme **polynucleotide phosphorylase** allowed artificial synthesis of RNA templates, which could be added to the cell-free system. This enzyme, isolated from bacteria, catalyzes the reaction shown in Figure 11.4. Discovered in 1955 by Marianne Grunberg-Manago and Severo Ochoa, the enzyme functions metabolically in bacterial cells to degrade RNA. However, *in vitro,* with high concentrations of ribonucleoside diphosphates, the reaction can be "forced" in the opposite direction to synthesize RNA.

In contrast to RNA polymerase, polynucleotide phosphorylase requires no DNA template. As a result, ribonucleotides are assembled at random, according to the relative concentration of the four ribonucleoside diphosphates added to the reaction mixture. **This point is absolutely critical to understanding the work of Nirenberg and others in the ensuing discussion**.

Taken together, the cell-free system for protein synthesis and the availability of synthetic mRNAs provided a means of deciphering the composition of various triplet codes.

Nirenberg and Matthaei's Homopolymer Codes

In their initial experiments, Nirenberg and Matthaei synthesized **RNA homopolymers,** each consisting of only one ribonucleotide. Therefore, the mRNA added to the *in vitro* system was either UUUUUU. . ., AAAAAA. . ., CCCCCC. . ., or GGGGGG. . . . In testing each mRNA, they were able to determine which, if any, amino acids were incorporated into newly synthesized proteins. They determined this by labeling one of the 20 amino acids added to the *in vitro* system and conducting a series of experiments, each with a different amino acid made radioactive.

For example, consider one of the initial experiments, where ^{14}C-phenylalanine was used (Table 11.1). From these and related experiments, Nirenberg and Matthaei concluded that the message poly U (polyuridylic acid) directs the incorporation of only phenylalanine into the homopolymer polyphenylalanine. Assuming a triplet code, they had determined the first specific codon assignment! UUU codes for phenylalanine.

In the same way, they quickly found that AAA codes for lysine and CCC codes for proline. Poly G did not serve as an adequate template, probably because the molecule folds back on itself. Thus, the assignment for GGG had to await other approaches. Note that the specific triplet codon assignments were possible only because of the use of homopolymers.

FIGURE 11.4
The reaction catalyzed by the enzyme polynucleotide phosphorylase. Note that the equilibrium of the reaction favors the degradation of RNA but can be "pushed" in favor of synthesis.

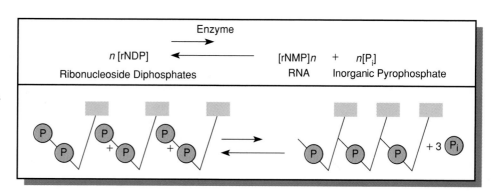

TABLE 11.1
Incorporation of ^{14}C-phenylalanine into protein.

Artificial mRNA	Radioactivity (counts/min)
None	44
Poly U	39,800
Poly A	50
Poly C	38
Poly inosinic acid	57
Poly AU	53
Poly U + Poly A	60

SOURCE: After Nirenberg and Matthaei, 1961, p. 1595.

This method yields only the composition of triplets, not the sequence. However, three U's, C's, or A's can have only one possible sequence (i.e., UUU, CCC, and AAA).

The Use of Mixed Copolymers

With these techniques in hand, Nirenberg and Matthaei, as well as Ochoa and coworkers, turned to the use of **RNA heteropolymers**. In this technique, two or more different ribonucleoside diphosphates are added in combination to form the message. These researchers reasoned that if the relative proportion of each type of ribonucleoside diphosphate is known, the frequency of any particular codon occurring in the synthetic mRNA can be predicted. If the mRNA is then added to the cell-free system and the percentage of any particular amino acid present in the new protein is ascertained, composition assignments may be predicted.

This concept is illustrated in Figure 11.5. Suppose that A and C are added in a ratio of 1A:5C. Now, the insertion of a ribonucleotide at any position along the RNA molecule during its synthesis is determined by the ratio of A:C. Therefore, there is a 1/6 possibility for an A and a 5/6 chance for a C to occupy each

FIGURE 11.5
Results and interpretation of a mixed copolymer experiment.

Possible Compositions	Probability of Occurrence of Any Triplet	Possible Triplets	Final %
3A	$(1/6)^3 = 1/216 = 0.4\%$	AAA	0.4
2A:1C	$(1/6)^2(5/6) = 5/216 = 2.3\%$	AAC ACA CAA	$3 \times 2.3 = 6.9$
1A:2C	$(1/6)(5/6)^2 = 25/216 = 11.6\%$	ACC CAC CCA	$3 \times 11.6 = 34.8$
3C	$(5/6)^3 = 128/216 = 57.9\%$	CCC	57.9
			100.0%

Transcription of Synthetic Message

CCCCCCCCACCCCCCAACCACCCCCACCCCCACCCAAACCCCCACCCCCC   RNA

Translation of Message

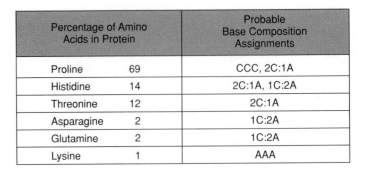

Percentage of Amino Acids in Protein		Probable Base Composition Assignments
Proline	69	CCC, 2C:1A
Histidine	14	2C:1A, 1C:2A
Threonine	12	2C:1A
Asparagine	2	1C:2A
Glutamine	2	1C:2A
Lysine	1	AAA

position. On this basis, we can calculate the frequency of any given triplet appearing in the message.

For AAA, the frequency is $(1/6)^3$, or about 0.4 percent. For AAC, ACA, and CAA, the frequencies are identical—that is, $(1/6)^2(5/6)$, or about 2.3 percent. Together, all three 2A:1C triplets account for 6.9 percent of the total three-letter sequences. In the same way, each of three 1A:2C triplets accounts for $(1/6)(5/6)^2$, or 11.6 percent (or a total of 34.8 percent). CCC is represented by $(5/6)^3$, or 57.9 percent of the triplets.

By examining the percentages of any given amino acid incorporated into the protein synthesized under the direction of this message, we may make tentative composition assignments. Since proline appears 69 percent of the time (Figure 11.5) and since 69 percent is close to 57.9 percent + 11.6 percent, we can deduce that proline is coded by CCC and by one triplet of the 2C:1A variety. Histidine, at 14 percent, is probably coded by one 2C:1A (11 percent) and one 1C:2A (2 percent). Threonine, at 12 percent, is likely coded by only one 2C:1A. Asparagine and glutamine appear each to be coded by one of the 1A:2C triplets, and lysine appears to be coded by AAA.

In similar experiments, some using as many as all four ribonucleotides to construct the mRNA, many possible combinations were studied. Thus, composition assignments were determined for all amino acids. Although this represented a very significant breakthrough, specific sequences of triplets were still unknown. Their determination awaited still other approaches.

The Triplet Binding Technique

It was not long before more advanced techniques were developed. In 1964 Nirenberg and Philip Leder developed the **triplet binding assay,** which led to specific assignments of triplets. The technique took advantage of the observation that ribosomes, when presented with an RNA sequence as short as three ribonucleotides, will bind to it and attract the correct charged tRNA. For example, if ribosomes are presented with an RNA triplet UUU and tRNA^phe, a complex will form that is similar to what actually occurs *in vivo*. The triplet acts like a codon in mRNA and it attracts the complementary sequence called the anticodon of tRNA (Figure 11.6). While it was not yet feasible to chemically synthesize long stretches of RNA, triplets of known sequence could be constructed in the laboratory.

All that was needed now was a method to determine which tRNA-amino acid bound to the mRNA-ribosome complex. The test system devised

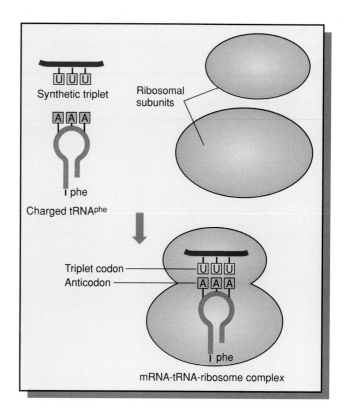

FIGURE 11.6
Molecular components used in the triplet binding assay.

was quite simple. The amino acid to be tested is made radioactive, and a charged tRNA (where the amino acid is bonded to tRNA) is produced. Since code compositions were known, it was possible to narrow the decision as to which amino acids should be tested for each specific triplet.

The radioactive charged tRNA, the RNA triplet, and ribosomes are incubated together on a nitrocellulose filter, which will retain the ribosomes but not the other components, such as charged tRNA. If radioactivity is not retained on the filter, an incorrect amino acid has been tested. If radioactivity remains on the filter, it is retained because the charged tRNA has bound to the triplet associated with the ribosome. In such a case, a specific codon assignment may be made.

Work proceeded in several laboratories, and in many cases clearcut, unambiguous results were obtained. Table 11.2, for example, shows 25 triplets assigned to 10 amino acids. However, in some cases the degree of binding was insufficient, and unambiguous assignments were not possible. Eventually, about 50 of the 64 triplets were assigned. The binding technique was a major innovation in deciphering the code. Based on these specific assignments of triplets to amino acids, two major conclusions were drawn.

TABLE 11.2
Amino acid assignments to specific trinucleotides derived
from the triplet binding assay.

Trinucleotide	Amino Acid
UGU UGC	Cysteine
GAA GAG	Glutamic acid
AAU AUC AUA	Isoleucine
UUA UUG CUU	Leucine
CUC CUA CUG	Leucine
AAA AAG	Lysine
AUG	Methionine
UUU UUC	Phenylalanine
CCU CCC	Proline
CCG CCA	Proline
UCU UCC	Serine
UCA UCG	Serine
UCU UCC	Valine
UCA UCG	Valine

The genetic code is **degenerate;** that is, one amino
acid may be specified by more than one triplet. The
code is also **unambiguous;** that is, a single triplet
specifies only one amino acid. As we shall see later in
this chapter, these conclusions have been upheld
with only minor exceptions.

The Use of Repeating Copolymers

Still another innovative technique used to decipher
the genetic code was developed by Gobind Khorana.
He was able to chemically synthesize long repeating
RNA sequences, which could be used in the cell-free
protein synthesizing system. First, he created shorter
sequences (e.g., di-, tri-, and tetranucleotides), which
were then replicated many times and finally joined
enzymatically to form the long polynucleotides.

As illustrated in Figure 11.7, a dinucleotide made
in this way is converted to a message with two
repeating triplets. A trinucleotide is converted to one
with three potential triplets, depending on the point
at which initiation occurs. Similarly, a tetranucleotide
creates four repeating triplets.

When these synthetic mRNAs were added to a
cell-free system, the predicted number of different
amino acids incorporated was upheld. Several examples are shown in Table 11.3. When such data were
combined with conclusions drawn from other approaches (composition assignment, triplet bindings),
specific assignments were possible.

One example of a specific assignment made from
such data will illustrate the value of Khorana's
approach. The repeating trinucleotide sequence
UUCUUCUUC. . . produces three possible triplets:
UUC, UCU, and CUU, depending on the first
nucleotide to initiate reading. When placed in a
cell-free translation system, the polypeptides contain-

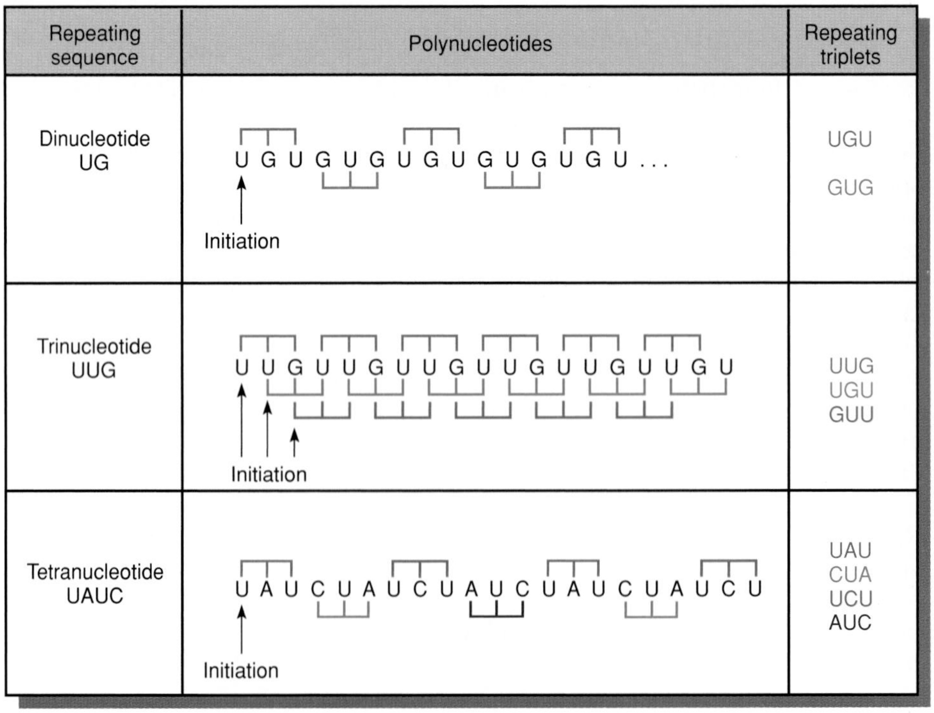

FIGURE 11.7
The conversion of di-, tri-, and
tetranucleotides into repeating
copolymers. The triplet codons
produced in each case are
shown.

Repeating sequence	Polynucleotides	Repeating triplets
Dinucleotide UG	U G U G U G U G U G U G U G U . . . Initiation	UGU GUG
Trinucleotide UUG	U U G U U G U U G U U G U U G U Initiation	UUG UGU GUU
Tetranucleotide UAUC	U A U C U A U C U A U C U A U C U A U C U Initiation	UAU CUA UCU AUC

TABLE 11.3
Amino acids incorporated using synthetic copolymers of RNA.

Copolymer	Codons Produced	Amino Acids in Polypeptide
UG	UGU	Cysteine
	GUG	Valine
AC	ACA	Threonine
	CAC	Histidine
UUC	UUC	Phenylalanine
	UCU	Serine
	CUU	Leucine
AUC	AUC	Isoleucine
	UCA	Serine
	CAU	Histidine
UAUC	UAU	Tyrosine
	CAU	Leucine
	UCU	Serine
	AUC	Isoleucine
GAUA	GAU	None
	AGA	
	UAG	
	AUA	

ing phenylalanine (phe), serine (ser), and leucine (leu) are produced. On the other hand, the dinucleotide sequence UCUCUCUC. . . produces the triplets UCU and CUC with the incorporation of leucine and serine. Therefore, the triplets UCU and CUC specify leucine and serine, but it cannot be determined which is which. By a process of elimination, the remaining trinucleotide UUC is assigned to phenylalanine.

Of UCU and CUU, which codes for serine and which for leucine? To answer this question, we must examine the results of using the repeating tetranucleotide sequence UUAC, which produces the triplets UUA, UAC, ACU, and CUU. The CUU triplet is one of the two in which we are interested. Three amino acids are incorporated: leucine, threonine, and tyrosine. Because CUU must specify only serine or leucine, and because, of these two, only leucine appears, we may conclude that CUU specifies leucine. By the process of elimination, UCU would seem to code for serine.

From such interpretations, Khorana reaffirmed triplets already deciphered and filled in gaps left from other approaches. For example, the use of two tetranucleotide sequences, GAUA and GUAA, suggested that at least two triplets were termination signals. This conclusion was drawn because neither of these sequences directed the incorporation of any amino acids into a polypeptide (Table 11.3). Since there are no triplets common to both messages, it was predicted that each repeating sequence contains at least one triplet that terminates protein synthesis. As we shall see, there are three such triplets, two of which are included in poly-(GAUA) and poly-(GUAA).

THE CODING DICTIONARY

The various techniques applied to decipher the genetic code have yielded a dictionary of 61 triplet codon-amino acid assignments. The remaining three triplets are termination signals, not specifying any amino acid. Figure 11.8 designates the assignments in a particularly illustrative form first suggested by Crick.

Degeneracy and Wobble

The degenerate nature of the code is immediately apparent when we inspect this presentation of the genetic code. While the amino acids tryptophan and methionine are encoded by only single triplets, most amino acids are specified by two, three, or four triplets. Three amino acids (serine, arginine, and leucine) are coded by six triplets each.

What also becomes evident is the pattern of degeneracy. Most often in a set of codons specifying the same amino acid, the first two letters are the

Second Position

First Position (5′ end)	U	C	A	G	Third Position (3′ end)
U	UUU phe	UCU ser	UAU tyr	UGU cys	U
	UUC phe	UCC ser	UAC tyr	UGC cys	C
	UUA leu	UCA ser	UAA Stop	UGA Stop	A
	UUG leu	UCG ser	UAG Stop	UGG trp	G
C	CUU leu	CCU pro	CAU his	CGU arg	U
	CUC leu	CCC pro	CAC his	CGC arg	C
	CUA leu	CCA pro	CAA gln	CGA arg	A
	CUG leu	CCG pro	CAG gln	CGG arg	G
A	AUU ile	ACU thr	AAU asn	AGU ser	U
	AUC ile	ACC thr	AAC asn	AGC ser	C
	AUA ile	ACA thr	AAA lys	AGA arg	A
	AUG met	ACG thr	AAG lys	AGG arg	G
G	GUU val	GCU ala	GAU asp	GGU gly	U
	GUC val	GCC ala	GAC asp	GGC gly	C
	GUA val	GCA ala	GAA glu	GGA gly	A
	GUG val	GCG ala	GAG glu	GGG gly	G

FIGURE 11.8
The coding dictionary. AUG encodes methionine, which initiates most polypeptide chains. All other amino acids except tryptophan, which is encoded by UGG, are represented by two to six triplets.

same, with only the third differing. This interesting pattern prompted Crick in 1966 to postulate the **wobble hypothesis**.

It was apparent that the first two ribonucleotides were more critical than the third member of the triplet code. Thus, Crick proposed that, in the anticodon of tRNA, less specific hydrogen bonding is required of the third ribonucleotide with its complement. This has the effect of resulting in a new set of base pairing rules at the third position of the codon. The anticodon is the triplet sequence of each tRNA that is complementary to the codon. The correct codon-anticodon interaction is essential to each step of translation, so that the amino acid corresponding to the codon is properly inserted into the growing polypeptide chain.

Such a "wobble" would allow the anticodon of a single tRNA species to pair with more than one triplet in mRNA. The code's degeneracy would often allow this to occur without changing the amino acid. Consistent with the coding assignments, it appears that U at the third position of the anticodon of tRNA may pair with A or G at the third position of the triplet in mRNA, and that G may likewise pair with U or C. Inosine, one of the modified bases found in tRNA, may pair with C, U, or A. Applying the "wobble" rules, a minimum of about 30 different tRNA species is necessary to accommodate the 61 triplets specifying an amino acid. If nothing more, wobble can be considered an economy measure, provided that fidelity of translation is not compromised.

Initiation, Termination, and Suppression

Initiation of protein synthesis is a highly specific process. In bacteria, the initial amino acid inserted into all polypeptide chains is a modified form of methionine—**N-formylmethionine (fmet)**. Only one codon, AUG, codes for methionine, and it is sometimes called the **initiator codon**. However, when AUG appears internally in mRNA, unformylated methionine is inserted into the polypeptide chain. Rarely, still another triplet, GUG, specifies methionine during initiation. It is not clear why this occurs, since GUG normally encodes valine.

In bacteria, either the formyl group is removed from the initial methionine upon completion of synthesis of a protein, or the entire formylmethionine residue is removed. In eukaryotes, methionine is also the initial amino acid during polypeptide synthesis. However, it is not formylated.

As mentioned in the preceding section, three other triplets (UAA, UAG, and UGA) serve as punctuation signals and do not code for any amino acid. They are not recognized by a tRNA molecule, and termination of translation occurs when they are encountered. Mutations occurring internally in a gene that produce any of the three triplets also result in termination. As a result, only a partial polypeptide is synthesized, since it is prematurely released from the ribosome. When such a change occurs, it is called a **nonsense mutation**. The terms **amber** (UAG), **ochre** (UAA), and **opal** (UGA) have been used to distinguish the three possibilities.

Interestingly, a distinct mutation in a second gene may cause **suppression** of premature termination. These mutations cause the chain-termination signal to be read as a sense codon. The "correction" usually inserts an amino acid other than that found in the wild-type protein. However, if the protein's structure is not altered drastically, it may function almost normally. Therefore, this second mutation has "suppressed" the mutant character resulting from the initial change to the termination codon.

Some suppressor mutations occur in genes specifying tRNAs. If the mutation results in a change in the anticodon such that it becomes complementary to a termination code, there is the potential for insertion of an amino acid and suppression.

Other types of suppression involve mutations in genes coding for aminoacyl synthetases, which are responsible for attaching the amino acid to tRNA (the process called *charging*), and in genes coding for ribosomal proteins. In both cases, suppression occurs as a result of misreading the mutant codon. That a ribosomal protein can cause ambiguity of translation demonstrates the intimate relationship between the ribosome, mRNA, and tRNA during translation.

CONFIRMATION OF CODE STUDIES: PHAGE MS2

All aspects of the genetic code discussed so far yield a fairly complete picture. The code is triplet in nature, degenerate, unambiguous, and comma-free, but contains punctuation with respect to start and stop signals. These individual principles have been confirmed by the detailed analysis of the RNA-containing bacteriophage MS2 by Walter Fiers and his coworkers.

MS2 is a bacteriophage that infects *E. coli*. Its nucleic acid (RNA) contains only about 3500 ribonucleotides, making up only three genes. These genes specify a coat protein, an RNA-directed replicase, and a maturation protein (the A protein). This simple system of a small genome and few gene products allowed Fiers and his colleagues to sequence the genes and their products. The amino acid sequence of the coat protein was completed in 1970, and the nucleotide sequence of the gene and a number of nucleotides on each end of it were reported in 1972. By 1976, this same research effort led to the complete sequencing of all three genes.

The coat protein contains 129 amino acids, and the gene contains 387 nucleotides, as expected for a triplet code. Each amino acid and triplet corresponds in linear sequence to the correct codon in the RNA code word dictionary, providing direct proof of the **colinear relationship** between nucleotide sequence and amino acid sequence. The codon for the first amino acid is preceded by AUG, the common initiator codon; and the codon for the last amino acid is succeeded by two consecutive termination codons, UAA and UAG.

The analysis clearly shows that the genetic code as established in bacterial systems is identical in this virus. We shall now briefly consider other evidence suggesting that the code is also identical in eukaryotes.

UNIVERSALITY OF THE CODE?

Between 1960 and 1978, it was generally assumed that the genetic code would be found to be universal, applying equally to viruses, bacteria, and eukaryotes. Certainly, the nature of mRNA and the translation machinery seemed to be very similar in these organisms. For example, cell-free systems derived from bacteria could translate eukaryotic mRNAs. Poly U was shown to stimulate translation of polyphenylalanine in cell-free systems when the components were derived from eukaryotes. Many recent studies involving recombinant DNA technology (see Chapter 16) have revealed that eukaryotic genes can be inserted into bacterial cells and transcribed and

translated. Within eukaryotes, mRNAs from mice and rabbits have been injected into amphibian eggs and efficiently translated. For those eukaryotic genes that have been sequenced, notably those for hemoglobin molecules, the amino acid sequence of the encoded proteins adheres to the coding dictionary established from bacterial studies.

However, several 1979 reports on the coding properties of DNA derived from yeast and human mitochondria (mtDNA) altered the principle of universality of the genetic language. Since then, mtDNA has been examined in other organisms, including the fungus *Neurospora*.

Not only do mitochondria contain DNA but transcription and translation occur within these organelles. Cloned mtDNA fragments were sequenced and compared with the amino acid sequences of various mitochondrial proteins. Several exceptions to the coding dictionary were revealed. Most surprising was that the codon UGA, normally causing termination, specifies the insertion of tryptophan during translation in yeast and human mitochondria. In human mitochondria, AUA, which normally specifies isoleucine, directs the internal insertion of methionine. In yeast mitochondria, threonine is inserted instead of leucine when CUA is encountered in mRNA.

More recently, in 1985, several other code alterations have been discovered. These and prior aberrant codes are summarized in Table 11.4. Such changes have been observed in the bacterium *Mycoplasma capricolum*, and in the protozoan ciliates *Paramecium*, *Tetrahymena*, and *Stylonychia*. As shown, each change converts one of the termination codons to glutamine (gln) or tryptophan (trp). These changes are significant because both a prokaryote and several eukaryotes are involved, representing distinct species that have evolved over a long period of time.

TABLE 11.4
Exceptions to the universal code.

Triplet	Normal Code Word	Altered Code Word	Source
UGA	Termination	trp	Human and yeast mitochondria Mycoplasma
CUA	leu	thr	Yeast mitochondria
AUA	ileu	met	Human mitochondria
AGA AGG	arg	Termination	Human mitochondria
UAA	Termination	gln	*Paramecium* *Tetrahymena* *Stylonychia*
UAG	Termination	gln	*Paramecium*

Note the apparent pattern in several of the altered codon assignments. The change in coding capacity involves only a shift in recognition of the third, or wobble, position. For example, AUA specifies isoleucine during translation in the cytoplasm and methionine in the mitochondrion. In cytoplasmic translation, methionine is specified by AUG. In a similar way, UGA calls for termination in the cytoplasmic system but tryptophan in the mitochondrion. In the cytoplasm, tryptophan is specified by UGG. Although it has been suggested that such changes in codon recognition may represent an evolutionary trend toward reducing the number of tRNAs needed in mitochondria, the significance of these findings is not yet clear. It is clear that only 22 tRNA species are encoded in human mitochondria. However, until still other examples are revealed, the differences must be considered as exceptions to the previously established general coding rules.

READING THE CODE: THE CASE OF OVERLAPPING GENES

In this chapter we established that the genetic code is nonoverlapping. This means that each ribonucleotide of an mRNA which specifies a polypeptide chain is part of only one triplet. However, this characteristic of the code does not rule out the possibility that a single mRNA may have multiple initiation points for translation. If so, these points could theoretically create several different frames of reading within the same mRNA, thus specifying more than one polypeptide. This concept, which would create **overlapping genes,** is illustrated in Figure 11.9(a).

That this might actually occur in some viruses was suspected when phage ϕX174 was carefully investi-

gated. The circular DNA chromosome consists of 5386 nucleotides, which should encode a maximum of 1795 amino acids, sufficient for five or six proteins. However, it was realized that this small virus in fact synthesizes 11 proteins consisting of more than 2300 amino acids! Comparison of the nucleotide sequence of the DNA and the amino acid sequences of the polypeptides synthesized has clarified this paradox. At least four cases of multiple initiation have been discovered, creating overlapping genes [Figure 11.9(b)].

The sequences specifying the K and B polypeptides are initiated with separate reading frames within the sequence specifying the A polypeptide. The K sequence overlaps into the adjacent sequence specifying the C polypeptide. The E sequence is out of frame with, but initiated in, that of the D polypeptide. Finally, the A' sequence, while in frame, begins in the middle of the A sequence. They both terminate at the identical point. In all, seven different polypeptides are created from a DNA sequence that might otherwise have specified only three (A, C, and D).

A similar situation has been observed in other viruses, including phage G4 and the animal virus SV40. Like ϕX174, phage G4 contains a circular, single-stranded DNA molecule. The use of overlapping reading frames optimizes the use of a limited amount of DNA present in these small viruses. However, such an approach to storing information has the distinct disadvantage that a single mutation may affect more than one protein and thus increase the chances that the change will be deleterious or lethal. In the case discussed above, a single mutation at the junction of genes *A* and *C* could affect three proteins (the A, C, and K proteins). It may be for this reason that such an approach has not become common in other organisms.

FIGURE 11.9

Illustration of the concept of overlapping genes. (a) An mRNA sequence initiated at two different AUG positions out of frame with one another will give rise to two distinct amino acid sequences. (b) The relative positions of the sequences encoding seven polypeptides of the phage ϕX174. Those encoding B, K, and E are out of frame.

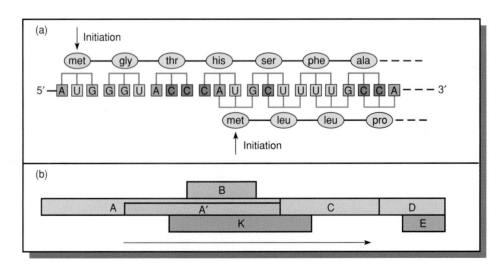

CHAPTER SUMMARY

1 The genetic code, stored in DNA, is transferred to RNA, where it is used to direct the synthesis of polypeptide chains.

2 The code is a triplet and has been demonstrated to be degenerate, unambiguous, nonoverlapping, and comma-free.

3 Around 1960, work performed by Nirenberg, Matthaei, Ochoa, Khorana, and others deciphered the code, assigning triplet ribonucleotide compositions and sequences to individual amino acids. These studies relied on cell-free, protein-synthesizing systems and synthetic mRNAs. The triplet binding assay provided the most specific assignments.

4 The complete coding dictionary determined that of the 64 possible codons, 61 encode the 20 amino acids found in proteins, while three triplets terminate translation. One of these 61 is the initiation codon and specifies methionine.

5 The observed pattern of degeneracy often involves only the third letter of a triplet series, and led Crick to propose the wobble hypothesis. This suggests that base pairing between the third letter of triplet codes in mRNA and their complementary anticodons in tRNA is less specific than that between the first two nucleotides of each code.

6 A mutation occurring internally in a gene may inadvertently produce one of the termination codons, causing the premature release of the polypeptide chain during its translation. Several modes of suppressing this and other types of mutations have been discovered.

7 Confirmation for the coding dictionary, including codons for initiation and termination, was obtained by comparing the complete nucleotide sequence of phage MS2 with the amino acid sequence of a corresponding protein. Other findings support the belief that, with only minor exceptions in mitochondria, the code is universal for all organisms.

8 In some bacteriophages multiple initiation points may occur during the translation of RNA, resulting in multiple reading frames.

Insights and Solutions

1 In 1990, J. Piccirilli, S. A. Benner, and colleagues reported findings showing that the nitrogenous base xanthine forms hydrogen bonds with a new base called kappa. When placed in a replicating system, DNA polymerase faithfully copies xanthine and kappa during synthesis. Had evolution seized on 6 bases rather than 4 within the structure of DNA, how many triplet codons would be possible? Would 6 bases accommodate a two-letter code, assuming 20 amino acids and start and stop codons?

ANSWER: Six things taken three at a time will produce $(6)^3$ or 216 triplet codes. If the code was a doublet, there would be $(6)^2$ or 36 two-letter codes, more than enough to accommodate 20 amino acids and start-stop punctuation.

2 In a heteropolymer experiment using 1/2C:1/4A:1/4G, how many different triplets will occur in the synthetic RNA molecule? How frequently will the most frequent triplet occur?

ANSWER: There will be $(3)^3$ or 27 triplets produced. The most frequent will be CCC, present $(1/2)^3$ or 1/8 of the time.

3 In a regular copolymer experiment, where UUAC is repeated over and over, how many different triplets will occur in the synthetic RNA, and how many amino acids will occur in the polypeptide when this RNA is translated? Be sure to consult Figure 11.8.

ANSWER: The synthetic RNA will repeat four triplets—UUA, UAC, ACU, and CUU—over and over.

Since both UUA and CUU encode leucine, while ACU and UAC encode threonine and tyrosine, respectively, polypeptides synthesized under the direction of such an RNA contain three amino acids in the repeating sequence leu-leu-thr-tyr.

PROBLEMS AND DISCUSSION QUESTIONS

1 In considering Gamow's diamond code, Brenner argued that if it were overlapping, only 16 different tripeptides involving any central amino acid encoded by a specific triplet could exist in proteins. Try to visualize why. Assuming 20 different amino acids, how many tripeptides containing a central amino acid can actually occur since the code is nonoverlapping?

2 In Gamow's diamond code, either the base pairs A-T or C-G forms the top and bottom of each diamond, while each side can be occupied by any of the four bases, creating 20 different conformations. Gamow believed each of these formed a pocket for a specific amino acid. Using the format below, draw the 20 possible diamonds.

 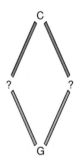

3 Crick, Barnett, Brenner, and Watts-Tobin, in their studies of frameshift mutations, found that either 3 (+)'s or 3 (−)'s restored the correct reading frame. If the code were a sextuplet (consisting of six nucleotides), would the reading frame be restored by either of the above combinations?

4 If, in Problem 3, 6 (+)'s or 6 (−)'s restored the original reading frame, could you decide on this basis if the genetic code was a triplet or a sextuplet? Why?

5 In a mixed copolymer experiment using polynucleotide phosphorylase, 3/4 G:1/4 C was added to form the synthetic message. The resulting amino acid composition of the ensuing protein was determined:

Glycine	36/64 (56%)
Alanine	12/64 (19%)
Arginine	12/64 (19%)
Proline	4/64 (7%)

From this information,
(a) Indicate the percentage (or fraction) of the time each possible triplet will occur in the message.
(b) Determine one consistent base composition assignment for the amino acids present.
(c) Considering the wobble hypothesis, predict as many specific triplet assignments as possible.

6 In a mixed copolymer experiment, messengers were created with either 4/5 C:1/5 A or 4/5 A:1/5 C. These messages yielded proteins with the following amino acid compositions. Using these data, predict the most specific coding composition for each amino acid.

4/5 C:1/5 A	Proline	63.0%	4/5 A:1/5 C	Proline	3.5%
	Histidine	13.0%		Histidine	3.0%
	Threonine	16.0%		Threonine	16.0%
	Glutamine	3.0%		Glutamine	13.0%
	Asparagine	3.0%		Asparagine	13.0%
	Lysine	0.5%		Lysine	50.0%
		98.5%			98.5%

7　When the amino acid sequences of insulin isolated from different organisms were determined, some differences were noted. For example, alanine was substituted for threonine, serine was substituted for glycine, and valine was substituted for isoleucine at corresponding positions in the protein. List the single base changes that could occur in triplets to produce these amino acid changes.

8　When repeating copolymers are used to form synthetic mRNAs, dinucleotides produce a single type of polypeptide that contains only two different amino acids. On the other hand, using a trinucleotide sequence produces three different polypeptides, each consisting of only a single amino acid. Why? What will be produced when a repeating tetranucleotide is used?

9　The mRNA formed from the repeating tetranucleotide UUAC incorporates only three amino acids, but the use of UAUC incorporates four amino acids. Why?

10　In studies using repeating copolymers, ACAC. . . . . incorporates threonine and histidine, and CAACAA. . . . . incorporates glutamine, asparagine, and threonine. What triplet code can definitely be assigned to threonine?

11　In a coding experiment using copolymers (as shown in Table 11.3), the following data were obtained:

Copolymer	Codons Produced	Amino Acids in Polypeptide
AG	AGA, GAG	Arg, Glu
AAG	AGA, AAG, GAA	Lys, Arg, Glu

AGG is known to code for arginine. Taking into account the wobble hypothesis, assign each of the four remaining different triplet codes to its correct amino acid.

12　Why doesn't polynucleotide phosphorylase synthesize RNA *in vivo*?

13　In the triplet binding technique (Figure 11.6), radioactivity remains on the filter when the amino acid corresponding to the triplet is labeled. Explain the basis of this technique.

14　Differentiate between the cause of suppression of a frameshift mutation and suppression of an amber mutation.

15　Of the changes noted in the coding dictionary for mitochondrial mRNAs, which represents the most surprising alteration?

16　In studies of the amino acid sequence of wild-type and mutant forms of tryptophan synthetase in *E. coli*, the following changes have been observed:

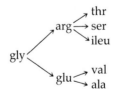

Determine a set of triplet codes in which a single nucleotide change produces each amino acid change.

17　Refer to Table 11.1. Can you hypothesize why Poly U + Poly A does not stimulate incorporation of ^{14}C-phenylalanine into protein?

18　What does this hypothesis (Problem 17) indicate about the RNA directing protein synthesis?

19　Predict the amino acid sequence produced during translation by the following short theoretical mRNA sequences. Note that the second sequence was formed from the first by a deletion of only one nucleotide.

Sequence 1:   AUGCCGGAUUAUAGUUGA
Sequence 2:   AUGCCGGAUUAAGUUGA

What type of mutation gave rise to Sequence 2?

20 A glycine residue exists at position 210 of the tryptophan synthetase enzyme of wild-type *E. coli*. If the codon specifying glycine is GGA, how many single base substitutions will result in an amino acid substitution at position 210? What are they? How many will result if the wild-type codon is GGU?

21 (a) Shown below is a theoretical viral mRNA sequence. Assuming that it could arise from overlapping genes, how many different polypeptide sequences can be produced? Using Figure 11.8, what are the sequences?

5'-AUGCAUACCUAUGAGACCCUUGGGA-3'

(b) A mutation at one position in the DNA giving rise to the above sequence occurred, eliminating the synthesis of all but one polypeptide. The base substitution of the mutant mRNA is shown below. Using Figure 11.8, determine why.

5'-AUGCAUACCUAUGUGACCCUUGGGA-3'

22 Most proteins have more leucine than histidine residues, but more histidine than tryptophan residues. Correlate the number of codons for these three amino acids with the above information.

SELECTED READINGS

BARRELL, B. G., et al. 1980. Different pattern of codon recognition by mammalian mitochondrial tRNAs. *Proc. Natl. Acad. Sci.* 77: 3164–66.

BARRELL, B. G., AIR, G., and HUTCHINSON, C. 1976. Overlapping genes in bacteriophage φX174. *Nature* 264: 34–40.

BARRELL, B. G., BANKIER, A. T., and DROUIN, J. 1979. A different genetic code in human mitochondria. *Nature* 282: 189–94.

BONITZ, S. G., et al. 1980. Codon recognition rules in yeast mitochondria. *Proc. Natl. Acad. Sci.* 77: 3167–70.

BRENNER, S. 1957. On the impossibility of all overlapping triplet codes in information transfer from nucleic acids to proteins. *Proc. Natl. Acad. Sci.* 43: 687–94.

BRENNER, S., STRETTON, A. O. W., and KAPLAN, D. 1965. Genetic code: The nonsense triplets for chain termination and their suppression. *Nature* 206: 994–98.

CALDWELL, P. C., and HINSHELWOOD, C. N. 1950. Some considerations on autosynthesis in bacteria. *J. Chem. Soc.*, pp. 3156–59.

COLD SPRING HARBOR LABORATORY. 1966. The genetic code. *Cold Spr. Harb. Symp.*, Vol. 31.

CRICK, F. H. C. 1962. The genetic code. *Scient. Amer.* (Oct.) 207: 66–77.

———. 1966. Codon-anticodon pairing: The wobble hypothesis. *J. Mol. Biol.* 19: 548–55.

———. 1966. The genetic code: III. *Scient. Amer.* (Oct.) 215: 55–63.

———. 1967. The Croonian lecture: The genetic code. *Proc. Roy. Soc. Biol.* 167: 331.

CRICK, F. H. C., BARNETT, L., BRENNER, S., and WATTS-TOBIN, R. J. 1961. General nature of the genetic code for proteins. *Nature* 192: 1227–32.

DICKERSON, R. E. 1983. The DNA helix and how it is read. *Scient. Amer.* (Dec.) 249: 94–111.

FIERS, W., et al. 1976. Complete nucleotide sequence of bacteriophage MS2 RNA: Primary and secondary structure of the replicase gene. *Nature* 260: 500–507.

GAMOW, G. 1954. Possible relation between DNA and protein structures. *Nature* 173: 318.

GAREN, A. 1968. Sense and nonsense in the genetic code. *Science* 160: 149–59.

JUDSON, H. F. 1979. *The eighth day of creation.* New York: Simon and Schuster.

JUKES, T. H. 1963. The genetic code. *Amer. Scient.* 51: 227–45.

KHORANA, H. G. 1967. Polynucleotide synthesis and the genetic code. *Harvey Lectures* 62: 79–105.

MIN JOU, W., HAGEMAN, G., YSEBART, M., and FIERS, W. 1972. Nucleotide sequence of the gene coding for bacteriophage MS2 coat protein. *Nature* 237: 82–88.

NIRENBERG, M. W. 1963. The genetic code: II. *Scient. Amer.* (March) 190: 80–94.

NIRENBERG, M. W., and LEDER, P. 1964. RNA codewords and protein synthesis. *Science* 145: 1399–1407.

NIRENBERG, M. W., and MATTHAEI, H. 1961. The dependence of cell-free protein synthesis in *E. coli* upon naturally occurring or synthetic polyribosomes. *Proc. Natl. Acad. Sci.* 47: 1588–1602.

PICCIRILLI, J. A., et al. 1990. Enzymatic incorporation of a new base pair into DNA and RNA extends the genetic alphabet. *Nature* 343: 33–37.

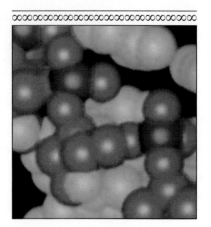

12

Synthesis of RNA and Protein: Transcription and Translation

TRANSCRIPTION: RNA SYNTHESIS
Experimental Evidence for the Existence of mRNA | RNA Polymerase |
The Sigma Subunit, Promoters, and Template Binding | The Synthesis of RNA |
Visualization of Transcription

TRANSLATION: PROTEIN SYNTHESIS
Ribosomal Structure | tRNA Structure | Charging tRNA | Initiation | Elongation |
Termination | Polyribosomes

TRANSCRIPTION AND TRANSLATION IN EUKARYOTES
Eukaryotic Template Binding | Heterogeneous RNA and its Processing | Intervening
Sequences and Split Genes | Splicing Mechanisms

CHAPTER CONCEPTS

Through the process of transcription, genetic information stored in
DNA is enzymatically transferred to an RNA molecule complementary
to one of the two DNA strands. Through the process of translation,
that RNA may be translated into a polypeptide chain as the result of
its association with a ribosome complex that includes adaptor
molecules called transfer RNAs. In bacteria, both processes are similar
to, but less complex than, those in eukaryotes.

Even while the genetic code was being studied, it was quite clear that proteins were the end products of many genes. Thus, while some geneticists were attempting to elucidate the code, other research efforts were directed toward the nature of genetic expression. The central question was how DNA, a nucleic acid, is able to specify a protein, composed of amino acids. Put still another way, How is information transferred between DNA and protein? We shall return in the next chapter to examine the evidence supporting the conclusion that proteins are specified by genes, as well as to discuss protein structure and function. Here we shall emphasize the concept of **information flow** as it occurs between DNA and protein. As you might imagine, this topic was extremely interesting and exciting in the early 1960s and it remains no less so in the 1990s!

Genetic information, stored in DNA, was shown to be transferred to RNA during the initial stage of gene expression. The process by which RNA molecules are synthesized on a DNA template is called **transcription**. The ribonucleotide sequence of RNA, written in a genetic code, is then capable of directing the process of **translation**. During translation, polypeptide chains—the precursors of proteins—are synthesized. Protein synthesis is dependent on a series of **transfer RNA (tRNA)** molecules, which serve as adaptors between the codons of mRNA and the amino acids specified by them. In addition, the process occurs only in conjunction with an intricate cellular organelle, the **ribosome**.

The processes of transcription and translation are complex molecular events. Like the replication of DNA, both rely heavily on base-pairing affinities between complementary nucleotides. The initial transfer from DNA to mRNA produces a molecule complementary to the gene sequence of one of the two strands of the double helix. Then, each triplet

codon is complementary to the anticodon region of tRNA as the corresponding amino acid is correctly inserted into the polypeptide chain during translation. In this chapter, we will describe in detail how these processes were discovered and how they are executed.

TRANSCRIPTION: RNA SYNTHESIS

The idea that an intermediate molecule is involved in the process of information flow between DNA and protein is suggested by the following observations:

1 DNA is, for the most part, associated with chromosomes in the nucleus of the eukaryotic cell. However, protein synthesis occurs in association with ribosomes located outside the nucleus in the cytoplasm. Therefore, DNA does not participate in protein synthesis.

The concept that this intermediate molecule is RNA is suggested by still other observations:

2 RNA is synthesized in the nucleus of eukaryotic cells, where DNA is found, and is chemically similar to DNA.
3 Following its synthesis, most RNA migrates to the cytoplasm, where protein synthesis occurs.
4 The amount of RNA is generally proportional to the amount of protein in a cell.

Collectively, these observations suggest that genetic information, stored in DNA, is transferred to an RNA intermediate which directs the synthesis of proteins. As with most new ideas in molecular genetics, the initial supporting evidence was derived from microbial experiments.

Experimental Evidence for the Existence of mRNA

In two papers published in 1956 and 1958, Elliot Volkin and his colleagues reported their analysis of RNA produced immediately after bacteriophage infection of *E. coli*. Using the isotope ^{32}P to follow newly synthesized RNA, they found that its base composition closely resembled that of the phage DNA but was different from that of bacterial RNA (Table 12.1). Although this newly synthesized RNA was unstable, or short-lived, its production was shown to precede the synthesis of new phage proteins. Thus, they considered the possibility that synthesis of RNA is a preliminary step in the process of protein synthesis.

Although ribosomes were known to participate in protein synthesis, their role in this process was not clear. One possible role was that each ribosome is specific for the protein synthesized in association with it. That is, perhaps genetic information in DNA is transferred to the RNA of a ribosome during its synthesis so that each class of ribosome specifies a particular protein. The alternative hypothesis was that ribosomes are nonspecific "workbenches" for protein synthesis and that specific genetic information rests with a "messenger" RNA.

In an elegant experiment using the *E. coli*–phage system, the results of which were reported in 1961, Sidney Brenner, François Jacob, and Matthew Meselson clarified this question. They labeled uninfected *E. coli* ribosomes with "heavy" isotopes and then allowed phage infection to occur in the presence of radioactive RNA precursors. They demonstrated that phage proteins were synthesized on bacterial ribosomes that were present prior to infection. Therefore, the ribosomes appeared to be nonspecific, strengthening the case that another type of RNA serves as an intermediary in the process of protein synthesis.

That same year, Sol Spiegelman and his colleagues isolated ^{32}P-labeled phage RNA following infection of bacteria and used it in molecular hybridization studies. They tried hybridizing this RNA to the DNA of both phages and bacteria in separate experiments. The RNA hybridized only with the phage DNA, showing that it was complementary in base sequence to the viral genetic information.

The results of all these experiments agree with the concept of a **messenger RNA (mRNA)** being made on a DNA template and then directing the synthesis of specific proteins in association with ribosomes. This concept was formally proposed by François Jacob and Jacques Monod in 1961 as part of a model for gene regulation in bacteria. Since then, mRNA has been isolated and thoroughly studied. There is no longer any question about its role in genetic processes.

RNA Polymerase

In order to prove that mRNA, as well as rRNA and tRNA, may be synthesized on a DNA template, it was necessary to demonstrate that there is an enzyme capable of directing this synthesis. By 1959, several investigators, including Samuel Weiss, had independently discovered such a molecule from rat liver. The enzyme, called **RNA polymerase,** directs the synthesis of the three types of RNA—messenger, ribosomal, and transfer—that are complementary to DNA templates. RNA polymerase has the same general requirements as DNA polymerase, the major exception being that the ribose rather than the deoxyribose form of the sugar is present in each nucleotide. The overall reaction that summarizes the synthesis of RNA on a DNA template may be expressed as:

$$n(\text{NTP}) + \text{DNA} \xrightarrow[\text{enzyme}]{\text{Mg}^{++}} (\text{NMP})_n + \text{DNA} + n(PP_i)$$

As the equation reveals, nucleoside triphosphates (NTPs) serve as substrates for the enzyme. It catalyzes the polymerization of nucleoside monophos-

TABLE 12.1

Base compositions (in mole percents) of RNA produced immediately following infection of *E. coli* by the bacteriophages T2 and T7 in contrast to the composition of RNA of uninfected *E. coli*.

	Adenine	Thymine	Uracil	Cytosine	Guanine
Postinfection RNA in T2-infected cells	33	—	32	18	18
T2 DNA	32	32	—	17*	18
Postinfection RNA in T7-infected cells	27	—	28	24	22
T7 DNA	26	26	—	24	22
E. coli RNA	23	—	22	18	17

*5-hydroxymethyl cytosine.
SOURCE: From Volkin and Astrachan, 1956; and Volkin, Astrachan, and Countryman, 1958.

phates (NMPs), or nucleotides, into a polynucleotide chain (NMP)$_n$. Nucleotides are linked during synthesis by 3'-5' phosphodiester bonds. The energy created by cleaving the triphosphate precursor into the monophosphate form drives the reaction and inorganic phosphates are produced (PP$_i$).

A second equation summarizes the sequential addition of each ribonucleotide as the process of transcription progresses:

$$NTP + (NMP)_n \xrightarrow[\substack{\text{DNA} \\ \text{and} \\ \text{enzyme}}]{Mg^{++}} (NMP)_{n+1} + PP_i$$

As this equation shows, each step of transcription involves the addition of one ribonucleotide (NMP) to the growing polyribonucleotide chain (NMP)$_{n+1}$, using a nucleoside triphosphate (NTP) as the precursor.

The RNA polymerase *holoenzyme* from *E. coli* has been extensively characterized and shown to consist of subunits designated α, β, β', and σ. The active form of the enzyme α$_2$ββ'σ has a molecular weight of almost 500,000 daltons. Little is known about the precise function of most subunits. However, it is the β polypeptide that provides the catalytic basis and active site for transcription. The **σ (sigma) subunit** can be removed from the complex without loss of catalytic activity to the remaining core enzyme. The sigma component plays a regulatory function and is involved in recognition of the points along the DNA template, called promoters, where RNA transcription is initiated.

While there is but a single form of the enzyme in *E. coli*, separate polymerases are involved in the transcription of the three types of RNA in eukaryotes. The nomenclature used in describing these is summarized in Table 12.2.

The three eukaryotic polymerases all consist of a greater number of polypeptide subunits than the bacterial form of the enzyme. One way in which the three forms may be distinguished is by their differential sensitivity to the mushroom poison, α-amanitin. Polymerase I, which catalyzes the tran-

scription leading to most rRNAs, is insensitive to the poison; polymerase II, which catalyzes the transcription producing mRNAs, is extremely sensitive (inhibited); polymerase III, which catalyzes transcription yielding tRNA and a small 5*S* rRNA component, is of intermediate sensitivity, being inhibited only at much higher concentrations. It is not immediately clear what functional differences exist among the three forms of this complex enzyme, but it is clear that they recognize different promoter sequences, thus providing the specificity of their transcriptive activities.

The Sigma Subunit, Promoters, and Template Binding

Transcription results in the synthesis of a single-stranded RNA molecule complementary to a region along one of the two strands of the DNA double helix. The initial step is referred to as **template binding,** where RNA polymerase interacts physically with DNA [Figure 12.1 (a)]. As mentioned above, the accuracy of this initial binding is achieved as a result of the recognition of specific DNA sequences called **promoters** by the sigma subunit (σ) of the holoenzyme.

These regions are located upstream (to the left in the illustration) from the point of initial transcription of a gene. Most likely, the holoenzyme "explores" a length of DNA until the promoter region is recognized, and a tightly bound complex results. Once this occurs, the enzyme then denatures or unwinds the helix locally prior to the initiation of transcription.

A great deal is now known about promoters and template binding. The enzyme is a complex molecule, large enough to bind about 60 nucleotide pairs of the helix, 40 of which are upstream from the point of initial transcription. The promoter regions for many genes have been isolated and their binding capacities analyzed in an ingenious way, called **DNA footprinting**. DNA representing isolated genes is incubated with copies of the RNA polymerase holoenzyme, but no ribonucleoside triphosphates are added to the reaction mixture. The polymerases bind to the promoter regions, but no transcription occurs. Then, endonucleases are added that cleave DNA except where it is protected by the polymerase. Those protected regions are isolated, and the DNA is separated from the polymerase and sequenced. In this way, the nucleotide sequences of promoter regions from various genes have been ascertained.

What this process has revealed is extremely interesting. Two sequences, each six nucleotides long, are strikingly similar in over 100 bacterial promoters thus far analyzed. The initial sequence, first discovered by

TABLE 12.2
RNA polymerases in eukaryotes.

Type	Product	Location
I	rRNA	Nucleolus
II	mRNAs	Nucleoplasm
III	5*S* RNA	Nucleoplasm
	tRNAs	Nucleoplasm

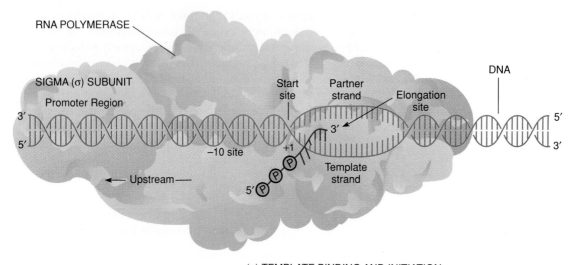

(a) TEMPLATE BINDING AND INITIATION
OF TRANSCRIPTION

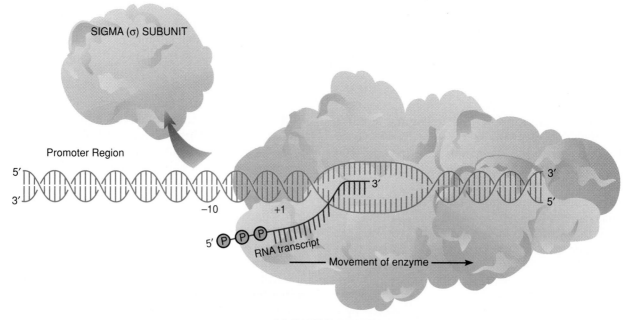

(b) CHAIN ELONGATION

FIGURE 12.1
Schematic representation of the early stages of transcription, including: (a) template binding and initiation involving the sigma subunit of RNA polymerase; and (b) chain elongation, after sigma has dissociated from the transcription complex.

David Pribnow, occurs 10 nucleotides upstream from the start of transcription; it is thus referred to as the **−10 sequence,** or **Pribnow box.** While it is not invariant, the sequence represented by each nucleotide present most often in all promoters analyzed is called the **consensus sequence**. By convention this is always depicted as that present on the nontemplate strand (the so-called partner or sense strand):

(−10)

5′ . . . TATAAT . . . 3′

The second consensus sequence is found about 35 nucleotides upstream (−35) from the start site:

(−35)

5′ . . . TTGACA . . . 3′

The importance of promoter sequences cannot be overemphasized. They govern the efficiency of initiation of transcription. Both strong promoters and weak promoters are recognized, leading to a variation of initiation from once every 1 to 2 seconds to only once every 10 to 20 minutes. In fact, mutations in the two consensus sequences described above may have the effect of severely reducing the initiation of gene expression. And, as we shall see later in this chapter, a consensus sequence rich in adenine and thymine residues (called the TATA box) is also present, but farther upstream, in almost all eukaryotic genes.

The high degree of conservation through evolution attests to the critical nature of these consensus sequences and the important role they play in transcription.

The Synthesis of RNA

Once the promoter has been recognized and bound by the enzyme complex, RNA polymerase catalyzes the insertion of the first 5'-ribonucleoside triphosphate, which is complementary to the first nucleotide at the start site of the DNA template strand. Subsequent ribonucleotide complements are inserted and linked together by phosphodiester bonds as RNA polymerization proceeds. This process, called **chain elongation** [Figure 12.1(a)], continues in the 5'-to-3' mode, creating a temporary DNA/RNA duplex whose chains run antiparallel to one another.

After a few ribonucleotides have been added to the growing RNA chain, the σ subunit is dissociated from the holoenzyme, and elongation proceeds under the direction of the core enzyme [Figure 12.1(b)]. In *E. coli* this process proceeds at the rate of about 50 nucleotides/second at 37°C.

Eventually, the enzyme traverses the entire gene and encounters a termination signal, a specific nucleotide sequence. Synthesis is completed usually in conjunction with the termination factor, rho (ρ). At that point, the transcribed RNA molecule is released from the DNA template, and the core enzyme dissociates. Under the direction of RNA polymerase, an RNA molecule is synthesized that is precisely complementary to a DNA sequence representing a gene. Wherever an A, T, C, or G residue existed on the DNA template strand, a corresponding U, A, G, or C residue has been incorporated into the RNA molecule, respectively. The significance of this synthesis is enormous, for it is the initial step in the process of information flow within the cell.

Visualization of Transcription

Electron microscope studies by Oscar Miller, Jr., Barbara Hamkalo, and Charles Thomas have provided striking visual demonstrations of the transcription process. Figure 12.2 shows micrographs and interpretive drawings of the DNA from two organisms, the newt, *Notophthalmus viridescens* and the bacterium *E. coli*. In both cases, the micrographs capture the process of RNA transcription on DNA templates. In the case of the newt, the segment of DNA is derived from oocytes that are known to produce an enormous amount of rRNA, which then becomes part of the ribosome. To accomplish this synthesis, the genes specific for RNA (**rDNA**) are replicated many times in the oocyte. This process is called **gene amplification**. The micrograph in Figure 12.2(a) shows many of these genes in tandem, each being transcribed simultaneously numerous times. For each rRNA gene, simultaneous transcription results in progressively longer and longer strands of incomplete rRNA molecules as the enzymes move along the DNA strand.

A different picture emerges from the study of *E. coli*, as seen in Figure 12.2(b). Because prokaryotes lack nuclei, cytoplasmic ribosomes are not separated from the bacterial chromosome. As a result, ribosomes are able to attach to partially transcribed mRNA molecules even before transcription is complete. Again, RNA molecules become progressively longer as transcription proceeds, with more and more ribosomes attaching to the longer strands. Visualizing the transcription process confirms what has previously been deduced from biochemical studies.

TRANSLATION: PROTEIN SYNTHESIS

Translation is the biological polymerization of amino acids into polypeptide chains. The process occurs only in association with ribosomes. The central question in translation is how triplet ribonucleotides of mRNA direct specific amino acids into their correct position in the polypeptide. This question was answered once transfer RNA (tRNA) was discovered. This class of molecules adapts specific triplet codons in mRNA to their correct amino acids. This adaptor role of tRNA was postulated by Crick in 1957.

In association with a ribosome, mRNA presents a triplet codon that calls for a specific amino acid. Because a specific tRNA molecule contains in its composition three consecutive ribonucleotides complementary to the **codon,** it is thus called the **anticodon** and can base pair with the codon. This tRNA is

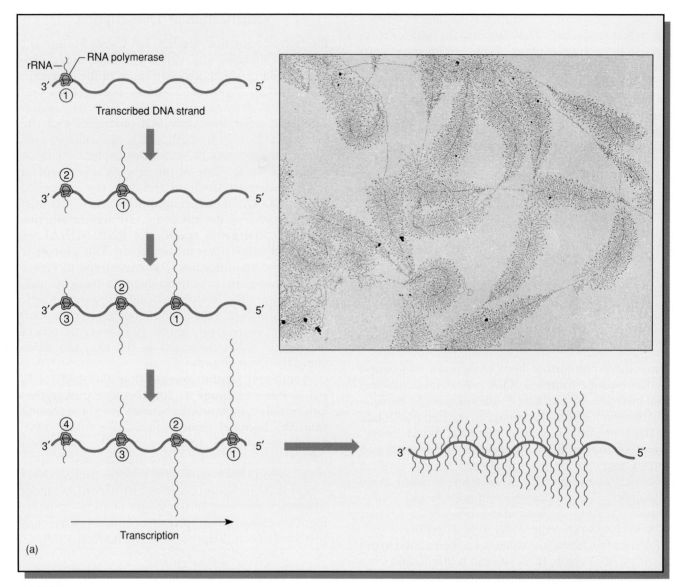

FIGURE 12.2
(a) Direct visualization of simultaneous transcription of multiple copies of the gene coding for rRNA in the oocyte of the newt *Notophthalmus* (Triturus) *viridescens*. The DNA was derived from one of many small nucleoli present in the oocyte. The diagram illustrates one transcription event. (b) Direct visualization of transcription of an unidentified gene of *E. coli* and the simultaneous translation of each mRNA transcript by multiple ribosomes. The diagram illustrates the sequence of transcription and translation events that lead to what is seen in the micrograph.

covalently bonded to the amino acid called for. As this process occurs over and over as mRNA runs through the ribosome, amino acids are polymerized into a polypeptide. The aspect most critical to the fidelity of translation is the codon-anticodon recognition.

In our discussion of translation, we will first consider the structure of the ribosome and transfer RNA. We will then subdivide the translation process into the following four phases and discuss each

separately: **tRNA charging, chain initiation, chain elongation,** and **chain termination**.

Ribosomal Structure

Because of its essential role in the expression of genetic information, the ribosome has been extensively analyzed. Bacterial cells contain about 10,000 of these structures, while eukaryotic cells contain many times more. Electron microscopy has revealed that

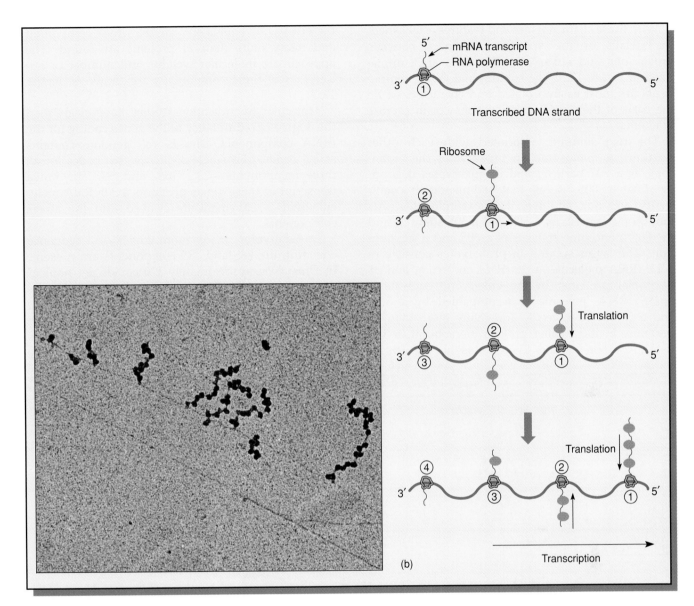

5′ ——— mRNA transcript
——— RNA polymerase
3′ ① Transcribed DNA strand 5′

Ribosome

②
3′ ① 5′

Translation
3′ ③ ② ① 5′

Translation
④ ②
3′ ③ ① 5′

Transcription

(b)

FIGURE 12.2
(continued)

the ribosome is about 250 Å in its largest diameter and consists of a larger and a smaller granule or particle intimately associated with each other. Called **subunits,** these particles can be separated and analyzed biochemically. Although the specific details differ among organisms, particularly between prokaryotes and eukaryotes, all ribosomes share certain characteristics and features:

1. Both subunits consist of one or more molecules of rRNA and an array of **ribosomal proteins**.
2. The large subunit contains two rRNA molecules, the larger of which is always a longer rRNA molecule than that found in the small subunit.

3. The number of proteins is always greater in the large subunit than in the small subunit.
4. The genes coding for the rRNA molecules are redundant, falling into the **moderately repetitive** category of DNA in eukaryotes.
5. The initial transcript of rRNA is larger than the final products incorporated into the subunits. The initial transcript is processed by endonuclease action, presumably in the nucleolus, in the case of eukaryotes. Specific enzymatic digestion leads to the final rRNA components of each subunit.
6. The functional rRNA molecules contain numerous **modified bases,** particularly ones that have been methylated.

With these characteristics and features established, the details of the specific differences between prokaryotic and eukaryotic ribosomes are summarized in Figure 12.3. The subunit and rRNA components are most easily isolated and characterized on the basis of their sedimentation behavior in sucrose gradients.

The two subunits associated with each other constitute a **monosome**. In prokaryotes the monosome is a 70S particle, and in eukaryotes it is approximately 80S. Recall that sedimentation coefficients are not additive. For example, the 70S monosome consists of a 50S and a 30S subunit, and the 80S monosome consists of a 60S and a 40S subunit. The larger subunit in prokaryotes consists of a 23S RNA molecule, a 5S rRNA molecule, and 31 ribosomal proteins. In the eukaryotic equivalent, a 28S rRNA molecule is accompanied by a 5.8S and a 5S rRNA molecule and many more than 34 proteins. In the smaller prokaryotic subunit, a 16S rRNA component and 21 proteins are found. In the eukaryotic equivalent, an 18S rRNA component and many more than 21 proteins are found. The approximate molecular weights and number of nucleotides of these components are shown in Figure 12.3.

Molecular hybridization studies have established the degree of redundancy of the genes coding for the rRNA components. The *E. coli* genome contains seven copies of a single sequence that codes for all three components—23S, 16S, and 5S. The initial transcript of these genes produces a 30S RNA molecule that is enzymatically cleaved into the above components.

In eukaryotes, many more copies of a sequence encoding the 28S and 18S components are present. In *Drosophila*, approximately 120 copies per haploid genome transcribe a molecule of about 34S. This is processed to the 28S, 18S, and 5.8S rRNA species. In *X. laevis*, over 500 copies per haploid genome are present. In mammalian cells, the initial transcript is 45S. The rRNA genes are part of the moderately

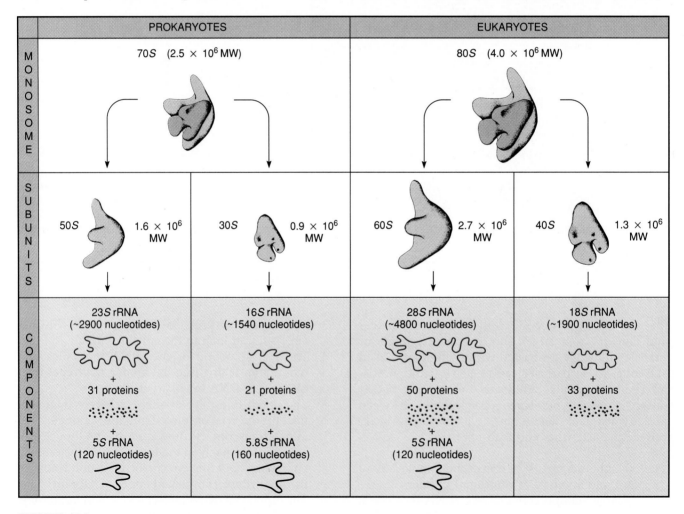

FIGURE 12.3
A comparison of the components in the prokaryotic and eukaryotic ribosomes.

repetitive DNA fraction and are present in clusters at various chromosomal sites. Each cluster consists of **tandem repeats,** with each unit separated by a noncoding **spacer DNA** sequence [Figure 12.2(b)]. In humans, these gene clusters have been localized on the ends of chromosomes 13, 14, 15, 21, and 22.

The 5S rRNA component of eukaryotes is not part of the larger transcript, as it is in *E. coli.* Instead genes coding for this ribosomal component are distinct and located separately. In humans, a gene cluster encoding them has been located on chromosome 1.

In spite of the detailed knowledge available on the structure and genetic origin of the ribosomal components, a complete understanding of the function of these components has eluded geneticists. This is not surprising, because the ribosome is perhaps the most complex of all cellular organelles. Together in bacteria, the 3 rRNA components and 55 proteins are precisely organized into a structure with a combined molecular weight of 2.5 million daltons, about 65 percent of which is RNA. The study of assembly mapping of the proteins, particularly by Masayasu Nomura, has provided some insights into the construction of the components. While no single protein is responsible for the binding of other proteins during assembly, groups of specific proteins facilitate the addition of others as the structure is constructed.

Once organized into functional subunits, it is clear that a precise association of ribosomes is essential to the translation process. Some insights to this organization have been gained from studies of the prokaryotic ribosome, the three-dimensional structure of which is illustrated in Figure 12.4.

The rRNA molecules are primarily confined to the interior of each ribosomal subunit, while proteins occupy most of the surface. However, studies using immune electron microscopy by James Lake and others show that the 3' ends of all three rRNAs as well as the 5' end of the 16S rRNA are found on the

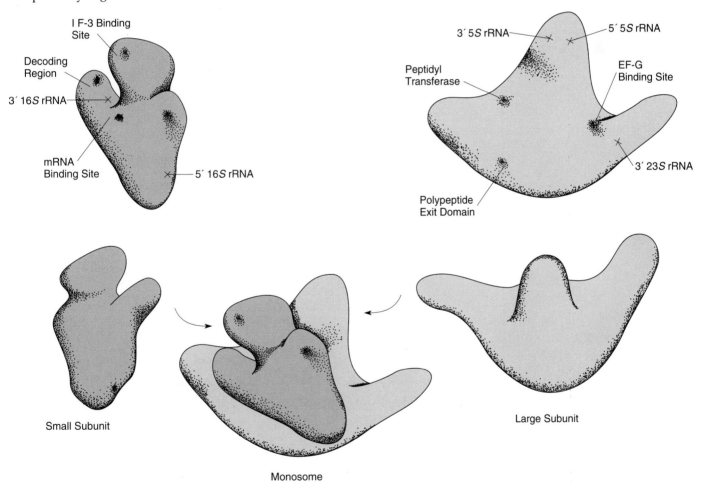

FIGURE 12.4
Three-dimensional models of the small and large prokaryotic ribosomal subunits and their interaction forming the monosome in *E. coli*. Several views of both subunits are provided. Numerous functional sites, regions, and domains that have been identified on the surface of each subunit are shown.

surfaces of their respective subunits. Certain functional domains have also been identified. The small subunit contains the mRNA binding site, the decoding region that binds the 3′ end of tRNA, and binding sites for the initiation factors. The large subunit contains regions for peptidyltransferase activity, the binding site for one of the elongation factors, and the exit point for the newly formed polypeptide chain. When the two subunits are combined into the functional monosome, each undergoes a conformational change and a groove is created between them where mRNA fits. The role of each of the above-mentioned components will soon become clear as we proceed with our discussion of translation.

tRNA Structure

Because of its small size and stability in the cell, tRNA has been investigated extensively. In fact, it is the best characterized RNA molecule. It is composed of only 75 to 90 nucleotides, displaying a nearly identical structure in bacteria and eukaryotes. In both types of organisms, tRNAs are transcribed as larger precursors, which are cleaved into mature 4S tRNA molecules. In *E. coli*, for example, tRNA^{tyr} is composed of 77 nucleotides, yet its precursor contains 126 nucleotides.

In 1965, Robert Holley and his coworkers reported the complete sequence of tRNA^{ala} (the superscript identifies the specific amino acid that binds to the tRNA) isolated from yeast. Their findings were the result of seven years of research, during which the techniques necessary to purify and separate specific tRNA species were developed. Methods to sequence moderate-sized RNA molecules were also worked out during that time. Their efforts paid handsome dividends, as a number of remarkable discoveries were soon revealed.

Of great interest was the finding that there are a number of nucleotides unique to tRNA. Each is a

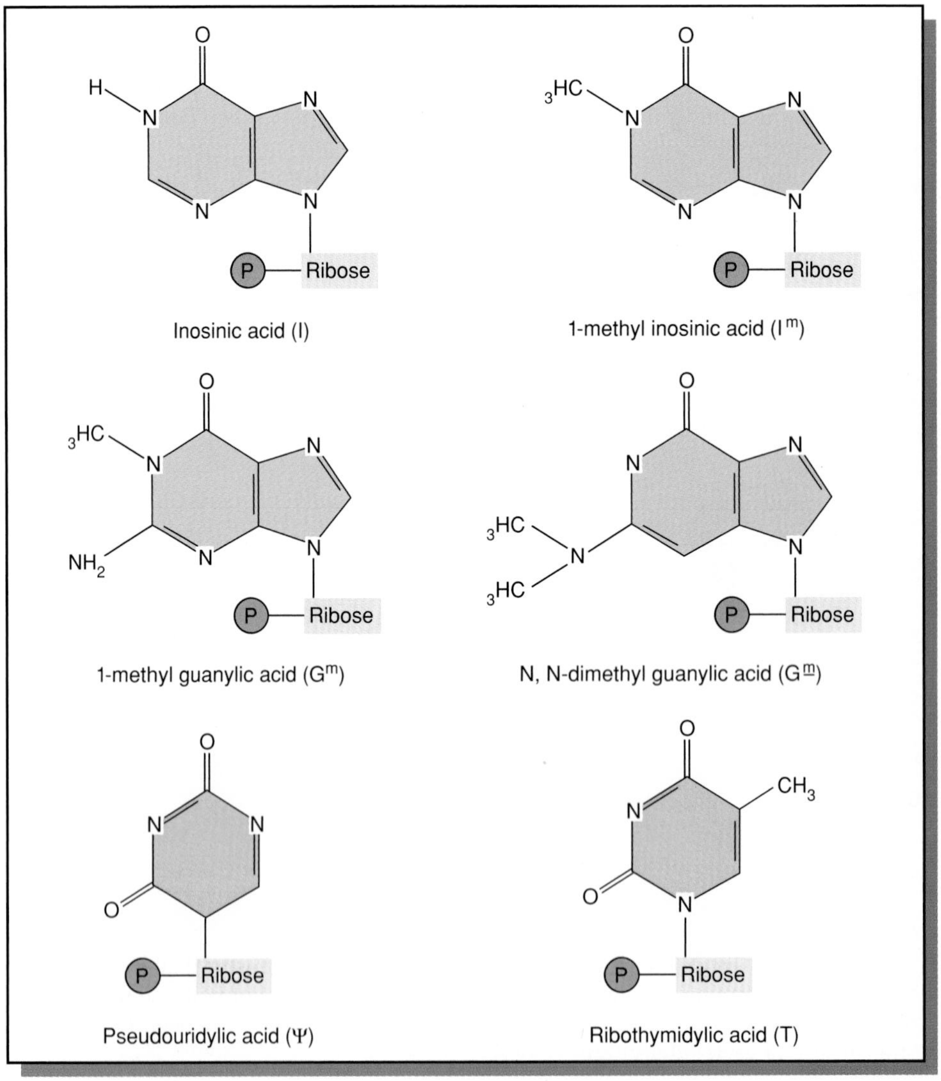

FIGURE 12.5
Unusual nitrogenous bases found in transfer RNA.

Inosinic acid (I)

1-methyl inosinic acid (I^m)

1-methyl guanylic acid (G^m)

N, N-dimethyl guanylic acid (G^m)

Pseudouridylic acid (Ψ)

Ribothymidylic acid (T)

modification of one of the four nitrogenous bases expected in RNA (G, C, A, and U). The structures of some of these are shown in Figure 12.5. **Inosinic acid (I)** contains the purine **hypoxanthine,** which differs from adenine only at position 6 of the heterocyclic ring. Others, **1-methyl inosinic acid (I^m), 1-methyl guanylic acid (G^m),** and **N,N-dimethyl guanylic acid ($G^{\underline{m}}$),** vary only in the additional methyl groups. **Pseudouridine (ψ),** on the other hand, has the uracil moiety attached by its 5-C rather than its normal 3-N to ribose. **Ribothymidylic acid (T)** contains thymine attached to ribose, creating an exception to the normal absence of thymidine in RNA. Finally, **dihydrouridylic acid (U^h)** contains additional hydrogens at the 5 and 6 carbon atoms of the pyrimidine ring.

These modified structures, sometimes referred to as unusual, rare, or odd bases, are created **post-transcriptionally.** That is, the unmodified base is produced during transcription, and then enzymatic reactions catalyze the modifications.

Holley's sequence analysis led him to propose the two-dimensional **cloverleaf model** of transfer RNA. It had been known that tRNA demonstrates a hyperchromic shift when heated, indicative of secondary structure due to base pairing. Holley discovered that he could arrange the linear model in such a way that stretches of base pairing resulted. This arrangement created a series of paired stems and unpaired loops resembling the shape of a four-leaf clover. Loops consistently contained modified bases, which do not generally form base pairs. Holley's model is shown in Figure 12.6.

Since the triplets GCU, GCC, and GCA specify alanine, Holley looked for the theoretical anticodon of his tRNA^{ala} molecule. He found it in the form of CGI, at the bottom loop of the cloverleaf. Recall from Crick's wobble hypothesis that I is predicted to pair with U, C, or A. Thus, the anticodon loop was established.

As other tRNA species were examined, numerous constant features were observed. First, at the 3' end, all tRNAs contain the sequence . . . . . **pCpCpA.** It is to the terminal adenosine residue that the amino acid is joined covalently during charging. Interestingly, it appears that the CCA sequence is added post-transcriptionally in some tRNAs. At the 5' terminus, all tRNAs contain . . . **pG.**

Additionally, the lengths of various stems and loops are very similar. All tRNAs examined also contain an anticodon complementary to the known amino acid code for which it is specific. These anticodon loops are present in the same position of the cloverleaf as well.

The cloverleaf model was predicted strictly on the basis of nucleotide sequence. Thus, there was great interest in X-ray crystallographic examination of tRNA, which reveals three-dimensional structure. By 1974, Alexander Rich and his coworkers in the United States, and J. Roberts, B. Clark, Aaron Klug, and their colleagues in England had been successful in crystallizing tRNA and performing X-ray crystallography at a resolution of 3 Å. At such resolution, the pattern formed by individual nucleotides is discernible.

FIGURE 12.6
The two-dimensional cloverleaf model of transfer RNA.

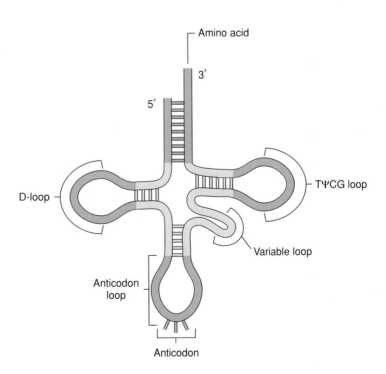

As a result of these studies, a complete three-dimensional model is now available (Figure 12.7). The model reveals tRNA to be L-shaped. At one end of the L is the anticodon loop and stem, and at the other end is the acceptor stem.

As different tRNAs were so examined, it became apparent that the stem areas were relatively constant. The loop areas, however, showed greater sequence variation. Thus, the areas contributing to the stems provide a rather constant structure for tRNAs. The loops, on the other hand, vary in shape and thus provide specificity to the individual tRNA species. For example, the T-loop in the initiator tRNA has a different sequence and shape from T-loops of other tRNAs. It has been speculated that the loop shapes may be recognized by specific **aminoacyl tRNA synthetases,** the enzymes responsible for adding the amino acid to tRNA. A model illustrating this interaction is portrayed in Figure 12.7.

Charging tRNA

Before translation can proceed, the tRNA molecules must be chemically linked to their respective amino acids. This activation process, called **charging,** occurs under the direction of enzymes called **aminoacyl synthetases**. Because there are 20 different amino acids, there must be at least 20 different tRNA molecules and as many different enzymes. In theory, since there are 61 triplet codes, there could be the same number of specific tRNAs and enzymes. Because of the ability of the third member of a triplet code to "wobble," however, it is now thought that there are about 30 different tRNAs; it is also believed that there are only 20 synthetases, one for each amino acid, regardless of the number of corresponding tRNAs.

The charging process is outlined in Figure 12.8. In the initial step, the amino acid is converted to an **activated form,** reacting with ATP to form an **aminoacyladenylic acid**. A covalent linkage is formed between the 5′ phosphate group and the carboxyl end of the amino acid. This molecule remains associated with the enzyme, forming an activated complex which then reacts with a specific tRNA molecule. In this second step, the amino acid is transferred to the appropriate tRNA and bonded covalently. The charged tRNA may participate directly in protein

FIGURE 12.7
A three-dimensional model of transfer RNA as it associates with its corresponding aminoacyl synthetase, which catalyzes the addition of the appropriate amino acid.

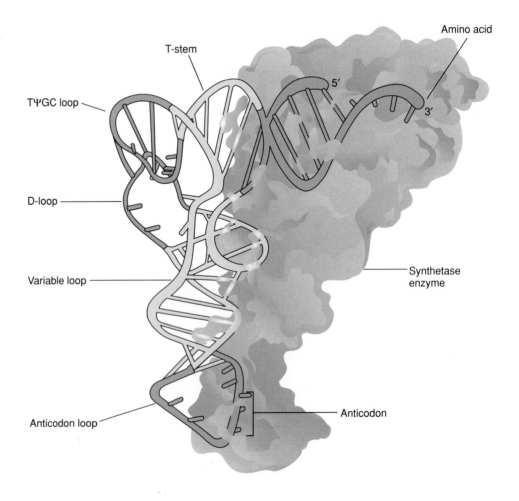

FIGURE 12.8
(a) Steps involved in charging tRNA. The X denotes that for each amino acid, only a specific tRNA and a specific aminoacyl synthetase enzyme are involved in the charging process. (b) Charged tRNA structure. The enlargement shows the attachment of the amino acid to the 3' end of tRNA.

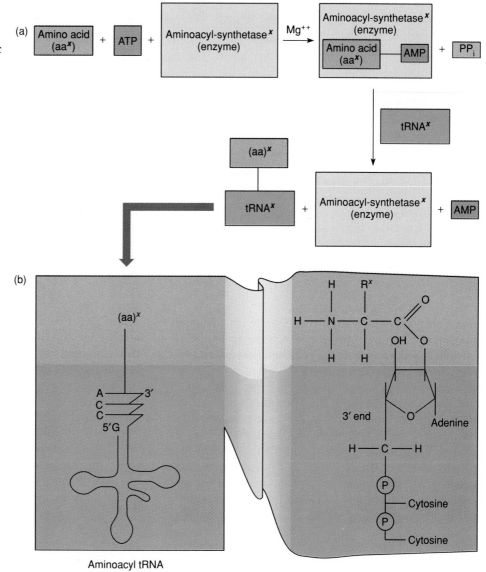

Aminoacyl tRNA

synthesis. Aminoacyl-tRNA synthetases are highly specific enzymes because they recognize only one amino acid and only a corresponding tRNA. This is a crucial point if fidelity of translation is to be maintained. The basis for this binding has sometimes been referred to as the "second genetic code," although this is not a particularly apt descriptor.

Initiation

Recall that ribosomes serve as a nonspecific workbench for the translation process. Many ribosomes that are uninvolved in translation are dissociated into their large and small subunits. Initiation of translation of *E. coli* involves these subunits, an mRNA molecule, a specific initiator tRNA, GTP, Mg^{++}, and at least three protein **initiation factors**. The protein

initiator molecules are not part of the ribosome, but are required to enhance the binding affinity of the various translational components (Table 12.3). In prokaryotes, the initiation code of mRNA calls for the modified amino acid formylmethionine.

The overall scheme of initiation events is outlined in Figure 12.9. The small ribosomal subunit binds to an initiation protein, and this complex in turn binds to mRNA. In bacteria, this binding involves a sequence of up to six ribonucleotides which precedes the initial AUG codon. It is this sequence (which is purine-rich and called the **Shine-Dalgarno sequence**) that base pairs with a region of the 16S rRNA of the small ribosomal subunit.

Several other initiation proteins then facilitate the binding of charged formylmethionyl-tRNA to the small subunit. This aggregate represents the **initia-**

SYNTHESIS OF RNA AND PROTEIN: TRANSCRIPTION AND TRANSLATION   325

TABLE 12.3
Various protein factors
involved during translation in
E coli.

Process	Factor	Role
Initiation	IF1	Stabilizes 30S subunit
	IF2	Binds fmet-tRNA to 30S-mRNA complex; Binds to and stimulates GTP hydrolysis
	IF3	Binds 30S subunit to mRNA; Dissociates monosomes into subunits following termination
Elongation	EF-Tu	Binds GTP; Brings aminoacyl-tRNA to the A site
	EF-Ts	Generates active EF-Tu
	EF-G	Stimulates translocation; GTP-dependent
Termination	RF1	Catalyzes release of the polypeptide chain from tRNA and dissociation of the translation complex; Specific for UAA and UAG termination codons.
	RF2	Behaves like RF1; Specific for UGA and UAA codons
	RF3	Stimulates RF1 and RF2

tion complex, which then combines with the large ribosomal subunit. In this process, a molecule of GTP is hydrolyzed to provide the required energy, and the initiation factors are released.

It is of interest to note that the synthesis of every *E. coli* protein begins with the modified amino acid **N-formylmethionine.** There are two tRNA species that may be charged with methionine. Only one, $tRNA^{fmet}$, serves as a substrate for the addition of a formyl group at the N-terminus of methionine. It is this charged tRNA that will bind to the small subunit initiation complex when AUG is present. The other, $tRNA^{met}$, is utilized when AUG appears internally in mRNA. It has no affinity for binding to the small ribosomal subunit during the formation of the initiation complex.

Elongation

As illustrated in Figure 12.9, the ribosomal subunits contain two binding sites for charged tRNA molecules; these are labeled the **P,** or **peptidyl,** and the **A,** or **aminoacyl, sites.** The initiation tRNA binds to the P site, provided the AUG triplet is in the corresponding position of the small subunit. The sequence of the second triplet in mRNA dictates which charged tRNA molecule will become positioned at the A site. Once it is present, **peptidyl transferase** catalyzes the formation of the peptide bond, which links the two amino acids together (Figure 12.10). This enzyme is part of the large subunit of the ribosome. At the same time, the covalent bond between the amino acid and the tRNA occupying the P site is hydrolyzed or broken.

The product of this reaction is a dipeptide, which is attached to the tRNA at the A site. The step in which the growing polypeptide chain has increased in length by one amino acid is called **elongation.**

Before elongation can be repeated, **translocation** must occur. That tRNA attached to the P site, which is now uncharged, is released from the large subunit. The entire **mRNA–tRNA–aa_2–aa_1** complex now shifts in the direction of the P site by a distance of three nucleotides. This translocation event requires several protein elongation factors as well as the energy derived from hydrolysis of GTP (Table 12.3). The result is that the third triplet of mRNA is now in a position to direct another specific charged tRNA into the A site. One simple way to distinguish the two sites in your mind is to remember that, following translocation, the P site contains a tRNA attached to a peptide chain (*P* for peptide) while the A site contains a tRNA with an amino acid attached (*A* for amino acid).

The sequence of elongation is repeated over and over. An additional amino acid is added to the growing polypeptide chain each time the mRNA advances through the ribosome. The efficiency of the process is remarkably high; the observed error rate is only 10^{-4}, so that an incorrect amino acid occurs once in every 20 polypeptides of an average length of 500 amino acids! In *E. coli* elongation occurs at a rate of about 15 amino acids per second at 37°C. The process can be likened to a tape moving through a tape recorder. As the tape moves, sequential sound is emitted from the recorder. Likewise, as mRNA moves, a growing polypeptide is produced by the ribosome.

FIGURE 12.9
Initiation events in the translation of an mRNA into a polypeptide chain.

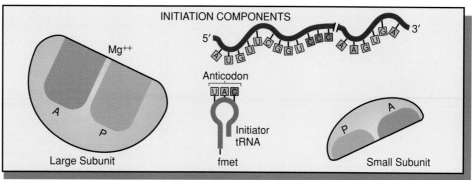

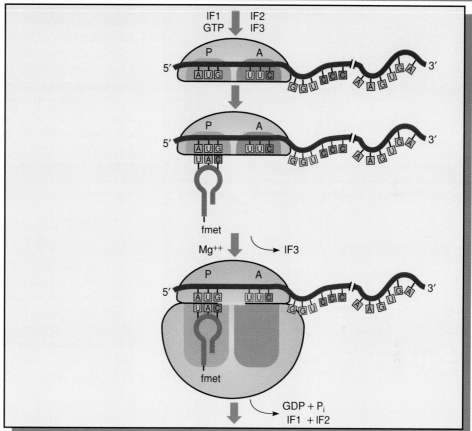

Termination

The termination of protein synthesis (Figure 12.11) is signaled by one of three triplet codes: UAG, UAA, or UGA. These codons do not specify an amino acid, nor do they direct tRNA into the A site. The finished polypeptide is therefore still attached to the terminal tRNA at the P site. The termination codon signals the action of **GTP-dependent release factors** (Table 12.3), which cleave the polypeptide chain from the terminal tRNA. Once this cleavage occurs, the tRNA is released from the ribosome, which then dissociates into its subunits. If a termination codon should appear in the middle of an mRNA molecule as a result of mutation, the same process occurs and the polypeptide chain is prematurely terminated.

Polyribosomes

As elongation proceeds and the initial portion of mRNA has passed through the ribosome, the message is free to associate with another small subunit to form another initiation complex. This process can be repeated several times with a single mRNA and

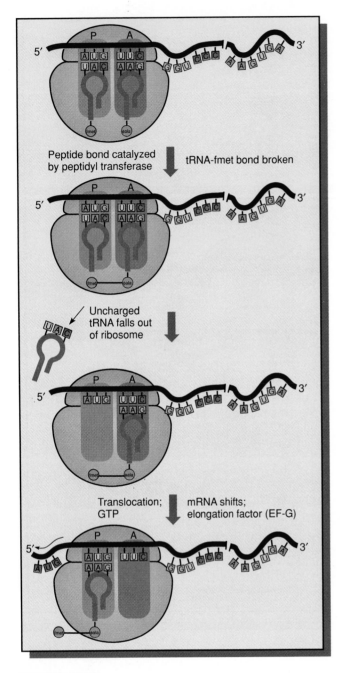

FIGURE 12.10
Elongation of the growing polypeptide chain during translation. After each amino acid is added, translocation of mRNA occurs, allowing the process to be repeated over and over.

results in what are called **polyribosomes** or just **polysomes**.

Polyribosomes can be isolated and analyzed following a gentle lysis of cells. Figures 12.12(a) and (b) illustrate these complexes as seen under the electron microscope. Note the presence of mRNA between the individual ribosomes in the micrograph on the right.

In Figure 12.12(c), the sedimentation properties of polyribosomes are shown. In this experiment, the cells were exposed briefly to radioactive RNA precursors prior to their lysis. As a result, the mRNA fraction was radioactive. After the ribosome fraction was isolated, it was centrifuged in a sucrose gradient, which was then eluted into successive fractions. Each fraction was then measured for absorbance at 260 nm and for radioactivity. Four absorbance peaks are apparent, each heavier than the one to the right of it. The three heaviest peaks represent polyribosome complexes consisting of four, three, and two ribosomes, respectively. The lightest peak is the single ribosome or monosome fraction. The radioactive mRNA is associated only with the polyribosome peaks, confirming that these complexes are the functional units of protein synthesis.

To complete the analogy with tapes (mRNA) and tape recorders (ribosomes), in polysome complexes one tape would be played simultaneously, but the transcripts (polypeptides) would all be at different stages of completion.

TRANSCRIPTION AND TRANSLATION IN EUKARYOTES

Much of our knowledge of transcription and translation has been derived from studies of prokaryotes. The general aspects of the mechanics of these processes are believed to be similar in eukaryotes. There are, however, numerous notable differences, several of which have been discussed earlier in this chapter. We will first summarize the differences and then expand on several yet to be mentioned:

1. Transcription in eukaryotes occurs within the nucleus under the direction of three separate forms of RNA polymerase (Table 12.2); for the mRNA to be translated, it must move out into the cytoplasm.
2. The initiation and regulation of transcription are under the control of extensive nucleotide sequences found in DNA upstream from the point of initial transcription.
3. Translation occurs on ribosomes that are larger and whose rRNA and proteins are more complex than those present in prokaryotes.
4. Protein factors similar to those in prokaryotes guide initiation, elongation, and termination of translation in eukaryotes. However, there appear to be more factors required during each of these steps.
5. Initiation of eukaryotic translation does not require the amino acid formylmethionine. How-

FIGURE 12.11
Termination of translation. The triplet UGA calls for no amino acid, but is instead recognized by release factors, leading to the dissociation of the translation complex.

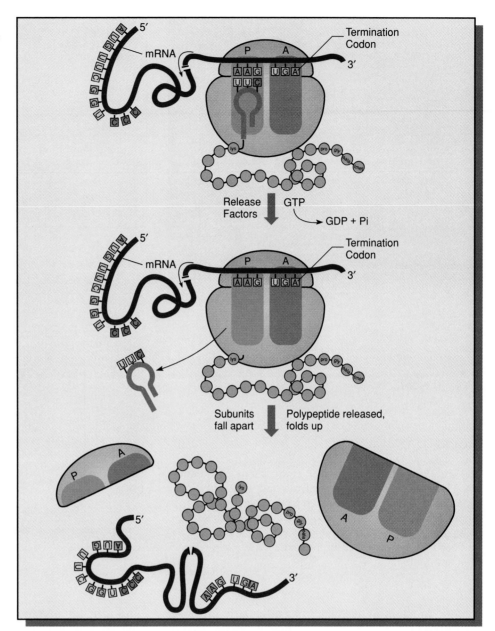

ever, the AUG triplet is essential to the formation of the translational complex and a unique transfer RNA ($tRNA_i^{met}$) is used for its initiation.

6 Most notably, extensive modifications occur to eukaryotic RNA transcripts that eventually serve as mRNAs. The initial transcripts are much larger than those that are eventually translated. Thus, they are called **pre-mRNAs** and are thought to constitute a group of molecules found only in the nucleus—a group referred to generally as **heterogeneous RNA (hnRNA)**. Only about 25 percent of hnRNA molecules are converted to mRNA. Those that are have substantial amounts of their ribonucleotide sequence excised, while the remaining segments are spliced back together prior to translation. This phenomenon has given rise to the concept of so-called **split genes** in eukaryotes.

7 Prior to the processing of an mRNA transcript, a cap and a tail are added to the molecule. These modifications are essential to efficient processing and subsequently to translation.

8 Eukaryotic mRNAs are much longer-lived than their bacterial counterparts. Most exist for hours, rather than minutes, prior to their degradation in the cell.

Eukaryotic Template Binding

The recognition of certain highly specific DNA regions by RNA polymerase is at the heart of orderly

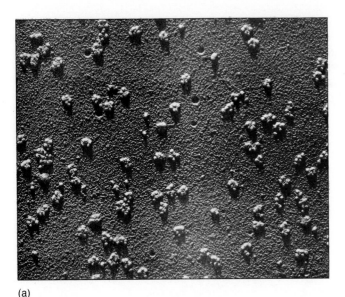

(a)

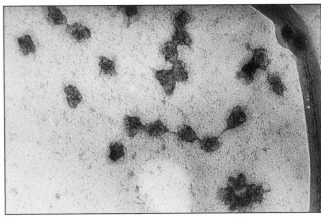

(b)

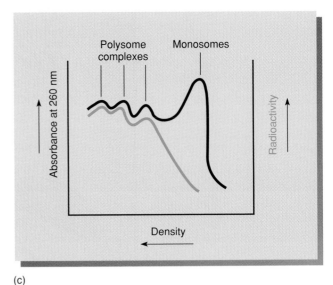

(c)

FIGURE 12.12
Polyribosomes visualized under the electron microscope [(a) and (b)] and in a graph analyzing a sucrose gradient (c). Those viewed by electron microscopy were derived from rabbit reticulocytes engaged in the translation of hemoglobin mRNA. In (c), ribosome complexes have been analyzed for their absorbance at 260 nm and for radioactivity, as described in the text.

genetic function in all cells, but particularly in those of differentiated eukaryotes. Thus it comes as no surprise that the nature of both the polymerases and promoters leading to template binding were found to be more complex in these more advanced organisms. RNA polymerase of eukaryotes, as earlier discussed, exists in three forms and each is more complex than the prokaryotic counterpart of the enzyme. In regard to the initial template binding step and promoter regions, most is known about polymerase II, which transcribes all mRNAs in eukaryotes.

There are at least three areas that are essential to the efficient initiation of transcription. The first is called the **Goldberg-Hogness** or **TATA box,** which is found about 25 nucleotide pairs upstream (−25) from the start point of transcription of most genes investigated. The consensus sequence is a heptanucleotide consisting solely of A and T residues. The sequence

and function are analogous to that found in the −10 region of prokaryotic genes. While mutations in this sequence reduce or eliminate transcription of a variety of genes, the TATA box may be fairly nonspecific and simply have the responsibility for fixing the site of initiation of transcription in a general way.

The second set of noncoding regions of interest are found farther upstream from the TATA box. In different genes studied, regions anywhere from 50 to 500 nucleotides upstream appear also to modulate transcription. These regions have been located on the basis of the effects of their deletion on transcription. The loss of various noncoding regions appears to drastically reduce *in vivo* transcription. Some noncoding regions, such as those associated with globin and SV40 genes, are about 50 to 100 nucleotides from the TATA sequence. Others, such as those associated with the sea urchin H2A gene and the *Drosophila* glue protein gene, are 200 to 500 nucleotides upstream. Frequently, GC-rich sequences as well as the sequence *CCAAT* are part of this noncoding region.

The third noncoding region is represented by elements called **enhancers**. While their location may vary, they often may be found even farther upstream. They have the effect of modulating transcription from a distance. Thus, they may not participate directly in template binding, even though they are essential to it. We will return to a discussion of these elements in Chapter 20.

Aside from the more extensive regulatory sequences present upstream in eukaryotes, a second major difference exists in the facilitation of template binding: the presence of protein transcription factors. It is generally believed that RNA polymerase II, unlike its prokaryotic counterpart, cannot bind directly to promoter sites and initiate transcription.

Instead, the enzyme appears to recognize a complex of DNA bound by specific proteins. For example, there is a **TATA-factor** (also called **TFIID**) which binds to that promoter sequence and which is critical to transcription. Still other factors specific to GC boxes and the CCAAT box have been discovered and studied. It may be that these transcription factors supplant the role of the sigma factor in the prokaryotic enzyme, thus providing a much greater degree of specific control of transcription. We will return in Chapter 20 to a consideration of the role of some of these factors in eukaryotic gene regulation.

Heterogeneous RNA and its Processing

Still other insights into regions of DNA that do not directly encode proteins have come from the study of RNA. This research has provided detailed knowledge of eukaryotic gene structure. The genetic code, as deciphered and shown in Figure 11.8, is written in the ribonucleotide sequence of mRNA. This information originated, of course, in the sense strand of DNA, where complementary sequences of deoxyribonucleotides exist. In bacteria, the relationship between DNA and RNA appears to be quite direct. The DNA base sequence is transcribed into an mRNA sequence, which is then translated into an amino acid sequence according to the genetic code.

However, in eukaryotes the situation is much more complex than in bacteria. It has been found that many internal base sequences of a gene may never appear in the mature mRNA that is translated. Other modifications occur at the beginning and the end of the mRNA prior to translation. These findings have made it clear that in eukaryotes, complex processing of mRNA occurs before it participates in translation.

By 1970, accumulating evidence showed that eukaryotic mRNA is initially transcribed as a much larger precursor molecule than that which is translated. This notion was based on the observation by James Darnell and his coworkers of **heterogeneous RNA (hnRNA)** in mammalian cells. Heterogeneous RNA is of large but variable size (up to 10^7 daltons), is found only in the nucleus, and is rapidly degraded. Nevertheless, hnRNA was found to contain nucleotide sequences common to the smaller mRNA molecules of the cytoplasm. Thus, it was proposed that the initial transcript of a gene results in a large RNA molecule which must first be processed in the nucleus before it appears in the cytoplasm as a mature mRNA molecule.

A subsequent discovery provided further evidence for this proposal. Both hnRNAs and mRNAs were found to contain at the 3′ end a stretch of up to 200 adenylic acid residues. Such **poly A** fragments are added **post-transcriptionally** to the initial gene transcript. In higher eukaryotes, for example, transcription is terminated, and the 3′ end of the initial transcript is reduced in length close to an AAUAAA sequence and then polyadenylated. Subsequent investigation has shown poly A at the 3′ end of almost all mRNAs studied in a variety of eukaryotic organisms. The exceptions seem to be the products of histone genes and some yeast genes.

As illustrated in Figure 12.13, Darnell proposed that poly A is added to the initial RNA transcript, which is then processed before its transport to the cytoplasm. It appears that only the segment of nucleotides nearest the 3′ end is retained as mRNA. The majority of the RNA transcript, cleaved from the 3′ poly A fragment, is then rapidly degraded by nucleases. Nonsurviving fragments contain both unique and repetitive sequences. In 1974, Darnell proposed that this process is a means of discriminating between nuclear RNA transcripts that survive to be translated and those that do not.

Still another modification of eukaryotic mRNA has been discovered (Figure 12.14). At the 5′ end of these molecules is found a **7-methyl guanosine (7mG) residue,** or cap. This cap is also added post-transcriptionally and is bonded in a unique way. Instead of the traditional 5′-to-3′ phosphodiester bond, 7mG is linked in a 5′-to-5′ configuration to three phosphates of the first nucleotide. It appears that G is first added and then methylated. Additionally, the initial nucleotides (N^1 and N^2) may also be methylated, yielding the sequence

$$7mGpppN^1mpN^2mpN^3pN^4p \ldots$$

where N^1 and N^2 are the first two nucleosides at the 5′ end of the message.

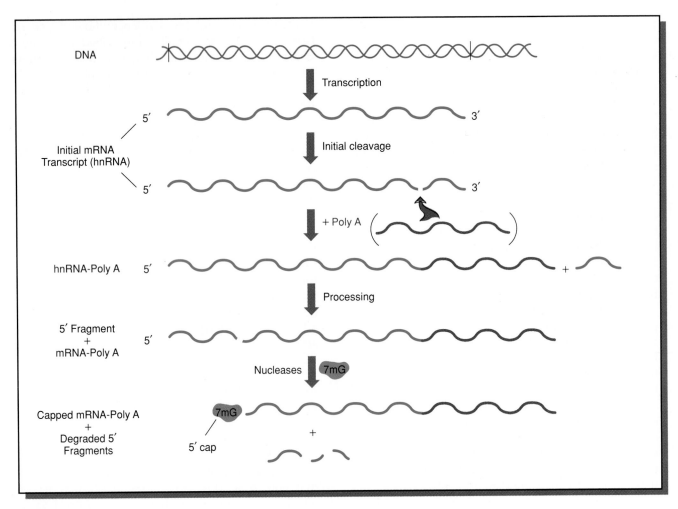

FIGURE 12.13
The conversion of heterogeneous RNA (hnRNA) into messenger (mRNA) in eukaryotes.

The significance of these 5' modifications of eukaryotic mRNAs is not yet clear but they may be essential to processing mRNA in the nucleus. As an analogue to the short Shine-Dalgarno sequence preceding the initiator codon in prokaryotic mRNAs, the cap very likely plays a role in initiation of translation. It may also stabilize mRNA from degradation during transcription.

Intervening Sequences and Split Genes

One of the most exciting discoveries in the history of molecular genetics occurred in 1977. At this time, direct evidence was provided by Susan Berget, Philip Sharp, and others that animal viruses as well as eukaryotic genes contain internal nucleotide sequences that are not expressed in the amino acid sequence of the proteins for which the genes code. That is, certain internal sequences in DNA do not always appear in the mature mRNA that is translated into a protein.

Detailed investigation has revealed numerous examples in which more than one noncontiguous DNA sequence fails to appear in the final mRNA. Such nucleotide segments have been called **intervening sequences,** contained within so-called **split genes**. Those DNA sequences not present in the final mRNA product are also called **introns** (*"int"* for intervening), and those retained and expressed are called **exons** (*"ex"* for expressed).

Intervening sequences were first suggested by studies of the animal viruses **adenovirus 2 (Ad2)** and **SV40**. Viral mRNAs were discovered to contain ribonucleotide sequences derived from different regions along the viral genome. This finding suggested that, in some manner, viral mRNAs were the product of an excision and rejoining process referred to as **splicing**.

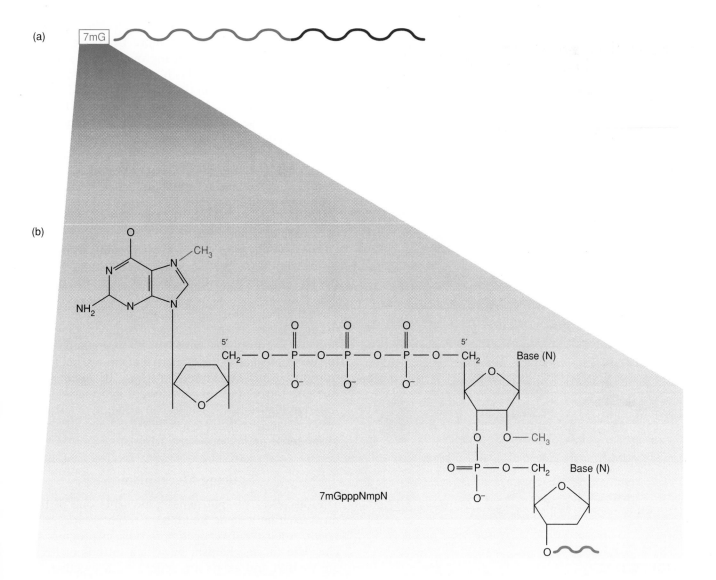

FIGURE 12.14
The 7-methyl guanosine cap of eukaryotic mRNA.

Similar discoveries were soon to be made in eukaryotes. Two approaches have been most fruitful. The first involves molecular hybridization of purified, functionally mature mRNAs with DNA containing the gene specifying that message. When hybridization occurs between nucleic acids that are not perfectly complementary, **heteroduplexes** are formed that may be visualized with the electron microscope. As illustrated in Figure 12.15, introns present in DNA but absent in mRNA must loop out and remain unpaired. This figure shows an electron micrograph and an interpretive drawing derived from hybridization between the template strand of the gene encoding chicken ovalbumin and the mature RNA prior to its translation into that protein. There are seven introns (A–G) whose sequences are present in DNA but not in the final mRNA.

The second approach provides more specific information. It involves a comparison of nucleotide sequences of DNA with those of mRNA and amino acid sequences. Such an approach allows the precise identification of all intervening sequences.

So far, a large number of genes from diverse eukaryotes have been shown to contain introns. One of the first so identified was the **beta-globin gene** in mice and rabbits, as studied independently by Philip Leder and Richard Flavell. The mouse gene contains an intron 550 nucleotides long, beginning immediately after the codon specifying the 104th amino acid. In the rabbit (Figure 12.16), there is an intron of 580

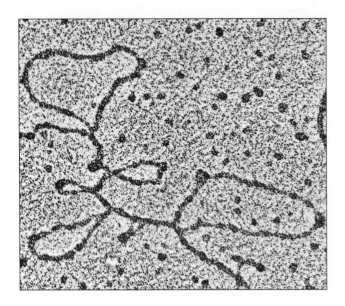

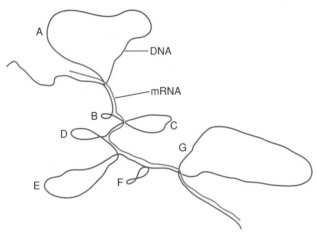

FIGURE 12.15
An electron micrograph and an interpretive drawing of the hybrid molecule formed between the template DNA strand of the ovalbumin gene and the mature ovalbumin mRNA. Seven DNA introns, A–G, produce unpaired loops.

base pairs near the codon for the 110th amino acid. Additionally, a second intron of about 120 nucleotides exists earlier in both genes. Similar introns have been found in the beta-globin gene in all mammals examined.

Several genes, notably those coding for histones and interferon, appear to contain no introns. However, intervening sequences have been identified in immunoglobulin genes of the mouse, some tRNA genes in yeast, and rRNA genes in *Drosophila,* among many others. The case of tRNA is interesting. Four different tRNA^{tyr} genes have been sequenced. In each case, an intron of 14 or 15 nucleotide pairs has been found immediately adjacent to the anticodon sequence.

As pointed out above, a more extensive set of introns has been located in the **ovalbumin gene** of chickens. The gene has been extensively characterized by Bert O'Malley in the United States and Pierre Chambon in France. As shown in Figure 12.16, the gene contains seven introns. Notice that the majority of the gene's DNA sequence is "silent," being composed of introns. The initial RNA transcript is four times the length of the mature mRNA. You should compare the information on the ovalbumin gene presented in Figures 12.15 and 12.16. Can you match the unpaired loops in 12.15 with the sequence of introns specified in 12.16?

The list of genes containing intervening sequences is growing rapidly. An extreme example of the number of introns in a single gene is that found in one of the chicken genes, *pro-α-2(I) collagen.* One of several genes coding for a subunit of this connective tissue protein, *pro-α-2(I) collagen* contains about 50 introns. The precision with which cutting and splicing occur must be extraordinary if errors are not to be introduced into the mature mRNA. The removal of just one intron and the subsequent splicing is illustrated in Figure 12.17. The loss or addition of a single ribonucleotide will create a shift in reading and result in missense and/or nonsense triplets.

Splicing Mechanisms

The discovery of split genes represents one of the most exciting genetic findings in recent years. As a result, intensive investigation is now in progress to elucidate the mechanism by which introns of RNA are excised and exons are spliced back together. A great deal of progress has already been made. Interestingly, it appears that somewhat different mechanisms are utilized for each of the three types of RNA as well as for RNAs produced in mitochondria (Figure 12.18). As each case is discussed below, refer to this figure.

The simplest mechanism appears to be used to process **tRNA molecules**. As you will recall, these are small RNAs consisting of approximately 80 nucleotides. Some of them contain a short internal intron of about 15 nucleotides. In such cases, initial folding of the molecule creates a **pre-tRNA** with an extra loop representing the intron. This is enzymatically removed. John Abelson has demonstrated this enzymatic process *in vitro.* The reactions can easily be visualized [Figure 12.18(a)]; a nuclease makes appropriate cuts in the RNA, leaving the ends that are to be joined in juxtaposition to one another. A ligase then links them together.

A second splicing mechanism has been discovered in the processing of **rRNA** by the protozoan,

FIGURE 12.16
Intervening sequences in various eukaryotic genes. The numbers indicate nucleotides present in various intron and exon regions.

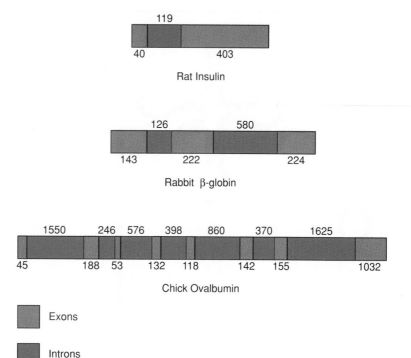

Rat Insulin

Rabbit β-globin

Chick Ovalbumin

Exons

Introns

Tetrahymena thermophila. We have earlier indicated that 18*S* and 28*S* eukaryotic rRNA molecules are derived from a 45*S* precursor that must be extensively altered. Working with *Tetrahymena,* Thomas Cech, a recent Nobel Prize winner, and his colleagues have demonstrated that in one of the larger 26*S* precursors **self-excision** of an intervening sequence occurs. The reaction involves no external catalyst and appears to require only a monovalent and a divalent cation (NH_4^+ and Mg^{++}) as well as guanosine as a cofactor. As illustrated in Figure 12.18(b), the RNA chain is cleaved and a guanosine residue (pG) is added to the 5'-linear portion of the intron. Then

a second step occurs where the intron is excised and the flanking exons are ligated. The intron, which is 414 nucleotides in length, is converted into a covalently closed circular form. The entire process is mediated by the RNA itself, and no enzymes are required. Thus, the RNA that undergoes this process behaves catalytically; it has been referred to as a **ribozyme**. A similar process has now been demonstrated in the splicing of rRNA introns in yeast and other fungi.

Another interesting splicing mechanism has been discovered for an RNA molecule transcribed on mitochondrial DNA (**mtDNA**). One of the introns of

FIGURE 12.17
Splicing of the initial RNA transcript to remove a single intervening sequence.

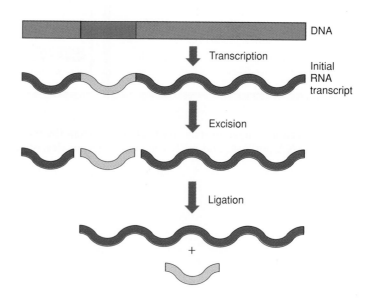

DNA

Transcription

Initial RNA transcript

Excision

Ligation

+

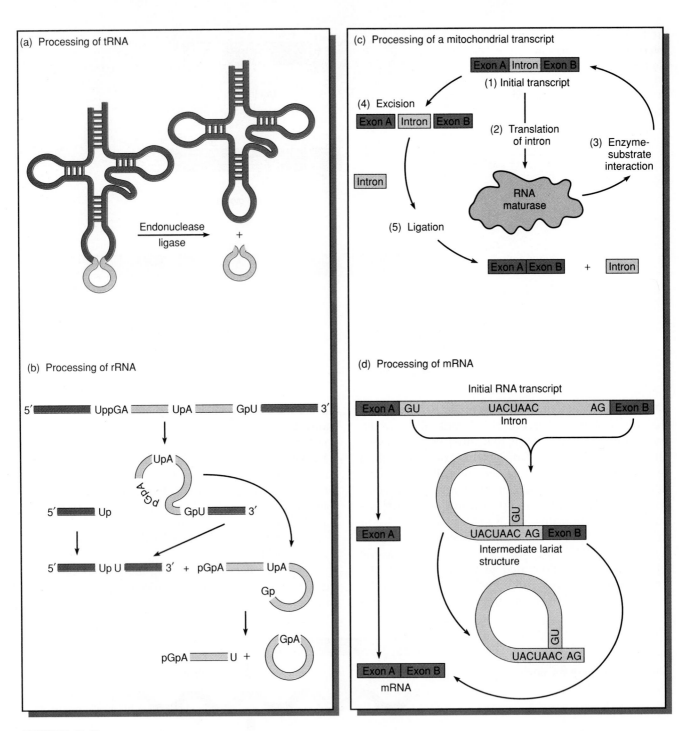

FIGURE 12.18
Four mechanisms utilized to remove introns and ligate exons from the initial RNA transcripts of tRNA, rRNA, mtRNA, and mRNA in eukaryotes, as described in the text.

the gene specifying the respiratory molecule cytochrome b encodes a separate enzyme. Once synthesized, this enzyme, called **RNA maturase,** directs the excision and splicing of the initial RNA transcript from which it was derived! Once processed, the resultant RNA is then translated into cytochrome b. This mechanism was proposed originally on genetic grounds, where mutations in an intron inhibited the production of cytochrome b. These steps are illustrated in Figure 12.18(c).

The final mechanism to be discussed is the least understood and involves eukaryotic **mRNAs.** Introns in mRNA, in comparison to other RNAs thus far discussed, can be much larger—up to 20,000 nucleotides—and they are more plentiful. Thus, their removal requires a much more complex mechanism, which until recently has been very difficult to define. Many clues are now emerging, however. First, the nucleotide sequence around different introns is often similar. Most begin at the 5′ end with a GU dinucleotide sequence and terminate at the 3′ end with an AG dinucleotide sequence. These, as well as other consensus sequences shared by introns, may attract a molecular complex essential to ligation and splicing. Such a complex has been identified in extracts of yeast as well as mammalian cells. Called a **spliceosome,** it is very large, being 40S in yeast and 60S in mammals. One group of components of these complexes consists of a unique set of small RNAs (usually 250 nucleotides or less) each complexed with proteins. Found only in the nucleus, these are called **small nuclear ribonucleoproteins** (or *snRNAs* or *SNURPs*). They have been arbitrarily designated U1, U2, U3, . . . U6, and the list continues to grow. Some SNURPs share one or more common proteins. Other proteins are unique to a single SNURP. Different complexes are thought to play different roles in splicing RNAs.

Intermediate products of processing have now been isolated during *in vitro* studies. An interesting structure called a **lariat** (or closed loop) is formed, as illustrated in Figure 12.18(d). First, the 5′ site adjacent to the first exon is cleaved, whereupon the intron folds back on itself, forming a covalent bond such that the lariat is still attached to the second exon. Then, the lariat structure is removed as the two exons are ligated. In many introns the sequence UACUAAC is located between 18 and 37 nucleotides proximal to the 3′ end. It seems to be critical to the lariat formation, which is thought to be essential to the formation of the spliceosome.

Even with the above knowledge available, the complexity of the splicing complex renders its analysis most difficult. While we have discussed the removal of one intron in a single RNA transcript, large numbers must be cleaved out and the many exons spliced back together with absolute precision. To ensure such precision an extremely efficient mechanism must be at work.

The finding of "genes in pieces," as split genes have been described, raises many interesting questions and has provided great insights into the organization of eukaryotic genes. Thus, we will return to this topic again in Chapter 18.

CHAPTER SUMMARY

1   Transcription and translation—RNA and protein biosynthesis, respectively—are the fundamental processes essential to the expression of genetic information. All of the various proteins in any cell can be traced back to the polynucleotide sequences in DNA representing genes, which are expressed by these two processes.

2   The processes of transcription and translation, like DNA replication, can be subdivided into the stages of initiation, elongation, and termination. Both processes rely on base-pairing affinities between complementary nucleotides.

3   Transcription describes the synthesis, under the direction of RNA polymerase, of a strand of RNA complementary to a DNA template. RNA polymerase, composed of the sigma subunit and the core enzyme, binds at specific initiation sites along the DNA called promotors. Following the initiation of RNA synthesis along the sense strand, sigma is released and elongation proceeds under the direction of the core enzyme.

4   Translation allows the genetic code present in mRNA to be deciphered, resulting in protein synthesis. This complex energy-requiring process involves charged tRNA molecules, numerous proteins, ribosomes, and mRNA. Transfer RNA (tRNA) serves as the adaptor molecule between an mRNA triplet and the appropriate amino acid. Its efficient function depends on anticodons complementary to specific triplets.

5   Ribosomes, in association with mRNA, facilitate the assembly of amino acids into polypeptide chains. One mRNA may initiate transcription with more than one ribosome, forming polyribosomes or polysomes.

6   The processes of transcription and translation are more complex in eukaryotes than in prokaryotes. The primary transcript in eukaryotes must be modified in various ways, including the addition of a cap and tail and the removal, through splicing, of intervening sequences, or introns.

Insights and Solutions

1 Actinomycin D inhibits DNA-dependent RNA synthesis. This antibiotic is added to a bacterial culture where a specific protein is being monitored. Compared to a control culture, translation of the protein declines over a period of 20 minutes, until no further protein is made. Explain these results.

ANSWER: The mRNA that is the basis for the translation of the protein has a lifetime of about 20 minutes. When actinomycin D is added, transcription is inhibited and no new mRNAs are made. Those already present support the translation of the protein for up to 20 minutes.

2 DNA and RNA base compositions were analyzed from a hypothetical bacterial species with the following results:

	(A + G)/(T + C)	(A + T)/(C + G)	(A + G)/(U + C)	(A + U)/(C + G)
DNA	1.0	1.2		
RNA			1.3	1.2

On the basis of these data, what can you conclude about the DNA and RNA of the organism? Are the data consistent with the Watson-Crick model of DNA? Is the RNA single-stranded or double-stranded, or can't we tell? If we assume that the entire length of DNA has been transcribed, do the data suggest that RNA has been derived from the transcription of one or both DNA strands, or can't we tell from these data?

ANSWER: This problem is a theoretical exercise designed to get you to look at the consequences of base complementarity as it affects base composition of DNA and RNA. The base composition of DNA is consistent with the Watson-Crick double helix. In a double helix, we expect A + G to equal T + C (the number of purines should equal the number of pyrimidines). In this case, there is a preponderance of A/T base pairs (120 A and T pairs to every 100 G and C pairs).

Given what we know about RNA, there is no reason to expect the RNA to be double-stranded, but if it were double-stranded, then we would expect that A = U and C = G. If so, then (A + G)/(U + C) = 1. Since it doesn't equal unity, we can conclude that the RNA is not double-stranded.

If all of the DNA is transcribed, from either one or both strands, the ratio of (A + U)/(C + G) in RNA should be 1.2, and as predicted, it is. Note that this ratio will not change, whether only one or both of the strands are transcribed. This is the case because for every A:T pair in DNA, for example, transcription of RNA will yield one A *and* one U if both strands are transcribed. If just one strand is transcribed, transcription will yield one A *or* one U. In either case, the (A + U)/(C + G) ratio in RNA will reflect the (A + T)/(C + G) ratio in the DNA from which it was transcribed. To prove this to yourself, draw out a DNA molecule with 12 A:T pairs and 10 C:G pairs and transcribe *both* strands and then transcribe *either* strand. Count the bases in the RNAs produced in both cases and calculate the ratios. Thus, we cannot determine whether just one or both strands are transcribed from the (A + U)/(C + G) ratio.

However, if both strands are transcribed, then the ratio of (A + G)/(U + C) should equal 1.0, and it doesn't. It equals 1.3. To verify this conclusion, examine the theoretical data you drew out on paper as you were asked to do above. One explanation for the observed ratio of 1.3 is that only one of the two strands is transcribed. If this is the case, then the (A + G)/(U + C) will reflect the proportion of A/T pairs that are A and the proportion of the G/C pairs that are G *on the DNA strand that is transcribed*. Another explanation is that transcription occurs only on one strand at any given point (for example, for one gene), but on the other strand at other points (for other genes).

PROBLEMS AND DISCUSSION QUESTIONS

1   Define and differentiate between transcription and translation. Where do these processes fit into the central dogma of molecular genetics?

2   What was the initial evidence for the existence of mRNA?

3   Describe the structure of RNA polymerase. What is the core enzyme? What is the role of the sigma subunit?

4   In a written paragraph, describe the abbreviated chemical reactions shown on pages 314–15 that summarize RNA polymerase-directed transcription.

5   What differences exist between the "visualization of transcription" studies of *E. coli* and *Notophthalmus*? Why?

6   List all of the molecular constituents present in a functional polyribosome.

7   How do the three main steps of translation fit the above model? Diagram each step.

8   Contrast the roles of tRNA and mRNA during translation.

9   List all enzymes that participate in the transcription and translation process.

10   Francis Crick proposed the "adaptor hypothesis" for the function of tRNA. Why did he choose that description?

11   What molecule bears the codon? the anticodon?

12   Contrast the roles of the small and large ribosomal subunits from a point prior to initiation to a point following termination of translation.

13   Compare the components of ribosomes in prokaryotes and eukaryotes.

14   What is meant by "fidelity" of transcription and translation? Why is it important, and how is it maintained?

15   The α chain of eukaryotic hemoglobin is composed of 141 amino acids. What is the minimum number of nucleotides in an mRNA coding for this protein chain? Assuming that each nucleotide is 0.34 nm long in the mRNA, how many triplet codes can at one time occupy space in a ribosome that is 20 nm in diameter?

16   Messenger RNA molecules are very difficult to isolate in prokaryotes because they are rather quickly degraded in the cell. Can you suggest a reason why this occurs? Eukaryotic mRNAs are more stable and exist longer in the cell than do prokaryotic mRNAs. Is this an advantage or disadvantage for a pancreatic cell making large quantities of insulin?

17   Summarize the steps involved in charging tRNAs with their appropriate amino acids.

18   In order to carry out its role, each transfer RNA requires at least four specific recognition sites that must be inherent in its tertiary structure. What are they?

19   In 1962, F. Chapeville and others reported an experiment where they isolated radioactive ^{14}C-cysteinyl-tRNA^{cys} (charged tRNA^{cys} + cysteine). They then removed the sulfur group from the cysteine, creating alanyl-tRNA^{cys} (charged tRNA^{cys} + alanine). When alanyl-tRNA^{cys} was added to a synthetic mRNA calling for cysteine but not alanine, a polypeptide chain was synthesized containing alanine. What can you conclude from this experiment?

20   A short RNA molecule was isolated that demonstrated a hyperchromic shift indicating secondary structure. Its sequence was determined to be AGGCGCCGACUCUACU.

(a)   Predict a two-dimensional model for this molecule.

(b) What DNA sequence would give rise to this RNA molecule through transcription?

(c) If the molecule were a tRNA fragment containing a CGA anticodon, what would the corresponding codon be?

(d) If the molecule were part of a message, what amino acid sequence would result from it following translation? (Refer to the code chart in Figure 11.8.)

21 The following represent deoxyribonucleotide sequences derived from the template strand of DNA:

Sequence 1   CTTTTTTGCCAT
Sequence 2   ACATCAATAACT
Sequence 3   TACAAGGGTTCT

(a) For each strand, determine the mRNA sequence that would be derived from transcription.

(b) Using Figure 11.8, determine the amino acid sequence that would result from translation of these mRNAs.

(c) If we assume that each sequence has been derived from the same gene, which represents the initial part? the middle region? the terminal portion?

22 For Sequence 1 in Problem 21, what is the sequence of the partner strand?

23 For Sequence 3, draw the structural formula of the polypeptide sequence resulting from translation (see Chapter 13).

24 During translation, there is only one case where a charged tRNA enters and occupies the peptidyl (P) site of the ribosome. Which is it?

25 Diagram and label the results found in an electron micrograph when the mature beta-globin mRNA from rabbits is hybridized under appropriate conditions with the template strand of DNA containing the beta-globin gene (see Figure 12.16).

SELECTED READINGS

ABELSON, J. 1979. RNA processing and the intervening sequence problem. *Ann. Rev. Biochem.* 48: 1035–70.

ALBERTS, B., et al. 1989. *Molecular biology of the cell.* 2nd ed. New York: Garland.

BARALLE, F. E. 1983. The functional significance of leader and trailer sequences in eukaryotic mRNAs. *Int. Rev. Cytol.* 81: 71–106.

BIRNSTIEL, M., BUSSLINGER, M., and STRUB, K. 1985. Transcription termination and 3′ processing: The end is in site. *Cell* 41: 349–59.

BRENNER, S. 1989. *Molecular biology: A selection of papers.* Orlando: Academic Press.

BRENNER, S., JACOB, F., and MESELSON, M. 1961. An unstable intermediate carrying information from genes to ribosomes for protein synthesis. *Nature* 190: 575–80.

BRIMACOMBE, R., and STIEGE, W. 1985. Structure and function of ribosomal RNA. *Biochem. J.* 229: 1–17.

CECH, T. R. 1986. The generality of self-splicing RNA: Relationship to nuclear mRNA splicing. *Cell* 44: 207–10.

———. 1986. RNA as an enzyme. *Scient. Amer.* (Nov.) 255, 5: 64–75.

CECH, T. R., ZAUG, A. J., and GRABOWSKI, P. J. 1981. *In vitro* splicing of the ribosomal RNA precursor of *Tetrahymena.* Involvement of a guanosine nucleotide in the excision of the intervening sequence. *Cell* 27: 487–96.

CHAMBON, P. 1975. Eucaryotic nuclear RNA polymerases. *Ann. Rev. Biochem.* 44: 613–38.

———. 1981. Split genes. *Scient. Amer.* (May) 244: 60–71.

CHAPEVILLE, F., et al. 1962. On the role of soluble ribonucleic acid in coding for amino acids. *Proc. Natl. Acad. Sci.* 48: 1086–93.

CIGAN, A. M., FENG, L., DONAHUE, T. F. 1988. tRNA^{met} functions in directing the scanning ribosome to the start site of translation. *Science* 242: 93–98.

CRICK, F. 1979. Split genes and RNA splicing. *Science* 204: 264–71.

DAHLBERG, A. E. 1989. The functional role of ribosomal RNA in protein synthesis. *Cell* 57: 525–29.

DANCHIN, A., and SLONIMSKI, P. 1985. Split genes. *Endeavour* 9: 18–27.

DARNELL, J. E. 1978. Implications of RNA: RNA splicing in the evolution of eukaryotic cells. *Science* 202: 1257–60.

———. 1983. The processing of RNA. *Scient. Amer.* (Oct.) 249: 90–100.

———. 1985. RNA. *Scient. Amer.* (Oct.) 253: 68–87.

DICKERSON, R. E. 1983. The DNA helix and how it is read. *Scient. Amer.* 249(6): 94–111.

DREYFUSS, G., PHILLIPSON, L., and MATTAJ, I. W. 1988. Ribonucleoprotein particles in cellular processes. *J. Cell Biol.* 106: 1419–25.

DUGAICZK, A., et al. 1978. The natural ovalbumin gene contains seven intervening sequences. *Nature* 274: 328–33.

GUTHRIE, C., and PATTERSON, B. 1988. Spliceosomal snRNAs. *Ann. Rev. Genet.* 22: 387–419.

HALL, B. D., and SPIEGELMAN, S. 1961. Sequence complementarity of T2-DNA and T2-specific RNA. *Proc. Natl. Acad. Sci.* 47: 137–46.

HAMKALO, B. 1985. Visualizing transcription in chromosomes. *Trends in Genet.* 1: 255–60.

HELMAN, J. D., and CHAMBERLIN, M. J. 1988. Structure and function of bacterial sigma factors. *Ann. Rev. Biochem.* 57: 839–72.

HOLLEY, R. W., et al. 1965. Structure of a ribonucleic acid. *Science* 147: 1462–65.

HUMPHREY, T., and PROUDFOOT, N. J. 1988. A beginning to the biochemistry of polyadenylation. *TIG* 4: 243–45.

KINNIBURGH, A. J., and ROSS, J. 1979. Processing of the mouse β-globin mRNA precursor: At least two cleavage-ligation reactions are necessary to excise the larger intervening sequence. *Cell* 17: 915–21.

KISH, V., and PEDERSON, T. 1975. Ribonucleoprotein organization of polyadenylate sequences in HeLa cell heterogeneous nuclear RNA. *J. Mol. Biol.* 95: 227–38.

LAKE, J. A. 1981. The ribosome. *Scient. Amer.* (Aug.) 245: 84–97.

LERNER, M. R., and STEITZ, J. A. 1981. Snurps and scyrps. *Cell* 25: 298–300.

MANIATIS, T., and REED, R. 1987. The role of small nuclear ribonucleoprotein particles in pre-mRNA splicing. *Nature* 325: 673–78.

MILLER, O. L., and BEATTY, B. R. 1969. Portrait of a gene. *J. Cell Physiol.* 74 (Supplement 1): 225–32.

MILLER, O. L., HAMKALO, B., and THOMAS, C. 1970. Visualization of bacterial genes in action. *Science* 169: 392–95.

MOORE, P. B. 1988. The ribosome returns. *Nature* 331: 223–27.

NOLLER, H. F. 1984. Structure of ribosomal RNA. *Ann. Rev. Biochem.* 53: 119–62.

———. 1973. Assembly of bacterial ribosomes. *Science* 179: 864–73.

NOMURA, M. 1984. The control of ribosome synthesis. *Scient. Amer.* (Jan.) 250: 102–14.

OHNO, S. 1980. Origin of intervening sequences within mammalian genes and the universal signal for their removal. *Differentiation* 17: 1–15.

O'MALLEY, B., et al. 1979. A comparison of the sequence organization of the chicken ovalbumin and ovomucoid genes. In *Eucaryotic Gene Regulation*, Axel, R., et al., 281–99. Orlando, Fla: Academic Press.

PADGETT, R. A., GRABOWSKI, P. J., KONARSKA, M. M., SEILER, S., and SHARP, P. A. 1986. Splicing of messenger RNA precursors. *Ann. Rev. Biochem.* 55: 1119–50.

PATWARDHAN, S., et al. 1985. Splicing of messenger RNA precursors. *BioEssays* 2: 205–08.

PEDERSON, T. 1981. Messenger RNA biosynthesis and nuclear structure. *Amer. Scient.* 69: 76–84.

REED, R., and MANIATIS, T. 1985. Intron sequences involved in lariat formation during pre-mRNA splicing. *Cell* 41: 95–105.

RICH, A., and HOUKIM, S. 1978. The three-dimensional structure of transfer RNA. *Scient. Amer.* (Jan.) 238: 52–62.

RICH, A., WARNER, J. R., and GOODMAN, H. M. 1963. The structure and function of polyribosomes. *Cold Spr. Harb. Symp.* 28: 269–85.

ROULD, M. A., et al. 1989. Structure of *E. coli* glutaminyl-tRNA synthetase complexed with tRNA^{gln} and ATP at 2.8 Å resolution. *Science* 246: 1135–42.

RUBY, S. W., and ABELSON, J. 1988. An early hierarchic role of U1 small nuclear ribonucleoprotein in spliceosome assembly. *Science* 242: 1028–35.

SCHIMMEL, P. 1987. Aminoacyl tRNA synthetases: General scheme of structure-function relationship in the polypeptides and recognition of transfer RNAs. *Ann. Rev. Biochem.* 56: 125–58.

SCHLEIF, R. 1988. DNA binding by proteins. *Science* 241: 1182–87.

SENTENAC, A. 1985. Eucaryotic RNA polymerases. *CRC Crit. Rev. Biochem.* 18: 31–91.

SHARP, P. 1987. Splicing of messenger RNA precursors. *Science* 235: 766–71.

SHARP, P. A., and EISENBERG, D. 1987. The evolution of catalytic function. *Science* 238: 729–30.

SHATKIN, A. J. 1985. mRNA cap binding proteins: Essential factors for initiating translation. *Cell* 40: 223–24.

SMITH, J. D. 1972. Genetics of tRNA. *Ann. Rev. Genet.* 6: 235–56.

STEITZ, J. A. 1988. Snurps. *Scient. Amer.* (June) 258(6): 56–63.

STRYER, L. 1988. *Biochemistry.* 3rd ed. New York: W. H. Freeman.

SUNG-HOU, K., and CECH, T. R. 1987. Three-dimensional model of the active site of the self-splicing rRNA precursor of Tetrahymena. *Proc. Natl. Acad. Sci.* 84: 8788–92.

VOLKIN, E., and ASTRACHAN, L. 1956. Phosphorus incorporation in *E. coli* ribonucleic acids after infection with bacteriophage T2. *Virology* 2: 149–61.

VOLKIN, E., ASTRACHAN, L., and COUNTRYMAN, J. L. 1958. Metabolism of RNA phosphorus in *E. coli* infected with bacteriophage T7. *Virology* 6: 545–55.

WARNER, J., and RICH, A. 1964. The number of soluble RNA molecules on reticulocyte polyribosomes. *Proc. Natl. Acad. Sci.* 51: 1134–41.

WATSON, J. D. 1963. Involvement of RNA in the synthesis of proteins. *Science* 140: 17–26.

WATSON, J. D., HOPKINS, N. H., ROBERTS, J. W., STEITZ, J. A., and WEINER, A. M. 1987. *Molecular Biology of the Gene.* 4th ed. Menlo Park, CA: Benjamin-Cummings.

WITTMAN, H. G. 1983. Architecture of prokaryotic ribosomes. *Ann. Rev. Biochem.* 52: 35–65.

ZUBAY, G. L., and MARMUR, J. 1973. *Papers in biochemical genetics.* 2nd ed. New York: Holt, Rinehart and Winston.

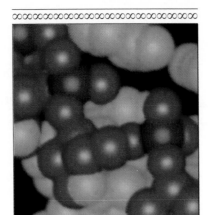

13
Genes and Proteins

∞∞∞∞∞∞∞∞∞∞∞∞∞∞∞∞∞∞∞∞∞∞∞∞∞∞∞∞∞∞

CHAPTER CONCEPTS

The end products of most genes are polypeptide chains. They achieve a three-dimensional conformation based on their primary amino acid sequences, often interacting with other such chains to create functional protein molecules. The function of any protein is closely tied to its structure, which can be disrupted by mutation, leading to a distinctive phenotypic effect.

In the two previous chapters, we have established that there is a rather straightforward genetic code that stores information in the form of triplet nucleotides in DNA, and that this information can be expressed through the orderly processes of transcription and translation. The final product of gene expression, in almost all instances, is a polypeptide chain consisting of a linear series of amino acids, whose sequence has been prescribed by the genetic code. In this chapter, we will review the evidence that confirmed that proteins are the end products of genes. Then we will discuss briefly the various levels of protein structure, diversity, and function. This information provides an important foundation in our understanding of how mutations, which arise in DNA, can result in the variety of phenotypic effects observed in organisms.

The first insight into the role of proteins in genetic processes was provided by observations made by Sir Archibald Garrod and William Bateson early in this century. Garrod established a correlation between certain inherited human disorders and abnormal metabolism of amino acids, the building blocks of proteins. Bateson was able to relate his own studies to Garrod's findings and link enzymes with genetic processes. However, it was not until around 1940 that solid evidence about the genetic role of proteins appeared. In a series of experiments, George Beadle and his coworkers were able to correlate genes and enzymes with the control of phenotypes and thus developed the **one-gene:one-enzyme theory**. Subsequently, Vernon Ingram and Linus Pauling showed that mutations alter the amino acid sequences of human hemoglobin molecules. Their work was particularly significant because it showed that nonenzymatic proteins are also produced as a result of transcription and translation, thus extending the one-gene:one-enzyme theory. These findings are the

foundation of what has more traditionally been called the field of **biochemical genetics**.

GARROD AND BATESON: INBORN ERRORS OF METABOLISM AND ENZYMES

Archibald Garrod was born into an English family of medical scientists. His father was a physician with a strong interest in the chemical basis of rheumatoid arthritis, and his eldest brother was a leading zoologist in London. Thus, it is not surprising that, as a practicing physician, Garrod became interested in several human disorders that seemed to be inherited. Although he also studied albinism and cystinuria, we will describe his investigation of the disorder **alkaptonuria**. Individuals afflicted with this disorder cannot metabolize the alkapton 2,5-dihydroxyphenylacetic acid, also known as **homogentisic acid**. As a result, an important metabolic pathway (Figure 13.1) is blocked. Homogentisic acid accumulates in cells and tissues and is excreted in the urine. The molecule's oxidation products are black and thus are easily detectable in the diapers of newborns and the urine of individuals. The products tend to accumulate in cartilaginous areas, causing a darkening of the ears and nose. In joints, this deposition leads to a benign arthritic condition. This rare disease is not a serious one, but it persists throughout an individual's life.

Garrod studied alkaptonuria by increasing dietary protein or adding the amino acids phenylalanine or tyrosine, which are chemically related to homogentisic acid, to the diet. Under such conditions, homogentisic acid levels increase in the urine of alkaptonurics but not in unaffected individuals. He concluded that normal individuals break down, or

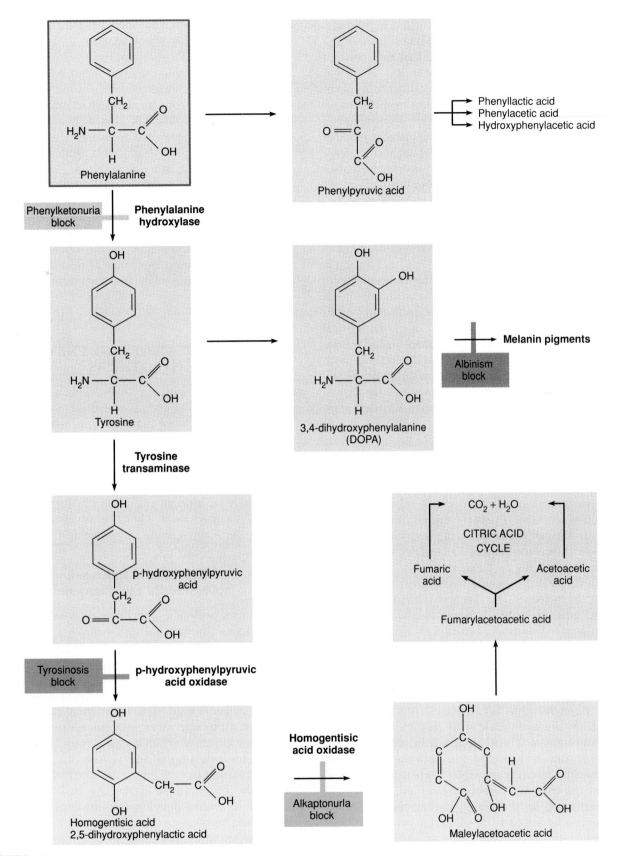

FIGURE 13.1

Metabolic pathway involving phenylalanine and tyrosine. Various metabolic blocks resulting from mutations lead to the disorders phenylketonuria, tyrosinosis, alkaptonuria, and albinism.

catabolize, this alkapton but that afflicted individuals cannot. By studying the patterns of inheritance of the disorder, Garrod further concluded that alkaptonuria was inherited as a simple recessive trait.

On the basis of these conclusions, Garrod hypothesized that the hereditary information controls chemical reactions in the body and that the inherited disorders he studied are the result of alternate modes of metabolism. While the terms *genes* and *enzymes* were not familiar during Garrod's work, the corresponding concepts of unit factors and ferments were. Ferments, now known to be enzymes, were recognized by 1900, but their chemical nature was not determined until 1926. Garrod published his initial observations in 1902 and his classic work *Inborn Errors of Metabolism* in 1909.

Only a few geneticists, including Bateson, were familiar with or referred to Garrod's work. His ideas fit nicely with Bateson's belief that inherited conditions were caused by the lack of some critical substance. In 1909, Bateson published *Mendel's Principles of Heredity*, in which he linked ferments with heredity. However, most geneticists failed to see the relationship between genes and enzymes for almost thirty years. Garrod and Bateson, like Mendel, were ahead of their time.

Phenylketonuria

Described first in 1934, **phenylketonuria (PKU)** may result in mental retardation and is inherited as an autosomal recessive disease. Afflicted individuals are unable to convert the amino acid phenylalanine to the amino acid tyrosine (Figure 13.1). If we examine the chemical structures of these molecules, we see that they differ by only a single hydroxyl group (—OH), present in tyrosine. The reaction is catalyzed by the enzyme **phenylalanine hydroxylase,** which is inactive in affected individuals and active at about a 30 percent level in heterozygotes. The enzyme functions in the liver. While the normal blood level of phenylalanine is about 1 mg/100 ml, phenylketonurics show a level as high as 50 mg/100 ml.

As phenylalanine accumulates, it may be converted in phenylpyruvic acid and, subsequently, other derivatives. These are less efficiently resorbed by the kidney and tend to spill into the urine more quickly than phenylalanine. Both phenylalanine and derivatives enter the cerebrospinal fluid, resulting in elevated levels in the brain. The presence of these substances during early development is thought to be responsible for retardation.

Retardation can be prevented by PKU screening of newborns. When the condition is detected in the analysis of an infant's blood, a strict dietary regimen is instituted. A low-phenylalanine diet can reduce such byproducts as phenylpyruvic acid, and abnormalities characterizing the disease can be diminished. Screening of newborns occurs routinely in almost every state in this country. Phenylketonuria occurs in approximately 1 in 11,000 births.

Knowledge of inherited metabolic disorders such as alkaptonuria, phenylketonuria, and albinism has caused a revolution in medical thinking and practice. No longer is human disease attributed solely to the action of invading microorganisms, viruses, or parasites. We know now that literally thousands of abnormal physiological conditions are caused by errors in metabolism that are the result of mutant genes. These human biochemical disorders are far-ranging and include all classes of organic biomolecules.

THE ONE-GENE : ONE-ENZYME HYPOTHESIS

In two separate investigations beginning in 1933, George Beadle was to provide the first convincing experimental evidence that genes are responsible for the synthesis of enzymes. The first investigation, conducted in collaboration with Boris Ephrussi, involved *Drosophila* eye pigments. Encouraged by these findings, Beadle then joined with Edward Tatum to investigate nutritional mutations in the pink bread mold *Neurospora crassa.* The latter investigation led to the one-gene : one-enzyme hypothesis.

Beadle and Ephrussi: *Drosophila* Eye Pigments

The studies of Beadle and Ephrussi involved **imaginal disks** in *Drosophila.* These embryonic cells found in the larvae of insects, upon metamorphosis, differentiate into a variety of adult structures. Among others, adult eyes, legs, antennae, and genital structures are derived from these groups of cells. Beadle and Ephrussi found that if an imaginal disk was removed from one larva and transplanted into the abdomen of another, that disk would differentiate during metamorphosis and its corresponding adult structure could be recovered from the abdomen of the adult fly. For example, if an eye disk were transplanted, it would develop into an eyelike structure that could be recovered and analyzed.

These investigators wondered whether a mutant eye disk, transplanted into a wild-type larva, would be altered by its new environment. Would it appear normally pigmented, or would it develop its characteristic mutant color? The first mutant used was *vermilion*, a sex-linked, bright red eye color mutation. Mutant development was reversed! The *vermilion* eye disk was altered to produce a normally pigmented wild-type eye. The normal wild-type color is brick red, produced as a combination of brown and bright red pigments. The *vermilion* mutant makes the bright red pigment, but lacks the brown pigment, indicating that its synthesis is inhibited. However, in the wild-type abdomen, synthesis of brown pigment also occurred normally in the mutant disk, producing wild-type color.

Beadle and Ephrussi then proceeded to perform similar experimentation with 25 other mutants. Only one, *cinnabar*, another bright red eye mutation missing the brown pigment, behaved like *vermilion* when transplanted; it developed a normal eye color. The remaining 24 behaved "autonomously" and exhibited their respective mutant color when differentiated.

They concluded that the flies with either *vermilion* or *cinnabar* phenotypes are mutant due to the absence of some *diffusible substance* in their disks during development. Because this substance was diffusible, it was present in the wild-type abdomen and thus accessible to the transplanted *v* or *cn* disks. Thus, it enabled full pigment production in the transplant. However, in the case of the 24 mutations that developed autonomously, the substance that might have permitted normal pigment production was not diffusible but confined to the host cells forming the eye. Thus, no "correction" was observed.

Beadle and Ephrussi then asked whether the same substance was lacking in both the *vermilion* and *cinnabar* mutants. To answer this question, they transplanted *vermilion* eye disks into *cinnabar* larvae, and vice versa. What they found was intriguing. The *vermilion* disks were "cured" or converted to wild-type color in *cn* abdomens, but *cinnabar* disks developed autonomously in *vermilion* hosts and retained their mutant phenotype. These experiments are diagramed in Figure 13.2 on page 348.

To explain such results, they proposed that two sequential biochemical reactions are involved in normal synthesis of the brown pigment. One is inactive in flies with the *vermilion* mutation, and the other is inactive in flies with the *cinnabar* mutation. Each reaction produces a product that is diffusible in the tissues of wild type. As shown in Figure 13.2, the substance produced under the direction of the wild-type allele of the *vermilion* locus (substance Y) occurs first in the pathway. It is then converted to a second substance in the pathway. This conversion is con-

trolled by the wild-type allele of the *cinnabar* locus (substance Z).

The order of this pathway explains the results of the reciprocal transplantations. Consider first the *vermilion* (*v*) disk in the *cinnabar* (*cn*) host. The *vermilion* disk lacks the v^+ enzyme but has the cn^+ enzyme. As the host, the *cinnabar* tissue can make substance Y because it has the wild-type allele of the *vermilion* gene (v^+). Substance Y is diffusible and enters the *vermilion* disk where the cn^+ enzyme converts it to substance Z. Substance Z is then converted to the brown pigment, restoring the wild-type eye color.

Now consider the *cinnabar* disk in the *vermilion* host. The host tissue, lacking the v^+ gene, cannot make the v^+ enzyme, and so cannot make substance Y. Even though the cn^+ enzyme is present, because no Y is available to be converted to Z, no Z is produced either. Thus, the transplanted *cn* disk is still blocked. It can make substance Y but cannot complete the transition through Z to the brown pigment. Thus, it remains bright red!

Within a few years, other workers had investigated the biochemical pathway leading to the brown xanthommatin pigment in *Drosophila*. As shown in Figure 13.3, the brown pigment is a derivative of the amino acid tryptophan. The reactions controlled by the specific enzymes altered by the *vermilion* and *cinnabar* mutations have been identified. Other autosomal recessive mutations, including *scarlet* and *cardinal*, have also been studied and shown to affect enzymes in this pathway. Most relevant to our discussion was the confirmation in the 1940s that mutant genes controlling distinctive morphological phenotypes were linked directly to biochemical "errors," most likely due to a lack of enzyme function.

Beadle and Tatum: *Neurospora* Mutants

In the early 1940s, Beadle joined Tatum to seek more substantial evidence about the nature of gene products. They chose to work with the organism *Neurospora crassa* because much was known about its biochemistry, and mutations could be induced and isolated with relative ease. By inducing mutations, they produced genetic blocks in reactions essential to the growth of the organism. The life cycle of this organism was presented in Figure 5.3.

Beadle and Tatum knew that *Neurospora* could manufacture nearly everything necessary for normal development. For example, using rudimentary carbon and nitrogen sources, this organism can synthesize 9 water-soluble vitamins, 20 amino acids, numerous carotenoid pigments, and purines and

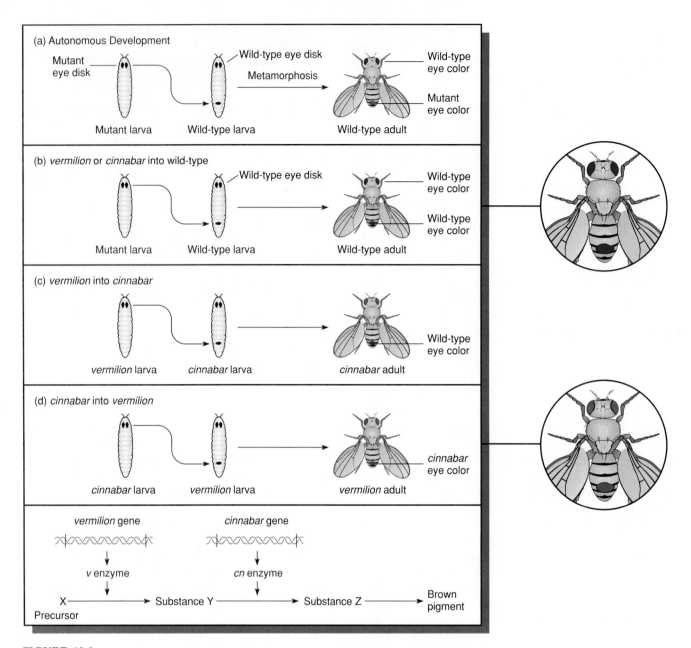

FIGURE 13.2
Beadle and Ephrussi's transplantation experiments involving imaginal disks that develop into eyes. A mutant disk, transplanted into a wild-type larval abdomen, either develops autonomously, resulting in a mutant eye color, or the mutation is "cured" and normal eye color results, as illustrated in (a) and (b), respectively. When a *vermilion* disk is transplanted into a *cinnabar* host (c), normal wild-type eye color develops in the transplanted disk. The reciprocal experiment (d) results in autonomous development of the mutant eye color in the transplanted disk. On the basis of these results, they concluded that the biosynthetic step controlled by the v^+ gene product occurs prior to that controlled by the cn^+ gene product within the same pathway.

pyrimidines. Beadle and Tatum irradiated conidia (asexually formed spores) with X-rays to increase the frequency of mutations and used the treated spores to produce mycelia and sexual structures used in fertilization. After the completion of a sexual cycle,

they picked out individual ascospores and grew them on "complete" medium containing vitamins, amino acids, etc. Under such growth conditions, a mutant strain would be able to grow by virtue of supplements present in the enriched complete medium. All

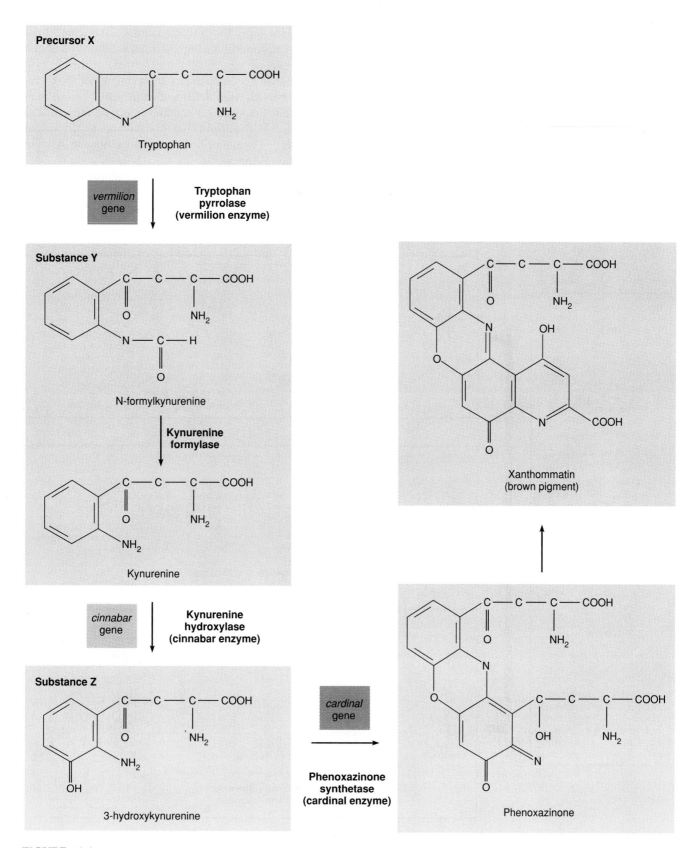

FIGURE 13.3

The biosynthetic pathway leading to the conversion of tryptophan to the brown eye pigment xanthommatin. Three different mutant blocks are shown, each interrupting synthesis at a different place in the pathway. In the absence of xanthommatin, all three mutations, when homozygous, result in the identical phenotype, bright red eyes.

single-spore cultures were then transferred to minimal medium. If growth occurred on minimal medium, then the single-spore culture did not contain a mutation. If no growth occurred, then it contained a mutation, and the only task remaining was to determine its type.

Literally thousands of individual spores derived by this procedure were isolated and grown on complete medium. In subsequent tests on minimal medium, many cultures failed to grow, indicating that a nutritional mutation had been induced. To identify the mutant type, the mutant strains were tested on a series of tubes containing minimal medium plus one of a series of vitamins, amino acids, purines, and pyrimidines until some single supplement that permitted growth was found. Beadle and Tatum reasoned that the supplement that restored growth was the molecule that the mutant strain could no longer synthesize. Details of how these mutants are isolated are presented in Figure 13.4.

The first mutant strain isolated required vitamin B-6 (pyridoxin), and the second one required vitamin

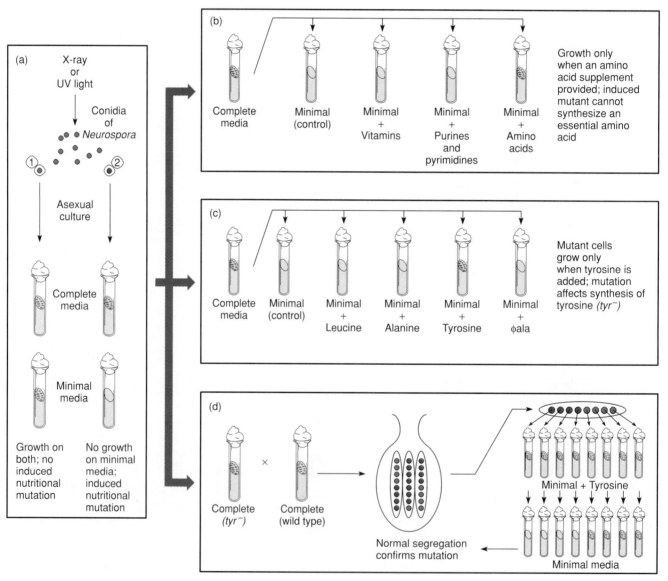

FIGURE 13.4

Induction, isolation, and characterization of a nutritional auxotrophic mutation in *Neurospora*. In (a), conidia 1 is not affected, but conidia 2 contains such a mutation. In (b) and (c), the precise nature of the mutation is determined to involve the biosynthesis of tyrosine. In (d), normal segregation in a cross between mutant and wild-type cells confirms that the *tyr⁻* mutation is nuclear in origin.

B-1 (thiamine). Using the same procedure, Beadle and Tatum eventually isolated and studied hundreds of mutants.

The next step was to demonstrate that each mutation was of nuclear origin. This was easily ascertained by crossing a given mutant with a wild type and observing Mendelian segregation in the ascospores following meiosis (Figure 13.4). A positive test for demonstrating the nuclear origin occurs when one-half of the meiotic products gives rise to wild-type cultures, which can grow on minimal medium, and one-half gives rise to cultures that cannot grow on minimal medium.

The findings derived from testing over 80,000 spores convinced Beadle and Tatum that genetics and biochemistry had much in common. It seemed likely that each nutritional mutation caused the loss of the enzymatic activity that facilitates an essential reaction in wild-type organisms. It also appeared that a mutation could be found for nearly any enzymatically controlled reaction. Beadle and Tatum had thus provided sound experimental evidence that **one gene specifies one enzyme,** a hypothesis alluded to over thirty years earlier by Garrod and Bateson. With modifications, this concept was to become a major principle of genetics.

Genes and Enzymes: Analysis of Metabolic Pathways

The one-gene: one-enzyme concept and its attendant methods have been used over the years to work out many details of metabolism in *Neurospora, Escherichia coli,* and a number of other microorganisms. One of the first metabolic pathways to be investigated in detail was that leading to the synthesis of the amino acid arginine in *Neurospora.* By studying seven mutant strains, each requiring arginine for growth, Adrian Srb and Norman Horowitz were able to ascertain a partial biochemical pathway leading to the synthesis of this molecule. The rationale followed in their work illustrates how genetic analysis can be used to establish biochemical information.

Srb and Horowitz tested each strain's ability to reestablish growth if either **citrulline** or **ornithine,** two compounds with close chemical similarity to arginine, was used as a supplement to minimal medium. If either was able to substitute for arginine, they reasoned that it must be involved in the biosynthetic pathway of arginine. In fact, both molecules could be substituted in one or more strains.

Of the seven mutant strains, four of them (*arg 1–4*) grew if supplied with either citrulline, ornithine, or arginine. Two of them (*arg 5* and *6*) grew if supplied with citrulline or arginine. One strain (*arg 7*) would grow only if arginine was supplied. Neither citrulline nor ornithine could substitute for it. From these experimental observations, the following pathway and metabolic blocks for each mutation were deduced:

$$\underset{\text{Enzyme A}}{\text{Precursor} \xrightarrow{\quad\overset{arg\ 1-4}{|}\quad} \text{Ornithine}} \underset{\text{Enzyme B}}{\xrightarrow{\quad\overset{arg\ 5-6}{|}\quad} \text{Citrulline}} \underset{\text{Enzyme C}}{\xrightarrow{\quad\overset{arg\ 7}{|}\quad} \text{Arginine}}$$

The reasoning supporting these conclusions was based on the following logic. If mutants *arg 1* through *4* can grow regardless of which of the three molecules is supplied as a supplement to minimal medium, the mutations preventing growth must cause a metabolic block that occurs prior to the involvement of ornithine, citrulline, or arginine in the pathway. When any of these three molecules is added, its presence bypasses the block. As a result, it can be concluded that both citrulline and ornithine are involved in the biosynthesis of arginine. However, the sequence of their participation in the pathway cannot be determined on the basis of these data.

On the other hand, the *arg 5* and *6* mutations grow if supplied citrulline but not ornithine. Therefore, ornithine must occur in the pathway prior to the block. Its presence will not overcome the block. Citrulline, however, does overcome the block, so it must be involved beyond the point of blockage. Therefore, the conversion of ornithine to citrulline represents the correct sequence in the pathway.

Finally, it can be concluded that *arg 7* represents a mutation preventing the conversion of citrulline to arginine. Neither ornithine nor citrulline can overcome the metabolic block because both participate earlier in the pathway.

Taken together, these reasons support the sequence of biosynthesis outlined here. Since Srb and Horowitz's work in 1944, the detailed pathway has been worked out and the enzymes controlling each step characterized. The chemical structures and specific enzymes are shown in Figure 13.5.

The concept of one-gene:one-enzyme developed in the early 1940s was not immediately accepted by all geneticists. This is not surprising, because it was not yet clear how mutant enzymes could cause variation in many phenotypic traits. For example, *Drosophila* mutants demonstrated altered eye size, wing shape, wing vein pattern, and so on. Plants exhibited mutant varieties of seed texture, height, and fruit size. How an inactive, mutant enzyme could result in such phenotypes was puzzling to many geneticists. Another reason for their reluctance to accept this concept was the paucity of information then available in molecular genetics. It was not until 1944 that

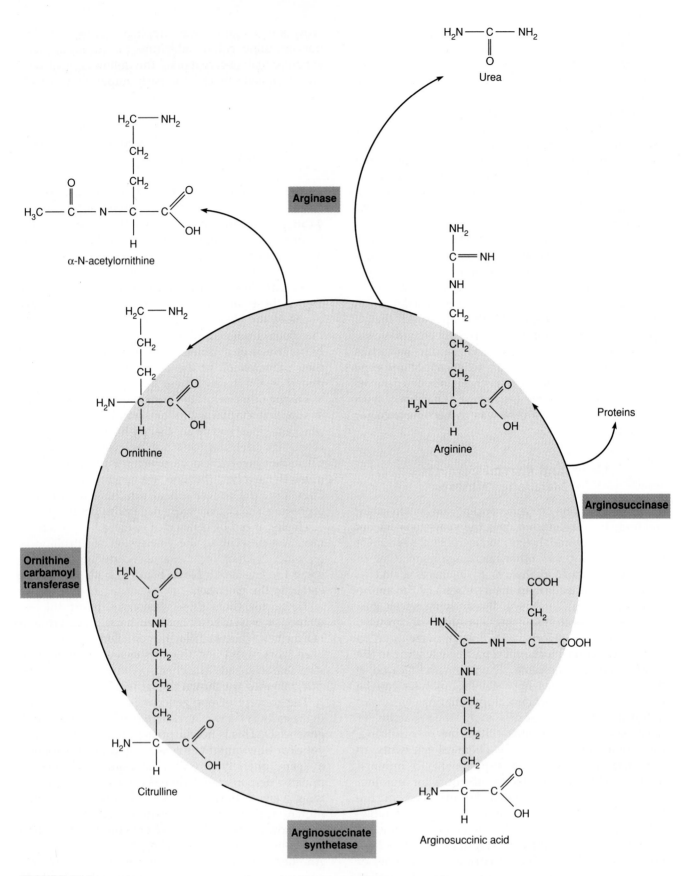

FIGURE 13.5
Arginine biosynthesis in *Neurospora*.

Avery, MacLeod, and McCarty showed DNA to be the transforming factor and not until the early 1950s that most geneticists believed that DNA serves as the genetic material. However, by this time, the evidence in support of the concept of enzymes as gene products was overwhelming, and the concept was accepted as valid. Nevertheless, the question of how DNA specifies the structure of enzymes remained unanswered.

ONE-GENE:ONE-PROTEIN/ONE-GENE: ONE-POLYPEPTIDE

Two factors soon modified the one-gene:one-enzyme hypothesis. First, while nearly all enzymes are proteins, not all proteins are enzymes. As the study of biochemical genetics proceeded, it became clear that all proteins are specified by the information stored in genes, leading to the more accurate phraseology **one-gene:one-protein**. Second, proteins were often shown to have a subunit structure consisting of two or more **polypeptide chains**. This is the basis of the **quaternary structure** of proteins, which we will discuss later in this chapter. Because each distinct polypeptide chain is encoded by a separate gene, a more modern statement of Beadle and Tatum's basic principle is **one-gene:one-polypeptide chain**. These modifications of the original hypothesis became apparent during the analysis of hemoglobin structure in individuals afflicted with sickle-cell anemia.

Sickle-Cell Anemia

The first direct evidence that genes specify proteins other than enzymes came from the work on mutant hemoglobin molecules derived from humans afflicted with the disorder **sickle-cell anemia**. Affected individuals contain erythrocytes which, under low oxygen tension, become elongated and curved because of the polymerization of hemoglobin. The "sickle" shape of these erythrocytes is in contrast to the biconcave disc shape characteristic of normal individuals (Figure 13.6). Individuals with the disease suffer attacks when red blood cells aggregate in the venous side of capillary systems, where oxygen tension is very low. As a result, a variety of tissues may be deprived of oxygen and suffer severe damage. When this occurs, an individual is said to experience a sickle-cell crisis. If untreated, a crisis is often fatal. The kidneys, muscles, joints, brain, gastrointestinal tract, and lungs may be affected.

In addition to suffering crises, these individuals are anemic because their erythrocytes are destroyed more rapidly than normal red blood cells. Compensatory physiological mechanisms include increased red cell production by bone marrow and accentuated heart action. These mechanisms lead to abnormal bone size and shape as well as dilation of the heart.

In 1949, James Neel and E. A. Beet demonstrated that the disease is inherited as a Mendelian trait. Pedigree analysis revealed three genotypes and phenotypes controlled by a single pair of alleles, A and S. Normal and affected individuals result from the homozygous genotypes AA and SS, respectively. The red blood cells of the AS heterozygote, which exhibits the **sickle-cell trait** but not the disease, undergo much less sickling since over half of their hemoglobin is normal. Although largely unaffected, such persons are carriers of the disorder.

In the same year, Linus Pauling and his coworkers provided the first insight into the molecular basis of the disease. They showed that hemoglobins isolated from diseased and normal individuals differed in

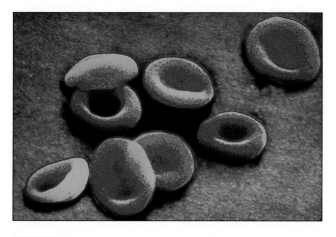 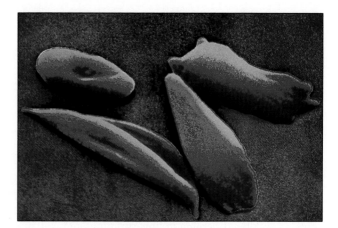

FIGURE 13.6
A comparison of erythrocytes derived from HbA (left) and HbS (right).

their rates of electrophoretic migration. In this technique (see Appendix A), charged molecules migrate in an electric field. If the net charge of two molecules is different, the rate of migration will vary. On this basis, Pauling and his colleagues concluded that a chemical difference existed between the two types of hemoglobin. The two molecules are now designated **HbA** and **HbS**.

Figure 13.7 illustrates the migration pattern of hemoglobin derived from individuals of all three possible genotypes when subjected to **starch gel electrophoresis**. The gel provides the supporting medium for the molecules during migration. In this experiment, samples are placed at a point of origin between the cathode (−) and the anode (+), and an electric field is applied. The migration pattern reveals that all molecules move toward the anode, indicating a net negative charge. However, HbA migrates farther than HbS, suggesting that its net negative charge is greater. The electrophoretic pattern of

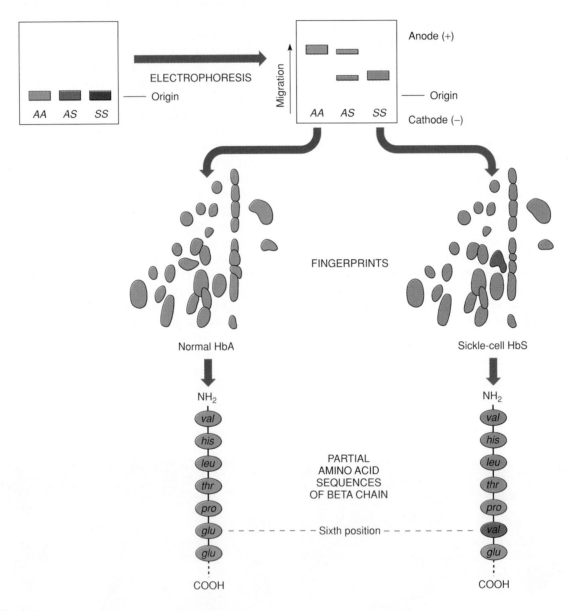

FIGURE 13.7
Investigation of hemoglobin derived from *AA* and *SS* individuals using electrophoresis, fingerprinting, and amino acids analysis. Hemoglobin from individuals with sickle-cell anemia (*SS*) migrates differently in an electrophoretic field, contains an altered peptide in fingerprinting, and contains an altered amino acid, valine, at the sixth position in the β chain. During electrophoresis, heterozygotes (*AS*) reveal both forms of hemoglobin.

hemoglobin derived from carriers reveals the presence of both HbA and HbS, confirming their heterozygous genotype.

Pauling's findings suggested two possibilities. It was known that hemoglobin consists of four nonproteinaceous, iron-containing **heme groups** and a **globin portion** that contains four polypeptide chains. The alteration in net charge in HbS could be due, theoretically, to a chemical change in either component.

Work carried out between 1954 and 1957 by Vernon Ingram resolved this question. He demonstrated that the chemical change occurs in the primary structure of the globin portion of the hemoglobin molecule. Using the **fingerprinting technique** (Figure 13.7), he showed that HbS differs in amino acid composition compared to HbA. Human adult hemoglobin contains two identical alpha (α) chains of 141 amino acids and two identical beta (β) chains of 146 amino acids in its quaternary structure.

The fingerprinting technique involves enzymatic digestion of the protein into peptide fragments. The mixture is then placed on absorbent paper and exposed to an electric field, where migration occurs according to net charge. The paper is then turned at a right angle and placed in a solvent, where chromatographic action causes migration of the peptides in the second direction. The end result is a two-dimensional separation of the peptide fragments into a distinctive pattern of spots or "fingerprints." Ingram's work revealed that HbS and HbA differed by only a single peptide fragment (Figure 13.7). Further analysis then revealed a single amino acid change: Valine was substituted for glutamic acid at the sixth position of the β chain, accounting for the peptide difference.

The significance of this discovery has been multifaceted. It clearly establishes that a single gene provides the genetic information for a single polypeptide chain. Studies of HbS also demonstrate that a mutation can affect the phenotype by directing a single amino acid substitution. Also, by providing the explanation for sickle-cell anemia, the concept of **molecular disease** was firmly established. Finally, this work led to a thorough study of human hemoglobins, which has provided valuable genetic insights.

In the United States, sickle-cell anemia is found almost exclusively in the black population. Extensive research is now underway to determine modes of treatment for those with the disease. It affects about one in every 625 black infants born in this country. Currently, about 50,000 to 75,000 individuals are afflicted. In about one of every 145 black married couples, both partners are heterozygous carriers, with the result that each of their children has a 25 percent chance of having the disease. Other estimates suggest that the frequency of such couples is even higher, perhaps one in 100.

Human Hemoglobins

Molecular analysis has revealed that a variety of hemoglobin molecules are produced in humans. All are tetramers consisting of numerous combinations of seven distinct polypeptide chains, each encoded by a separate gene. In our discussion of sickle-cell anemia, we learned that **HbA** contains two **alpha (α)** and two **beta (β)** chains. HbA represents about 98 percent of all hemoglobin found in an individual's erythrocytes after the age of six months. The remaining 2 percent consists of **HbA$_2$**, a minor adult component. This molecule contains two alpha and two **delta (δ)** chains. The latter chain is very similar to the beta chain, consisting of 146 amino acids.

During embryonic and fetal development, quite a different set of hemoglobins is found. The earliest set to develop is called **Gower 1** and contains two **zeta (ζ)** chains, which are alphalike, and two **epsilon (ϵ)** chains, which are betalike. By eight weeks of gesta-

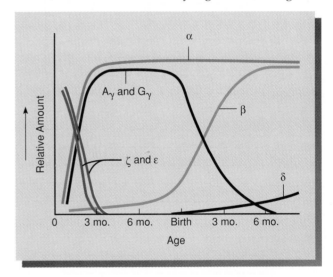

Hemoglobin Type	Chain Composition
Embryonic	$\zeta_2 \epsilon_2$
Fetal	$\alpha_2{}^A \gamma_2$
	$\alpha_2{}^G \gamma_2$
Adult	$\alpha_2 \beta_2$
Minor adult	$\alpha_2 \delta_2$

FIGURE 13.8
Sequence of appearance and chain compositions of human hemoglobins from conception until after birth.

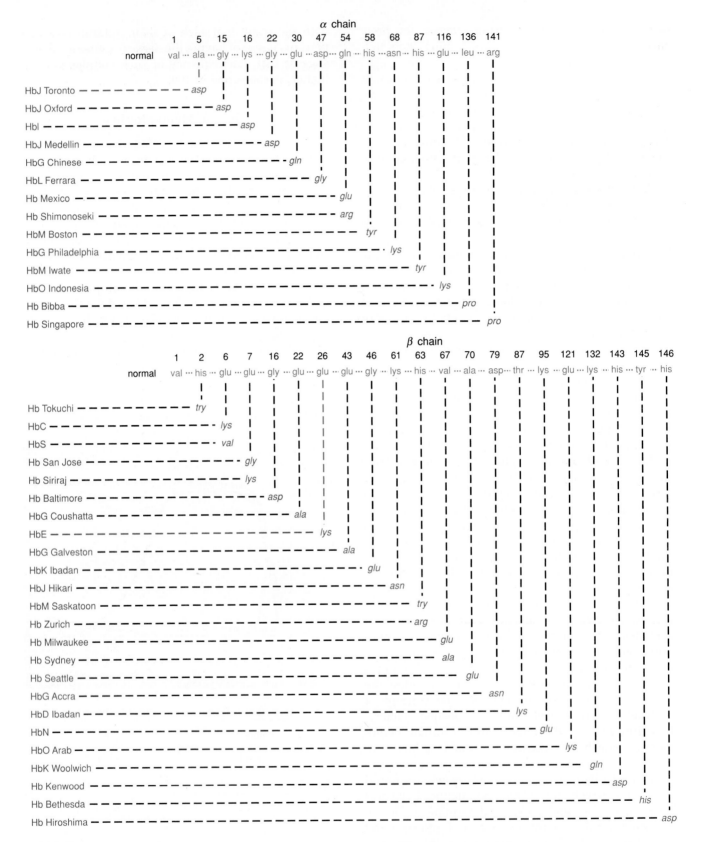

FIGURE 13.9
Hemoglobin mutations in humans. The normal amino acid at each affected position is shown across the top for both the α and β chains. The number indicates the position of the amino acid in the chain from the N-terminus. The amino acid substitution in each mutation is determined by following down each vertical line.

tion, this embryonic form is gradually replaced by still another hemoglobin molecule with different chains. This molecule is called **HbF,** or fetal hemoglobin, and consists of two alpha chains and two **gamma (γ)** chains. These gamma chains are of nearly identical types and are designated $^G\gamma$ and $^A\gamma$. Both are betalike and differ from each other by only a single amino acid.

The nomenclature and sequence of appearance of the five tetramers described so far are summarized in Figure 13.8. The genes coding for each of the seven chains have been mapped. Those coding for α and ζ are located on chromosome 16, while those coding for ε, $^G\gamma$ $^A\gamma$, δ, and β are located on chromosome 11. In each case, they are clustered together, constituting **gene families**, which we will discuss in detail in Chapter 18.

Many mutant varieties of human hemoglobin have been discovered. Most demonstrate a single amino acid change in either the α or β chains. One of the first mutant hemoglobins found was HbC. As in HbS, an amino acid change at the sixth position of the β chain has occurred. However, instead of a valine substitution for glutamic acid, lysine is found in HbC. Figure 13.9 illustrates some of more than 100 rare variants now known. Most have been discovered in the heterozygous condition and thus produce no noticeable effect or only a minor anemia. Each of the changes reflects a nucleotide alteration in the respective gene.

Hemoglobin is one of the most carefully and extensively studied proteins, both in humans and in other organisms. For example, human hemoglobin genes have been cloned through recombinant DNA technology. Their analysis has added to our knowledge of gene structure. The temporal sequence of gene activation and repression has been of great interest in studies of genetic regulation of development in eukaryotes. The homologies of amino acid sequences between Hb polypeptide chains of different organisms have served as one approach in molecular evolutionary investigations. Related findings from these studies will be described in subsequent chapters.

COLINEARITY

Once it was established that genes specify the synthesis of polypeptide chains, the next logical question was how genetic information contained in the nucleotide sequence of a gene can be transferred to the amino acid sequence of a polypeptide chain. It seemed most likely that a **colinear relationship** would exist between the two molecules. That is, the order of nucleotides in the DNA of a gene would correlate directly with the order of amino acids in the corresponding polypeptide.

While we have introduced this concept earlier in Chapter 12, it is useful to examine experimental evidence in its support. In studies of the A subunit of the enzyme **tryptophan synthetase** in *E. coli*, Charles Yanofsky sought to demonstrate colinearity. He isolated many independent mutants that had lost activity of the enzyme. He was able to map these mutations and establish their location with respect to one another within the gene. Then, he determined where the amino acid substitution had occurred in each mutant protein. When the two sets of data were compared, the colinear relationship was apparent. The location of each mutation in the *trp* A gene correlated with the position of the altered amino acid

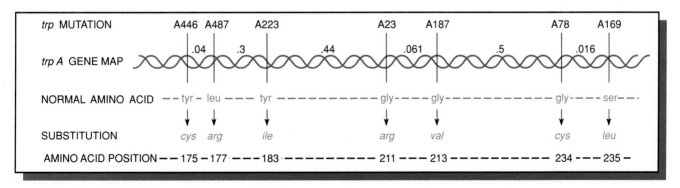

FIGURE 13.10
Demonstration of colinearity between the genetic map of various *trp A* mutations in *E. coli* and the affected amino acids in the protein product. The values shown between mutations represent linkage distances.

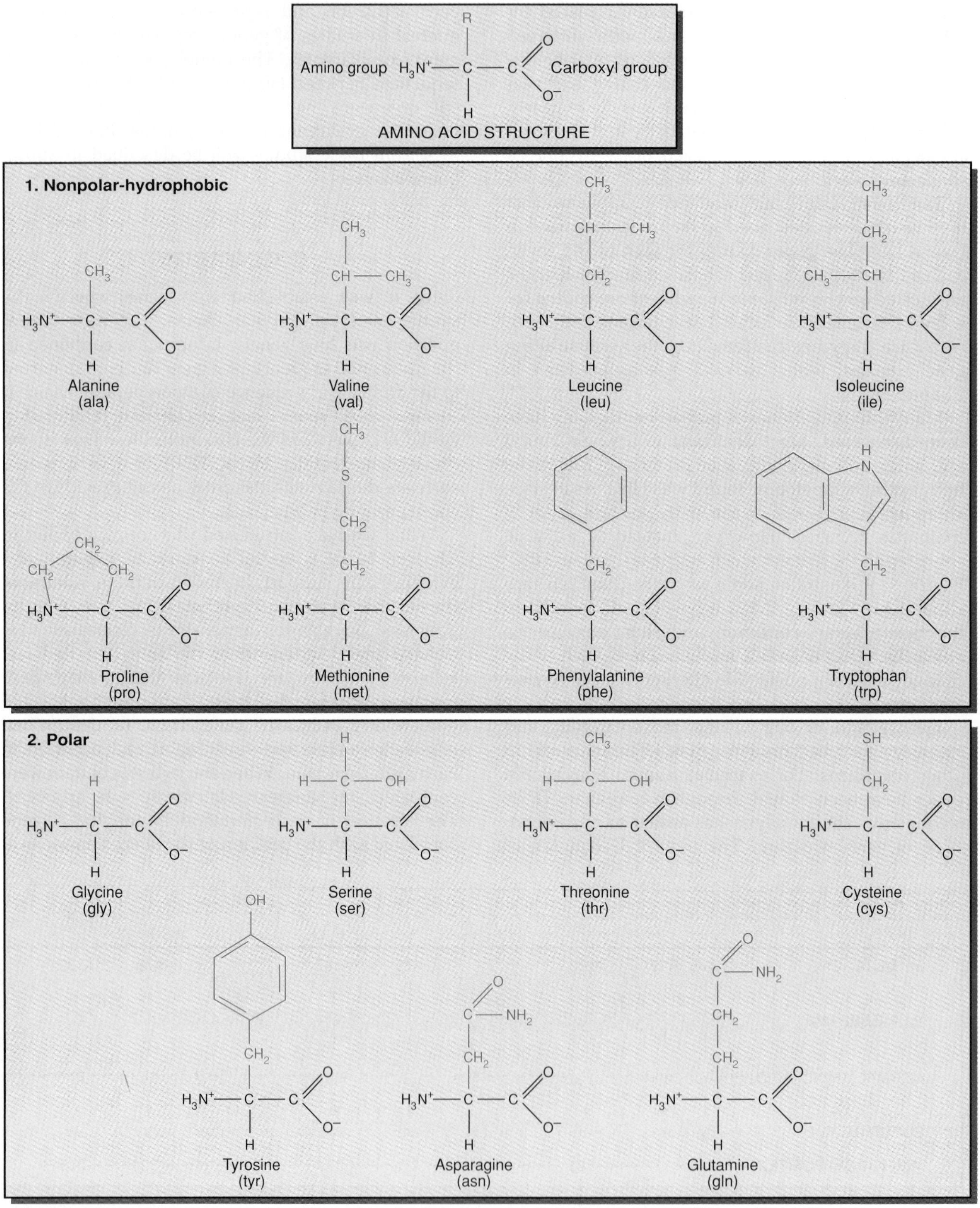

FIGURE 13.11
Structural formulae of amino acids, divided into four categories based on their chemical properties.

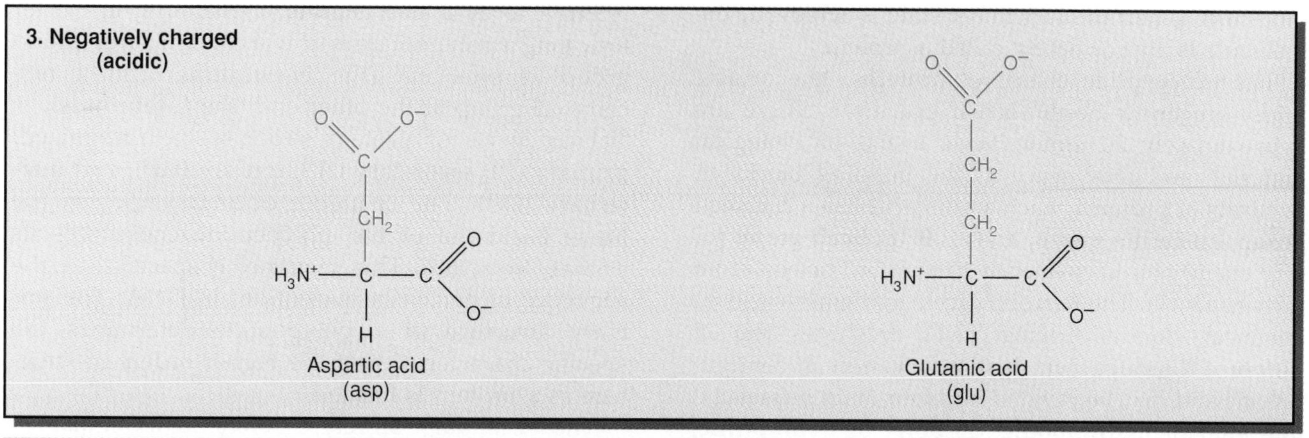

3. Negatively charged (acidic)

Aspartic acid (asp)

Glutamic acid (glu)

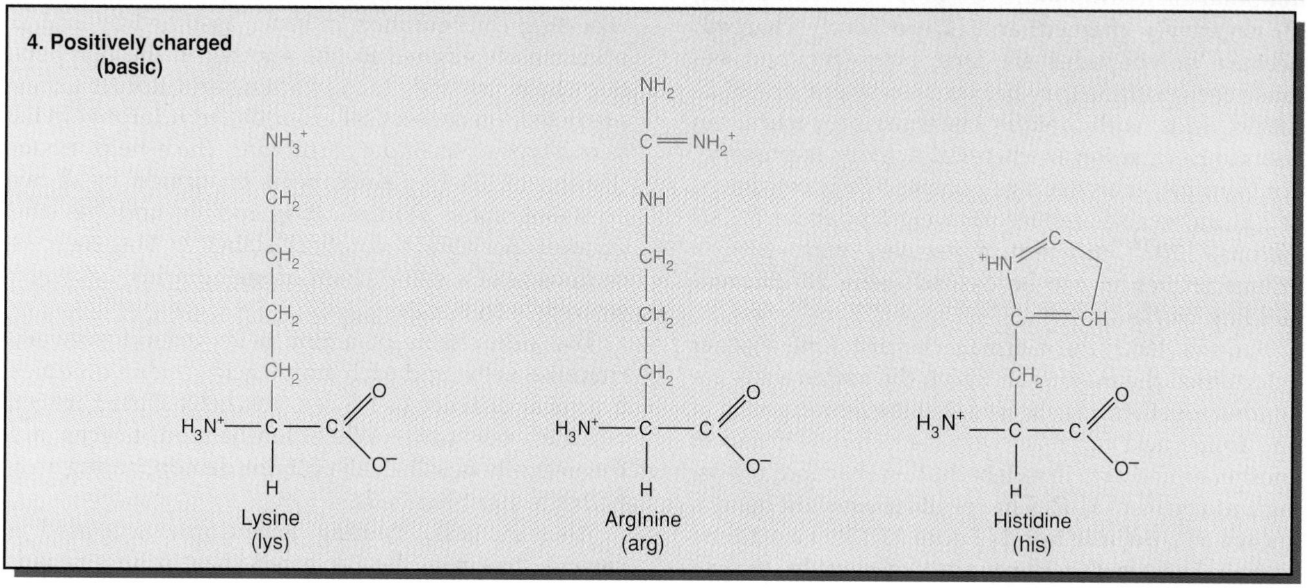

4. Positively charged (basic)

Lysine (lys)

Arginine (arg)

Histidine (his)

FIGURE 13.11
(continued)

in the A polypeptide of tryptophan synthetase. This comparison is illustrated in Figure 13.10. Recall that we have already discussed the details of how information is transferred from DNA to protein (transcription and translation) in Chapter 12.

PROTEIN STRUCTURE AND FUNCTION

Having established that the genetic information is stored in DNA, yet influences cellular activities through the proteins it encodes, we turn now to a discussion of protein structure and function. How is it that these molecules play such a critical role in determining the complexity of cellular activities? As we will see, the structure of proteins is intimately related to the functional diversity of these molecules.

Protein Structure

First, we should differentiate between the terms **polypeptides** and **proteins**. Both describe a molecule composed of amino acids. The molecules differ, however, in their state of assembly and functional capacity.

Polypeptides are the precursors of proteins. As assembled on the ribosome during translation, the molecule is called a polypeptide. When released from the ribosome following translation, a polypeptide folds up and assumes a higher order of structure. When this occurs, a three-dimensional conformation in space is produced. In many cases, several polypeptides interact to produce this conformation. Whether or not several polypeptides interact, the three-dimensional conformation is essential to the function of the molecule. When the functional state is achieved, the molecule is appropriately called a protein.

The polypeptide chains of proteins, like nucleic acids, are linear nonbranched polymers. There are approximately 20 amino acids found in biological material, and they serve as the building blocks or subunits of proteins. Each amino acid has a **carboxyl group**, an **amino group**, and an **R (radical) group** (or side chain) bound covalently to a central carbon atom (Figure 13.11). The R group gives each amino acid its chemical identity. Figure 13.11 illustrates the 20 different R groups, which show a variety of configurations and may be divided into four main classes: (1) **nonpolar** or **hydrophobic;** (2) **polar** or **hydrophilic;** (3) **negatively charged;** and (4) **positively charged.** Because polypeptides are long polymers, and because each position may be occupied by any one of 20 amino acids with unique chemical properties, an enormous variation in chemical activity is possible. For example, if an average polypeptide is composed of 200 amino acids (molecular weight of about 20,000 daltons), 20^{200} different molecules, each with a unique sequence, can be created using 20 different building blocks.

Around 1900, the German chemist Emil Fischer determined the manner in which the amino acids are bonded together. He showed that the amino group of one amino acid can react with the carboxyl group of another amino acid in a dehydration reaction, releasing a molecule of H_2O. The resulting covalent bond is known as a **peptide bond** (Figure 13.12). Two amino acids linked together constitute a **dipeptide,** three a **tripeptide,** and so on. When more than ten amino acids are linked by peptide bonds, the chain is referred to as a **polypeptide**. Generally, no matter how long a polypeptide is, it will contain a free amino group at one end (the **N-terminus**) and a free carboxyl group at the other end (the **C-terminus**).

Four levels of protein structure are recognized: **primary (I°); secondary (II°); tertiary (III°);** and **quaternary (IV°).** The sequence of amino acids in the linear backbone of the polypeptide constitutes its primary structure. This sequence is specified by the sequence of deoxyribonucleotides in DNA. The primary structure of a polypeptide determines the specific characteristics of the higher orders of structure as a protein is formed.

The secondary structure refers to a regular or repeating configuration in space assumed by amino acids closely aligned to one another in the polypeptide chain. In 1951, Linus Pauling and Robert Corey predicted, on theoretical grounds, an **α (alpha) helix** as one type of secondary structure. The α-helix model (Figure 13.13) has since been confirmed by X-ray crystallographic studies. It is rodlike and has the greatest possible theoretical stability. The helix is composed of a spiral chain of amino acids stabilized by hydrogen bonds.

The side chains of amino acids extend outward from the helix, and each amino acid residue occupies a vertical distance of 1.5 Å in the helix. There are 3.6 residues per turn. While left-handed helices are theoretically possible, all proteins demonstrating an α helix are right-handed.

Also in 1951, Pauling and Corey proposed a second structure, the β-pleated-sheet configuration. In this model, a single polypeptide chain folds back on itself, or several chains run in either parallel or

FIGURE 13.12
Peptide bond formation between two amino acids, resulting from a dehydration reaction.

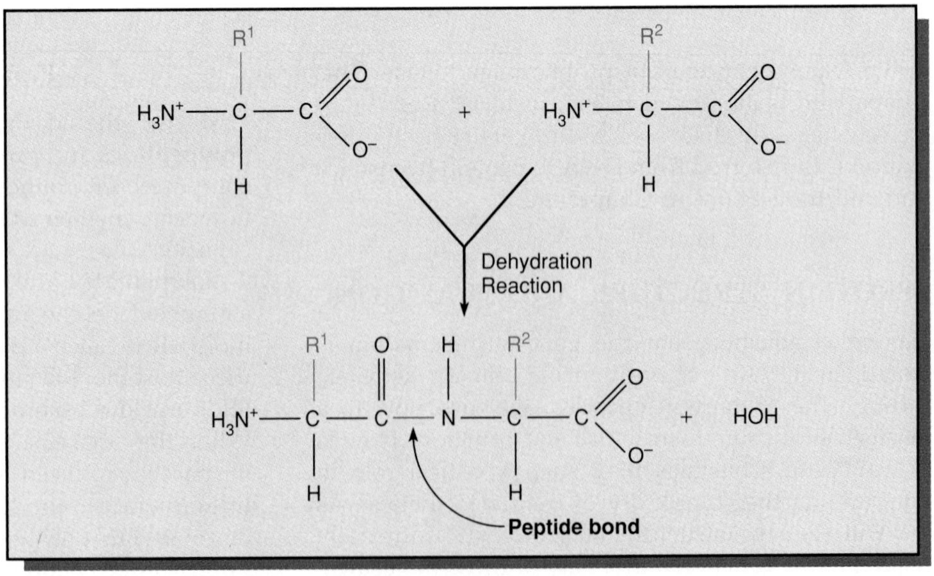

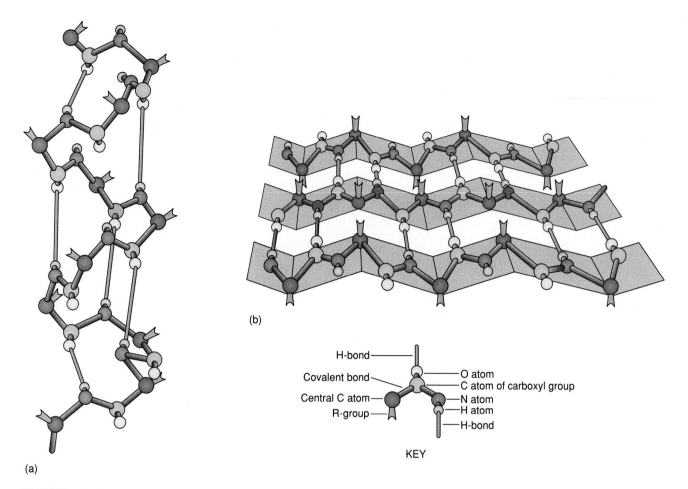

(b)

H-bond

Covalent bond

Central C atom

R-group

O atom

C atom of carboxyl group

N atom

H atom

H-bond

KEY

(a)

FIGURE 13.13

(a) The right-handed α helix, which represents one form of secondary structure of a polypeptide chain. In this simplified diagram, the R groups and hydrogen atoms not involved in hydrogen bonding are not shown. (b) The β-pleated-sheet configuration, an alternate form of secondary structure of polypeptide chains. As in part (a), some atoms are not shown for the sake of clarity.

antiparallel fashion next to one another. Each such structure is stabilized by hydrogen bonds formed between atoms present on adjacent chains [Figure 13.13(b)]. A single zigzagging plane is formed in space with adjacent amino acids 3.5 Å apart.

As a general rule, most proteins demonstrate a mixture of α and β structure. Globular proteins, most of which are round in shape and water soluble, usually contain a core of β-pleated-sheet structure as well as many areas demonstrating α helical structure. The more rigid structural proteins, many of which are water insoluble, rely on more extensive β-pleated-sheet regions for their rigidity. For example, **fibroin,** the protein made by the silk moth, depends extensively on this form of secondary structure.

While the secondary structure describes the arrangement of amino acids within certain areas of a polypeptide chain, tertiary protein structure defines the three-dimensional conformation of the entire chain in space. The molecule twists and turns and loops around itself in a very specific fashion, characteristic of the specific protein. Three aspects of this III° structure are most important in determining this conformation and in stabilizing the molecule.

1 Covalent disulfide bonds form between closely aligned cysteine residues to form the unique amino acid cystine.
2 Nearly all of the polar, hydrophilic R groups are located on the surface, where they interact with water.
3 The nonpolar, hydrophobic R groups are usually located on the inside of the molecule, where they interact with one another.

GENES AND PROTEINS **361**

FIGURE 13.14
A drawing illustrating the tertiary (III°) level of protein structure in myoglobin.

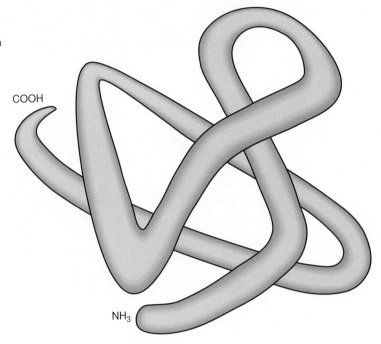

COOH

NH₃

It is important to emphasize that the three-dimensional conformation achieved by any protein is the direct result of the primary (I°) structure of the polypeptide. Thus, the genetic code need only specify the sequence of amino acids in order to encode information leading to the complete structure of proteins. The three stabilizing factors listed above

β chain   β chain

α chain   α chain

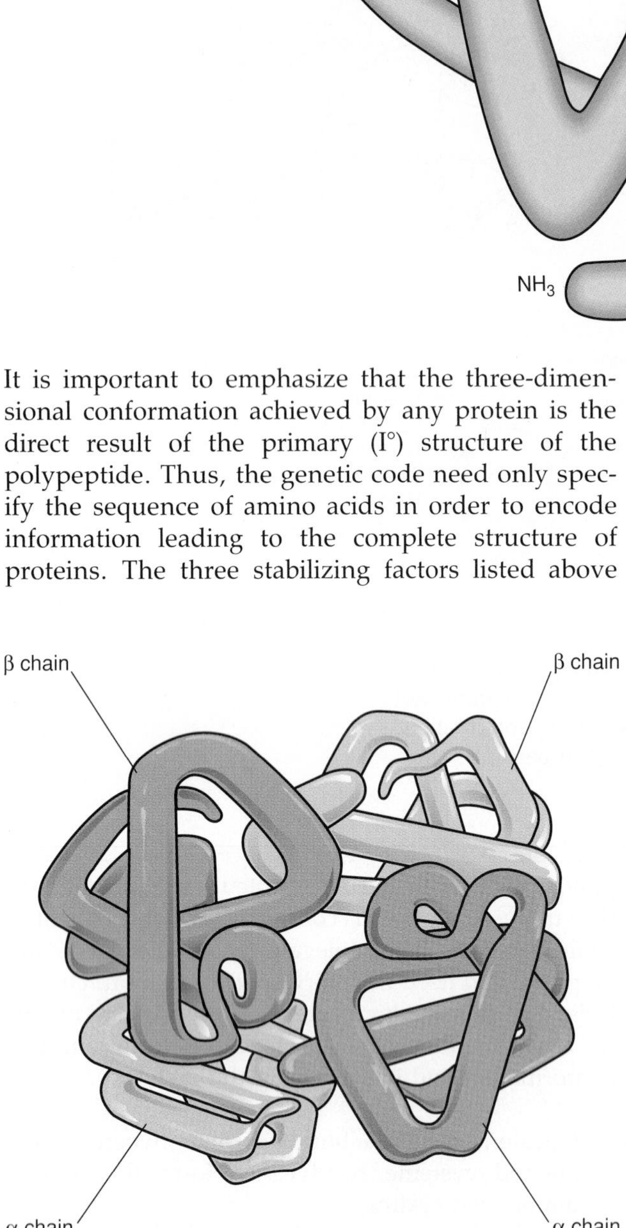

FIGURE 13.15
Schematic drawing illustrating the quaternary level (IV°) of protein structure as seen in hemoglobin. Four chains (2 alpha and 2 beta) interact with four heme groups (not shown) to form the functional molecule.

depend on the location of each amino acid relative to all others in the chain. As folding occurs, the most thermodynamically stable conformation possible results.

The three-dimensional structures of the globular protein **myoglobin** is diagramed in Figure 13.14. This III° level of organization is extremely important because the specific function of any protein is the direct result of its three-dimensional conformation.

The quaternary level of organization is characteristic of proteins composed of more than one polypeptide chain. The IV° structure indicates the conformation of the various chains in relation to one another. This type of protein is called **oligomeric,** and each chain is called a **protomer**. The individual protomers have native conformations that fit together in a specific complementary fashion. Hemoglobin, an oligomeric protein of four polypeptide chains, has been studied in great detail. Its IV° structure is shown in Figure 13.15. Many enzymes, including DNA and RNA polymerase, demonstrate IV° structure.

Protein Function

Proteins are the most abundant macromolecules found in cells. As the end products of genes, they play many diverse roles. For example, the respiratory pigments **hemoglobin** and **myoglobin** transport oxygen, which is essential for cellular metabolism. **Collagen** and **keratin** are examples of structural proteins associated with the skin, connective tissue, and hair of organisms. **Actin** and **myosin** are contractile proteins, found in abundance in muscle tissue.

Still other examples are the **immunoglobins,** which function in the immune system of vertebrates; **transport proteins,** involved in movement of molecules across membranes; **hormones,** which regulate various types of chemical activity; and **histones,** which bind to DNA in eukaryotic organisms.

The largest group of proteins with a related function are the **enzymes.** These molecules specialize in catalyzing biological reactions. Enzymes increase the rate at which a chemical reaction reaches equilibrium, but they do not alter the end point of the chemical equilibrium. Their remarkable, highly specific catalytic properties largely determine the biomolecular nature of any cell type. The specific functions of many enzymes involved in the genetic and cellular processes of cells have been described throughout the text.

Catalysis is a process whereby the **energy of activation** for a given reaction is lowered (Figure 13.16). The energy of activation is the increased kinetic energy state that molecules must usually reach before they react with one another. While this state can be attained as a result of elevated temperatures, enzymes allow biological reactions to occur at lower physiological temperatures. In this way, enzymes make possible life as we know it.

The catalytic properties and specificity of an enzyme are determined by the chemical configuration of the molecule's **active site.** This site is associated with a crevice, a cleft, or a pit on the surface of the enzyme which binds the reactants, or substrates,

enhancing their interaction. Enzymatically catalyzed reactions control metabolic activities in the cell. Each reaction is either **catabolic** or **anabolic.** Catabolism is the degradation of large molecules into smaller, simpler ones with the release of chemical energy. Anabolism is the synthetic phase of metabolism, yielding nucleic acids, proteins, lipids, and carbohydrates. Metabolic pathways that serve the dual function of anabolism and catabolism are **amphibolic.**

Protein Structure and Function: The Collagen Fiber

In order to provide a more in-depth example of the relationship between protein structure and function, we shall consider one of the most interesting proteins found in vertebrates: collagen. In mammals, it is the most abundant protein, constituting up to 25 percent of the total in an individual. It has been extensively studied and clearly illustrates this relationship.

Collagen is found in many places in the body, including tendons, ligaments, bone, connective tissue, skin, blood vessels, teeth, and the lens and cornea of the eye. In each case, the presence of collagen increases the tensile strength of, and provides support to, the tissue. An inspection of the above list makes it evident that collagen must be a versatile protein, providing a variable degree of flexibility and support to these tissues.

While there are several types of collagen produced in vertebrates, we will restrict our discussion to just the major one, called type I. Its main subunit or building block is called **tropocollagen.** This consists of three polypeptide chains wrapped around one another in a triple helix. This unit is extremely large, being 15 Å in diameter and 3000 Å long. Each polypeptide contains about 1000 amino acids, and the three interacting chains of the helix are stabilized by hydrogen bonds between them. There is no hydrogen bonding between amino acids within a single chain. Thus, no α-helical or β-pleated-sheet secondary structure exists in collagen.

The maturation of the polypeptides into tropocollagen and the subsequent condensation of these helical structures into the more densely coiled collagen fibers is a fascinating tale (Figure 13.17). The polypeptides are first synthesized by fibroblasts as even longer units called procollagen. These are secreted into extracellular spaces, where they are cleaved and shortened at both N-terminus and C-terminus by specific enzymes called **procollagen peptidases.**

The amino acid composition of the chains is rather unusual. Almost one-third of the amino acid residues

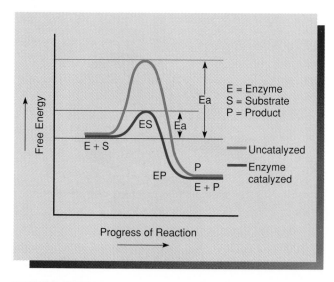

FIGURE 13.16
Energy requirements of an uncatalyzed versus an enzymatically catalyzed chemical reaction. The energy of activation necessary (E_a) to initiate the reaction is substantially lower as a result of catalysis.

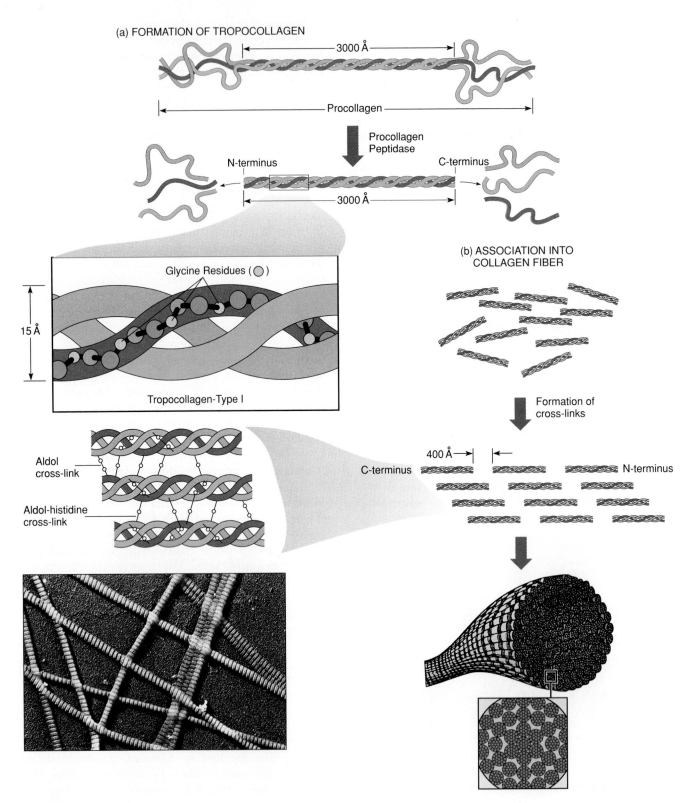

(a) FORMATION OF TROPOCOLLAGEN

3000 Å

Procollagen

Procollagen
Peptidase

N-terminus

C-terminus

3000 Å

(b) ASSOCIATION INTO
COLLAGEN FIBER

Glycine Residues (⃝)

15 Å

Tropocollagen-Type I

Formation of
cross-links

Aldol
cross-link

Aldol-histidine
cross-link

400 Å

C-terminus

N-terminus

FIGURE 13.17
Various steps and intermediate structures involved in the formation of collagen.
(a) The process of post-translational modification of polypeptides during the formation
of tropocollagen, the precursor of collagen. (b) The formation of cross-links leading to
the assembly of the collagen fiber. An electron micrograph illustrates such fibers.

are glycine and a high percentage of them are proline. Additionally, two modified amino acids, **hydroxyproline** and **hydroxylysine,** are abundant. The tripeptide sequences glycine-X-proline or glycine-X-hydroxyproline occur frequently. The abundant presence of glycine is important in the triple helical structure because of the minimal size of its R group. Anything larger would present great difficulties in such a helix.

When the procollagen polypeptides are first synthesized, no hydroxyproline or hydroxylysine residues are part of them. Rather, the hydroxyl groups are added to the amino acids proline and lysine after they have become part of the polypeptide chains. This is an example of the more general phenomenon known as **post-translational modification.** This occurs as a result of two enzymes, **prolyl hydroxylase** and **lysyl hydroxylase.** Each enzyme acts on its specific substrate amino acid (proline or lysine) only if that amino acid is on the amino side of an adjacent glycine residue. The enzymatic reactions occur prior to the formation of the triple helix.

Some of the former amino acids are further modified by the addition of a disaccharide containing one molecule each of glucose and galactose. The number of carbohydrate units per tropocollagen molecule varies in different tissues. For example, the lens of the eye contains many more such units than tendons or ligaments.

Once tropocollagen has been formed, these triple-helical rods spontaneously associate to form dense collagen fibers. The association follows an orderly pattern where rows of end-to-end molecules line up in a staggered fashion next to one another [Figure 13.17(b)]. In each row a gap of approximately 400 Å exists between each tropocollagen unit. This complex structure is stabilized by covalent bonds formed within and between the tropocollagen units. First, the free terminal amino groups (called the ϵ-amino termini) of two lysine residues within the α helix of a tropocollagen molecule may react to form a covalent linkage. This structure, called an aldol cross-link, can react with a histidine residue from another tropocollagen chain, forming another covalent bond. This modified structure can further interact with other amino acids forming still other cross-links. Many such linkages can form, creating great stability in the collagen fibers.

This complex structural arrangement creates protein fibers that strengthen and support a variety of tissues. While it is not yet completely clear how certain tissues are strengthened or rendered flexible to a greater or lesser extent by collagen, the degree of cross-linkage between tropocollagen units is one determining factor. The frequency of carbohydrate side chains may be another. The fibers are usually embedded in an extracellular matrix specific to a given tissue. This undoubtedly adds to each unique environment in which collagen finds itself.

While we have concentrated our discussion on type I collagen, there are about 20 different genes, each encoding a slightly different primary chain. At least ten different combinations of chains have been discovered, each constituting a variant collagen molecule. It is likely that each imparts slightly different properties and characteristics to the different forms of collagen.

The Genetics of Collagen

We began this chapter with a discussion of an inherited biochemical disorder first studied early in this century by Garrod. It is appropriate that we conclude this chapter with a short discussion of more modern discoveries of biochemical disorders, this time related to collagen.

The hydroxylation of proline is very important to normal collagen structure and function. The enzyme that hydroxylates this amino acid, prolyl hydroxylase, contains an iron atom near its active site which must be in the reduced ferrous state. It is interesting to note that individuals with the vitamin-deficiency disease **scurvy,** where inadequate dietary amounts of ascorbic acid (vitamin C) are ingested, show symptoms related to defective collagen production. These include loss of strength, skin lesions, fragile blood vessels, and degeneration of gum tissue. It is now clear that in individuals with normal diets, ascorbic acid serves as a reducing agent for the iron atom of the prolyl hydroxylase enzyme, maintaining it in its active state. Collagen synthesized in the absence of ascorbic acid is under-hydroxylated and less stable, causing the various maladies associated with scurvy.

While this response to biochemically abnormal collagen is not inherited, it certainly could be, simply as the result of mutations in any number of the genes associated with collagen synthesis. At least two inherited disorders known to involve defects in collagen synthesis and assembly are fairly well characterized genetically. For example, the inability to adequately transform procollagen to collagen leads to the human connective tissue disorder, **Ehlers-Danlos syndrome.** Individuals exhibiting this disorder have fragile, stretchable skin and are loose-jointed, providing hypermobility. Some forms of the syndrome render individuals susceptible to arterial and colon ruptures as well as periodontal problems. They exhibit elevated levels of procollagen and decreased activity of the enzyme procollagen peptidase.

Our second example involves the lethal human disorder **osteogenesis imperfecta**. There are many variations, but all affect long bone formation and result from some defect in collagen structure. In its most severe form (type II), widespread bone defects occur in the fetus and newborn; fractures occur spontaneously and death often occurs soon after birth. An intact skeletal system is obviously critical to survival.

Type I, in comparison, is fairly mild, with an onset sometimes as late as age 35 to 45. However, at that point, declining health is associated with the degeneration of collagen-rich tissues: blood vessels weaken, bones become fragile, and they fracture frequently. A stroke or heart attack usually causes premature death.

In one instance, the specific genetic defect has been uncovered and found to involve but a single glycine residue in the primary procollagen chain. At position 988, a glycine residue has been altered to the amino acid cysteine by a mutation of a single base change in the gene. This alteration of the primary structure inhibits the formation of highly ordered fibers that make up the bundles of collagen and leads to all of the aberrant phenotypic effects associated with the disease.

Osteogenesis imperfecta type I is inherited as a dominant disorder and provides an excellent insight into the concept of **genetic dominance.** Envision a heterozygote with one copy of the normal allele of the gene encoding the primary collagen chain and one copy of the mutant allele. Assume that this individual synthesizes, through transcription and translation, 50 percent normal chains and 50 percent mutant chains, which then are randomly assembled into triple helices. Only one out of eight triple helices $(1/2)^3$ will have all normal chains. Seven out of eight will have one to three abnormal chains. As a result, there is no way for the normal allele to compensate adequately for the mutant allele; most collagen molecules and all collagen fibers will contain mutant polypeptides. Clearly, once the biochemistry of an inherited dominant disorder is clarified, the basis for its dominant expression is more easily comprehended.

Collagen disorders are quite prevalent in the human population. It is estimated that over 100,000 individuals worldwide suffer from osteogenesis imperfecta. While Ehlers-Danlos syndrome is much rarer, there are about 20,000 individuals who suffer from still another collagen-based disorder called **Marfan syndrome.** Long, thin arms, legs, and digits of the hands and feet characterize the appearance of affected individuals. Aortic heart disease most frequently is the cause of premature death. Abraham Lincoln is suspected to have suffered from this disorder. Given the complexity of the biochemistry of collagen and its wide distribution throughout the body, it is not surprising that these and other genetic disorders exist.

CHAPTER SUMMARY

1 Basic to Garrod's studies in the early 1900s of inborn, human metabolic disorders such as cystinuria, albinism, and alkaptonuria, was the concept that genes control the synthesis of specific metabolic products. Mutations in genes may produce blocks in metabolic pathways. Subsequent accumulation of byproducts from failed biochemical conversions results in metabolic disorders.

2 The investigation of eye pigments in *Drosophila* and nutritional requirements in *Neurospora* by Beadle and his colleagues made it clear that mutations cause the loss of enzyme activity. Their work led to the concept of one gene:one enzyme.

3 The concept of the one gene:one enzyme hypothesis was later revised. Pauling and Ingram's investigations of hemoglobins from patients with sickle-cell anemia led to the discovery that one enzyme directs the synthesis of only one polypeptide chain.

4 Thorough investigations have revealed the existence of several major types of human hemoglobin molecules, found in the embryo, fetus, and the adult. Specific genes control each polypeptide chain constituting these various hemoglobin molecules. Mutations resulting in amino acid substitutions in these chains are rare, but well documented, in the human population.

5 The proposal suggesting that a gene's nucleotide sequence specifies in a colinear way the amino acids in a polypeptide chain was confirmed by Yanofsky's experiments involving mutations in the tryptophan synthetase gene in *E. coli.*

6 Proteins, the end products of genes, demonstrate four levels of structural organization. The primary structure is the linear amino acid sequences of a polypeptide chain. The linear polypeptide chain will assume a secondary structure as an α helix or a β pleated sheet, imparting stability to the chain. Tertiary structure represents the three-dimensional conformation of polypeptide chains. This conformation is directly related to the functional capacity of the molecule. Quaternary structure refers to the combination of two or more polypeptide chains composing a single protein molecule, as occurs in hemoglobin.

7 Of the myriad functions performed by proteins, the most influential role is assumed by enzymes. These highly specific, cellular catalysts play a central role in the production of all classes of molecules in living systems.

8 Collagen is an abundant, specialized protein in vertebrates, serving a structural role in a variety of tissues. There are about 20 different genes for collagen, each coding for slightly different primary chains, which are embedded in a matrix unique to the given tissue.

9 Ehlers-Danlos syndrome and osteogenesis imperfecta are examples of inherited human disorders resulting from mutations that alter collagen structure and its function.

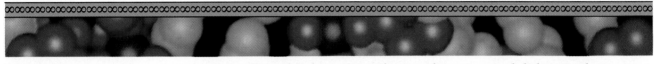

Insights and Solutions

1 In Beadle and Ephrussi's studies involving imaginal disk transplants, most eye color mutants behaved autonomously (i.e., they were not cured, or converted to wild-type color) when allowed to develop in a wild-type abdominal environment. One concludes that in such cases, the product influenced by the gene under study is *not* diffusible. The *vermilion* (*v*) and *cinnabar* (*cn*) eye disks were the exception, developing into wild-type eyes, because the products controlled by these genes *are* diffusible and supplied by the host. Recall that *cn* disks developing in *v* hosts remain as mutant *cn* eyes, but that *v* disks developing in *cn* hosts are cured and develop wild-type color. This is the basis of determining that the v^+ gene product acts at a point in the biosynthetic pathway prior to the cn^+ gene product.

Suppose that a new bright red mutation is discovered and the product controlled in this gene is not diffusible. However, the gene product acts prior to the steps controlled by both the v^+ and cn^+ gene products in the biosynthesis of the brown xanthommatin pigment. Predict the outcome of a transplant experiment of the new mutant eye disk into a wild-type larva.

ANSWER: Since the block in the new mutation precedes either the cn^+ or the v^+ steps, the wild-type host will make both of these diffusible products. As a result, the mutant disk will be cured and develop wild-type color. This happens because these diffusible products can enter the disk and allow the block in the new mutation to be bypassed.

2 The following growth responses were obtained using four mutant strains of *Neurospora* and the related compounds A, B, C, and D. None of the mutations grow on minimal medium. Draw all possible conclusions.

	Growth Product			
Mutation	C	D	B	A
1	−	−	−	−
2	−	+	+	+
3	−	−	+	+
4	−	−	+	−

ANSWER: First, nothing can be concluded about mutation 1, except that it is lacking some essential growth factor, perhaps even unrelated to the biochemical pathway represented by mutations 2–4. Nor can anything be concluded about compound C. If it is involved in the pathway, it is a product synthesized prior to the synthesis of A, B, and D.

We must now analyze these three compounds and the control of their synthesis by the enzymes encoded by genes 2, 3, and 4. Since product B allows growth in all three cases, it may be considered the "end product." It bypasses the block in all three instances. Using similar reasoning, product A precedes B in the pathway since it allows a bypass in two of the three steps. Product D precedes B, yielding the final pathway:

$$C(?) \longrightarrow D \longrightarrow A \longrightarrow B$$

Now determine which mutations control which steps. Since mutation 2 can be alleviated by products D, B, and A, it must control a step prior to all three products, perhaps the direct conversion to D, although we can't be certain. Mutation 3 is alleviated by B and A, so its effect must precede them in the pathway. Thus, we will assign it as controlling the conversion of D to A. Likewise, we can assign mutation 4 to the conversion of A to B, leading to the final pathway:

$$C(?) \xrightarrow{2(?)} D \xrightarrow{3} A \xrightarrow{4} B$$

1 Discuss the potential difficulties involved in designing a diet to alleviate the symptoms of phenylketonuria.
2 Phenylketonurics cannot convert phenylalanine to tyrosine. Why don't these individuals exhibit a deficiency of tyrosine?
3 Phenylketonurics are often more lightly pigmented than normal individuals. Can you suggest a reason why this is so?
4 Consider the following hypothetical circumstances involving reciprocal transplantations of three mutant imaginal eye disks. The three mutations, a, b, and c, are responsible for the enzymatic pathway shown below. However, it is not known which mutant gene controls which step in the reactions.

$$\underset{\downarrow}{?} \quad \underset{\downarrow}{?} \quad \underset{\downarrow}{?}$$
$$w \longrightarrow x \longrightarrow y \longrightarrow z$$

The results of the transplantations are shown below. Assuming that the w, x, y, and z substances are diffusible, determine which reactions are controlled by which genes.

Donor	Host	Transplant Eye Pigment	Development
a	c	Mutant	Autonomous
a	b	Mutant	Autonomous
c	b	Mutant	Autonomous
c	a	Wild type	Nonautonomous
b	c	Wild type	Nonautonomous
b	a	Wild type	Nonautonomous

5 What F_1 and F_2 ratios will occur in a cross between *vermilion* females and *cinnabar* males? Recall that v is sex-linked and cn is autosomal, and that these mutants produce identical bright red phenotypes. What ratios will occur in the reciprocal cross (v males and cn females)?

6 The synthesis of flower pigments is known to be dependent upon enzymatically controlled biosynthetic pathways. In the crosses shown below, postulate the role of mutant genes and their products in producing the observed phenotypes.

(a) P_1: white strain A × white strain B

F₁: all purple

F_2: 9/16 purple: 7/16 white

(b) P_1: white × pink

F₁: all purple

F_2: 9/16 purple: 3/16 pink: 4/16 white

7 A series of mutations in the bacterium *Salmonella typhimurium* result in the requirement of either tryptophan or some related molecule in order for growth to occur. From the data shown below, suggest a biosynthetic pathway for tryptophan.

Mutation	Growth Response Supplement				
	Minimal Medium	Anthranilic Acid	Indole Glycerol Phosphate	Indole	Tryptophan
trp-8	−	+	+	+	+
trp-2	−	−	+	+	+
trp-3	−	−	−	+	+
trp-1	−	−	−	−	+

8 The study of biochemical mutants in organisms such as *Neurospora* has demonstrated that some pathways are branched. The data shown below illustrate the branched nature of the pathway resulting in the synthesis of thiamine. Why don't the data support a linear pathway? Can you postulate a pathway for the synthesis of thiamine in *Neurospora*?

Mutation	Growth Response Supplement			
	Minimal Medium	Pyrimidine	Thiazole	Thiamine
thi-1	−	−	+	+
thi-2	−	+	−	+
thi-3	−	−	−	+

9 Explain why the one-gene : one-enzyme concept is not considered accurate today.

10 Why is an alteration of electrophoretic mobility interpreted as a change in the primary structure of the protein under study?

11 Contrast the polypeptide chain components of each of the hemoglobin molecules found in humans.

12 Using sickle-cell anemia as a basis, describe what is meant by a molecular or genetic disease. What are the similarities and dissimilarities between this type of a disorder and a disease caused by an invading microorganism?

13 Contrast the contributions of Pauling and Ingram to our understanding of the genetic basis for sickle-cell anemia.

14 Hemoglobins from two individuals are compared by electrophoresis and by fingerprinting. Electrophoresis reveals no difference in migration, but fingerprinting shows an amino acid difference. How is this possible?

15 When the amino acid sequences of the polypeptide Hb chains are compared, many homologous sequences are detected, but the degree of homology varies between any two chains. For example, a greater degree of homology exists between the β and γ chains than between the α and β chains. Relate this information to Ohno's theory of gene duplication (discussed in Chapter 7).

16 Describe what *colinearity* means.

17 Certain mutations called *amber* in bacteria and viruses result in premature termination of polypeptide chains during translation. Many *amber* mutations found at different points along the gene coding for a head protein in phage T4 have been detected. How might this system be further investigated to demonstrate and support the concept of colinearity?

18 Define and compare the four levels of protein organization.

19 List as many different protein functions as you can, with an example of each.

20 How does an enzyme function? Why are enzymes essential for living organisms on earth?

21 Discuss the role of the following in the production of the collagen fiber: (a) glycine residues in the primary chain, (b) procollagen peptidase, (c) prolyl hydroxylase, (d) aldol cross-links, (e) hydrogen bonds.

22 Why is Fiers' work with phage MS2, discussed in Chapter 11, a more direct evidence in support of colinearity than Yanofsky's work with the *trp A* locus in *E. coli*?

23 Using the amino acid substitutions shown in Figure 13.9 for the α and β chains of human hemoglobin and the code table (Figure 11.8), determine how many of them can occur as a result of a single nucleotide change.

SELECTED READINGS

BARTHOLOME, K. 1979. Genetics and biochemistry of phenylketonuria—Present state. *Hum. Genet.* 51: 241–45.

BATESON, W. 1909. *Mendel's principles of heredity.* Cambridge, England: Cambridge University Press.

BEADLE, G. W. 1945. Genetics and metabolism in *Neurospora. Physiol. Rev.* 25: 643.

———. 1946. Genes and the chemistry of the organism. *Amer. Scient.* 34: 31–53.

BEADLE, G. W., and EPHRUSSI, B. 1937. Development of eye colors in *Drosophila*: Diffusible substances and their interrelations. *Genetics* 22: 76–86.

BEADLE, G. W., and TATUM, E. L. 1941. Genetic control of biochemical reactions in *Neurospora. Proc. Natl. Acad. Sci.* 27: 499–506.

BEET, E. A. 1949. The genetics of the sickle-cell trait in a Bantu tribe. *Ann. Eugenics* 14: 279–84.

BOYER, P. D., ed. 1974. *The enzymes.* 3rd ed. Vol. 10. Orlando: Academic Press.

BRENNER, S. 1955. Tryptophan biosynthesis in *Salmonella typhimurium. Proc. Natl. Acad. Sci.* 41: 862–63.

BYERS, P. H. 1989. Inherited disorders of collagen gene structure and expression. *Am. J. Med. Genet.* 34: 72–80.

COHN, D. H., BYERS, P. H., STEINMANN, B., and GELINAS, R. E. 1986. Lethal osteogenesis imperfecta resulting from a single nucleotide change in one human pro-α 1(I) collagen allele. *Proc. Natl. Acad. Sci.* 83: 6045–47.

DICKERSON, R. E. 1964. X-ray analysis and protein structure. In *The proteins*, 2nd ed., ed. H. Neurath, vol. 2. Orlando: Academic Press.

DICKERSON, R. E., and GEIS, I. 1983. *Hemoglobin: Structure, function, evolution, and pathology.* Menlo Park, Calif.: Benjamin/Cummings.

DOOLITTLE, R. F. 1985. Proteins. *Scient. Amer.* (Oct.) 253: 88–99.

EPHRUSSI, B. 1942. Chemistry of eye color hormones of *Drosophila. Quart. Rev. Biol.* 17: 327–38.

GARROD, A. E. 1902. The incidence of alkaptonuria: A study in chemical individuality. *Lancet* 2: 1616–20.

———. 1909. *Inborn errors of metabolism.* London: Oxford University Press. (Reprinted 1963, Oxford University Press, London.)

GARROD, S. C. 1989. Family influences on A. E. Garrod's thinking. *J. Inher. Metab. Dis.* 12: 2–8.

HOLLISTER, D. W., BYERS, P. H., and HOLBROOK, K. A. 1982. Genetic disorders of collagen metabolism. *Adv. Human Genet.* 12: 1–88.

INGRAM, V. M. 1957. Gene mutations in human hemoglobin: The chemical difference between normal and sickle cell hemoglobin. *Nature* 180: 326–28.

———. 1963. *The hemoglobins in genetics and evolution*. New York: Columbia University Press.

KOSHLAND, D. E. 1973. Protein shape and control. *Scient. Amer.* (Oct.) 229: 52–64.

LaDU, B. N., ZANNONI, V. G., LASTER, L., and SEEGMILLER, J. E. 1958. The nature of the defect in tyrosine metabolism in alkaptonuria. *J. Biol. Chem.* 230: 251.

LESK, A. M., and HARDMAN, K. D. 1982. Computer-generated schematic diagrams of protein structure. *Science* 216: 539–40.

MANIATIS, T., et al. 1980. The molecular genetics of human hemoglobins. *Ann. Rev. Genet.* 14: 145–78.

MECHANIC, G. 1972. Cross-linking of collagen in a heritable disorder of connective tissue: Ehlers-Danlos syndrome. *Biochem. Biophys. Res. Commun.* 47: 267–72.

MURAYAMA, M. 1966. Molecular mechanism of red cell sickling. *Science* 153: 145–49.

NEEL, J. V. 1949. The inheritance of sickle-cell anemia. *Science* 110: 64–66.

PAULING, L., ITANO, H. A., SINGER, S. J., and WELLS, I. C. 1949. Sickle cell anemia, a molecular disease. *Science* 110: 543–48.

PROCKOP, D. J., and KIVIRIKKO, K. I. 1984. Heritable diseases of collagen. *New Engl. J. Med.* 311: 376–86.

SARABHAI, A., STRETTON, A., BRENNER, S., and BOLLE, A. 1964. Colinearity of the gene with the polypeptide chain. *Nature* 201: 13–17.

SCOTT-MONCRIEFF, R. 1936. A biochemical survey of some Mendelian factors for flower colour. *J. Genet.* 32: 117–70.

SCRIVER, C. R., and CLOW, C. L. 1980. Phenylketonuria and other phenylalanine hydroxylation mutants in man. *Ann. Rev. Genet.* 14: 179–202.

SRB, A. M., and HOROWITZ, N. H. 1944. The ornithine cycle in *Neurospora* and its genetic control. *J. Biol. Chem.* 154: 129–39.

STRYER, L. 1988. *Biochemistry*. 3rd ed. New York: W. H. Freeman.

SYKES, B. 1985. The molecular genetics of collagen. *BioEssays* 3:112–17.

WAGNER, R. P., and MITCHELL, H. K. 1964. *Genetics and metabolism*. 2nd ed. New York: Wiley.

YANOFSKY, C., DRAPEAU, G., GUEST, J., and CARLTON, B. 1967. The complete amino acid sequence of the tryptophan synthetase A protein and its colinear relationship with the genetic map of the *A* gene. *Proc. Natl. Acad. Sci.* 57: 296–98.

ZIEGLER, I. 1961. Genetic aspects of ommochrome and pterin pigments. *Adv. in Genet.* 10: 349–403.

ZUBAY, G. L. 1988. *Biochemistry*. 2nd ed. New York: Macmillan.

ZUBAY, G. L., and MARMUR, J., eds. 1973. *Papers in biochemical genetics*. 2nd ed. New York: Holt, Rinehart and Winston.

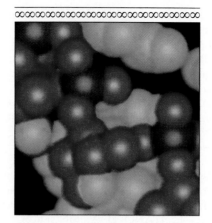

14
Mutation and Mutagenesis

CHAPTER CONCEPTS

Aside from chromosomal mutations, the major basis of diversity among organisms, even those that are closely related, is genetic variation at the level of the gene. The origin of such variation is the process of mutation, whereby groups of coding sequences are altered as a result of base substitution or the addition or deletion of one or more bases within those sequences. Gene mutations are the working tools of the geneticist and the raw material upon which evolution relies.

In Chapter 8, we defined the four characteristics or functions ascribed to the genetic information: **replication, storage, expression,** and **variation by mutation**. In a sense, mutation is a failure to store the genetic information faithfully. If a change occurs in the stored information, it may be reflected in the expression of that information and will be propagated following replication. Historically, the term **mutation** includes both chromosomal changes and changes within single genes. We have discussed the former alterations in Chapter 7, referring to them collectively as chromosomal aberrations. In this chapter, we will be concerned with the so-called **gene mutations**. As we will see, a change may be a simple substitution of a single nucleotide or involve the addition or deletion of one or more nucleotides within the normal sequence of DNA.

The term *mutation* was coined in 1901 by Hugo DeVries to explain the variation he observed in crosses involving the evening primrose, *Oenothera lamarckiana*. Most of the variation was actually due to multiple translocations, but two cases were subsequently shown to be caused by gene mutations. As the studies of mutations progressed, it soon became clear that mutations serve as the source of most alleles and thus are the origin of much of the existing genetic variability within populations. As new alleles arise, whether they are detrimental, neutral, or beneficial, they constitute "raw material" to be tested by the evolutionary process of natural selection.

Mutations are the working tool of the geneticist. The resulting phenotypic variability allows the geneticist to identify and study the genes that control the modified traits. Without the phenotypic variability that mutations provide, all genetic crosses would be meaningless. For example, if all pea plants displayed a uniform phenotype, Mendel would have had no basis for his experimentation. Because of the impor-

tance of mutations, great attention has been given to their origin, induction, and classification.

Certain organisms lend themselves to induction of mutations that can be easily detected and studied throughout reasonably short life cycles. Viruses, bacteria, fungi, fruit flies, certain plants, and mice fit these criteria to various degrees. Thus, these organisms have often been used in studying mutation and mutagenesis and through other studies have also contributed to more general aspects of genetic knowledge.

In this chapter we shall consider the classification, detection, and origin of mutations. Additionally, we will discuss the potential dangers posed to human populations by environmental factors that increase mutation rates. These same environmental sources are also believed to induce a wide variety of human cancers.

CLASSIFICATION OF MUTATIONS

There are various schemes by which mutations are classified. They are not mutually exclusive, but instead depend simply on which aspects of mutation are being investigated or discussed. In this section we will describe three sets of distinctions concerning mutations.

Spontaneous versus Induced Mutations

All mutations are described as either **spontaneous** or **induced**. While these two categories overlap to some degree, **spontaneous mutations** are considered as those that arise in nature. No specific agent—other than natural forces—is associated with their occurrence, and they are assumed to arise randomly as changes in the nucleotide sequences of genes.

What causes spontaneous mutations? Although their origin is not fully understood, many such mutations can be linked to normal chemical processes or phenomena that result in rare errors. Such errors may cause alterations in the chemical structure of nitrogenous bases that are part of existing genes; more often, they may occur during the enzymatic process of DNA replication. We will explore examples of such errors later in this chapter. It is generally agreed that any natural phenomenon that heightens chemical reactivity in cells will lead to more errors. For example, background radiation from cosmic and mineral sources and ultraviolet light from the sun are energy sources to which most organisms are exposed. As such, they may be factors leading to spontaneous mutations. Once an error is present in the genetic code, it may be reflected in the amino acid composition of the specified protein. If the changed amino acid is present in a part of the molecule critical to its biochemical activity, a functional alteration may result.

In contrast to such spontaneous events, those that arise as a result of the influence of any artificial factor are considered to be **induced mutations**. The earliest demonstration of the induction of mutation occurred in 1927, when Herman J. Muller reported that X-rays could cause mutations in *Drosophila*. In 1928, Lewis J. Stadler reported the same finding in barley. In addition to various forms of radiation sources, a wide spectrum of chemical agents is also known to be mutagenic. We will examine specific aspects of mutagenic agents later in this chapter.

Gametic versus Somatic Mutations

When considering the effects of mutation in eukaryotic organisms, it is important to distinguish whether the change occurs in somatic cells or gametes. Mutations arising in somatic cells are not transmitted to future generations. In tissues of an adult organism, thousands upon thousands of cells may be performing a similar function. Thus, a mutation in a single cell may not impair the organism, even if the mutation is detrimental. First, a random mutation might occur in a gene that is not active or essential to the function of that cell. Second, even if a critical gene is affected, there are still thousands of unaffected cells to perform the function of the tissue in question.

Somatic mutations have their greatest impact if they are dominant or if they are sex-linked recessives, and when they occur early in development. If the mutation occurs in an undifferentiated cell whose descendants are likely to give rise to several differentiated tissues or organs, the impact may be great.

However, if the mutation is in a differentiated cell serving as the progenitor to only a small number of cells, the effect will be much less substantial, since there are so many nonmutant cells that can assume the responsibility for normal function.

There is a very important exception to the above description. If a mutation occurs, even in a cell that would not *normally* serve as a progenitor to other cells, that disrupts the control of proliferation, such a mutation may be responsible for a malignant state, where cells divide uncontrollably. Such a condition is, of course, of great concern.

Mutations in gametes or gamete-forming tissue are part of the germ line and are of greater concern because they have the potential of being expressed in all cells of an offspring. Such a mutation may also be transmitted to future generations, gradually increasing in frequency within the population. **Dominant autosomal mutations** will be expressed phenotypically in the first generation. **Sex-linked recessive mutations** arising in the gametes of a heterogametic female may be expressed in hemizygous male offspring. This will occur provided the male offspring receives the affected X chromosome. Because of heterozygosity, the occurrence of an **autosomal recessive mutation** in the gametes of either males or females, even one resulting in a lethal allele, may go unnoticed for many generations until it has become widespread in the population. The new allele will become evident only when a chance mating brings two copies of it together in the homozygous condition.

Categories of Mutation

Various types of mutations are classified on the basis of their effect on the organism. Even though some of these categories were introduced earlier, we will briefly review several of them again. A single mutation may well fall into more than one category.

The most obvious mutations are those affecting a **morphological trait**. For example, all of Mendel's pea characters and most genetic variations encountered in the study of *Drosophila* fit this designation. All such morphological variations deviate from the normal or wild-type phenotype.

A second broad category includes mutations that are **nutritional** or **biochemical variations** from the norm. In bacteria and fungi, the inability to synthesize an amino acid or vitamin is an example of a typical nutritional mutation. In humans, sickle-cell anemia and hemophilia are examples of biochemical mutations. While such mutations in these organisms are not visible and do not usually affect specific morphological characters, they can have a more general effect on the well-being and survival of the affected individual.

A third category consists of mutations that affect behavior patterns of an organism. For example, mating behavior or circadian rhythms of animals may be altered. The primary effect of **behavior mutations** is often difficult to discern. For example, the mating behavior of a fruit fly may be impaired if it cannot beat its wings. However, the defect may be in: (1) the flight muscles, (2) the nerves leading to them, or (3) the brain, where the nerve impulses that initiate wing movements originate. The study of behavior and the genetic factors influencing it has benefited immensely from investigations of behavior mutations.

Still another type of mutation may affect the regulation of genes. A regulatory gene may produce a product that controls the transcription of another gene. In other instances, a region of DNA either close to or far away from a gene may modulate its activity. In either case, **regulatory mutations** may disrupt this process and permanently activate or inactivate a gene. Our knowledge of genetic regulation has been dependent on the study of mutations that disrupt this process.

Another group consists of **lethal mutations**. Nutritional and biochemical mutations may also fall into this category. A mutant bacterium that cannot synthesize a specific amino acid it needs will cease to grow if plated on a medium lacking that amino acid. Various human biochemical disorders, such as Tay-Sachs disease and Huntington disease, are lethal at different points in the life cycle of humans.

Finally, any of the above groups can exist as **conditional mutations**. Even though a mutation is present in the genome of an organism, it may not be evident under certain conditions. The best examples are the **temperature-sensitive mutations,** found in a variety of organisms. At certain permissive temperatures, a mutant gene product functions normally, only to lose its functional capability at a different restrictive temperature. When shifted to this temperature, the impact of the mutation becomes apparent, even lethal, and is amenable to investigation. The study of conditional mutations has been extremely important in experimental genetics, particularly in understanding the function of genes essential to the viability of organisms.

DETECTION OF MUTATION

Before geneticists can study directly the mutational process or obtain mutant organisms for genetic investigations, they must be able to detect mutations. The ease and efficiency of detecting mutations in a particular organism has by and large determined the organism's usefulness in genetic studies. In this section, we will use several examples to illustrate how mutations are detected.

Detection in Bacteria and Fungi

Detection of mutations is most efficient in haploid microorganisms such as bacteria and fungi. Detection depends on a selection system where mutant cells are isolated easily from nonmutant cells. The general principles are similar in bacteria and fungi. To illustrate, we will describe how nutritional mutations in the fungus *Neurospora crassa* (see Figure 5.3) are detected.

Neurospora is a pink mold that normally grows on bread. It may also be cultured in the laboratory. This eukaryotic mold is haploid in the vegetative phase of its life cycle. Thus, mutations may be detected without the complications generated by heterozygosity in diploid organisms. Pioneering biochemical genetic studies with *Neurospora* were performed by George Beadle and Edward Tatum around 1940.

Visible mutants such as *albino* have been well studied, but the full potential of *Neurospora* genetics was attained during the investigation of nutritional mutants. Wild-type *Neurospora* grows on a **minimal culture medium** of glucose, a few inorganic acids and salts, a nitrogen source such as ammonium nitrate, and the vitamin biotin. Induced nutritional mutants will not grow on minimal medium, but will grow on a supplemented or **complete medium** that also contains numerous amino acids, vitamins, nucleic acid derivatives, and so forth. Microorganisms that are nutritional wild types (requiring only minimal medium) are called **prototrophs,** while those mutants that require a specific supplement to the minimal medium are called **auxotrophs.**

The procedural details for the detection of nutritional mutants were illustrated in Figure 13.4. Nutritional mutants may be detected and isolated by their failure to grow on minimal medium and their ability to grow on complete medium. The mutant cells can no longer synthesize some essential compound absent in minimal medium but present in complete medium. Once a nutritional mutant is detected and isolated, the missing compound is determined by attempts to grow the mutant strain in a series of tubes, each containing minimal medium supplemented with a single compound. Once the missing compound is found, crosses with the wild type are made, and individual ascospores are dissected out for testing on supplemented and minimal media. Normal segregation yields a ratio of 1 wild type:1

nutritional mutant, confirming that the mutant is of nuclear origin.

In Chapter 15, where the topics of bacterial and viral genetics are introduced, we will return to the topic of mutation. There, we will see that it was not until 1943 that spontaneous mutations were believed to occur in bacteria. Techniques for the detection and study of such mutations will be discussed. Analysis of mutations in microorganisms has been particularly important to the study of molecular genetics.

Detection in *Drosophila*

Muller, in his studies demonstrating that X-rays are mutagenic, developed a number of detection systems in *Drosophila melanogaster*. These systems allow the estimation of the spontaneous and induced rates of sex-linked and autosomal recessive lethal mutations. We will consider two techniques: the **CIB** and the **attached-X procedures,** which Muller devised. The *CIB* system (Figure 14.1) detects the rate of induction of sex-linked recessive lethal mutations. The *CIB* stock involves *C*, an inversion that suppresses the recovery of crossover products; *l*, a recessive lethal mutation; and *B*, the dominant gene duplication causing *Bar* eye, described in Chapter 7. All of these traits are located on the X chromosome.

In this technique, wild-type P_1 males are treated with a mutagenic agent—in this case, X-rays—and mated to untreated, heterozygous *CIB* females. The F_1 offspring are of four types. One group contains the *CIB* males, which die because they contain the recessive lethal *l*, which is expressed in the hemizygous condition. The remaining F_1 males are wild type. They are mated with those F_1 females with *Bar* eyes. The F_1 wild-type females are not utilized.

The F_1 *Bar*-eyed females receive the *CIB* chromosome from their mothers and the X-irradiated chromosome from their fathers. Following mating, the females are placed individually in culture bottles and allowed to lay eggs, which will yield an F_2 generation. If a lethal mutation has been produced in the P_1 generation sperm, the F_2 generation will not include any males. One-half of the males die because of the *l* allele on the *CIB* chromosome, and the other half die because of the induced lethal allele *l**. By inspecting the culture bottles for female populations only, the number of induced, sex-linked, recessive lethal mutations is found. If 500 such F_2 culture bottles were inspected, and 25 contained only females, the induced rate of such mutations would be 5 percent. This method also permits the detection of sex-linked, morphological mutations. If a mutation of this type has been induced, all surviving F_2 males will show the trait.

The second technique, using females with attached-X chromosomes, is even simpler to use in detecting recessive morphological mutations because

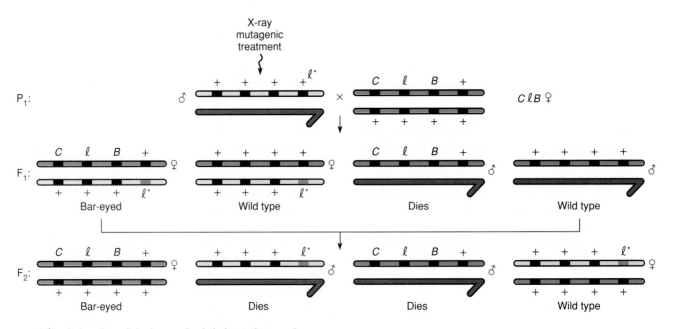

(ℓ^* = induced sex-linked recessive lethal mutation = ▬)

FIGURE 14.1
Muller's classical *CIB* technique for the detection of induced, sex-linked recessive lethal mutations in *Drosophila*.

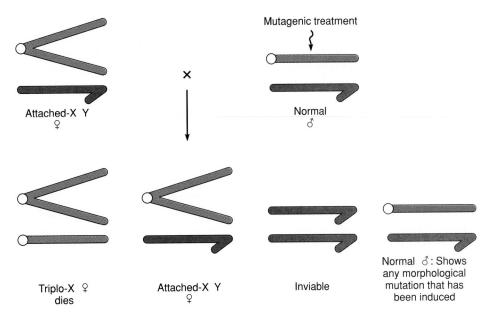

FIGURE 14.2
The attached-X method for detection of induced morphological mutations in *Drosophila*.

it requires only one generation. These females have two X chromosomes attached to a single centromere and one Y chromosome, in addition to the normal diploid complement of autosomes. When attached-X females are mated to males with normal sex chromosomes (XY), four types of progeny result: triplo-X females that die; viable attached-X females; YY males that also die; and viable XY males. Figure 14.2 illustrates how P_1 males that have been treated with a mutagenic agent produce F_1 male offspring that express any induced sex-linked recessive mutation. In contrast to the *ClB* technique, which tests only a single X chromosome, the attached-X method tests numerous X chromosomes at one time in a single cross.

Detection techniques have also been devised for recessive autosomal lethals in *Drosophila*. In these techniques, dominant marker mutations are followed through a series of three generations. While more cumbersome to perform, these techniques are also fairly efficient.

Detection in Plants

Genetic variation in plants is extensive. Mendel's peas, for example, were the basis for the fundamental postulates of transmission genetics. Studies of plants have also enhanced our understanding of gene interaction, polygenic inheritance, linkage, sex determination, chromosome rearrangements, and poly-

FIGURE 14.3
A hypothetical pedigree of inherited cataract of the eye in humans.

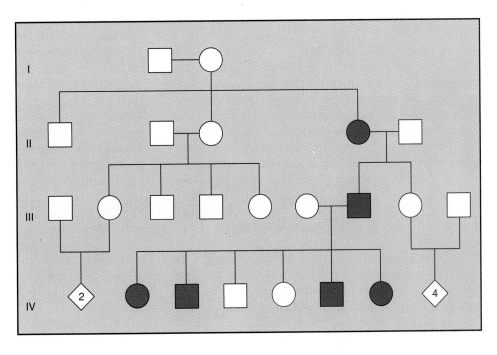

ploidy. Most variations are detected simply by visual observation. However, there are also techniques for the detection of biochemical mutations in plants. The first is the analysis of biochemical composition of plants. For example, the isolation of proteins from maize endosperm, hydrolysis of the proteins, and determination of the amino acid composition have revealed that the *opaque*-2 **mutant strain** contains significantly more lysine than do other, nonmutant lines. As a result of this discovery, plant geneticists and other specialists have begun to analyze the amino acid compositions of various strains of several grain crops, including corn, rice, wheat, barley, and millet. When the results of these analyses are completed and catalogued, the information will be useful in combating malnutrition diseases resulting from inadequate protein or the lack of the essential amino acids in the diet.

The second detection technique involves tissue culture of plant cell lines in defined medium. The plant cells are handled as microorganisms, and biochemical requirements may be determined by adding or deleting nutrients in the culture medium. There are other advantages to this method. Techniques associated with conditional lethal mutants can be used on plant cells in tissue culture and then applied to the genetics of higher plants. Also, this method provides a detection system that is generally not useful with the intact plant. Temperature-sensitive mutations in plants are beginning to be explored, particularly in tobacco. These studies may add significantly to our understanding of plant growth, metabolism, and genetics.

Detection in Humans

Since designed matings are not possible or desirable in humans, many techniques that have been developed for the detection of mutations in organisms such as *Drosophila* are not available to human geneticists. To determine the mutational basis for any human characteristic or disorder, geneticists must first analyze a pedigree that traces the family history as far back as possible. Once any trait has been shown to be inherited, it is possible to predict whether the mutant allele is behaving as a dominant or a recessive and whether it is sex-linked or autosomal.

Dominant mutations are the simplest to detect. If they are present on the X chromosome, affected fathers pass the phenotypic trait to all their daughters. If dominant mutations are autosomal, approxi-

mately 50 percent of the offspring of an affected heterozygous individual are expected to show the trait. Figure 14.3 shows a hypothetical pedigree illustrating the initial occurrence of an autosomal dominant allele for cataracts of the eye. The parents in generation I were unaffected, but one of three offspring (Generation II) developed cataracts. This female, the proband, produced two children, of which the male child was affected. Of his six offspring (IV), four of six were affected (Generation III). These observations are consistent with, but do not prove, an autosomal dominant mode of inheritance. However, the high percentage of affected offspring in generation IV favors this conclusion. Also, the unaffected daughter in this generation argues against sex linkage because she received her X chromosome from her affected father. Provided that the mutant allele is completely penetrant, such a conclusion is soundly based.

Sex-linked recessive mutations may also be detected by pedigree analysis, as discussed in Chapter 5. The most famous case of a sex-linked mutation in humans is that of **hemophilia,** which was found in the descendants of Queen Victoria. The recessive mutation for hemophilia has occurred many times in human populations, but the political consequences of the mutation that occurred in the royal family have been sweeping. Inspection of the pedigree in Figure 14.4 leaves little doubt that Victoria was heterozygous (*Hh*) for the trait. Her father was not affected, and there is no reason to believe that her mother was a carrier, as was Victoria. Robert Massie's *Nicholas and Alexandra* and Robert and Suzanne Massie's *Journey* (see the list of selected readings at the end of this chapter) provide fascinating reading on the topic of hemophilia.

In a similar manner, it is possible to detect autosomal recessive alleles. Because this type of mutation is "hidden" when heterozygous, it is not unusual for the trait to appear only intermittently through a number of generations. An affected individual and a homozygous normal individual will produce unaffected carrier children. Matings between two carriers will produce, on the average, one-fourth affected offspring.

In addition to pedigree analysis, human cells may now be routinely cultured *in vitro*. This procedure has allowed the detection of many more mutations than any other form of analysis. Analysis of enzyme activity, protein migration in electrophoretic fields, and direct sequencing of DNA and proteins are among the techniques that have demonstrated wide genetic variation between individuals in human populations.

SPONTANEOUS MUTATION RATE

The types of detection systems just described allow geneticists to estimate mutation rates. It is of considerable interest to determine the rate of spontaneous mutation. Not only does such information provide insights into evolution, but it also provides the baseline for measuring the rate of experimentally induced mutation. Induction of mutation can only be ascertained when the induced rate clearly exceeds the spontaneous rate for the organism under study.

Table 14.1 shows the spontaneous mutation rate determined in a variety of organisms. There are three particularly striking features evident in these data. First, the rate is exceedingly low for all organisms studied. Second, the rate is seen to vary considerably in different organisms. Third, even within the same species, the spontaneous mutation rate varies from gene to gene.

Viral and bacterial genes undergo spontaneous mutation on an average of about 1 in 100 million (10^{-8}) cell divisions. While *Neurospora* exhibits a

TABLE 14.1
Rates of spontaneous mutations at various loci in different organisms.

Organism	Character	Gene	Rate	Units
Bacteriophage T2	Lysis inhibition	$r \rightarrow r^+$	1×10^{-8}	Per gene replication
	Host range	$h^+ \rightarrow h$	3×10^{-9}	
E. coli	Lactose fermentation	$lac^- \rightarrow lac^+$	2×10^{-7}	
	Lactose fermentation	$lac^+ \rightarrow lac^-$	2×10^{-6}	
	Phage T1 resistance	$T1\text{-}s \rightarrow T1\text{-}r$	2×10^{-8}	
	Histidine requirement	$his^+ \rightarrow his^-$	2×10^{-6}	
	Histidine independence	$his^- \rightarrow his^+$	4×10^{-8}	Per cell division
	Streptomycin dependence	$str\text{-}s \rightarrow str\text{-}d$	1×10^{-9}	
	Streptomycin sensitivity	$str\text{-}d \rightarrow str\text{-}s$	1×10^{-8}	
	Radiation resistance	$rad\text{-}s \rightarrow rad\text{-}r$	1×10^{-5}	
	Leucine independence	$leu^- \rightarrow leu^+$	7×10^{-10}	
	Arginine independence	$arg^- \rightarrow arg^+$	4×10^{-9}	
	Tryptophan independence	$try^- \rightarrow try^+$	6×10^{-8}	
Salmonella typhimurium	Trytophan independence	$try^- \rightarrow try^+$	5×10^{-8}	Per cell division
Diplococcus pneumoniae	Penicillin resistance	$pen^s \rightarrow pen^r$	1×10^{-7}	Per cell division
Chlamydomonas reinhardi	Streptomycin sensitivity	$str^r \rightarrow str^s$	1×10^{-6}	Per cell division
Neurospora crassa	Inositol requirement	$inos^- \rightarrow inos^+$	8×10^{-8}	Mutant frequency among asexual spores
	Adenine independence	$ade^- \rightarrow ade^+$	4×10^{-8}	
Zea mays	Shrunken seeds	$sh^+ \rightarrow sh^-$	1×10^{-6}	
	Purple	$pr^+ \rightarrow pr^-$	1×10^{-5}	Per gamete per generation
	Colorless	$c^+ \rightarrow c$	2×10^{-6}	
	Sugary	$su^+ \rightarrow su$	2×10^{-6}	
Drosophila melanogaster	Yellow body	$y^+ \rightarrow y$	1.2×10^{-6}	
	White eye	$w^+ \rightarrow w$	4×10^{-5}	
	Brown eye	$bw^+ \rightarrow bw$	3×10^{-5}	Per gamete per generation
	Ebony body	$e^+ \rightarrow e$	2×10^{-5}	
	Eyeless	$ey^+ \rightarrow ey$	6×10^{-5}	
Mus musculus	Piebald coat	$s^+ \rightarrow s$	3×10^{-5}	
	Dilute coat color	$d^+ \rightarrow d$	3×10^{-5}	Per gamete per generation
	Brown coat	$b^+ \rightarrow b$	8.5×10^{-4}	
	Pink eye	$p^+ \rightarrow p$	8.5×10^{-4}	
Homo sapiens	Hemophilia	$h^+ \rightarrow h$	2×10^{-5}	
	Huntington disease	$Hu^+ \rightarrow Hu$	5×10^{-6}	
	Retinoblastoma	$R^+ \rightarrow R$	2×10^{-5}	Per gamete per generation
	Epiloia	$Ep^+ \rightarrow Ep$	1×10^{-5}	
	Aniridia	$An^+ \rightarrow An$	5×10^{-6}	
	Achondroplasia	$A^+ \rightarrow A$	5×10^{-5}	

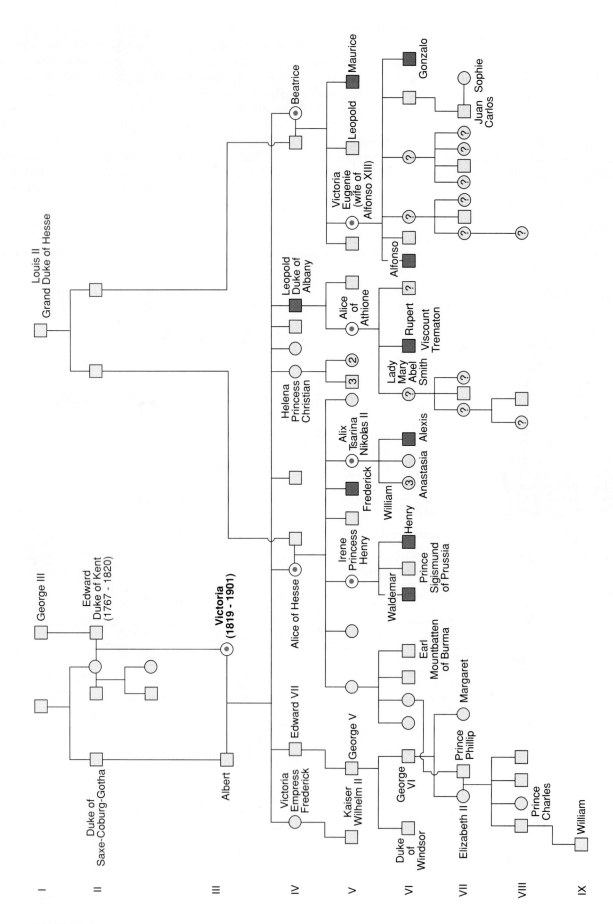

FIGURE 14.4
Pedigree of hemophilia in the royal family descended from Queen Victoria (opposite).
The pedigree is typical of the transmission of sex-linked recessive traits. Circles with a
dot in them indicate presumed female carriers heterozygous for the trait. Circles with
a question mark indicate females whose status is uncertain. The photograph shows
Queen Victoria (seated front center) and some of her immediate family. Standing to
the left of Victoria (wearing a feather boa) is Alexandra, the future Tsarina of Russia.
To her left is Nicholas II, who became the last Tsar of Russia.

similar rate, maize, *Drosophila,* and humans demonstrate a rate several orders of magnitude higher. The genes studied in these groups average between 1/1,000,000 to 1/100,000 (10^{-6} to 10^{-5}) mutations per gamete formed. Mouse genes are still another order of magnitude higher in their spontaneous mutation rate, 1/100,000 to 1/10,000 (10^{-5} to 10^{-4}). It is not clear why such a large variation occurs in mutation rate. The variation might reflect the relative efficiency of enzyme systems whose function is to repair errors created during replication. Repair systems will be discussed later in this chapter.

THE MOLECULAR BASIS OF MUTATION

For the purposes of the following discussion, we will use a *simplified* definition of a gene in the description of the molecular basis of mutation. In this context, it is easiest to consider a gene as a linear sequence of nucleotide pairs representing stored chemical information. Since the genetic code is a triplet, each sequence of three nucleotides specifies a single amino acid in the corresponding polypeptide. Any change that disrupts the coded information provides sufficient basis for a mutation. The simplest change would be the substitution of a single nucleotide. In Figure 14.5, such a change is compared with our own written language, using three-letter words to be consistent with the genetic code. As you can see, a single change can alter the meaning of the sentence "THE CAT SAW THE DOG," creating what is called

missense. These are analogies to what are most appropriately referred to as **base substitutions** or **point mutations**. The mutation has turned information that makes sense into various forms of missense. Two terms are often used to describe nucleotide substitutions. If a pyrimidine replaces a pyrimidine or a purine replaces a purine, a **transition** has occurred. If a purine and a pyrimidine are interchanged, a **transversion** has occurred.

A second type of change that could occur is the insertion or deletion of a single nucleotide at any point along the gene. As also shown in Figure 14.5, wherever this occurs, the remainder of the information becomes garbled, again with missense, when one letter is inserted or deleted. These examples are called **frameshift mutations** because the frame of reading has become altered. In either the point or frameshift mutations, the end result is a change in the amino acid sequence of the protein encoded by the altered gene.

Tautomeric Shifts

In 1953, after they had proposed a molecular structure of DNA, Watson and Crick published a paper in which they discussed the genetic implications of this structure. They recognized that the purines and pyrimidines found in DNA could exist in **tautomeric forms;** that is, each can exist in several chemical forms, differing by only a single proton shift in the molecule. Watson and Crick suggested that **tautomeric shifts** could result in base pair changes or mutations. The biologically important tautomers in-

FIGURE 14.5
The impact of the substitution, addition, or deletion of one letter in a sentence composed of three-letter words, creating either missense or nonsense.

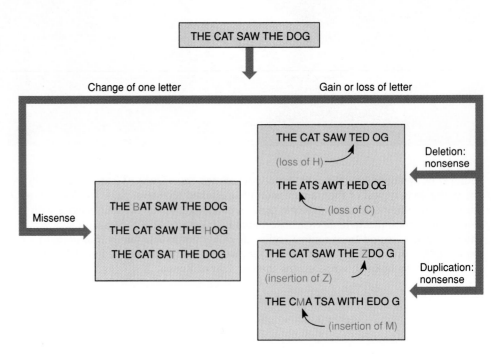

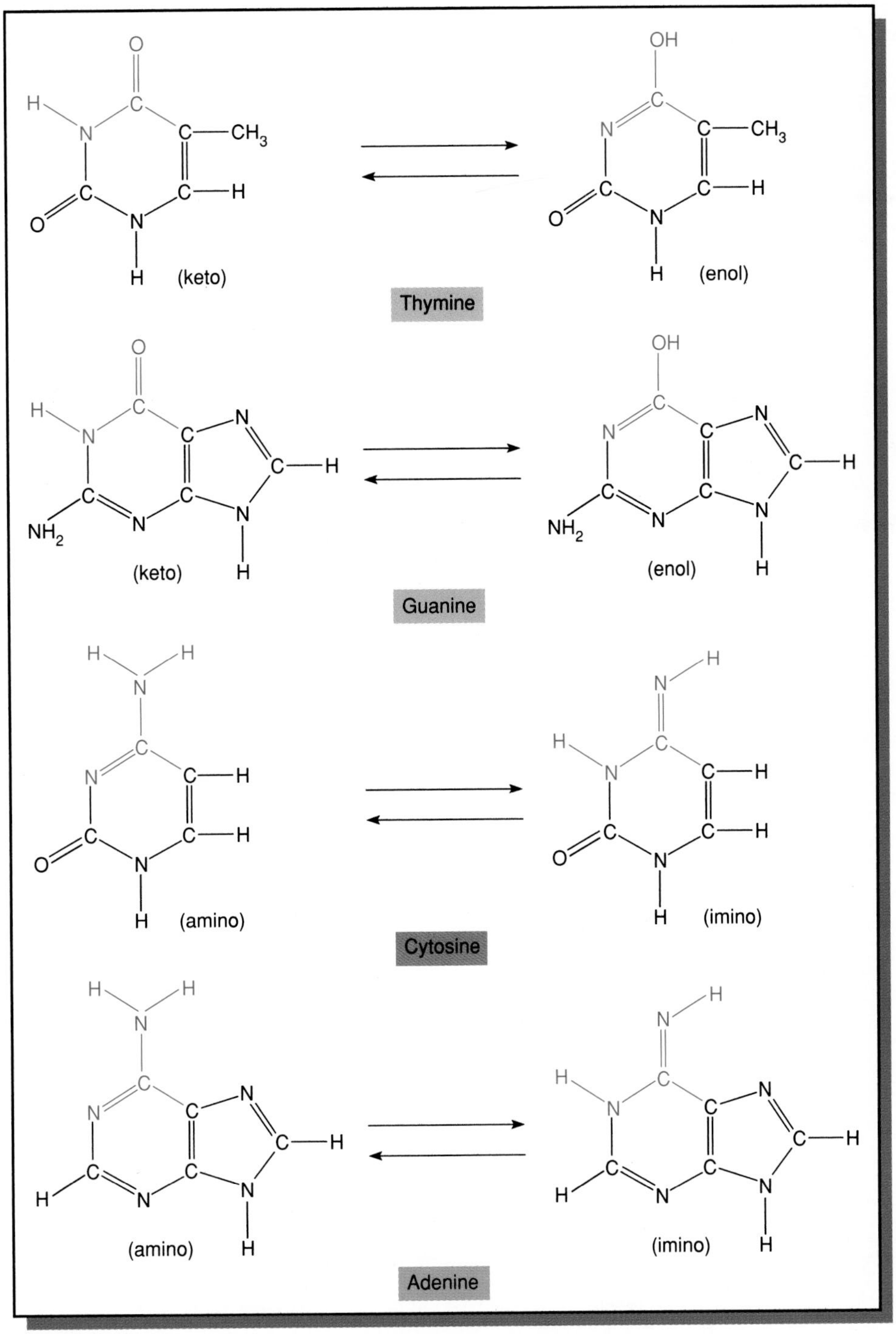

FIGURE 14.6
Rate tautomeric shifts that occur in the chemical structure of the four nitrogenous bases of DNA.

volve keto-enol pairs for thymine and guanine, and amino-imino pairs for cytosine and adenine (Figure 14.6).

The infrequent tautomer is capable of hydrogen bonding with a normally noncomplementary base. However, the pairing is always between a pyrimidine and a purine. The end result is the transition where an A=T pair is replaced by a G≡C pair, or vice versa. Figure 14.7 compares the normal base-pairing relationships with transition pairings. The effect of the rare tautomers comes at the time of DNA replication when a rare tautomer in the template strand matches with a noncomplementary base. In the next round of replication, the "mismatched" members of the base

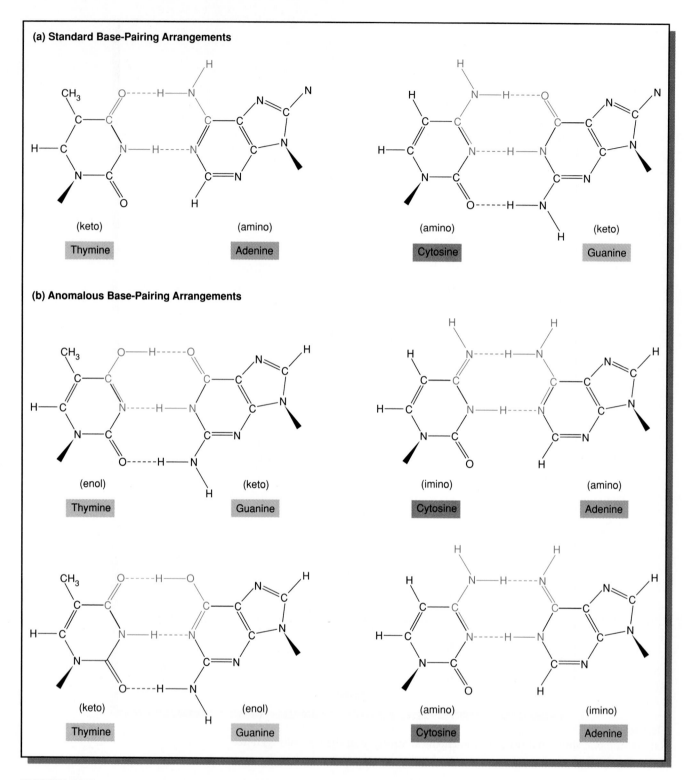

FIGURE 14.7

The standard base-pairing relationships compared with anomalous arrangements occurring as a result of tautomeric shifts. The dense arrow indicates the point of bonding to the pentose sugar.

FIGURE 14.8
Formation of an A=T to a G≡C transition mutation as a result of a tautomeric shift in adenine.

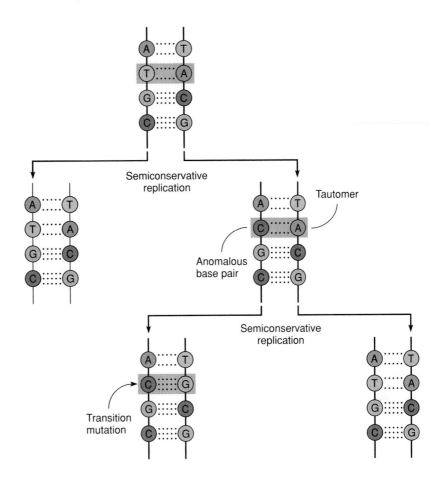

pair are separated, and each specifies its normal complementary base. As shown in Figure 14.8, the end result is a mutation.

Base Analogues

The **base analogues,** another group of mutagenic chemicals, are molecules that may substitute for purines or pyrimidines during nucleotide and DNA biosynthesis. They were recognized in 1956 to be efficiently incorporated into the DNA of the bacteriophage T4. The analogues were halogenated derivatives of uracil in the number-5 position of the pyrimidine ring: 5-bromouracil, 5-chlorouracil, and 5-iodouracil. Figure 14.9 compares the structure of a thymine analogue, **5-bromouracil** (5-BU),* with the structure of thymine. The presence of the bromine atom in place of the methyl group increases the probability that a tautomeric shift will occur. If 5-BU is incorporated into DNA in place of thymine and then a tautomeric shift occurs, the result is an A=T to G≡C transition. If the tautomeric shift to the enol

---

*If 5-BU is chemically linked to d-ribose, the nucleoside analogue bromodeoxyuridine (BUdR) is formed.

form occurs before the analogue is incorporated into DNA and 5-BU is mistaken for C, the transition is from G≡C to A=T.

There are other base analogues that are mutagenic. One, **2-amino purine (2-AP),** can serve successfully as an analogue of adenine. In addition to its base-pairing affinity with thymine, 2-AP can also base pair with cytosine. As such, transitions from A=T to G≡C may result following replication.

Because of the specificity by which base analogues such as 2-AP induce transition mutations, base analogues may also be used to induce reversion to the wild-type nucleotide sequence. This alteration is called **reverse mutation.** The process may also occur spontaneously, but at a much lower rate.

Nitrous Acid

Nitrous acid is one of a number of chemical compounds known to be mutagenic. It is capable of oxidatively deaminating purines and pyrimidines. In this reaction, an amino group is converted to a keto group in cytosine and adenine (Figure 14.10). In these two cases, cytosine is converted to uracil, and adenine is changed to hypoxanthine.

FIGURE 14.9
Similarity of 5-bromouracil (5-BU) structure to thymine structure. In the common keto form, 5-BU pairs normally with adenine, behaving as an analogue. In the rare enol form, it pairs anomalously with guanine.

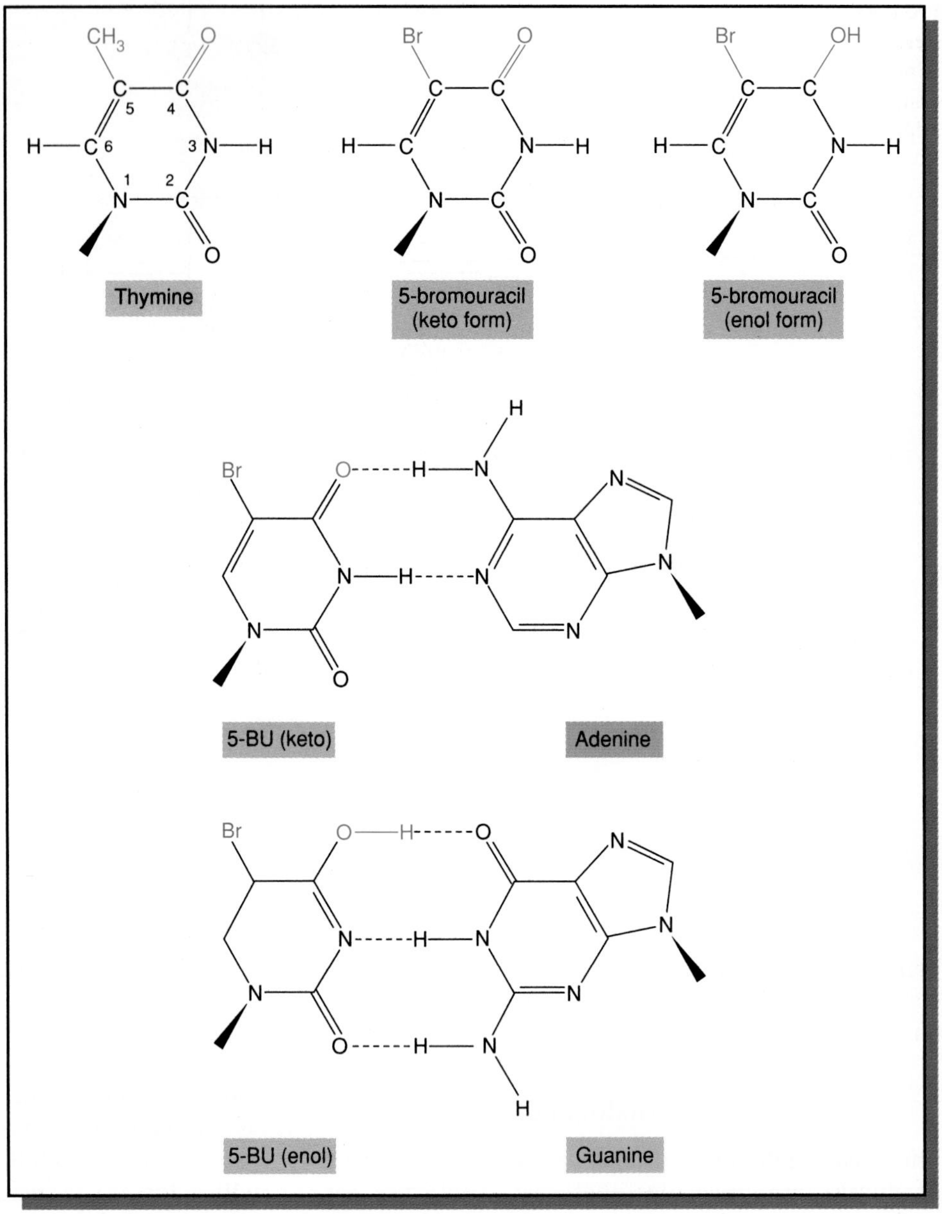

The major effect of these changes is to alter the base-pairing specificities of these two molecules during DNA replication. For example, cytosine normally pairs with guanine. Following conversion to uracil, which pairs with adenine, the original $G\equiv C$ pair converted to an $A=U$ pair and, following an additional replication, to an $A=T$ pair. When adenine is deaminated, an original $A=T$ pair is converted to a $G\equiv C$ pair because hypoxanthine pairs naturally with cytosine.

Hydroxylamine

Hydroxylamine (NH_2OH) reacts specifically with cytosine by adding a hydroxyl group to its existing amino group (Figure 14.11). The new product, hy-

droxylaminocytosine, may now undergo a tautomeric shift, allowing it to pair with adenine. Following two replications, a $G\equiv C$ pair will be converted to an $A=T$ pair. In this example, a specific, predictable mutation is induced.

Alkylating Agents

The sulfur-containing **mustard gases** were one of the first groups of chemical mutagens discovered. This discovery was made as a result of studies concerned with chemical warfare during World War II. Mustard gases are **alkylating agents;** that is, they donate an alkyl group such as CH_3- or CH_3-CH_2- to amino or keto groups in nucleotides. **Ethylmethane sulfonate**

FIGURE 14.10
Deaminations caused by nitrous acid (HNO$_2$) leading to new base-pairing arrangements and mutations. G≡C to A=T and A=T to G≡C transition mutations result.

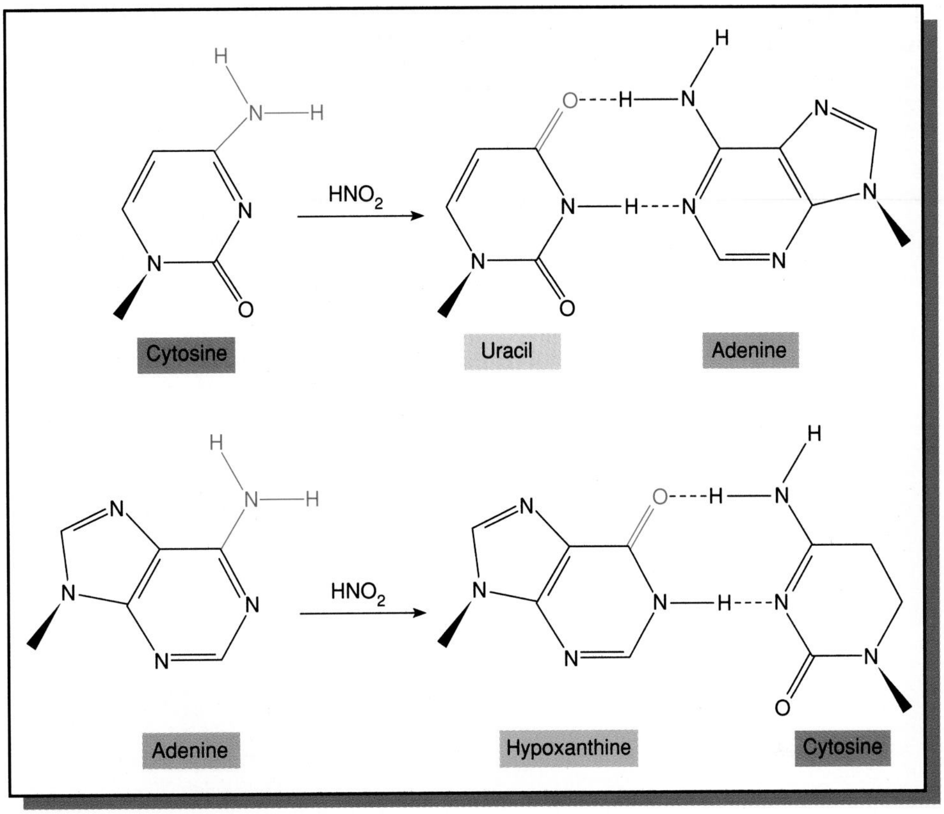

FIGURE 14.11
Reaction of hydroxylamine with cytosine, converting it to a form that base pairs with adenine. The result is the G≡C to A=T transition mutation.

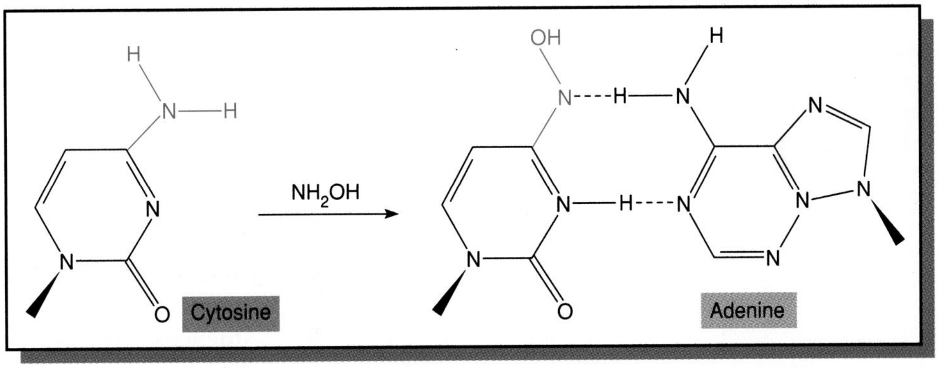

FIGURE 14.12
Conversion of guanine to 6-ethylguanine by the alkylating agent ethylmethane sulfonate (EMS). 6-ethylguanine base pairs with thymine.

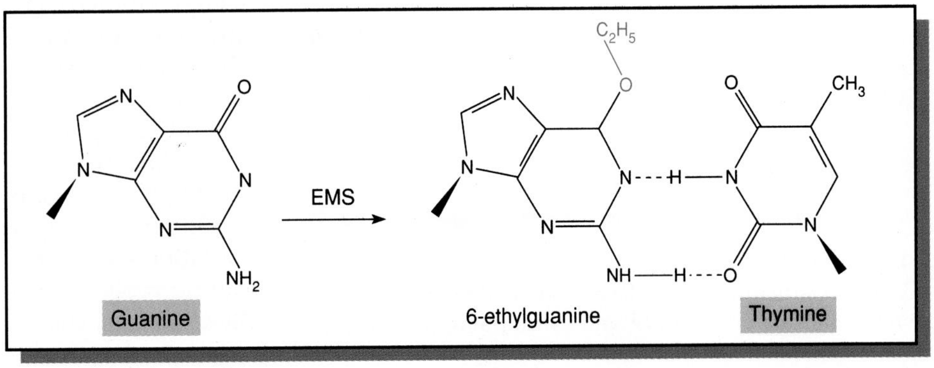

TABLE 14.2
Alkylating agents.

Common Name or Symbol	Chemical Name	Chemical Structure
Mustard gas (sulfur)	Di-(2-chloroethyl)sulfide	Cl—CH₂—CH₂—S—CH₂—CH₂—Cl
EMS	Ethylmethane sulfonate	CH₃—CH₂—O—S(=O)(O)—CH₃
EES	Ethylethane sulfonate	CH₃—CH₂—O—S(=O)(O)—CH₂—CH₃
MMS	Methylmethane sulfonate	CH₃—O—S(=O)(O)—CH₃
DES	Diethylsulfate	CH₃—CH₂—O—S(=O)(O)—O—CH₂—CH₃
NG	*N*-methyl-*N'*-nitro-*N*-nitrosoguanidine	HN=C—NH—NO₂ ; O=N—N—CH₃

(EMS), for example, alkylates the keto groups in the number-6 position of guanine and in the number-4 position of thymine (Figure 14.12). As with nitrous acid and hydroxylamine, base-pairing affinities are altered and transition mutations result. In the case of 6-ethyl guanine, this molecule acts like a base analogue of adenine, causing it to pair with thymine. Table 14.2 lists the chemical names and structures of several frequently used alkylating agents known to be mutagenic.

Acridine Dyes and Frameshift Mutations

Other chemical mutagens cause **frameshift mutations**. These result from the addition or removal of one or more base pairs in the polynucleotide sequence of the gene. Inductions of frameshift mutations have been studied in detail with a group of aromatic molecules known as **acridine dyes** or **acridines**. The structures of **proflavin,** the most widely studied acridine mutagen, and **acridine orange** are shown in Figure 14.13. Acridine dyes are of about the same dimension as a nitrogenous base pair and are known to intercalate or wedge between purines and pyrimidines of intact DNA. Intercalation of acridine dyes is considered to induce contortions in the DNA helix, causing deletions and additions.

One model suggests that the resultant frameshift mutations are generated at gaps produced in DNA during replication, repair, or recombination. During these events, there is the possibility of slippage and improper base pairing of one strand with the other.

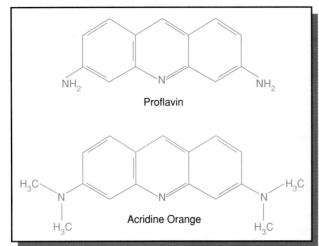

FIGURE 14.13
Chemical structures of proflavin and acridine orange, which intercalate with DNA and cause frameshift mutations.

The model suggests that intercalation of the acridine into an improperly base-paired region may prolong the time during which these slippage structures exist. If so, the probability increases that the mispaired configuration will exist at the time that synthesis and rejoining occurs, thereby resulting in an addition or deletion of one or more bases from one of the strands.

MUTATIONS IN HUMANS: TWO CASE STUDIES

The preceding section provides an important overview of the molecular basis of mutation. The information presented relies almost exclusively on the study of nucleic acid chemistry and has described numerous ways in which one nucleotide pair can be converted to another, either spontaneously or as a result of an induced change. We know that such mutations actually occur, primarily because analysis of amino acid sequences of many proteins within populations of a species show substantial diversity. This diversity has arisen during evolution and is a reflection of changes in the genetic code following substitution of one or more nucleotides in the DNA sequences constituting genes.

As our ability to analyze DNA more directly has increased, we are now able to look specifically at the actual nucleotide sequence of genes and gain even greater insights into the process of mutation. Two techniques capable of accurate, rapid sequencing of DNA are described in Appendix A. These and other approaches involving the analysis of DNA have greatly extended our knowledge in molecular genetics.

In this brief section, we will examine the results of two studies that have investigated the actual gene sequence of various mutations affecting humans. The first will provide an interesting insight into the molecular basis of the **ABO antigens,** originally presented in Chapter 4, as an example of multiple alleles. The second case involves the nature of the mutations that have led to the devastating sex-linked disorder **muscular dystrophy.**

The ABO system is based on a series of antigenic determinants found on erythrocytes as well as other cells, particularly epithelial types. As previously shown in Figure 4.2, three alleles of a single gene exist, the product of which is designed to modify the H substance. The modification involves glycosyltransferase activity, converting the H substance to either the A or B antigen, as a result of the product of the I^A or I^B allele, respectively, or failing to modify the H substance, as a result of the I^O allele.

Using recombinant DNA technology, the responsible gene has been examined in 14 cases of varying ABO status. The results are rather straightforward. Four consistent nucleotide substitutions were found when the DNAs of the I^A and I^B alleles were compared. It is assumed that the changes in the amino acid sequence of the glycosyltransferase gene product resulting from these substitutions lead to the different modifications of the H substance. In one case, N-acetyl-α-D-galactosamine is added, producing the A antigen, while in the other case, α-D-galactose is added, producing the B antigen. An individual carrying both alleles contains both forms of the enzyme and produces both antigens.

The situation with the I^O allele is quite different and very interesting. Individuals homozygous for this allele are type O, lack glycosyltransferase activity, and fail to modify the H substance. Analysis of the DNA of this allele shows one consistent change compared to that of the other alleles: the deletion of a single nucleotide early in the coding sequence, causing a frameshift mutation. A complete messenger RNA is transcribed, but upon its translation, the frame of reading shifts at the point of deletion and continues out of frame for about one hundred nucleotides before a "stop" codon is encountered. At this point, premature termination of the resulting polypeptide chain occurs, producing a nonfunctional product.

These results provide a direct molecular explanation of the ABO allele system and the basis for the biosynthesis of the corresponding antigens. The molecular basis for the antigenic phenotypes is clearly the result of structural alterations, or mutations, of the nucleotide sequence of the gene encoding the glycosyltransferase enzyme.

The second case of mutational analysis involves the severe disorder muscular dystrophy. It is characterized by progressive muscle degeneration, or myopathy, resulting in the death of affected individuals by early adulthood. Because the condition is recessive and sex-linked, and since affected males do not reproduce, females are never affected by the disorder. The incidence of 1/3500 live male births makes muscular dystrophy one of the most common life-shortening hereditary disorders known. Two forms exist: Duchenne muscular dystrophy (DMD) is more common and more severe than the allelic form called Becker muscular dystrophy (BMD).

The region containing the gene has been extensively analyzed and consists of over two million base pairs. In unaffected individuals, transcription results in a messenger RNA containing about 14,000 bases (14kb) that is translated into a protein, **dystrophin,** consisting of 3685 amino acids. This protein can be detected in most cases of the less severe BMD, but is not usually found in the DMD form of the disorder. This has led to the hypothesis that most mutations causing BMD do not alter the reading frame, but that most DMD mutations change the reading frame early in the gene, resulting in premature termination of dystrophin translation. This hypothesis is consistent with the observed differences in severity of the two forms of the disease.

In an extensive analysis of the DNA of 194 patients (160 DMD and 34 BMD), J. T. Den Dunnen and associates have made a remarkable observation. They found that 128 of these mutations (65 percent) consisted of substantial deletions or duplications. Of 115 deletions, 17 occurred in BMD, and of 13 duplications, 1 was in BMD, with the remainder being found in DMD cases. In most cases, the above results were consistent with the "reading frame" hypothesis. With just a few exceptions, DMD mutations changed the frame of reading of exon areas. In most instances, BMD mutations did not alter the reading frame.

Perhaps the most noteworthy finding is the high percentage of deleterious mutations studied that represent the deletion or duplication of nucleotides within the gene. This observation no doubt reflects the fact that a mutation caused by a random single nucleotide substitution within a gene is more likely to be tolerated without the devastating effect seen in muscular dystrophy than the addition or loss of numerous nucleotides that may alter the frame of reading. There are three reasons for this:

1 A nucleotide substitution may not change the code, since it is degenerate.
2 If an amino acid substitution does result, the change may not be present at a location within the protein that is critical to its function.
3 Even if the altered amino acid is present at a critical region, it may still have little or no effect on the function of the protein. For example, an amino acid might be changed to another with nearly identical chemical properties or to one with very similar recognition properties, such as shape.

As a result, single base substitutions may have little or no effect on protein function, or may simply reduce the overall efficiency of the gene product. Thus, it may well be that most severe hereditary disorders will reveal much more substantial genetic changes, as is the case with muscular dystrophy. As a result of this analysis, it is clear that we cannot look at mutations with the oversimplified expectation that most of them are single base substitutions. As more of them are analyzed directly, our picture of mutation will become increasingly clear.

DETECTION OF MUTAGENICITY: THE AMES TEST

There is particular concern about the possible mutagenic properties of any chemical that enters the human body, whether through the skin, digestive tract, or respiratory tract. For example, great attention has been given to residual materials of air and water pollution, food preservatives and additives, artificial sweeteners, herbicides, pesticides, and pharmaceutical products. While mutagenicity may be tested in various organisms, including *Drosophila* and mice, as well as in cultured mammalian cells, the most common test involves bacteria and was devised by Bruce Ames.

The Ames system utilizes four tester strains of the bacterium *Salmonella typhimurium* that were selected for sensitivity and specificity for mutagenesis. One strain is used to detect base-pair substitutions, and the other three detect various frameshift mutations. Each mutant strain requires histidine for growth (*his*⁻). The assay measures the frequency of reverse mutation, which yields wild-type (*his*⁺) bacteria. Greater sensitivity to mutagens occurs because these strains bear other mutations that eliminate the DNA excision repair system, discussed later in this chapter, and the lipopolysaccharide barrier that coats and protects the surface of the bacteria.

It is of considerable interest to note that many substances that enter the human body are relatively innocuous until activated metabolically to a more chemically reactive electrophilic product. This usually occurs in the liver. Thus, the **Ames test,** which is performed *in vitro*, includes a step in which the test compound is incubated in the presence of a mammalian liver extract. Or, test compounds are actually injected into the mouse, which is later sacrificed and the liver removed. Extracts are then tested.

In the initial use of Ames test in the 1970s, a large number of known carcinogens were tested. Over 80 percent of these were shown to be strong mutagens! This is not so surprising, since transformation of cells to the malignant state undoubtedly occurs as a result of some alteration of DNA. While a positive response as a mutagen does not prove the carcinogenic nature of a test compound, the Ames test is useful as a preliminary screening device. As such, it is used extensively in conjunction with the industrial and pharmaceutical development of chemical compounds.

RADIATION, DNA DAMAGE, AND ITS REPAIR

In addition to chemicals, other components of the natural and man-made environment are capable of inducing mutations. The following sections consider ultraviolet radiation (UV light) and high-energy radiation as two such components. Both are part of the electromagnetic spectrum and have shorter wavelengths than visible light (Figure 14.14). X-rays and gamma radiation are characterized by exceedingly short wavelengths. Since the inherent energy associated with components of this spectrum is inversely proportional to wavelength, both X-rays and gamma radiation are considered together as **high-energy radiation**.

As we shall see, the study of the mutagenicity of UV light paved the way for the discovery of DNA damage; a great deal is now known about natural repair of many forms of DNA damage. The ability to perform such repair is essential to the survival of all organisms, including humans.

Since we live in the nuclear age, concern about high-energy radiation is particularly keen. We shall consider not only the mutagenic effects of high-energy radiation but also some of the concerns involving radiation safety. The relevance is obvious as we approach the end of the twentieth century.

UV Radiation, Thymine Dimers, and Photoreactivation

In Chapters 8 and 9, we emphasized the fact that purines and pyrimidines absorb **ultraviolet light (UV)** most intensely at a wavelength around 260 nm. This property has been used extensively in the detection and analysis of nucleic acids. In 1934, as a result of studies involving *Drosophila* eggs, it was discovered that ultraviolet light is mutagenic. By 1960, several studies concerning the *in vitro* effect of UV light on the components of nucleic acids had been completed with the following conclusions. The major effect of UV light is on pyrimidines at a point where the addition of water to the ring structure is initiated. Subsequently, pyrimidine dimers are formed, particularly between two thymine residues (Figure 14.15). While cytosine-cytosine and thymine-cytosine dimers may also be formed, they are less prevalent. It is believed that the dimers distort the DNA conformation and inhibit normal replication. This inhibition seems to be responsible for the killing effects of UV light on microorganisms.

Other studies have provided more complete information on the relationship of UV light to genetic phenomena. These studies have demonstrated that an intricate set of repair processes function to counteract the lesions produced in DNA as a result of UV

FIGURE 14.14
The components of the electromagnetic spectrum and their associated wavelengths.

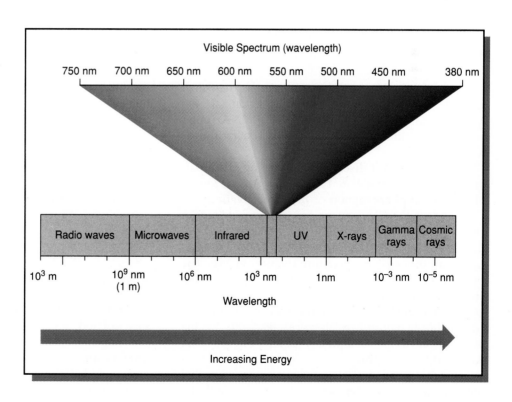

FIGURE 14.15
Formation of thymine dimers
induced by ultraviolet light.

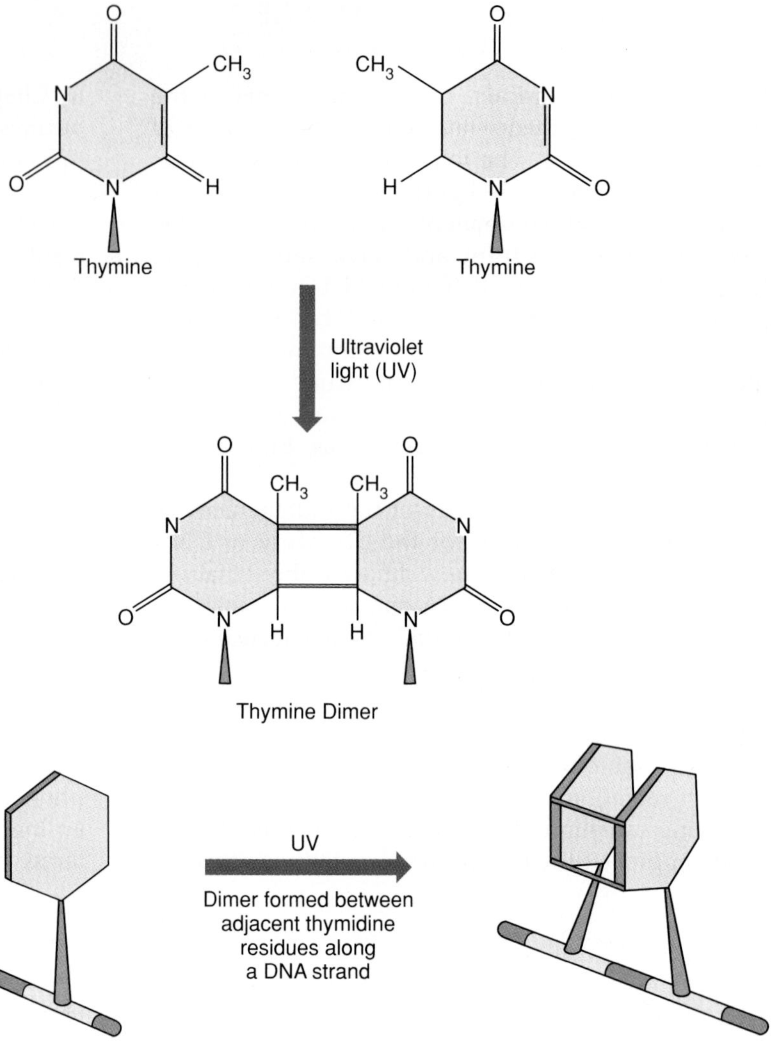

irradiation. It is during the repair process that most errors leading to UV-induced mutations occur. In the following discussion, we will examine DNA repair systems, as initially elucidated in the bacterium *E. coli*.

The first relevant discovery concerning UV repair in bacteria was made in 1949 when Albert Kelner observed the phenomenon of **photoreactivation**. He showed that the UV-induced damage to *E. coli* DNA could be partially reversed if, following irradiation, the cells were briefly exposed to light in the blue range of the visible spectrum (Figure 14.14). The photoreactivation process has subsequently been shown to be temperature-dependent, thereby suggesting that the light-induced mechanism involves an enzymatically controlled chemical reaction. Visible light appears to induce a process of repair in the DNA damaged by UV light.

Further studies of photoreactivation have revealed that the process is due to an enzyme, called the **photoreactivation enzyme (PRE)**. This molecule may be isolated from extracts of *E. coli* cells, and its repair activity is destroyed by heat. The enzyme's mode of action is to cleave the bonds between thymine dimers, thus reversing the effect of UV light on DNA [Figure 14.16(a)]. While the enzyme will associate with a dimer in the dark, it must absorb a photon of light to cleave the dimer.

Excision Repair of DNA

Investigations in the early 1960s suggested that a repair system or systems that do not require light also existed in *E. coli*. Paul Howard-Flanders and coworkers isolated several independent mutants demonstrating increased sensitivity to ultraviolet radiation. One group was designated *uvr* (ultraviolet repair) and included the *uvrA*, *uvrB*, and *uvrC* mutations. These genes and their protein products were subse-

FIGURE 14.16
Contrasting diagrams of photoreactivation and excision repair of UV-induced thymine dimers.

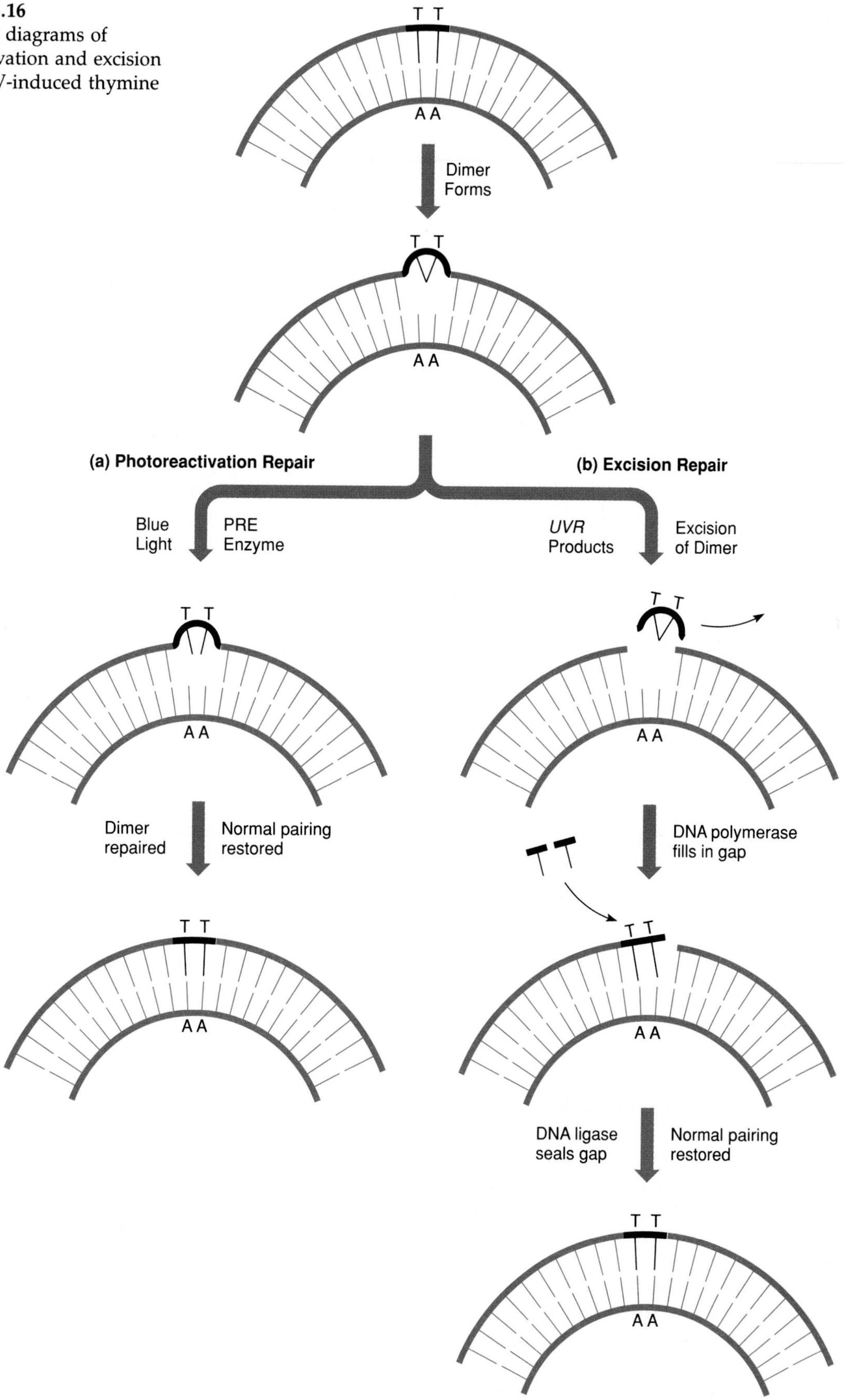

T T
A A
Dimer
Forms
T T
A A
(a) Photoreactivation Repair
(b) Excision Repair
Blue
Light
PRE
Enzyme
UVR
Products
Excision
of Dimer
T T
A A
T T
A A
Dimer
repaired
Normal pairing
restored
DNA polymerase
fills in gap
T T
A A
T T
A A
DNA ligase
seals gap
Normal pairing
restored
T T
A A

quently shown to be involved in a process called **excision repair**. During this process, three steps have been shown to occur, as illustrated in Figure 14.16(b):

1 The portion of the strand containing the UV-induced dimer is recognized and enzymatically clipped out by a nuclease that cleaves the phosphodiester bonds. This "excision" leaves a gap in the helix. The *uvr* gene products operate at this step.
2 **DNA polymerase I** fills this gap by inserting d-ribonucleotides complementary to those on the intact strand. The enzyme adds these bases to the 3'-OH end of the clipped DNA.
3 The joining enzyme, **DNA ligase**, seals the final "nick" that remains at the 3'-OH end of the last base inserted, closing the gap.

Recall from Chapter 10 that DNA polymerase I is the enzyme discovered by Kornberg and long assumed to be the universal DNA-replication enzyme. The discovery of the *pol A1* mutation demonstrated that this was not the case. *E. coli* cells carrying the *pol A1* mutation lack functional polymerase I. However, replication of DNA takes place normally. Cells with this mutation are unusually sensitive to UV light. Apparently, such cells are unable to fill in the gap created by the excision of the thymine dimers. This finding demonstrated the importance of excision repair mechanisms in counteracting the effects of UV light.

It has since been shown that the process of excision repair can be activated in response to any damaged DNA that distorts the helix. For example, the loss of a purine from d-ribose of one strand creates what is called an **apurinic site**. Such a loss can occur spontaneously or be induced by a powerful carcinogen, aflatoxin B (derived from an *Aspergillus* mold). The complementary pyrimidine on the opposite strand has nothing with which to form hydrogen bonds. Such a sugar with a missing base is recognized by an enzyme called **AP endonuclease** (AP for *apurinic*), which clips out the DNA at that site. After its removal, repair occurs using DNA polymerase I and DNA ligase. Other enzymes, called **DNA glycosylases** recognize other altered bases and catalyze their excision, leading to repair.

The ability to repair such errors is extremely important to maintaining DNA sequences with high fidelity. The importance is underscored when it is realized that approximately 5000 spontaneous depurinations occur daily from the DNA of each human cell. These chemical events are attributable to random thermal fluctuations.

Proofreading and Mismatch Repair

Since we have chosen to introduce DNA repair mechanisms during our consideration of mutations, it is logical to complete our discussion of this topic here. Spontaneous errors that occur during replication are introduced at a rate of approximately one in 100,000 (10^{-5}). Each error, involving a **mismatched base pair,** is immediately subject to proofreading associated with DNA polymerase III, as discussed in Chapter 10. Proofreading is thought to increase fidelity by two orders of magnitude, leaving only one in 10,000,000 (10^{-7}) mismatches.

Once proofreading has occurred, still another mechanism, called **mismatch repair,** may be activated. Proposed over 20 years ago by Robin Holliday, the molecular basis of this process is now well established. Like other DNA lesions, (1) the alteration or mismatch must be detected, (2) the incorrect nucleotide must be removed, and (3) replacement with the correct base must occur. But a special problem exists with correction of a mismatch. There is no physical distortion of the helix to "signal" that a problem exists as there is with UV damage. In a mismatch, one of the two strands, the parental strand, which served as the template during replication, contains the "correct base." The newly synthesized strand, however, bears the mismatch. How the repair system discriminates and recognizes the nonparental strand puzzled geneticists for decades. If the mismatch is recognized, but no discrimination occurred and excision was random, half the time the strand bearing the correct base would be clipped out. The concept of strand discrimination by a repair enzyme is thus the critical step.

At least in some bacteria, including *E. coli*, this process has been elucidated and is based on the process of **DNA methylation**. These bacteria contain methylases which recognize the DNA sequence

$$5' \ldots GATC \ldots 3'$$
$$3' \ldots CTAG \ldots 5'$$

as a substrate. Upon recognition, a methyl group is added to each of the adenine residues. This modification is stable throughout the cell cycle.

Following methylation and a further round of replication, the newly synthesized strands remain temporarily unmethylated. It is at this point that the repair enzyme recognizes the mismatch and preferentially binds to the unmethylated strand. It is excised and the correct complement is inserted! Interestingly, the GATC sequence need only be within several thousand base pairs of the mismatch. A series of *E. coli* gene products, *mutH, L, S,* and *U* are involved in the discrimination step. Mutations in each result in strains quite deficient in mismatch repair.

While it is agreed that mismatch repair undoubtedly also occurs in the DNA of higher organisms, the question of strand discrimination remains speculative in the absence of GATC methylation.

Recombinational Repair

The final mode of repair to be discussed was discovered in an excision-defective strain of *E. coli*. Called **recombinational repair,** this system is thought to respond when damaged DNA has escaped repair and the damage disrupts the process of replication. The cells that show this phenomenon are dependent on the product of a gene, *rec A*, which is involved in several types of recombination in *E. coli*.

When such DNA is being replicated, DNA polymerase at first stalls at and then skips over the distortion, creating a gap on one of the newly synthesized strands. To counteract this, the *rec A* protein directs a recombinational exchange process whereby this gap is filled as a result of the insertion of a segment initially present on the intact homologous strand. The resulting gap now present on the "donor" strand is filled by repair synthesis as replication proceeds.

This mode of repair is a "rescue operation" that allows replication to create accurate and complete DNA copies from a template that is damaged. Phil Hanawalt and Paul Howard-Flanders, among others, have established that many different gene products are involved. Of particular interest is the *lex A* product, which serves to repress the function of the *rec A* and *uvr* genes. However, when a *rec A* protein binds to DNA in the area of a gap, this binding somehow activates a second function of the *rec A* protein—the ability to cleave the *lex A* repressor molecule. The absence of a functional repressor molecule allows the activation of the *rec A* and *uvr* genes, leading to an increased production of the proteins for which they code. The activation of the repair system in this way has been called the **SOS response,** which counteracts most of the dimers through the process of excision and recombinational repair.

One disadvantage of the repair process associated with the SOS response is a diminution of an important function performed by DNA polymerase III. Normally, the enzyme **proofreads** as it proceeds along the DNA strand. This editing function is possible because of 3'-exonuclease function of the enzyme. Proofreading occurs at the 3' terminus of the growing DNA chain; a nucleotide is excised immediately if it is not correctly base paired with its corresponding nucleotide present in the template strand. However, during the SOS response, the proofreading function is relaxed so that polymerization does not halt when it encounters a dimer that has distorted the shape of the double helix. While the mechanism by which the editing system is shut down is not yet understood, the effect is to prevent the correction of spontaneous errors at sites other than dimers. Thus, the fidelity of DNA replication is diminished. As a result, the SOS response to the repair of DNA is said to be an **error-prone system.** Fortunately, the SOS repair system is not constantly engaged; instead it only becomes active when induced by UV-light damage to DNA.

UV Light and Human Skin Cancer: Xeroderma Pigmentosum

Xeroderma pigmentosum (XP), a rare autosomal recessive disorder in humans, predisposes individuals to epidermal pigment abnormalities. Exposure to sunlight results in malignant growth of the skin.

Figure 14.17 contrasts two XP individuals, one of whom has been detected early and protected from sunlight. The condition is very severe and may be lethal. Since sunlight contains UV radiation, scientists have suspected a causal relationship between thymine dimer production and XP. They have also been trying to determine which of the three forms of repair processes counteracting the effects of UV-induced damage to DNA, if any, are operating in humans. Furthermore, the question has been raised whether XP individuals might lack one or more repair systems, which may cause them to be susceptible to UV-induced skin damage.

The various modes of repair of UV-induced lesions have been investigated in human fibroblast cultures derived from XP and normal individuals. Fibroblasts are undifferentiated connective tissue cells. The results, most of which have been obtained since 1968, are varied and suggest that the XP phenotype may be caused by more than one mutant gene, as the following summary shows:

1   In 1968, James Cleaver showed that cells from XP patients were deficient in the **unscheduled DNA synthesis** elicited in normal cells by UV light. This assay is thought to represent activity of the excision repair system, suggesting that XP cells are deficient in this form of repair.
2   There are two clinical forms of XP, the classical disease and the **DeSanctis-Cacchione syndrome**

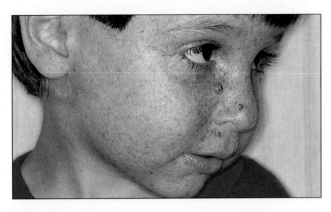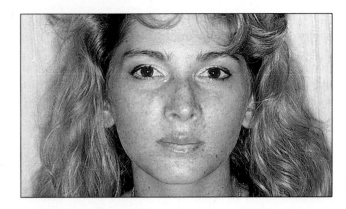

FIGURE 14.17

Xeroderma pigmentosum. The four-year-old boy on the left shows marked skin lesions induced by sunlight. Mottled redness (erythema) and irregular pigment changes that are a response to cellular injury are apparent. Two nodular cancers are present on his nose. The eighteen-year-old girl on the right has been carefully protected from sunlight since the diagnosis of xeroderma pigmentosum made in infancy. Several cancers have been removed and she works as a successful model.

characterized by the additional involvement of neurological symptoms. Both types have been shown to be deficient in the excision repair process.

3 In 1974, the presence of a **photoreactivation enzyme** (PRE) was established in normal human cells. Betsy Sutherland identified the enzyme first in leukocytes and subsequently in fibroblast cells.

4 Sutherland further demonstrated that some XP cultures contained a lower PRE activity than control cultures. The activity in various XP cell strains ranged from zero to 50 percent in cultures established from different patients.

5 Cultured cells from any two XP variants may be induced to fuse together, forming a **heterokaryon** where the two nuclei share a common cytoplasm. This process is one aspect of **somatic cell genetics** (see Chapter 7). When fusion is performed with cells of different XP variants, excision repair, as assayed by unscheduled DNA synthesis, is sometimes reestablished. When this occurs, the two variants are said to undergo **complementation**. Alone, neither demonstrates excision repair, but together in a heterokaryon the process occurs. In genetic terms, this is taken as evidence that different genes are involved in the defect of each variant. Complementation occurs because the heterokaryon has at least one normal copy of each gene. In many studies, variants have been divided into seven complementation groups, suggesting that as many as seven different genes may be involved in causing XP.

These findings are of great interest for two reasons. First, they suggest that XP results if any of the three repair systems is inactivated as a result of mutation. Second, they establish that normal individuals are probably susceptible to UV-induced damage of DNA by exposure to sunlight. However, this damage activates the repair systems that counteract it. We can expect that future work will clarify the precise role and mechanism of repair of UV-induced damage in humans.

High-Energy Radiation

Within the electromagnetic spectrum, energy varies inversely with wavelength. Recall that we earlier compared the relative wavelengths of the various portions of the electromagnetic spectrum (see Figure 14.14). **X-rays, gamma rays,** and **cosmic rays** have even shorter wavelengths and are therefore more energetic than ultraviolet light. As a result, they are strong enough to penetrate deeply into tissues, causing **ionization** of the molecules encountered along the way. It might be anticipated that these sources of **ionizing radiation** would be mutagenic. Indeed, Herman Muller detected such effects in 1927 while working with *Drosophila*. Since that time, the effects of ionizing radiation, particularly X-rays, have been studied intensely.

As X-rays penetrate cells, electrons are ejected from the atoms of molecules encountered by the radiation. Thus, stable molecules and atoms are transformed into free radicals and reactive ions. Along the path of a high-energy ray, a trail of ions is left that can initiate a variety of chemical reactions. These reactions can directly or indirectly affect the genetic material, altering the purines and pyrimidines in DNA and resulting in point mutations. Such ionizing radiation is also capable of breaking chromosomes, resulting in a variety of aberrations.

Figure 14.18 shows a plot of induced sex-linked recessive lethals versus the dose of X-rays administered. The graph shows a straight line that, if extrapolated, intersects near the zero axis. A linear relationship is evident between X-ray doses and the induction of mutation. For each doubling of dose, twice as many mutations are induced. Because the line intersects near the zero axis, this graph suggests that there is no minimal dose of irradiation that is not mutagenic.

These observations may be interpreted in the form of the **target theory,** first proposed in 1924 by J. A. Crowther and F. Dessauer. The theory proposes that there are one or more sites, or targets, within cells and that a single event of irradiation at one site will bring about a damaging effect, or mutation. In a simple form, the target theory says that one "hit" of irradiation will cause one "event" or mutation, suggesting that the X-rays interact directly with the genetic material.

The target theory is not completely applicable because it has been shown that indirect effects of irradiation may also cause mutations. For example, if prior to inoculation of a bacterial culture the medium upon which the cell will grow is treated with X-rays, the bacteria subsequently demonstrate an induced mutation rate. Thus, reactive molecules in the medium are produced that can secondarily interact with DNA of bacterial cells. It has also been observed in

several organisms, including *Drosophila*, that reduced oxygen concentration in tissue lowers the induced mutation rate. Oxygen is believed to be critical in the formation of highly reactive peroxides in tissues. Peroxides may be formed in the presence of oxygen following irradiation of water, which is split into its hydrogen and hydroxyl groups. These findings support the concept that the indirect effects of high-energy radiation may be extremely important in the observed mutagenic effect.

Two other observations concerning irradiation effects are of particular interest. In general, the **intensity of the dose** (dose-rate) administered seems to make little difference in mutagenic effect for most organisms studied. That is, a 2000-roentgen exposure (a **roentgen** is a measure of dose equal to the production of two billion ionizations per cubic centimeter), whether occurring in a single acute dose or cumulatively in many smaller chronic doses, seems to produce the same mutagenic effect. *Drosophila*, for example, shows this response. Mice, however, are an important exception to this general rule. In this organism, and presumably in other mammals, repair of the damage seems to occur during the intervals between irradiation. Thus, several smaller doses are not so potent as a single large dose.

The second observation is that certain portions of the cell cycle are more susceptible to irradiation effects. As mentioned above, in addition to a mutagenic effect, X-rays will also break chromosomes, resulting in terminal or intercalary deletions, translocations, and general chromosome fragmentation. Damage occurs most readily when chromosomes are greatly condensed in mitosis. This property constitutes one of the reasons why radiation is used to treat human malignancy. Since tumorous cells are more often undergoing division than their nonmalignant counterparts, they are more susceptible to the immobilizing effect of radiation.

RADIATION SAFETY

Technological advances during the past 50 years have led to widespread applications of radiation in our society. Largely due to the pioneering efforts of Muller, the potential dangers of radiation to humans are known and radiation safety has received increasing attention. In addition to natural sources of radiation, therapeutic and diagnostic use, industrial application, and atomic and nuclear fallout represent the sources of major concern.

It is now known that a dose of radiation equivalent to 50 to 150 roentgens (r) is lethal to 50 percent of human cells in culture. Whole-body irradiation of

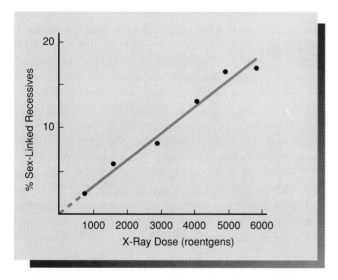

FIGURE 14.18
Plot of sex-linked recessive mutations induced by increasing doses of X-rays. If extrapolated, the graph intersects the zero axis (dashed line).

400 r is usually fatal to individuals, and 100 r is sufficient to cause radiation sickness. Sublethal doses of radiation may also significantly shorten the human life span and increase the occurrence of various forms of cancer, particularly leukemia.

Since there seems to be no lower limit for the amount of radiation that will induce mutation and since the effects of ionizing radiation can be cumulative, precautionary measures are essential to the human population. In 1970, a National Academy of Sciences panel was commissioned to investigate radiation safety standards. At that time, the federal guidelines stated that the general population should not receive more than 170 millirems of artificial radiation per year, exclusive of medical sources. A **rem** is a "roentgen equivalent in man," and a millirem (mrem) is 1/1000 of a rem. The committee's report included the following statement:

> The present guides of 170 millirems per year grew out of an effort to balance societal needs against genetic risks. It appears that these needs can be met with far lower average exposures and lower genetic and somatic risk than permitted by the current radiation protection guide. To this extent, the current guide is unnecessarily high.

The committee strongly admonished the medical profession and urged it to improve the efficiency of X-ray equipment; increase shielding, particularly of the reproductive organs; and avoid unnecessary mass screening for many diseases. The panel estimated that an exposure of 170 millirems per year increases the death rate in the United States by between 3000 and 15,000 people each year. For each additional 100-millirem increase, 3500 added deaths from cancer were estimated per year. Table 14.3 summarizes the report's estimate of annual exposure from various sources of radiation.

TABLE 14.3
Estimates of annual whole-body exposure to radiation from various sources in the United States (1970).

Source	Average Dose Rate (mrem/yr)
Natural	102
Global fallout	4
Nuclear power	0.003
Diagnostic	72
Radiopharmaceuticals	1
Occupational	0.8
Miscellaneous	2
TOTAL	182

The information presented by the National Academy of Sciences and more recently by the International Commission on Radiological Protection (ICRP) is sufficient to cause concern about the future of the human species. Many scientists have projected that the human population will continue to accumulate deleterious mutations as a result of exposure to mutagens such as radiation. The concept of **genetic burden** refers to deleterious alleles that are already present in the human gene pool. It is estimated that the average individual carries two to ten recessive lethal alleles in the heterozygous condition. Before absolute conclusions concerning the dangers of radiation to the genetic well-being of future human populations can be drawn, more direct evidence is needed. Tragically, the atomic bomb attacks on Japan in 1945 have allowed us to investigate this question.

Immediately following World War II, the **Atomic Bomb Casualty Commission (ABCC)** was established. Its large-scale study began in 1946 as a joint venture between the U.S. National Research Council and the Japanese National Institute of Health. Conducted at the Radiation Effects Research Foundation in Hiroshima, this scientific project is still active in the 1990s. The overall study considered various parameters of radiation effects on somatic and germinal tissue.

The results of massive radiation exposure to the human population were devastating. Between 100,000 and 250,000 fatalities were estimated. Somatic effects on survivors were severe. Sublethal radiation doses led to chromosome aberrations, brain damage and retardation, decreased lifespans, and an elevation of various cancers, particularly leukemia.

In spite of these dire consequences, one aspect of the study was more encouraging and came from investigations to assess the impact of radiation exposure on germinal tissue and thus on future generations. Scientists studied children conceived after August, 1945, who had at least one parent exposed in Hiroshima or Nagasaki. From 1948 to 1953, over 70,000 such cases were located. All newborns were examined, and 40 percent of the children were reexamined when they were 8 to 10 months old.

The results of the study through 1953 revealed no statistically significant effect of increasing levels of radiation exposure. Increases in congenital malformations, stillbirths, altered birth weights, or abnormal physical development at age 8 to 10 months could not be detected. In an ongoing investigation, members of this group as well as other children born after 1953 to exposed parents are being examined. The focus of this study has been on mortality, chromosome abnormalities, and alterations in specific proteins. The findings parallel the results of the examination of newborns. No statistically significant increases have been observed as a result of radiation exposure. While the development of cancer has been the major delayed effect on those individuals exposed directly, their children, thus far in the early 1990s, fail to exhibit such an effect. All indicators suggest that humans are less sensitive to the genetic effects of ionizing radiation than expected. We can only hope that this conclusion continues to be upheld as this important study progresses.

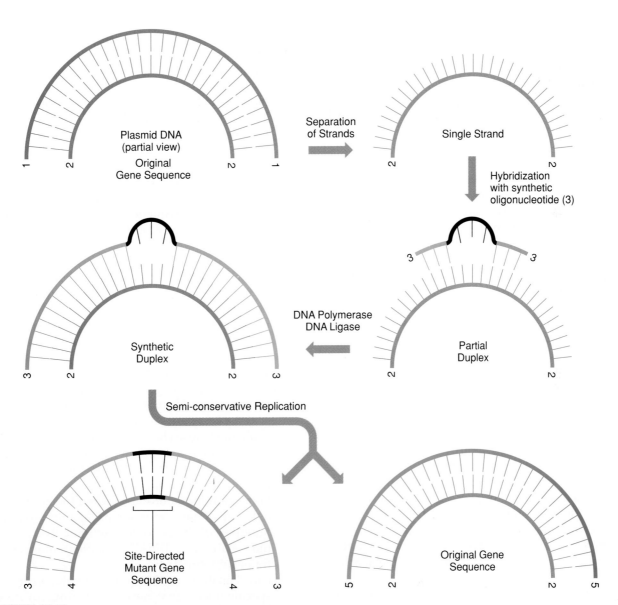

FIGURE 14.19
Site-directed mutagenesis, achieved by first obtaining a single strand of DNA from a gene of interest. This is hybridized with a synthetic oligonucleotide containing a triplet altered so as to encode an amino acid of choice. The partial duplex is completed under the direction of DNA polymerase and DNA ligase. Following semiconservative replication, a different complementary base pair is present in one of the new duplexes. Upon transcription and translation, a mutant protein, designed in the laboratory, will be produced. Only part of the plasmid DNA is shown.

SITE-DIRECTED MUTAGENESIS

We conclude this chapter by introducing a recently developed, sophisticated technique that allows tailor-made changes within genes: **site-directed mutagenesis**. The technique relies on a number of manipulations involving recombinant DNA technology, which will be introduced in Chapter 16. However, the underlying principles are based on information presented previously. This technique is of great interest here because it illustrates the application of basic knowledge of a topic such as mutation to current research investigations.

The goal of the technique is to specifically alter one or more nucleotides within a gene in order to change the genetic code. Upon transcription and translation, this change will cause the insertion of a "mutant" amino acid in the protein encoded by the original gene. Such designed mutations are particularly useful in studying the effects of specific mutations on protein function.

The first step is to determine the nucleotide sequence of the gene being studied. This can be accomplished by direct sequencing techniques (see Appendix A), or it can be predicted if the amino acid sequence of the protein is known by utilizing our knowledge of the genetic code.

The next step is to isolate the DNA of known sequence and obtain from it one of the two complementary strands. A decision is then made as to which nucleotides are to be changed. Then, a small piece of DNA complementary to that region is chemically synthesized. It is complementary at all points, except in the triplet sequence (or sequences) that is to be altered. This triplet encodes the amino acid that will change in the protein.

As illustrated in Figure 14.19, this short piece of DNA will hybridize with the original parent strand because of its complementarity along most of its length. If DNA polymerase and DNA ligase are then added to this hybrid complex, the short sequence will be extended so that a duplex of the entire gene is formed. The two strands are perfectly complementary except at the point of alteration.

When this DNA undergoes semiconservative replication, two types of duplexes are formed: one is like the original, unaltered gene and the other contains the newly designed sequence. Using recombinant DNA technology, it is possible not only to complete the above manipulations, but also to allow the altered gene to be expressed so that large amounts of the desired protein are available for study. Such an approach is useful in investigating the role of specific amino acids during protein function, one of the topics of the next chapter.

CHAPTER SUMMARY

1 The phenomenon of mutation not only provides the basis for most of the inherent variation present in living organisms, but also serves as the working tool of the geneticist in studying and understanding the nature of genes and the mechanisms governing genetic processes.

2 Mutations are distinguished by the tissues affected. Somatic mutations are those that may affect the individual, but are not heritable. Mutations arising in the gametic tissues may produce new alleles that can be passed on to offspring and enter the gene pool.

3 Another classification of mutations relies on their effect. Morphological mutations, for example, may be detected visibly. Other types include biochemical, lethal, conditional, and regulatory mutations, groups that are not mutually exclusive.

4 Organisms in which mutations can be easily induced and detected are most often used in genetic studies. Viruses, bacteria, fungi, and *Drosophila* are frequently used because of these properties, as well as the fact that they have short life cycles.

5 Spontaneous mutations may arise naturally as a result of rare chemical rearrangements of atoms, or tautomeric shifts, and as the result of errors occurring during DNA replication. While spontaneous mutations are very rare, their rate of occurrence may be increased experimentally by a variety of mutagenic agents.

6 The elucidation of the molecular mechanisms involved in mutagenesis is an area of active investigation. Mutagenic agents such as nitrous acid, hydroxylamine, alkylating agents, and base analogues cause chemical changes in nucleotides that alter their base pairing affinities. As a result, base substitution mutations arise following DNA replication.

7 Frameshift mutations arise when the addition or deletion of one or more nucleotides (but not multiples of three) occurs. Acridine dyes, which intercalate with DNA, are potent inducers of frameshift mutations.

8 Direct analysis of DNA from individuals carrying specific alleles that have arisen through mutations, including the ABO blood types and muscular dystrophy, have been informative. When complete loss of function occurs, as is the case in blood type O and the Duchenne form of muscular dystrophy, deletions or duplications of nucleotides have resulted in a shift in reading frame,

eventually causing premature termination of translation of the resulting messenger RNA.

9 Ultraviolet light and high-energy radiation from gamma, cosmic, or X-ray sources are another form of potent mutagenic agents. UV light induces the formation of pyrimidine dimers in DNA, while high-energy radiation is able to penetrate tissues, causing the ionization of molecules in its path.

10 Various forms of genetically controlled repair of DNA lesions, such as pyrimidine dimers, have been discovered, including photoreactivation, excision, mismatch, and recombinational repair. Loss of repair function by mutation in humans results in the severe disorder xeroderma pigmentosum.

11 Safety reports from the NAS and ABCC have increased our awareness of and concern for the potential mutagenicity and carcinogenicity of ionizing radiation. It appears that there is no minimal dose of radiation that is without effect. Massive exposure, as occurred at Hiroshima and Nagasaki at the end of World War II, caused widespread death and somatic damage to survivors. Thus far, the genetic effect has been much less than expected, as studied in children conceived by exposed parents.

12 Site-directed mutagenesis is a technique allowing researchers to create specific alterations in the nucleotide sequence of the DNA of genes. In this way, geneticists may investigate the functional role of various amino acids, and the importance of their positions within proteins.

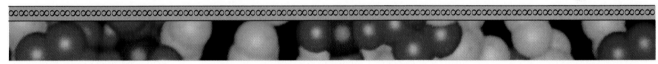

Insights and Solutions

1 How could you isolate a mutant strain of bacterial cells that is resistant to penicillin, an antibiotic that inhibits cell wall synthesis?

ANSWER: Grow a culture of bacterial cells in liquid medium and plate the cells on agar medium to which penicillin has been added. Only penicillin-resistant cells will reproduce and form colonies. Each colony will, in all likelihood, represent a cloned group of cells with the identical mutation. Isolate members of each colony. To enhance the chance of such a mutation arising, you might want to add a mutagen to the liquid culture.

2 The base analogue 2-amino purine (2-AP) substitutes for adenine during DNA replication, but it may base pair with cytosine. The base analogue 5-bromouracil (5-BU) substitutes for thymidine, but it may base pair with guanine. Follow the double-stranded trinucleotide sequence shown below through three rounds of replication, assuming that in the first round, both analogues are present and become incorporated wherever possible. In the second and third round of replication, they are removed. What final sequences occur?

ANSWER: The answer to this problem is illustrated on the following page.

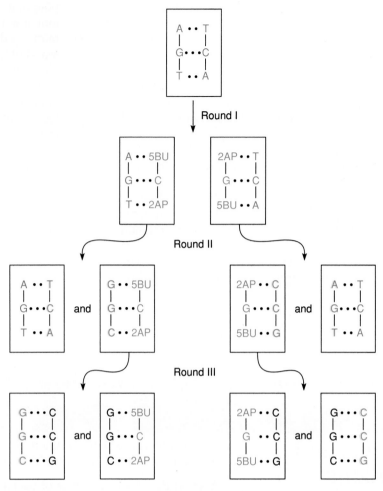

3 Assuming that in Problem 2 the initial trinucleotide TCA is the sense strand, consult the code table and determine what initial amino acid is encoded. Also determine what mutant amino acids might subsequently occur from the trinucleotide sequences produced after three rounds of replication in the presence of the base analogues.

ANSWER: Assuming that TCA is transcribed into the mRNA 5′-AGU, then the wild-type amino acid encoded by this sequence is serine. Subsequently, the DNA sequences GGC and CCG are produced, which upon transcription produce the RNA complements of CCG and GGC, respectively. These encode the amino acids proline and glycine.

PROBLEMS AND DISCUSSION QUESTIONS

1 What is the difference between a chromosomal aberration and a point mutation?
2 Discuss the importance of mutations to the successful study of genetics.
3 Describe the technique for detection of nutritional mutants in *Neurospora*.
4 Most mutations are thought to be deleterious. Why, then, is it reasonable to state that mutations are essential to the evolutionary process?
5 Why do you suppose that a random mutation is more likely to be deleterious than beneficial?
6 Most mutations in a diploid organism are recessive. Why?
7 What is meant by a conditional lethal mutation?
8 Contrast the concerns about mutation in somatic and gametic tissue.

9   In the *ClB* technique for detection of mutation in *Drosophila*:
    (a)   What type of mutation may be detected?
    (b)   What is the importance of the *l* gene?
    (c)   Why is it necessary to have the crossover suppressor (C) in the genome?
    (d)   What is C?

10 In an experiment with the *ClB* system, a student irradiated wild-type males and subsequently scored 500 F_2 cultures, finding all of them to have both males and females present. What conclusions can you draw about the induced mutation rate?

11 In *Drosophila*, induced mutations on chromosome 2 that are recessive lethals may be detected using a second chromosomal stock, *Curly, Lobe/Plum (Cy L/Pm)*. These alleles are all dominant and lethal in the homozygous condition. Detection is performed by crossing *Cy L/Pm* females to wild-type males that have been subjected to a mutagen. Three generations are required. In the F_1, *Cy L* males are selected and individually backcrossed to *Cy L/Pm* females. In the F_2 of each cross, flies expressing *Cy* and *L* are mated to produce a series of F_3 generations. Diagram these crosses and predict how an F_3 culture will vary if a recessive lethal was induced in the original test male compared to the case where no lethal mutation resulted. In which F_3 flies would a recessive morphological mutation be expressed?

12 Describe tautomerism and the way in which this chemical event may lead to mutation.

13 Contrast and compare the mutagenic effects of nitrous acid, hydroxylamine, alkylating agents, and base analogues.

14 Acridine dyes induce frameshift mutations. Is such a mutation likely to be more detrimental than a point mutation where a single pyrimidine or purine has been substituted?

15 Why is X-radiation a more potent mutagen than UV light?

16 Contrast the induction of mutations by UV light and X-radiation.

17 Contrast the various types of DNA repair mechanisms known to counteract the effects of UV light and other DNA damage. What is the role of visible light in photoreactivation?

18 Mammography is an accurate screening technique for the early detection of breast cancer in humans. Because this technique uses X-rays diagnostically, it has been highly controversial. Can you explain why?

19 The ABCC study of the radiation effects on humans examined various forms of somatic and gametic damage. What type of genetic effects could not be examined immediately, but are of great concern? When might they become "visible?"

20 Describe the Ames assay for screening potential environmental mutagens. Why is it thought that a compound that tests positively in the Ames assay may also be carcinogenic?

21 Describe the methodology used in site-directed mutagenesis.

22 Describe the disorder xeroderma pigmentosum in humans.

23 What is known about repair mechanisms of UV-induced lesions in humans? Relate this information to what has been learned about the genetic basis of xeroderma pigmentosum.

24 Presented below are theoretical findings from studies of heterokaryons formed from human xeroderma pigmentosum cell strains:

	XP1	XP2	XP3	XP4	XP5	XP6	XP7
XP1	0						
XP2	0	0					
XP3	0	0	0				
XP4	+	+	+	0			
XP5	+	+	+	+	0		
XP6	+	+	+	+	0	0	
XP7	+	+	+	+	0	0	0

NOTE: + = complementing; 0 = noncomplementing.

These data represent the occurrence of unscheduled DNA synthesis in the fused heterokaryon when neither of the strains alone showed synthesis. What does unscheduled DNA synthesis represent? Which strains fall into the same complementation groups? How many groups are revealed based on these limited data? How does one interpret the presence of these complementation groups?

25 If the human genome contains 100,000 genes and the mutation rate at each of these loci is 5×10^{-5} per gamete formed, what is the average number of new mutations that exist in each individual? If the current population is 4.3 billion people, how many newly arisen mutations exist in the current populace?

26 In a bacterial culture where all cells were unable to synthesize leucine (*leu*$^-$) a potent mutagen was added, and the cells were allowed to undergo one round of replication. At that point, samples were taken and a series of dilutions was made prior to plating the cells on either minimal medium or minimal medium to which leucine was added. The first culture condition (minimal medium) allows the detection of mutations from *leu*$^-$ to *leu*$^+$, while the second culture condition allows the determination of total cells, since all bacteria can grow. From the following results, determine the frequency of mutant cells. What is the rate of mutation at the locus involved with leucine biosynthesis?

Culture Condition	Dilution	Colonies
minimal medium	10^{-1}	18
minimal + leucine	10^{-7}	6

∞∞∞∞∞∞∞∞∞∞∞∞∞∞∞∞∞∞∞∞∞∞∞∞∞∞∞∞

SELECTED READINGS

AMES, B. N., McCANN, J., and YAMASAKI, E. 1975. Method for detecting carcinogens and mutagens with the *Salmonella*/mammalian microsome mutagenicity test. *Mut. Res*: 31: 347–64.

AUERBACH, C. 1978. Forty years of mutation research: A pilgrim's progress. *Heredity* 40: 177–87.

AUERBACH, C., and KILBEY, B. J. 1971. Mutations in eukaryotes. *Ann. Rev. Genet.* 5: 163–218.

BALTIMORE, D. 1981. Somatic mutation gains its place among the generators of diversity. *Cell* 26: 295–96.

BEADLE, G. W., and TATUM, E. L. 1945. Neurospora II. Methods of producing and detecting mutations concerned with nutritional requirements. *Amer. J. Bot.* 32: 678–86.

BECKER, M. M., and WANG, Z. 1989. Origin of ultraviolet damage in DNA. *J. Mol. Biol.* 210: 429–38.

BRENNER, S. L., BARNETT, F. H. C., and ORGEL, L. 1961. The theory of mutagenesis. *J. Mol. Biol.* 3: 121–24.

CARLSON, E. A. 1981. *Genes, radiation, and society: The life and work of H. J. Muller.* Ithaca, NY: Cornell University Press.

CARTER, P. 1986. Site-directed mutagenesis. *Biochem. J.* 237: 1–7.

CLEAVER, J. E. 1968. Defective repair replication of DNA in xeroderma pigmentosum. *Nature* 218: 652–56.

————. 1990. Do we know the cause of xeroderma pigmentosum? *Carcinogenesis* 11: 875–82.

CLEAVER, J. E., and KARENTZ, D. 1986. DNA repair in man: Regulation by a multiple gene family and its association with human disease. *Bioessays* 6: 122–27.

CROW J. F., and DENNISTON, C. 1985. Mutation in human populations. *Adv. Hum. Genet.* 14: 59–216.

DEERING, R. A. 1962. Ultraviolet radiation and nucleic acids. *Scient. Amer.* (Dec.) 207: 135–44.

DEN DUNNEN, J. T., et al. 1989. Topography of the Duchenne muscular dystrophy (DMD) gene. *Am. J. Hum. Genet.* 45: 835–47.

DEVORET, R. 1979. Bacterial tests for potential carcinogens. *Scient. Amer.* (Aug.) 241: 40–49.

DRAKE, J. W. 1970. *Molecular basis of mutation.* San Francisco: Holden-Day.

DRAKE, J. W.,. et al. 1975. Environmental mutagenic hazards. *Science* 187: 505–14.

DRAKE, J. W., GLICKMAN, B. W., and RIPLEY, L. S. 1983. Updating the theory of mutation. *Amer. Scient.* 71: 621–30.

FRIEDBERG, E.C. 1985. *DNA repair*. New York: W.H. Freeman.

HANAWALT, P. C., and HAYNES, R. H. 1967. The repair of DNA. *Scient. Amer.* (Feb.) 216: 36–43.

HARRIS, M. 1971. Mutagenicity of chemicals and drugs. *Science* 171: 51–52.

HASELTINE, W. A., 1983. Ultraviolet light repair and mutagenesis revisited. *Cell* 33: 13–17.

HOWARD-FLANDERS, P. 1981. Inducible repair of DNA. *Scient. Amer.* (Nov.) 245: 72–80.

KELNER, A. 1951. Revival by light. *Scient. Amer.* (May) 184: 22–25.

KNUDSON, A. G. 1979. Our load of mutations and its burden of disease. *Am. J. Hum. Genet.* 31: 401–13.

KRAEMER, F. H., et al. 1975. Genetic heterogeneity in xeroderma pigmentosum: Complementation groups and their relationship to DNA repair rates. *Proc. Natl. Acad. Sci.* 72: 59–63.

LITTLE, J. W., and MOUNT, D. W. 1982. The SOS regulatory system of *E. coli*. *Cell* 29: 11–22.

MASSIE, R. 1967. *Nicholas and Alexandra*. New York: Atheneum.

MASSIE, R., and MASSIE, S. 1975. *Journey*. New York: Knopf.

McCANN, J., CHOI, E., YAMASAKI, E., and AMES, B. 1975. Detection of carcinogens as mutagens in the *Salmonella*/microsome test: Assay of 300 chemicals. *Proc. Natl. Acad. Sci.* 72: 5135–39.

McKUSICK, V. A. 1965. The royal hemophilia. *Scient. Amer.* (Aug.) 213: 88–95.

MODRICH, P. 1987. DNA mismatch correction. *Ann. Rev. Biochem.* 56: 435–66.

MULLER, H. J. 1927. Artificial transmutation of the gene. *Science* 66: 84–87.

———. 1955. Radiation and human mutation. *Scient. Amer.* (Nov.) 193: 58–68.

NEEL, J. V., et al. 1980. Search for mutations affecting protein structure in children of atomic bomb survivors: Preliminary report. *Proc. Natl. Acad. Sci.* 77: 4221–25.

NEWCOMBE, H. B. 1971. The genetic effects of ionizing radiation. *Adv. in Genet.* 16: 239–303.

OSSANA, N., PETERSON, K. R., and MOUNT, D. W. 1986. Genetics of DNA repair in bacteria. *Trends Genet.* 2: 55–58.

RADMAN, M., and WAGNER, R. 1988. The high fidelity of DNA duplication. *Scient. Amer.* (August) 259(2): 40–46.

SANCAR, A. Z., and SANCAR, G. B. 1988. DNA repair enzymes. *Ann. Rev. Biochem.* 57: 29–67.

SCHULL, W. J., OTAKE, M., and NEEL, J. V. 1981. Genetic effects of the atomic bombs: A reappraisal. *Science* 213: 1220–27.

SHORTLE, D., DiMARIO, D., and NATHANS, D. 1981. Directed mutagenesis. *Ann. Rev. Genet.* 15: 265–94.

SIGURBJORHSSON, B. 1971. Induced mutations in plants. *Scient. Amer.* (Jan.) 224: 86–95.

SINGER, B., and KUSMIEREK, J. T. 1982. Chemical mutagenesis. *Ann. Rev. Biochem.* 51: 665–94.

STADLER, L. J. 1928. Mutations in barley induced by X-rays and radium. *Science* 66: 84–87.

SUTHERLAND, B. M. 1981. Photoreactivation. *Bioscience* 31: 439–44.

TOMLIN, N. V., and APRELIKOVA, O. N. 1989. Uracil DNA glycosylases and DNA uracil repair. *Intern. Rev. Cytol.* 114: 81–124.

TOPAL, M. D., and FRESCO, J. R. 1976a. Complementary base pairing and the origin of substitution mutations. *Nature* 263: 285–89.

WILLS, C. 1970. Genetic load. *Scient. Amer.* (March) 222: 98–107.

WOLFF, S. 1967. Radiation genetics. *Ann. Rev. Genet.* 1: 221–44.

YAMAMOTO, F., et al. 1990. Molecular genetic basis of the histo-blood group ABO system. *Nature* 345: 229–33.

YOSHIMOTO, Y. 1990. Cancer risk among children of atomic bomb survivors. *JAMA* 264:596–600.

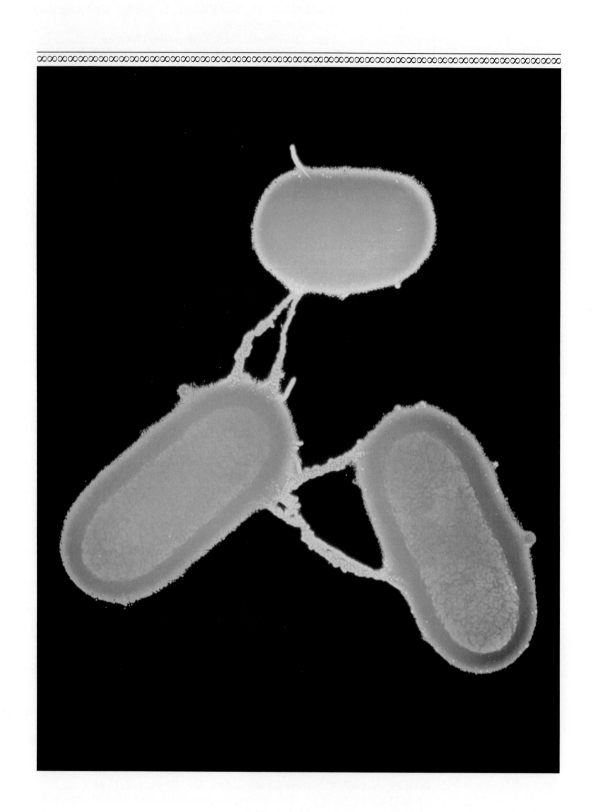

PART THREE

Organization and Regulation of Genetic Information

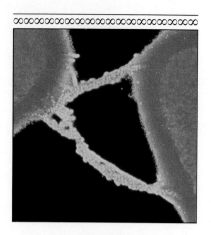

15
Genetics of Bacteria and Viruses

CHAPTER CONCEPTS

Both bacteria and viruses that have bacteria as their hosts

(bacteriophages) have been the subject of extensive genetic analysis and

demonstrate mechanisms by which genetic exchange occurs. These

processes serve as the basis for genetic mapping. Bacteria may contain

extrachromosomal DNA in the form of plasmids, which, in addition to

bacteriophage DNA, can be integrated into the bacterial chromosome.

The use of bacteria and viruses has been essential to the accumulation of knowledge in many areas of genetic study. For example, much of what is known about molecular genetics, recombinational phenomena, and gene structure has been initially derived from experimental work with these organisms. Their successful use in genetic studies is due to numerous factors. Both bacteria and viruses have extremely short reproductive cycles. Literally hundreds of generations, giving rise to billions of organisms, can be produced in short periods of time. Importantly, they can be studied in pure cultures. That is, a single species or mutant strain of bacteria or one type of virus can be investigated independently of other similar organisms. Pure culture techniques for the study of microorganisms were developed in the latter part of the nineteenth century.

It was not until the 1940s and thereafter that genetic studies with microorganisms became productive. In 1943, Salvador Luria and Max Delbruck established that bacteria, like eukaryotic organisms, undergo spontaneous mutation. These mutations were demonstrated to be the source of variation observed in bacteria, and this finding paved the way for genetic investigation. Mutant cells that arise in an otherwise pure culture can be isolated and established independently from the parent strain by using efficient selection techniques. As a result, mutations for almost any desired characteristic can now be induced and isolated in pure culture. Since bacteria and viruses are haploid, all mutations are expressed directly in the descendants of mutant cells, adding to the ease with which these microorganisms are studied.

In this chapter, we will review a number of the historical developments leading to the extensive use of bacteria and viruses in genetics. Additionally, we will examine the processes by which bacteria and viruses exchange genetic information.

BACTERIAL GROWTH

Bacteria can be grown in either a liquid culture medium or in a Petri dish on a semisolid agar surface. If the nutrient components of the growth medium are very simple and consist only of an organic carbon source (such as a glucose or lactose) and a variety of ions, including Na^+, K^+, Mg^{++}, Ca^{++}, and NH_3^+ present as inorganic salts, it is called **minimal medium**. In order to grow on such a medium, a bacterium must be able to synthesize all essential organic compounds, e.g., amino acids, purines, pyrimidines, sugars, vitamins, fatty acids, etc. A bacterium that can accomplish this remarkable biosynthetic feat—one that we ourselves cannot duplicate—is termed a **prototroph**. It is said to be wild type for all growth requirements. On the other hand, if a bacterium loses the ability to synthesize one or more organic components, it is said to be an **auxotroph**. For example, if it loses the ability to make histidine, then this amino acid must be added as a supplement to the minimal medium in order for growth to occur. Such a loss of ability can occur as the result of a mutation, and the resulting bacterium would be designated as a *his*⁻ auxotroph, as opposed to its prototrophic *his*⁺ counterpart.

When an inoculum of bacteria is placed in liquid culture medium, a characteristic growth pattern is exhibited, as illustrated in Figure 15.1. Initially, during the **lag phase,** growth is slow. Then, a period of rapid growth ensues called the **log phase**. During this phase, cells divide many times with a fixed time interval between cell divisions, resulting in logarithmic growth. When a cell density of about 10^9 cells per ml is reached, nutrients and oxygen become limiting and cells enter the **stationary phase**. Since the doubling time during the log phase may be as short as 20 minutes, an initial inoculum of a few thousand cells can easily achieve a maximum cell density in an overnight culture.

We can determine the approximate number of cells present in each milliliter by performing serial dilutions of an aliquot of the liquid culture and then plating a 0.1-ml sample of each dilution on nutrient agar medium (Figure 15.2). Each bacterium so plated will, following incubation, give rise to a colony of progeny cells. If the dilution is too low, a **lawn of cells** will occur because the many colonies will fuse together during their growth. With greater dilution, the colonies do not fuse, and they can be counted. When this count and the dilution factor are known, the original cell density can be easily calculated. In Figure 15.2, one milliliter of medium is first removed from an overnight culture. It is serially diluted, ultimately attaining a final dilution of one in ten million. In other terms, a 10^7 dilution factor or a 10^{-7} dilution has been achieved. Two colonies grew at this dilution, indicating that two bacteria per 0.1 milliliter were present. Therefore, there were an estimated 20 bacteria per ml at that dilution. The original density is then calculated as

$$(\text{colony number/ml}) \times (\text{dilution factor})$$
$$\text{or}$$
$$20\,\frac{\text{bacteria}}{\text{ml}} \times 10^7$$
$$\text{which equals } 20 \times 10^7 \text{ bacteria/ml}$$

If we examine the plate derived from the 10^{-5} dilution, we expect one hundred times the number of colonies present in the 10^{-7} dilution. The observation of 203 colonies fulfills this expectation. Since 203 colonies were obtained from a 0.1 ml sample, 2030 colonies are estimated per ml. From this plate, the original density is calculated as:

$$(2030)\,(10^5) = 20.3 \times 10^7 \text{ cells/ml}$$

This value is close to that obtained using the 10^{-7} dilution.

Calculations such as these make possible the quantitative analysis of a variety of parameters of a given bacterial population. As we will see, such analysis is particularly useful in mutational studies designed to determine mutation rates. Additionally, serial dilutions allow for the isolation of individual colonies. Since each colony contains only genetic clones of the original cell, true-breeding or pure strains can be isolated, propagated, and studied.

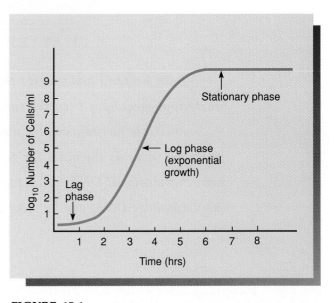

FIGURE 15.1
A typical bacterial population growth curve illustrating the initial lag phase, the log phase where exponential growth occurs, and the stationary phase that results as nutrients are exhausted.

MUTATION IN BACTERIA

While it was known well before 1943 that pure cultures of bacteria could give rise to small numbers of cells exhibiting heritable variation, particularly with respect to survival under different environmental conditions, the source of the variation was hotly debated. The majority of bacteriologists believed that environmental factors induced changes in certain bacteria that led to their survival or adaptation to the new conditions. For example, strains of *E. coli* are known to be *sensitive* to infection by the bacteriophage T1. Infection by the bacteriophage leads to the reproduction of the virus at the expense of the bacterial cell, which is lysed or destroyed (see Figure 8.5). If a plate of *E. coli* is homogeneously sprayed with T1, almost all cells are lysed. Rare *E. coli* cells, however, survive infection and are not lysed. If these cells are isolated and established in pure culture, all descendants are *resistant* to T1 infection. The **adaptation hypothesis,** put forth to explain this type of observation, implied that the interaction of the phage and bacterium was essential to the acquisition of immunity. In other words, the phage had "induced" resistance in the bacteria.

The occurrence of **spontaneous mutations** provided an alternative model to explain the origin of T1 resistance in *E. coli*. In 1943, Luria and Delbruck presented the first direct evidence that bacteria, like eukaryotic organisms, are capable of spontaneous

FIGURE 15.2
Diagrammatic illustration of the serial dilution technique and subsequent culture of bacteria from several of the dilutions. The dilution value for each tube indicates how diluted the initial 1-ml sample has become after each transfer. For example, a dilution of 10^{-3} means that the original 1 ml is now equivalent to 1 ml in 1000 ml. The dilution factor indicates the degree of dilution; e.g., 10^{-3} equals a 1000× dilution. When 0.1 ml of the 10^{-3} dilution is plated, so many bacteria are still present that a lawn of cells is formed as the individual colonies fuse. Plating of 10^{-5} and 10^{-7} dilutions yields 203 and 2 colonies, respectively. The original cell density may be calculated from this information, as discussed in the text. Each colony represents a clone of genetically identical bacteria arising from a single bacterium.

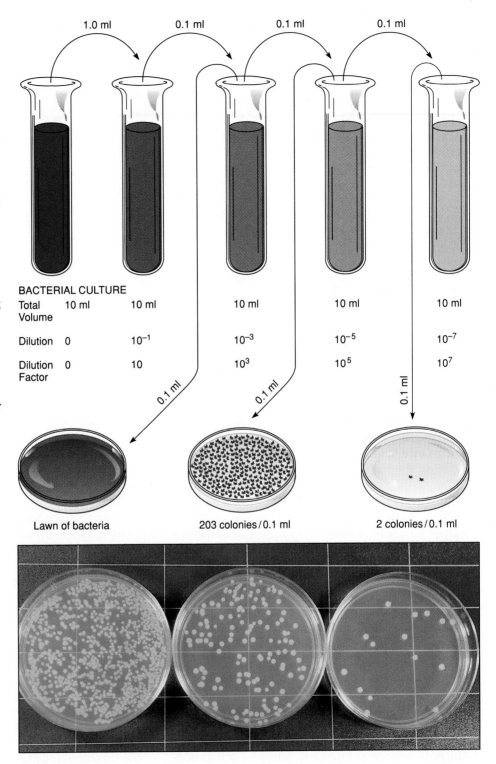

mutation. This experiment marked the initiation of modern bacterial genetic study.

The Luria-Delbruck Experiment

In a beautiful example of analytical and theoretical work, Luria and Delbruck carried out experiments to differentiate between the adaptation and spontaneous mutation hypotheses. In their work, they used the *E. coli*/T1 systems just described. Many small individual liquid cultures of phage-sensitive *E. coli* were grown up. Then, numerous aliquots from each culture were added to Petri dishes containing agar medium to which T1 bacteriophages had been previously added. Following incubation, each plate was scored for the number of phage-resistant bacterial colonies. This was easy to ascertain because only mutant cells were not lysed and survived to be counted. The precise number of total bacteria added to each plate prior to incubation was also determined so that quantitative data could be obtained.

The experimental rationale for distinguishing between the two hypotheses was as follows:

1 **Adaptation**: Every bacterium has a small but constant probability of acquiring resistance as a result of contact with the phages on the Petri dish. Therefore, the number of resistant cells will depend only on the number of bacteria and phages added to each plate. The final results should be independent of all other experimental conditions.

The adaptation hypothesis predicts, therefore, that if a constant number of bacteria and phages is used for each culture and if incubation time is constant, then little fluctuation in the number of resistant cells should be noted on different plates and from experiment to experiment.

2 **Spontaneous Mutation**: On the other hand, if resistance is acquired as a result of mutation, resistance will occur at a low but constant rate during the incubation in liquid medium prior to plating. If mutations occur early during incubation, which they may do randomly, subsequent reproduction of the mutant bacteria will produce large numbers of resistant cells. If the mutations occur relatively late during incubation, far fewer resistant cells will be present.

The mutation hypothesis predicts, therefore, that significant fluctuation in the number of resistant cells will be observed from experiment to experiment, depending on the time when most spontaneous mutations occurred while in liquid culture.

Table 15.1 shows a representative set of data from the Luria-Delbruck experiments. The middle column shows the number of mutants recovered from a series of aliquots derived from one large individual liquid culture. In a large culture, experimental differences are evened out because the culture is constantly mixed. As a result, these data serve as a control because the number in each aliquot should be nearly identical. As predicted, little fluctuation is observed. The right-hand column, however, shows the number of resistant mutants recovered from a single aliquot from each of ten smaller liquid cultures. The amount

TABLE 15.1
The Luria-Delbruck experiment demonstrating that spontaneous mutations are the source of phage-resistant bacteria.

| Sample No. | Number of T1-Resistant Bacteria | |
	Same Culture (Control)	Different Cultures
1	14	6
2	15	5
3	13	10
4	21	8
5	15	24
6	14	13
7	26	165
8	16	15
9	20	6
10	13	10
Mean	16.7	26.2
Variance	15.0	2178.0

SOURCE: After Luria and Delbruck, 1943.

of fluctuation in the data will support only one of the two alternate hypotheses. For this reason, the experiment has been designated the **fluctuation test**. A great fluctuation *is* observed between cultures, thus supporting the spontaneous mutation theory. Fluctuation is measured by the amount of variance, a statistical calculation.

The conclusion reached in the Luria-Delbruck experiment received further support from other experimentation. Nevertheless, until the 1950s, staunch supporters of the adaptation theory were not immediately convinced of the spontaneous mutation origin of variation. At that time, Joshua and Esther Lederberg and their coworkers developed the **replica plating technique**.

Replica Plating: Selection of Bacterial Mutants

In the replica plating technique, a dilute suspension of bacterial cells is plated on a Petri dish containing agar and then incubated until discernible colonies appear. An absorbent material, usually velvet, in the circular shape of the Petri dish is then pressed onto the surface of the growing bacterial culture, as shown in Figure 15.3. A few cells from each colony are picked up on the absorbent surface, which is then pressed lightly onto a second fresh agar plate. The transferred cells produce colonies on the replica plate in a pattern of growth similar to that of the original Petri dish. Consecutive replicas may be prepared in the same way.

FIGURE 15.3
Replica plating technique, which allows new cultures to be "seeded" with bacterial colonies at identical positions on successive plates.

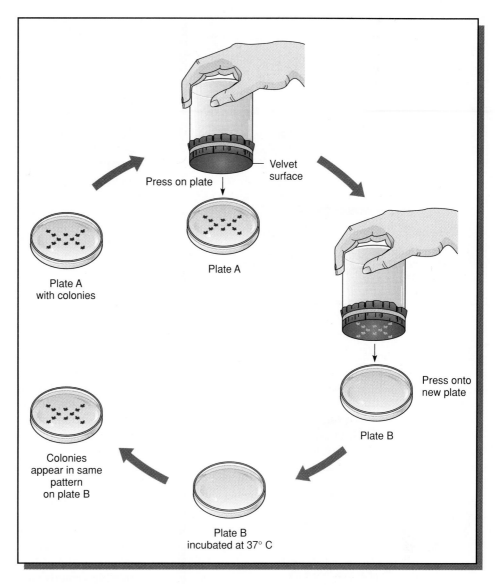

Press on plate

Velvet surface

Plate A

Plate A with colonies

Press onto new plate

Plate B

Plate B incubated at 37° C

Colonies appear in same pattern on plate B

In their original publication in 1956, the Lederbergs described work using the replica plating technique that provided strong support for Luria and Delbruck's conclusion. In one experiment, bacteria were added to agar medium and allowed to reproduce until a dense lawn of cells covered the plate surface. Several replicas were then made on agar medium to which T1 bacteriophages had been added. As in the Luria-Delbruck experiment, only phage-resistant mutant cells could grow on these plates. As shown in Figure 15.4, many phage-resistant cells appeared at the same places on all replica plates. This observation suggests that all resistant cells were present at congruent spots on the original lawn of bacteria, prior to any exposure to the bacteriophage. This conclusion was directly verified by returning to the original plate and "picking" individual colonies from the locations revealed to contain resistant cells

on the replica plates. Such cells were expected to be phage-resistant mutants, even though no phage had been in contact with them at any time in the process. They were.

The replica plating technique is now routinely used in microbial genetic studies. It has been particularly useful in isolating spontaneous and induced mutations. For example, the Lederbergs showed that if streptomycin-sensitive bacteria were used and if streptomycin was added to the replica plates, cells resistant to this antibiotic could be isolated. The technique has also been used to isolate **auxotrophic** mutations, which interfere with the biosynthesis of an essential nutrient molecule. For example, the mutant strain may be unable to synthesize an amino acid or a vitamin. The bacteria harboring the mutation will grow only if minimal medium is supplemented with the missing component. Prototrophs,

GENETICS OF BACTERIA AND VIRUSES   **413**

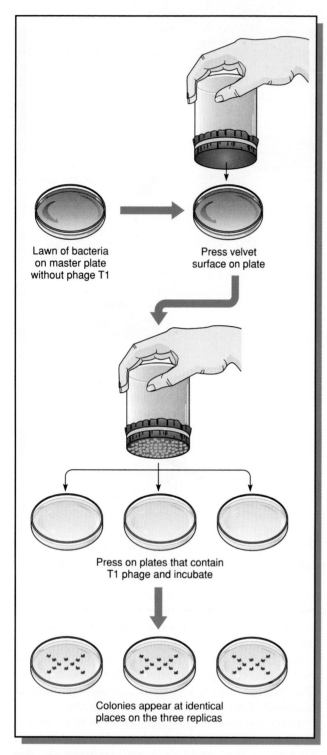

Lawn of bacteria
on master plate
without phage T1

Press velvet
surface on plate

Press on plates that contain
T1 phage and incubate

Colonies appear at identical
places on the three replicas

FIGURE 15.4
Demonstration of phage T1-resistant bacteria using the
replica plating technique. While a large number of
bacterial cells are transferred, only a few colonies, which
are phage-resistant, appear. Replica plating reveals that
the resistant cells giving rise to each colony were present
on the original plate prior to exposure to phage T1.

on the other hand, are wild-type strains of bacteria
that will grow on minimal medium.

Phenotypic and Segregation Lag

Two phenomena, **phenotypic lag** and **segregation
lag,** operate to delay the immediate expression and
detection of newly arisen mutations. In phenotypic
lag, the mutant trait may be expressed only several
generations after the mutation actually occurred. The
reason for this can be illustrated by considering
phage resistance.

Phage-sensitive bacteria contain specific receptors
in the cell wall where phage particles are adsorbed
prior to infection. Phage-resistant mutations eliminate the ability to establish any new functional
receptor sites. However, those sites already present
still allow phage attachment, and as the mutant
bacterium divides, these existing sites are gradually
diluted in the descendant cells. Thus, phenotypic
expression of resistance occurs only after descendant
cells are produced containing too few receptors to
permit significant phage adsorption.

Segregation lag, a related phenomenon, is explained in the following way. Bacteria in the log
phase of exponential growth divide more slowly than
chromosome replication occurs. As a result, a single
cell may contain two, four, or even eight chromosomes. If a spontaneous auxotrophic mutation arises,
for example, in one of the two, four, or eight
chromosomes present, the mutation will not be
expressed until the nonmutant chromosomes are
subsequently distributed to daughter cells (Figure
15.5). As long as one chromosome is still present
containing the nonmutant gene in question, prototrophic growth will occur.

Observations such as these, indicative of either
phenotypic or segregation lag, must be taken into
account when quantitative studies of mutation are
performed. Phenotypic lag clearly affected the Luria
and Delbrück experiment. Other studies attempting
to estimate spontaneous mutation rates in bacteria
are likely to be underestimates because of phenotypic
or segregation lag.

GENETIC EXCHANGE AND
RECOMBINATION IN BACTERIA

The development of the isolation techniques just
described and the information derived from the
subsequent study of bacterial mutagenesis led to
detailed investigations of the arrangement of genes

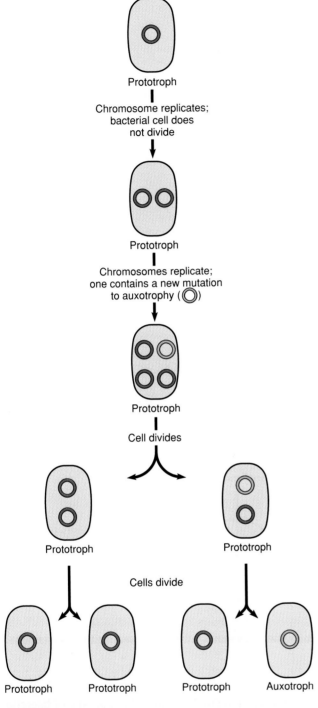

FIGURE 15.5
Explanation of segregation lag, where the chromosome replicates more quickly than the cell divides. If an auxotrophic mutation arises, it may not be expressed immediately if a second chromosome containing the wild-type gene is present. Eventually, cell division catches up with replication, the auxotroph-bearing chromosome is present alone in the cytoplasm, and the mutation is expressed.

Prototroph

Chromosome replicates; bacterial cell does not divide

Prototroph

Chromosomes replicate; one contains a new mutation to auxotrophy (⊙)

Prototroph

Cell divides

Prototroph

Prototroph

Cells divide

Prototroph

Prototroph

Prototroph

Auxotroph

on the bacterial chromosome. These studies were initiated in 1946 by Joshua Lederberg and Edward Tatum. They showed that bacteria undergo **conjugation,** a parasexual process in which the genetic information from one bacterium is transferred to and recombined with that of another bacterium. Like meiotic crossing over in eukaryotes, genetic recombination in bacteria led to methodology for chromosome mapping.

Two other phenomena, **transformation** and **transduction,** also result in the transfer of genetic information from one bacterium to another. These processes also have served as a basis for determining the arrangement of genes on the bacterial chromosome. Transformation (see Chapter 8) involves the entry and integration of a piece of DNA from one bacterium into the chromosome of another intact organism. Transduction, on the other hand, is a phage-mediated transfer of small pieces of DNA from one bacterial cell to another.

BACTERIAL CONJUGATION

Lederberg and Tatum's initial experiments were performed with two multiple auxotroph strains of *E. coli* K12. Strain A required methionine and biotin in order to grow, while strain B required threonine, leucine, and thiamine (Figure 15.6). Neither strain would grow on minimal medium. The two strains were first grown separately in supplemented media, and then cells from both were mixed and grown together for several more generations. They were then plated on minimal medium. Any bacterial cells that grew on minimal medium would be prototrophs. It was highly improbable that any of the cells that contained two or three mutant genes would undergo spontaneous mutation simultaneously at two or three locations. Therefore, any prototrophs recovered must have arisen as a result of some form of genetic exchange and recombination.

In this experiment, prototrophs were recovered at a rate of $1/10^7$ (10^{-7}) cells plated. The controls for this experiment involved separate plating of cells from the strains A and B on minimal medium. No prototrophs were recovered. Lederberg and Tatum therefore concluded that genetic exchange had occurred!

In the following decade, many experiments designed to elucidate the mechanism of bacterial recombination were performed. Because the description of these experiments is lengthy and the results and conclusions complex, they will be discussed in a stepwise fashion.

FIGURE 15.6
Genetic exchange between two auxotrophic strains resulting in prototrophs. Neither auxotroph will grow on minimal medium, but prototrophs will, allowing their recovery.

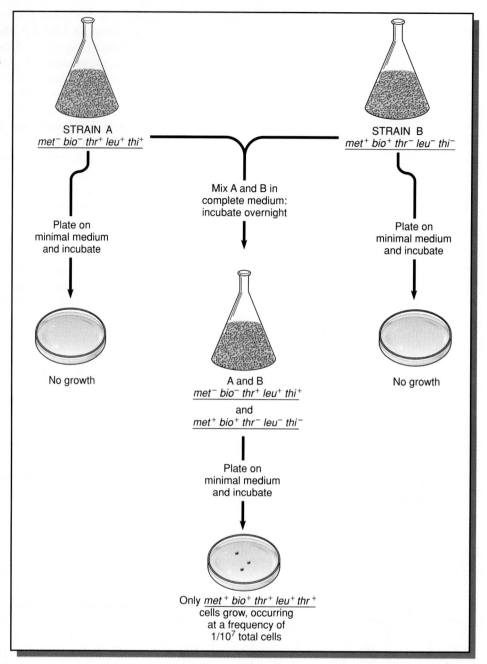

STRAIN A
met⁻ bio⁻ thr⁺ leu⁺ thi⁺

STRAIN B
met⁺ bio⁺ thr⁻ leu⁻ thi⁻

Mix A and B in complete medium: incubate overnight

Plate on minimal medium and incubate

Plate on minimal medium and incubate

No growth

No growth

A and B
met⁻ bio⁻ thr⁺ leu⁺ thi⁺
and
met⁺ bio⁺ thr⁻ leu⁻ thi⁻

Plate on minimal medium and incubate

Only *met⁺ bio⁺ thr⁺ leu⁺ thr⁺* cells grow, occurring at a frequency of $1/10^7$ total cells

F^+ and F^- Bacteria: Observations and Conclusions

1 Lederberg and Tatum showed that filtrates of lysed cultures of either strain A or B would not lead to the production of prototrophs when added to intact cells of the other strain. Therefore, it appeared that the cells must interact directly for genetic recombination to occur. This idea received strong support from an experiment by Bernard Davis using a U-tube (Figure 15.7). At the base of the tube was a sintered glass filter with a pore size that allowed the passage of the liquid medium, but was too small to allow the passage of bacteria. Strain A was placed on one side and strain B on the other side of the filter. The medium was pulled back and forth by suction so that the cells essentially shared a common medium during bacterial incubation. Samples from both sides of the tube were then plated on minimal medium, but no prototrophs were found. Thus, it was concluded that **physical contact is essential to genetic recombination**. This physical interaction is the initial stage of the process of conjugation

FIGURE 15.7
When strain A and B auxotrophs are grown in a common medium but are separated by a filter, no genetic exchange occurs and no prototrophs are produced. This apparatus is called a Davis U-tube.

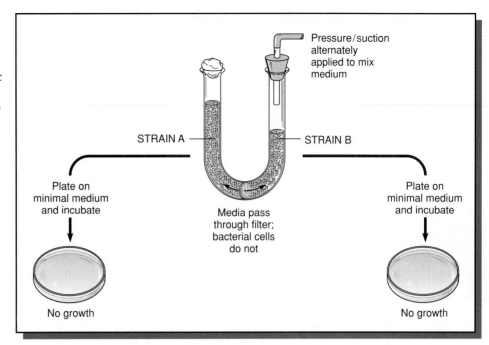

Pressure/suction alternately applied to mix medium

STRAIN A

STRAIN B

Plate on minimal medium and incubate

Plate on minimal medium and incubate

Media pass through filter; bacterial cells do not

No growth

No growth

and is mediated through a structure called the **F sex pilus.** Bacteria often have many pili, which are microscopic extensions of the cell. After contact has been initiated between mating pairs (Figure 15.8), transfer of DNA begins.

2   In 1952, William Hayes performed experiments with auxotrophs similar to those used by Lederberg and Tatum. His findings established that **transfer of the genetic material is unidirectional.** Hayes found that the number of prototrophs was not diminished greatly if strain A cells were inactivated, or sterilized, by exposing them to high concentrations of streptomycin prior to the cross. Streptomycin is a potent antibiotic that inhibits protein synthesis and prevents subsequent cell divisions. However, no prototrophs were recovered when strain B was similarly treated with streptomycin.

The conclusion drawn was that conjugation involves a **donor** and a **recipient** cell. In Hayes' experiment, cells of strain A were serving as donors, and cells of strain B were the recipients. The reasoning used in drawing this conclusion was that streptomycin, while inhibiting bacterial reproduction of the donor cells, will not prevent them from transferring the genetic material to the recipient. On the other hand, by inhibiting the growth and division of cells serving as recipients, streptomycin effectively terminates the process leading to the recovery of prototrophs. Therefore, these observations support the concept of the occurrence of a nonreciprocal transfer of the bacterial chromosome where cells of strain A

serve as "male" donors and those of strain B serve as "female" recipients.

3   In subsequent independent experiments performed by Hayes, the Lederbergs, and Luca Cavalli-Sforza, it was shown that certain conditions could eliminate donor ability in otherwise fertile cells. However, if these cells were then grown with fertile donor cells, fertility was reestablished. The conclusion drawn was that a **fertility factor,** or **F factor,** controls donor ability. It can

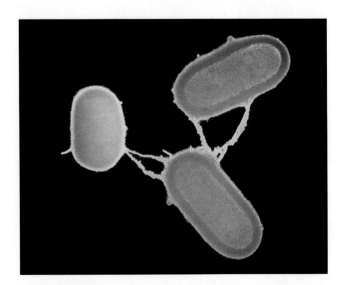

FIGURE 15.8
A false-color electron micrograph of conjugation between an F^+ *E. coli* cell and two F^- cells. The F-pili are visible because they are covered with attacking MS2 bacteriophages.

be lost, converting the donor cell to a recipient, and can also be regained during conjugation. Cells containing the F factor are designated **F⁺**, and those lacking it are designated **F⁻**. On this basis, the initial crosses of Lederberg and Tatum can be clarified:

Strain A   Strain B
F⁺   ×   F⁻
Donor   Recipient

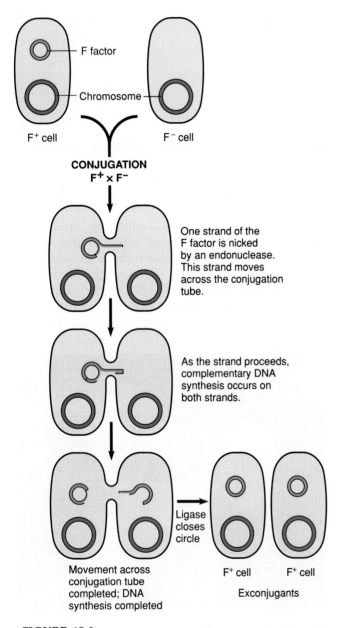

One strand of the F factor is nicked by an endonuclease. This strand moves across the conjugation tube.

As the strand proceeds, complementary DNA synthesis occurs on both strands.

Ligase closes circle

Movement across conjugation tube completed; DNA synthesis completed

F⁺ cell   F⁺ cell

Exconjugants

FIGURE 15.9
An F⁺ × F⁻ mating demonstrating how the recipient F⁻ cell is converted to F⁺. The F factor is transferred across the conjugation tube and then replicated inside the recipient cell.

4 **The F factor has been subsequently shown to consist of a circular, double-stranded DNA molecule.** It exists independently from the bacterial chromosome and contains about 100,000 nucleotide base pairs (100 kb). This amount of DNA is equivalent to about 2 percent of that making up the bacterial chromosome. Contained in this DNA, among others, are 19 genes (*tra*) involved in the transfer of genetic information, including those essential to the formation of the F pilus.

5 Following conjugation, **virtually every recipient cell becomes F⁺,** suggesting that a copy of the F factor is somehow passed through the conjugation tube during each mating, as diagramed in Figure 15.9. It is believed that the transfer involves strand separation of the F factor DNA and the movement of one of the two strands into the recipient. Both parental strands, one remaining in the donor cell, complete semiconservative replication, resulting in two F⁺ cells.

F⁺ and F⁻: Summary

In *E. coli*, cells either contain the F factor or not. When this factor is present, the cell is able to form a sex pilus and potentially serve as a donor of genetic information. During conjugation, a copy of the F factor is always transferred from the F⁺ cell to the F⁻ recipient, converting it to the F⁺ state.

Hfr Bacteria and Chromosome Mapping: Observations and Conclusions

Subsequent discoveries revealed the circumstances surrounding genetic exchange during conjugation. Again, we will approach this topic by listing a series of observations and conclusions, followed by a summary.

1 In 1950, Cavalli-Sforza treated an F⁺ strain of *E. coli* K12 with nitrogen mustard, a potent mutagen. From these treated cells he recovered a strain of donor bacteria that underwent recombination at a rate of 1/10⁴ (10⁻⁴), 1000 times more frequently than the original F⁺ strains. In 1953, Hayes isolated a similar strain that yielded recombinants at a similar frequency. Both strains were designated **Hfr,** or **high-frequency recombination.** Since Hfr cells behave as donors, they are a special class of F⁺ cells.

2 Another important difference was noted between Hfr strains and the original F⁺ strains. The recipient cells, while sometimes displaying genetic recombination, rarely became Hfr; they remained F⁻. In comparison, then,

$F^+ \times F^- \rightarrow F^+$ (low rate of recombination)
$Hfr \times F^- \rightarrow F^-$ (higher rate of recombination)

3 Perhaps the most significant characteristic of Hfr strains is the nature of recombination. In any given strain, certain genes are more frequently recombined than others, and some not at all. This **nonrandom pattern** was shown to vary from Hfr strain to Hfr strain.

4 While these results were puzzling, Hayes interpreted them to mean that some physiological alteration of the F factor had occurred, resulting in the production of Hfr strains of *E. coli*. In the mid-1950s, experimentation by Ellie Wollman and François Jacob clarified the basis of Hfr and showed how these strains allowed genetic mapping of the *E. coli* chromosome. In their experiments, Hfr and F⁻ strains with suitable marker genes were mixed and recombination of specific genes assayed with time. To accomplish this, cultures of an Hfr and an F⁻ strain were first incubated together. At various intervals, samples were removed and placed in a blender. The shear forces created in the blender separated conjugating bacteria so that the transfer of the chromosome was effectively terminated. The cells were then assayed for genetic recombination. This process, called the **interrupted mating technique,** demonstrated that specific genes of a given Hfr strain were transferred and recombined sooner than others. Figure 15.10 illustrates this point. During the first 8 minutes after the two strains were initially mixed, no genetic recombination could be detected. At about 10 minutes, recombination of the *azi* gene could be detected, but no transfer of the *tonˢ*, *lac⁺*, or *gal⁺* genes was noted. By 15 minutes, 70 percent of the recombinants were *azi⁺*; 30 percent were now also *tonˢ*; but none was *lac⁺* or *gal⁺*. By 20 minutes, the *lac⁺* gene was found among the recombinants; and by 30 minutes, *gal⁺* was also being transferred. Therefore, Wollman and Jacob had demonstrated an **oriented transfer** of genes that was correlated with the length of time conjugation was allowed to proceed.

It appeared that the chromosome of the Hfr bacterium was transferred linearly and that the sequence and distance between the genes, as measured in minutes, could be predicted from such experiments (Figure 15.11). This information served as the basis for the first genetic map of the *E. coli* chromosome. Apparently, conjugation does not usually last long enough to allow the entire chromosome to pass across to the recipient cell.

Wollman and Jacob then repeated the same type of experimentation with other Hfr strains, obtaining similar results with one important dif-

FIGURE 15.10
The progressive transfer during conjugation of various genes from a specific Hfr strain of *E. coli* to an F⁻ strain. Certain genes (*azi* and *ton*) are transferred sooner than others and recombine more frequently. Others (*lac* and *gal*) take longer to be transferred and recombine with a lower frequency.

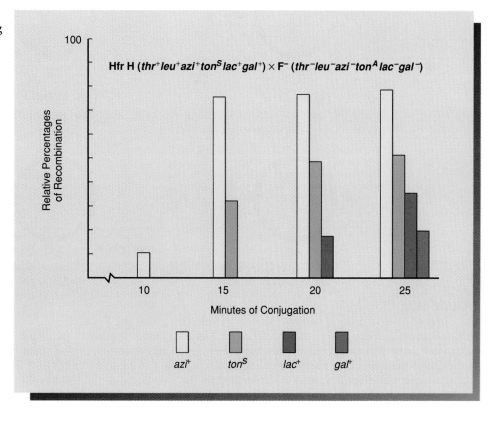

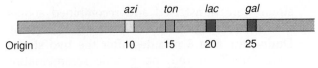

azi   *ton*   *lac*   *gal*

Origin   10   15   20   25

Time in Minutes

FIGURE 15.11
A time map of the genes studied in the experiment
depicted in Figure 15.10.

ference. While genes were always transferred
linearly with time, as in their original experiment,
which genes entered first and which followed
later seemed to vary from Hfr strain to Hfr strain
[Figure 15.12(a)]. When they reexamined the rate
of entry of genes, and thus the different genetic
maps for each strain, a definite pattern emerged.
The major differences between all strains were
simply the point of the origin (*O*) and the direction
in which entry proceeded from that point [Figure
15.12(b)].

(a)

Hfr Strain	Order of Transfer
H	*thr* – *leu* – *azi* – *ton* – *pro* – *lac* – *gal* – *thi*
1	*leu* – *thr* – *thi* – *gal* – *lac* – *pro* – *ton* – *azi*
2	*pro* – *ton* – *azi* – *leu* – *thr* – *thi* – *gal* – *lac*
7	*ton* – *azi* – *leu* – *thr* – *thi* – *gal* – *lac* – *pro*

(b)

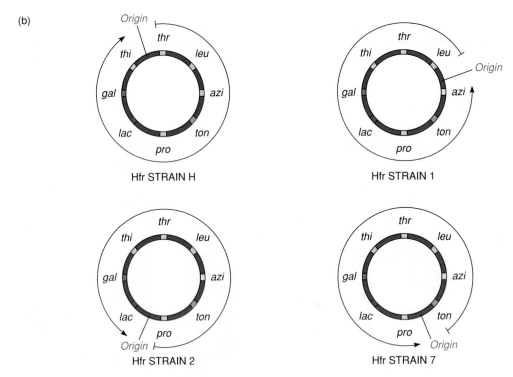

FIGURE 15.12
The order of gene transfer in various Hfr strains, suggesting that the *E. coli*
chromosome is circular. The position designated *O* in each strain represents the origin
of transfer. Note that transfer can proceed in either direction, depending on the strain.
The origin is determined by the point of integration into the chromosome of the F
factor, and the direction of transfer is determined by the orientation of the F factor as
it integrates.

5 In order to explain these results, Wollman and Jacob postulated that **the *E. coli* chromosome is circular**. If the point of origin (*O*) varied from strain to strain, a different sequence of genes would be transferred in each case. But what determines *O*? They proposed that in various Hfr strains, **the F factor integrates into the chromosome at different points**. Its position determines the *O* site. One such case is shown in Figure 15.13. During conjugation between an Hfr and F⁻ cell, the position of the F factor determines the initial point of transfer. Those genes adjacent to *O* are transferred first. The F factor becomes the last part to be transferred. Apparently, conjugation rarely, if ever, lasts long enough to allow the entire chromosome to pass across the conjugation tube. This proposal explains why recipient cells, when mated with Hfr cells, remain F⁻.

The use of the interrupted mating technique with different Hfr strains has provided the basis for mapping the entire *E. coli* chromosome. Mapped in time units, strain K12 (or *E. coli* K12) is 90 minutes long. Over 900 genes have now been placed on the map. In most instances, only a

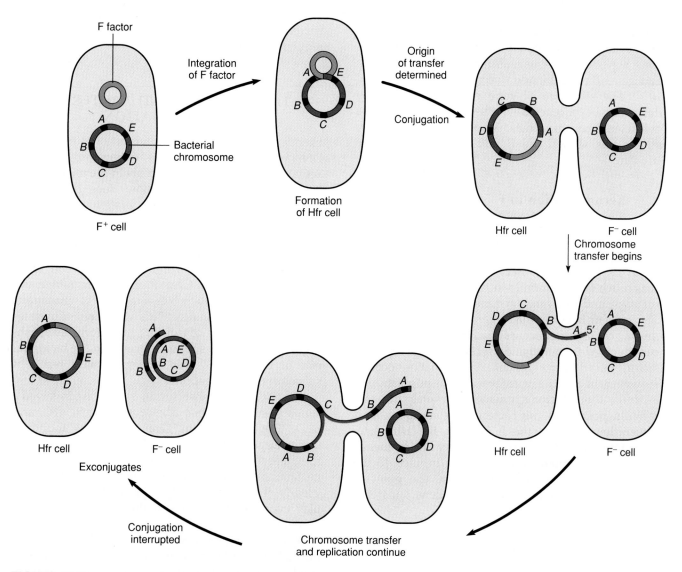

FIGURE 15.13

Conversion of F⁺ to an Hfr state occurs by the integration of the F factor into the bacterial chromosome. During conjugation, the F factor is nicked by an enzyme, creating the origin of transfer of the chromosome. The major portion of the F factor is now on the end of the chromosome opposite the origin and is the last part to be transferred. Conjugation is usually interrupted prior to complete transfer. Above, only the A and B genes were transferred.

single copy of each gene exists. Another interesting observation apparent from the map is that genes whose products share related functions are often closely linked. As we shall see in Chapter 19, this observation is the basis of **operons,** where groups of related genes are under common regulatory control.

Hfr Bacteria and Chromosome Mapping: Summary

Hfr bacteria are produced by the spontaneous and random integration of the cytoplasmic F factor into the bacterial chromosome. In the Hfr condition, conjugation is initiated with F^- bacteria. Depending on the site of integration of the F factor, the chromosome is passed to the recipient cell in a linear fashion, and genetic recombination occurs at a rate of about 10^{-4}. The sequence of gene entry, which can be determined experimentally, is used to construct maps of bacterial chromosomes. Mapping of various Hfr strains has shown the *E. coli* chromosome to be circular.

Recombination in $F^+ \times F^-$ Matings: A Reexamination

The above model has helped geneticists to understand better how genetic recombination occurs during the $F^+ \times F^-$ matings. Recall that recombination occurs much less frequently in them than in Hfr $\times F^-$ matings, and that random gene transfer is involved. The current belief is that when F^+ and F^- cells are mixed, conjugation occurs readily and that each F^- cell involved in conjugation receives a copy of the F factor, but that no genetic recombination occurs. However, in a very low frequency in a population of F^+ cells, the F factor spontaneously integrates at a random point into the bacterial chromosome. This integration converts the F^+ cell to the Hfr state. Therefore, in $F^+ \times F^-$ crosses, the very low frequency of genetic recombination (10^{-7}) is attributed to the rare, newly formed Hfr cells. Since the point of integration is random, a nonspecific gene transfer is noted.

The F' State and Merozygotes

In 1959 during experiments with Hfr strains of *E. coli,* Edward Adelberg discovered that the F factor could lose its integrated status, causing reversion to the F^+ state (Figure 15.14). When this occurs, the F factor frequently carries several adjacent bacterial genes along with it. He labeled this condition **F'** to distin-

guish it from F^+ and Hfr. F' is thus a special case of F^+.

The presence of bacterial genes within a cytoplasmic F factor creates an interesting situation. An F' bacterium behaves as an F^+ cell, initiating conjugation with F^- cells. When this occurs, the F factor, along with the chromosomal genes, is always transferred to the F^- cell. As a result, whatever chromosomal genes are part of the F factor are now present in duplicate in the recipient cell because they are also present in the recipient cell's chromosome. This creates a partially diploid cell called a **merozygote.** Pure cultures of F' merozygotes may be established. They have been extremely useful in the study of bacterial genetics, particularly in the area of genetic regulation (see Chapter 19).

THE *rec* GENES AND BACTERIAL RECOMBINATION

Having established that a unidirectional transfer of DNA occurs between bacteria, determining how the actual recombination event occurs in the recipient cell has been of interest. Just how does the genetic exchange occur? As with many systems, the mechanism by which a biochemical event occurs is often deciphered as a result of genetic studies. Such an approach has been particularly fruitful in the case of the phenomenon of recombination between DNA molecules in bacteria. Great insights were gained as a result of the isolation of a group of mutations involving the *rec* genes. This discovery illustrates quite well the value of isolating mutations, establishing their phenotypes, and determining the biological role of the normal, wild-type gene as a result of subsequent investigation.

The fist relevant observation in this case involved a series of mutant genes labeled *recA, recB, recC,* and *recD.* The first mutant gene, *recA,* was found to diminish genetic exchange in bacteria one-thousand-fold, nearly eliminating it altogether. The other *rec* mutations reduced exchange by about 100 times! Clearly, the normal wild-type products of these genes play some essential role in the process of genetic exchange.

By looking for a functional gene product present in normal cells, but missing in mutant cells, two *rec* gene products were subsequently isolated and studied. They have been found to play a direct role in the process of DNA recombination in bacterial cells. The first is a protein called the *recA* gene product. The second is a more complex protein called the *recBCD* product, consisting of polypeptide subunits derived

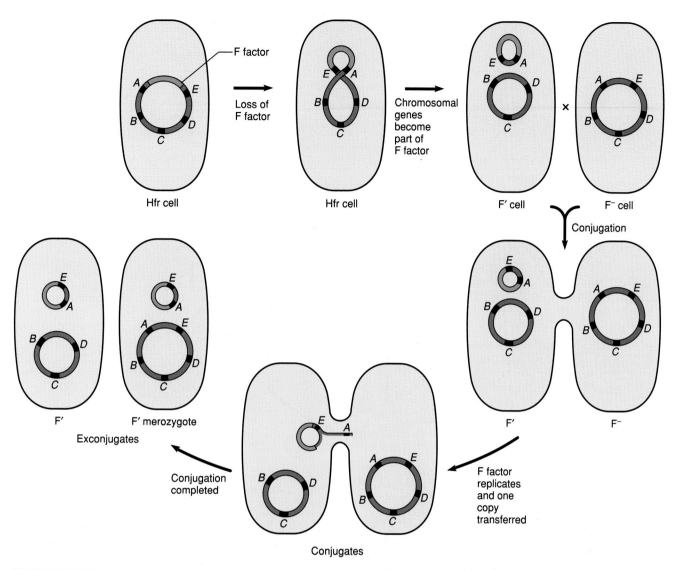

FIGURE 15.14
Conversion of an Hfr bacterium to F' and its subsequent mating with an F⁻ cell. The conversion occurs when the F factor loses its integrated status and carries with it one or more chromosomal genes (A and E). Following conjugation with such an F' cell, an F⁻ recipient becomes partially diploid and is called a merozygote. It also becomes F'.

from the *B, C,* and *D* genes. The role of these proteins has now been elucidated. As a result of this genetic research, our knowledge of the process of recombination has been extended considerably.

The *recA* protein has a strong affinity *in vitro* for single-stranded DNA (ssDNA). The binding creates a DNA-protein complex which invades and probes an independent duplex of DNA. The complex migrates along until the ssDNA locates its homologous region within the duplex. When the homologous region is encountered, the invading ssDNA, which is complementary to one of the two strands, base pairs with it, replacing its counterpart in the initial duplex. Once

the hybrid duplex is formed, the *recA* protein is released. The displaced region of DNA juts out of the duplex, forming what is called a **D-loop**. This general scheme is illustrated in Figure 15.15.

The *recBCD* protein, on the other hand, has an affinity *in vitro* for double-stranded DNA (dsDNA). It binds to and migrates along the duplex, separating the helix as it goes. The separated strands reassociate, but more slowly than they separated. This leaves behind loops of single-stranded DNA (ssDNA). Since mutations in any of these genes diminish recombination, we can speculate that separation of strands, as described above, may be an essential step in allowing

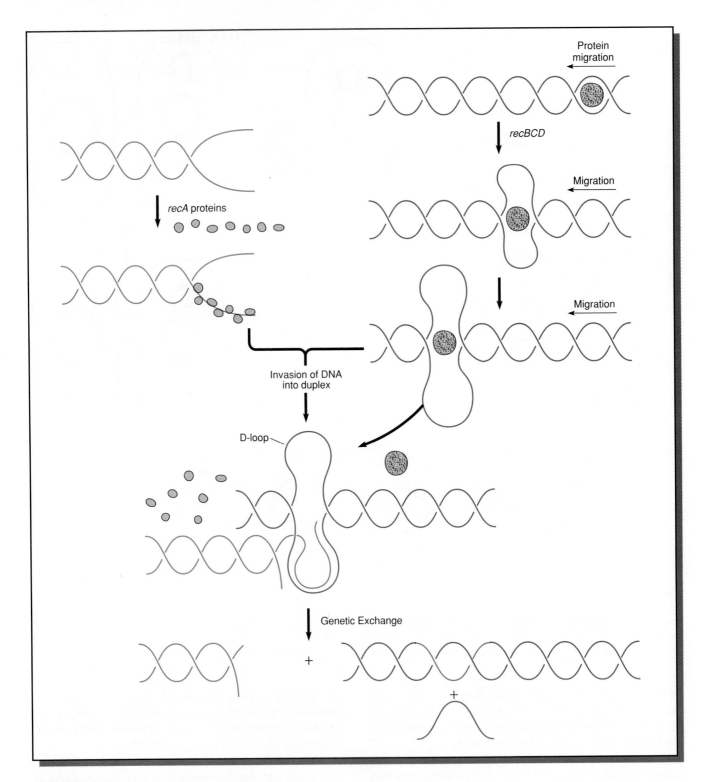

FIGURE 15.15
A theoretical model illustrating the possible role of the *recA* and *recBCD* proteins in the process of genetic recombination between bacterial DNA molecules. The *recA* protein has an affinity to bind to single-stranded regions of DNA and to facilitate their invasion of double-stranded DNA. The *recBCD* protein has an affinity for double-stranded DNA. This protein migrates along the helix, causing the complementary strands to dissociate. Once the invasive region finds its homologous DNA sequence, the *recA* proteins are shed and a hybrid complement is formed. The displaced strand forms a D-loop and genetic exchange occurs.

the invading *recA*–ssDNA complex to locate its homologous region within the duplex.

Together, the *recA* and *recBCD* products appear to have the effect of positioning one of the two strands of donor DNA from the F⁺ cells at the appropriate location within the F⁻ chromosome in order to facilitate genetic recombination. What remains to be accomplished, of course, is the actual insertion of the F⁺ DNA. This may be accomplished by the *recBCD* protein, which has been shown to also be able to make cuts in DNA. DNA ligase might then complete the task of recombination by enzymatically sealing the ends of the invading DNA with those of the DNA of the recipient cell. This picture will undoubtedly be clarified as further research is conducted. Figure 15.15 illustrates the general role of the *recBCD* product as well as a way in which both gene products might work in an integrated fashion.

PLASMIDS

In the above sections, we introduced and discussed extensively the extrachromosomal heredity unit called the F factor. When existing autonomously in the bacterial cytoplasm, this unit takes the form of a double-stranded closed circle of DNA. These characteristics place the F factor in the more general category of the genetic structure called a **plasmid**. These structures contain one or more genes and often, quite a few. Their replication depends on the host cell's enzymes and they are distributed to daughter cells along with the host chromosome during cell division.

Plasmids are generally classified according to the genetic information specified by their DNA. The F factor confers fertility and contains genes essential for sex pilus formation. Other examples include the **R** and the **Col plasmids**.

Most R plasmids consist of two components: the **RTF (resistance transfer factor)** and one or more **r-determinants**. The RTF contains information essential to transfer of the plasmid between bacteria, and r-determinants are genes conferring antibiotic resistance. While the DNA of RTFs is quite similar in a variety of plasmids from different bacteria species, there is a wide variation in r-determinants. Each is specific for resistance to one antibiotic. If a bacterial cell contains r-determinant plasmids but no RTF is present, the cell is resistant but not infectious; that is, resistance cannot be transferred. However, the most commonly studied plasmids consist of the RTF as well as one or more r-determinants. Resistances to tetracycline, streptomycin, ampicillin, sulfonamide,

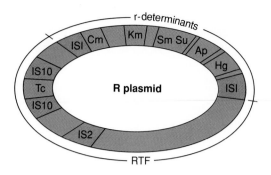

FIGURE 15.16

R plasmid containing resistance transfer factors (RTFs) and multiple r-determinants (Tc—tetracycline; Km—kanamycin; Sm—streptomycin; Su—sulfonamide; Ap—ampicillin; Hg—mercury).

kanamycin, and chloramphenicol are most frequently encountered.

As we shall see in the ensuing section, still other DNA sequences may associate with an R plasmid. As a result, r-determinants from different R plasmids may combine into single units, conferring multiple resistance to several antibiotics (Figure 15.16).

The ColE1 plasmid, derived from *E. coli*, encodes one or more proteins which are highly toxic to bacterial strains that do not harbor the same plasmid. These proteins, called **colicins**, may kill neighboring bacteria. Those that carry the plasmid are said to be colicinogenic. Present 10 to 20 times per cell, the plasmid also contains a gene encoding an immunity protein which protects the host cell from the toxin. Unlike the F and R plasmids, the Col plasmid is not transmissible to other cells unless one of the former structures is present to confer the ability to engage in conjugation.

Interest in plasmids has increased dramatically because of their role in the genetic technology referred to as recombinant DNA research. Specific genes from any source may be inserted into a plasmid, which may be inserted into a bacterial cell. As the cell replicates its DNA and undergoes division, the foreign gene is treated like one of the cell's own. This is the major topic of the next chapter.

INSERTION SEQUENCES AND TRANSPOSONS IN BACTERIA

In the early 1970s, a number of independent workers, including Peter Starlinger and James Shapiro, discovered a unique class of mutations in *E. coli*. These mutations affected several genes in different strains of bacteria. For example, transcription of a cluster of genes (or **operon**) controlling galactose metabolism in

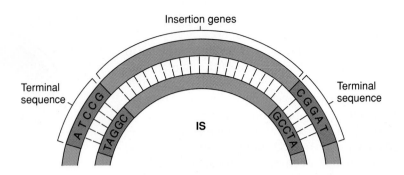

FIGURE 15.17
Diagrammatic representation of an insertion sequence (IS).

E. coli was repressed or shut off as a result of one of these mutations. This phenotypic effect was heritable, but was found not to be caused by a nucleotide change characteristic of conventional mutations. Instead, it was shown that a short, specific DNA segment had been inserted into the bacterial chromosome at the beginning of the galactose gene cluster. When this segment was spontaneously excised from the bacterial chromosome, wild-type function was restored!

It was subsequently revealed that several other distinct DNA segments could behave in a similar fashion, inserting into the chromosome and affecting gene function. These DNA segments are relatively short, not exceeding 2000 base pairs (2 kb, or 2 kilobases), and are now called **insertion sequences (IS)**. The genetic information contained within an IS unit appears to be involved only in the insertion process itself.

Analysis of the DNA sequences of most IS units has revealed a most interesting feature. At each end of any given double-stranded unit, the nucleotide sequence consists of a **perfect inverted repeat** of the other end. While Figure 15.17 shows this terminal repeating unit to consist of only a few nucleotides, many more are actually involved. For example, in E. coli, the IS1 termini contain about 30 nucleotide pairs, IS2 about 40 pairs, and IS4 about 18 pairs. It seems likely that these terminal sequences are involved closely in the mechanism of insertion of IS units into DNA. This contention is supported by the fact that insertion of IS units is not random but instead is more likely to occur at certain DNA regions. This suggests that the termini may be involved in recognition sites during the insertion process.

Careful investigation has revealed that IS units are present in the wild-type E. coli chromosome as well as in various plasmids. Thus, their presence does not always result in mutation. In the E. coli chromosome, five or more copies of IS1 and IS2 are known to be present, but the number of IS3 and IS4 is yet to be determined. The position of IS units is not fixed; rather, they move from point to point, inserting at preferential positions. The F factor also contains IS

units, as shown in Figure 15.18. Two copies of IS3, one IS2, and one designated γδ are present. It appears that these IS sites are involved in the integration of the F factor into the E. coli chromosome during Hfr formation. A possible mode of integration and excision of the F factor into and out of the bacterial chromosome is illustrated in Figure 15.19.

In addition to its mutational effects, the insertion of IS units has a significant role in the formation and movement of the larger **transposon (Tn) elements**. Transposons in bacteria consist of IS units that are covalently joined to other genes whose functions are unrelated to the insertion process. Like IS units, Tn elements can insert into bacterial and viral chromosomes as well as plasmids. This property is undoubtedly related to the presence of IS units in each transposon. Thus, Tn elements provide a mechanism for movement of genetic information from place to place both within and between organisms.

Transposons were first discovered to move between DNA molecules as a result of observations of

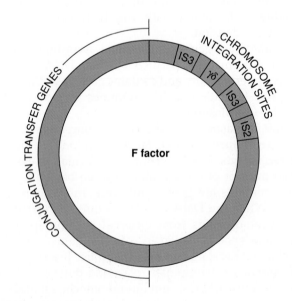

FIGURE 15.18
Diagram of the F factor in E. coli showing four insertion sequences that serve as integration sites into the bacterial chromosome.

426   CHAPTER 15

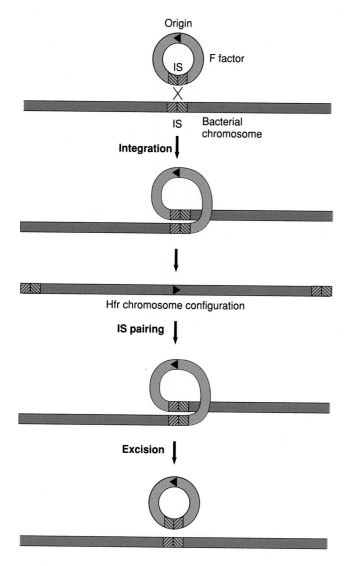

FIGURE 15.19
A model depicting the role of insertion sequences (IS) in the integration and excision of the F factor into and out of the bacterial chromosome. Insertion leads to the conversion of an F⁺ bacterium to an Hfr bacterium.

Labels within figure:
Origin
IS — F factor
IS — Bacterial chromosome
Integration
Hfr chromosome configuration
IS pairing
Excision

repeat units remain single stranded and form loops on either end of the double-stranded stem.

One of the most significant aspects of these findings is that transposition does not appear to require homologous pairing in order for insertion to occur. Recombination of a more conventional nature—conjugation and transformation, for example—is thought to rely on homologous pairing prior to the recombination event. In *E. coli*, such events are dependent upon the product of the *recA* locus. In *recA* mutants, as discussed above, such recombination is abolished. However, transposon movement can occur in such mutants. Therefore, transposition as a form of recombination would seem not to involve or be dependent on the homologous pairing-breakage-rejoining mechanism. Because of this characteristic, transposition is described as a form of **illegitimate recombination**. However, the precise mechanism of insertion remains unknown.

Nevertheless, several aspects of the insertion process are quite well understood. First, insertion can be either conservative or replicative. In the former case, the transposon is excised from one location and reinserted into another DNA sequence. The new location may be in the same genetic unit or another within the cell; that is, a transposon may move from a plasmid to the host chromosome or vice versa. The original transposon is conserved and it moves to its new location. In the latter case of replicative transposition, the initial copy of the genetic unit is left behind and a new copy, presumably the result of selective replication, moves and is inserted into another DNA sequence.

One consequence of the transposition process is the generation of a short duplication of an existing DNA sequence at the point of insertion. For example, the integration of IS1 in *E. coli* results in the repetition of nine base pairs at the target site. Both nine and five base-pair repeats are common during transposition in bacteria. The repetition is thought to arise as a result of the illegitimate form of recombination mentioned above. If the two strands of the target DNA are not cleaved at the same point, the integration of an insertion sequence or transposon creates gaps on both ends which must be filled to reclose the double helical structure. As the gaps are filled enzymatically, the duplicated sequences are produced. One model of transposition postulates that IS and transposon units encode a **transposase enzyme** which is responsible for the type of uneven cut which ultimately leads to these duplications.

Transposons have become the focus of increased interest, particularly because they have been found not to be unique to bacteria. Bacteriophages that demonstrate the ability to insert their genetic material

antibiotic-resistant bacteria. In the mid-1960s, Susumu Mitsuhashi suggested that an R plasmid containing a gene responsible for resistance to antibiotic chloramphenicol had been transferred to the bacterial chromosome. Since then, it has been demonstrated that antibiotic-resistant genes can move from plasmid to plasmid as well as between plasmids and chromosomes. Electron microscopic studies have confirmed that the two terminal inverted repeat sequences are indeed present within plasmids harboring transposons. As shown in Figure 15.20, when the strands of DNA from such a plasmid are separated and allowed to reanneal separately, the inverted repeat units are complementary, and a heteroduplex is formed as expected. All areas other than the

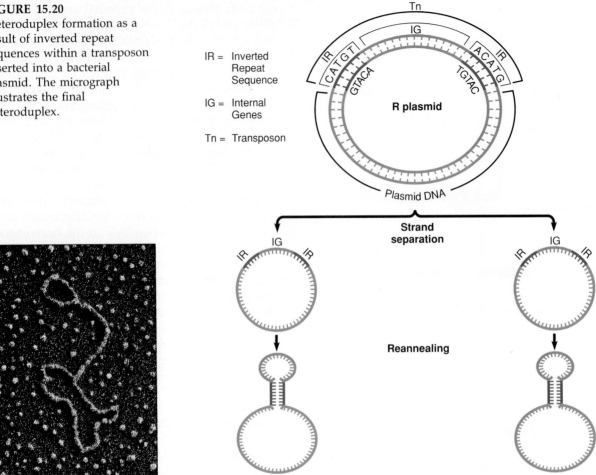

FIGURE 15.20
Heteroduplex formation as a result of inverted repeat sequences within a transposon inserted into a bacterial plasmid. The micrograph illustrates the final heteroduplex.

IR = Inverted Repeat Sequence

IG = Internal Genes

Tn = Transposon

into the host chromosome behave in a similar fashion. (See the discussion of this topic later in this chapter.) The **bacteriophage mu,** consisting of over 35,000 nucleotides, can insert its DNA anywhere in the *E. coli* chromosome! And like IS units, insertion into a gene results in mutant behavior at that locus. Transposons have also been discovered in higher organisms, including yeast, corn, and *Drosophila*. Thus, we shall return to this important topic again, but at a more appropriate time when we discuss genetic control mechanisms in eukaryotes in Chapter 20.

BACTERIAL TRANSFORMATION

In Chapter 8 we described how the study of transformation provided direct evidence that DNA is the genetic material. Transformation, like conjugation, provides a mechanism for the recombination of genetic information in some bacteria. Transformation can also be used to map bacterial genes, although in a more limited way.

This process consists of numerous steps that can be divided into two main categories: (1) entry of exogenous DNA into a recipient cell, and (2) recombination of the donor DNA with its homologous region in the recipient chromosome. In a population of cells, only those in a particular physiological state, referred to as **competence,** take up DNA. Entry apparently occurs at a limited number of receptor sites on the surface of the bacterial cell. The efficient length of transforming DNA is about 10,000 to 20,000 base pairs, an amount equal to about 1/200 of the *E. coli* chromosome. Passage across the cell wall and membrane is an active process requiring energy and specific transport molecules. This concept is supported by the fact that substances that inhibit energy production or protein synthesis in the recipient cell also inhibit the transformation process.

During the process of entry, one of the two strands of the double helix is digested by nucleases, leaving only a single strand to participate in transformation. This DNA segment aligns with its complementary region of the bacterial chromosome. In a process involving several enzymes, the segment replaces its counterpart in the chromosome, which is excised and degraded (Figure 15.21).

In order for recombination to be detected, the transforming DNA must be derived from a genetically different strain of bacteria. Once it is integrated into the chromosome, that region contains one strand originally present and one mutant strand. Since these strands are not genetically identical, this helical region is referred to as a **heteroduplex**. Following one round of semiconservative replication, one chromosome is identical to that of the original recipient, and the other contains the mutant gene. Following cell division, one nonmutant and one mutant cell are produced.

Transformation and Linked Genes

Because transforming DNA is usually about 10,000 to 20,000 base pairs long, it contains a sufficient number of nucleotides to encode several genes. Genes that are adjacent or very close to one another on the bacterial chromosome may be carried on a single piece of DNA of this size. Because of this fact, a single event may result in the **cotransformation** of several genes simultaneously. Genes that are close enough to each other to be cotransformed are said to be **linked**. Note that here *linkage* refers to the proximity of genes, in contrast to the use of this term in eukaryotes to indicate all genes on a single chromosome.

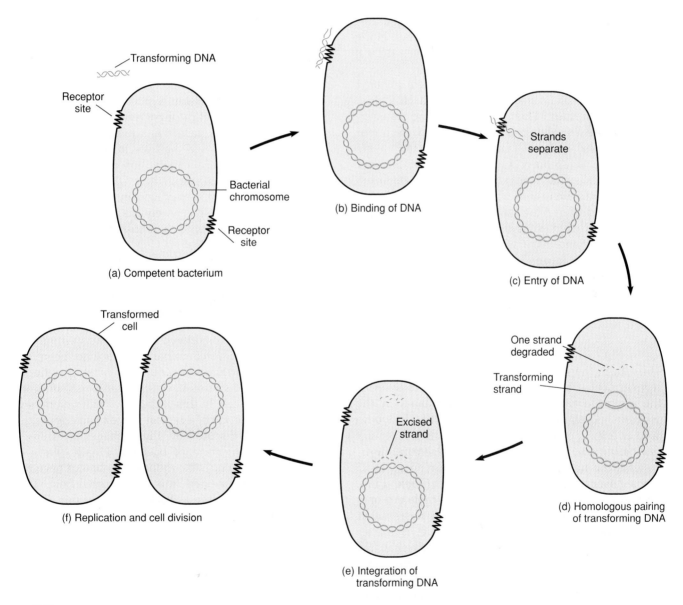

FIGURE 15.21
Proposed steps leading to transformation of a bacterial cell by exogenous DNA.

TABLE 15.2
The results of several transformation experiments, which establish linkage between the *str* and *mtl* loci in *Pneumococcus*.

Donor DNA	Recipient Cell Genotype	Percentage of Transformed Genotypes		
		str^r mtl^-	*str^s mtl^+*	*str^r mtl^+*
str^r mtl^+	*str^s mtl^-*	4.3	0.40	0.17
str^r mtl^- and *str^s mtl^+*		2.8	0.85	0.0066

SOURCE: Data from Hotchkiss and Marmur, 1954, p. 55.

If two genes are not linked, simultaneous transformation can occur only as a result of two independent events involving two distinct segments of DNA. As in double crossing over in eukaryotes, the probability of two independent events occurring simultaneously is equal to the product of the individual probabilities. Thus, the frequency of two unlinked genes being transformed simultaneously is much lower than if they are linked.

Linked genes were first demonstrated in 1954 during studies of *Pneumococcus* by Rollin Hotchkiss and Julius Marmur. They were examining transformation at the *streptomycin* and *mannitol* loci. Recipient cells were *str^s* and *mtl^-*, meaning that they were sensitive to streptomycin and could not ferment mannitol. Cells that were *str^r mtl^+*, which are resistant to streptomycin and able to ferment mannitol, were used to derive the transforming DNA. If these genes are unlinked, simultaneous transformation for both genes would occur with such a minimal probability as to be nearly undetectable. However, as shown in Table 15.2, a low but detectable rate (0.17%) of double transformation did occur.

In order to confirm that the genes were indeed linked, a second experiment was performed. Instead of using *str^r mtl^+* DNA, a mixture of *str^r mtl^-* and *str^s mtl^+* DNA simultaneously served as the donor DNA. Thus, both the *str^r* and *mtl^+* genes were available to recipient cells, but on separate segments of DNA. If the observed rate of double transformants in the previous experiment had been due to two independent events and the genes unlinked, then a similar rate would also be observed under these conditions. Instead, the number of double transformants was 25 times fewer, thus confirming linkage (Table 15.2). Careful examination of the data in this table not only confirms the linkage of these loci, but raises other interesting questions. Why, for example, is the *str* locus always transformed more frequently than the *mtl* locus? This observation suggests that entry into the cell is not totally random.

Subsequent studies of transformation have shown that a variety of bacteria readily undergo this process of recombination (e.g., *Bacillus subtilis*,

Shigella paradysenteriae, and, to a lesser extent, *E. coli*). Under certain conditions, studies have shown that relative distances between linked genes can be determined from the recombination data provided by transformation experiments. While analysis is more complex, such data are interpreted in a manner analogous to chromosome mapping in eukaryotes.

There is still a third mode involving genetic transfer between bacteria, called **transduction**. We shall delay our discussion temporarily until we have considered the genetics of bacteriophages and other viruses, since transduction is a process mediated by bacteriophages. As we shall see in the next section, bacterial viruses are capable of either lysing the bacterial host or inserting their DNA into the host chromosome where it may remain genetically quiescent. In some cases, it is when this integrated DNA is extricated from the host chromosome that transduction becomes possible.

THE GENETIC STUDY OF BACTERIOPHAGES

So far, we have considered recombination involving bacterial genes. Recombination has also been observed in bacterial viruses, but only when genetically distinguishable strains have simultaneously infected the same host cell. In this section, we will first review briefly the life cycle of a typical T-even bacteriophage. Then, we will discuss how these phages are studied during their infection of bacteria. Finally, we will contrast the two possible modes of behavior once the phage DNA is injected into the bacterial host. This information will serve as background for our discussion of transduction. We will then entertain more esoteric topics involving viral genetics.

The T4 Phage Life Cycle

In Chapter 8, we discussed the Hershey-Chase experiment, which utilized bacteriophage T2 in providing support for the concept that DNA is the genetic material. Phage T4 is very similar and typical of the family of T-even phages. It contains double-stranded DNA in a quantity sufficient to encode more than 150 average-sized genes. The genetic material is enclosed by an icosahedral protein coat (a polyhedron with 20 faces) making up the head of the virus. This is connected to a tail that contains a collar and a contractile sheath that surrounds a central core. Six tail fibers protrude from the tail; their tips contain binding sites that specifically recognize unique areas of the external surface of *E. coli* cell wall.

Binding of tail fibers to these unique areas of the cell wall is the first step of infection. Then, an ATP-driven contraction of the tail sheath causes the central core to penetrate the cell wall. The DNA in the head is extruded where it then moves across the cell membrane into the bacterial cytoplasm. Within minutes, all bacterial DNA, RNA, and protein synthesis is inhibited, and synthesis of viral macromolecules begins. This initiates the **latent** or **eclipse period,** in which the production of viral components occurs prior to the assembly of virus particles.

RNA synthesis is initiated by the transcription of viral genes by the host cell RNA polymerase. The viral genes are divided into three groups: (1) **immediate early genes,** (2) **delayed early genes,** and (3) **late genes**. The products of the first two groups are made prior to the initiation of T4 DNA synthesis. One of the earliest gene products is a DNase enzyme that degrades the *E. coli* chromosome. The T4 DNA molecule is protected from digestion by this enzyme because its cytosine residues are modified to form **5-hydroxymethylcytosine** (Figure 15.22); as a result, this DNase is unable to use T4 DNA as a substrate.

A plethora of early and late gene products are subsequently synthesized using the host cell ribosomes. For example, about 20 different proteins are part of the protein capsid of the head. Many others are structural components of the tail and its fibers. Additionally, numerous enzymes participate in assembly of mature bacteriophages, but are not themselves structural components. One late gene product, **lysozyme,** digests the bacterial cell wall, leading to its rupture and release of mature phages once they are assembled.

Prior to the synthesis of most viral gene products, semiconservative DNA replication begins, leading to a pool of viral DNA molecules that are available to be packaged into phage heads. It is at this time that recombination may occur between DNA molecules.

Assembly of mature viruses is a complex process that has been well studied by William Wood, Robert Edgar, and others. Three major independent pathways occur, leading to: (1) DNA packaged into heads; (2) construction of tails; and (3) synthesis and assembly of tail fibers. A mutation that interrupts any one of the three pathways does not inhibit the other two. As shown in Figure 15.23, the head is assembled and DNA is packaged into it. This complex then combines with the tail, and only then are the tail fibers added. Total construction is a combination of self-assembly and enzyme-directed processes. When approximately 200 viruses are assembled, the bacterial cell is ruptured by the action of lysozyme and the mature phages are released. If this occurs on a lawn of bacteria, the 200 phages will infect other bacterial cells and the process will be repeated over and over again. On the lawn, this leads to a clear area where all bacteria have been lysed. This area is called a **plaque** and represents multiple clones of the single infecting T4 bacteriophage.

The Plaque Assay

The experimental study of bacteriophages and other viruses has played a critical role in our understanding of molecular genetics. During infection of bacteria, enormous quantities of bacteriophages may be obtained for investigation. Often, over 10^{10} viruses per milliliter of culture medium are produced. Many genetic studies have relied on the ability to quantitate the number of phage produced following infection under specific culture conditions. The technique that works nicely is called the **plaque assay**.

This assay is illustrated in Figure 15.24, where actual plaque morphologies are also shown. A serial dilution of the original virally infected bacterial culture is first performed. Then, a 0.1-ml aliquot from one or more dilutions is added to a small volume of melted nutrient agar (about 3 ml) to which a few drops of a healthy bacterial culture have been mixed. The solution is then poured evenly over a base of solid nutrient agar in a Petri dish and

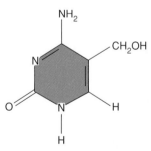

5-hydroxymethylcytosine

FIGURE 15.22
The chemical structure of 5-hydroxymethylcytosine, which is present in some bacteriophage DNA. This modified nitrogenous base protects the viral DNA from the enzymatic degradation that affects the bacterial host DNA during infection.

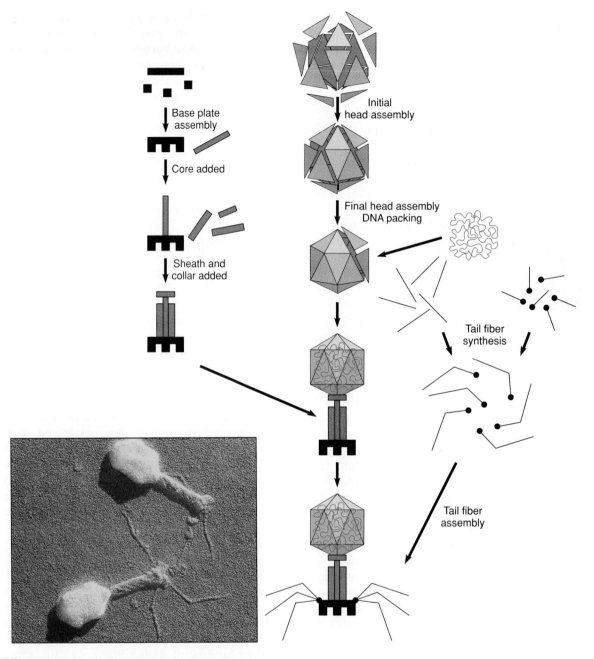

FIGURE 15.23
The assembly of bacteriophage T4 following infection of *E. coli*. Three separate
pathways lead to the formation of an icosahedral head filled with DNA, a tail
containing a central core, and tail fibers. The tail is added to the head and then tail
fibers are added. Each step of each pathway is influenced by one or more phage
genes.

allowed to solidify prior to incubation. As described
in the above section and illustrated in Figure 15.24,
viral plaques occur at each place where a single
virus has initially infected one bacterium in the lawn
that has grown up during incubation. If the dilution
factor is too low, the plaques are plentiful and will
fuse, lysing the entire lawn. This has occurred in the
10^{-3} dilution in Figure 15.24. On the other hand, if

the dilution factor is increased, plaques can be
counted and a density of viruses in the initial culture
can be estimated. The calculation is the same type as
that used for determining bacterial density by counting colonies following serial dilution of an initial
culture:

$$(\text{plaque number/ml}) \times (\text{dilution factor})$$

FIGURE 15.24
Diagrammatic illustration of the plaque assay for bacteriophage analysis. Serial dilutions of a bacterial culture infected with bacteriophage are first made. Then, three of the dilutions $(10^{-3}, 10^{-5}, \text{ and } 10^{-7})$ are analyzed using the plaque assay technique. In each case 0.1 ml of the bacterial-viral culture is mixed with a few drops of a healthy bacterial culture in nutrient agar. This mixture is spread over and allowed to solidify on a base of agar. Each clear area that arises is a plaque and represents the initial infection of one bacterial cell by one bacteriophage. Each generation of viral reproduction involves more bacteria, creating a visible plaque. In the 10^{-3} dilution, so many phage are present that all bacteria are lysed. In the 10^{-5} dilution, 23 plaques are produced. In the 10^{-7} dilution, the dilution factor is so great that no phage are present in the 0.1-ml sample, and thus no plaques form. From the 0.1-ml aliquot of the 10^{-5} dilution, the original bacteriophage density can be calculated as $23 \times 10 \times 10^5$ phage/ml, as described in the text. The photo illustrates plaque morphology of phage T2 (upper left), T4 (upper right), and lambda (bottom), all grown on *E. coli*. Note that plaques appear as "clear" areas in the drawing, but as purple areas in the color-enhanced photograph.

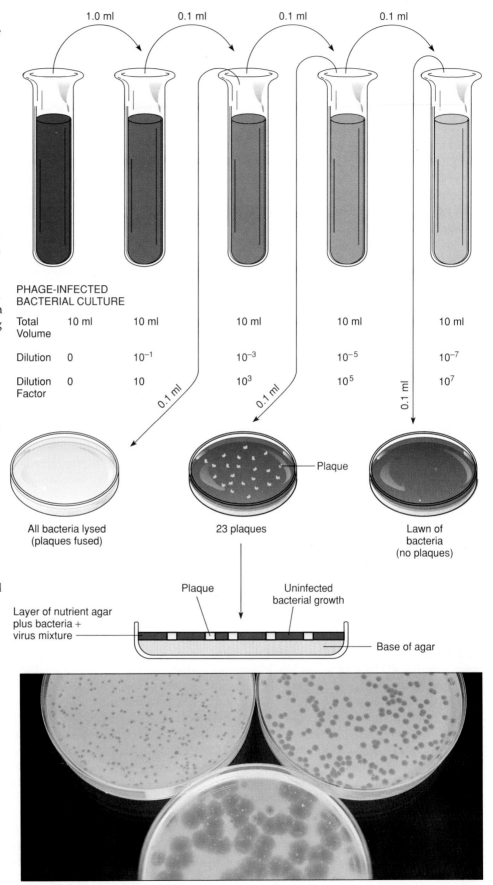

Using the results shown in Figure 15.24, it is observed that there are 23 phage plaques derived from the 0.1-ml aliquot of the 10^{-5} dilution. Therefore, we can estimate that there are 230 phage per ml at this dilution. The initial viral density in the undiluted sample, where 23 plaques are observed from 0.1 ml, is calculated as:

$$(230/\text{ml}) \times (10^5) = 230 \times 10^5/\text{ml}$$

Since this figure is derived from the 10^{-5} dilution, we can estimate that there will be only 0.23 phage per 0.1 ml in the 10^{-7} dilution. As a result, when 0.1 ml from this tube is assayed, it is not unexpected that no phage particles will be present. This possibility is borne out in Figure 15.24, where an intact lawn of bacteria exists. The dilution factor is simply too great.

The use of the plaque assay has been invaluable in mutational and recombinational studies of bacteriophages. We will apply the above technique more directly in Chapter 18 when Seymour Benzer's elegant genetic analysis of a single gene in phage T4 is discussed.

Lysis and Lysogeny

The relationship between virus and bacterium does not always result in viral reproduction and lysis. As early as the 1920s it was known that a virus could enter a bacterial cell and establish a symbiotic relationship with it. The precise molecular basis of this symbiosis is now well understood. Upon entry, the viral DNA, instead of replicating in the bacterial cytoplasm, is integrated into the bacterial chromosome, a step that characterizes the **lysogenic pathway**. Subsequently, each time the bacterial chromosome is replicated, the viral DNA is also replicated and passed to daughter bacterial cells following division. No new viruses are produced and no lysis of the bacterial cell occurs. However, under certain stimuli, such as chemical or ultraviolet light treatment, the viral DNA may lose its integrated status and initiate replication, phage reproduction, and lysis of the bacterium.

Several terms are used to describe this relationship. The viral DNA, integrated in the bacterial chromosome, is called a **prophage**. Viruses that can either lyse the cell or behave as a prophage are called **temperate**. Those that can only lyse the cell are referred to as **virulent**. A bacterium harboring a prophage is said to be **lysogenic;** that is, it is capable of being lysed as a result of induced viral reproduction. The viral DNA, which can replicate either in the bacterial cytoplasm or as part of the bacterial chromosome, may be classified as an **episome**.

The relationship of a prophage and its host cell is symbiotic because lysogenic bacteria are immune to viral attacks by the same phage whose genetic information it harbors. As we shall see in Chapter 19, this immunity is due to the synthesis of a prophage repressor molecule that regulates the expression of the viral DNA.

TRANSDUCTION: VIRUS-MEDIATED BACTERIAL DNA TRANSFER

In 1952, Joshua Lederberg and Norton Zinder were investigating possible recombination in the bacterium *Salmonella typhimurium*. Although they recovered prototrophs from mixed cultures of two different auxotrophic strains, subsequent investigations revealed that recombination was occurring in a manner different from that attributable to the presence of an F factor, as in *E. coli*. What they were to discover was still a third mode of bacterial recombination, one mediated by bacteriophages and now called **transduction**.

The Lederberg-Zinder Experiment

Lederberg and Zinder mixed the *Salmonella* auxotrophic strains LA-22 and LA-2 together and recovered prototroph cells when the mixture was plated on minimal medium. LA-22 was unable to synthesize the amino acids phenylalanine and tryptophan (phe^- trp^-), and LA-2 could not synthesize the amino acids methionine and histidine (met^- his^-). Prototrophs (phe^+ trp^+ met^+ his^+) were recovered at a rate of about $1/10^5$ (10^{-5}) cells.

Although these observations at first appeared to suggest that the type of recombination involved was the kind observed earlier in *E. coli*, experiments using the Davis U-tube soon showed otherwise (Figure 15.25). When the two auxotrophic strains were separated by a glass-sintered filter, thus preventing cell contact but allowing growth to occur in a common medium, a startling observation was made. When samples were removed from both sides of the filter and plated independently on minimal medium, prototrophs were recovered, but only from one side of the tube. Recall that if conjugation were responsible, the conditions in the Davis U-tube would have prevented recombination.

Prototrophs were recovered only when cells from the side of the tube containing LA-22 bacteria were plated. Obviously, the presence of LA-2 cells on the other side of the tube was essential for recombination, since LA-2 cells were the source of the new genetic information. Because the genetic information

FIGURE 15.25
The Lederberg-Zinder
experiment using *Salmonella*.
After mixing two auxotrophic
strains in a Davis U-tube,
Lederberg and Zinder
recovered prototrophs from the
side containing LA-22 but not
from the side containing LA-2.
These initial observations led to
the discovery of the
phenomenon called
transduction.

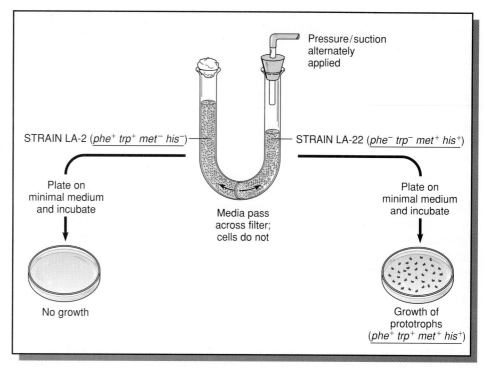

STRAIN LA-2 (phe^+ trp^+ met^- his^-)

STRAIN LA-22 (phe^- trp^- met^+ his^+)

Pressure/suction
alternately
applied

Plate on
minimal medium
and incubate

Plate on
minimal medium
and incubate

Media pass
across filter;
cells do not

No growth

Growth of
prototrophs
(phe^+ trp^+ met^+ his^+)

responsible for recombination had somehow to pass across the filter, but in an unknown form, it was initially designated simply as a **filterable agent (FA)**.

Three subsequent observations made it clear that this recombination was quite distinct from any other form of recombination.

1 The FA would not pass across filters with a pore diameter of less than 100 nm, a size that normally allows passage of small DNA molecules.

2 Testing the FA in the presence of DNase, which will enzymatically digest DNA, showed that the FA was not destroyed. These first two observations demonstrate that FA is not naked DNA.

3 The third observation was particularly important. It was observed that FA was produced by the LA-2 cells only when they were grown in association with LA-22 cells. If LA-2 cells were grown independently and that culture medium was then added to LA-22 cells, recombination was not observed. Therefore, LA-22 cells play some role in the production of FA by LA-2 cells and do so only when sharing common growth medium. This was a key observation in determining the basis of genetic exchange.

These observations were explained by the presence of a prophage (P22) present in the LA-22 *Salmonella* cells. Rarely, P22 prophages entered the vegetative or lytic phase, reproduced, and lysed some of the LA-22 cells. This phage, being much smaller than a bacterium, was then able to cross the filter and lyse some of the LA-2 cells, because this strain was not immune to attack. In the process of lysis, the P22 phages produced in LA-2 often acquired a region of the LA-2 chromosome along with their own genetic material. If this region contained the phe^+, and trp^+ genes present in LA-2, and if the phages subsequently passed back across the filter and reinfected LA-22 cells, prototrophs were produced. The exact nature of how this occurred was not immediately clear.

The Nature of Transduction

Further studies revealed the existence of transducing phages in other species of bacteria. For example, *E. coli*, *Bacillus subtilis*, and *Pseudomonas aeruginosa* can be transduced by the phages P1, SP10, and F116, respectively. The precise mode of transfer of DNA during transduction has also been established. The process most often begins when a prophage enters the lytic cycle and progeny viruses subsequently infect and lyse other bacteria. During infection, the bacterial DNA is degraded into small fragments and the viral DNA is replicated. As packaging of the viral chromosomes in the protein head of the phage occurs, errors are sometimes made. Rarely, instead of viral DNA, segments of bacterial DNA are packaged.

Even though the initial discovery of transduction involved lysogenic bacteria, the same process can occur during the normal lytic cycle. Following infection, the bacterial chromosome is degraded into small pieces. If, during bacteriophage assembly, a small piece of bacterial DNA is packaged along with the viral chromosome, subsequent transduction is possible.

Sometimes, only bacterial DNA is packaged! Regions as large as 1 percent of the bacterial chromosome may become enclosed randomly in the viral head. Following lysis, these aberrant phages, lacking their own genetic material, are released in the culture medium. Because the ability to infect is a property of the protein coat, they can initiate infection of other unlysed bacteria. When this occurs, bacterial rather than viral DNA is injected into the bacterium and can either remain in the cytoplasm or recombine with its homologous region of the bacterial chromosome. If the bacterial DNA remains in the cytoplasm, it does not replicate but may remain in one of the progeny cells following each division. When this happens, only a single cell, partially diploid for the transduced genes, is produced—a phenomenon called **abortive transduction**. If the bacterial DNA recombines with its homologous region of the bacterial chromosome, the transduced genes are replicated as part of the chromosome and passed to all daughter cells. This process is called **complete transduction**. Both abortive and complete transduction are subclasses of the broader category of **generalized transduction**. As described above, transduction is characterized by the random nature of DNA fragments and genes transduced. Each fragment has a finite but small chance of being packaged in the phage head. Most cases of generalized transduction are of the abortive type; some data suggest that complete transduction occurs 10 to 20 times less frequently. This finding may be related to the fact that double-stranded DNA is involved. In comparison with single-stranded DNA, which is integrated during transformation, it may be much more difficult for double-stranded DNA to become integrated.

Mapping and Specialized Transduction

As with transformation, transduction has been used in linkage studies and mapping of the bacterial chromosome. Cotransduced genes must be aligned closely to one another along the chromosome. By concentrating on two or three linked genes, transduction studies can also determine the precise order of genes. Such an analysis is predicated on the same rationale underlying other mapping techniques, where outcomes resulting from single events occur much more frequently than those relying on two or more independent events occurring simultaneously.

Another aspect of virally mediated bacterial recombination involves **specialized transduction**. Com-pared with generalized transduction, where all genes have an equal probability of being transduced, specialized transduction is restricted to certain genes. One of the best examples involves transduction of E. coli by the prophage λ. In this case, only the gal (galactose) or bio (biotin) genes are transduced.

The reason why specialized transduction occurs became clear when it was learned how the λ DNA integrates into the E. coli chromosome during lysogeny. A region of λ DNA, designated att, is some 15 nucleotides long and is responsible for integration. A precisely homologous region exists on the E. coli chromosome, which is flanked by the gal and bio loci on either end. Therefore, λ DNA always integrates at a location in the chromosome between these genes (Figure 15.26).

Phage λ DNA can be induced to detach from the chromosome and cause lysis. Sometimes the detachment process occurs incorrectly and carries either the gal or bio E. coli genes in place of part of the viral DNA. This happens when the recombinational event leading to detachment occurs incorrectly, outside the att region of the λ chromosome. The resulting chromosome is defective because it has lost some of its genetic information, but it is nevertheless replicated and packaged during the formation of mature phage particles. Once the previously lysogenized cell is lysed, the virus can inject the defective chromosome into another bacterial cell.

In a process involving lysogeny by a nondefective chromosome, the defective viral chromosome may also be integrated into the bacterial chromosome and is replicated along with it. Such cells are diploid for the gal or bio genes. If the recipient cells are gal⁻ and are unable to use galactose as a carbon source, the presence of the transducing gal⁺ DNA will cause them to revert to a gal⁺ phenotype, where they can use this carbohydrate. In a similar way, bio⁻ cells can be transduced to a bio⁺ phenotype.

As is evident from this discussion, specialized transduction occurs in quite a different way from generalized transduction. Because it is limited to certain genes, it is not useful in linkage or mapping studies.

MUTATION AND RECOMBINATION IN VIRUSES

Much of what is known concerning viral genetics has been derived from studies of bacteriophages. Phage mutations often affect the morphology of the plaques formed following lysis of the bacterial cells. For example, in 1946 Alfred Hershey observed unusual

FIGURE 15.26
Production of a defective phage λ, leading to specialized transduction.

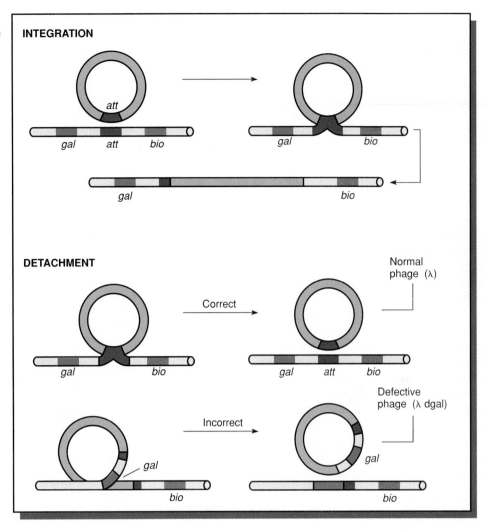

INTEGRATION

att

gal att bio

gal bio

gal bio

DETACHMENT

gal bio

Correct

Normal phage (λ)

gal att bio

gal

bio

Incorrect

Defective phage (λ dgal)

gal

bio

T2 plaques on plates of *E. coli* strain B. While the normal T2 plaques are small and have a clear center surrounded by a turbid, diffuse halo, the unusual plaques were larger and possessed a sharp outer perimeter. When the viruses were isolated from these plaques and replated on *E. coli* B cells, an identical plaque appearance was noted. Thus, the plaque phenotype was an inherited trait resulting from the reproduction of mutant phages. Hershey named the mutant *rapid lysis (r)* because the plaques were larger, apparently resulting from a more rapid or more efficient life cycle of the phage. It is now known that wild-type phages undergo an inhibition of reproduction once a particular-sized plaque has been formed. The *r* mutant T2 phages are able to overcome this inhibition, producing larger plaques with a sharp perimeter.

Another bacteriophage mutation, *host range (h)*, was discovered by Luria. This mutation extends the range of bacterial hosts that the phage can infect. Although wild-type T2 phages can infect *E. coli* B,

they cannot normally attach or be absorbed to the surface of *E. coli* B-2. The *h* mutation, however, provides the basis for adsorption and subsequent infection.

Table 15.3 lists other representative types of mutations that have been isolated and studied in the T-even series of bacteriophages (T2, T4, T6, etc.). These mutations are important to the study of genetic phenomena in bacteriophages. Conditional, temperature-sensitive mutations have been particularly valuable in the study of essential genes where mutations eliminate the ability of the phage to reproduce.

Genetic Exchange between Viruses

Around 1947, several research teams demonstrated that recombination occurs between viruses. This discovery of genetic exchange was made during experiments in which two mutant strains of bacteriophage were allowed to simultaneously infect the

TABLE 15.3
Some mutant types of T-even phages.

Name	Description
minute	Small plaques
turbid	Turbid plaques on *E. coli* B
star	Irregular plaques
uv-sensitive	Alters UV sensitivity
acriflavin-resistant	Forms plaques on acriflavin agar
osmotic shock	Withstands rapid dilution into distilled water
lysozyme	Does not produce lysozyme
amber	Grows in *E. coli* K12, but not B
temperature-sensitive	Grows at 25°C, but not at 42°C

same bacterial culture. These **mixed infection experiments** were designed so that the number of viral particles sufficiently exceeded the number of bacterial cells so as to ensure simultaneous infection of most cells by both viral strains.

For example, in one study using the T2/*E. coli* system, the viruses were of either the h^+r or hr^+ genotype. If no recombination occurred, these two parental genotypes (wild-type host range restriction, rapid lysis; and extended host range, normal lysis) would be the only expected phage progeny. However, the recombinant h^+r^+ and hr were detected in addition to the parental genotypes (Figure 15.27). As with eukaryotes, the percentage of recombinant plaques divided by the total number of plaques reflects the relative distance between the genes:

$$\frac{(h^+r^+) + (hr)}{\text{total plaques}} \times 100 = \text{recombinational frequency}$$

Sample data for the *h* and *r* loci are shown in Table 15.4.

Similar recombinational studies have been performed with large numbers of mutant genes in a variety of bacteriophages. Data are analyzed in much the same way as they are in eukaryotic mapping experiments. Two- and three-point mapping crosses are possible, and the percentage of recombinants in the total number of phage progeny is calculated. This value is proportional to the relative distance between two genes along the chromosome.

An interesting observation in phage crosses is **negative interference**. Recall that in eukaryotic mapping crosses, positive interference is the rule. Fewer than expected double-crossover events are observed. In phage crosses, often just the reverse often occurs. In three-point analysis, a greater than expected frequency of double exchanges is observed. Negative interference is explained on the basis of the dynamics of the conditions leading to recombination within the bacterial cell. The available evidence supports the

concept that recombination between phage chromosomes involves a breakage and reunion process similar to that of eukaryotic crossing over. The process is facilitated by nucleases that nick and reseal

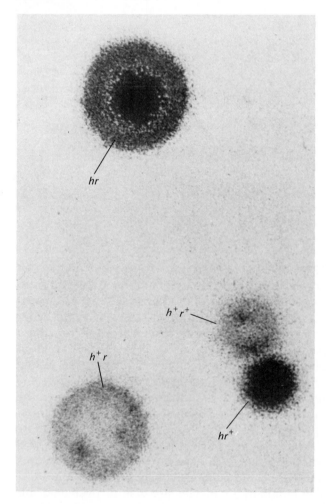

FIGURE 15.27
Actual plaque phenotypes observed following simultaneous infection of *E. coli* by two strains of phage T2, h^+r and hr^+. In addition to the parental genotypes, recombinant plaques hr and h^+r^+ were also recovered.

TABLE 15.4
Results of a cross involving the *h* and *r* genes in phage T2.

Genotype	Plaques	Designation
h r⁺	42	Parental Progeny 76%
h⁺*r*	34	
h⁺*r*⁺	12	Recombinants 24%
h r	12	

SOURCE: Data derived from Hershey and Rotman, 1949.

DNA strands. A fairly clear picture of the dynamics of viral recombination has emerged.

Following the early phase of infection, the chromosomes of each phage begin replication. As this stage progresses, a pool of chromosomes accumulates in the bacterial cytoplasm. If double infection by phages of two genotypes has occurred, then the pool of chromosomes initially consists of the two parental types. Genetic exchange between these two types will occur, producing recombinant chromosomes.

In the case of the *h*⁺*r* and *hr*⁺ example just discussed, *h*⁺*r*⁺ and *hr* chromosomes are produced. Each of these recombinant chromosomes may undergo replication and is also free to undergo new exchange events with the other and with the parental types. Furthermore, recombination is not restricted to exchange between two chromosomes—three or more may be involved. As phage development progresses, chromosomes are randomly removed from the pool and packed into the phage head, forming mature phage particles. Thus, parental and recombinant genotypes are produced.

Recombination events in viruses are not restricted to intergenic exchanges. With powerful selection systems, it is possible to detect intragenic recombination, where exchanges occur within a single gene. Such studies have led to the fine-structure analysis of the gene (see Chapter 18).

Packaging and Concatemer Formation

Once a sufficient pool of phage chromosomes is available as a result of replication, they are individually packaged into the head of the virus particle. Most often, this packaging involves a complex of phage chromosomes called a **concatemer** (Figure 15.28). This structure represents a linear series of chromosomes covalently bonded to one another. A concatemer results from a series of recombinational events between the terminally redundant portions of individual chromosomes. This process may be illustrated by considering the events that occur during the reproduction of the T-even series of phages.

Following several rounds of replication of the phage chromosome in the bacterial cytoplasm, numerous individual chromosomes are produced. Each one contains redundant terminal regions that are identical to one another. Recombination between these ends results in concatemer formation. In this group of phages, the concatemer is further replicated, and it is in this form that intergenic and intragenic recombination occurs.

When replication is completed, packaging is initiated. The phage head is just large enough to hold a length of DNA equivalent to one chromosome, including its terminally redundant ends. Packaging is accomplished by what is referred to as the **headful mechanism**. The end of a concatemer enters and is condensed into the phage head until it is filled. Then, an endonuclease cleaves the molecule, producing a concatemer shorter by one chromosome length. The process is repeated until the entire complex has been cleaved and the individual units packaged.

An interesting consequence of this process is the production of **circularly permuted** molecules of DNA (Figure 15.28). Assuming that the endonuclease cuts precise lengths of the chromosome, each subsequent molecule will contain a different region at its terminal ends. Until the mechanism of packaging was clarified, circular permutation presented great difficulty in mapping the chromosome. However, it is now clear that each chromosome contains a full set of phage genes plus the terminally redundant sequences.

In phage λ, a similar mechanism involving concatemers is utilized. The ends of the chromosomes are marked by specific sequences called *cos* **sites,** which define the perimeters of the DNA to be packaged by serving as markers for cleavage of the concatemers. Up to a point, all DNA between *cos* sites is packaged in the phage head. This is the basis of the use of **cosmids** in DNA cloning techniques (Chapter 16).

OTHER STRATEGIES FOR VIRAL REPRODUCTION

In contrast to the T-even phages, other viruses have evolved many different strategies in order to reproduce. For example, during the infection of *E. coli*, the smaller **bacteriophage φX174** contains only a circular single strand of DNA, called the plus (+) strand (see

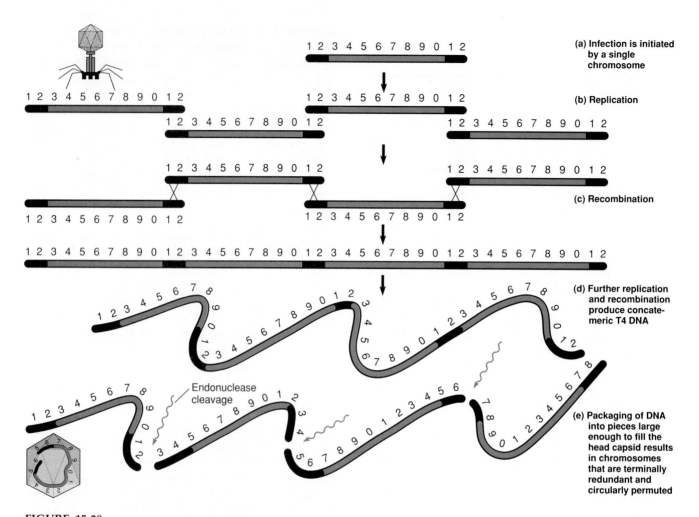

FIGURE 15.28
Concatemer formation and the production of circularly permuted chromosomes as packaging occurs during bacteriophage maturation.

Chapter 10). It infects the host and immediately serves as the template for the synthesis of the complementary negative (−) strand. This results in a circular double-stranded molecule called the **replicative form (RF)**. The RF is itself replicated semiconservatively, producing about 50 progeny molecules. These then serve as templates for the **rolling circle mode** of production of only (+) strands (see Figure 10.14). As hundreds of these DNA strands are produced, they are packaged into protein capsids synthesized under the direction of the viral genetic information.

While DNA viruses can pirate the use of the host DNA polymerase in order to replicate their nucleic acid, no comparable enzyme exists when RNA-containing viruses invade their host cells. Some RNA viruses require an RNA-dependent RNA polymerase,

which is sometimes more simply called **RNA replicase**. For example, when the small **bacteriophage Qβ** infects *E. coli* as its host cell, the infective single-stranded RNA molecule first serves as a messenger RNA (mRNA), encoding three proteins. Two are components of the coat protein, while one is a subunit of RNA replicase. This subunit combines with three other subunits that are bacterial in origin. The resultant enzyme is specific for the replication of the phage RNA (as compared with the bacterial RNAs that are present). First, complementary RNA virus strands (−) are created. These serve as the template for the synthesis of many (+) strands. Ultimately, the (+) strands are encapsulated by coat proteins as mature viruses are formed. A similar mode of replication occurs in the eukaryotic, RNA-containing **poliovirus**.

Reverse Transcriptase

The final type of viral reproduction to be discussed was the basis of an extraordinary finding made in the early 1960s. A long series of investigations by Howard Temin and David Baltimore led to the discovery of **reverse transcriptase,** an enzyme capable of synthesizing DNA on an RNA template. This enzyme is also called **RNA-directed DNA polymerase.** Temin proposed that the RNA that is the genetic material of certain RNA-containing animal tumor viruses is transcribed reversibly into DNA, which then serves as the template for replication and transcription of RNA during viral infection. These viruses are called **oncogenic** because they induce malignant growth.

Temin's proposed reverse transcriptase was largely ignored because it contradicted the previously accepted notion that genetic information flows only from DNA to RNA. In addition, other RNA-containing viruses (polio virus and double- and single-stranded RNA bacteriophages), whose RNA is replicated by the enzyme RNA replicase, did not demonstrate reverse transcriptase activity. Thus, molecular biologists questioned why the tumor viruses should require this enzyme.

Working with the RNA-containing oncogenic **Rous sarcoma virus (RSV),** which produces cancer in chickens, Temin added the antibiotic actinomycin D to cultures of cells that had been infected. He found that the antibiotic inhibited viral multiplication. Actinomycin D is known to specifically inhibit DNA-directed RNA synthesis and not to affect RNA-directed RNA synthesis. Temin also found that inhibitors of DNA synthesis such as 5-fluorodeoxyuridine also blocked infection. Thus, he reasoned that DNA must therefore serve as an essential intermediate molecule during infection.

Temin subsequently demonstrated directly that DNA is synthesized by RSV during infection. He announced these results in 1970 and found that other workers, using a mouse leukemia virus, had arrived at the same conclusions. These reports stimulated a tremendous amount of work on the enzyme and its role in tumor production. Reverse transcriptase was subsequently isolated independently by Temin and Baltimore. It has since been found in all RNA oncogenic viruses. Because they "reverse" the flow of genetic information they are called **retroviruses.**

The enzyme uses the infecting (+) RNA strand as a template, synthesizing DNA in the 5'-to-3' direction. An RNA primer is first synthesized. Then, the enzyme directs synthesis of the complementary DNA strand, creating an RNA/DNA hybrid molecule.

Reverse transcriptase possesses RNase activity, and as a result, it degrades the RNA portion of the hybrid molecule. The enzyme then uses the remaining DNA strand as a template and synthesizes the complementary strand, creating a double-stranded DNA molecule. This is capable of integrating into the genome of the host cell, forming what is called a **provirus.** When the host genome is transcribed into RNA, so is the proviral DNA. This produces (+) RNA strands, some of which are translated into viral proteins. Packaging of other newly formed RNA molecules into the viral capsid occurs to form mature viruses. Such a host cell is said to be **transformed.** A similar life cycle occurs in all retroviruses.

Acquired Immunodeficiency Syndrome (AIDS)

Interest in retroviruses has heightened considerably with the discovery of the **human immunodeficiency virus (HIV),** considered to be the causative agent in **acquired immunodeficiency syndrome (AIDS).** HIV is a member of a subgroup of retroviruses transmitted by sexual contact or the exchange of body fluids. The genetic material consists of single-stranded RNA containing about 9000 ribonucleotides, surrounded by a protein capsid. Genetically programmed surface glycoproteins that are associated with this capsid are essential to the infection process. One set of these, called gp120 (a glycoprotein of 120,000 molecular weight), interacts with a surface receptor on the host cell called CD4, which is found mainly on a subset of T lymphocytes, the T helper cells.

Viral infection and reproduction is dependent on DNA produced under the direction of reverse transcriptase. Following assembly, the viruses bud through the cell membrane of the infected cell (Figure 15.29). In addition to the production of viral particles, an excess amount of gp120 is produced that is inserted into the surface of the infected cell. Its presence facilitates fusion with other uninfected T helper cells, leading to the death of all involved cells. The depletion of the T lymphocyte helper cell population has a catastrophic effect on the immune system, eventually immobilizing it almost completely.

Infected individuals are, as a result, susceptible to so-called opportunistic infections that healthy immune systems fend off fairly easily. Such diseases, including *Pneumocystitis carini* pneumonia and Kaposi's sarcoma, are often the cause of death in AIDS patients.

HIV has an unusually high error rate during its replication, in comparison to many other viruses. It is

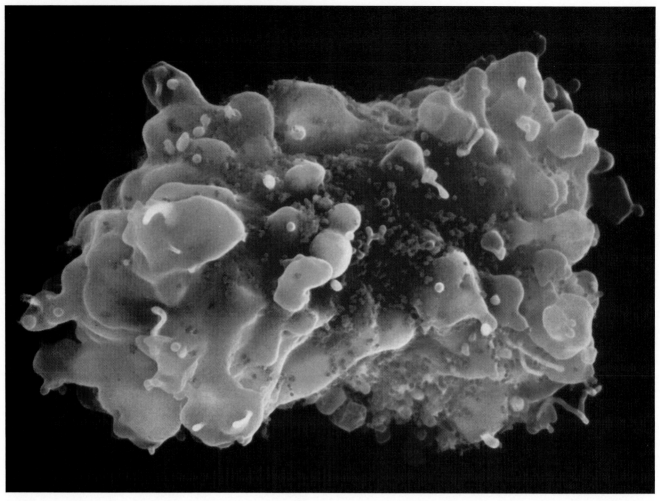

FIGURE 15.29
A scanning electron micrograph of a T lymphocyte infected by HIV virus. The
projections from the surface are microvilli. The virus particles are present on the cell
surface and are seen as small, blue, rounded dots.

this characteristic that leads to the production of virus
particle populations with substantial genetic diversity, particularly in regard to the chemical specificity
of the envelope protein gp120. As a result, the
development of a vaccine against this devastating
virus has been extremely difficult.

THE STUDY OF ESSENTIAL
GENES IN VIRUSES:
COMPLEMENTATION ANALYSIS

If a viral gene is responsible for an essential function
during reproduction, a mutation eliminating that
function is lethal. In order to be studied, such

mutations must be conditional; that is, they must be
expressed only under certain conditions. **Temperature-sensitive conditional lethals** have been particularly useful in the study of essential phage genes. At
a permissive temperature, the gene under study is
expressed normally, but at a restrictive temperature,
mutant expression occurs and wild-type function is
lost.

It is possible to simultaneously infect a bacterium
with two viral strains, each containing a separate
conditional lethal mutation. If the mutations are
contained in different genes, they will demonstrate
complementation under restrictive conditions. The
chromosome of each strain contains the normal allele
of the gene that is mutant in the other strain. As a

result, a pool of all normal gene products will be produced in the bacterial cytoplasm during simultaneous infection, normal phage assembly will occur, and the bacterium will be lysed. The concept of complementation is illustrated in part (b) of Figure 15.30.

On the other hand, if two mutations present in the same gene are studied simultaneously, complementation will not occur under restrictive conditions. Each type of chromosome will produce the same defective gene product. As a result, reproduction cannot be completed since one essential gene product is always defective. This case is illustrated in part (c) of Figure 15.30.

If many conditional lethal mutations are isolated, it is possible to saturate the chromosome so that all regions contain at least one mutation. When mutations are studied in pairs, complementation shows that the two mutations are in separate genes. Non-complementation, on the other hand, shows that they are in the same gene.

About 40 conditional lethal mutations in the phage ϕX174 have been investigated in this way. They were shown to fall into eight complementation groups. This finding suggests that there are eight essential genes in the small genome of this phage. In the larger genome of phage λ, 25 complementation groups or genes are recognized. These include genes required for head and tail production, DNA replication, cell lysis, and regulation of gene expression. Complementation analysis thus allows a determination of the number of essential genes in certain bacteriophages.

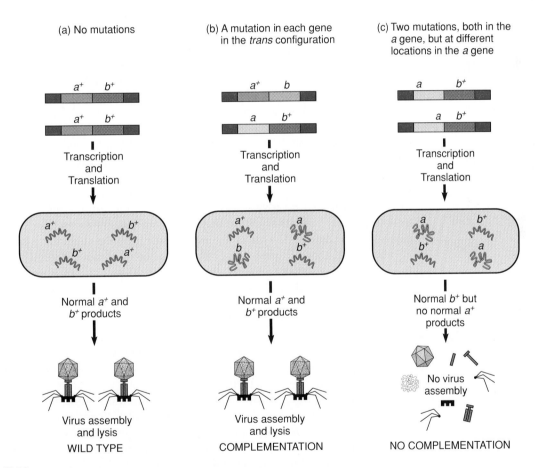

FIGURE 15.30
Diagrammatic illustration of (a) normal wild-type phage infection, compared with (b), showing how mutations in different genes in different viral strains will complement one another during simultaneous infection, resulting in the production of normal-appearing phage. (c) Simultaneous infection by two viral strains with different mutations in the same gene fail to complement one another.

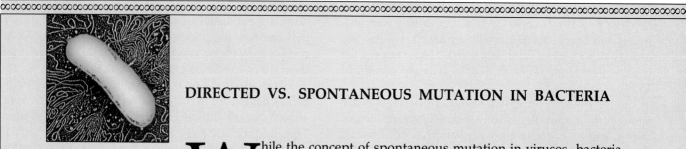

DIRECTED VS. SPONTANEOUS MUTATION IN BACTERIA

While the concept of spontaneous mutation in viruses, bacteria, and even higher organisms is no longer disputed, the possibility that organisms such as bacteria might also be capable of "selecting" a specific set of mutations occurring as a result of environmental pressures has long intrigued geneticists. Two independent investigations published in 1988, by John Cairns and Barry Hall and their colleagues, have provided preliminary evidence that this might be the case. Cairns devised an experimental protocol that improves on that used in the fluctuation test of Luria and Delbruck, discussed at the beginning of this chapter. The procedures were designed to detect nonrandom mutations arising in response to factors in the environment in which bacteria are cultured. The results of this work suggest that some bacteria may select mutations that are "adaptive" to the environment.

Instead of using the characteristic of phage T1 resistance, where the only surviving cells are those which have mutated, the Cairns work involved a strain of bacteria that contained a lactose mutation (lac^-). Such bacteria cannot use lactose as a carbon source. Cells first were grown in a rich liquid medium that provided an adequate carbon source other than lactose. Thus, the lac^- cells as well as any spontaneous lac^+ mutants were able to grow and reproduce quite well. This gene was subsequently followed by plating aliquots of cells on Petri plates containing minimal medium to which lactose had been added. The lac^- cells, present in the vast majority, will survive on the plates, but since they lack a carbon source that they can metabolize, they cannot proliferate and form colonies. On the other hand, cells that have mutated to lac^+, whether in the original liquid culture or as they sit on the plate, will be detected. This is the major experimental difference between this approach and that of Luria and Delbruck some 45 years earlier. Those cells that have not mutated to lac^+ in liquid medium have the opportunity to do so after they are plated, and they may also be detected, since they will then form colonies.

Using a slightly more sophisticated mathematical analysis than Luria and Delbruck, Cairns attempted to identify the same sort of "fluctuation," or the lack of it, as an indication of either spontaneous or adaptive mutations, respectively. Cairns found both types of data! The composite distribution was calculated for the case in which both spontaneous mutations occurred in the liquid cultures at random times (creating "fluctuation") and directed mutations arose as a response to the presence of lactose on the Petri plates. Indeed, such a composite distribution was observed.

But, how were they to prove that, in fact, the elevated frequency of mutations occurred on the plates in response to lactose? To answer this question, the Cairns group asked whether another mutation, unrelated to the metabolism of lactose, had also occurred on the plates. They chose to assay for the production of Val^R mutations, which confer resistance to high concentrations of the amino acid valine. The assay was easily accomplished by overlaying the plates with agar containing valine and glucose. Only Val^R mutants will grow. Cairns reasoned that if the lac^+ mutations were not necessarily in response to the lactose, then Val^R mutations should arise with a similar distribution as earlier observed for the lac mutations. The result was that plates accumulating lac^+ mutants were not, at the same time, accumulating Val^R mutations. They concluded that the increased frequency of mutations affecting lactose metabolism was the result of the presence of lactose in the medium. This is, of course, an indirect inference.

Cairns' work suggests that the bacterial cell's genetic machinery can profit from and respond to its environment by producing adaptive mutations. This conclusion is contrary to the current thinking that mutations are spontaneous events, many of which occur as random errors during replication of DNA. As Cairns has pointed out, these findings, at first glance, seem to support the romantic notion that life is mysterious and that certain observations cannot be explained in simple, straightforward terms based on the laws of physics. Whatever the case may be, Cairns' observations are not isolated examples.

Barry Hall's work demonstrated a similar adaptive response by *E. coli* to growth on salicin. In this case, interestingly, a two-step genetic change was required. The first occurred in response to salicin even though that mutation offered no selective advantage without the second genetic alteration, the removal, or deletion, of a segment of nucleotides called an insertion sequence (IS). Further, the changes must occur sequentially. They do so at a much higher frequency than could be predicted, provided salicin is present in the growth medium. The observed frequency is several orders of magnitude higher!

What possible explanation can account for observations such as these? As Hall has pointed out, our failure to adequately explain such phenomena may be attributable to our ignorance of fundamental mechanisms and rates of mutations in nongrowing cells. Such information and knowledge is extremely important because this physiological condition is much more akin to a natural environment as opposed to that found with cells grown under laboratory conditions. One suggestion is that under stressful nutritional conditions (starvation), bacteria may be capable of activating mechanisms that create a hypermutable state in genes that will enhance survival.

In the context of a textbook such as this, an important idea to be realized from this discussion is that knowledge may often be derived from observations that initially cannot be explained. This, and the ensuing debates and surrounding controversy, are what constantly maintain intrigue and interest in science!

CHAPTER SUMMARY

1 Genetic analysis of bacteria and viruses has been pursued successfully since the 1940s. The ease of obtaining large quantities of pure cultures of bacteria and mutant strains of viruses as experimental material and the efficient selection and isolation of naturally occurring and induced mutations have made these the organisms of choice in numerous types of genetic study.

2 Although the concept of spontaneous mutation is supported experimentally and is widely upheld, recent evidence suggests that some mutations that are adaptive for the survival of the organism may arise in response to selective pressure from the environment.

3 Genetic exchange of genetic information between bacteria involves three different modes of recombination: conjugation, transformation, and transduction.

4 Conjugation is initiated by a bacterium harboring an F factor, a bacterial plasmid. Genetic information flows unidirectionally from an F^+ cell and upon its entry into the recipient F^- cell, recombination with the recipient chromosome may occur.

5 The integration of the F factor into the donor chromosome creates Hfr cells and provides for more efficient transfer of the donor DNA. Time mapping, using Hfr donor cells, is based upon the orientation of transfer that is determined by the point and polarity of integration.

6 The group designated as the *rec* genes have been found to be either directly involved in the process of recombination or in the regulation of this process. These genes were discovered as a result of isolating mutations that affected the products of these genes.

7 Insertion sequences, unique DNA sequences present in both plasmids and the bacterial chromosome, facilitate movement of genetic information such as the insertion and excision of the F factor into and out of the chromosome. These sequences may combine with genes, producing transposons, which also move from one genetic molecule or position to another.

8 The phenomenon of transformation, which does not require cell contact, involves the entry of single-stranded exogenous DNA into the host chromosome of a recipient bacterial cell. Linkage mapping of closely aligned genes may be performed using this process.

9 Transduction, or virus-mediated bacterial DNA transfer, requires the formation of a symbiotic relationship between bacteria and bacteriophages that infect them. The viral genetic material may become integrated into the bacterial chromosome as a prophage and lysogenize the host cell.

10 In the process of generalized transduction, prophages lose their integrated status and the resultant viruses mediate the transfer of a random part of the bacterial chromosome between organisms. In specialized transduction, only specific genes adjacent to the point of insertion of the prophage into the bacterial chromosome are transferred. Transduction may also be used for bacterial linkage and mapping studies.

11 Phage are studied using the plaque assay. The discovery of mutations in plaques, such as those causing variation in plaque morphology and the alteration of the host range of infectivity, have allowed the analysis of recombination between viruses. Discovery of such genetic exchange occurred during simultaneous or mixed infection of bacterial cells by two distinct mutant viral strains. While the viral nucleic acid molecules are present in the bacterial cytoplasm, recombination may occur. Observations of this recombination may be used to map the circular viral chromosome.

12 Retroviruses are RNA oncogenic viruses that depend on reverse transcriptase activity for infection. This enzyme reverses the flow of genetic information from RNA to DNA, which most often integrates into the genome of the host chromosome. HIV, a subgroup of the retroviruses, debilitates the immune system by infecting T lymphocyte helper cells, consequently increasing the patients' susceptibility to diseases characteristic of AIDS.

13 Complementation analysis of conditional, lethal mutations has elucidated the number of genes essential to the reproduction of bacteriophages.

Insights and Solutions

1 Three strains of bacteria, each bearing a separate mutation, a^-, b^-, or c^-, were used as the sources of donor DNA in a transformation experiment involving recipient cells that were wild type for those genes, but expressed the mutant d^-.

(a) Based on the following data, and assuming that the location of the d gene precedes the a, b, and c genes, propose a linkage map for the four genes:

Donor DNA	Recipient	Transformants	Frequency of ++ Transformants
$a^- d^+$	$a^+ d^-$	$a^+ d^+$	0.21
$b^- d^+$	$b^+ d^-$	$b^+ d^+$	0.18
$c^- d^+$	$c^+ d^-$	$c^+ d$	0.63

ANSWER: These data reflect the relative distances between each of the a, b, c genes and the d gene. The a and b genes are about the same distance away from the d gene and are thus tightly linked to one another. The c gene is more distant. Assuming that the d gene precedes the others, the map looks like this:

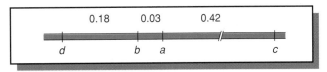

(b) If the donor DNA was wild type and recipient cells were either $a^- b^-$, $a^- c^-$ or $b^- c^-$, in which case would wild-type transformants be expected most frequently?

ANSWER: Since the a and b genes are closely linked, they may be cotransformed in a single event. Thus, recipient cells of $a^- b^-$ are most likely to be converted to wild type.

2 Time mapping was performed in a cross involving the genes *his*, *leu*, *mal*, and *xyl*. After 25 minutes, mating was interrupted with the following results in recipient cells.

90% were *xyl*
80% were *mal*
20% were *his*
none were *leu*

What are the positions of these genes relative to the origin (O) of the F factor and to one another?

ANSWER: Since the *xyl* gene was transferred most frequently, it is closest to O (*very* close). The *mal* gene is next and reasonably close to *xyl*, followed by the *his* gene. The *leu* gene is well beyond these three, since no recombinants are recovered that include it.

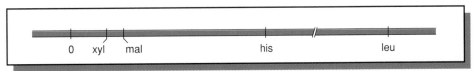

1 Contrast the two major hypotheses explaining the origin of heritable variation in bacteria.

2 Discuss the experiments performed by Luria and Delbruck that clarified the origin of heritable variation in bacteria. What was the rationale for their conclusions? Why were these experiments called the "fluctuation tests"?

3 Describe the replica plating technique. Why is this technique so valuable experimentally?

4 How did the use of replica plating strengthen the conclusions drawn in the Luria-Delbruck experiment?

5 Differentiate between phenotypic lag and segregation lag. Do these phenomena lead to overestimates or underestimates of mutation rates?

6 Distinguish between the three modes of recombination in bacteria.

7 With respect to F^+ and F^- bacterial matings, answer the following questions:
   (a) How was it established that physical contact was necessary?
   (b) How was it established that chromosome transfer was unidirectional?
   (c) What is the physical-chemical basis of a bacterium being F^+?

8 List all major differences between (a) the $F^+ \times F^-$ and the Hfr $\times F^-$ bacterial crosses and (b) F^+, F^-, Hfr, and F' bacteria.

9 Describe the basis for chromosome mapping in the Hfr $\times F^-$ crosses.

10 When the interrupted mating technique was used with five different strains of Hfr bacteria, the following order of gene entry and recombination was observed. On the basis of these data, draw a map of the bacterial chromosome. Do the data support the concept of circularity?

Hfr Strain	Order
1	T C H R O
2	H R O M B
3	M O R H C
4	M B A K T
5	C T K A B

11 Why are the recombinants produced from an Hfr $\times F^-$ cross almost never F^+?

12 Describe the origin of *F' bacteria* and *merozygotes*.

13 Describe what is known about the mechanisms of the transformation process.

14 In a transformation experiment involving a recipient bacterial strain of genotype $a^- b^-$, the following results were obtained:

	% Transformants		
Transforming DNA	$a^+ b^-$	$a^- b^+$	$a^+ b^+$
$a^+ b^+$	3.1	1.2	0.04
$a^+ b^-$ and $a^- b^+$	2.4	1.4	0.03

What can you conclude about the location of the *a* and *b* genes relative to each other?

15 In a transformation experiment, donor DNA was obtained from a prototroph bacterial strain ($a^+ b^+ c^+$), and the recipient was a triple auxotroph ($a^- b^- c^-$). The following transformant classes were recovered:

$$
\begin{array}{ll}
a^+ b^- c^- & 180 \\
a^- b^+ c^- & 150 \\
a^+ b^+ c^- & 210 \\
a^- b^- c^+ & 179 \\
a^+ b^- c^+ & 2 \\
a^- b^+ c^+ & 1 \\
a^+ b^+ c^+ & 3 \\
\end{array}
$$

What general conclusions can you draw about the linkage relationships among the three genes?

16 Explain the observations that led Zinder and Lederberg to conclude that the prototrophs recovered in their transduction experiments were not the result of F-mediated conjugation.

17 Define *plaque, lysogeny,* and *prophage.*

18 Differentiate between generalized and restricted transduction. Which can be used in mapping and why?

19 Two theoretical genetic strains of a virus ($a^-\ b^-\ c^-$ and $a^+\ b^+\ c^+$) are used to simultaneously infect a culture of host bacteria. Of 10,000 plaques scored, the following genotypes were observed:

$a^+\ b^+\ c^+$	4100	$a^-\ b^+\ c^-$	160
$a^-\ b^-\ c^-$	3990	$a^+\ b^-\ c^+$	140
$a^+\ b^-\ c^-$	740	$a^-\ b^-\ c^+$	90
$a^-\ b^+\ c^+$	670	$a^+\ b^+\ c^-$	110

Determine the genetic map of these three genes on the viral chromosome. Determine whether interference was positive or negative.

20 Describe the conditions under which viral recombination may occur.

21 Describe concatemer formation and its role in the headful packaging mechanism.

22 In complementation studies in phage T4, six mutations were tested in three separate experiments. In the first two experiments, the results were misplaced for one of the crosses. For these experiments, predict the results of the missing cross. From an analysis of all three experiments, determine which mutations are in the same gene. (Note: plus (+) = complementation; minus (−) = no complementation.)

Expt. 1	Expt. 2	Expt. 3
$d \times e \rightarrow +$	$g \times h \rightarrow -$	$d \times g \rightarrow +$
$d \times f \rightarrow -$	$g \times i \rightarrow -$	$e \times i \rightarrow -$
$e \times f \rightarrow ?$	$h \times i \rightarrow ?$	

23 A culture of an auxotrophic *leu*⁻ strain of bacteria was irradiated and incubated until it reached the stationary phase. A control culture, which was not irradiated, was also studied. These cultures were then serially diluted, and 0.1 ml of various dilutions was plated on minimal medium plus leucine and on minimal medium. The results, shown below, were used to determine the spontaneous and X-ray induced mutation rate of *leu*⁻ to *leu*⁺.

Culture Condition	Culture Medium	Dilution	Number of Colonies
Irradiated	(1) Minimal medium plus leucine	10^{-9}	24
	(2) Minimal medium	10^{-2}	12
Control	(3) Minimal medium plus leucine	10^{-9}	12
	(4) Minimal medium	10^{-1}	3

(a) Describe what is represented by each value obtained. Which values should be approximately equal? Are they?

(b) Determine the induced and spontaneous mutation rate leading to prototrophic growth (*leu*⁻ to *leu*⁺).

24 Distinguish between insertion sequences and transposons in bacteria.

25 If a single bacteriophage infects one *E. coli* cell present on a lawn of bacteria, and upon lysis yields 200 viable viruses, how many phage will exist in a single plaque if only three more lytic cycles occur?

SELECTED READINGS

∞∞∞∞∞∞∞∞∞∞∞∞∞∞∞∞∞∞∞∞∞∞

ADELBERG, E. A. 1960. *Papers on bacterial genetics*. Boston: Little, Brown.

ADELBERG, E. A., and PITTARD, J. 1965. Chromosome transfer in bacterial conjugation. *Bacteriol. Rev.* 29: 161–72.

BIRGE, E. A. 1988. *Bacterial and bacteriophage genetics—An introduction*. New York: Springer-Verlag.

BRODA, P. 1979. *Plasmids*. New York: W. H. Freeman.

BUKHARI, A. I., SHAPIRO, J. A., and ADHYA, S. L., eds. 1977. *DNA insertion elements, plasmids, and episomes*. Cold Spring Harbor, NY: Cold Spring Harbor Laboratory.

CAIRNS, J., OVERBAUGH, J., and MILLER, S. 1988. The origin of mutants. *Nature* 335: 142–45.

CAIRNS, J., STENT, G. S., and WATSON, J. D., eds. 1966. *Phage and the origins of molecular biology*. Cold Spring Harbor, NY: Cold Spring Harbor Laboratory.

CAMPBELL, A. M. 1976. How viruses insert their DNA into the DNA of the host cell. *Scient. Amer.* (Dec.) 235: 102–13.

CAVALLI-SFORZA, L. L., and LEDERBERG, J. 1956. Isolation of pre-adaptive mutants in bacteria by sib selection. *Genetics* 41: 367–81.

COHEN, S. 1976. Transposable genetic elements and plasmid evolution. *Nature* 263: 731–38.

COHEN, S., et al. 1973. Construction of biologically functional bacterial plasmids *in vitro*. *Proc. Natl. Acad. Sci.* 70: 3240–44.

COHEN, S. N., and SHAPIRO, J. A. 1980. Transposable genetic elements. *Scient. Amer.* (Feb.) 242: 40–49.

COLD SPRING HARBOR LABORATORY. 1981. Movable genetic elements. *Cold Spr. Harb. Symp.*, vol. 45, parts 1 and 2.

EARNSHAW, W. C., and CASJENS, S. R. 1980. DNA packaging by the double-stranded DNA bacteriophages. *Cell* 21: 319–31.

EDGAR, R. 1982. Max Delbruck. *Ann. Rev. Genet.* 16: 501–6.

EISENSTARK, A. 1977. Genetic recombination in bacteria. *Ann. Rev. Genet.* 11: 369–96.

FOX, M. S. 1966. On the mechanism of integration of transforming deoxyribonucleate. *J. Gen. Physiol.* 49: 183–96.

HALL, B. G. 1988. Adaptive evolution that requires multiple spontaneous mutations. I. Mutations involving an insertion sequence. *Genetics* 120: 887–97.

———. 1990. Spontaneous point mutations that occur more often when advantageous than when neutral. *Genetics* 126: 5–16.

HAYES, W. 1953. The mechanisms of genetic recombination in *Escherichia coli*. *Cold Spr. Harb. Symp.* 18: 75–93.

———. 1968. *The genetics of bacteria and their viruses*. 2nd ed. New York: Wiley.

HERSHEY, A. D. 1946. Spontaneous mutations in bacterial virus. *Cold Spr. Harb. Symp.* 11: 67–76.

HERSHEY, A. D., and CHASE, M. 1951. Genetic recombination and heterozygosis in bacteriophage. *Cold Spr. Harb. Symp.* 16: 471–79.

HERSHEY, A. D., and ROTMAN, R. 1949. Genetic recombination between host range and plaque-type mutants of bacteriophage in single cells. *Genetics* 34: 44–71.

HOTCHKISS, R. D., and MARMUR, J. 1954. Double marker transformations as evidence of linked factors in deoxyribonucleate transforming agents. *Proc. Natl. Acad. Sci.* 40: 55–60.

JACOB, F., and WOLLMAN, E. L. 1961a. *Sexuality and the genetics of bacteria*. Orlando: Academic Press.

———. 1961b. Viruses and genes. *Scient. Amer.* (June) 204: 92–106.

LANDY, A., and ROSS, W. 1977. Viral integration and excision: Structure of the lambda *att* sites. *Science* 197: 1147–60.

LEDERBERG, J. 1986. Forty years of genetic recombination in bacteria: A fortieth anniversary reminiscence. *Nature* 324: 627–28.

———. 1989. Replica plating and indirect selection of bacterial mutants: Isolation of preadaptive mutants in bacteria by sib selection. *Genetics* 121: 395–99.

LEDERBERG, J., and LEDERBERG, E. M. 1952. Replica plating and indirect selection of bacterial mutants. *J. Bacteriol.* 63: 399–406.

LOW, B., and PORTER, R. 1978. Modes of genetic transfer and recombination in bacteria. *Ann. Rev. Genet.* 12: 249–87.

LURIA, S. E., and DELBRUCK, M. 1943. Mutations of bacteria from virus sensitivity to virus resistance. *Genetics* 28: 491–511.

LWOFF, A. 1953. Lysogeny. *Bacteriol. Rev.* 17: 269–337.

MESELSON, M., and WEIGLE, J. J. 1961. Chromosome breakage accompanying genetic recombination in bacteriophage. *Proc. Natl. Acad. Sci.* 47: 857–68.

MORSE, M. L., LEDERBERG, E. M., and LEDERBERG, J. 1956. Transduction in *Escherichia coli* K12. *Genetics* 41: 141–56.

NOVICK, R. P. 1980. Plasmids. *Scient. Amer.* (Dec.) 243: 102–27.

OZEKI, H., and IKEDA, H. 1968. Transduction mechanisms. *Ann. Rev. Genet.* 2: 245–78.

PETERS, J. A. 1969. *Classic papers in genetics.* Englewood Cliffs, NJ: Prentice-Hall.

SMITH, H. O., DANNER, D. B., and DEICH, R. A. 1981. Genetic transformation. *Ann. Rev. Biochem.* 50: 41–68.

SMITH-KEARY, P. F. 1989. *Molecular genetics of Escherichia coli.* New York: Guilford Press.

STAHL, F. W. 1979. *Genetic recombination: Thinking about it in phage and fungi.* New York: W. H. Freeman.

———. 1987. Genetic recombination. *Scient. Amer.* (Nov.) 256: 91–101.

STARLINGER, P. 1980. IS elements and transposons. *Plasmid* 3: 241–59.

STENT, G. S. 1963. *Molecular biology of bacterial viruses.* New York: W. H. Freeman.

———. 1966. *Papers on bacterial viruses.* 2nd ed. Boston: Little, Brown.

STENT, G. S., and CALENDAR, R. 1978. *Molecular genetics: An introductory narrative.* 2nd ed. New York: W. H. Freeman.

TEMIN, H. 1972. RNA-directed DNA synthesis. *Scient. Amer.* (Jan.) 24–33.

TEMIN, H. M., and MIZUTANI, S. 1970. RNA-dependent DNA polymerase in virions of Rous sarcoma virus. *Nature* 226: 1211–13.

VARMIS, H. 1987. Reverse transcription. *Scient. Amer.* (Sept.) 257: 56–64.

VISCONTI, N., and DELBRUCK, M. 1953. The mechanism of genetic recombination in phage. *Genetics* 38: 5–33.

WOLLMAN, E. L., JACOB, F., and HAYES, W. 1956. Conjugation and genetic recombination in *Escherichia coli* K-12. *Cold Spr. Harb. Symp.* 21: 141–62.

ZINDER, N. D. 1953. Infective heredity in bacteria. *Cold. Spr. Harb. Symp.* 18: 261–69.

———. 1958. Transduction in bacteria. *Scient. Amer.* (Nov.) 199: 38.

ZINDER, N. D., and LEDERBERG, J. 1952. Genetic exchange in *Salmonella. J. Bacteriol.* 64: 679–99.

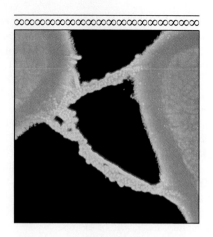

16

Recombinant DNA

RECOMBINANT DNA TECHNOLOGY: AN OVERVIEW
Restriction Enzymes | Vectors

LIBRARY CONSTRUCTION
Genomic Libraries | Chromosome-Specific Libraries | cDNA Libraries

SELECTION OF RECOMBINANT CLONES
Probes to Identify Specific Clones | Colony and Plaque Hybridization | PCR Analysis

CHARACTERIZATION OF CLONED SEQUENCES
Restriction Mapping | Nucleic Acid Blotting | DNA Sequencing

CLONING IN EUKARYOTIC SYSTEMS
Cloning in Yeast | Cloning in Higher Plant Cells | Cloning in Mammalian Cells

APPLICATIONS OF RECOMBINANT DNA TECHNOLOGY
Gene Mapping | Disease Diagnosis | Commercial Applications of
Recombinant DNA Technology

CHAPTER CONCEPTS

Recombinant DNA technology depends in part on the ability to cleave and rejoin DNA segments at specific base sequences. Using this methodology, individual DNA segments can be transferred to viruses or bacteria and amplified, isolated, and identified. The use of this technology has brought about significant advances in gene mapping, disease diagnosis, the commercial production of human gene products, and the trans-specific transfer of genes in plants and animals.

The independent rediscovery of Mendel's work by de Vries, Correns, and von Tschermak in 1900 marks the beginning of genetics as an organized discipline. During the course of its growth and development, several key discoveries have served as landmarks or turning points, each accelerating the rate at which our knowledge of genetics has grown, and, in turn, opening new fields of investigation.

One of the first turning points was the concept that chromosomes are the repository for genes. This concept, proposed by Sutton and Boveri in 1902, was developed by Morgan and his colleagues using the fruit fly, *Drosophila*. From these studies came our understanding of transmission genetics, sex determination, linkage, and the use of polytene chromosomes to map genes to specific cytological loci.

A second landmark was the discovery by Avery, MacLeod, and McCarty that DNA is the macromolecular carrier of genetic information. This work stimulated the use of viruses and bacteria as organisms for genetic research and led to the Watson-Crick model for the structure of DNA. From this model has come knowledge of the molecular basis for genetic coding, transcription, translation, and gene regulation.

We are now in the beginning stages of another and perhaps the most profound transition in the history of genetics—the development and application of **recombinant DNA technology,** also known as **gene splicing** or **genetic engineering**. This technology is being used to generate new knowledge and to develop new commercial products such as vaccines and drugs. It has also raised fears about epidemics or widespread ecological changes that might result from the release of genetically engineered organisms into the environment. In this chapter we will review some of the methods used in recombinant DNA technology to isolate, replicate, and analyze genes, and we will

discuss some of the applications of this technology to agriculture, medicine, and industry.

RECOMBINANT DNA TECHNOLOGY: AN OVERVIEW

The term **recombinant DNA** refers to the creation of a new association between DNA molecules or segments of DNA molecules that are not found together naturally. Although genetic mechanisms such as crossing over technically produce recombinant DNA, the term is generally reserved for DNA molecules produced by joining segments derived from different biological sources.

Recombinant DNA technology uses techniques derived from the biochemistry of nucleic acids coupled with genetic methodology originally developed for the study of bacteria and viruses. The basic procedures involve a series of steps:

1 DNA fragments are generated by using enzymes that recognize and cut DNA molecules at specific sequences.
2 These segments are joined to other DNA molecules that serve as **vectors**. Vectors facilitate the manipulation and identification of the newly created recombinant DNA molecule.
3 The vector, carrying an inserted DNA segment, is transferred to a host cell. Within this cell, the recombinant DNA molecule is replicated, producing copies of the inserted DNA segment known as **clones**.
4 The cloned DNA segments can be recovered from the host cell, purified, and analyzed.
5 Potentially, the cloned DNA can be transcribed, its mRNA translated, and the gene product isolated and studied.

The development of genetic engineering techniques affords new opportunities for research, making it easier to obtain large amounts of DNA encoding specific genes and facilitating studies of gene organization, structure, and expression. This methodology has also given impetus to the development of a burgeoning biotechnology industry that has already delivered a number of products to the marketplace.

Restriction Enzymes

The cornerstone of recombinant DNA technology is a class of enzymes called **restriction endonucleases**. These enzymes, isolated from bacteria, received their name because they restrict viral infection by degrading the invading nucleic acid. Two classes of restriction endonucleases have been discovered, and more than 80 enzymes of both types are now recognized. Type I restriction enzymes recognize a specific nucleotide sequence and cut the DNA at a random location at some distance from the recognition site. Type II enzymes recognize a specific sequence and produce a double-stranded break precisely within the sequence. In Type II enzymes, the recognition site is **palindromic** (that is, it reads the same in both directions). The DNA of bacteria producing a specific restriction enzyme is protected by **modification enzymes** that methylate the nucleotides contained in sequences vulnerable to digestion.

Of the two classes, type II enzymes are most often used in genetic engineering. Because of their specificity, some of them produce DNA fragments with single-stranded complementary tails. For example, the first such enzyme discovered was from *E. coli* and was designated *Eco*RI. Its palindromic recognition site and points of cleavage are shown in the following diagram:

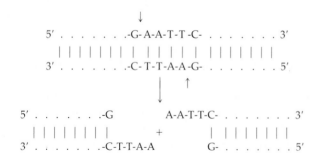

The complementary tails are said to be "sticky" because under hybridization conditions, they can reanneal with each other. If DNAs from different sources share the same palindromic recognition sites, then both will contain complementary single-stranded tails when treated with a restriction endo-

nuclease. If placed together under proper conditions, the DNA fragments from these two sources can form recombinant molecules by annealing of their sticky ends. The enzyme DNA ligase is used *in vitro* to chemically join these fragments (Figure 16.1).

Other restriction enzymes such as *Sma*I, however, cleave DNA in a manner that produces blunt-end fragments:

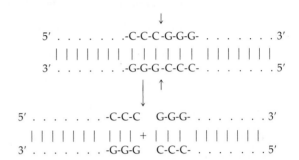

Ligation of DNA fragments with blunt ends is less efficient but can also be used to create recombinant molecules. Commonly, the enzyme **terminal deoxynucleotidyl transferase** may be used to enhance annealing. This enzyme will extend single-stranded ends by the addition of nucleotide "tails." If a poly dA tail is added to DNA fragments from one source (such as plasmid DNA), and poly dT is added to eukaryotic DNA, complementary tails are created, and the fragments can anneal. Recombinant molecules can be created by ligation. Table 16.1 lists some common restriction enzymes and their recognition sequences, some of which yield cohesive or "sticky" ends, while others generate blunt ends. After its formation, the recombinant molecule must be inserted into a host organism for replication.

Vectors

By using a vector or cloning vehicle, a DNA segment can gain entry into a host cell and be replicated or cloned. Vectors are, in essence, carrier DNA molecules. To serve as a vector, a DNA molecule must have several properties: (a) it must be able to replicate itself and inserted DNA segments independently of the chromosome(s) of the host cell, (b) it should contain a number of unique restriction enzyme cleavage sites for insertion of DNA segments, (c) it should carry a selectable marker to allow identification of host cells that contain it, and (d) it should be easy to recover from the host cell. There are a number of vectors currently in use, including **plasmids, bacteriophage,** and **cosmids.** Other vectors, used in specific applications, will be described in following sections.

FIGURE 16.1
Construction of a hybrid plasmid using the restriction endonuclease *Eco*RI. The locations of the *Eco*RI recognition and cutting sites are indicated by the small arrows.

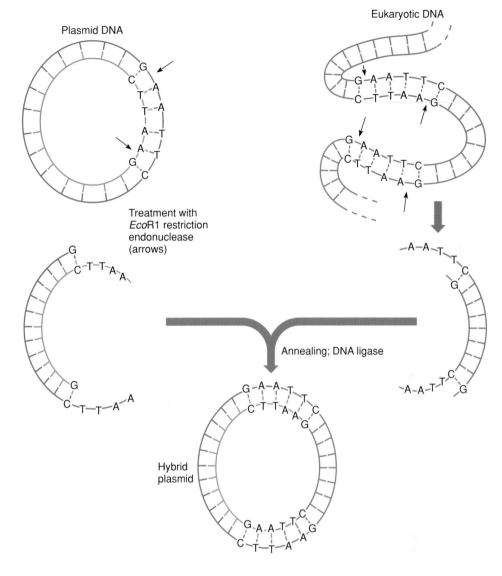

Plasmid DNA

Eukaryotic DNA

Treatment with
*Eco*R1 restriction
endonuclease
(arrows)

Annealing; DNA ligase

Hybrid
plasmid

TABLE 16.1
Some restriction enzymes, cleavage pattern and source.

Enzyme	Recognition, Cleavage Sequence	Source	Enzyme	Recognition, Cleavage Sequence	Source
*Eco*RI	↓ GAATTC CTTAAG ↑	*E. coli*	*Alu*I	↓ AGCT TCGA ↑	*Arthrobacter luteus*
*Hind*III	↓ AAGCTT TTCGAA ↑	*Hemophilus influenza*	*Hae*III	↓ GGCC CCGG ↑	*Hemophilus aegypticus*
*Bam*HI	↓ GGATCC CCTAGG ↑	*Bacillus amyloliquefaciens*	*Bal*I	↓ TGGCCA ACCGGT ↑	*Brevibacterium albidum*
*Taq*I	↓ TCGA AGCT ↑	*Thermus aquaticus*	*Sau*3A	↓ GATC CTAG ↑	*Staphylococcus aureus*

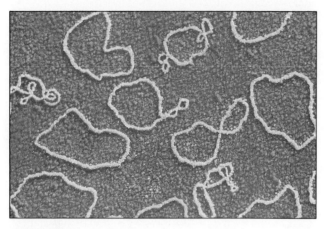

FIGURE 16.2
Circular plasmids isolated from the bacterium *E. coli*. Genetically engineered plasmids are used as vectors for cloning DNA segments.

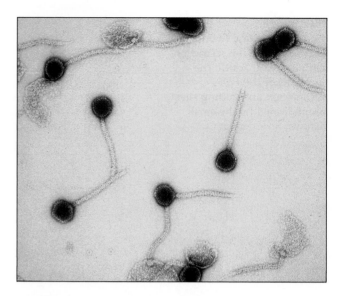

FIGURE 16.4
An electron micrograph showing a cluster of the bacteriophage lambda, widely used as a vector in recombinant DNA work.

As discussed in Chapter 15, plasmids are naturally occurring, extrachromosomal double-stranded DNA molecules that replicate autonomously within bacterial cells (Figure 16.2). Some plasmids can also be amplified within cells, increasing their copy number from 10–20 to almost 1000, allowing more copies of cloned DNA to be produced. Many plasmids have been modified or engineered to contain a limited number of restriction sites and specific antibiotic

resistance genes. The plasmid pBR322, shown in Figure 16.3, carries single cleavage sites for several restriction enzymes and genes for resistance to ampicillin and tetracycline. If a DNA segment is inserted into a cleavage site in the tetracycline gene (such as

FIGURE 16.3
Restriction map of the plasmid pBR322, showing the locations of the restriction enzyme sites that cleave the plasmid only once. Also shown are the locations of the genes for antibiotic resistance.

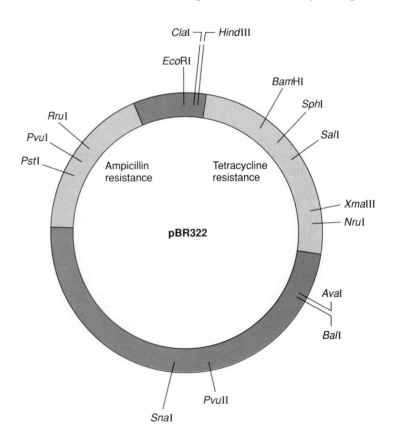

the *Bam*HI site), bacterial cells that acquire the recombinant plasmid will become resistant to ampicillin. Bacterial cells that contain pBR322 plasmids with inserted DNA segments can be selected because they are sensitive to tetracycline.

A bacteriophage widely used in recombinant DNA work is the phage lambda (Figure 16.4). Its genes have all been identified and mapped, and the DNA sequence of the entire genome is known. The middle third of the lambda chromosome contains a cluster of genes necessary for the lysogenic phase of its life cycle, but not for the lytic phase. As a result, this region can be replaced with foreign DNA without affecting the ability of the phage to infect cells and form plaques (Figure 16.5). Over one hundred vectors based on lambda phage have been derived by removing various portions of the central gene cluster,

and all but one of a particular restriction enzyme cleavage site. Lambda vectors containing inserted DNA can be introduced into bacterial host cells in a process called transfection, in which the host cells have been made permeable by a chemical treatment. Each phage vector can be amplified by growth on plated bacterial cells to form plaques.

Other bacteriophage are also used as vectors, including the single-stranded filamentous phage known as M13 (Figure 16.6). When M13 infects a bacterial cell, the single strand (+ strand) replicates to produce a double-stranded molecule known as the **replicative form (RF)**. RF molecules can be regarded as similar to plasmids, and foreign DNA can be inserted into single restriction enzyme cleavage sites present in the genome. When reinserted into bacterial cells, the progeny of RF molecules consists of a

FIGURE 16.5

A diagrammatic representation of the lambda chromosome, showing how the middle section can be removed and replaced with foreign DNA. The recombinant chromosome is then packaged into phage proteins to form a recombinant virus. This virus is able to infect bacterial cells and replicate its chromosome, including the foreign DNA insert.

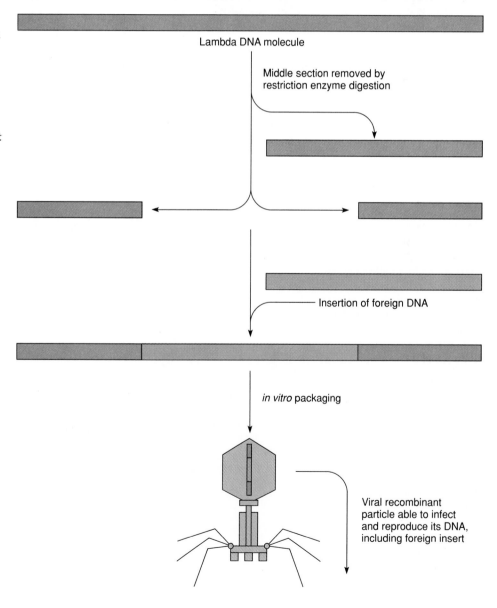

Lambda DNA molecule

Middle section removed by restriction enzyme digestion

Insertion of foreign DNA

in vitro packaging

Viral recombinant particle able to infect and reproduce its DNA, including foreign insert

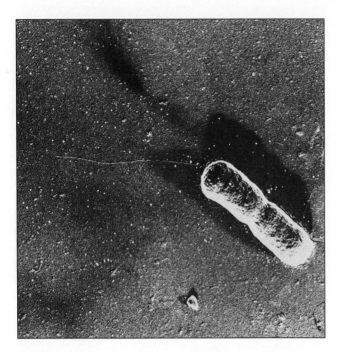

FIGURE 16.6
An electron micrograph of the single-stranded, filamentous phage M13. Foreign DNA inserted into the M13 chromosome is replicated as a single strand within bacterial cells and extruded into the medium. The recovered inserts are used for DNA sequencing or as templates for mutagenic experiments.

single (+) strand, containing one or the other strand of the inserted DNA segment. The single-stranded DNA cloned into M13 can be used for DNA sequencing or as a template for mutational alteration of the cloned sequence.

Cosmids are vectors constructed using the terminal *cos* sequence of phage lambda, necessary for the assembly of chromosomes into phage heads, and plasmid (the *mid* in cosmid) sequences for antibiotic resistance and replication (Figure 16.7). Cosmid DNA containing DNA inserts is packaged into lambda protein heads. These constructs will infect bacterial cells, and replicate as plasmids within host cells. Because almost the entire lambda genome has been deleted, cosmids allow the cloning of large segments of DNA, up to 50 kb in length. Phage vectors, on the other hand, can accommodate DNA inserts of about 20 kb, while plasmids are usually limited to inserts of 10 kb of foreign DNA.

Other hybrid vectors constructed with replication sequences derived from different sources (e.g., plasmids and animal viruses such as SV40) are able to function in more than one host cell and are known as **shuttle vectors**. These vectors usually contain genetic markers that are selectable in both host systems, and can be used to shuttle cloned sequences back and forth between hosts. Often, such vectors are employed in studying gene expression. Other vectors,

FIGURE 16.7
The cosmid pJB8 contains a bacterial origin of replication (*ori*), a single *cos* site (*cos*), an ampicillin resistance gene (*amp*), and a region containing four restriction sites for cloning (*Bam*HI, *Eco*RI, *Cla*I, and *Hind*III). Because the vector is only 5.4 kb long, it can accept foreign DNA segments between 33 and 46 kb in length. The *cos* site allows cosmids carrying large inserts to be packaged into lambda viral coat proteins as if they were viral chromosomes. The viral coats carrying the cosmid can be used to infect a suitable bacterial host, and the vector, carrying a foreign DNA insert, will be transferred into the host cell. Once inside, the *ori* sequence allows the cosmid to replicate as a bacterial plasmid.

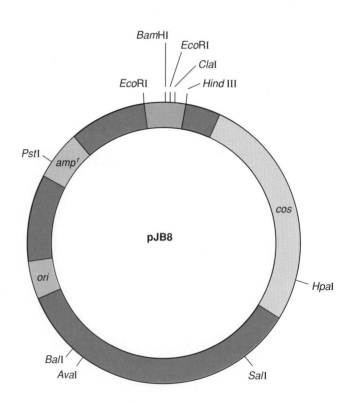

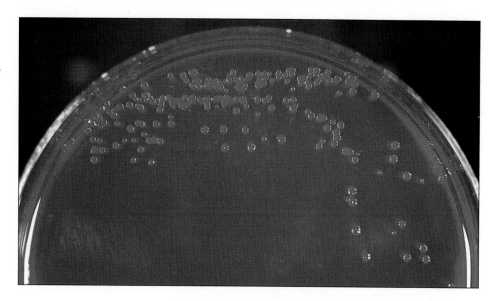

FIGURE 16.8
A petri plate containing bacterial colonies. Each colony is composed of millions of cells, all derived from a single ancestor; each colony, therefore, can be regarded as a clone.

used in specific applications, will be described in following sections.

A variety of cell types can be used as hosts for replication of DNA segments in vectors. The most commonly used host is a laboratory strain of *E. coli* known as K12. *E. coli* strains are genetically well characterized, and can serve as hosts for a wide range of plasmid, phage, and cosmid vectors. Plasmids containing foreign DNA inserts can be transferred to bacterial cells by transformation. The bacteria are then plated on a solid medium and form discrete colonies (Figure 16.8). Since each colony consists of members derived from a single ancestor, all members of the colony are genetically identical, or **clones**. If these cells contain a recombinant plasmid, many copies of the foreign DNA have been replicated or cloned. Similarly, phage containing foreign DNA can be used to transfect *E. coli*, and when plated on solid medium, the resulting plaques each represent a cloned descendant of a single ancestral phage, and are regarded as clones. Cosmids behave as phage and infect bacterial cells, but then replicate as plasmids inside the host cell, and can be identified and recovered in the same way as plasmids. In a subsequent section, we will consider the use of eukaryotic vectors and hosts in cloning DNA segments.

LIBRARY CONSTRUCTION

Since each cloned DNA segment is relatively small, many separate clones must be constructed in order to include even a small portion of the total genome of an organism. A set of cloned DNA segments derived from a single individual is called a **library**. Cloned libraries can represent the entire genome of an individual, the DNA from a single chromosome, or the set of genes that are actively transcribed in a single cell type.

Genomic Libraries

A genomic library is usually constructed to provide a reference source for clones from a specific organism. Ideally, a genomic library contains at least one copy of all sequences represented in the genome. Since each vector molecule can contain only a relatively few kilobases of foreign DNA, the task in preparing a genomic library is the selection of the proper vector to contain all sequences in the smallest number of clones. The number of clones required to contain all the sequences in a genome is dependent on the average size of the cloned inserts and the size of the genome to be cloned. This number can be represented by the following formula:

$$N = \frac{\ln(1 - P)}{\ln(1 - f)}$$

where N is the number of required clones, P is the probability of recovering a given sequence, and f represents the fraction of the genome present in each clone. Suppose we wish to prepare a library of the human genome using a lambda phage vector. The genome is 3.0×10^6 kb, and if the average size of cloned inserts in the vector is 17 kb, a library of about 8.1×10^5 phage would be required (where $f = 1.7 \times 10^4/3.0 \times 10^9$) to have a 99 percent ($P = 0.99$) probability that any human gene is present in at least one copy. If we had selected pBR322 as the vector, with an average insert of about 5 kb, several million clones would be needed to contain the library.

Chromosome-Specific Libraries

A library made from a subgenomic fraction such as a single chromosome can be of great value in the selection of specific genes and the study of chromosome organization. In an early attempt to prepare a library from a specific region of the genome, DNA from a small segment of the X chromosome of *Drosophila* corresponding to a region of about 50 polytene bands was isolated by microdissection. DNA was extracted from this chromosome, cut with a restriction endonuclease and cloned into a lambda vector. This region of the chromosome contains the genes *white*, *zeste*, and *Notch*, as well as the original site of a transposing element that can translocate a chromosomal segment to more than a hundred chromosomal sites. Though technically difficult, this procedure produces a library that contains only the genes of interest and their adjacent sequences.

Libraries derived from individual human chromosomes have been prepared using a technique known as **flow cytometry**. In this procedure, chromosomes from mitotic cells are stained with two fluorescent dyes, one that binds to A-T pairs, the other to G-C pairs. The stained chromosomes flow past a laser beam that stimulates them to fluoresce, and a photometer sorts and fractionates the chromosomes by differences in dye binding and light scattering (Figure 16.9). Using this technique, the National Laboratories at Los Alamos and Lawrence-Berkley have prepared cloned libraries for each of the human chromosomes and made them available for distribution to research scientists all over the world.

Recently, using a modification of electrophoresis known as **pulse field gel electrophoresis**, yeast chromosomes have been separated from each other and used to construct chromosome-specific libraries (Figure 16.10). Libraries constructed using these techniques can be used to gain access to genetic loci for which there are no convenient selection systems using DNA, mRNA, or proteins, or where the gene product is unknown. In addition, chromosome-specific libraries provide a means for studying the molecular organization and even the nucleotide sequence in a defined region of the genome.

cDNA Libraries

A library representing the structural genes active in a specific cell type at a specific time can be constructed by first synthesizing **cDNA (complementary DNA) molecules**. If we begin with an mRNA molecule

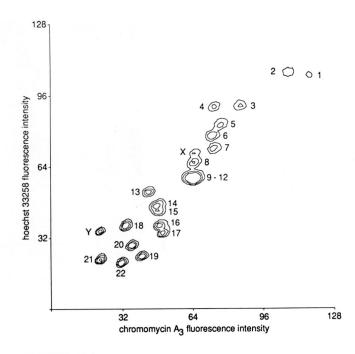

FIGURE 16.9
A fraction of human chromosomes after separation by flow sorting.

containing a 3′ poly A tail, poly dT can be used as a primer to pair with the poly A residues (Figure 16.11). This poly dT serves as the starting point for the synthesis of a DNA strand using the enzyme **reverse transcriptase**. The result is an RNA-DNA double-stranded duplex. The RNA strand can be removed by treatment with alkali or the enzyme **ribonuclease H**. The single strand of DNA is then used as a template to synthesize a complementary strand of DNA using the enzyme **DNA polymerase I**. Because the 3′ end of the single strand of DNA often loops back upon itself, it can serve as the primer for the synthesis of the second strand. The result is a DNA duplex with the strands joined together at one end. The hairpin loop can be opened using the enzyme **S₁ nuclease**. The result is a double-stranded DNA molecule that can be cloned by attaching sequences to both ends to allow it to be inserted into a plasmid or phage vector.

SELECTION OF RECOMBINANT CLONES

As we calculated above, a library can contain several thousand or several hundred thousand clones. The problem is to identify and select only the clone or clones that contain a DNA sequence or gene of interest. Several techniques can be employed to accomplish this task, and the choice often depends on circumstances and available information. The

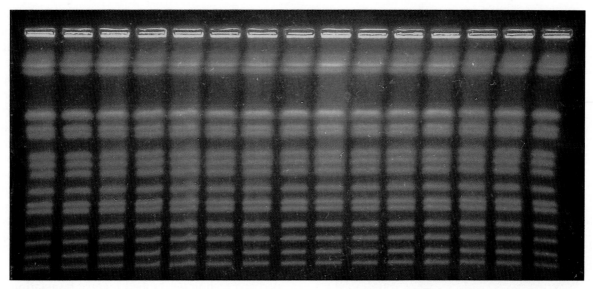

FIGURE 16.10
Intact yeast chromosomes separated using a method of electrophoresis employing contour-clamped homogeneous electric field (CHEF). In each lane, 15 of the 16 yeast chromosomes are visible, separated by size, with the largest chromosomes at the top.

methods described below utilize somewhat different approaches to this problem.

Probes to Identify Specific Clones

Most procedures to select specific clones from a library employ a radioactive polynucleotide that contains a base sequence complementary to all or part of the gene of interest. This radioactively labeled nucleic acid is commonly referred to as a **probe**. Probes can be derived from a variety of sources; even related genes isolated from other species can serve as probes if enough of the DNA sequence has been

FIGURE 16.11
The production of cDNA from mRNA. Since many eukaryotic mRNAs have a polyadenylated tail of variable length (A_n) at their 3' end, a short oligonucleotide sequence composed of thymidine nucleotides can be annealed to this tail. This oligonucleotide acts as a primer for the enzyme reverse transcriptase, which uses the mRNA as a template to synthesize a complementary DNA strand. A characteristic hairpin loop is often formed as synthesis is terminated on the template. The mRNA can be removed by alkaline treatment of the complex, and DNA polymerase is used to synthesize the second DNA strand. S_1 nuclease is used to open the hairpin loop, and the result is a double-stranded cDNA molecule that can be cloned into a suitable vector, or used as a probe in library screening.

conserved. For example, extrachromosomal copies of the ribosomal genes of the clawed frog *Xenopus laevis* were isolated by centrifugation (Figure 16.12) and

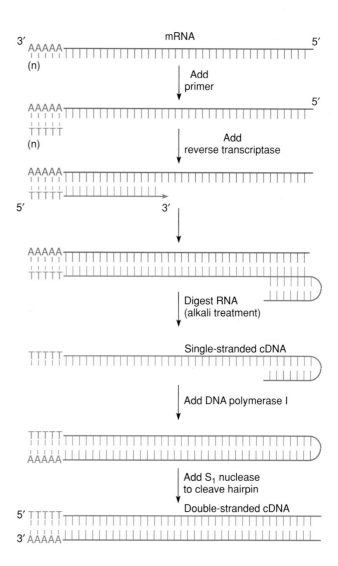

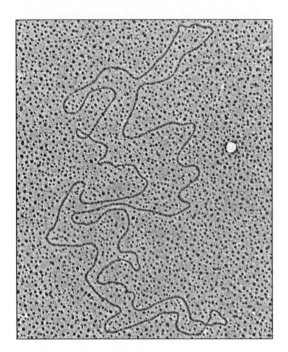

FIGURE 16.12
Electron micrograph showing extrachromosomal circles of *Xenopus* ribosomal DNA recovered from oocytes by centrifugation. These circles can be treated with restriction enzymes, and the resulting fragments cloned into plasmid vectors.

cloned into plasmid vectors. Because ribosomal gene sequences have been conserved during evolution, the *Xenopus* clones were used as probes to isolate the human ribosomal genes from a cloned library.

If the gene to be selected from a genomic library is transcriptionally active in certain cell types, a cDNA probe can be prepared. This technique is particularly useful when purified mRNA species can be obtained. For example, hemoglobin is the predominant mRNA produced at a certain stage of red blood cell development. Similarly, ovalbumin mRNA can be obtained from hormonally stimulated avian uterine tissue. Purified mRNA is copied by reverse transcriptase into a cDNA molecule that can be used as a probe to recover the structural gene from a cloned genomic library.

Still another approach to obtaining a probe is the use of a technique dubbed **reverse translation**. If the protein product of the gene is available, and at least part of its amino acid sequence is known, a polynucleotide probe can be constructed with this information. Beginning with the amino acid sequence, the nucleotide sequence encoding these amino acids can be deduced (Figure 16.13). Note that because of degeneracy in the genetic code, several combinations of nucleotide sequences are possible. In the example

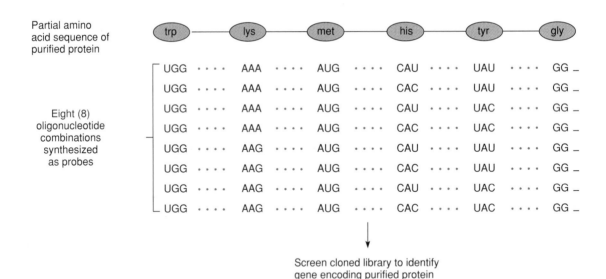

FIGURE 16.13
The process of reverse translation. From the partial amino acid sequence of a purified protein, the coding nucleotide sequence can be deduced. In this case, two amino acids (*trp* and *met*) have only one codon each. Three others (*lys, his,* and *tyr*) are encoded by two codons. The sixth amino acid (*gly*) is encoded by a degenerate series of codons that encompasses all base combinations at the third base position. Consequently, the eight coding sequences shown encompass all possible combinations. The exact coding sequence of the gene must be one of the eight. Using a radioactive mixture of these sequences as probes, a cloned library can be screened to isolate the structural gene for the complete protein.

given, eight different oligonucleotides could encode the amino acid sequence

UGGAAAUAGCAUUAUCU__
UGGAAGUAGCAUUAUCU__
etc.

Each of these are chemically synthesized using radioactive nucleotides and are used as probes to identify the clone encoding the gene of interest. Note that it is not necessary to know the entire amino acid sequence of the protein or to synthesize the nucleotide sequence of the entire gene. A fragment corresponding to 6–7 amino acids and the corresponding 18–21 nucleotides is usually sufficient.

Colony and Plaque Hybridization

Once DNA segments have been produced, ligated into a vector, and transferred into a host cell, the next task is to ensure that the host cells contain vectors that have indeed incorporated foreign DNA and can be used as a viable source of recombinant DNA clones. This step is necessary, because during ligation of vector with foreign DNA inserts, a mixture of ligation products is present, some of which may represent head-to-tail associated vectors, or vectors that have religated without incorporating foreign DNA. In the case of plasmids such as pBR322 or its derivatives, this can be accomplished by transferring bacterial cells from a given colony or collection of colonies to plates containing ampicillin or tetracycline. If, for example, foreign DNA is inserted at the *Bam*HI site of pBR322 (Figure 16.3), the gene for tetracycline resistance will be inactivated. Colonies formed from the cell carrying this vector will not grow in the presence of tetracycline, but will grow on plates containing ampicillin. If the cell had incorporated a vector with no insert, it would grow on both tetracycline and ampicillin.

Other plasmid and phage vectors have been engineered to produce blue plaques when plated on medium containing a compound called Xgal. The restriction sites for the insertion of foreign DNA occur within the gene responsible for the formation of blue plaques, and consequently, phage carrying insertions produce clear plaques (Figure 16.14). These methods represent the initial screening steps in the isolation of suitable clones.

The **colony hybridization** method was developed by Grunstein and Hogness for screening a plasmid library (Figure 16.15). The bacterial colonies to be screened are grown on plates, and a replica is made by gently pressing a filter made of nitrocellulose or other DNA-binding material to the plate to transfer bacterial cells from the colonies to the filter. The colonies on the filter are lysed, and the DNA from the bacterial cells is denatured. The filter is then placed in

FIGURE 16.14
A petri plate showing colonies derived from a cloning experiment. The medium on the plate contains a compound called Xgal. The vector has been engineered so that foreign DNA inserts disrupt the gene responsible for the formation of blue colonies. As a result, blue colonies do not carry any cloned foreign DNA, while clear colonies contain cloned vectors carrying foreign DNA inserts.

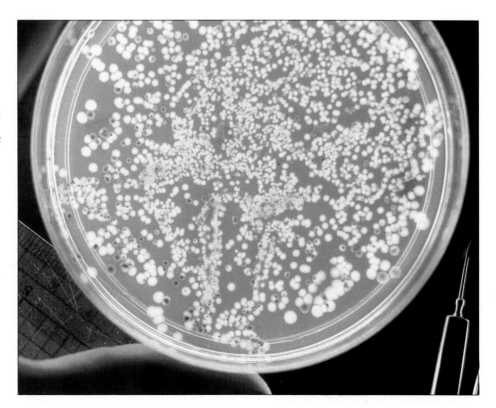

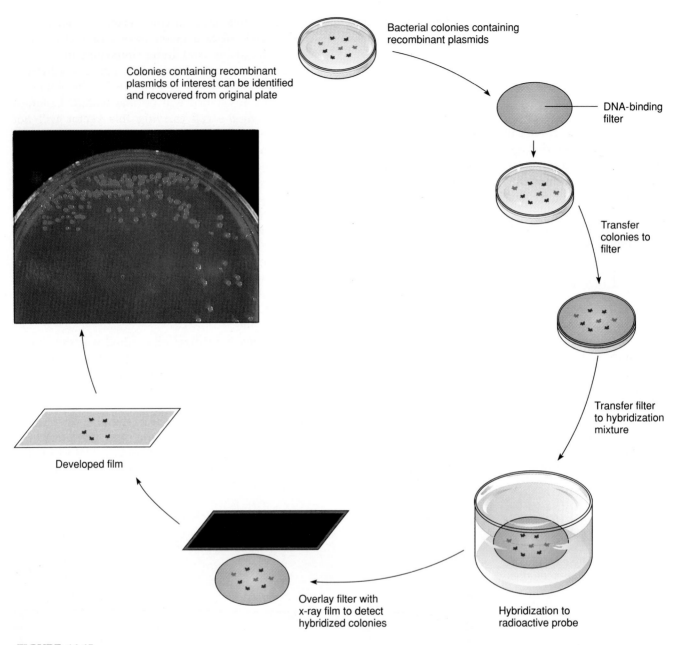

Colonies containing recombinant plasmids of interest can be identified and recovered from original plate

Bacterial colonies containing recombinant plasmids

DNA-binding filter

Transfer colonies to filter

Transfer filter to hybridization mixture

Developed film

Overlay filter with x-ray film to detect hybridized colonies

Hybridization to radioactive probe

FIGURE 16.15
The colony screening method for the detection and recovery of cloned DNA inserts of interest.

a solution with a radioactive probe to allow hybridization between the probe and complementary sequences present on the filter. After excess probe is washed away, the filter is overlaid with a piece of photographic film. If the probe is complementary to the DNA of one of the lysed colonies, the hybridized colony will be identified as a dark spot on the developed film. That colony can be recovered from the initial plate, and used in further experiments.

A similar technique known as **plaque hybridization** has been developed by Benton and Davis to

screen phage vectors carrying DNA inserts. Because many more phage plaques than bacterial colonies can be screened on a single filter, this method is more efficient in screening large genomic libraries.

PCR Analysis

Recombinant DNA techniques were developed in the early 1970s, and in the subsequent years they revolutionized the way geneticists and molecular biologists conduct research. In 1986, another technique, called the **polymerase chain reaction (PCR),** was developed, and has extended the power of recombinant DNA research, and has even replaced some of the earlier methods. PCR analysis has found applications in a wide range of disciplines, including molecular biology, human genetics, evolution, development, and even forensics.

One of the prerequisites for many recombinant DNA techniques is the availability of large quantities of a specific DNA segment. Often such quantities are obtained through tedious and labor-intensive efforts in cloning and recloning. PCR allows a more direct amplification of specific DNA segments, and can be used on fragments of DNA that are initially present in infinitesimally small quantities. The PCR method is based on the amplification of a DNA segment using DNA polymerase and oligonucleotide primers that hybridize to opposite strands of the sequence to be amplified (Figure 16.16). There are three basic steps in the PCR reaction, and the amount of amplified DNA produced is limited only by the number of times this series of steps is repeated.

The first step is the denaturation of the DNA to be amplified. The DNA to be amplified does not have to be purified, and can come from any number of sources, including genomic DNA, forensic samples such as dried blood or semen, dried samples stored as part of medical records, single hairs, mummified remains, and fossils. The double-stranded DNA is denatured by heating until it dissociates into single strands.

The second step is the annealing of primer sequences to the single strands of DNA. These primers are synthetic oligonucleotides that anneal to sequences flanking the segment to be amplified. Two different primers are most often used in PCR. Each primer has a sequence complementary to only one of the two strands of DNA. The primers are said to align themselves with their 3' ends facing each other since they anneal to opposite strands (Figure 16.16).

The third step in the PCR is the actual amplification event, carried out by extension of the primers using the single-stranded DNA as a template (Figure 16.16). This is carried out by adding DNA polymerase to the reaction mixture. The polymerase extends the oligonucleotide primers in the 5'-to-3' direction, using the single-stranded DNA bound to the primer as a template. The product is a double-stranded DNA molecule with the primers incorporated into the final product (Figure 16.16).

Each set of three steps—denaturation of the double-stranded product, annealing of primers, and extension by polymerase—is referred to as a cycle. The cycle can be repeated by carrying out each of the steps again. Twenty-five cycles of amplification re-

sults in a one million-fold increase in the amount of target DNA. The process has been automated through the development of instruments that can be programmed to carry out a predetermined number of cycles, yielding large amounts of amplified DNA segments that can be used in other procedures such as cloning, sequencing, diagnosis of infectious disease, and genetic screening.

CHARACTERIZATION OF CLONED SEQUENCES

Restriction Mapping

A restriction map is a compilation of the number and order of, and distance between, restriction enzyme cutting sites along a cloned segment of DNA. The map units are expressed in **base pairs (bp)** or kilobases (kb). The fragments generated by cutting with restriction enzymes can be separated by **gel electrophoresis,** a method that separates fragments by size, with the smallest pieces moving farthest (see Appendix A for a description of electrophoresis). The fragments appear as a series of bands that can be visualized by staining the DNA with ethidium bromide and viewing under ultraviolet illumination (Figure 16.17). The size of individual fragments can be determined by running a set of fragments of known size on the same gel in another lane. Figure 16.18 shows the construction of a restriction map from a cloned DNA segment that contains cutting sites for two restriction enzymes, designated A and B. Digestion with enzyme A alone produces a single fragment, with terminal A sites (Figure 16.18). Digestion with enzyme B alone produces a different fragment with terminal B sites (Figure 16.18). To construct a map, we must make use of fragments produced by each enzyme individually and both enzymes simultaneously, as well as fragments produced by one enzyme that are then digested by the other enzyme (Figure 16.18). When each of these digestion experiments has been completed and analyzed, it is possible to arrange the fragments and enzyme cutting sites to form a restriction map (Figure 16.18) that shows the location of four restriction enzyme cutting sites contained in this DNA segment. Additional sites on this segment can be mapped through the use of other restriction enzymes.

Restriction maps provide an important way of characterizing a DNA segment, and can be constructed in the absence of any information about the genetic content or function of the mapped DNA. In conjunction with other techniques, restriction mapping can be used to define the boundaries of a gene

FIGURE 16.16

PCR amplification. In the polymerase chain reaction (PCR) method, the DNA to be amplified is first denatured into single strands, and then each strand is annealed to a different primer. The primers are synthetic oligonucleotides complementary to sequences flanking the region to be amplified. DNA polymerase is added along with nucleotides to extend the primers in the 3' direction, resulting in a double-stranded DNA molecule with the primers incorporated into the newly synthesized strand. In a second PCR cycle, the products of the first cycle are denatured into single strands, primers are added, and they are annealed and extended by DNA polymerase. Repeated cycles can amplify the original DNA sequence by more than a million times.

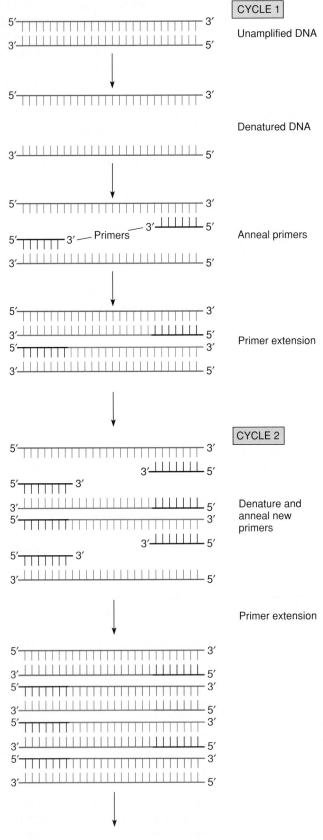

CYCLE 1

Unamplified DNA

Denatured DNA

Anneal primers

Primers

Primer extension

CYCLE 2

Denature and anneal new primers

Primer extension

Cycles 3-25 for a 10^6-fold increase in DNA

FIGURE 16.17
An agarose gel showing DNA fragments separated on the basis of size. Smaller fragments migrate faster and farther than larger fragments, resulting in the distribution shown.

and provides a way of dissecting the molecular organization within a gene and its flanking regions. Mapping can also serve as a starting point for the isolation of an intact gene from cloned segments of DNA, and provides a means for locating mutational sites within genes.

Restriction maps can also be used to construct and refine genetic maps. To a large extent, the accuracy of genetic maps is based on the frequency of recombination between genetic markers, and the number of genes or genetic markers used in construction of the map. If there is a large distance between markers and variation in recombination frequency, the genetic map may not correspond to the physical map of the chromosome or chromosome region. In humans, for example, the genome size is large (3.2×10^9 bp), and the number of mapped genes is small (a few thousand), meaning that each map unit is composed of millions of base pairs of DNA. The result is a poor correlation between the genetic and physical maps of chromosomes. Restriction enzyme cutting sites can be used as genetic markers, reducing the distance between sites for map construction, increasing the accuracy of genetic maps, and providing reference points for the correlation of genetic and physical maps.

Restriction sites have played an important role in mapping genes to specific human chromosomes and to defined regions of individual chromosomes. In addition, if a restriction site maps close to a mutant gene, it can be used as a marker in a diagnostic test for detecting the disease. This has proven especially useful because the mutant genes underlying many human genetic diseases are poorly characterized at the molecular level, and closely linked restriction sites have been successfully used in the detection of afflicted individuals and heterozygotes at risk for having affected children.

Nucleic Acid Blotting

DNA fragments cloned into vectors can be used in hybridization reactions to characterize the identity of specific genes, to locate coding regions or flanking regulatory regions within cloned sequences, and to study the molecular organization of genomic sequences.

In this procedure, the DNA to be probed is first cut into fragments with one or more restriction enzymes, and the fragments separated by gel electrophoresis (Figure 16.19). The DNA in the gel is denatured into single-stranded fragments by treatment with alkali. These fragments are transferred to a sheet of DNA-binding material, usually nitrocellulose or a nylon derivative. The transfer is effected by placing the transfer membrane on top of the gel and causing buffer to flow through the gel and the nitrocellulose or nylon sheet. The buffer flows through both, pulling the DNA out of the gel, and immobilizing it in the nitrocellulose. The denatured DNA is then hybridized with a radioactive probe. Those single-stranded DNA fragments imbedded in the membrane that are complementary to the base sequence of the probe will form hybrids. The unbound probe is washed away, and the hybridized fragments are visualized by autoradiography. (See Appendix A for a detailed discussion of autoradiography).

This technique, developed by Edward Southern, has become known as a **Southern blot,** and is a procedure with many applications in genetics. The Southern blot is widely used to identify cloned genes that correspond to particular probes, to identify related genes in genomes of other organisms, and to find related sequences or members of a gene family within a single genome.

A related technique, developed by James Alwine and colleagues, can be used to determine whether a cloned gene is transcriptionally active in a given cell or tissue type. This characterization of gene activity is accomplished by extracting RNA from one or several

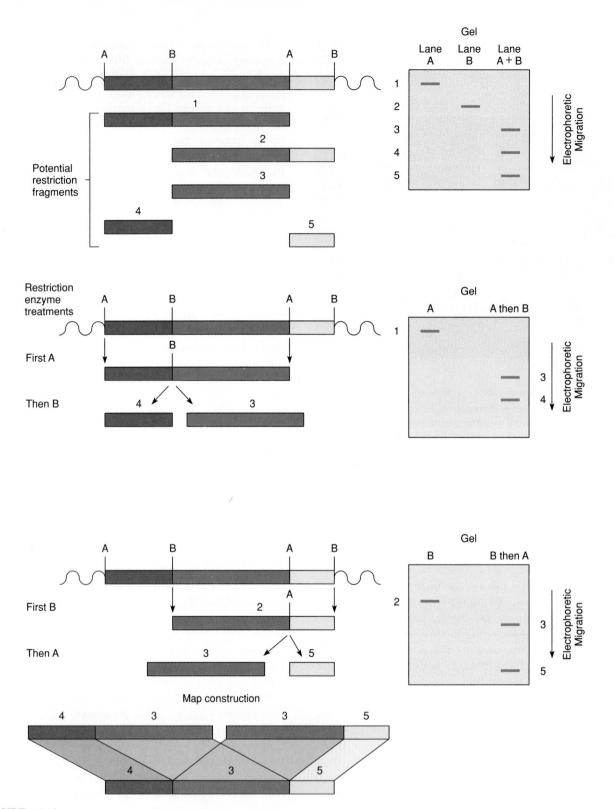

FIGURE 16.18

The process of restriction mapping. A cloned DNA segment contains four restriction sites, two for enzyme A, and two for enzyme B. Vector sequences are shown as wavy lines. Cutting with enzyme A generates fragment 1, shown on the gel (at right) in lane A. Cutting with enzyme B generates fragment 2, shown on the gel in lane B. Cutting with enzyme A + enzyme B generates fragments 3, 4, and 5, as shown in lane A + B. For the sake of clarity, bands produced by vector molecules and by cloned fragments attached to vector molecules are not shown on the gel. If the cloned DNA segment is first cut with A, and fragment 1 is recovered from the gel and cut with enzyme B, two smaller fragments, 3 and 4, are generated (middle gel, lane A + B). If the clone is first cut with enzyme B, and fragment 2 is recovered from the gel and cut with A, fragments 3 and 5 are generated (lower gel, lane B + A). With this information in hand, the order and spatial relationships of the fragments can be assembled into a restriction map.

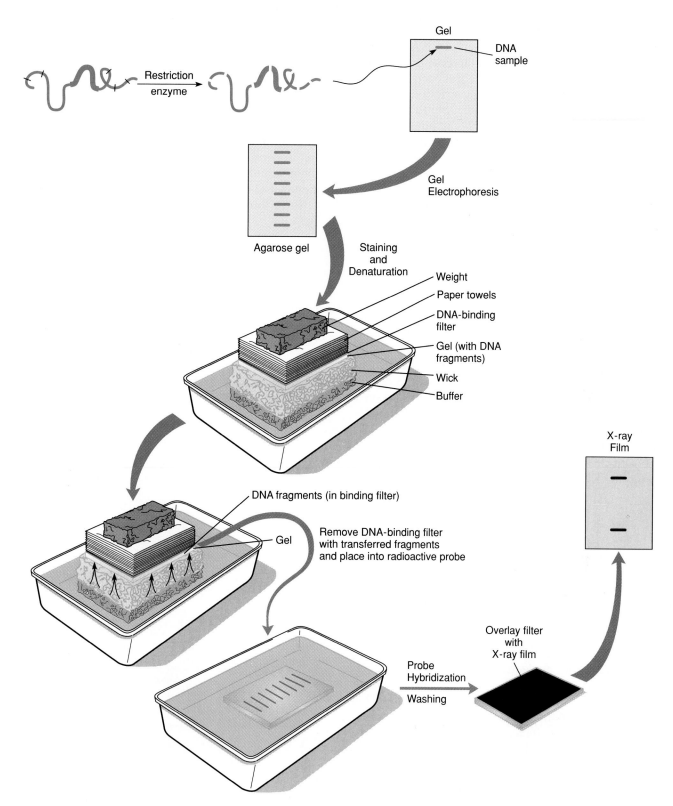

FIGURE 16.19

The Southern blotting technique. DNA is first cut with one or more restriction enzymes and the fragments separated by gel electrophoresis. The pattern of fragments can be visualized and photographed under ultraviolet illumination by staining the gel with ethidium bromide. The gel is then placed on a filter paper wick in contact with a buffer solution and covered with a sheet of a DNA binding filter. Layers of paper towels or blotting paper are placed on top of the DNA binding filter and held in place with a weight. Capillary action draws the buffer through the gel, transferring the pattern of DNA fragments from the gel to the DNA binding filter. The DNA fragments on the binding filter are then denatured and hybridized with a radioactive probe, washed, and overlaid with a piece of X-ray film for autoradiography. The hybridized fragments show up as bands on the X-ray film.

cell and/or tissue types. The RNA is fractionated by gel electrophoresis and the pattern of RNA bands is transferred to a sheet of RNA-binding membrane. The sheet is then hybridized to a radioactive single-stranded DNA probe, derived from cloned genomic DNA or cDNA. If RNA complementary to the DNA probe is present, it will be detected by autoradiography as a band on the film. Because the original procedure using DNA bound to a filter became known as a Southern blot, this procedure using RNA bound to a filter was called a **Northern blot**. (Following this somewhat perverse logic, another procedure involving proteins bound to a filter is known as a **Western blot**.)

Northern blots provide information about whether RNA transcripts complementary to cloned DNA segments are present in a cell or tissue, and also provide two additional bits of information. If marker RNAs are run in an adjacent lane, the size and molecular weight of the transcribed RNA can be calculated. Secondly, the amount of transcribed RNA present in the cell or tissue is reflected by the density of the RNA band on the autoradiograph. This can be quantified by measuring the density of the band on a densitometer, providing a measurement of the degree of transcriptional activity. Thus, Northern blotting can be used to characterize and quantify the transcriptional activity of a cloned DNA segment in different cells and tissues.

DNA Sequencing

In a sense, the ultimate characterization of a cloned DNA segment is the determination of its nucleotide sequence. The ability to sequence cloned DNA has added immensely to our understanding of gene structure and the mechanisms of gene regulation. Although techniques were available in the 1940s to determine the base composition of DNA (see Chapter 9), it was not until the 1960s that methods for the determination of nucleotide sequences were developed. In 1965 Robert Holley determined the sequence of a tRNA molecule consisting of 74 nucleotides. This work required about one year of concentrated effort. In the 1970s, more efficient and direct methods were developed, and it is now possible to conveniently sequence several hundred or a thousand bases in a week's work. In the near future, with widespread automation of this process, it will be commonplace to sequence thousands of nucleotides in a single day.

The application of DNA sequencing depends on the ability of cloning methods to provide large amounts of specific DNA fragments. The use of specific vectors such as M13 or modifications of the

PCR technique are also important in providing single-stranded DNA segments that can be directly sequenced.

Two methods of DNA sequencing are in general use. One was developed by Alan Maxam and Walter Gilbert; the second was created as part of an effort by Fredrick Sanger and his colleagues to sequence the genome of the virus ΦX174. Both sequencing strategies depend on methods that generate a series of single-stranded DNA molecules of the same sequence, but of differing lengths. Each method uses a series of four reactions, each in a separate tube, to

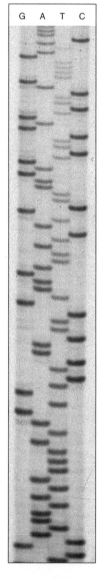

FIGURE 16.20
Photograph of a DNA sequencing gel, showing the separation of bases that are analyzed to reveal the base sequence of the DNA segment.

determine the sequence of the four bases in a DNA segment. These sequences, differing in length by as little as one base, are separated from each other by gel electrophoresis in four adjacent lanes. The result is a series of bands that form a ladderlike pattern. The sequence can be read directly from the band pattern in the four lanes (Figure 16.20). Details of each of the methods of DNA sequencing are presented in Appendix A.

DNA sequencing has provided information about the organization of genes and the nature of mutational events that alter both genes and gene products, confirming the genetic conclusion that genes and proteins are colinear molecules. Sequencing has also provided information about the organization of regulatory regions that flank prokaryotic and eukaryotic genes, and has been used to infer the structure of proteins. The gene for cystic fibrosis, an autosomal recessive human genetic disorder, was first identified by the fact that it was a DNA sequence with an open reading frame, since the nature of the protein product was unknown. The DNA sequence was then used to infer that the gene encodes a protein of 1480 amino acids with properties similar to membrane proteins that function in ion transport. This discovery serves to illustrate how the use of DNA sequencing now plays an important role in genetic analysis.

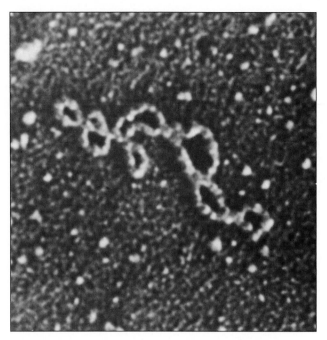

FIGURE 16.21

An electron micrograph showing the yeast 2-micron circle. This independently replicating, double-stranded plasmid is present in 50–100 copies in the cytoplasm of many different strains of yeast and has been modified to serve as the basis for the construction of cloning vectors.

CLONING IN EUKARYOTIC SYSTEMS

We have described in detail the use of cloning systems based on *E. coli*. Other bacterial hosts can also be used for cloning, including *B. subtilis* and *Streptomyces*. These bacterial host systems use methodologies similar to those we have already described for *E. coli*. To study expression and regulation, it is convenient and even necessary to use eukaryotic hosts for some cloned eukaryotic genes. Several such systems have been developed, and in this section we will describe three gene cloning systems that employ eukaryotic cells as hosts.

Cloning in Yeast

Although yeast is a eukaryotic organism, it can be manipulated and grown in much the same way as bacterial cells. Further, the genetics of yeast has been intensively studied, providing a large catalog of mutations and highly developed genetic maps of the yeast chromosomes. Finally, a naturally occurring plasmid named the **2-micron plasmid** is present in yeast (Figure 16.21), and forms the basis for a number of vector systems. By combining bacterial plasmid

sequences with those of the yeast 2-micron plasmid, vectors with different properties can be produced. One such vector is called a **yeast integrative plasmid (YIP)** (Figure 16.22). Such plasmids are valuable in the study of yeast genes that cannot be expressed or selected when cloned into a bacterial host. An integrative plasmid carrying the normal allele of a yeast gene can be transformed into a yeast host cell carrying a mutant allele of the same gene. The result is an integration event, during which the normal allele is inserted into a host chromosome at the site of the mutant allele. This event can result in the integration of the entire plasmid, or the replacement of the chromosomal (in this case the mutant) allele with the (normal or wild-type) allele carried on the plasmid. The converse can also occur, providing a powerful way to study the effect of specific mutations on gene function. An allele mutated *in vitro* can be used to replace the normal allele by integrative transformation in YIP to study the effect of specific mutations on gene function. In other configurations, integrative plasmids can be used to map the location of genes and mutational sites within genes.

A second type of yeast vector is the **yeast artificial chromosome (YAC)**. This vector is linear, has yeast

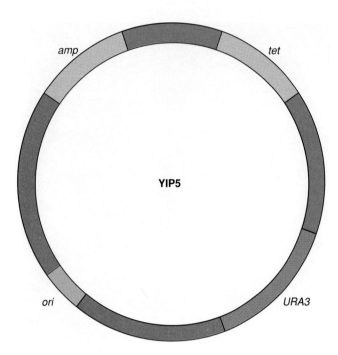

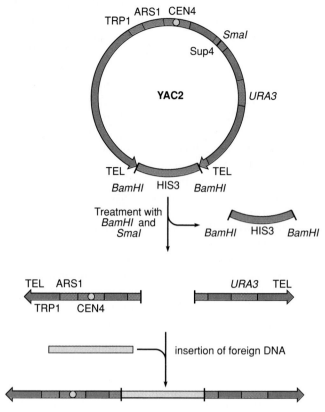

FIGURE 16.22

A yeast integrative plasmid (YIP) that is used to introduce cloned DNA sequences into yeast chromosomes. The plasmid carries genes for tetracycline and ampicillin resistance derived from the bacterial plasmid pBR322, and an origin of replication *(ori)* that allows the plasmid to replicate when grown in a bacterial host. It also contains the *URA3* gene from yeast that serves as a selectable marker when the plasmid is introduced into yeast cells deficient for this gene. This gene can also serve as a region of homology for recombination of the plasmid into a yeast chromosome. The YIP cannot replicate when present in yeast cells, and is carried by integration into a yeast chromosome.

FIGURE 16.23

A summary of the yeast artificial chromosome (YAC) cloning system. YAC2 is constructed from pBR322 sequences and the *SUP4, TRP1, HIS3,* and *URA3* genes from yeast. The *ARS1* and *CEN4* sequences represent a chromosome segment adjacent to the *TRP1* gene on yeast chromosome 4. The *TEL* sequences act as telomeres and are derived from *Tetrahymena* ribosomal gene clusters. The centromere (CEN4), telomere, and ARS (autonomously replicating sequence, in effect an origin of replication) confer the basic properties of yeast chromosomes on the construct. Cleavage with *Bam*HI and *Sma*I results in the production of two arms that can be ligated to foreign DNA, followed by transformation into yeast host cells.

telomeric sequences at each end, and contains a yeast origin of replication, which confers the ability to replicate autonomously. It has a yeast centromere to allow distribution of replicated YACs at cell division, a yeast selectable marker to allow identification of cells carrying YACs, and restriction sites for the integration of foreign DNA (Figure 16.23). YACs have proven to be very useful in the cloning of human genes, where the size range exceeds the cloning capacity of lambda vectors and cosmids. A human genomic library has been cloned using YAC vectors and yeast as the host cells. YACs have also been used to clone DNA from large regions of human chromosome 21, and a complete library of human chromosome 21 sequences cloned in YACs is being prepared. If the average size of the insert in each YAC is 440 kb, the entire chromosome would be represented with some redundancy in only 330 YACs.

Although cloning in yeast is at present the most advanced eukaryotic host system, other fungal hosts are being developed, and other eukaryotes such as plant cells and mammalian tissue culture cells show great promise for use as host systems.

Cloning in Higher Plant Cells

As strange as it might seem, cloning using higher plants as hosts has been achieved using a bacterial plasmid. The infectious soil bacterium *Agrobacterium tumifaciens* produces tumors or galls in many species of dicotyledonous plants. The ability to cause tumor formation is associated with the presence of a tumor-inducing (Ti) plasmid in the bacteria (Figure 16.24). When bacteria infect the plant cells, a segment of the Ti plasmid, known as T-DNA, is transferred into the genome of the host cell. The T-DNA controls tumor formation and the synthesis of small molecules known as opines that are required for growth of the infecting bacteria. Foreign genes can be inserted into the T-DNA segment of Ti, and the altered Ti can be transferred into a plant cell by the infecting bacteria. The foreign DNA is inserted into the plant genome when the T-DNA is integrated into a host cell chromosome. Such genetically altered single cells can be induced to grow in tissue culture to form a cell mass known as a **callus**. By manipulating the culture medium, the callus can be induced to form roots and shoots, and eventually develop into mature plants carrying a foreign gene. Such plants are said to be **transgenic**. In a later section, we will see how this system has been used to genetically alter agriculturally important crop plants.

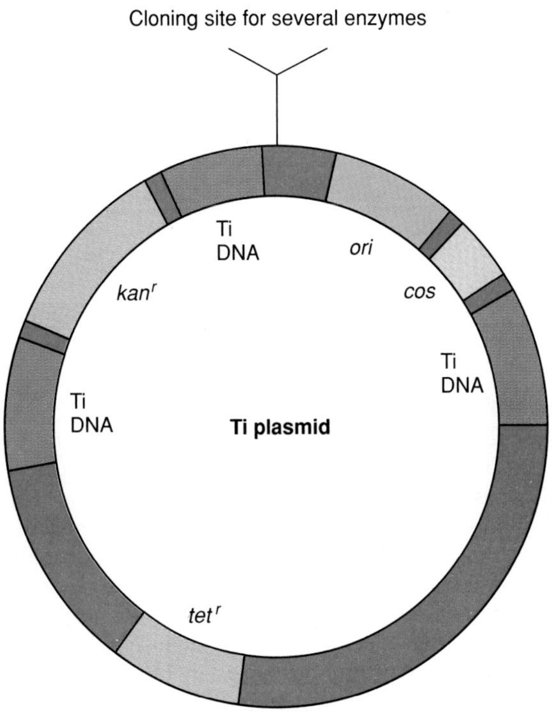

Cloning in Mammalian Cells

DNA can be taken up into mammalian cells by several methods, including coprecipitation with calcium phosphate and endocytosis, direct microinjection, exposure of cells to short pulses of high voltage (electroporation), encapsulation of DNA into artificial membranes (liposomes) followed by fusion with cell membranes, or the use of a variety of vectors based on viruses. DNA introduced into a mammalian cell by any of these methods usually results in the stable integration of the foreign DNA into the host genome at a relatively high frequency. This range of DNA transfer methods has been used successfully to transfer genes to fertilized eggs, producing transgenic animals; to study molecular aspects of gene expression; and to replicate cloned genes using mammalian cells as hosts.

Two approaches have been taken to the development of viral vectors: the direct cloning of DNA sequences using mammalian cells as hosts, and the transfer of single genes to replace defective alleles in genetic disease. For replicating cloned genes, vectors based on the bovine papilloma virus have been constructed. These vectors employ a portion of the bovine papilloma genome and sites for the insertion of foreign DNA. When transferred into mammalian cell hosts, the vectors containing foreign DNA replicate extrachromosomally, making it possible to clone genes in mammalian cells.

Vectors for the delivery of functional genes into mammalian cells have been developed using avian and mouse retroviruses. The genome of such retro-

Cloning site for several enzymes

FIGURE 16.24
A Ti plasmid designed for cloning in plants. Segments of Ti DNA, including those necessary for opine synthesis and integration, are combined with bacterial segments that incorporate cloning sites and antibiotic resistance genes (kan^r and tet^r). The vector also contains an origin of replication (*ori*) and a lambda *cos* site that permits recovery of cloned inserts from the host plant cell.

viruses is a single-stranded RNA that is transcribed by reverse transcriptase into a double-stranded DNA. This DNA is integrated into the host genome and passed on to daughter cells as part of the host chromosome. The retroviral genome can be engineered to remove some viral protein genes to create vectors that can carry foreign DNA, including human structural genes (Figure 16.25).

APPLICATIONS OF RECOMBINANT DNA TECHNOLOGY

Gene Mapping

Although an increasing number of the 50,000 to 100,000 human genes have been localized to chromosomal sites, in the majority of human genetic disorders the primary defect is unknown, and without a marker the gene cannot be mapped. A defective gene product has been identified in just under 400 single-

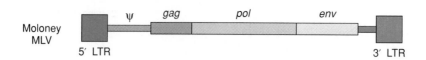

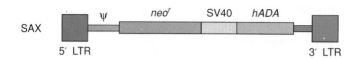

FIGURE 16.25

Retroviral vectors constructed from the Moloney murine leukemia virus (Moloney MLV). The native MLV genome contains a *psi* sequence required for encapsulation and genes that encode viral coat proteins (*gag*), an RNA-dependent DNA polymerase (*pol*), and surface glycoproteins (*env*). At each end, the genome is flanked by long terminal repeat (LTR) sequences that control transcription and integration into the host genome. The SAX vector retains the LTR and *psi* sequences, and includes a bacterial neomycin resistance (*NEOʳ*) gene that can be used as a selective marker. As shown, the vector carries a cloned human adenosine deaminase (*hADA*) gene, fused to an SV40 early region promoter/enhancer. The SAX construct is typical of retroviral vectors that will be used in human gene therapy.

gene disorders, and the nature of the mutation is known in only about 40 of these. In the majority of cases, the mutant gene product responsible for a genetic disease has not been identified, and as a result, classical genetic methods cannot be used to identify the chromosomal locus of the disorder.

The development of recombinant DNA technology has provided a new method for mapping allelic variants without the need to identify the gene product. This method makes use of the fact that single base changes in the recognition sequence of restriction enzymes can alter the pattern of cuts made in DNA. This gives rise to a detectable variation in DNA fragment length that is inherited in a Mendelian codominant fashion (Figure 16.26). These allelic variants or polymorphisms are called **restriction fragment length polymorphisms** or **RFLPs**.

The use of RFLP markers for gene mapping requires a collection of cloned DNA segments that recognize variations in enzyme cutting sites, and the mapping of these sequences to specific human chromosomes. Currently, over 200 such RFLP markers have been identified, and at least one marker is available for each human chromosome. The second requirement is large, multigenerational families in which an RFLP marker and a disease to be mapped are inherited.

The value of RFLP mapping in human genetics can be illustrated by considering the search for the chromosomal locus of type 1 neurofibromatosis

(NF1). This disorder is an autosomal dominant condition with an incidence of about 1 in 3000. NF1 is associated with a range of nervous system disorders, including tumors and an increased incidence of learning disorders. The spontaneous mutation rate at this locus is very high, and it is estimated that almost 50 percent of all cases represent new mutations. To map this autosomal gene by conventional methods in a systematic fashion would require the identification of large families carrying genes mapped to each of the 22 autosomes, and also carrying neurofibromatosis. The difficulty of this task can be illustrated by recalling that despite intense efforts, the first case of autosomal linkage was not discovered until 1959, and by 1969, only five cases of linkage were well documented.

The mapping isolation of the NF1 gene was accomplished in several steps. First, a number of laboratories compared the inheritance of NF1 with that of dozens of RFLP markers, each representing a specific human chromosome or chromosome region. This produced an exclusion map, indicating the chromosomes and chromosome regions where NF1 was not located. A second step used a collection of RFLP markers to determine that the disorder did segregate with an RFLP that spanned the centromere of chromosome 17 (Figure 16.27). Further efforts using over 30 RFLP markers from the pericentromeric region of chromosome 17 and 13,000 genotypings refined the location of this gene to a small region on

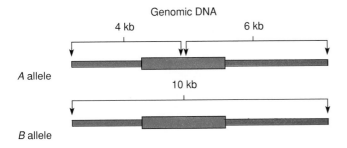

Genomic DNA

4 kb   6 kb

A allele

10 kb

B allele

1,1 1,2

1,2 1,2 1,2 1,2

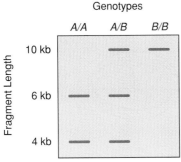

Genotypes

A/A   A/B   B/B

Fragment Length

10 kb

6 kb

4 kb

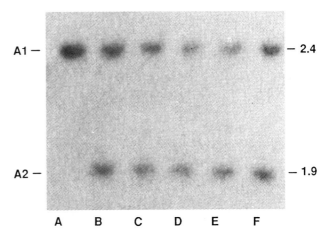

A1 —

A2 —

— 2.4

— 1.9

A   B   C   D   E   F

FIGURE 16.26

Restriction fragment length polymorphisms (RFLPs). The A and B alleles represent DNA segments from homologous chromosomes. The thick region represents a segment that can be detected with a DNA probe. Arrows indicate the pattern of restriction enzyme cutting sites that define the alleles. In allele A, three cutting sites generate fragments of 6 and 4 kb. In allele B, only two cutting sites are present, generating a fragment 10 kb in length. The absence of the cutting site in B could be the result of a single base mutation within the enzyme recognition/cutting site. Because the alleles are codominant, three genotypes (AA, AB and BB) can be generated. The allele combination carried by any individual can be detected by restriction digestion of genomic DNA (obtained from leukocytes or skin fibroblasts), followed by gel electrophoresis, transfer to a DNA binding filter, and hybridization to the appropriate probe. The fragment patterns for the three possible genotypes are shown as they would appear on a Southern blot.

FIGURE 16.27

The segregation of RFLP allele A2 with type 1 neurofibromatosis (NF1) in each of four affected offspring and their father. This RFLP is detected by probe pA10-41, which is known to be a DNA segment derived from human chromosome 17. On the basis of this and results from the use of other probes, the locus for NF1 was assigned to chromosome 17.

the long arm designated as 17q11.2. Finally, using a collection of cloned DNA sequences that spanned a subregion of 17q11.2, the gene for NF1 was identified in 1990. Note that this feat was accomplished in a little over three years with no direct knowledge of either the nature of the gene product or the mutational events that result in the NF1 phenotype.

Once the DNA sequence of the NF1 gene is known, it will be possible to construct probes to study the patterns of expression specific to tissues and development. It will also be possible to reconstruct the amino acid sequence of the gene product

from the nucleotide sequence, and deduce its functional properties. This information will in turn allow the isolation of the gene product, and the construction of antibodies to study its pattern of distribution in the cell. It is hoped that the sum of this knowledge will allow the development of therapeutic strategies to effectively treat NF1.

The use of RFLPs in gene mapping is a significant breakthrough in human genetics and has been used to map an increasing number of human genes. Coupled with other techniques, RFLP analysis has revolutionized human gene mapping, and promises to produce a human gene map with a level of resolution approaching that of other eukaryotic organisms such as *Drosophila* and the mouse.

Disease Diagnosis

The most widely used methods for prenatal detection of genetic diseases are **amniocentesis** and **chorionic villus sampling (CVS)**. In amniocentesis, a needle is

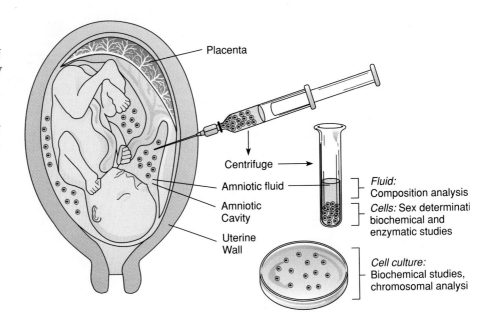

FIGURE 16.28
The technique of amniocentesis. The position of the fetus is first determined by ultrasound, and then a needle is inserted through the abdominal and uterine wall to recover fluid and fetal cells for cytogenetic and/or biochemical analysis.

Placenta

Centrifuge

Amniotic fluid

Amniotic Cavity

Uterine Wall

Fluid: Composition analysis

Cells: Sex determinati biochemical and enzymatic studies

Cell culture: Biochemical studies, chromosomal analysi

used to withdraw amniotic fluid (Figure 16.28). The fluid and the cells it contains can be analyzed for chromosomal or single-gene disorders. In CVS, a catheter is inserted into the uterus and used to retrieve a small tissue sample of the fetal chorion. This tissue is used for cytogenetic and biochemical prenatal diagnosis.

Coupled with these sampling methods, recombinant DNA technology has proven to be a highly sensitive and accurate tool for the detection of genetic disorders. The use of cloned DNA sequences has the advantage of allowing direct examination of the genotype rather than a potentially unknown or unexpressed gene product. The prenatal detection of two different defects in the beta globin gene serves to illustrate the application of recombinant DNA technology to disease detection. These two disorders are difficult to detect prenatally using methods that rely on detection of gene products because the beta globin gene is not active until a few days after birth. The first disorder is beta thalassemia, an autosomal recessive condition associated with lowered or absent production of beta globin. At the molecular level, this condition can be caused by a variety of deletions or point mutations. One such deletion covers about 600 nucleotides, and includes the third exon of the gene (Figure 16.29). Restriction enzyme digestion and the use of a Southern blot and hybridization to a cloned beta globin probe can be used to diagnose the genetic status of a fetus and other members of a family (Figure 16.29). Because the Southern blot detects fragments separated on the basis of size, this method is useful in studying mutations that involve deletions, insertions, and rearrangements.

Deletions, however, are relatively rare mutational events, and account for only a small fraction (less than 5 percent) of all mutations. More commonly, mutations are associated with changes in one or a small number of nucleotides (point mutations). Even changes in a single nucleotide can have devastating clinical and phenotypic effects. One such condition is sickle-cell anemia, an autosomal recessive condition common in those with family origins in areas of West Africa and the Mediterranean basin. Recall from Chapter 13 that sickle-cell anemia is caused by a single nucleotide substitution in codon 6 (changing the codon from GAG, encoding glutamine, to GTC, encoding valine). While the mutation does not change the length of the beta globin gene, it does destroy a cutting site for several restriction enzymes, including *Mst*II and *Cvn*I. This results in changes in the length of restriction fragments, allowing diagnosis of genotypes by Southern blot analysis. The standard method of detecting the mutant gene is by DNA digestion with a restriction enzyme that recognizes the normal nucleotide sequence at codon 6 (such as *Mst*II), followed by gel electrophoresis to separate fragments by size, and Southern blot hybridization with a radioactive cloned probe. With this procedure, it is possible to determine fetal genotypes as well as those of the parents (Figure 16.30).

Naturally, not every point mutation involving a single base change alters a restriction site. In fact, it is estimated that only about 5 to 10 percent of all such mutations can be detected by restriction analysis. Alternately, if the mutant gene is well characterized, it is possible to use synthetic oligonucleotides as probes. Under the proper conditions, such a probe,

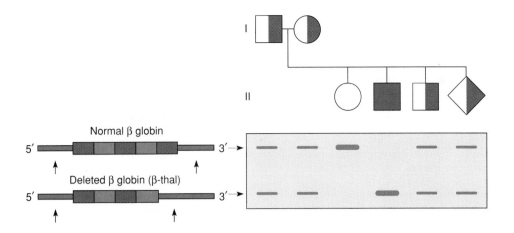

FIGURE 16.29

Diagnosis of beta thalassemia cause by a partial deletion of the beta-globin gene. The family pedigree (□ = male, ○ = female, ◇ = fetus) is shown positioned above the individuals' respective genotypes on a Southern blot. The normal and deleted beta-globin genes that produce the genotypic band patterns are shown at the right. Arrows indicate the cutting sites for restriction enzymes used in this analysis. The normal gene produces larger fragments shown as the top row of fragments on the gel; the smaller fragments produced by the deleted gene are represented at the bottom of the gel. Thin lines on the gel represent one copy of an allele, thick lines represent two copies.

known as an **allele-specific oligonucleotide (ASO),** will hybridize only with its homologous sequence, and not with other sequences that might vary by only a single nucleotide. When coupled with PCR amplification, the method can be used to detect situations in which one or a small number of point mutants are responsible for causing a disorder. A newer method based on ASOs and PCR analysis has been developed to screen for genotypes related to sickle-cell anemia. In this method, white blood cells are lysed and heated to denature the DNA, and the beta globin gene in the crude cell lysate is amplified by PCR. An aliquot of the amplified DNA is spotted on nitrocel-

lulose filters, and each filter is hybridized to an ASO (Figure 16.31). The genotype can be read directly from the filters. This rapid, inexpensive, and highly accurate technique will probably be the method of choice for diagnosis of a wide range of genetic disorders caused by point mutations.

Commercial Applications of Recombinant DNA Technology

Although recombinant DNA techniques were originally developed to facilitate basic research into gene organization and regulation of expression, scientists

FIGURE 16.30

Southern blot diagnosis of sickle-cell anemia. Arrows below the genes represent restriction enzyme cutting sites. In the mutant (β^S) globin gene, a point mutation (GAG → GTG) has destroyed a restriction enzyme cutting site, resulting in an altered pattern of DNA fragments. In the pedigree, the family has one unaffected homozygous normal daughter (II-1), an affected son (II-2), and an unaffected fetus (II-3).

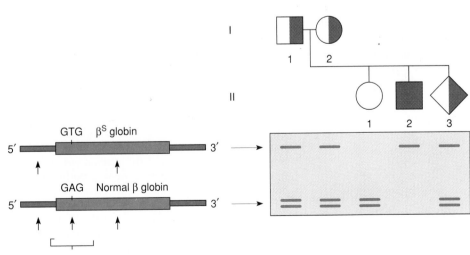

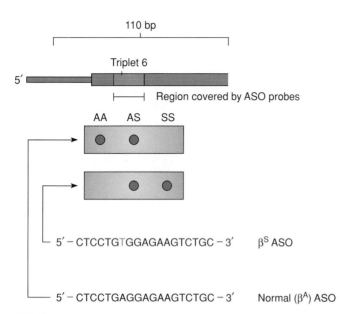

110 bp

Triplet 6

5′

Region covered by ASO probes

AA   AS   SS

5′ – CTCCTGTGGAGAAGTCTGC – 3′ β^S ASO

5′ – CTCCTGAGGAGAAGTCTGC – 3′   Normal (β^A) ASO

FIGURE 16.31
Genotype determination using allele-specific oligonucleotides (ASO). In this technique, DNA extracted from leukocytes is denatured and amplified by PCR. Aliquots of the amplified DNA are spotted onto strips of DNA binding filters. Each strip is hybridized to a specific ASO. The patterns expected for each genotype can be read directly from the filter. *AA*—homozygous normal hemoglobin; *AS*—heterozygous sickle-cell anemia; and *SS*—homozygous for sickle-cell anemia.

TABLE 16.2
Genetically engineered pharmaceutical products available or in clinical testing.

Gene Product	Condition Being Treated
Atrial natriuretic factor	Heart failure, hypertension
Epidermal growth factor	Burns, skin transplants
Erythropoiten	Anemia
Factor VIII	Hemophilia
Gamma interferon	Cancer
Granulocyte colony-stimulating factor	Cancer
Hepatitis B vaccine	Hepatitis
Human growth hormone	Dwarfism
Insulin	Diabetes
Interleukin-2	Cancer
Superoxide dismutase	Transplants
Tissue plasminogen activator	Heart attacks

have not been blind to the commercial possibilities of this technology. As a result, scientists have participated in the formation of biotechnology companies to exploit recombinant DNA technology. The commercial production of human gene products, such as hormones and clotting factors, from sequences cloned into bacterial cells are examples of the commercial use of recombinant DNA. The biotechnology industry is expected to grow into a multibillion dollar industry in the next decade.

Such commercial applications often represent ingenious solutions to problems of producing eukaryotic gene products in prokaryotic cells, or animal proteins in plant cells. For example, the hormone somatostatin is a 14-amino-acid hormone secreted by the hypothalamus. Working back from the amino acid sequence, Herbert Boyer and his colleagues synthesized a DNA segment encoding the gene and fused it to the 3′ end of a bacterial beta-galactosidase gene. The junction codon joining the prokaryotic and eukaryotic coding sequence directed the insertion of the amino acid methionine. The gene for this hybrid protein was inserted into the plasmid pBR322 and transformed into *E. coli*. The synthetic gene produced about 10,000 hybrid protein molecules per cell, which were recovered from the bacteria cells. These were treated with cyanogen bromide, a chemical that cleaves the active somatostatin hormone from the beta-galactosidase fragment by breaking the chain at the methionine residue (Figure 16.32). Several other clinically important hormones, such as insulin and human growth hormone, are produced using similar methods (Table 16.2).

Antibodies are proteins involved in the immune response, and are composed of two polypeptide chains, an H chain and an L chain. Many antibodies are used in clinical treatment, including cancer ther-

FIGURE 16.32 (opposite)
Production of human somatostatin in bacteria. The gene for the human growth hormone somatostatin was synthesized by reverse translation from the amino acid sequence. At the 5′ end of the gene, a *met* codon (AUG) was added to act as a bridge in order to fuse the human gene to a portion of the bacterial gene for beta-galactosidase. This construction was inserted into a plasmid and transformed into a bacterial cell. Within the bacterial cell, the transcription of the fusion gene produces a protein composed of a portion of the bacterial beta-galactosidase enzyme fused to the complete human somatostatin protein. This chimeric protein can be recovered from the bacterial cell and cleaved *in vitro* at the *met* site, bridging the two polypeptides to yield human somatostatin.

Amino acid sequence for somatostatin

Ala–Gly–Cys–Lys–Asn–Phe–Phe–Trp–Lys–Thr–Phe–Thr–Ser–Cys

↓

Synthetic gene derived from amino acid sequence

GCTGGTTGTAAGAACTTCTTTTGGAAGACTTTCACTTCGTGT
CGACCAACATTCTTGAAGAAAACCTTCTGAAAGTGAAGCACA

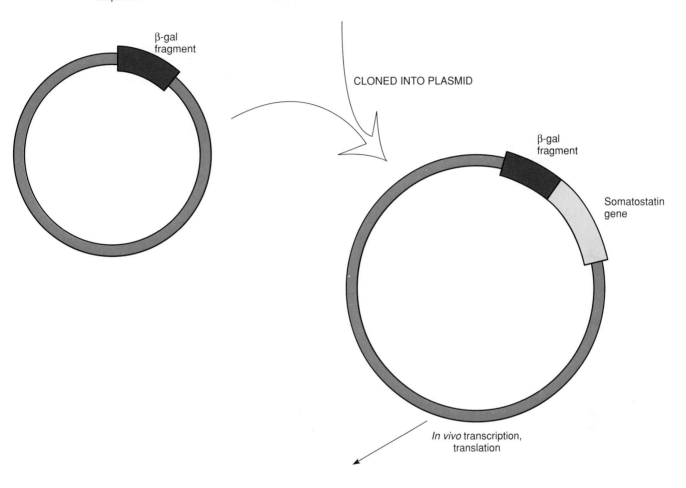

β-gal fragment

CLONED INTO PLASMID

β-gal fragment

Somatostatin gene

In vivo transcription, translation

〰〰〰–Met–Ala–Gly–Cys–Lys–Asn–Phe–Phe–Trp–Lys–Thr–Phe–Thr–Ser–Cys

β-gal fragment   Met linker   Somatostatin Sequence

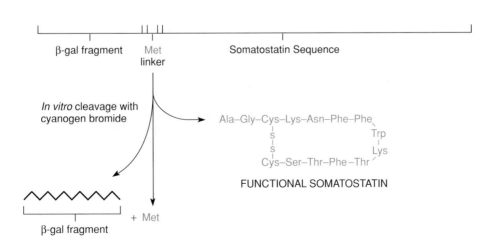

In vitro cleavage with cyanogen bromide

Ala–Gly–Cys–Lys–Asn–Phe–Phe
      |s   Trp
      |s   Lys
Cys–Ser–Thr–Phe–Thr

FUNCTIONAL SOMATOSTATIN

β-gal fragment   + Met

apy. At present, antibodies are commercially produced through a process of cell hybridization that is inefficient, time-consuming, and expensive. To improve the commercial production of antibodies, tobacco plants have been converted into antibody factories. Genes for the H and L chains of an antibody known as IgG were cloned into separate vectors, and inserted separately into different plants. Tobacco plants expressing H chains were crossed with plants expressing L chains to produce progeny that expressed complete antibody chains. Not only were the genes expressed, the antibodies amounted to 1.5 percent of the protein in the tobacco leaves. Although this has been accomplished only on an experimental basis, it is easy to see that this method can be adapted to produce commercial quantities of antibodies, such as those that attack tumor cells, from an acre or two of genetically modified plants.

CHAPTER SUMMARY

1   The independent confirmation of Mendel's work by de Vries, Correns, and von Tschermak in 1900 marked the beginning of genetics as an organized discipline. The development of the chromosome theory of inheritance, coupled with genetic studies of *Drosophila* and other organisms, provided the basis for our understanding of transmission genetics.

2   The discovery of DNA as the macromolecular carrier of genetic information and the development of the Watson-Crick model for the structure of DNA provided the basis for the development of molecular genetics, leading to our understanding of the mechanisms of genetic coding, transcription, and translation.

3   Recombinant DNA refers to the creation of a new association between DNA molecules or segments not usually found together. The cornerstone of recombinant DNA technology is a class of enzymes called restriction endonucleases that cut DNA at specific recognition sites. The fragments produced are joined with DNA vector segments using the enzyme DNA ligase to form recombinant DNA molecules. Vectors have been constructed from many sources, including bacterial plasmids, viruses, and artificial yeast chromosomes.

4   Recombinant DNA molecules are transferred into a host, and cloned copies are produced during host cell replication. A variety of host cells may be used for replication, including yeast, bacteria, cells of higher plants, and mammalian cells in tissue culture. However, the most common host is *E. coli*. Cloned copies of foreign DNA sequences can be recovered, purified, and analyzed.

5   Cloned libraries have been constructed containing DNA sequences from entire genomes, single chromosomes, or chromosome segments. Researchers use libraries to select probes complementary to all or part of a gene being surveyed.

6   Cloned DNA segments are used in a variety of applications, including gene mapping, disease diagnosis, and forensic analysis. In addition, cloned human genes inserted into bacterial hosts are used for the commercial production of human gene products for therapeutic treatment of a range of disorders including diabetes and growth defects.

7   Gene transfer in higher plants can be accomplished by using the Ti plasmid carrying foreign genes to infect a host plant and become incorporated into the host genome. This transgenic technique has been used to produce altered strains of agricultural crops that are resistant to disease and pesticides.

8   The correction of human genetic defects may be possible using retroviruses. Some of the viral genes can be removed from the retroviral genome to create a vector capable of transferring human structural genes into human target tissues, where the human genes become activated.

Insights and Solutions

The recognition site for the restriction enzyme *Sau*3A is GATC (Table 16.1). The recognition site for the enzyme *Bam*HI is GGATCC, where the four internal bases are identical to the *Sau*3A site. This means that the single-stranded ends produced by the two enzymes are identical. Suppose you have a cloning vector containing a *Bam*HI site, and foreign DNA that you have cut with *Sau*3A.

1 Can this DNA be ligated into the *Bam*HI site of the vector? Why?

ANSWER: DNA cut with *Sau*3A can be ligated into the *Bam*HI site of the vector because the single-stranded ends generated by the two enzymes are identical.

2 Can the DNA segment cloned into this site be cut from the vector with *Sau*3A? With *Bam*HI? What potential problems do you see with the use of *Bam*HI?

ANSWER: The DNA can be cut from the vector with *Sau*3A since the recognition site for this enzyme is maintained. The DNA may be recovered from the vector by *Bam*HI, but there is a potential problem with the use of this enzyme. When the *Sau*3A fragment is ligated into the *Bam*HI site, a base at each end of the *Bam*HI site must be added (arrows). Even though a template is present on the opposite strand, insertion of the wrong base will occur in a fraction of cases, destroying the *Bam*HI site, making it impossible to remove the insert with *Bam*HI.

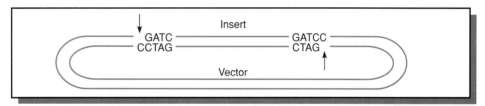

3 Calculate the average size of the fragments generated when cutting DNA with *Sau*3A and with *Bam*HI.

ANSWER: According to the Watson-Crick model of DNA, the sequence of bases in a DNA molecule can be at random. Therefore, if an A is present in a strand of DNA, the next base in that strand can be any base, even another A. Thus, the chance that any given base in DNA can be an A is 1/4, and the chance that the next base is a T is also 1/4. As a result, the chance that the four-base sequence GATC will occur is 4^4, or 1 in every 256 bases. The chance of encountering the six-base sequence GGATCC is 4^6, or 1 in every 4096 bases. Accordingly, cuts with *Sau*3A should generate fragments that average 256 bp in length, while cuts with *Bam*HI should generate fragments that average 4096 base pairs in length. This, of course, is only an approximation, and assumes that each base is present about 25 percent of the time. In reality, however, many genomes depart from this ideal, and may have only 40 or 42 percent G + C content, instead of the 50 percent ideal.

PROBLEMS AND DISCUSSION QUESTIONS

1 In recombinant DNA studies, what is the role of each of the following: restriction endonucleases, terminal transferase, vectors, calcium chloride, and host cells?

2 Why is poly dT an effective primer for reverse transcriptase?

3 An ampicillin-resistant, tetracycline-resistant plasmid is cleaved with *Eco*RI, which cuts within the ampicillin gene. The cut plasmid is ligated with *Eco*RI-digested *Drosophila* DNA to prepare a genomic library. The mixture is used to transform *E. coli* K12.

(a) Which antibiotic should be added to the medium to select cells that have incorporated a plasmid?

(b) What antibiotic resistance pattern should be selected to obtain plasmids containing *Drosophila* inserts?

(c) How can you explain the presence of colonies that are resistant to both antibiotics?

4 Clones from the previous question are found to have an average length of 5 kb. Given that the *Drosophila* genome is 1.5×10^5 kb long, how many clones would be necessary to give a 99 percent probability that this library contains all genomic sequences?

5 The human insulin gene contains a number of introns. In spite of the fact that bacterial cells will not excise introns from mRNA, how can a gene like this be cloned into a bacterial cell and produce insulin?

6 Type II restriction enzymes recognize palindromic sequences in intact DNA molecules and cleave the double-stranded helix at these sites. Inasmuch as the bases are internal in a DNA double helix, how is this recognition accomplished?

7 In a control experiment, a plasmid containing a *Hind*III site within a kanamycin-resistant gene is cut with *Hind*III, religated, and used to transform *E. coli* K12 cells. Kanamycin-resistant colonies are selected, and plasmid DNA from these colonies subjected to electrophoresis. Most of the colonies contain plasmids that produce single bands that migrate at the same rate as the original, intact plasmid. A few colonies, however, produce two bands, one of original size, and one that migrates much higher in the gel. Diagram the origin of this slow band during the relegation process.

8 In mice transfected with the rabbit beta-globin gene, the rabbit gene is active in a number of tissues including spleen, brain, and kidney. In addition, some mice suffer from thalassemia, caused by an imbalance in the coordinate production of alpha and beta globins. What problems associated with gene therapy are illustrated by these findings?

9 What facts should you consider in deciding which vector to use in constructing a genomic library of eukaryotic DNA?

10 Although the potential benefits of cloning in higher plants are obvious, the development of this field has lagged behind cloning in bacteria, yeast, and mammalian cells. Can you think of any reason for this?

SELECTED READINGS

ALWINE, J. C., KEMP, D. J., and STARK, G. R. 1977. Method for detection of specific RNAs in agarose gels by transfer to diazobenzyloxymethyl paper and hybridization with DNA probes. *Proc. Natl. Acad. Sci.* 74: 5350–54.

ANDERSON, W. F., and DIACUMAKOS, E. G. 1981. Genetic engineering in mammalian cells. *Scient. Amer.* (July) 245: 106–21.

ANTONARAKIS, S. 1989. Diagnosis of genetic disorders at the DNA level. *New Engl. J. Med.* 320: 153–163.

ANTONARAKIS, S. E., et al. 1985. Hemophilia A: Detection of molecular defects and carriers by DNA analysis. *New Engl. J. Med.* 313: 842–48.

BARKER, D., et al. 1987. Gene for von Recklinghausen neurofibromatosis is in the pericentromeric region of chromosome 17. *Science* 236: 1001–1102.

BERG, P. 1981. Dissections and reconstructions of genes and chromosomes. *Science* 213: 296–303.

BOYER, H. W. 1971. DNA restriction and modification mechanisms in bacteria. *Ann. Rev. Microbiol.* 25: 153–76.

CAWTHON, R. M., et al. 1990. A major segment of the neurofibromatosis type 1 gene: cDNA sequence, genomic structure, and point mutations. *Cell* 62: 193–201.

CHANG, J. C., and KAN, Y. W. 1981. Antenatal diagnosis of sickle-cell anemia by direct analysis of the sickle mutation. *Lancet*, pp. 1127–29.

COHEN, S., et al. 1973. Construction of biologically functional bacterial plasmids *in vitro*. *Proc. Natl. Acad. Sci.* 70: 3240–44.

COLLINS, J., and HOHN, B. 1978. Cosmids: A type of plasmid gene cloning vector that is packageable *in vitro* in bacteriophage heads. *Proc. Natl. Acad. Sci.* 75: 4242–46.

DAVIES, K. E. 1981. The applications of DNA recombinant technology to the analysis of the human genome and human disease. *Hum. Genet.* 58: 351–57.

———. 1982. A comprehensive list of cloned eukaryotic genes. In *Genetic engineering—3*, ed. R. Williamson. Orlando: Academic Press.

DAVIES, K. E., et al. 1981. Cloning of a representative genomic library of the human X chromosome after sorting by flow cytometry. *Nature* 293: 374–76.

ELLIS, K. P., and DAVIES, K. E. 1985. An appraisal of the application of recombinant DNA techniques to chromosomal defects. *Biochem. J.* 226: 1–11.

FISHER, E. M. C., CAVANNA, J. S., and BROWN, S. D. M. 1985 Microdissection and microcloning of the mouse X chromosome. *Proc. Natl. Acad. Sci.* 82: 5846–49.

FRIEDMANN, T. 1989. Progress toward human gene therapy. *Science* 244: 1275–81.

GEEVER, R. F., et al. 1981. Direct identification of sickle-cell anemia by blot hybridization. *Proc. Natl. Acad. Sci.* 78: 5081–85.

GLOVER, D. M. 1984. *Gene cloning: The mechanism of DNA manipulation.* London: Chapman and Hall.

GRAF, L. H. 1982. Gene transformation. *Amer. Scient.* 70: 496–505.

GRIESBACH, R. J., KOIVUNIEMI, P. J. and CARLSON, P. S. 1981. Extending the range of plant genetic manipulation. *Bioscience* 31: 754–56.

HOPWOOD, D. A. 1981. Genetic programming of industrial microorganisms. *Scient. Amer.* (Sept.) 245: 91–102.

KNORR, D., and SINSKEY, A. J. 1985. Biotechnology in food production and processing. *Science* 229: 1224–29.

KRUMLAUFF, R., JEANPIERRE, M., and YOUNG, B. D. 1982. Construction and characterization of genomic libraries from specific human chromosomes. *Proc. Natl. Acad. Sci.* 79: 2971–75.

MANIATIS, T., FRITSCH, E. F., and SAMBROOK, J. 1988. *Molecular cloning: A laboratory manual.* 2nd ed. Cold Spring Harbor, NY: Cold Spring Harbor Laboratory.

McKUSICK, V. A. 1988. The new genetics and clinical medicine. *Hospital Practice* 23: 177–91.

MOLLIS, K.B. 1990. The unusual origin of the polymerase chain reacion. *Scient. Amer.* (April) 262:56–65.

OLD, R. W., and PRIMROSE, S. B. 1985. *Principles of genetic manipulation: An introduction to genetic engineering.* Palo Alto, CA: Blackwell Scientific.

OSTE, C. 1988. Polymerase chain reaction. *BioTechniques* 6: 162–67.

REISS, J., and COOPER, D. N. 1990. Application of the polymerase chain reaction to the diagnosis of human genetic disease. *Human Genetics* 85:1–8.

SAIKI, R. K., et al. 1986. Analysis of enzymatically amplified beta-globin and HLA-DQ alpha DNA with allele-specific oligonucleotide probes. *Nature* 324: 163–66.

SCALENGHE, F., et al. 1981. Microdissection and cloning of DNA from a specific region of *Drosophila melanogaster* polytene chromosomes. *Chromosoma* 82: 205–16.

SOUTHERN, E. M. 1975. Detection of specific sequences among DNA fragments separated by gel electrophoresis. *J. Mol. Biol.* 98: 503–17.

THOMAS, T. L. and HALL T. C. 1985. Gene transfer and expression in plants: Implications and potential. *BioEssays* 3: 149–53.

TORREY, J. G. 1985. The development of plant biotechnology. *Amer. Scient.* 73: 354–63.

VILLA-KOMAROFF, L., et al. 1978. A bacterial clone synthesizing proinsulin. *Proc. Natl. Acad. Sci.* 75: 3727–31.

WAGNER, T. E., et al. 1981. Microinjection of a rabbit beta-globin gene into zygotes and its subsequent expression in adult mice and their offspring. *Proc. Natl. Acad. Sci.* 78: 6376–80.

WALLACE, M., et al. 1990. Type 1 neurofibromatosis gene: Identification of a large transcript disrupted in three NF1 patients. *Science* 249: 181–86.

WHITE, R. 1985. DNA sequence polymorphisms revitalize linkage approaches in human genetics. *Trends in Genetics.* 1: 177–80.

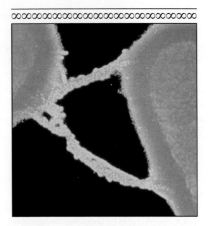

17
Organization of DNA in Chromosomes

CHAPTER CONCEPTS

DNA is organized in a manner consistent with the complexity of the host structure with which it is associated. Viruses, bacteria, mitochondria, and chloroplasts contain a shorter, often circular, DNA molecule relatively free of proteins. Eukaryotic cells contain greater amounts of DNA organized into nucleosomes and present as chromatin fibers. This increase in complexity is related to the greater amount of genetic information present as well as the greater complexity associated with their genetic function.

The elucidation of how DNA is organized in chromosomes is currently an extremely active area of research. There has been much interest in this topic because it is believed that the determination of the spatial arrangement of the genetic material and associated molecules will provide valuable insights into other aspects of genetics. For example, how the genetic information is stored, expressed, and regulated must be related to the organization of the genetic molecule, DNA. In eukaryotes, how the chromatin fibers characteristic of interphase are condensed into chromosome structures visible during mitosis and meiosis is also of great interest.

Investigations have involved viruses, bacteria, and a variety of eukaryotes. The genetic material has been studied using numerous approaches, including molecular analysis and direct visualization by light and electron microscopy. Despite these sophisticated approaches, we do not yet completely understand the organization of all molecular components directly associated with genetic material.

In this chapter, we provide a survey of the various forms of chromosome organization. Wherever possible, the physical arrangement of the genetic material will be related to its function. In a later chapter, we will focus directly on the organization of individual genes.

VIRAL AND BACTERIAL CHROMOSOMES

In comparison with higher forms, the chromosomes of viruses and bacteria are much less complicated. They consist of a single nucleic acid molecule that is often, but not always, circular in form. Such chromosomes are largely devoid of associated proteins, relative to the extensive, intimate association characteristic of eukaryotes. Contained within the single

chromosome of viruses and bacteria is much less genetic information than in the multiple units of higher forms. These various characteristics have greatly simplified analysis, which has now provided a fairly comprehensive view of viral and bacterial chromosomes.

Organization of Viral DNA and RNA

The chromosomes of viruses consist of a single nucleic acid molecule—either DNA or RNA—which is either single- or double-stranded. Viruses cannot reproduce in the absence of their host cell, because they depend on the host's biosynthetic machinery to duplicate their genetic material and to synthesize their protein coat. The DNA or RNA contained within the protein coat, or **capsid,** becomes functional only after its entrance into a host cell. In bacteriophages, the nucleic acid is literally injected into the host.

Viral nucleic acid molecules may exist as ring structures, or they may take the form of linear molecules. The single-stranded DNA of the **ΦX174 bacteriophage** and the double-stranded DNA of the **polyoma virus** are ring-shaped nucleic acid molecules both within the protein coat of the mature virus and within the host cell. The **bacteriophage lambda (λ),** on the other hand, possesses a linear double-stranded DNA molecule prior to infection, which closes to form a ring upon infection of the host cell. The two ends of the DNA molecule are single-stranded and described as being "sticky" because they are composed of complementary base sequences; thus, they have a chemical affinity for forming a ring. Following infection, an enzymatic reaction involving **polynucleotide ligase** occurs, forming a covalent bond and effecting closure of the ring. This process is illustrated in Figure 17.1. Circu-

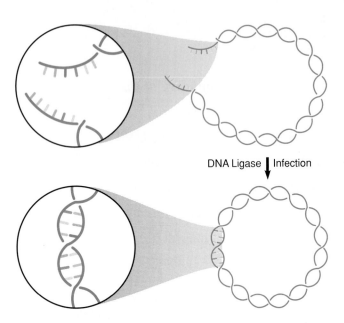

FIGURE 17.1
Schematic diagram of the complementary single-stranded tails at the ends of the chromosome of bacteriophage λ. Under the action of DNA ligase, the linear molecule is converted to a covalently sealed circular structure.

larization of the nucleic acid after invasion of the host is thought to be related to the mode of replication of the molecule. Still other viruses, such as the **T-even series of bacteriophages,** have linear, double-stranded chromosomes of DNA, which do not form circles inside the bacterial host. Thus, circularity is not an absolute requirement for viral replication.

Viral nucleic acid molecules have been visualized under the electron microscope. Figure 17.2 shows a mature bacteriophage lambda with its double-stranded DNA molecule in the circular configuration. One constant feature shared by viruses, bacteria, and eukaryotic cells is the ability to package an exceedingly long DNA molecule into a relatively small volume. In lambda, the DNA is 17 μm long and must fit into the phage head, which is less than 0.1 μm on any side. In Figure 17.2, the phage head is enlarged six times more than the DNA molecule.

In Table 17.1, the measured lengths of the genetic molecules are compared with the size of the head in several viruses. In each case a similar packaging feat must be accomplished. The dimensions given for phage T2 may be compared with the micrograph of both the DNA and viral particle shown in Figure 17.3.

When the diameter as well as the length of the nucleic acid molecule is taken into account, the minimum volume of the entire chromosome can be estimated and compared to the volume of the viral head. Seldom does the space available in the head exceed the chromosome volume by more than a factor of two. In many cases, almost all space is filled, indicating nearly perfect packing. In single-stranded RNA viruses like the tobacco mosaic virus (TMV), the capsid is assembled around the chromosome. The protein subunits of the capsid interact with the RNA molecule, playing a role in the packing phenomena. On the other hand, DNA viruses with nearly spherical heads, such as the T-even group, first assemble the capsid. Then, the chromosome is inserted into the

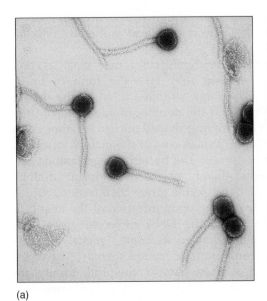

(a)

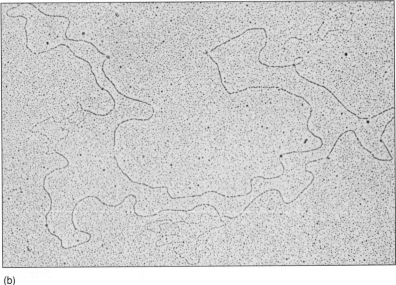

(b)

FIGURE 17.2
(a) Electron micrograph of phage λ and (b) the DNA isolated from it.

TABLE 17.1
The genetic material of representative viruses and bacteria.

| Organism | Nucleic Acid | | | Overall Size of Virus or Bacteria (μm) |
	Type	SS or DS	Length (μm)	
VIRUSES				
φX174	DNA	SS	2.0	0.025 × 0.025
Tobacco mosaic virus	RNA	SS	3.3	0.30 × 0.02
Lambda phage	DNA	DS	17.0	0.07 × 0.07
T2 phage	DNA	DS	52.0	0.07 × 0.10
BACTERIA				
Hemophilus influenzae	DNA	DS	832.0	1.00 × 0.30
Escherichia coli	DNA	DS	1200.0	2.00 × 0.50

SS = single stranded; DS = double stranded.

head which, when filled, is sealed by addition of a tail. Once packed within the head, the genetic material is functionally inert until released into a host cell.

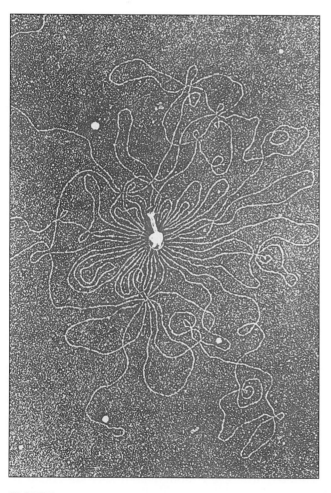

FIGURE 17.3
Electron micrograph of bacteriophage T2, which has had its DNA released by osmotic shock.

Organization of Bacterial DNA

Bacterial chromosomes are also relatively simple in form compared with those of eukaryotic cells. They always consist of a double-stranded DNA molecule, compacted into a structure sometimes referred to as the **nucleoid**. *Escherichia coli*, the most extensively studied bacterium, has a large circular chromosome, measuring approximately 1200 μm (1.2 mm) in length. When the cell is gently lysed and the chromosome released, it can be visualized under the electron microscope (Figure 17.4).

The DNA is found to be associated with several types of **DNA-binding proteins,** including those called **HU** and **H**. They are small but abundant in the cell and contain a high percentage of positively charged amino acids that can bond ionically to the negative charges of the phosphate groups in DNA. As we will soon see, these proteins resemble structurally similar molecules called histones that are found associated with eukaryotic DNA. Whether the bacterial counterparts serve precisely the same role remains to be seen. Unlike the tightly packed chromosome of a virus, the bacterial chromosome is not functionally inert. In spite of the compacted condition of the bacterial chromosome, transcription readily occurs.

DNA OF MITOCHONDRIA AND CHLOROPLASTS

The results of a number of experiments have demonstrated that both **mitochondria** and **chloroplasts** contain their own genetic information. It was observed that certain mutations in yeast, other fungi, and plants correlated with altered functioning of mitochondria or chloroplasts. These mutations were not always inherited in the ratios expected by Mendelian principles. Often these traits seemed to follow

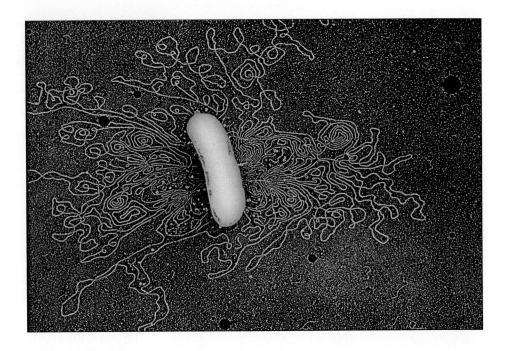

FIGURE 17.4
Electron micrograph of the bacterium *Escherichia coli*, which has had its DNA released by osmotic shock. The chromosome is approximately 1200 μm long.

a **maternal mode of inheritance;** that is, the mutation would be passed on only if it were possessed by the female parent. When it became clear that both mitochondria and chloroplasts are self-replicating, it seemed logical that these organelles might contain their own genetic information.

When autoradiographic studies showed that chloroplasts and mitochondria could incorporate radioactive DNA precursors, geneticists set out to look for DNA in these organelles. Electron microscopists not only documented the presence of DNA in both organelles, but they also saw DNA in a form quite unlike that seen in the chromosomes of eukaryotic cells. Instead, this DNA looked remarkably similar to that seen in viruses and bacteria. As we shall see, this similarity did not occur by chance as unrelated events. Both mitochondria and chloroplasts were once primitive, free-living bacteria-like organisms.

Molecular Organization and Function of Mitochondrial DNA

Extensive information is now available about the molecular aspects of mitochondrial DNA and related gene function. In most eukaryotes, mtDNA (Figure 17.5) is a circular duplex that replicates semiconservatively and is free of proteins. In size, mtDNA differs among organisms. This can be seen by examining the information presented in Table 17.2. In a variety of animals, mtDNA consists of about 16 to 18 kb. In vertebrates, there are 5 to 10 molecules per organelle. There is a considerably greater amount of

DNA present in plant mitochondria, where 100 kb is not unusual.

Several general statements can now be made concerning mtDNA. There appear to be few or no gene repetitions, and replication is dependent upon enzymes encoded by nuclear DNA. Genes have been identified that code for the ribosomal RNAs, over 20 tRNAs, and numerous products essential to the cellular respiratory functions of the organelles.

The protein-synthesizing apparatus as well as the molecular components for cellular respiration are jointly derived from nuclear and mitochondrial DNA. Ribosomes found in the organelle are different from those present in the cytoplasm. The sedimentation coefficient of these particles is variable in different organisms. The data in Table 17.3 show that mitochondrial ribosomes can be either lower in their coefficients than in prokaryotes ($55S$ to $60S$ in vertebrates), the same general coefficient (about $70S$ in some algae and fungi), or as high as eukaryotic cytoplasmic ribosomes ($80S$ in certain protozoans and fungi).

Many nuclear-coded gene products are essential to biological activity within mitochondria: DNA and RNA polymerases, initiation and elongation factors essential for translation, ribosomal proteins, aminoacyl tRNA synthetases, and some tRNA species. As in chloroplasts, these imported components are generally regarded as distinct from their cytoplasmic counterparts, even though both sets are coded by nuclear genes. For example, the synthetases essential to charging tRNA molecules (a process essential to

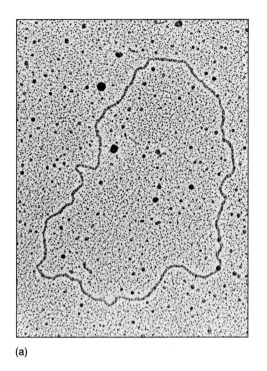

(a)

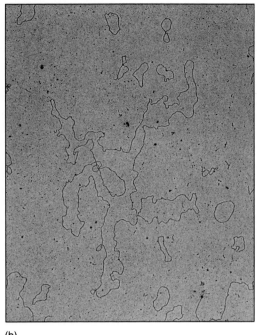

(b)

FIGURE 17.5
(a) Electron micrograph of mitochondrial DNA (mtDNA). (b) Electron micrograph of chloroplast DNA derived from corn.

translation) show a distinct affinity for the mitochondrial tRNA species as compared with the cytoplasmic tRNAs. Similar affinity has been shown for the initiation and elongation factors. Furthermore, while bacterial and nuclear RNA polymerases are known to be composed of numerous subunits, the mitochondrial variety consists of only one polypeptide chain. This polymerase is generally susceptible to bacterial antibiotics specific to RNA synthesis but not to eukaryotic inhibitors. The relative contributions of nuclear and mitochondrial gene products are illustrated in Figure 17.6.

There is a surprising diversity between mtDNA derived from mammalian (including human) and yeast mitochondria. The mammalian DNA is compact, is replicated in a single orderly process, and is free of intergenic (noncoding) regions. Yeast mtDNA,

on the other hand, demonstrates multiple origins of replication and extensive intergenic DNA sequences. Human mtDNA is free of **introns**, while yeast mtDNA contains such intragenic sequences. To some extent, the presence of introns accounts for the greater amount of DNA in mitochondria of plants compared with animals.

That genes have been exchanged between mitochondrial and nuclear DNA during evolution is established by the case of the ATPase subunit 9 gene. In yeast this gene is found in the nucleus, and in *Neurospora* it is present in the mitochondrion. It has been observed that the genetic code used in mitochondria contains several minor variations compared

TABLE 17.2
The size of mtDNA in different organisms.

Organism	Size in Kilobases
Human	16.6
Mouse	16.2
Xenopus (frog)	18.4
Drosophila (fruit fly)	18.4
Saccharomyces (yeast)	84.0
Pisum sativum (pea)	110.0

TABLE 17.3
Sedimentation coefficients of mitochondrial ribosomes.

Kingdom	Examples	Sedimentation Coefficient (S)
Animals	Vertebrates	50-60
	Insects	60-71
Protists	*Euglena*	71
	Tetrahymena	80
Fungi	*Neurospora*	73-80
	Saccharomyces	72-80
Plants	Maize	77

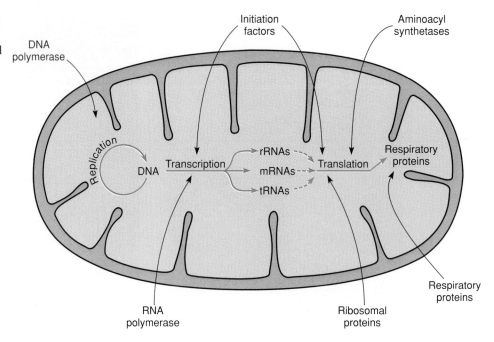

FIGURE 17.6
A comparison of the origin of gene products that are essential to mitochondrial function. Those shown entering the organelle are derived from the cytoplasm and encoded by the nucleus.

with that used in the cytoplasmic translation of mRNA derived from the nucleus (see Chapter 11).

Molecular Organization and Function of Chloroplast DNA

There is now available a substantial amount of molecular information about the DNA and genetic function of chloroplasts. The chloroplast, like the mitochondrion, contains a complete genetic system distinct from that found in the nucleus and cytoplasm. This system includes DNA (**cpDNA**) as a source of genetic information and a complete protein-synthesizing apparatus. However, the molecular components of the translation apparatus are jointly derived from both nuclear and chloroplast genetic information. As seen in Figure 17.5, chloroplast DNA is much larger than mitochondrial DNA, but it is nevertheless very similar to that found in prokaryotic cells.

DNA isolated from chloroplasts is found to be circular, double-stranded, replicated semiconservatively, and free of the associated proteins characteristic of eukaryotic DNA. Compared with nuclear DNA of the same organism, it invariably shows a different buoyant density, base composition, and nearest-neighbor frequency. Furthermore, cpDNA lacks 5-methylcytosine, a modified base present in nuclear DNA of many plants.

In *Chlamydomonas*, there are about 75 copies of the chloroplast DNA molecule per organelle. Each copy consists of a length of DNA that contains 195,000 bases (195 kb), almost twice the size of a T-even bacteriophage genome. In higher plants such as the sweet pea, multiple copies of the DNA molecule are present in each organelle, but the molecule is considerably smaller than that in *Chlamydomonas*, consisting of 134 kb.

Molecular hybridization studies have demonstrated that some chloroplast gene products function to synthesize proteins. In a variety of higher plants (beans, lettuce, spinach, maize, and oats), such studies reveal two sets of the genes coding for the ribosomal RNAs—5*S*, 16*S*, and 23*S* rRNA. These are present in two large DNA segments (greater than 20 kb), which are of opposite polarity and separated by a large spacer region (greater than 10 kb). Additionally, certain genes in chloroplast DNA code for at least 25 tRNA species and a number of ribosomal proteins specific to the chloroplast ribosomes. These ribosomes have a sedimentation coefficient slightly less than 70*S*, similar to that of bacteria.

Still other chloroplast genes have been identified that are specific to photosynthetic function. One of the major chloroplast gene products is the large subunit of the photosynthetic enzyme **ribulose-1-5-biphosphate carboxylase (RuBP)**.* Interestingly, the small subunit of this enzyme is encoded by a nuclear gene. The gene for the small subunit is thus transcribed in the nucleus, the corresponding mRNA translated in the cytoplasm, and the resultant small subunit protein transported into the chloroplast before it can function in photosynthesis. The large subunit, like all other chloroplast-derived gene products, is synthesized and remains within the organelle.

---

*The enzyme is also named ribulose-1-5-diphosphate carboxylase.

Chloroplast ribosomes are sensitive to the same protein-synthesis-inhibiting antibiotics as bacterial ribosomes: chloramphenicol, erythromycin, streptomycin, and spectinomycin. Even though ribosomal proteins are derived from nuclear and chloroplast DNA, most, if not all, are distinct from the equivalent proteins of cytoplasmic ribosomes.

Given this information, it is not surprising that experiments have revealed mutations relating to chloroplast function. Since products essential to photosynthesis are coded by chloroplast DNA and synthesized in the chloroplast, critical mutations in the chloroplast DNA can lead to inactivation of photosynthesis. If such chloroplasts are transmitted to progeny through the cytoplasm, offspring may similarly lack photosynthetic activity.

Based on the observation that mitochondrial and chloroplast DNA and their genetic apparatus are similar to their counterparts in bacteria, a theory of the origin of these organelles has been proposed. Championed by Lynn Margulis and others, this hypothesis, called the **endosymbiont theory,** states that mitochondria and chloroplasts may have originated as distinct bacteria-like particles that became incorporated into primitive eukaryotic cells. In the evolution of this symbiotic relationship, the particles lost their ability to function independently and underwent an extensive evolution as the eukaryotic host cell became dependent on them. While there are many unanswered questions, the basic tenets of this theory are difficult to dispute.

EUKARYOTIC CHROMOSOMES

The structure and organization of the genetic material in **eukaryotic cells** is much more intricate than in viruses or bacteria. This complexity is due to the greater amount of DNA per chromosome and the presence of large numbers of proteins associated with DNA in eukaryotes. For example, while DNA in the *E. coli* chromosome is 1200 μm long, the DNA in human chromosomes ranges from 14,000 to 73,000 μm in length as a linear duplex. In a single human nucleus, all 46 chromosomes contain sufficient DNA to extend almost 2 meters! This genetic material, along with its associated proteins, is contained within a nucleus that usually measures about 5 μm in diameter.

For many reasons this intricacy is to be expected. It parallels the structural and biochemical diversity of the many types of eukaryotic cells making up a single organism. In a single multicellular organism, cells assume specific functions by "division of labor." It is assumed that any functional capability is based upon highly specific biochemical activity. Since this activity is under the direction of the genetic information, a highly ordered regulatory system governing the readout of this information must exist if dissimilar cells are to perform a variety of different functions. Such a system must in some way be imposed on or related to the molecular structure of the genetic material.

While bacteria can reproduce themselves, they never exhibit a detailed process similar to mitosis. As was pointed out in Chapter 2, eukaryotic cells exhibit a highly organized cell cycle. During interphase, the genetic material is finely dispersed throughout the nucleus in chromatin form. As mitosis begins, the chromatin condenses greatly, and during prophase it is compressed into recognizable chromosomes. This condensation represents a contraction in length of some 10,000 times for each chromatin fiber. This highly regular condensation-uncoiling cycle poses special organizational problems in eukaryotic genetic material.

Mitotic Chromosomes

Initially, biologists knew of eukaryotic chromosome structure only from observations made with the light microscope. Most information came from studies of mitotic chromosomes. By observing preparations of mitotic chromosomes from different species, geneticists and cytologists have been able to determine the diploid number characteristic of any species and to note the varying sizes and gross morphology of chromosome pairs.

By the metaphase stage of mitosis it becomes apparent that each chromosome is really a double structure consisting of two **sister chromatids.** Sister chromatids are held together at a single point, the **centromere,** which is the area of attachment to the spindle fibers. Chromosomes for any given species are classified by the location of the centromere and the overall size of the chromosome. The **karyotype** consists of a micrograph of the chromosome pairs in metaphase arranged by size and centromere location. The number of chromosome pairs in the karyotype is equal to the haploid number. In species with a low haploid number, each pair of chromosomes may be distinct in gross morphology from all other pairs.

In humans, however, there is considerable morphological similarity between some pairs. In an initial classification scheme, human mitotic chromosome pairs were divided into seven categories (A–G), determined by size and by centromere position. The centromere position establishes the arm lengths on either side of the centromere. This scheme is called the **Denver classification system** because it was established at a conference held in that city in 1960. Since then, revised procedures for staining mitotic chromosomes have been devised to produce **chromosome banding**. The various patterns observed allow all 23 human chromosomes to be distinguished from one another. We will discuss this topic later in this chapter.

The study of mitotic chromosomes and karyotypes has been important in several genetics disciplines. The way in which chromatin folds up into mitotic chromosomes is of interest to students of structure. **Chromosome mutations** or **aberrations,** discussed in Chapter 7, may be detected in karyotypes and linked to human disorders. The evolutionary relationships between different species may be gauged by similarities between chromosome number and morphology (see Chapter 26).

The Folded-Fiber Model for Mitotic Chromosomes

We have so far considered what has been learned about mitotic chromosomes from studies involving light microscopy. In this section, we will discuss findings derived from electron microscopic observations. During the transition from interphase to mitosis, chromatin must undergo considerable compression and condensation. It has been estimated that a 5000-fold contraction in the length of DNA occurs. This process must be extremely precise, given the highly ordered nature and consistent appearance of the mitotic chromosomes in all organisms. Electron microscopic observations of mitotic chromosomes have provided an excellent overview of the intact chromosome following condensation. A whole-mount electron micrograph of a human mitotic chromosome is shown in Figure 17.7(a). In areas of greatest spreading, individual fibers similar to those seen in interphase chromatin are apparent. Very few fiber ends seem to be present. In some cases, none can be seen. Instead, individual fibers always seem to loop back into the interior. Such fibers are obviously twisted and coiled around one another, forming the regular pattern of the mitotic chromosome. The fibers are so tightly packed that, unless uncoiled slightly, the chromosome has a uniform or homogenous appearance throughout much of its length. You should compare this illustration with the light microscopic view of human chromosomes shown in Figures 2.4 and 17.20.

Observations of mitotic chromosomes in varying states of coiling led Ernest DuPraw to postulate the **folded-fiber model,** illustrated in Figure 17.7(b). During metaphase, each chromosome consists of two sister chromatids joined at the centromeric region.

Each arm of the chromatid appears to consist of a single fiber wound up much like a skein of yarn. The fiber is composed of double-stranded DNA and protein tightly coiled together. When these intact chromosomes are viewed under the scanning electron microscope, a pattern very similar to the model emerges [Figure 17.7(c)]. An orderly coiling-twisting-condensing process appears to be involved in the transition of the interphase chromatin to the more condensed, genetically inert mitotic chromosomes. As we shall see, more recent findings involving nucleosome and solenoid structures provide a basis for understanding the initial coiling and packing.

Synaptonemal Complex Organization

The electron microscope has also been used to visualize an additional ultrastructural component of the chromosome seen only in cells undergoing meiosis. This structure, mentioned in Chapter 2, is intimately associated with synapsed homologues during the pachytene stage of the first meiotic prophase and is called the **synaptonemal complex.*** In 1956, Montrose Moses observed this complex in spermatocytes of crayfish, and Don Fawcett saw it in pigeon and human spermatocytes. Because there was not yet any satisfactory explanation of the mechanism of synapsis or of crossing over and chiasmata formation, many researchers became interested in this structure. With few exceptions, the ensuing studies revealed the synaptonemal complex to be present in most plant and animal cells visualized during meiosis.

Figure 2.10 is an electron micrograph of the synaptonemal complex. It is composed primarily of a triplet set of parallel strands. The **central element** of this tripartite structure is usually less dense and thinner (100–150 Å) than the two identical outer elements (500 Å). The outer structures, called **axial** or **lateral elements,** are intimately associated with the chromatin of the synapsed homologues on either side. Selective staining has revealed that these axial elements contain protein, DNA, and perhaps RNA. Some DNA fibrils traverse the axial elements, making connections with the central element, which is composed primarily of protein. Figure 17.8 provides a diagrammatic interpretation of the electron micrograph and the above description.

The formation of the complex is initiated prior to the pachytene stage. As early as leptonema of the first meiotic prophase, lateral elements are seen in association with sister chromatids. Homologues have yet to associate with one another and are randomly

---

*An alternate spelling of this term is *synaptinemal complex.*

FIGURE 17.7
(a) Whole-mount transmission electron micrograph of a human mitotic chromosome. (b) Depiction of the coiling and folding of the chromatin fiber into the structure characteristic of a mitotic chromosome. (c) Scanning electron micrograph of human mitotic chromosomes.

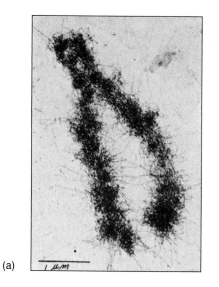

(a)

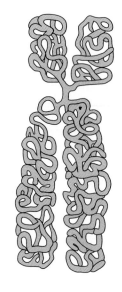

(b)

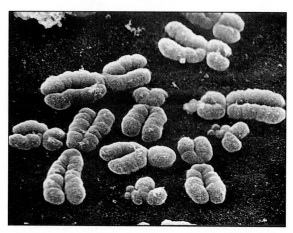

(c)

FIGURE 17.8
(a) Drawing of the synaptonemal complex (SC) observed during meiosis. Part (b) shows a schematic model of the interaction of theoretical components of this organelle with the chromatin fiber of the bivalents. (ce, central element; le, lateral element; c, chromatin.)

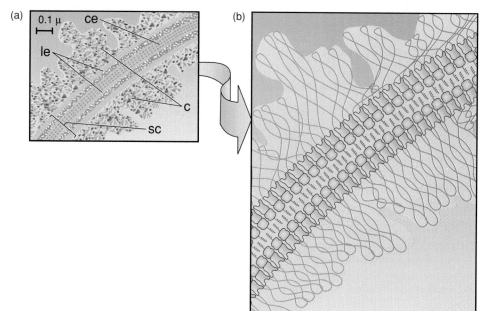

dispersed in the nucleus. By the next stage, zygonema, homologous chromosomes begin to align with one another, but remain distinctly apart by some 300 nm. Then, during pachynema, an intimate association between homologues occurs as formation of the complex is completed. In some diploid organisms, this occurs in a zipperlike fashion beginning with the telomeres, which may be attached to the nuclear envelope.

It is now agreed that the synaptonemal complex is the vehicle for the pairing of homologues and their subsequent segregation during meiosis. However, synapsis can occur in certain cases where no synaptonemal complexes are formed. Thus, it is possible that the function of this structure may be more extensive than just its involvement in the formation of bivalents.

In certain instances where no synaptonemal complexes are formed during meiosis, synapsis is not so complete and crossing over is reduced or eliminated.

For example, in *Drosophila melanogaster* meiotic crossing over rarely, if ever, occurs in males, and it is not observed in females homozygous for the third chromosomal mutant gene *c(3)G*. Synaptonemal complexes are not usually observed in either case. This information suggests that the synaptonemal complex must be present in order for chiasmata to form and crossing over to occur. In addition, it has been hypothesized that the genetic product coded for at the *c(3)G* locus may in some way be involved in the biosynthesis of the complex.

There may be other, yet undiscovered molecular components that are essential to the function of the synaptonemal complex. Even so, the discovery of the complex is significant to the field of genetics. If the complex is critical to synapsis during meiosis, then all sexual reproduction is dependent on it. If the complex is essential to crossing over, then it is largely responsible for eukaryotic genetic recombination, enhancing the role of that process in evolution.

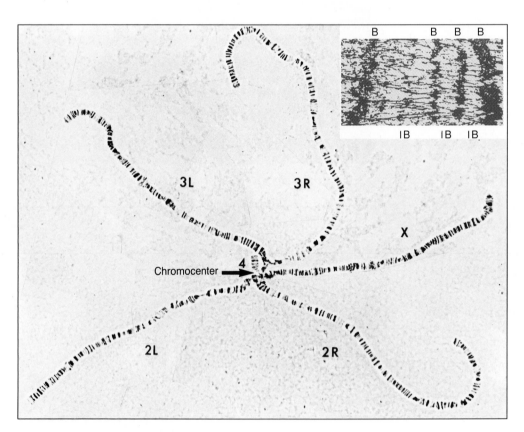

FIGURE 17.9
Giant polytene chromosomes derived from salivary gland cells of *Drosophila melanogaster*. The X chromosome, the right and left arms of chromosomes 2 and 3, and the small chromosome 4 are seen projecting from the chromocenter, where the centromeres of all chromosomes are aggregated. The insert depicts bands and interband regions along the axis of each chromosome.

Polytene Chromosomes

Studies of other unique chromosomes with the light microscope have yielded additional information about gross chromosome morphology. Interphase cells from a variety of organisms contain giant **polytene chromosomes**. They are found in various dipteran larval cells (salivary, midgut, rectal, and malpighian excretory tubules) and in several species of protozoans and plants. The large amount of information obtained from studies of these chromosomes has provided a model system for more recent investigations. Such structures were first observed by E. G. Balbiani in 1881 and are illustrated in Figure 17.9.

The large size and distinctiveness of such chromosomes result from the many DNA strands that compose them. Many replication cycles of paired homologues occur without strand separation or cytoplasmic division. As replication proceeds, chromosomes are created having 1000 to 5000 DNA strands remaining in parallel register with one another. In conjunction with associated chromosomal proteins, giant chromosomes result. They may be more easily studied than haploid interphase chromosomes because of their dramatic visibility.

When polytene chromosomes are observed under the light microscope, each is seen to be composed of a linear series of alternating **bands** and **interbands** (see insert in Figure 17.9). The banding pattern is distinctive for each chromosome in any given species. Individual bands are sometimes called **chromomeres,** a more generalized term describing lateral condensations of material along the axis of a chromosome. Each polytene chromosome is 200 to 600 μm long. Each of the 1000 or more fibers is a continuous DNA double helix.

The genetic information is contained in these fibers. In order for any particular gene that is contained within a band to become active in transcription, a localized uncoiling usually occurs. As the DNA unreels, the enzymes essential to transcription have access to it. A micrograph and a diagrammatic interpretation of this description are shown in Figure 17.10. Such localized sites of genetic activity are referred to as **puffs** because of their appearance.

That puffs are visible manifestations of transcription is evidenced by their high rate of incorporation of radioactively labeled uridine into RNA as assayed by autoradiography. Bands that are not extended into puffs incorporate much less uridine or none at all. During development in insects such as *Drosophila* and the midge fly *Chironomus*, the study of bands reveals differential gene activity. A characteristic pattern of band formation, equated with gene activation, occurs during development. This topic is pursued in more detail in Chapter 21.

Until recently, it has been unclear what, if anything, is different about the DNA composing bands compared with interbands. The DNA appears to be more tightly packed and more condensed in the band areas, as revealed in Figure 17.10. Studies at the molecular level have established that this condensation is not the result of a greater number of DNA strands in band areas. The degree of polyteny is identical in the band and interband areas of a given chromosome. Then why is the DNA arranged in these alternate patterns?

For several decades, it was thought that the bands were the site of genes, while the interbands were intergenic areas of the chromosome. The bands are

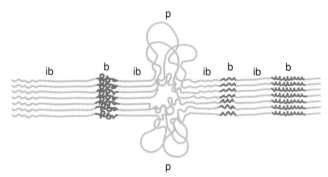

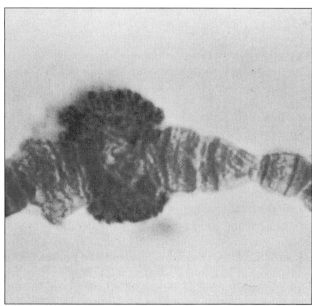

FIGURE 17.10

Photograph of a puff within a polytene chromosome. The schematic representation depicts the uncoiling of strands within a band (b) region to produce a puff (p) in polytene chromosomes. Interband regions (ib) are also labeled.

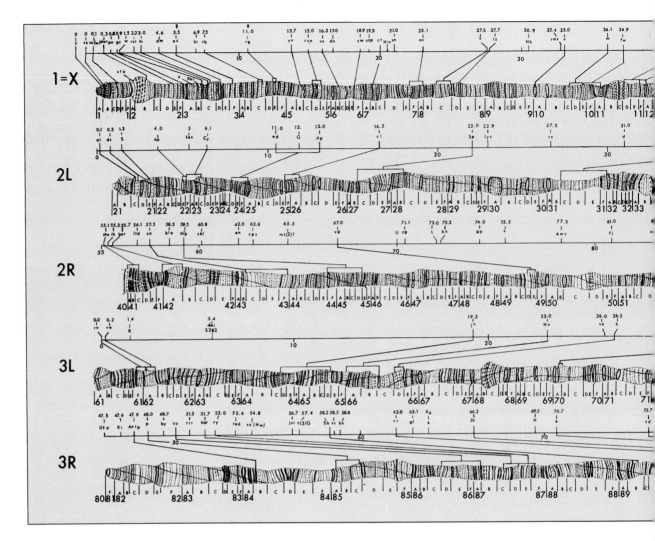

FIGURE 17.11

Comparison of the cytological map and the genetic map of the chromosomes of *Drosophila melanogaster*. The large designations at the left indicate the chromosome number and the right or left arm. The cytological map has been devised from the polytene chromosomes of larval salivary gland cells. It is divided into 102 major divisions, each of which is subdivided into six lettered subdivisions (A–F). The genetic map and many gene sites are shown above each cytological map.

obvious sources of genetic expression since puffs emanate from them. And many indirect observations suggested that each band might be the site of just a single gene. For example, an extremely supportive observation was the correlation made in the 1930s between the estimated number of genes present on the X chromosome of *Drosophila melanogaster* and the bands that could be counted along that chromosome (968). A similar correlation between the estimated number of genes in the entire genome and the total number of bands on all chromosomes (5000) has also been cited in support of this concept.

More recent work, by Burke Judd in the 1970s and by Pierre Spierer and others in the 1980s, has

analyzed the number of mutations that could be induced and isolated along limited regions of *Drosophila* chromosomes. Careful analysis revealed that the mutations fell into **complementation groups,** the number of which was about equivalent to the number of bands in the region studied. A complementation group is considered the equivalent of one functional genetic unit, or one gene. These observations also support the **one band–one gene paradigm**.

Nevertheless, other more direct observations suggest that this idea simply isn't true. These are the result of very convincing molecular studies of DNA along lengths of chromosomes spanning numerous contiguous bands and interbands. One inconsistency

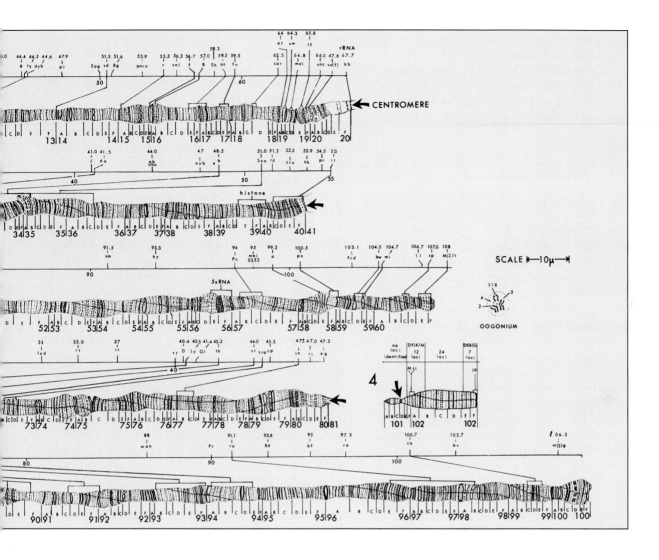

in the one band–one gene paradigm involves the amount of DNA present in single bands. Estimates range from 3,000 to 100,000 nucleotide pairs with an average of about 30,000! This is at least ten times more DNA than is needed for an average-sized gene.

Much more convincing are studies utilizing cloned DNA fragments. Such analyses have allowed the identification of mRNAs that are transcribed from a known region of a given chromosome. The results confirm that in several instances, the DNA of a single band encodes several rather than just a single gene. More than three times the number of distinctive mRNAs are produced than bands present along a given length of DNA. Furthermore, there are known arrays of genes and gene families that have been localized to single bands. For example, the tandem array of 150 repeated 5S RNA genes of *Drosophila* are all localized to a single band on chromosome 2, while the tandem array of 100 histone gene clusters (each with 5 genes) has been localized to a region of just four bands on that same chromosome. Thus, we can be nearly certain that other bands also contain more than one gene.

These assorted observations have prompted the reassessment of our classical view of polytene chromosomes. While we now know much more about them, we must conclude that we still do not understand the significance of alternating bands and interbands. This organizational format may somehow be related to the functional organization and regulation of the genes that reside within each of these regions. Whatever the final disposition, polytene chromosomes have remained, even in the age of molecular genetics, a popular and important resource for genetic study.

Cytological Mapping

By the early 1930s **genetic** or **linkage maps,** which gave the sequence and distances between many genes on the chromosomes of *Drosophila,* had been worked out (see Chapter 6). In that decade, these linkage maps were successfully correlated with the order of bands seen on the polytene chromosomes of *Drosophila.* This was accomplished by studying the altered banding patterns produced by flies with **chromosomal aberrations**. Aberrations (see Chapter 7) include duplications, deletions, and other rearrangements of the chromosomal material. In flies that demonstrated both a mutant phenotype and an aberration detectable in polytene chromosomes, the mutation was assigned to a region or band along the chromosome. In this way, a **cytological map** of many mutations on the *Drosophila* polytene chromosomes was compiled. A part of the *Drosophila* cytological map and its correlation with genetic loci identified by linkage studies are shown in Figure 17.11. Both the linkage and cytological maps are in agreement on the sequence of genes. However, the relative distance between any two loci as determined by linkage mapping and cytological techniques differs considerably in many instances.

Lampbrush Chromosomes

Another type of chromosome that has provided insights into chromosomal structure is the **lampbrush chromosome**. It was given this name because it was imagined to be similar in appearance to the brushes used to clean lamp chimneys in centuries past. Lampbrush chromosomes were first discovered in 1892 in sharks and are now known to be characteristic of vertebrate oocytes as well as spermatocytes of some insects. Most experimental work has been done with material taken from amphibian oocytes.

These unique chromosomes are easily isolated from oocytes in the diplotene stage of the first prophase of meiosis, where they are active in directing the metabolic activities of the developing cell. The homologues are seen as synapsed pairs held together by chiasmata, but instead of condensing, as do most meiotic chromosomes, the lampbrush chromosomes

are often extended to lengths of 500 to 800 μm. Later in meiosis they revert to their normal length of 15 to 20 μm. Thus, lampbrush chromosomes are interpreted as uncoiled and unfolded versions of the normal meiotic chromosomes.

Figure 17.12 shows several views of these structures. Collectively, they provide significant insights into the morphology of these chromosomes. In part (a), the meiotic configuration is visualized in a photomicrograph. The linear axis of each structure contains large numbers of repeating condensations. As in polytene chromosomes, the more general term, chromomere, describes each condensation. Each chromomere appears to support a pair of **lateral loops,** which give the chromosome its distinctive appearance. In part (b), an extended axis and adjacent loops are seen using the scanning electron microscope (SEM). Molecular analysis of the DNA present in loops has shown an average of about 100,000 nucleotide pairs/loop. This is obviously in excess of the amount of DNA needed to encode a single gene.

The structure and the functional activities of lampbrush chromosomes have been studied with enzymatic digestion and radioactive tracer techniques. Although considerable amounts of RNA and protein are associated with the chromosome, neither RNase nor protease enzymes disrupt the linear integrity of either the axis or the loops. It has been concluded, therefore, that DNA provides the skeletal structure of the chromosome, which is associated with protein and RNA.

Part (c) of Figure 17.12 presents another SEM, but at a much higher magnification. This micrograph provides a detailed view of one chromomere and the loop emanating from it. Clearly, the chromomere is a coiled-up chromosome fiber containing primarily a DNA molecule and associated protein, referred to as DNP (for deoxyribonucleoprotein). Each loop is thought to be composed of one double helix, while the central axis is made up of two DNA helices. This hypothesis is consistent with the belief that each chromosome is composed of a pair of sister chromatids. When oocytes are incubated in the presence of radioactive RNA precursors, the sites of transcription may be determined by autoradiographic analysis. Such studies reveal that the loops are active in the synthesis of RNA. The lampbrush loops, similar in a way to puffs in polytene chromosomes, represent DNA that has been reeled out from the central chromomere axis during transcription. As with polytene chromosomes, the study of lampbrush chromosomes has provided many insights into the arrangement and function of the genetic information.

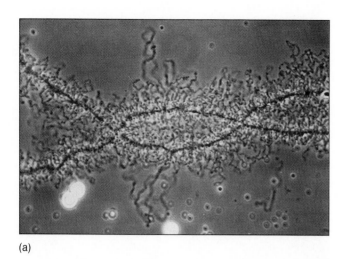
(a)

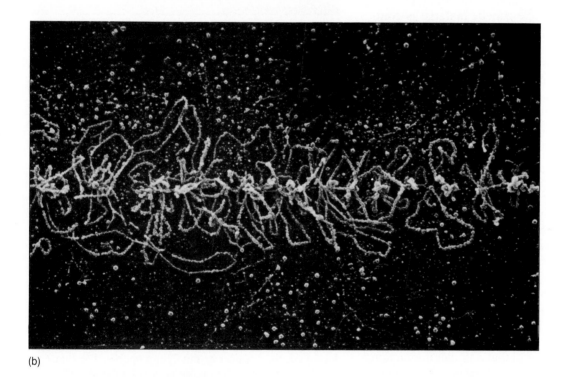
(b)

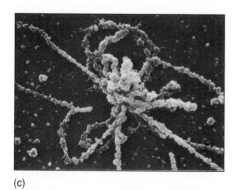
(c)

FIGURE 17.12
Lampbrush chromosomes derived from amphibian oocytes. Part (a) is a photomicrograph, while parts (b) and (c) are scanning electron micrographs.

MOLECULAR ORGANIZATION
OF CHROMATIN

Early studies of the structure of eukaryotic genetic material concentrated on intact chromosomes, preferably large ones, because of the limitation of light microscopy. The development of techniques for biochemical analysis, as well as the examination of relatively intact eukaryotic chromatin and mitotic figures under the electron microscope, have greatly enhanced our understanding of chromosome structure.

As established earlier, the genetic material of viruses and bacteria consists of essentially naked strands of DNA or RNA. In eukaryotes, the structure of the chromosome is much more complex. A substantial amount of protein is associated with the chromosomal DNA in all phases of the eukaryotic cell cycle. Thus, the eukaryotic genetic material is composed of **nucleoprotein;** such material is referred to as **chromatin** during interphase, when it is uncoiled. The associated proteins are divided into basic, positively charged **histones** and less positively charged **nonhistones**. The histones are intimately associated with chromatin structure, while the nonhistone proteins play other roles, including genetic regulation.

Nucleosome Structure

The general model for chromatin structure (Figure 17.13) is based on the assumption that chromatin fibers, composed of DNA and protein, undergo extensive coiling and folding in order to fit into the cell nucleus.

Of the proteins associated with DNA, the histones clearly play an essential role. Histones contain large amounts of the positively charged amino acids lysine and arginine, making it possible for them to bond electrostatically to the negatively charged phosphate groups of nucleotides. Recall that a similar interaction has been proposed for several bacterial proteins. There are five main types of histones (Table 17.4) that together exist in a 1:1 mass ratio with DNA.

X-ray diffraction studies confirm that histones play an important role in chromatin structure. Chromatin produces regularly spaced diffraction rings, suggesting that repeating structural units occur along the chromatin axis. If the histone molecules are chemically removed from chromatin, the regularity of this diffraction pattern is altered.

The basic model for chromatin structure was worked out between 1973 and 1977. Observations from a large number of workers in many laboratories served as the basis for the development of the proposed structure of the nucleoprotein fiber.

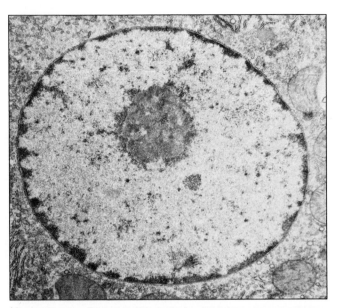

(a)

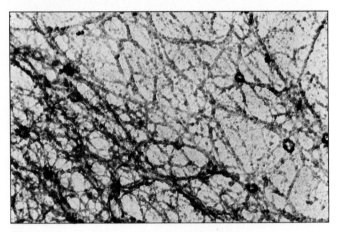

(b)

FIGURE 17.13
Eukaryotic chromatin seen in two different ways. (a) Electron micrograph of a thin section cut through a mouse liver nucleus. Because the fibers have been sectioned, they appear as the lighter specks. The dense circular structure is the nucleolus. (b) Whole-mount electron micrograph of chromatin fibers derived from a mouse liver nucleus. The circular structures are nuclear pores.

TABLE 17.4
Categories and properties of histone proteins.

Histone Type	Lysine-Arginine Content	Molecular Weight (daltons)
H1	Lysine-rich	21,000
H2A	Slightly lysine-rich	14,500
H2B	Slightly lysine-rich	13,700
H3	Arginine-rich	15,300
H4	Arginine-rich	11,300

1 Digestion of chromatin by certain endonucleases, such as **micrococcal nuclease,** yields DNA fragments that are approximately 200 base pairs in length or multiples thereof. Such work was reported by Dean Hewish and Leigh Burgoyne in 1973 and subsequently by others. These data demonstrate that enzymatic digestion is not random, for if it were, we would expect a wide range of fragment sizes. This observation suggests that chromatin consists of repeating units, each of which is protected from enzymatic cleavage except where any two units are joined. It is the area between all units that is attacked and cleaved by the nuclease. Such information is consistent with the results of X-ray diffraction studies, which suggest regular spacing arrangements in chromatin.

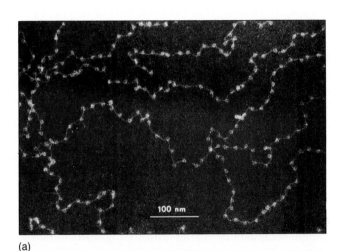

(a)

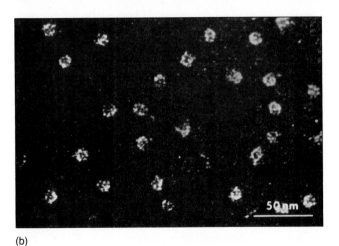

(b)

FIGURE 17.14
(a) Dark-field electron micrograph of nucleosomes present in chromatin derived from a chicken erythrocyte nucleus.
(b) Dark-field electron micrograph of nucleosomes produced by micrococcal nuclease digestion.

2 In 1974, Ada and Donald Olins reported on electron microscopic observations of chromatin prepared by methods different from those of most earlier work. Two such micrographs appear in Figure 17.14; they show chromatin fibers composed of linear arrays of spherical particles. The particles occur regularly along the axis of a chromatin strand and resemble beads on a string. These particles are now referred to as **ν-bodies** or **nucleosomes** (ν is the Greek letter *nu*). These findings conform nicely to the earlier proposal, which suggested the existence of repeating units.

3 In 1975, Roger Kornberg published the results of his study of the precise interactions of histone molecules and DNA in chromatin. His work showed that histones H2A, H2B, H3, and H4 occur as two types of tetramers: $(H2A)_2 \cdot (H2B)_2$, and $(H3)_2 \cdot (H4)_2$. He suggested that each repeating nucleosome unit consists of one of each tetramer, which together interact with about 200 base pairs of DNA. These data correlate well with the two previous observations and provide the basis for a model that explains the interaction of histone and DNA in chromatin.

4 Between 1975 and 1977, a more accurate understanding of the nucleosome emerged. Nuclease digestion of chromatin from diverse tissues and species was shown to yield wide variation in the number of DNA base pairs in the nucleosome unit. However, when digestion time is extended, DNA is removed from both the entering and exiting strands, creating a **nucleosome core particle** consisting of 146 base pairs. This number is consistent in all organisms studied.

The DNA lost in this prolonged digestion—that is, the difference in length between the nucleosome and the core particle—is responsible for linking nucleosomes together. This **linker DNA** is associated with histone H1. Extended nuclease digestion of the nucleosome first creates an intermediate structure containing 166 base pairs. Then, following further digestion, during which the conversion of this intermediate form to the 146-base-pair core particle occurs, histone H1 falls off. This observation provides evidence that histone H1 is located very near the initial portion of the DNA of the next core particle but is not an integral part of it.

5 In 1977 John T. Finch, Aaron Klug, and other investigators, who had performed X-ray and neutron scattering analysis of crystallized core particles, proposed a model of this core particle

(Figure 17.15). In this model, the 146-base-pair length of DNA is a secondary helix surrounding the core of eight histone molecules. The coiled DNA does not quite complete two full turns, yielding a flattened wedge shape of the dimensions $57 \times 110 \times 110$ Å. Because of this shape, the core particle of the nucleosome was called a **platysome**. Although this model does not establish the precise arrangement of histones within the helix, it provides a close approximation of the basic unit making up chromatin.

6 The extensive investigation of nucleosomes now provides the basis for predicting how the initial stages of packaging of chromatin occur (Figure 17.16). In its most extended state under the electron microscope, the chromatin fiber is about 100 Å in diameter, a size consistent with the longer dimension of the ellipsoidal nucleosome. The nucleosome itself condenses the DNA of chromatin at least tenfold. This **packing ratio** is calculated by dividing the length of DNA in 200 base pairs ($3.4 \text{ Å} \times 200 = 6800 \text{ Å}$) by the height of the nucleosome (60 Å).

The 100-Å fiber of chromatin is apparently further condensed into a fiber 300 Å in diameter, as viewed under the electron microscope. This transition has been studied carefully and can be achieved *in vitro* by increasing the ionic strength of the surrounding medium. The H1 histone is essential to this condensation since, in its absence, the transition cannot occur. The 300-Å fiber appears to consist of five or six nucleosomes coiled closely together. This structure has been called a **solenoid** and its formation increases the initial condensation by another factor of five, yielding an overall packing ratio of about 50.

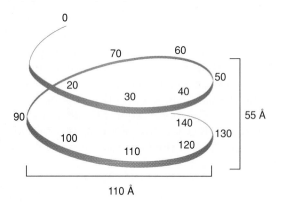

FIGURE 17.15
Representation of the DNA core of the nucleosome, called a platysome. The numbers indicate the relative positions of the 146 nucleotides contained in the core particle.

These 100- and 300-Å fibers are characteristic of interphase chromatin. The larger fiber can undoubtedly be further bent, coiled, and packed as even greater condensation occurs during the formation of mitotic chromosomes. There, a packing ratio of approximately 5000 must be achieved.

HETEROCHROMATIN

All recent evidence supports the concept that the DNA of each eukaryotic chromosome consists of one continuous double-helical fiber along its entire length. A continuous fiber is the basis of the **unineme model** of DNA within a chromosome. This finding, along with our knowledge of nucleosomes, might lead one to suspect that the chromatin fiber of each unit would demonstrate structural uniformity along its length. However, in the early part of this century, it was observed that some parts of the fiber remain condensed and stain deeply during interphase, but most do not. In 1928, E. Heitz coined the terms **heterochromatin** and **euchromatin** to describe the parts of chromosomes that remain condensed and those that are uncoiled, respectively.

Subsequent investigation has revealed a number of characteristics of heterochromatin that distinguish it from euchromatin. Heterochromatic areas are genetically inactive either because they lack genes or contain genes that are repressed. Also, heterochromatin replicates later during the S phase of the cell cycle than does euchromatin. This characteristic has been established using autoradiography and is thought to be a secondary feature of the extreme condensation. The discovery of heterochromatin provided the first clues that parts of eukaryotic chromosomes are not related to storage of genetic information. Instead, some chromosome regions are clearly involved in the maintenance of the chromosome's structural integrity.

Heterochromatin is characteristic of the genetic material of eukaryotes. Early cytological studies showed that areas closely adjacent to the centromeres were composed of heterochromatin. The ends of chromosomes, called **telomeres**, are also heterochromatic. In certain cells, notably the polytene chromosome-containing salivary gland cells of *Drosophila*, the heterochromatin of centromeres may clump together to form a **chromocenter** (see Figure 17.9). This observation suggests that some molecular affinity involved in heterochromatin is responsible for this adherence.

Heterochromatin has been subdivided into two types. **Constitutive heterochromatin** is present at

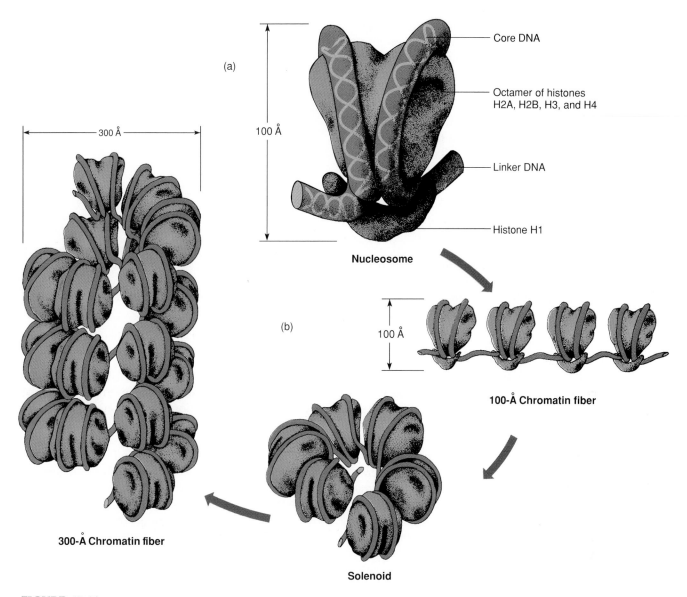

(a)

100 Å

Core DNA

Octamer of histones
H2A, H2B, H3, and H4

Linker DNA

Histone H1

Nucleosome

300 Å

(b)

100 Å

100-Å Chromatin fiber

300-Å Chromatin fiber

Solenoid

FIGURE 17.16
(a) General model of the association of histones and DNA in the nucleosome.
(b) Schematic illustration of the way in which the chromatin fiber may be coiled into a
more condensed structure.

homologous sites on pairs of chromosomes and
always exists in a genetically inert state. This form
has been shown to contain DNA sequences that are
not organized into genes. Telomeres and centromeric
regions are examples of constitutive heterochromatin.

The second type of heterochromatin, called **facultative,** is chromatin with the faculty, or potential, to
become heterochromatic. For example, one of the two
X chromosomes in mammalian females exhibits this
property when a **Barr body** is formed (see Chapter 5).
In certain species, such as the mealy bug, one entire
haploid set of chromosomes becomes heterochromatic. In both of these examples, only one member of

a homologous pair or set of chromosomes becomes
heterochromatic, in contrast with constitutive heterochromatin. The Y chromosome of many species,
lacking a true homologue, is also classified as facultative heterochromatin. It remains condensed during
interphase, and the majority of its length is genetically inert.

Facultative heterochromatin may contain genetic
information, but it is not used by the organism once
it becomes condensed. The extreme condensation
and coiling of heterochromatic material is thought to
physically preclude genetic activity in the same way
that the condensation of mitotic and meiotic chromosomes renders them inactive.

When certain heterochromatic areas from one chromosome are translocated to a new site on the same or another nonhomologous chromosome, genetically active areas sometimes become genetically inert if they now lie adjacent to the translocated heterochromatin. This influence on existing euchromatin is one example of what is more generally referred to as a **position effect**. That is, the position of a gene or groups of genes relative to all other genetic material may affect their expression. We first introduced this term in Chapter 7.

Heterochromatin, Satellite DNA, and Repetitive DNA

Chapter 9 and Appendix A describe techniques useful in the analysis of DNA. Two of these—sedimentation equilibrium centrifugation and reassociation kinetics—have provided important information about the nature of DNA in chromatin. Specifically, the DNA of heterochromatin and euchromatin exhibits different molecular characteristics. The nucleotide composition of the DNA of a particular species is reflected in its density, which can be measured with sedimentation equilibrium centrifugation. When the DNA of any eukaryotic species is analyzed in this way, the majority of it shows up as a single main peak or band of uniform density. However, one or more satellite peaks represent DNA that differs slightly in nucleotide composition and density from main-band DNA. This component, called **satellite DNA,** represents a variable proportion of the total DNA, depending on the species. A profile of main-band and satellite DNA from the mouse is shown in Figure 17.17. It should be noted that bacteria, as representative prokaryotes, contain only main-band DNA.

The significance of satellite DNA remained an enigma until the mid 1960s, when the techniques for measuring the reassociation kinetics of DNA became available. These techniques, developed by Roy Britten and David Kohne, allowed geneticists to determine the rate of reassociation of fragmented DNA. Britten and Kohne showed that when complementary strands of DNA fragments were separated by heating and then reassociated by cooling, certain portions were capable of reannealing more rapidly than others. They concluded that rapid reannealing was characteristic of multiple DNA fragments composed of identical or nearly identical nucleotide sequences—the basis of **repetitive DNA.**

As discussed in Chapter 9, repeating nucleotide sequences are classified as either highly or moderately repetitive DNA. Evidence has now accumulated linking satellite DNA to repetitive DNA. Further, this

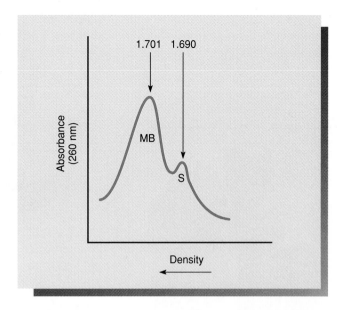

FIGURE 17.17
Separation of main-band (MB) and satellite (S) DNA from the mouse using ultracentrifugation in a CsCl gradient.

specific DNA fraction has been shown to be present in heterochromatic regions of chromosomes. The following paragraphs review information supporting and clarifying these conclusions.

When satellite DNA is subjected to analysis by reassociation kinetics, it falls into the category of highly repetitive DNA. It therefore consists of short sequences repeated a large number of times. The available evidence further suggests that these sequences are present as tandem repeats clustered at various positions in the genome.

Satellite DNA is found in very specific chromosome areas known to be heterochromatic—the **centromeric regions**. This was discovered in 1969 when several workers, including Mary Lou Pardue and Joe Gall, applied the technique of *in situ* **hybridization** to the study of satellite DNA. This technique (see Figure 9.16 and Appendix A) involves the molecular hybridization between an isolated fraction of radioactive DNA or RNA and the DNA contained in the chromosomes of a cytological preparation. Following the hybridization procedure, autoradiography is performed to locate the chromosomal areas complementary to the fraction of DNA or RNA.

In their work, Pardue and Gall demonstrated that RNA transcribed by mouse satellite DNA hybridizes with DNA of centromeric regions of mouse mitotic chromosomes (Figure 17.18). Thus, several conclusions can be drawn. Satellite DNA differs from main-band DNA in its molecular composition, as established by buoyant density studies. It is also composed of short repetitive sequences. Finally,

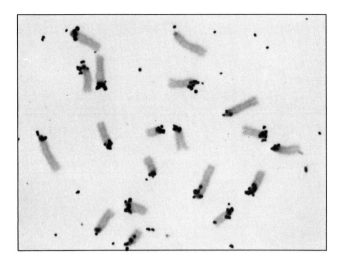

FIGURE 17.18

In situ hybridization between RNA transcribed by mouse satellite DNA and mitotic chromosomes. The grains in the autoradiograph localize the chromosome regions (the centromeres) containing satellite DNA sequences.

satellite DNA is found in the heterochromatic centromeric regions of chromosomes.

These conclusions establish one basis of the structural and functional differences between heterochromatin and euchromatin. They further reflect the complexity of the genetic material in eukaryotes. Instead of viewing the chromosome as a linear series of genes, we must now see it as quite variable in its molecular organization and functional capacity. For example, there are nucleotide sequences that do not specify gene products. Are these genetically inert

DNA sequences of heterochromatin related to the maintenance of the structural integrity of chromatin? Or, might they be involved in a regulatory role of transcription? Such noninformational sequences may play more than one role. There is no doubt that more extensive investigation will continually expand and modify our concept of eukaryotic genetic organization.

Heterochromatin Identification: Chromosome Banding

Until about 1970, mitotic chromosomes viewed under the light microscope could be distinguished only by their relative sizes and the positions of their centromeres. Even in organisms with a low haploid number, two or more chromosomes are often indistinguishable from one another. However, around 1970, differential staining along the longitudinal axis of mitotic chromosomes was made possible by new cytological procedures. These methods are now called **chromosome banding techniques** because the staining patterns resemble the bands of polytene chromosomes.

One of the first chromosome banding techniques was devised by Pardue and Gall during the development of the *in situ* hybridization procedure. They recognized that if chromosome preparations were heat denatured and then treated with Giemsa stain, a unique staining pattern emerged. The centromeric regions of mitotic chromosomes preferentially took up the stain! Thus, this cytological technique stained a specific area of the chromosome composed of

FIGURE 17.19

Karyotypes of a male (a) and female (b) mouse where chromosome preparations were processed and stained to demonstrate C-bands.

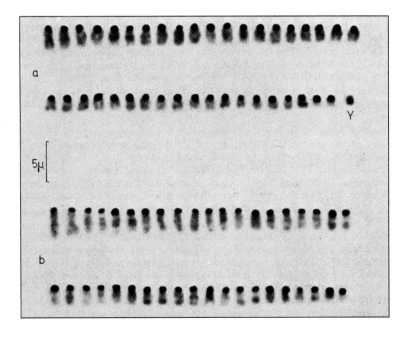

constitutive heterochromatin. A diagram of the mouse karyotype treated in this way is shown in Figure 17.19. The staining pattern is referred to as **C-banding**.

Still other chromosome banding techniques were developed about the same time. A group of Swedish researchers, led by Tobjorn Caspersson, used a technique that provided even greater staining differentiation of metaphase chromosomes. They used fluorescent dyes that bind to nucleoprotein complexes and produce unique banding patterns. When the chromosomes are treated with the fluorochrome **quinacrine mustard** and viewed under a fluorescent microscope, precise patterns of differential brightness are seen. Each of the 23 human chromosome pairs can be distinguished by this technique. The bands produced by this method are called **Q-bands**.

Another banding technique, which produces a staining pattern nearly identical to the Q-bands, has been developed by Tau-Chiuh Hsu and Frances E. Arrighi. This method, producing **G-bands** (Figures 17.20 and 17.21), involves the digestion of the mitotic chromosomes in the cytological preparations with the proteolytic enzyme **trypsin** followed by Giemsa staining. Another technique results in the reverse G-band staining pattern, called an **R-band** pattern. In 1971, a meeting was held in Paris to establish the nomenclature for these various patterns. Still other banding techniques are now available.

Intense efforts are currently underway to elucidate the molecular mechanisms involved in producing these banding patterns. The variety of staining reactions under different conditions reflects the heterogeneity and complexity of chromosome composition. The variations in nucleic acid composition that exist along the longitudinal axis of any given chromosome, with its alternating sections of heterochromatin and euchromatin, suggest functional diversity. For example, the following section describes the molecular organization of two chromosome components, the centromere and the telomere. As additional information becomes available, a clearer and more unified concept of genetic organization is expected.

Centromeres, Telomeres, and DNA Sequences

Linear chromosomes of most eukaryotes contain two regions that have essential structural roles. The first is the **centromere,** the region of the chromosome that attaches to one or more spindle fibers and that is pulled to one of the poles during the anaphase stages of mitosis and meiosis. In sister chromatids during mitosis, the common centromere must divide and the daughter structures must always move to opposite

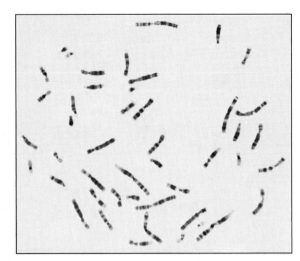

(a)

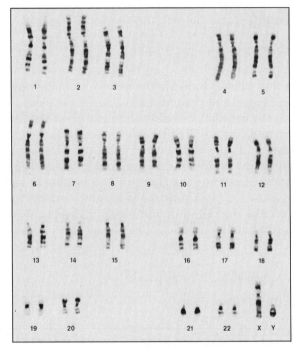

(b)

FIGURE 17.20
(a) G-banded human metaphase preparation.
(b) G-banded karyotype of a normal male showing approximately 400 bands per haploid set of chromosomes. Chromosomes were derived from cells in metaphase.

poles. In the first meiotic division, on the other hand, the common centromere of each pair of sister chromatids must remain a single structure. But, the two structures from each pair of homologous chromosomes must align during synapsis, attach properly to spindle fibers, and then move in opposite directions during anaphase I. During the second meiotic division, the centromeres of sister chromatids must behave as they do subsequently in mitosis.

Maintenance of these patterns is essential to the fidelity of chromosome distribution during mitosis and meiosis. Most estimates of infidelity during the transmission of chromosomes are exceedingly low: 10^{-5} to 10^{-6}, or one error per 100,000 to 1,000,000 cell divisions. As a result, it has been generally assumed that analysis of the DNA sequence of centromeric regions will provide insights into the rather remarkable features of this chromosomal region.

Before pursuing a discussion of this information, several distinctions will be helpful. The region referred to as the centromere, since we are considering eukaryotes, consists of both DNA and associated proteins. The DNA sequence itself has been designated **CEN** (for centromere). It is assumed that this specific sequence somehow is responsible for organizing the formation of the granular structure within the centromeric region called the **kinetochore**. You may recall from Chapter 2 that it is the kinetochore that attaches to one or more microtubules that compose the spindle fibers. Thus, the kinetochore most likely consists of both the CEN and surrounding proteins, which are specific to centromeric function.

The unicellular eukaryote *Saccharomyces cerevisiae* (a yeast) is ideal for such an analysis, primarily because of its rather uncomplicated arrangement of kinetochore-microtubule association. While human kinetochores may each bind to as many as a hundred microtubules during spindle attachment, the comparable yeast structures each bind to only a single microtubule.

The CEN regions of all 17 yeast chromosomes have now been isolated and sequenced. Since each centromere really serves an identical function, all CENs are remarkably similar. The region consists of about 225 base pairs, which can be divided into three regions (Figure 17.22). The first and third (I and III) are relatively short and highly conserved, consisting of only 8 and 25 base pairs, respectively. Region II, which is larger (80–85 base pairs) and extremely A-T rich, varies in sequence between different chromosomes. Up to 95 percent of the sequences are adenine or thymine, making it highly unlikely that this region might be transcribed into a meaningful RNA molecule.

The sequence requirements within the CEN region have now been investigated using the technique generally called mutational analysis. Site-specific mutations or deletions are induced and the CEN function assayed. Regions I and II are much more capable of absorbing sequence alterations without loss of function than region III. Base pair deletions in region

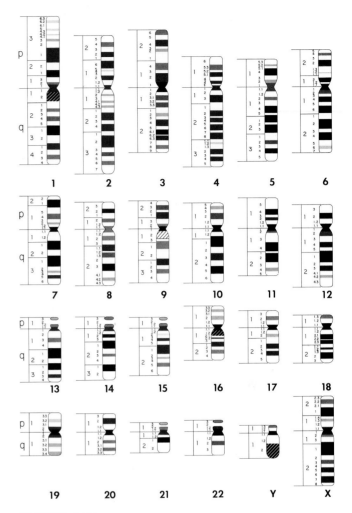

FIGURE 17.21
Schematic representation and nomenclature of human chromosomes at the 400-band stage using G-banding. Different shades represent varying intensities of bands.

I and changes in A-T composition and overall size in region I are tolerated. However, a single base change in region III disrupts centromere function. Thus the integrity of this sequence appears to be essential to proper kinetochore organization and function. Furthermore, it is now clear that CENs do not encode mRNAs or regulatory molecules, since transcripts complementary to these DNA sequences are not detected. Transcripts from adjoining regions have been detected, but their significance is not yet clear.

The above findings strongly suggest that at least part of the CEN contains a DNA sequence essential to the proper binding and organization of the proteins that form the kinetochore which provides the ability to attach to the spindle fiber. Even the DNA regions that can tolerate genetic change without loss of centromere function must be quite important to the

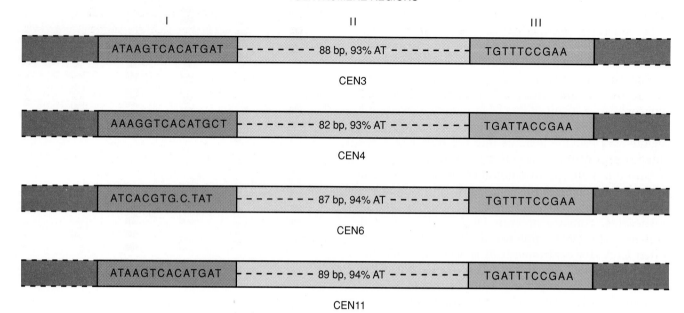

FIGURE 17.22
Nucleotide sequence information derived from DNA of the centromere regions of chromosomes 3, 4, 6, and 11 of yeast.

overall process to have been conserved in all 17 chromosomes. And, even though these findings have not centered on higher eukaryotes, they serve as excellent starting points; we can certainly anticipate parallel information to be forthcoming from the study of more advanced organisms.

The second important structure is that portion of the linear chromosome constituting each end, called the **telomere**. Its major function is to provide stability, aiding in the maintenance of individual integrity of the chromosome. This role is accomplished by virtue of rendering chromosome ends generally inert in interactions with other chromosome ends. In contrast to broken chromosomes, whose ends are "sticky" and readily rejoin other such ends, telomere regions do not fuse with one another or with broken ends. Thus, it has been thought that some aspect of the DNA sequence of chromosome ends must be unique compared with most other chromosome regions.

As with centromeres, the problem has been approached by investigating the smaller chromosomes of less advanced eukaryotes, such as protozoans and yeast. The idea that all telomeres in a given species might share a common nucleotide sequence has now been borne out.

Two types of sequences have been discovered. The first type, called simply **telomeric DNA sequences,** consists of short tandem repeats. It is this group that contributes to the stability and integrity of the chromosome. In a variety of lower eukaryotes, these take the form of

$$5'-\left[C_{1-8}\begin{smallmatrix}A\\T\end{smallmatrix}_{1-4}\right]-3'$$

For example, in the ciliate *Tetrahymena*, over fifty tandem repeats of the hexanucleotide sequence TTGGGG occur. In another ciliate, *Stylonychia*, TTTGGGG is repeated many times. In the slime mold *Physarum*, the tandem repeat TAGGG occurs.

It is not yet clear how such sequences might relate to the creation of inert chromosome ends. Several interesting observations may eventually provide clues to the explanation. Within the tandem repeats of telomeric DNA sequences, several single-stranded nucleotide gaps are present. The structure created is apparently resistant to the ligase action that would normally seal the gaps. A second observation is that the number of tandem repeats at the end of each chromosome may increase during each cell generation. For example, in the flagellate parasite *Trypanosoma*, seven to ten base pairs are added during each generation of a cell clone. However, there is apparently some mechanism limiting the absolute length of the terminus of the chromosome. Although we do not yet comprehend the significance of these observations, they are intriguing.

In addition to the above satellitelike sequences made up of repetitive DNA, a number of examples of what are called **telomere-associated sequences** are known. These have been identified by the *in situ* molecular hybridization technique described earlier in this chapter (see Figure 17.18). For example, in the amphibian *Xenopus laevis*, the gene encoding a small ribosomal RNA component (5S RNA) is tandemly repeated a varying number of times at the end of the long arms of all chromosomes. In *Drosophila*, a tandem repeat of a 3-kb sequence has been identified at the telomere of all polytene chromosomes. In yeast, a 6.7-kb repeat has been found.

One final observation involving the various repeated sequences associated with the telomere is related to the process of DNA replication at the ends of linear chromosomes. While replication can proceed normally to the end of the leading strand, a problem is encountered on the lagging strand. The 3′ end of the lagging strand is apparently an inadequate template for the synthesis of the RNA primer essential to replication. The result is that normal replication procedures will leave a gap on the end of the complement of the lagging strand following each replication cycle [Figure 17.23(a)].

In bacteria and viruses that contain circular DNA molecules this is not a problem, since no free ends are encountered during replication. In eukaryotes, the discovery of a unique enzyme, **telomerase,** has helped us understand how some organisms solve this problem. In the ciliated protozoan *Tetrahymena*, the many telomeres terminate in the sequence 5′–TTGGGG–3′. The enzyme is capable of adding repeats of TTGGGG to the ends of molecules already containing this sequence. This process, now known to occur in other organisms under the direction of a similar enzyme, prevents the telomeric ends from shortening following each replication [Figure 17.23(b)].

Further investigation of the *Tetrahymena* telomerase enzyme, isolated and studied extensively by Elizabeth Blackburn and Carol Greider, has yielded an extraordinary finding. This enzyme adds the same TTGGGG sequence to other DNA termini. Thus, it is not using a specific terminus as a template to somehow repeat the sequence. Blackburn and Greider have now established how the enzyme works. They have discovered that the enzyme contains a short piece of RNA that is essential to its catalytic activity! The functional enzyme is thus a ribonucleoprotein. The RNA component encodes the sequences added by the enzyme, serving as its own template. The RNA contains 129 bases, including the sequence 5′–CCCCAA–3′, which is complementary to the sequence whose synthesis it directs. Analogous enzyme functions have now been found in still other organisms, including the ciliate *Euplotes* and cultured human cells. In both cases, the RNA-containing telomerase enzyme behaves like the retroviral enzyme, reverse transcriptase, in that it synthesizes the DNA complement of an RNA template. In this case, the template is much shorter, and the enzyme supplies its own!

The analysis of telomeric DNA sequences has shown them to be highly conserved throughout evolution. Such conservation reflects not only the critical function of these structures but also their unusual mode of synthesis.

FIGURE 17.23
Replication of the telomere with and without the telomerase enzyme. New DNA synthesis is shown in blue. The dotted lines indicate the nonterminus portion of the chromosome, while the 3′ and 5′ designations identify the terminus of the chromosome occupied by the telomere. Without the new synthesis provided by telomerase, a gap is created on the lagging strand following each round of replication.

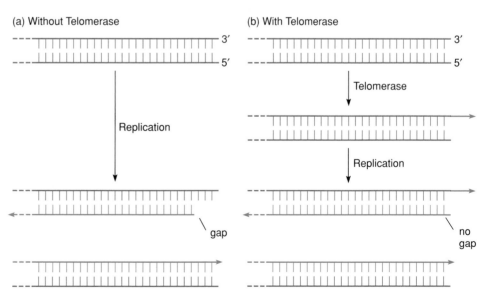

CHAPTER SUMMARY

1 Knowledge of the organization of the molecular components forming chromosomes is essential to the understanding of the function of the genetic material. Largely devoid of associated proteins, bacteriophage and bacterial chromosomes are double helical DNA molecules present in a form equivalent to the Watson-Crick model.

2 Mitochondria and chloroplasts contain DNA that encodes products essential to their biological function. This DNA is remarkably similar in form and appearance to viral and bacterial DNA, lending support to the endosymbiont theory stating that these organelles were once free-living organisms.

3 Classical light microscopy studies of polytene chromosomes and lampbrush chromosomes have provided valuable insights regarding chromosome morphology. These chromosomes have served as model systems for studying the relation of chromosome structure to genetic function.

4 Mitotic chromosomes, which serve as the basis for constructing karyotypes, are composed of long chromatin fibers that are folded into structures characteristic of the division stages.

5 The eukaryotic chromatin fiber is a nucleoprotein organized into repeating units called nucleosomes. Composed of about 200 base pairs of DNA and an octamer of four types of histones, the nucleosome is important in facilitating the conversion of the extensive chromatin fiber characteristic of interphase into the highly condensed chromosome seen in mitosis.

6 The structural heterogeneity of the chromosome axis has been established as a result of both biochemical and cytological investigation. The chromosomal banding techniques demonstrate that the axis of the mitotic chromosome is not chemically homogeneous. Genetically inert heterochromatin, often composed of repetitive nucleotide sequences, may be distinguished from the potentially active euchromatin both cytologically and biochemically.

7 DNA analysis has revealed highly conserved, unique nucleotide sequences in both the heterochromatic centromere and telomere regions. These sequences distinguish these structures from other parts of the chromosome, and provide the basis for the unique, critical features imparted to the chromosome by these regions.

Insights and Solutions

A previously undiscovered single-celled organism was found living at a great depth on the ocean floor. Its nucleus contained only a single linear chromosome containing 7×10^6 nucleotide pairs of DNA coalesced with three types of histonelike proteins.

1 A short micrococcal nuclease digestion yielded DNA fractions consisting of 700, 1400, and 2100 base pairs. What do these fractions represent? What conclusions can be drawn?

ANSWER: The chromatin fiber may consist of a variation of nucleosomes containing 700 base pairs of DNA. The 1400 and 2100 bp fractions represent two and three nucleosomes, respectively, linked together. Enzymatic digestion has been incomplete in the latter two fractions.

2 Analysis of individual nucleosomes revealed that each unit contained one copy of each protein, and that the short linker DNA contained no protein bound to it. If the entire chromosome consists of nucleosomes, how many are there, and how many total proteins are needed to form them?

ANSWER: Since the chromosome contains 7×10^6 base pairs of DNA, the number of nucleosomes, each containing 7×10^2 base pairs, is equal to

$$\frac{7 \times 10^6}{7 \times 10^2} = 10^4 \text{ nucleosomes}$$

The chromosome thus contains 10^4 copies of each of the three proteins, for a total of 3×10^4 molecules.

3   Analysis then revealed DNA to be a double helix similar to the Watson-Crick model, but containing 20 base pairs per complete turn of the right-handed helix. The physical size of the nucleosome was exactly double the volume occupied by that found in all other known eukaryotes, by virtue of increasing the distance along the fiber axis by a factor of two. Compare the degree of compaction of this organism's nucleosome to that found in other eukaryotes.

ANSWER: The unique organism compacts a length of DNA consisting of 35 complete turns of the helix (700 base pairs per nucleosome/20 base pairs per turn) into each nucleosome. The normal eukaryote compacts a length of DNA consisting of 20 complete turns of the helix (200 base pairs per nucleosome/10 base pairs per turn) into a nucleosome one-half the volume of that in the unique organism. The degree of compaction is therefore less in the unique organism.

4   No further coiling and compaction of this unique chromosome occurs in the unique organism. Compare this to a eukaryotic chromosome. Do you think an interphase human chromosome 7×10^6 base pairs in length would be a shorter or longer chromatin fiber?

ANSWER: The eukaryotic chromosome contains still another level of condensation in the form of "solenoids," which are dependent on the H1 histone molecule associated with linker DNA. Solenoids condense the eukaryotic fiber by still another factor of five.

The length of the unique chromosome is compacted into 10^4 nucleosomes, each containing an axis length twice that of the eukaryotic fiber. The eukaryotic fiber consists of $7 \times 10^6/2 \times 10^2 = 3.5 \times 10^4$ nucleosomes, 3.5 more than the unique chromosome organism. However, they are compacted by the factor of 5 in each solenoid. Therefore, the chromosome of the unique organism is a longer chromatin fiber.

PROBLEMS AND DISCUSSION QUESTIONS

1   Compare and contrast the chemical nature, size, and form assumed by the genetic material of viruses and bacteria.

2   Contrast the DNA associated with mitochondria and chloroplasts.

3   Why might it be predicted that the organization of eukaryotic genetic material would be more complex than that of viruses or bacteria?

4   Describe the formation of polytene chromosomes. What do the terms *band, interband, chromomere,* and *puff* signify?

5   Describe the sequence of research findings leading to the model of chromatin structure. What is the molecular composition and arrangement of the nucleosome? What is the solenoid?

6   Speculate as to why a structure such as the synaptonemal complex facilitates the precision associated with the processes of synapsis and crossing over.

7   When chloroplasts and mitochondria are isolated, they are found to contain ribosomes that are similar to prokaryotic cells. How does this observation relate to the endosymbiont theory?

8   Provide a comprehensive definition of heterochromatin.

9   List examples of constitutive and facultative heterochromatin.

10   Why were the cytological banding techniques considered to be significant to the study of genetics in the 1930s?

11   Mammals contain a diploid genome consisting of at least 10^9 base pairs. If this amount of DNA is present as chromatin fibers where each group of 200 base pairs of DNA is combined with 9 histones into a nucleosome, and each group of 5

nucleosomes is combined into a solenoid, achieving a final packing ratio of 50, determine:

(a) the total number of nucleosomes in all fibers.

(b) the total number of solenoids in all fibers.

(c) the total number of histone molecules combined with DNA in the diploid genome.

(d) the combined length of all fibers.

12 Assume that a viral DNA molecule is in the form of a 50 μm long circular rod of a uniform 20-Å diameter. If this molecule is contained within a viral head that is a sphere with a diameter of 0.08 μm, will the DNA molecule fit into the viral head, assuming complete flexibility of the molecule? Justify your answer mathematically.

SELECTED READINGS

ANGELIER, N., et al. 1984. Scanning electron microscopy of amphibian lampbrush chromosomes. *Chromosoma* 89: 243–53.

BEERMAN, W., and CLEVER, U. 1964. Chromosome puffs. *Scient. Amer.* (April) 210: 50–58.

BLACKBURN, E. H. 1990. Telomeres: Structure and synthesis. *J. Biol. Chem.* 265:5919–21.

BLACKBURN, E. H., and SZOSTAK, J. W. 1984. The molecular structure of centromeres and telomeres. *Ann. Rev. Biochem.* 53: 163–94.

BLOOM, K., HILL, A., and YEH, E. 1986. Structural analysis of a yeast centromere. *BioEssays* 4: 100–105.

BRITTEN, R. J., and KOHNE, D. 1968. Repeated sequences in DNA. *Science* 161: 529–40.

CALLAN, H. G. 1986. *Lampbrush chromosomes*. New York: Springer-Verlag.

CARBON, J. 1984. Yeast centromeres: Structure and function. *Cell* 37: 351–53.

CHEN, T. R., and RUDDLE, F. H. 1971. Karyotype analysis utilizing differentially stained constitutive heterochromatin of human and murine chromosomes. *Chromosoma* 34: 51–72.

CORNEO, G., et al. 1968. Isolation and characterization of mouse and guinea pig satellite DNA. *Biochemistry* 7: 4373–79.

DuPRAW, E. J. 1970. *DNA and chromosomes*. New York: Holt, Rinehart and Winston.

GALL, J. G. 1963. Kinetics of deoxyribonuclease on chromosomes. *Nature* 198: 36–38.

GOLOMB, H. M., and BAHR, G. F. 1971. Scanning electron microscopic observations of surface structure of isolated human chromosomes. *Science* 171: 1024–26.

GREEN, B. R., and BURTON, H. 1970. *Acetabularia* chloroplast DNA: Electron microscopic visualization. *Science* 168: 981–82.

HEWISH, D. R., and BURGOYNE, L. 1973. Chromatin sub-structure. The digestion of chromatin DNA at regularly spaced sites by a nuclear deoxyribonuclease. *Biochem. Biophys. Res. Comm.* 52: 504–10.

HILL, R. J., and RUDKIN, G. T. 1987. Polytene chromosomes: The status of the band-interband question. *BioEssays* 7: 35–40.

HSU, T. C. 1973. Longitudinal differentiation of chromosomes. *Ann. Rev. Genet.* 7: 153–77.

JUDD, B. H., SHEN, M., and KAUFMAN, T. 1972. The anatomy and function of a segment of the X chromosome of *Drosophila melanogaster*. *Genetics* 71: 139–56.

KING, R. C. 1970. The meiotic behavior of the *Drosophila* oocyte. *Int. Rev. Cytol.* 28: 125–68.

KORNBERG, R. D. 1975. Chromatin structure: A repeating unit of histones and DNA. *Science* 184: 868–71.

KORNBERG, R. D., and KLUG, A. 1981. The nucleosome. *Scient. Amer.* (Feb.) 244: 52–64.

MILLER, O. L. 1965. Fine structure of lampbrush chromosomes. *Natl. Canc. Inst. Monogr.* 18: 79–99.

OLINS, A. L., and OLINS, D. E. 1974. Spheroid chromatin units (ν bodies). *Science* 183: 330–32.

————. 1978. Nucleosomes: The structural quantum in chromosomes. *Amer. Scient.* 66: 704–11.

PARDUE, M. L., and GALL, J. G. 1970. Chromosomal localization of mouse satellite DNA. *Science* 168: 1356–58.

van HOLDE, K. E. 1989. *Chromatin*. New York: Springer-Verlag.

VERMA, R. S., ed. 1988. *Heterochromatin: Molecular and structural aspects*. Cambridge, England: Cambridge University Press.

WISCHNITZER, S. 1976. The lampbrush chromosomes: Their morphology and physiological importance. *Endeavour* 35: 27–31.

YUNIS, J. J. 1981. Chromosomes and cancer: New nomenclature and future directions. *Hum. Pathol.* 12: 494–503.

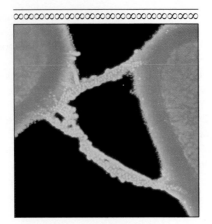

18
Organization and Structure of Genes

FINE STRUCTURE ANALYSIS OF THE GENE
Allelism and Complementation Testing | The *rII* Locus of Phage T4 | Deletion Testing of the *rII* Locus | Genetic Analysis of Eukaryotic Genes

THE MOLECULAR ORGANIZATION OF THE EUKARYOTIC GENE
Eukaryotic Genomes and the *C* Value Paradox | Eukaryotic Gene Structure | Exon Shuffling and Protein Domains

MULTIGENE FAMILIES
The Alpha- and Beta-Globin Families | The Histone Gene Family | Ribosomal RNA Genes: Tandem Repeats of a Gene Family

∞∞∞∞∞∞∞∞∞∞∞∞∞∞∞∞∞∞∞∞∞∞∞∞∞∞∞∞∞∞

CHAPTER CONCEPTS

Studies designed to elucidate the organization of genes within the
genome have included both genetic and molecular analyses and
have revealed diverse forms of organization, ranging from single
copies of genes to large numbers of tandem repeats of families of related
genes. A most striking feature is the degree to which noncoding
DNA is associated with the organization of genes throughout
the eukaryotic genome.

In the previous chapter, we discussed in detail how DNA is organized into chromosomes. That discussion illustrated the various forms and structures that have been observed as well as the chemical basis for them. For example, we observed and discussed bacterial and viral chromosomes, mitochondrial and chloroplast structures, and polytene, lampbrush, and mitotic chromosomes. In eukaryotes, we saw how DNA combines with protein to form a chromatin fiber, which is organized into nucleosomes and can exist as either euchromatin or heterochromatin. This information provides a comprehensive overview and framework for the investigation of the organization of the more basic genetic units, the genes.

We have, of course, known about the existence of genes for the better part of this century. And, for some time, we have been aware of the nature of the genetic code and thus, how information is stored in DNA. In this chapter, we expand this discussion by first describing studies designed to elucidate the precise nature of the gene. Sometimes referred to as **fine structure analysis of the gene,** this topic has been pursued continuously since Watson and Crick presented their model of DNA structure in 1953. At the forefront of genetics, such investigations will continue to be important throughout this century and beyond.

Studies performed by Seymour Benzer in the 1950s resulted in the now classical genetic analysis of the *rII* locus in bacteriophage T4. His findings suggested that a gene is an inclusive unit consisting of a linear series of nucleotides, all of which encode information specifying a polypeptide. His work supported the notion that the chromosome consists of a linear series of contiguous genes. To a large extent, these concepts

are still considered fairly accurate in bacteriophages and bacteria.

Detailed genetic analysis of individual loci in eukaryotic organisms has also been performed for several genes in *Drosophila melanogaster*. However, it was not until the development of recombinant DNA and nucleotide sequencing technologies in the 1970s that we extensively modified our concept of the genes of animal viruses and eukaryotic organisms. These technologies have allowed the detailed biochemical analysis of gene structure.

For example, as we have learned in previous chapters, a large portion of many eukaryotic genes consists of noncoding intron and flanking sequences that do not become part of the final mRNA transcript. Further, we now know that a large portion of the total DNA comprising the genome consists of intergenic areas, some of which are occupied by repetitive sequences. Beyond this information, we are learning much more about the organization of genes within the genome. We will review some of these topics and expand upon them with particular emphasis on the arrangement of multigene families. As we do, we will attempt to shed some light on an intriguing paradox involving the tremendous increase in DNA content in the genome of many eukaryotic organisms.

These latter topics represent some of the most exciting and unexpected discoveries made in the field of genetics. For all geneticists, but particularly those trained prior to the 1970s, the findings concerning eukaryotic gene structure serve as a source of wonderment. Perhaps we would have all been better prepared to learn of the much greater complexity of eukaryotic genes and genomes compared with those of prokaryotes and viruses if we had paid more attention to the quotation from Lewis Carroll's *Through the Looking Glass*, which was insightfully included by Richard B. Goldschmidt in his Presidential Address to the IX International Congress of Genetics in 1954:

> "I can't believe that," said Alice. "Can't you?" the Queen said in a pitying tone. "Try again: draw a long breath, and shut your eyes."
>
> Alice laughed. "There's no use trying," she said, "one can't believe impossible things." "I dare say you haven't had much practice," said the Queen. "When I was your age I did it for half-an-hour a day. Why, sometimes I've believed as many as six impossible things before breakfast."

For all of us interested in genetics, it is certain that the next "impossible thing" is just around the corner!

FINE STRUCTURE ANALYSIS OF THE GENE

The experimentation described in the next few sections represents the approach referred to as **genetic analysis,** as applied to probing the detailed structure of the gene. This is in clear contrast to the **biochemical analysis** of DNA and associated molecules making up the gene. While genetic analysis provides results that are less direct and often more cumbersome to obtain than biochemical data, this general approach is often the initial one used in the study of the role genes play in directing life processes. Thus, it is interesting that one of the most ingenious cases of genetic analysis was applied to the analysis of the structure of the gene itself. Before we describe these experiments, we will review the concepts of **allelism** and **complementation testing**.

Allelism and Complementation Testing

When two recessive mutations cause similar phenotypes, they may be suspected to be alleles of the same gene. If it is possible to map them and they are found at approximately the same chromosome location, the case for allelism is further strengthened. In order to prove this, however, a **complementation test** must be performed.

In a diploid organism, a complementation test is conducted by examining the phenotype of an individual carrying one copy of each recessive mutation. There are two configurations by which this can occur. In the *cis* state, both mutations are present on the same chromosome. In the *trans* state, one mutation is on one chromosome and the other mutation is present on its homologue. Since the mutations can

either be alleles or not be alleles, four possible conditions are created (Figure 18.1).

As shown in case 2, when two independent mutations (m_1 and m_2) are actually alleles and are present in the *cis* arrangement, they are both present within one of the two copies of the same gene. As a result, the wild-type phenotype will be expressed. This will occur because the homologue bearing the gene with the wild-type versions of these mutations (m_1^+ and m_2^+) will produce the normal gene product. However, if these alleles were present in the *trans* condition (case b), one mutation would disrupt the function of the gene on one homologue while the other mutation would disrupt the gene on the other homologue. No normal gene product would be made, and a mutant phenotype would be expressed.

Now consider the cases where the two mutations m_1 and m_2 are not alleles of one another and thus affect separate gene products. In both the *cis* and *trans* arrangement, normal products of the two genes can be produced, and the wild-type phenotype will be expressed. In the case of the *cis* arrangement (case c), both normal products result from the two genes on the same homologue. In the case of the *trans* arrangement (case d), one normal product is produced from a gene on one homologue, while the other normal product is produced by a gene on the other homologue. In such a situation, **complementation** is said to occur. The genetic information on one chromosome complements that present on another chromosome.

The experimental examination of these possibilities distinguishes between mutations that affect the same gene product and are thus alleles and those that are not. This is called a **complementation test,** or sometimes just an **allelism test**. Barring minor exceptions such as intragenic recombination and intragenic complementation, two mutations that are alleles will always fail to complement one another when present in the *trans* arrangement. As such, they establish a **complementation group,** all the mutations of which affect the same genetic product. On the other hand, mutations that do complement one another in the *trans* arrangement are in separate complementation groups (or genes) and are not alleles.

The same underlying principles apply to organisms other than eukaryotes. Thus, complementation testing need not be restricted to diploid organisms. For example, if viruses from two different genetic strains are used to infect a single bacterium simultaneously, complementation analysis can be performed. While this approach was described earlier (see Figure 15.30), we shall review it again here, since it serves as the basis of the analysis that will be discussed in the next section.

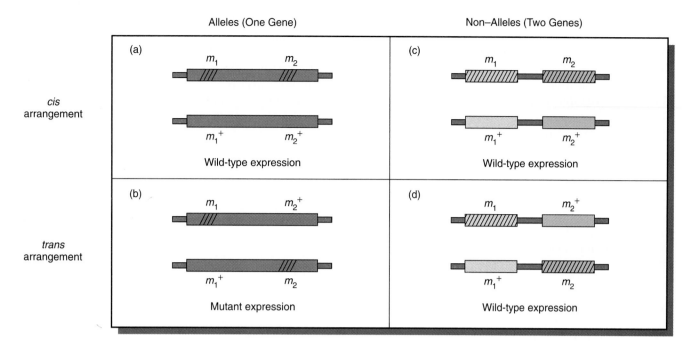

| Alleles (One Gene) | Non–Alleles (Two Genes) |

cis arrangement

(a)

m_1 m_2

m_1^+ m_2^+

Wild-type expression

(c)

m_1 m_2

m_1^+ m_2^+

Wild-type expression

trans arrangement

(b)

m_1 m_2^+

m_1^+ m_2

Mutant expression

(d)

m_1 m_2^+

m_1^+ m_2

Wild-type expression

FIGURE 18.1

A comparison of two mutations, m_1 and m_2, present heterozygously in *cis* and *trans* arrangements. The left column depicts both arrangements for two mutations within the same gene. The right column depicts both arrangements found in separate genes.

Let us consider a hypothetical case in which each genetic strain carries a separate mutation that is lethal, but both are alleles of one another. Following simultaneous infection, no viral progeny will be produced when tested, because no complementation occurs. If, however, the two lethal mutations are not alleles, but instead are present in separate genes, complementation *will* occur, and normal viral progeny will be produced. Extensive testing of many mutations can, in fact, establish all complementation groups. All mutations within the same group fail to complement one another.

The description and definition of a complementation group will aid you in understanding the above discussion and what is soon to follow. In general terms, a complementation group is often equivalent to a single gene. Since there is some ambiguity, however, another term, **cistron,** was coined by Benzer. A cistron, as determined by the complementation test, is a functional genetic unit specifying the formation of a single polypeptide chain. Alleles associated with one cistron are most often the result of mutations at various sites within that cistron.

The *rII* Locus of Phage T4

A detailed analysis of a single locus, *rII* in phage T4, was initiated in 1953 by Seymour Benzer. Mutants at this locus produce distinctive plaques when plated on *E. coli* strain B (see Figure 15.27). Benzer's approach was to isolate a large number of independent *rII* mutants—he eventually obtained about 20,000—and to attempt to perform recombinational studies in order to produce a genetic map of this locus. While the mapping techniques discussed in Chapters 6 and 15 were of the **intergenic** type, Benzer wished to study **intragenic** recombination; that is, he wanted to establish within the *rII* locus the relative positions of all mutations.

The key to his analysis was that *rII* mutant phages, while capable of infecting and lysing *E. coli* B, could not successfully lyse a second related strain, *E. coli* K12 (λ). However, any wild-type phages could lyse both the B and the K12 strains. Benzer realized that he had the potential for a highly sensitive screening system. If he analyzed two *rII* mutations present in separate viral strains, neither alone could successfully lyse strain K12. However, simultaneous infection could result in lysis in all cases where either complementation occurred or where a recombinational event occurred between the mutations being studied. In the latter case, genetically wild-type phages are produced that result in lysis of strain K12. If no complementation or recombination occurred, all phages produced would still express the *rII* mutation and thus would not be able to lyse K12. When recombination is investigated, the resolution of this system is phenomenal. As few as one in one hundred million recombinant phages (1 in 10^8, or at a rate of 10^{-8}) may be recovered! We will illustrate the application of this screening system as we proceed with our discussion.

First the question of complementation was addressed. It was discovered that many mutations did complement one another. A thorough analysis revealed that all *rII* mutations fell into one of two complementation groups. All mutations in the same group failed to complement one another, but complemented all those in the other group. On the basis of these results, the *rII* locus was divided into two cistrons, A and B.

The comparison of complementation and no complementation is illustrated in Figure 18.2. When complementation occurs, the functional product of

FIGURE 18.2
Comparison of two *rII* mutations that either complement one another (a), or do not (b). Complementation occurs when each mutation is in a separate cistron and results in lysis of the cell. Failure to complement occurs if the two mutations are in the same cistron and results in no lysis.

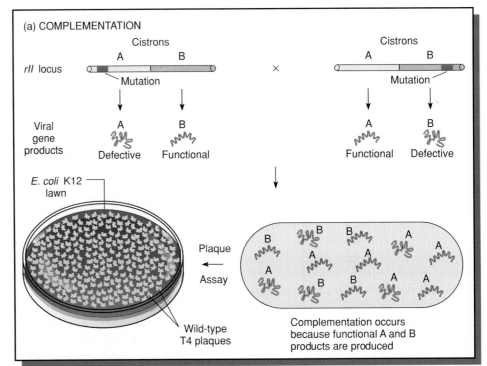

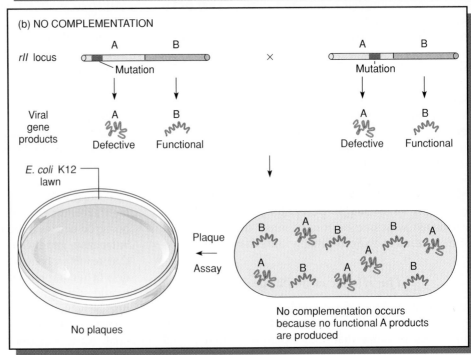

the A cistron is specified by one mutant DNA molecule, and the functional product of the B cistron is specified by the other mutant DNA molecule. Recall that during simultaneous infection, DNA from both viral strains is injected into the bacterium and replicated, forming a mixed pool of viral chromosomes. Provided that both the A and B functional products can be produced, lysis of *E. coli* K12 will occur.

Of the approximately 20,000 *rII* mutations, roughly half fell into each cistron. Benzer then set about to map the mutations within each one. For example, if two *rIIA* mutants were first allowed to infect *E. coli* B in a liquid culture, and if a recombination event occurred between the mutational sites in the A cistron (Figure 18.3), then rare wild-type progeny viruses would be produced. If samples of the progeny viruses from such an experiment were then plated on *E. coli* K12, only the recombinants would lyse the bacteria and produce plaques. The total number of nonrecombinant progeny viruses could also be determined by plating samples on *E. coli* B. The percentage of recombinants can be determined using these two figures. As in eukaryotic mapping experiments, the frequency of recombination can be taken as an estimate of the distance between the two mutations within the cistron.

For example, if the number of recombinants is equal to 4×10^3/ml, and the total number of progeny is 8×10^9/ml, then the frequency of recombination between the two mutants is

$$2\left(\frac{4 \times 10^3}{8 \times 10^9}\right) = 2(0.5 \times 10^{-6})$$
$$= 10^{-6}$$
$$= 0.000001$$

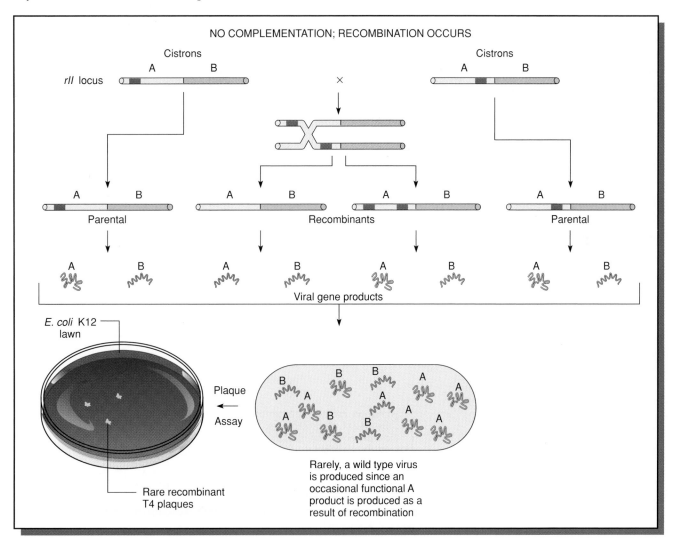

FIGURE 18.3
Production of rare wild-type phages as a result of recombination between the chromosomes of two strains, each with a mutation in the A cistron.

Multiplying by 2 is necessary because if each recombinant event yields reciprocal products, only one of them—the wild type—is detected.

Deletion Testing of the *rII* Locus

While the selective system of recombination just described was available to map mutations within each cistron, testing 1000 mutants two at a time in all combinations would have required millions of experiments. Fortunately, Benzer eliminated this step because he discovered that some of the *rII* mutations were in reality **deletions** of small parts of each cistron. That is, the genetic changes giving rise to the *rII* properties were not point mutations involving a single nucleotide change, but instead were due to the loss or deletion of a variable number of nucleotides within the cistron. These deletions could be identified by their failure to revert to wild type. Doing so is a characteristic of point mutations. Furthermore, and this is the key point, it was observed that a deletion, when tested during simultaneous infection with a point mutation located in the deleted part of the same cistron, never yielded any wild-type recombinants. The basis for the failure to do so is illustrated in Figure 18.4. Thus, a method was available that could quickly but roughly localize any mutation, provided it was contained within a region covered by a deletion.

As shown in Figure 18.5, this second property served as the basis for localization of each mutation.

Seven overlapping deletions (shown in color) covered variable portions of the A and B regions and were used for initial screening of the point mutations. Each point mutation was ultimately assigned to an area of the cistron corresponding to one specific deletion. Then, further deletions within each of the seven areas were used to localize or map each *rII* point mutation more specifically. Remember that in each case, a point mutation is localized in the area of a deletion when it fails to give rise to any wild-type recombinants.

After several years of work, Benzer produced a genetic map of the two cistrons composing the *rII* locus of phage T4 (Figure 18.6). Of the 20,000 mutations analyzed, 307 distinct sites within this gene had been mapped in relation to one another. Many mutations fell into the same site. An area containing many mutations was designated as a **hot spot**. It appeared that such an area was more susceptible to mutation than areas where only one or a few mutations are found. Additionally, Benzer found areas within the cistrons where no mutations were localized. He estimated that as many as 200 recombinational units had not been localized by his studies.

The significance of Benzer's work is his application of genetic technologies to the analysis of what had previously been considered an abstract unit—the gene. Benzer demonstrated that a gene is not an indivisible particle but instead consists of mutational and recombinational units. His research was per-

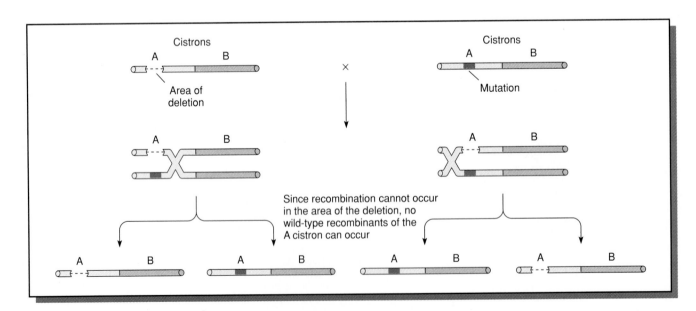

FIGURE 18.4
Demonstration that recombination between a phage chromosome with a deletion in the A cistron and another with a mutation overlapped by that deletion cannot yield a chromosome with wild-type A and B cistrons.

formed shortly after the publication of Watson and Crick's work on DNA and some time before the genetic code was unraveled in the early 1960s. Thus, his analysis is considered one of the classic examples of genetic experimentation.

Genetic Analysis of Eukaryotic Genes

Fine structure analysis of eukaryotic genes is much more difficult to perform than that of prokaryotic genes. Screening procedures to detect rare recombinants make such analysis much more complex. Nevertheless, studies even before Benzer's work had demonstrated that recombination occurs within regions thought to represent single genes. These studies, performed in *Drosophila*, took advantage of the large number of progeny produced by this organism in relatively short periods of time.

Although many genes in *Drosophila* have been studied, we will focus on three loci that have yielded somewhat different results. The first, the X-linked *lozenge* (*lz*) locus, has been thoroughly investigated by Melvin M. Green and K. C. Green. About 20 recessive alleles were examined. The most characteristic phenotype produced by alleles involves eye morphology and pigmentation.

In a complementation test, most alleles fail to complement one another in the *trans* position, as expected. However, when various combinations of two alleles present in the *trans* position are tested for recombination, certain pairs recombine at a low but discernible frequency, while others do not. When the

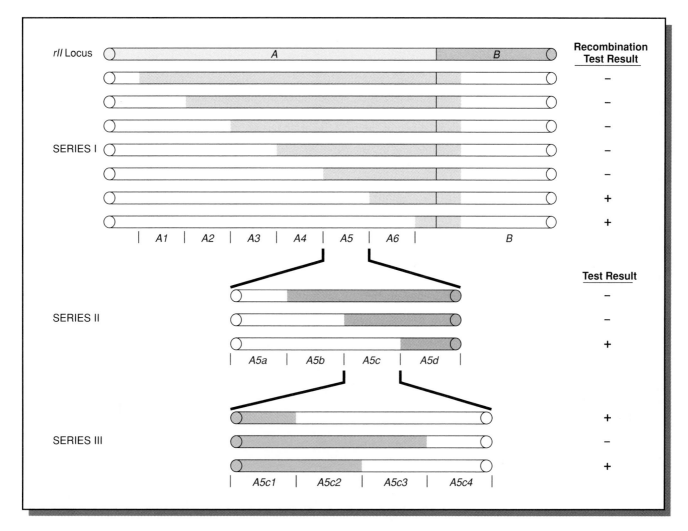

FIGURE 18.5
Three series of overlapping deletions in the *rII* locus used to localize the position of an *r* mutation. For example, if a mutant strain tested against each deletion (shown in color) in Series I for the production of recombinant wild-type progeny shows the results at the right (+ or −), the mutation must be in segment A5. In Series II, the mutation is further narrowed to segment A5c, and in Series III to segment A5c3.

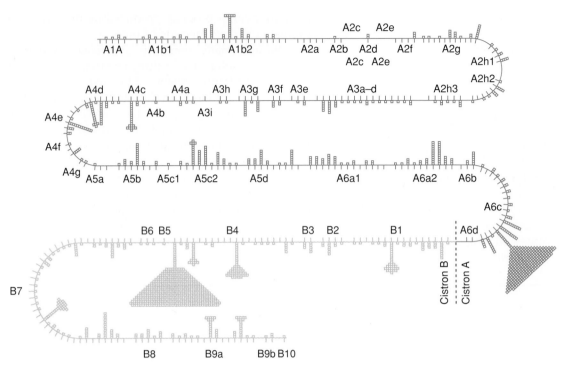

FIGURE 18.6

A partial map of mutations in the A and B cistrons of the *rII* locus of phage T4. Each square represents an independently isolated mutation.

results of testing all combinations were sorted out, each allele could be placed in one of four groups. Alleles would not recombine with others in the same group, but would recombine with those in the other groups (Figure 18.7).

On this basis, the *lozenge* locus was subdivided into four subloci and was deemed a **complex locus,** much the way in which the *rII* locus in phage T4 was subdivided into two genetically distinct regions. Recombination, illustrated in Figure 18.8, occurred between 0.03 and 0.09 percent of the time, establishing the relative distances between the subloci. Findings such as these led to the concept of **pseudoalleles,** which test positively for allelism in complementation studies but between which recombination

can occur. All those in the same sublocus are considered true alleles, but any two in different subloci are considered pseudoalleles. Edward B. Lewis has suggested that in complex loci, each sublocus is responsible for an essential step in the synthesis of the final gene product and that these steps must occur sequentially in the order established by the subloci along the chromosome. Whether or not this is true must await more detailed biochemical analyses of the DNA making up complex loci.

The second locus to be discussed is *rosy (ry)*, located on chromosome 3 of *Drosophila melanogaster*. It has been studied extensively by Arthur Chovnick and his many colleagues. Unlike *lozenge, rosy* is not considered a complex locus. Detailed analysis of *rosy*

FIGURE 18.7

Various alleles contained in the subloci of the complex *lozenge* locus of *Drosophila melanogaster*.

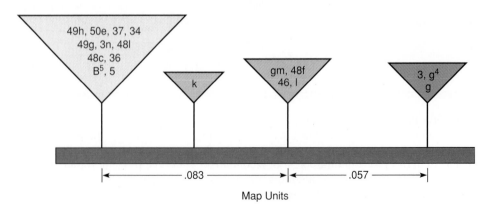

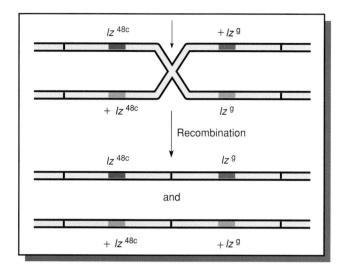

FIGURE 18.8
Intragenic recombination between subloci within the *lozenge* locus.

mutants is possible because a unique screening technique is available to detect wild-type recombinants. The locus contains the structural gene coding for the enzyme **xanthine dehydrogenase (XDH)**. Recessive mutant flies demonstrate a rosy eye color and are missing the pigment isoxanthopterin. Mutant larvae are sensitive to the toxic effects of excess purines, which they cannot metabolize in the absence of XDH activity.

Thus, if a suitable cross has been made involving two different *ry* alleles in the *trans* position—where only recombinants are wild type—the recombinants may be detected by adding excess purines to the

medium during larval development. Larvae containing two *ry* alleles fail to survive, but recombinants with at least one *ry$^+$* gene sequence can counteract the lethal effects of the purines because of partial XDH activity. Such a cross is illustrated in Figure 18.9.

Recombinational analysis of many *ry* alleles has established that the *ry* locus is considerably smaller than the *lz* locus. While *lozenge* consists of 0.140 recombinational unit, *rosy* consists of only 0.005 unit (Figure 18.10). It has been estimated that the resolution of fine structure analysis in this investigation is nearly comparable to that attained in the *rII* locus. The organization of the *ry* locus appears to be similar to that of each *rII* cistron.

The final locus of interest is that called the *bithorax* gene complex in *Drosophila*. Mutations of this gene complex alter the development of the thorax and abdominal regions of the fly's body. Those affecting the thorax are considered to be part of the *Ultrabithorax* domain of this complex. This domain has been well studied by Edward Lewis and others. Genetic testing has established several complementation groups. Mutations in each group display a similar phenotype, affecting a specific compartment or segment of the thorax. For example, the *bithorax (bx)* allele converts the thoracic segment T2 to T3 (there are three such segments). Because the wings originate in T3, a fly homozygous for this mutation has two T3 segments and forms two pairs of wings (see Figure 21.17). The *postbithorax (pbx)* allele behaves in the opposite fashion; segment T3 is converted to T2 and no wings develop.

The map of this complex is shown in Figure 18.11. Because this complex locus is of great interest in the

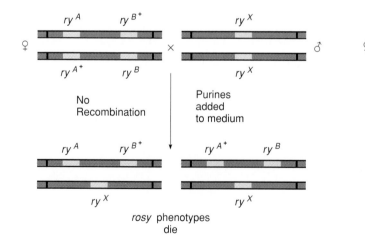
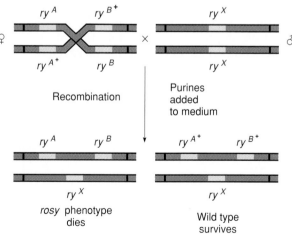

FIGURE 18.9
Selection for wild-type recombinants from a cross between two *rosy* mutants in *Drosophila melanogaster*.

FIGURE 18.10
Approximate map location of
numerous *rosy* alleles, as
determined by recombinant
mapping.

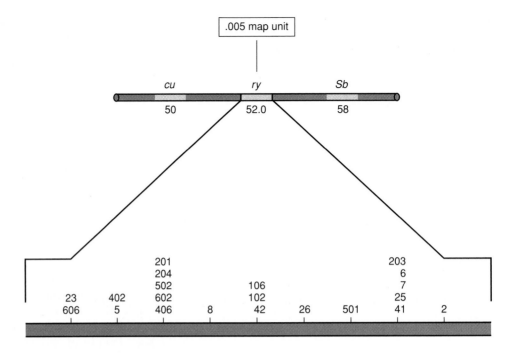

THE MOLECULAR ORGANIZATION OF THE EUKARYOTIC GENE

Prior to 1970, there was little evidence at the molecular level involving eukaryotic gene structure and organization. Nevertheless, many geneticists believed that when the eukaryotic gene could be examined directly, it might well be quite different from that of viruses and bacteria. Eukaryotes, it was reasoned, have several unique requirements distinguishing them from phages and bacteria. Primarily, cell differentiation, tissue organization, and coordinated development and function depend on genetic expression and its regulation. Do these requirements

study of developmental genetics, we will discuss it again in Chapter 22.

also depend on different modes of gene structure and organization?

When the technologies of recombinant DNA and rapid nucleotide sequencing were developed (Chapter 16), the answer to this question began to emerge. The eukaryotic gene is more complex than we could ever have imagined. In this section, we will review many findings related to gene structure and organization.

Eukaryotic Genomes and the *C* Value Paradox

The amount of DNA contained in the haploid genome of a species is called the **C value**. When such values were determined for a large variety of eukaryotic organisms (Figure 18.12), several trends were apparent. The most notable trends are that:

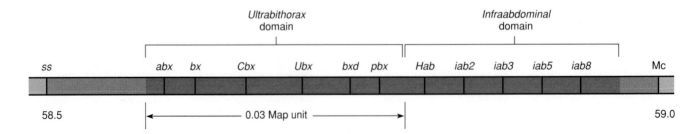

FIGURE 18.11
The *bithorax* complex in *Drosophila* consists of the *Ultrabithorax* and *Infraabdominal* domains. This complex is located on chromosome 3 and is flanked by the unrelated mutations *spineless (ss)* at locus 58.5 and *microcephalus (Mc)* at locus 59.0.

FIGURE 18.12

The range of DNA content of the haploid genome of many representative groups of organisms.

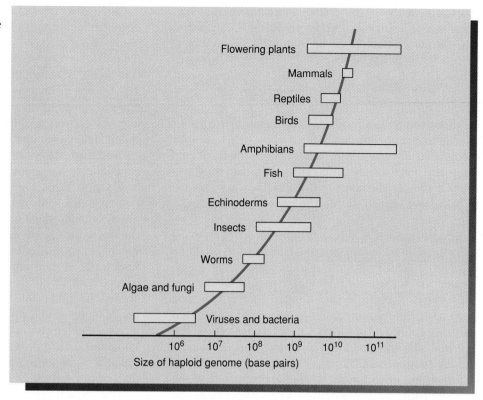

1 Eukaryotes contain substantially more DNA in their genomes than viruses or prokaryotes and exhibit a wide variation between organisms.

2 Evolutionary divergence has been accompanied by increased amounts of DNA. In many major phylogenetic groups, more advanced forms generally contain more DNA than less advanced forms.

It might be argued that increasing C values are simply the result of a greater need for increased amounts and varieties of gene products in more advanced organisms. However, several observations make this explanation unacceptable. First, the increases are very dramatic. While viruses and bacteria contain 10^4 to 10^6 nucleotide pairs in their single chromosomes, eukaryotes contain 10^7 to 10^{11} nucleotide pairs in each set of chromosomes. Does a eukaryote require 100,000 more genes, as these figures might suggest? Second, closely related organisms with the same degree of complexity in body form and tissue and organ types often vary tenfold in DNA content. Third, amphibians and flowering plants, which vary as much as 100-fold within their taxonomic classes, often contain much more DNA than other more recently evolved eukaryotes. In spite of this observation, there is no correlation between increased genome size and the morphological complexity of these groups.

Therefore, it is doubtful that the development of greater complexity during evolution can account for the amount of DNA found in eukaryotic genomes. This conclusion is the basis of what has been called the **C value paradox**. Excess DNA is present that does not seem to be essential to the development or evolutionary divergence of eukaryotes. The paradox raises the question of whether all or most of the excess DNA codes for proteins. If it does not, as the preceding arguments suggest, are eukaryotic genomes composed mainly of noncoding nucleotide sequences? Is such DNA vital to the organisms carrying it in their genomes, or is it simply excess DNA that has somehow accumulated during evolution and has no function?

Recent studies of gene structure are beginning to unravel these mysteries. Some insights have been gained from findings presented earlier in the text. For example, we have already established that noncoding sequences are indeed present in the genomes of higher organisms in the form of **repetitive DNA** (Chapters 9 and 17). Recall that there are two major classes—highly repetitive sequences and moderately or middle repetitive sequences. Can these account for the excess DNA?

To answer this question, consider the human genome. Only about 5 percent of human DNA is highly repetitive. Some of this is in the form of satellite DNA located at heterochromatic regions associated with the centromere and telomeres. Another major source of highly repetitive DNA is the **Alu family** characteristic of mammals. In humans, there are about 300,000 copies of a repeated sequence of about 300 base pairs. These are interspersed throughout the genome and constitute roughly 3 percent of the total DNA.

These sequences received their name by virtue of the presence of a common nucleotide sequence recognized by the *Alu I* restriction endonuclease. They are particularly interesting because they are **mobile genetic elements** and, thus, they are capable of being transposed within the genome (see Chapter 20).

Middle repetitive DNA comprises only 15 to 30 percent of the human genome. This fraction contains duplicated genes as well as nontranscribable DNA sequences interspersed between unique or single-copy genes. Finally, it is estimated that only 1 to 2 percent of the human genome consists of genes that encode proteins. This leaves over 60 percent of the genome still unaccounted for, much of which is unique sequence DNA.

This is not uncommon. While the proportion of the genome consisting of repetitive DNA varies between organisms, one feature seems to be shared. Only a small part of the genome codes for proteins. For example, sea urchins contain an estimated 20,000 to 30,000 genes coding for proteins, which occupy less than 10 percent of the genome. In *Drosophila*, only 5 to 10 percent of the genome is occupied by genes coding for proteins. Similar estimates have been made for other organisms. We must conclude that while there is sufficient DNA to code for over one-half million genes in many eukaryotic organisms, the majority of DNA does not serve this function.

Eukaryotic Gene Structure

One of the first insights into regions of DNA that are neither repetitive nor directly encode proteins has come from the study of RNA. This research has provided detailed knowledge of eukaryotic gene structure.

The genetic code, as deciphered and shown in Figure 11.8, is written in the ribonucleotide sequence of mRNA. This information originated, of course, in the sense strand of DNA, where complementary sequences of deoxyribonucleotides exist. In bacteria, the relationship between DNA and RNA appears to be quite direct. The DNA base sequence is transcribed into an mRNA sequence, which is then translated into an amino acid sequence according to the genetic code.

However, in eukaryotes the situation is very different from that in bacteria. It has been found that many internal base sequences of genes, called **introns,** never appear in the mature mRNAs that are translated. The remaining areas of each gene, which are reflected in the amino acid sequence of the encoded protein, are called **exons**. While the entire gene is transcribed into a large precursor mRNA, called **heterogeneous RNA (hnRNA),** the introns are excised and the exons spliced back together prior to translation. These findings were presented in detail in Chapter 12.

Given this information about internal gene sequences, two questions can now be asked. First, do the nucleotide sequences found in introns solve the C value paradox? That is, do introns account for the remainder of the noncoding sequences beyond those of repetitive DNA? The answer is emphatically no! As shown in Table 18.1, the presence of introns may increase by four or five times our estimate of the nucleotide length occupied by single copy genes. However, since this estimate is seldom above 10 percent of the genome and usually much less, a great deal of noncoding DNA is still unaccounted for.

The second question concerns our concept of the gene. Can we any longer provide an accurate definition for it? In his 1980 book *Gene Expression—2, The Eukaryotic Chromosome*, Benjamin Lewin has summarized beautifully this dilemma:

> More than a hundred years of work has led to the concept that the gene is a contiguous region of DNA that is colinear with its protein product. This definition encompasses the complementation test, which formally defines the gene as a unit all of whose parts must be present on one chromosome; and it has found fulfillment in the detailed molecular studies of the past two decades demonstrating that DNA first is transcribed into RNA and then is translated into protein by reading the triplets of the genetic code. Colinearity appeared to be preserved through the stages of gene expression. The recent discovery that eukaryotic genes contain intervening sequences that are not represented in the mRNA from which the protein product is synthesized has caused this concept now to be discarded. . . .This has made it difficult, if not impossible, to arrive at a single satisfactory definition of the gene.

It is clear then that the presence of introns does not entirely resolve the C value paradox and, furthermore, has created a dilemma in arriving at a universal definition of the gene. Nevertheless, studies of introns and exons have expanded considerably our knowledge of eukaryotic gene structure.

Beyond consideration of introns, we next must examine the noncoding flanking regions in order to

TABLE 18.1
A comparison of mRNAs and the initial transcripts from which they are derived.

Gene	mRNA Size	Minimum Transcript Length	Ratio
Rabbit β-globin	589	1295	2.2
Mouse β^{maj}-globin	620	1382	2.2
Mouse β^{min}-globin	575	1275	2.2
Mouse α-globin	585	850	1.5
Rat insulin I	443	562	1.3
Rat insulin II	443	1061	2.4
Chick ovalbumin	1859	7500	4.0
Chick ovomucoid	883	5600	6.3
Chick lysozyme	620	3700	6.0

SOURCE: From Lewin, 1980, p. 825.

complete our review of the eukaryotic gene. In the discussion of transcription in Chapter 12, we introduced the concept of the **promoter region**. Found "upstream" from the coding sequence that initiates the 5' end of mRNA during transcription, the promoter region is responsible for the initial RNA polymerase binding in prokaryotes. A similar but more extensive region, called the **5' flanking region**, precedes eukaryotic genes. While we will return to these topics in Chapter 20, we shall consider these regions here as we complete the picture of the eukaryotic gene, summarized in Figure 18.13.

Three areas of the 5' flanking region appear to be essential to efficient transcription and have been investigated extensively. The first is called the **Goldberg-Hogness** or **TATA box**. Found at a point about 30 nucleotides upstream from the point where transcription begins, this short region consists of the sequence TATXAX, where X is either T or A. The sequence has been found in most but not all eukaryotic genes examined. The TATA box is analogous to the Pribnow sequence found associated with prokaryotic gene promoters.

The TATA box has generated much excitement because of its possible involvement in the initiation of transcription. In many instances, both mutations and deletions in this region severely reduce or eliminate the *in vitro* transcription by RNA polymerase II of the associated gene. This is the case in studies of the conalbumin and ovalbumin genes of chickens, the beta-globin gene in rabbits, and some genes in animal viruses. Other evidence suggests that the region is only responsible for fixing the site of initiation of transcription *in vivo*.

The second noncoding region of interest is found farther upstream from the TATA box. In different genes studies, regions anywhere from 50 to 500 nucleotides upstream appear also to modulate transcription. These regions have been located on the basis of the effects on transcription of their deletion. The loss of various noncoding regions appears to drastically reduce *in vivo* transcription. Some noncoding regions, such as those found associated with globin and SV40 genes, are about 50 to 100 nucleotides from the TATA sequence. Others, such as those associated with the sea urchin H2A gene and the *Drosophila* glue protein gene, are 200 to 500 nucleotides upstream. Frequently, the sequence **CCAAT** (called the CAAT box) is part of this noncoding region.

The third noncoding region is represented by elements called **enhancers**. While their location may vary, they often may be found even farther upstream. We will return to a discussion of these elements in Chapter 20.

Since these regulatory regions are not transcribed, but instead seem to facilitate template binding, they are most likely important to the conformation of chromatin structure during gene activity. Such configurations are thought to be essential if DNA is to

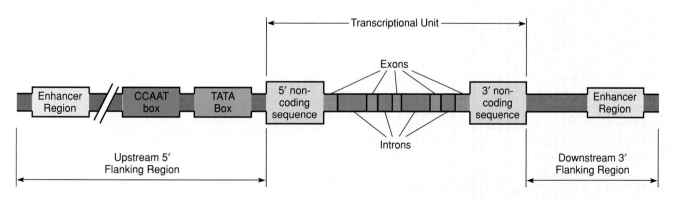

FIGURE 18.13
Modern concept of the eukaryotic gene.

become accessible to the polymerase enzyme and initiation proteins involved in transcription.

The discovery of the TATA box and other flanking sequences has extended our knowledge of gene structure considerably. As shown in Figure 18.13, a more concise view of the typical eukaryotic gene is gradually emerging. This model can be expected to be modified and further extended as research continues.

Exon Shuffling and Protein Domains

The discovery of introns and exons is, at first glance, most puzzling. Why, during the evolution of eukaryotes, would their genes acquire noncoding introns between the coding portions of genes whose presence requires precision splicing following transcription and whose replication requires a substantial expenditure of energy? While we do not yet have a solid answer, important clues are emerging primarily as a result of our ability to rapidly determine the nucleotide sequence of many genes. With the knowledge of the sequence of the exons of genes, the amino acid sequences of the corresponding proteins can be accurately predicted. When composite data bases are established and compared by computer analysis, the uniqueness of genes can be examined. As a result of this approach, some very interesting findings have been forthcoming, as first recognized by Walter Gilbert in 1977. He suggested that genes in higher organisms may consist of collections of exons present in ancestral genes that are brought together through recombination during the course of evolution. He proposed that exons are modular in the sense that each might encode a functional domain in the structure of a protein. For example, a particular amino acid sequence encoded by a single exon might always create a specific type of fold in a protein. This **protein domain** may impart a unique but characteristic function to all molecules containing such a fold. Serving as the basis of useful parts of a protein, many similar domains might be functional in a variety of proteins. In Gilbert's proposal, many exons could be "mixed and matched" to form unique genes in eukaryotes.

Three observations lend immediate support to this proposal. First, introns and exons are nearly universal in vertebrate organisms, rare in lower eukaryotes such as yeast, and with few exceptions, absent in prokaryotes. This pattern is to be expected if introns and exons are evolutionary products. Second, most exons are fairly small, averaging about 100 to 150 nucleotides and encoding 30 to 50 amino acids. This size is consistent with the production of functional domains in proteins. Third, recombinational events that lead to **exon shuffling** would be expected to occur within areas of genes represented by introns. Since introns are free to accumulate mutations without harm to the organism, recombinational events would tend to further randomize their nucleotide sequences. Over extended evolutionary periods, diverse sequences would tend to accumulate. This is, in fact, what is observed. Introns range from 50 to 20,000 bases and exhibit fairly random base sequences.

Since 1977, a vast research effort has been directed toward the analysis of split genes. In 1985, direct evidence in favor of Gilbert's proposal of exon modules was presented. Analysis was made of the human gene encoding the membrane receptor for low density lipoproteins (LDL). This **LDL receptor protein** is essential to the transport of plasma cholesterol into the cell. It mediates **endocytosis** and is expected to have numerous functional domains. These include the capability of this protein to bind specifically to the LDL substrates and to interact with other proteins at different levels of the membrane during transport across it. In addition, this receptor molecule is modified post-translationally by the addition of a carbohydrate; a domain must exist that links to this carbohydrate.

Detailed analysis of the gene encoding this protein supports the concept of exon modules and their shuffling during evolution. The gene is quite large—45,000 nucleotides—and contains 18 exons.

FIGURE 18.14

A comparison of the 18 exons making up the LDL receptor protein gene and their organization into five functional domains and one signal sequence.

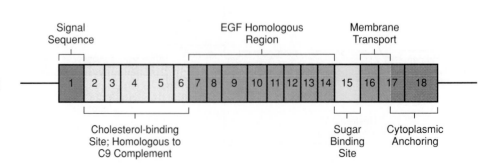

LDL RECEPTOR PROTEIN GENE

These represent only slightly less than 2600 nucleotides. These exons are related to the functional domains of the protein *and* appear to have been recruited from other genes during evolution.

Figure 18.14 illustrates these relationships. The first exon encodes a signal sequence that is removed before the LDL receptor becomes part of the membrane. The next five exons represent the domain specifying the binding site for cholesterol. This domain is made up of a 40-amino-acid sequence repeated seven times. This amino acid sequence closely resembles one found in the C9 component of the immunological molecule complement, which is found in human plasma. The next domain consists of a sequence of 400 amino acids bearing a striking homology to the mouse peptide hormone **epidermal growth factor (EGF)**. This region is encoded by eight exons and contains three repetitive sequences of 40 amino acids. A similar sequence is also found in three blood-clotting proteins. The fifteenth exon specifies the domain for the post-translational addition of the carbohydrate, while the remaining two exons specify regions of the protein that are part of the membrane, anchoring the receptor to specific sites called coated pits on the cell surface.

These observations are fairly compelling in favor of the theory of exon shuffling during evolution. Certainly, there is no disagreement concerning the concept of protein domains providing three-dimensional surfaces responsible for specific interactions with a variety of substrate molecules. For example, the central exon of both the myoglobin and hemoglobin genes specifies a domain that binds to the iron-containing heme portion of both molecules. It is this interaction that creates the molecular conformation conferring the ability to reversibly bind oxygen. We will also see an example of a related phenomenon in our discussion of immunoglobulin diversity in Chapter 20.

The phenomenon of shared protein domains is not restricted to eukaryotes. Examination of the amino acid sequences of a series of regulatory molecules (to be discussed in the next chapter) illustrates this well. The bacterial catabolite activator protein (CAP), the *lac* repressor protein from *E. coli*, and the *cro* repressor molecule encoded by the bacteriophage lambda (λ) all share a common property: they regulate genetic activities by binding to DNA. All three molecules share a protein domain responsible for this interaction.

Further research will undoubtedly clarify this theory. We conclude our discussion of this section by raising a very interesting question. We know that the presence of introns is primarily a eukaryotic phenomenon. In prokaryotes, could introns at one time have been present, but subsequently lost during evolution? We know that these organisms have evolved strategies that minimize expenditure of energy and emphasize rapid rates of reproduction. Could they have streamlined their genome by eliminating introns? This is an intriguing, but difficult inquiry that is likely to be debated for years to come.

MULTIGENE FAMILIES

The study of still another level of eukaryotic gene organization has contributed to our understanding of the distribution of genetic units within the genome. This information further provides valuable insights into the *C* value paradox. This level consists of the organization of **multigene families**. Two types of families of genes will be considered.

First, we will examine cases where groups of genes encode very similar, but not identical, polypeptide chains that become part of proteins which are very closely related in function. The globin and the histone gene families are examples that provide many insights into the organization of the genome. Then, we will consider the case of identical, tandemly repeated genes, as represented by ribosomal RNA genes.

While here we will stress the organization of the members of the families as well as their distribution in the genome, we will return to a consideration of the evolutionary origin of gene families in Chapter 26. Usually, nonidentical members of the same family contain sufficient nucleotide sequence homology to suggest that they arose through the process of gene duplication and then diverged during evolution.

The Alpha- and Beta-Globin Families

The human **alpha- and beta-globin gene families** are two of the most extensive and best characterized groups. The alpha family resides on the short arm of chromosome 16 and contains five genes, while the beta family resides on the short arm of chromosome 11 and contains six genes. Members of both families show homology to one another, but not nearly to the extent members within the same family exhibit. Members of both families encode globin polypeptides that combine into a single tetrameric molecule, which interacts with a heme group to reversibly bind to oxygen. And, within each group of genes, members are coordinately turned on and off during the embryonic, fetal, and adult stages of development in the precise order in which they occur along the chromosome within each family.

The alpha family [Figure 18.15(a)] spans more than 30,000 nucleotide pairs (30 kb) and contains five

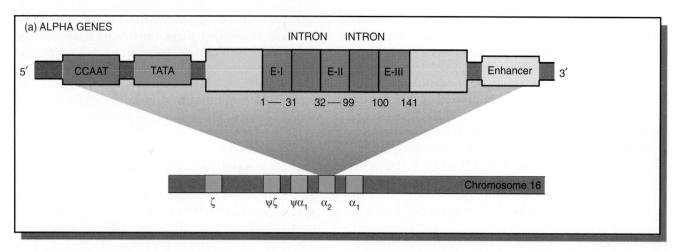

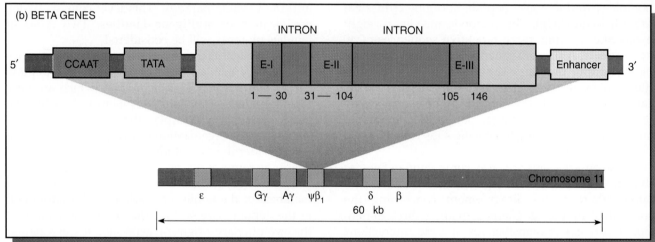

FIGURE 18.15

The comparative arrangement of the human alpha and beta gene families, depicted in parts (a) and (b), respectively. The functional globin genes are shown in red, while the pseudogenes are shaded gray. The internal organization of the functional genes from each family are also shown. All genes contain three exons (E–I, II, and III) and two introns, as well as 5′ and 3′ noncoding regions. The arabic numbers under each gene designate the amino acid positions encoded by each exon.

genes: the zeta gene (ζ), expressed only in the embryonic stage, two nonfunctional **pseudogenes,** and two copies of the alpha (α) gene, expressed during the fetal and adult stages. Pseudogenes, found in both families, are similar in sequence to the other genes in their family, but contain significant nucleotide substitutions, deletions, and duplications that prevent their transcription. The first in the alpha family shows greater homology to the zeta gene, while the second shows greater homology to the alpha gene. Pseudogenes are designated by the prefix ψ (the Greek letter psi), followed by the symbol of the gene they most resemble. Thus, the designation ψα1 indicates a pseudogene of the first alpha gene.

While we will do a more exact analysis of the beta family, examination of the organization of the mem-

bers of the alpha family as well as their intragenic distribution of introns and exons, as shown in Figure 18.15(a), reveals several interesting features. First, only a small portion of the chromosomal region housing the entire family is occupied by the three functional genes. The vast majority of the region consists of intergenic regions. Second, all genes in this family contain two introns, at precisely the same positions within the genes. The nucleotide sequences within corresponding exons is nearly identical when the zeta and alpha genes are compared. Both encode polypeptide chains that are 141 amino acids long. However, the intron sequences are extremely divergent, even though they are about the same size. Significantly, a high percentage of the nucleotide sequence within each gene is contained in these noncoding introns. The above information bears on both the structural organization of a gene family as well as on our earlier discussion of the C value paradox.

The beta-globin family in humans is even more extensive, containing six genes and occupying a 60-kb sequence of DNA [Figure 18.15(b)]. As with the alpha family, sequence analysis has been dependent on restriction endonuclease digestion studies as well as cDNA cloning technology. And, as with the alpha family, the sequence of genes along the chromosome parallels the order of expression during development. A pseudogene is also present.

Of the six genes, three are expressed prior to birth: the epsilon (ϵ) gene is embryonic, while two nearly identical gamma genes ($^G\gamma$ and $^A\gamma$) are fetal. The two gamma gene products vary by only a single amino acid. The two remaining functional genes, delta (δ) and beta (β), are expressed following birth. Finally, a single pseudogene, $\psi\beta1$, is present within the family.

All five functional genes encode products that are 146 amino acids in length and contain two similarly sized introns at exactly the same positions. The second intron is significantly larger than its counterpart in the functional alpha genes. These introns and their locations within the gene are contrasted in Figure 18.15.

The 3' and 5' flanking regions of each gene have also been studied. In all functional genes, both a TATA box (29 or 30 base pairs upstream) and a CCAAT region (70 to 78 base pairs upstream from the initial codon) have been observed in the 3' region. In the 5' region, globin enhancers have been identified.

Several observations of these genes pertain to the C value paradox. Only about 5 percent of the 60-kb region consists of coding sequences. The remaining 95 percent includes the introns, flanking regions, and spacer DNA found between genes. Of this percentage, only about 11 percent consists of introns. The remainder of the DNA serves no known function and consists of sequences not found elsewhere in the genome. Because it represents the majority of DNA in the cluster, we must conclude that it is either excess nonfunctional DNA *or* DNA whose function has yet to be discovered.

The Histone Gene Family

A variation on the theme established in the globin gene families is illustrated by the **histone genes**. Here, a cluster of five related but nonidentical genes, separated from each other by highly divergent nontranscribed spacer sequences, is tandemly repeated many times. As you recall, histones are positively charged (basic) proteins that interact with the nega-

tively charged phosphate groups of DNA to form the nucleosomes characteristic of the chromatin fiber. The need for histone proteins is extremely great because of the large amount of DNA in eukaryotic chromosomes. In rapidly dividing cells of some organisms, not only is the need great, but the necessary quantity must appear rapidly as one cell cycle runs into the next, time and time again, without pause to catch up with the cell's biochemical needs. To accommodate this need, evolution appears to have seized upon the duplication of the entire cluster in order to increase the number of copies of each histone gene.

Many interesting correlations exist that support this contention. Yeast cells, with much less DNA than other eukaryotes, have only two copies of each of four histone genes (they lack histone H1). Contrast this with the number of times the entire cluster is tandemly repeated in many evolutionarily advanced organisms, as displayed in Table 18.2. From 10 to 600 repeats occur. Sea urchins, which display the largest number of duplications of the cluster, are noted for their extremely rapid rate of cell division during development.

Beyond the repeat nature of the histone gene family, there are several other differences in comparison to the globin families. First, almost all histone genes lack introns, as well as the flanking region that encodes the poly A tail of the corresponding mRNA. Related to these observations, the entire cluster of genes is considerably smaller than either globin cluster. While some variation exists between organisms, due primarily to the different sizes of the noncoding spacer sequences, most clusters are less than 10 kb in size. The other interesting difference is the observation that the individual genes within the cluster are often oriented in opposite directions.

As shown in Figure 18.16, the arrangement of genes and their polarity of transcription vary in the sea urchin and *Drosophila*, two organisms where this cluster has been well studied. Polarity differences exist in other organisms studied, as well.

In mice and humans, members of this gene family

TABLE 18.2
The number of tandemly repeated histone gene clusters found in a variety of eukaryotic organisms.

Organism	Repeats
Chickens	10
Mammals	22
Xenopus	40
Drosophila	100
Sea urchins	300–600

FIGURE 18.16

The histone gene clusters in *Drosophila* and the sea urchin. The arrows indicate the polarity of transcription.

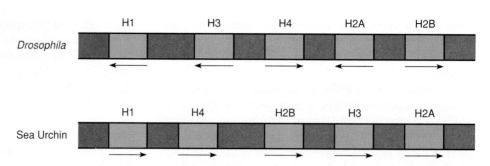

appear to be clustered but not present as a tandem array of repeat units. For example, a region of the distal tip of the long arm of the human chromosome 7 contains all members of the histone gene family, but they may be interspersed with other non-histone genes. Therefore, no single pattern of organization applies to all organisms.

Ribosomal RNA Genes: Tandem Repeats of a Gene Family

We conclude our discussion of the organization of multigene families by considering the example of **tandem repeats** of the gene family encoding the rRNA molecules present in ribosomes. This gene family, like the histone cluster, encodes separate gene products. In this case, three molecules are encoded in the following order along the chromosome: 18*S*, 5.8*S*, and 28*S* rRNAs (Figure 18.17). Unlike the histone family, the rRNA family contains substantial regions between each gene that are part of the initial transcript but that are not found in the final rRNA gene products. Thus, the initial transcript must be processed in a manner analogous to the removal of introns from the primary transcript of most mRNAs. In addition, there is a nontranscribed spacer region between each gene family unit.

The basic repeating unit that is transcribed varies in length in different organisms from almost 7 kb in yeast to 13 kb in humans (Figure 18.17). The nontranscribed spacer sequence of DNA between each repeating unit also varies in length in different organisms. In humans, this is quite large, extending some 30 kb. These spacer regions are clearly visible in Figure 12.2a, where transcription of tandem repeats has been visualized under the electron microscope. Together, each gene family unit and its accompanying spacer regions occupy a substantial portion of DNA. In the human rRNA gene family, a length of about 43 kb of DNA is utilized, over and over in the tandem repeats. This amount of DNA is as large as the smaller globin gene cluster.

Multiple copies of a single transcriptional unit encoding the rRNAs also exist in bacteria. In *E. coli*, seven dispersed copies are present, presumably to accommodate the need for the rapid production of the approximately 15,000 ribosomes found in each cell. It is not surprising, therefore, to find many more than seven copies in the larger, more complex eukaryotic cells. Yeast contains over 100, while *Xenopus*, *Drosophila*, and humans have up to 400 copies per genome. A single eukaryotic cell may have a million or more ribosomes if it is actively producing proteins.

FIGURE 18.17

Comparison of the rRNA gene family in yeast (*Saccharomyces cerevisiae*), the fruit fly (*Drosophila melanogaster*), the frog (*Xenopus laevis*), and humans. Each unit is tandemly repeated many times and is connected by a nontranscribed spacer sequence that varies in length in each organism.

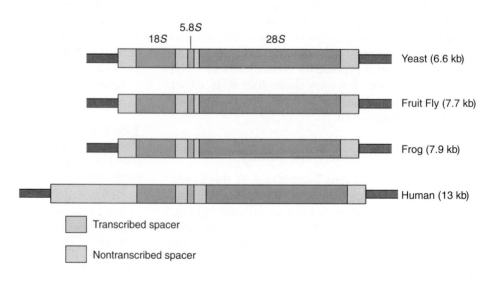

532 CHAPTER 18

If all of these genes were present in a single cluster in an organism such as humans, a substantial part of one chromosome would be occupied by them (400×45 kb = 18,000 kb or 18 million base pairs). This amount of DNA is almost five times greater than the entire *E. coli* chromosome (4×10^6 base pairs). As it turns out, they are present in clusters dispersed throughout the genome. In humans, for example, the clusters are found near the ends of chromosomes 13, 14, 15, 21, and 22. In *Drosophila*, they are present on both the X and Y chromosomes. Within the nucleus, these various gene regions are collectively referred to as the **nucleolus organizer regions,** or simply the **NORs,** because of their association with the production of ribosomes in the nucleolus.

The above information extends our knowledge and overall picture of how some genes are organized within the genome. It also provides insights into the vast amount of DNA that may be utilized in order to provide cells with one set of gene products. Examination of the same gene family in different organisms also provides information valuable to the study of evolution at the molecular level.

CHAPTER SUMMARY

1 The complementation test, which functionally defines allelism, was one of the first experimental approaches designed to elucidate gene structure and organization. Two heterozygous mutations present in the *trans* configuration that produce a mutant phenotype are considered to be alleles and to belong to the same complementation group, or cistron, a more accurate term for the smallest functional genetic unit.

2 Genetic analysis of the *rII* locus in bacteriophage T4 allowed Benzer in 1957 to study intragenic recombination. By isolating *rII* mutants, performing recombinational studies, and relying on deletion mapping, Benzer was able to locate 20,000 mutations at 307 distinct sites within the two cistrons of the *rII* locus.

3 Fine structure analysis has been undertaken in eukaryotes, but with much greater difficulty. Studies involving *Drosophila* have enhanced such analysis, and the following three loci serve as examples: *lozenge,* a complex locus divided into four subunits; *rosy,* a simple locus whose resolution is nearly as great as the *rII* locus when mapped; and *bithorax,* a complex locus divided into distinct regions, each of which specifies a genetic product affecting a specific body segment during development.

4 Pseudoalleles refer to mutations in different subunits which test positively for allelism in complementation studies, but can be separated by crossing over.

5 The *C* value paradox, made apparent during the study of the organization of eukaryotic DNA, suggests that eukaryotic organisms contain much more DNA than necessary to encode only gene products essential to the development and normal functions of an organism.

6 Detailed analysis of eukaryotic gene structure has revealed numerous categories of noncoding DNA sequences. Introns, internal noncoding sequences, are represented in the initial mRNA molecules or heterogeneous RNA but never appear in the mature mRNAs from which proteins are synthesized. This variation is accounted for by the process of splicing, whereby noncoding introns are excised while coding exons are spliced back together, giving rise to the mature mRNA transcript that may be translated.

7 The flanking regions of genes, particularly those giving rise to the 5' initiation point of mRNA, have also been analyzed. Three "upstream" regions that appear to be essential to efficient transcription are the TATA box, found about 30 nucleotides from the point of initiation of transcription; a second region located 100 to 500 nucleotides upstream; and an enhancer element.

8 Gene families, representing the final level of eukaryotic gene organization, are structurally related genes, usually clustered together, encoding functionally similar products.

9 The human alpha-globin gene family, contained in a 30-kb region of DNA, encodes five genes, two of which are nonfunctional pseudogenes. The beta-globin family occupies 60 kb and encodes six genes, including one pseudogene. In both families, the majority of the DNA consists of intergenic sequences, which include introns, flanking regions, and spacer DNA.

10 The histone family occupies a smaller DNA region and contains five genes whose polarity varies in different organisms. Unlike the globin family, the histone cluster of genes is tandemly repeated a variable number of times in different organisms.

11 Ribosomal RNA genes provide an example of tandem repeats of identical genes with the same polarity. The repeating transcribed unit as well as the nontranscribed spacer sequence between repeating units vary in length in different organisms. The repeating unit varies from almost 7 kb in yeast to 13 kb in humans.

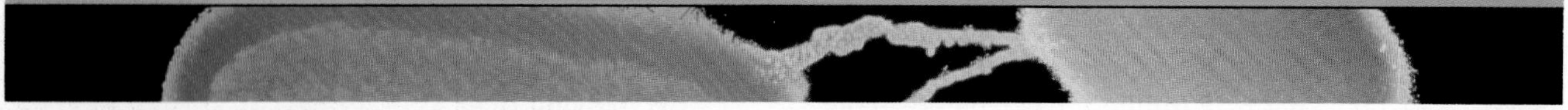

Insights and Solutions

1 In Benzer's fine structure analysis of the *rII* locus in phage T4, he was able to perform complementation testing of any pair of mutations once it was clear that the locus contained two cistrons. Complementation was assayed by simultaneously infecting *E. coli* K12 with two phage strains, each with an independent mutation, neither of which could alone lyse K12. From the following data, determine which mutations are in which cistron, assuming that mutation 1 (M-1) is in the A cistron and mutation 2 (M-2) is in the B cistron. Are there any cases where the mutation cannot be properly assigned?

Test Pair	Results
1, 2	+
1, 3	−
1, 4	−
1, 5	+
2, 3	−
2, 4	+
2, 5	−

ANSWER: M-1 and M-5 complement one another and are therefore not in the same cistron. Thus, M-5 must be in the B cistron. M-2 and M-4 complement one another. Using the same reasoning, M-4 is not with M-2 and therefore is in the A cistron.

M-3 fails to complement either M-1 or M-2 and would not seem to be in either cistron. One explanation is that the physical basis of M-3 somehow overlaps both the A and B cistron. It might be a double mutation with one in each cistron. It might also be a deletion that overlaps both cistrons, making it impossible for it to complement either M-1 or M-2.

2 Another mutation, M-6, was tested with the results shown below. Draw all possible conclusions about M-6.

Test Pair	Results
1, 6	+
2, 6	−
3, 6	−
4, 6	+
5, 6	−

ANSWER: These results are consistent with assigning M-6 to the B cistron.

3 Recombination testing was then performed for M-2, M-5, and M-6 in order to map the B cistron. Recombination analysis using both *E. coli* B and K12 showed that recombination occurred between M-2 and M-5 and between M-5 and M-6, but not between M-2 and M-6. Why not?

ANSWER: Either M-2 and M-6 represent identical mutations, or one of them may be a deletion that overlaps the other, but not M-5. Furthermore, the data cannot rule out the possibility that both are deletions.

4 In recombination studies, what is the significance of the value determined by calculating growth on the K12 vs. the B strains of *E. coli* following simultaneous infection in *E. coli*? Which value is always greater?

ANSWER: By performing plaque analysis on *E. coli* B, where wild-type and mutant phage are both lytic, the total number of phage per ml can be determined. Since almost all cells are *rII* mutants of one type or another, this value is much larger. To avoid total lysis of the plate, extensive dilutions are necessary. On K12, *rII* mutations will not grow, but wild-type phage will. Since wild-type phage are the rare recombinants, there are relatively few of them and extensive dilution is not required.

PROBLEMS AND DISCUSSION QUESTIONS

1 Many independent mutants of a haploid organism that cannot synthesize substance X are isolated. In its life cycle, the organism passes through a brief diploid phase as a result of fusion of haploid cells. During this stage, two mutants can be tested for complementation because the mutations are present in the *trans* condition. The following data result when five mutations are tested. Determine the number of complementation groups and which mutations are in each one.

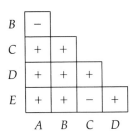

2 Three mutations were discovered with a similar phenotype. In complementation testing, mutants 1 and 2 were shown to be allelic, but neither was allelic to mutant 3. Complete the following table indicating + for positive complementation and − for negative complementation.

	Condition	
Mutants	cis	trans
1, 2		
1, 3		
2, 3		

3 In complementation studies of the *rII* locus of phage T4, three groups of three different mutations were tested. For each group, only two combinations were tested. On the basis of each set of data, predict the results of the third experiment.

Group A	Group B	Group C
d × *e*—lysis	*g* × *h*—no lysis	*j* × *k*—lysis
d × *f*—no lysis	*g* × *i*—no lysis	*j* × *l*—lysis
e × *f*—?	*h* × *i*—?	*k* × *l*—?

4 In an analysis of other *rII* mutants, complementation testing yielded the following results:

Mutants	Results (lysis)
1, 2	+
1, 3	+
1, 4	−
1, 5	−

(a) Predict the results of testing 2 and 3, 2 and 4, and 3 and 4 together.

(b) If further testing yielded the following results, what would you conclude about mutant 5?

Mutants	Results
2, 5	−
3, 5	−
4, 5	−

(c) Following mixed infection of mutants 2 and 3 on *E. coli* B, progeny virus were plated in a series of dilutions on both *E. coli* B and K12 with the following results. What is the recombination frequency between the two mutants?

Strain plated	Dilution	Colonies
E. coli B	10^{-5}	2
E. coli K12	10^{-1}	5

(d) Another mutation, 6, was then tested in relation to mutations 1 through 5. In initial testing, mutant 6 complemented mutants 2 and 3. In recombination testing with 1, 4, and 5, mutant 6 yielded recombinants with 1 and 5, but not with 4. What can you conclude about mutation 6?

5 Define *pseudoalleles*. Why was the analysis of the *rosy* locus easier to accomplish than that of the *lozenge* locus in *Drosophila melanogaster*?

6 Describe what is meant by exon shuffling.

7 The β-globin gene family consists of 60 kb of DNA yet only 5 percent of the DNA encodes β-globin gene products. Account for the remaining 95 percent of the DNA.

8 Compare the histone gene family with the globin gene families.

9 Describe the rRNA gene family in eukaryotes. What accounts for the major difference in the size of tandem repeats in different organisms?

10 "In the 1990s, the C value paradox is not so paradoxical." Agree or disagree with this statement, and support your position.

SELECTED READINGS

BENZER, S. 1955. Fine structure of a genetic region in bacteriophage. *Proc. Natl. Acad. Sci.* 42: 344–54.

———. 1961. On the topography of the genetic fine structure. *Proc. Natl. Acad. Sci.* 47: 403–15.

CHAMBON, P. 1981. Split genes. *Scient. Amer.* (May) 244: 60–71.

CHOVNICK, A. 1989. Perspectives: Intragenic recombination in *Drosophila*: The *rosy* locus. *Genetics* 123: 621–24.

CHOVNICK, A., GELBART, W., McCARRON, M. 1977. Organization of the *rosy* locus in *Drosophila melanogaster*. *Cell* 11: 1–10.

CRICK, F. 1979. Split genes and RNA splicing. *Science* 204: 264–71.

DARNELL, J. E. 1978. Implications of RNA: RNA splicing in the evolution of eukaryotic cells. *Science* 202: 1257–60.

FRITSCH, E. F., LAWN, R. M., and MANIATIS, T. 1980. Molecular cloning and characterization of the human β-like globin gene cluster. *Cell* 19: 959–72.

GELBART, W., et al. 1976. Extension of the limits of the XDH structural element in *Drosophila melanogaster*. *Genetics* 84: 211–32.

GREEN, M. M., and GREEN, K. C. 1949. Crossing-over between alleles at the *lozenge* locus in *Drosophila melanogaster*. *Proc. Natl. Acad. Sci.* 48: 586–91.

HENTSCHEL, C. C., and BIRNSTIEL, M. L. 1981. The organization and expression of histone gene families. *Cell* 25: 301–13.

LEWIS, E. B. 1948. Pseudoallelism and gene evolution. *Cold Spr. Harb. Symp.* 16: 159–74.

MANIATIS, T., et al. 1980. The molecular genetics of human hemoglobins. *Ann. Rev. Genet.* 14: 145–78.

SCHMID, C. W., and JELINEK, W. R. 1982. The Alu family of dispersed repetitive sequences. *Science* 216: 1065–70.

SÜDHOF, T. C., et al. 1985. The LDL receptor gene: A mosaic of exons shared with different proteins. *Science* 228: 815–22.

TESSMAN, I. 1965. Genetic ultrafine structure in the T4 *rII* region. *Genetics* 51: 63–75.

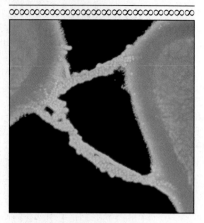

19

Genetic Regulation in Bacteria and Bacteriophages

∞∞∞∞∞∞∞∞∞∞∞∞∞∞∞∞∞∞∞∞∞∞∞∞∞∞∞∞∞∞

CHAPTER CONCEPTS

Efficient expression of genetic information is dependent on regulatory mechanisms that either activate or repress gene activity. In bacteria, mechanisms exist that regulate genes in ways that meet metabolic needs of the cell. Gene regulation in bacteriophages determines the viruses' mode of existence within the bacterial host. In bacteria and their phages, these mechanisms are responses to the prevailing cellular and extracellular conditions.

In earlier chapters, we established how DNA is organized into genes, how genes store information, and how this information can be expressed. But, is the information in any gene always expressed, or can genes be turned on and off, depending on the needs of the cell? On theoretical grounds, it is rather easy to argue that the answer to the latter question is yes. Bacterial cells in varying environments may or may not require certain gene products in order to metabolize varying substrates. In mammals, cells of the pancreas do not synthesize hemoglobin and retinal cells do not synthesize insulin. If every cell fully expressed every gene, cellular phenotypes would not vary. Furthermore, organisms would consume phenomenal amounts of energy in the processes of transcription and translation to produce unnecessary gene products.

Thus, it seems clear that some form of **genetic regulation** must exist in organisms. Detailed analysis of proteins in *Escherichia coli* has shown that for 4000 or so polypeptide chains encoded by the genome, there is a vast range of concentration of gene products. Some proteins may be present in as few as 5 to 10 molecules per cell, whereas others, such as ribosomal proteins and the many proteins involved in the glycolytic pathway, are present in as many as 100,000 copies per cell. While a basal level of most gene products exists, it is clear that this level can be increased and subsequently decreased in prokaryotes. Thus, fundamental regulatory mechanisms to control the expression of the genetic information must exist.

In this chapter, we will explore what is known about the regulation of genetic expression in bacteria and bacteriophages. Highly detailed information has now been obtained. The following chapter will focus on the regulation of gene expression in eukaryotes.

GENETIC REGULATION IN PROKARYOTES: AN OVERVIEW

The regulation of genetic expression has been most extensively studied in prokaryotes, particularly in *E. coli*. What has been found is that highly efficient genetic mechanisms have evolved that turn genes on and off, depending on the cell's metabolic need for the respective gene products. The activity of enzymes may also be regulated once they have been synthesized in the cell, but we will focus primarily on what is known about regulation at the level of gene transcription. It is important to remember that it is the resulting proteins ultimately present or absent that are critical to efficient cell function under varying environmental conditions.

It is not a particularly new concept that microorganisms regulate the synthesis of gene products. As early as 1900 it was shown that when the galactose-glucose-containing disaccharide **lactose** is present in the growth medium of yeast, enzymes specific to lactose metabolism are produced. When this substrate is absent, the enzymes are not manufactured. It was soon shown that bacteria also "adapt" to their chemical environment, producing certain enzymes only when specific substrates are present. Such enzymes were thus referred to as **adaptive**. In contrast, those produced continuously regardless of the chemical makeup of the environment were called **constitutive**. Since then, the term *adaptive* in descriptive enzymology has been replaced with a more accurate term. We now call them **inducible** enzymes, reflecting the role of the substrate, or **inducer,** in their production.

More recent investigation has revealed other cases where the presence of a specific molecule causes inhibition of genetic expression. This is usually true

GENETIC REGULATION IN BACTERIA AND BACTERIOPHAGES   **539** for molecules that are end products of biosynthetic pathways. For example, an amino acid such as tryptophan can be synthesized by bacterial cells. If an exogenous supply of this amino acid is present in the environment or culture medium, it is energetically inefficient to synthesize the enzymes necessary for tryptophan production. As a result, a mechanism exists whereby tryptophan plays a role in repressing transcription of RNA essential to the translation of the appropriate biosynthetic enzymes. In contrast to the inducible system controlling lactose metabolism, that governing tryptophan is said to be **repressible**.

As we will soon see, instances of regulation, whether inducible or repressible, may be under either **negative** or **positive control**. Under negative control, genetic expression occurs unless it is shut off by some form of regulator. This is in contrast to positive control, where transcription does not occur unless a regulator molecule directly stimulates RNA production. In theory each type of control can govern inducible or repressible systems. Examples discussed in the ensuing sections of this chapter will help distinguish between these possible mechanisms.

LACTOSE METABOLISM IN *E. COLI*: AN INDUCIBLE GENE SYSTEM

The most extensively studied system of gene regulation has involved the metabolism of lactose in *E. coli*. Beginning in 1946 with the studies of Jacques Monod and continuing through the next decade with significant contributions by Joshua Lederberg, François Jacob, and Andre L'woff, genetic and biochemical evidence was amassed. This research provided clear insights into the way in which the genes responsible for lactose metabolism are turned off, or repressed, when lactose is absent but activated or induced when it is available. In the presence of lactose, the concentration of the enzymes responsible for its metabolism increases rapidly from 5 to 10 molecules to thousands per cell. Thus, the enzymes are **inducible**, and lactose serves as the **inducer**.

Paramount to the understanding of genetic regulation in this system was the discovery of two genes that serve strictly in a regulatory capacity. They do not code for enzymes necessary for lactose metabolism. Three other genes are responsible for the production of enzymes involved in lactose metabolism. Together, the five genes function in an integrated fashion and provide a rapid response to the presence or absence of lactose.

Structural Genes

Genes coding for the primary structure of the enzymes are called **structural genes**. The so-called *lac z* gene specifies the amino acid sequence of the **β-galactosidase enzyme,** which converts lactose to glucose and galactose (Figure 19.1). This conversion is essential if lactose is to serve as the primary energy source in glycolysis. The second gene, *lac y*, specifies the primary structure of **β-galactoside permease,** which facilitates the entry of lactose into the bacterial cell. The third gene, *lac a,* codes for a **transacetylase** enzyme whose physiological role is unrelated to our discussion.

Studies of the genes coding for these three enzymes relied on the isolation of numerous mutations that eliminated the function of one or the other enzyme. Such *lac⁻* mutants were isolated and studied by Lederberg. Mutant cells fail to produce either active β-galactosidase or permease molecules and so are unable to utilize lactose as an energy source. Mutations also were found in the transacetylase gene. Mapping studies by Lederberg established that all three genes are closely linked or contiguous to one another in the order *z-y-a* (Figure 19.2).

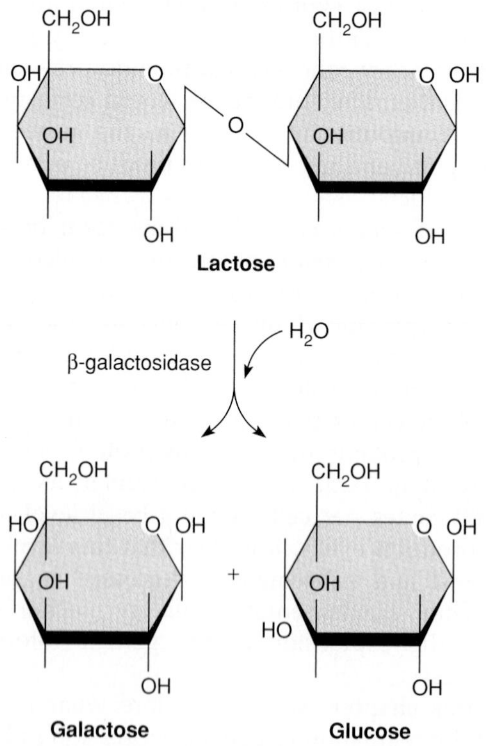

FIGURE 19.1
The catabolic conversion of the disaccharide lactose into its monosaccharide units, galactose and glucose.

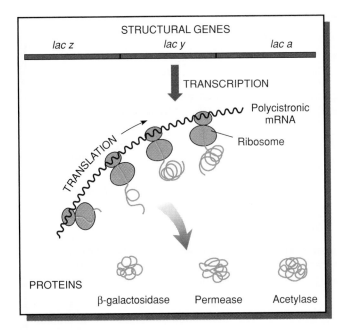

STRUCTURAL GENES

lac z   *lac y*   *lac a*

TRANSCRIPTION

Polycistronic mRNA

Ribosome

TRANSLATION

PROTEINS

β-galactosidase   Permease   Acetylase

FIGURE 19.2
The structural genes of the *lac* operon in *E. coli*. The three genes are transcribed into a single polycistronic mRNA, which is simultaneously translated into the three enzymes encoded by the operon.

Two other observations are relevant to what became known about the structural genes. First, knowledge of their close linkage led to the discovery that all three genes are transcribed together, resulting in a single **polycistronic message** or **mRNA** (Figure 19.2). Additionally, it has been shown that upon induction by lactose, the rapid appearance of the enzymes results from the *de novo* synthesis or translation of this mRNA. Although this finding might seem obvious or expected, it is in contrast to the proposal that induced enzyme activity may result from the activation of existing but inactive forms of the enzymes.

The Discovery of Regulatory Mutations

How, then, can lactose activate structural genes and induce the synthesis of the related enzymes? The discovery and study of **gratuitous inducers** ruled out one possibility. These molecules, which are chemical analogues of lactose, serve as inducers but not as substrates for the enzymatic reaction. One such gratuitous inducer is the sulfur analogue **isopropyl-thiogalactoside (IPTG),** shown in Figure 19.3. The discovery of such molecules is strong evidence that the primary induction event is not the result of the interaction between the inducer and the enzyme. What, then, is the role of lactose in induction?

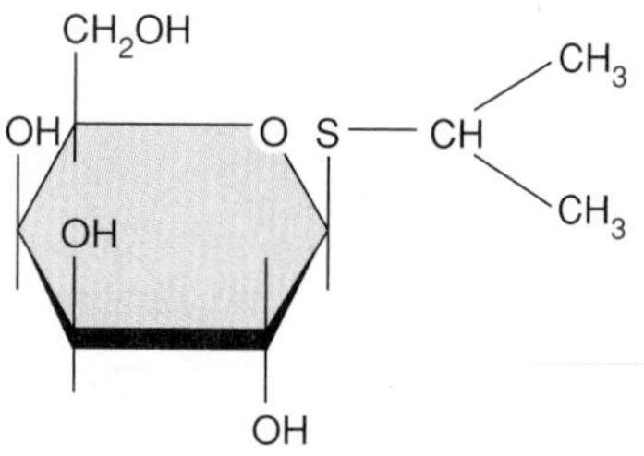

FIGURE 19.3
The gratuitous inducer isopropylthiogalactoside (IPTG).

The answer to this question required the study of a second class of mutation, the **constitutive mutants**. In this type of mutant, the enzymes are produced regardless of the presence or absence of lactose. These mutations served as the basis of studies that defined the regulatory scheme for lactose metabolism.

Maps of the first type of constitutive mutation, *lac i⁻*, showed that it is located at a site on the DNA close to, but distinct from, the structural genes. As we will soon see, the *i* gene is appropriately called a **repressor gene**. A second set of mutations produced identical effects but was found in a region immediately adjacent to the structural genes. This class is designated *lac oᶜ* and represents the **operator region**. Because the enzymes are continually produced for both types of constitutive mutation, they clearly represent regulatory units.

The Operon Model: Negative Control

In 1961, Jacob and Monod proposed a scheme of negative control of regulation, which they called the **operon model** (Figure 19.4). In this model, the **operon** consists of the structural genes as well as the adjacent region of DNA represented by the *lac oᶜ* mutation. They proposed that the *lac i* gene regulates the transcription of the structural genes by producing a **repressor molecule**. The repressor was hypothesized to be **allosteric**. This concept is applied particularly to proteins that reversibly interact with another smaller molecule, causing a conformational change in three-dimensional shape.

Jacob and Monod suggested that the repressor normally interacts with the DNA sequence of the operator region. When it does so, it inhibits the action of RNA polymerase, effectively repressing the transcription of the structural genes [Figure 19.4(b)]. However, in the presence of lactose, this disaccharide binds to the repressor, causing a conformational change. This change alters the binding site of the repressor capable of interacting with operator DNA

FIGURE 19.4
The components involved in the regulation of the *lac* operon and their interaction under various genotypic conditions, as described in the text.

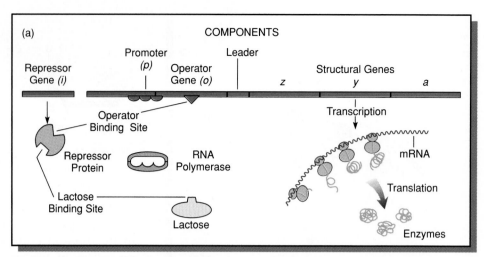

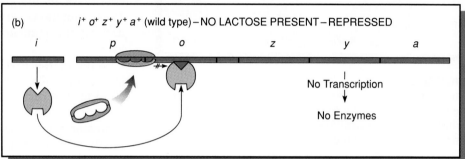

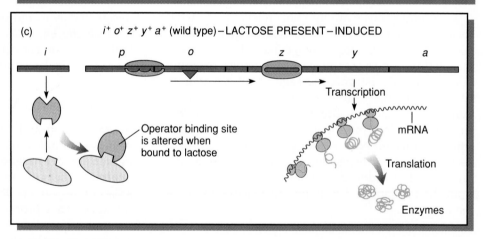

[Figure 19.4(c)]. In the absence of the repressor/operator interaction, RNA polymerase transcribes the structural genes, and the enzymes necessary for lactose metabolism are translated. *Since transcription occurs only in the absence of the repressor*, **negative control** is exerted.

The operon model uses these potential molecular interactions to explain the efficient regulation of the structural genes. In the absence of lactose, the enzymes encoded by the genes are not needed and so are repressed. When lactose is present, it indirectly induces the activation of the genes by binding with the repressor.* If all lactose is metabolized, none is available to bind to the repressor, which is again free to bind to operator DNA and repress transcription.

Both the i^- and o^c constitutive mutations interfere with these molecular interactions, allowing continuous transcription of the structural genes. In the case of the i^- mutant, the repressor product is altered and cannot bind to the operator region, so the structural genes are always turned on. In the case of the o^c

---

*Technically, allolactose, an isomer of lactose, is the inducer. Allolactose is produced during the metabolism of lactose.

FIGURE 19.4
continued

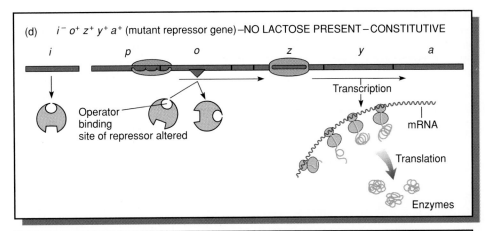

(d) $i^-\ o^+\ z^+\ y^+\ a^+$ (mutant repressor gene)–NO LACTOSE PRESENT–CONSTITUTIVE

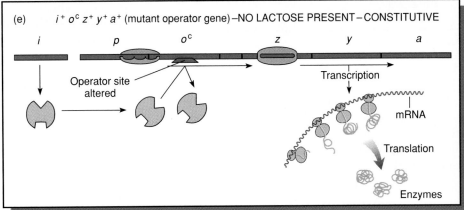

(e) $i^+\ o^c\ z^+\ y^+\ a^+$ (mutant operator gene)–NO LACTOSE PRESENT–CONSTITUTIVE

mutant, the nucleotide sequence of the operator DNA is altered and will not bind with a normal repressor molecule. The result is the same: Structural genes are always transcribed. Both types of constitutive mutations are illustrated diagrammatically in Figure 19.4(d) and (e).

Genetic Proof of the Operon Model

The operon model is a good one because there are predictions derived from it that can be tested to determine its validity. The major assumptions to be tested are (1) the *i* gene produces a diffusible cellular product; (2) the *o* region does not; and (3) the *o* region must be adjacent to the structural genes in order to regulate transcription.

The construction of partial diploid bacteria (see Chapter 15) allows an assessment of these assumptions. For example, it is possible to construct genotypes where an i^+ gene has been introduced into an i^- host, or where an o^+ region is added to an o^c host. The Jacob-Monod operon model predicts that adding an i^+ gene to an i^- cell should restore inducibility,

because a normal repressor would again be produced. Adding an o^+ region to an o^c cell should have no effect on constitutive enzyme production, since regulation depends on an o^+ region immediately adjacent to the structural genes.

The results of these experiments are shown in Table 19.1, where *z* represents the structural genes. In both cases just described, the Jacob-Monod model is upheld (part B of Table 19.1). Part C shows the reverse experiments, where either an i^- gene or an o^c region is added to cells of normal inducible genotypes. The model predicts that inducibility will be maintained in these partial diploids, and it is.

Another prediction of the operon model is that certain mutations in the *i* gene should have the opposite effect of i^-. That is, instead of being constitutive by failing to interact with the operator, mutant repressor molecules should be produced, which cannot interact with the inducer, lactose. As a result, the repressor would always bind to the operator sequence, and the structural genes would be permanently repressed. If this were the case, the presence of an additional i^+ gene would have little or no effect on repression.

TABLE 19.1
Genotypes and enzyme activity for the *lac* operon of *E. coli*.

Genotype	Presence of β-galactosidase Activity	
	Lactose Present	Lactose Absent
A. $i^+o^+z^+$	+	−
$i^+o^+z^-$	−	−
$i^-o^+z^+$	+	+
$i^+o^cz^+$	+	+
B. $i^-o^+z^+/F'i^+$	+	−
$i^+o^cz^+/F'o^+$	+	+
C. $i^+o^+z^+/F'i^-$	+	+
$i^+o^+z^+/F'o^c$	+	−
D. $i^so^+z^+$	−	−
$i^so^+z^+/F'i^+$	−	−

In fact, as shown in Table 19.1D, such a mutation, i^s, was discovered. An additional i^+ gene does not effectively relieve repression of gene activity. These observations again support the operon model for gene regulation.

Isolation of the Repressor

Although the operon theory of Jacob and Monod succeeded in explaining many aspects of genetic regulation in prokaryotes, the nature of the repressor molecule was not known when their landmark paper was published in 1961. While they had assumed that the allosteric repressor was a protein, RNA was also a candidate, since the molecule required binding to DNA. Despite many attempts to isolate and characterize the hypothetical repressor molecule, no direct chemical evidence was immediately forthcoming. A single *E. coli* cell contains no more than ten or so copies of the *lac* operon repressor. Direct chemical identification of ten molecules in a population of millions of proteins and RNAs in a single cell presented a tremendous challenge.

In 1966, Walter Gilbert and Benno Müller-Hill reported the isolation of the *lac* repressor in partially purified form. To achieve the isolation, they used a *regulator quantity* (i^q) mutant strain that contains about ten times as much repressor as wild-type *E. coli* cells. Also instrumental in their success were the use of the gratuitous inducer, IPTG, which binds to the repressor, and the technique of **equilibrium dialysis**. Extracts of i^q cells were placed in a dialysis bag and allowed to attain equilibrium with an external solution of radioactive IPTG, which is small enough to diffuse freely in and out of the bag. At equilibrium,

the concentration of IPTG was higher inside the bag than in the external solution, indicating that an IPTG-binding material was present in the cell extract and that it was too large to diffuse across the wall of the bag. Ultimately the binding material was purified; it was shown to be heat labile and to have other characteristics of a protein as well. Extracts of i^- constitutive cells having no *lac* repressor activity did not exhibit IPTG-binding activity, strongly suggesting that the isolated protein was the repressor molecule.

To demonstrate this further, Gilbert and Müller-Hill grew *E. coli* cells on a medium containing radioactive sulfur and isolated the IPTG-binding protein, which was labeled in its sulfur-containing amino acids. This protein was mixed with DNA from a strain of phage lambda (λ) which carries the *lac* o^+ gene. The DNA sediments at 40S, while the IPTG-binding protein sediments at 7S. The DNA and protein were mixed and sedimented in a glycerol gradient in an ultracentrifuge. If the radioactive protein binds to the DNA, then it should sediment at a much faster rate, moving in a band with the DNA. This was found to be the case. Further experiments showed that the IPTG-binding or repressor protein binds only to DNA containing the *lac* region. It was also shown not to bind to *lac* DNA containing an operator constitutive (o^c) mutation.

Subsequent study of the repressor protein has revealed that it consists of four identical subunits of 38,000 daltons. Each can bind to one molecule of IPTG. The complete protein has a low binding to DNA in general, but has an affinity about 1000-fold greater for the operator region, which consists of about 30 nucleotide pairs.

The Catabolite Activating Protein (CAP): Positive Control of the *lac* Operon

When the *lac* repressor is bound to the inducer, the *lac* operon is activated and transcription of the structural genes proceeds under the direction of RNA polymerase. As we have learned in earlier chapters, this process is initiated as a result of the binding that occurs between this enzyme and a nucleotide sequence constituting the **promoter region,** found upstream from the initial coding sequences. Within the *lac* operon, the promoter is found between the i gene and the operator region. Careful examination has revealed that, in this case, polymerase binding is not very efficient unless another protein is present to facilitate the process. This protein is called the **catabolite activating protein (CAP)**.

The mechanism by which CAP works is an interesting one. In contrast to the negative control of the operon by the repressor, CAP exerts **positive control**. Unless CAP is bound to a region of DNA to the left of the promoter, transcription is severely diminished. When bound, transcription is greatly facilitated, provided, of course, that the repressor is not also bound to the operator region. Under what circumstances does CAP bind?

The initial observation that drew attention to this phenomenon was that transcription of the operon (and thus, binding of CAP) is inhibited in the presence of glucose, even though the system is under inducible or constitutive conditions. Since glucose is a catabolic product of the reaction catalyzed by the gene product β-galactosidase, and since its presence inhibits transcription of the operon, the phenomenon is designated **catabolite repression**. Such regulation results in efficient energy utilization, because the presence of glucose will override the need for the metabolism of lactose, should it also be available to the cell. Catabolite repression has also been observed for other inducible operons, including those controlling the metabolism of galactose and arabinose.

The details of how catabolite repression is achieved have now been elucidated. In addition to CAP, there are two other components: a region to the left of the promoter, the so-called **CAP binding site,** and a modified nucleotide, **cyclic adenosine monophosphate (cAMP)**. These interact to repress gene activity, as summarized in Figure 19.5.

The promoter contains two regions. The region closest to the operator contains the nucleotide sequence that interacts with the sigma subunit of RNA polymerase. The opposite end, closest to the *i* gene, contains the CAP site. Before RNA polymerase can effectively interact with its promoter region, the CAP site must be bound by CAP. However, this interaction requires that CAP be linked to cAMP.

In the absence of glucose, the enzyme **adenyl cyclase** readily converts ATP to cAMP (Figure 19.5). The availability of cAMP completes the chain of events just described, and, if the *lac* system is induced or constitutive, transcription of the structural genes proceeds.

The presence of glucose, however, drastically reduces the activity of adenyl cyclase, interrupting the events essential to transcription. The concentration of cAMP is also reduced. Under this condition, the CAP-cAMP complex is unavailable to bind to the CAP site, inhibiting transcription.

Unlike regulation by the repressor, the CAP interaction exerts a **positive control**. When bound to cAMP, CAP facilitates transcription. The repressor, on the other hand, exerts a **negative control**. When it is not bound to lactose, the operon is repressed. Taken together, both systems involve the direct interaction of DNA with three proteins. The operator binds the repressor, and the promoter binds CAP-cAMP and RNA polymerase.

THE *ara* REGULATOR PROTEIN: POSITIVE AND NEGATIVE CONTROL

Before turning to a consideration of repressible systems, we want to make mention of a rather unique inducible operon in *E. coli* that appears to be under both positive and negative control.

In *E. coli*, the metabolism of the sugar arabinose is under the direction of the enzymatic products of the *ara B, A,* and *D* genes. These genes are regulated by the **ara C protein** encoded by the *ara C* gene. In the absence of arabinose, the protein behaves as a repressor, binding to three regulatory sites within the operon. In the presence of arabinose, the *ara C* protein binds to it and this complex becomes an activator. Attaching to a separate initiator site on the operon stimulates the transcription of the *B, A,* and *D* genes. Thus, the *ara* regulatory protein can alternately behave as a repressor or an activator, depending on the absence or presence of arabinose, respectively. In the former case, negative control is exerted, while in the latter case, positive control occurs.

A further complication is of interest in this system: it is, like the *lac* operon, sensitive to glucose and under the positive control of the CAP protein. When glucose is present and the level of cyclic AMP is low, the CAP binding site is empty. Under such a condition the *ara C* protein binds to two of the three regulatory sites, drawing DNA into a loop. This creates a repressive condition, preventing the *ara C*–arabinose complex from activating the *BAD* gene promoter. Thus, both cyclic AMP and arabinose must be present for the expression of the *BAD* genes in this operon.

TRYPTOPHAN METABOLISM IN *E. COLI*: A REPRESSIBLE GENE SYSTEM

Although induction had been known for some time, it was not until 1953 that Monod and his coworkers discovered **enzyme repression**. Wild-type *E. coli* are capable of producing the necessary enzymes essential to the biosynthesis of amino acids as well as other essential macromolecules. Monod focused his studies on the amino acid tryptophan and the enzyme **tryptophan synthetase**. He discovered that if tryptophan is present in sufficient quantity in the growth medium, the enzymes necessary for its synthesis are **repressed**. Energetically, such enzyme repression is highly economical to the cell, because synthesis is

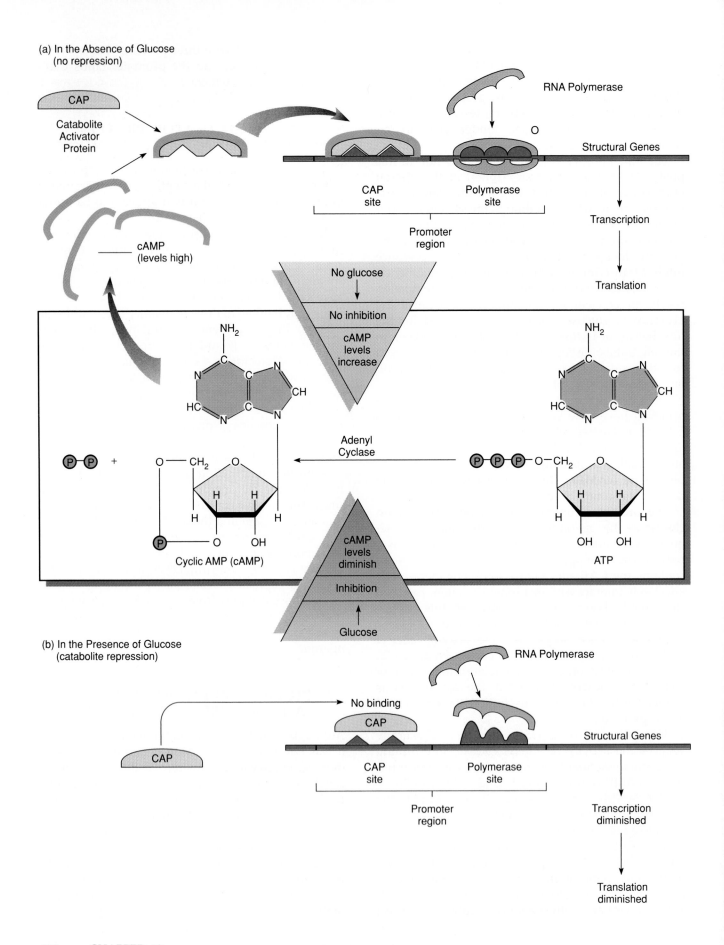

(a) In the Absence of Glucose
(no repression)

CAP

Catabolite
Activator
Protein

RNA Polymerase

Structural Genes

CAP
site

Polymerase
site

Promoter
region

Transcription

Translation

cAMP
(levels high)

No glucose

No inhibition

cAMP
levels
increase

NH_2

C

N   N

C   CH

HC   C

N   N

O — CH_2 O

H   H

H   H

P   O   OH

Cyclic AMP (cAMP)

Adenyl
Cyclase

NH_2

C

N   N

C   CH

HC   C

N   N

O — CH_2 O

H   H

H   H

OH   OH

ATP

cAMP
levels
diminish

Inhibition

Glucose

(b) In the Presence of Glucose
(catabolite repression)

No binding

CAP

RNA Polymerase

Structural Genes

CAP

CAP
site

Polymerase
site

Promoter
region

Transcription
diminished

Translation
diminished

unnecessary in the presence of exogenous tryptophan.

Further investigation showed that a series of enzymes encoded by five contiguous genes on the *E. coli* chromosome is involved in tryptophan synthesis. These genes are part of an operon, and in the presence of tryptophan, all genes are coordinately repressed. As a result, none of the enzymes are produced. Because of the great similarity between this repression and the induction of enzymes for lactose metabolism, Jacob and Monod proposed a model of gene regulation analogous to the *lac* system.

To account for repression, they suggested that an inactive repressor is normally made that alone cannot interact with the operator region of the operon. However, it is allosteric, interacting as well with tryptophan, if present. The resulting interaction activates the repressor that may now bind to the operator, repressing transcription. Thus, when the end product of this anabolic biosynthesis pathway is present, enzymes are not made. Since the regulatory complex inhibits transcription of the operon, this **repressible system** is under **negative control**. Because tryptophan participates in repression, it is referred to as a **corepressor** in this regulatory scheme.

Evidence for and Concerning the *trp* Operon

Support for the concept of a repressible operon was soon forthcoming. Primarily, two distinct categories of constitutive mutations were isolated. The first class, *trp R⁻*, maps at a considerable distance from the structural genes. This locus represents genetic information coding for the repressor. Presumably, the R^- mutation either inhibits the interaction of a mutant repressor with tryptophan or inhibits repressor formation entirely. Thus, no repression ever occurs. The presence of an R^+ gene restores repressibility.

The second constitutive mutant is analogous to that of the operator of the lactose operon because it maps immediately adjacent to the structural genes. Furthermore, the addition of a wild-type operator gene into mutant cells does not restore enzyme repression. This is predictable if the mutant operator can no longer interact with the repressor-tryptophan complex.

The entire *trp* operon has now been well defined, as shown in Figure 19.6. Five contiguous structural genes (*trp E, D, C, B,* and *A*) are transcribed as a polycistronic message that directs translation of the polypeptide components. These components catalyze the biosynthesis of tryptophan. As in the *lac* operon, a promotor region (*trp P*), representing the binding site for RNA polymerase, and an operator region (*trp O*), which binds the repressor, have been demonstrated. In the absence of binding, transcription is initiated within the overlapping *trp P-trp O* region and proceeds along a **leader sequence**. Transcription is initiated 162 nucleotides prior to the first structural gene (*E*). Within this leader sequence, still another regulatory site has been demonstrated. This component, called an **attenuator,** has been investigated extensively by Charles Yanofsky and his colleagues. As we shall see, this regulatory unit is an integral part of the control mechanism of this operon.

The Attenuator

Yanofsky observed that, whether tryptophan is present or not, initiation of transcription of this leader sequence usually occurs. Thus, the activated repressor, even when bound to the operator region, does not strongly inhibit expression of the operon. As transcription of the leader sequence proceeds, if tryptophan is present in high concentration, mRNA synthesis is terminated at a point about 140 nucleotides along the transcript. It is this region that is called the **attenuator**. If, however, tryptophan is absent or present in low concentrations, this process of **attenuation** is overcome, and transcription of the entire operon occurs, with the subsequent production of the enzymes essential to the biosynthesis of tryptophan.

The identification of the site of the attenuator was made possible by the isolation of deletion mutations in the region 123 to 150 nucleotides into the leader sequence. Such mutations abolish attenuation. The phenomenon of attenuation appears to be common to other bacterial operons, including those regulating the biosynthesis of histidine and leucine.

An explanation of how attenuation occurs and how it is overcome has been put forward by Yanofsky and is summarized in Figure 19.7. Present in the

FIGURE 19.5 (opposite)
Catabolite repression. (a) In the absence of glucose, cAMP levels are high. When cAMP binds to CAP, a complex is formed that binds to a promoter region and facilitates transcription. (b) In the presence of glucose, the activity of adenyl cyclase is inhibited and cAMP levels are reduced. CAP, when not bound to cAMP, cannot efficiently bind to the promoter region, and transcription is reduced.

FIGURE 19.6
(a) The components involved in the regulation of the tryptophan operon. (b) Regulatory conditions involving either activation (top) or repression (bottom) of the structural genes. In the absence of tryptophan, an inactive repressor is made that cannot bind to the operator (*O*), thus allowing transcription to proceed. In the presence of tryptophan, it binds to the repressor, causing an allosteric transition to occur. This complex binds to the operator region, resulting in repression of the operon.

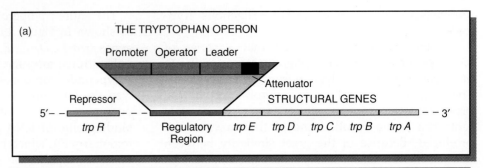

(a) THE TRYPTOPHAN OPERON

Promoter Operator Leader

Attenuator

Repressor

STRUCTURAL GENES

5′ — — — — —   3′

trp R   Regulatory Region   *trp E*   *trp D*   *trp C*   *trp B*   *trp A*

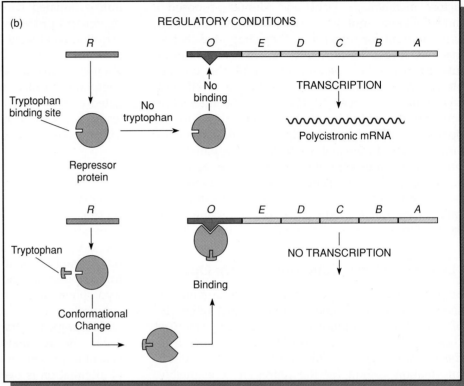

(b) REGULATORY CONDITIONS

Tryptophan binding site

No tryptophan

No binding

TRANSCRIPTION

Polycistronic mRNA

Repressor protein

Tryptophan

Conformational Change

Binding

NO TRANSCRIPTION

region of the attenuator is a DNA sequence that, when transcribed, gives rise to an RNA molecule that immediately folds back on itself and forms a **hairpin loop**. This is then followed by eight uridine residues. Such a structure is typical of that found at the 3′ end of RNA transcripts, leading to the termination phase of their transcription. Thus, even when the operon should be repressed, transcription begins, but the process is immediately terminated.

The question then remains as to how the absence (or a low concentration) of tryptophan allows attenuation somehow to be bypassed. Yanofsky's proposal is based on the discovery that the leader transcript contains an AUG initiator codon closely followed by two triplets encoding tryptophan. Even though this leader sequence is not part of the structural genes in the operon, following its transcription, the AUG sequence prompts the initiation of its translation by

ribosomes. Recall that in bacteria, translation of an mRNA is initiated well before transcription is completed.

If cells are starved of tryptophan (which is a rather rare amino acid in proteins anyway), the ribosome "stalls" during the translation of triplets calling for charged tRNA^{trp}. This event influences the secondary structure of the transcript such that the termination hairpin does not form [Figure 19.7(b)], attenuation is bypassed, and simultaneous transcription and translation of the entire set of structural genes then proceeds. However, when adequate tryptophan is present, charged tRNA^{trp} is present, no stalling occurs during translation, and the termination hairpin is formed [Figure 19.7(a)].

The details of the secondary structures of the transcript have now been worked out and described. They satisfy conditions leading to either continued

FIGURE 19.7
Diagrammatic representation of the leader sequence of the RNA transcript of the *trp* operon of *E. coli* and its role in attenuation. Shown at the top are the *trp* codons, the potential base-pairing sites, and the hairpin termination configuration in the leader sequence (shaded). In the presence of tryptophan (a), rapid movement of the ribosome across the *trp* codons occurs during translation, and a hairpin configuration forms at the attenuator site. Termination of transcription usually occurs, leading to attenuation. When cells are starved of tryptophan (b), the ribosome stalls at the *trp* codons, no hairpin configuration forms at the attenuation site, and transcription proceeds across it. The model assumes that transcription by RNA polymerase and translation by a ribosome are occurring simultaneously, as is known to happen in bacteria.

(a) mRNA regions/selected base sequences

(b) No translation

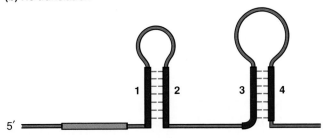

(c) High concentrations of tryptophan: no stalling occurs. Transcription terminates at termination hairpin.

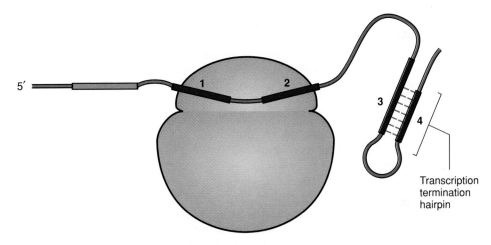

(d) Low concentrations of tryptophan: ribosome stalls. Transcription proceeds.

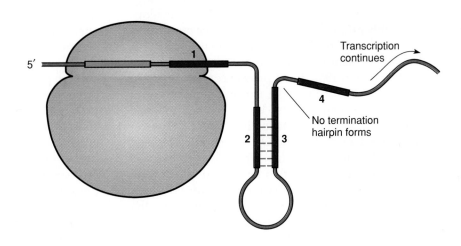

transcription or termination (attenuation). Additionally, various mutations in the leader sequence predicted to alter the secondary structure of the transcript and its impact on attenuation have been isolated. In each case, the predicted result has been upheld. The model, while complex, has significantly extended our knowledge of genetic regulation. Furthermore, the supporting evidence represents an integrated approach involving genetic and biochemical analysis, which is becoming commonplace in molecular genetic research.

GENETIC REGULATION IN PHAGE LAMBDA: LYSOGENY OR LYSIS?

Our understanding of genetic regulation at the transcriptional level has benefitted from studies of bacteriophage lambda as well as from studies of operons in bacteria. Lambda DNA contains about 45,000 base pairs, enough bioinformation for about 35 to 40 average-sized genes. After injection of lambda DNA into *E. coli*, either the **lysogenic** or **lytic** pathways may be followed (see Chapter 15). In the first pathway, the phage DNA is integrated into the host genome and is almost totally repressed; in the second, the phage DNA is expressed and viral reproduction ensues.

If the lysogenic pathway is followed, the genes responsible for phage reproduction and lysis are turned off by the **λ repressor protein,** which is produced by one of the virus's own genes, *cI*. If the lytic pathway is followed, the expression of the *cI* gene is repressed by a second protein, **cro,** which is produced by a distinct viral gene of the same name. Both the *cI* repressor (sometimes called the λ repressor) and the cro protein have been isolated and characterized. Their interaction with λ DNA has also been determined. The various sites of the regulatory system are diagramed in Figure 19.8. The development of this information, which enhances our understanding of gene regulation, is a fascinating story.

The λ repressor was isolated and characterized by Mark Ptashne in 1967. It is a protein consisting of 236 amino acids. The cro protein, isolated more recently, consists of 66 amino acids. When the repressor is produced, it recognizes two different operator regions in the λ DNA, O_L and O_R, found on either side (left or right) of the *cI* gene. When the repressor is bound to these regions, two sets of so-called early genes are repressed, causing the remainder of the λ genes to be turned off as well. In this case, the repressor behaves as a negative control element. It is now clear that during this same binding, the cI repressor also stimulates transcription of its own *cI*

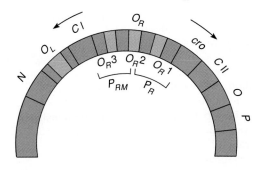

FIGURE 19.8
The regulatory sites controlling lysogeny and lysis in phage λ.

gene. Thus, in this role, it behaves as a positive control element and enhances transcription by a factor of ten.

When the repressor binding to O_L and O_R is absent, initiation at the respective promoters (P_{RM} and P_R) proceeds, resulting in the production of two proteins, N and cro. The N protein functions as an antiterminator, allowing completion of transcription of genes essential to reproduction and lysis. The cro protein, as stated earlier, serves as a repressor of *cI* gene transcription. Mutational analysis supports this model. Mutations in either O_L or O_R prevent repressor binding and abolish the potential for lysogeny. *cro*⁻ mutations abolish the potential for lysis.

The operator-repressor interactions have also been worked out in detail. The interaction occurring at O_R will be used as an illustration. The O_R region consists of 80 DNA base pairs and is located between the *cI* gene and the *cro* gene. There are three binding sites in O_R, each with a distinctive nucleotide sequence. The three regions are labeled O_R1, O_R2, and O_R3, and each consists of 17 nucleotides. Overlapping this region are the promoters for the *cI* and *cro* genes, called P_{RM} and P_R, respectively. Transcription of these genes occurs in opposite directions, using alternate strands of the helix to achieve 5'-to-3' RNA synthesis.

The greatest surprise was that both the λ repressor and cro protein demonstrate a binding affinity to the three sites of O_R. However, the binding of the two proteins to the three sites varies, depending on the concentration of the proteins. In addition, when either protein is bound to these sites, the other cannot bind. The cI protein binds initially to O_R1 and O_R2 coordinately and then to O_R3, but only when present at very high concentrations. The cro protein behaves in the opposite fashion. It initially binds O_R3 and subsequently binds O_R1 and O_R2.

Following initial infection, the determination of lysogeny versus lysis appears to depend on which operator region is bound first. If O_R1 and O_R2 are bound, P_R is inhibited and the *cro* gene is not transcribed. P_{RM}, on the other hand, is available to RNA polymerase and the *cI* gene is transcribed, resulting in increased amounts of the repressor. Lysogeny is now established. However, if O_R3 is bound by the cro protein, P_{RM} is inhibited and the *cI* gene is repressed. This allows initiation at the P_R site, and transcription essential to lysis proceeds.

While this description explains how the determination is made, it does not address why one pathway is favored over the other. Still another gene and its product are critical to this decision: *cII*. Both the cII protein and the cro protein accumulate during early infection. The cII protein affects a special promotor that activates the *cI* gene, causing the production of the repressor. This favors lysogenesis. Apparently, the activity of the cII protein changes under certain cellular conditions. For example, if glucose is plentiful, the activity of cII is reduced, thus favoring the lytic cycle. The cII protein, which is unstable, may serve to monitor growth conditions. When it fails to prevail, the cro protein successfully represses the *cI* gene and viral reproduction commences.

To pursue this question further, Ptashne and others have found it useful to examine the molecular events that occur when an induction event transforms the lysogenic pathway to the lytic sequence. Discovered in 1946 by L'woff, treatment of lysogenized *E. coli* by various agents (including UV light and mutagens) induces a lytic response. While such an event occurs spontaneously about once in a million bacterial cell divisions, all treated bacteria are induced to enter the lytic cycle.

Induction events have one common property: They all involve interaction with DNA. In so doing, they stimulate what has been called the SOS response (see Chapter 14). During this response, the recA protein, normally essential to recombination, is synthesized. Under inductive conditions, it behaves as a proteolytic enzyme and plays an essential role in the induction of lysis.

The λ repressor protein contains two structural domains and may exist in monomer or dimer form. It is the dimer form that binds strongly to O_R. The recA protein cleaves the monomers at a sensitive region between the two domains. Cleavage of monomers prevents their dimerization; as a result, the repressor protein can no longer bind to O_R, stimulating the expression of the *cro* gene. The cro product, also active in dimer form, then binds to O_R, inhibiting further transcription of the *cI* gene. Genetic events essential to viral reproduction then proceed, new viruses are constructed, and lysis of the bacterial cell occurs.

The research leading to the information just described is remarkable. It illustrates the depth of analysis attainable with regard to what at first may appear as a simple question: How is lysogeny regulated? Important insights have also been gained regarding protein/nucleic acid interactions during genetic regulation. Molecular interactions such as these undoubtedly play major roles in many other genetic phenomena.

PHAGE TRANSCRIPTION DURING LYSIS

When a bacteriophage invades its host bacterium and initiates the lytic cycle, it faces a novel problem. How can its genes be transcribed without the presence of RNA polymerase, itself a product of transcription? The strategy that has evolved is an interesting example of parasite-host interaction. You may wish to review the discussion of the phage life cycle and strategies for replication of DNA discussed in Chapter 15.

The study of phage T7, which invades *E. coli*, has provided some answers to this question. This phage has evolved DNA sequences that are recognized as promoter sites by the *E. coli* RNA polymerase. As a result, upon the phage's entry to the host cell, a set of so-called **early genes** is transcribed into mRNAs, which are translated on bacterial ribosomes. One of these gene products is a viral-specific RNA polymerase that recognizes the promoters for the remaining T7 genes. A second early gene product, a protein kinase, inhibits bacterial transcription. The site of inhibitory interaction is presumably the bacterial RNA polymerase.

With the host cell's genetic apparatus secured, the remaining viral genes are expressed. The major products are DNA-replicating enzymes, head and tail components, enzymes involved in DNA packaging, and, finally, a lytic enzyme to break open the host cell.

Studies of the phage **SPO1,** which invades *Bacillus subtilis*, have revealed an even finer control or regulation of transcription during lysis (Figure 19.9). As in T7, **early genes** are read by the host RNA polymerase. One of the early gene products is a protein subunit that associates with the host polymerase. This complex can now interact with the promoter regions of the **middle genes**. This association between the early gene subunit and the bacterial polymerase curbs transcription of the early genes, and the middle genes associated with DNA replication are expressed. Two of the middle genes produce additional polypeptides that can interact with the host polymerase, again altering its specificity such

FIGURE 19.9
Initiation and regulation of phage SPO1 gene expression during lysis, as described in the text.

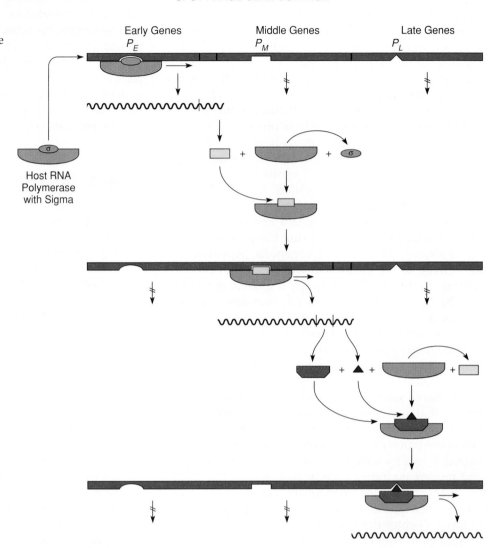

that it directs the expression of the **late genes**. Thus, sequential gene action results, providing a finer mechanism for the regulation of gene expression during the lytic cycle.

CHAPTER SUMMARY

1 A system of genetic regulation must exist if the complete genome is not to be continuously active in transcription throughout the life of every cell of all species. Highly refined mechanisms have evolved to regulate transcription of the genetic information maintaining genetic efficiency in prokaryotes and the specialization associated with differentiation in eukaryotes.

2 Inducible and repressible operons, both of which have been documented and studied in bacteria, involve genes of a regulatory nature in addition to structural genes that code for the enzymes of the system.

3 The operon model, proposed by Jacob and Monod, demonstrates an inducible mode of regulation under negative control. Studies of many types of mutations that altered regulation of lactose metabolism served as the basis for the model. This model predicted the presence of a repressor molecule, which has been successfully isolated and shown to be a protein with quaternary structure.

4 Inducible systems control catabolic processes, producing enzymes to metabolize the substrate only when the inducer is present. In the inducible *lac* operon, the regulatory gene produces an allosteric repressor molecule. In the absence of

the inducer, lactose, the repressor binds to the operator DNA adjacent to the structural genes, inhibiting transcription. If lactose is present, it binds to the repressor, causing a conformational change that prevents binding of the repressor to the operator. As a result, the structural genes are transcribed.

5 The catabolite activating protein (CAP) facilitates the binding of RNA polymerase to the promoter in the *lac* operon. Catabolite repression, a mechanism that represses the operon when glucose is present, has evolved since glucose can be utilized more efficiently than lactose.

6 The *ara* regulator protein exerts both positive and negative control over expression of the genes specifying the enzymes that metabolize the sugar arabinose.

7 Repressible systems control anabolic processes such as tryptophan synthesis. When a high concentration of tryptophan is present, transcription is repressed as the allosteric repressor becomes complexed with this end product and binds to the operator. If tryptophan is not present, the uncomplexed repressor cannot bind to the operator, resulting in transcription.

8 The attenuator, a controlling element in the *trp* operon, consists of a sequence of nucleotides located in a leader sequence of DNA occurring prior to the structural genes. When tryptophan is present in high concentration, transcription is terminated by the formation of a hairpin loop. This process of attenuation is overcome when tryptophan is absent or present in low concentration.

9 Repression of transcription in phage lambda has been shown to be due to one of its own gene products *cI*, which behaves as a repressor. A second gene product, *cII*, stimulates the production of the repressor which favors the lysogenic pathway. Another gene product, *cro*, acts to inhibit the transcription of *cI* and promotes the lytic pathway when the activity of the *cII* protein is reduced.

10 During lysis, bacteriophages must rely on the host RNA polymerase to initiate transcription of their genome. Certain viral promoter sequences are recognized by the host polymerase. One of the early gene products in phage T7 is the RNA polymerase that transcribes the remainder of the genome.

Insights and Solutions

1 A theoretical operon (*theo*) in *E. coli* contains several structural genes encoding enzymes that are involved sequentially in the biosynthesis of an amino acid. Unlike the *lac* operon, where the repressor gene is separate from the operon, the gene encoding the regulator molecule is contained within the *theo* operon. When the end product (the amino acid) is present, it combines with the regulator molecule and this complex binds to the operator, repressing the operon. In the absence of the amino acid, the regulatory molecule fails to bind to the operator and transcription proceeds.

Characterize this operon; then consider the following mutations as well as the situation in which the wild-type gene is present along with the mutant gene in partially diploid cells (F'). In each case, will the operon be active or inactive in transcription, assuming that the mutation affects the regulation of the *theo* operon? Compare each response to the equivalent situation in the *lac* operon.

(a) Mutation in the operator gene
(b) Mutation in the promoter region
(c) Mutation in the regulator gene

ANSWER: The operon is under negative control and is repressible. The regulatory molecule, when bound to the amino acid, binds to the operator region and inhibits gene expression.

(a) As in the *lac* operon, a mutation in the *theo* operator gene inhibits binding with the repressor complex, and transcription occurs constitutively. The presence of an F′ plasmid bearing the wild-type allele would have no effect.

(b) A mutation in the *theo* promoter region would no doubt inhibit binding to RNA polymerase and therefore inhibit transcription. This would also happen in the *lac* operon. A wild-type allele present in an F′ plasmid would have no effect.

(c) A mutation in the *theo* regulator gene, as in the *lac* system, may either inhibit its binding to the repressor or its binding to the operator gene. In both cases, transcription will be constitutive because the *theo* system is repressible. Both cases result in the failure of the regulator to bind to the operator, allowing transcription to proceed. In the *lac* system, failure to bind the corepressor lactose would permanently repress the system. The addition of a wild-type allele would restore repressibility, provided that this gene was transcribed constitutively.

PROBLEMS AND DISCUSSION QUESTIONS

1. Contrast the need for the enzymes involved in the metabolism of lactose and tryptophan in bacteria in the presence and absence of lactose and tryptophan, respectively.

2. Contrast positive and negative control systems.

3. Contrast the role of the repressor in an inducible system and in a repressible system.

4. Describe how the *lac* repressor was isolated. What properties demonstrate it to be a protein? Describe the evidence that it indeed serves as a repressor within the operon scheme.

5. Even though the *z*, *y*, and *a* structural genes are transcribed as a single polycistronic mRNA, each gene contains the appropriate initiation and termination signals essential for translation. Predict what will happen when a cell growing in the presence of lactose contains a deletion of one nucleotide early in the *z* gene.

6. For the following *lac* genotypes, predict whether the structural genes (*z*) are constitutive, permanently repressed, or inducible in the presence of lactose.

Genotype	Constitutive	Repressed	Inducible
$i^+o^+z^+$			×
$i^-o^+z^+$			
$i^+o^cz^+$			
$i^-o^+z^+/\text{F}'i^+$			
$i^+o^cz^+/\text{F}'o^+$			
$i^so^+z^+$			
$i^so^+z^+/\text{F}'i^+$			

7. For the following genotypes and condition (lactose present or absent), predict whether functional enzymes are made, nonfunctional enzymes are made, or no enzymes are made.

Genotype	Condition	Functional Enzyme Made	Nonfunctional Enzyme Made	No Enzyme Made
$i^+o^+z^+$	No lactose	—	—	×
$i^+o^cz^-$	Lactose			
$i^-o^+z^-$	No lactose			
$i^-o^+z^-$	Lactose			
$i^-o^+z^+/F'i^+$	No lactose			
$i^+o^cz^+/F'o^+$	Lactose			
$i^+o^+z^-/F'i^+o^+z^+$	Lactose			
$i^-o^+z^-/F'i^+o^+z^+$	No lactose			
$i^so^+z^+/F'o^+$	No lactose			
$i^+o^cz^-/F'o^+z^+$	Lactose			

8   In a theoretical operon, genes a, b, c, and d represent the repressor gene, the promoter sequence, the operator gene, and the structural gene, but not necessarily in that order. This operon is concerned with the metabolism of a theoretic molecule (tm). From the data given below, first decide if the operon is inducible or repressible. Then assign a, b, c, and d to the four parts of the operon. (AE = active enzyme; IE = inactive enzyme; NE = no enzyme)

Genotype	tm Present	tm Absent
$a^+b^+c^+d^+$	AE	NE
$a^-b^+c^+d^+$	AE	AE
$a^+b^-c^+d^+$	NE	NE
$a^+b^+c^-d^+$	IE	NE
$a^+b^+c^+d^-$	AE	AE
$a^-b^+c^+d^+/F'a^+b^+c^+d^+$	AE	AE
$a^+b^-c^+d^+/F'^+b^+$	AE	NE
$a^+b^+c^-d^+/F'a^+b^+c^+d^+$	AE + IE	NE
$a^+b^+c^+d^-/F'a^+b^+c^+d^+$	AE	NE

9   If the cI gene of phage λ contained a mutation with effects similar to the i^- mutation in *E. coli*, what would be the result?

10   Describe the level of genetic activity of the *lac* operon as well as the status of the *lac* repressor and the CAP protein under the following cellular conditions:

	lactose	glucose
(a)	−	−
(b)	+	−
(c)	−	+
(d)	+	+

11 A bacterial operon is responsible for the production of the biosynthetic enzymes needed to make a theoretical amino acid tisophane (tis). The operon is regulated by a separate gene, *r*. Deletion of the *r* gene causes the loss of synthesis of the enzymes. In the wild-type condition, when tis is present, no enzymes are made. In the absence of tis, the enzymes are made. Mutations in the operator gene (O^-) result in repression regardless of the presence of tis.

Is the operon under positive or negative control? Propose a model for (1) repression of the genes in the presence of tis in wild-type cells; and (2) the O^- mutations.

SELECTED READINGS

ANDERSON, J. E., PTASHNE, M., and HARRISON, S. C. 1985. A phage repressor-operator complex at 7 Å resolution. *Nature* 316: 596–601.

BECKWITH, J., and ROSSOW, P. 1974. Analysis of genetic regulatory mechanisms. *Ann. Rev. Genet.* 8: 1–13.

BECKWITH, J. R., and ZIPSER, D., eds. 1970. *The lactose operon.* Cold Spring Harbor, NY: Cold Spring Harbor Laboratory.

BERTRAND, K., et al. 1975. New features of the regulation of the tryptophan operon. *Science* 189: 22–26.

ENGLESBERG, E., and WILCOX, G. 1974. Regulation: Positive control. *Ann. Rev. Genet.* 8: 219–42.

GILBERT, W., and MÜLLER-HILL, B. 1966. Isolation of the *lac* repressor. *Proc. Natl. Acad. Sci.* 56: 1891–98.

———. 1967. The *lac* operator in DNA. *Proc. Natl. Acad. Sci.* 58: 2415–21.

GILBERT, W., and PTASHNE, M. 1970. Genetic repressors. *Scient. Amer.* (June) 222: 36–44.

GOLDBERGER, R. F., DEELEY, R. G., and MULLINIX, K. P. 1976. Regulation of gene expression in prokaryotic organisms. *Adv. in Genet.* 18: 1–67.

GOTTESMAN, S. 1984. Bacterial regulation: Global regulatory networks. *Ann. Rev. Genet* 18: 415–41.

HERSHEY, A. D., ed. 1971. *The bacteriophage lambda.* Cold Spring Harbor, NY: Cold Spring Harbor Laboratory.

JACOB, F., and MONOD, J. 1961. Genetic regulatory mechanisms in the synthesis of proteins. *J. Mol. Biol.* 3: 318–56.

JOHNSON, A. D., et al. 1981. λ repressor and *cro*—Components of an efficient molecular switch. *Nature* 294: 217–23.

KELLER, E. B., and CALVO, J. M. 1979. Alternative secondary structures of leader RNAs and the regulation of the *trp, phe, his, thr,* and *leu* operons. *Proc. Natl. Acad. Sci.* 76: 6186–90.

LEWIN, B. 1974. Interaction of regulator proteins with recognition sequences of DNA. *Cell* 2: 1–7.

MANIATIS, T., and PTASHNE, M. 1976. A DNA operator-repressor system. *Scient. Amer.* (Jan.) 234: 64–76.

MILLER, J. H., and REZNIKOFF, W. S. 1978. *The operon.* Cold Spring Harbor, NY: Cold Spring Harbor Laboratory.

PETRUSEK, R. H., DUFFY, J., and GEIDUSCHEK, E. 1976. Control of gene action in phage SPO1 development: Phage-specific modifications of RNA polymerase and a mechanism of positive control. In *RNA polymerase,* eds. R. Losick and M. Chamberlin, pp. 567–85. Cold Spring Harbor, NY: Cold Spring Harbor Laboratory.

PLATT, T. 1981. Termination of transcription and its regulation in the tryptophan operon of *E. coli. Cell* 24: 10–23.

PTASHNE, M. 1986. *A genetic switch.* Palo Alto: Blackwell Scientific Publ.

PTASHNE, M., et al. 1980. How the λ repressor and *cro* work. *Cell* 19: 1–11.

PTASHNE, M., JOHNSON, A. D., and PABO, C. O. 1982. A genetic switch in a bacterial virus. *Scient. Amer.* (Nov.) 247: 128–40.

RAIBAUD, O., and SCHWARTZ, M. 1984. Positive control of transcription initiation in bacteria. *Ann. Rev. Genet.* 18: 173–206.

REZNIKOFF, W. S. 1972. The operon revisited. *Ann. Rev. Genet.* 6: 133–56.

STROYNOWSKI, I., and YANOFSKY, C. 1982. Transcript secondary structures regulate transcription termination at the attenuator of *S. marcescens* tryptophan operon. *Nature* 298: 34–38.

STUDIER, F. W. 1972. Bacteriophage T7. *Science* 176: 367–76.

UMBARGER, H. E. 1978. Amino acid biosynthesis and its regulation. *Ann. Rev. Biochem.* 47: 533–606.

YANOFSKY, C. 1981. Attenuation in the control of expression of bacterial operons. *Nature* 289: 751–58.

YANOFSKY, C., and KOLTER, R. 1982. Attenuation in amino acid biosynthetic operons. *Ann. Rev. Genet.* 16: 113–134.

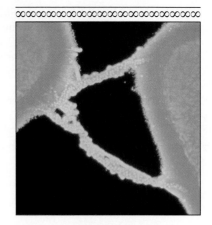

20

Genetic Regulation in Eukaryotes

∞

CHAPTER CONCEPTS

Genetic regulation in eukaryotes occurs at the level of transcription and during post-transcriptional events. Transcription is modulated by the interaction of various regulatory molecules with short DNA sequences most often located upstream from affected genes. Post-transcriptional mechanisms involve the selection of alternate products from a single transcript and the control of mRNA stability.

EUKARYOTIC GENE REGULATION: AN OVERVIEW

The discovery of the operon in *E. coli* by Jacob and Monod in 1960 was the first step in unraveling the mechanisms that govern gene expression. In the elation that followed this breakthrough, operons were invoked to explain everything from bacterial gene expression to the developmental expression of hemoglobin synthesis in humans. Only after a few years was it realized that even prokaryotes employ other mechanisms of regulation, and that eukaryotes may regulate gene action in fundamentally different and more complex ways. The discovery of introns and splicing of eukaryotic messages in 1977 confirmed this suspicion and accelerated the reconsideration of gene regulation in eukaryotes.

What the early proponents of operonlike controls in eukaryotes had overlooked is the fundamental difference in the evolutionary history of prokaryotes and eukaryotes. Specifically, each eukaryotic cell contains a much greater amount of genetic information, and this DNA is complexed with histones and other proteins to form chromatin, while in prokaryotic organisms, the DNA does not form into chromatin. The genetic information in eukaryotes is carried on many chromosomes, rather than one, and these chromosomes are enclosed within a nuclear membrane. In eukaryotes, the process of transcription is spatially and temporally separated from that of translation. The transcripts of eukaryotic genes are processed, cleaved, and realigned before transport to the cytoplasm, and many transcribed sequences never leave the nucleus. In addition, eukaryotes for the most part are multicellular. One consequence of this is that highly differentiated cells often synthesize large amounts of a restricted number of gene products even though they contain a complete set of

genetic information. In addition, during the development of multicellular organisms, changes in gene expression are often directed by external signals such as inducers and hormones.

Without belaboring the point, it seems obvious that these properties of eukaryotes provide opportunities for the regulation of genetic expression in many ways. For example, the level and type of gene products may be altered by amplifying or rearranging the genetic material. Regulation of gene expression may occur at many levels, including transcription, processing, transport, and translation. Although the details of gene regulation in eukaryotes have not yet been described for many genes, enough is known to allow some general features to be presented.

GENE AMPLIFICATION

One way to regulate the amount of a given gene product that is produced over a limited time period is to increase the number of copies of the structural gene that is being transcribed. This process, termed **gene amplification,** plays a role in the developmentally regulated expression of some gene sets, such as the ribosomal RNA genes of *Xenopus* (described in Chapter 12) and the chorion genes in *Drosophila,* and results in a high rate of gene expression over a limited period of time. Controlling the number of copies of a gene can be an effective way to speed the accumulation of a gene product. For example, the amplification of the ribosomal genes during oogenesis in *Xenopus* allows the developing oocyte to accumulate 10^{12} ribosomes in about 60 to 80 days, at an average rate of 300,000 per second, instead of the several thousand days the diploid number of genes would require.

FIGURE 20.1

The onionskin model of DNA amplification. Multiple initiations of DNA replication at an origin of replication (marked by small arrows) gives rise to a nested complex of replicating DNA molecules. Multiple recombination events within this complex can generate an intrachromosomal linear array, while recombination events within the same duplex can generate circular, extrachromosomal DNA.

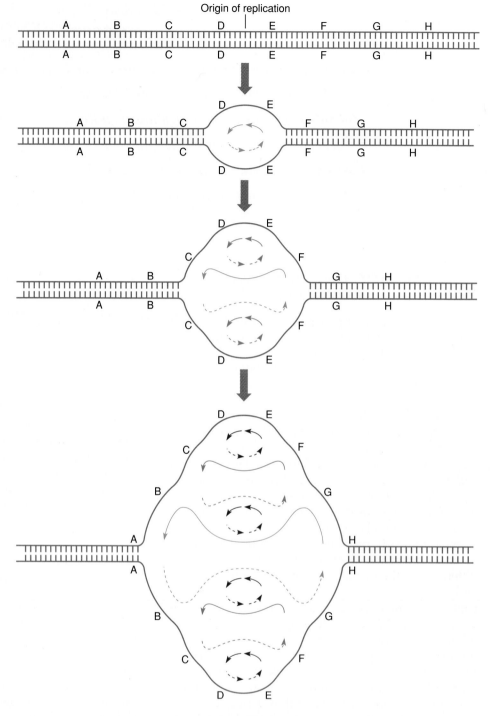

Origin of replication

In mammals, gene amplification does not appear to play a developmental role, but is often associated with the appearance of drug resistance in tumors during chemotherapy, and the multiple copies of tumor-associated genes known as oncogenes that are characteristic of the progressive changes associated with tumor growth. Amplified DNA sequences are often visible in mammalian cells as alterations in chromosome structure known as **extended chromosome regions (ECRs)** and **homogeneously staining**

regions (HSRs), or as extrachromosomal elements known as **double minute chromosomes (DMs)**.

Several mechanisms have been proposed to explain gene amplification, and they have been grouped into two types: mechanisms associated with DNA replication and those associated with recombination and segregation (Table 20.1)

One of the replication-based mechanisms, the onionskin model, associated with the amplification of chorion genes during oogenesis in *Drosophila*, is

TABLE 20.1
Molecular mechanisms for gene amplification.

Replication-Based Models	Segregation and Recombination-Based Models
Onionskin model	Deletion plus episome model
Extrachromosomal double rolling circle model	Sister chromatid exchange model
Chromosomal spiral model.	

shown in Figure 20.1. In this model, amplification is the result of multiple reinitiations of DNA synthesis within a single cell cycle. This causes the growth of nested replication complexes resembling the layers of an onion. Coupled with recombination events that

occur during replication, this model predicts a wide range of outcomes, including the origin of extrachromosomal DNA that might grow into DM structures after further rounds of replication.

Perhaps the most straightforward example of recombination- and segregation-driven models of gene amplification is the sister chromatid exchange model (Figure 20.2). If two sister chromatids engage in an unequal exchange, segments of DNA will be duplicated on one chromatid, and lost from the other. If this process is repeated for several cell divisions, the outcome will be a cell that contains one chromosome containing an ESR or HSR with many amplified copies of a gene, and another homologue with a single copy of the gene. Cytogenetic analysis of cultured cells that have undergone amplification in

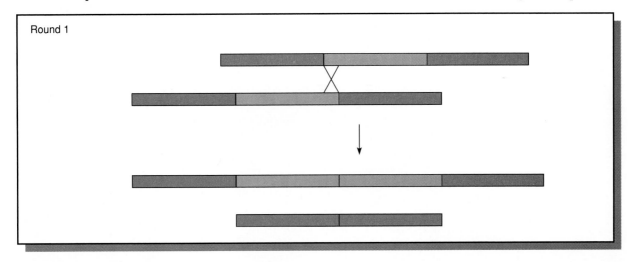

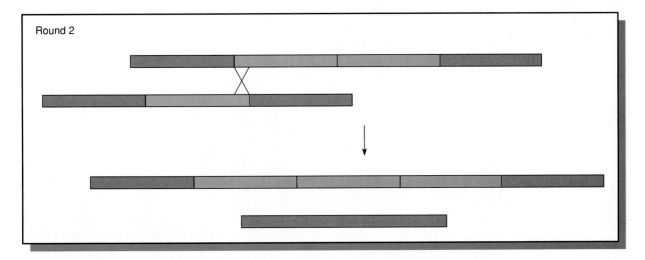

FIGURE 20.2
Sister chromatid recombination model of gene amplification. Misalignment and recombination between sister chromatids results in one chromatid with a duplicated segment (green portion) and a sister chromatid with a deleted segment. Further rounds of misalignment and recombination involving the chromosome carrying the duplication will result in a linear amplification of the shaded sequence.

response to drug selection confirm this expectation. When drug selection is relaxed, there is often a reduction in the copy number of the amplified gene, and a loss or reduction of the modified chromosomal region. Thus the sister chromatid model provides for a reversible means of altering the copy number of structural genes. It is hoped that future work on amplification in tumors and cultured cells will provide information about the organization of chromosomes and the chromosomal loci that serve as origins of DNA replication.

GENE REARRANGEMENT AND ANTIBODY DIVERSITY

A second general mechanism of gene regulation in higher eukaryotes does not involve increasing the number of loci that are transcribed, but reshuffling the organization of gene segments in a group of related genes to generate new combinations, resulting in the appearance of novel gene products. One excellent example of this mechanism is found in the production of antibody variability. One of the most complex families of genes is that specifying the proteins known as antibodies that participate in the **immune response** of vertebrates. These molecules are responsible for the highly specific response mounted when foreign substances called **antigens** enter the blood stream of an organism. The antigen may be an independent molecule, a part of the surface coat of an invading bacterium, virus, or organ transplant, or any other foreign substance. Two types of cells respond specifically to antigens. **B lymphocytes,** produced in bone marrow, react by synthesizing **antibodies** that are secreted into plasma and react with soluble antigens. These antibodies are also called **immunoglobulins**. **T lymphocytes,** white blood cells that are produced in bone marrow and

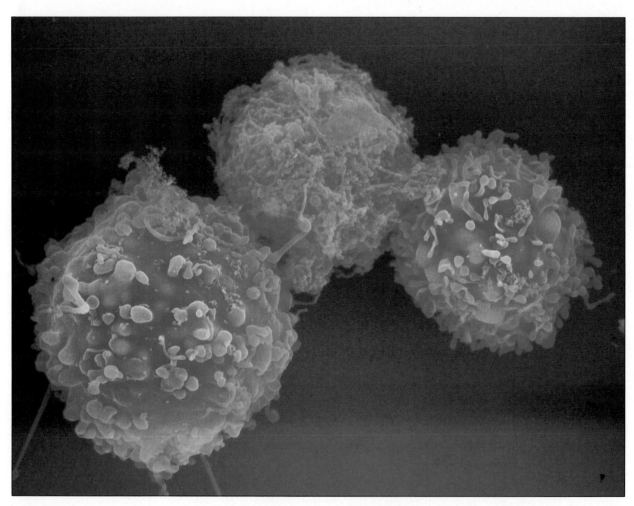

FIGURE 20.3
A scanning electron micrograph showing the T and B cells of the immune system.

mature in the thymus, react with antigens present on cell surfaces. They produce a counterpart to antibodies called the **T-cell receptors,** which mediate their specific immune response (Figure 20.3).

Mammals can produce millions of different antibodies, each responding to a different antigen. It is now clear that the basis for this molecular diversity lies in the various amino acid sequences composing the chains of the antibodies. In humans, antibodies, or immunoglobulins (Ig), have been divided into five classes: **IgG, IgA, IgM, IgD,** and **IgE**. The first class, IgG, represents about 80 percent of the antibodies found in blood and is the most extensively characterized.

Each antibody molecule consists of two different types of polypeptide chains, each present in duplicate. The larger or **heavy chain (H)** consists of approximately 440 amino acids, and the smaller or **light chain (L)** contains approximately 220 amino

acids. The first 110 amino acids of the N-terminus of each chain vary in sequence in different immunoglobulin molecules and are thus known as the **variable region (V)**. The remaining C-terminal amino acid sequence of each chain is invariable and is called the **constant region (C)**. The variable regions are responsible for antibody specificity and contain the **antibody combining sites** that bind to the antigens. This general scheme, illustrated in Figure 20.4, was first determined by Rodney Porter and Gerald Edelman in the early 1960s.

If a foreign organism invades the body, a small part of a surface protein or glycoprotein will serve as an antigen. Once specific antibodies are produced, they combine with the antigenic determinants in a lock-and-key configuration. Since each antibody has two combining sites (Figure 20.4), large complexes are formed that can be engulfed by white blood cells and effectively removed from the body.

FIGURE 20.4
(a) General scheme of the IgG immunoglobulin showing the constant, variable, and hypervariable regions of the heavy (H) and light (L) chains.
(b) Antibody-binding sites of the variable regions and complexes that may be formed.

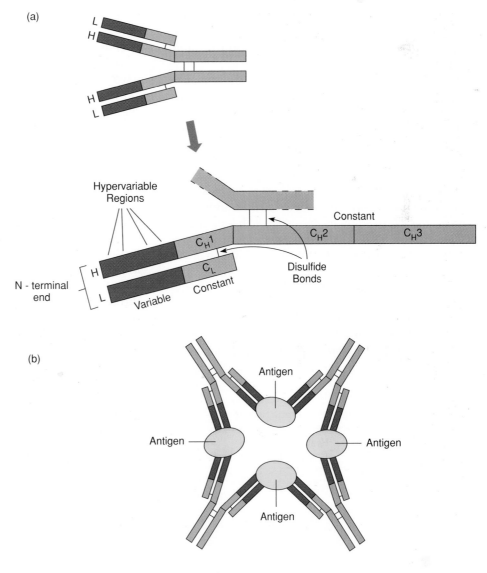

Within the variable region are areas that vary more than others. There are three such **hypervariable regions** in both light chains and heavy chains (Figure 20.4). They are most important to the three-dimensional configuration of the combining site.

The question of major genetic interest is how the vast array of molecular variability demonstrated by antibodies is encoded in the genome. Three hypotheses have been proposed and supported to one degree or another over the past several decades: the **germline theory,** the **somatic mutation theory,** and the **recombination theory**. The latter has received the greatest experimental support. Basically, this theory proposes that there are a limited number of genes that encode each of the various portions of the different types of H and L chains. As antibody-forming B lymphocytes develop, a mechanism involving DNA recombination brings various combinations of these genes together so that each lymphocyte comes to encode only one specific type of antibody. In the presence of a specific antigen, the corresponding lymphocyte is stimulated to differentiate and proliferate further. As a result, antibodies containing the appropriate combining sites are produced and interact with the antigen.

To comprehend how the diversity of antibodies is generated, we must delve into the structure of immunoglobulins and the genes encoding the regions of their polypeptide components. Recall that there are light and heavy polypeptide chains, each containing variable and constant regions (V_L, C_L, V_H, and C_H). Also, there are five classes of immunoglobulins. As shown in Table 20.2, each class contains one of two types of light chains, λ or κ. Furthermore, each class is characterized by its own specific heavy chain, designated α, γ, δ, ε, or μ. Thus, there are ten different general classes of immunoglobulins that can be formed.

Although there is some variation found in the C region of different chains, the major variation is found in the V region (Figure 20.5). The V_L region appears to be encoded by two sets of genes, designated V and J. **V genes** specify the N-terminal

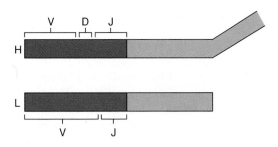

FIGURE 20.5
Schematic diagram of the V and J regions of light chains and the V, D, and J regions of heavy chains.

portion, including the first two hypervariable regions and part of the third. The **J genes** specify the remainder of the V region, including the latter part of the third hypervariable sequence. When an antibody-producing cell is formed, different combinations of V and J genes are fused together through recombination. If numerous V and J genes combine at random, an extremely large number of V_L sequences can be encoded.

Does this really happen? In 1976, Susumo Tonegawa applied the techniques of recombinant DNA cloning and nucleotide sequencing to demonstrate that it does. He compared a cloned fragment specifying a λ V_L chain derived from embryonic germline DNA with that derived from DNA of an antibody-producing cell. In the embryonic DNA, the region coding for the first 95 amino acids of the V_L region was separated by a 4.5-kb region from that coding for the remaining 13 amino acids (the J region) and the C region. However, in the DNA of the antibody-producing cell, the V and J segments were joined contiguously to form a single gene encoding a specific λ V_L chain. This observation strongly supports the hypothesis that immunoglobulin genes are formed through a process of recombination. Similar processes lead to the formation of DNA encoding the κ light chain, illustrated in Figure 20.6.

Even more diversity is possible in the V_H region because it is divided into three segments, each thought to be represented by distinct genetic segments (see Figure 20.5). In V_H chains, a V region containing the first two hypervariable segments occurs. Additionally, a D region containing the third hypervariable segment and a J segment comprising the rest of the V_H region have been described. The DNA encoding heavy chains is formed by joining D- and J-coding regions. The combined DJ segment is then combined with a V-coding region. The diagram in Figure 20.7 illustrates the diversity of heavy chains that may be formed in this way.

TABLE 20.2
Categories and components of immunoglobins.

Ig Class	Light Chain	Heavy Chain	Tetramers	
IgA		α	$\kappa_2\alpha_2$	$\lambda_2\alpha_2$
IgD		δ	$\kappa_2\delta_2$	$\lambda_2\delta_2$
IgE	κ or λ	ε	$\kappa_2\epsilon_2$	$\lambda_2\epsilon_2$
IgG		γ	$\kappa_2\gamma_2$	$\lambda_2\gamma_2$
IgM		μ	$\kappa_2\mu_2$	$\lambda_2\mu_2$

FIGURE 20.6

The formation of the DNA segments encoding a human κ light chain and the subsequent transcription, mRNA splicing, and translation leading to the final polypeptide chain. In germline DNA, up to 150 different L-V (Leader-Variable) segments are present. These are separated from the J regions by a long noncoding sequence. The J regions are separated from a single C gene by an intervening sequence (intron) that must be spliced out of the initial mRNA transcript. Following translation, the amino acid sequence derived from leader RNA is cleaved off as the mature polypeptide chain passes across the cell membrane.

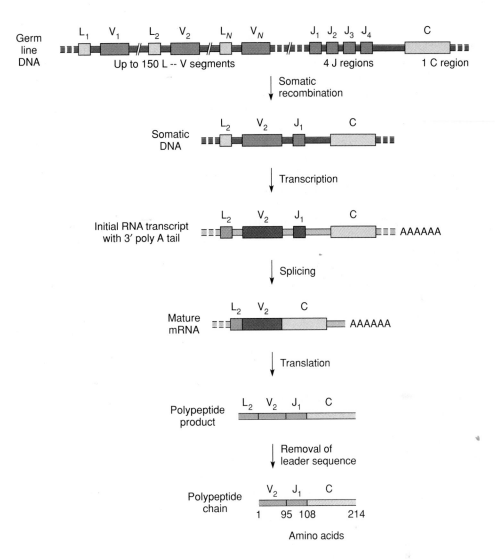

Based on this information, a general explanation of the origin of antibody diversity is emerging. Certain clusters or gene families encode the various segments of heavy and light chains. Recombination brings together various gene fragments, and the newly formed composite gene encodes a specific chain.

This shuffling of DNA segments allows the production of literally millions of different antibodies from a restricted amount of genetic information. The larger question is, what controls the shuffling and reshuffling of genes in cells of the immune system to generate the multitudes of antibodies? One candidate gene that might regulate this activity, called **RAG-1** for **recombination-activating gene,** has been isolated and characterized. This gene is expressed only in lymphocytes, and only during those stages when immune recombination is taking place. When transferred in an active form to cultured fibroblasts, this gene confers the ability to bring about the assembly of variable (V), joining (J) and diversity (D) segments, a property not normally found in fibroblasts.

A second candidate for the VDJ recombinase is RBP-2, which encodes a protein that binds to the recognition sequences that flank the J region before recombination. The protein is similar in several ways to bacterial integrases, proteins involved in splicing

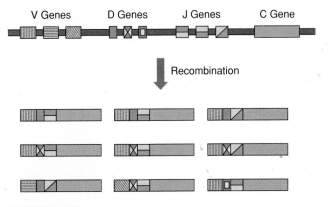

FIGURE 20.7

The theoretical formation of various heavy chain genes during the maturation of an antibody-forming cell.

DNA into bacterial chromosomes. It is not clear whether these genes work independently or are part of a coordinated pattern of activity involving a number of regulatory genes. Further work may reveal whether there is yet another level of regulation that switches on the activity of the recombinases.

TRANSPOSABLE GENETIC UNITS

With our knowledge of insertion sequences and transposons in bacteria (Chapter 15), it comes as no surprise that mobile genetic units also exist in eukaryotes. They are, in fact, more widespread, and they often have the effect of altering the expression of genetic information. In the following sections, we will review a variety of cases where transposable genetic units affect gene expression in a variety of organisms, including humans.

The *Ac-Ds* System in Maize

About twenty years before the discovery of transposons in prokaryotic organisms, Barbara McClintock analyzed the genetic behavior of two mutations, *Dissociation (Ds)* and *Activator (Ac)*, in corn plants. Initially, she determined that *Ds* was located on chromosome 9. Provided that *Ac* was also present in the genome, *Ds* had the effect of inducing breakage at a point on the chromosome adjacent to its location. If breakage occurred in somatic cells during their development, progeny cells often lost part of chromosome 9, causing a variety of phenotypic effects.

Subsequent analysis showed that both the *Ds* and *Ac* genes were sometimes transposed to different chromosomal locations. While *Ds* moved only if *Ac* was also present, *Ac* was capable of autonomous movement. Where *Ds* came to reside determined its genetic effect. Although it might cause chromosome breakage, it might instead inhibit gene expression. In cells where expression was inhibited, *Ds* might move again, releasing this inhibition. In these cases, the *Ds* element is believed to insert into a gene and subsequently to depart from it, causing changes in gene expression. Often, analysis involved an examination of the phenotypes of the kernels of maize resulting from genes expressed in either the endosperm or aleurone layers (Figure 20.8). Figure 20.9 illustrates the sort of movements and effects of the *Ds* and *Ac* elements described above. McClintock concluded that the *Ds* and *Ac* genes were **transposable controlling elements**.

Until many years later, nothing comparable to the controlling elements in maize was recognized in other organisms. When the behavior of certain prophages and plasmids was revealed and insertion sequences and transposons were discovered, many parallels were evident. Transposons and insertion sequences were seen to move into and out of chromosomes, to insert at different positions, and to affect gene expression at the point of insertion.

Several *Ac* and *Ds* elements have now been isolated and carefully analyzed (Figure 20.10). As a result of this information, the relationship between the two elements has been clarified. The first *Ac* element sequence is 4563 bases long and strikingly similar to the bacterial transposon **Tn3**. This sequence

FIGURE 20.8
Corn kernel showing spots of colored aleurone produced by genetic transposition involving the *Ac/Ds* system.

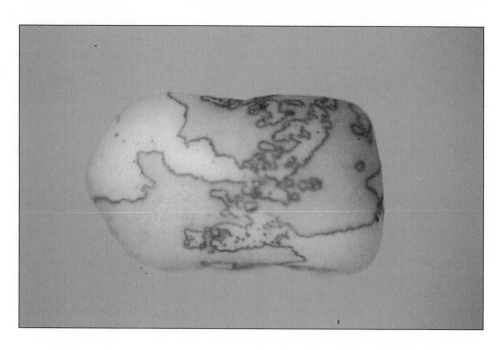

FIGURE 20.9

Two consequences of the influence of the *Activator* (*Ac*) element on the *Dissociation* (*Ds*) element. In (a), *Ds* is transposed to a region adjacent to a theoretical gene *W*. Subsequent chromosomal breakage is induced, the *W*-bearing segment is lost, and mutant gene expression occurs. In (b), *Ds* is transposed to a region within the *W* gene, causing immediate mutant expression. *Ds* may also "jump" out of the *W* gene with the accompanying restoration of *W* gene activity and its wild-type expression.

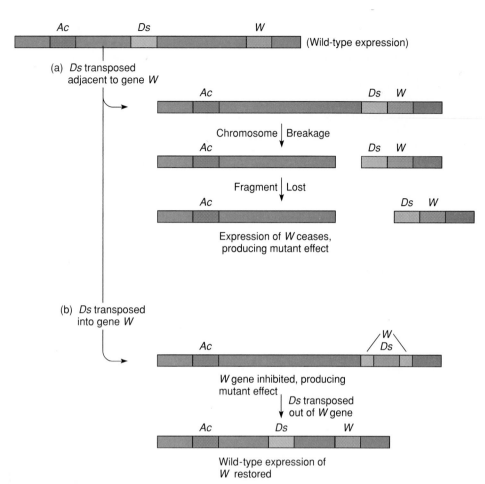

contains two 11-base-pair imperfect terminal inverted repeats, two open reading frames (ORFs), and three noncoding regions. The first *Ds* element studied (*Ds-a*) is nearly identical in structure to *Ac* except for a 194-base segment that has been deleted from the larger open reading frame (Figure 20.10). There is some evidence that this gene encodes a **transposase** enzyme, essential to transposition of both the *Ac* and *Ds* elements. The deletion in this gene in the *Ds* element explains its dependence on the *Ac* element for transposition. Several other *Ds* elements have also been sequenced and each reveals an even larger deletion in the same region. In each case, however, the terminal repeats are retained and seem to be essential for transposition, provided that a functional transposase enzyme is supplied by the gene in the *Ac* element.

Other Mobile Genetic Elements

More recent work on transposable elements in plants has led us full circle to the union of Mendel's own observations with molecular genetics. Recall that one of the first phenotypes investigated by Mendel involved the inheritance of round and wrinkled peas.

The two phenotypes are produced by alleles of a single gene, *rugosus*. It is now known that the wrinkled phenotype is associated with the absence of an enzyme, starch-branching enzyme (SBEI), that controls the formation of branch points in starch molecules. The lack of starch synthesis leads to the accumulation of sucrose and a higher water content and osmotic pressure in the developing seeds. As the seeds mature, the wrinkled seeds (genotype *rr*) lose more water than the smooth seeds (*RR* or *Rr*), producing the wrinkled phenotype (Figure 20.11).

The structural gene for SBEI has now been cloned and characterized in both wild type and mutant genotypes. In the *rr* genotype, the SBEI protein is nonfunctional, and the SBEI gene is interrupted by a 0.8-kb insertion in an exon, resulting in the production of an abnormal RNA transcript. The inserted DNA has 12-bp inverted repeats at each end that are highly homologous to the terminal sequences in the transposable element *Ac* from maize, and other *Ac*-like elements from snapdragons and parsley. As we will see below, the terminal sequences and the genetic information encoding a transposase enzyme are universal components of transposons in all organisms studied.

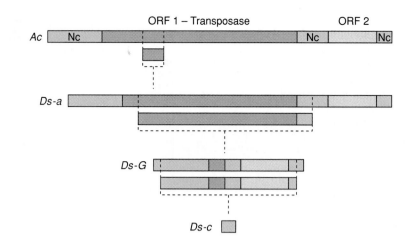

FIGURE 20.10
A comparison of the structure of an *Ac* element with three *Ds* elements, all of which have been isolated and sequenced. The imperfect and inverted repeats are at the ends of the *Ac* element. The transposase gene is in open reading frame (ORF) 1. No function has yet been assigned to open reading frame 2. Noncoding regions are designated by NC. The dotted lines indicate the area that has been deleted, leading to each successive *Ds* element. As this scheme shows, *Ds-a* appears to be simply an *Ac* element containing a small deletion in the gene encoding the transposase enzyme.

Transposable elements have been discovered in other eukaryotic organisms, notably in yeast, *Drosophila*, and primates, including humans. In 1975, David Hogness and his colleagues David Finnigan, Gerald Rubin, and Michael Young identified a class of genes in *Drosophila melanogaster*, which they desig-

nated as *copia*. These genes transcribe "copious" amounts of RNA (thus, their name), which can be isolated in the polyadenylated mRNA fraction. Present up to 30 times in the genome of cells, *copia* genes are nearly identical in nucleotide sequence. Mapping studies show that they are transposable to

FIGURE 20.11
The *wrinkled* trait in garden peas, studied by Gregor Mendel, is caused by the insertion of a transposable element into the structural gene for starch-branching enzyme.

different chromosomal locations and are dispersed throughout the genome.

Copia appears to be only one of the approximately 30 families of transposable elements in *Drosophila*, each of which is present 20 to 50 times in the genome. Since the discovery of *copia*, many other families of transposable elements have been recognized, all of which are referred to as *copia*-like. Together, the many families constitute about 5 percent of the *Drosophila* genome and over half of the middle repetitive DNA of this organism.

Despite the variability in DNA sequence between the members of different families, they share a common structural organization thought to be related to the insertion and excision processes of transposition. Each *copia* gene consists of approximately 5000 base pairs of DNA, including a long family-specific **direct terminal repeat** sequence **(DTR)** of 276 base pairs at each end. Within each repeat is a short **inverted terminal repeat (ITR)** of 17 base pairs. These features are illustrated in Figure 20.12. The DTR sequences are found in other transposons in other organisms but are not universal. However, the shorter ITR sequences are considered universal.

Insertion of *copia* and other transposable elements appears to be dependent on ITR sequences and occurs at specific target sites in the genome. This is inferred by the observation that certain regions of chromosomes are more disposed to transposon-induced changes. Upon insertion, short direct repeats are usually generated at points of insertion. In the case of *copia*, this repeat sequence is 5 nucleotides long. For *Ds* in maize an 8-nucleotide sequence is generated. In general, eukaryotic transposons are strikingly similar to one another and share many features with those in bacteria.

The *copia*-like elements demonstrate regulatory effects at the point of their insertion in the chromosome. Certain mutations, including ones affecting eye color and segment formation, have been found to be due to insertions within genes. For example, the eye color mutation *white-apricot* (*w^a*), an allele of the *white* (*w*) mutation, contains a *copia* insertion element. Removing the transposable element restores the wild-type allele.

The P and I elements are associated with a phenomenon in *Drosophila* referred to as **hybrid dysgenesis**. When P or I insertions are contributed paternally to an offspring whose mother lacks them, increased levels of sterility, male recombination, chromosomal rearrangements, and mutations may result. The offspring is said to be "dysgenic" and thus frequently fails to reproduce. The transposability of these elements, which induce most of these problems, is under the control of the maternal

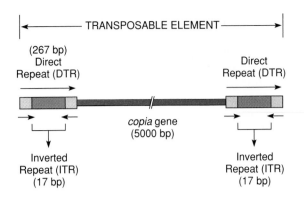

FIGURE 20.12
Structural organization of a *copia* transposable element in *Drosophila melanogaster*.

cytoplasm in the zygote. One model suggests that P elements encode a transposase enzyme that is active in maternal cytoplasm derived from females lacking them. In such cases, P insertions then induce hybrid dysgenesis. Females already carrying P elements apparently contain cytoplasm that renders the enzyme inactive.

The final class of mobile eukaryotic genetic units that we will discuss is the **Alu family**. Found in mammals, the *Alu* family consists of large numbers of highly repetitive repeat sequences, originally detected using reassociation kinetics. In humans, there are about 300,000 copies found interspersed throughout the genome. These sequences represent most of the highly repetitive DNA in humans and constitute 3 percent of the total genome. The family has received its name because a high percentage of these sequences are cleaved by the restriction endonuclease *Alu* I.

The significant role of the *Alu* family in genetic processes is supported by the following observations:

1 These sequences are found in all primate and rodent DNA.
2 There is conserved within the sequences in mammals a specific 40-base-pair unit.
3 These sequences are reflected in nuclear transcripts generated by RNA polymerase II and III.

These sequences are considered transposable based on several lines of evidence. Most important is the fact that the 300-bp sequences are flanked on either side by direct repeat sequences consisting of 7 to 20 bp. This observation is similar to that of bacterial insertion sequences. These flanking regions are related to the insertion process during transposition. The second line of evidence involves the observation that clustered regions of *Alu* sequences vary in the DNA of normal and diseased individuals and in different tissues of the same individual. Additionally, the sequences have been found extrachromosomally.

Potential mobility of such a large number of short nucleotide sequences is a most intriguing concept. Do these units play a significant role in gene expression and regulation? Have they been important in the evolutionary process of primates? Future research will determine the answers to these questions.

CELL FUSION AND GENE ACTIVITY

Although most differentiated eukaryotic cells contain a complete set of genetic information, most genes are repressed or inactive. Along with genes needed for basic housekeeping functions such as nutrient metabolism, only a small tissue-specific subset of genes are transcriptionally active. This pattern of tissue-specific synthesis is itself evidence that eukaryotic genes are transcriptionally regulated. One type of evidence for factors that control transcription has been gathered through the analysis of gene activity following cell fusion.

When two genetically distinct cell lines are grown together in culture, cell fusion is a spontaneous and rare event. The new cell produced as a result of the fusion contains two nuclei and is called a *heterokaryon*. These heterokaryons remain viable for some period of time, but do not replicate and thus die off after a short period in culture. Although they are a transient phenomenon, heterokaryons have been useful in providing evidence for cytoplasmic control of nuclear transcription.

Rabbit white blood cells normally synthesize RNA but not DNA. The hen erythrocyte (which, unlike that of mammals, retains its nucleus) is highly differentiated and synthesizes almost no RNA or

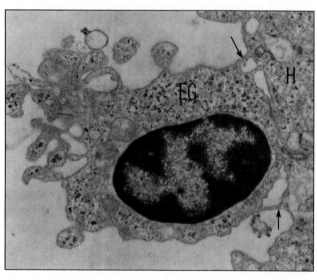

(a)

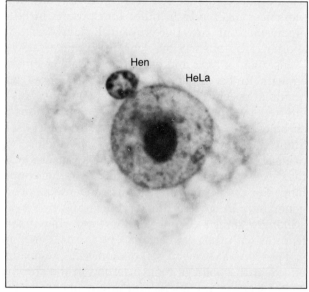

(b)

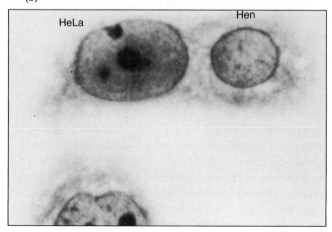

(c)

FIGURE 20.13
Stages in heterokaryon formation between HeLa cells and hen erythrocytes. (a) Cytoplasmic bridges between hen erythrocyte ghost (EG) and HeLa cell (H). (b) Heterokaryon contains a large HeLa cell nucleus and a small hen erythrocyte nucleus. (c) Enlargement and dispersion of chromatin in hen nucleus following incorporation into HeLa cytoplasm.

DNA. The human-derived HeLa cell, on the other hand, synthesizes both RNA and DNA. When various combinations of these three cell types are used two at a time as the source of heterokaryons, several general observations can be made:

1 If one parental cell synthesizes RNA prior to fusion, both nuclei of the heterokaryon synthesize RNA.
2 If one cell normally synthesizes DNA prior to fusion, both nuclei of the heterokaryon synthesize DNA.
3 If neither cell normally synthesizes RNA or DNA prior to fusion, neither of the nuclei synthesizes RNA or DNA, respectively, following heterokaryon formation.

Since the three cells above are from quite diverse sources (rabbit, chicken, and human), a very significant conclusion may be drawn. The signals initiating either RNA or DNA synthesis in an inactive nucleus are shared by the three organisms and are not at all species-specific.

When hen erythrocytes are used as a source for heterokaryon formation, the cells, prior to fusion, contain a small nucleus with highly condensed chromatin. Using Sendäi virus on these erythrocytes causes hemolysis, rupturing the cell membrane and dispersing the cytoplasmic contents. As shown in Figure 20.13, what remains in the intact nucleus and parts of the cell membrane, referred to as an **erythrocyte ghost (EG)**. These EGs attach to whatever cells are being utilized in the fusion study and are ultimately engulfed into the second cell's cytoplasm. Of importance here is that the heterokaryon that is formed does not contain hen erythrocyte cytoplasm, but only cytoplasm of the other cell type used in the experiment. Thus, any initiating signals for reactivation of DNA and RNA synthesis or morphological change of the hen nucleus must originate from the foreign cytoplasm.

When HeLa cells are fused with hen EGs, synthesis of both RNA and DNA is detected in the hen nucleus. Even more striking are the morphological changes that may be observed in the hen nucleus following heterokaryon formation. Not only does the erythrocyte nucleus proceed to enlarge its volume 20 to 30 times, but the highly condensed chromatin, characteristic of genetic inactivity, becomes greatly dispersed (Figure 20.13). Presumably, this dispersion makes more of the DNA accessible to transcription and is related to the increasing amount of RNA that is synthesized as the nucleus enlarges. Further studies by Henry Harris and colleagues have shown that hen-specific antigens and enzymes are synthesized in

the heterokaryon, and HeLa cell proteins can be detected entering the hen nucleus prior to gene action.

From this investigation it is apparent that the nucleus from a highly specialized cell, upon entering a foreign cytoplasm that is actively metabolizing, reverses its differentiated and genetically inert state. The original genetic control of the intact erythrocyte is lost when its cytoplasm is removed, and the regulation of genetic activity in the nucleus is strongly influenced by the new cytoplasmic environment following fusion. In the next section, we will investigate the factors that regulate transcriptional events in one nucleus, including factors that move from the cytoplasm into the nucleus.

REGULATION OF TRANSCRIPTION

As in prokaryotes, transcriptional control in eukaryotes involves the interaction of DNA sequences adjacent to the regulated genes and DNA-binding proteins known as **transcription factors**. As outlined in Chapter 18, the regulatory sequences adjacent to a wide range of genes have been identified, mapped, and sequenced. The development of DNA protection and gel retardation assays have allowed the detection of DNA-binding proteins in cell extracts, and new affinity chromatography techniques have been used to isolate transcriptional factors present in very low concentrations. The result has been a virtual explosion of information about eukaryotic transcriptional factors. In this section, we will review some of this new information, including the upstream organization of eukaryotic genes, the characteristics of transcription factors, their interaction with DNA sequences, other regulatory factors, and the initiation of transcription by RNA polymerase.

Promoters and Enhancers in Eukaryotes

Eukaryotic genes are divided into three classes, depending on whether they are transcribed by RNA polymerase I (ribosomal RNA), RNA polymerase II (messenger RNA) or by RNA polymerase III (transfer RNA, 5S RNA, and other small RNAs). Each class requires a different set of regulatory elements. In addition, while the regulatory sequences of prokaryotic genes are generally organized according to a fairly inflexible plan, there is much more variation in the organization and location of regulatory elements within or adjacent to eukaryotic genes. Two general classes of such elements can be distinguished: **promoters** and **enhancers**. Promoters that are recognized by RNA polymerase II are composed of short modular sequences of DNA usually located within 100 bp upstream (in the 5′ direction) of the gene. In the promoter closest to the start of most genes is a sequence called the **TATA box**. Located some 25 to 30 bases upstream from the initial point of transcription (symbolized as −25 to −30), it consists of an 8-base-pair consensus sequence composed only of A-T pairs, often flanked on either side by GC-rich regions. Mutations in the TATA box severely restrict transcription, and deletions often alter the initiation point of transcription. A second modular component of many promoters is called the **CAAT box**. Its consensus sequence is CAAT or CCAAT, it frequently appears in the region −70 or −80 bp from the start site, and can function in either orientation. Mutational analysis suggests that the CAAT box determines the efficiency of the promoter or promoter strength. A third modular element of some promoters is called the **GC box,** which has the consensus sequence GGGCGG and is often found at about position −110. This module is found in either orientation, and often occurs in multiple copies. Frequently, other modules, with the sequences GCCACACCC, ATGCAAAT, etc., are present as upstream regulatory sequences.

The variation in the organization of the upstream regulatory elements in several eukaryotic genes is shown in Figure 20.14. Note that in different genes there are great differences in the number and orientation of promoter elements and the distance between them. Just how a functional promoter can be composed of different elements scattered along a DNA molecule is not clear. One suggestion is that RNA polymerase II initially makes contact with the DNA at an upstream region and then moves along the DNA to the start site of transcription (Figure 20.15). In this model, the spacing of upstream elements is relatively unimportant, as long as transcription factors bind to the regulatory sequences, and provide recognition for RNA polymerase. Alternately, the DNA molecule may fold or loop to bring distant upstream sites into juxtaposition (Figure 20.15), implying that the spacing between elements is more important than the identity of the promoter modules.

In addition to promoters, transcription of most if not all eukaryotic genes is regulated by additional *cis*-acting sequences called **enhancers**. Enhancers greatly stimulate the transcriptional activity of a promoter and can be distinguished from promoters by several characteristics:

1   The position of the enhancer need not be fixed; it can be placed upstream, downstream, or within the gene it regulates.
2   Its orientation can be inverted without significant effect on its action.
3   Enhancers are not restricted to specific genes. If an enhancer is moved to another location in the genome, or if an unrelated gene is placed near an enhancer, transcription of the adjacent gene is enhanced.

The enhancer of the SV40 virus consists of two tandemly repeated 72-bp sequences located some 200

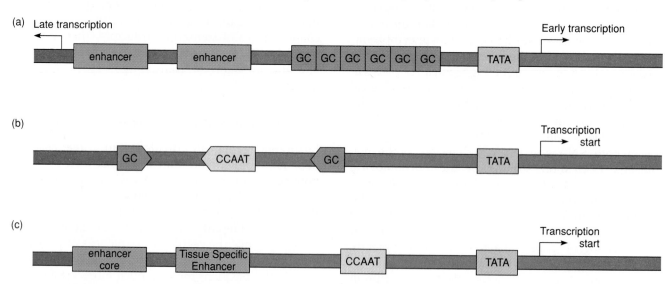

FIGURE 20.14
Upstream organization of several eukaryotic genes, illustrating the variable nature, number, and arrangement of controlling elements. (a) SV40 control region, with early genes to right, late genes to left. (b) Thymidine kinase. (c) Insulin gene.

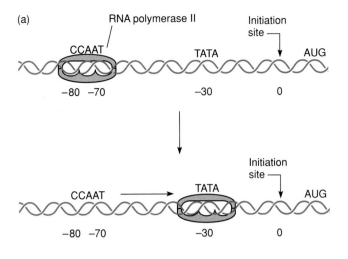

(a)

RNA polymerase II

CCAAT   TATA   Initiation site

AUG

−80 −70   −30   0

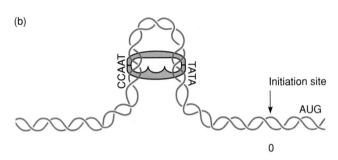

Initiation site

CCAAT   TATA

AUG

−80 −70   −30   0

(b)

CCAAT   TATA

Initiation site

AUG

0

FIGURE 20.15
Two models to explain the interaction of polymerase molecules with promoters. (a) Polymerase binds upstream, then moves along DNA to initiation site. (b) DNA folds to bring upstream region (CCAAT) in proximity to TATA box.

bp upstream from a transcriptional start point (Figure 20.16). If one or the other of these sequences is deleted, there is no effect on transcription, but if both are deleted, *in vivo* transcription is greatly reduced.

Most eukaryotic genes are also under the control of enhancers. In the immunoglobulin heavy chain genes, an enhancer is located in the intron between

the V-D-J coding region and the C region of the rearranged DNA in B lymphocytes. In this case, the enhancer is located *within* the gene it regulates. It is also active only in B cells in which the immunoglobulin genes are being expressed, indicating that tissue-specific gene expression can be modulated through enhancers.

Internal enhancers have been discovered in other eukaryotic genes, including the immunoglobulin light chain gene and the *Xenopus* 5S RNA genes. Downstream enhancers are found in the human beta globin gene and the chicken thymidine kinase gene. In chickens, an enhancer is located between the beta globin and the epsilon globin genes and apparently works in one direction to control the epsilon gene during embryonic development and in the opposite direction to regulate the beta gene during adult life. In yeast, regulatory sequences similar to enhancers, called **upstream activator sequences (UAS),** can function upstream at variable distances and in either orientation. They differ from enhancers in that they cannot function downstream of the transcription start point.

Like promoters, enhancers contain modules composed of short DNA sequences. The SV40 enhancer contains sequences that when mutated reduce the efficiency of the enhancer by 40 to 80 percent. This and other enhancers contains a copy of an octamer sequence (ATGCAAATNA) that is found in both promoters and enhancers, but so far, it has proven difficult to determine which sequences constitute the basic units of enhancer structure. These postulated units, called **enhansons,** are probably short sequences that represent the binding sites for transcriptional factors, and may have sequence-specific properties for the binding of different factors.

Overall, the distinction between promoters and enhancers is really an operational one. Enhancers are able to stimulate transcription at a distance, while this is not necessarily true of promoters. As a group, promoters contain some modules (the TATA box) that must be located at a fixed site upstream of the

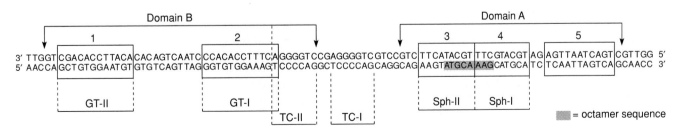

FIGURE 20.16
DNA sequence for the SV40 enhancer. Below the sequence are the various control sequences contained within this region.

transcription start point, while enhancers do not contain these modules. Although enhancers and promoters share some modules, the modules in enhancers are most often contiguous, while in promoters, they are spaced apart. On the other hand, both enhancers and promoters have some striking similarities; they both stimulate transcription, they both have modular organization, with some modular sequences in common, and they both bind transcriptional factors.

The most intriguing question about enhancers is how they are able to exert control over transcription at a great distance from either promoters or the transcriptional start site. While there is no clear answer yet, some intriguing possibilities should be considered.

1   The transcription factors that bind to the enhancer may alter the structure of chromatin and/or DNA to allow the binding of RNA polymerase. For example, in SV40, the DNA within and just outside the 72-bp repeat of the enhancer is coiled to form Z-DNA. This alteration in DNA conformation may influence transcriptional activity.

2   The binding of transcriptional factors to enhancers may alter the configuration of chromatin and/or DNA by bending or looping the DNA to bring distant enhancers and promoters into direct contact.

While the general features of enhancer activity are known, there are several important issues about the function of enhancers that are unresolved. It is not known what fraction of all genes require the action of an enhancer, nor do we understand what role enhancers might play in tissue-specific gene expression, nor how enhancers work at the molecular level. It is clear that such regulatory sequences work by binding protein transcription factors, and that these factors, in turn, control transcription.

Transcription Factors

Proteins that are not part of the RNA polymerase molecule itself, but are needed for the initiation of transcription, are called **transcription factors**. These proteins have at least two functional domains: one that binds to DNA sequences present in promoters and enhancers, and another that activates transcription via protein-protein interaction by binding to RNA polymerase or to other transcription factors. These domains of eukaryotic transcription factors show several structural adaptations to their functional requirements.

The **DNA binding domains** or **structural motifs** of eukaryotic transcription factors analyzed to date have been classified into three major types: **zinc fingers, the homeodomain (HD)** or **helix-turn-helix,** and the **leucine zipper** (Figure 20.17). This classification is not exhaustive, and other new groups will undoubtedly be established as new factors are characterized.

Zinc fingers consist of amino acid sequences containing two cysteine and two histidine residues at repeating intervals. The consensus amino acid repeat is: $Cys-N_{2-4}-Cys-N_{12-14}-His-N_3-His$. The interspersed cysteine and histidine residues covalently bind zinc atoms, folding the amino acids into loops known as zinc fingers (Figure 20.17). Each finger consists of approximately 23 amino acids, with a loop of 12 to 14 amino acids between the Cys and His residues, and a linker between loops consisting of 7 or 8 amino acids. The amino acids in the loop interact with and bind to specific DNA sequences. The number of fingers in a transcription factor varies from 2 to 13, as does the length of the DNA binding sequence. The mammalian transcription factor Sp1, for example, contains three zinc-finger regions at its carboxyl terminus, and selectively binds to GC-box modules in promoters. Similar proteins with zinc-finger motifs have been isolated from yeast, *Drosophila, Xenopus,* and humans.

FIGURE 20.17 (opposite)
Three structural motifs of DNA binding proteins. (a) In the helix-turn-helix configuration, one helix of the protein recognizes and binds to the DNA, and the other helix helps lock the DNA-recognition helix into place. (b) The zinc-finger motif shown here consists of stretches of about 30 amino acids containing two cysteine and two histidine residues that bind to a Zn^{2+} ion. These stretches are separated by regions of about 12 amino acids. A second class of zinc fingers uses two pairs of cysteine residues rather than the cysteine-histidine pair to bind the zinc ion. (c) Proteins with leucine zippers contain a conserved stretch of about 30 amino acids with a net basic charge followed immediately by a region containing four leucine residues spaced at intervals of seven amino acids. This region allows the protein to form functional dimers that bind to DNA.

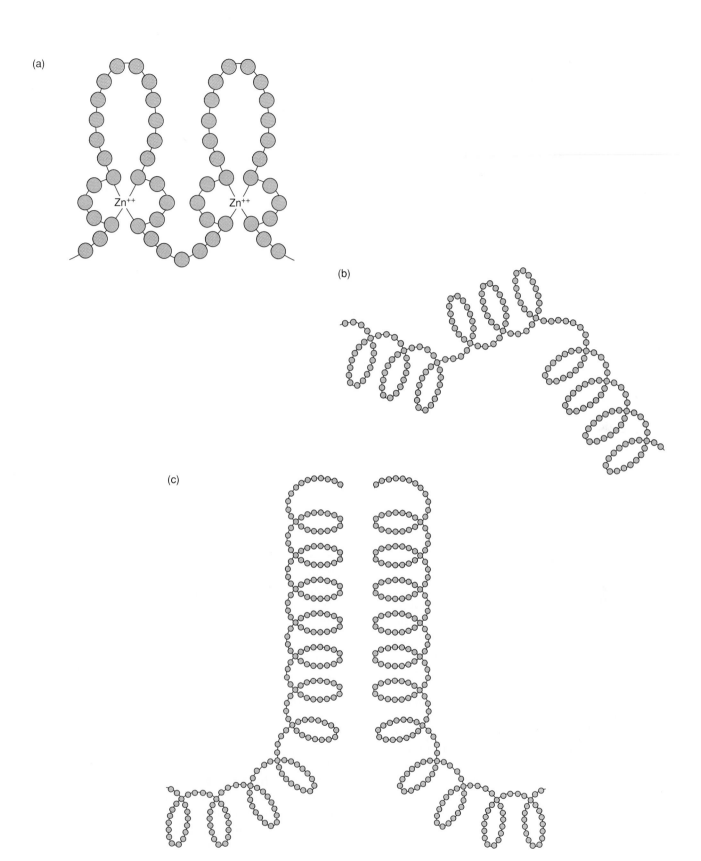

(a)

Zn++   Zn++

(b)

(c)

The second motif, known as a homeodomain (HD) or helix-turn-helix motif, consists of 60 amino acids named after its discovery in homeotic loci (see Chapter 22) of *Drosophila* (Figure 20.17). Proteins with this conserved sequence are found in organisms ranging from nematodes to humans, and related sequences are present in bacteria and yeast. A subset of HD factors also contain another domain of about 160 amino acids near the amino terminus, called the POU domain (because it is found in *p*ituitary cells, *o*ctamer factors, and *u*nc-86, a transcription factor from *C. elegans*). Together, the HD and POU regions are termed the POU box. In this subset of HD factors, the DNA target sequences are AT-rich octamers. Mutations in the HD region abolish DNA binding capacity, but it is not yet clear which segments are responsible for the separate functions of DNA binding and transcription activation.

The rat liver nuclear protein C/EPB was isolated and characterized because of its properties as a DNA binding protein. The gene was cloned and sequenced, and its DNA binding domain identified as a stretch of approximately 30 to 35 amino acids containing 4 leucine residues spaced 7 amino acids apart, flanked by a region of amino acids with basic charges. The leucine-rich region forms a helix with leucine residues protruding at every other turn. When two such molecules dimerize, the leucine residues "zip" together to form a domain required for DNA binding (Figure 20.17). Other proteins that may act as transcription factors, including AP1, and several oncogene products, including FOS and JUN, contain leucine zippers.

In addition to domains that bind DNA, transcription factors contain domains that activate transcription. These regions can occupy from 30 to 100 amino acids, and are distinct from the DNA binding domains. Activator domains were first identified in the yeast transcription factors *GAL4* and *GCN4* as short stretches of acidic (negatively charged) amino acids that can form alpha helices, and have been identified in other eukaryotic transcription factors. These stretches of acidic amino acids are proposed to function by interacting with other transcription factors (such as those that bind to the TATA sequence) or directly to the RNA polymerase. Two other domains, one glutamine-rich and the other proline-rich, have been identified in transcription factors from insect and mammalian systems. In all cases, these domains are likely to function by contacting or interacting with other proteins. The nature of the interaction is still unknown, but since RNA polymerase II requires 4 or 5 transcription factors for activation, the existence of a number of different protein-protein interaction mechanisms seems likely.

Overall, the picture of transcriptional regulation in eukaryotes is somewhat complex, but a number of generalizations can be drawn. First, regulation of transcription by protein factors is largely positive. The binding of one or more factors is a prerequisite to transcriptional activation of a locus. Promoter and enhancer sequences are recognized and bound by transcription factors, rather than by RNA polymerase. Because of the modular nature of recognition sequences, different transcription factors may compete for binding to a given DNA sequence, or the same site may bind different factors in a cell-specific fashion. For example, the TATA box or CAAT box can bind and transcription factor that can recognize these modules. Finally, because of the variability in the location and modular nature of promoters and enhancers, and the variety of transcription factors, an initiation complex (consisting of transcription factors and RNA polymerase II) can be constructed in a number of ways, all involving protein-protein interaction. Because of this variability, it is impossible to provide a general transcriptional activation scheme for all genes. However, to illustrate how eukaryotic transcriptional factors operate, we will consider the details of transcriptional regulation in a yeast gene, and the related phenomenon of steroid hormone activation of transcription.

Transcriptional Activation in Galactose Metabolism

The activity of four loci are required for the fermentation of galactose in the yeast *Saccharomyces cerevisiae*. Three of these loci (*GAL1*, *GAL7*, *GAL10*) are clustered on chromosome 2, and encode a *kinase*, a *transferase*, and an *epimerase*, respectively. Expression of these genes is positively regulated by the presence of galactose, and constitutive mutants have been isolated, indicating the existence of a regulatory gene or genes. Further analysis has revealed the existence of two regulatory loci, *GAL4* (located on chromosome 16) and *GAL80* (chromosome 7). *GAL4* is required for expression of the *GAL1, 7,* and *10* genes, and *GAL80* inhibits the function of *GAL4*. Between the *GAL1* and *GAL10* genes (a distance of about 600 bp) is an **upstream activating sequence (UAS)**, called UAS_G, located within a DNase hypersensitive chromatin site that is the site of regulation (Figure 20.18). The *GAL4* gene product (made constitutively) is a transcription factor that binds to sites within UAS_G and activates transcription of *GAL1* and *GAL10* in a bidirectional fashion. In the absence of galactose, the *GAL80* gene product binds to the *GAL4* transcription factor. The *GAL4/GAL80* complex binds to UAS_G but cannot activate transcription. When galactose is present, the

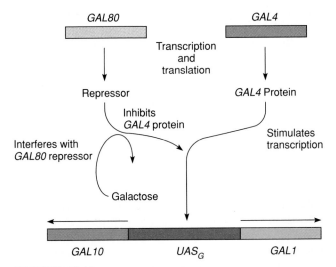

FIGURE 20.18

A representation of the interactions of the products of the GAL80 and GAL4 genes with galactose during regulation of the GAL1 and GAL10 genes in yeast. In the absence of galactose, the GAL80 repressor inhibits the positive control exerted by the GAL4 protein, and transcription is not stimulated. If galactose is added to the medium, it interferes with the action of the GAL80 repressor, and transcription is stimulated by the GAL4 regulatory protein, which interacts with UAS_G, the upstream activating sequence.

GAL80 protein binds to galactose, preventing its association with GAL4, allowing transcriptional activation of GAL1 and GAL10.

The GAL4 gene product is a protein 881 amino acids long, with three functions: binding to DNA sequences in the UAS_G, interacting with the GAL80 protein, and activating transcription. Mark Ptashne and his colleagues at Harvard and others have intensively studied the GAL4 transcription factor in order to dissect its functional properties. By cloning truncated GAL4 genes and assaying their ability to bind DNA and activate transcription, a region at the N-terminus of the gene product consisting of the first 98 or the first 147 amino acids was found to be the location of the DNA binding site (Figure 20.19). This derived protein binds to DNA in the UAS_G, but fails to activate transcription and cannot bind the GAL80 protein. Analysis of the amino acid composition of this region has identified a single zinc finger responsible for the DNA binding function. Similar experiments have identified two regions of the protein involved in activation of transcription: region I, consisting of amino acids 149–196, and region II, encompassing amino acids 768–881 (Figure 20.19). Constructs that consist of the DNA binding domain and region I (amino acids 1–96) or the DNA binding

FIGURE 20.19

The GAL4 protein has three domains that participate in the activation of the GAL1 and GAL10 loci. Amino acids 1–98 or 1–147 bind to DNA recognition sites in UAS_G. Amino acids 148–196 and 768–881 are required for transcriptional activation. GAL80 binding activity resides in the amino acids region 851–881.

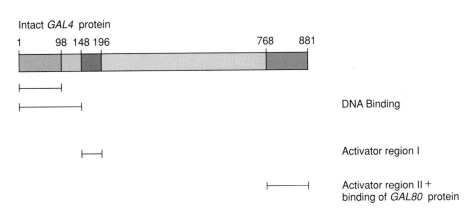

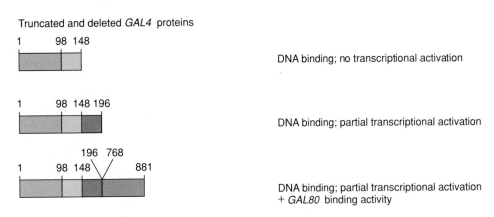

domain and region II (amino acids 1–96 and 768–881) have reduced transcriptional activity. Deletions of amino acids 852–881 destroy the ability to bind *GAL80,* and cause the truncated protein to constitutively activate the *GAL1* and *GAL10* loci.

The nature of the activating regions was explored by Ptashne's group by starting with a protein that contains the DNA binding region and activating region I (*GAL4* 1–238) and isolating point mutations in region I that result in increased activation. Most single mutations that increase activation represent amino acid substitutions that increase the negative charge of region I, and most exert a threefold increase on transcription. A multiple mutant that contains four more negative charges than the starting protein has a ninefold increase in transcriptional activation, and has almost 80 percent of the activity of the complete *GAL4* protein. However, negative charge alone is not responsible for the activating function, since some mutants with decreased activity do not have fewer negative charges than the parental *GAL4* 1–238 molecule.

Further experiments by Ptashne and his colleagues suggest that an amphipathic alpha helix is an essential part of an activating region. Amphipathic helices are those that carry negative charges along one surface, and hydrophobic residues along the other. To test this, a synthetic oligonucleotide (AH) was constructed that encoded a 15-amino-acid peptide that would form an amphipathic alpha helix. When attached to DNA encoding *GAL4* 1–147, the resulting protein activated transcription. A second construct, encoding the same 15 amino acids but in scrambled order (SH), did not activate transcription.

These results strongly suggest that transcriptional activation involves interaction between the activating domain of the transcription factor and other proteins. These other proteins must normally be components of transcription complexes, and might include the polymerase molecule itself, or any of the 4 or 5

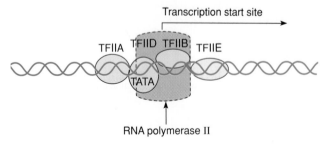

FIGURE 20.20
The array of transcription factors that bind to DNA elements clustered around the transcriptional start site. RNA polymerase II binds to the transcription factors through protein-protein interaction.

accessory proteins required for transcription (Figure 20.20). An attractive candidate for the target protein is TFIID, a factor that binds to the TATA sequence and is required for transcription. This idea is supported by the observation that the *GAL4* 1–147 + AH protein stimulates transcription in mammalian cell extracts, and changes the conformation of DNA-bound TFIID.

According to a general model formulated by Ptashne and Gann, *GAL4* represents a class of transcriptional activators that contain two functional regions: a DNA-binding region, and another region that interacts with a target protein, resulting in activation of the transcriptional complex. A second class of activators contains only a single functional site (either a DNA-binding site or a target activator site). These activators require the action of an additional factor to bring about transcriptional activation. Examples of this class of activators include the DNA binding protein OCT-1, and the herpes virus protein VP16, which contains only a target protein activating site. The Ptashne/Gann model proposes that all transcriptional activators work through an interaction with TFIID, either directly as in the case of *GAL4,* or indirectly, through a second factor, like OCT-1 or VP16 (Figure 20.21). Further experiments on TATA-binding proteins should reveal the nature and organization of their functional domains and shed more light on the molecular mechanisms involved in transcriptional activation.

Transcriptional Regulation by Steroid Hormones

Evidence that gene expression in eukaryotes is regulated by effector molecules originating outside the cell was elegantly provided by Ulrich Clever and Peter Karlson with their discovery that the steroid hormone **ecdysone** induces specific changes in the puffing pattern in polytene chromosomes (recall from Chapter 17 that puffs represent the transcription of specific genes). In eukaryotes, steroid hormones are used to regulate growth and development and to maintain homeostasis. The major sex hormones are all steroids, as is vitamin D, and the homeostatic adrenal hormones (more than 30 different types) that regulate glucose metabolism and mineral utilization are also steroids (Figure 20.22).

Steroid hormones enter the cell by passing through the plasma membrane and binding to a specific receptor protein in the cytoplasm. The receptor-hormone complex is translocated to the nucleus and activates transcription of one or more specific genes. Hormone receptors and the DNA sequences to which they bind have been intensively studied

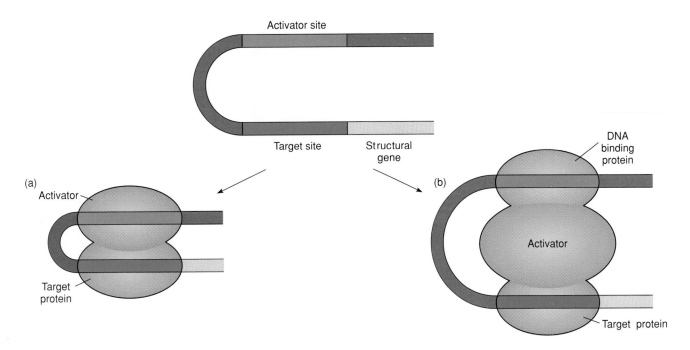

FIGURE 20.21
A model for gene activation by two classes of proteins. In (a), the activator contains a
DNA binding site and an acidic activating region that interacts with a target protein,
activating the transcriptional complex. A second class of activators (b) contains only a
single functional site, either DNA binding or target activating, requiring the action of
two factors to activate the target protein and stimulate transcription.

over the last decade. The structural genes encoding
virtually every known hormone receptor protein
have been cloned, and enough is now known to
provide a general picture of how these receptors
work. All receptors analyzed to date have three
functional domains: a variable N-terminus region; a
short, highly conserved central region; and a fairly
well-conserved C-terminus (Figure 20.23). The central region contains two zinc fingers. The zinc fingers
in hormone receptors bind to specific DNA sequences
known as **hormone-responsive elements** or **HREs**.

In steroid receptors, the finger closer to the
N-terminus binds in a base-specific way to the HRE
and determines the specificity of binding to a particular target gene. The C-terminus finger contacts the
sugar-phosphate residues adjacent to the HRE and
may have other functions. In general, HREs share
some characteristics with promoters and enhancers.
They are composed of short consensus sequences
that are related but not always identical (Table 20.3).
HREs are often located several hundred bases upstream from the transcription start site, and may be
present in multiple copies. Often, HREs are present
within promoter or enhancer sequences.

Binding of the receptor to the HRE is necessary for
activation of a specific gene, but it may not be
sufficient. Binding of the receptor to the HRE serves

to place regions of both the C-terminus and N-
terminus in position to activate a specific promoter,
but activation may require the participation of other
transcription factors, and, as described below, involves the alteration of chromatin structure in the
HRE and adjacent regions.

In addition to activating the regulated gene, the
C-terminal domain of steroid receptors interacts with
and binds to the steroid hormone. (It has been
suggested that in the absence of bound hormone, the
receptor is unable to recognize the HRE). The C-
terminal region may also be partially responsible for
controlling the translocation of the hormone-receptor
complex from the cytoplasm to the nucleus. The
N-terminal region of the receptor is not as well
understood, but it apparently modulates the degree
of gene activation.

It is important to note that gene activation by
steroid hormones requires alteration of chromatin
structure near the regulated gene. Genes in chromatin that are actively transcribed or are able to be
transcribed are hypersensitive to digestion with the
enzyme DNase I, while inactive or repressed genes
are relatively resistant to DNase I. *In vitro* studies of
steroid activation of genes in chromatin indicate that
following hormone administration, DNase I sensitivity greatly increases in the regulated gene and its

FIGURE 20.22

Structure of several classes of steroid hormones that activate transcription by binding to cytoplasmic receptor proteins and, after transport to the nucleus, bind to specific DNA sequences.

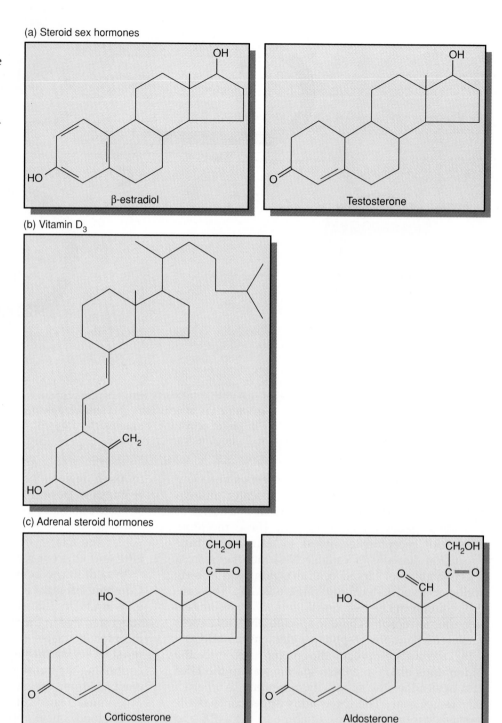

(a) Steroid sex hormones

β-estradiol

Testosterone

(b) Vitamin D₃

(c) Adrenal steroid hormones

Corticosterone

Aldosterone

flanking regions, including the hormone receptor binding site. The interpretation of these results is consistent with the idea that the binding of one or more receptor-hormone complexes (receptor may need to act as dimers or tetramers to activate genes) to the HRE causes a change or shift in the nucleosome structure of chromatin at the binding site, making other binding sites, including the promoter, available

to transcription factors and RNA polymerase II, resulting in the initiation of transcription.

In some ways, then, the biology of steroid hormone regulation of gene transcription in eukaryotes is similar to the *lac* operon system in prokaryotes in that an external effector binds to a cytoplasmic receptor, changes the configuration of the receptor, and moves the receptor to the DNA, where it acts as

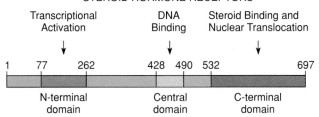

STEROID HORMONE RECEPTORS

| Transcriptional Activation | DNA Binding | Steroid Binding and Nuclear Translocation |

1   77   262   428 490 532   697

N-terminal domain   Central domain   C-terminal domain

FIGURE 20.23
Steroid hormone receptors have three functional domains, as exemplified by the glucocorticoid receptor: the N-terminal domain (amino acids 77–262 in the glucocorticoid receptor), the central domain (amino acids 428–490), and the C-terminal domain (amino acids 532–697). The central domain binds DNA sequences known as the glucocorticoid response element (GRE) by means of two zinc fingers. The C-terminal domain binds to the steroid hormone and may be responsible for allowing the hormone-receptor complex to be translocated to the nucleus. This region is also responsible for regulating the activity of the receptor.

a transcription factor. In other ways, however, the regulation of gene expression is quite different, in that the DNA binding sequence is often at a great distance from the regulated gene, and that alterations in chromatin structure mediate the action of the receptor.

POST-TRANSCRIPTIONAL REGULATION OF GENE EXPRESSION

As outlined above, there are many opportunities in the processes of transcription and translation where regulation of genetic expression can occur. Although transcriptional control is perhaps the most obvious and widely used mode of regulation in eukaryotes, post-transcriptional modes of regulation are also employed in many organisms. For example, nuclear RNA transcripts are modified prior to translation, noncoding **introns** are removed, and the remaining **exons** are precisely spliced together. This processing step offers several possibilities for regulation, two of which will be described below.

Processing Transcripts

In discussing gene regulation through processing of transcripts, we will first consider the case in which a single gene is transcribed, but following processing, two mRNAs, differing in their untranslated leader sequences, are produced. In mice, the enzyme alpha-amylase is produced in both salivary glands and liver. The alpha-amylase gene contains two promoters, one

TABLE 20.3
DNA receptor sequences for hormone binding proteins.

Hormone	Consensus Sequence	Element Name
Glucocorticoid	GGTACANNNTGTTCT	GRE
Estrogen	AGGTCANNNTGACCT	ERE
Thyroid	AGGTCA . . . TGACCT	TRE

used in salivary glands, and the other in liver (Figure 20.24). The salivary gland promoter is about 2850 bp upstream from the liver promoter, and two different exons are adjacent to the promoters. The use of the two different promoters creates two different pre-mRNAs that are differentially spliced to yield tissue-specific mRNAs (see Figure 20.23). It is important to emphasize that the protein product in both tissues is identical. The differences lie in the exons at the 5' end of the mRNA, a region that constitutes part of the 5' untranslated leader sequence.

Researchers are especially interested in cases where splicing variations might generate different mRNAs responsible for directing the synthesis of different polypeptides. Several instances have now been documented demonstrating that splicing may include or exclude an entire exon. This leads to distinctive mRNAs that specify different polypeptides. This is the case in several systems, including the expression of the **myosin** gene in *Drosophila*, the **α-crystallin** gene in rodents, and the immunoglobulin genes discussed earlier.

Figure 20.25 illustrates still a third example of this phenomenon where the polypeptide products are more clearly distinct from one another. Shown is a diagram of the initial bovine RNA transcript that is processed into one of two **preprotachykinin mRNAS (PPT mRNA)**. This mRNA molecule potentially includes the genetic information specifying two neuropeptides called **P** and **K**. These two peptides are members of the family of sensory neurotransmitters referred to as **tachykinins,** and are believed to play different physiological roles. While the P neuropeptide is largely restricted to tissues of the nervous system, the K neuropeptide is found more predominantly in the intestine and thyroid.

The RNA sequences for both neuropeptides are derived from the same gene. However, processing of the initial RNA transcript can occur in two different ways. In one case, exclusion of the K-exon during processing results in the α-PPT mRNA, which upon translation yields neuropeptide P but not K. On the other hand, processing that includes both the P and K exons yields β-PPT mRNA, which upon translation results in the synthesis of both the P and K neuropeptides. Analysis of the relative levels of the two types

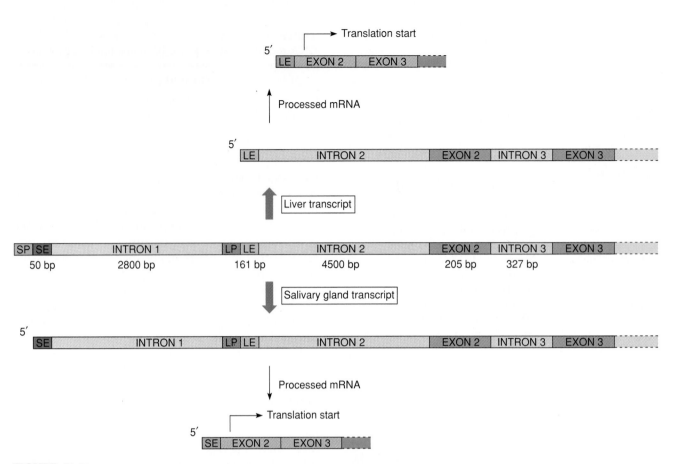

FIGURE 20.24
Organization of the mouse alpha-amylase gene. The salivary promoter (SP) and its adjacent exon (SE) are located approximately 3000 bp upstream from the liver promoter (LP) and the liver exon (LE). The use of alternate promoters results in the transcription of different pre-mRNAs in liver and salivary glands. Processing of these transcripts results in mRNAs that start with different exons in the 5′ untranslated region.

FIGURE 20.25
Diagrammatic representation of the alternate splicing of the initial RNA transcript of the preprotachykinin gene (PPT). Introns are unshaded and are labeled I. Exons are either numbered or designated by the letters P or K. Inclusion of the P and K exons leads to β-PPT mRNA, which upon translation yields both the P and K tachykinin neuropeptides. When the K exon is excluded, α-PPT mRNA is produced, and only the P neuropeptide is synthesized.

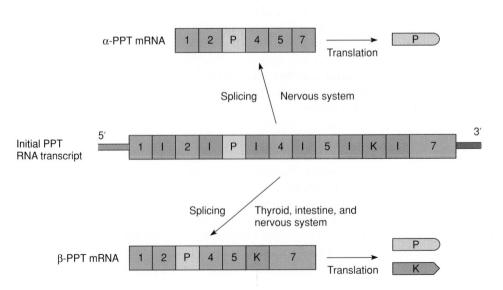

of RNA has demonstrated striking differences between tissues. In nervous system tissues, α-PPT mRNA predominates by as much as a threefold factor, while β-PPT mRNA is the predominant type in the thyroid and intestine.

Alternate splicing of transcripts from a single gene, leading to two distinct products, is governed by tissue-specific alteration of the initial RNA transcript. Further investigations will very likely uncover still other examples of gene regulation at the post-transcriptional level.

Controlling mRNA Stability

After mRNA precursors are processed and transported, they enter the population of cytoplasmic mRNA molecules, from which messages are recruited for translation. Although regulation at the level of translation might be unexpected and seem inefficient, a growing body of evidence suggests that, in fact, the stability of mRNA engaged in translation may be a significant control point for gene regulation in eukaryotes. One of the best-studied examples of what is termed **autoregulation** is the synthesis of alpha and beta tubulins, the subunit components of eukaryotic microtubules. Treatment of a cell with the drug colchicine leads to a rapid disassembly of microtubules, and an increase in the concentration of the alpha and beta subunits. Under these conditions, the synthesis of alpha and beta tubulins drops dramatically. However, when cells are treated with vinblastine, a drug which also causes microtubule disassembly, the synthesis of tubulins is increased. The difference between the two drugs is that in addition to causing microtubule disassembly, vinblastine precipitates the subunits, lowering their concentration of the alpha and beta subunits. At low concentrations, synthesis of tubulins is stimulated, while at higher concentrations, synthesis is inhibited.

Further work by Don Cleveland and his colleagues has described the conditions necessary for tubulin mRNA to be regulated at the level of translation. Fusion of tubulin gene fragments to a cloned thymidine kinase gene show that the first 13 nucleotides at the 5' end of the messenger RNA following the transcription start site (encoding the amino terminal region of the protein) are necessary and sufficient to cause the hybrid mRNA to be regulated as efficiently as intact tubulin mRNA. These nucleotides encode the amino acids Met-Arg-Glu-Lys (abbreviated as MREI). Deletions, translocations, or point mutations in this 13-nucleotide segment abolish regulation. Examination of the cytoplasmic distribution of mRNAs demonstrate that only tubulin mRNAs in poly-

somes were depleted by drug treatment. Copies of tubulin mRNAs not bound to polysomes were protected from degradation. Third, translation must proceed to at least codon 41 of the mRNA in order for regulation to occur.

A model, shown in Figure 20.26, has been proposed to explain these observations. The model proposes that the first four amino acids of the tubulin gene product (in this case, beta-tubulin) constitute a recognition element to which the regulatory factors bind. The concentration of alpha and beta tubulin subunits in the cytoplasm may serve this regulatory function. This protein-protein interaction invokes the action of an RNase that may be a ribosomal component or a nonspecific cytoplasmic RNase. Action by the RNase degrades the tubulin mRNA in the act of translation, shutting down tubulin biosynthesis. As an alternative, binding of the regulatory element may cause the ribosome to stall in its translocation along the mRNA, leaving it exposed to the action of RNase. Translation must proceed to at least codon 41, because the first 30 to 40 amino acids translated occupy a tunnel within the large ribosomal subunit, and only the translation of this number of amino acids will make the MREI recognition sequence accessible for binding.

RNA instability has been proposed as a regulatory mechanism for other genes, including histones, some transcription factors, lymphokines, and cytokines. This suggests the existence of a regulatory mechanism that acts on mRNAs in the act of translation, probably by protein-protein interaction between a regulatory element (most likely the gene product) and the nascent polypeptide chain, resulting in the activation of RNases.

ONCOGENES AND REGULATION

Oncogenes are genes that induce or maintain uncontrolled cellular proliferation associated with cancer. The existence of specific genes associated with the transformation of normal cells into cancerous cells was inferred by Peyton Rous in 1910–1911. Using cells from a tumor of chickens known as a **sarcoma,** Rous demonstrated that injection of cell-free extracts from these tumors could induce sarcoma formation in healthy chickens. He postulated the existence of a "filterable agent" that was responsible for transmitting the disease. His work was criticized by his colleagues who believed that his extracts were really not cell-free, and he was simply transferring the disease by injecting cancer cells into healthy chickens.

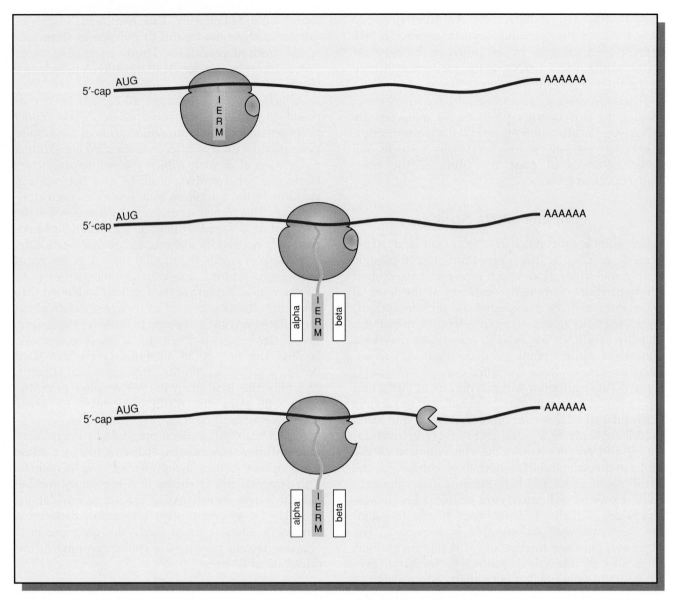

FIGURE 20.26
Model of post-translational regulation of tubulin synthesis. Tubulin subunits (or another effector) bind to the amino terminal tubulin tetrapeptide as it emerges from the large ribosomal subunit. Binding activates an RNase (in this case, a ribosomal RNase) that breaks down the polysomal tubulin mRNA.

Decades later, his filterable agent was shown by other investigators to be a virus, now known as the Rous sarcoma virus (RSV). This virus belongs to a group known as **retroviruses** because the RNA genome must be re-transcribed by an enzyme, **reverse transcriptase,** into single-stranded DNA. The single-stranded DNA in converted to double-stranded DNA and integrated into the genome of the infected cell, where it can be transcribed to produce an infectious RNA. In the case of RSV, the tumor-forming ability results from the presence of a single gene, called the *src* gene in the viral genome. This single gene, responsible for the induction of sarcoma formation in chicken cells, is designated as an oncogene, and retroviruses that carry oncogenes are known as **acute transforming viruses**. Other retroviruses that can cause tumor formation do not carry oncogenes, but are able to induce the activity of cellular genes that bring about tumor formation. They are designated as **nonacute** or **nondefective viruses**.

FIGURE 20.27
A transforming retrovirus has acquired a copy of a gene from the host genome, converting it from a *c-onc* or proto-oncogene into an oncogene that confers on the virus the ability to transform a specific type of host cell into a cancerous cell.

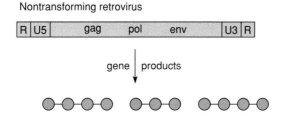
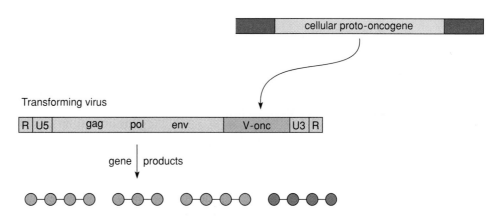

The oncogenes carried by acute transforming viruses are acquired from the cellular genome during the process of infection, when a portion of the viral genome is exchanged for a cellular gene (Figure 20.27), as in bacterial transduction. *Onc* genes carried by retroviruses are called *v-onc,* and the cellular version of the gene is called a *c-onc* gene or **proto-oncogene**. Retroviruses that carry a *v-onc* gene are able to infect and transform a specific type of host cell. For RSV, the oncogene captured from the chicken genome is called *v-src,* and it confers the ability to infect chicken cells and transform them into tumorous growths known as sarcomas. The cellular version of the same gene, found in the chicken genome, is called *c-src.* More than 20 oncogenes have been identified by their presence in retroviral genomes, and over 50 oncogenes have been identified overall. Some of these are listed in Table 20.4.

Two questions about oncogenes come to mind. First, are *v-onc* and *c-onc* different, and second, what mechanisms allow oncogenes to bring about cellular transformation and tumor formation? Comparison of the DNA sequence of *v-onc* genes with the corresponding *c-onc* sequence reveals several examples

TABLE 20.4
Oncogenes.

c-onc	Origin	Species	Human Chromosome Location
A. Cellular Oncogenes with Viral Equivalents			
src	Rous sarcoma virus	Chicken	20
fos	FBJ osteosarcoma virus	Mouse	2
sis	Simian sarcoma virus	Monkey	22
fes	ST feline virus	Cat	15
abl	Abelson murine leukemia virus	Mouse	9
erb-B	Avian erythroblastosis virus	Chicken	7
Ha-ras-1	Harvey murine sarcoma virus	Mouse	11
myc	Avian MC29 myelocytomatosis virus	Chicken	8
B. Oncogenes without Viral Equivalents			
N-ras	Neuroblastoma, leukemia	Human	1
N-myc	Neuroblastoma	Human	12
neu	Neuroglioblastoma	Human, rat	17
man	Mammary carcinoma	Human, mouse	?

(such as *v-ras, v-mos*) in which there are only minimal differences, probably generated by point mutations. In other oncogenes, portions of the *c-onc* sequence have been replaced or lost. In *v-src*, 19 amino acids of the *c-onc* sequence have been replaced with 12 different amino acids. In *v-erb*, the *v-onc* version retains only the C-terminus half of the gene. Mutations, therefore, play a role in differentiating some *v-onc* sequences from their *c-onc* counterparts. However, not all oncogenes are mutant versions of cellular genes carried by retroviruses. We must, therefore, consider both qualitative and quantitative mechanisms in oncogenetic events.

Oncogenes and Gene Expression

At least three mechanisms can be invoked to explain the conversion of proto-oncogenes into oncogenes. These mechanisms, which are related to the range of functions carried out by the proto-oncogenes in the cellular genome, include **point mutations, translocations,** and **overexpression**. While some of these are mediated by viruses, others operate in the absence of retroviruses, and are generated solely by intracellular events.

The *ras* gene family, encoding a protein of 189 amino acids involved with the transduction of external signals across the cell membrane, illustrates how point mutations as small as single base changes

underlie the differences between the normal *c-onc* sequence and the oncogenic version of the gene. Comparison of the amino acid sequence of *ras* proteins from a number of different human carcinomas reveals that all *ras* mutations involve a single amino acid substitution at position 12 or 61 that is caused by a single base change (Figure 20.28). The normal *ras* protein functions as a molecular switch, alternating between an "on" and "off" state (Figure 20.29). In some cases, the mutational event causes the *ras* protein to be stuck in the "on" position, stimulating cell growth. In the human tumors, this event was not mediated by a retrovirus, and is presumed to be the result of a somatic mutation.

One of the well-characterized examples of oncogene activation by chromosome rearrangement is the translocation of *c-abl* associated with chronic myelogenous leukemia (CML). In 96 percent of all patients with this condition, there is a translocation between chromosome 9 and 22, with the *c-abl* proto-oncogene moving to chromosome 22 (Figure 20.30). Molecular analysis of the breakpoint indicates that the *c-abl* gene and a DNA segment on chromosome 22 known as the *breakpoint cluster region (bcr)* form a fused gene (*bcr/abl*). The hybrid protein generated by translocation resembles the retroviral *v-abl* gene, and causes tumor formation. In Burkitt's lymphoma, (a B cell tumor primarily affecting children) a *c-myc* gene on chromosome 8 is translocated to an immunoglobulin heavy-

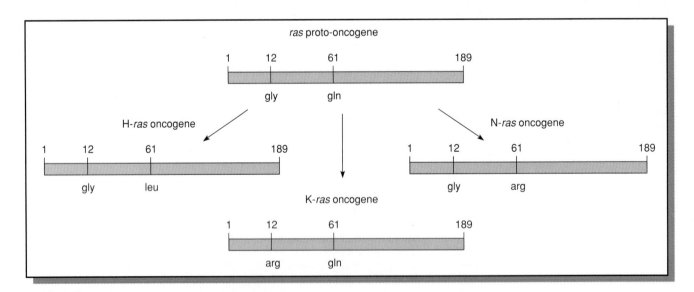

FIGURE 20.28
The *ras* proto-oncogene encodes a protein of 189 amino acids. In the normal protein, glycine is encoded at position 12, and glutamine at position 61. Analysis of *ras* oncogene proteins from several tumors shows a single amino acid substitution at one of these positions. A single amino acid substitution resulting from a single base change can convert a proto-oncogene into a tumor-promoting oncogene.

chain locus on chromosome 14, causing activation of the *c-myc* gene as a result of its acquiring upstream regulatory sequences that normally control immunoglobulin production.

At least three separate mechanisms of protooncogene activation are associated with overexpression. First, the *c-onc* may acquire a new promoter, causing an increase in the level of transcript production, or activating a silent locus. This is the case in avian leukosis, where strong viral promoters become integrated upstream from a proto-oncogene, causing an increase in transcript production and an increase in the amount of the gene product. The second is overexpression by acquisition of new upstream regulatory sequences including enhancers. The *c-myc*

gene in Burkitt's lymphoma, discussed above, involves this form of activation, associated with a translocation. The third mechanism involves the amplification of the proto-oncogene. In human tumors, members of the *myc* family of oncogenes are frequently amplified. The *c-myc* proto-oncogene is found amplified up to several hundred copies in human solid tumors, and are often associated with HSR regions or double minute chromosomes described earlier in this chapter.

Oncogenes and Transcription Factors

The protein products of proto-oncogenes are found in and associated with the plasma membrane, cyto-

FIGURE 20.29
A three-dimensional computer-generated image of *ras* proteins in two different conformations. Normal *ras* proteins act as molecular switches controlling cell growth and differentiation. The switch is "on" when GTP binds to the protein, and "off" when the GTP is hydrolyzed to GDP. Switching the protein between states alters the conformation of the protein in the two regions shown in blue and yellow. Oncogenic mutations of *ras* are stuck in the "on" state, continuously signalling for cell growth.

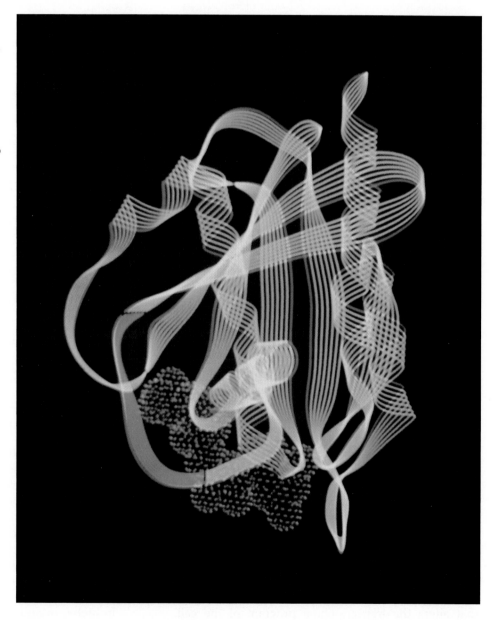

FIGURE 20.30

The standard Philadelphia chromosome (Ph) resulting from a reciprocal translocation between one member of a chromosome 9 pair and one member of a chromosome 22 pair. The bands shown schematically are derived from preparations of prophase chromosomes.

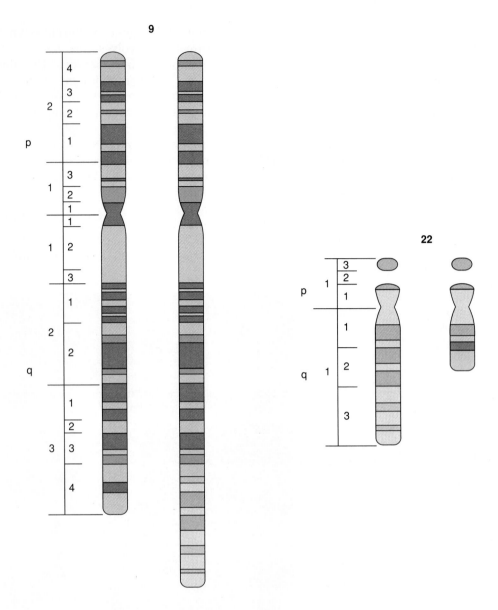

plasm, and nucleus. While their specific functions vary widely, all products characterized to date alter gene expression in a direct or an indirect manner. Specifically, some of the oncogene products that are found in the nucleus have characteristics of transcription factors. The human gene *c-jun* and its viral counterpart *v-jun* encode a protein that binds to DNA and activates transcription. There is no difference between the *c-jun* and *v-jun* proteins in their ability to bind DNA, so presumably the difference lies in the ability to activate transcription. The *jun* protein normally forms a dimer with the product of the *c-fos* gene, and this protein-protein complex binds to the TGACTCA box present in several enhancers. While the cascade of events that leads to uncontrolled cellular growth is not yet known, it is clear that these events can be triggered by the action of a single gene that alters the transcriptional program of the cell.

Thus, the fundamental problem of cancer can be defined and perhaps understood in terms of gene expression and its regulation.

CHAPTER SUMMARY

1   Mechanisms controlling gene regulation in eukaryotes are constrained by the properties of multicellularity, expanded genome size, and the spatial and temporal separation between transcription and translation.

2   The roles played by DNA organization and structure, transcriptional control, and signal transduction, as well as the expression of specific gene sets such as hormone receptors and oncogenes, are studied as models for gene regulation in higher organisms.

3 Gene amplification, a variant of DNA organization, maximizes the accumulation of gene product per unit of time by increasing the number of gene copies necessary during times of cell growth and differentiation.

4 One of the most complex families of genes specifying proteins are those participating in the immune response of vertebrates. Human antibodies, or immunoglobulins, are characterized by amino acid sequence diversity. DNA shuffling during maturation of the immune system produces the genetic variation required to match the millions of different antigens with corresponding antibodies.

5 Eukaryotic transposons and insertion sequences have the ability to move in and out of chromosomal sequences, thus affecting gene expression throughout the genome.

6 Transcription in eukaryotes is controlled by regulatory DNA sequences known as promoters and enhancers. Regulatory sequences, including the TATA box and the CCAAT box, are elements of promoters found near the transcriptional starting point. Enhancer elements, which appear to control the degree of transcription, can be located before, after, or within the gene expressed.

7 Transcription factors are proteins that bind to DNA recognition sequences within the promoters and enhancers and activate transcription through protein-protein interaction.

8 Several types of post-transcriptional control of gene expression are possible in eukaryotes. One such mechanism is alternate processing of a single class of pre-mRNAs to generate different mRNA species. Another mechanism provides a feedback system to mRNA regulation in response to cytoplasmic concentration of the gene product, controlling mRNA stability.

9 Gene expression in eukaryotes can be regulated by effector molecules originating outside the cell. Such signals are transduced by proteins that shuttle from the cytoplasm to the nucleus and regulate patterns of transcription.

10 The discovery of oncogenes represents a turning point in our understanding of the molecular mechanisms of cancer. Their normal counterparts, proto-oncogenes, function to regulate cell growth. Proto-oncogenes are converted to oncogenes by alterations in activity or by mutations that produce an altered gene product. Since all human tumors examined to date express oncogenes, it is possible that mutations in a limited number of genes is all that is required for the development of malignancies.

Insights and Solutions

Regulatory sites for eukaryotic genes are usually located within a few hundred nucleotides of the transcriptional starting site, but can be located up to several kilobases away. DNA sequence-specific binding assays have been used to detect and isolate protein factors present at low concentrations in nuclear extracts. In these experiments, DNA binding sequences are bound to material on a column, and nuclear extracts are passed over the column. The idea is that if proteins that specifically bind to the sequence represented on the column are present, they will bind to the DNA, and can be recovered from the column after all other nonbinding material has been washed away. Once a DNA binding protein has been identified and isolated, the problem is to devise a general method for screening cloned libraries for genes encoding these and other DNA binding factors. Determining the amino acid sequence of the protein and constructing synthetic oligonucleotide probes is time-consuming and useful only for one factor at a time. Knowing of the strong affinity for binding between the protein and its DNA recognition sequence, how would you screen for genes encoding binding factors?

ANSWER: Several general strategies have been developed, and one of the most promising has been devised by McKnight's laboratory. cDNA isolated from cells expressing the binding factor is cloned into the lambda vector, gt11. Plaques of this library containing proteins derived from expression of cDNA inserts is adsorbed onto nitrocellulose filters and probed with double-stranded, radioactive DNA corresponding to the binding site. If a fusion protein corresponding to the binding factor is present, it will bind DNA probe. After washing off unbound probe, the filter is subjected to autoradiography, and plaques corresponding to the DNA binding signals can be identified. An added advantage of this strategy is recycling, by washing the bound DNA from the filters for reuse. This ingenious procedure is similar to the colony hybridization and plaque hybridization procedures described for library screening in Chapter 16, and provides a general method for isolating genes encoding DNA binding factors.

PROBLEMS AND DISCUSSION QUESTIONS

1 Why is gene regulation assumed to be more complex in a multicellular eukaryote than in a prokaryote? Why is the study of this phenomenon in eukaryotes more difficult?

2 List and define the levels of gene regulation discussed in this chapter.

3 What is the evidence from heterokaryons that suggests that cytoplasmic factors play an important role in regulating genetic activity or inactivity of the nucleus of a cell?

4 Distinguish between the regulatory elements referred to as promoters and enhancers.

5 Is the binding of a transcription factor to its DNA recognition sequence necessary and sufficient for an initiation of transcription at a regulated gene? What else plays a role in this process?

6 In the autoregulation of tubulin synthesis, two models for the mechanism were proposed: first, that the tubulin subunits bind to the mRNA, or that the subunits interact with the nascent tubulin polypeptides. To distinguish between these two models, Cleveland and colleagues introduced mutations into the 13-base regulatory element of the beta-tubulin gene. Some of the mutations resulted in amino acid substitution, while others did not. In addition, they shifted the reading frame of the intact 13-base sequence. Below are results from a mutagenesis study of the mRNA. Which of the two models is supported by the results? What experiments would you do to confirm this?

					Autoregulation
Wild type	met AUG	arg AGG	glu GAA	lys ATC	+
		UGG			−
Second codon mutations		GGG			−
		CGG			+
		AGA			+
		AGC			−
Third codon mutations			GAC		+
			AAC		−
			UAU		−
			UAC		−

7 Assuming your answer to problem 6 is correct, predict the results of the following experiment. Site-directed mutagenesis is used to insert four bases just in front of the AUG initiating codon of the 13-base sequence, keeping the sequence intact, but shifting its position in the mRNA by four bases, and shifting the reading frame to an alternate sequence different from the normal *met-arg-glu-lys* sequence. Will tubulin subunits regulate translation of this mRNA? Why or why not?

8 What are the possible ways that a proto-oncogene can differ from an oncogene?

9 In humans, one of the ways that oncogenes are activated involves chromosome translocations. Many translocations place proto-oncogenes near loci that are involved with the immune system. What may genes of the immune system have to do with oncogene activation?

∞∞∞∞∞∞∞∞∞∞∞∞∞∞∞∞∞∞∞∞∞∞∞∞∞∞∞∞∞∞∞

SELECTED READINGS

BAKER, B. S. 1989. Sex in flies: The stuff of life. *Nature* 340: 521–24.

BEATO, M. 1989. Gene regulation by steroid hormones. *Cell* 56: 335–44.

BHATTACHARYYA, M. K., SMITH, A. M., ELLIS, T. H. N., HEDLEY, C. and MARTIN, C. 1990. The wrinkled seed character of pea described by Mendel is caused by a transposon-like insertion in a gene encoding starch-branching enzyme. *Cell* 60: 115–22.

BRAWERMAN, G. 1989. mRNA decay: Finding the right targets. *Cell* 57: 9–10.

BURCH, J. B. E., EVANS, M. I., FRIEDMAN, T. M., and O'MALLEY, P. J. 1988. Two functional estrogen response elements are located upstream of the major chicken vitellogen gene. *Mol. Cell. Biol.* 9: 1123–31.

BUSCH, S. J. and SASSONE-CORSI, P. 1990. Dimers, leucine zippers and DNA-binding domains. *Trends in Genet.* 6: 36–40.

CLEVELAND, D. W. 1988. Autoregulated instability of tubulin mRNAs: A novel eukaryotic regulatory mechanism. *Trends in Biochem. Sci.* 13: 339–43.

DYNAN, W. S. 1988. Modularity in promoters and enhancers. *Cell* 58: 1–4.

HUNTER, T. 1989. The function of oncogene products. *Prog. Clin. Biol. Res.* 288: 25–34.

JOHNSON, P. F. and McKNIGHT, S. L. 1989. Eukaryotic transcriptional regulatory proteins. *Ann. Rev. Biochem.* 58: 799–839.

KAKIDANI, H., and PTASHNE, M. 1988. GAL4 activates gene expression in mammalian cells. *Cell* 52: 161–67.

KARIN, M., CASTRILLO, J-L., and THEILL, L. E. 1990. Growth hormone regulation: A paradigm for cell type-specific gene activation. *Trends in Genet.* 6: 92–96.

LEVINE, M., and MANLEY, J. L. 1989. Transcriptional repression of eukaryotic promoters. *Cell* 59: 405–8.

MANIATIS, T., GOODBOURN, S., and FISCHER, J. A. 1987. Regulation of inducible and tissue-specific expression. *Science* 236: 1237–45.

MILBURN, M. V., TONG, L., deVOS, A. M., BRUNGER, A., YAMAIZUMI, Z., NISHIMURA, S., and KIM, S. H. 1990. Molecular switch for signal transduction: Structural differences between active and inactive forms of proto-oncogenic *ras* proteins. *Science* 247: 939–45.

MITCHELL, P. J., and TJIAN, R. 1989. Transcriptional regulation in mammalian cells by sequence-specific DNA binding proteins. *Science* 245: 371–78.

PIATORGORSKY, J., and WISTOW, G. J. 1989. Enzyme/crystallins: Gene sharing as an evolutionary strategy. *Cell* 57: 197–99.

PTASHNE, M. 1988. How eukaryotic transcriptional activators work. *Nature* 335: 683–89.

PTASHNE, M., and GANN, A. A. F. 1990. Activators and targets. *Nature* 346: 329–31.

SCHIMKE, R. T. 1989. The discovery of gene amplification in mammalian cells: To be in the right place at the right time. *Bioessays* 11: 69–73.

SIBLEY, E., KASTELIC, T., KELLY, T. J., and LANE, M. D. 1989. Characterization of the mouse insulin receptor gene promoter. *Proc. Natl. Acad. Sci.* 86: 9732–36.

STARK, G. R., DEBATISSE, M., GIULOTTO, E., and WAHL, G. 1989. Recent progress in understanding mechanisms of gene amplification. *Cell* 57: 901–8.

STRINGER, K. F., INGLES, C. J., and GREENBLATT, J. 1990. Direct and selective binding of an acidic transcriptional activation domain to the TATA-box factor TFIID. *Nature* 345: 783–86.

SUAREZ, H. G. 1989. Activated oncogenes in human tumors. *Anticancer Res.* 9: 1331–44.

YEN, T. J., GAY, D. A., PACHTER, J. S., and CLEVELAND, D. W. 1988. Autoregulated changes in stability of polyribosome-bound beta-tubulin mRNAs are specified by the first 13 translated nucleotides. *Mol. Cell. Biol.* 8: 1224–35.

PART FOUR

Genetics of Organisms and Populations

21
Developmental Genetics

DEVELOPMENTAL CONCEPTS

THE VARIABLE GENE ACTIVITY THEORY
Genomic Equivalence

EVIDENCE OF DIFFERENTIAL TRANSCRIPTION IN SPECIALIZED CELLS
Chromosome Puffs | Isozymes | Growth Hormone

DIFFERENTIAL TRANSCRIPTION DURING DEVELOPMENT
Bacterial Sporulation | Slime Mold Development | The Effects of the Cellular and
Extracellular Environment | Overview of *Drosophila* Development | Gene Expression
During Early *Drosophila* Development | Organization and Regulation of
Homoeotic Genes

THE STABILITY OF THE DIFFERENTIATED STATE
Transdetermination | Regeneration

CHAPTER CONCEPTS

The genetic basis of development is differential gene action. Gene actions vary over time in the same tissue, and between different tissues present at the same developmental stage. Gene action is required to bring about and maintain adult structures. Processes such as regeneration involve a reiteration of the embryonic program of gene action.

In multicellular plants and animals, a fertilized egg, without further stimulus, begins a cycle of developmental events that ultimately give rise to an adult member of the species from which the egg and sperm were derived. Thousands, millions, or even billions of cells are organized into a cohesive and coordinated unit that we perceive as a living organism. The heterogeneous series of events whereby organisms attain their final adult form is studied by developmental biologists. This area of study is perhaps the most intriguing in biology because comprehension of developmental processes requires knowledge of many biological disciplines.

In the past one hundred years, investigations in embryology, genetics, biochemistry, molecular biology, cell physiology, and biophysics have contributed to the study of development. Largely, the findings have pointed out the tremendous complexity of developmental processes. Unfortunately, this description of what actually happens does not answer the "why" and "how" of development. Nevertheless, further hypotheses and experimentation will be based on this knowledge and will lead us closer to a comprehensive understanding of development.

In this chapter the primary emphasis will be on the role of genetic information during development. Because genetic information directs cellular function and determines the cellular phenotype, the genetics of development is being actively studied using a wide range of experimental organisms and techniques.

DEVELOPMENTAL CONCEPTS

Studies of development in a wide range of organisms have made it clear that common processes govern the formation of embryonic and adult structures from the fertilized egg. These phenomena include *determination, differentiation,* and *cell-to-cell interaction.* Determination can be defined as one or more regulatory events that establish a specific pattern of gene activity and developmental fate for a given cell. In other words, through determination, an undifferentiated embryonic cell is programmed to develop into, for example, a muscle cell, a liver cell, or an epidermal cell. Differentiation is the process of genetic and morphological changes by which cells attain their adult form and function. It is the mechanism by which the determined state becomes expressed. Cell-to-cell interactions represent a form of intercellular communication that provides the basis for cell migration and the formation of gradients and patterns. Cell-to-cell communication can be involved in both processes of determination and differentiation.

The current thrust in developmental genetics is to provide a molecular explanation of these phenomena in order to establish a continuous relationship between the presence or absence of molecules, receptors, transcriptional events, and cell and tissue interactions, and the observable morphological events that accompany the process of development. To explain the sequence of developmental events that take place in any organism, geneticists rely on certain heuristic concepts. These include the theory of variable gene activity, and the principles of differential transcription and the stability of the differentiated state. These concepts have been derived from the study of a wide range of organisms used as model systems, and these ideas form the theoretical framework that is used to explain how the developmental potential present in a single cell becomes elaborated into a recognizable yet individual form with a certain behavioral repertoire we identify as an adult organism.

THE VARIABLE GENE ACTIVITY THEORY

From a genetic perspective, development may be described as the attainment of a differentiated state. For example, an erythrocyte active in hemoglobin synthesis is differentiated, but a cell in a blastula-stage embryo is undifferentiated. In order to accomplish this specialization, certain genes are actively transcribed but many other genes are not. Because most eukaryotic organisms are composed of a large number of cell types, differential transcriptive capacities characterize functionally diverse cells.

The concept of differential transcription has led to the **variable gene activity hypothesis** of differentiation. This theory was first entertained by Thomas H. Morgan in 1934, later proposed by Edgar and Ellen Stedman as well as Alfred Mirsky in the 1950s, and articulated in modern form by Eric Davidson in his book, *Gene Activity in Early Development*. The theory holds that the forms of differentiation assumed by any specific cell type are qualitatively determined by those genes that are actively transcribed. Its underlying assumption is that each cell contains the entire diploid genome in the nucleus and that only certain gene products characteristic of the cell type are produced. The rest of the genome is actively shut down and not transcribed.

Given our knowledge of molecular biology, the variable gene activity theory is readily acceptable. Every cell is a biochemical entity composed of macromolecules that are either informational, structural, catalytic, or metabolic in nature. In an organized fashion, these molecules constitute the cell. The presence or absence of any molecule in the cell is thus directly or indirectly influenced by gene activity or inactivity. Therefore, a cell is what its active genes direct it to be.

The variable gene activity theory is a very useful model system for experimental design. The remainder of this chapter examines the validity of this theory and its premises and provides examples that offer experimental support. It is important to remember, however, that this type of approach initially tells us what happens. We still do not know why certain genes are active and others inactive and why qualitatively different gene activity occurs between different cells.

Genomic Equivalence

The variable gene activity hypothesis is based on the premise that in multicellular organisms, somatic cells all contain a complete diploid set of genetic information. While biochemical and cytophotometric analysis has shown the quantity of DNA to be equivalent and equal to the diploid content in cells of most species studied, other investigations have approached the question differently. Is it possible to show that differentiated cells are qualitatively equivalent? In other words, is a complete set of genes present in each cell?

One approach is to show that a differentiated cell is **totipotent** and thus capable of giving rise to a complete organism under the proper conditions. In the early part of this century, Hans Spemann demonstrated that a nucleus derived from the 16-cell stage of a newt embryo was capable of supporting total development. By constricting a newly fertilized egg prior to the first cell division, Spemann partitioned the zygote nucleus into one half of the cell. The half containing the nucleus proceeded to undergo division and produce 16 cells. At that point, the constriction was loosened and one of the 16 nuclei was allowed to pass back into the non-nucleated cytoplasm on the other side. Both halves subsequently produced complete but separate embryos. Therefore, at the 16-cell stage in this organism, genomic equivalence of the 16 nuclei was demonstrated.

Beginning in the 1950s, more sophisticated experiments were performed by Robert Briggs and Thomas King using the grass frog *Rana pipiens*, and by John Gurdon in the 1960s using an African frog, *Xenopus laevis*. After inactivating or surgically removing the nucleus from an egg, these investigators were able to test the developmental capacity of somatic nuclei by injecting nuclei isolated from cells at various stages of differentiation. Nuclei derived from the blastula stage of development were capable of supporting the development of complete and normal adults when transplanted into enucleated oocytes. In *Rana* (Figure 21.1), transplanted nuclei derived from later stages such as the gastrula and neurula usually allowed only partial development. In *Xenopus*, however, Gurdon's experiments showed that epithelial gut nuclei of tadpoles were able to support the development of an adult frog when transplanted into enucleated oocytes. In both cases, it is clear that under normal developmental conditions, progressive restrictions that may prevent the expression of totipotency are placed on differentiating nuclei.

To test whether the nucleus of a highly differentiated adult cell is irreversibly specialized or can support the development of a normal embryo, Gurdon used nuclei from adult frog skin cells in serial transplant experiments (Figure 21.2). In these experiments, a donor nucleus was transplanted into an enucleated egg, and the recipient was allowed to develop for a short time, for example, to the blastula stage. The blastula cells were then dissociated and a

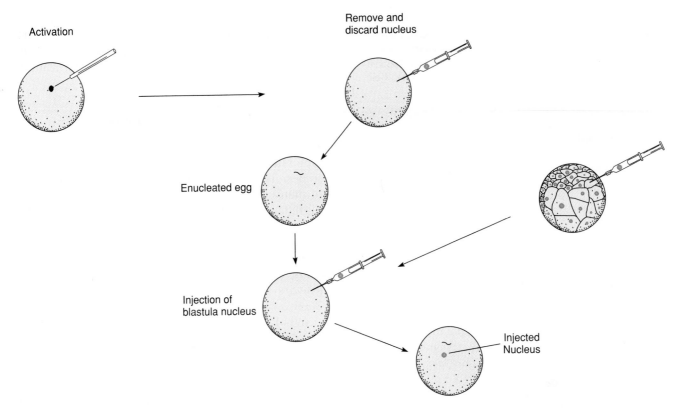

FIGURE 21.1
The process of nuclear transplantation in the frog.

FIGURE 21.2
Serial transplantation experiments using adult frog skin cell nuclei. Serial transplantation dramatically increases the percentage (from 5 to 30 percent) of recipient eggs that undergo complete cleavage.

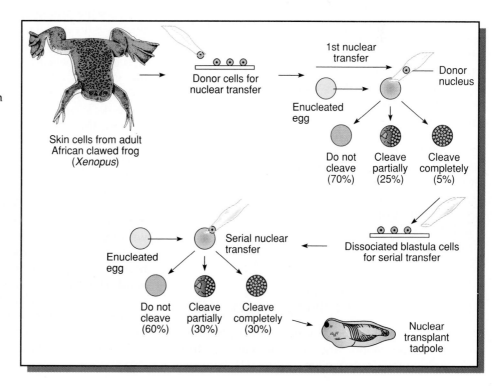

TABLE 21.1

First transfer vs. serial transplantation in *Xenopus laevis*.

Donor Cells	Percentage Reaching Tadpole Stage with Differentiated Cell Types	
	1st Transfer	Serial Transfer
Intestinal epithelial cells from tadpoles	1.5	7
Cells from adult skin	0.037	8
Blastula or gastrula	36	57

SOURCE: From Gurdon, 1974, p. 24.

nucleus removed from one of them. This nucleus was transplanted into still another enucleated egg and development was allowed to occur. Such serial transfers were repeated a number of times. Subsequently, the blastula was not dissociated, but instead was allowed to continue development as far as it would go. After several such transfers, Gurdon found that 30 percent of the nuclear transplants resulted in the formation of advanced tadpoles (Table 21.1). Because nuclei from specialized epidermal cells can eventually direct the synthesis of gene products such as myosin, hemoglobin, and crystallin and promote the organization of cells and tissues into a tadpole, we are certain that such cells contain a complete copy of the genome. Under the proper circumstances, genes not expressed in the most specialized cells can become reactivated.

In plants, the totipotency of terminally differentiated nonmeristematic cells has been demonstrated using the carrot. Fredrick Steward observed that when individual phloem cells are explanted into a liquid culture medium, each cell divides and eventually forms a mass called a callus. Under appropriate conditions, this cell mass will differentiate into a mature plant.

These studies all convincingly argue that differentiated adult cells have not lost any of the genetic information present in the zygote. Instead, the majority of genes in any given cell type are shut off or repressed, but can be reprogrammed to direct normal development. Unfortunately, we do not yet understand the nature of the processes controlling nuclear differentiation during cellular specialization.

EVIDENCE OF DIFFERENTIAL TRANSCRIPTION IN SPECIALIZED CELLS

The entire thrust of the variable gene activity hypothesis of differentiation rests on the concept that differential gene transcription occurs (1) between

different cells of an organism at the same time; and (2) within the same cell at different times during development. Evidence that supports the role of differential transcription is available from the cellular, biochemical, and molecular levels.

Chromosome Puffs

As pointed out in Chapter 17, several larval tissues in dipteran insects contain cells housing giant chromosomes. These polytene chromosomes are composed of many DNA strands in parallel register, are complexed with protein, and show distinctive banding patterns (see Figure 17.11). Each band represents one or several genes. Puffing at particular banded regions is taken as cytological evidence of transcription. Thus, a comprehensive study of puffing patterns in salivary gland cells during larval development has been very informative.

What is seen in such studies is that polytene chromosomes undergo a very specific puffing pattern with time. With great consistency, specific bands show puffing at precise times. As development proceeds, these regions show a regression of puffs, and new puffs are seen associated with different sets of bands [Figure 21.3(b)]. These changing patterns continue until pupation and are interpreted as representing differential synthesis of RNA and thus the transcription of specific but different genes over time.

Of particular interest are the studies performed by Wolfgang Beerman and Ulrich Clever using the fly *Chironomus*. They identified a genetic locus that specifies the production of proteinaceous granules in certain salivary gland cells. In one species, *Ch. pallidivittatus*, a granular secretion is seen. In a second species, *Ch. tentans*, the secretion is clear and lacks the granules. When granules are present (in *Ch. pallidivittatus*), specific puffing occurs on chromosome 4. When no granules are produced (in *Ch. tentans*), this specific puffing is absent.

To provide a firm correlation between the puffing (representing gene transcription) and the granular product, Beerman and Clever produced hybrids between the two species and examined the larval salivary glands. In hybrid cells, the amount of granular secretion is considerably reduced compared with that in *Ch. pallidivittatus*. Since a single salivary chromosome is composed of both members of a homologous pair of chromosomes, only part of the band should puff in the hybrid if it indeed is responsible for the granular product, and this is what is observed. This work, summarized in Figure 21.3(a), provides strong support for the concept of differential transcription of the same gene between two species.

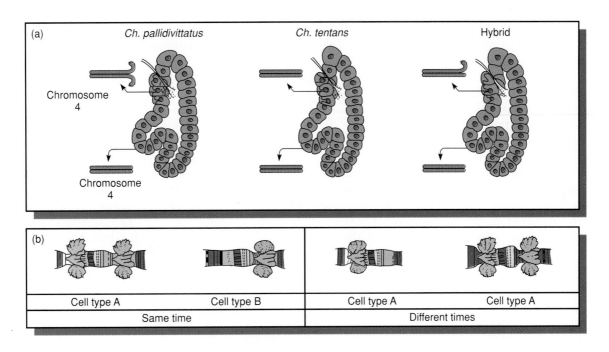

FIGURE 21.3
(a) Puffing patterns in parental species and hybrids of *Chironomus*. Salivary glands of *Ch. pallidivittatus* (left) produce granules. Chromosome 4 from the granule-producing cells has a puff at one end, but chromosome 4 from other cells of the same gland (lower left) has no puff at that location. *Ch. tentans* (center) makes no granules. No puffs at that location are produced in *Ch. tentans*. Hybrids of the two species (right) have a puff only on the chromosome from the *pallidivittatus* parent in the granule-producing cells, and these cells make far fewer granules. (b) Differential puffing patterns are evidence of differential gene activity.

Isozymes

A direct means of demonstrating differential transcription is to show that different gene products are present in the cells of different tissues. One approach has involved studies of **isozymes**. Isozymes are multiple molecular forms of the same enzyme that may be distinguished electrophoretically because of differing net charge. In other words, an enzyme that catalyzes a single reaction may exist in more than one form. Different forms of the same enzyme are possible when it is composed of separate polypeptide chains or subunits. If the enzyme is composed of a combination of nonidentical subunits, isozymes are observed.

This is best illustrated by examining the isozymic forms of the enzyme **lactate dehydrogenase (LDH)**. Clement Markert and his colleagues have shown that this enzyme is a tetrameric molecule composed of four subunits. Each subunit in the molecule may be composed of either an "A" polypeptide or a "B" polypeptide; thus, there are five different isozymes possible (Figure 21.4). LDH is ubiquitous among vertebrates and found in many invertebrates as well.

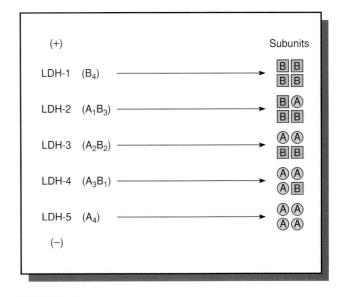

FIGURE 21.4
Zymogram of the multiple forms of the enzyme lactate dehydrogenase.

This is to be expected because of its catalytic capacity for the interconversion of pyruvic and lactic acid, essential reactions in glycolysis. Even though both the A and B subunits have a molecular weight of 35,000 daltons (and thus the weight of the tetrameric form is 140,000 daltons), they differ slightly in amino acid composition and net electrical charge. Thus, any of the five isozymes will migrate differently when placed in an electric field.

When various tissues of the same species or the same tissues of different species are examined, LDH isozyme patterns are found to differ considerably. The isozyme patterns in Figure 21.5(a) show that either the A subunit (LDH-3,4,5) or the B subunit

(LDH-1,2,3) seems to predominate in each of eight tissues of the rat. Since the A and B subunits are presumably the products of separate, distinct genes, these patterns directly reflect differential transcription.

In Figure 21.5(b), LDH isozymes are shown at various stages during the development of the rat heart. Prior to birth (−9, −5 days) LDH-5 predominates, indicative of A gene activity. Just before birth (−1 day) a binomial distribution is seen, indicating that both the A and B genes are equally transcribed. Following birth, there is a gradual shift toward the preponderance of LDH-1 and the B subunit, most notably in the adult. Thus, in the same tissue during

FIGURE 21.5
(a) Tissue-specific pattern of expression of LDH activity in the rat. (b) Developmental sequence of LDH activity in the rat heart.

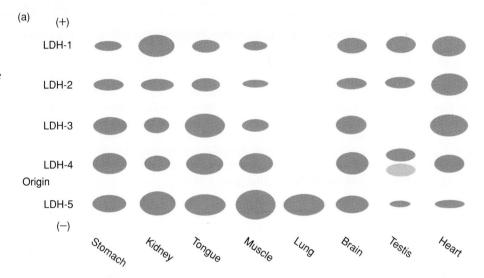

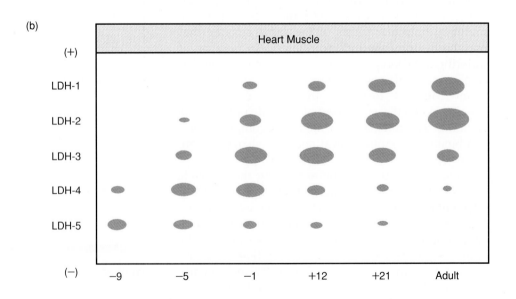

development, the A and B genes are differentially transcribed. Similar examples are found in other tissues and between related species.

An even more remarkable variation of LDH gene activity is found in the pigeon. In primary spermatocytes of this species, a third subunit (designated C and clearly distinguishable from A and B) is found. The C-containing tetramer (C_4) is found only in primary spermatocytes and no other cells of the body. Following division and secondary spermatocyte formation, the C gene is turned off. The physiological advantage conferred by one isozymic form or another is not clear. However, isozymes do have slightly different kinetic and substrate specificities, as in the case of the C_4 tetramer.

Growth Hormone

In mammals, synthesis and secretion of growth hormone (GH) take place only in a small cluster of cells in the anterior pituitary gland. No other cells of the body are known to express GH. The GH gene is activated by a transcription factor called GHF-1 (Figure 21.6). GHF-1 is 229 amino acids long, and contains a 60-amino-acid homeodomain near its carboxy terminus, and an adjacent POU domain. Recall from the discussion in Chapter 20 that homeodomains are DNA binding regions, and that POU regions contribute indirectly to DNA binding. Within the first 72 amino acids the amino terminal portion of the molecule contains a region that is responsible for transcriptional activation (Figure 21.6). From the structure of the transcription factor, it appears that GHF-1 positively regulates expression of GH by binding to one or more promoter elements and activating GH transcription.

Studies using *in situ* hybridization and immunochemistry in mouse embryos have shown that GHF-1 is responsible for the developmental regulation of GH and for maintaining GH expression in adult, differentiated cells of the anterior pituitary. In the mouse, GHF-1 transcripts are detectable only in anterior pituitary cells beginning on day 13 of embryonic development, and GH transcripts appear shortly thereafter. This observation and other experimental evidence provides strong support for the idea that GHF-1 is the only factor necessary for the activation of the GH gene.

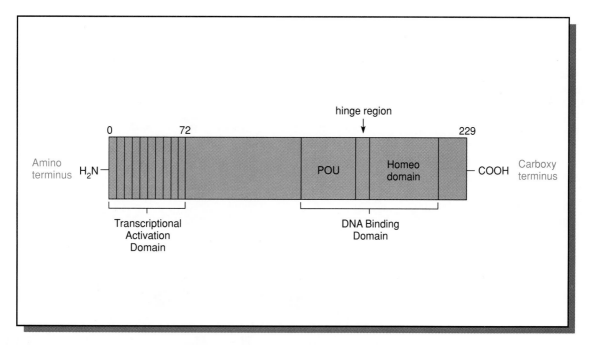

FIGURE 21.6
Organization of the GHF-1 transcription factor. The amino terminus includes a 72-amino-acid serine/threonine-rich region necessary for activation of the GH gene. Near the carboxy terminus is a 60-amino-acid homeodomain adjacent to a POU domain. These two domains are separated by a short sequence that may act as a hinge, folding the molecule.

To explain how a single factor can be responsible for the tissue-specific activation of a given gene, we need to consider the organization of the 5′ flanking region of the GH gene, and the role of GHF-1 in expression of GH. The 5′ flanking region of the GH gene contains several blocks of DNA sequences that can function as promoters, all located in the first 300 bp of DNA (Figure 21.7). Directly upstream from the TATA box is a region (at about −60 to −100 bp) that binds a factor mediating the role of cAMP in the transcriptional induction of GH. GHF-1 binds to two adjacent sites, one centered at −80, and the other at −120. Each site functions as an important promoter element. Further upstream is a glucocorticoid receptor (GRE) that acts to mediate the response to GHF-1. In addition, a thyroid hormone response element (ENH) is also present within the 5′ flanking region (Figure 21.7). With the exception of GHF-1, factors that bind to these promoter elements are distributed in many different cell types, including anterior pituitary cells, indicating that specificity for GH transcriptional activation must reside in the GHF-1 factor. A model has been proposed, however, to explain the role of GHF-1 and other factors in the developmental regulation of the GH gene in the mouse.

According to this model [Figure 21.7(a)], the GH promoter region is unresponsive to the presence of all other factors such as the glucocorticoid factor, the thyroid factor, and the cAMP mediating factor. In addition, the 5′ flanking promoter region may be involved in nucleosome formation, making it more inaccessible to activation. At about day 13 of embryonic development, GHF-1 transcripts appear, and as the GHF-1 factor accumulates, it binds to the promoter region of the GH gene, initiating transcription of GH mRNA [Figure 21.7(b)]. The transcriptional activation by GHF-1 alone results in low level synthesis of GH. Binding at the GHF-1 receptor may enhance the accessibility of other promoter sites, resulting in binding of glucocorticoid factors, thyroid factors, and the cAMP mediating factors. The net result of this cooperative binding is a large increase in GH transcription [Figure 21.7(c)]. Thus, the appearance of GHF-1 is the key step in the developmental

FIGURE 21.7
Model for cell-specific GHF-1 activation of the mouse GH gene. (a) Prior to day 13 of embryonic development, the GH gene is inactive, even though factors that can bind to the promoter region are present (with the exception of GHF-1). (b) After day 13, GHF-1 (black squares) appears, and binds to sites in the GH promoter, activating transcription of GH. (c) Transcription is enhanced by alterations of DNA conformation in the promoter region, allowing other factors to bind.

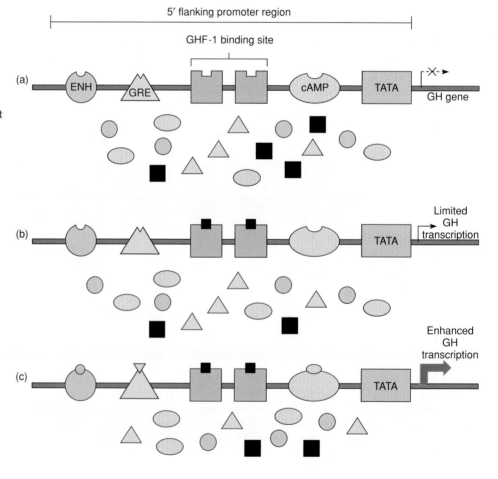

pathway leading to the cell-specific expression of the GH gene. If this model can be confirmed, future experiments will be directed at determining how the activation of GHF-1 at a specific stage of embryonic development is controlled.

DIFFERENTIAL TRANSCRIPTION DURING DEVELOPMENT

We have now come to the central issue of developmental biology. In the preceding discussion, we established what is responsible for making cells different from one another (i.e., differential transcription). However, why one cell's fate is different from another's during development is a broader and more critical issue. This question is distinct from that discussed in Chapter 20, where we asked what actually turns any given gene on or off at the cellular level. We consider here the question of how cells eventually acquire one differentiated state or another.

As alluded to earlier, there is no simple answer to this question. However, information derived from the study of many different organisms provides a starting point. We shall examine several of these examples, which together reflect the direction and scope of our knowledge in this area.

Studies of viruses, bacteria, and the less complex eukaryotes have yielded interesting information pertinent to development. These organisms are used because they are more amenable to experimental approaches. In contrast to complex multicellular eukaryotes, these organisms may be cultured under controlled conditions. In addition, their biochemical and developmental responses are less dependent on the interactions of many differentiated cell and tissue types. Thus, before describing more detailed organisms, we shall briefly discuss two model systems that have been studied in some detail. In this discussion we will also illustrate that developmental events in apparently simple systems are in reality complex mechanisms requiring the coordinated action of many gene sets.

Bacterial Sporulation

Certain gram-positive bacilli have developed a mechanism of survival under adverse conditions—the formation of spores. These structures are highly resistant to heat, desiccation, and toxic conditions, and are induced to form in nutritionally depleted environments. Although the efficiency of sporulation may depend on cell density, the process is usually

regarded as an example of unicellular differentiation that requires no interaction with other cells. Spores are formed within the bacterial cell over a period of several hours and the process has been divided into seven morphologically distinct stages (Figure 21.8). When maturation is complete, spores possess no metabolic activity; thus, they resemble the seeds of plants. Under favorable conditions, dormancy may be broken and germination will give rise to a vegetative bacterial cell.

As vegetative cells (stage 0) enter sporulation, the replicated bacterial chromosome condenses to form an axial filament. In stage II, a septum divides the cell into a larger mother cell and a smaller forespore or prespore. In subsequent stages the prespore is enclosed in a second membrane layer, and a proteoglycan cell wall is laid down between the membranes. The spore coat is formed in stage V, and after maturation (stage VI), the mature spore is released by lysis of the mother cell in stage VII.

A number of biochemical events also occur during the sporulation process, including those that are spore-specific and those that take place in both vegetative and prespore compartments. These biochemical events are the result of programmed gene action, under the control of regulatory genes. Significant to developmental studies are the shifts in genetic activity that take place during sporulation. In the early stages of development at least 20 loci—all of which are spore-specific—are switched on, and transcription of the genes normally used in vegetative growth is shut down. Although the exact sequence of the biochemical events is not yet known, some of the steps are shown in Figure 21.8. Many of the products appearing in a stage-specific fashion, such as spore coat proteins, are used directly in spore production. Others may serve as triggers for the next series of events. If RNA synthesis is inhibited experimentally during sporulation, the entire process is interrupted, thus documenting that the sporulation process depends on the transcription of a distinct set of genes. As sporulation moves beyond stage II, different genes are expressed in the prespore and the mother cell. As differentiation proceeds, the spore retains only a small number of ribosomes, the enzymes necessary for germination, and a copy of the bacterial chromosome.

Recent studies have provided insight into the genetic regulation of sporulation. These studies have attempted to categorize sporulation genes into temporally discrete classes, and to define the regulatory elements that activate each class at a specific developmental stage. Most of the mutants recovered in this work block sporulation at a specific time in

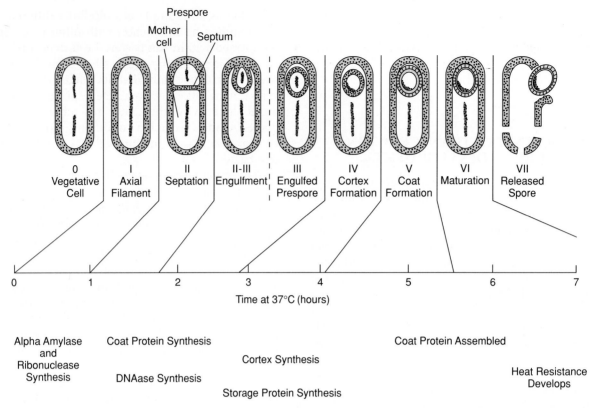

FIGURE 21.8
Morphological stages and some of the biochemical events during sporulation in species of the bacterium *Bacillus*.

development and are classified accordingly. Mutations that block the initiation of sporulation are called stage 0 or *spoO* mutations, those that block septum formation (stage II) are called *spoII* mutations, and so on through stage VI. Hundreds of developmental mutations have been isolated, and approximately 80 loci have been mapped. The total number of genes involved in the sporulation process is unknown, but probably involves well over 100 loci.

A start has also been made in identifying regulatory elements that are involved in controlling sporulation. These include different forms of the enzyme RNA polymerase. Recall that prokaryotic RNA polymerase is a large molecule composed of a core containing subunits and a sigma (σ) factor. The sigma factor allows polymerase to bind and initiate transcription from a promoter site adjacent to a gene. In spore-forming bacteria, at least six forms of sigma factor have been identified. Some, like σ^{43} (σ^a) are associated with vegetative growth, while others, such as σ^{30} (σ^h), encoded by the *spoOH* gene, control the full expression of several *spoO* genes. Evidence suggests that other, more complex regulatory elements also participate in controlling gene expression during sporulation.

Thus, the differentiation of a spore within a single cell over a period of eight to ten hours has proven to be a genetically complex event. This process involves at least 80 genes within at least six classes, the activation of several different sigma factors, and a carefully regulated temporal sequence of gene expression. This example serves to illustrate the intricate nature of gene regulation and the interaction between genes and gene products in the process of development.

Slime Mold Development

The eukaryotic cellular slime mold *Dictyostelium discoideum* is a unique organism in studies of development. As shown in Figure 21.9, the slime mold exists in one of two alternating forms during its life cycle. Germinating spores give rise to amoeboid cells called **myxamoebae**. These cells feed on soil bacteria and decaying vegetation, grow, and multiply in numbers. However, if nutrients become depleted, free-living cells aggregate together and form a multicellular body or slug consisting of about 10^5 cells. The process of aggregation and further differentiation is mediated by cyclic AMP. As cells cluster and form aggregation

FIGURE 21.9
Stages in the life cycle of the slime mold, *Dictyostelium discoideum*.

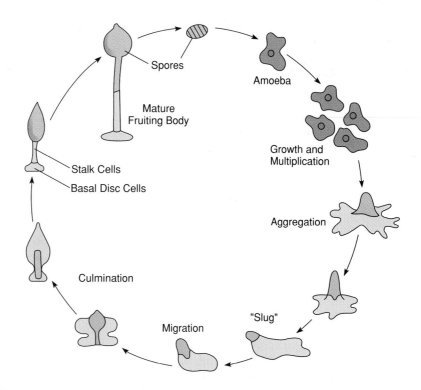

Aggregation stage

Slug stage

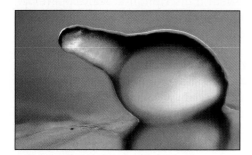

Upright stage following culmination

Immature fruiting body

FIGURE 21.10
Cycle of gene expression during development in *Dictyostelium*. The signals that induce gene response are shown at the top. Aside from the first, which is environmental, the signals are produced by the cells themselves during development. Each signal causes a class of genes to become active during a developmental stage, and the products of each class are required for transition to the next stage.

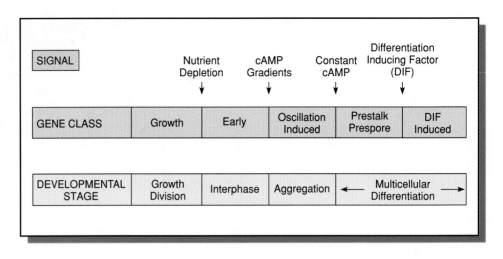

centers, they secrete waves of cAMP that provide a series of chemotactic gradients to signal and direct other cells to the aggregation center. Two distinct cell types form in the slug, depending on their position in the aggregate. Cells in the posterior 75 to 80 percent of the slug form prespore cells, while the cells at the anterior tip become prestalk cells. The slug, which is initially migratory, differentiates into vacuolated stalks and spores. When the spores mature and are released, they germinate into myxamoebae, and the cycle is repeated.

The slime mold life cycle is an ideal model for investigation of differentiation and cell-to-cell interactions because only two cell types are involved. Cellular structure and function are completely transformed during the conversion of a myxamoeba into a portion of the fruiting body. Proteins present in individual cells are broken down, and synthesis of different gene products occurs. For example, the posterior prespore cells express a characteristic set of protein markers that are absent from the anterior prestalk cells. Because the fate of any given amoeba is determined solely by its position in the aggregate, all cells must have the genetic potential for becoming any part of the fruiting body. In addition, the anterior tip of the slug has properties similar to the embryonic organizer found in vertebrates. If the tip (anterior 25% of the slug, containing the prestalk cells) is cut off, the posterior fragment mounds up, and a new tip is formed before morphogenesis is restarted. If a tip is grafted to another slug, it forms a new axis, causing the slug to separate into two slugs, each with an anterior tip.

Mutations that affect each step in the process of aggregation and differentiation are known. For example, over 100 complementation groups exhibit an aggregation-minus phenotype. In this group, *frigid A*

mutants show no chemotactic or signalling response and cannot be induced to differentiate. On the other hand, *synag* mutants can be induced to differentiate if supplied with cAMP. A general scheme for the timing of gene expression is shown in Figure 21.10. It is apparent that a programmed sequence of gene activation leading to the production of differentiation-inducing factor (DIF) is essential for the complete life cycle of *Dictyostelium*. Because this organism is less complex than higher eukaryotic forms, further study may help decipher the role of cell-to-cell interaction in development.

The Effects of the Cellular and Extracellular Environment

Eukaryotic organisms develop from a *zygote,* a cell formed by the fusion of sperm and oocyte. The oocyte cytoplasm is complex, heterogeneous, and nonuniform in distribution within the cell. Following fertilization and early cell division, the nuclei of progeny cells will therefore find themselves in different environments as the maternal cytoplasm is distributed into the new cells. Evidence suggests that these different forms of cytoplasm exert variable influences on the genetic material of different cells, causing differential transcription at some point during development. Recall that when a specific developmental fate for cells becomes fixed prior to the actual events of differentiation or specialization, the cells are said to have undergone determination. Although such cells show no immediate evidence of structural or functional specialization, their position in the developing embryo or organism seems to determine the ultimate differentiated form they will assume. It is as if their fate has been programmed prior to the actual events leading to specialization.

Early gene products further alter the total cytoplasmic content of each cell, producing a still different cellular environment that may in turn lead to the activation of other genes, and so on. In this way, cells embark on different pathways as development proceeds. As the number of cells increases, they influence one another. The total environment acting on the genetic material, therefore, now includes the cell's individual cytoplasm as well as the influence of other cells. Thus, as the developing organism becomes more complex, so does the total environment. In this way, different forms of determination and differentiation occur during development.

We have already discussed several examples that lend credence to the concept of cytoplasmic influence on nuclear activity. For example, Gurdon's nuclear transplantation work with *Xenopus* shows that nuclei from differentiated cell types are capable of reversing their role and serving as zygote nuclei. It may be concluded that the cytoplasmic environment of the enucleated egg exerts a profound influence on nuclear activity.

Similar conclusions may also be drawn from studies of somatic cell hybridization, discussed in detail in Chapter 6. When cultured somatic cells are experimentally fused, two nuclei may exist in a common cytoplasm, forming a heterokaryon. One of the two nuclei usually conforms to the cytoplasm native to the other nucleus so that both nuclei become uniform in their degree of genetic activity.

On a more general environmental level, consider the examples of bacterial sporulation and of the aggregation and differentiation of slime mold myxamoebae discussed earlier in this chapter. In both cases, extracellular conditions trigger specific transcriptional events.

Overview of *Drosophila* Development

The formation of the body plan of the larva and adult of *Drosophila melanogaster* is an elegant example of the interaction between cytoplasm and nucleus. The body of *Drosophila* is composed of a number of head, thoracic, and abdominal segments. During embryogenesis, a series of nuclear divisions occurs immediately after fertilization. When nine divisions have occurred, the approximately 512 nuclei migrate to the egg's outer surface or cortex, where further divisions take place. The nuclei become enclosed in membranes, forming a single layer of cells (blastoderm) over the embryo surface (Figure 21.11). Each segment of the adult body is formed from the descendants of cells set aside at the blastoderm stage into discrete structures called **imaginal disks** (Figure 21.12). These disks, formed from hollow sacs of cells, are deter-

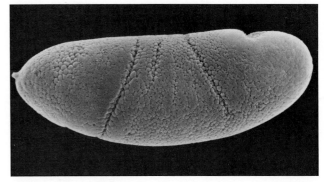
(a)

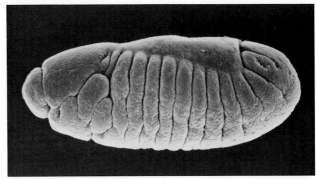
(b)

(c)

FIGURE 21.11
Stages of development in the *Drosophila* embryo. (a) During gastrulation, the single cell layer of the embryo begins to invaginate and migrate to form the principal layers of the embryo. (b) At a later stage, the regions that will form the body segments of the larva begin to emerge as distinct landmarks. (c) A few hours later, the segmentation pattern of the body is clearly established.

mined to form specific parts of the adult body. There are 12 bilaterally paired disks and one genital disk. For example, there are eye-antennal disks, leg disks, wing disks, and so forth. Other cells in the blastoderm will form the internal and external structures of the larva. During the metamorphosis, which occurs during pupation, most of the body parts of the larva

FIGURE 21.12
Imaginal disks of the *Drosophila*
larva and the adult structures
that are derived from them.

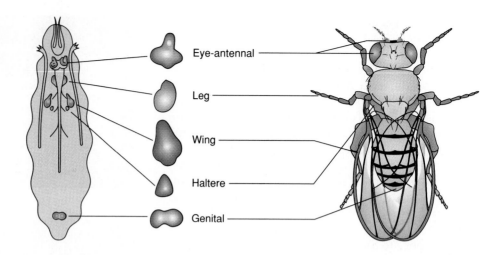

histolyze, or break down, and the imaginal disks
differentiate into adult structures of the head, thorax,
and abdomen.

Studies using a combination of physical and genetic techniques have revealed that the major features of the larval and adult body plans are present by
the blastoderm stage of embryonic development in
Drosophila. The two-dimensional projections of the
external adult body parts on the surface of the
blastoderm are known as **fate maps** (Figure 21.13).

One of the most striking features of fate maps is
that the embryo is organized along an anterior-
posterior axis that reflects the body axis of the larva,
pupa, and adult (Figure 21.13). The existence of a fate
map implies that developmental fate is governed by
the location to which nuclei migrate during blastoderm formation, and that positional information of an
unknown nature directs the cleavage nuclei to adopt
developmental programs that result in the morphologically distinct segments of the adult body.

At a slightly later stage of development, the
precursors of the adult body structures are present as
distinct morphological landmarks in the embryo
(Figure 21.14). Even at this stage of development,
cells in each segment are determined to form either
an anterior or a posterior half or **compartment** of the
segment. Progressive restrictions continue to occur
throughout development so that cells in each compartment are committed to forming a limited number
of adult structures.

Gene Expression During Early *Drosophila* Development

The shape of the *Drosophila* egg provides a set of
morphological landmarks that define the anterior-
posterior and dorsal-ventral axes of the egg (Figure
21.15). The anterior pole of the egg contains the
micropyle, a specialized structure for the entrance of
sperm into the egg, while the posterior pole is

FIGURE 21.13
Diagram showing the origin of
some *Drosophila* structures from
the embryonic blastoderm.

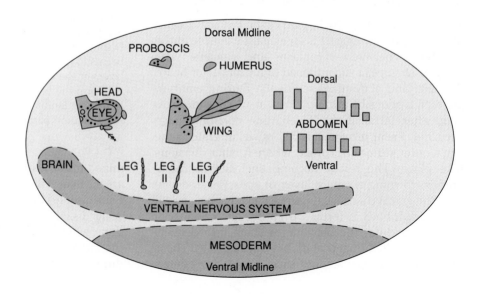

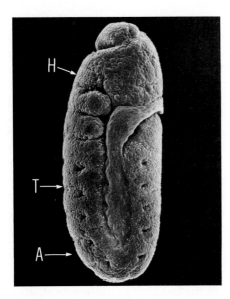

FIGURE 21.14
Scanning electron micrographs of stages of the *Drosophila* embryo. Head (H), thorax (T), and abdominal (A) structures are shown.

rounded and marked by a series of aeropyles that are openings for gas exchange during development. The dorsal side of the egg is flattened and contains the chorionic appendages, while the ventral side is curved. These axes are associated with the distribution of molecular gradients within the egg that play a

key role in directing the development of nuclei that migrate into a specific region of the embryo.

As zygotic nuclei migrate into different regions of the cytoplasm, maternal transcripts and gene products localized in these regions initiate a program of gene expression in the nuclei that results in the formation of the anterior-posterior axis of the embryo. Table 21.2 lists the maternal genes that are required for formation of the anterior-posterior embryonic axis. One of the three maternal genes required for the formation of the anterior pattern of body segments is *bicoid*. Embryos of *bicoid* mutants lack head and thoracic structures and also have abnormalities in the first four segments of the abdomen. In the egg and early embryo, *bicoid* mRNA is localized in a small region at the anterior end. After fertilization, *bicoid* protein appears in a gradient in the anterior two-thirds of the embryo. The *bicoid* protein contains a homeodomain (see Chapter 16 for a description of such domains), suggesting it may be a transcription control factor. In fact, the *bicoid* protein is known to bind to the upstream region of at least one gene involved in the formation of body segments. Whether other gene products involved in anterior specification act as transcription factors remains to be demonstrated, but *bicoid* clearly acts as a positive transcription regulator on nuclei within the anterior portion of the embryo.

Nine posterior genes are known (Table 21.2), and the available genetic and molecular evidence suggests that this group may act on the formation of abdominal segments by repression of a maternal gene product or transcript. Thus, in contrast to the action of anterior genes, the posterior group of genes works by a negative effect on a maternal gene product.

Cloning of two of the terminal genes, *torso* and *fs(1)polehole*, suggests that this group may work by encoding genes that synthesize transmembrane signal receptors associated with the reception and transduction of an extracellular signal.

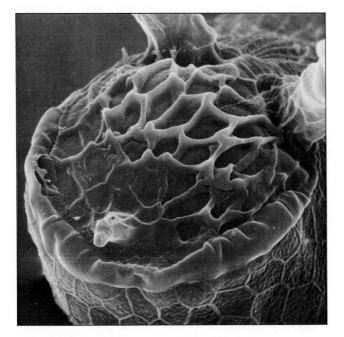

FIGURE 21.15
Scanning electron micrograph of the anterior pole of the newly oviposited *Drosophila* egg.

TABLE 21.2
Maternal effect genes involved in anterior-posterior axis formation in *Drosophila*.

Anterior Group	Posterior Group	Terminal Group
bicoid	*nanos*	*torso*
swallow	*pumilio*	*l(l)polehole*
exuperantia	*vasa*	*torsolike*
	staufer	*trunk*
	cappuccino	*fs(l)nasrat*
	spire	*fs(l)polehole*
	valois	
	oskar	
	tudor	

These three groups of maternal genes establish a spatial pattern of segment-organizing activities during oogenesis. Following fertilization and the appearance of embryonic nuclei, these genes regulate the transcriptional activity of nuclei along the anterior-posterior axis of the egg.

The transcription of the dorsal-ventral segmentation genes begins at the blastoderm stage, and is restricted to a number of narrow bands that extend circumferentially around the embryo. The temporal expression of this gene set establishes the boundaries of the body segments and compartments, and the developmental fate of the cells within each segment.

Many of the segmentation genes belong to two gene complexes located on chromosome 3: the *Antennapedia* complex (*Ant-C*) and the *bithorax* complex (*BX-C*). It has been postulated that genes in these complexes represent *selector* genes. Such genes are cued by positional information in the embryo cytoplasm and determine a particular developmental pathway for cells in a restricted portion of the embryo. In different segments and compartments, the activation of unique combinations of selector genes produces the differentiated segmental structures of the larva and adult bodies.

Mutant selector genes called **homoeotic mutants** shift the determined state of a group of cells in a compartment or segment to form structures normally seen in another segment. Thus, the mutant *tumorous head* transforms the head into the last abdominal segment and genitalia. Studies on the *Antennapedia* complex (*ANT-C*) and the *bithorax* complex (*BX-C*) suggest that *Antennapedia* controls segmentation of the head and anterior regions of the fly, while *bithorax*

acts to regulate the development of the middle and posterior regions of the body. The genes of the *bithorax* complex consist of eight or more genes arranged on the chromosome in an order that corresponds to the anterior-posterior arrangement of body segments in which they are active (Figure 21.16). These genes normally act to direct the correct developmental pathway for thoracic and abdominal segments (Figure 21.17). According to a model developed by E. B. Lewis, who has studied these genes for more than thirty years, loci in the *bithorax* complex act to specify segments by controlling the number of genes active in each segment. Thus, none of the genes are active in the second thoracic segment (mesothorax). But development of the third thoracic segment requires the first gene to be activated, and in each subsequent segment one more gene is switched on until, in the last abdominal segment, all the genes in the complex are switched on. In the mutant condition, each of these genes transforms a segment into structures normally formed by other segments, presumably by activating the wrong number of loci in the complex (Figure 21.16).

Organization and Regulation of Homoeotic Genes

Because of their important role in controlling development, *ANT-C* and *BX-C* genes have been analyzed using recombinant DNA techniques and molecular cloning. Some 260 kilobases in the distal region of the *ANT-C* complex have been cloned, and the entire *BX-C* region, spanning almost 300 kb of DNA, has been cloned. The results to date have been somewhat

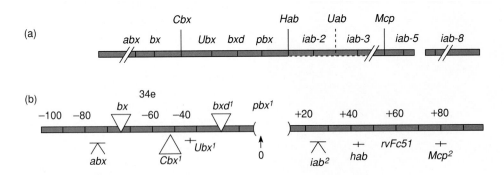

FIGURE 21.16
(a) Map of the *bithorax* locus in *Drosophila* derived from genetic studies. Mutations in the upper row (*Cbx, Hab, Uab, Mcp*) transform anterior segments into posterior ones. Mutations in the lower row cause posterior segments to become more like anterior ones. (b) Map derived from studies using recombinant DNA. The zero and the arrow indicate the location of the first clone recovered. Triangles and inverted Vs are insertions; lines are inversions; and *pbx* is a deletion.

FIGURE 21.17
Wild-type fly (left) compared to a fly with two pairs of wings (right) produced by a combination of *bithorax* mutants.

surprising in several respects. The recombinant DNA studies have largely confirmed the order of genes in each complex that were derived from genetic studies, but they show that the complexes are very large—on the order of 200 to 300 kilobases of DNA, instead of the 8 to 10 kb predicted from genetic studies. An average gene with no introns is usually thought to be about 1 kb in length, so these regions have enormous coding capacities. Many of the genes in these complexes are also very large. The *Antennapedia* gene covers more than 100 kb of DNA, and includes two promoters, eight exons, and two terminator processing regions. One of the introns in this gene covers almost 60 kb. Similarly, in *BX-C*, the *Ultrabithorax (Ubx)* gene covers some 70 kb.

Another unexpected and important discovery is that most of the mutations in this region are not point mutations involving a change in one or even a few bases, but instead are insertions or deletions covering several thousand base pairs of DNA. The ends of these insertions have characteristics in common with the transposable elements found in the genomes of other higher organisms.

Not only is there evidence that some DNA regions in the complex may be transposable, but it appears that changing the position of a section of DNA alters its pattern of expression. The pbx^1 mutation is caused by the removal of a 17-kb DNA segment, and Cbx^1 is caused by the insertion of the same segment some 40 kb upstream from its original position. While pbx^1 transforms the rear of the haltere into the rear of the wing, Cbx^1 produces the opposite effect. The conclusion is that the 17-kb piece encodes the information for the rear half of the haltere (or, more generally, the

rear half of the third thoracic segment), and the segment in which this information is expressed depends on the location of the DNA within the *bithorax* complex.

Molecular studies of the transcription and translation of genes in the *bithorax* cluster have provided evidence for the mechanism of action of homeotic genes. *In situ* hybridization experiments show that homeotic genes are expressed in order, as proposed by Lewis, and are maximally expressed in the segments which they specify. The results of such experiments have also led Walter Gehring and his colleagues to propose a modification of Lewis's model of homoeotic gene action. Their observations indicate that the action of a homoeotic gene that is active in a more anterior segment is repressed in more posterior segments by the activity of genes located more posteriorly in the complex. This idea is supported by genetic experiments demonstrating that anterior genes are strongly expressed in more posterior segments when the next (more posterior) gene in the complex is deleted.

All of the steps in determining the segmental arrangement of the body as well as the activation of the correct combination of selector genes depend on gene products produced by the maternal genome and arranged into a positional array in the cortex of the egg. Some of these genes must act to control the action of the homeotic selector genes by providing positional cues in the early embryo. Such genes should be active in the maternal genome and, in the mutant condition, have some effect on the expression of homeotic genes. A number of genes fulfilling these criteria have been identified, including *extra sex combs* (esc), *Polycomb (Pc), super sex combs (sxc),* and *trithorax (trx)*. In the second chromosomal mutant *extra sex combs (esc)*, some of the head and all of the thoracic and abdominal segments develop as posterior abdominal segments (Figure 21.18), indicating that this gene normally controls the expression of *BX-C* genes in all body segments. The mutation does not affect either the number or the polarity of segments but only their developmental fate, indicating that the *esc*⁺ gene product that is synthesized and stored in the egg by the maternal genome may be required for

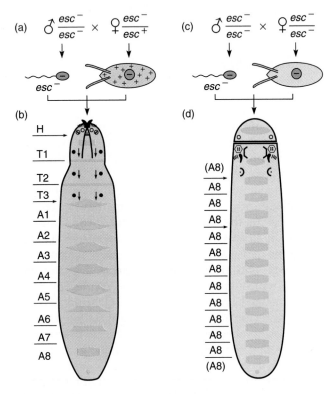

FIGURE 21.18
Action of the *extra sex combs* mutation in *Drosophila*. (a) Heterozygous females form wild-type *esc* gene product and store it in the oocyte. (b) Fertilization of *esc*⁻ egg formed at meiosis by the heterozygous female by *esc*⁻ sperm still produces wild-type larva with normal segmentation pattern (maternal rescue). (c) Homozygous *esc*⁻ females produce defective eggs which, when fertilized by *esc*⁻ sperm, (d) produce a larva in which most of the segments of the head, thorax, and abdomen are transformed into the eighth abdominal segment. Maternal rescue demonstrates that the *esc*⁻ gene product is produced by the maternal genome and is stored in the oocyte for use in the embryo. (H, head; T1–3, thoracic segments; A1–8, abdominal segments). Borderlines between the head and thorax and between thorax and abdomen are marked with arrows.

the correct interpretation of the information gradient in the egg cortex.

Lewis has proposed that genes in the *bithorax* complex, and perhaps other genes involved in segmentation, arose from a common ancestral gene by tandem duplication and subsequent divergence of function. In fact, recent evidence has shown that genes of the *BX-C* and *ANT-C* share a highly conserved DNA sequence of about 200 bp called a *homeo box* or *homeodomain*. Similar sequences have been found in the genomes of other eukaryotes with segmented body plans, including *Xenopus*, chicken, mice, and humans. Homeodomains from all organisms examined to date are very similar in DNA sequence and encode a protein associated with the transcriptional regulation of a specific gene set. This suggests that the metameric or segmented body plan may have evolved only once.

The examples cited represent the striking influence of the cellular and extracellular environment on the fate of cells during development. In many other organisms, determination is not demonstrated as early as it is in *Drosophila*, and it is not as easy to pinpoint and document. This does not imply, however, that determination does not occur. Rather, it seems just a matter of timing during development.

THE STABILITY OF THE DIFFERENTIATED STATE

Once a cell has taken on a specific structural and functional capacity, under normal circumstances in an adult organism the cell maintains that biological status. That is, kidney epithelial cells, red blood cells, muscle cells, and so on are not normally converted into other cell types. Such observations have led to the question of whether or not differentiation can be reversed under experimental conditions. Differentiation is a two-step process, beginning with determination, and followed by the biochemical and morphological changes that lead to the differentiated state. Thus the question must be addressed in two steps: can we reverse determination, and can we reverse differentiation?

Recall that we have already established that nuclear differentiation in specialized cell types is not necessarily stable. This conclusion was drawn based on the nuclear transplantation experiments discussed earlier in this chapter. As we will see in the following examples, there are certain cases where differentiated cells, and not just nuclei, may alter their developmental status.

Transdetermination

Ernst Hadorn demonstrated that if imaginal disks are explanted from a larva and implanted into the abdomen of an adult, these primordia will continue to proliferate by cell division but will not differentiate. The disk subsequently may be recovered from the adult, cut in half, and reimplanted—part to another adult abdomen and part back into a larva (Figure 21.19). The half implanted into a second adult continues to proliferate and can be serially propagated in this way. The part placed into the larva will undergo metamorphosis and differentiate into an adult structure.

FIGURE 21.19
In vitro culturing of imaginal disk fragments over four generations. In each generation, a disk fragment is implanted into a larva to test its developmental potential and fate.

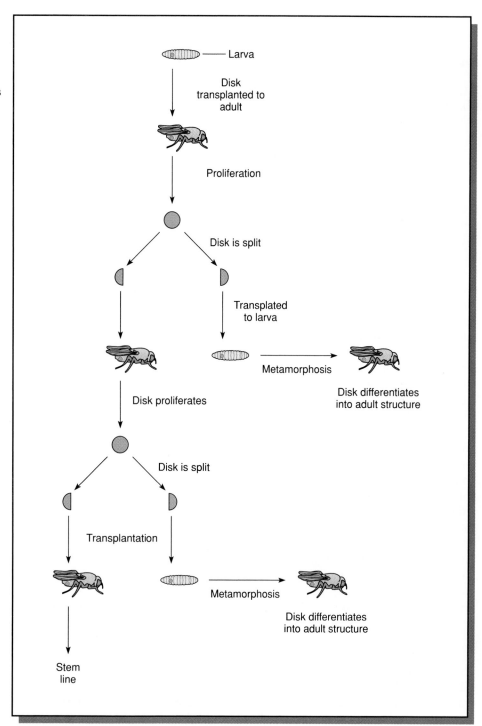

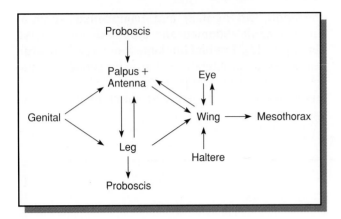

FIGURE 21.20
Observed sequence of transdeterminations from each imaginal disk.

Hadorn showed that during the early transfers, the disk primordium always maintained its determined state (i.e., a leg disk always differentiated into leg structures, etc.) following experimental manipulation. After several passages through adult abdomens, atypical structures were occasionally discovered; that is, an antennal disk sometimes produced wing structures, a genital disk gave rise to leg structures, and so forth. This shift in determination is referred to as **transdetermination**.

The major forms of transdetermination are summarized in Figure 21.20. All disks can be transdetermined into one or another disk type, and certain sequences of transdetermination occur; a disk can give rise only to a limited number of other disks.

Hadorn's experiments showed that developmental programming or determination can be altered by serially "subculturing" disk cells in adult abdomens. This transdetermination involves complete cells, not just the nucleus (as in nuclear transplantation experiments). Thus, it may be concluded that irreversible changes do not occur during preadult development of imaginal disks. Homoeotic mutants show many of the properties of transdetermination; in fact, most of the transdeterminations are also known as homoeotic mutations, emphasizing that these developmental programs are under genetic control.

Regeneration

There are still other examples in which differentiated cells show a lack of developmental stability. Regeneration of amputated limbs occurs in the tadpoles of many frogs and toads and in some larval and adult

newts and salamanders. Since replacement of the missing structure often follows the normal developmental process, regeneration has been used to study the stability of the differentiated state as well as normal developmental pathways.

Following limb amputation, a sequence of morphogenetic events follows:

1 The cut end of the stump is covered by a distal migration of epidermal cells, forming a thin, transparent sheet over the wound.
2 Degenerative changes occur in the cells at the cut tip, and phagocytes remove cellular debris.
3 Cells from all or most tissues at the cut tip undergo a dedifferentiation and accumulate under the epidermal covering to form a **blastema,** from which all regenerating tissues will be derived.

The blastema grows by cell division into a cone projecting from the stump and then flattens into a paddle or palette (Figure 21.21). Blastemal mesenchyme cells condense into ridges, forming digits that elongate from the palette. Muscle and nerve differentiation confers movement on the regenerate limb, and growth continues until it reaches full size.

Experiments involving transplantation of ^{3}H-thymidine labeled or triploid blastema tissue onto unlabeled or diploid amputated stumps have demonstrated that the undifferentiated blastema cells (mesenchyme) accumulate from most of the internal tissues of the adjacent stump. Cells from the cartilage, muscle, and connective tissues all contribute to

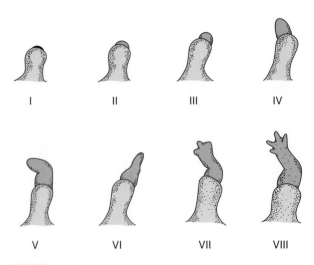

FIGURE 21.21
Stages of limb regeneration in the newt.

the formation of the blastema. While all of the cells in the blastema have undergone dedifferentiation, some apparently retain their determined state and can redifferentiate only in one direction—a process known as **modulation**. Other cells can transform into several tissue types during differentiation. Cartilage cells, for example, can dedifferentiate into blastemal cells, but undergo modulation and regenerate into cartilage. Muscle cells, on the other hand, can redifferentiate into other cell types, including cartilage. These observations again support the concept that even though cells have attained a highly specialized state where differential transcription is occurring, the potential still exists for inactive genes to be reactivated following cell proliferation.

CHAPTER SUMMARY

1   The role of genetic information during development and differentiation has been studied extensively using two simple model systems: bacterial sporulation and slime mold development. In both systems, environmental conditions induce specific gene activity, which, in turn, leads to altered development. By isolating developmental mutations, researchers have elucidated the regulatory elements involved in controlling these two processes.

2   The variable gene activity theory, which applies to higher organisms, assumes that all somatic cells in an organism contain equivalent genetic information, but that it is differentially expressed. Studies conducted on genomic equivalence have demonstrated that somatic cells undergoing differentiation retain the entire gene set, which can be reprogrammed to direct the development of the entire organism.

3   The concept of differential gene activity accounts for the structural and functional differences that exist among specialized cells. Examples that support this concept are chromosome puffs, isozymes, and human hemoglobin synthesis.

4   During embryogenesis, specific gene activity appears to be affected by the total environment of the cell, which includes the cell's individual cytoplasm as well as the maternal cytoplasm. As development proceeds, the cell's environment becomes further altered by the presence of other cells and early gene products.

5   In *Drosophila*, genetic and molecular studies have confirmed that the egg contains information that specifies the body plan of the larva and adult, and that interaction of embryonic nuclei with the maternal cytoplasm initiates a transcriptional program characteristic of specific developmental pathway.

6   Determination is the regulatory event whereby cells become progressively restricted in developmental capacity during early development. Determination becomes more specific with time and precedes the actual differentiation or specialization of distinctive cell types.

7   The differentiated state of a cell may be stable, as in eukaryotes, or reversed, as demonstrated by transdetermination in *Drosophila* and regeneration in amphibians.

Insights and Solutions

In *Dictyostelium*, the experimental evidence suggests that cyclic AMP (cAMP) plays a central role in the developmental program leading to spore formation. The genes encoding the cAMP cell surface receptor have been cloned and the amino acid sequence of the protein components is known. Aggregating cells secrete waves or oscillations of cAMP to foster aggregation of cells, and then continuously secrete cAMP to activate genes at later stages of development. Because of the proposed central role of cAMP in cell-to-cell interaction and gene expression in this developmental program, it is important to test this hypothesis by using several different experimental techniques. What different approaches can you devise to test this hypothesis, and what specific experimental systems would you employ to test them?

ANSWER: Two of the most powerful forms of analysis in biology involve the use of biochemical analogs or inhibitors to block gene transcription or the action of gene products in a predictable way, and the use of mutations to alter the gene and its products. These two approaches can be used to study the role of cAMP in the developmental program of *Dictyostelium*. First, compounds chemically related to cAMP, such as GTP and GDP, can be used to test whether they have any effect on the processes controlled by cAMP. In fact, both GTP and GDP lower the affinity of cell surface receptors for cAMP, effectively blocking the action of cAMP. To inhibit the synthesis of the cAMP receptor, it is possible to construct a vector that contains a DNA sequence that transcribes an antisense RNA (a molecule that has a base sequence complementary to the mRNA). Antisense RNA forms a double-stranded structure with the mRNA, preventing it from being transcribed. If normal cells are transformed with vector that expresses antisense RNA, no cAMP receptors will be produced. It is possible to predict that such cells will fail to respond to a gradient of cAMP and consequently will not migrate to an aggregation center. In fact that is what happens. Such cells remain dispersed and nonmigratory in the presence of cAMP. Similarly, it is possible to determine whether this response to cAMP is necessary to trigger changes in the transcriptional program by assaying for the expression of developmentally regulated genes in cells expressing this antisense RNA.

Mutational analysis can be used to dissect components of the cAMP receptor system. One way to approach this is to use transformation with wild type genes to restore mutant function. Similarly, since the genes for the receptor proteins have been cloned, it is possible to construct mutants with known alterations in the component proteins and transform them into cells to assess their effects.

∞∞∞∞∞∞∞∞∞∞∞∞∞∞∞∞∞∞∞∞∞∞∞∞∞∞∞

PROBLEMS AND DISCUSSION QUESTIONS

1 Carefully distinguish between the terms *differentiation* and *determination*. Which phenomenon occurs initially during development?

2 The *Drosophila* mutant *spineless aristapedia* (ss^a) results in the formation of a miniature tarsal structure (normally part of the leg) on the end of the antenna. Such a mutation is referred to as *homoeotic*. From your knowledge of imaginal disks and Hadorn's transdetermination studies, what insight is provided by ss^a concerning the role of genes during determination?

3 How many isozymic forms are possible for a molecule that is a dimer and may contain any combination of three nonidentical polypeptide subunits?

4 How does the concept of isozymes relate to the theory of gene duplication discussed in Chapter 7? How would you test for the molecular homology between two genes coding for two polypeptide subunits, as in LDH?

5 In the sea urchin, early development up to gastrulation may occur even in the presence of actinomycin D, which inhibits RNA synthesis. However, if actinomycin D is removed at the end of blastula formation, gastrulation does not proceed. In fact, if actinomycin D is present only between the 6th and 11th hours of development, gastrulation (normally occurring at the 15th hour) is arrested. What conclusions can be drawn concerning the role of gene transcription between hours 6 and 15?

6 In this chapter, we have pointed out several examples of a specific molecule being synthesized by only one specific differentiated cell type (e.g., skin cells and keratin; erythrocytes and hemoglobin). From your past training in biology, what similar examples can you cite?

7 In the slime mold, the enzyme UDPG-pyrophosphorylase is essential to carbohydrate metabolism necessary for formation of the fruiting body. At a specific time during fruiting-body formation, mRNA specific to the enzyme is produced, translation of the enzyme ensues, and transcription ends. This is an example of specific gene activity during development. After transcription of UDPG-pyrophosphorylase mRNA has stopped, the cells of the fruiting body may be dissociated and allowed to reaggregate. Again, at the same stage in the newly formed fruiting body, transcription and translation begin on schedule. What conclusions can you draw concerning the activation of the UDPG-pyrophosphorylase gene during development?

8 Nuclei from almost any source may be injected into frog oocytes. Studies have shown that these nuclei remain active in transcription and translation. How can such an experimental system be useful in developmental genetic studies?

9 The concept of epigenesis indicates that an organism develops by forming cells that acquire new structures and functions, which become greater in number and complexity as development proceeds. This theory is in contrast to the preformationist doctrine that miniature adult entities are contained in the egg that must merely unfold and grow to give rise to a mature organism. What sorts of isolated evidence presented in this chapter might have led to the preformation doctrine? Why is the epigenetic theory held as correct today?

SELECTED READINGS

BEACHY, P. A., HELFAND, S. L., and HOGNESS, D. S. 1985. Segmental distribution of bithorax complex proteins during *Drosophila* development. *Nature* 313: 545–51.

BEERMAN, W., and CLEVER, U. 1964. Chromosome puffs. *Scient. Amer.* (April) 210: 50–58.

BENDER, W., AKAM, M., KARCH, F., BEACHY, P., PEIFER, M., SPIERER, P., LEWIS, E. B., and HOGNESS, D. 1983. Molecular genetics of the *bithorax* complex in *Drosophila melanogaster*. *Science* 221: 23–29.

BRIGGS, R., and KING, T. 1952. Transplantation of living nuclei from blastula cells into enucleated frog eggs. *Proc. Natl. Acad. Sci.* 38: 455–63.

BROTHERS, A. J. 1976. Stable nuclear activation dependent upon a protein synthesized during oogenesis. *Nature* 260: 112–15.

BROWER, D. L. 1985. The sequential compartmentalization of *Drosophila* segments revisited. *Cell* 41: 361–64.

DAVIDSON, E. H. 1976. *Gene activity in early development.* 2nd ed. Orlando: Academic Press.

DEVREOTES, P. 1989. Cell-cell interactions in *Dictyostelium* development. *Trends Genet.* 5: 242–45.

EBERT, J. 1965. *Interacting systems in development.* New York: Holt, Rinehart and Winston.

FOSTER, S. J., and JOHNSTONE, K. 1990. Pulling the trigger: The mechanism of bacterial spore germination. *Mol. Microbiol.* 4: 137–41.

FULTON, C., and KLEIN, A. 1976. *Explorations in developmental biology.* Cambridge, MA: Harvard University Press.

GEHRING, W. 1968. The stability of the differentiated state in cultures of imaginal disks in *Drosophila.* In *The stability of the differentiated state,* ed. H. Unsprung. New York: Springer-Verlag.

GODFREY, S., and SUSSMAN, M. 1982. The genetics of development in *Dictyostelium discoideum. Ann. Rev. Genet.* 16: 385–404.

GURDON, J. B. 1968. Transplanted nuclei and cell differentiation. *Scient. Amer.* (Dec.) 219: 24–35.

———. 1974. *The control of gene expression in animal development.* Cambridge, MA: Harvard University Press.

GURDON, J. B., LASKEY, R. A., and REEVES, O. R. 1975. The developmental capacity of nuclei transplanted from keratinized skin cells of adult frogs. *J. Embryol. Exp. Morphol.* 34: 93–112.

HADORN, E. 1968. Transdetermination in cells. *Scient. Amer.* (Nov.) 219: 110–20.

HOTTA, Y., and BENZER, S. 1973. Mapping of behavior in *Drosophila* mosaics. In *Genetic mechanisms of development*, ed. F. H. Ruddle, pp. 129–67. Orlando: Academic Press.

KAPPEN, C., SCHUGHART, K., and RUDDLE, F. H. 1989. Organization and expression of homeobox genes in mouse and man. *Ann. N. Y. Acad. Sci.* 567: 243–52.

KING, T. J. 1966. Nuclear transplantation in amphibia. *Methods in Cell Physiology*, ed. D. Prescott, vol. 2. Orlando: Academic Press.

LEIGHTON, T., and LOOMIS, W. F., eds. 1980. *The molecular genetics of development*. Orlando: Academic Press.

LEVINE, M., RUBIN, G., and TJIAN, R. 1984. Human DNA sequences homologous to a protein coding region conserved between homoeotic genes of *Drosophila*. *Cell* 38: 667–73.

LEWIS, E. B. 1976. A gene complex controlling segmentation in *Drosophila*. *Nature* 276: 565–70.

LOSICK, R. 1973. The question of gene regulation in sporulating bacteria. In *Genetic mechanisms of development*, ed. F. H. Ruddle, pp. 15–28. Orlando: Academic Press.

LOSICK, R., and PERO, J. 1981. Cascades of sigma factors. *Cell* 25: 582–84.

MARKERT, C. 1975. *Isozymes: Genetics and evolution*. Orlando: Academic Press.

MANSEAU, L. J., and SCHÜPBACH, T. 1989. The egg came first, of course! Anterior-posterior pattern formation in *Drosophila* embryogenesis and oogenesis. *Trends Genet.* 5: 400–405.

MARX, J. L., 1981. Genes that control development. *Science* 213: 1485–88.

McGINNIS, W., HART, C. P., GEHRING, W. J., and RUDDLE, F. H. 1984. Molecular cloning and chromosome mapping of a mouse DNA sequence homologous to homoeotic genes of *Drosophila*. *Cell* 38: 675–80.

REINERT, J., and HOLTZER, H., eds. 1975. *The cell cycle and cell differentiation*. New York: Springer-Verlag.

RUDDLE, F. H., HART, C. P., and McGINNIS, W. 1985. Structural and functional aspects of the mammalian homeo-box sequences. *Trends Genet.* 1: 46–50.

SANCHEZ-HERRERO, E., and AKAM, M. 1989. Spatially ordered transcription of regulatory DNA in the bithorax complex of Drosophila. *Development* 107: 321–29.

SETLOW, P. 1988. Small, acid-soluble spore proteins of *Bacillus* species: Structure, synthesis, genetics, function, and degradation. *Ann. Rev. Microbiol.* 42: 319–38.

SINGER, M. 1952. Influence of the nerve in regeneration of the amphibian extremity. *Quart. Rev. Biol.* 27: 169–200.

STRUHL, G. 1981. A gene product required for correct initiation of segmental determination in *Drosophila*. *Nature* 293: 36–41.

STRUHL, G., and BROWER, D. 1982. Early role of the esc^+ gene product in the determination of segments in *Drosophila*. *Cell* 31: 285–92.

SUBTELNY, S., and GREEN, P. B., eds. 1982. *Developmental order: Its origin and regulation*. New York: Alan R. Liss.

SUSSMAN, M., ed. 1972. *Molecular genetics and developmental biology*. Englewood Cliffs, NJ: Prentice-Hall.

———. 1973. *Developmental biology, its cellular and molecular foundations*. Englewood Cliffs, NJ: Prentice-Hall.

TSONIS, P. A. 1990. Molecular approaches in limb development and regeneration. *Trends in Biochem. Sci.* 15: 82–83.

TYSON, J. J., and MURRAY, J. D. 1989. Cyclic AMP waves during aggregation of *Dictyostelium* amoebae. *Development* 106: 421–26.

WATSON, J. D. 1976. *Molecular biology of the gene*. 3rd ed. Menlo Park, CA: Benjamin/Cummings.

WEBER, R. 1975. *Biochemistry of animal development: Molecular aspects of animal development*. Vol. 3. Orlando: Academic Press.

WESTPHAL, H., and GRUSS, P. 1989. Molecular genetics of development studied in the transgenic mouse. *Ann. Rev. Cell Biol.* 5: 181–96.

YOUNGMAN, P., et al. 1985. New ways to study developmental genes in spore-forming bacteria. *Science* 228: 285–91.

22

Extrachromosomal Inheritance

MATERNAL EFFECT
Ephestia Pigmentation | *Limnaea* Coiling | *Drosophila* Embryogenesis

ORGANELLE HEREDITY: MATERNAL INHERITANCE
Chloroplasts: Variegation in Four O'Clock Plants | *Iojap* in Maize | *Chlamydomonas* Mutations | Mitochondria: *poky* in *Neurospora* | *Petite* in *Saccharomyces* | Mitochondrial DNA and Human Disease

INFECTIOUS HEREDITY
Kappa in *Paramecium* | Infective Particles in *Drosophila*

∞∞∞∞∞∞∞∞∞∞∞∞∞∞∞∞∞∞∞∞∞∞∞∞∞∞∞∞∞∞

CHAPTER CONCEPTS

Inheritance patterns displayed for some traits show them to be controlled by genes that appear to be located outside the nucleus. In these cases, transmission is usually through the female parent, whose gametes either contain maternal gene products that influence development, or are the sole source of chloroplasts and mitochondria that affect the offspring's phenotype.

Throughout the history of genetics, occasional reports have challenged the basic tenets of transmission genetics—the production of the phenotype through the transmission of genes located on chromosomes carried in the nucleus. These reports have indicated an apparent extranuclear or extrachromosomal influence on the phenotype. Such reports were often regarded with skepticism. However, with the increasing knowledge of molecular genetics and the discovery of DNA in mitochondria and chloroplasts, extrachromosomal inheritance is now recognized as an important aspect of genetics.

There are many diverse examples of these unusual modes of inheritance. The expression of such information can influence the phenotype in many ways. In this chapter we will focus on three general types of extrachromosomal genetic phenomena: (1) maternal influence resulting from the effect of stored products of nuclear genes of the female parent during early development; (2) organelle heredity resulting from the expression of DNA contained in mitochondria and chloroplasts; and (3) infectious heredity resulting from the symbiotic association of microorganisms with eukaryotic cells. Each has the effect of producing inheritance patterns that vary from those predicted by the concepts of Mendelian and neo-Mendelian genetics.

MATERNAL EFFECT

Maternal effect, also referred to as maternal influence, implies that an offspring's phenotype for a particular trait is strongly influenced by the nuclear genotype of the maternal parent. This is in contrast to most cases, where inheritance of traits is biparental. Crosses involving traits inherited due to a maternal effect produce results that do not adhere to Mende-

lian or neo-Mendelian patterns. In such cases, the genetic information of the female gamete is transcribed and these genetic products are present in the egg cytoplasm. Following fertilization, these products influence patterns or traits established during early development. Three examples will illustrate the influences of the maternal genome on particular traits.

Ephestia Pigmentation

In the Mediterranean meal moth, *Ephestia kuehniella,* the wild-type larva has a pigmented skin and brown eyes as a result of the dominant gene *A*. The pigment is derived from a precursor molecule, kynurenine, which is in turn a derivative of the amino acid tryptophan. A mutation, *a*, results in red eyes and little pigmentation in larvae when homozygous. As illustrated in Figure 22.1, different results are obtained in the cross *Aa* × *aa*, depending on which parent carries the dominant gene. When the male is the heterozygous parent, a 1:1 brown/red-eyed ratio is observed in larvae as predicted by Mendelian segregation. However, when the female is heterozygous for the *A* gene, all larvae are pigmented and have brown eyes. As these larvae develop into adults, one-half of them gradually develop red eyes.

One explanation of these results is that the *Aa* oocytes synthesize kynurenine or an enzyme necessary for its synthesis and accumulate it in the ooplasm prior to the completion of meiosis. Even in *aa* progeny (whose mothers were *Aa*), this pigment is distributed in the cytoplasm of the cells of the developing larvae—thus, they develop pigmentation and brown eyes. Eventually, the pigment is diluted among many cells and used up, resulting in the conversion to red eyes as adults. The *Ephestia* example demonstrates the maternal effect in which a

FIGURE 22.1
Illustration of maternal
influence in the inheritance of
eye pigment in the meal moth
Ephestia kuehniella. Multiple
light receptor structures (eyes)
are present on each side of the
anterior portion of larvae.

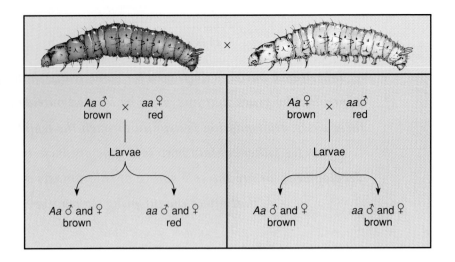

cytoplasmically stored nuclear gene product influences the larval phenotype and at least temporarily overrides the genotype of the progeny.

Limnaea Coiling

Shell coiling in the snail *Limnaea peregra* represents a permanent rather than a transitory maternal effect. Some strains of this snail have left-handed or sinistrally coiled shells (*dd*), while others have right-handed or dextrally coiled shells (*DD* or *Dd*). These snails are hermaphroditic and may undergo either cross- or self-fertilization, thus providing a variety of types of matings.

As shown in Figure 22.2, the coiling pattern of the progeny snails is determined by **the genotype of the parent producing the female gamete, regardless of the parental phenotype**. Investigation of the developmental events in these snails reveals that the orientation of the spindle in the first cleavage division after fertilization apparently determines the direction of coiling. Spindle orientation is thought to be controlled by maternal genes acting on the developing eggs in the ovary. The orientation of the spindle, in turn, influences cell divisions following fertilization and quickly establishes the permanent adult coiling pattern. Therefore, the progeny phenotypes are determined by the maternal genotype.

The dextral allele (*D*) produces an active gene product that causes right-handed coiling. If ooplasm from dextral eggs is injected into uncleaved sinistral eggs, they cleave in a dextral pattern. However, in the converse experiment, sinistral ooplasm has no effect when injected into dextral eggs. Apparently, the sinistral allele is the result of a classical recessive mutation inactivating the gene product.

Drosophila Embryogenesis

A third example illustrates that the absence of certain maternal gene products may have a lethal effect following fertilization. In *Drosophila*, the sex-linked recessive gene *fused* (*fu*) causes partial sterility as well as the fusion of two longitudinal wing veins. As illustrated in Figure 22.3, at least one wild-type allele of the *fused* locus must be present in either the maternal parent or in the genome of the progeny in order for embryogenesis to proceed normally. As a result of this effect, progeny with identical genotypes (*fu/fu* or *fu/ /*) may either die or develop to maturity, depending on the genotype of their mother. Thus, during oogenesis in a +/+ or *fu/*+ female, an essential gene product appears to be made and stored in the egg, assuring complete development. This may occur even though the genotype of the progeny lacks the wild-type allele of *fused*.

ORGANELLE HEREDITY: MATERNAL INHERITANCE

In this section we will examine examples of inheritance patterns of phenotypes related to chloroplast and mitochondrial function. Prior to the discovery of DNA in these organelles and extensive characterization of genetic processes occurring within them, these patterns were grouped under the category of cytoplasmic inheritance. That is, certain mutant phenotypes seemed to be inherited through the cytoplasm rather than through the genetic information of chromosomes. Transmission was most often from the maternal parent through the ooplasm, and thus the results of reciprocal crosses (as in Figure 22.1) varied.

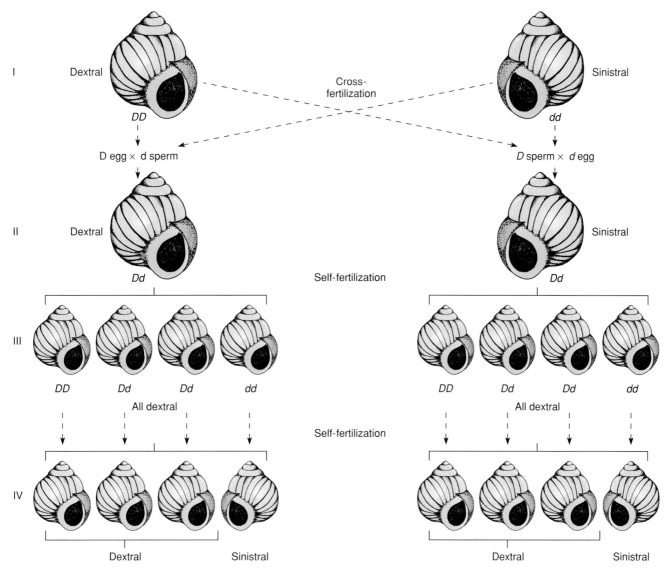

FIGURE 22.2
Inheritance of coiling in the snail *Limnaea peregra*. Coiling is either dextral
(right-handed) or sinistral (left-handed). A maternal effect is evident in generations II
and III, where the genotype of the maternal parent controls the phenotype of the
offspring rather than the offspring's own genotype.

Such patterns are now more appropriately considered examples of **maternal inheritance**.

While such results are similar to the examples of maternal effect presented in the preceding section, a major difference exists. Maternal effects (due to the presence of nuclear gene products) are transitory in the sense that they are not necessarily heritable. With maternal inheritance, however, the phenotype is stable and is continually passed to future generations.

Analysis of the hereditary transmission of mutant alleles of chloroplast and mitochondrial DNA has

been difficult owing to a number of factors. First, the function of these organelles is dependent upon gene products of both nuclear and organelle DNA. Second, the transmission of mitochondrial or chloroplast DNA to progeny is necessarily through the transmission of the organelle itself. Sometimes, it is difficult to determine the origin (maternal vs. paternal) of the organelle contributed to the zygote. This pattern often varies from organism to organism. Third, the number of organelles contributed to each progeny often exceeds one. If many chloroplasts and/or mitochondria are contributed and only one or a few of

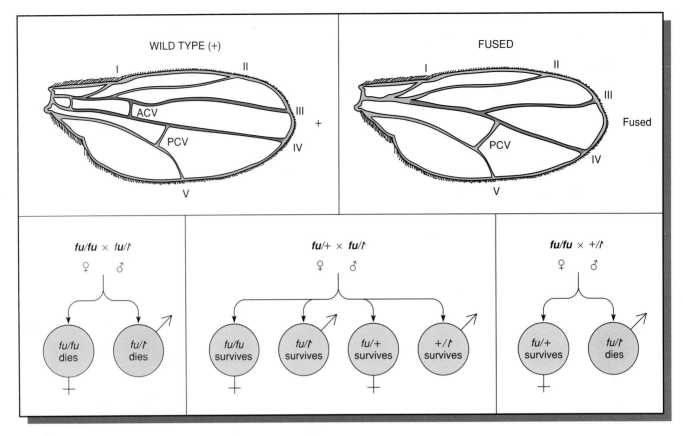

FIGURE 22.3
Results of various crosses in *Drosophila* involving the *fused* allele. Survival of offspring often depends on the genotype of the maternal parent. The *fused* phenotype is contrasted with the normal wild type wing vein pattern. ACV and PCV designate the anterior and posterior cross veins, respectively. The latter is absent in the surviving *fused* flies.

them contain a mutant gene, the corresponding mutant phenotype may not be revealed. Taken together, these complexities have made analysis much more involved than for Mendelian characters.

In this section, for both chloroplasts and mitochondria, we shall discuss examples of inheritance patterns seemingly related to the respective organelle. We have previously provided a fairly detailed analysis of the molecular genetics of these organelles in Chapter 17. You should review and relate this information to the discussion that follows.

Chloroplasts: Variegation in Four O'Clock Plants

In 1908, Carl Correns (one of the rediscoverers of Mendel's work) provided the earliest example of inheritance linked to chloroplast transmission. Correns discovered a variety of the four o'clock plant, *Mirabilis jalapa*, which had branches with either white, green, or variegated leaves. As shown in Table

22.1, inheritance in all possible combinations of crosses is strictly determined by the phenotype of the ovule source. For example, if the seeds (representing the progeny) were derived from ovules on branches with green leaves, all progeny plants bore only green leaves regardless of the phenotype of the source of pollen.

Correns concluded that inheritance was through the cytoplasm of the maternal parent because the pollen, which contributes little or no cytoplasm to the zygote, had no influence on the progeny phenotypes. Since the leaf coloration involves the chloroplast, either genetic information in that organelle or in the cytoplasm and influencing the chloroplast could be responsible for this inheritance pattern.

Iojap In Maize

A phenotype similar to *Mirabilis* but with a different pattern of inheritance has been analyzed in maize by Marcus M. Rhoades. In this case, green, colorless, or

TABLE 22.1
Crosses between flowers from various branches of variegated four o'clock plants.

Source of Pollen	Source of Ovule		
	White Branch	Green Branch	Variegated Branch
White branch	White	Green	White, green, or variegated
Green branch	White	Green	White, green, or variegated
Variegated branch	White	Green	White, green, or variegated

green-and-colorless striped leaves are under the influence not only of the cytoplasm but, in addition, of a nuclear gene located on the seventh linkage group. This locus is called *iojap*, after the recessive mutation *ij* located there. The wild-type allele is designated *Ij*. Plants homozygous for the mutation *ij/ij* may have green-and-white striped leaves. However, when reciprocal crosses are made between plants with striped leaves and green leaves, the results are seen to vary, depending upon which parent is mutant (Figure 22.4). If the female is striped (*ij/ij*) and the male is green (*Ij/Ij*), plants with colorless, striped, and green leaves are observed as progeny. If the male parent is striped and the female green, only green plants are produced! In both types of cross, all offspring have identical genotypes (*Ij/ij*). We may conclude that although a nuclear gene is somehow involved, the inheritance pattern is influenced maternally.

We can understand this pattern better by examining the offspring resulting from self-fertilization of these heterozygous plants (Figure 22.4). The striped plant gives rise to progeny with colorless, striped, and green leaves, regardless of their genotype. The green plants give rise to both green and striped progeny in a 3:1 ratio. The results of these self-fertilizations substantiate that the mutant phenotype is controlled solely through the female cytoplasm, regardless of the plant's genotype.

The role of the nucleus in the origin of the mutant phenotype is unclear. However, the colorless areas of the leaf are due to the lack of the green chlorophyll pigment in chloroplasts. Once acquired, colorless chloroplasts are transmitted through the egg cytoplasm, establishing the phenotypes of leaves of progeny plants.

Chlamydomonas Mutations

The unicellular green alga *Chlamydomonas reinhardi* has provided an excellent system for the investigation of maternal inheritance. The organism is eukaryotic, contains a single large chloroplast as well as numerous mitochondria, undergoes matings which are

followed by meiosis, yet can easily be cultured in the laboratory. The first cytoplasmic mutant, *streptomycin resistance (sr)*, was reported in 1954 by Ruth Sager. Although *Chlamydomonas'* two mating types (see Chapter 5)—*mt+* and *mt−*—appear to make equal cytoplasmic contributions to the zygote, Sager determined that the *sr* phenotype was transmitted only through the *mt+* parent.

Since this discovery, a number of other *Chlamydomonas* mutations (including an acetate requirement as well as resistance to or dependence on a variety of bacterial antibiotics) have been discovered that show a similar maternal inheritance pattern. These mutations have been linked to the transmission of the chloroplast, and their study has extended our knowledge of chloroplast inheritance.

Following fertilization, the single chloroplasts of the two mating types fuse. After the resulting zygote has undergone meiosis, it is apparent that the genetic information of the chloroplasts of progeny cells is derived only from the *mt+* parent.

Further, studies suggest that these chloroplast mutations, representing numerous gene sites, form a single circular linkage group—the first such extrachromosomal unit to be established in a eukaryotic organism. It is also apparent that the linkage groups from the two mating types are capable of undergoing recombination following zygote formation. All available evidence supports the hypothesis that this linkage group consists of DNA residing in the chloroplast.

Mitochondria: *poky* in *Neurospora*

Mitochondria, like chloroplasts, play a critical role in cellular bioenergetics and contain a distinctive genetic system. Mutants affecting mitochondrial function have been discovered and studied. As with chloroplast mutants, these are transmitted through the cytoplasm and result in non-Mendelian inheritance patterns. Additionally, the mitochondrial genetic system has now been extensively characterized.

In 1952, Mary B. and Hershel K. Mitchell discovered a slow-growing mutant strain of the mold *Neurospora crassa* and called it *poky*. (It is also desig-

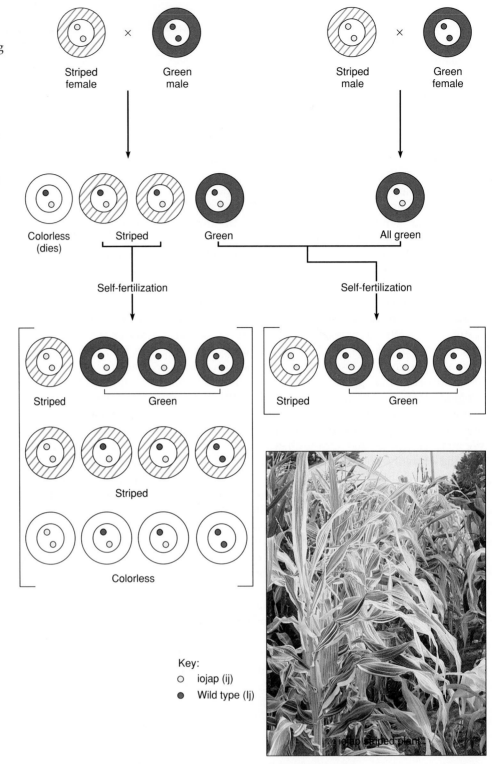

FIGURE 22.4
Maternal inheritance of striping in maize. Regardless of the genotype of the *iojap* genes, offspring reflect the maternal phenotypes in their appearance. If the maternal parent has green leaves, so do all progeny plants. If the maternal parent is striped (containing green *and* colorless areas), progeny are either colorless, green, or striped. Since color is due to chloroplasts in the leaves, inheritance is controlled by these organelles as they are passed through the maternal cytoplasm.

Striped female × Green male

Striped male × Green female

Colorless (dies)   Striped   Green

All green

Self-fertilization

Self-fertilization

Striped   Green

Striped   Green

Striped

Colorless

Key:
○ iojap (ij)
● Wild type (Ij)

nated *mi-1* for maternal inheritance.) Studies have shown slow growth to be associated with impaired mitochondrial function specifically related to certain cytochromes essential to electron transport. The results of genetic crosses between wild-type and *poky* strains suggest that the trait is maternally inherited. If the female parent is *poky* and the male parent is wild type, all progeny colonies are *poky*. The reciprocal cross produces normal colonies.

Studies with *poky* mutants illustrate the use of **heterokaryon** formation during the investigation of maternal inheritance in fungi. Occasionally, hyphae from separate mycelia fuse with one another, giving rise to structures containing two or more nuclei in a common cytoplasm. If the hyphae contain nuclei of different genotypes, the structure is called a heterokaryon. The cytoplasm will contain mitochondria derived from both initial mycelia. A heterokaryon may give rise to haploid spores, or **conidia,** that produce new mycelia. The phenotypes of these structures may be determined.

Heterokaryons produced by the fusion of *poky* and wild-type hyphae initially show normal rates of growth and respiration. However, mycelia produced through conidia formation become progressively more abnormal until they show the *poky* phenotype. This occurs in spite of the presumed presence of both wild-type and *poky* mitochondria in the cytoplasm of the hyphae.

To explain the initial growth and respiration pattern, it is assumed that the wild-type mitochondria support the respiratory needs of the hyphae. The subsequent expression of the *poky* phenotype suggests that the presence of the *poky* mitochondria may somehow prevent or depress the function of these wild-type mitochondria. Perhaps the *poky* mitochondria replicate more rapidly and "wash out" or dilute wild-type mitochondria numerically. Another possibility is that *poky* mitochondria produce a substance that inactivates the wild-type organelle or interferes with the replication of its DNA (mtDNA). As a result of this type of interaction, mutations such as *poky* are labeled **suppressive**. This general phenomenon is characteristic of many other suspected mitochondrial mutations of *Neurospora* and yeast.

Petite in Saccharomyces

Another extensive study of mitochondrial mutations has been performed with the yeast *Saccharomyces cerevisiae*. The first such mutation, *petite*, was described by Boris Ephrussi and his coworkers in 1956. The mutant was so named because of the small size of the yeast colonies. Many independent *petite* mutations have since been discovered and studied. They all have a common characteristic: deficiency in cellular respiration involving abnormal electron transport. Fortunately, this organism is a facultative anaerobe and can grow by fermenting glucose through glycolysis. Thus, although colonies are small, the organism may survive the loss of mitochondrial function by generating energy anaerobically.

The complex genetics of *petite* mutations is diagramed in Figure 22.5. A small proportion of these mutants exhibit Mendelian inheritance and are called **segregational petites,** indicating that they are the result of nuclear mutations. The remainder demonstrate cytoplasmic transmission, producing one of

two effects in matings. The **neutral petites,** when crossed to wild type, produce ascospores that give rise only to wild-type or normal colonies. The same pattern continues if progeny of this cross are backcrossed to *neutral petites*. These results are due to the fact that the majority of neutrals lack mtDNA completely or have lost a substantial portion of it. Thus, the wild-type gamete is the effective source of normal mitochondria capable of reproduction. Neutral petites have now been shown to lack most, if not all, mitochondrial DNA (mtDNA).

A third type, the **suppressive petites,** behaves similarly to *poky* in *Neurospora*. Crosses between mutant and wild type give rise to mutant diploid zygotes which, upon undergoing meiosis, immediately yield all mutant cells. Under these conditions, the *petite* mutation behaves "dominantly" and seems to suppress the function of the wild-type mitochondria. *Suppressive petites* also represent deletions of mtDNA, but not nearly to the extent of the *neutral petites*. Buoyant density studies show a lower G-C content in suppressive mtDNA compared with normal mtDNA.

Suppressiveness remains unexplained. Two major hypotheses have been advanced. One explanation suggests that the mutant (or deleted) mtDNA replicates more rapidly, and thus mutant mitochondria "take over" or dominate the phenotype by numbers alone. The second explanation suggests that recombination occurs between the mutant and wild-type mtDNA, introducing errors into or disrupting the normal mtDNA. It is not yet clear which, if either, of these explanations is correct.

Mitochondrial DNA and Human Diseases

As you may recall from Chapter 17, the DNA found in mitochondria (**mtDNA**) has been extensively characterized in a variety of organisms, including humans. Human mtDNA, which is circular, contains 16,569 base pairs and is strictly inherited maternally. The mitochondrial gene products encoded include:

a. 13 proteins, required for oxidative respiration
b. 22 tRNAs, required for translation
c. 2 rRNAs, required for translation

Unlike the nuclear genome, mitochondria contain little extraneous DNA that fails to code for gene products. Thus, mtDNA would seem particularly vulnerable to mutations, since most genetic alterations can potentially result in a disruption of either translation or oxidative respiration within the organelle.

FIGURE 22.5
The outcome of crosses involving the three types of *petite* mutations affecting mitochondrial function in the yeast *Saccharomyces cerevisiae*. The photograph shows normal yeast colonies.

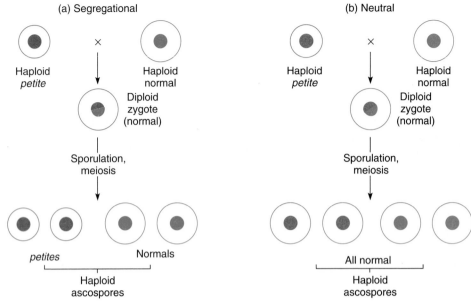

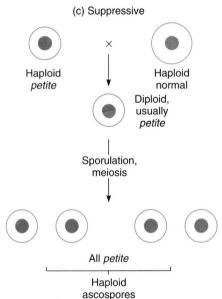

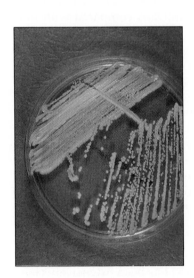

On the other hand, a zygote receives a large number of organelles through the egg, so if only one of them contains a mutation, its impact is diluted because there will be many more mitochondria that will function normally. During early development, cell division disperses the initial population of mitochondria present in the zygote, and in newly formed cells, these organelles reproduce autonomously. Therefore, adults will exhibit cells with a variable mixture of normal and abnormal organelles, should a deleterious mutation arise or already be present in the initial population of organelles. This is a condition called **heteroplasmy**.

In order for a human disorder to be attributable to genetically altered mitochondria, several criteria must be met:

1 Inheritance must exhibit a maternal rather than a Mendelian pattern.
2 The disorder must reflect a deficiency in the bioenergetic function of the organelle.
3 A specific genetic mutation in one of the mitochondrial genes must be documented.

Thus far, several such cases are known that demonstrate these characteristics.

Myoclonic epilepsy demonstrates a pedigree consistent with maternal inheritance. Individuals with this rare disorder express deafness and dementia in addition to seizures, which are associated with ragged red muscle fibers and the abnormal appearance of mitochondria (Figure 22.6). Upon analysis of mtDNA, a single nucleotide substitution has been

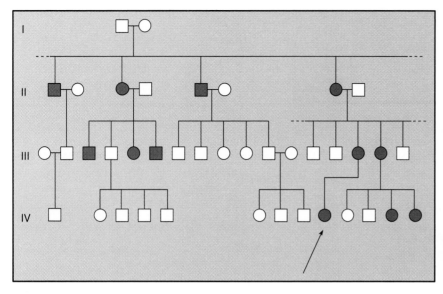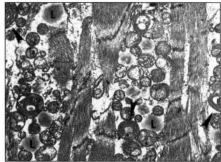

FIGURE 22.6
A partial pedigree illustrating maternal inheritance of myoclonic epilepsy. Some of the individuals were not available for analysis, including the great-grandparents (I) of the proband (arrow). The photograph demonstrates abnormal mitochondria derived from cells of an afflicted individual. Note the abnormal concentric cristae. L = lipid droplets.

found in tRNA^{lys} that apparently interferes with translation within the organelle. Presumably, the deficiency in efficient mitochondrial function is related to the various manifestations of the disorder.

A second disorder, **Leber's hereditary optic neuropathy (LHON)** also exhibits maternal inheritance as well as mtDNA lesions. In this case, a missense mutation has been identified in the gene encoding subunit 4 of NADH:ubiquinone reductase (ND4). Some patients display still other nucleotide changes. In a third disorder, **Kearns-Sayre syndrome,** patients develop encephalomyopathy and bear large deletions at various positions within their mtDNA. It is felt that these heterogeneous deletions in different patients lead to a common phenotype due to the disruption of the translational machinery within these organelles.

The documentation of hereditary, mitochondrial-based diseases extends our knowledge of the impact of genetics on normal development. Additionally, the discovery of the basis of the above diseases paves the way for the exploration of the relationships between mitochondrial function and other neuromuscular disorders.

INFECTIOUS HEREDITY

There are numerous examples of cytoplasmically transmitted phenotypes in eukaryotes that are due to an invading microorganism or particle. The foreign invader coexists in a symbiotic relationship, is usually passed through the maternal ooplasm to progeny cells or organisms, and confers a specific phenotype that may be studied. We shall consider several examples illustrating this phenomenon.

Kappa in *Paramecium*

First described by Tracy Sonneborn, certain strains of *Paramecium aurelia* are called **Killers** because they release a cytoplasmic substance called **paramecin** that is toxic and sometimes lethal to sensitive strains. This substance is produced by particles called **kappa** that replicate in the Killer cytoplasm, contain DNA and protein, and depend for their maintenance on a dominant nuclear gene, *K*. One cell may contain 100 to 200 such particles.

Paramecia are diploid protozoans that can undergo sexual exchange of genetic information through the process of **conjugation**. In some instances, cytoplasmic exchange also occurs. Thus, there is a variety of ways in which the *K* gene and kappa can be transmitted.

The genetic events occurring during conjugation in *Paramecium aurelia* are shown in Figure 22.7. Paramecia contain two diploid micronuclei, which are involved in cross-fertilization. Prior to conjugation, both micronuclei in each mating pair undergo meiosis, resulting in eight haploid nuclei. However, seven of these degenerate, and the remaining one undergoes a single mitotic division. During conjugation, each cell donates one of the two haploid nuclei to the

FIGURE 22.7
Genetic events occurring
during conjugation in
Paramecium. The photographic
insert shows a pair of
organisms undergoing
conjugation.

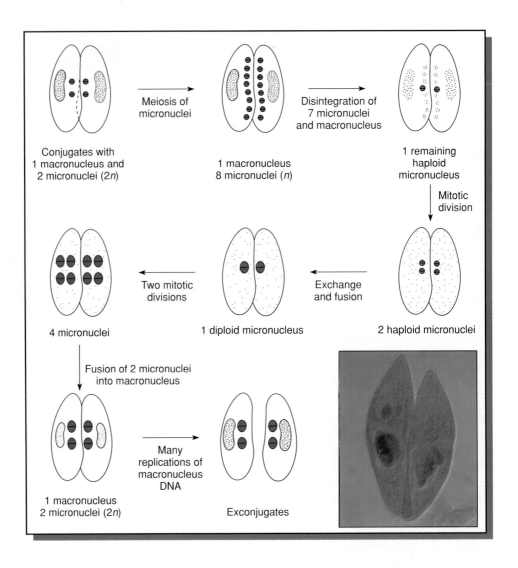

Conjugates with
1 macronucleus and
2 micronuclei (2*n*)

Meiosis of
micronuclei

1 macronucleus
8 micronuclei (*n*)

Disintegration of
7 micronuclei
and macronucleus

1 remaining
haploid
micronucleus

Mitotic
division

4 micronuclei

Two mitotic
divisions

1 diploid micronucleus

Exchange
and fusion

2 haploid micronuclei

Fusion of 2 micronuclei
into macronucleus

1 macronucleus
2 micronuclei (2*n*)

Many
replications of
macronucleus
DNA

Exconjugates

other, recreating the diploid condition in both cells. As a result, exconjugates are of identical genotypes.

In a similar process involving only a single cell, **autogamy** occurs. Following meiosis of both micronuclei, seven products degenerate and one survives. This nucleus divides, and the resulting nuclei fuse to recreate the diploid condition. If the original cell was heterozygous, autogamy results in homozygosity because the newly formed diploid nucleus was derived solely from a single haploid meiotic product. In a population of cells that were originally heterozygous, half of the new cells express one allele and half express the other allele.

Figure 22.8 illustrates the results of crosses between *KK* and *kk* cells, without and with cytoplasmic exchange. When no cytoplasmic exchange occurs, even though the resultant cells may be *Kk* (or *KK* following autogamy), they remain sensitive if no kappa particles are transmitted. When exchange occurs, the cells become Killers provided the kappa

particles are supported by at least one dominant *K* allele.

Kappa particles are bacterialike and may contain temperate bacteriophages. One theory holds that these viruses of kappa may become vegetative; during this multiplication, they produce the toxic products that are released and kill sensitive strains.

Infective Particles in *Drosophila*

Two examples of similar phenomena are known in *Drosophila*: **CO_2 sensitivity** and **sex-ratio**. In the former, flies that would normally recover from carbon dioxide anesthetization instead become permanently paralyzed and are killed by CO_2. Sensitive mothers pass this trait to all offspring. Furthermore, extracts of sensitive flies induce the trait when injected into resistant flies. Phillip L'Heritier has postulated that sensitivity is due to the presence of a virus, **sigma**. The particle has been visualized and is

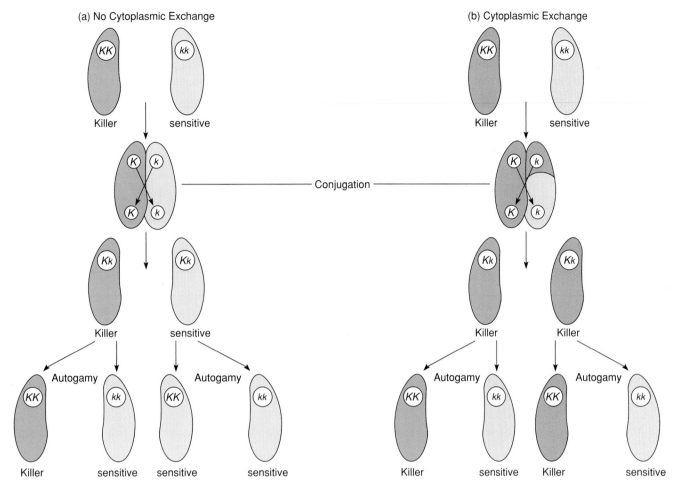

FIGURE 22.8
Results of crosses between *Killer (KK)* and *sensitive (kk)* strains of *Paramecium,* with and without cytoplasmic exchange during conjugation.

smaller than kappa. Attempts to transfer the virus to other insects have been unsuccessful, demonstrating that specific nuclear genes support the presence of sigma in *Drosophila*.

A second example of infective particles comes from the study of *Drosophila bifasciata*. A small number of these flies were found to produce predominantly female offspring if reared at 21°C or lower. This condition, designated **sex-ratio,** was shown to be transmitted to daughters but not to the low percentage of males produced. This phenomenon was subsequently investigated in *Drosophila willistoni*. In these flies, the injection of ooplasm from sex-ratio females into normal females induced the condition. This observation suggested that an extrachromosomal element is responsible for the sex-ratio phenotype. The agent has now been isolated and shown to be a protozoan. While the protozoan has been found in both males and females, it is lethal primarily to developing male larvae. There is now some evidence that a virus harbored by the protozoan may be responsible for producing a male-lethal toxin.

CHAPTER SUMMARY

1 Patterns of inheritance sometimes vary from that expected of nuclear genes. In such instances, phenotypes most often appear to result from genetic information transmitted through the egg.
2 Maternal-effect patterns result when nuclear gene products controlled by the maternal genotype of the egg influence early development. *Ephestia* pigmentation and coiling in snails are examples.
3 Maternal inheritance patterns result from the genotypes of chloroplast and mitochondrial DNA as these organelles are transmitted through the egg. These are cases of true extrachromosomal inheritance.

4 Chloroplast mutations affect the photosynthetic capabilities of plants, while mitochondrial mutations affect cells highly dependent on ATP generated through cellular respiration. The resulting mutants display phenotypes related to the loss of function of these organelles.

5 Another form of extrachromosomal inheritance is due to the transmission of infectious microorganisms, which establish symbiotic relationships with their host cells. Kappa particles and CO_2-sensitivity and sex-ratio determinants are examples.

Insights and Solutions

1 Analyze the following theoretical pedigree and determine the most consistent interpretation of how the trait is inherited. Given your interpretation, are there any inconsistencies in the various individuals?

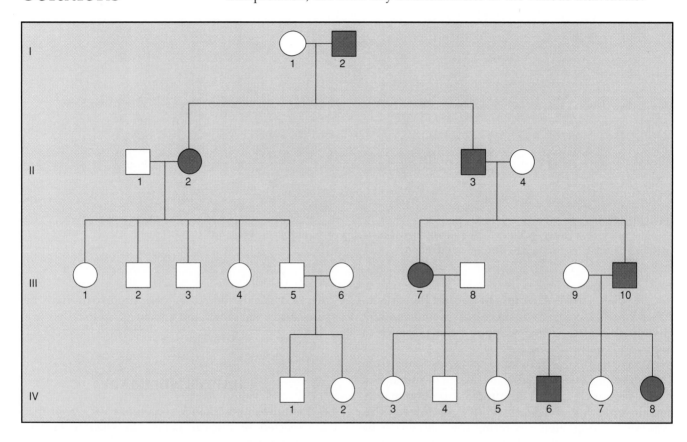

ANSWER: The trait is passed from all male parents to all but one offspring, and *never* passed maternally. Individual IV-7 (a female) is the only exception.

2 Can the above explanation be attributed to a gene on the Y chromosome? Defend your answer.

ANSWER: No, since male parents pass the trait to their daughters as well as to their sons.

3 Is the above case an example of a paternal effect or of paternal inheritance?

ANSWER: It has all the earmarks of paternal inheritance since males pass the trait to all of their offspring. To assess whether the trait is due to a paternal effect (resulting from a nuclear gene in the male gamete), analysis of further matings would be needed.

∞∞∞∞∞∞∞∞∞∞∞∞∞∞∞∞∞∞∞∞∞∞∞∞∞∞∞∞∞∞

**PROBLEMS AND
DISCUSSION QUESTIONS**

1 What genetic criteria distinguish a case of extrachromosomal inheritance from a case of Mendelian autosomal inheritance? from a case of sex-linked inheritance?

2 In *Limnaea*, what results would be expected in a cross between a *Dd* dextrally coiled and a *Dd* sinistrally coiled snail, assuming cross-fertilization occurs as shown in Figure 22.2? What results would occur if the *Dd* dextral produced only eggs and the *Dd* sinistral produced only sperm?

3 Streptomycin resistance in *Chlamydomonas* may result from a mutation in a chloroplast gene or in a nuclear gene. What phenotypic results would occur in a cross between a member of an mt^+ strain resistant in both genes and a member of an mt^- strain sensitive to the antibiotic? What results would occur in the reciprocal cross?

4 A plant may have green, white, or green and white (variegated) leaves on its branches owing to a mutation in the chloroplast (which produces the white leaves). Predict the results of the following crosses.

	Ovule Source		Pollen Source
(a)	Green branch	×	White branch
(b)	White branch	×	Green branch
(c)	Variegated branch	×	Green branch
(d)	Green branch	×	Variegated branch

5 In diploid yeast strains, sporulation and subsequent meiosis can produce haploid ascospores. These may fuse to reestablish diploid cells. When ascospores from a *segregational petite* strain fuse with those of a normal wild-type strain, the diploid zygotes are all normal. However, following meiosis, ascospores are 1/2 petite and 1/2 normal. Is the *segregational petite* phenotype inherited as a dominant or recessive gene?

6 Predict the results of a cross between ascospores from a *segregational petite* strain and a *neutral petite* strain. Indicate the phenotype of the zygote and the ascospores it may subsequently produce.

7 Described below are the results of three crosses between strains of *Paramecium*. Determine the genotypes of the parental strains.
(a) Killer × sensitive → ½ Killer: ½ sensitive
(b) Killer × sensitive → all Killer
(c) Killer × sensitive → ¾ Killer: ¼ sensitive

8 *Chlamydomonas*, a eukaryotic green alga, is sensitive to the antibiotic erythromycin that inhibits protein synthesis in prokaryotes.
(a) Explain why.
(b) There are two mating types in this alga, mt^+ and mt^-. If an mt^+ cell sensitive to the antibiotic is crossed with an mt^- cell that is resistant, all progeny cells are sensitive. The reciprocal cross (mt^+ resistant and mt^- sensitive) yields all resistant progeny cells. Assuming that the mutation for resistance is in cpDNA, what can be concluded?

9 In *Limnaea*, a cross where the snail contributing the eggs was dextral but of unknown genotype mated with another snail of unknown genotype and phenotype. All F_1 offspring exhibited dextral coiling. Ten of the F_1 snails were allowed to undergo self-fertilization. One-half produced only dextrally coiled offspring, while the other half produced only sinistrally coiled offspring. What were the genotypes of the original parents?

10 In *Drosophila subobscura*, the presence of a recessive gene called *grandchildless* (*gs*) causes the offspring of homozygous females, but not homozygous males, to be sterile. Can you offer an explanation as to why females but not males are affected by the mutant gene?

SELECTED READINGS

BOGORAD, L. 1981. Chloroplasts. *J. Cell Biol.* 91: 256s–70s.

COHEN, S. 1973. Mitochondria and chloroplasts revisited. *Amer. Sci.* 61: 437–45.

FREEMAN, G., and LUNDELIUS, J. W. 1982. The developmental genetics of dextrality and sinistrality in the gastropod *Lymnaea peregra*. *Wilhelm Roux Arch.* 191: 69–83.

GILLHAM, N. W. 1978. *Organelle hereditary.* New York: Raven Press.

GOODENOUGH, U., and LEVINE, R. P. 1970. The genetic activity of mitochondria and chloroplasts. *Scient. Amer.* (Nov.) 223: 22–29.

GRIVELL, L. A. 1983. Mitochondrial DNA. *Scient. Amer.* (March) 248: 78–89.

LANDER, E. S., et al. 1990. Mitochondrial diseases: Gene mapping and gene therapy. *Cell* 61: 925–26.

LEVINE, R. P., and GOODENOUGH, U. 1970. The genetics of photosynthesis and of the chloroplast in *Chlamydomonas reinhardi. Ann. Rev. Genet.* 4: 397–408.

MARGULIS, L. 1970. *Origin of eukaryotic cells.* New Haven, Conn.: Yale University Press.

MITCHELL, M. B., and MITCHELL, H. K. 1952. A case of maternal inheritance in *Neurospora crassa. Proc. Natl. Acad. Sci.* 38: 442–49.

PREER, J. R. 1971. Extrachromosomal inheritance: Hereditary symbionts, mitochondria, chloroplasts. *Ann. Rev. Genet.* 5: 361–406.

ROSING, H. S., et al. 1985. Maternally inherited mitochondrial myopathy and myoclonic epilepsy. *Ann. Neurol.* 17: 228–37.

SAGER, R. 1965. Genes outside the chromosomes. *Scient. Amer.* (Jan.) 212: 70–79.

———. 1985. Chloroplast genetics. *BioEssays* 3: 180–84.

SCHWARTZ, R. M., and DAYHOFF, M. O. 1978. Origins of prokaryotes, eukaryotes, mitochondria and chloroplasts. *Science* 199: 395–403.

SLONIMSKI, P. 1982. *Mitochondrial genes.* Cold Spring Harbor, N. Y.: Cold Spring Harbor Laboratory.

SONNEBORN, T. M. 1959. Kappa and related particles in *Paramecium. Adv. in Virus Res.* 6: 229–356.

STRATHERN, J. N., et al., eds. 1982. *The molecular biology of the yeast* Saccharomyces: *Life cycle and inheritance.* Cold Spring Harbor, N.Y.: Cold Spring Harbor Laboratory.

STURTEVANT, A. H. 1923. Inheritance of the direction of coiling in *Limnaea. Science* 58: 269–70.

TZAGOLOFF, A. 1982. *Mitochondria.* New York: Plenum Press.

WALLACE, D. C., et al. 1988. Familial mitochondrial encephalomyopathy (MERRF): Genetic, pathophysiological and biochemical characterization of a mitochondrial DNA disease. *Cell* 55: 601–10.

23
Quantitative Genetics

CONTINUOUS VARIATION AND POLYGENES
Continuous vs. Discontinuous Variation | Polygenic Traits

ANALYSIS OF POLYGENIC TRAITS
The Mean | Variance | Standard Deviation | Standard Error of the Mean |
Analysis of a Quantitative Character

PHENOTYPIC EXPRESSION
Penetrance and Expressivity | Genetic Background: Suppression and Position
Effects | Temperature Effects | Nutritional Effects | Onset of Genetic Expression |
Heredity vs. Environment

CHAPTER CONCEPTS

Some traits are controlled by a number of genes and are classified as polygenic traits. Such quantitative traits are studied using biometric and statistical methods. Statistical methods are also used to determine how genetic variation interacts with the environment to produce the phenotype. That fraction of phenotypic variation that is due to genetic variation is its heritability.

In Chapter 4 we considered how the classical patterns of Mendelian inheritance (the 3:1 and 9:3:3:1 F_2 ratios) are modified because of gene interaction or the action of multiple genes on a single trait. At that time, we discussed examples of polygenic traits such as the inheritance of grain color in wheat, a trait that is controlled by three gene pairs. These traits were discussed to illustrate that even in complex modes of inheritance, the fundamental principles of segregation and independent assortment discovered by Mendel are operational.

Having considered the patterns of inheritance that are characteristic of polygenic traits in that earlier discussion, in this chapter we will describe the methods used by geneticists to study traits that are controlled by several genes. These methods are often statistical, and involve analysis of traits using mathematical tools in addition to those of biochemistry or molecular biology. We will also discuss the effects of some extrinsic factors such as temperature, nutrition, and age on phenotypic expression. Finally, we will examine the concept of **heritability,** which is used by geneticists to estimate the degree to which phenotypic variation is a reflection of genetic differences, and to estimate indirectly the degree of environmental influence on the expression of complex traits.

Polygenic traits are at the heart of several disciplines in genetics, including plant breeding, livestock breeding, and wildlife management. Polygenic traits are also an important part of human genetics; traits such as intelligence, skin color, obesity, and predisposition to certain diseases are thought to be under polygenic control. In the following discussion, remember that in polygenic inheritance, as in monogenic inheritance, barring accidents, the genotype is fixed at the moment of fertilization, while the pheno-

type is more flexible and changes over the lifespan of the organism as the genotype interacts with the environment.

CONTINUOUS VARIATION AND POLYGENES

Continuous vs. Discontinuous Variation

The traits studied by Gregor Mendel in the garden pea could each be separated into two classes: for height, the plants were either tall or short; for color, either green or yellow, and so forth. Such traits are known as **discontinuous traits** because they fall into a limited number of distinct or discontinuous groups (Figure 23.1). Other investigators working on the mechanisms of inheritance in the nineteenth century studied traits that showed a continuous gradation of phenotypes. Over a period of years, Sir Francis Galton used the sweet pea in a series of genetic crosses, and measured the diameter of the pea as a phenotypic trait. When he crossed plants with large peas to plants with small peas, the F_1 peas were all intermediate in size to the parents. When the F_1 were crossed among themselves, the resulting F_2 generation contained peas of many different diameters. Some of the F_2 peas were as large or as small as the parents, but most were intermediate in size when compared to the parents (Figure 23.1).

Galton devised statistical methods to study these **continuous traits,** and the resulting field became known as **biometry.** While these mathematical techniques provided an important tool for the analysis of experimental data, they offered no explanation for the underlying mechanisms of inheritance. Not surprisingly, this led to many erroneous conclusions

FIGURE 23.1
Histograms comparing phenotypic distributions in two crosses involving pea plants. (a) The distribution of phenotypes in the F_1 and F_2 for a trait exhibiting discontinuous variation. (b) The distributions of phenotypes in the F_1 and F_2 for a trait showing continuous variation.

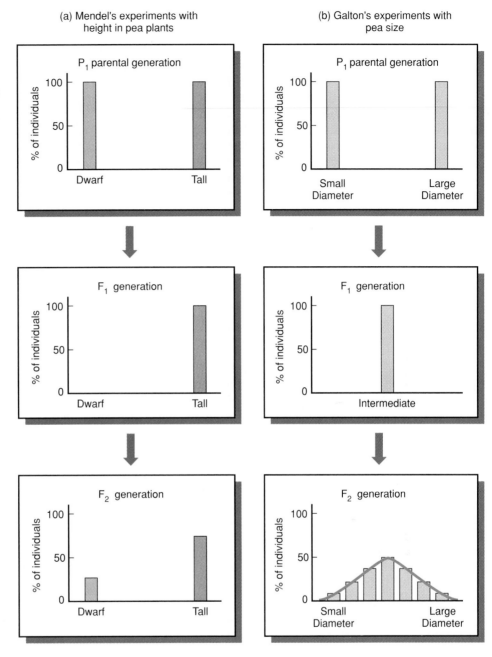

(a) Mendel's experiments with height in pea plants

(b) Galton's experiments with pea size

about the characteristics of continuous variation, and to a subsequent debate as to whether Mendelian principles could explain this form of inheritance. As outlined in Chapter 4, this debate was settled in a series of experiments that demonstrated the role of Mendelian factors in continuous variation. The resulting **multiple-factor hypothesis** proposed that many genes can contribute to the phenotype in a cumulative fashion. Recall that in multiple-factor inheritance, the traits are quantified in some way, and that, for simplicity, it is usually assumed that each gene pair consists of one additive and one

nonadditive allele. Again for simplicity, the total effect of each additive allele is small and roughly equal to all others, and is influenced by phenotypic and environmental variation. When two or more genes contribute to a phenotypic trait in an additive way, the trait is said to be **polygenic**.

Polygenic Traits

In discussing polygenic traits, it is useful to know how to approximate the number of genes that control a given trait. In a given cross, if the parents represent highly inbred and homozygous extremes of the phenotypic range, the phenotypic trait measured in the F_2 generation may show the same mean as the F_1 generation, but the F_2 generation will show greater phenotypic variability. The result is that the phenotypes in the F_2 will vary so greatly that they will overlap with those of the parental generation. The multiple-factor hypothesis suggests that the number of allelic pairs of genes controlling a trait is inversely proportional to the number of F_2 individuals with the most extreme phenotype. That relationship is expressed as: $1/4^n$, where n equals the number of gene pairs that control the trait.

While the number of genes involved in a polygenic trait can be estimated from observation of extreme phenotypes, such calculations provide no information about the physical location and relationships of these genes. Because polygenes act on a single trait, it is not unreasonable to ask whether such genes might be physically clustered on a contiguous portion of a single chromosome, or scattered throughout the genome.

In *Drosophila*, resistance to the insecticide DDT has been shown to be a polygenic trait. To find the loci responsible for resistance, strains selected for resistance to DDT and strains selected for sensitivity were crossed to a stock carrying dominant markers on each chromosome (Figure 23.2). Backcrosses and $F_1 \times F_1$

matings were used to create offspring carrying combinations of marker chromosomes and chromosomes from resistant strains. These combinations were then tested for resistance by exposure to DDT. The results indicate that each of the chromosomes in the *Drosophila* karyotype contains genes that contribute to resistance. In other words, the loci that control DDT resistance are not clustered together on a single chromosome, but instead are scattered throughout the genome. This does not mean that in all cases polygenes controlling a trait are scattered throughout the genome, but it does show that polygenes need not be clustered in order to affect a single trait.

It is possible to map genes responsible for DDT resistance in *Drosophila* because identified genetic markers are present on each chromosome. In many other organisms, especially those of agricultural importance, genetic markers are not available for each chromosome, so that systematic mapping of quantitative traits is not feasible. However, the development of restriction fragment length polymorphisms (RFLPs) as genetic markers provides a new approach to the problem of enumerating and mapping the loci responsible for quantitative traits.

RFLPs were first used in mapping human genes (described in Chapter 16), but, in principle, can be used in almost any organism. For example, an RFLP map covering all 12 chromosomes of the tomato has

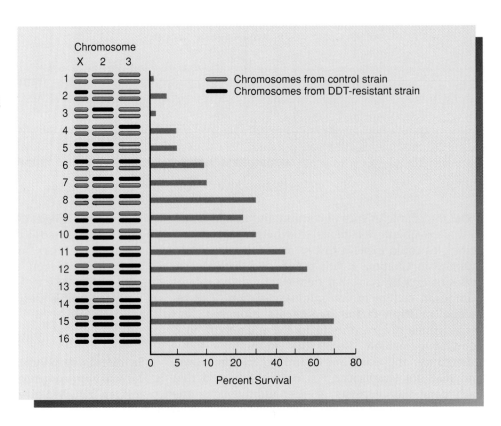

FIGURE 23.2
The differential survival of *Drosophila* carrying combinations of chromosomes from DDT-resistant and DDT-sensitive strains when exposed to DDT. The results indicate that DDT resistance is polygenic, with genes on each of the chromosomes making a major contribution.

been constructed, using some 300 RFLP markers. This provides a map of the tomato genome with genetic markers spaced at an average distance of 5 cM (recall that 1 cM, or centimorgan, is the distance that allows 1 percent recombination between markers). Using this map and some newly derived analytical methods, quantitative trait loci (QTLs) for fruit weight and soluble solids have been mapped in the tomato. Genes for fruit weight were found on six chromosomes (1, 4, 6, 7, 9, and 11), and four loci for soluble solids were identified (on chromosomes 3, 4, 6, and 7). Further work using RFLP markers on each of these chromosomes has allowed further localization of several loci to relatively short chromosomal segments.

The use of chromosome-specific RFLP markers and identified loci for a desirable trait will make it possible to transfer traits such as disease and heat resistance or improved flavor from wild type or other domestic strains into agriculturally important strains of domesticated plants. Eventually such methods can be used to isolate and characterize the genes that control quantitative traits in the same way that RFLP analysis has led to the identification and isolation of genes associated with specific human genetic disorders such as cystic fibrosis or neurofibromatosis.

ANALYSIS OF POLYGENIC TRAITS

Because traits inherited in a quantitative fashion cannot be described in the same way as Mendelian traits, it is necessary to provide a way of evaluating the inheritance of quantitative characters. In most cases, the outcome of quantitative genetic experiments is a set of measurements that can be expressed as a frequency diagram (Figure 23.3). These measurements, made from one or a series of crosses, are taken as a sample of all possible crosses that could have been made in the experimental population. If we had made an infinite number of such measurements, the frequency distribution would take on the shape of the curve in Figure 23.3, which is a normal, or bell-shaped, distribution curve. Obviously, in many cases, it is inconvenient or impossible to count all the individuals in an experimental population or to collect exhaustive amounts of data. But gathering a smaller data sample that is unbiased, and yet representative of the measured group, is often a problem. In addition, individual variation is a universal attribute of living systems, and this variability is reflected in the data gathered from observations of organisms. How can we distinguish between normal variation that is strictly due to chance and that due to an experimental variable?

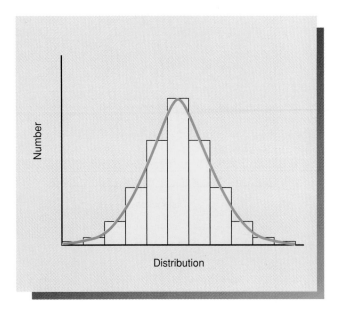

FIGURE 23.3
A normal frequency distribution.

The solution to these and other problems with data analysis has been to employ the techniques of **statistics**. Statistical analysis serves three purposes:

1 Data can be mathematically reduced to provide a **descriptive summary** of the sample.
2 Data from a small but random sample can be utilized to infer information about groups larger than those from which the original data were obtained **(statistical inference)**. The sample values are called **statistics** (symbolized by Roman letters) and are used as an estimate of the value for the population. These population values are called **parameters** (symbolized by Greek letters).
3 Two or more sets of experimental data may be compared to determine whether they represent significantly different populations of measurements.

Several statistical methods are useful in the analysis of traits that exhibit a normal distribution, including the mean, variance, standard deviation, and standard error of the mean.

The Mean

The distribution of phenotypic values in Figure 23.4(a) tends to cluster around a central value. This clustering is called a **central tendency,** the most common measurement of which is the mean ($\overline{X}$). The mean is simply the arithmetic average of a set of measurements or data and is calculated as

$$\overline{X} = \frac{\Sigma X_i}{n} \qquad (23.1)$$

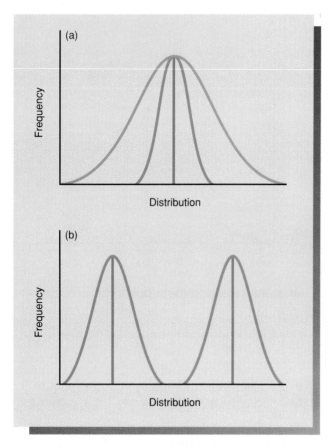

FIGURE 23.4
(a) Two normal frequency distributions with the same mean but different amounts of variation. (b) Two normal distributions with different means but the same amount of variation.

where $\overline{X}$ is the mean, ΣX_i represents the sum of all individual values in the sample, and n is the number of individual values.

Although the mean provides a descriptive summary of the sample, it is in itself of limited value. All values in the sample may be clustered near the mean, or they may be distributed widely around it. Figure 23.4 shows normal or symmetrical distributions with identical means, but a widely different range of values. These contrasting conditions represent a different type and amount of variation within the sample. Whether due to chance or to one or more experimental variables, such variation creates the need for methods to describe variation around the mean.

Variance

As seen in Figure 23.4(a), the range of values on either side of the mean will determine the width of the distribution curve. Measurement of this variation

in a sample is called the sample variance (s^2 or V) and is used as an estimate of the variation present in an infinitely large population (σ^2). The variance for a sample is calculated as:

$$s^2 = V = \frac{\Sigma(X_i - \overline{X})^2}{n - 1} \tag{23.2}$$

where the sum (Σ) of the squared differences between each measured value (X_i) and the mean ($\overline{X}$) is divided by one less than the total sample size ($n - 1$). To avoid the numerous subtraction functions necessary in calculating s^2 for a large sample, we can convert the equation to its algebraic equivalent:

$$s^2 = V = \frac{\Sigma X_i^2 - n\overline{X}^2}{n - 1} \tag{23.3}$$

The variance is a valuable measure of sample variability. Two distributions may have identical means ($\overline{X}$), yet vary considerably in their concentration around the mean. The variance represents the average squared deviation of the measurements from the mean. Obviously, the larger the variance, the greater the range of values on either side of the mean. The estimation of variance has been particularly valuable in determining the degree of genetic control of traits where the immediate environment also influences the phenotype.

Standard Deviation

Since the variance is a squared value, its unit of measurement is also squared (cm^2, mg^2, etc.). To express variation around the mean in the original units of measurement, it is necessary to calculate the square root of the variance, a term called the **standard deviation (s)**:

$$s = \sqrt{s^2} \tag{23.4}$$

Table 23.1 shows what fraction of the individual values within a normal or bell-shaped distribution is included with different multiples of the standard deviation. The mean plus or minus one standard deviation ($\overline{X} \pm 1s$) includes 68 percent of all values in

TABLE 23.1
Sample inclusion for various s values.

Multiples of s	% of Sample Included
$\overline{X} \pm 1s$	68.3
$\overline{X} \pm 1.96s$	95
$\overline{X} \pm 2s$	95.5
$\overline{X} \pm 3s$	99.7

the sample. Over 95 percent of all values are found within two standard deviations ($\overline{X} = 2s$). As such, the standard deviation provides an important descriptive summary of a set of data.

Standard Error of the Mean

To estimate how much the means of other similar samples drawn from the same population might vary, we can calculate the **standard error of the mean** **($S_{\overline{X}}$)**:

$$S_{\overline{X}} = \frac{s}{\sqrt{n}} \qquad (23.5)$$

where s is the standard deviation and $\sqrt{n}$ is the square root of the sample size. The standard error of the mean is a measure of the accuracy of the sample mean, that is, the variation of sample mean in replications of the experiment. Because it can be expected that the standard deviation of mean values will reflect less variance than the standard deviation of a set of individual measurements, the standard error is always less than the standard deviation. Since the standard error of the mean is computed by dividing s by $\sqrt{n}$, it is always a smaller value.

Analysis of a Quantitative Character

Many characteristics of importance in livestock and crop plants are controlled by polygenic systems. In an earlier chapter we discussed such phenotypes as ear length in corn and kernel color in wheat. To illustrate how biometric methods are used in the analysis of quantitative characters, we will consider a simplified example involving fruit weight in tomatoes. Let us assume that fruit weight is a quantitative character, and that one highly inbred strain produces tomatoes averaging 18 oz. in weight, and another highly inbred strain produces fruit averaging 6 oz. in weight. These two varieties are crossed, and produce an F_1 generation with weights ranging from 10 oz. to 14 oz. (Table 23.2). The F_2 population contains individuals that produce fruit ranging from 6 oz. to 18 oz.

The mean value for the fruit weight in the F_1 generation can be calculated as:

$$\overline{X} = \frac{\Sigma X_i}{n} = \frac{626}{52} = 12.04$$

Similarly, the mean value for fruit weight in the F_2 generation is calculated as:

$$\overline{X} = \frac{\Sigma X_i}{n} = \frac{872}{72} = 12.11$$

Thus, average fruit weight is 12.04 oz. in the F_1 generation, and 12.11 oz. in the F_2 generation. While these mean values are similar, it is apparent that there is more variation present in the F_2 generation, because fruit weight ranges from 6 to 18 oz. in the F_2 generation, but only from 10 to 14 oz. in the F_1 generation.

To assist in quantitating the amount of variation present in each generation, we can construct a frequency table (Table 23.3). The first column (x) represents the midpoint of a class interval. For example, all tomatoes weighing between 5.5 and 6.5 oz. are classified as 6 oz. The second column (marked f) lists the number of tomatoes that fall into that category. The last column, $f(x)$, computes the product of the first two columns ($x \times f$), and lists the total weight in oz. for each class. The sum of the f column (Σf) equals the total number of tomatoes counted, and the sum of the $f(x)$ column [$\Sigma f(x)$] equals the total weight of all the tomatoes in the sample. The mean value for fruit weight in the F_1 and F_2 generations taken from the frequency table can be calculated as:

$$F_1: \frac{\Sigma f(x)}{\Sigma f} = \frac{626}{52} = 12.03 \qquad F_2: \frac{\Sigma f(x)}{\Sigma f} = \frac{842}{72} = 12.11$$

As noted above, the sample variance can be calculated as the sum of the squared differences between each value and the mean, divided by one less than the total number of observations (Equation 23.2). However, in the case where a number of observations (f) have been grouped into representa-

TABLE 23.2
Distribution of F_1 and F_2 progeny.

		Weight in oz.												
		6	7	8	9	10	11	12	13	14	15	16	17	18
Number of	F_1:					4	14	16	12	6				
Individuals	F_2:	1	1	2		9	13	17	14	7	4	3		1

TABLE 23.3
Frequency distribution in F_1 and F_2.

F_1:	x	f	f(x)		F_2:	x	f	f(x)
	6					6	1	6
	7					7	1	7
	8					8	2	16
	9					9		
	10	4	40			10	9	90
	11	14	154			11	13	143
	12	16	192			12	17	204
	13	12	156			13	14	182
	14	6	84			14	7	98
	15					15	4	60
	16					16	3	48
	17					17		
	18					18	1	18
		$\Sigma f = \overline{52}$	$\Sigma f(x) = \overline{626}$				$\Sigma f = \overline{72}$	$\Sigma f(x) = \overline{872}$

TABLE 23.4
Calculation of variance.

F_1:	x	f	f(x)	f(x²)		F_2:	x	f	f(x)	f(x²)
	6						6	1	6	36
	7						7	1	7	49
	8						8	2	16	128
	9						9			
	10	4	40	400			10	9	90	900
	11	14	154	1,694			11	13	143	1,573
	12	16	192	2,304			12	17	204	2,448
	13	12	156	2,028			13	14	182	2,366
	14	6	84	1,176			14	7	98	1,372
	15						15	4	60	900
	16						16	3	48	768
	17						17			
	18						18	1	18	324
		$52 = n$	$626 = \Sigma f(x)$	$7,602 = \Sigma f(x^2)$				$72 = n$	$872 = \Sigma f(x)$	$10,864 = \Sigma f(x^2)$

$$s^2 = \frac{n\Sigma f(x^2) - (\Sigma fx)^2}{n(n-1)}$$

For F_1:

$$s^2 = \frac{52 \times 7,602 - (626)^2}{52(52-1)}$$

$$= \frac{395,304 - 391,876}{2,652}$$

$$= 1.29$$

For F_2:

$$s^2 = \frac{72 \times 10,864 - (872)^2}{72(72-1)}$$

$$= \frac{782,208 - 760,304}{5,112}$$

$$= 4.28$$

tive classes (x), the variance can be calculated according to the formula:

$$s^2 = V = \frac{n\Sigma f(x^2) - (\Sigma fx)^2}{n(n-1)}$$

As shown in Table 23.4, the value for the F_1 generation is 1.29, and for the F_2 generation, it is 4.28. When converted to the standard deviation ($s = \sqrt{s^2}$), the values become 1.13 and 2.06 respectively. Thus, the distribution of tomato weight in the F_1

generation can be described as 12.04 ± 1.13, and in the F_2 as 12.11 ± 2.06. This analysis indicates that the mean fruit weight of the F_1 is identical to that of the F_2, but that the F_2 generation showed greater variability in the distribution of weights than the F_1.

The observations about the inheritance of fruit weight in crosses between these two strains of tomatoes meet the expectations for polygenic traits (review these in Chapter 7). For the sake of this example, if we assume that each parental strain represents a phenotypic extreme (i.e., is homozygous for the genes that control fruit weight), we can calculate the number of gene pairs involved in controlling fruit weight in these two strains of tomatoes. The number of gene pairs can be calculated as $1/4^n$, and since 1/72 of the F_2 exhibit a phenotype as extreme as one of the parental strains, we can conclude that in this example, 3 or 4 gene pairs are responsible for the inheritance of fruit weight.

PHENOTYPIC EXPRESSION

Gene expression is often discussed as if the genes operate in a closed, "black box" system in which the presence or absence of functional products directly determines the collective phenotype of an individual. The situation is actually much more complex. Most gene products function within the internal milieu of the cell, cells interact with one another in various ways, and the organism must survive under diverse environmental influences. Thus, gene expression and the resultant phenotype are often modified through the interaction between an individual's particular genotype and the internal and external environment.

The degree of environmental influence may vary from inconsequential to subtle to very strong. Subtle interactions are the most difficult to detect and document, and have led to unresolvable "nature-nurture" conflicts in which scientists debate the relative importance of genes versus environment. How easily such conflicts are resolved depends on the characteristic being investigated, and even then it is sometimes impossible to provide definitive information. In this section we will deal with some of the variables known to modify gene expression.

Penetrance and Expressivity

Some mutant genotypes are always expressed by a distinct phenotype, while others produce a proportion of individuals whose phenotypes cannot be distinguished from wild type. The variable expression of a particular trait may be quantitatively studied by determining the degree of penetrance and expressivity. The percentage of individuals that show at least some degree of expression of a mutant genotype defines the **penetrance** of the mutation. For example, many mutant alleles in *Drosophila* are said to overlap with wild type. Flies homozygous for the recessive mutant gene *eyeless* yield phenotypes that range from the presence of normal eyes to the complete absence of one or both eyes (Figure 23.5). If 15 percent of mutant flies show the wild-type appearance, the mutant gene is said to have a penetrance of 85

FIGURE 23.5
Gradations in phenotype, ranging from wild-type to eyeless, associated with the *eyeless* gene in *Drosophila*.

percent. On the other hand, the range of expression of the genotype defines the **expressivity**. In the case of *eyeless*, although the average reduction of eye size is one-fourth to one-half, the range of expressivity is from complete loss of both eyes to completely normal eyes.

Examples such as the expression of the *eyeless* phenotype have provided the basis for experiments designed to determine the causes of phenotypic variation. If the laboratory environment is held constant and extensive variation is still observed, the genetic background may be investigated. It is possible that other genes are influencing or modifying the *eyeless* phenotype. On the other hand, if it is experimentally proven that the genetic background is not the cause for the phenotypic variation, environmental variables such as temperature, humidity, and nutrition may be examined. In the case of the *eyeless* phenotype, it has been determined experimentally that both genetic background and environmental factors influence its expression.

Genetic Background: Suppression and Position Effects

With only certain exceptions, it is difficult to assess the specific relationship between the **genetic background** and the expression of a gene responsible for determining the potential phenotype.

The influence of genetic background may occur in two ways. First, the expression of other genes throughout the genome may have an effect on the phenotype produced by the gene in question. The phenomenon of **genetic suppression** is an example. Mutant genes such as *suppressor of vermilion (su-v)*, *suppressor of forked (su-f)*, and *suppressor of Hairy-wing (su-Hw)* in *Drosophila* completely or partially restore the normal phenotype to an organism that is homozygous (or hemizygous) for these recessive genes. For example, flies homozygous for both *vermilion* (a bright red eye color mutation) and *su-v* have eyes with wild-type color.

The phenomenon of suppression occurs in a wide variety of organisms. In microorganisms, where molecular studies are easier to perform, some suppressor gene products are molecules that function in the genetic translation process by correcting certain errors produced by the mutation of other genes. Transfer RNA in mutant form has been shown to have such an effect. It has also been hypothesized that a suppressor gene product might provide an alternate metabolic route to bypass a block in a biosynthetic pathway caused by the primary mutation. Thus, suppressor genes are excellent examples of the genetic background modifying primary gene effects.

Second, the physical location of a gene in relation to other genetic material may influence its expression. Such a situation is called a **position effect**. For example, if a gene is included in a **translocation** or **inversion** event (in which a region of a chromosome is relocated), the expression of the gene may be affected. This is particularly true if the gene is translocated to or near an area of the chromosome

composed of **heterochromatin**. These regions are genetically different and often appear condensed in interphase.

An example of a position effect involves female *Drosophila* heterozygous for the sex-linked recessive eye-color mutant *white (w)*. If the region of the X chromosome containing the wild-type w^+ allele is translocated so that it is close to heterochromatin, expression of the w^+ allele is modified (Figure 23.6). Before translocation, the w^+/w genotype results in a wild-type brick red eye color. After translocation, the dominant effect of the w^+ allele is reduced. Instead of having a red color, the eyes are variegated, or mottled with red and white patches. A similar position effect

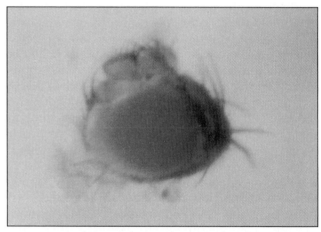

(a)

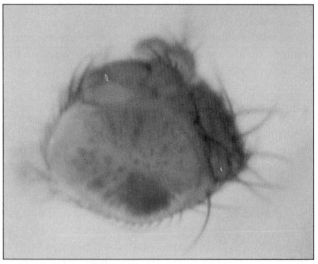

(b)

FIGURE 23.6
Eye phenotype in two female *Drosophila* heterozygous for the gene *white*. (a) Normal dominant phenotype showing brick red eye color. (b) Variegated color of an eye caused by transposition of the *white* gene to another location in the genome.

is produced if a heterochromatic region is relocated next to the *white* locus on the X chromosome. Apparently, heterochromatic regions inhibit the expression of adjacent genes. Loci in many other organisms also exhibit position effects, providing proof that alteration of the normal arrangement of genetic information can modify its expression.

Temperature Effects

Because chemical activity depends on the kinetic energy of the reacting substances, which in turn depends on the surrounding temperature, we can expect temperature to influence phenotypes. For example, the evening primrose produces red flowers at 23°C and white flowers at 18°C. Siamese cats and Himalayan rabbits exhibit dark fur in regions of the nose, ears, and paws because the body temperatures of these extremities are slightly cooler (Figure 23.7). In these cases, it appears that an enzyme that is functional in pigment production at the lower temperatures present in the extremities loses its catalytic function at the slightly higher temperatures throughout the rest of the body.

The most striking effect of temperature is the complete inhibition of mutant expression. Organisms carrying a certain mutant allele may show the mutant phenotype at one temperature, but if they are grown under a second temperature, they exhibit a wild-type phenotype. Mutations of this type are said to be **conditional** and are called **temperature sensitive;** they have been discovered in numerous organisms, including viruses, fungi, and *Drosophila.*

The effects of temperature are used in studying mutations that interrupt essential processes during development and are thus normally lethal to the

FIGURE 23.7
A Himalayan rabbit, showing dark fur color at the extremities of the ears, nose, and paws. These patches are due to expression of a temperature-sensitive allele for dark fur at the lower temperatures found in the extremities.

organism. If viruses carrying a **temperature-sensitive lethal mutation** are allowed to infect bacteria cultured at 42°C, infection progresses until the essential gene product is needed and then arrests, because this is the **restrictive temperature**. If cultured at the **permissive temperature** of 25°C, infection occurs normally, new viruses are produced, and the bacteria are lysed. The use of temperature-sensitive mutations, which may be induced and isolated, has added immensely to the study of viral genetics.

Similarly, many temperature-sensitive mutations have been discovered in *Drosophila*. Not only have dominant and recessive conditional lethal mutations been recovered in great numbers, but nonlethal temperature-sensitive mutations affecting development, morphology, and behavior have also been recovered. These nonlethal mutants have proven to be valuable in the dissection of genetic control mechanisms. We will discuss the use of such mutants in behavioral genetics in the next chapter.

Nutritional Effects

Another example of conditional mutations involves nutrition. In microorganisms, mutations that prevent synthesis of nutrient molecules are quite common. These **nutritional mutants** arise when an enzyme essential to a biosynthetic pathway becomes inactive. If the end product of a biochemical pathway can no longer be synthesized, and if that molecule is essential to normal growth and development, the mutation is lethal. If, for example, the bread mold *Neurospora* can no longer synthesize the amino acid leucine, proteins cannot be synthesized unless leucine is added to the growth medium. If leucine is added, the lethal effect is overcome. Nutritional mutants have been crucial to molecular genetic studies and also served as the basis for George Beadle and Edward Tatum's proposal, in the early 1940s, that one gene functions to produce one enzyme.

In humans, a slightly different set of circumstances is known. The presence or absence of certain dietary substances, which normal individuals may consume without harm, may adversely affect individuals with abnormal genetic constitutions. Often, a mutation may prevent an individual from metabolizing some substance commonly found in normal diets. For example, those afflicted with the genetic disorder **phenylketonuria** cannot metabolize the amino acid phenylalanine. Those with **galactosemia** cannot metabolize galactose. Other individuals are intolerant of the milk sugar lactose. Those with **diabetes** or **hypoglycemia** cannot adequately metabolize glucose.

In each case, excessive amounts of the molecule accumulate in the body and become toxic, and a characteristic phenotype results. However, if the dietary intake of the molecule is drastically reduced or eliminated, the associated phenotype may be reversed.

The case of **lactose intolerance** illustrates the general principles involved. Lactose is a disaccharide consisting of a molecule of glucose and a molecule of galactose. It is present as 7 percent of human milk and 4 percent of cow's milk. To metabolize lactose, humans require the enzyme **lactase,** which cleaves the disaccharide. Adequate amounts of lactase are produced during the first few years after birth. However, in many racial and ethnic groups, the levels of this enzyme soon drop drastically, and adults become intolerant of milk. The major phenotypic effect involves severe intestinal diarrhea, including flatulence and cramps. This condition is particularly prevalent in Eskimos, Africans, Asiatics, and Americans with these heritages. In these cultures, milk is usually used for food in different forms, including cheese, butter, and yogurt. In these forms, the amount of lactose is reduced significantly and adverse effects are largely eliminated. Thus, by altering the diet, phenotypic expression of a genetic trait can be modified.

Onset of Genetic Expression

Not all genetic traits are expressed at the same time during an organism's life span. In most cases, the age at which a gene is expressed corresponds to the normal sequence of growth and development. In humans, the prenatal, infant, preadult, and adult phases require different kinds of genetic information. In a similar way, many genetic disorders can be expected to manifest themselves at different stages of life. Lethal genes account for many of the frequent spontaneous abortions and miscarriages occurring in the human population. These mutations are believed to alter genetic products essential to prenatal development. **Tay-Sachs disease** and the **Lesch-Nyhan syndrome** (see Chapter 24) are severe disorders of lipid and nucleic acid metabolism, respectively, but do not cause lethality until early in childhood. **Huntington disease** (see Chapter 4) demonstrates a wide range of age of onset, but not usually before the age of 20, and most often after age 40. Observations of the development of these conditions support the concepts that the need for certain gene products may be more essential at certain times and that the internal environment of an organism changes with age. In the broadest sense, aging must be thought of as a process that begins following conception and continues until death.

Heredity vs. Environment

Many genetically determined traits are influenced by the environment and have considerable impact on the organism's phenotype. How do geneticists attempt to define the relative importance of heredity compared with that of the environment? Some traits demonstrate minimal variation under normal environmental circumstances, but other traits show distinct variability under the same environmental conditions. The latter case is particularly true for traits controlled by polygenic inheritance. Often, polygenic traits show such a large degree of variation in a population that geneticists can only speculate that a large but unknown number of genes provide the genetic potential, which is then modified by the environment.

Provided that a trait can be quantitatively measured, we can approach analytically the question of genetics versus environment. Experiments on plants and animals other than humans can test the causes of variation. Inbred strains containing individuals of a relatively homogeneous or constant genetic background can be generated, thus producing an **isogenic** population. Experiments are then designed to test the effects of the range of prevailing environmental conditions on phenotypic variability.

In studies with such isogenic lines, the **heritability index** or **ratio (H^2)** for a given trait can be calculated. Also called the **broad heritability** of the character being studied, H^2 measures the degree to which phenotypic variation (V_P) is due to genetic factors for a single population under the limits of environmental variability during the study. It is important to emphasize here that H^2 *does not* measure the proportion of the total phenotype attributed to genetic factors, but only the observed variation in the phenotype. Technically, the term H^2 is an expression of phenotypic variance, which is due to the sum of three components: environmental variance (V_E), genetic variance (V_G), and variance resulting from the interaction of genetics and environment (V_{GE}). Therefore, phenotypic variance (V_P) is theoretically expressed as

$$V_P = V_E + V_G + V_{GE} \qquad (23.6)$$

Since V_{GE} is nearly impossible to analyze, it is usually omitted. Therefore, the simpler equation is generally used:

$$V_P = V_E + V_G \qquad (23.7)$$

Broad heritability expresses that proportion of variance due to the genetic component:

$$H^2 = \frac{V_G}{V_P} \qquad (23.8)$$

A very high H^2 value indicates that the environmental conditions have had little impact on phenotypic variance in the population studied. A very low H^2 value indicates that the variation in the environment has been almost solely responsible for the observed phenotypic variation. Table 23.5 lists estimates of heritability for traits in a variety of organisms.

It is not possible to obtain an absolute H^2 value for any given character. If measured in a different population under a greater or lesser degree of environmental variability, H^2 might well change for that character.

Practically, heritability is most useful in animal and plant breeding as a measure of potential selection. Breeders are most interested in the improvement of economically important characters. From the standpoint of selection, V_G must be examined in a more complex way:

$$V_G = V_D + V_A + V_I \qquad (23.9)$$

where V_D is the genetic effect of dominant genes, V_A is due to the effect of additive genes, and V_I is the effect of interactive or epistatic genes. Of these components, V_A is most important in selection. This more limited estimate has been called **narrow heritability (h^2)**. Therefore, heritability in breeding experiments becomes effectively

$$h^2 = \frac{V_A}{V_P} \qquad (23.10)$$

TABLE 23.5
Estimates of heritability for several traits.

Trait	Percent Heritability (H^2)
Mouse	
Tail length	60
Litter size	15
Drosophila	
Abdominal bristle number	52
Wing length	45
Egg production	18
Humans	
Asthma	80
Diabetes (late onset)	70
Ulcers	37

Narrow heritability estimates, while characterizing a particular population, have been very useful in agricultural programs. Based on these estimates, selection techniques have led to vast improvements in the quality and quantity of plant and animal products.

In humans, isogenic strains are obviously not available for study, nor can the environment be controlled for individuals or populations. However, twins are very useful subjects for studying the heredity versus environment question in humans. **Monozygotic** or **identical twins,** derived from the division and splitting of a single egg following fertilization, are identical in their genetic compositions. Although most identical twins are reared together and are exposed to very similar environments, some pairs are separated and raised in different settings. For any particular trait, average similarities or differences can be investigated. Such an analysis is particularly useful because characteristics that remain similar in different environments are believed to be inherited. These data can then be compared with a similar analysis of **dizygotic** or **fraternal twins,** who originate from two separate fertilization events. Dizygotic twins are thus no more genetically similar than any two siblings.

A form of quantitative analysis of characteristics of twins reared together may also be pursued. Twins are said to be **concordant** for a given trait if both express it or neither expresses it and **discordant** if one shows the trait and the other does not. Table 23.6 lists concordance values for various traits in both types of twins.

These data must be examined very carefully before any conclusions are drawn. If the concordance value approaches 90 to 100 percent with monozygotic twins, we might be inclined to interpret this value as indicating a large genetic contribution to the expression of the trait. In some cases—blood types and eye color, for example—we know this is indeed true. In the case of measles, however, a high concordance value merely indicates that the trait is almost always induced by a factor in the environment—in this case, a virus.

Therefore, it is more meaningful to compare the difference between the concordance values of monozygotic and dizygotic twins. If these values are significantly higher for monozygotic twins than dizygotic twins, we suspect that there is a genetic component involved in the determination of the trait. We reach this conclusion because monozygotic twins, with identical genotypes, would be expected to show a greater concordance than genetically related, but not genetically identical, dizygotic twins. In the case of measles, where concordance is high in both types of twins, the environment is assumed to contribute significantly.

Even though a particular trait may be determined to be influenced substantially by genetic factors, it is often difficult to formulate a precise mode of inheritance based on available data. In many cases the trait is considered to be controlled by multiple-factor inheritance. However, when the environment is also exerting a partial influence, such a conclusion is particularly difficult to prove.

CHAPTER SUMMARY

1 Discontinuous variation is represented by those traits that fall into discrete phenotypic categories. Traits that demonstrate considerably more variation and are not easily categorized into distinct classes are examples of continuous variation. Polygenic inheritance, which controls characters that can be measured, involves genes with additive alleles that influence a character in a quantitative way.

2 Polygenic characteristics can be analyzed using statistical methods, which include the mean, the variance, the standard deviation, and the standard error of the mean. Such statistical analysis can be descriptive, can be used to make inferences about a population, or can be used to compare sets of data.

3 The penetrance of a mutant allele is measured by the percentage of a population that exhibits evidence of the mutant phenotype. Expressivity, on the other hand, measures the range of expression within a population.

TABLE 23.6
A comparison of concordance for various traits between monozygotic (MZ) and dizygotic (DZ) twins.

Trait	Concordance	
	MZ	DZ
Blood types	100%	66%
Eye color	99	28
Mental retardation	97	37
Measles	95	87
Idiopathic epilepsy	72	15
Schizophrenia	69	10
Diabetes	65	18
Identical allergy	59	5
Tuberculosis	57	23
Cleft lip	42	5
Club foot	32	3
Mammary cancer	6	3

4 Phenotypic expression may be modified by factors such as genetic background, temperature, nutrition, and other environmental factors. The phenomena of genetic suppression and position effects have been used to illustrate the existence of genetic background factors. Together, all such factors constitute the total environment in which genetic information is expressed.

5 The time of onset of gene expression varies as the need for certain gene products occurs at different periods during the processes of development, growth, and aging. This observation is particularly evident in the study of human disorders.

6 The degree of impact of genetic and environmental factors in the establishment of a given phenotype is difficult to ascertain. Heritability can be calculated for many characters but is especially useful in selective breeding of commercially valuable plants and animals.

7 Studies involving twins are aimed at resolving the question of heredity versus environment in human traits. The degree of concordance of a trait may be compared in monozygotic (identical) and dizygotic (fraternal) twins raised together or apart.

Insights and Solutions

The following results were recorded for ear length in corn:

Length of ear in cm.

	5	6	7	8	9	10	11	12	13	14	15	16	17	18	19	20	21
Parent A	4	21	24	8													
Parent B									3	11	12	15	26	15	10	7	2
F_1					1	12	12	14	17	9	4						
F_2				1	10	19	26	47	73	68	68	39	25	15	9	1	

1 For each of the parental strains, and the F_1 and F_2, calculate the mean values for ear length.

ANSWER: The mean values can be calculated from equation 23.1.

$$\overline{X} = \frac{\Sigma X_i}{n}$$

$$P_A: \quad \overline{X} = \frac{\Sigma X_i}{n} = \frac{378}{57} = 6.63$$

$$P_B: \quad \overline{X} = \frac{\Sigma X_i}{n} = \frac{1697}{101} = 16.80$$

$$F_1: \quad \overline{X} = \frac{\Sigma X_i}{n} = \frac{836}{69} = 12.11$$

$$F_2: \quad \overline{X} = \frac{\Sigma X_i}{n} = \underline{\quad} =$$

2 Compare the mean of the F_1 with that of each parental strain. What does this tell you about the type of gene action involved?

ANSWER: The F_1 mean (12.11) is almost midway between the parental means of 6.63 and 16.80. This indicates that the genes in question are probably additive in effect.

3 If the F_1 mean was skewed toward one or the other parental mean value, what genetic factor or factors might be at work?

ANSWER: If the mean values are skewed (especially upwards), it may indicate that the F_1 mean approximates the geometric ($X = \sqrt{\overline{X}_A \times \overline{X}_B}$)

rather than the arithmetic mean of the parental strains. In this case, it would suggest that there is a multiplying rather than an additive effect of some genes on the trait in question.

1 Distinguish between epistasis and polygenic inheritance, and between discontinuous and continuous variation.
2 List as many human traits as you can that are likely to be under the control of a polygenic mode of inheritance.
3 Describe the difference between penetrance and expressivity.
4 Define and discuss the significance of the following terms: (a) position effect, (b) suppressor genes, (c) monozygotic and dizygotic twins, (d) concordance and discordance, and (e) heritability.
5 In the following table, average differences of height and weight between monozygotic twins (reared together and apart), dizygotic twins, and siblings are compared. Draw as many conclusions as you can concerning the effects of genetics and the environment in influencing these human traits.

Trait	MZ Reared Together	MZ Reared Apart	DZ Reared Together	Sibs Reared Together
Height (cm)	1.7	1.8	4.4	4.5
Weight (kg)	1.9	4.5	4.5	4.7

SOURCE: Newman, Freeman, and Holzinger, 1937.

6 In *Drosophila*, the sex-linked recessive mutation *vermilion (v)* causes bright red eyes, which is in contrast to brick red eyes of wild type. A separate autosomal recessive mutation, *suppressor of vermilion (su-v)*, causes flies homozygous or hemizygous for *v* to have wild-type eyes. In the absence of vermilion alleles, *su-v* has no effect on eye color. Determine the F_1 and F_2 phenotypic ratios from a cross between a female with wild-type alleles at the *vermilion* locus, but who is homozygous for *su-v*, with a *vermilion* male who has wild-type alleles at the *su-v* locus.
7 Corn plants from a test plot are measured, and the distribution of heights at 10-cm intervals is recorded below:

Height (cm)	100	110	120	130	140	150	160	170	180
Plants (no.)	20	60	90	130	180	120	70	50	40

Calculate (a) the mean height, (b) the variance, (c) the standard deviation, and (d) the standard error of the mean.
8 A dark red strain and a white strain of wheat are crossed to produce an intermediate or medium red F_1. The F_1 plants are self-crossed to produce an F_2 in a ratio of 1 dark red:4 medium-dark red:6 medium red:4 light red:1 white. Further crosses reveal that the dark red and white F_2 plants are true breeding.
    (a) Based on the ratio of offspring in the F_2, how many loci are involved in the production of color? How many alleles at each locus?
    (b) Are the allele effects additive?
    (c) How many units of color are produced by each allele?
    (d) How many genotypes are represented by each phenotypic class? What are they?
9 Height in humans depends on the additive action of genes. Assume that this trait is controlled by four loci, *R, S, T, U*, and that environmental effects are negligible. Dominant alleles contribute two units and recessive alleles contribute one unit to height.

(a) Can two individuals of moderate height produce offspring that are much taller or shorter than either parent? How?

(b) If an individual with the minimum height specified by these genes marries an individual of intermediate or moderate height, will any of their children be taller than the tall parent? Why?

10 The mean length and variance of corolla length in two true-breeding strains of *Nicotiana* and their progeny are as shown below.

Strain	Mean length (mm)	Variance
P_1 short	40.47	3.124
P_2 long	93.75	3.876
$F_1(P_1 \times P_2)$	63.90	4.743
$F_2(F_1 \times F_1)$	68.72	47.708

Calculate heritability of flower length in three strains.

SELECTED READINGS

BRINK, R. A., ed. 1967. *Heritage from Mendel*. Madison: University of Wisconsin Press.

CHAPMAN, A. B. 1985. *General and quantitative genetics*. Amsterdam: Elsevier.

CORWIN, H. O., and JENKINS, J. B. 1976. *Conceptual foundations of genetics: Selected readings*. Boston: Houghton-Mifflin.

CROW, J. F. 1966. *Genetics notes*. 6th ed. Minneapolis: Burgess.

DUNN, L. C. 1966. *A short history of genetics*. New York: McGraw-Hill.

FALCONER, D. 1989. *Introduction to quantitative genetics*. 3rd ed. Harlow, England: Longman Scientific and Technical Pub.

FARBER, S. L. 1980. *Identical twins reared apart*. New York: Basic Books.

FELDMAN, M. W., and LEWONTIN, R. C. 1975. The heritability hang-up. *Science* 190: 1163–68.

FOSTER, M. 1965. Mammalian pigment genetics. *Adv. in Genet.* 13: 311–39.

LANDER, E. S. and BOTSTEIN, D. 1989. Mapping Mendelian factors underlying quantitative traits using RFLP linkage maps. *Genetics* 121: 185–99.

LEWONTIN, R. C. 1974. The analysis of variance and the analysis of causes. *Amer. J. Hum. Genet.* 26: 400–11.

LINDSLEY, D. C., and GRELL, E. H. 1967. *Genetic variations of* Drosophila melanogaster. Washington, D.C.: Carnegie Institute of Washington.

MATHER, K. 1965. *Statistical analysis in biology*. London: Methuen.

NEWMAN, H. H., FREEMAN, F. N., and HOLZINGER, K. J. 1937. *Twins: A study of heredity and environment*. Chicago: The University of Chicago Press.

NOLTE, D. J. 1959. The eye-pigmentary system of *Drosophila*. *Heredity* 13: 233–41.

PATERSON, A. H., DeVERNA, J. W., LANINI, B., and TANKSLEY, S. D. 1990. Fine mapping of quantitative train loci using selected overlapping recombinant chromosomes, in an interspecific cross of tomato. *Genetics* 124: 735–42.

PATERSON, A. H., LANDER, E. S., HEWITT, J. D., PETERSON, S., LINCOLN, S. E., and TANKSLEY, S. D. 1988. Resolution of quantitative traits into Mendelian factors by using a complete linkage map of restriction fragment length polymorphisms. *Nature* 335: 721–26.

PAWELEK, J. M., and KÖRNER, A. M. 1982. The biosynthesis of mammalian melanin. *Amer. Scient.* 70: 136–45.

RANSON, R., ed. 1982. *A handbook of* Drosophila *development*. New York: Elsevier Biomedical Press.

VOELLER, B. R., ed. 1968. *The chromosome theory of inheritance—Classic papers in development and heredity*. New York: Appleton-Century-Crofts.

ZIEGLER, I. 1961. Genetic aspects of ommochrome and pterin pigments. *Adv. in Genet.* 10: 349–403.

24

Behavior Genetics and Neurogenetics

CHAPTER CONCEPTS

Behavior in many organisms (both prokaryotic and eukaryotic) is controlled by both single genes and polygenic systems. At the biochemical level, these genes work by controlling cascades of metabolic reactions such as phosphorylation, methylation, and ion flux.

Behavior is defined generally as reaction to stimuli or environment. In broad terms, every action, reaction, and response represents a type of behavior. Animals run, remain still, or counterattack in the presence of a predator; birds build complex and distinctive nests; fruit flies execute intricate courtship rituals; plants bend toward light; and humans reflexively avoid painful stimuli as well as "behave" in a variety of ways as guided by their intellect, emotions, and culture.

Even though clearcut cases of genetic influence on behavior were known in the early 1900s, the study of behavior was of greater interest to psychologists, who were concerned with learning and conditioning. While some traits were recognized as innate or instinctive, behavior that could be modified by prior experience received the most attention. Such traits or patterns of behavior were thought to reflect the previous environmental setting to the exclusion of the organism's genotype. This philosophy served as the basis of the **behaviorist school**.

Such thinking provided a somewhat distorted view of the nature of behavioral patterns. It is logical that a genotype can be expressed within a series of environmental levels (e.g., cell, tissue, organ, organism, population, surrounding environment) and that a behavioral pattern must rely on the expression of the individual genotype for its execution. Nevertheless, the so-called **nature-nurture controversy** flourished well into the 1950s. By that time it became clear that while certain behavioral patterns, particularly in less complex animals, seemed to be innate, others were the result of environmental modifications limited by genetic influences. The latter condition is particularly true in organisms with more complicated nervous systems.

Since about 1950, studies of the genetic component of behavioral patterns have intensified, and support

for the importance of genetics in understanding behavior has increased. The prevailing view is that all behavior patterns are influenced both genetically and environmentally. The genotype provides the physical basis and/or mental ability essential to execute the behavior and further determines the limitations of environmental influences.

Behavior genetics has blossomed into a distinct specialty within the larger field of genetics as more and more behaviors have been found to be under genetic control. This chapter provides an overview of the role of genes in behavior. More extensive treatment can be found in the list of selected readings at the end of the chapter. There seems to be no question that this topic will be one of the most exciting areas of genetics in the years to come.

THE STUDY OF BEHAVIOR GENETICS: METHODOLOGY

Historically, three major approaches have been used in studying behavior genetics. The first involves the determination of behavioral differences between genetic strains of the same species or between closely related species. If such closely related organisms exist in similar environments and their survival needs are identical, any observed behavioral differences may be correlated with genetic differences. In the second approach, a modified behavioral trait is selected from a heterozygous population. If such a modified trait can be established in a new genetic strain, the positive influence of the genotype is established. The first two approaches identify behavioral patterns as being under the control of genes. Genetic investigation through controlled breeding experiments can then be performed in an attempt to establish inheritance patterns. In most cases where these approaches have been used successfully, it has only been shown that the inheritance is not due to simple Mendelian patterns. Instead, behavioral traits have been attributed to polygenic inheritance (i.e., many genes controlling one trait).

The third approach, used extensively today, is to study the effects of single genes on behavior. Within inbred strains of organisms with relatively constant genotypes, spontaneous mutants with behavioral deviations may arise. Even more productive is the induction and isolation through selection of mutations altering behavior. While a single complex behavioral trait may be controlled by many genes, disruption of the trait by a single mutation may occur and be analyzed. This approach is the most attractive because it offers the most objective information concerning the role of genes in behavior. In selected organisms, defining a gene controlling behavior is a prelude to the isolation, cloning, and molecular characterization of such genes. As we proceed through this chapter, all three approaches will be illustrated.

COMPARATIVE APPROACHES IN STUDYING BEHAVIOR GENETICS

Before turning to specific examples of genetic influence on behavior, we shall present several case studies to illustrate two of the approaches used in behavior genetics. In the first approach, behavior is compared in closely related strains. Alcohol preference in mice, open-field behavior in mice, and mate selection in baboons are case studies representative of this method. The second approach involves selection within heterozygous populations for a modification of behavior. To illustrate this mode of investigation, we will describe studies of learning in rats and geotaxis in *Drosophila*.

Alcohol Preference in Mice

Many studies of **alcohol preference** in mice have been reported. Different inbred strains have been compared for preference or aversion to ethanol, with all studies indicating genetic control of this response.

For example, D. A. Rogers and Gerald E. McClearn compared four strains of inbred mice over a period of three weeks. Each strain was presented with seven vessels containing either pure water or alcohol varying in strength from 2.5 to 15.0 percent. Daily consumption was measured. Table 24.1 shows the proportion of absolute alcohol to total liquid consumed on a weekly basis.

Examination of these data shows clearly that the C57BL and C3H/2 strains exhibit a preference for

alcohol, while BALB/c and A/3 demonstrate an aversion toward it. Since the environment in raising the mice over many generations has been constant, the preference differences are attributed to the genotypes of each strain. Presumably, these strains contain identical sets of genes and vary only by the fixation of different alleles at these loci.

It has been suggested that differences in alcohol preference, metabolism, and severity of withdrawal symptoms are related to differences in the major enzymes of alcohol metabolism, particularly **alcohol dehydrogenase (ADH)** and **acetaldehyde dehydrogenase (AHD)**. The major enzymes and their isozymes have been isolated and characterized with respect to biochemical and kinetic properties, and their distribution and activities in different mouse strains have been catalogued. Genetic variants for these enzymes present in different strains of inbred mice have been used to try to establish an association between the enzyme and a form of alcohol-related behavior. Unfortunately, to date, no correlation between alcohol preference or metabolism and these biochemical markers has been established. It may be that other genetic markers, perhaps controlling the neuropharmacological effects of alcohol, will be needed to establish a link between specific genes and alcohol-related behavior in mice.

Open-Field Behavior in Mice

First used in 1934, an **open-field test** was designed to study exploratory and emotional behavior in mice.

TABLE 24.1
Alcohol consumption in mice.

Strain	Week	Proportion of Absolute Alcohol to Total Liquids	$\bar{x}$
C57BL	1	0.085	
	2	0.093	9.4% alcohol
	3	0.104	
C3H/2	1	0.065	
	2	0.066	6.9% alcohol
	3	0.075	
BALB/c	1	0.024	
	2	0.019	2.0% alcohol
	3	0.018	
A/3	1	0.021	
	2	0.016	1.7% alcohol
	3	0.015	

SOURCE: Modified from Rogers and McClearn, 1962. Reprinted by permission from *Quarterly Journal of Studies on Alcohol*, Vol. 23, pp. 26-33, 1962. Copyright by Journal of Studies on Alcohol, Inc. New Brunswick, NJ 08903.

When placed in a new environment, mice normally explore the surroundings, but they are a bit cautious or "nervous" about the new setting. The latter response is evidenced by their elevated rate of defecation and urination. To study these behaviors in the laboratory, an enclosed, brightly illuminated box was devised with the floor marked into squares. Exploration is measured by counting the number of movements into different sections, and emotion is measured by counting the number of defecations. These measurements thus attempt to quantify the two behavioral patterns.

As in the study of alcohol preference, different inbred strains of mice have been shown to vary significantly in their response to the open-field setting. Of greatest interest is the experimentation of John C. DeFries and his associates in 1966. DeFries concentrated his work on two strains. BALB/cJ and C57BL/6j. The BALB strain is homozygous for a coat color allele, *c*, and is albino, while the C57 strain has normal pigmentation (*CC*). BALB demonstrates low exploratory activity and is very emotional, while C57 is field-active and nonemotional.

DeFries proceeded to cross the two strains and then to interbreed each generation, creating F_2, F_3, F_4, etc., generations. Each generation beyond the F_1 contained albino and nonalbino mice, and these were tested. Regardless of generation, pigmented mice behaved as strain C57, while albino mice behaved as BALB. The general conclusion may be drawn that the *c* allele behaves **pleiotropically,** affecting both coat color and behavior.

Detailed heritability and variance analysis has been performed to assess the degree of input of the *c* gene to these behavioral patterns. Such analysis has shown that this locus accounts for 12 percent of the additive genetic variance in open-field activity and 26 percent of defecation-related emotion. Thus, these behaviors are under polygenic control, presumably in an additive way.

What might be the relation between albinism and behavior? DeFries sought an answer by testing albinos and nonalbinos under both white and red light. Red light provides little visual stimulation to mice. The behavioral differences between mice with the two types of coat pigmentation disappeared under red light. Thus, albino open-field responses are visually mediated. This is not surprising, since albino mice are photophobic, and lack of pigmentation in albinos extends to the iris as well as to the coat.

Baboon Mate Selection

Baboons are large, quadrupedal terrestrial monkeys found in Africa (Figure 24.1). Because they are primates and exhibit a variety of social behavioral patterns, they have been investigated rather thoroughly. They are classified into one genus, *Papio*, and several species. *Papio anubis* is a savanna-dwelling group, while *Papio hamadryas* is a desert dweller. Their ranges sometimes overlap.

P. hamadryas has a social system that includes harems. Each male is associated permanently with one or more adult females. *P. anubis* displays the more common social structure of baboons, in which permanent associations are not formed. Instead, matings are promiscuous but affected by dominance ranking of males.

In the overlapping range, which is an arid region resembling that of the hamadryas more than that of

FIGURE 24.1
Male and female Gelada baboons.

the anubis, hamadryas males have been observed to kidnap juvenile anubis females. These females learned to cooperate with the herding behavior of males in harem life. Hybrid males resulting from such matings were not so efficient in herding females as the pure hamadryas. Since both hamadryas and hybrid males appear to have the same opportunity to learn the herding behavior, it has been concluded that herding and harem formation are at least partly genetically determined.

Selection: Maze Learning in Rats

The second major approach in behavior genetics—selection for behavior modification from a heterozygous population—leads to the production of inbred lines with differences in behavior. The end result is similar to the previous approach of initially comparing inbred lines. The classical illustration of the second approach involves measurement of **maze learning** in rats.

The first experiment of this kind was reported by E. C. Tolman in 1924. He began with 82 white rats of heterozygous ancestry and measured their ability to "learn" to obtain food at the end of a multiple T-maze (Figure 24.2) by recording the number of errors and trials. When first exposed to the maze, a rat explores all alleys and eventually arrives at the end, to be rewarded with food. In succeeding trials, fewer and fewer mistakes are made as the rat learns the correct

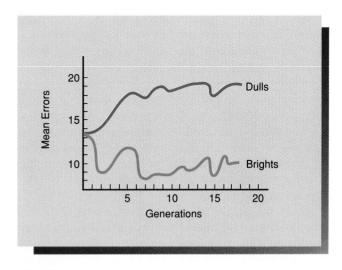

FIGURE 24.3
Selection for the ability and inability of rats to learn to negotiate a maze.

route. Eventually, a hungry rat may proceed to the food with no errors.

From the initial 82 rats, nine pairs each of the "brightest" and "dullest" rats were selected and mated to produce two lines. In each generation, selection was continued. Even in the first generation, Tolman demonstrated that he could select and breed rats whose offspring performed more efficiently in the maze. Subsequently, his approach was pursued by others, notably R. C. Tryon, who in 1942 published results of 18 generations of selection.

As shown in Figure 24.3, two lines, one clearly superior and one clearly inferior in maze learning, were established. There is some variation around the mean, but by the eighth generation there was no overlap between lines even in this variation. That is, the dullest of the bright rats were then superior to the brightest of the dull rats.

In other studies, bright and dull rats were tested for other related traits with varying results. Brights were found to be better in solving hunger-motivation problems but inferior in escape-from-water tests. Brights were also found to be more emotional in open-field experiments. Thus, it was concluded that selection of genetic strains superior in certain traits is possible, but care must be taken not to generalize such studies to overall intelligence, which is composed of many learning parameters. Tryon seems to have selected for specific capacities rather than for general intelligence.

Geotaxis in *Drosophila*

A **taxis** is the movement of a free-living organism toward or away from the source of an external

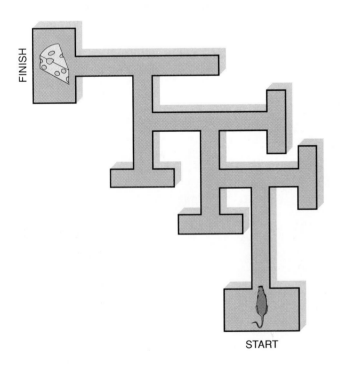

FIGURE 24.2
A multiple T-maze used in learning studies with rodents.

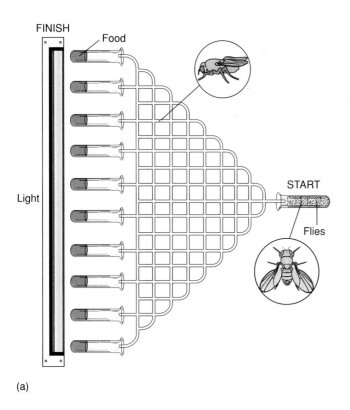

(a)

FIGURE 24.4
(a) Schematic drawing and (b) photograph of a maze used to study geotaxis in *Drosophila*.

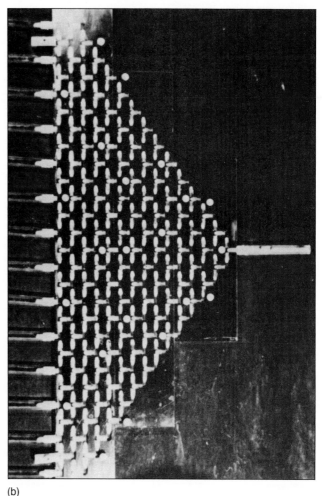

(b)

stimulation. The response may be positive or negative, and the sources of stimulation may include chemicals, gravity, light, and so on. The investigation of **geotaxis** (response to gravity) in *Drosophila* illustrates this general behavior response as well as the selection technique. In addition, this investigation used a novel approach that determines the genetic influences of specific chromosomes on geotaxis.

Jerry Hirsch and his colleagues designed a mass screening device that allows about 200 flies to be tested per trial, as shown in Figure 24.4. The maze is placed vertically, a fluorescent light illuminates the side opposite the entry point, and flies are added. Those that continue to turn up at each junction will finally arrive at the top; those that always turn down will arrive at the bottom; and those making both "up" and "down" decisions along the way will ultimately reside somewhere in between. It was found that flies could be selected for both positive and negative geotropism, establishing the genetic influence on this behavioral response.

As shown in Figure 24.5, mean scores may vary from +4.0 to −6.0, corresponding to the number of

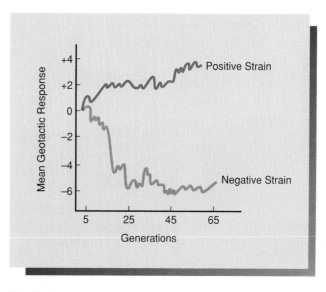

FIGURE 24.5
Selection for positive and negative geotaxis in *Drosophila* over many generations.

T-junctions the fly encounters going up or down. These data show that the extreme of negative geotropism is stronger than the extreme of positive geotropism. The two lines have now undergone selection for almost 30 years, encompassing over 500 generations, and the testing of more than 80,000 flies. Throughout the experiment, clearcut but fluctuating differences were observed. Such results indicate the additive effects of polygenic inheritance.

Hirsch and his colleagues also performed an analysis of the importance of genes located on different chromosomes to geotaxis. They were able to differentiate between genes located on chromosomes 2 and 3 and the X in a rather ingenious way. They carried out a set of crosses that produced flies that were either heterozygous or homozygous for a given chromosome.

Figure 24.6 shows how this is accomplished. From either a selected or unselected line, a male is crossed to a "tester" female whose chromosomes are each suitably marked with a dominant mutation. Each marked chromosome carries an inversion to suppress the recovery of any crossover products. As shown, one of the F_1 females heterozygous for each chromo-

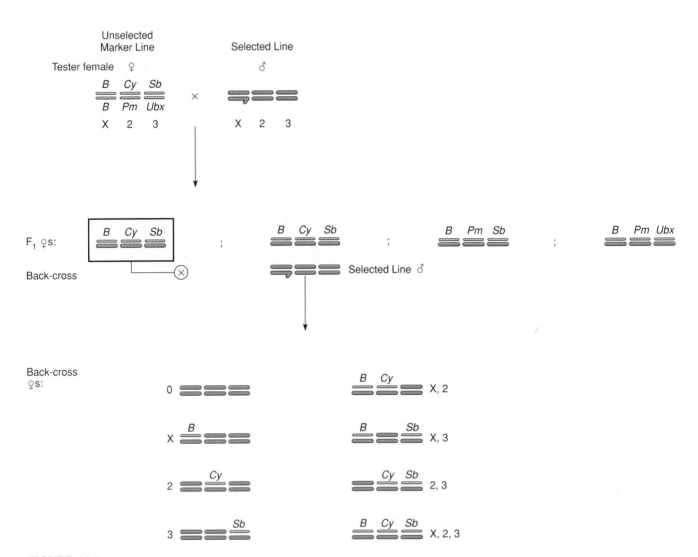

FIGURE 24.6

In this mating scheme in *Drosophila*, the effect of genes located on specific chromosomes that contribute to geotaxis can be assessed. The progeny produced by back-crossing the F_1 female contain all combinations of chromosomes. Examination of the phenotype of these flies makes it possible to determine which chromosomes from the selected strain are present. Subsequent testing for geotaxis is then performed (*B*–Bar eyes; *Cy*–Curly wing; *Sb*–Stubble bristles). The designations alongside each genotype (X, 2, 3, etc.) indicate which chromosomes are heterozygous.

some is back-crossed to a male from the original line. The resulting female offspring contain all combinations of chromosomes. The dominant mutations make it possible to recognize which chromosomes from the selected lines are present in homozygous or heterozygous configurations. For example, combinations O and X (Figure 24.6) differ by being homozygous and heterozygous, respectively, for the X chromosome. Similar combinations exist for chromosome 2 (0 and 2) and chromosome 3 (0 and 3). Thus, by subjecting these flies to the geotaxic maze, it is possible to assess the influence of genes on any given chromosome on the behavioral response.

Many flies of each genotype have been tested, and the chromosomal effects have been compiled individually and in all combinations. For the negatively geotactic lines (flies that go up in the maze), genes on the second chromosome are the most important, followed by loci on chromosome 3 and the X (2 > 3 > X). For the positive line (flies that go down in the maze), the reverse arrangement (X > 3 > 2) indicates the relative effects of chromosomes in controlling this trait. Thus, the overall results indicate that geotaxis is under polygenic control, and that the responsible loci are distributed on all three major chromosomes of *Drosophila*.

Further genetic testing in which each chromosome from a selected line has been isolated in homozygous form in an unselected background has been used to estimate the number of genes that control the geotaxic response in *Drosophila*. While this work has not yet produced definitive results, it does indicate that a small number of genes, perhaps two to four loci, are responsible for this behavior.

SINGLE-GENE EFFECTS ON BEHAVIOR

By far the most definitive information about the genetic influence on behavior has come from the study of the effects of single genes on behavior. In this more recent approach, either spontaneous or, more often, induced mutations are analyzed in order to infer general principles about how normal behavior is created and regulated.

There are obvious advantages to this approach. By and large, in these laboratory studies the environmental effect on the behavioral response is minimized or eliminated. Thus, the genetic influence is more straightforward and easier to define than in studies where the environment is a major factor. As a result, it is theoretically possible to dissect a behavioral pattern into its components.

In this section we shall look at a number of single-gene influences. However, because of the large amount of information resulting from such studies, we shall be selective in our discussion.

Nest-Cleaning Behavior in Honeybees

Honeybee nests are frequently infected with *Bacillus larvae*, the agent causing American foulbrood disease. The disease may be counteracted by what is called **hygienic behavior** by worker bees. The cells of infected combs containing afflicted larvae are opened, and the diseased organisms are removed from the hive. Hygienic hives are resistant to infection, while hives containing strains that do not display removal behavior are susceptible to the disease.

In 1964, Walter Rothenbuhler published results of his cross between a hygienic (Brown) line with a nonhygienic (Van Scoy) line. This work strongly favors the hypothesis that two recessive independently assorting genes (*u* and *r*) or a gene complex are responsible for hygienic behavior.

The F_1 hybrids were all nonhygienic. However, when F_1 drones were back-crossed to hygienic queens, four phenotypes were produced in roughly equal proportions, as shown in Figure 24.7. While one group was hygienic and one group nonhygienic, the other two groups were most interesting. One could uncap cells but not remove infected larvae. The fourth group, which at first appeared nonhygienic, was shown to be able to remove larvae if the cells were artificially uncapped. Thus, they were not able to uncap.

It appears that one gene pair (*u/u*) or a linked complex of genes determines uncapping behavior, and a second gene pair (*r/r*) or complex determines removal ability. We have begun with this example to illustrate the way a genetic study has allowed components of a more complex behavior to be dissected. Nest cleaning in honeybees is also one of the most striking examples of the far-reaching effects of genes on behavioral responses.

Taxes in Bacteria

Behavioral responses exist even in single-celled organisms such as bacteria. These responses are in the form of taxes (plural of *taxis*) and are mediated by flagella or cilia. Bacteria demonstrate **chemotaxis** and are attracted to or avoid a variety of stimuli. These responses have now been carefully analyzed, and mutants have been isolated that disrupt normal behavior.

Motile bacteria such as *E. coli* and *Salmonella* are capable of monitoring gradients of chemicals and moving along the gradient by controlling flagellar

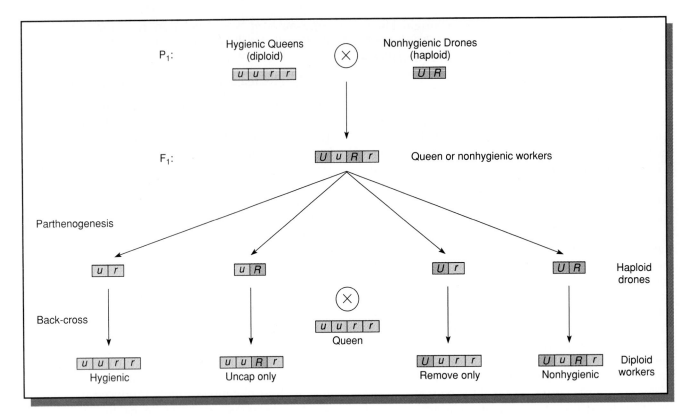

FIGURE 24.7
Results of a honeybee cross between hygienic diploid females and nonhygienic
haploid males.

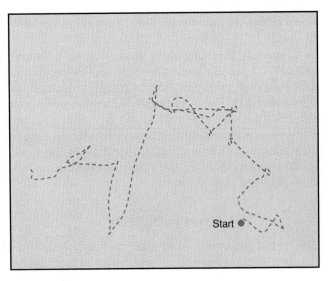

FIGURE 24.8
Runs and tumbles executed by an *E. coli* cell in
homogenous medium. Tracking of the cell began at the
large dot at lower right.

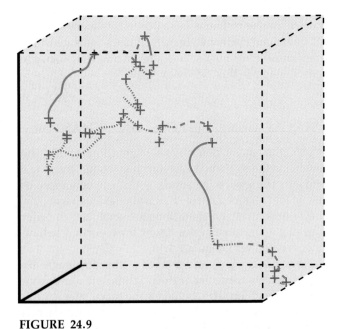

FIGURE 24.9
Runs and tumbles executed by a single *E. coli* cell in an
exponential gradient of attractant (highest concentration
at top). Movement up the gradient occurred in 35 runs,
34 tumbles. Two extended runs represent most of the
vertical distance traveled.

action. Cells exhibit periods of smooth swimming called *runs* caused by counterclockwise (CCW) rotation of flagella, which act much like propellers, driving the bacteria through the medium. During runs, the flagella form a cohesive bundle, leading to movement in a single direction. Runs, which have an average duration of 1 second, alternate with events called *tumbles* during which the flagella switch direction and rotate clockwise (CW). In clockwise rotation, the flagella become dispersed, and each acts independently. As a result, during tumbles, the cell achieves little net displacement (Figure 24.8). Tumbles, which last only about 0.1 second, serve to change the direction of the cell. In the presence of an attractant, runs that carry the cell up the gradient are extended, while those that move the cell down the gradient are not (Figure 24.9). In the presence of a repellent, the parameters are inverted.

Although molecular details of the chemotactic response in bacteria are not yet available, the major components of the system have been identified, and the role of several gene products have been described. *E. coli* contains at least four different transmembrane receptor proteins that bind to effectors and initiate an intracellular response. These proteins, called **transducers,** are members of a gene family that contains divergent periplasmic domains that bind chemical effectors, and conserved cytoplasmic domains involved in transmitting signals to the flagella. The intracellular pathways involve a set of cytoplas-

mic chemotaxis proteins that includes *CheA, CheB, CheR, CheW, CheY,* and *CheZ.*

The excitation pathway involves interaction of the transducer protein with *CheW* and *CheA* (Figure 24.10), resulting in phosphorylation of *CheA*. Signal transmission from *CheA* to *CheB* and *CheY* involves transfer of phosphate groups to these gene products. *CheA* apparently integrates the response of the cell to chemical stimuli, as *CheA* mutants are nonchemotactic. It is thought that activated *CheY* binds directly to the base of the flagellar apparatus and favors CW rotation by changing the rates of transition between CW and CCW rotation. *CheZ* acts to dephosphorylate *CheY* and inactivate its signal. Mutants of *CheZ* are apparently unable to inactivate the signal, with the result that runs are extended by a factor of three. A second pathway, called the adaptation pathway, involves *CheR* and *CheW*. This pathway apparently enhances the sensitivity of the transducer protein to the changing environmental conditions by methylation of the transducer by *CheR*. This is reversed through demethylation mediated by *CheB*. Thus, the chemotactic behavior of *E. coli* involves a number of genes that receive and process signals and generate a response through a metabolic network involving protein phosphorylation and methylation. At the present time, the analysis of chemotaxis in bacteria represents the best-known example of the relationship between a behavioral response and its underlying molecular mechanisms.

FIGURE 24.10
Sensory transduction during chemotaxis. Stimulation of the transmembrane transducer leads to the phosphorylation of *CheA* with the aid of *CheW*. Activated *CheA* transfers the phosphate group to *CheY*. Phosphorylated *CheY* interacts with the flagellar switch to alter direction of flagellar rotation. *CheZ* inactivates *CheY* by removing the phosphate group. In a second level of response, the cytoplasmic domain of the transducer is methylated by *CheR* and demethylated by *CheB* to adjust transducer sensitivity to changes in environmental conditions.

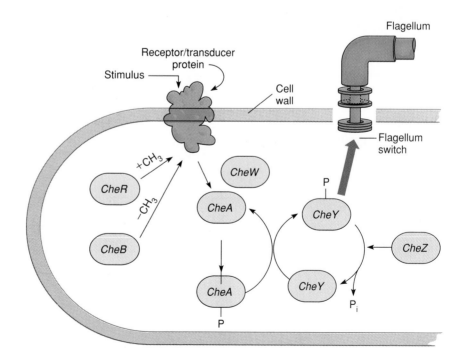

Avoidance Reactions in *Paramecium*

A great deal is known about the neurophysiology of ciliary action in *Paramecium*. This organism swims in a more coordinated fashion than a bacterium. Upon encountering a physical barrier or noxious stimulus, it reverses the direction in which all cilia are beating, causing it to back up. A reversal again occurs, and it swims off in a new direction.

The close coordination of ciliary direction is controlled by changes in electrical potential associated with the outer membrane. This depolarization passes along the membrane across the cell, leading to a calcium influx. The increased Ca^{++} inside the cell is thought to initiate the reversal of direction of beating cilia.

Several mutants unable to execute this avoidance reaction have been identified. The *Pawn* mutant, named after its chess counterpart, cannot swim backward under normal conditions. However, if the membrane is disrupted with detergent and the calcium ion concentration is increased sufficiently, backward swimming is observed. Thus, the ciliary apparatus is completely functional. The pattern shown in Figure 24.11 confirms that the mutation has eliminated electrical potential changes in the membrane and thus affects the electrical conductance essential to reversal. A second mutation, *Paranoic*, is also aptly named. Its electrical discharge patterns are compared with those of *Pawn* and wild type in Figure 24.11. A *Paranoic* mutant shows random reversals of ciliary direction and backing up in the presence of increased sodium ions, even in the absence of barriers.

In these and other mutants, behavioral responses are mediated by changes in the electrical potential of the cell membrane. The transducers are a class of integral membrane proteins known as **ion channels**. Because ions will not pass through lipid bilayer in the plasma membrane, they move through ion channels in order to flow into and out of the cell. Ion channels can be open, allowing ions to move passively in response to a gradient, or they can be closed, preventing the flow of ions and electrical impulses.

Such channels act as switches, transducing chemical or mechanical signals into electrical impulses that control the direction of ciliary beat.

In *Paramecium*, mutants in seven different complementation groups affect Ca^{++} current. In one such mutant, *pantophobiac* (*pnt*), the action potential is reversed and the cell exhibits a prolonged period of backward swimming when exposed to various stimuli. The defective protein responsible for this mutation has been identified by microinjecting mutant cells with wild-type cytoplasm. Fractionation of wild-type cytoplasm before injection has identified **calmodulin**, a Ca^{++} binding protein as the defective gene product in this mutant. Calmodulin binds to the Ca^{++} ion, and this complex in turn binds to and activates a number of enzymes. In the *pnt A* mutant, there is an amino acid substitution of serine for phenylalanine at position 101, and the mutant protein binds Ca^{++} weakly or not at all. Because ion channels in *Paramecium* and humans are identical in structure and function, further work on ion channels in *Paramecium* will not only shed light on genetic control of behavior in this organism, but also on the mechanisms that underlie taste, hearing, smelling, and the action of neurotransmitters in humans and other higher organisms.

Behavior Genetics of a Nematode

In 1968, in one of the boldest attempts to define the total genetic influence on the behavior of a single organism, Sidney Brenner began an investigation of the nematode *Caenorhabditis elegans*. Brenner had previously made valuable contributions in the field of molecular genetics, including studies of DNA replication, F factors, mRNA, and the genetic code. When he turned to the study of behavior, he hoped that it would be possible to dissect genetically the nervous system of *C. elegans* using techniques previously applied successfully to other organisms.

He chose this nematode because it was possible to determine the complete structure of the nervous system. Adult worms are about 1 mm long and are

FIGURE 24.11
Recordings of electrical potential from wild-type and mutant *Paramecium* in response to barium.

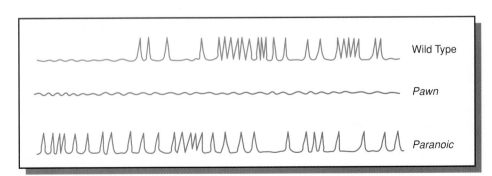

Wild Type

Pawn

Paranoic

FIGURE 24.12
Chemotactic response to ammonium chloride (NH₄Cl) of wild-type and mutant *Caenorhabditis*.

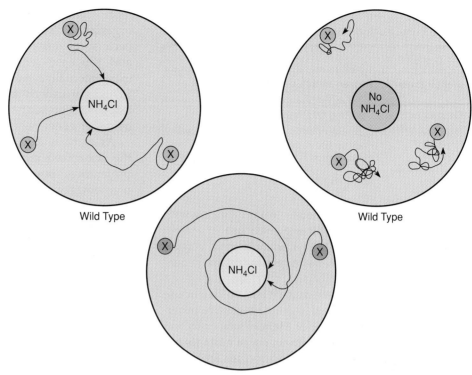

Wild Type

Wild Type

Bent-headed Mutants

composed of only 959 cells, about 300 of which are neurons. As a result, it is possible to cut serial sections of an embedded organism and reconstruct the entire organism and its nervous system in a three-dimensional model. Brenner then hoped to induce large numbers of behavioral mutations and to correlate aberrant behavior with structural and biochemical alterations in the nervous system. Because of the vast scope of this research endeavor, work is still underway. Some progress has been made, particularly in isolating large numbers of mutations. Three types of behavioral mutants have been characterized. The worms are positively chemotactic to a variety of stimuli (cyclic AMP and GMP; anions such as Cl^-, Br^-, and I^-; and cations such as Na^+, Li^+, K^+, and Mg^{++}). As shown in Figure 24.12, positive attraction can be tracked in gradients on agar plates. The study of mutants has shown that sensory receptors in the head alone mediate the orientation responses to attractants.

A second class of behavior studied involves *thermotaxis*. Cryophilic mutants move toward cooler temperatures, and thermophilic mutants move toward warmer temperatures. However, this behavior has not yet been correlated with the responsible component of the nervous system.

The third class of behavior involves generalized movement on the surface of an agar plate. Of 300 induced mutations, 77 affected the movement of the animal. While wild-type worms move with a smooth, sinuous pattern, mutants are either **uncoordinated** (*unc*) or **rollers** (*rol*). Those that are uncoordinated vary from the display of partial paralysis to small aberrations of movement, including twitching. Rollers move by rotating along their long axis, creating circular tracks on an agar surface. Of these mutants, some have been correlated with defects in the dorsal or ventral nerve cord or in the body musculature.

Linkage mapping has also begun. All mutants are distributed on six linkage groups, corresponding to the haploid number of chromosomes characteristic of this organism. Numerous *unc* mutants are found on each of the six chromosomes, indicating extensive genetic control of nervous system development.

Brenner's work is being extended in many laboratories throughout the world. In fact, an international congress now meets regularly to discuss work on *C. elegans*. Research now includes genetic analysis of development, and nongenetic problems are also being approached. This organism promises to rank with *Drosophila* in the amount of information acquired about it. It is hoped that study of *C. elegans* will unlock the mysteries of how genes control the structure of the nervous system.

Single-Gene Effects in Mice

Surely the oldest recorded behavior mutant is the *waltzer* mutation in the mouse, recorded in 80 B.C. in

TABLE 24.2
Inherited neurological defects in mice.

Group	Name
Waltzer-shaker	*Waltzer, shaker, pirouette, jerker, fidget, twirler, zig-zag*
Convulsive	*Trembler, tottering, spastic*
Incoordination	*Jumping, quaking, reeler, staggerer, leaner, Purkinje cell degeneration, cerebral degeneration*

China and described as a mouse "found dancing with its tail in its mouth." Waltzing mice run in a tight circle and demonstrate both horizontal and vertical head shaking as well as hyperirritability. Some waltzers are also deaf.

Genetic crosses between mutant and normal house mice (*Mus musculus*) reveal a simple recessive inheritance pattern characteristic of Mendel's monohybrid matings. Investigation of the inner ear of mutants has revealed degeneration of both the cochlea and semicircular canals, accounting for the deafness and circling behavior. This is an example of a mutation causing a structural anomaly which, in turn, alters behavior.

Many other single-gene behavior effects are known in the mouse, an organism particularly well characterized genetically. Of over 300 mutants discovered representing about 250 loci, over 90 are neurological in nature and alter behavior. These are classified into three groups of syndromes, as shown in Table 24.2. The neurological basis of many of these has now been determined.

The incoordination mutants *quaking* and *jumping* are due to faulty myelination of nervous tissue. *Quaking* is an autosomal recessive mutation, and *jumping* is inherited as a sex-linked recessive. The former cannot synthesize adequate myelin, while the latter demonstrates a degenerative process involving myelin. *Cerebral degeneration* is a mutant exhibiting progressive deterioration of behavior. As its name implies, part of the brain deteriorates.

The study of such mutants clearly establishes the genetic basis of normal development. Additionally, and more germane to this chapter, abnormal development is linked directly to altered behavior by these findings.

DROSOPHILA BEHAVIOR GENETICS

Genes and Mating Behavior in *Drosophila*

We have already discussed geotaxis in *Drosophila* as an example of the selection methodology, but much more information on the behavior genetics of this organism is available. This is not at all surprising because of our extensive knowledge of its genetics and the ease with which it can be manipulated experimentally. Advances made beginning in the 1970s, particularly by Seymour Benzer and his colleagues, represent significant strides in the field of behavior genetics.

As early as 1915, Alfred Sturtevant observed that the sex-linked recessive gene *yellow* affects mating preference, besides conferring the more obvious pigmentation difference. Recall that such a situation is an example of **pleiotropic** gene expression (see Chapter 4). Sturtevant found that both wild-type and *yellow* females, when given the choice of wild-type or *yellow* males, prefer to mate with wild type. Wild-type and *yellow* males prefer to mate with *yellow* females. These conclusions were based on quantitative measurements of success in mating between all combinations of *yellow* and wild-type (gray-bodied) males and females.

In 1956, Margaret Bastock extended these observations by investigating which, if any, component of courtship behavior was affected by the *yellow* mutant gene. Courtship in wild-type *Drosophila* is a complex ritual. The male first undergoes **orientation,** where he follows the female, perhaps circles her, and then orients usually at right angles and taps her on the abdomen. Once he has her attention, male wing display or **vibration** occurs. The wing closest to the female is raised and vibrates rapidly for several seconds. He then moves behind her, and contact is made between the male proboscis and female genitalia, described as "licking." Following this phase, if she has signaled acceptance by remaining in place, he mounts her and copulation occurs.

Bastock compared wild-type and *yellow* males for courtship rituals. What she observed was that *yellow* males prolong orientation but spend much less time in the vibrating and licking phases. Courtship patterns displayed by wild-type and *yellow* males are represented in Figure 24.13.

It appears that the *yellow* mutation has disrupted the intricate sequences of male courtship. Differences in the finer aspects of courtship have also been noted between related species of *Drosophila*. In these instances, the behavioral differences are thought to serve as possible isolating mechanisms during evolution.

The Genetic Dissection of Behavior in *Drosophila*

In 1967, Seymour Benzer and his colleagues initiated a comprehensive study of behavior genetics in *Droso-*

FIGURE 24.13

Courtship patterns of wild-type and *yellow* males in *Drosophila*. The duration of the behavioral patterns of licking, vibration, and orientation are recorded over time.

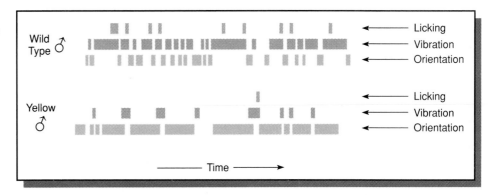

phila. Benzer's approach is an excellent example of the use of genetic techniques to "dissect" a complex biological phenomenon into its simpler components. Furthermore, the use of such techniques allows geneticists to study the underlying basis of the phenomenon—in this case, behavior. Benzer's goals in this research illustrate the "genetic dissection" approach:

1 To discern the genetic components of behavioral responses by the isolation of mutations that disrupt normal behavior.
2 To identify the mutant genes by chromosome localization and mapping.
3 To determine the actual site within the organism at which the gene expression influences the behavioral response.
4 To learn, if possible, how the particular gene expression influences behavior.

All four steps are illustrated in a discussion of **phototaxis,** one of the first behaviors studied by Benzer. Normal flies are positively phototactic; that is, they move toward a light source. Mutations are induced by feeding male flies sugar water containing **EMS** (ethylmethanesulfonate, a potent mutagen) and

mating them to attached-X virgin females. As shown in Figure 24.14, the F_1 males receive their X chromosome from their fathers. Because they are hemizygous, any sex-linked recessive mutations are expressed.

The F_1 males must be tested for their response to light, and those that are not attracted to it are isolated. Benzer found *runner* mutants, which move quickly to and from light; *negatively phototactic* mutants, which move away from light; and *nonphototactic* mutants, which show no preference for light or darkness. He established that the behavior changes were due to mutations by mating these F_1 males to attached-X virgin females. Male progeny of this cross also showed the abnormal phototactic responses, confirming them as products of sex-linked recessive mutations.

We shall now delve into further work concerning just one group, the *nonphototactic* mutants. Such flies behave in the light as normal flies do in the dark. They can walk normally, but respond to light as if they were blind. Benzer and Yoshiki Hotta tested the electrical activity at the surface of mutant eyes in response to a flash of light. The pattern of electrical activity was recorded as an **electroretinogram**. Vari-

FIGURE 24.14

Genetic cross in *Drosophila* that facilitates the recovery of X-linked induced mutations. The female parent contains two X chromosomes that are attached, in addition to a Y chromosome. In a cross between this female and a normal male that has been fed the mutagen ethylmethanesulfonate, all surviving males receive their X chromosome from their father and express all mutations induced on that chromosome.

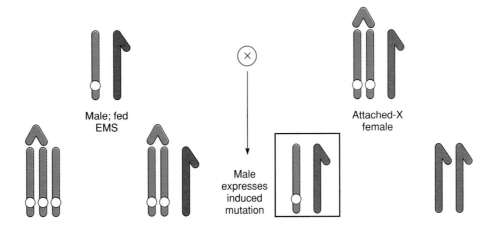

FIGURE 24.15
Production of a mosaic fruit fly as a result of fertilization by a gamete carrying an unstable ring-X chromosome (shown in color). If this chromosome is lost in one of the two cells following the first mitotic division, the body of the fly will consist of one part which is male (XO) and the other part which is female (XX). The male side will express all mutations contained on the X chromosome.

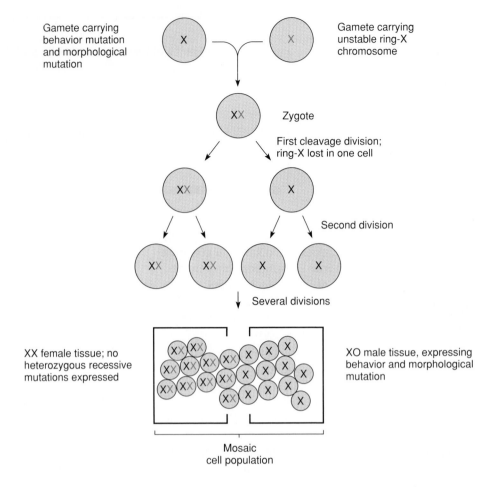

FIGURE 24.16
The effect of spindle orientation on the production of mosaic flies as shown in Figure 24.15.

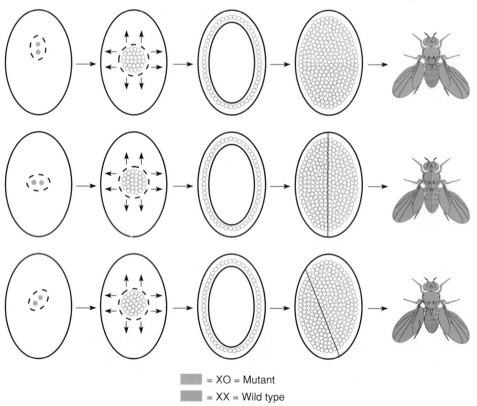

ous types of abnormal responses were detected, and in none of the mutants was a normal pattern observed. When these mutations were mapped, they were not all allelic; instead, they were shown to occupy several loci on the X chromosome. Thus, it can be concluded that several gene products contribute to the formation of the behavioral response to light.

Where, within the fly, must adequate gene expression occur to yield a normal pattern? In an ingenious approach aimed at answering this question, Benzer turned to the use of mosaics. In mosaic flies, some tissues are mutant and others are wild type. If it can be ascertained which part must be mutant in order to yield the abnormal behavior, the **primary focus** of the genetic alteration can be determined.

To facilitate the production of mosaic flies, Benzer used a strain that has one of its X chromosomes in an unstable ring shape. When present in a zygote undergoing cell division, the ring-X is frequently lost by nondisjunction. If the zygote is female and has two X chromosomes (one normal and one ring-X), loss of the ring-X at the first mitotic division will result in two cells—one with a single X (normal X) and one with two X chromosomes (one normal and one ring-X). The former cell goes on to produce male tissue (XO) and expresses all alleles on the remaining X, while the latter produces female tissue and does not express heterozygous recessive X-linked genes. Such an occurrence is illustrated in Figure 24.15. One can see that loss of the ring-X will produce mosaic flies with male and female parts that express or do not express sex-linked recessive genes, respectively.

In the embryo, when and where the ring-X is lost determines the pattern of mosaicism. The loss usually occurs early in development, before the cells migrate to the surface of the blastula to form the **blastoderm**. As shown in Figure 24.16, depending on the orientation of the spindle when the loss occurs, different types of mosaics will be created. If the stable X chromosome contains the behavior mutation and an obvious mutant gene (*yellow*, for example), the pat-

TABLE 24.3
Some behavioral mutants of *Drosophila*.

Class	Name	Characteristics
Locomotor	*sluggish*	Moves slowly
	hyperkinetic	High consumption of O_2; shaking of legs; early death
	wings up	Wings perpendicular to body
	flightless	Does not fly well, although wings are well developed
	uncoordinated	Lacks coordinated movements
	nonclimbing	Fails to climb
Response to stress	*easily shocked*	Mechanical shock induces "coma"
	stoned	Stagger induced by mechanical shock
	shaker	Vibrates all legs while etherized
	freaked out	Grotesque, random gyrations under the influence of ether
	paralyzed	Collapses above a critical temperature
	parched	Dies quickly in low humidity conditions
	tko	Epilepticlike response
	comatose	Paralyzed by cool temperature
	out-cold	Similar to comatose
Circadian rhythm	*period^o*	Eclosion at any time; locomotor activity spread randomly over the day
	period^s	19-hour cycle rather than 24-hour cycle
	period^l	28-hour cycle rather than 24-hour cycle
Sexual	*savoir-faire*	Males unsuccessful in courtship
	fruity	Males pursue each other
	stuck	Male is often unable to withdraw after copulation
	coitus interruptus	Males disengage in about half the normal time
Visual	*nonphototactic*	Blind
	negatively phototactic	Moves away from light
Learning	*dunce*	Fails to learn conditioned response

tern of mosaicism will be readily apparent. For example, the fly may consist of a mutant head on a wild-type body, a normal head on a mutant body, one normal and one mutant eye on a normal or mutant body, and so on.

When such mosaics for *nonphototactic* mutants were studied, it was found that the **focus** of the genetic defect was in the eye itself. In mosaics where every part of the fly except the eye was normal, abnormal behavior was still detected. Even if only one eye was mutant, a modified abnormal behavior was observed. Instead of crawling straight up toward light as the normal fly does, the single mutant-eyed fly crawls upward to light in a spiral pattern. In the dark, such a fly will move in a straight line. Thus, mosaic studies have established that the focus of *nonphototactic* gene expression is in the eye itself. Additionally, the abnormal behavior is due to altered electrical conductance of the cells of the eye.

Benzer and other workers in the field have identified a large number of genes affecting behavior in *Drosophila*. As shown in Table 24.3, mutants have been isolated that affect locomotion, response to stress, circadian rhythm, sexual behavior, visual behavior, and even learning. The mutations have received very descriptive and often humorous names.

Many mutations have been analyzed with the mosaic technique in order to localize the focus of gene expression. While it was easy to predict that the focus of the *nonphototactic* mutant would be in the eye, other mutants are not so predictable. For example, the focus of mutants affecting circadian rhythms has been located in the head, presumably in the brain. The *wings up* mutant might have a defect in the wings, articulation with the thorax, the thorax musculature, or the nervous system. Mosaic studies have pinpointed the indirect flight muscles of the thorax as the focus. Cytological studies have confirmed this finding, showing a complete lack of myofibrils in these muscles. Temperature-sensitive *paralytic* mutants are paralyzed at a raised temperature (29°C) but recover rapidly if the temperature is lowered. Mosaic studies have revealed that both the brain and thoracic ganglia represent the focus causing this abnormal behavior. More recent work has shown that mutant flies have defective sodium channels, and that the *paralytic* locus encodes a protein that controls the movement of sodium across the membrane of nerve cells.

The *drop-dead* mutant is most interesting. Such mutant flies appear completely normal for the first few days of adult life. Then they begin to stagger, fall over, and die. The cause of death could be related to

any vital function anywhere in the body. However, mosaic study revealed that the defect is in the head. Most flies with mutant heads and normal bodies drop dead, while most with normal heads and mutant bodies do not. As shown in Figure 24.17, the brain of the *drop-dead* mutant appears to be full of holes. Examination of the brain of mutants before death shows them to be normal.

The mosaic technique has also been used to determine which regions of the brain are associated with sex-specific aspects of courtship and mating behavior. Jeffrey Hall and his associates have shown that mosaics with male cells in the most dorsal region of the brain, the protocerebrum, exhibit the initial stages of male courtship toward females. Later stages of male sexual behavior including wing vibrations and attempted copulations require male cells in the thoracic ganglion. Similar studies of female-specific sexual behavior have shown that the ability of a mosaic to induce courtship by a male depends on female cells in the posterior thorax or abdominal region. A region of the brain within the protocerebrum must be female for receptivity to copulation. Anatomical studies have confirmed that there are fine structural differences in the brains of male and female

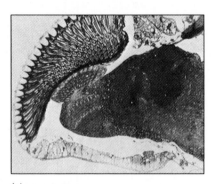

(a)

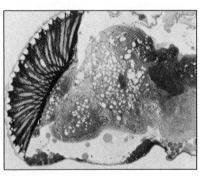

(b)

FIGURE 24.17
Photomicrographs of sections cut through the brains of (a) wild-type and (b) *drop-dead* mutant of *Drosophila melanogaster*.

Drosophila, indicating that some forms of behavior may be dependent on the development and maturation of specific parts of the nervous system.

Learning in *Drosophila*

To study the genetics of behavior such as learning, it would be advantageous to use an organism like *Drosophila*, in which methods of genetic analysis are highly advanced. However, the first question is: Can *Drosophila* learn? Recent work from a number of laboratories indicates that organisms such as *Drosophila* are, in fact, capable of learning. Using a simple apparatus, flies are presented with a pair of olfactory cues, one of which is associated with an electrical shock. Flies quickly learn to avoid the odor associated with the shock. That this response is learned is indicated by a number of factors: first, performance is associated with the pairing of a stimulus/response with a reinforcer; second, the response is reversible. Flies can be trained to select an odor that they previously avoided, and flies exhibit short-term memory for the training they have received.

Other evidence suggests that more innate forms of behavior, such as courtship, are also associated with learning. Richard Siegel and his colleagues have found that following unsuccessful courtship of sexually unreceptive females, males show a reduction of courtship behavior for about 3 hours, even in the presence of sexually receptive females. Memory-deficient mutants, such as *amnesiac,* resume active courtship in about 1 hour, presumably because the previous experience has been forgotten.

The demonstration that *Drosophila* can learn opens the way to selecting mutants that are defective in learning and memory. To accomplish this, males

from an inbred wild-type strain are mutagenized and mated to females from the same strain. Their progeny are recovered and mated to produce populations of flies, each of which carries a mutagenized X chromosome. Mutants that affect learning are selected by testing the population for response in the olfactory/shock apparatus. A number of learning-deficient mutants including *dunce, turnip, rutabaga,* and *cabbage* have been recovered. In addition, a memory-deficient mutant, *amnesiac* (mentioned previously), that learns normally but forgets four times faster than normal. Each of these mutations represents a single gene defect that affects a specific form of behavior. Because of the method used to recover them, all the mutants found so far are X-linked genes. Presumably similar genes controlling behavior are also located on autosomes.

Molecular Biology of Behavior

Since in many cases, mutation results in the alteration or abolition of a single protein, biochemical study of the learning mutants described above can provide a link between behavior and molecular biology. The *dunce* mutation is one of the first for which this link has been established. The locus for *dunce* has been shown to encode the structural gene for the enzyme cyclic AMP phosphodiesterase. It appears that *rutabaga* is a mutation in the structural gene for adenylate cyclase and that the *turnip* mutation is in the gene for a GTP-binding protein associated with adenylate cyclase. The unexpected clustering of these independently derived mutations in the biochemical pathway of the adenyl cyclase system suggests a role for cyclic nucleotides in learning. This conclusion is consistent with work in the sea hare, *Aplysia,* indicating that

TABLE 24.4
Behavioral mutants of *Drosophila* affecting nerve impulse transmission.

Mutation	Map Location	Ion Channel Affected	Phenotype
nap^ts	2–56.2	sodium	Adults, larvae paralyzed at 37.5° C. Reversible at 25° C.
para^ts	1–53.9	sodium	Adults paralyzed at 29° C, larvae at 37° C. Reversible at 25° C.
tip-E	3–13.5	sodium	Adults, larvae paralyzed at 39–40° C. Reversible at 25° C.
sei^ts	2–106	sodium	Adults paralyzed at 38° C, larvae unaffected. Adults recover at 25° C.
Sh	1–57.7	potassium	Aberrant leg shaking in adults exposed to ether.
eag	1–50.0	potassium	Aberrant leg shaking in adults exposed to ether.
Hk	1–30.0	potassium	Ether-induced leg shaking.
slo	3–85.0	potassium	At 22° C, adults are weak fliers; at 38° C, adults are weak, uncoordinated.

short-term memory is associated with an increase in cyclic nucleotides within neurons. Further work will be necessary to clarify the role of such molecules in behavior.

Because behavior is controlled and modified by cells of the nervous system, recent studies in *Drosophila* have centered on the isolation and characterization of mutations that affect the generation and propagation of electrical signals in the nervous system. These mutations have been identified by screening behavioral mutants for electrophysiological abnormalities. Two general classes of mutations have been isolated: those affecting sodium channels, and those affecting potassium channels (Table 24.4). Using recombinant DNA techniques, the genes responsible for these mutations are being cloned; the *Shaker* mutation, originally identified over 40 years ago as a behavioral mutant, has been cloned and found to encode a potassium channel structural protein. These cloned sequences can be used to answer fundamental questions about the function of the nervous system and the molecular basis of memory and learning. Certainly, work undertaken in this field in the next few years will be important and exciting.

HUMAN BEHAVIOR GENETICS

The genetic input to behavior in humans is much more difficult to characterize than that of other organisms. Not only are humans unavailable as experimental subjects in genetic investigations, but the types of responses considered to be interesting behavior are extremely difficult to study. The most popular forms of studied behavior all include some aspects of **intelligence, language, personality,** or **emotion**. There are two problems in examining such traits. First, all are difficult to define objectively and to measure quantitatively. Second, they are the traits most affected by the environment. In each case, while there is undoubtedly a genetic basis, it is a complex one. Furthermore, the environment is extremely important in shaping, limiting, or facilitating the final phenotype for each trait.

The study of human behavior genetics has also been hampered by two other factors. Many studies of human behavior have been performed by psychologists without adequate input from the biologist or geneticist. Second, traits involving intelligence, personality, and emotion have the greatest social and political significance. As such, these traits are more likely to be the subject of sensationalism when reported to the lay public. Because their study comes closest to infringing upon individual liberties such as the right to privacy, these traits are the basis of the most controversial investigations.

In lamenting the gulf between psychology and genetics in explaining human behavior genetics, C. C. Darlington in 1963 wrote, "Human behavior has thus become a happy hunting ground for literary amateurs. And the reason is that psychology and genetics, whose business it is to explain behavior, have failed to face the task together." Since 1963, some progress has been made in bridging this gap, but the genetics of human behavior remains a controversial area.

Mental Disorders with a Clearcut Genetic Basis

Many genetic disorders in humans result in some behavioral abnormality. One of the most prominent examples is **Huntington disease (HD)**. Inherited as an autosomal dominant disorder, it affects the nervous system, including the brain. Symptoms of HD usually appear in the fifth decade of life with a gradual loss of motor function and coordination. Degeneration of the nervous system is progressive, and personality changes occur. The affected individual soon is unable to care for himself. Most victims die within 10 to 15 years after onset of the disease. Since onset is usually after a family has been started, all children of an affected person must live with the knowledge that they face a 50 percent probability of developing the disorder. The gene for HD maps to the tip of the short arm of chromosome 4 and is associated with elevated brain levels of quinolinic acid, a naturally occurring neurotoxin. Although the gene has not yet been isolated, nearby molecular markers can be used in RFLP analysis (Chapter 16) to diagnose those carrying the dominant allele before symptoms appear.

The **Lesch-Nyhan syndrome** is inherited as a sex-linked recessive disorder. Onset is within the first year, and the disease is most often fatal early in childhood. The disorder is of a metabolic nature, involving purine biosynthesis. Affected individuals lack hypoxanthine-guanine phosphoribosyltransferase (HGPRT), and accumulate high levels of uric acid. Mental and physical retardation occurs, and these individuals demonstrate uncontrolled self-mutilation. They also strike out at individuals attempting to care for them.

Other metabolic disorders are also known to affect mental function. For example, **Tay-Sachs disease,** an autosomal recessive disorder, involves severe mental retardation among other phenotypic characteristics. The disease is apparent soon after birth and is fatal.

Porphyria, under the control of an autosomal dominant allele, usually has a much later onset and is marked by recurring periods of dementia. The autosomal recessive disease **phenylketonuria,** unless detected and treated early, results in mental retardation. All of these disorders alter the normal biochemistry of the affected individuals and are inherited in a Mendelian fashion.

Chromosome abnormalities also produce syndromes with behavioral components. **Down syndrome** (trisomy 21) results in mental retardation. While there is a wide range of variability, the mean IQ of affected individuals is estimated to be 25 to 50. The onset of walking and talking is often delayed until four to five years of age. Both Klinefelter syndrome (XXY) and Turner syndrome (XO) may also result in diminished mental capacity.

Human Behavior Traits with Less-Defined Genetic Bases

Other aspects of human behavior, notably **schizophrenia** and **manic-depressive illness,** have been the subject of extensive study. Investigations have sought to relate the development of the mental disorder or the display of intelligence to the closeness of family relationships or twin studies. In all cases, it has been concluded that a genetic component influences the trait, but that environment also plays a substantial role.

A discussion of schizophrenia may serve to illustrate the methodology used. This mental disorder is characterized by withdrawn, bizarre, and sometimes delusional behavior. Those affected by the disease are unable to lead organized lives and are periodically disabled by the condition. It is clearly a family disease, with relatives of schizophrenics having a much higher incidence of this disorder than the general population. Furthermore, the closer the relationship to the index case or proband, the greater is the probability of the disorder occurring.

The concordance of schizophrenia in monozygotic and dizygotic twins has been the subject of many studies. In almost every investigation, concordance has been higher in monozygotic twins than in dizygotic twins reared together. Although these results suggest that a genetic component exists, they do not reveal the precise genetic basis of schizophrenia. Simple monohybrid and dihybrid inheritance as well as multiple gene control have been proposed for schizophrenia. However, it seems unlikely that only one or two loci are involved, nor is it likely that the control is strictly quantitative, as in polygenic inheritance. In both schizophrenia and manic-depressive

illness, it is most sound to conclude that each individual is endowed with a genetic predisposition for normal or abnormal behavior and that environmental factors can serve to alter the final phenotype.

There is a long-standing controversy about genetic differences in intelligence between races. While IQ testing has established intelligence differences in populations of different races, there is currently no strong evidence to support the conclusion that this is due to a genetic component. With regard to intelligence, genetic factors may provide an upper and lower potential range, but an individual's environment may modify the development of intelligence. The environment undoubtedly has a profound effect on the type of intelligence measured by the various forms of IQ tests. It seems likely that the genetic component of intelligence does not differ any more significantly between races than it does between individuals within the same race.

CHAPTER SUMMARY

1   Behavioral genetics has emerged as an important specialty within the field of genetics because both genotype and environment have been found to have an impact in determining an organism's behavioral response.

2   Since both genotype and environment play a role in the expression of behavioral traits, the production of purely objective data in this area is particularly difficult.

3   Three areas of genetically influenced behavior research are being pursued: the behavior of closely related organisms from similar environments whose survival needs appear to be identical; the laboratory modification of behavioral traits for evidence of heritability; and the specific effects of a single gene on behavior patterns.

4   The studies of alcohol preference and open-field behavior in mice and mate selection in baboons illustrate behaviors strongly influenced by the genotype.

5   Studies of maze learning in rats and geotaxis in *Drosophila* have successfully established both bright and dull lines of rats and either positively or negatively geotaxic *Drosophila*. These results have led researchers to ascribe the relative contributions of genes on a specific chromosome to these behaviors.

6   By isolating mutations that cause deviations from normal behavior, the role of a corresponding wild-type allele in the respective response is established. Numerous examples have been examined, including honeybee hygenic behavior;

chemotaxis, thermotaxis, and general movement in nematodes; neurological mutations in mice; and a variety of behaviors in *Drosophila*.

7 Any aspect of human behavior is difficult to study because the individual's environment makes an important contribution towards trait develop-

ment. In humans, studies using twins have shown that while a family may have a predisposition to schizophrenia, the expression of this disorder may be modified by the environment. General intelligence is also a product of both genetic and environmental influences.

Insights and Solutions

Manic depression is an affective disorder associated with recurring mood changes. It is estimated that one in four individuals will suffer from some form of affective disorder at least once in his lifetime. Genetic studies indicate that manic depression is familial, and that single genes may play a major role in controlling this behavioral disorder. In 1987 two separate studies using RFLP analysis and other genetic markers reported linkage between manic depression and markers on the X chromosome and to the short arm of chromosome 11. At the time, these reports were hailed as landmark discoveries, opening the way to the isolation and characterization of genes that control specific forms of behavior, and the development of therapeutic strategies based on knowledge of the nature of the gene product and its action. However, further work on the same populations reported in 1989 and 1990 demonstrated that the original results were invalid, and concluded that no linkage to markers on the X chromosome and to markers on chromosome 11 could be established. These findings do not exclude the role of major genes on the X chromosome and autosomes in manic depression, but do exclude linkage to the markers used in the original reports.

This setback not only caused embarrassment and confusion, but it also forced a reexamination of the validity of the methods used in mapping human genes, and the analysis of data from linkage studies involving complex behavioral traits. Several factors have been proposed to explain the flawed conclusions reported in the original studies. What do you suppose some of these factors to be?

ANSWER: While some of the criticisms were directed at the choice of markers, most of the factors at work in this situation appear to be related to the phenotype of manic depression. At least three confounding elements have been identified. One is age of onset. There is a positive correlation between age and the appearance of manic depression. Therefore, at the time of a pedigree study, younger individuals who will be affected later in life may not show any signs of manic depression. Another confusing factor is that in the populations studied, there may in fact be more than one major sex-linked gene and more than one major autosomal gene that cause manic depression. A third factor relates to the diagnosis of manic depression itself. The phenotype is complex and not as easily quantified as height or weight. In addition, swings in mood are a universal part of everyday life, and it is not always easy to distinguish transient mood alterations and the role of environmental factors from affective disorders having a biological and/or genetic basis.

These factors point up the difficulty in researching the genetic basis of complex behavioral traits. Individually or in combinations, the factors described above might skew the results enough so that guidelines for proof that are adequate for other traits are not stringent enough for behavioral traits with complex phenotypes and complex underlying causes.

PROBLEMS AND DISCUSSION QUESTIONS

1 Contrast the methodologies used in studying behavior genetics. What are the advantages of each method with respect to the type of information gained?

2 Contrast the advantages of using *Drosophila* versus *Caenorhabditis* for studying behavior genetics.

3 Theoretical data concerning the genetic effect of *Drosophila* chromosomes on geotaxis are shown below. What conclusions can you draw? See Table 24.2 for comparative data.

	Chromosome		
	X	2	3
Positive geotaxis	+0.2	+0.1	+3.0
Unselected	+0.1	−0.2	+1.0
Negative geotaxis	−0.1	−2.6	+0.1

4 If a haploid honeybee drone of the genotype *uR* is mated to a queen of genotype *UuRR*, what ratio of behavioral phenotypes will be observed in the hive in the offspring with respect to American foulbrood disease? For *Ur × UURr*? For *uR × Uurr*?

5 Assume that you discovered a fruit fly that walked with a limp and was continually off balance as it moved. Describe how you would determine if this behavior was due to an injury (induced in the environment) or was an inherited trait. Assuming that it is inherited, what are the various possibilities for the focus of gene expression causing the imbalance? Describe how you would locate the focus experimentally if it were X-linked.

6 In humans, the chemical phenylthiocarbamide (PTC) is either tasted or not. When the offspring of various combinations of taster and nontaster parents are examined, the following data are obtained:

PARENTS	Both tasters	Both tasters	Both tasters	One taster One nontaster	One taster One nontaster	Both nontasters
OFFSPRING	All tasters	½ tasters ½ nontasters	¾ tasters ¼ nontasters	All tasters	½ tasters ½ nontasters	All nontasters

Based on these data, how is PTC tasting behavior inherited?

7 Discuss why the study of human behavior genetics has lagged behind that of other organisms.

8 J. P. Scott and J. L. Fuller studied 50 traits in five pure breeds of dog. Almost all traits varied significantly in the five breeds, but very few bred true in crosses. What can you conclude with respect to the genetic control of these behavioral traits?

BASTOCK, M. 1956. A gene which changes a behavior pattern. *Evolution* 10: 421–39.

———. 1967. *Courtship—An ethological study*. Chicago: Aldine.

BENZER, S. 1973. Genetic dissection of behavior. *Scient. Amer.* (Dec.) 229: 24–37.

BERG, H. C. 1988. A physicist looks at bacterial chemotaxis. *Cold Spring Harb. Symp. Quant. Biol.* 53: (Part 1) 1–9.

BODMER, W. F., and CAVALLI-SFORZA, L. L. 1976. *Genetics, evolution and man*. New York: W. H. Freeman.

BRENNER, S. 1974. The genetics of *Caenorhabditis elegans*. *Genetics* 77: 71–94.

BYERS, D., DAVID, R. L., and KIGER, J. A. Jr. 1981. Defect in the cyclic AMP phosphodiesterase due to the *dunce* mutation of learning in *Drosophila*. *Nature* 289: 79–81.

DEVOR, E. J., and CLONINGER, C. R. 1989. Genetics of alcoholism. *Ann. Rev. Genet.* 23: 19–36.

EHRMAN, L., and PARSONS, P. A. 1981. *Behavior genetics and evolution*. New York: McGraw-Hill.

FARBER, S. L. 1980. *Identical twins reared apart*. New York: Basic Books.

FULLER, J. C., and THOMPSON, W. R. 1978. *Foundations of behavior genetics*. St. Louis: C. V. Mosby.

FULLER, J. L. 1960. Behavior genetics. *Ann. Rev. Psychol.* 11: 41–63.

GAILEY, D. A., HALL, J. C., and SIEGEL, R. W. 1985. Reduced reproductive success for a conditioning mutant in experimental populations of *Drosophila melanogaster*. *Genetics* 111: 795–804.

GANTEZKY, B. 1989. There's a whole lot of shaking going on. *Genetics* 121: 201–4.

GANTEZKY, B., and WU, C. F. 1986. Neurogenetics of membrane excitability in *Drosophila*. *Ann. Rev. Genet.* 20: 13–44.

GOODMAN, R. M., ed. 1970. *Genetic disorders of man*. Boston: Little, Brown.

GOTTESMAN, I. I., and SHIELDS, J. 1972. *Schizophrenia and genetics: A twin study vantage point*. Orlando: Academic Press.

HALL, J. C., GREENSPAN, R. J., and HARRIS, W. A. 1982. *Genetic neurobiology*. Cambridge, MA: The MIT Press.

HARRIS, W. A. 1985. Genetics and development of the nervous system. *J. Neurogenet.* 2: 179–96.

HAY, D. A. 1985. *Essentials of behaviour genetics*. Palo Alto, CA: Blackwell Scientific Publ.

HESTON, L. L. 1970. The genetics of schizophrenia and schizoid disease. *Science* 167: 249–56.

HIRSCH, J., ed. 1967. *Behavior-genetic analysis*. New York: McGraw-Hill.

HOTTA, Y., and BENZER, S. 1972. Mapping behavior of *Drosophila* mosaics. *Nature* 240: 527–35.

KAPLAN, A. R. 1976. *Human behavior genetics*. Springfield, IL: Charles C. Thomas.

KIDD, K. K., and CAVALLI-SFORZA, L. L. 1973. An analysis of the genetics of schizophrenia. *Social Biol.* 20: 254–65.

KUNG, C., and NAITOH, Y. 1973. Calcium-induced ciliary reversal in the extracted models of *pawn*, a behavioral mutant of *Paramecium*. *Science* 179: 195–96.

KUNG, G., CHANG, S. Y., SATOW, Y., VanHOUTEN, J., and HANSMA, H. 1975. Genetic dissection of behavior in *Paramecium*. *Science* 188: 898–904.

LEWONTIN, R. C. 1975. Genetic aspects of intelligence. *Ann. Rev. Genet.* 9: 387–405.

LINDZEY, G., LOEHLIN, J., MANOSEVITZ, M., and THIESSEN, D. 1971. Behavioral genetics. *Ann. Rev. Psychol.* 22: 39–94.

LINDZEY, G., and THIESSEN, D. D. 1970. *Contributions to behavior-genetic analysis: The mouse as a prototype*. New York: Appleton-Century-Crofts.

LIVINGSTONE, M. S. 1985. Genetic dissection of *Drosophila* adenylate cyclase. *Proc. Natl. Acad. Sci.* 82: 5795–99.

LOUGHNEY, K., KREBER, R., and GANTEZKY, B. 1989. Molecular analysis of the *para* locus, a sodium channel gene in *Drosophila*. *Cell* 58: 1143–54.

McCLEARN, G. E. 1970. Behavioral genetics. *Ann. Rev. Genet.* 4: 437–68.

MERRELL, D. J. 1965. Methodology in behavior genetics. *J. Hered.* 56: 263–66.

PLOMIN, R., DeFRIES, J. C., and McCLEARN, G. E. 1980. *Behavioral genetics*. 2nd ed. New York: W. H. Freeman.

QUINN, W. B., and GREENSPAN, R. J. 1984. Learning and courtship in *Drosophila*: Two stories with mutants. *Ann. Rev. Neurosci.* 7: 67–93.

QUINN, W. G., and GOULD, J. L. 1979. Nerves and genes. *Nature* 278: 19–23.

RICKER, J. P., and HIRSCH, J. 1988. Genetic changes occurring over 500 generations in lines of *Drosophila melanogaster* selected divergently for geotaxis. *Behavior Genet.* 18: 13–24.

RICKER, J. P., and HIRSCH, J. 1988. Reversal of genetic homeostasis in laboratory populations of *Drosophila melanogaster* under long-term selection for geotaxis and estimates of gene correlates: Evolution of behavior-genetic systems. *J. Comp. Psychol.* 102: 203–14.

RIDDLE, D. L. 1978. The genetics of development and behavior in *Caenorhabditis elegans*. *J. Nematol.* 10: 1–15.

ROGERS, D. A., and McCLEARN, G. E. 1962. Mouse strain differences in preference for various concentrations of alcohol. *Quart. J. Stud. Alc.* 23: 26–33.

ROTHENBUHLER, W. C., KULINCEVIC, J. M., and KERR, W. E. 1968. Bee genetics. *Ann. Rev. Genet.* 2: 413–38.

SAIMI, Y., MARTINAC, B., GUSTIN, M. C., CULBERTSON, M. R., ADLER, J., and KUNG, C. 1988. Ion channels in *Paramecium*, yeast, and *Escherichia coli*. *Cold Spring Harb. Symp. Quant. Biol.* 53: (Part 2) 667–73.

SATOW, Y., and KUNG, C. 1974. Genetic dissection of active electrogenesis in *Paramecium aurelia*. *Nature* 247: 69–71.

SCOTT, J. P., and FULLER, J. L., eds. 1974. *Dog behavior: The genetic basis*. Chicago: The University of Chicago Press.

SIEGEL, R. W., HALL, J. C., GAILEY, D. A., and KYRIACOU, C. P. 1984. Genetic elements of courtship in *Drosophila*: Mosaics and learning mutants. *Behav. Genet.* 14: 383–410.

SPRINGER, M. S., GOY, M. F., and ADLER, J. 1979. Protein methylation in behavioral control mechanisms and in signal transduction. *Nature* 208: 279–84.

STANSBURY, J. B., WYNGAARDEN, J. B., and FREDERICKSON, D. S., eds. 1972. *The metabolic basis of inherited disease*. 3rd ed. New York: McGraw-Hill.

TRYON, R. C. 1942. Individual differences. In *Comparative psychology*, 2nd ed., ed. F. A. Moss. Englewood Cliffs, NJ: Prentice-Hall.

TULLY, T. 1984. *Drosophila* learning: Behavior and biochemistry. *Behav. Genet.* 14: 527–57.

VALE, J. R. 1980. *Genes, environment, and behavior: An interactionist approach*. New York: Harper & Row.

WARD, S. 1973. Chemotaxis by the nematode, *Caenorhabditis elegans*: Identification of attractants and analysis of the response by use of mutants. *Proc. Natl. Acad. Sci.* 70: 817–21.

25
Population Genetics

POPULATIONS, GENE POOLS, AND GENE FREQUENCIES

THE HARDY-WEINBERG LAW
Sex-Linked Genes | Linkage Disequilibrium | Multiple Alleles | Heterozygote
Frequency | Demonstrating Equilibrium

FACTORS THAT ALTER GENE FREQUENCIES
Mutation | Migration | Selection | Genetic Drift | Inbreeding and Heterosis

CHAPTER CONCEPTS

Individuals carry only two different alleles of a given gene. A group of individuals can carry a larger number of different alleles, giving rise to a reservoir of genetic diversity. The diversity contained in the population can be measured by the Hardy-Weinberg equation. Mutation is the ultimate source of genetic variation, and other factors such as drift, migration, and selection can alter the amount of genetic variation in populations.

When Charles Darwin published *The Origin of Species* in 1859 following his briefer statement, coauthored with Alfred Russel Wallace, on the likely mechanism of natural selection, the foundation for the modern interpretation of evolution was established. Although organisms are capable of reproducing in a geometric fashion, Darwin observed that this growth potential of species is not realized. Instead, population numbers remain relatively constant in nature. He deduced that some form of competitive struggle for survival must therefore occur. He also observed that there is a natural variation between individuals within a species; on this observation, he based his theory of natural selection: ". . .that any being, if it vary however slightly in any manner profitable to itself. . .will have a better chance of surviving." Darwin included in his concept of survival ". . .not only the life of the individual, but success in leaving progeny."

While Mendel was familiar with Darwin's work, Darwin was unaware of any underlying mechanism to account for the morphological variation he had observed. However, with the development of the concept of genes and alleles, the genetic basis of inherited variation was established.

As others pursued the study of evolution, it became apparent that the population rather than the individual was the functional unit in this process. In order to study the role of genetics in the process of evolution, therefore, it was necessary to consider gene frequencies in populations rather than offspring from individual matings. Thus arose the discipline of **population genetics**.

Early in the twentieth century various workers, including Gudny Yule, William Castle, Godfrey Hardy, and Wilhelm Weinberg, formulated ideas that became the basic principles of this field of investigation. In the earlier years of population genetics, the emphasis was on theory and mathematical models used to describe the genetic structure of a population. Only in the past thirty years have experimentalists and field workers attempted to test these theories. Gene frequencies and the forces that alter these frequencies, such as mutation, migration, selection, and random genetic drift, have been and are being examined. In this chapter, we shall consider some general aspects of population genetics and also discuss other areas of genetics relating to evolution.

POPULATIONS, GENE POOLS, AND GENE FREQUENCIES

Members of a species are often distributed over a wide geographic range. A **population,** however, is a local group belonging to a single species, within which mating is actually or potentially occurring. Within the population, the set of genetic information carried by all interbreeding members is called the **gene pool**. For a given gene, this pool includes all the alleles of that gene present in the population. In population genetics the focus is on groups rather than on individuals, and on the measurement of gene (allelic) and genotype frequencies from generation to generation rather than on the distribution of genotypes resulting from a single mating. The term **allelic frequency** will be used throughout the chapter and will represent the frequency of alleles in contrast to genotype frequencies. Gametes produced by one generation represent withdrawals from the gene pool to form the zygotes of the next generation. This new generation has a reconstituted gene pool that may differ from that of the preceding generation.

Obviously, then, populations are dynamic; they may grow and expand or diminish and contract through changes in birth or death rates, by migration, or by merging with other populations.

One approach used to study a population's genetic structure is to measure the frequency of a given gene controlling a known trait. This is possible once the mode of inheritance and the number of different alleles present in the population have been established. Gene frequencies cannot always be determined directly because in many cases only phenotypes, and not genotypes, can be observed. However, if alleles expressed in a codominant fashion are considered, phenotypes are equivalent to genotypes. Such is the case with the autosomally inherited **MN blood group** in humans.

In this case, the gene L on chromosome 2 has two alleles, L^M and L^N (often referred to as M and N, respectively).* Each controls the production of a distinct antigen on the surface of red blood cells. Thus, an individual may be type M ($L^M L^M$), N ($L^N L^N$), or MN ($L^M L^N$).

The genotypes, phenotypes, and immunological reactions of the MN blood group are shown in Table 25.1. Because they are codominant, the frequency of M and N in a population can be determined simply by counting the number of alleles for each phenotype. For example, consider a population of 100 individuals of which 36 are MM, 48 are MN, and 16 are NN. The 36 MM individuals represent 72 M alleles, and the 48 MN heterozygotes represent 48 additional M alleles. There are 72 + 48 or 120 M alleles out of a total of 200 alleles in the population (100 individuals, each with two alleles at this locus). Therefore, the frequency of the M allele in this population is 0.6 (120/200 = 60% = 0.6). The frequency of the N allele can be calculated in a similar fashion (80/200 = 40%) and is 0.4. Table 25.2 illustrates two methods for computing the frequency of M and N alleles in a hypothetical population of 100 individuals, and Table 25.3 lists the frequencies of M and N alleles actually measured in several human populations.

THE HARDY-WEINBERG LAW

In the example of the MN blood group, M and N are codominant. If, on the other hand, one allele had been recessive, the heterozygotes would have been phenotypically identical to the homozygous dominant individuals, and the frequency of the alleles could not have been directly determined. However, a mathematical model developed independently by the British mathematician Godfrey H. Hardy and a German physician, Wilhelm Weinberg, can be used

*L stands for Karl Landsteiner, the geneticist for whom the locus is named.

TABLE 25.1
MN blood groups.

		Reaction with Antibodies	
Genotype	Blood Type	Anti-M	Anti-N
$L^M L^M$	M	+	−
$L^M L^N$	MN	+	+
$L^N L^N$	N	−	+

to calculate allele frequencies in this case. The **Hardy-Weinberg law (HWL)** is one of the fundamental concepts in population genetics. As is the case with most mathematical models, certain assumptions must be made. In the Hardy-Weinberg law, the following conditions are presumed:

1 The population is infinitely large, or large enough that sampling error is negligible.
2 Mating within the population occurs at random.
3 There is no selective advantage for any genotype; that is, all genotypes produced by random mating are equally viable and fertile.
4 There is an absence of other factors such as mutation, migration, and random genetic drift.

To illustrate the application of the Hardy-Weinberg law in an ideal population, suppose that a locus has two alleles, A and a. The frequency of the dominant allele A in both eggs and sperm is represented by p, and the frequency of the recessive allele in gametes is represented by q. Because the sum of p and q represents 100 percent of the alleles for that gene in the population, then $p + q = 1$. A Punnett square can be used to represent the random combination of gametes containing these alleles and the resulting phenotypes:

	Sperm	
	$A(p)$	$a(q)$
$A(p)$	AA (p^2)	Aa (pq)
$a(q)$	Aa (pq)	aa (q^2)

Eggs

In the random combination of gametes in the population, the probability that sperm and egg both contain the A allele is $p \times p = p^2$. Similarly, the chance that gametes will carry unlike alleles is $(p \times q) + (p \times q) = 2pq$, and the chance that a homozygous recessive individual will result is $q \times q = q^2$. Note that

TABLE 25.2
Methods of determining allele frequencies for codominant alleles.

A. Counting Alleles

Genotype/Phenotype	MM	MN	NN	Total
No. of individuals	36	48	16	100
No. of M alleles	72	48	0	120
No. of N alleles	0	48	32	80
Total no. of alleles	72	96	32	200

Frequency of M in population: $\frac{120}{200} = 0.6 = 60\%$

Frequency of N in population: $\frac{80}{200} = 0.4 = 40\%$

B. From Genotypes

Genotype/Phenotype	MM	MN	NN	Total
No. of individuals	36	48	16	100
Genotype frequency	36/100 = 0.36	48/100 = 0.48	16/100 = 0.16	1.00

Frequency of M in population: $36 + (1/2)48 = 0.36 + 0.24 = 0.60 = 60\%$

Frequency of N in population: $16 + (1/2)48 = 0.16 + 0.24 = 0.40 = 40\%$

these terms also describe an important aspect of the HWL: allelic frequencies determine genotype frequencies. In other words, while the value p^2 is the probability that both gametes in a fertilization event will carry the A allele, it is also a measure of the frequency of AA homozygotes in the ensuing generation. In a similar way, $2pq$ describes the frequency of Aa heterozygotes, and q^2 is a measure of the frequency of homozygous recessive (aa) zygotes. Thus, the distribution of genotypes in the next generation can be expressed as

$$p^2 + 2pq + q^2 = 1 \qquad (25.1)$$

Let us consider a population in which 70 percent of the alleles for a given gene are A, and 30 percent are a. Thus, $p = 0.7$ and $q = 0.3$, and $p(0.7) + q(0.3) = 1$.

The distribution of genotypes produced by random mating is shown in the following Punnett square.

Sperm

	A (p = 0.7)	a (q = 0.3)
A (p = 0.7)	AA p^2 (0.49)	Aa pq (0.21)
a (q = 0.3)	Aa pq (0.21)	aa q^2 (0.09)

Eggs

In this new generation, 49 percent (p^2) of the individuals will be homozygous dominant, 42 percent ($2pq$)

TABLE 25.3
Frequencies of M and N alleles in various populations.

Population	Genotype Frequency (%)			Allele Frequency	
	MM	MN	NN	M	N
Eskimos (Greenland)	83.48	15.64	0.88	0.913	0.087
U.S. Indians	60.00	35.12	4.88	0.776	0.224
U.S. whites	29.16	49.38	21.26	0.540	0.460
U.S. blacks	28.42	49.64	21.94	0.532	0.468
Ainus (Japan)	17.86	50.20	31.94	0.430	0.570
Aborigines (Australia)	3.00	29.60	67.40	0.178	0.822

will be heterozygous, and 9 percent (q^2) will be homozygous recessive. By inspection of the Punnett square, the frequency of the *A* allele in the new generation can be calculated as

$$p^2 + \frac{1}{2} 2pq \qquad (25.2)$$
$$0.49 + \frac{1}{2} (0.42)$$
$$0.49 + 0.21 = 0.70$$

For *a*, the frequency is

$$q^2 + \frac{1}{2} 2pq \qquad (25.3)$$
$$0.09 + \frac{1}{2} (0.42)$$
$$0.09 + 0.21 = 0.30$$

Since $p + q = 1$, the value for *a* could have been calculated as

$$q = 1 - p \qquad (25.4)$$
$$q = 1 - 0.70 = 0.30$$

The values of *A* and *a* in the new generation are the same as in the previous generation.

A population in which the frequency of a given gene remains constant from generation to generation is said to be in state of **genetic equilibrium** for that gene. In this case, since the frequencies of *A* and *a* remain constant, we can conclude that the conditions presumed for the Hardy-Weinberg law hold true in this population.

Several important points are relevant to the preceding example. First, while the hypothetical genes we have considered were at equilibrium, not all genes in a population are. This is particularly true when the assumptions made in the Hardy-Weinberg law do not hold. Second, the examples illustrate why dominant traits do not tend to increase in frequency as new generations are produced. Finally, the examples demonstrate that genetic equilibrium maintains a state of **genetic variability** in a population. Once fixed in a population, allelic frequencies remain unchanged during equilibrium—a factor important to the evolutionary process.

Sex-Linked Genes

In considering genotype and gene frequencies for autosomal recessive traits using the Hardy-Weinberg equation, we assumed that the frequency of *A* is the same in both sperm and eggs. What about sex-linked genes? Since human females have two X chromosomes, one maternal and the other paternal in origin, they carry two copies of all genes on the X chromosome. Males, on the other hand, receive an X chromosome only from the mother and carry only one copy of all X-linked genes. Genes on the X chromosome are therefore distributed unequally in the population, with females carrying two-thirds and males one-third of the total number. Can the Hardy-Weinberg law be applied to calculate allelic frequencies and genotypic frequencies in such circumstances?

It is easy to determine frequencies for X-linked genes in males. Because they have only one copy of all genes on this chromosome, the phenotype of both dominant and recessive alleles is expressed. Therefore, the frequency of an X-linked allele of a gene is the same as the phenotypic frequency. In Western Europe, a form of sex-linked color blindness occurs with a frequency of 8 percent in males ($q = 0.08$). Since females have two doses of all genes on the X chromosome, color-blind females exhibit the phenotypic frequencies expected for complete dominance, and the genotype and frequencies can be calculated by using the standard Hardy-Weinberg equation. For example, since color blindness in males has a frequency of 0.08, the expected frequency in females is q^2, or 0.0064. This means that 800 out of 10,000 males surveyed would be expected to be color-blind, but only 64 of 10,000 females would show this trait. Expected values of the frequency of X-linked traits in males and females are compared in Table 25.4.

Linkage Disequilibrium

If the frequency of an X-linked allele differs between males and females, then the population is not in equilibrium. In contrast to autosomal recessive genes, equilibrium in each group will not be reached in a single generation, but will be approached over a series of succeeding generations. Because males inherit their X chromosome maternally, the allelic frequency in females will determine the frequency in males of the next generation. Daughters will inherit both a maternal and a paternal X, and their gene frequency is the average of that found in the parents.

TABLE 25.4
Expected relative frequency of X-linked traits in males and females.

Frequency of Males with Trait	Expected Frequency in Females
90/100	81/100
50/100	25/100
10/100	1/100
1/100	1/10,000
1/1000	1/1,000,000
1/10,000	1/100,000,000

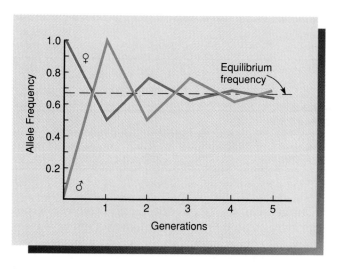

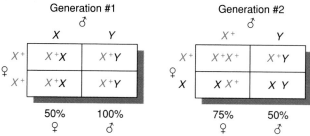

Generation #1		
♂		
	X	Y
♀ X⁺	X⁺X	X⁺Y
X⁺	X⁺X	X⁺Y

FIGURE 25.1
Approach to equilibrium for an X-linked trait with an initial frequency of 1.0. The fluctuations in frequency in the first two generations are shown in the Punnett square. In the parents of generation 1, the X-linked trait (x^+) is present in 100% of the females and 0% of the males. In the progeny, the trait is present in 50% of all female X chromosomes and in 100% of all male X chromosomes. Thus, as the graph shows, the frequency in females drops from 1.0 to 0.5 in a generation, while the frequency in males rises from 0.0 to 1.0. In generation 2, the frequency in females is 0.75, and in males 0.5. This oscillation continues until eventually an equilibrium is reached.

Even though the population's overall gene frequency remains constant, there is an oscillation in gene frequencies for the two sexes in each generation, with the differences being halved in each succeeding generation until an equilibrium is reached. This concept is illustrated in Figure 25.1 for the simple case where an allele has an initial frequency of 1.0 in females and 0.0 in males.

Multiple Alleles

In addition to autosomal recessive and sex-linked genes, it is common to find several alleles of a single locus in a population. The ABO blood group in humans is such an example. The single locus I

(isoagglutinin) has three alleles (I^A, I^B, and I^O), yielding six possible genotypic combinations (I^AI^A, I^BI^B, I^OI^O, I^AI^B, I^AI^O, I^BI^O). Recall that in this case A and B are codominant alleles, and both of these are dominant to O. The result is that homozygous AA and heterozygous AO individuals are phenotypically identical, as are BB and BO individuals, so we can distinguish only four phenotypic combinations.

By adding another term to the Hardy-Weinberg equation, we can calculate both genotype and gene frequencies for the situation involving three alleles. In an equilibrium population, the frequency of the three genes can be described by

$$p(A) + q(B) + r(O) = 1 \qquad (25.5)$$

and the distribution of genotypes will be given by

$$(p + q + r)^2 \qquad (25.6)$$

In our hypothetical population, the genotypes AA, AB, AO, BB, BO, and OO will be found in the ratio

$$p^2(AA) + 2\,pq(AB) + 2pr(AO) + q^2(BB) \\ + 2qr(BO) + r^2(OO) = 1 \quad (25.7)$$

Knowing the frequencies of A, B, and O for a population, we can then calculate both the genotypic and phenotypic frequencies for all combinations of these three alleles. For example, in an Armenian population, the frequency of A (p) is 0.38; B (q) is 0.11; and O (r) is 0.51. By using the formula $(p + q + r)^2 = 1$, we can calculate the genotypic and phenotypic frequencies for this population as shown in Table 25.5, combining phenotypically identical genotypes to obtain the phenotypic frequencies.

Heterozygote Frequency

One of the practical applications of the Hardy-Weinberg law, mentioned briefly in the discussion of multiple alleles, is the calculation of heterozygote frequency in a population. When a deleterious recessive trait is investigated, the frequency of the recessive phenotype usually can be determined by counting such individuals in a sample of the population. This information and the Hardy-Weinberg equation are used to calculate the gene and genotype frequencies.

Albinism, an autosomal recessive trait, has an incidence of about 1/10,000 (0.0001) in some populations. Albinos are easily distinguished from the population at large by a lack of pigment in skin, hair, and iris. Since this is a recessive trait, albino individuals must be homozygous. Their frequency in a population is represented by q^2, provided that mating

TABLE 25.5
Calculating genotypic and phenotypic frequencies for multiple alleles where the frequency of $A(p)$ is 0.38, $B(q)$ is 0.11, and $O(r)$ is 0.51.

Genotype	Genotypic Frequency	Phenotype	Phenotypic Frequency
AA AO	$p^2 = (0.38)^2 = 0.14$ $2pr = 2(0.38 \times 0.51) = 0.39$	A	0.53
BB BO	$q^2 = (0.11)^2 = 0.01$ $2qr = 2(0.11 \times 0.51) = 0.11$	B	0.12
AB OO	$2pq = 2(0.38 \times 0.11) = 0.082$ $r^2 = (0.51)^2 = 0.26$	AB O	0.08 0.26

has been at random and all Hardy-Weinberg conditions have been met in the previous generation.

The frequency of the recessive allele therefore is

$$\sqrt{q^2} = \sqrt{0.0001} \qquad (25.8)$$
$$q = 0.01, \text{ or } 1/100$$

Since $p + q = 1$, then the frequency of p is

$$p = 1 - q \qquad (25.9)$$
$$= 1 - 0.01$$
$$= 0.99, \text{ or } 99/100$$

In the Hardy-Weinberg equation, the frequency of heterozygotes is given as $2pq$. Therefore, we have

$$\text{Heterozygote frequency} = 2pq \qquad (25.10)$$
$$= 2[(0.99)(0.01)]$$
$$= 0.02, \text{ or } 2\%, \text{ or } 1/50$$

Thus, heterozygotes for albinism are rather common in the population (2%), even though the incidence of homozygous recessives is only 1/10,000.

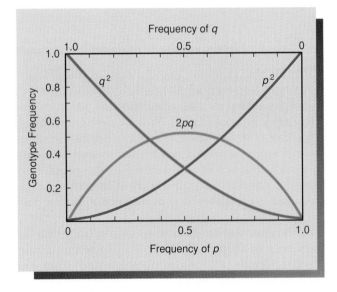

FIGURE 25.2
Relationship between genotype frequency and gene frequencies derived from the Hardy-Weinberg equation.

In general, the frequencies of all three genotypes can be calculated once the frequency of either allele is known. The relationship between genotype frequency and gene frequency is shown in Figure 25.2. It is important to note how fast heterozygotes increase in a population as the values of p and q move away from zero. This observation confirms our conclusion that when a recessive trait like albinism is rare, the majority of those carrying the allele are heterozygotes. In some cases, where the frequencies of p and q are between 0.33 and 0.67, heterozygotes actually constitute the major class in the population.

Demonstrating Equilibrium

The Hardy-Weinberg law can also be used to determine whether a given population is in equilibrium or is evolving. In order to do this, one must be able to phenotypically identify heterozygotes. If this is possible, then it must be ascertained whether the existing population fits a $p^2 + 2pq + q^2 = 1$ relationship. If so, the population has probably reached equilibrium. If not, some factor, presumably either selection, mutation, migration, or some other force, is causing gene frequencies to shift with each successive generation.

The MN blood group in one human population, the Australian aborigines, is a good example of this application of the Hardy-Weinberg law. From the values of allele frequencies in Table 25.3 (0.178 for M and 0.822 for N), the expected frequencies of blood types M, MN, and N can be calculated to determine if the population is in equilibrium:

$$\text{Expected frequency of type M} = p^2 \qquad (25.11)$$
$$= (0.178)^2 = 0.032 = 3.2\%$$

$$\text{Expected frequency of type MN} = 2pq$$
$$= 2(0.178)(0.822) = 0.292 = 29.2\%$$

$$\text{Expected frequency of type N} = q^2$$
$$= (0.822)^2 = 0.676 = 67.6\%$$

These **expected frequencies** are nearly identical to the **observed frequencies** shown in Table 25.3, confirming that the population is in equilibrium. If there

TABLE 25.6
Frequencies of *M* and *N* alleles
and genotypes in natural and
artificial populations.

Population	Genotype Frequencies			Allele Frequencies	
	MM	*MN*	*NN*	*p*	*q*
Australian aborigines	0.03	0.296	0.674	0.178	0.822
American Indians	0.60	0.351	0.049	0.776	0.224
Mixed population (500 + 500):					
Observed	0.315	0.324	0.361	0.477	0.523
Expected	0.228	0.498	0.274		

were a question as to whether the observed frequencies varied significantly from the expected frequencies, a chi-square analysis could be performed (see Chapter 3). This calculation could ascertain whether the values agree statistically.

In these calculations, the values are so close that we shall accept them as being in agreement. In addition to demonstrating that equilibrium has been reached, we may conclude that all assumptions of the HWL are being met: mating between individuals of these genotypes is random in this population; there is no selective advantage to any of the three genotypes; the population is fairly large; there is no net change of gene frequency as a result of mutation; and there is insignificant migration between this and any other population.

On the other hand, if the Hardy-Weinberg test demonstrates that the population is not in equilibrium, one or more of these necessary conditions are not being met. Using the values listed for *M* and *N* frequencies in Table 25.3, we can construct a hypothetical mixed population of 500 Australian aborigines and 500 American Indians. The frequencies for *M* and *N* for this hypothetical population are shown in Table 25.6. In such a mixed population, the frequency of *M* (*p*) would be

$$0.315 + 1/2(0.324) = 0.477 \qquad (25.12)$$

and that of *N* (*q*) would be

$$0.361 + 1/2(0.324) = 0.523 \qquad (25.13)$$

If these allelic frequencies had been derived by random mating, then the proportions of phenotypes we should expect are MM (p^2), 22.8 percent; MN (2*pq*), 49.8 percent; and NN (q^2), 27.4 percent. Since the expected genotype frequencies do not fit those observed (and a chi-square test would confirm this), we would conclude that the population is in a state of nonequilibrium brought about, obviously, by a lack of random mating in this hypothetical population.

What would happen, however, if such a mixed population were to be established on an island with no previous inhabitants and mating did take place and occurred at random? How long would it take for equilibrium to be established? Applying the Hardy-Weinberg equation, we can predict that after one generation the observed and expected genotype frequencies would converge. Using the *p* and *q* values just calculated, and remembering that the gene frequencies represent those of the total gene pool of the first generation mixed population, we can construct the following Punnett square:

	♂ *M*(*p* = 0.477)	*N*(*q* = 0.523)
M (*p* = 0.477)	*MM* p^2 0.228	*MN* *pq* 0.249
N (*q* = 0.523)	*MN* *pq* 0.249	*NN* q^2 0.274

We can see that after one generation, the observed genotype frequencies of 22.8 percent for MM (p^2); 49.8 percent for MN (2*pq*); and 27.4 percent for NN (q^2) are those expected for an equilibrium population. The gene frequencies of *M* = 0.477 and *N* = 0.523 are also those expected for an equilibrium population. However, to rigorously demonstrate HWL equilibrium, we would need to calculate the frequency of matings.

FACTORS THAT ALTER GENE FREQUENCIES

The Hardy-Weinberg law allows us to compare gene frequencies and genotype frequencies in populations in which our initial assumptions about random mating, absence of selection and mutation, and equal viability and fecundity hold true. Obviously, it is difficult—if not impossible—to find natural populations in which all these conditions are met. Populations in nature are dynamic, and changes in size and structure are part of their life cycles. This does not mean that the Hardy-Weinberg relationship is invalid; rather, it provides an opportunity to study those forces that introduce and maintain genetic diversity in populations. In this section, we will discuss factors that prevent populations from reaching equilibrium and the relative contribution of these factors to evolutionary change.

Mutation

Within a population, the gene pool is reshuffled each generation to produce new combinations in the genotypes of the offspring. Because the number of possible combinations is so large, the members of a population alive at any given time represent only a small fraction of all possible genotypes. Although an enormous genetic reserve is present in the gene pool, genetic variability is produced by Mendelian assortment and recombination, and these processes do not produce any new alleles. **Mutation** alone acts to create new alleles and is a force in increasing genetic variation. But in the absence of other forces, mainly selection, mutation is a negligible force in changing gene frequencies. Details of the chemical and molecular aspects of mutation have been presented in Chapter 13.

For our purposes, we need only consider that mutational events occur at random—that is, without regard for any possible benefit or disadvantage to the organism. Here we will discuss only the generation of mutant alleles and in a later section will consider the spread and distribution of new alleles through the population.

To know whether mutation is a significant force in changing gene frequencies, we must measure the rate at which mutations are produced. Since most mutations are recessive, it is difficult to observe mutation rates directly in diploid organisms. Indirect methods using probability and statistics or large-scale screening programs must be employed. For certain dominant mutations, however, a direct method of measurement can be used. To ensure accuracy, several conditions must be met:

1 The trait must produce a distinctive phenotype that can be distinguished from similar traits produced by recessive alleles.
2 The trait must be fully expressed or completely penetrant so that mutant individuals can be identified.
3 An identical phenotype must never be produced by nongenetic agents such as drugs or chemicals.

Mutation rates can be expressed as the number of new mutant genes per given number of gametes. Suppose for a given gene that undergoes mutation to a dominant allele, 2 out of 100,000 births exhibit a mutant phenotype. Because the zygotes that produced these births each carry two copies of the gene, we have actually surveyed 200,000 copies of the gene (or 200,000 gametes). If we assume that the affected births are each heterozygous, we have uncovered 2 mutant alleles out of 200,000. Thus the mutation rate is 2/200,000 or 1/100,000, which in scientific notation would be written as 1×10^{-5}.

In humans, a dominant form of dwarfism known as **achondroplasia** fulfills the requirements outlined above for the measurement of mutation rates. Individuals with this skeletal disorder have an enlarged skull, short arms and legs, and can be diagnosed by radiologic examination at birth. In a survey of almost 250,000 births, the mutation rate (μ) for achondroplasia has been calculated as

$$1.4 \times 10^{-5} \pm 0.5 \times 10^{-5} \qquad (25.14)$$

Knowing the rate of mutation, we can then employ the Hardy-Weinberg law to calculate the change in frequency for each generation. If initially only the normal d allele exists (i.e., all individuals are homozygous dd), the frequency of d (q_0) is 1.0 and the frequency of D (dwarf) is $p_0 = 0$. If the rate of mutation from d to D is μ, then in each successive generation,

$$\text{Frequency of } D = p_1 = q_0\mu \qquad (25.15)$$

The new frequency for d, which will be lowered by the rate of mutation, can be expressed as

$$\text{Frequency of } d = q_1 = 1 - p_0\mu \qquad (25.16)$$

Although the rate of mutation is constant from generation to generation, the rate of change in the mutant allele is initially high, but declines to zero at mutational equilibrium.

As a more general example, let us assume that the mutation rate for genes in the human genome is 1.0×10^{-5}. With such a low mutation rate, changes in gene frequency brought about by mutation alone are very small. In fact, Figure 25.3 shows that if A is the only allele of the a locus in the population ($p = 1$), with a mutation rate of 1.0×10^{-5} for A to a, it would require about 70,000 generations to reduce the frequency of A to 0.5. Even if the rate of mutation were to increase through exposure to higher levels of radioactivity or chemical mutagens, the impact of mutation on gene frequencies would be extremely weak. This example

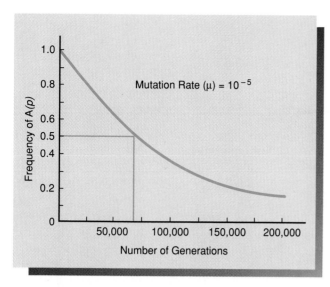

FIGURE 25.3
Rate of replacement of an allele by mutation alone, assuming an average mutation rate of 1.0×10^{-5}.

once again emphasizes that mutation is a major force in generating genetic variability, but by itself has an insignificant role in changing gene frequencies.

Migration

Frequently, a species of plant or animal becomes divided into subpopulations that to some extent are geographically separated. Differences in mutation rate and selective pressures can establish different allele frequencies in the subpopulations. **Migration** occurs when individuals move between these populations. Consider a single pair of alleles, A (p) and a (q). The change in the frequency of A in one generation can be expressed as

$$\Delta p = m(p_m - p) \qquad (25.17)$$

where

p = frequency of A in existing population
p_m = frequency of A in immigrants
Δp = change in one generation
m = coefficient of migration (proportion of migrant genes entering the population per generation)

For example, assume that $p = 0.4$ and $p_m = 0.6$, and that ten percent of the parents giving rise to the next generation are immigrants ($m = 0.1$). Then, the change in the frequency of A in one generation is

$$\begin{aligned}
\Delta p &= m(p_m - p) \\
&= 0.1 \ (0.6 - 0.4) \\
&= 0.1 \ (0.2) \\
&= 0.02
\end{aligned}$$

In the next generation, the frequency of A (p_1) will increase as follows:

$$\begin{aligned}
p_1 &= p + \Delta p \qquad (25.18) \\
&= 0.40 + 0.02 \\
&= 0.42
\end{aligned}$$

If either m is large and/or if p is much larger or smaller than p_m, then a rather large change in the frequency of A will occur in a single generation. All other factors being equal (including migration), an equilibrium will be attained when $p = p_m$. These calculations reveal that the change in gene frequency attributable to migration is proportional to the differences in frequency between the donor and recipient populations. Since the migration coefficient can have a wide range of values, the effect of migration can substantially alter gene frequencies in populations. Although migration can be somewhat difficult to quantify, it can often be estimated.

Migration can also be regarded as the flow of genes between two populations that were once but are no longer geographically isolated. For example, West African populations were the source of most of the blacks brought to the United States as slaves. In Africa, the frequency of the Duffy blood group allele Fy^0 is almost 100 percent. In Europe, the source of U.S. whites, the frequency is close to zero percent, and almost all individuals are Fy^a or Fy^b. By measuring the frequency of Fy^a or Fy^b among U.S. blacks, we can estimate the amount of white gene flow into the black population. Figure 25.4 shows the frequency of the Fy^a allele among blacks in three African nations and in several regions of the United States. Using the average frequency and assuming an equal rate of gene flow in each generation, we calculate the migration of the Fy^a allele at 5 percent per generation.

Selection

Mutations and migration introduce new alleles into populations. **Natural selection,** on the other hand, is the principal force that shifts gene frequencies within large populations and is one of the most important factors in evolutionary change.

Darwin's main contribution to the study of evolution was his recognition that natural selection is the mechanism that leads to the divergence and eventual separation of populations into distinct species. In any

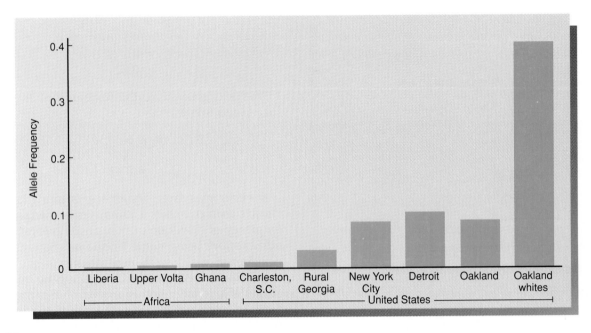

FIGURE 25.4
Frequencies of the Fy^a allele in African and U.S. black populations

population at a given time, there are individuals with different genotypes. Because of these inherent genetic differences, some of the individuals in this population will be better adapted to the environment than others, leading to the differential survival and reproduction of some genotypes over others. Natural selection is thus the differential reproduction of genotypes. Therefore, allelic frequencies will change over time. Natural selection, then, is a major force in changing gene frequencies and therefore represents a departure from the Hardy-Weinberg assumption that all genotypes have equal viability and fertility.

When a particular genotype/phenotype combination confers an advantage to organisms in competition with others harboring an alternate combination, selection occurs. The relative strength of selection varies with the amount of advantage provided. The probability that a particular phenotype will survive and leave offspring is a measure of its **fitness**. Thus, fitness refers to a total reproductive potential or efficiency. The concept of fitness is usually expressed in relative terms by comparing a particular genotypic/phenotypic combination with one regarded as optimal. Fitness is a relative concept because as environmental conditions change, so does the advantage conferred by a particular genotype.

Mathematically, the difference between the fitness of a given genotype and another regarded as optimal is called the **selection coefficient (s)**. For a phenotype conferred by the genotype aa, when only 99 of every

100 organisms successfully reproduce, $s = 0.01$. If the aa genotype is a homozygous lethal, then $s = 1.0$, and the a allele is propagated only in the heterozygous carrier state. Starting with any original frequencies of A (p_0) and a (q_0) in a population, when $s = 1.0$, the effect of selection on any successive generation can be calculated using the formula

$$q_n = \frac{q_0}{1 + nq_0} \qquad (25.19)$$

where n is the number of generations elapsing since p_0 and q_0.

Figure 25.5 demonstrates the change in allele frequencies when A (p_0) = a (q_0) = 0.5. Initially, because of the high percentage of aa genotypes, the frequency of the a allele is reduced rapidly. The frequency of a is halved (0.25) in only two generations. By the sixth generation, the frequency of a is again reduced twofold (0.12). By now, however, the majority of a alleles are carried by heterozygotes. Since selection operates on phenotypes and not on components of the genotype, this means that heterozygotes are not selected against. Subsequent reductions in frequency occur very slowly in successive generations and depend on the rate at which heterozygotes are removed from the population. For this reason, it is difficult to eliminate a recessive gene from the population by selection as long as heterozygotes continue to mate.

FIGURE 25.5
Changes in allele frequency under selection where $s = 1.0$. The frequency of a is halved in two generations, and halved again by the sixth generation. Subsequent reductions occur slowly because the majority of a alleles are carried by heterozygotes.

Generation	p	q	p^2	$2pq$	q^2
0	0.50	0.50	0.25	0.50	0.25
1	0.67	0.33	0.44	0.44	0.12
2	0.75	0.25	0.56	0.38	0.06
3	0.80	0.20	0.64	0.32	0.04
4	0.83	0.17	0.69	0.28	0.03
5	0.86	0.14	0.73	0.25	0.02
6	0.88	0.12	0.77	0.21	0.01
10	0.91	0.09	0.84	0.15	0.01
20	0.95	0.05	0.91	0.09	<0.01
40	0.98	0.02	0.95	0.05	<0.01
70	0.99	0.01	0.98	0.02	<0.01
100	0.99	0.01	0.98	0.02	<0.01

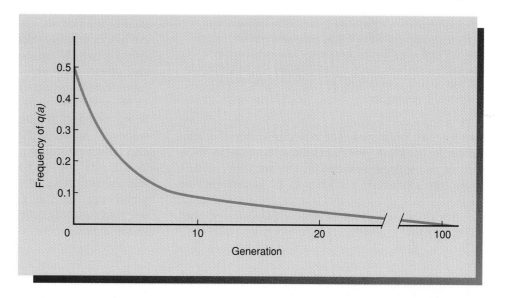

When s is less than 1.0, it is possible to calculate the effects of selection on each successive generation with any s, p_0, and q_0 values using the formula

$$q_1 = \frac{q_0 - sq_0^2}{1 - sq_0^2} \qquad (25.20)$$

Figure 25.6 demonstrates a variety of initial conditions and the change of the frequency of q (a) through a large number of generations. If q_0 is initially 0.5 and $s = 0.5$, five generations must elapse before q is halved. Then, another nine generations must elapse

to cut q in half again. Figure 25.6 shows the relative trends as q_0 and s change.

There has been much study of the action of selection on both laboratory and natural populations. A classic example of selection in natural populations is that of the peppered moth, *Biston betularia*, in England. Before 1850, 99 percent of the moth population was light colored, allowing the moths to fit well into their surroundings. As toxic gases produced by industry killed the lichens and mosses growing on trees and buildings, and soot deposits darkened the landscape, the light-colored moths became easy tar-

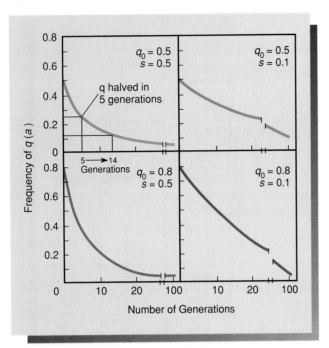

FIGURE 25.6

Changes in allele frequencies under different selection coefficients and different initial gene frequencies. When $q_0 = 0.5$ and $s = 0.5$, the frequency of q is halved in 5 generations. When $q_0 = 0.8$ and $s = 0.5$, the frequency of q is halved in 5 generations and halved again in an additional 5 generations.

gets for their predators. The rare dark-colored moths suddenly gained a great advantage because of their natural camouflage. A rapid shift in frequency of this phenotype occurred, probably in 50 generations or less.

Laboratory experiments have demonstrated that the dark form is due to a single dominant allele, C.

Figure 25.7 demonstrates the protective advantage from predators provided by the dark phenotype. Following the adoption of laws to restrict environmental pollution in 1964, the frequency of nonmelanic forms of the moth has started to increase around the Manchester area, showing the close relationship between the environment and selection.

Because selection acts on the organism's genotypic/phenotypic combination, it also acts on polygenic or quantitative traits—those that are controlled by a number of genes and may be susceptible to environmental influences. Such quantitative traits, including adult body height and weight in humans, often demonstrate a continuously varying distribution resembling a bell-shaped curve. Selection for such traits can be classified as (1) directional, (2) stabilizing, or (3) disruptive.

In **directional selection,** which is important to plant and animal breeders, desirable traits, often representing phenotypic extremes, are selected for. In fact, if the trait is polygenic, the most extreme phenotypes that the genotype can express will appear in the population only after prolonged selection. An example is the selection for oil content in domestic corn kernels. In upward selection, C. M. Woodworth and others were able to raise the oil content of corn more than threefold, from about 4 percent to just over 15 percent in 50 generations, with no sign of a plateau being reached (Figure 25.8). Similarly, downward selection reduced the content to about 1 percent. Directional selection favors an extreme phenotype and tends to produce genetic uniformity in the population. In nature, directional selection can occur when one of the phenotypic extremes becomes selected for or against, usually as a result of changes in the environment.

FIGURE 25.7

The speckled and melanic forms of the peppered moth, *Biston,* seen on a normal, lichen-covered tree trunk (left) and on a darkened, soot-covered trunk (right).

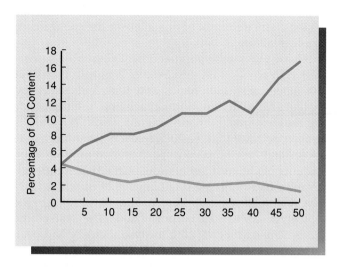

FIGURE 25.8
Long-term selection to alter the oil content of corn kernels.

Stabilizing selection, on the other hand, tends to favor intermediate types, with both extreme phenotypes being selected against. One of the clearest demonstrations of stabilizing selection is the data of Mary Karn and Sheldon Penrose on human birth weight and survival. Figure 25.9 shows the distribution of birth weight for 13,730 children born over an 11-year period and the percentage of mortality at four weeks of age. Infant mortality increases dramatically on either side of the optimal birth weight of 7.5 pounds. At the genetic level, stabilizing selection represents a situation where a population is adapted to its environment.

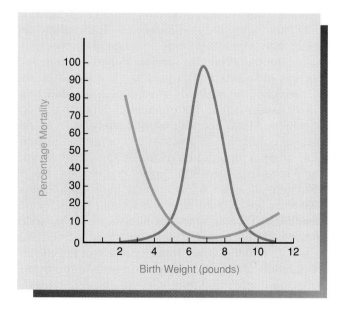

FIGURE 25.9
Relationship between birth weight and mortality in humans.

Disruptive selection is selection against intermediates and for both phenotypic extremes. It may be viewed as the opposite of stabilizing selection. John Thoday applied disruptive selection to progeny of *Drosophila* strains with high and low bristle number. Matings were carried out using females from one strain and males from the other. Progeny were selected for high and low bristle number, and matings were carried out for a number of generations. Despite the fact that opportunities for gene flow between the two lines through random mating were experimentally provided at each generation, the two lines diverged rapidly. In natural populations, such a situation may exist for a population in a heterogeneous environment. Gene flow may occur between organisms in two niches in which selection is favoring opposite extremes.

The types and effects of selection are summarized in Figure 25.10.

Genetic Drift

In laboratory crosses, one condition essential to realizing theoretical genetic ratios (i.e., 3:1, 1:2:1, 9:3:3:1) is a fairly large sample size. This condition is equally important to the study of population genetics as gene and genotype frequencies are examined or predicted. For example, if a population consists of 1000 randomly mating heterozygotes (*Aa*), the next generation will consist of approximately 25 percent *AA*, 50 percent *Aa*, and 25 percent *aa* genotypes. Provided that the initial and subsequent populations are large, there will be at most only minor deviations from this mathematical ratio, and the frequency of *A* and *a* will remain about equal (0.50).

However, if a population is formed from only one set of heterozygous parents and they produce only two offspring, gene frequency can change drastically. In this case, the probability of the occurrence of genotypes and gene frequencies in the subsequent generation can be predicted as summarized in Table 25.7. As shown, in 10 of 16 times such a cross is made, the new allele frequencies would be altered dramatically. In 2 of 16 times, either the *A* or *a* allele would be eliminated in a single generation.

The conditions used to generate these calculations are extreme, but they illustrate why large interbreeding populations are essential to the Hardy-Weinberg equilibrium. In small populations, significant random fluctuations in allelic frequencies are possible strictly on the basis of chance deviation. The degree of fluctuation will increase as the population size decreases. Such changes illustrate the concept of **genetic drift**. In the extreme case, genetic drift may lead

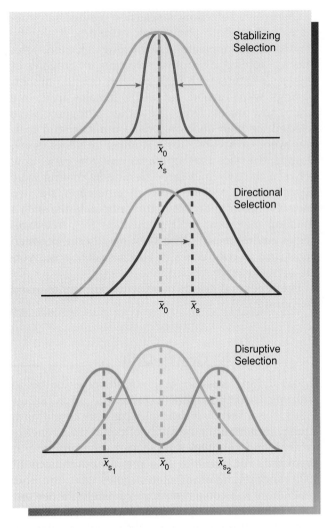

FIGURE 25.10

Comparison of the impact of stabilizing, directional, and disruptive selection. In each case, $\bar{x}_0$ is the mean of an original population, and $\bar{x}_s$ is the mean of the population following selection.

TABLE 25.7

All possible pairs of offspring produced by two heterozygous parents ($Aa \times Aa$).

Possible Genotypes of Two Offspring	Probability	Allele Frequency	
		A	a
AA and AA	$(1/4)(1/4) = 1/16$	1.00	0.00
aa and aa	$(1/4)(1/4) = 1/16$	0.00	1.00
AA and aa	$2(1/4)(1/4) = 2/16$	0.50	0.50
Aa and Aa	$(2/4)(2/4) = 4/16$	0.50	0.50
AA and Aa	$2(2/4)(1/4) = 4/16$	0.75	0.25
aa and Aa	$2(1/4)(2/4) = 4/16$	0.25	0.75

to the chance fixation of one allele to the exclusion of another allele.

Using laboratory populations of *Drosophila melanogaster*, Warwick Kerr and Sewall Wright set up over 100 lines, with four males and four females as the parents for each line. Within each line, the frequency of the sex-linked bristle mutant *forked* (*f*) and its wild-type allele (*f⁺*) was 0.5. In each generation, four males and four females were chosen at random to be parents of the next generation. After 16 generations, fixation had occurred in 70 lines—29 in which only the *forked* allele was present and 41 in which the wild-type allele had become fixed. The rest of the lines were still segregating the two alleles or had been lost. If fixation had occurred randomly, then an equal number of lines should have become fixed for each allele. In fact, the experimental results do not differ statistically from the expected ratio of 35:35, demonstrating that alleles can spread through a population and eliminate other alleles by chance alone.

How are small populations created in nature? In one instance, a large group of organisms may be split by some natural event, creating a small, isolated subpopulation. A natural disaster such as an epidemic might also occur, leaving a small number of survivors to constitute the breeding population. Third, a small group might emigrate from the larger population, as founders, to a new environment, such as a volcanically created island.

Allelic frequencies in certain human isolates best support the role of drift as an evolutionary force in natural populations. The Pingelap atoll in the western Pacific Ocean (lat. 6° N, long. 160° E) has in the past been devastated by typhoons and famine, and in 1775 there were only 30 survivors. Today there are fewer than 2000 inhabitants, and their ancestry can be traced to the typhoon survivors. Approximately 5 percent of the current population is affected by the recessively determined disorder **achromatopsia,** which causes ocular disturbances and a form of color blindness. This disorder is extremely rare in the human population as a whole. However, the responsible allele is present at a relatively high frequency in the Pingelap population. From genealogical reconstruction, it was found that one of the original survivors was a chief who was heterozygous for the condition. If we assume that he was the only carrier in the founding population, the initial gene frequency was 1/60, or 0.016. Since some 5 percent of the current population is affected, they represent homozygous recessives, and the frequency of the gene in the population is calculated as 0.23.

Another example of genetic drift involves the Dunkers, a small religious isolate living in Pennsylvania who emigrated from the German Rhineland.

Since their religious beliefs do not permit marriage with outsiders, the population has grown only through intermarriage. When the frequencies of the ABO and MN blood group alleles are compared, significant differences are found among the Dunkers, the German population, and the U.S. population. The frequency of blood group A in the Dunkers is about 60 percent, in contrast to 45 percent in the U.S. and German populations. The I^B allele is nearly absent in the Dunkers. Type M blood is found in about 45 percent of the Dunkers, compared with about 30 percent in the U.S. and German populations. Since there is no evidence for a selective advantage of these alleles, it is apparent that their observed frequencies are the result of chance events in a relatively small isolated population.

Inbreeding and Heterosis

One of the assumptions of the Hardy-Weinberg equilibrium is random mating within the population. This is an ideal condition and as such does not often occur in nature. Within a widespread population, individuals tend to mate with those in physical proximity rather than with those separated by some distance. The result is the restriction of gene flow within that population.

Nonrandom mating occurs in many species, including humans. One form of nonrandom mating is called **assortative**. In humans, pair bonds may be established by religious practices, physical characteristics, professional interests, and so forth on a nonrandom basis. In nature, phenotypic similarity plays the same role.

Another form of nonrandom mating is **inbreeding,** where mating occurs between relatives. Inbreeding results in increased homozygosity. To illustrate this concept, we shall consider the most extreme form of inbreeding, **self-fertilization**.

Figure 25.11 shows the consequences of four generations of self-fertilization starting with a single individual heterozygous for one pair of alleles. By the fourth generation, only about 6 percent of the individuals are still heterozygous. Note that the frequencies of A and a still remain at 50 percent.

In humans, inbreeding (called **consanguineous marriages**) is related to population size, mobility, and social customs governing marriages among relatives. To determine the probability that two alleles at the same locus in an individual are derived from a common ancestor, Sewall Wright devised the **coefficient of inbreeding**. Expressed as F, the coefficient can be defined as the probability that two alleles of a given gene in an individual are derived from a common allele in an ancestor. If $F = 1$, all genotypes

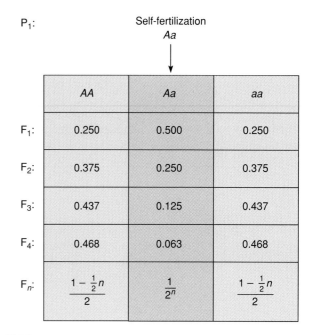

FIGURE 25.11
Reduction in heterozygote frequency brought about by self-fertilization. After n generations, the frequencies of the genotypes can be calculated according to the formulas in the bottom row.

are homozygous and derived from a common ancestor. If $F = 0$, no alleles present are derived from a common ancestor. Shown in Figure 25.12 are pedigrees of first- and second-cousin marriages. A lethal allele, a, is assumed to be present in one heterozygous parent. For each relative, the probability of the presence of this lethal allele is shown. The pedigrees then culminate in the final probability, F, that the lethal allele will become homozygous in the offspring of first or second cousins. Since every population carries as its genetic burden a number of heterozygous recessive alleles that are lethal or deleterious in homozygotes, mortality will rise in matings between relatives.

Thus, from an evolutionary standpoint, if a population is split into smaller subpopulations, the effects of inbreeding will become evident. **Inbreeding depression,** an expression of reduced fitness in a population caused by inbreeding, may occur; and because an increase in homozygous individuals results, genetic variability will decrease. Both favorable and unfavorable alleles will become more frequently homozygous. From the standpoint of Darwinian fitness, the adaptive nature of the population will decrease as well.

This knowledge has been applied in the breeding of domesticated plant and animal species. First, an inbreeding program is initiated. As homozygosity increases, some populations will become fixed for

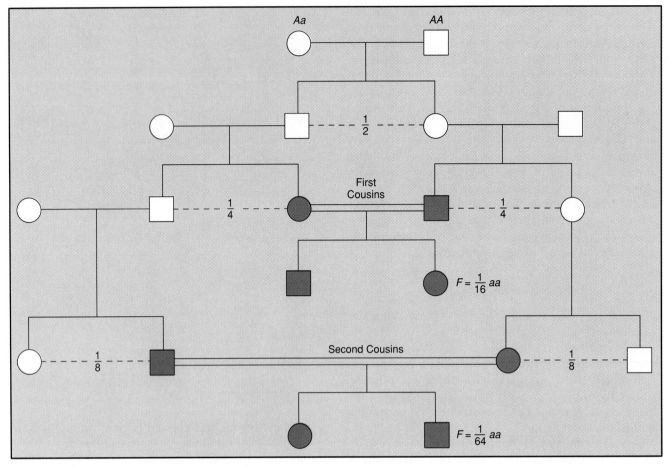

FIGURE 25.12
Changes in the coefficient of inbreeding (F) brought about by consanguineous matings. Numbers indicate the probability of carrying the recessive allele *a*.

favorable alleles and others for unfavorable alleles. By selecting the more viable and vigorous plants or animals, the proportion of individuals carrying desirable traits can be increased.

If members of two favorable inbred lines are mated, hybrid offspring are often more vigorous in desirable traits than is either of the parental lines. This phenomenon has been called **hybrid vigor** or **heterosis.** Such an approach was used in the breeding programs established for corn, and crop yields increased tremendously. Unfortunately, the hybrid vigor extends only through the first generation. Many hybrid lines are sterile, and those that are fertile show subsequent declines in yield. Consequently, the hybrids must be regenerated each time by crossing the inbred parental lines.

Heterosis has been explained in two ways. The first hypothesis, the **dominance hypothesis,** incorporates the obvious reversal of inbreeding depression,

which inevitably must occur in out-crossing. Consider a cross between two strains of corn with the following genotypes:

$$\text{Strain A} \qquad \text{Strain B}$$
$$aaBBCCddee \times AAbbccDDEE$$
$$\downarrow$$
$$AaBbCcDdEe$$

The F_1 hybrids are heterozygotes at all loci shown. Deleterious recessive alleles present in the homozygous form in the parents will be masked by the more favorable dominant alleles in the hybrids. Such masking is thought to cause hybrid vigor or heterosis.

The second theory, **overdominance,** holds that in many cases the heterozygote is superior to either homozygote. This may be related to the fact that in the heterozygote, there may be two forms of a protein present, providing a form of biochemical diversity.

Thus, the cumulative effect of heterozygosity at many loci accounts for the hybrid vigor. Most likely, such vigor results from a combination of phenomena explained by both hypotheses.

CHAPTER SUMMARY

1 The idea that the population rather than the individual is the effective unit in the process of evolution has led to the study of gene frequencies and to the emergence of the discipline of population genetics.

2 The Hardy-Weinberg law established the conditions for genetic equilibrium within a population. These conditions include a large, randomly breeding population, and the absence of selection, mutation, and migration. If any of these conditions are not met, gene or genotype frequencies will change from generation to generation.

3 The Hardy-Weinberg formula is also useful in demonstrating whether or not a population is in equilibrium and in calculating the degree of heterozygosity within a population for any particular locus.

4 The process of evolution requires conditions leading to change, such as changing gene frequencies that provide genetic variation. The mathematical derivation of the Hardy-Weinberg equilibrium has served to assess the forces of evolution: selection, mutation, migration, genetic drift, and inbreeding.

5 Mutation and migration introduce new alleles into a population, but the retention or loss of these alleles depends on the degree of fitness conferred and the action of selection.

6 Natural selection is the most powerful force altering a population's genetic constitution. In small populations, however, inbreeding and genetic drift may be important factors.

Insights and Solutions

1 Color blindness is caused by a recessive allele on the X chromosome. In a survey, 16 women out of 20,000 were found to be color blind. What is the expected frequency of color-blind males in this population?

   ANSWER: Remember that for X-linked traits, females exhibit the phenotypic frequencies expected for complete dominance. This means that the frequency of color blindness in women ($16/20,000 = 0.0008$) is equal to q^2. If $0.0008 = q^2$, then $q = 0.0282$, the value for color blindness in males. The expected frequency in males would be 2.82% or about 1 in 34.

2 Eugenics is the term employed for the selective breeding of humans to bring about improvements in the species. As a eugenic measure, it has been suggested (sometimes by force of law) that individuals suffering from serious genetic disorders should be prevented from reproducing (by sterilization, if necessary) in order to reduce the frequency of the disorder in future generations. Suppose that such a trait were present in the population at a frequency of 1 in 40,000, and that affected individuals did not reproduce. In 100 generations, or about 2,500 years, what would be the frequency of the condition? What are your assumptions about this condition? Are the eugenic measures effective in this case?

   ANSWER: Let us assume that the trait is recessive, and that homozygous recessive individuals are prevented from reproducing. In this case, selection against the homozygous recessive phenotype is equal to 1. Using formula 25.19, we can calculate the frequency after 100 generations as follows:

$$q_n = \frac{q_0}{1 + nq_0}$$

$$q_n = \frac{0.000025}{1 + 100\,(0.000025)}$$

$$= \frac{0.000025}{1.0025}$$

where

$$q_n = 1/41{,}666$$
$$s = 1$$
$$n = 100$$
$$q_0 = 0.000025$$

The frequency of the condition has been reduced from 1 in 40,000 to 1 in 41,666 in 100 generations, indicating that this eugenic measure is ineffective.

PROBLEMS AND DISCUSSION QUESTIONS

1 The ability to taste the compound PTC is controlled by a dominant allele T, while individuals homozygous for the recessive allele t are unable to taste this compound. In a genetics class of 125 students, 88 were able to taste PTC, 37 could not. Calculate the frequency of the T and t alleles in this population, and the frequency of the genotypes.

2 Calculate the frequencies of the AA, Aa, and aa genotypes after one generation if the initial population consists of 0.2 AA, 0.6 Aa, and 0.2 aa genotypes. What frequencies will occur after a second generation?

3 Consider rare disorders in a population caused by an autosomal recessive mutation. From the following frequencies of the disorder in the population, calculate the percentage of heterozygous carriers:
(a) 0.0064
(b) 0.000081
(c) 0.09
(d) 0.01
(e) 0.10

4 What must be assumed in order to validate the answers in Problem 3?

5 In a population where only the total number of individuals with the dominant phenotype is known, how can you calculate the percentage of carriers and homozygous recessives?

6 For the following two sets of data, determine whether they represent populations that are in equilibrium. Use chi-square analysis if necessary.
(a) MN blood groups: MM, 60%; MN, 35.1%; NN, 4.9%.
(b) Sickle-cell hemoglobin: AA, 75.6%; AS, 24.2%; SS, 0.2%.

7 If 4 percent of the population in equilibrium expresses a recessive trait, what is the probability that the offspring of two individuals who do not express the trait will express it?

8 Consider the gene frequencies $p = 0.7$ and $q = 0.3$ where selection occurs against the homozygous recessive genotype. What will be the gene frequencies after one generation if
(a) $s = 1.0$
(b) $s = 0.5$
(c) $s = 0.1$
(d) $s = 0.01$?

9 If initial gene frequencies are $p = 0.5$ and $q = 0.5$, and $s = 1.0$, what will be the gene frequencies after 1, 5, 10, 25, 100, and 1000 generations?

10 Determine the frequency of allele A after one generation under the following conditions of migration:

(a) $p = 0.6$; $p_m = 0.1$; $m = 0.2$
(b) $p = 0.2$; $p_m = 0.7$; $m = 0.3$
(c) $p = 0.1$; $p_m = 0.2$; $m = 0.1$

11 Assume that a recessive autosomal disorder occurs in one of 10,000 individuals (0.0001) in the general population. Assume that in this population about 2 percent (0.02) of the individuals are carriers for the disorder. Estimate the probability of this disorder occurring in the offspring of a marriage between first cousins and between second cousins. Compare these probabilities to the population at large.

12 What is the basis of inbreeding depression?

13 Does inbreeding cause an increase in frequency of recessive alleles within a population?

14 Describe how inbreeding may be used in the domestication of plants and animals. Discuss the theories underlying these techniques.

15 Evaluate the following statement: inbreeding increases the frequency of recessive alleles in a population.

16 In a breeding program to improve crop plants, which of the following mating systems should be employed to produce a homozygous line in the shortest possible time?
(a) self-fertilization
(b) brother-sister matings
(c) first cousin matings
(d) random matings
Illustrate your choice with pedigree diagrams.

17 If the recessive trait albinism (a) is present in 1/10,000 individuals, calculate the frequency of:
(a) the recessive mutant allele
(b) the normal dominant allele
(c) heterozygotes in the population
(d) matings between heterozygotes

18 One of the first Mendelian traits identified in man was a dominant condition known as *brachydactyly*. This gene causes an abnormal shortening of the fingers and/or toes. At the time, it was thought by some that this dominant trait would spread until 75 percent of the population would be affected (since the phenotypic ratio of dominant to recessive is 3:1). Show how this line of reasoning is incorrect.

19 Achondroplasia is a dominant trait that causes a characteristic form of dwarfism. In a survey of 50,000 births, 5 infants with achondroplasia were identified. Three of the affected infants had affected parents, while the rest had normal parents. Calculate the mutation rate for achondroplasia and express the rate as the number of mutant genes per given number of gametes.

SELECTED READINGS

BARTON, N. H., and TURELLI, M. 1989. Evolutionary quantitative genetics: How little do we know? *Ann. Rev. Genet.* 23: 337–70.

BODMER, W. F., and CAVALLI-SFORZA, L. L. 1976. *Genetics, evolution and man.* New York: W. H. Freeman.

CAVALLI-SFORZA, L. L., and BODMER, W. F. 1971. *The genetics of human populations.* New York: W. H. Freeman.

COOK, L. M., ASKEW, R. R., and BISHOP, J. A. 1970. Increasing frequency of the typical form of the peppered moth in Manchester. *Nature* 227: 1155.

CROW, J. F. 1986. *Basic concepts in population, quantitative and evolutionary genetics.* New York: W. H. Freeman.

CROW, J. F., and KIMURA, M. 1970. *An introduction to population genetic theory.* New York: Harper and Row.

DAVID, J. R., and CAPY, P. 1988. Genetic variation of *Drosophila melanogaster* natural populations. *Trends Genet.* 4: 106–11.

DOBZHANSKY, T. 1955. A review of some fundamental concepts and problems of population genetics. *Cold Spr. Harb. Symp.* 20: 1–15.

EDWARDS, J. H. 1989. Familiarity, recessivity and germline mosaicism. *Ann. Hum. Genet.* 53: 33–47.

FISHER, R. A. 1930. *The genetical theory of natural selection.* Oxford, England: Clarendon Press. (Reprinted in 1958 by Dover Press.)

FREIRE-MAIA, N. 1990. Five landmarks in inbreeding studies. *Am. J. Med. Genet.* 35: 118–20.

HARTL, D. L. 1981. *A primer of population genetics.* Sunderland, MA: Sinauer Associates.

JONES, J. S. 1981. How different are human races? *Nature* 293: 188–90.

KARN, M. N., and PENROSE, L. S. 1951. Birth weight and gestation time in relation to maternal age, parity and infant survival. *Ann. Eugen.* 16: 147–64.

KERR, W. E., and WRIGHT, S. 1954. Experimental studies of the distribution of gene frequencies in very small populations of *Drosophila melanogaster.* I. *Forked. Evolution* 8: 172–77.

KETTLEWELL, H. B. D. 1961. The phenomenon of industrial melanism in *Lepidoptera. Ann. Rev. Entomol.* 6: 245–62.

———. 1973. *The evolution of melanism: The study of a recurring necessity, with special reference to industrial melanism in the Lepidoptera.* London: Oxford University Press.

KHOURY, M. J., and FLANDERS, W. D. 1989. On the measurement of susceptibility to genetic factors. *Genet. Epidemiol.* 6: 699–701.

LI, C. C. 1977. *First course in population genetics.* Pacific Groves, CA: Boxwood Press.

METTLER, L. E., GREGG, T., and SCHAFFER, H. E. 1988. *Population genetics and evolution.* 2nd ed. Englewood Cliffs, NJ: Prentice-Hall.

NEEL, J. V., and ROTHMAN, E. 1981. Is there a difference among human populations in the rate with which mutation produces electrophoretic variants? *Proc. Natl. Acad. Sci.* 78: 3108–12.

REED, T. E. 1969. Caucasian genes in American Negroes. *Science* 165: 762–68.

ROBERTS, D. F. 1988. Migration and genetic change. *Hum. Biol.* 60: 521–39.

SPIESS, E. B. 1989. *Genes in populations.* 2nd ed. New York: Wiley.

WALLACE, B. 1981. *Basic population genetics.* New York: Columbia University Press.

———. 1989. One selectionist's perspective. *Quart. Rev. Biol.* 64: 127–45.

WOODWORTH, C. M., LENG, E. R., and JUGENHEIMER, R. W. 1952. Fifty generations of selection for protein and oil in corn. *Agron. J.* 44: 60–66.

26
Evolutionary Genetics

GENETIC DIVERSITY AND SPECIATION
Inbreeding Depression | Protein Polymorphisms | Chromosomal Polymorphisms | DNA
Sequence Polymorphisms | The Adaptive Norm | Speciation | Formation of Races and
Species | Quantum Speciation

MOLECULAR EVOLUTION
Amino Acid Sequence Phylogeny | Nucleotide Sequence Phylogeny | RFLP Analysis
of Nucleotide Phylogenies | Evolution of Genome Size | Evolution of Gene Families
by Duplication

THE ROLE OF MUTATION IN EVOLUTION
Selectionists versus Neutralists | Recent Trends

〇〇〇〇〇〇〇〇〇〇〇〇〇〇〇〇〇〇〇〇〇〇〇〇〇〇〇〇〇〇〇〇〇〇

CHAPTER CONCEPTS

Evolution is concerned with the effects of gene frequency changes within the gene pool of a population. Changes in gene frequency lead to changes in genetic diversity and the ability to undergo evolutionary divergence. The process of speciation divides an originally homogeneous gene pool into two or more reproductively separate gene pools. This division may be accompanied by changes in morphology, physiology, and adaptation to the environment.

In Chapter 25, we described populations in terms of allele frequencies and genetic equilibria and outlined the forces that alter the genetic constitution of populations. Mutation, migration, selection, and drift are some of the forces that individually and collectively alter gene frequencies and lead to evolutionary divergence and species formation. This process depends not only upon genetic divergence but also upon environmental or ecological diversity. If a population is spread over a geographic range encompassing a number of subenvironments or **niches,** the subpopulations occupying these niches will adapt and become genetically differentiated. This process may lead to the formation of **races** (ecotypes) or subspecies. Because the formation and maintenance of these races depend ultimately on the interaction of the genotype and environment, races are dynamic entities that may remain in existence, become extinct, merge with the parental population, or continue to separate from the parental population until they become reproductively isolated and form new species.

It is difficult to define the exact moment of transition between a subspecies and species. A **species** can be defined as one or more groups of interbreeding or potentially interbreeding organisms that are reproductively isolated in nature from all other organisms. The process of **speciation** depends on changing gene frequencies and involves the division of a homogeneous population (gene pool) into two or more reproductively isolated subunits (separate gene pools). Changes in morphology, physiology, and adaptation to an ecological niche may also occur, but are not necessary components of the speciation event. Speciation can take place gradually over a long time period or within a generation or two. Throughout this process, the degree of genetic diver-

sity is the key factor. Individuals of a species are members of a common gene pool that distinguishes them from members of other species. Therefore, it is clear that genetic divergence should be the starting point for a discussion of speciation and evolution.

In this chapter, we will focus on three central topics: first, the extent of genetic diversity in closely related organisms and its potential contribution to speciation; second, analysis of the evolution of diverse organisms at the molecular level; and third, the role of mutation in evolution, a matter of current controversy.

GENETIC DIVERSITY AND SPECIATION

If a population of organisms constituting a species is well suited to inhabit and reproduce successfully in its environment, how much genetic diversity does it possess? We might assume that members of a population are fit because the most favorable allele at each locus has become fixed homozygously. Certainly, an examination of most populations of plants and animals reveals them to be phenotypically similar, if not identical. However, current evidence supports the opinion that a high percentage of heterozygosity is maintained within the genome of diploid individuals in a population. This built-in diversity is concealed, so to speak, because it is not necessarily apparent in the phenotype. Such diversity may better adapt a population to the inevitable changes in the environment and may promote the survival of the species.

The detection of this concealed genetic variation is not a simple task. Nevertheless, such investigation has been successful using several techniques.

Inbreeding Depression

One method of measuring genetic diversity is to determine what fraction of the alleles carried by individuals of a population are deleterious. Deleterious alleles in a population can be detected by monitoring increasing homozygosity resulting from inbreeding. Because inbreeding involves mating between relatives, mates have genotypic similarities that extend across all segregating loci. As a result of inbreeding, recessive alleles previously present in the heterozygous condition become homozygous, and if these alleles are deleterious, a reduction in viability called **inbreeding depression** is evident in successive generations.

Theodosius Dobzhansky and his colleagues have succeeded in making groups of alleles and sometimes entire chromosomes of *Drosophila* species completely homozygous. When this is accomplished, viability is depressed. As shown in Table 26.1, such studies reveal lethal, semilethal, subvital, and male and female sterility alleles to be prominent in natural populations of different species. In *Drosophila pseudoobscura*, almost 100 percent of chromosomes 2 and 3 from a sample of the population cause a depression in viability when made homozygous. In *Drosophila willistoni*, almost 70 percent of chromosomes 2 and 3 contain male sterile alleles when made homozygous.

Although these and other studies reveal the substantial genetic burden carried by organisms, we discuss them here as examples of genetic variation maintained in natural populations. If detrimental alleles are maintained heterozygously, certainly alleles that are effectively neutral or potentially advantageous must also be retained in the heterozygous

condition. As recessive alleles, they represent concealed genetic variation within the "normal" heterozygous genotypes characterizing the population.

Protein Polymorphisms

The use of Mendelian genetics to estimate genetic diversity is both laborious and slow. Gel electrophoresis (see Appendix A), on the other hand, is a simpler technique that can be used to separate protein molecules on the basis of differences in electrical charge (Figure 26.1). If variation in a structural gene results in the substitution of a charged amino acid such as glutamic acid for an uncharged amino acid such as glycine, the net charge on the protein will be altered. This change in charge can be detected by electrophoresis. In the mid-1960s, John Hubby and Richard Lewontin introduced the use of gel electrophoresis to measure protein variation in natural populations of *Drosophila*. Since then, genetic variation has been studied using gel electrophoresis in a wide range of organisms.

When two alleles from one locus produce slightly different electrophoretic forms, yet perform identical functions, the resultant proteins are designated as **allozymes**. As shown in Table 26.2, a surprisingly large percentage of loci examined from diverse species produce allozymes. Of the populations shown in Table 26.2, approximately 30 loci per species were examined, and about 30 percent of the loci were polymorphic (i.e., they revealed allozymes). An approximate average of 10 percent heterozygosity per diploid genome was revealed.

These values apply only to genetic variation detectable by altered protein migration in an electric

TABLE 26.1

Percentage of homozygous chromosomes tested that reveal recessive modifiers of viability and fertility.

Drosophila Species	Chromosome	Lethals and Sublethals	Subvitals	Supervitals	Female Sterility	Male Sterility
D. pseudoobscura	2	33.0	62.6	<0.1	10.6	8.3
(Mather, Calif.)	3	25.0	58.7	<0.1	13.6	10.5
	4	22.7	51.8	<0.1	4.3	11.8
D. persimilis	2	25.5	49.8	0.2	18.3	13.2
(Mather, Calif.)	3	22.7	61.7	2.1	14.3	15.7
	4	28.1	70.7	0.3	18.3	8.4
D. prosaltans	2	32.6	33.4	<0.1	9.2	11.0
(Brazil)	3	9.5	14.5	3.0	6.6	4.2
D. willistoni	2	38.8	57.5	<0.1	40.5	64.8
(Venezuela, British Guiana, and Trinidad)	3	34.7	47.1	1.0	40.5	66.7

SOURCE: From Spiess, 1977, p. 288.

FIGURE 26.1
Photograph of gel
electrophoretic separation of
isozymes of the enzyme alcohol
dehydrogenase.

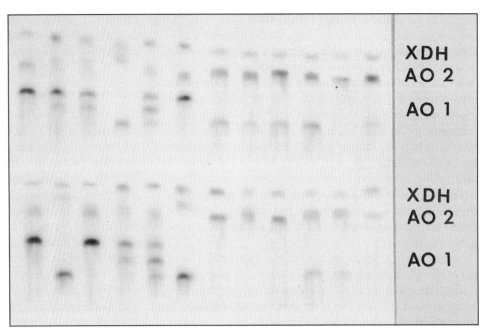

field. Electrophoresis probably detects only about 30 percent of the actual variation that is due to amino acid substitutions because many substitutions do not change the net electric charge on the molecule. Richard Lewontin has thus estimated that about two-thirds of all loci are polymorphic in a population, and that in any individual within that group about one-third of the loci exhibit genetic variation in the form of heterozygosity.

There is some controversy over the significance of genetic variation as detected by electrophoresis. Some argue that alleles producing allozymes are effectively neutral and therefore do not play any role in evolution. We shall address this argument later in this chapter.

Chromosomal Polymorphisms

While the chromosome number is usually invariant for a given species, the arrangement of chromosomal material is often polymorphic as a result of chromosomal inversions and translocations. These chromosomal aberrations usually have little direct effect on the phenotype because gene content is rearranged but not altered.

Inversions are produced by two breaks in a chromosome followed by rejoining of the broken fragment rotated 180° (see Chapter 7). The presence of an inversion does not imply that any new genes are present, nor does the presence of a variety of specific inversions prove that gene frequencies vary between any two arrangements. However, we have learned in Chapter 7 that inversion heterozygotes only rarely produce viable crossover products involving genes contained within the inversion. Therefore, an inversion suppresses the recovery of crossover gametes. In so doing, inversions tend to preserve specific allele arrangements along the inverted stretch of the chromosome involved. The net effect of inversions in populations, therefore, is a reduction in

TABLE 26.2
Heterozygosity at the molecular level.

Species	Number of Populations Studied	Number of Loci Examined	Loci per Population	Heterozygosity per Locus
Homo sapiens (humans)	1	71	28	6.7
Mus musculus (house mouse)	4	41	29	9.1
Drosophila pseudoobscura (fruit fly)	10	24	43	12.8
Limulus polyphemus (horseshoe crab)	4	25	25	6.1

SOURCE: From Lewontin, 1974, p. 117.

available, realized genetic variability. In addition, inversions reduce heterozygote fitness because of the production of inviable crossover gametes. Therefore, if inversions are present in a species, it implies a causal relationship with selective factors that favor inversion heterozygotes. In other words, perhaps the inversion has preserved a favorable gene arrangement, which at least partially accounts for the fitness of the population in a specific habitat. Linkage then becomes a possible criterion for selection. The role of inversions in natural populations has been intensively studied in *Drosophila* and will be discussed later in this section.

Translocations, in which chromosome segments move to other, nonhomologous chromosomes, are also found in populations. In simple translocations, a segment of one chromosome becomes joined to a nonhomologous chromosome. In reciprocal translocations, chromosome parts are exchanged between two nonhomologous chromosomes. Translocations also reduce realized genetic variability in populations because translocation heterozygotes are usually subfertile, producing 50 percent chromosomally abnormal gametes. Although somewhat rare in animal populations, translocation heterozygotes are found in many plants. In a population of the flowering plant *Clarkia elegans*, Harlan Lewis found that 13 percent of the individuals carried a translocation. In the extreme case, the angiosperm *Rhoeo discolor* has all of its chromosomes in translocation heterozygotes.

DNA Sequence Polymorphisms

Obviously, the most direct way to estimate genetic variation is by examination of the actual nucleotide sequence diversity between individuals of a population. With the development of techniques for cloning and sequencing DNA, nucleotide sequence variations have been cataloged for an increasing number of gene systems. Using restriction endonucleases to detect polymorphisms (see Chapter 16 for the methodology), Alec Jeffrey surveyed 60 unrelated individuals to estimate the total number of DNA sequence variants in humans. His results show that within the genes of the beta-globin cluster, 1 in 100 base pairs shows polymorphic variation. If this region is representative of the genome, this indicates that at least 3×10^7 nucleotide variants per genome are possible. Data from other organisms such as *Drosophila*, the rat, and the mouse have produced similar estimates of nucleotide diversity, indicating that there is an enormous reservoir of genetic variability within a population, and that at the level of DNA, most and perhaps all genes exhibit diversity from individual to individual.

The Adaptive Norm

A population of interbreeding organisms is more or less adapted to its surrounding environment. Adaptive phenotypes are thought to be a consequence of the array of genotypes possessed by the individuals constituting the group. These genotypes are present as a result of the evolutionary history of the population.

Dobzhansky has called the array of genotypes present in the population the **adaptive norm**. Ideally, each individual should possess a genotype and phenotype best suited to the immediate environment. However, the preceding discussion on genetic variability in natural populations suggests that there is a wide variation in genotypes within many populations. Therefore, the adaptive norm tends to be one of balanced heterozygosity. Because numerous recessive alleles are concealed and do not alter the phenotype, it follows that a variety of genotypes can be well adapted to the same environment. As Dobzhansky has pointed out, the variety making up the adaptive norm constitutes a coadapted set of genotypes that must be able to meet the environmental diversity and stresses over space and time.

Alleles that at one point in time do not seem to be of major significance to individual fitness may be of great value to the population in future generations. Under changing environmental conditions, previously insignificant alleles may become essential to the maintenance of fitness. The concept of **preadaptation** describes the so-called hidden or concealed genetic variation as a storehouse of genetic information available to enhance survival under new environmental conditions.

Speciation

In the classical sense, the process of splitting a genetically homogeneous population into two or more populations that undergo genetic differentiation and eventual reproductive isolation is called **speciation**. According to Ernst Mayr, species originate in two predominant ways. In the first mode, often called **phyletic evolution,** species A over a long period of time becomes, through genetic change, transformed into species B. In the second method, one species gives rise to one or more derived species, bringing about multiplication of species. This second process can occur over a long period of time or abruptly in a generation or two. Table 26.3 summarizes the principal methods of speciation.

The most developed classical model of speciation is **geographic** or **allopatric speciation,** first proposed

TABLE 26.3
Modes of speciation.

I. Transformation of Species (phyletic evolution)
    1. Autogenous speciation
II. Reduction in Number of Species (fusion of two species)
III. Multiplication of Species (true species)
    A. Instant speciation (through individuals)
        1. Genetic
            (a) Single mutation in asexual species
            (b) Macrogenesis
        2. Cytological
            (a) Chromosomal mutation (translocations, etc.)
            (b) Autopolyploidy
            (c) Amphidiploidy
    B. Gradual speciation (through populations)
        1. Sympatric speciation
        2. Semigeographic speciation
        3. Geographic speciation (allopatric)

SOURCE: From Mayr, 1963, Table 15.1.

by Moritz Wagner in 1868. According to Wagner, physical isolation of populations by geographic features such as lakes, rivers, or mountains that act as barriers to gene flow is the first step toward species formation. In a second step, these isolated populations undergo independent evolution and may continue to diverge to produce two distinct species.

Twentieth-century workers such as Mayr and Dobzhansky have refined and updated this model but have retained the general features of Wagner's hypothesis. In the refined model, the first step requires that populations become separated; that is, gene flow must be interrupted. The absence or interruption of gene flow is a prerequisite for the development of genetic differences by adaptation to local conditions. Genetic diversity brought about by natural selection, or by random genetic drift, will be reflected in the presence of new alleles, changes in allele frequency, or the presence of new chromosomal arrangements. Eventually, a point will be reached when the populations have enough differences that they can be identified as distinct races or semispecies. This process may continue uninterrupted until two or more species are present.

If at any time during the process of genetic divergence, the conditions that prevent gene flow between the populations are removed, two outcomes are possible: (1) the two populations may fuse into a single gene pool, because hybridization does not reduce fertility or viability; or (2) the gene pools of the populations may have diverged to the point where isolating mechanisms may have arisen.

The various biological and behavioral properties of organisms that act to prevent or reduce interbreeding are called **reproductive isolating mechanisms**. These

mechanisms are classified in Table 26.4. For example, genetic divergence may have reached the stage where the viability or fertility of hybrids is reduced. Hybrid zygotes may be formed, but all or most may be inviable. Alternately, the hybrids may be viable but have reduced fertility or be sterile. In another possibility, the hybrids themselves may be fertile, but their progeny may have lowered viability or fertility. These mechanisms, called **postzygotic,** are a byproduct of genetic divergence. Such isolating mechanisms waste gametes and zygotes and lower the fitness of hybrid survivors. Selection will therefore favor the spread of alleles that will reduce the formation of hybrids, leading to the development of **prezygotic isolating mechanisms**. Speciation and the development of isolating mechanisms may gradually occur in the absence of selection, as when populations remain permanently isolated on two or more islands. Selection accelerates speciation in those cases where some reproductive isolation has resulted from genetic diversity, but not all isolating mechanisms are represented in each speciation event. Usually at least two isolating mechanisms are developed by natural selec-

TABLE 26.4
Reproductive isolating mechanisms.

Prezygotic Mechanisms (prevent fertilization and zygote formation)

1. Geographic or ecological: The populations live in the same regions but occupy different habitats.
2. Seasonal or temporal: The populations live in the same regions but are sexually mature at different times.
3. Behavioral (only in animals): The populations are isolated by different and incompatible behavior before mating.
4. Mechanical: Cross-fertilization is prevented or restricted by differences in reproductive structures (genitalia in animals, flowers in plants).
5. Physiological: Gametes fail to survive in alien reproductive tracts.

Postzygotic Mechanisms (fertilization takes place and hybrid zygotes are formed, but these are nonviable or give rise to weak or sterile hybrids)

1. Hybrid nonviability or weakness.
2. Developmental hybrid sterility: Hybrids are sterile because gonads develop abnormally or meiosis breaks down before completion.
3. Segregational hybrid sterility: Hybrids are sterile because of abnormal segregation to the gametes of whole chromosomes, chromosome segments, or combinations of genes.
4. F_2 breakdown: F_1 hybrids are normal, vigorous, and fertile, but F_2 contains many weak or sterile individuals.

SOURCE: From G. Ledyard Stebbins, *Processes of Organic Evolution*, 3rd edition, © 1977, p. 143. Reprinted by permission of Prentice-Hall, Inc., Englewood Cliffs, N.J.

tion drawing on the genetic variability present in the evolving populations.

Formation of Races and Species

Although the theory of geographic speciation is well developed, it has been more difficult to observe directly the processes involved. Diversification of isolated populations occurs gradually over thousands or hundreds of thousands of years. In addition, the geographic changes that paralleled this divergence may be complex and completely unknown. In most cases, therefore, the formation of species is a historical event, and biologists studying this process must rely on the present-day distribution of races, subspecies, and sibling species to reconstruct stages in the evolutionary process.

To study speciation, therefore, biologists must first find examples in nature where all or most of the stages of race formation and speciation can be documented. The intensive studies carried out on natural populations of *Drosophila* provide good examples of the stages involved in **geographic** or **allopatric** speciation.

To illustrate the first step, the formation of races, we shall consider the case of *Drosophila pseudoobscura* as studied by Dobzhansky and his colleagues. This species is found over a wide range of environmental habitats, including the western and southwestern United States. Although the flies throughout this range are morphologically similar, Dobzhansky discovered that populations from different locations varied substantially in the arrangement of genes on chromosome 3. He found a variety of different inversions in this chromosome that could be detected by loop formations in the salivary chromosomes. Each particular inversion sequence was named after the locale in which it was first detected (e.g., AR = Arrowhead, British Columbia; CH = Chiricahua Mountains, etc.) and was compared with one standard sequence, designated ST.

For example, Figure 26.2 shows a comparison of three arrangements detected in populations found at three different elevations in the Sierra Nevada in California's Yosemite region. The ST arrangement is most common at low elevations but declines in frequency as elevation increases. At 8000 feet, AR is most common and ST least common. In these same populations, the frequency of the CH arrangement gradually increases with elevation. The gradual change in inversion frequencies is probably the result of natural selection and thus parallels the gradual environmental changes occurring at ascending elevations. Since the populations along this gradient show a continuous and gradual change in inversion frequencies, it is difficult to classify a fly into any one racial group.

Dobzhansky also found that if populations were collected at a single site throughout the year, inversion frequencies also changed. During the different seasons cyclic variation occurred, as shown in Figure 26.3. Such variation was observed to be repeated over a period of several years. ST was always observed to decline during the spring, with a concomitant increase in CH during the same period.

To test the hypothesis that this cyclic change is a response to natural selection, Dobzhansky devised the following laboratory experiment. He constructed a large population cage in which samples could be periodically removed and studied. He began with a

FIGURE 26.2
Inversions in chromosome 3 of *D. pseudoobscura* found at different elevations in the Sierra Nevada near Yosemite National Park.

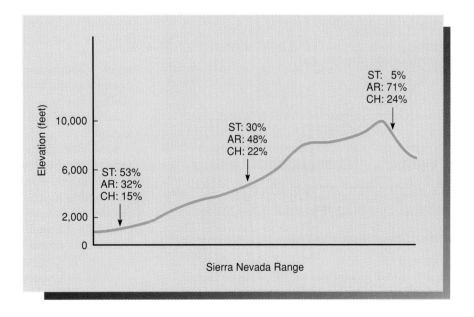

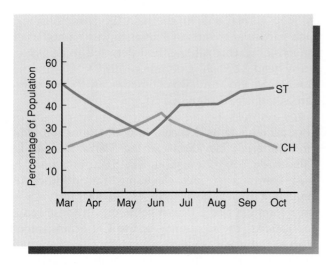

FIGURE 26.3
Changes in the ST and CH arrangements in *D. pseudoobscura* throughout the year.

population of a known inversion frequency, 88 percent CH and 12 percent ST. He reared it at 25°C and sampled it over a one-year period. As shown in Figure 26.4, the frequency of ST increased gradually until it was present at a level of 70 percent. At this point, an equilibrium between ST and CH was reached. When the same experiment was performed at 16°C, no change in inversion frequency occurred. It can be concluded that the equilibrium reached at 25°C was in response to the elevated temperature, the only variable in the experiment.

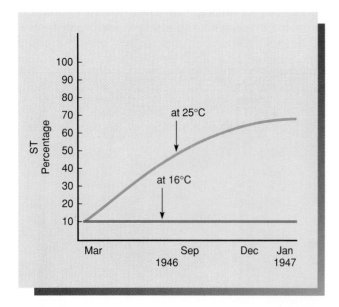

FIGURE 26.4
Increase in the ST arrangement of *D. pseudoobscura* in population cages under laboratory conditions.

The results of Dobzhansky's experiment are strong evidence that a balanced condition of the two inversions and their respective gene arrangements is superior to either inversion by itself. The equilibrium attained presumably represents the greatest degree of fitness in the population under varying laboratory conditions. This interpretation of the experiment suggests that natural selection is the driving force toward equilibrium. If so, then we may conclude that the existence of various inversions is representative of genetic variation.

In a more extensive study, Dobzhansky went on to sample populations over a much broader geographic range. Twenty-two different gene arrangements were found in populations from 12 locations. In Figure 26.5, the frequencies of five inversions are shown according to geographic location. The differences are largely quantitative, with most populations differing only in relative frequencies of inversions. However, in some cases, there are qualitative differences. For example, PP is absent in all four California locales where ST is predominant. In the Texas locations, ST is absent or present in very low frequency and PP is predominant. Some general trends are also apparent. ST increases in frequency from east to west, while just the opposite is true of PP. AR is least common in the east-west extremes, yet predominant in Arizona and New Mexico.

Because these locations represent varied environments and the inversions preserve different gene combinations, we can conclude that numerous races of *D. pseudoobscura* have been formed. Not only do the studies of *D. pseudoobscura* illustrate the principles of the initial step in speciation, but they also strongly support the concept that the adaptive norm consists of balanced genotypes within a population.

For speciation to occur, the development of races must be followed by a second step, reproductive isolation. We might then ask whether the evolution of *D. pseudoobscura* has gone beyond the formation of races. Dobzhansky investigated this question by examining the chromosome structure of other related **sibling species**. Sibling species have become reproductively isolated but remain very similar morphologically.

One sibling species, *Drosophila persimilis*, has provided very interesting information. *D. persimilis*, like *pseudoobscura*, contains five pairs of chromosomes and carries inversions within chromosome 3. When Dobzhansky examined the chromosome 3 inversions carefully, 11 such arrangements were found. One, ST, is also found in *D. pseudoobscura*. Based on the common arrangement, it was possible to construct a phylogenetic sequence for all arrangements found in both species. The phylogeny was based on the

FIGURE 26.5
Frequencies of five chromosomal inversions in *D. pseudoobscura* in different geographic regions.

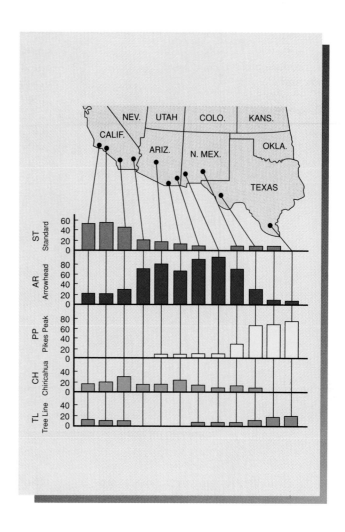

similarity of any two inversions. If a fly is heterozygous for different inversions (each member of the pair of chromosome 3 contains a different inversion), only one inversion loop may be necessary to achieve homologous pairing. If this criterion is met, the populations bearing the inversions are on a direct line; that is, one can give rise directly to the other. However, while the sequence of arrangements constitutes a phylogeny, we cannot be absolutely sure of its direction.

As shown in Figure 26.6, ST is shared by both species. Only one hypothetical arrangement is necessary to complete the continuity of the tree. It appears that an ancestral population with the ST arrangement gave rise to many different inversions. Some were incorporated into races as members of the *pseudoobscura* species, and others gave rise to the *persimilis* species.

Today, even when the geographic distributions of these sibling species overlap, several isolating mechanisms keep them from interbreeding. They are isolated by prezygotic mechanisms such as habitat selection, with *persimilis* preferring higher elevations and cooler temperatures. Differences in courtship rituals allow females to distinguish between males of the two species and choose only males of their own species for mating. Even the time of day when courtship and mating occur is different in the species, with *persimilis* tending to court in the morning and *pseudoobscura* more active in the evening. Postzygotic mechanisms also maintain reproductive isolation in these species. When cross-fertilized in the laboratory, the species produce sterile F_1 hybrid males, with male sterility being associated with interactions between the X chromosome and chromosome 2. Back-crosses between F_1 females and parental males exhibit hybrid breakdown through lowered viability of the offspring.

Quantum Speciation

At the heart of the classical theory of speciation is the concept of **gradualism**. According to this concept, speciation is a microevolutionary event resulting from the accumulation of many minute gene differences over a long period of time under the influence

FIGURE 26.6
Inversion phylogeny for arrangements of chromosome 3 in *D. pseudoobscura* and *D. persimilis*. The ST arrangement is shared by both species.

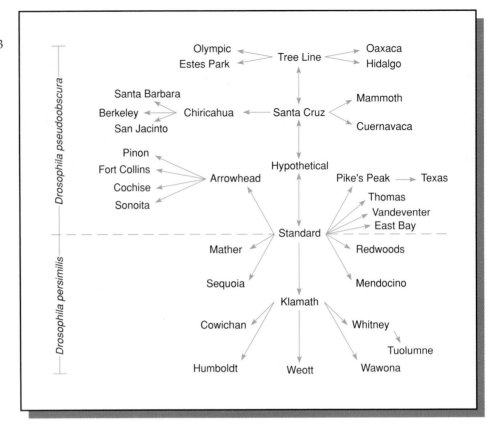

of natural selection. Recently, however, more **stochastic** or **catastrophic models of speciation** have been proposed. These models emphasize the role of evolutionary events that occur suddenly and intermittently and are therefore referred to as **quantum speciation**. We will briefly discuss one model derived from the fossil record, and two examples that depend on chromosome rearrangements as isolating mechanisms. Finally, we will consider a mechanism that has long been known to exist in angiosperms—speciation through polyploidy.

Drawing on evidence from the fossil record, Niles Eldredge and Stephen Jay Gould have proposed a mechanism of species formation known as punctuated equilibria. According to this model, evolutionary changes are not gradual and continuous, but occur intermittently and rapidly as events that punctuate or interrupt long periods of evolutionary equilibrium during which little or no change occurs. In paleontological terms, *rapidly* means within thousands or even hundreds of thousands of years. The results of such evolutionary alterations are thought to cause abrupt changes in the fossil record, since relatively few fossils will be formed during such changes, compared to the large numbers of fossils formed during long periods (extending into millions of years) of evolutionary equilibrium.

The evolutionary spurts that punctuate periods of stasis are thought to be events that produce mass extinctions, or events that take place in small populations located at the fringes of the species range. In these locales, environmental stresses disrupt evolutionary stability, eventually causing the emergence of new species. Proponents of this hypothesis point out that such changes leading to speciation take place through natural selection of individual variations, and do not invoke any unusual genetic mechanisms. On the other hand, some population geneticists argue that the mechanism of punctuated equilibria unnecessarily separates the process of species formation from other evolutionary events that generate diversity between members of a species. Much more evidence and interpretation of the fossil record are needed to evaluate this idea.

In his study of the evolution of Hawaiian *Drosophila*, Hampton Carson has proposed a **founder-flush speciation theory**. According to this theory, populations and even species can be started by a single individual. The founder principle is not new, as this idea was advanced much earlier by Mayr. In Carson's proposal, however, gradualism is not a necessary component of speciation, and more importantly, reproductive isolation *precedes* adaptation rather than being a consequence of genetic diversification. In other words, Carson has changed the order of steps in the classical model of speciation. According to the founder-flush theory, a single fertilized female can colonize an isolated territory previously unoccupied by members of this species. If conditions are favorable, the population founded by this individual will undergo a **flush,** or rapid expansion. After several generations, it is likely that the population growth will outstrip the environment's carrying capacity, causing a **population crash**. The crash, a completely random event, causes the death or dispersal of almost the entire population. A single survivor, or at most a few survivors, may rebuild the population, which eventually undergoes several flush-crash cycles before coming to equilibrium with the environment. This cycle is diagrammed in Figure 26.7. Through the genetic revolution it has undergone, the colony will, in all probability, have acquired adaptations making it unable to interbreed with the parental population.

According to Carson's theory, genetic changes are brought about in two ways. First, the original founding of the population by one or a very few individuals can establish allele frequencies different from those in the ancestral population. Second and most important, selection is relaxed during a population flush. If descendants of the founder can invade a new niche, they expand their numbers in a flush. Carson proposes that normally some blocks of genes on chromo-somes remain tightly linked or "closed" to recombination because of some selective advantage conferred by their configuration. Any new genotypes produced by recombination within these closed areas have reduced fitness. In fact, the advantage conferred by balanced polymorphisms for inversion heterozygotes may derive from the protection of such closed gene complexes. A diagram of open and closed chromosomal regions is shown in Figure 26.8.

During the flush period, genotypes produced by recombination in closed regions of the genome may survive under the relaxed conditions of selection. After a crash, the survivor's reshuffled genome is acted upon by selection to produce a new combination of open and closed gene groups adapted to the environment. Several such cycles can produce enough genetic differences so that crosses between the progenitor and colony populations produce hybrids with lowered fitness commonly displayed by interspecific hybrids. Carson views the cycles of disorganization and reorganization of the genomes as the essence of speciation rather than as the gradual genetic divergence of isolated populations over long periods of time.

The evolution of *Drosophila* species in the Hawaiian Islands appears to have followed such a founder-flush cycle. Geologic evidence indicates that the northwesternmost islands are the oldest, and that the southeastern island of Hawaii (produced by volcanic action) is the youngest at about 700,000 years old. The relationships between species of *Drosophila* can be traced by mapping the location and frequency of inversions in the banded polytene chromosomes of larval salivary glands. One group, the *planitibia* complex, has three species with the same basic set of chromosome inversions: *D. planitibia, D. heteroneura,* and *D. silvestris. D. planitibia* is found on the older island of Maui, and the other two are found on Hawaii. From this evidence, it is postulated that an

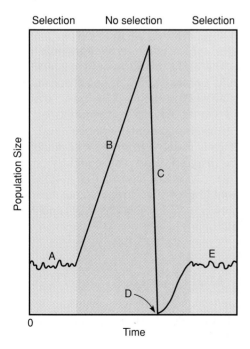

FIGURE 26.7
The flush-crash cycle. Small populations descended from a single founder (A) undergo a population flush (B), followed by a crash (C). A single survivor in the form of a fertilized female (D) builds a new population (E).

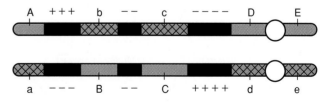

FIGURE 26.8
Model for open and closed regions of chromosomes. Products of crossing over anywhere within the open system (blue) result in offspring with high fitness, while crossovers in the closed regions (black) produce zygotes with low fitness. The letters represent genes or polygenes, and the pluses and minuses represent internally balanced gene complexes. Recombination within these blocks produces unfit gene combinations.

immigrant fertilized female belonging to an ancestral stock on Maui, chromosomally related to the present-day *D. planitibia*, crossed the Alenuihaha Channel to Hawaii. Subsequent flush-crash cycles led to the reconstruction of the colony's genotype, giving rise to the two species found on Hawaii—*D. heteroneura* and *D. silvestris* (Figure 26.9). An alternate theory proposes that one of the species on Hawaii could have arisen as the result of a second colonization from Maui. Similar evidence indicates that two other groups on Maui have given rise to a total of five species on Hawaii.

In laboratory tests of this theory, Jeffrey Powell has found that 15 generations after several flush-crash cycles, laboratory strains of *Drosophila* species showed significant prezygotic (behavioral) isolation from other strains. Subsequent testing several months later showed that these differences were not transitory. These experimental results are important in establishing that the first stages of speciation can occur rapidly under certain circumstances.

The third model of quantum speciation that we will discuss, Michael White's **statispatric speciation,** was originally derived to account for the evolution of flightless grasshoppers in Australia. In this model, a chromosomal aberration such as a translocation arises by chance in a small population. If heterozygotes for the translocation have a slightly reduced fitness, perhaps caused by abnormal meiosis, selection will favor either homokaryotype (two copies of the translocation or two normal chromosomes). The translocation homokaryotype, with a chromosome number

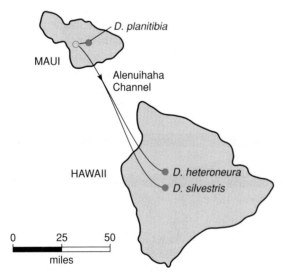

FIGURE 26.9
Proposed pathway of colonization of Hawaii by members of the *D. planitibia* species complex. Open circles represent a population ancestral to the three present-day species.

different from the original, may spread and partially displace the ancestral population. Semispecies would then be reproductively isolated because hybrids would have an unbalanced chromosome complement and a lowered fitness. Thus, in White's as well as Carson's model, reproductive isolation precedes the development of genetic variability. The statispatric model has also been applied to the development of closely related species of mole rats differing in chromosome number, the *Spalax ehrenbergi* complex.

The final example illustrating quantum speciation involves polyploidy in plants. The formation of species by polyploidy in animals is rare, but has been an important factor in the evolution of plants. It is estimated that one-half of all flowering plants have evolved by polyploidy. One such form of polyploidy is **allopolyploidy** (see Chapter 7), produced by doubling the chromosome number in an interspecific hybrid. If two species of related plants have the genetic constitution *SS* and *TT,* where *S* and *T* represent the haploid set of chromosome in each species, then the F_1 hybrid would have the chromosome constitution *ST*. Normally such a plant would be sterile because there are few or no homologous chromosome pairs, and aberrations would arise during meiosis. If the hybrid undergoes a spontaneous doubling of chromosome number, however, a tetraploid *SSTT* would be produced. This might occur in somatic tissue, giving rise to a partially tetraploid plant that would produce some tetraploid flowers. Alternatively, aberrant meiotic events may produce *ST* gametes, which when fertilized would yield *SSTT* zygotes. The *SSTT* plants would be fertile because they would possess homologous chromosomes producing viable *ST* gametes. This new, true-breeding tetraploid would have a combination of characters derived from the parental species, and would be reproductively isolated from them because F_1 hybrids would be triploids and consequently sterile. The tobacco plant *Nicotiana tabacum* ($2n = 48$) is the result of the doubling of the chromosome number in the hybrid between *N. otophora* ($2n = 24$) and *N. silvestris* ($2n = 24$).

MOLECULAR EVOLUTION

The diversity of life forms inhabiting the earth is overwhelming. Nearly two million species of plants and animals have been described, and surely there are many yet to be classified. Nevertheless, all organisms share the same molecular features: they are composed principally of carbon, hydrogen, nitrogen, and oxygen atoms; they use nucleic acids to store and transfer chemical information; and proteins are, for them, the indispensable products of the stored genetic information.

The recent development of techniques to analyze and sequence proteins and nucleic acids has allowed biologists to infer relatedness of organisms and to construct molecular phylogenetic trees. In this section we shall review some of the findings in this area of evolutionary study and examine the genetic variability demonstrated at the molecular level within populations.

Amino Acid Sequence Phylogeny

Evolutionary relatedness or divergence can be measured by comparing the amino acid sequences of proteins common to various organisms. The first protein to be sequenced was **insulin,** composed of only 51 amino acids. In the early 1950s, the amino acid sequence of insulin was examined in a variety of mammals, including cattle, pigs, horses, sperm whales, and sheep. With the exception of a stretch of three amino acids, the protein was shown to contain an identical sequence in each of these mammals.

Cytochrome c is another commonly investigated protein. It is a respiratory pigment found in the mitochondria of eukaryotes. The molecule consists of 104 amino acids in many vertebrates and a slightly higher number in most other organisms. Cytochrome c has changed very slowly during evolution. For example, the amino acid sequence in humans and chimpanzees is identical; between humans and rhesus monkeys only one amino acid is different. This is remarkable considering that lines leading to humans and monkeys diverged from a common ancestral form approximately 20 million years ago.

Table 26.5 shows the number of amino acid differences between a variety of organisms, using human cytochrome c as a standard. Even as distantly removed from humans as yeasts are, only 38 amino acids are different. In a comparison of a large number of organisms, more than 15 percent of the sequences remain unchanged.

In addition to the number of amino acid differences, it is possible to assess the minimum number of nucleotide changes that must have occurred during the evolution of a protein. This assessment requires knowledge of the genetic code and is a more refined analysis. For example, more than one nucleotide change may have been necessary to establish any given amino acid change found between two organisms. Thus, over evolutionary time, two or more independent mutations may have been essential in order to produce the observed change. When all nucleotide changes necessary for all amino acid differences are totaled, the **minimal mutational dis-**

TABLE 26.5
A comparison of the number of amino acid differences and the minimal mutational distance in cytochrome c.

Organism	Number of Amino Acid Differences	Minimal Mutational Distance
Human	0	0
Chimpanzee	0	0
Rhesus monkey	1	1
Rabbit	9	12
Pig	10	13
Dog	10	13
Horse	12	17
Penguin	11	18
Moth	24	36
Yeast	38	56

SOURCE: From W. M. Fitch and E. Margoliash. "Construction of Phylogenetic Trees," *Science* 155: 279–84, 20 January 1967. Copyright 1967 by the American Association for the Advancement of Science.

tance between any two species is established. Table 26.5 shows such an analysis of the genes coding for cytochrome c in ten organisms. One can see that, as expected, these values are larger than the corresponding number of amino acids separating humans from the other nine organisms.

Minimal mutational distance provides an estimation of evolutionary divergence between organisms. Furthermore, on the basis of sequence data, the order of derivation of species from common ancestors can be deduced. Thus, it is possible to construct **divergence dendrograms** or **phylogenetic trees** based on the analysis of amino acid sequences of a single protein from diverse organisms. This is perhaps the most fascinating application of this information to evolutionary study.

The underlying assumption of this analysis is that all present-day sequences from different species represent gene products that diverged from common ancestral sequences at various points in evolutionary time. By determining minimal mutational distances between all species under analysis and in which specific amino acids have changed, the computer program can establish the most likely relationships between the species. It can also determine the point in these relationships at which a now extinct ancestral sequence must have existed in order to lead to evolutionary divergence.

This information may be summarized in the form of a phylogenetic tree. The phylogenetic tree shown in Figure 26.10 is based on the sequences used to derive the data of Table 26.5. The number along each line represents the minimal mutational distance between any two points. The tree is plotted so that the ordinate represents proportional amounts of dis-

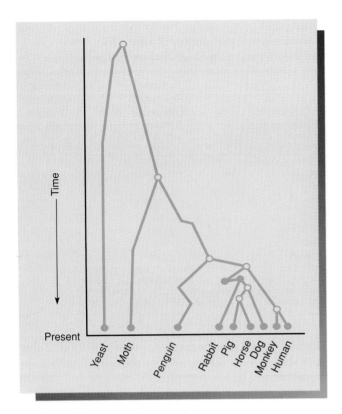

FIGURE 26.10
Phylogenetic sequence constructed by comparison of homologies in cytochrome c amino acid sequences.

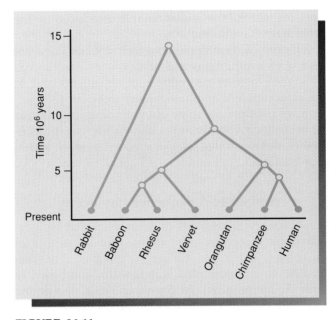

FIGURE 26.11
Phylogenetic sequence of carbonic anhydrase amino acid substitutions.

cule and all hemoglobin chains are proposed to have arisen from a common ancestral sequence. Furthermore, all four hemoglobin chains have a common origin.

Single chains such as the alpha and beta globins can also be compared between species. The differ-

tance. If a constant mutation rate is assumed, the ordinate represents a relative estimate of geologic time.

When phylogenetic trees are constructed in this way, they agree remarkably well with trees constructed using more conventional approaches such as morphological and paleontological evidence. Once a number of proteins from a variety of organisms are sequenced and analyzed simultaneously, even more accurate trees may be constructed.

When closely related species are to be examined in this way, proteins that have evolved more rapidly than cytochrome c are more useful. The 115-amino-acid protein carbonic anhydrase has been used to analyze more accurately the relationship between humans and several other primates (Figure 26.11).

This phylogenetic approach may also be applied to related molecules that have arisen during evolution through gene duplication. (The importance of duplication in producing genetic variation has been discussed in Chapter 7.) The various hemoglobin chains and myoglobin have been analyzed in this way. As in the comparison of species, each protein chain may be studied using comparative sequences. As shown in Figure 26.12, the oxygen-carrying myoglobin mole-

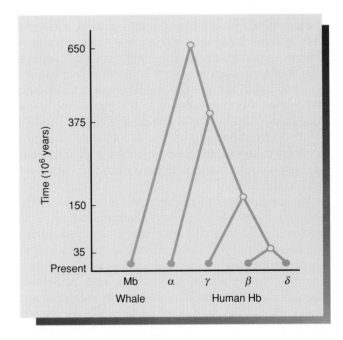

FIGURE 26.12
Phylogenetic sequence of myoglobin and hemoglobin proteins. These genes arose by duplication and subsequently diverged in amino acid sequence.

ences between humans and their close relatives are noteworthy. Human and chimpanzee alpha and beta chains are identical in sequence, but the gorilla differs from the human in one amino acid in each chain. Table 26.6 shows the number of amino acid differences between humans and a number of species.

All in all, amino acid sequence data have been extremely useful in evolutionary studies. These independent analyses complement other types of evolutionary evidence. Perhaps their greatest value is that these studies are performed directly at the level of genetic variation, which is the underlying resource supporting evolutionary change.

Nucleotide Sequence Phylogeny

The molecular hybridization of DNA from different sources has been a valuable technique in evolutionary studies. Although it has been discussed in Chapter 9 and is described in Appendix A, we shall briefly review it here.

Hybridization of nucleic acids is based on nucleotide sequence homology. Molecules are heated until they dissociate (melt) into single strands and are then allowed to reanneal by slow cooling of a given mixture. In reannealing, complementary sequences join together in a stable double-stranded form at the lower temperature. Molecular hybridization will occur between mixtures of DNA strands (DNA:DNA) or between mixtures of DNA and RNA strands (DNA:RNA).

In studies of sequence diversity between species, radioactive DNA from one species is prepared and

those sequences present once in the haploid genome (single copy fraction) are isolated. This single copy tracer is dissociated into individual strands and can be reassociated with melted single copy DNA from the same species (homologous reaction) or from other species (heterologous reaction). The difference in thermal stability (ΔT_m) between homologous and heterologous duplex molecules is a measure of the nucleotide sequence divergence between the two species. For example, a ΔT_m of 1°C roughly corresponds to 1 percent mismatching in nucleotide sequences. The results of experiments with two species of sea urchin are shown in Figure 26.13. From differences in thermal stability, it can be calculated that there is a sequence divergence between *Strongylocentrotus purpuratus* and *S. franciscanus* of about 19 percent. A similar experiment shows about 7 percent sequence diversity between *S. purpuratus* and another species, *S. drobachiensis*. From the data on nucleotide diversity and what evidence is available in the fossil record, it has been proposed that *S. purpuratus* diverged from *S. franciscanus* some 15 to 20 million years ago.

Where the nucleotide sequence divergence can be compared with other indicators of genetic variability, such as protein polymorphisms and chromosomal rearrangements, the evidence indicates that nucleotide sequence diversity is a more sensitive indicator

TABLE 26.6

A comparison of the alpha and beta hemoglobin chains between humans and other organisms.

Organism	Amino Acid Differences Between Humans and Various Organisms	
	α Chains	β Chains
Chimpanzee	0	0
Gorilla	1	1
Macaque	5	10
Mouse	19	31
Sheep	26	33
Pig	20	28
Horse	22	30
Rabbit	28	16
Chicken	45	—
Kangaroo	—	54
Carp	93	—
Lamprey	113	—

SOURCE: Data from various sources.

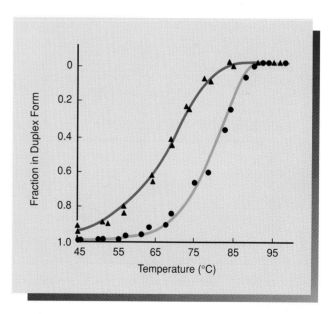

FIGURE 26.13

Measurement of nucleotide sequence diversity in the sea urchin. The thermal stability of *S. purpuratus* single copy DNA reassociated with *S. purpuratus* (circles) and *S. franciscanus* DNA (triangles) is measured. Lowered stability with *S. franciscanus* DNA is a measurement of nucleotide sequence divergence.

of evolutionary divergence than amino acid replacements. *Drosophila heteroneura* and *D. silvestris*, found only on the island of Hawaii, are estimated to have diverged only about 300,000 years ago, but it is difficult to demonstrate significant differences between these species in chromosomal inversion patterns or protein polymorphisms (Figure 26.14). Figure 26.15 shows the results obtained when labeled single copy DNA from *D. silvestris* is hybridized with itself and with DNA from *D. heteroneura* and *D. picticornis*. (*D. picticornis* is another member of the *planitibia* group and is found only on Kauai, a much older island.) The sequence diversity between the two species from Hawaii is about 0.55 percent, but *D. silvestris* and *D. picticornis* show a 2.1 percent difference in nucleotide sequences. Thus, nucleotide sequence diversity may precede the development of protein or chromosomal polymorphisms.

The fact that *D. heteroneura* and *D. sylvestris* share identical chromosome arrangements, have almost no detectable protein differences, are 99 percent homologous in DNA sequence, and yet are classified as separate species may seem paradoxical. However, the two species are clearly separated from each other by different and incompatible courtship and mating behaviors (a prezygotic isolating mechanism), by morphology, and by pigmentation of the body and wings (Figure 26.16). The available evidence suggests that these differences are controlled by a relatively small number of genes. In a genetic analysis of the differences in head shape and pigmentation, F. C. Val has estimated that only 15 to 19 loci are responsible for the morphological differences between these species, demonstrating that the process of speciation need only involve a small number of genes.

Problems similar to those encountered in measuring genetic differences in the Hawaiian *Drosophila* are also encountered in studies of higher primates. The hominoid primates include the chimpanzee, gorilla, orangutan, gibbon, and human. Past chromosomal studies and data on protein and isozyme differences have failed to accurately resolve the taxonomic relationships among the chimpanzee, gorilla, and humans, because they are so closely related. Using hybridization of single copy DNA sequences from a large number of individuals and using calibrations derived from the fossil record, Charles Sibley and Jon Ahlquist have clarified the evolutionary branching pattern in the hominoids and have estimated the times at which divergence occurred (Figure 26.17).

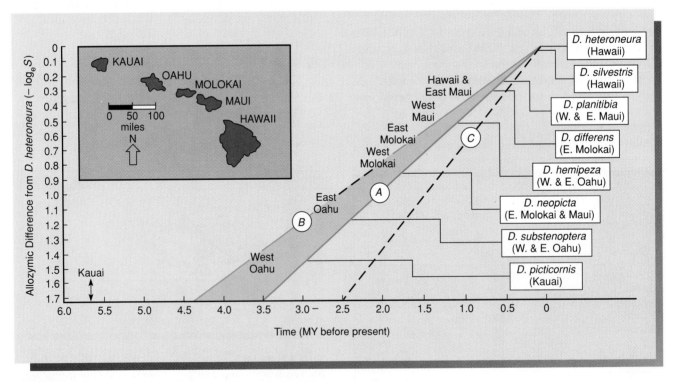

FIGURE 26.14
Detectable allozyme differences and proposed time of species origin in some Hawaiian *Drosophila* (line A). Line B assumes a slower accumulation of genetic differences and C is a faster accumulation. In any case, genetic differences based on enzymes in the most recently evolved species (*planitibia, sylvestris, heteroneura*) are difficult to quantify.

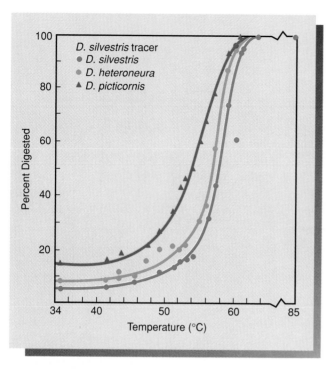

FIGURE 26.15
Nucleotide sequence diversity in the *planitibia* species complex. The degree of shift to the left by the heterologous hybrids is an indication of the degree of nucleotide diversity.

According to their data, humans and chimpanzees are more closely related than either is to the gorilla. The measure of relatedness is called the $\Delta T_{50}H$ and is related to the T_m discussed earlier. In man and chimpanzee, the $\Delta T_{50}H$ is 1.9, somewhat less than the value of 2.1 to 2.3 for the distance between the gorilla line and the chimpanzee/human line. Their

calculation of divergence from the chimpanzee line is 8 to 10 million years (MY) for the gorilla, 6.3 to 7.7 MY for humans, and 2.4 to 3.0 MY for the pygmy chimpanzee. The high degree of nucleotide homology, similarity in chromosome patterns, and protein homology are reminiscent of the situation in *D. heteroneura* and *D. sylvestris* and lend support to the notion that distinct species are not necessarily separated by large numbers of gene differences.

RFLP Analysis of Nucleotide Phylogenies

The question of where the human species originated has spawned several theories. Africa, Asia, Europe, and the Americas have been suggested as the site of human origin, and another theory proposes that modern humans arose simultaneously in several locations. The available fossil record indicates that *Homo sapiens* was present in Southern Africa more than 100,000 years ago, and was found in Asia at least 50,000 years ago, and it appears that *Homo sapiens* replaced Neanderthals in Europe about 30,000 to 40,000 years ago. The bulk of evidence from fossils supports the origin of humans in Africa, probably by phyletic evolution from *Homo erectus,* with a rapid spread to non-African regions. Over the past decade, molecular techniques, including RFLP analysis, have been used to provide additional evidence for the origin and spread of the human species across the globe.

In Chapter 16, we discussed the use of restriction fragment length polymorphisms (RFLPs) to map genes to specific chromosomes and/or chromosomal regions. RFLPs can be generated by single nucleotide base changes in a DNA sequence that is recognized by a restriction enzyme. Since restriction enzymes

(a)

(b)

FIGURE 26.16
(a) Differences in pigmentation patterns in *D. sylvestris* (left) and *D. heteroneura* (right).
(b) Head morphology in *D. sylvestris* (left) and *D. heteroneura* (right).

FIGURE 26.17

Phylogenetic sequence of the hominoid primates and the Old World monkeys. The numbers at the branch points are $\Delta T_{50}H$ measurements. The evolutionary branch points are dated by reference to the fossil record and nucleotide divergence.

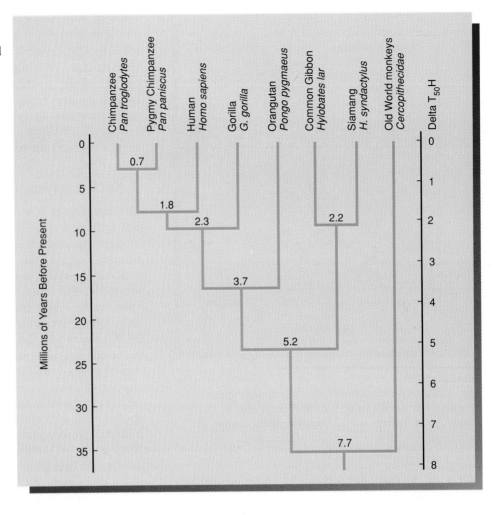

make double-stranded cuts in DNA, alterations in recognition sequences result in changes in the pattern of cuts made in the DNA. These detectable changes or polymorphisms are inherited in a codominant Mendelian fashion. RFLP analysis has been used to construct evolutionary branching patterns known as dendrograms and to reconstruct evolutionary events.

A study of restriction polymorphisms in mitochondrial DNA (which is maternally inherited) by Cavalli-Sforza and his colleagues first suggested that African populations are ancestral to others. More recent work on mitochondrial RFLPs has suggested that Caucasoid populations may have diversified from the ancestral African population and served as the source of all other groups.

Recently, RFLP analysis of nuclear genes has been used to study the evolutionary relationships among various human populations. Using five different restriction sites clustered within the beta globin complex, as markers, J. S. Wainscoat and colleagues studied the distribution of these sites in eight population groups. In all, 32 combinations of these five

restriction sites are possible, and 14 of these were actually observed in the populations surveyed. Three combinations of restriction sites (known as haplotypes) are common in non-African populations, and are rare in African populations. These haplotypes are represented as $+ - - - -$, $- + - + +$, and $- + + - +$. The African populations studied have high frequencies of two haplotypes: $- - - - +$ and $- + - - +$, both of which are rare or absent in all other populations studied.

Using the fossil evidence and the RFLP data, it is possible to construct a dendrogram for the eight populations studied (Figure 26.18). The model is consistent with the suggestion that *Homo sapiens* originated in Africa, and that a small founder population migrated from Africa and later gave rise to all non-African populations. Further RFLP screening of other nuclear genes will help to resolve the issues raised in this study, and to answer other questions about evolutionary events, such as the size of the founding population, and the time scale over which these evolutionary events may have transpired.

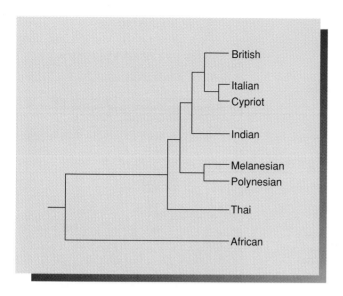

FIGURE 26.18
Evolutionary dendrogram of human populations constructed from RFLP analysis of alleles adjacent to the beta globin gene.

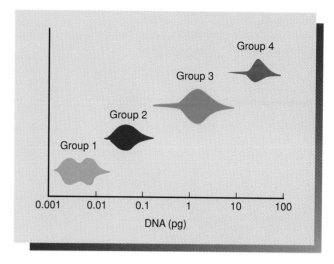

FIGURE 26.19
Classes of organisms grouped according to their DNA content. Group 1, bacteria; Group 2, fungi; Group 3, most animals and some plants; Group 4, many plants, salamanders, and some fish.

Evolution of Genome Size

An early study on the relationship between cellular DNA content and evolution by Alfred Mirsky and Hans Ris in 1951 concluded that

1 Nuclear DNA content increased from the lower to higher invertebrates.
2 Closely related organisms usually have similar amounts of DNA.
3 The development of terrestrial vertebrates has been accompanied by reductions in nuclear DNA content.

DNA content measurements from a wide range of species are now available, and organisms can be classified into four somewhat overlapping groups, as shown in Figure 26.19. The lowest DNA content in free-living organisms is found in bacteria (Class 1), with a range between 0.003 and 0.01 picogram (pg) per cell. Fungi (Class 2) average less than 0.1 pg. Class 3 includes most animals and plants, including sea urchins, reptiles, birds, and mammals. Class 4 is composed of organisms whose genomes are larger than 10 pg, including salamanders, some fish, and many plant species. The clustering of haploid DNA content in higher organisms and the clustering of related organisms are shown in Figure 26.20.

Although the relationship between DNA content and evolution is somewhat complex, several general statements can be made. A large increase in DNA content occurred during evolution from bacteria to higher plants. Few higher eukaryotes have genomes smaller than 0.1 pg, suggesting that a minimum DNA content is necessary to support this level of organization. Interestingly, while there is no necessary correlation between DNA content and chromosome number, there is a direct relationship between DNA content and nuclear and cellular size. In fact, a rough estimate of DNA content can be made from measurements of nuclear size.

Several explanations have been offered for the trend toward increases in DNA content per haploid genome during evolution, including the development of control system redundancy, generation of repetitive DNA sequences, and slower cell division and development. Unfortunately, we are not sure whether increases in DNA content are the raw material for selection or are produced as a byproduct. An examination of genome size in closely related organisms suggests that changes in DNA content have occurred mainly by duplication of small segments of DNA rather than by saltational increases associated with polyploidy. Figure 26.21 shows the distribution of DNA content in teleost fishes. If changes had arisen through polyploidy, a distribution into discrete classes reflecting multiples of a basic genome size would be expected. However, the differences in DNA content are seen to be the result of small incremental increases. Such studies confirm the role of gene duplication in evolution.

FIGURE 26.20
Haploid genome size clusters in higher organisms.

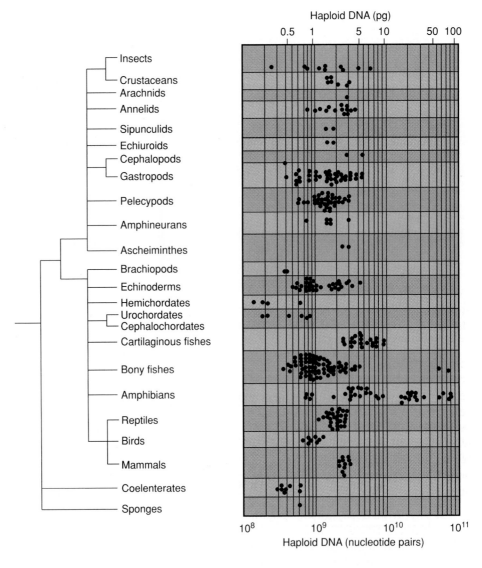

Haploid DNA (pg)

Insects
Crustaceans
Arachnids
Annelids
Sipunculids
Echiuroids
Cephalopods
Gastropods
Pelecypods
Amphineurans
Ascheiminthes
Brachiopods
Echinoderms
Hemichordates
Urochordates
Cephalochordates
Cartilaginous fishes
Bony fishes
Amphibians
Reptiles
Birds
Mammals
Coelenterates
Sponges

Haploid DNA (nucleotide pairs)

Evolution of Gene Families by Duplication

By examining functionally distinct but sequence-related genes that arose by duplication, we can study the evolution of gene function and the development of regulatory systems. The origin of the globin genes from an ancestral oxygen-carrying molecule has already been mentioned. Cloning and nucleotide sequencing have produced a great deal of information about the evolution of this gene family. Figure 26.22 summarizes what is known about the organization of the human alpha- and beta-globin gene families. Several features of these gene clusters are worth noting:

1 Both families consist of several genes packed together over a relatively short distance, all of which are oriented in the same direction with 5′ ends to the left and 3′ ends to the right.

2 In addition to sequences that code for all known globin polypeptides, several related gene sequences called **pseudogenes** are present.
3 The genes in each family are arranged in the order in which they are expressed in development.

In the beta-globin cluster, the embryonic epsilon form is first, followed by the fetal gamma genes, and the adult beta and delta genes. It is not known whether this order is related to mechanisms regulating gene expression or is a reflection of evolutionary history.

The complete DNA sequences for all these genes are now available. A phylogenetic tree based on these sequences is shown in Figure 26.23. According to this tree, which was constructed from differences in nucleotide composition, the original duplication that gave rise to the alpha and beta gene clusters occurred some 500 million years ago. The adult sequences diverged from the prenatal forms about 200 MY ago,

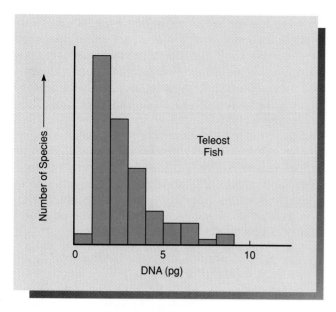

FIGURE 26.21
Haploid genome size in teleost fish.

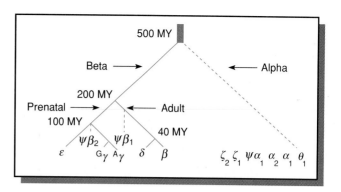

FIGURE 26.23
Phylogenetic sequence of globin genes based on nucleotide sequence of coding regions. Dotted line indicates relationships not yet clarified.

sometime during the evolution of reptiles, which are ancestral to mammals. Separate embryonic and fetal sequences diverged at the beginning of the mammalian radiation some 100 MY ago, and this event may be linked to the development of placental mammals with their unique problems of oxygen transport. The adult beta and delta genes diverged some 40 MY ago, before the separation of higher primates into two lines—one leading to the New World monkeys and

the other to the Old World monkeys, including great apes and humans.

Comparison of the nucleotide sequences in the coding and noncoding regions of the genes indicates that these regions have continued to diverge by base substitutions as well as by duplication and deletion of short regions. An unexpected outcome of this analysis has been the detection of pseudogenes in each gene family. **Pseudogenes** are stretches of DNA that have some sequence homology with protein-coding regions but are themselves inactive. Such sequences can arise as a result of increased crossing over between homologous chromosome regions, or by reverse transcription from mRNA. These extra copies—free from selection because they have no phenotypic effect—can accumulate mutations that would cause the gene to become inactive, but retain many of the features associated with active genes such as promoter sites, splicing sequences, and poly-A addition sites. In the human globin gene families, pseudogenes separate the embryonic and adult alpha genes, and the prenatal and adult beta genes. The significance and function of such pseudogenes are matters of intense speculation. Oliver Smithies has proposed that because of their location in human globin families, pseudogenes modulate the expression of the protein-producing alpha and beta genes. Philip Leder and his colleagues, on the other hand, have shown that pseudogenes of the alpha-globin complex in the mouse are located on different chromosomes from the active genes. Dispersed single pseudogenes (called **orphons**) derived from tandemly repeated families, including histone and ribosomal genes, have recently been found in a wide range of organisms. Drawing on these results, Leder suggests that the genomes of higher eukaryotes are mosaics of

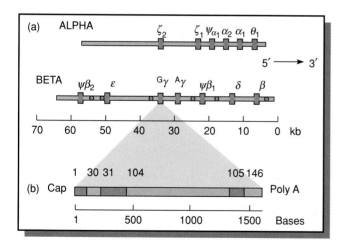

FIGURE 26.22
(a) Organization of the alpha- and beta-globin genes in humans. The genes are represented as rectangles, and repeated DNA sequences within the beta-globin gene cluster are shown as circles. (b) The coding (solid) and noncoding regions of the $^G\gamma$-globin gene. The top numbers are amino acid residues.

dispersed pseudogenes, which serve as a reservoir of sequences that can evolve new function. This interpretation implies that gene duplication leading to the formation and dispersal of pseudogenes has been a critical factor in the evolution of higher organisms.

THE ROLE OF MUTATION IN EVOLUTION

Selectionists versus Neutralists

When we discussed the concept of the adaptive norm in an earlier section, it was pointed out that there is a controversy over the importance of extensive genetic variation within a population. This controversy centers particularly on the variation detected at the molecular level. On the one hand, the **classical hypothesis** proposes that natural selection favors the fixation of the most favorable alleles at each locus. Homozygosity therefore should be the rule rather than the exception. On the other hand, the **balance hypothesis** favors the maintenance through natural selection of a high degree of heterozygosity in populations.

The classical hypothesis is supported by the observation that induced mutations almost invariably lower fitness. Therefore, if most mutations are detrimental, the accumulation of genetic variability will result in a substantial genetic burden to the fitness of a population. According to the classical theory, this detrimental variation will be purged by natural selection.

In contrast, Dobzhansky and his colleagues have provided the initial support for the balance hypothesis. Having observed a high degree of heterozygosity in adapted *Drosophila* populations, they postulated that populations displaying genetic polymorphism will have added fitness. The observation of protein polymorphisms by Lewontin in *Drosophila* and by Henry Harris in humans, and the evidence for numerous amino acid substitutions in single proteins throughout evolution, attest to a high degree of heterozygosity, particularly at the molecular level, in many diploid and polyploid populations. These findings do not confirm the balance hypothesis, however. Instead, they lend support to it and argue against the classical theory.

However, the existence of a high degree of genetic variation does not prove that it is the basis of evolutionary fitness. A third theory, the **neutralist hypothesis** proposed by James Crow and Motoo Kimura, argues that mutations leading to amino acid substitutions are rarely favorable. They are more often detrimental, but most often neutral or genetically equivalent to the allele that is replaced. Those that are favorable or detrimental are either preserved or removed from the population, respectively, by natural selection. However, neutral genetic changes will not be affected by selection and will instead become randomly fixed in the population. Their frequency will be determined by the rate of mutation and the principles of random genetic drift.

The neutralist interpretation, as most recently articulated by Kimura, is based on several observations and theoretical considerations. These ideas focus on several important aspects of genetic variation as an evolutionary phenomenon. The neutralist theory is based on the following points:

1 The rate of amino acid substitution in a given protein is about the same in organisms with diverse lineages.
2 The types of substitutions do not seem to demonstrate any particular pattern, but are instead random.

With regard to these observations, the neutralists believe that if selection favored the amino acid changes even slightly, the rate would vary in different organisms exposed to different selective pressures. The substitutions would therefore not be random.

3 The overall rate of mutation leading to amino acid substitutions is relatively high (it is about equal to the substitution of one nucleotide per genome every two years in mammals) and has remained relatively constant for a long period of time.

For example, when nucleotide substitution as estimated from amino acid changes in seven proteins from 16 pairs of mammals is estimated over the last 150 MY, the points fall close to a straight line. A constant rate of change is indicated. The neutralists argue that if natural selection were acting on all such mutations, the rate of fixation would not be constant in fluctuating environments over time.

4 The number of amino acid substitutions is higher in the less important parts of molecules (those parts not critical to tertiary structure and the active site of proteins) and in molecules less critical to the organism.

This argument centers on **functional constraint**. That is, those regions that determine whether the molecule functions or not are less able to tolerate change. The regions of lesser importance—those determining only how well the molecule functions—can tolerate greater amounts of change. It is argued that if such changes in less constrained regions of proteins are neutral, the rate of amino acid substitution in these regions will be substantially higher than in constrained regions. The same argument is applicable to total proteins of lesser importance to organisms. Examination of less-constrained regions of proteins and less-critical proteins does show a higher amino acid substitution rate. This is interpreted by the neutralists as support for their contentions.

5 The rate of amino acid substitution is much too high to be accounted for by selection.

If all heterozygosity within a population represented alleles with even a slight advantage, the neutralists argue that the **cost of selection** necessary to fix and maintain them in populations would be astronomically high. The cost of selection is the genetic death of those organisms that must be replaced by those bearing the more favorable change. Mathematical calculations can be made to support this theory.

The neutralist theory runs counter to the theory of natural selection as set down by Darwin and as expanded by modern knowledge of genetics. In the 1960s there was general agreement that all biological characteristics could be interpreted as arising by adaptive evolution through natural selection. In this sense, most mutations that have survived through evolution should be at least slightly adaptive in homozygotes, heterozygotes, or both.

Those who oppose the neutralist theory have been called **selectionists**. They point out examples where enzyme or protein polymorphism is associated with adaptation to certain environmental conditions. The well-known advantage of sickle-cell anemia carriers in malarial regions is such an example.

Selectionists also stress that enzyme polymorphism may often appear to offer no advantage, but exists in such a frequency that it is impossible to explain as a random occurrence. Thus, even though no currently available analytical technique can detect any physiological difference, it cannot be proved that some slight advantage associated with any given amino acid substitution does not exist.

On many specific points of contention, selectionists are able to offer persuasive theoretical arguments and calculations. For example, although the nucleotide substitution rate appears on the average to be constant, localized variation is sometimes as much as 2.5 times higher than expected by chance.

Even though this controversy is highly theoretical and esoteric, it is important that we not lose sight of several factors. For example, the neutralists do not discount natural selection as a guiding force in evolution. Rather, they suggest that some features of organisms' genotypes are nonadaptive, fluctuate randomly, and may have been fixed by genetic drift. On the other hand, the selectionists certainly do not deny that genetic drift is an important factor in establishing gene frequencies.

Finally, it should be pointed out that the two theories are not mutually exclusive. It is difficult to argue against the notion that some genetic variation must be neutral. The difference between the two theories is in the degree of neutrality that exists. While current data are insufficient to resolve the problem one way or the other, one important point has emerged from the arguments. In considering natural selection, there are clearly two levels that must be examined: the phenotypic level, including all morphological and physiological characteristics imparted by the genotype, and the molecular level, represented by the precise nucleotide and amino acid sequences of DNA and proteins. There is no question about selection acting at the phenotypic level. The extent to which selection occurs at the molecular level, however, is in question.

Recent Trends

One of the major premises of the classical theory of speciation is that mutations (along with recombination) are the source of all variation. In this context, new alleles are inherited in Mendelian fashion and may gradually increase in frequency if they confer greater fitness in accord with the principles of population genetics. Speciation involves a large number of allelic substitutions sequentially incorporated into the gene pool. We have already seen that the concept of gradualism is being challenged by chromosomal theories of quantum speciation. Work in molecular biology, particularly the discovery of intervening sequences, transposable elements, and pseudogenes, is causing fundamental changes in our concepts of genome structure, stability, and mutation.

The genome of higher organisms has been regarded as a largely static entity, with low rates of mutation and chromosome aberration. We now know that the genome is highly dynamic and in constant flux, with families of repeated sequences being created, fixed, deleted, and turned over. Previously, mutation has been thought of in terms of nucleotide substitution or, at most, small duplications or deletions. However, nucleotide sequence analysis of the *bithorax* gene complex in *Drosophila* has shown that most mutations in these genes involve the insertion or deletion of pieces of DNA several thousand base pairs in length. Insertion of transposable elements into genes produces mutant phenotypes, and their removal produces reversion to wild type. In addition, since many transposable elements contain start and stop signals for transcription, the dispersal and mobility of such elements in the genome can create new patterns of gene expression in a single generation.

The evolutionary role of regulatory mutations in development is also receiving increased attention. Many workers now believe that mutations that change the timing of gene expression in embryogenesis can have a significant impact on the rates of speciation. Such mutations would not change the frequency of protein polymorphisms in a population, but rather the order or duration of gene expression during embryonic or preadult development, producing instant morphological variation. Such a mutant has already been described in the *bithorax* region of *Drosophila*. By moving a gene some 40 kb within the complex, the mutant changes the gene's time of expression during development. This gene, *Cbx*, produces a fly with a greatly altered phenotype.

Finally, pseudogenes are seen as possible reservoirs of inactive genes that can be rapidly mobilized to perform new functions. In sum, the classical theory of speciation appears ready to undergo a new and dramatic synthesis, and it is likely that this synthesis will arise from within molecular biology.

CHAPTER SUMMARY

1   Speciation, which depends on genetic variation and forces controlling its distribution (e.g., natural selection), is initiated when a population becomes separated into smaller, reproductively semi-isolated breeding groups.

2   The partitioning of a population's gene pool results in a better fitness for each group in its new environment. Races form as each group acquires its own unique set of genetic variations. Races may become reproductively isolated from one another and form new species as evolution proceeds. The formation of races and speciation are illustrated by several studies of *Drosophila*.

3   Two approaches have been especially fruitful in evolutionary study at the molecular level: (1) comparison of complementary sequences present in the DNA of different organisms, and (2) comparison of amino acid substitutions in proteins common to a variety of organisms. Such comparisons provide not only a measure of evolutionary relatedness, but further allow the assessment of genetic variation during evolution.

4   The central question concerning the importance of variation at the molecular level is whether or not all nucleotide and amino acid changes preserved through evolution in DNA and proteins are adaptive to the population. Some believe that the majority of these changes are neutral or genetically equivalent, while others adhere to the theory of adaptive evolution through natural selection.

Insights and Solutions

1   Sequence analysis of DNA can be accomplished by a number of techniques (see Appendix A for a detailed description). Protein sequencing, on the other hand, is made more complex by the fact that twenty different subunits need to be unambiguously identified and enumerated, rather than the four nucleotides of DNA. Because of their unique properties, the N-terminal and C-terminal amino acids in a protein are easy to identify, but the array in between offer a difficult challenge, since many proteins contain hundreds of amino acids. How is it that protein sequencing is accomplished?

ANSWER: The overall strategy for protein sequencing is the same as for DNA sequencing: divide and conquer. To accomplish this, specific enzymes are used that reproducibly cleave proteins between certain amino acids. The use of different enzymes produces overlapping fragments. Each fragment is isolated and its amino acid sequence is determined by chemical means. The sequences from overlapping fragments are then assembled into the amino acid sequence for the entire protein.

2 A single plant twice the size of others in the same population suddenly appears. Normally, plants of this species reproduce by self-fertilization and by cross-fertilization. Is this new giant plant simply a variant, or could it be a new species? How would you determine this?

ANSWER: One of the most widespread mechanisms of speciation in higher plants is polyploidy, the multiplication of entire chromosome sets. The result of polyploidy is usually a larger plant with larger flowers and seeds. There are two ways of testing this new variant to determine whether it is a new species. First, the giant plant should be crossed with a normal-sized plant to see if it produces viable, fertile offspring. If it does not, the two different types of plants would appear to be reproductively isolated. Secondly, the giant plant should be cytogenetically screened to examine its chromosome complement. If it has twice the number of its normal-sized neighbors, it is a tetraploid that may have arisen spontaneously. If the chromosome number differs by a factor of two, and the new plant is reproductively isolated from its normal-sized neighbors, it is a new species.

PROBLEMS AND DISCUSSION QUESTIONS

1 Discuss the rationale behind the statement that inversions in chromosome 3 of *Drosophila pseudoobscura* represent genetic variation.

2 Describe the process of race formation. What is the role of natural selection?

3 Contrast the classical and balance hypotheses as they relate to the adaptive norm.

4 What types of nucleotide substitutions will not be detected by electrophoretic studies of a gene's protein product?

5 In a sequencing experiment (using the numbers 1 through 6 to represent amino acids), the following two sets of peptide fragments were obtained in independent experiments with different proteolytic enzymes:

Proteins	Fragments
Enzyme I	624 24635 135
Enzyme II	13 6356 524 24

Determine the sequence of fragments and amino acids in the protein.

6 Shown below are two homologous lengths of the alpha and beta chains of human hemoglobin. Consult the genetic code dictionary (Figure 11.8) and determine how many amino acid substitutions may have occurred as a result of a single nucleotide substitution. For any that cannot occur as the result of a single change, determine the minimal mutational distance.

Alpha:	Ala	Val	Ala	His	Val	Asp	Asp	Met	Pro
Beta:	Gly	Leu	Ala	His	Leu	Asp	Asn	Leu	Lys

7 Determine the minimal mutational distances between the following amino acid sequences of cytochrome c from various organisms. Compare the distance between humans and each organism.

Human:	Lys	Glu	Glu	Arg	Ala	Asp
Horse:	Lys	Thr	Glu	Arg	Glu	Asp
Pig:	Lys	Gly	Glu	Arg	Glu	Asp
Dog:	Thr	Gly	Glu	Arg	Glu	Asp
Chicken:	Lys	Ser	Glu	Arg	Val	Asp
Bullfrog:	Lys	Gly	Glu	Arg	Glu	Asp
Fungus:	Ala	Lys	Asp	Arg	Asn	Asp

8 The genetic difference between *D. heteroneura* and *D. sylvestris* as measured by nucleotide diversity is about 1.8 percent. The difference between chimpanzees

(*Pan troglodytes*) and humans (*Homo sapiens*) is about the same, yet the latter species are classified in different genera. In your opinion, is this valid? If so, why; if not, why not?

9 As an extension of the previous question, consider the following: In sorting out the complex taxonomic relationships among birds, species with $\Delta T_{50}H$ values of 4.0 are placed in the same genus, even by traditional taxonomy based on morphology. Using the data in Figure 26.17, construct a phylogeny that obeys this rule, using any of the appropriate genus names (*Pongo, Pan, Homo*), or constructing new ones.

10 The use of nucleotide sequence data to measure genetic variability is complicated by the fact that the genes of higher eukaryotes are complex in organization and contain 5' and 3' flanking regions as well as introns. Slightom and colleagues have compared the nucleotide sequence of two cloned alleles of the gamma-globin gene from a single individual and found a variation of 1 percent. Those differences include 13 substitutions of one nucleotide for another, and three short DNA segments that have been inserted in one allele or deleted in the other. None of the changes take place in the exons (coding regions) of the gene. Why do you think this is so, and should it change the concept of genetic variation?

11 Discuss the arguments supporting the neutral mutation theory. What counterarguments are proposed by the selectionists?

12 Of what value to our understanding of genetic variation and evolution is the debate concerning the neutral mutation theory?

SELECTED READINGS

ANDERSON, W., et al. 1975. Genetics of natural populations, XLII. Three decades of genetic change in *Drosophila pseudoobscura*. *Evolution* 29: 24–36.

AYALA, F. J. 1976. *Molecular evolution*. Sunderland, MA: Sinauer Associates.

———. 1984. Molecular polymorphism: How much is there, and why is there so much? *Dev. Genet.* 4: 379–91.

BACHMAN, K. O. B., et al. 1972. Nuclear DNA amounts in vertebrates. In *Evolution of genetic systems*, ed. H. B. Smith. Brookhaven Symposium in Biology, vol. 23.

BARTON, N. H., and HEWITT, G. M. 1989. Adaptation, speciation and hybrid zones. *Nature* 341: 497–503.

BERNARDI, G., MOUCHIROUD, D., GAUTIER, C., and BERNARDI, G. 1988. Compositional patterns in vertebrate genomes: Conservation and change in evolution. *J. Mol. Evol.* 28: 7–18.

BRITTEN, R. J., and DAVIDSON, E. H. 1971. Repetitive and non-repetitive DNA sequences and a speculation on the origin of evolutionary novelty. *Quart. Rev. Biol.* 46: 111–38.

CARSON, H. 1970. Chromosome tracers of the origin of species. *Science* 168: 1414–18.

———. 1975. The genetics of speciation at the diploid level. *Amer. Natur.* 109: 83–92.

DAYHOFF, M. O. 1969. Computer analysis of protein evolution. *Scient. Amer.* (July) 221: 86–95.

DOBZHANSKY, T. 1947. Adaptive changes induced by natural selection in wild populations of *Drosophila*. *Evolution* 1: 1–16.

———. 1948. Genetics of natural populations, XVI. Altitudinal and seasonal changes produced by natural selection in certain populations of *Drosophila pseudoobscura* and *Drosophila persimilis*. *Genetics* 33: 158–76.

———. 1955. *Genetics of the evolutionary process*. New York: Columbia University Press.

DOBZHANSKY, T., et al. 1966. Genetics of natural populations, XXXVIII. Continuity and change in populations of *Drosophila pseudoobscura* in Western United States. *Evolution* 20: 418–27.

DOVER, G. 1982. Molecular drive: A cohesive mode of species evolution. *Nature* 299: 11–17.

DOVER, G. A., and FLAVELL, R. B., eds. 1982. *Genome evolution*. Orlando: Academic Press.

EFSTRATIADIS, A., et al. 1980. The structure and evolution of the human beta-globin gene family. *Cell* 21: 653–68.

FITCH, W. M. 1973. Aspects of molecular evolution. *Ann. Rev. Genet.* 7: 343–80.

FITCH, W. M., and MARGOLIASH, E. 1967. Construction of phylogenetic trees. *Science* 155: 279–84.

———. 1970. The usefulness of amino acid and nucleotide sequences in evolutionary studies. *Evol. Biol.* 4: 67–109.

GOULD, S. J. 1982. Darwinism and the expansion of evolutionary theory. *Science* 216: 380–87.

HALL, T., GRULA, J., DAVIDSON, E. H., and BRITTEN, R. J. 1980. Evolution of sea urchin non-repetitive DNA. *J. Mol. Evol.* 16: 95–110.

HINEGARDNER, R. 1976. Evolution of genome size. In *Molecular evolution*, ed. F. J. Ayala, pp. 179–99. Sunderland, MA: Sinauer Associates.

HUNT, J., et al. 1981. Evolution distance in Hawaiian *Drosophila*. *J. Mol. Evol.* 17: 361–67.

JEFFREY, A. 1979. DNA sequence variation in the $^{G}\gamma$-, $^{A}\gamma$-, δ- and B-globin genes of man. *Cell* 18: 1–10.

JENKINS, N., COPELAND, N. G., TAYLOR, B. A., and LEE, B. K. 1981. Dilute(d) coat colour mutation of DBA.2 J mice is associated with the site of integration of an ecotropic MuLV genome. *Nature* 293: 370–74.

JUKES, T. H. 1966. *Molecules and evolution*. New York: Columbia University Press.

KIMURA, M. 1979a. Model of effectively neutral mutations in which selective constraint is incorporated. *Proc. Natl. Acad. Sci.* 76: 3440–44.

———. 1979b. The neutral theory of molecular evolution. *Scient. Amer.* (Nov.) 241: 98–126.

———. 1989. The neutral theory of molecular evolution and the world view of the neutralists. *Genome* 31: 24–31.

KING, M. C., and WILSON, A. C. 1975. Evolution at two levels: Molecular similarities and biological differences between humans and chimpanzees. *Science* 188: 107–16.

KOHNE, D. E., CHISCON, J. A., and HOYER, B. H. 1972. Evolution of primate DNA sequences. *J. Hum. Evol.* 1: 627–44.

LANDE, R. 1989. Fisherian and Wrightian theories of speciation. *Genome* 31: 221–27.

LEDER, A., SWAN, D., RUDDLE, F. H., D'EUSTACHIO, P., and LEDER, P. 1981. Dispersion of alpha-like globin genes of the mouse to three different chromosomes. *Nature* 293: 196–200.

LEWONTIN, R. C., and HUBBY, J. L. 1966. A molecular approach to the study of genic heterozygosity in natural populations. II. Amount of variation and degree of heterozygosity in natural populations of *Drosophila pseudoobscura*. *Genetics* 54: 595–609.

MAYR, E. 1963. *Animal species and evolution*. Cambridge, MA: Harvard University Press.

NAGEL, R. L., and LABIE, D. 1989. DNA haplotypes and the beta s globin gene. *Prog. Clin. Biol. Res.* 316B: 371–93.

PAGEL, M. D., and HARVEY, P. H. 1989. Comparative methods for examining adaptation depend on evolutionary models. *Folia Primatol.* 53: 203–20.

POWELL, J. 1978. The founder-flush speciation theory: An experimental approach. *Evolution* 32: 465–74.

ROBERTS, D. F. 1988. Migration and genetic change. *Hum. Biol.* 60: 521–39.

SIBLEY, C., and AHLQUIST, J. 1984. The phylogeny of the hominoid primates, as indicated by DNA-DNA hybridization. *J. Mol. Evol.* 20: 2–15.

SMITH, J. M., ed. 1982. *Evolution now: A century after Darwin*. New York: W. H. Freeman.

SMITHIES, O., BLECHL, A. E., SHEN, S., SLIGHTOM, J. L., and VANIN, E. F. 1981. Co-evolution and control of globin genes. In *Levels of genetic control in development*, eds. S. Subtelny and U. Abbot, pp. 185–200. New York: Alan R. Liss.

STEBBINS, G. L. 1977. *Processes of organic evolution*. 3rd ed. Englewood Cliffs, NJ: Prentice-Hall.

STEBBINS, G. L., and AYALA, F. J. 1981. Is a new evolutionary synthesis necessary? *Science* 213: 967–71.

TASHIAN, R. E., and CARTER, N. D. 1976. Biochemical genetics of carbonic anhydrase. In *Advances in human genetics*, eds. H. Harris and K. Hirschhorn, pp. 1–56. New York: Plenum Press.

TEMPLETON, A. R. 1985. Phylogeny of the hominoid primates: A statistical analysis of the DNA-RNA hybridization data. *Mol. Biol. Evol.* 2: 420–33.

VAL, F. C. 1977. Genetic analysis of the morphological differences between two interfertile species of Hawaiian *Drosophila*. *Evolution* 31: 611–29.

WHITE, M. J. D. 1977. *Modes of speciation*. New York: W. H. Freeman.

YUNIS, J. J., and PRAKASH, O. 1982. The origin of man: A chromosomal pictorial legacy. *Science* 215: 1525–30.

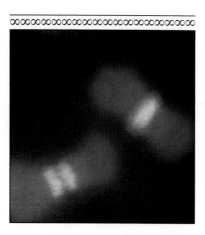

A
Experimental Methods

In addition to the techniques of genetic analysis, physical and chemical techniques for the separation and analysis of macromolecular components of the cell nucleus and cytoplasm have been instrumental in advancing our understanding of genetics at the molecular level. In this appendix we will describe the background and theoretical basis of some techniques that have been important in molecular genetics.

ISOTOPES

Isotopes are forms of an element that have the same number of protons and electrons but differ in the number of neutrons contained in the atomic nucleus. For example, the most common form of carbon has an atomic number of 6 (the number of protons in the nucleus) and an atomic weight of 12 (the sum of the protons and neutrons in the nucleus). In a very small percentage of carbon atoms, a seventh neutron is present, producing an atom with an atomic weight of 13. This is an example of a so-called **heavy isotope**. Since the number of protons and electrons, and thus the net charge, has not changed, the atom has the same chemical properties as **carbon-12** (^{12}C) and differs only in mass. **Carbon-13** (^{13}C) is thus a stable, heavy isotope of carbon.

Although the addition of neutrons does not alter the chemical properties of an atom, it can produce instabilities in the atomic nucleus. If another neutron is added to a carbon-13 atom, the isotope **carbon-14** (^{14}C) results. However, the presence of eight neutrons and six protons is an unstable condition, and the atom undergoes a nuclear reaction in which radiation is emitted during the transition to a more stable condition. Therefore, carbon-14 is a **radioactive isotope** of carbon.

The type of radiation emitted and the rate at which these nuclear events take place are characteristic of the element. Table A.1 lists types of radioactivity. The rate at which a radioactive isotope emits radiation is expressed as its **half-life,** which is the time required for a given amount of a radioactive substance to lose one-half of its radioactivity. Table A.2 lists some of the isotopes available for use in research.

DETECTION OF ISOTOPES

The choice of which isotope to use as a tracer in biological experiments depends on a combination of its physical and chemical properties, which enable the investigator to quantitate the amount of radioactivity or measure the ratio of heavy to light isotopes.

TABLE A.1
Properties of ionizing radiation.

Type	Relative Penetration	Relative Ionization	Range in Biological Tissue
Alpha particle (2 protons + 2 neutrons)	1	10,000	Microns
Beta particle (electron)	100	100	Microns−mm
Gamma ray	>1000	<1	∞

TABLE A.2
Some isotopes used in research.

Element	Isotope	Half-life	Radiation
H	^{2}H	—	Stable
	^{3}H	12.3 years	β
C	^{13}C	—	Stable
	^{14}C	5700 years	β
N	^{15}N	—	Stable
O	^{18}O	—	Stable
P	^{32}P	14 days	β
S	^{35}S	87 days	β
K	^{40}K	1.2×10^{9} years	β, gamma
Fe	^{59}Fe	45 days	β, gamma
I	^{125}I	60 days	β, gamma
	^{131}I	8 days	β, gamma

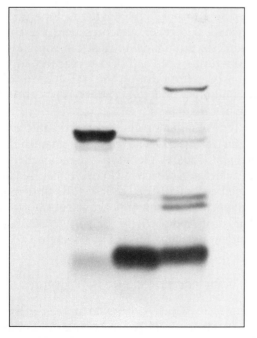

FIGURE A.1
Autoradiogram of radioactive bacterial proteins synthesized in a minicell system.

For the detection of heavy isotopes, two methods are commonly employed: **mass spectrometry** and **equilibrium density gradient centrifugation** (discussed in the following section). Although the use of heavy isotopes has been more restricted than that of radioisotopes, they have been instrumental in several basic advances in molecular biology, e.g., demonstrating the semiconservative nature of DNA replication and the existence of messenger RNA (mRNA).

There are a number of methods to detect radioisotopes, the foremost being **liquid scintillation spectrometry,** which provides quantitative information about the amount of radioactive isotope present in a sample, and **autoradiography,** which is used to demonstrate the cytological distribution and localization of radioactively labeled molecules.

In recording radioactivity by liquid scintillation counting, a small sample of the material to be counted is solubilized and immersed in a solution containing a **phosphor,** an organic compound that emits a flash of light after it absorbs energy released by decay of the radioactive compound. The counting chamber of the liquid scintillation spectrometer is equipped with very sensitive photomultiplier tubes that record the light flashes emitted by the phosphor. The data are recorded as counts of radioactivity per unit time and are displayed on a printout or can be fed into a computer for storage.

In autoradiography, a gel, chromatogram, or plant or animal part is placed against a sheet of photographic film. Radioactive decay from the incorporated isotope behaves just as light energy does and reduces silver grains in the emulsion. After exposure, the sheet or film is developed and fixed, revealing a deposit of silver grains corresponding to the location of the radioactive substance (Figure A.1).

Alternately, to record the subcellular localization of an incorporated labeled isotope, cells or chromosomal preparations that have been incubated with radioactively labeled compounds are affixed to microscope slides and covered with a thin layer of liquid photographic emulsion. After they are exposed, the slides are processed to develop and fix the reduced silver grains in the emulsion. After staining, microscopic examination reveals the location and extent of labeled isotope incorporation (Figure A.2).

CENTRIFUGATION TECHNIQUES

The centrifugation of biological macromolecules is widely employed to provide information about their physical characteristics (e.g., size, shape, density, and molecular weight) and to purify and concentrate cells, organelles, and their molecular components.

Differential centrifugation is commonly used to separate materials such as cell homogenates according to size. Initially, the homogenate is distributed uniformly in the centrifuge tube. After a period of centrifugation, the pellet obtained is enriched for the largest and most dense particles, such as nuclei, in the homogenate. After each step, the supernatant can be recentrifuged at higher speeds to pellet the next heavier component. A typical fractionation scheme for cell homogenates is shown in Figure A.3. Further fractionation using density gradient techniques can be used to purify any of the fractions obtained by differential centrifugation.

Rate zonal centrifugation is used to separate particles on the basis of differences in their sedimentation rates. It may be used to separate mixtures of macromolecules such as proteins or nucleic acids and cellular organelles such as mitochondria. In this technique, which employs a medium of increasing density, the rate at which particles sediment depends on size, shape, density, and the frictional resistance of the solvent.

In addition to the preparation and purification of macromolecules and cellular components, rate zonal centrifugation can be used to determine the **sedimentation coefficients** and **molecular weights** of biological macromolecules. If a purified molecule such as a protein is spun in a centrifugal field, the molecule will eventually sediment toward the bottom at a constant velocity. At this point, the molecular weight (M) can be calculated as

$$M = f \times v/\omega^2 r$$

where f is the frictional coefficient of the solvent system (which has been calculated from other measurements) and $v/\omega^2 r$ is the rate of sedimentation per

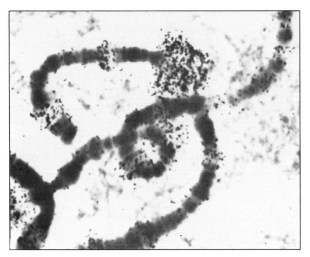

FIGURE A.2
Autoradiogram of RNA synthesis in salivary glands of *Drosophila* larva. Silver grains are deposited over sites of RNA synthesis at chromosome puffs.

FIGURE A.3
Fractionation scheme for cell homogenates.

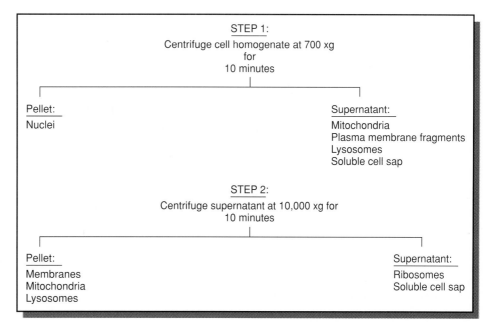

STEP 1:
Centrifuge cell homogenate at 700 xg
for
10 minutes

Pellet:
Nuclei

Supernatant:
Mitochondria
Plasma membrane fragments
Lysosomes
Soluble cell sap

STEP 2:
Centrifuge supernatant at 10,000 xg for
10 minutes

Pellet:
Membranes
Mitochondria
Lysosomes

Supernatant:
Ribosomes
Soluble cell sap

unit applied centrifugal field. The latter value is given the symbol *S*, or sedimentation coefficient. The *S* value for most proteins is between 1×10^{-13} sec and 2×10^{-11} sec. A sedimentation coefficient of 1×10^{-13} is defined as one **Svedberg unit (S);** this unit is named for The Svedberg, a pioneer in the field of centrifugation. Thus, a protein with a sedimentation value of 2×10^{-11} sec would have a value of 200*S*. Figure A.4 shows the *S* values of selected molecules and particles.

 Isopycnic centrifugation, or **equilibrium density gradient centrifugation,** is one of the most widely used techniques in genetics and molecular biology. In this technique, the solvent varies in density from one end of the tube to the other. A mixture of molecules layered on top and centrifuged through this gradient will migrate toward the bottom of the tube until each particle reaches its isopycnic point—that is, the place in the gradient where the density of the solvent equals the buoyant density of the particle. When each molecular species in the mixture migrates to its own characteristic isopycnic point, it no longer moves and is at equilibrium no matter how much longer the centrifugal field is applied. After separation by this method, components may be recovered by puncturing the bottom of the tube and collecting fractions.

Two methods of forming density gradients are commonly employed (Figure A.5), one using preformed gradients of sucrose, or soluble salts of heavy metals such as cesium chloride or cesium sulfate. In the second method, the gradient is formed by the action of the centrifugal field on the salt solution.

RENATURATION AND HYBRIDIZATION OF NUCLEIC ACIDS

The ability of separated complementary strands of nucleic acids to unite and form stable, double-stranded molecules has been used to measure the relatedness of nucleic acids from different parts of the cell, different organs, and even different species. If both strands are DNA, the process is known as **renaturation** or **reassociation**. If one strand is RNA and the other DNA, it is known as **hybridization**. Renaturation and hybridization involve two steps: (1) a rate-limiting step, in which collision or nucleation between two homologous strands initiates base pairing, and (2) a rapid pairing of complementary bases, or "zippering" of the strands, to form a double-stranded molecule. For DNA renaturation, the formation of double-stranded molecules can be assayed at any time during an experiment. A sample is passed

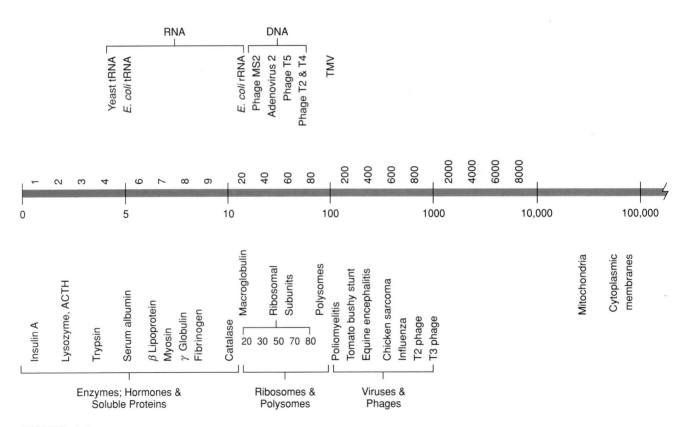

FIGURE A.4
S values of some common biological molecules and particles.

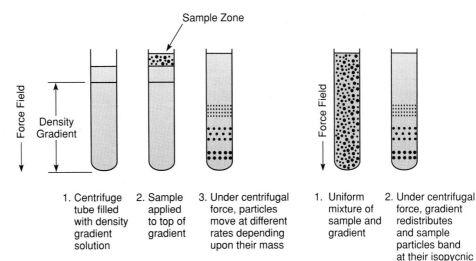

1. Centrifuge tube filled with density gradient solution
2. Sample applied to top of gradient
3. Under centrifugal force, particles move at different rates depending upon their mass

1. Uniform mixture of sample and gradient
2. Under centrifugal force, gradient redistributes and sample particles band at their isopycnic positions

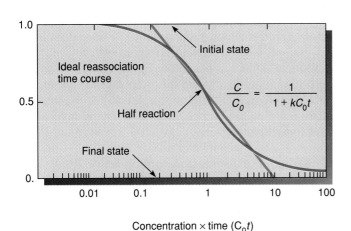

FIGURE A.6
Idealized C_0t curve.

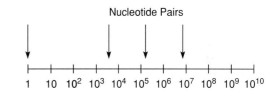

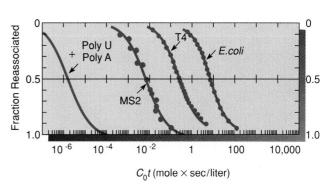

FIGURE A.7
Changes in C_0t value as genome size increases.

over a column of **hydroxyapatite,** which selectively binds double-stranded DNA but allows single-stranded molecules to pass through. The double-stranded molecules can be released from the column by raising the temperature or salt concentration.

The process of renaturation follows second-order kinetics according to the equation

$$\frac{C}{C_0} = \frac{1}{1 + kC_0t}$$

where C is the single-stranded DNA concentration at time t, C_0 is the total DNA concentration, and k is a second-order rate constant. It is usually convenient to express the data from renaturation experiments as the fraction of single-stranded DNA at any time t versus the product of total DNA concentration and time, as shown in Figure A.6.

In the process of renaturation, the DNA is initially single-stranded and is renatured to double-stranded structures in the final state. The time necessary to reassociate half of the DNA in a sample at a given concentration should be proportional to the number of different pieces of DNA present. Consequently, the half reassociation time should be proportional to the DNA content of the genome, with smaller genomes having shorter half-renaturation rates. Figure A.7 confirms this expectation and shows increases in C_0t values as the size of the genome increases. This proportional relationship between $C_0 t$ and genome size is valid only in cases where repetitive DNA sequences are absent from the DNA being studied. DNA from calf thymus (and many other eukaryotic sources) exhibits a complex pattern of reassociation, indicating that bovine DNA contains some sequences (in this case, 40 percent of the total DNA) that reassociate rapidly and others that reassociate more slowly. The rapidly reassociating fraction must therefore contain sequences that are present in many copies. The more slowly renaturing DNA, however, contains sequences present in only one copy per genome. The *E. coli* DNA shown in Figure A.8 renatures with a pattern close to that of an ideal second-order reaction, indicating the absence of repeated DNA sequences.

In the case of hybridization between DNA and RNA, two approaches can be used—either the RNA or the DNA can be in excess. RNA-excess hybridization is usually preferred because most double-stranded molecules that are formed are RNA:DNA hybrids. Since DNA is present in low concentration, and since single-stranded RNA molecules lack complementary RNA strands in the mixture, the number of RNA:RNA and DNA:DNA hybrids is negligible.

The extent of hybridization can be followed by using hydroxyapatite columns or by using radioactively labeled RNA or DNA. In practice, one of the components of the hybridization reaction is usually immobilized on a substrate such as nitrocellulose paper. For example, DNA may be sheared to a uniform size, denatured to single strands, immobi-

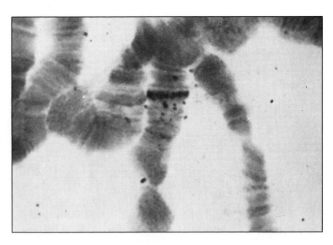

FIGURE A.9
Light micrograph of *in situ* hybridization of radioactive 5*S* RNA to a single band of polytene chromosome in the salivary gland of a *Drosophila* larva.

lized on nitrocellulose filters, and hybridized with an excess of labeled RNA. After hybridization, the unbound RNA is removed by washing, and the hybrids assayed by liquid scintillation counting. DNA:RNA hybridization can also be performed using cytological preparations, a technique called ***in situ* hybridization**. DNA, which is a part of an intact chromosome preparation fixed to a slide, can be denatured and hybridized to radioactive RNA. Hybrid formation is detected using autoradiography (Figure A.9).

ELECTROPHORESIS

Electrophoresis is a technique that measures the rate of migration of charged molecules in a liquid-containing medium when an electrical field is applied to the liquid. Negatively charged molecules (anions) will migrate toward the positive electrode (anode) and positively charged molecules (cations) will migrate toward the negative electrode (cathode). Several factors affect the rate of migration, including the strength of the electrical field and the molecular sieving action of the medium (paper, starch, or gel) in which migration takes place. Since proteins and nucleic acids are electrically charged, electrophoresis has been used extensively to provide information about the size, conformation, and net charge of these macromolecules. On a larger scale, electrophoresis provides a method of fractionation that can be used to isolate individual components in mixtures of proteins or nucleic acids.

More recently, the analytical separation of proteins has been enhanced with the development of two-dimensional electrophoresis techniques, in which

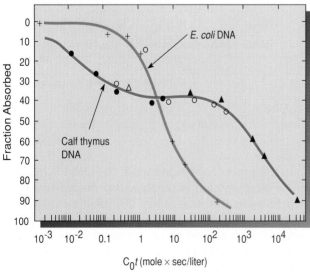

FIGURE A.8
Renaturation curve for *E. coli* DNA, containing no repetitive DNA sequences, and for calf thymus DNA, containing several families of repetitive DNA.

separation in the first dimension is by net charge, and in the second dimension by size and molecular weight.

In practice, electrophoresis employs a buffer system; a medium (paper, cellulose acetate, starch gel, agarose gel, or polyacrylamide gel); and a source of direct current. Samples are applied, and current is passed through the system for an appropriate time. Following migration of the molecules, the gel or paper may be treated with selective stains to reveal the location of the separated components (Figure A.10).

DNA SEQUENCING

The ability to directly determine the sequence of DNA segments has added immensely to our understanding of gene structure and the mechanisms of gene regulation. Although techniques were available in the 1940s to determine the base composition of DNA, it was not until the 1960s that methods providing direct chemical analysis of nucleotide sequence were developed and used. These first methods were based on those used in determining the amino acid sequence of proteins, and were time-consuming and laborious. For example, in 1965 Robert Holley determined the sequence of a tRNA molecule consisting of 74 nucleotides, a task that required almost a year of concentrated effort.

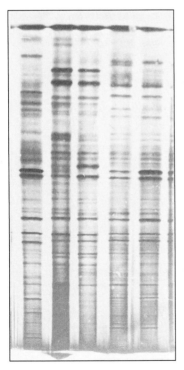

FIGURE A.10
Coomassie blue-stained protein slab gel.

In the 1970s more efficient and direct methods of analysis of nucleotide sequencing were developed. These methods developed in parallel with recombinant DNA techniques that allow the isolation of large quantities of purified DNA segments from any organism. A chemical method, developed by Allan Maxam and Walter Gilbert, cleaves DNA at specific bases. A second method, developed by Fred Sanger and colleagues, synthesizes a stretch of DNA that terminates at a given base. In both methods, the DNA to be sequenced is subjected to four individual reactions (one for each base). The products of the four reactions are a series of DNA fragments that differ in length by only one nucleotide. These reaction products are electrophoretically separated in four adjacent lanes on a gel. Each band on the gel corresponds to a base, and the sequence of the DNA segment can be read from the bands on the gel.

In the Maxam/Gilbert method, the complementary strands of the DNA fragment to be sequenced are separated and recovered. The strand to be sequenced is labeled at its 5' end with radioactive ^{32}P using the enzyme polynucleotide kinase. This provides a means of identifying specific DNA fragments after gel electrophoresis. In the next step, aliquots of the DNA are subjected to each of four chemical treatments that cleave the strand at a specific nucleotide. The reaction is carried out for a limited time so that any given molecule is cleaved at only a small number of the target nucleotides. The result is a collection of fragments, all labeled at the 5' end, but differing in length, depending on the point of cleavage.

Four different reactions cleave DNA at guanine (G > A), adenine (A > G), cytosine alone (C), or cytosine and thymine (C + T). In these reactions, purines are cleaved by using dimethylsulfate. This reagent methylates guanine far more efficiently than adenine, and when heat is applied, the strand is broken at the methylated site, producing a DNA fragment most frequently cleaved at a G residue (G > A). This can be reversed by cleaving the strand in acid, producing fragments most often cleaved at an A residue (A > G). The reaction for pyrimidines uses hydrazine, which cleaves both cytosine and thymine; however, in high salt (2M NaCl), only cytosine reacts. Thus, in the two reactions, one represents only cytosine (C), and the other represents cytosine and thymine (C + T).

For each set of reactions, the fragments produced are then subjected to gel electrophoresis in adjacent lanes on a gel. Each reaction contains a series of fragments in which strand breakage has occurred at one of the target bases, resulting in a series of fragments that differ in length. Under the influence of the electric field during electrophoresis, the different-sized fragments separate from each other, with the smallest fragments migrating the farthest. Since the DNA fragments contain a radioactive 5' end, the gel is subjected to autoradiography by placing a sheet of X-ray film over the gel for an appropriate exposure time. The DNA sequence is then analyzed by reading the bands on the gel from the bottom up, reading across all four lanes. In the example shown in Figure A.11, the first bases on the gel read GCGG.

In the Sanger method for DNA sequencing, which relies on an initial enzymatic treatment, the DNA strand to be sequenced is used as a template for the synthesis of a new DNA strand catalyzed by DNA polymerase I. This method employs a modified dideoxynucleoside triphosphate to generate a series of DNA fragments. The dideoxynucleoside triphosphates lack a 3' OH group. This allows them to be added to a DNA strand undergoing synthesis, but since they lack a 3' OH group, no nucleotide can be added to them, causing termination of strand synthesis, producing a DNA fragment. In use, four separate reaction mixtures are set up, each with a template DNA strand, a primer, all four radioactive nucleoside triphosphates, and a small amount of a single dideoxyribonucleoside triphosphate. Each of the four reactions contains a different dideoxynucleoside triphos-

phate that acts as a chain terminator. Because only a small amount of the modified nucleoside is used in each reaction, the newly synthesized strands are randomly terminated, producing a collection of fragments.

After synthesis, the radioactive fragments are electrophoresed in four adjacent lanes, one corresponding to each of the reactions. The fragments are visualized by autoradiography, and the gel is analyzed by reading the sequence from the bottom, as in the Maxam/Gilbert reaction. In the example shown in Figure A.12, the first band is in the lane with ddT, so it is a T residue, and the next few bands have the sequence GCAATCG.

DNA sequencing methods have generated a great deal of information about the structure and organization of many genes in a wide range of organisms. In the case of some viruses, the sequence of the entire genome is known, and large portions of the genome of other organisms, including *E. coli*, have been sequenced. To date, only a very small portion of the human genome has been sequenced, but the U.S. Department of Energy and the National Institutes of Health are coordinating efforts to develop the technology to sequence the more than 3 billion bases that constitute the haploid human genome.

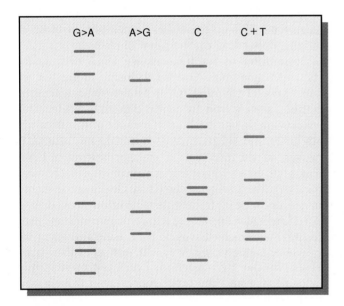

FIGURE A.11
Nucleotide sequence derived by the Maxam/Gilbert method.

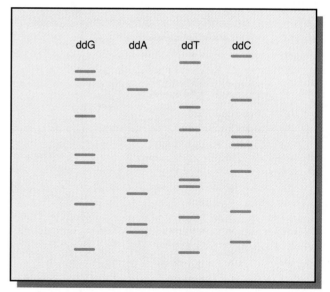

FIGURE A.12
Nucleotide sequence derived by the Sanger method.

Answers to Selected Problems

CHAPTER 2

3 The genetic loci along the length of the chromosome.

5 In mitosis, 16 chromosomes containing 32 chromatids are visible at the end of prophase. 16 chromosomes move to each pole during anaphase.

11 (a) 8 tetrads
    (b) 8 dyads
    (c) 8 monads

14 As shown in Figure 2.9, an alternate alignment at metaphase II would have resulted in a new combination of maternally and paternally derived chromosomal segments in each spermatid. Had the initial alignment at metaphase I been different than shown (one other alignment is possible), two other arrangements would have been possible at metaphase II. Therefore, three other unique combinations of four spermatids are possible in addition to the one shown.

15 (a) 7
    (b) 5
    (c) trisomy

17 (a) pachynema
    (b) zygonema
    (c) leptonema
    (d) diplonema

18 Meiosis results in the production of haploid microspores and megaspores, which themselves develop into haploid gametes. Following fertilization between male and female gametes, the diploid zygote develops into the prominent sporophyte.

CHAPTER 3

1 (a) P_1: $WW \times ww$
        F_1: $WW \times Ww$
        F_2: ¼ WW; ½ Ww; ¼ ww
    (b) All white
    (c) Cross 1: $WW \times WW$ or $WW \times Ww$
        Cross 2: $Ww \times Ww$

2 (a) $Aa \times Aa$
    (b) $AA \times aa$
    (c) $Aa \times aa$

3 Dominance/recessiveness; unit factors in pairs; segregation

7 P_1: $WWGG \times wwgg$
   F_1: $WwGg$
   F_2: 9/16 round, yellow (W–G–)
        3/16 round, green (W–gg)
        3/16 wrinkled, yellow (wwG–)
        1/16 wrinkled, green ($wwgg$)

9 (d) is a test cross.

10 Independent assortment

14 (a) F_1: All gray, long ($EeVv$)
        F_2: 9/16 gray, long (E–V–); 3/16 gray, vestigial (E–vv); 3/16 ebony, long (eeV–); 1/16 ebony, vestigial ($eevv$)

(b) Same answer as (a).
(c) F_1: All gray, long (*EEVv*)
F_2: ¾ gray, long (*EEV–*); ¼ gray, vestigial (*EEvv*)

17 P_1: $GG \times gg$
F_1: All *Gg*
F_2: ¼ *GG* (breed true); 2/4 *Gg* (produce ¾ yellow: ¼ green); ¼ *gg* (breed true)

19 ¼ *WwGg* (round, yellow); ¼ *Wwgg* (round, green); ¼ *wwGg* (wrinkled, yellow); ¼ *wwgg* (wrinkled, green)

21 (a) $\chi^2 = 0.07$; $p = 0.80$
(b) $\chi^2 = 0.38$; $p = 0.55$

23 For the 3:1 ratio, $\chi^2 = 33.33$. For the 1:1 ratio, $\chi^2 = 25.00$. In both cases, the probability of obtaining the data with the observed deviation strictly by chance is much less than 0.0001. Therefore, both null hypotheses are rejected.

25 The pedigree fits an autosomal recessive mode of inheritance. I-1 is *aa;* I-2 is *Aa;* I-3 is *Aa;* and I-4 is *Aa.*

27 No conclusion can be drawn. The trait could be recessive, in which case the female parent is a heterozygous carrier. Or, the trait could be due to a dominant allele, in which case the male parent is heterozygous.

29 $p = 0$; it is impossible.

CHAPTER 4

1 Incomplete dominance; roan is *Aa.*
$AA \times AA \rightarrow AA$; $aa \times aa \rightarrow aa$; $AA \times aa \rightarrow Aa$
$Aa \times Aa \rightarrow$ ¼ AA; ½ Aa; ¼ aa

3 ⅔ platinum; ⅓ silver. The *P* allele behaves as a recessive lethal, but as a dominant in its influence on coat color.

5 The I^A and I^B alleles are dominant to the recessive I^O allele. The I^A and I^B alleles are codominant to each other.

7 Blood type AB would exclude the male in question. If he were type A, B, or O, he could contribute an I^O allele, and could be the father. However, countless males are of these blood types, so no proof is provided.

11 (a) $c^kc^a \times c^dc^a$ = ½ sepia: ¼ cream: ¼ albino
(b) $c^kc^a \times c^dc^d$ = ½ sepia: ½ cream (or $c^kc^a \times c^dc^a$)
(c) $c^kc^k \times c^dc^d$ = all sepia (or $c^kc^k \times c^dc^a$)
(d) $c^kc^a \times c^dc^d$ = ½ sepia; ½ cream

13 As in Problem 12, pink is the heterozygote, and white and red are the two homozygotes. Personate is dominant to peloric. F_1 (a) × F_1 (b) cross: 3/16 red, personate; 1/16 red, peloric; 6/16 pink, personate; 2/16 pink, peloric; 3/16 white, personate; 1/16 white, peloric.

15 (a) F_1: All gray (*CcAa*)
F_2: 9/16 gray (*C–A–*); 3/16 black (*C–aa*); 4/16 albino (*ccA–* and *ccaa*)
(b) (1) *CcAA* (2) *CCAa* (3) *CcAa*

17 (a) All gray
(b) All gray
(c) 16/32 albino; 9/32 gray; 3/32 yellow; 3/32 black; 1/32 cream
(d) 9/16 gray; 3/16 black; 4/16 albino
(e) ⅜ gray; ⅛ yellow; 4/8 albino

21 (a) Two gene pairs;
A–B– tall
A–bb, *aaB–*, *aabb* } dwarf
(b) ¼ of the F_2 are true-breeding:
1/16 *AABB*
1/16 *AAbb*
1/16 *aaBB*
1/16 *aabb*

23 3/16 type A: 6/16 type AB: 3/16 type B: 4/16 type O. All of those that type as O actually lack the H substance.

24 1–c; 2–d; 3–b, 4–e, 5–a. No!

26 Two gene pairs are involved. The first determines black (*B*–) or brown (*bb*) color. The second is epistatic and when homozygous recessive (*gg*), pigment deposition is affected, resulting in the golden color. Dogs that are *GG* or *Gg* are black or brown. Based on this explanation, you should be able to determine the parental genotypes.

27 (a) 40 cm
(b) F_1: All *AaBb,* 30 cm
F_2: 1/16 *AABB,* 40 cm; 4/16 [*AABb*, *AaBB*] 35 cm;
4/16 [*aaBb*, *Aabb*], 25 cm; 6/16 [*AAbb*, *AaBb*, *aaBB*], 30 cm;
1/16 *aabb,* 20 cm

28 Only one possible genotype will have only two additive alleles: *AabbCc.* It will occur in ⅛ of all progeny.

30 If four gene pairs control height such that each of the eight potential additive alleles adds 2 in. to the base height of 4 in., then the following parental genotypes are consistent with the data shown:
(a) *aabbccdd* (4″) × *AABBCCDD* (20″)
(b) *AAbbccdd* (8″) × *AABBCCDD* (20″)
(c) *AABBccdd* (12″) × *AABBCCDD* (20″)
(d) *AABBCCdd* (16″) × *AABBCCDD* (20″)
Other genotypes will fit equally well for the shorter parents in (b), (c), and (d). For example, in (b), *aaBBccdd, aabbCCdd,* and *aabbccDD* will all yield offspring that are 14″ tall.

31 (a) polygenic inheritance
(b) 3 gene pairs
(c) 2 cm
(d) P_1: *AAbbcc* × *aaBBCC*
F_1: *AaBbCc*
(e) *AAbbcc, aaBBcc, aabbCC* are all 20 cm *and* true-breeding. *AaBbcc, AabbCc,* and *aaBbCc* are all 20 cm but are *not* true-breeding.

CHAPTER 5

2 (a) F_1: ½ wild ♀; ½ wild ♂
F_2: ¼ wild ♀; ¼ reduced ♀; ½ wild ♂
(b) F_1: ½ wild ♀; ½ reduced ♂
F_2: ¼ wild ♀; ¼ reduced ♀; ¼ wild ♂; ¼ reduced ♂
(c) No

5 ⅜ wild ♀; ⅛ wild ♂ (transformed); 4/8 singed ♂

7 The Y chromosome is essential for initial differentiation of testicular tissue. However, hormones subsequently produced by the testes are responsible for differentiation of the remainder of the male reproductive tract.

9 As a result of crossing over between the X and Y chromosomes during male meiosis, the X chromosome normally passed on to daughters would receive male-determining genes.

11 Klinefelter, one; Turner, none; 47, XYY, none; 47, XXX, two; 48, XXXX, three.

15 In *Drosophila*, the condition results in a normal, fertile female. In humans, the condition produces a male with Klinefelter syndrome.

17 If nondisjunction occurred and the parthenogenic egg retained two X chromosomes.

21 (a) ¼ (b) ½ (c) ¼ (d) 0

23 (a) F_1: ½ scalloped ♂; ½ normal ♀
    F_2: ¼ scalloped ♀; ¼ scalloped ♂; ¼ normal ♀; ¼ normal ♂
  (b) Assuming the female is homozygous, the F_1 consists of all normal males and females. In the F_2, ½ normal ♀; ¼ normal ♂; ¼ scalloped ♂.

25 F_1: ½ red, normal ♀; ½ red, normal ♂
    F_2: $^6/_{16}$ red, normal ♀; $^2/_{16}$ normal, dumpy ♀; $^3/_{16}$ red, normal ♂; $^3/_{16}$ white, normal ♂; $^1/_{16}$ red, dumpy ♂; $^1/_{16}$ white, dumpy ♂

26 F_1: All wild-type ♂ and ♀
    F_2: $^8/_{16}$ wild-type ♀; $^5/_{16}$ wild-type ♂; $^3/_{16}$ vermilion ♂

29 F_1: all hen-feathered males and females
    F_2: $^4/_8$ hen-feathered females
    $^3/_8$ hen-feathered males
    $^1/_8$ cock-feathered males

31 The condition is inherited as a sex-linked dominant trait.

CHAPTER 6

3 Because each crossover event occurs randomly along the length of the chromosome. The farther apart two loci are, the more frequently a random event will occur between them.

5 Because the probability of two independent events occurring simultaneously is equal to the product of the probability of each individual event.

7 One parent must be heterozygous for all genes to be mapped. Second, the phenotypes of the offspring must reveal the genotypes of the gametes formed by the crossover parent that gave rise to these progeny.

8 *dp–cl–ap*

11 Regardless of the percentage of offspring produced as a result of crossover vs. noncrossover gametes, both groups yield progeny in a 2 +:1 *ca*:1 *e* ratio. It is impossible to determine from the data the map distance between the *e* and *ca* loci.

13 *Female A:* 3 and 4 are NCO; 7 and 8 are SCO between the *d* and *b* loci; 1 and 2 are SCO between the *b* and *c* loci; 5 and 6 are DCO. These occur in the relative frequency such that 3, 4 > 1, 2 > 7, 8 > 5, 6.

Female B: 7 and 8 are NCO; 5 and 6 are SCO between

the *d* and *b* loci; 3 and 4 are SCO between the *b* and *c* loci; 1 and 2 are DCO. These occur in the relative frequency such that 7, 8 > 3, 4 > 5, 6 > 1, 2.

15 (a) $\dfrac{y\ w\ +}{+\ +\ ct} \times \dfrac{y\ w\ +}{\xrightarrow{\hspace{1cm}}}$

  (b) $\dfrac{0.0\qquad 1.5\qquad\quad 20.0}{y\qquad\ w\qquad\qquad ct}$

  (c) Yes; $(0.015)(0.185)(1000) = 2.775$

  (d) No; the phenotype of the females would not always reflect their genotypes.

16 (b) The F_1 heterozygote is $\dfrac{D\ +\ +}{+\ p\ e}$, with *pink* in the middle. The interlocus distances are 3.0 for *D* to *p* and 18.5 for *p* to *e*.

18 Since *Sb* is dominant and lethal when homozygous, *curled* male $\left(\dfrac{+\ cu}{+\ cu}\right)$ would be the suitable choice.

21 10 map units

25 (a) P = 1, 2; NP = 3; T = 4, 5, and 6
  (b) Category 2 will arise if a SCO occurs between one of the genes and the centromere.
  (c) $\dfrac{6\ +\ ½\ (22)}{71} = 0.239 = 24$ map units

27 No, in both cases.

30 (a) NP T P NP T P T
  (b) There is a significant excess of parental ditypes over nonparental. P(41 + 3):NP (1 + 1) = 44P:2 NP. Thus the null hypothesis that the ratio would be 1:1 if independent assortment occurred is not supported, and the few nonparental ditypes arose through double crossovers between the linked loci.
  (c) Second division segregation at the *c* locus is ½ (1 + 5 + 3 + 2)/70 = 0.078, or 7.8 cM. Second division segregation at the *d* locus is ½ (17 + 1 + 3 + 2)/70 = 16.4 cM.
  (d) (NP + ½T)/Total = 2 + ½ (17 + 5 + 2)/70 = 20 cM.
  (e)
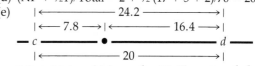
  7.8 + 16.4 = 24.2 vs. the 20. Two-strand double crossovers would not be detected by the method in (d), so the more correct map would be the one determined by adding the gene-centromere distances.
  (f) Requires 2 double-strand crossovers.

CHAPTER 7

8 The hybrid was an alloploid with one member of each chromosome pair from each species. Treatment with colchicine to double the chromosome number may create a fertile plant.

12 Haploid, 9; triploid, 27; tetraploid, 36; trisomic, 19; monosomic, 17.

16 No; they suppress *the recovery* of crossover products when the crossover occurs within the inverted region.

18 Crossover products between *d* and *e* are the only ones recovered. Genes *a*, *b*, and *c* are part of an inversion on one of the homologues. The distance between d and c can be mapped. They are 10 map units apart.

23 Since she received her single X chromosome from her father, her mother contributed the gamete devoid of sex chromosomes.

24 *P. kewensis* is an allotetraploid that arose from a cross between *P. floribunda* and *P. verticillata*. The initial hybrid was sterile and contained 18 chromosomes which then duplicated in a few reproductive cells, giving rise to fertile offspring containing 36 chromosomes.

25 The varieties are examples of autopolyploidy. Each contains an even ploidy value (e.g., 2, 4, 6, 8, and 10n). The variety with 27 chromosomes is triploid and produces unbalanced gametes.

26 The second crossover may negate the possible deleterious effects of the first crossover, which would have led to duplications and deletions as well as dicentric and acentric chromosomes if the inversion were paracentric.

28 No, at most, only one-fourth of the gametes will be normal because an acentric and a dicentric chromosome have been created.

CHAPTER 8

5 How does the transforming DNA enter the cell? Does it enter preferentially at any particular time of the host's cell cycle? In what form and size does it enter? How does it recombine with the host DNA?

7 Technically, ^{32}P will label RNA as well as DNA. However, since mature T2 phages lack RNA, DNA must be the genetic material of the virus that enters the bacterium.

CHAPTER 9

3 Guanine: 2-amino, 6-oxy purine
Cystosine: 2-oxy, 4-amino pyrimidine
Thymine: 2,4-dioxy, 5-methyl pyrimidine
Uracil: 2,4-dioxy pyrimidine

7 They might have concluded that adenine:guanine and thymine:cytosine base pairs form between the two strands. Since A and G are purines and T and C are pyrimidines, the diameter between strands would not be constant, as observed by Franklin and Wilkins.

11 The nitrogenous base is responsible for UV absorption.

21 Fraction A contains a mixture of types of DNA sequences. One group, which reanneals more quickly than *E. coli* DNA, is highly repetitive and possibly short. Another group reanneals more slowly and appears to be moderately repetitive. The third group consists of sequences that are single copies and of greater complexity than the *E. coli* single copy sequences. Fraction B consists of all single copy, nonrepetitive DNA sequences of greater complexity than those in either fraction A or *E. coli*.

22 The curve would be similar to that produced by fraction A DNA in Problem 21.

24 The bottom part of the helix is left-handed, while the top portions are right-handed, an obvious inconsistency. The molecule is replicating semiconservatively.

25 Since T4, MS-2, and *E. coli* do not contain repetitive sequences, a direct proportionality between $C_0t_{1/2}$ and X complexity exists. Thus, MS-2 DNA consists of 200 base pairs, and *E. coli* DNA consists of 2×10^6 base pairs.

CHAPTER 10

3 The conservative mode can be ruled out after the first round; the dispersive mode can be ruled out after the second round.

5 Semiconservative replication would be very difficult, if not impossible, for a triple helix. Conservative replication, while difficult, might work better.

7 Vast quantities of *E. coli* were necessary in order to obtain small amounts of the highly purified enzyme because purification methods were not as refined as those subsequently developed.

22 (a) No, nearest neighbor data reflects proportionality, not absolute amount.
(b) Yes, since nearest neighbor frequencies are dependent on a similar frequency of the bases.
(c) Not necessarily. A group of 10 unique sequences, each 20 nucleotides long, could be put together in a number of ways and still yield similar nearest neighbor data.

23

(a) Base		Nearest Neighbor	(b) Base		Nearest Neighbor
G	$\longrightarrow$	C, C, T	G	$\longrightarrow$	C, T, A
C	$\longrightarrow$	T, A	C	$\longrightarrow$	A, T, G
T	$\longrightarrow$	T, A, A, G	T	$\longrightarrow$	T, A, G
A	$\longrightarrow$	C, A, G	A	$\longrightarrow$	C, A, T

CHAPTER 11

1 If only a single central amino acid is considered, then the first and third amino acid can be any of the 20 possibilities. Therefore, 20^2 or 400 different tripeptides can be created. Since any of 20 amino acids can serve as the central one, a total of $20(20^2)$ or 20^3 different tripeptides can be formed using all amino acids.

3 No

4 No. Either a triplet or sextuplet code would allow restoration of the reading frame.

6 3C: proline
2C:1A: proline, histidine, threonine
1C:2A: threonine, glutamine, asparagine
3A: lysine

9 For the repeating UUAC sequence, two of the triplets (UUA and CUU) code for leucine. For the repeating UAUC sequence, all four possible triplets code for a different amino acid.

10 ACA

11 AGA: arginine; GAG and GAA: glutamic acid; AAG: lysine

12 *In vivo*, the enzyme degrades RNA.

15 The most surprising is that CUA, which normally codes for leucine, specifies threonine (normally coded by ACA, ACU, ACC, and ACG). The other changes involve triplets related by two of the three nucleotides.

17 Poly U and Poly A are complementary strands, forming a double-helical structure.

18 mRNA must be single-stranded.

19 Sequence 1: met-pro-asp-tyr-ser
Sequence 2: met-pro-asp
Sequence 2 contains a frameshift mutation, creating UAA (a termination codon) as the fourth triplet.

21 (a) Two initiation points (AUG triplets) exist in the mRNA sequence giving rise to:
met-his-thr-tyr-glu-thr-leu-gly . . .
met-arg-pro-leu-gly . . .
(b) The mutation has produced a termination codon directly after the second initiation triplet. This causes an amino acid substitution in the first polypeptide chain, but immediately terminates the synthesis of the second polypeptide chain.

22 Leucine is encoded by six triplets, histidine by two, and tryptophan by one, correlating directly with the frequency of these amino acids in proteins.

CHAPTER 12

15 426 nucleotides, including a termination codon. Each triplet must occupy 3(0.34) or about 1 nm; about 20 triplets could potentially occupy a single ribosome.

19 The recognition of the corrected charged tRNA involves the tRNA portion of this molecule and not the amino acid portion.

21 (a) Sequence 1: GAAAAAACGGUA
Sequence 2: UGUAGUUAUUGA
Sequence 3: AUGUUCCCAAGA
(b) Sequence 1: glu-lys-thr-val
Sequence 2: cys-ser-tyr-nonsense
Sequence 3: met-phe-pro-arg
(c) Initial: sequence 3
Middle: sequence 1
Terminal: sequence 2

22 Sequence 1: GAAAAAACGGTA

24 The initiator tRNA^{fmet}.

CHAPTER 13

4 The conversion of substance *w* to *x* is controlled by gene product **b;** of substance *x* to *y* by gene product **c;** of substance *y* to *z* by gene product **a.**

6 (a) Two independently assorting autosomal recessive genes are involved. The P_1 generation is *AAbb* × *aaBB*. The F_1 is *AaBb*, and in the F_2, the following genotypic ratios occur:

$$A - B - \quad 9/16$$
$$A - b\,b \quad 3/16$$
$$a\,a\,B - \quad 3/16$$
$$a\,a\,b\,b \quad 1/16$$

Only those with at least one dominant *A* and one dominant *B* allele are purple (9/16). All other genotypes (7/16) are white.

(b) The same general explanation as in (a) accounts for inheritance, except that

$$A - B - \quad \text{purple}$$
$$A - bb \quad \text{pink}$$
$$a\,a\,B - \quad \text{white}$$
$$a\,a\,b\,b \quad \text{white}$$

7 Precursor $\xrightarrow{trp\text{-}8}$ Anthranilic Acid $\xrightarrow{trp\text{-}2}$ IGP $\xrightarrow{trp\text{-}3}$ Indole $\xrightarrow{trp\text{-}1}$ Trp

8 Precursor $\xrightarrow{thi\text{-}2}$ Pyrimidine
Precursor $\xrightarrow{thi\text{-}1}$ Thiazole $\xrightarrow{thi\text{-}3}$ Thiamine

14 The amino acid difference in the hemoglobin of the two individuals involves no change in charge. Thus, the net charge is not different in the two molecules. Fingerprinting, which involves chromatography (in addition to electrophoresis) may detect an amino acid difference on the basis of peptide migration in the solvent used.

17 If the mutations are mapped within the gene, and if colinearity holds, each successive mutant should result in a slightly longer polypeptide chain following termination.

23 You should find that almost all of them can be related by a single nucleotide change.

CHAPTER 14

9 (a) Sex-linked necessive lethal mutations.
(b) It is critical to the screening of F_2 cultures lacking male offspring.
(c) To insure that both the *Bar* and lethal alleles remain together on the same homologue.
(d) An inversion, which suppresses the recovery of crossover products.

10 The induced mutation rate is probably less than 0.002. However, because of the small sample size, one cannot be certain. The experiment should be repeated and the data pooled.

11 In the F_3, any cultures lacking wild-type flies were derived from parents bearing an induced recessive lethal mutation on chromosome 2. Any induced morphological mutation will be expressed in flies that are not *Cy L*.

15 X-rays are derived from the portion of the electromagnetic spectrum with a shorter and more energetic wavelength.

18 Because of the dangers inherent in X-ray exposure.

19 Autosomal recessive mutations could not be assessed. Many generations will be required before such a morphological or lethal allele becomes frequent enough in the gene pool to become homozygous, at which point it will be expressed.

25 Since each individual is the product of two gametes, the average number of new mutations in each individual is

$$2(5 \times 10^{-5})(10^5) = 10 \text{ new mutations}$$

Thus, the current human population contains 43 billion (4.3×10^{10}) newly arisen mutations.

26 The frequency of mutant cells following one round of replication is

$$\frac{18 \times 10^1}{6 \times 10^7} = 3 \times 10^{-6}$$

The rate of mutation per cell replication must take into account the fact that the total number of cells replicating was only one-half the total assayed, since each replicating cell produces two progeny cells. Thus, the mutation rate per cell replication is 1.5×10^{-6}.

CHAPTER 15

10 A circular chromosome of sequence TCHROMBAK can be derived from the data. Note that recombination of various sequences runs in either direction.

11 Because the F factor is always the last part of the chromosome to enter the recipient.

14 The data are consistent with the fact that genes *a* and *b* are not linked (or not close enough to each other to be included in a single transformation event).

15 The *a* and *b* genes are linked. The *c* gene is not linked to either of them. As a result, single transformation events can produce $a^+b^+c^-$, $a^+b^-c^-$, $a^-b^+c^-$, and $a^-b^-c^+$. The $a^+b^+c^+$, $a^+b^-c^+$, and $a^-b^+c^+$ genotypes all require two simultaneous events. Consistent with our conclusion, these all occur at much lower frequencies.

19 In this theoretical mapping experiment, $a^-b^-c^+$ and $a^+b^+c^-$ appear as double exchanges in the lowest frequency. Therefore, the *c* gene is located between the *a* and *b* genes. The distance between *a* and *c* is calculated as

$$(740 + 670 + 90 + 110)/10{,}000 = 16.1\%$$

The distance between *c* and *b* is

$$(160 + 140 + 90 + 110)/10{,}000 = 5\%$$

The percentages represent map units.
Interference is negative. A double exchange produces the $a^-b^-c^+$ and $a^+b^+c^-$ genotypes, which are expected to occur about 80 times in each 10,000 viruses. 200 are observed.

22 Expt. 1: $e \times f \rightarrow +$
Expt. 2: $h \times i \rightarrow -$
Overall, *d* and *f* are in the same gene, and *e*, *g*, *h*, and *i* are located together in a separate gene.

CHAPTER 16

2 Because mRNA to be copied usually has a poly dA tail to pair with the poly dT.

3 (a) Cells that have incorporated a plasmid should be resistant to tetracycline.
(b) Growth on tetracycline but no growth on ampicillin.
(c) Colonies resistant to both antibiotics contain plasmids that religated without incorporating *Drosophila* DNA.

5 First, a full-length cDNA copy of the insulin gene is prepared from the mRNA. This will not contain introns. This cDNA is cloned through the use of linkers into a vector containing a prokaryotic promoter, and transformed into *E. coli* K12.

6 Although the bases are internal to the DNA molecule, the side groups attached to the bases extend into the major groove of the helix and serve as the basis of recognition.

7 The slow band is formed by the religation of two plasmids in opposite orientation.

9 The size of the genome to be cloned and the cloning capacity of the vector. Generally, it is better to clone large genomes in high capacity vectors so that fewer clones need to be screened to select a gene of interest.

10 There are very few suitable vectors for cloning in plants.

CHAPTER 17

11 (a) 5×10^6 nucleosomes
(b) 10^6 solenoids
(c) 4.5×10^7 histone molecules
(d) 6.8×10^6 nm

12 Yes; using the value of π as 3.14 and the formulas πr^2 for the area of a circle and $\frac{4}{3}\pi r^3$ for the volume of a sphere, the total volume of the chromosome is approximately $1.57 \times 10^8 \text{Å}^3$, while the viral head has a volume of $2.68 \times 10^8 \text{Å}^3$.

CHAPTER 18

1 There are three complementation groups. *A* and *B* are in one; *C* and *E* are in another; *D* is in the third group.

3 Group A: $e \times f \rightarrow$ lysis
Group B: $h \times i \rightarrow$ no lysis
Group C: $k \times l \rightarrow$ no lysis

CHAPTER 19

5 The remainder of the *z* gene and all of the *y* and *a* genes will be out of frame, creating missense. Likely, one of the out-of-frame triplets will be a nonsense codon causing premature termination of the garbled polypeptide during translation.

8 The operon is inducible. The data are consistent with *a* being the operator (a^- is constitutive and not corrected by a^+), *b* being the promoter (b^- is permanently repressed unless an entire operon is sexduced into the merozygote), *c* being the structural gene (c^- yields inactive enzymes), and *d* being the repressor (d^- is constitutive and corrected by d^+).

10 (a) Operon is off; both the repressor and CAP are bound to their respective locations on DNA.

(b) Operon is being transcribed; the repressor is bound to lactose, while CAP is bound to its site on DNA.

(c) Operon is off; the repressor is bound to the operator region, while CAP is not bound to DNA.

(d) Operon is off; the repressor is bound to lactose, but CAP is not bound to DNA.

11 The operon is under positive control. In the presence of tis, the activator produced by the r gene is bound to tis and inhibited from stimulating the operon. The O^- mutations prevent stimulation by the activator, causing repression of the genes.

CHAPTER 20

1. Gene regulation in eukaryotes is more complex than in prokaryotes because of several factors, including:

(a) Each eukaryotic cell contains much greater amounts of genetic information, and this DNA is complexed with histones to form chromatin.

(b) Genetic information in eukaryotes is carried on multiple chromosomes, making coordinate expression of genes more complex.

(c) In eukaryotes, the process of transcription is spatially and temporally separated from translation, making transport of messages necessary.

(d) Gene transcripts in eukaryotes are processed, cleaved, and realigned before transport, and many transcribed sequences never leave the nucleus.

(e) In eukaryotes, highly differentiated cells often synthesize large amounts of a single gene product even though they contain a complete set of genetic information.

3 Cell fusions in which a transcriptionally inactive nucleus is placed into a foreign cytoplasm results in the reactivation of DNA and mRNA synthesis, indicating that the signals for this reactivation originate from the foreign cytoplasm.

4 Both promoters and enhancers are DNA sequences that regulate gene transcription. Enhancers can be distinguished from promoters in several ways:

(a) The position of an enhancer is not fixed; it can be upstream, downstream, or within the gene it regulates. Promoters are usually located within 100 bp upstream of the genes they regulate.

(b) The orientation of an enhancer can be inverted without significant effect on its action.

(c) Enhancers are not limited to specific genes; movement to another location does not affect its ability to enhance transcription of a different gene.

5 Experiments with fragments of the *GAL4* protein have shown that binding of a transcription factor is necessary but not sufficient to activate transcription. Other factors are required to initiate transcription, and might include proteins associated with the RNA polymerase.

8 Oncogenes can differ in several ways from proto-oncogenes, including point mutations, overexpression, or gene rearrangement.

CHAPTER 21

1 Undifferentiated cells become determined preceding differentiation into specialized form.

3 Six; for A, B, and C polypeptides, AA, BB, CC, AB, AC, and BC

5 Gene products essential for gastrulation are synthesized between hours 6 and 15 during development. While blastula cells appear to be undifferentiated, biochemical events occurring well in advance of gastrulation are critical to subsequent development.

CHAPTER 22

2 In both cases, only dextrally coiled F_1 snails would result.

3 In the original cross, all progeny cells would be resistant to streptomycin. In the reciprocal cross, one-half of the progeny would be sensitive and one-half resistant, with the phenotypes dependent on the nuclear genotype.

5 *Segregational petites* are inherited as a recessive gene.

6 The diploid zygote would be normal because it carries a normal nuclear allele derived from the *neutral petite* parent. This zygote produces ½ normal: ½ petite ascospores.

8 (a) The protein synthesis capability in both chloroplasts and mitochondria is likely to be affected. As a result, aerobic respiration and photosynthesis will be inhibited.

(b) If the mutation is in cpDNA, then only the mt^+ strains contribute functional chloroplasts to progeny cells.

9 The maternal parent was *Dd* and the paternal parent was *dd*.

CHAPTER 23

5 Determination of height has a substantial genetic component since MZ twins, whether reared together or apart, are much more similar than DZ twins and sibs reared together. Determination of weight is under much weaker genetic control than height. Both height and weight are also influenced by the environment (diet).

6 F_1: All wild-type ♂ and ♀

F_2: $\frac{8}{16}$ wild-type ♀; $\frac{5}{16}$ wild-type ♂; $\frac{3}{16}$ vermilion ♂

7 (a) $\overline{X} = 140$ cm

(b) $s^2 = 374.176$

(c) $s = 19.343$

(d) $S_{\overline{X}} = 0.6892$

8 (a) Because the F_2 ratios are in sixteenths, 2-locus inheritance is indicated, with 2 alleles at each locus.

(b) Yes.

(c) White has lack of color, is homozygous recessive. Therefore, recessive alleles produce no color, dominants each produce an equal amount of color, and dosage of the dominant alleles determines color.

(d)

1 dark red :	4 medium dark red :	6 medium red :	4 light red :	1 white :
AABB	*AaBB*	*AaBb*	*Aabb*	*aabb*
	AABb	*AAbb*	*aaBb*	
		aaBB		

9 (a) Yes, individuals of moderate height are partially heterozygous. Since dominant alleles contribute more units of height than recessive alleles, offspring can be significantly taller or shorter. For example:

$$R/r\ S/s\ t/t\ u/u \times r/r\ s/s\ T/t\ U/u$$
$$\downarrow$$
$$R/r\ S/s\ T/t\ U/u\ \text{tall offspring}$$
or
$$r/r\ s/s\ t/t\ u/u\ \text{short offspring}$$

(b) No. Minimum height is represented by the homozygous recessive parent. Matings will not place any additional dominant alleles in the offspring, so they can never be taller than tall parent.

$$r/r\ s/s\ t/t\ u/u \times R/r\ S/s\ T/t\ U/u$$
$$\downarrow$$

tallest child: $R/r\ S/s\ T/t\ U/u$ same height as tall parent

10 Heritability is estimated as $H^2 = \dfrac{V_G}{V_P}$,

where $V_P = V_E + V_G$.

Because the parental strains are true breeding, we can assume that they are genetically homogeneous, and that variance in corolla length is due to environmental factors. The average V_E for the P_1 is therefore 3.50. If we further assume that the F_1 is genetically homogeneous, then the overall V_E is $= \dfrac{3.50 + 4.743}{2} = 4.1215$.

If we now define the phenotypic variance V_P as the variance shown by the F_2 ($V_P = 47.708$), then the genetic variance becomes

$$V_G = V_P - V_E$$
$$V_G = 47.708 - 4.1215$$
$$V_G = 43.5865$$

In this case, then,

$$H^2 = \frac{43.5865}{47.708} = 0.913$$

About 91% of the variance in corolla length is contributed by genetic factors.

CHAPTER 24

3 Genes on chromosome 2 influence negative geotropism, and genes on chromosome 3 have an impact on positive geotropism.

4 (a) $uR \times UuRR \rightarrow$ ½ $UuRR$ (uncap): ½ $UuRR$ (nonhygienic)

(b) $Ur \times UURr \rightarrow$ ½ $UURr$ (nonhygienic): ½ $UUrr$ (remove if uncapped)

(c) $uR \times Uurr \rightarrow$ ½ $UuRr$ (nonhygienic): ½ $uuRr$ (uncap)

6 If PTC tasting is inherited as a dominant allele such

that *TT* or *Tt* individuals show the trait, various genotypic combinations can be devised consistent with the data in this problem.

8 The traits are not under the control of single loci.

CHAPTER 25

2 After one generation, $AA = 0.25$, $Aa = 0.50$, and $aa = 0.25$. In the absence of evolutionary forces, the same genotypic frequencies will occur in subsequent generations.

3 (a) 14.72% (b) 1.78% (c) 42%
  (d) 36% (e) 44.2%

4 The population is in equilibrium.

6 (a) The population is in equilibrium.
  (b) The population is not in equilibrium.

7 ¼

10 (a) 0.5 (b) 0.35 (c) 0.11

13 Inbreeding causes a change in the distribution of recessive alleles by increasing the number of homozygous recessive individuals, but inbreeding does not increase the absolute number of recessive alleles in the population. Only mutation or migration can cause changes in the number of alleles.

16 (a) Self-fertilization. Figure 25.11 shows the rate at which self-fertilization creates homozygous lines.

19 This statement confuses the mode of inheritance (dominant) with the frequency of the gene. In a cross involving two heterozygous individuals, the offspring will show a phenotypic ratio of 3 brachydactyly : 1 normal. In the population, however, the frequency of the gene will depend on a host of factors that are independent of the mode of inheritance, and the frequency of the gene in the population cannot be predicted because it is dominant.

20 The 2 children with normal parents represent mutations. Therefore, the survey records 2/50,000 births as being the result of mutation. Each birth carries 2 copies of the gene, only 1 of which we will assume is mutant. Thus, the survey has actually covered 100,000 copies of the gene, and found 2/100,000 copies of the gene have mutated. Thus the mutation rate would be 2×10^{-5}.

CHAPTER 26

4 (a) Those that do not change the code.
  (b) Those that convert a positively charged amino acid to one of the same charge.
  (c) Those that convert a negatively charged amino acid to one of the same charge.
  (d) Those that convert one neutral amino acid to another neutral amino acid.

5 13524635624

7 The minimal mutational distance between humans and each organism shown is horse, 3; pig, 2; dog, 3; chicken, 3; bullfrog, 3; and fungus, 6.

Glossary

A-DNA An alternate form of the right-handed double-helical structure of DNA in which the helix is more tightly coiled, with 11 base pairs per full turn of the helix. In this form, the bases in the helix are displaced laterally and tilted in relation to the longitudinal axis. It is not yet clear whether this form has biological significance.

abortive transduction An event in which transducing DNA fails to be incorporated into the recipient chromosome. (See *transduction*.)

acentric chromosome Chromosome or chromosome fragment with no centromere.

acquired immunodeficiency syndrome (AIDS) An infectious disease caused by a retrovirus designated as human immunodeficiency virus (HIV). The disease is characterized by a gradual depletion of T lymphocytes, recurring fever, weight loss, multiple opportunistic infections, and rare forms of pneumonia and cancer associated with collapse of the immune system.

acrocentric chromosome Chromosome with the centromere located very close to one end. Human chromosomes 13, 14, 15, 21, and 22 are acrocentric.

active immunity Immunity gained by direct exposure to antigens followed by antibody production.

active site That portion of a protein, usually an enzyme, whose structural integrity is required for function (e.g., the substrate binding site of an enzyme).

adaptation A heritable component of the phenotype which confers an advantage in survival and reproductive success. The process by which organisms adapt to the current environmental conditions.

additive genes See *polygenic inheritance*.

albinism A condition caused by the lack of melanin production in the iris, hair, and skin. In humans, most often inherited as an autosomal recessive.

aleurone layer In seeds, the outer layer of the endosperm.

alkaptonuria An autosomal recessive condition in humans caused by the lack of an enzyme, homogentisic acid oxidase. Urine of homozygous individuals turns dark upon standing due to oxidation of excreted homogentisic acid. The cartilage of homozygous adults blackens from deposition of a pigment derived from homogentisic acid. Such individuals often develop arthritic conditions.

allele One of the possible mutational states of a gene, distinguished from other alleles by phenotypic effects.

allele frequency Measurement of the proportion of individuals in a population carrying a particular allele.

allelic exclusion In plasma cell heterozygous for an immunoglobulin gene, the selective action of only one allele.

allopatric speciation Process of speciation associated with geographic isolation.

allopolyploid Polyploid condition formed by the union of two or more distinct chromosome sets with a subsequent doubling of chromosome number.

allosteric effect Conformational change in the active site of a protein brought about by interaction with an effector molecule.

allotetraploid Diploid for two genomes derived from different species.

allozyme An allelic form of a protein that can be distinguished from other forms by electrophoresis.

alpha fetoprotein (AFP) A 70-kd glycoprotein synthesized in embryonic development by the yolk sac. High levels of this protein in the amniotic fluid are associated with nerural tube defects such as spina bifida. Lower than normal levels may be associated with Down syndrome.

***Alu* sequence** An interspersed DNA sequence of approximately 300 bp found in the genome of primates that is cleaved by the restriction enzyme *Alu* I. *Alu* sequences are composed of a head to tail dimer, with the first monomer approximately 140 bp and the second approximately 170 bp. In humans, they are dispersed throughout the genome and are present in 300,000 to 600,000 copies, constituting some 3 to 6 percent of the genome. See *SINES*.

amber codon The codon UAG, which does not code for an amino acid but for chain termination.

Ames test An assay developed by Bruce Ames to detect mutagenic and carcinogenic compounds using reversion to histidine independence in the bacterium *Salmonella typhimurium*.

amino acid Any of the subunit building blocks that are covalently linked to form proteins.

aminoacyl tRNA Covalently linked combination of an amino acid and a tRNA molecule.

amniocentesis A procedure used to test for fetal defects in which fluid and fetal cells are withdrawn from the amniotic layer surrounding the fetus.

amphidiploid See *allotetraploid*.

anabolism The metabolic synthesis of complex molecules from less complex precursors.

analogue A chemical compound structurally similar to another, but differing by a single functional group (e.g., 5-bromodeoxyuridine is an analogue of thymidine).

anaphase Stage of cell division in which chromosomes begin moving to opposite poles of the cell.

aneuploidy A condition in which the chromosome number is not an exact multiple of the haploid set.

angstrom Unit of length equal to 10^{-10} meter. Abbreviated Å.

antibody Protein (immunoglobulin) produced in response to an antigenic stimulus with the capacity to bind specifically to the antigen.

anticodon The nucleotide triplet in a tRNA molecule which is complementary to, and binds to, the codon triplet in a mRNA molecule.

antigen A molecule, often a cell surface protein, that is capable of eliciting the formation of antibodies.

antiparallel Describing molecules in parallel alignment, but running in opposite directions. Most commonly used to describe the opposite orientations of the two strands of a DNA molecule.

apoenzyme The protein portion of an enzyme that requires a cofactor or prosthetic group to be functional.

ascospore A meiotic spore produced in certain fungi.

ascus In fungi, the sac enclosing the four or eight ascospores.

asexual reproduction Production of offspring in the absence of any sexual process.

assortative mating Nonrandom mating between males and females of a species. Selection of mates with the same genotype is positive; selection of mates with opposite genotypes is negative.

ATP Adenosine triphosphate.

attached-X chromosome Two conjoined X chromosomes that share a single centromere.

attenuator A nucleotide sequence between the promoter and the structural gene of some operons that can act to regulate the transit of RNA polymerase and thus control transcription of the structural gene.

autogamy A process of self-fertilization resulting in homozygosis.

autoimmune disease The production of antibodies that results from an immune response to one's own molecules, cells, or tissues. Such a response results from the inability of the immune system to distinguish self from nonself. Diseases such as arthritis, scleroderma, systemic lupus erythematosus, and perhaps diabetes are considered to be autoimmune diseases.

autopolyploidy Polyploid condition resulting from the replication of one diploid set of chromosomes.

autoradiography Production of a photographic image by radioactive decay. Used to localize radioactively labeled compounds within cells and tissues.

autosomes Chromosomes other than the sex chromosomes. In humans, there are 22 pairs of autosomes.

auxotroph A mutant microorganism or cell line which requires a substance for growth that can be synthesized by wild-type strains.

B-DNA See *double helix*.

back-cross A cross involving an F₁ heterozygote and one of the P₁ parents (or an organism with a genotype identical to one of the parents).

bacteriophage A virus that infects bacteria (synonym is *phage*).

bacteriostatic A compound that inhibits the growth of bacteria, but does not kill them.

balanced lethals Recessive, nonallelic lethal genes, each carried on different homologous chromosomes. When organisms carrying balanced lethal genes are interbred, only organisms with genotypes identical to the parents (heterozygotes) survive.

balanced polymorphism Genetic polymorphism maintained in a population by natural selection.

Barr body Densely staining nuclear mass seen in the somatic nuclei of mammalian females. Discovered by Murray Barr, this body is thought to represent an inactivated X chromosome.

base analogue See *analogue*.

base substitution A single base change in a DNA molecule that produces a mutation. There are two types of substitutions: *transitions,* in which a purine is substituted for a purine or a pyrimidine for a pyrimidine; and *transversions,* in which a purine is substituted for a pyrimidine, or vice versa.

biotechnology Commercial and/or industrial processes that utilize biological organisms or products.

bivalents Synapsed homologous chromosomes in the first prophase of meiosis.

BrdU (5-bromodeoxyuridine) A mutagenically active analogue of thymidine in which the methyl group at the 5′ position in thymine is replaced by bromine.

buoyant density A property of particles (and molecules) that depends upon their actual density, as determined by partial specific volume and degree of hydration. Provides the basis for density gradient separation of molecules or particles.

CAAT box A highly conserved DNA sequence found about 75 base pairs 5′ to the site of transcription in eukaryotic genes.

canonical sequence See *consensus sequence.*

cAMP Cyclic adenosine monophosphate. An important regulatory molecule in both prokaryotic and eukaryotic organisms.

CAP Catabolite activator protein. A protein that binds cAMP and regulates the activation of inducible operons.

carcinogen A physical or chemical agent that causes cancer.

carrier An individual heterozygous for a recessive trait.

cassette model First proposed to explain mating type interconversion in yeast, this model proposes that both genes for mating types, *a* and *alpha* are present as silent or unexpressed genes in transposable DNA segments (cassettes) that are activated (played) by transposition to the mating type locus.

catabolism A metabolic reaction in which complex molecules are broken down into simpler forms, often accompanied by the release of energy.

catabolite activator protein See *CAP.*

catabolite repression The selective inactivation of an operon by a metabolic product of the enzymes encoded by the operon.

cdc mutation A class of mutations in yeast that affect the timing and progression through the cell cycle.

cDNA DNA synthesized from an RNA template by the enzyme reverse transcriptase.

cell cycle Sum of the phases of growth of an individual cell type; divided into G_1 (gap 1), S (DNA synthesis), G_2 (gap 2), and M (mitosis).

cell-free extract A preparation of the soluble fraction of cells, made by lysing cells and removing the particulate matter, such as nuclei, membranes, and organelles. Often used to carry out the synthesis of proteins by the addition of specific, exogenous mRNA molecules.

CEN In yeast, fragments of chromosomal DNA, about 120 bp in length, that when inserted into plasmids confer the ability to segregate during mitosis. These segments contain at least three types of sequence elements associated with centromere function.

centimeter A unit of length equal to 10^{-2} meter. Abbreviated cm.

centimorgan A unit of distance between genes on chromosomes. One centimorgan represents a value of 1 percent crossing over between two genes.

central dogma The concept that information flow progresses from DNA to RNA to proteins. Although exceptions are known, this idea is central to an understanding of gene function.

centric fusion See *Robertsonian translocation.*

centriole A cytoplasmic organelle composed of nine groups of microtubules, generally arranged in triplets. Centrioles function in the generation of cilia and flagella and serve as foci for the spindles in cell division.

centromere Specialized region of a chromosome to which the spindle fibers attach during cell division. Location of the centromere determines the shape of the chromosome during the anaphase portion of cell division. Also known as the primary constriction.

centrosome Region of the cytoplasm containing the centriole.

character An observable phenotypic attribute of an organism.

charon phage A group of genetically modified lambda phage designed to be used as vectors for cloning foreign DNA. Named after the ferryman in Greek mythology who carried the souls of the dead across the River Styx.

chemotaxis Negative or positive response to a chemical gradient.

chiasma (pl., chiasmata) The crossed strands of nonsister chromatids seen in diplotene of the first meiotic division. Regarded as the cytological evidence for exchange of chromosomal material, or crossing over.

chi-square (χ^2) analysis Statistical test to determine if an observed set of data fits a theoretical expectation.

chloroplast A cytoplasmic self-replicating organelle containing chlorophyll. The site of photosynthesis.

chorionic villus sampling (CVS) A technique of prenatal diagnosis that intravaginally retrieves fetal cells from the chorion and uses them to detect cytogenetic and biochemical defects in the embryo.

chromatid One of the longitudinal subunits of a replicated chromosome, joined to its sister chromatid at the centromere.

chromatin Term used to describe the complex of DNA, RNA, histones, and nonhistone proteins that make up chromosomes.

chromatography Technique for the separation of a mixture of solubilized molecules by their differential migration over a substrate.

chromocenter An aggregation of centromeres and heterochromatic elements of polytene chromosomes.

chromomere A coiled, beadlike region of a chromosome most easily visible during cell division. The aligned chromomeres of polytene chromosomes are responsible for their distinctive banding pattern.

chromosomal aberration Any change resulting in the duplication, deletion, or rearrangement of chromosomal material.

chromosomal mutation See *chromosomal aberration*.

chromosomal polymorphism Alternate structures or arrangements of a chromosome that are carried by members of a population.

chromosome In prokaryotes, an intact DNA molecule containing the genome; in eukaryotes, a DNA molecule complexed with RNA and proteins to form a threadlike structure containing genetic information arranged in a linear sequence.

chromosome banding Technique for the differential staining of mitotic or meiotic chromosomes to produce a characteristic banding pattern or selective staining of certain chromosomal regions such as centromeres, the nucleolus organizer regions, and GC- or AT-rich regions. Not to be confused with the banding pattern present in unstained polytene chromosomes, which is produced by the alignment of chromomeres.

chromosome map A diagram showing the location of genes on chromosomes.

chromosome puff A localized uncoiling and swelling in a polytene chromosome, usually regarded as a sign of active transcription.

cis configuration The arrangement of two mutant sites within a gene on the same homologue, such as

$$\frac{a^1 \; a^2}{+ \; +}$$

Contrasts with a *trans* arrangement, where the mutant alleles are located on opposite homologues.

cis dominance The ability of a gene to affect the expression of other genes adjacent to it on the chromosome.

cis-trans test A genetic test to determine whether two mutations are located within the same cistron.

cistron That portion of a DNA molecule that codes for a single polypeptide chain; defined by a genetic test as a region within which two mutations cannot complement each other.

cline A gradient of genotype or phenotype distributed over a geographic range.

clonal selection Theory of the immune system that proposes that antibody diversity precedes exposure to the antigen, and that the antigen functions to select the cells containing its specific antibody to undergo proliferation.

clone Genetically identical cells or organisms all derived from a single ancestor by asexual or parasexual methods. For example, a DNA segment that has been enzymatically inserted into a plasmid or chromosome of a phage or a bacterium and replicated to form many copies.

cloned library A collection of cloned DNA molecules representing all or part of an individual's genome.

code See *genetic code*.

codominance Condition in which the phenotypic effects of a gene's alleles are fully and simultaneously expressed in the heterozygote.

codon A triplet of bases in a DNA or RNA molecule which specifies or encodes the information for a single amino acid.

coefficient of coincidence A ratio of the observed number of double-crossovers divided by the expected number of such crossovers.

coefficient of selection A measurement of the reproductive disadvantage of a given genotype in a population. If for genotype *aa*, only 99 of 100 individuals reproduce, then the selection coefficient (*s*) is 0.1.

colchicine An alkaloid compound that inhibits spindle formation during cell division. Used in the preparation of karyotypes to collect a large population of cells inhibited at the metaphase stage of mitosis.

colicin A bacteriocidal protein produced by certain strains of *E. coli* and other closely related bacterial species.

colinearity The linear relationship between the nucleotide sequence in a gene (or the RNA transcribed from it) and the order of amino acids in the polypeptide chain specified by the gene.

competence In bacteria, the transient state or condition during which the cell can bind and internalize exogenous DNA molecules, making transformation possible.

complementarity Chemical affinity between nitrogenous bases as a result of hydrogen bonding. Responsible for the base pairing between the strands of the DNA double helix.

complementation test A genetic test to determine whether two mutations occur within the same gene. If two mutations are introduced into a cell simultaneously and produce a wild-type phenotype (i.e., they complement each other), they are often nonallelic. If a mutant phenotype is produced, the mutations are noncomplementing and are often allelic.

complete linkage A condition in which two genes are located so close to each other that no recombination occurs between them.

complexity The total number of nucleotides or nucleotide pairs in a population of nucleic acid molecules as determined by reassociation kinetics.

complex locus A gene within which a set of functionally related pseudoalleles can be identified by recombinational analysis (e.g., the *bithorax* locus in *Drosophila*).

concatemer A chain or linear series of subunits linked together. The process of forming a concatemer is called concatenation (e.g., multiple units of a phage genome produced during replication).

concordance Pairs or groups of individuals identical in their phenotype. In twin studies, a condition in which both twins exhibit or fail to exhibit a trait under investigation.

conditional mutation A mutation that expresses a wild-type phenotype under certain (permissive) conditions and a mutant phenotype under other (restrictive) conditions.

conjugation Temporary fusion of two single-celled organisms for the sexual exchange of genetic material.

consanguine Related by a common ancestor within the previous few generations.

consensus sequence A nucleotide sequence most often found in a defined segment of DNA.

cosmid A vector designed to allow cloning of large segments of foreign DNA. Cosmids are hybrids composed of the cos sites of lambda inserted into a plasmid. In cloning, the recombinant DNA molecules are packaged into phage protein coats, and after infection of bacterial cells, the recombinant molecule replicates and can be maintained as a plasmid.

coupling conformation See *cis configuration.*

covalent bond A nonionic chemical bond formed by the sharing of electrons.

cri-du-chat syndrome A clinical syndrome in humans produced by a deletion of a portion of the short arm of chromosome 5. Afflicted infants have a distinctive cry which sounds like that of a cat.

crossing over The exchange of chromosomal material (parts of chromosomal arms) between homologous chromosomes by breakage and reunion. The exchange of material between nonsister chromatid during meiosis is the basis of genetic recombination.

cross-reacting material (CRM) Nonfunctional form of an enzyme, produced by a mutant gene, which is recognized by antibodies made against the normal enzyme.

C-terminal amino acid The terminal amino acid in a peptide chain which carries a free carboxyl group.

cytogenetics A branch of biology in which the techniques of both cytology and genetics are used to study heredity.

cytokinesis The division or separation of the cytoplasm during mitosis or meiosis.

cytological map A diagram showing the location of genes at particular chromosomal sites.

cytoplasmic inheritance Non-Mendelian form of inheritance involving genetic information transmitted by self-replicating cytoplasmic organelles such as mitochondria, chloroplasts, etc.

dalton A unit of mass equal to that of the hydrogen atom, which is 1.67×10^{-24} gram. A unit used in designating molecular weights.

Darwinian fitness See *fitness.*

deficiency (deletion) A chromosomal mutation involving the loss or deletion of chromosomal material.

degenerate code Term used to describe the genetic code, in which a given amino acid may be represented by more than one codon.

deletion See *deficiency.*

deme A local interbreeding population.

denatured DNA DNA molecules that have been separated into single strands.

de novo Newly arising; synthesized from less complex precursors rather than having been produced by modification of an existing molecule.

density gradient centrifugation A method of separating macromolecular mixtures by the use of centrifugal force and solvents of varying density. In sedimentation velocity centrifugation, macromolecules are separated by the velocity of sedimentation through a preformed gradient such as sucrose. In density gradient equilibrium centrifugation, macromolecules in a cesium salt solution are centrifuged until the cesium solution establishes a gradient under the influence of the centrifugal field, and the macromolecules sediment until the density of the solvent equals their own.

deoxyribonuclease A class of enzymes that breaks down DNA into oligonucleotide fragments by introducing single-stranded breaks into the double helix.

dermatoglyphics The study of the surface ridges of the skin, especially of the hands and feet.

deoxyribonucleic acid (DNA) A macromolecule usually consisting of antiparallel polynucleotide chains held together by hydrogen bonds, in which the sugar residues are deoxyribose. The primary carrier of genetic information.

determination A regulatory event that establishes a specific pattern of gene activity and developmental fate for a given cell.

diakinesis The final stage of meiotic prophase I in which the chromosomes become tightly coiled and compacted and move toward the periphery of the nucleus.

dicentric chromosome A chromosome having two centromeres.

differentiation The process of complex changes by which cells and tissues attain their adult structure and functional capacity.

dihybrid cross A genetic cross involving two characters in which the parents possess different forms of each character (e.g., tall, round × short, wrinkled peas).

diploid A condition in which each chromosome exists in pairs; having two of each chromosome.

diplotene A stage of meiotic prophase I immediately after pachytene. In diplotene, one pair of sister chromatids begins separating from the other, and chiasmata become visible. These overlaps move laterally toward the ends of the chromatids (terminalization).

directional selection A selective force that changes the frequency of an allele in a given direction, either toward fixation or toward elimination.

discontinuous variation Phenotypic data that fall into two or more distinct classes that do not overlap.

discordance In twin studies, a situation where one twin shows a trait but the other does not.

disjunction The separation of chromosomes at the anaphase stage of cell division.

disruptive selection Simultaneous selection for phenotypic extremes in a population, usually resulting in the production of two discontinuous strains.

dizygotic twins Twins produced from separate fertilization events; two ova fertilized independently. Also known as fraternal twins.

DNA See *deoxyribonucleic acid.*

DNA footprinting See *footprinting.*

DNA gyrase One of the DNA topoisomerases that functions during DNA replication to reduce molecular tension caused by supercoiling. DNA gyrase produces, then seals, double-stranded breaks.

DNA ligase An enzyme that forms a covalent bond between the 5′ end of one polynucleotide chain and the

3' end of another polynucleotide chain. Also called polynucleotide-joining enzyme.

DNA polymerase An enzyme that catalyzes the synthesis of DNA from deoxyribonucleotides and a template DNA molecule.

DNase Deoxyribonucleosidase, an enzyme that degrades or breaks down DNA into fragments or constitutive nucleotides.

dominance The expression of a trait in the heterozygous condition.

dosage compensation A genetic mechanism that regulates the levels of gene products at certain autosomal loci; this results in homozygous dominants and heterozygotes having the same amount of a gene product. In mammals, random inactivation of one X chromosome in females leads to equal levels of X chromosome-coded gene products in males and females.

double-crossover Two separate events of chromosome breakage and exchange occurring within the same tetrad.

double helix The model for DNA structure proposed by James Watson and Francis Crick, involving two antiparallel, hydrogen-bonded polynucleotide chains wound into a right-handed helical configuration, with 10 base pairs per full turn of the double helix. Often called B-DNA.

duplication A chromosomal aberration in which a segment of the chromosome is repeated.

dyad The products of tetrad separation or disjunction at the first meiotic prophase. Consists of two sister chromatids joined at the centromere.

effector molecule Small, biologically active molecule that acts to regulate the activity of a protein by binding to a specific receptor site on the protein.

electrophoresis A technique used to separate a mixture of molecules by their differential migration through a stationary phase in an electrical field.

endogenote The segment of the chromosome in a partially diploid bacterial cell (merozygote) which is homologous to the chromosome transmitted by the donor cell.

endomitosis Chromosomal replication that is not accompanied by either nuclear or cytoplasmic division.

endonuclease An enzyme that hydrolyzes internal phosphodiester bonds in a polynucleotide chain or nucleic acid molecule.

endoplasmic reticulum A membraneous organelle system in the cytoplasm of eukaryotic cells. The outer surface of the membranes may be ribosome-studded (rough ER) or smooth ER.

endopolyploidy The increase in chromosome sets that results from endomitotic replication within somatic nuclei.

endosymbiont theory The proposal that self-replicating cellular organelles such as mitochondria and chloroplasts were originally free-living organisms that entered into a symbiotic relationship with nucleated cells.

enhancer Originally, a 72-bp sequence in the genome of the virus, SV40, that increases the transcriptional activity of nearby structural genes. Similar sequences that enhance transcription have been identified in the genomes of eukaryotic cells. Enhancers can act over a distance of thousands of base pairs and can be located 5' or 3' to the gene they affect, and thus are different from promoters.

enhanson The DNA sequence that represents the core sequence of an enhancer.

environment The complex of geographic, climatic, and biotic factors within which an organism lives.

enzyme A protein or complex of proteins that catalyzes a specific biochemical reaction.

epigenesis The idea that an organism develops by the appearance and growth of new structures. Opposed to preformationism, which holds that development is the growth of structures already present in the egg.

episome A circular genetic element in bacterial cells that can replicate independently of the bacterial chromosome or integrate and replicate as part of the chromosome.

epitope That portion of a macromolecule or cell that acts to elicit an antibody response; an antigenic determinant. A complex molecule or cell can contain several such sites.

epistasis Nonreciprocal interaction between genes such that one gene interferes with or prevents the expression of another gene. For example, in *Drosophila*, the recessive gene *eyeless*, when homozygous, prevents the expression of eye color genes.

equatorial plate See *metaphase plate*.

euchromatin Chromatin or chromosomal regions that are lightly staining and are relatively uncoiled during the interphase portion of the cell cycle. The region of the chromosomes thought to contain most of the structural genes.

eukaryotes Those organisms having true nuclei and membranous organelles and whose cells demonstrate mitosis and meiosis.

euploid Polyploid with a chromosome number that is an exact multiple of a basic chromosome set.

evolution The origin of plants and animals from preexisting types. Descent with modifications.

excision repair Repair of DNA lesions by removal of a polynucleotide segment and its replacement with a newly synthesized, corrected segment.

exogenote In merozygotes, the segment of the bacterial chromosome contributed by the donor cell.

exon (extron) The DNA segment(s) of a gene that are transcribed and translated into protein.

exonuclease An enzyme that breaks down nucleic acid molecules by breaking the phosphodiester bonds at the 3' or 5' terminal nucleotides.

expression vector Plasmids or phage carrying promoter regions designed to cause expression of cloned DNA sequences.

expressivity The degree or range in which a phenotype for a given trait is expressed.

extranuclear inheritance Transmission of traits by genetic information contained in cytoplasmic organelles such as mitochondria and chloroplasts.

F⁺ cell   A bacterial cell having a fertility (F) factor. Acts as a donor in bacterial conjugation.

F⁻ cell   A bacterial cell that does not contain a fertility (F) factor. Acts as a recipient in bacterial conjugation.

F factor   An episome in bacterial cells that confers the ability to act as a donor in conjugation.

F′ factor   A fertility (F) factor that contains a portion of the bacterial chromosome.

F_1 generation   First filial generation; the progeny resulting from the first cross in a series.

F_2 generation   Second filial generation; the progeny resulting from a cross of the F_1 generation.

F pilus   See *pilus*.

facultative heterochromatin   Chromatin that may alternate in form between euchromatic and heterochromatic. The Y chromosome of many species contains facultative heterochromatin.

familial trait   A trait transmitted through and expressed by members of a family.

fate map   A diagram or "map" of an embryo showing the location of cells whose developmental fate is known.

fertility (F) factor   See *F factor*.

filial generations   See F_1, F_2 *generations*.

fingerprint   The pattern of ridges and whorls on the tip of a finger. The pattern obtained by enzymatically cleaving a protein or nucleic acid and subjecting the digest to two-dimensional chromatography or electrophoresis.

fitness   A measure of the relative survival and reproductive success of a given individual or genotype.

fixation   In population genetics, a condition in which all members of a population are homozygous for a given allele.

fixity of species   The idea that members of a species can give rise only to other members of the species, thus implying that all species are independently created.

fluctuation test   A statistical test developed by Salvadore Luria and Max Delbruck to determine whether bacterial mutations arise spontaneously or are produced in response to selective agents.

fMet   See *formylmethionine*.

footprinting   A technique for identifying a DNA sequence that binds to a particular protein, based on the idea that the phosphodiester bonds in the region covered by the protein are protected from digestion by deoxyribonucleases.

formylmethionine (fMet)   A molecule derived from the amino acid methionine by attachment of a formyl group to its terminal amino group. This is the first amino acid inserted in all bacterial polypeptides. Also known as N-formyl methionine.

founder effect   A form of genetic drift. The establishment of a population by a small number of individuals whose genotypes carry only a fraction of the different kinds of alleles in the parental population.

fragile site   A heritable gap or nonstaining region of a chromosome that can be induced to generate chromosome breaks.

frameshift mutation   A mutational event leading to the insertion of one or more base pairs in a gene, shifting the codon reading frame in all codons following the mutational site.

fraternal twins   See *dizygotic twins*.

gamete   A specialized reproductive cell with a haploid number of chromosomes.

gene   The fundamental physical unit of heredity whose existence can be confirmed by allelic variants and which occupies a specific chromosomal locus. A DNA sequence coding for a single polypeptide.

gene amplification   The process by which gene sequences are selected for differential replication either extrachromosomally or intrachromosomally.

gene conversion   A directional process that changes one allele into another specific allele. In fungi, this process involves meiosis and a recombinational step, but in other organisms, such as trypanosomes, meiosis may not be required.

gene duplication   An event in replication leading to the production of a tandem repeat of a gene sequence.

gene flow   The gradual exchange of genes between two populations, brought about by the dispersal of gametes or the migration of individuals.

gene frequency   The percentage of alleles of a given type in a population.

gene interaction   Production of novel phenotypes by the interaction of alleles of different genes.

gene mutation   See *point mutation*.

gene pool   The total of all genes possessed by reproductive members of a population.

generalized transduction   The transduction of any gene in the bacterial genome by a phage.

genetic burden   Average number of recessive lethal genes carried in the heterozygous condition by an individual in a population. Also called genetic load.

genetic code   The nucleotide triplets that code for the 20 amino acids or for chain initiation or termination.

genetic counseling   Analysis of risk for genetic defects in a family and the presentation of options available to avoid or ameliorate possible risks.

genetic drift   Random variation in gene frequency from generation to generation. Most often observed in small populations.

genetic engineering   The technique of altering the genetic constitution of cells or individuals by the selective removal, insertion, or modification of individual genes or gene sets.

genetic equilibrium   Maintenance of allele frequencies at the same value in successive generations. A condition in which allele frequencies are neither increasing nor decreasing.

genetic fine structure   Intragenic recombinational analysis that provides mapping information at the level of individual nucleotides.

genetic load   See *genetic burden*.

genetic polymorphism   The stable coexistence of two or more discontinuous genotypes in a population. When the frequencies of two alleles are carried to an equilibrium, the condition is called balanced polymorphism.

genetics The branch of biology that deals with heredity and the expression of inherited traits.

genome The array of genes carried by an individual.

genotype The specific allelic or genetic constitution of an organism; often, the allelic composition of one or a limited number of genes under investigation.

Goldberg-Hogness box A short nucleotide sequence 20 to 30 bp 5' to the initiation site of eukaryotic genes to which RNA polymerase II binds. The consensus sequence is TATAAAA. Also known as TATA box.

graft versus host disease (GVHD) In transplants, reaction by immunologically competent cells of the donor against the antigens present on the cells of the host. In human bone marrow transplants, often a fatal condition.

gynandromorph An individual composed of cells with both male and female genotypes.

gyrase One of a class of enzymes known as topoisomerases. Gyrase converts closed circular DNA to a negatively supercoiled form prior to replication, transcription, or recombination.

H substance The carbohydrate group present on the surface of red blood cells. When unmodified, it results in blood type O; when modified by the addition of monosaccharides, it results in type A, B, and AB.

haploid A cell or organism having a single set of unpaired chromosomes. The gametic chromosome number.

haplotype The set of alleles from closely linked loci carried by an individual and usually inherited as a unit.

Hardy-Weinberg law The principle that both gene and genotype frequencies will remain in equilibrium in an infinitely large population in the absence of mutation, migration, selection, and nonrandom mating.

heat shock A transient response following exposure of cells or organisms to elevated temperatures. The response involves activation of a small number of loci, inactivation of previously active loci, and selective translation of heat shock mRNA. Appears to be a nearly universal phenomenon observed in organisms ranging from bacteria to humans.

helicase An enzyme that participates in DNA replication by unwinding the double helix near the replication fork.

hemizygous Conditions where a gene is present in a single dose. Usually applied to genes on the X chromosome in heterogametic males.

hemoglobin (Hb) An iron-containing, conjugated respiratory protein occurring chiefly in the red blood cells of vertebrates.

hemophilia A sex-linked trait in humans associated with defective blood-clotting mechanisms.

heredity Transmission of traits from one generation to another.

heritability A measure of the degree to which observed phenotypic differences for a trait are genetic.

heterochromatin The heavily staining, late replicating regions of chromosomes that are condensed in interphase. Thought to be devoid of structural genes.

heteroduplex A double-stranded nucleic acid molecule in which each polynucleotide chain has a different origin. These structures may be produced as intermediates in a recombinational event, or by the *in vitro* reannealing of single-stranded, complementary molecules.

heterogametic sex The sex that produces gametes containing unlike sex chromosomes.

heterogenote A bacterial merozygote in which the donor (exogenote) chromosome segment carries different alleles than does the chromosome of the recipient (endogenote). A heterozygous merozygote.

heterogeneous nuclear RNA (hnRNA) The collection of RNA transcripts in the nucleus, representing precursors and processing intermediates to rRNA, mRNA, and tRNA. Also represents RNA transcripts that will not be transported to the cytoplasm, such as snRNA (small nuclear RNAs).

heterokaryon A somatic cell containing nuclei from two different sources.

heterosis The superiority of a heterozygote over either homozygote for a given trait.

heterozygote An individual with different alleles at one or more loci. Such individuals will produce unlike gametes and therefore will not breed true.

Hfr A strain of bacteria exhibiting a high frequency of recombination. These stains have a chromosomally integrated F factor that is able to mobilize and transfer all or part of the chromosome to a recipient F$^-$ cell.

histocompatibility antigens See *HLA*.

histones Proteins complexed with DNA in the nucleus. They are rich in the basic amino acids arginine and lysine and function in the coiling of DNA to form nucleosomes.

HLA Cell surface proteins, produced by histocompatibility loci, which are involved in the acceptance or rejection of tissue and organ grafts and transplants.

hnRNA See *heterogeneous nuclear RNA*.

holandric A trait transmitted from males to males. In humans, genes on the Y chromosome are holandric.

homeotic mutation A mutation that causes a tissue normally determined to form a specific organ or body part to alter its differentiation and form another structure. Alternately spelled: homoeotic.

homogametic sex The sex that produces gametes that do not differ with respect to sex chromosome content; in mammals, the female is homogametic.

homogeneously staining regions (hsr) Segments of mammalian chromosomes that stain lightly with Giemsa following exposure of cells to a selective agent. These regions arise in conjunction with gene amplification and are regarded as the structural locus for the amplified gene.

homogenote A bacterial merozygote in which the donor (exogenote) chromosome carries the same alleles as the chromosome of the recipient (endogenote). A homozygous merozygote.

homologous chromosomes Chromosomes that synapse or pair during meiosis. Chromosomes that are identical with respect to their genetic loci and centromere placement.

homozygote An individual with identical alleles at one or more loci. Such individuals will produce identical gametes and will therefore breed true.

metafemale In *Drosophila*, a poorly developed female of low viability in which the ratio of X chromosomes to sets of autosomes exceeds 1.0. Previously called a superfemale.

metamale In *Drosophila*, a poorly developed male of low viability in which the ratio of X chromosomes to sets of autosomes is less than 0.5. Previously called a supermale.

metaphase The stage of cell division in which the condensed chromosomes lie in a central plane between the two poles of the cell, and in which the chromosomes become attached to the spindle fibers.

metaphase plate The arrangement of mitotic or meiotic chromosomes at the equator of the cell during metaphase.

MHC Major histocompatibility loci. In humans, the HLA complex; and in mice, the H2 complex.

micrometer A unit of length equal to 1×10^{-6} meter. Previously called a micron. Abbreviated μm.

micron See *micrometer*.

migration coefficient An expression of the proportion of migrant genes entering the population per generation.

millimeter A unit of length equal to 1×10^{-3} meter. Abbreviated mm.

minimal medium A medium containing only those nutrients that will support the growth and reproduction of wild-type strains of an organism.

missense mutation A mutation that alters a codon to that of another amino acid, causing an altered translation product to be made.

mitochondrion Found in the cells of eukaryotes, a cytoplasmic, self-reproducing organelle that is the site of ATP synthesis.

mitogen A substance that stimulates mitosis in nondividing cells; e.g., phytohemagglutinin.

mitosis A form of cell division resulting in the production of two cells, each with the same chromosome and genetic complement as the parent cell.

mode In a set of data, the value occurring in the greatest frequency.

monohybrid cross A genetic cross between two individuals involving only one character (e.g., $AA \times aa$).

monosomic An aneuploid condition in which one member of a chromosome pair is missing; having a chromosome number of $2n - 1$.

monozygotic twins Twins produced from a single fertilization event; the first division of the zygote produces two cells, each of which develops into an embryo. Also known as identical twins.

mRNA An RNA molecule transcribed from DNA and translated into the amino acid sequence of a polypeptide.

mtDNA Mitochondrial DNA.

multiple alleles Three or more alleles of the same gene.

multiple-factor inheritance See *polygenic inheritance*.

multiple infection Simultaneous infection of a bacterial cell by more than one bacteriophage, often of different genotypes.

mu phage A phage group in which the genetic material behaves like an insertion sequence, capable of insertion, excision, transposition, inactivation of host genes, and induction of chromosomal rearrangements.

mutagen Any agent that causes an increase in the rate of mutation.

mutant A cell or organism carrying an altered or mutant gene.

mutation The process which produces an alteration in DNA or chromosome structure; the source of most alleles.

mutation rate The frequency with which mutations take place at a given locus or in a population.

muton The smallest unit of mutation in a gene, corresponding to a single base change.

nanometer A unit of length equal to 1×10^{-9} meter. Abbreviated nm.

natural selection Differential reproduction of some members of a species resulting from variable fitness conferred by genotypic differences.

nearest-neighbor analysis A molecular technique used to determine the frequency with which nucleotides are adjacent to each other in polynucleotide chains.

neutral mutation A mutation with no immediate adaptive significance or phenotypic effect.

noncrossover gamete A gamete which contains no chromosomes that have undergone genetic recombination.

nondisjunction An accident of cell division in which the homologous chromosomes (in meiosis) or the sister chromatids (in mitosis) fail to separate and migrate to opposite poles; responsible for defects such as monosomy and trisomy.

nonsense codon The nucleotide triplet in an mRNA molecule that signals the termination of translation. Three such codons are known: UGA, UAG, and UAA.

nonsense mutation A mutation that alters a codon to one which encodes no amino acid; i.e., UAG (amber codon), UAA (ochre codon), or UGA (opal codon). Leads to premature termination during the translation of mRNA.

NOR See *nucleolar organizer region*.

normal distribution A probability function that approximates the distribution of random variables. The normal curve, also known as a Gaussian or bell-shaped curve, is the graphic display of the normal distribution.

np See *nucleotide pair*.

N-terminal amino acid The terminal amino acid in a peptide chain that carries a free amino group.

nu body See *nucleosome*.

nuclease An enzyme that breaks bonds in nucleic acid molecules.

nucleoid The DNA-containing region within the cytoplasm in prokaryotic cells.

nucleolar organizer region (NOR) A chromosomal region containing the genes for rRNA; most often found in physical association with the nucleolus.

nucleolus A nuclear organelle that is the site of ribosome biosynthesis; usually associated with or formed in association with the NOR.

nucleoside   A purine or pyrimidine base covalently linked to a ribose or deoxyribose sugar molecule.

nucleosome   A complex of four histone molecules, each present in duplicate, wrapped by two turns of a DNA molecule. One of the basic units of eukaryotic chromosome structure. Also known as a nu body.

nucleotide   A nucleoside covalently linked to a phosphate group. Nucleotides are the basic building blocks of nucleic acids. The nucleotides commonly found in DNA are deoxyadenylic acid, deoxycytidylic acid, deoxyguanylic acid, and deoxythymidylic acid. The nucleotides in RNA are adenylic acid, cytidylic acid, guanylic acid, and uridylic acid.

nucleotide pair   The pair of nucleotides (A and T, or G and C) in opposite strands of the DNA molecule that are hydrogen-bonded to each other.

nucleus   The membrane-bounded cytoplasmic organelle of eukaryotic cells that contains the chromosomes and nucleolus.

nullisomic   Describes an individual with a chromosomal aberration in which both members of a chromosome pair are missing.

ochre codon   A codon that does not code for the insertion of an amino acid into a polypeptide chain, but signals chain termination. The ochre codon is UAA.

Okazaki fragment   The small, discontinuous strands of DNA produced during DNA synthesis.

oncogene   A gene whose activity promotes uncontrolled proliferation in eukaryotic cells.

operator region   A region of a DNA molecule that interacts with a specific repressor protein to control the expression of an adjacent gene or gene set.

operon   A genetic unit that consists of one or more structural genes (that code for polypeptides) and an adjacent operator gene that controls the transcriptional activity of the structural gene or genes.

orphon   Single copies of tandemly repeated genes found dispersed in the genome. For example, histone genes are members of a multigene family present in several hundred copies clustered in a tandem array. A single copy of a histone gene found elsewhere in the genome is said to have lost its family and is regarded as an orphon.

overlapping code   A genetic code first proposed by George Gamow in which any given nucleotide is shared by two adjacent codons.

pachytene   The stage in meiotic prophase I when the synapsed homologous chromosomes split longitudinally (except at the centromere), producing a group of four chromatids called a tetrad.

palindrome   A word, number, verse, or sentence that reads the same backward or forward (e.g., *able was I ere I saw elba*). In nucleic acids, a sequence in which the base pairs read the same on complementary strands (5′→3′). For example: 5′GAATTC3′, 3′CTTAAG5′. These often occur as sites for restriction endonuclease recognition and cutting.

pangenesis   A discarded theory of development that postulated the existence of pangenes, small particles from all parts of the body that concentrated in the gametes, passing traits from generation to generation, blending the traits of the parents in the offspring.

paracentric inversion   A chromosomal inversion that does not include the centromere.

parasexual   Condition describing recombination of genes from different individuals which does not involve meiosis, gamete formation, or zygote production. The formation of somatic cell hybrids is an example.

parental gamete   See *noncrossover gamete*.

parthenogenesis   Development of an egg without fertilization.

partial diploids   See *merozygote*.

partial dominance   See *incomplete dominance*.

passive immunity   A form of immunity produced by receipt of antibodies synthesized by another individual.

patroclinous inheritance   A form of genetic transmission in which the offspring have the phenotype of the father.

pedigree   In human genetics, a diagram showing the ancestral relationships and transmission of genetic traits over several generations in a family.

penetrance   The frequency (expressed as a percentage) with which individuals of a given genotype manifest at least some degree of a specific mutant phenotype associated with a trait.

peptide bond   The covalent bond between the amino group of one amino acid and the carboxyl group of another amino acid.

pericentric inversion   A chromosomal inversion that involves both arms of the chromosome and thus involves the centromere.

permissive condition   Environmental conditions under which a conditional mutation (such as a temperature-sensitive mutant) expresses the wild-type phenotype.

phage   See *bacteriophage*.

phenocopy   An environmentally induced phenotype (nonheritable) which closely resembles the phenotype produced by a known gene.

phenotype   The observable properties of an organism that are genetically controlled.

phenylketonuria (PKU)   A hereditary condition in humans associated with the inability to metabolize the amino acid phenylalanine. The most common form is caused by the lack of the liver enzyme phenylalanine hydroxylase.

phosphodiester bond   In nucleic acids, the covalent bond between a phosphate group and adjacent nucleotides, extending from the 5′ carbon of one pentose (ribose or deoxyribose) to the 3′ carbon of the pentose in the neighboring nucleotide. Phosphodiester bonds form the backbone of nucleic acid molecules.

photoreactivation repair   Light-induced repair of damage caused by exposure to ultraviolet light. Associated with an intracellular enzyme system.

phyletic evolution   The gradual transformation of one species into another over time; vertical evolution.

pilus   A filamentlike projection from the surface of a bacterial cell. Often associated with cells possessing F factors.

plaque   A clear area on an otherwise opaque bacterial